Herwig Unger · Phayung Meesad
Editors

Advances in Natural Language Processing and Information Retrieval

Set 1

Springer

Editors
Herwig Unger
Lehrgebiet Kommunikationsnetze
FernUniversität Hagen
Hagen, Germany

Phayung Meesad
Bangkok, Thailand

ISSN 2367-3370 ISSN 2367-3389 (electronic)
Lecture Notes in Networks and Systems
ISBN 978-3-032-20896-5 ISBN 978-3-032-20897-2 (eBook)
https://doi.org/10.1007/978-3-032-20897-2

This Springer imprint is published by the registered company Springer Nature Switzerland AG
The registered company address is: Gewerbestrasse 11, 6330 Cham, Switzerland

The series "Lecture Notes in Networks and Systems" publishes the latest developments in Networks and Systems—quickly, informally and with high quality. Original research reported in proceedings and post-proceedings represents the core of LNNS.

Volumes published in LNNS embrace all aspects and subfields of, as well as new challenges in, Networks and Systems.

The series contains proceedings and edited volumes in systems and networks, spanning the areas of Cyber-Physical Systems, Autonomous Systems, Sensor Networks, Control Systems, Energy Systems, Automotive Systems, Biological Systems, Vehicular Networking and Connected Vehicles, Aerospace Systems, Automation, Manufacturing, Smart Grids, Nonlinear Systems, Power Systems, Robotics, Social Systems, Economic Systems and other. Of particular value to both the contributors and the readership are the short publication timeframe and the world-wide distribution and exposure which enable both a wide and rapid dissemination of research output.

The series covers the theory, applications, and perspectives on the state of the art and future developments relevant to systems and networks, decision making, control, complex processes and related areas, as embedded in the fields of interdisciplinary and applied sciences, engineering, computer science, physics, economics, social, and life sciences, as well as the paradigms and methodologies behind them.

Indexed by SCOPUS, EI Compendex, INSPEC, WTI Frankfurt eG, zbMATH, SCImago.

All books published in the series are submitted for consideration in Web of Science.

For proposals from Asia please contact Aninda Bose (aninda.bose@springer.com).

Lecture Notes in Networks and Systems 1904

Preface

Recent advances in artificial intelligence, particularly Large Language Models (LLMs), have significantly advanced natural language understanding, knowledge integration, and human–computer interaction. Since its inception, the International Conference on Natural Language Processing and Information Retrieval (NLPIR) has provided an important forum for presenting innovative research in NLP and IR, including conversational systems, semantic search, and related applications.

While LLMs have driven rapid progress, important challenges remain in efficiency, interpretability, multilingual adaptability, and domain-specific deployment. Addressing these issues requires continued research beyond model scaling, including low-resource learning, ethical and human-centered design, and interpretable architectures that bridge theory and real-world impact.

This volume contains 85 accepted papers of the 9th International Conference on Natural Language Processing and Information Retrieval (NLPIR 2025), held on December 12–14, 2025, in Fukuoka, Japan. Building on the success of previous events in locations such as Okayama, Seoul, Sanya, Tokushima, and Bangkok, the conference provides an international forum for researchers, scholars, and practitioners to exchange ideas and present recent advances in these dynamic fields. These papers were carefully selected through a rigorous review process to ensure high academic quality and relevance.

We would like to express our sincere gratitude to all authors for their valuable submissions and to the reviewers and program committee members for their time, expertise, and dedication. Their collective efforts have been essential to the success of the conference and the quality of this proceedings volume. We hope that these papers will inspire further research and foster collaboration within the community.

Herwig Unger
Conference General Chair

Organization

Committee

General Chair

Herwig Unger	FernUniversität in Hagen, Germany

General Co-chair

Phayung Meesad	King Mongkut's University of Technology North Bangkok, Thailand

Local Chair

Tsunenori Mine	Kyushu University, Japan

Local Co-chairs

Koichi Takeuchi	Okayama University, Japan
Kazuyuki Matsumoto	Tokushima University, Japan
Hiromitsu Shiina	Okayama University of Science, Japan

Program Chairs

Thomas Böhme	Technical University of Ilmenau, Germany
Janusz Kacprzyk	The Polish Academy of Sciences, Poland

Program Co-chairs

Siripong Potisuk	'The Citadel School of Engineering, United States
Minoru Yoshida	Tokushima University, Japan

Steering Committee

Herwig Unger	FernUniversität in Hagen, Germany
Phayung Meesad	King Mongkut's University of Technology North Bangkok, Thailand
James Z. Wang	Clemson University, United States

International Program Committees

Adrien Guille	University Lyon 2, France
Akinori Ito	Tohoku University, Japan
Alessio Faccia	University of Birmingham Dubai, UAE
Ann Franchesca Laguna	De La Salle University, Philippines
Au Yeung Siu Kei Jeff	Hong Kong Metropolitan University, China
Bahar Ilgen	Robert Koch Institute, Germany
Chisom Ezekannagha	Robert Koch Institute; Freie Universitat Berlin, Germany
Chunjie Guo	Nanjing University of Aeronautics and Astronautics, China
Dinh Dien	Ho Chi Minh University of Science (HCMUS), Vietnam
E. Hashmi	Norwegian University of Science and Technology, Norway
Endang Wahyu Pamungkas	Universitas Muhammadiyah Surakarta, Indonesia
Eric R. Harley	Toronto Metropolitan University, Canada
Ethel Ong	De La Salle University, Philippines
Florian Matthes	Technical University of Munich, Germany
Haemin Jung	Korea National University of Transportation, South Korea
Harinda Fernando	Sri Lanka Institute of Information Technology, Sri Lanka
Hector Rafael Morano Okuno	Tecnologico de Monterrey, Mexico
Hidekazu Yanagimoto	Osaka Metropolitan University, Japan
Hoi-Jeong Lim	Chonnam National University, South Korea
Hongwei Du	California State University, United States
Hosik Choi	University of Seoul, South Korea
Irene Weber	University of Applied Sciences Kempten, Germany
Jairo Cugliari	University Lyon 2, France
Javier Martin Arista	Universidad de La Rioja, Spain
Jerry Chih-Yuan Sun	National Yang Ming Chiao Tung University, Taiwan

Kang Xin	Tokushima University, Japan
Kevin Hirschi	University of Texas at San Antonio, United States
Khang Nhut Lam	Can Tho University, Vietnam
Long Nguyen	University of Science, Ho Chi Minh City, Vietnam
Lucia Nacinovic Prskalo	University of Rijeka, Croatia
Makoto P. Kato	University of Tsukuba and National Institute of Informatics, Japan
Maleerat Maliyaem	King Mongkut's University of Technology North Bangkok, Thailand
Marija Brkic Bakaric	University of Rijeka, Croatia
Mario Flores	University of Texas at San Antonio, United States
Mario Kubek	Georgia State University, United States
Martin Strecker	IRIT, University de Toulouse, France
Masaki Murata	Tottori University, Japan
Mehmet Gonen	Koc University, Turkey
Miguel Couceiro	Inria, France
Minoo Jafarlou	Georgia State University, United States
Moritz Marutschke	Ritsumeikan University, Japan
Mosima A. Masethe	Sefako Makgatho Health Sciences University, South Africa
Ngoc Hong Tran	Vietnamese-German University, Vietnam
Ondrej Bojar	Institute of Formal and Applied Linguistics, Czechia
Pythagoras Petratos	Coventry University, UK
Richa Sharma	PES University, India
Shylaja Ss	PES University, India
Soma Das	Indian Institute of Information Technology Kalyani, India
SUnited Statesn Gauch	University of Arkansas, United States
Swapna Gokhale	University of Connecticut, United States
Takashi Inui	University of Tsukuba, Japan
Tanioka, Hiroki	Tokushima University, Japan
Tee Hean Tan	Suwnay University, Malaysia
Tejas Pant	Accrete AI, United States
Thaweesak Yingthawornsuk	King Mongkut's University of Technology Thonburi, Thailand
Xiaoge Li	Xi'an University of Posts and Telecommunications, China
Xiaorui Jiang	Sheffield University, UK
Yves Lepage	Waseda University, Japan
Zichao Li	University of Waterloo, Canada

Contents

Multimodal NLP Recognition, Semantic Analysis and Modeling

Sentiment Analysis and Corpus Construction

Context-Aware Large Language Models and Parameter Adaptation

Language Representation and Information Retrieval in Multimodal Information Systems

Syntactic Analysis, Language Representation and MachineTranslation

Efficient Language Expansion for Machine Translation Using LoRA: A Case Study for Kazakh

Saida Mussakhojayeva and Huseyin Atakan Varol(✉)

Institute of Smart Systems and Artificial Intelligence, Nazarbayev University, 010000 Astana, Kazakhstan
{saida.mussakhojayeva,ahvarol}@nu.edu.kz

Abstract. Adding a new low-resource language to translation-oriented large language models (LLMs) often demands costly retraining and induces regressions elsewhere. We extend a multilingual machine translation (MT) LLM with Kazakh (KK) under a frozen backbone, thus preserving performance on directions not seen during training. Using Seed-X-Instruct-7B (28 languages, no KK), we attach a single LoRA adapter and train jointly on EN/RU/TR/ZH→KK with instruction-formatted prompts. On FLORES zero-shot, X→KK rises from sub-1 BLEU to 17.1 BLEU with the adapter alone, and to 18.03 BLEU when combined with KK-LM and synthetic data. A controlled full-model fine-tune on EN→KK attains similar single-direction quality but degrades KK→X. Despite updating only 4% of parameters, the adapter is competitive with a 13B MT-LLM on EN→KK and operationally simple: it is loaded only for X→KK requests and left unloaded otherwise, so non-Kazakh directions continue to run unchanged. We demonstrate a practical and reproducible recipe for extending MT-LLM coverage to new low-resource targets with minimal compute and no regressions on existing pairs.

Keywords: machine translation · low-resource languages · Kazakh MT · multilingual LLM · LoRA · parameter-efficient fine-tuning

1 Introduction

LLMs now form the foundation of modern natural language processing (NLP), with models scaling to hundreds of billions of parameters and achieving strong zero- and few-shot performance across a range of tasks, including question answering, reasoning, summarization, coding, and translation [22,28]. A key advancement in these models is their broad multilingual support. For example, Qwen3 [28] was trained on a corpus of approximately 36 trillion tokens spanning over 100 languages. Despite this breadth, the out-of-the-box translation remains uneven. Major evaluations have found that while state-of-the-art LLMs perform competitively on high-resource language pairs, they are considerably weaker on low-resource directions [6,34].

H. Unger and P. Meesad (Eds.): NLPIR 2025, LNNS 1904, pp. 3–18, 2026.
https://doi.org/10.1007/978-3-032-20897-2_1

In response, several LLMs are trained or aligned specifically for MT. ALMA [26] fine-tunes LLaMA with parallel data and preference optimization, closing much of the gap to supervised MT. Hunyuan-MT [33] and Seed-X [5] report strong results across tens of languages by combining mono/bitext pre-training with MT-oriented fine-tuning and alignment. However, language support remains a limited subset (often a few dozen), missing many real-world targets, including Kazakh. Adapting an LLM to a new low-resource language poses challenges. Directly fine-tuning all parameters is costly and risks degrading existing capabilities through interference and overwriting. In contrast, parameter-efficient fine-tuning (PEFT) methods like Low-Rank Adaptation (LoRA) [11] offer a lightweight alternative: small low-rank modules are inserted into selected layers while the base model is frozen, yielding large reductions in trainable parameters and memory. LoRA is effective in low-data settings, often mitigating catastrophic forgetting [13]. Moreover, recent studies show that adapter-style tuning can effectively teach a pre-trained model new languages, sometimes outperforming continued full-model pre-training or even full fine-tuning in those settings [4,32]. These properties make LoRA an appealing choice for adding Kazakh with minimal disruption.

We study how to add an unseen low-resource language to a multilingual LLM-based MT system using PEFT. We take Kazakh—absent from our base model (Seed-X-Instruct-7B; 28 languages)—and add support with four sources: English, Russian, Turkish, and Chinese (Simplified). Our contributions are:

- A practical recipe to add a new target language via LoRA without modifying the base model: a separable adapter is activated only for X→KK, so other directions run unchanged.
- Empirical evidence that this lightweight recipe lifts X→KK from near-zero to 13–23 BLEU (by source) while preserving KK→X, outperforming full-model fine-tuning on the stability/cost trade-off.
- Systematic ablations that identify effective choices—rank 64–128, MLP or MLP+attention placement, interleaving with Kazakh-only text batches, and a small EN→KK synthetic augmentation—yielding a compact, deployable configuration.

These results offer a reproducible, low-overhead path to extend MT-LLMs to new low-resource targets.

2 Related Work

2.1 General-Purpose LLMs for Machine Translation

General LLMs translate competitively on high-resource pairs but are less reliable on low-resource or distant directions. Early broad evaluations on GPT-class models report strong performance on well-resourced European pairs with clear gaps elsewhere [10]; a 102-language, 606-direction study finds GPT-4 surpasses NLLB-54B on 41% of directions yet is still behind Google Translate overall, especially on low-resource languages [34]. Fine-grained analyses likewise note weaknesses on agreement/morphology relative to supervised MT [18]. Recent results are consistent: a large-scale evaluation of open LLMs (e.g., Gemma2-9B) reports rapid gains yet still finds them behind supervised MT like NLLB-54.5B with the gap largest on low-resource pairs: Google Translate outperforms GPT-4 Turbo on FLORES and WMT-24 [6]. Prompting can help at inference time (e.g., pivot prompting and few-shot exemplars [12], domain prompts [23]), and injecting structured monolingual knowledge yields additional gains across low-resource languages [9].

2.2 Translation-Specialized LLMs

Several recent works built LLMs specifically for translation tasks. ALMA [26] trains LLaMA-2 with a two-stage pipeline of monolingual pre-training followed by supervised fine tuning on high quality bitext, gaining about 12 BLEU and 12 COMET over zero-shot on WMT'21–'22 and surpassing prior models. Its successor X-ALMA [27] further extends the framework to 50 languages with plug-and-play modules, achieving SOTA results on FLORES and WMT'23 by COMET. GemmaX2-28 (9B) is an MT-oriented LLM trained first on parallel bitext, then on monolingual data. It consistently beats open translation baselines (e.g., X-ALMA) and is competitive with Google Translate and GPT-4-Turbo across 28 languages [6]. More recently, Seed-X [5] introduced a 7B family (instruct, reasoning, and reward) built from multilingual mono+bitext pre-training across 28 languages, followed by translation-oriented instruction tuning (with chain-of-thought) and reinforcement learning (RL) alignment. It reports parity with closed models (Gemini-2.5 and GPT-4o) in its 28 language suite and substantial wins over larger open baselines. Similarly, Hunyuan-MT [33] is a 7B multilingual translation LLM for 33 languages, trained with MT pre-training, supervised fine-tuning, and RL, with a focus on Chinese↔minority language directions. The report claims SOTA on WMT'25, ranking first on 30 of 31 pairs. Notably, it includes Kazakh, but the Kazakh is the Arabic script variant used in China, rather than the standard Cyrillic Kazakh used in Kazakhstan.

2.3 Parameter-Efficient Fine-Tuning (PEFT)

To adapt LLMs to new languages or MT tasks without updating all weights, many works use PEFT methods. For instance, LoRA [11] injects small low-rank matrices into selected linear layers while freezing the base model, slashing trainable parameters and memory while retaining the base model's knowledge. In practice, LoRA and related adapters are often paired with quantization (e.g., QLoRA) to make training feasible on modest setups [7]. A growing set of case studies shows that language addition using LoRA is feasible in practice. A case study for Uyghur language shows that curating monolingual data, generating synthetic parallel data from it, and then applying parameter-efficient tuning yields sizable MT gains from small, targeted corpora [16]. A Persian pipeline combines monolingual continuation with bilingual alignment (including LoRA), improving Persian generation/classification without hurting English [17]. For

Icelandic, systematic ablations find LoRA in feed-forward blocks particularly effective for LM-style adaptation, with larger LoRA budgets yielding more robust language addition [15].

2.4 NLP and MT for Kazakh

Kazakh is a Kipchak-branch Turkic language with agglutinative morphology and vowel harmony; it is the state language of Kazakhstan and is also spoken in neighboring regions (China/Xinjiang, Russia, and Mongolia), with Cyrillic, Latin, and Arabic scripts in use. Public figures vary, but Kazakhstan's 2021 census reports over 13M Kazakh speakers domestically [3]. From an NLP perspective, Kazakh remains lower-resourced on the web; CC-100 style crawls list Kazakh among the smaller corpora, a common marker for low digital presence [19].

Several open datasets and toolkits beyond MT have been developed for Kazakh in recent years. On text, KazNERD (112k sentences, 25 types) provides a strong named entity recognition baseline with guidelines and code [29], and KazSAnDRA [31] (180k reviews) offers large-scale sentiment supervision. On speech, KSC [14] (~332h) and KSC2 [21] (~1.2k hours) deliver transcribed ASR corpora with speech recognition baselines. For speech synthesis, KazakhTTS2 [20] and KazEmoTTS [1] (~75h, multi-emotion) contribute parallel text–audio resources. These corpora accelerated Kazakh NLP and provide monolingual data useful for tokenization, morphology, and style.

For Kazakh MT, parallel data have been scattered across different collections and crawls. Quality and size vary and are often English-centric: acquisition and alignment translate from English, concentrating resources in KK–EN and underserving non-English pairs [8, 24]. WMT'19 ran EN↔KK as a shared task, where systems leaned on morphological segmentation, transfer from Turkish/Russian, and synthetic data to compensate for scarcity [25]. More recently, KazParC introduced a large-scale KK-EN-RU-TR parallel corpus with over 372k sentence pairs [30]. It emphasizes domain diversity and human-assisted curation.

3 Methodology

We extend a multilingual LLM with support for Kazakh using parameter-efficient LoRA adapters. Concretely, we train a single adapter for many-to-one translation from English (EN), Russian (RU), Turkish (TR), and Chinese (Simplified; ZH) into Kazakh: EN/RU/TR/ZH→KK.

3.1 Data and Preprocessing

We train on curated X→KK pairs, combining KazParC with internal sources. After preprocessing the totals are: EN→KK 602,600 pairs, RU→KK 578,855, TR→KK 569,091, and ZH→KK 484,382. ZH→KK portion is constructed by translating the English side of EN↔KK with GPT-4o to form ZH sources, then filtering for Han character coverage, removing refusal or boilerplate strings, and enforcing length-ratio bounds. We additionally use 2M monolingual KK sentences (Wikipedia) for a Kazakh-only LM objective, and in some runs include 953k filtered EN→KK synthetic pairs generated via back-translation using Google Translate. Pre-processing includes Unicode normalization, markup stripping, language-ID filtering, and length/ratio constraints. Unless noted otherwise, we use FLORES devtest for evaluation (2,009 segments per direction). For the closed-source comparisons (see Sect. 4.6), we additionally report results on the official KazParC test set (4,750 segments). To prevent leakage, we drop any training pair whose source or target matches a test segment after identical normalization (Unicode, lowercasing, punctuation/markup removal, and digit/whitespace normalization).

3.2 Model and Training

The base model is Seed-X-Instruct-7B, a Mistral-derived 7B decoder trained on 28 languages (excluding KK), with a tokenizer that already covers Cyrillic via Russian. Therefore, no vocabulary changes were required, though some KK-specific characters can be split into more tokens. Translation is framed as instruction following and adapted via parameter-efficient LoRA while keeping the backbone frozen. LoRA adapters are added into Transformer projections (`q_proj`, `k_proj`, `v_proj`, `o_proj`, `up_proj`, `gate_proj`, `down_proj`) with the default r=α=128 and dropout 0.05. Optimization uses AdamW with learning rate 1e-4, cosine decay, warmup ratio 0.08, gradient checkpointing, and sequence length 2048. Micro-batch size is 28 with sample packing enabled. We use a single autoregressive cross-entropy objective. Training runs on 4 H100 graphics processing units (GPUs) using `transformers+peft` via `Axolotl` [2].

3.3 Prompt Format and Examples

Translation batches are instruction-formatted. We standardize translation prompts to the following template (language tags in braces; the `<kk>` tag explicitly marks the target:

```
Translate the following {source} sentence into {target}:
{src_text}<kk>
```

KK-LM batches use a continuation instruction to reinforce fluency and morphology without parallel supervision. See Table 1 for examples of the instruction templates and sample completions.

3.4 Baselines and Variants

We compare the following systems on FLORES with greedy decoding, prioritizing speed and reproducibility over small potential gains from beam search. We report automatic evaluation metrics: BLEU (SacreBLEU) and chrF (word_order=2, i.e., chrF++).

Table 1. KK-LM prompts and example completions.

Instruction	Example KK completion
Produce one or two fluent sentences in Kazakh.`<kk>`	Бүгін ауа райы құбылмалы болды, түстен кейін жаңбыр жауды. Ертең таудағы жолдар жабылуы мүмкін.
Write a natural text in Kazakh.`<kk>`	Университет кітапханасы жаңа бөлім ашты, онда қазақ тіліндегі сирек еңбектер жинақталған. Келушілерге ашық.
Continue in Kazakh only.`<kk>`	Қала орталығындағы жөндеу жұмыстары аяқталуға жақын, қоғамдық көлік қозғалысы қалыпқа келеді.

Baseline (Zero-Shot). Unadapted `Seed-X-Instruct-7B` evaluated on all eight directions (EN/RU/TR/ZH ↔ KK). We use it to quantify how much headroom there is on X→KK before any adaptation, to verify whether KK→X is already usable without fine-tuning, and to establish a reference for monitoring performance drift across directions.

LoRA (Joint X →KK). We train a single LoRA adapter for X→KK jointly on EN/RU/TR/ZH→KK parallel data (KazParC + internal). We compare two learning rates: 1e-5 and 1e-4. Joint training tests whether one adapter can enable Kazakh across a number of sources while keeping deployment simple and encouraging cross-source generalization on the target-side.

Full Fine-Tune Baseline EN →KK vs. LoRA. To stress-test interference and efficiency, we unfreeze all weights of Seed-X and fully fine-tune on our best EN→KK mixture. We report EN→KK quality, regression (if any) on KK→{EN,RU,TR,ZH} and efficiency signals (wall-clock, tokens/sec, peak VRAM, and update size on full model versus a 336M parameter adapter).

Adapter Capacity and Placement. We vary two factors, one at a time: *(i) Adapter capacity.* Ranks $r \in \{16, 32, 64\}$ with $\alpha{=}r$. Trainable parameters scale roughly linearly with r (≈42M, 84M, 168M vs. 336M for $r{=}128$), exposing the size–quality trade-off. *(ii) Adapter placement.* With rank fixed to $r{=}128$, we attach LoRA either only to attention (`q,k,v,o`; ≈109M), only to MLP (`up,gate,down`; ≈226M), or to both stacks. This isolates where capacity helps most and whether a single-stack adapter suffices.

Kazakh-only Language-Model (KK-LM) Augmentation. To better capture Kazakh morphology, we inject target-side signal by adding KK-LM batches during LoRA training. Because Kazakh is absent from the Seed-X backbone, parallel alone may not provide sufficient coverage; KK-LM might improve fluency and stabilize learning across sources especially for typologically distant directions (e.g., ZH→KK). The implementation is minimal: every n-th batch is a KK-only continuation prompt. We vary the proportion of KK-LM batches by schedule:

(0%) none, ≈ 20% (every 5th), ≈ 33% (every 3rd), ≈ 50% (every 2nd), and ≈ 67%, i.e., two KK-LM batches per MT batch.

Synthetic EN →KK Augmentation. We augment our data mix with a filtered weakly supervised EN→KK pairs (back-translated from monolingual KK) and retrain the same adapter. This tests whether strengthening the target-side KK prior improves X→KK overall and whether gains, if any, transfer across sources.

4 Results and Discussion

4.1 Bidirectional Baseline

Table 2 shows the baseline Seed-X-Instruct-7B on KK↔X. Seed-X carries enough multilingual knowledge to translate *from* Kazakh into high-resource targets (KK→X avg. 26.94 BLEU / 44.55 chrF). However, the base model cannot translate *into* Kazakh without adaptation (X→KK avg. 0.70 BLEU / 3.48 chrF). The rest of this section focuses on X→KK: how much we can recover with LoRA, what capacity is needed, and which training parameters matter.

Table 2. Zero-shot Seed-X (no adaptation)

Into Kazakh (X→KK)			**From Kazakh (KK→X)**		
Source	BLEU	chrF	Target	BLEU	chrF
EN	0.78	3.95	EN	40.06	59.58
RU	0.60	3.05	RU	19.57	42.63
TR	0.75	3.41	TR	17.20	43.72
ZH	0.65	3.51	ZH	30.94	32.27
AVG	**0.70**	**3.48**	**AVG**	**26.94**	**44.55**

4.2 Overall X→KK Performance

Table 3 shows that LoRA training converts zero-shot performance into functional translation across all sources. A higher learning rate is important in this setting: LoRA-1e-4 substantially outperforms LoRA-1e-5 across the board. EN→KK shows the largest absolute gain; RU/TR→KK follow the same pattern. Among sources, ZH→KK remains the hardest. ZH and KK are not direct translations, introducing style and alignment noise. The typological gap (analytic Chinese versus agglutinative Kazakh) further requires the model to recover the case/possessive morphology absent in the source.

Table 3. Overall X→KK results

Pair	Seed-X		LoRA-1e-5		LoRA-1e-4	
	BLEU	chrF	BLEU	chrF	BLEU	chrF
EN→KK	0.78	3.95	19.48	50.95	**21.76**	**52.61**
RU→KK	0.60	3.05	15.08	46.80	**16.56**	**48.07**
TR→KK	0.75	3.41	15.03	46.41	**17.12**	**48.19**
ZH→KK	0.65	3.51	10.33	41.32	**12.95**	**43.34**
AVG	0.70	3.48	14.98	46.37	**17.10**	**48.05**

4.3 Full Fine-Tune Baseline (EN→KK) Versus LoRA

We compare a LoRA adapter trained on EN→KK (LR 1e-4) to full-model fine-tuning on the same EN→KK mix (all weights, LR 1e-5; higher LRs were unstable). LoRA delivers competitive EN→KK while leaving KK→X unchanged (identical to the frozen Seed-X baseline), whereas full FT improves EN→KK further but induces substantial regressions on reverse directions (see Table 4).

Table 4. EN→KK BLEU on *Seed-X*: Full FT (all weights) vs. LoRA (frozen backbone, adapter only). KK→X is evaluated zero-shot post-training; for LoRA, KK→X equals the original Seed-X.

Method	EN→KK	KK→EN	KK→RU	KK→TR	KK→ZH
Full FT	**23.98**	30.11	2.22	0.80	0.19
LoRA	23.14	**40.06**	**19.57**	**17.20**	**30.94**

Both runs used micro-batch 24 (full FT could not fit 28), making the comparison conservative for LoRA. Even so, full FT took 11.5 h at 79.1 GB/GPU, while LoRA finished in 10.1 h at 67.2 GB/GPU, updating only ~336 M parameters (Table 5). Extrapolating full FT from EN→KK to all four X→KK sources (EN, RU, TR, ZH; ~2.24 M pairs) implies ~3.7× wall-clock, i.e., ~43 h per two-epoch pass on the same node, with additional risk of overfitting the source and catastrophic forgetting; repeating per-source or multi-source FT only increases cost and operational risk.

Table 5. EN→KK efficiency and quality under the same schedule.

Method	Runtime	Tokens/s	VRAM (GB)	Trainable Params
Full FT	11:32:13	22998	79.1	8.0B (100%)
LoRA	10:04:58	26335	67.2	336M (4.2%)

External Reference. Released MT–LLMs with Kazakh are rare; to our knowledge, *X-ALMA (13B)* is the only public model that reports EN→KK results on FLORES (22.2 BLEU). On the same benchmark, our *Seed-X (7B) + LoRA* adapter (336M trainable, ~4.2% of weights) reaches 23.14 BLEU while *Full FT* reaches 23.98 BLEU. These figures are reference points rather than head-to-head comparisons (training/decoding differ).

4.4 Adapter Capacity and Placement

We probe both adapter size ($r \in \{16, 32, 64, 128\}$ with $\alpha = r$) and placement (MLP, attention, or both) and report the results in Table 6.

Table 6. Adapter ablations for X→KK: rank sweep (top) and placement (bottom). BLEU / chrF shown; *Ref* = r=128 with adapters on all projections (336M).

	(a) Rank sweep			
Pair	r=16 (42M)	r=32 (84M)	r=64 (168M)	*Ref*
EN→KK	20.96 / 51.75	21.67 / 52.43	21.62 / **52.64**	**21.76** / 52.61
RU→KK	15.48 / 47.39	16.63 / **48.14**	**16.67** / 48.11	16.56 / 48.07
TR→KK	16.02 / 47.08	16.60 / 47.77	16.89 / 47.90	**17.12** / **48.19**
ZH→KK	11.64 / 41.96	12.23 / 42.72	12.53 / 43.08	**12.95** / **43.34**
AVG	16.03 / 47.04	16.78 / 47.76	16.93 / 47.93	**17.10** / **48.05**
	(b) Placement			
Pair	MLP (226M)	Attn (109M)	*Ref*	
EN→KK	21.63 / 52.53	20.89 / 51.83	**21.76** / **52.61**	
RU→KK	**16.63** / **48.01**	16.21 / 47.60	16.56 / 48.07	
TR→KK	17.14 / 48.08	15.96 / 46.94	**17.12** / **48.19**	
ZH→KK	12.88 / 43.30	11.75 / 42.07	**12.95** / **43.34**	
AVG	17.07 / 47.98	16.20 / 47.11	**17.10** / **48.05**	

Table 7. Effect of interleaving KK-LM batches during LoRA training. Column names are the proportion of LM batches.

	0%		~20%		~33%		~50%		~67%	
Pair	BLEU	chrF	BLEU	chrF	BLEU	chrF	BLEU	chrF	BLEU	chrF
EN→KK	21.76	52.61	22.41	53.34	22.63	53.46	**22.97**	53.64	22.90	**53.71**
RU→KK	16.56	48.07	17.51	48.78	17.25	48.66	**17.56**	**48.90**	16.90	48.12
TR→KK	17.12	48.19	17.66	48.65	17.69	48.66	**17.55**	**48.82**	17.43	48.45
ZH→KK	12.95	43.34	13.24	43.84	13.41	**44.22**	13.61	44.14	13.19	44.15
AVG	17.10	48.05	17.70	48.65	17.75	48.75	**17.92**	**48.87**	17.58	48.61

For rank, diminishing returns appear by $r = 64$: halving parameters from r=128 to r=64 costs only 0.17 BLEU on average. At r=16 the drop is visible across all sources, suggesting a capacity bottleneck.

For placement, attaching adapters to both stacks yields the strongest average, but MLP-only is essentially tied while reducing parameters by one-third. Attention-only is weaker, losing ~0.9 BLEU on average despite two-thirds fewer parameters. Together, these results show that compact settings such as $r = 64$, MLP-only are viable; we keep $r = 128$ with both stacks as default.

4.5 Augmentation with KK-LM Batches

Table 7 shows a steady improvement as the share of KK-LM batches rises from 0% to 50%: average BLEU moves from 17.10 to 17.92 and chrF from 48.05 to 48.87. Each source language benefits: English→Kazakh sees the largest absolute gain, Russian and Turkish move by a solid margin, and Chinese—our hardest pair—also improves. Pushing beyond that to a 2:1 LM:MT schedule (≈ 67% LM) stops helping overall: English is essentially flat, Russian and Turkish slip a bit, and Chinese no longer improves compared to the 50% setting. The picture is consistent with the intended role of the LM batches. A moderate amount of target-only training seems to improve lexical/morphological realization and stabilize decoding, which shows up in both BLEU and chrF. When the balance tilts too far toward unconditional Kazakh, the model appears to spend less capacity on conditioning on the source, and payoffs diminish. On these results, the sweet spot is around 20–50% LM: 20–33% captures most of the gain with less overhead; 50% gives the best average score across sources.

4.6 Synthetic EN→KK, Final Model, and Closed-Source Comparison

Adding Synthetic Data. Adding synthetic EN→KK yields modest, consistent gains. The macro-average rises from 17.10→17.58 BLEU ($\Delta_{\text{rel}} = +2.8\%$) and 48.05→48.58 chrF (+1.1%). Improvements concentrate in EN→KK, from 21.76→22.67 BLEU (+4.2%) and 52.61→53.61 chrF (+1.9%). RU→KK and TR→KK increase slightly (RU: +0.75 BLEU, +0.54 chrF; TR: +0.42 BLEU, +0.38 chrF), consistent with target-side transfer. ZH→KK shows no meaningful change (12.95→12.81 BLEU; 43.34→43.54 chrF), likely due to source mismatch, since ZH inputs are machine-generated from EN (Table 8).

Final Combined Model. Finally, we combine moderate KK-LM interleaving (every third batch, ≈33%) with the filtered synthetic EN→KK pool. The mix is consistently additive: average BLEU/chrF rises to 18.03/49.34 (see Table 9), improving over either LM-only (17.75/48.75) or synthetic-only (17.58/48.58). EN→KK sees the largest absolute lift (+1.64 BLEU over the base; +0.73 over the better of the single-component runs), and RU/TR→KK gain modestly; ZH→KK also nudges up despite the pivoted source. We keep the LM share at 33% here—higher LM proportions that work well without synthetic made the combined

Table 8. Effect of adding synthetic EN→KK bitext (back-translation) to the X→KK training mix.

	Baseline		+Synthetic	
Pair	BLEU	chrF	BLEU	chrF
EN→KK	21.76	52.61	**22.67**	**53.61**
RU→KK	16.56	48.07	**17.31**	**48.61**
TR→KK	17.12	48.19	**17.54**	**48.57**
ZH→KK	**12.95**	43.34	12.81	**43.54**
AVG	17.10	48.05	**17.58**	**48.58**

setting more "causal" and increased hallucinations when noisy MT-generated pairs were present. To summarize, a small amount of target-side continuation plus a carefully filtered synthetic pool gives a reliable, low-cost bump without destabilizing decoding.

Table 9. Effect of target-side KK-LM mixing and synthetic EN→KK on X→KK. *Baseline* = parallel only; *+KK-LM* = interleave ≈33% KK-LM batches; *+Synth* = add filtered EN→KK bitext; *+KK-LM+Synth* = both.

	Baseline		+KK-LM		+Synth		+KK-LM+Synth	
Pair	BLEU	chrF	BLEU	chrF	BLEU	chrF	BLEU	chrF
EN→KK	21.76	52.61	22.63	53.46	22.67	53.61	23.40	54.23
RU→KK	16.56	48.07	17.25	48.66	17.31	48.61	17.57	49.19
TR→KK	17.12	48.19	17.69	48.66	17.54	48.57	17.70	49.12
ZH→KK	12.95	43.34	13.41	44.22	12.81	43.53	13.46	44.83
AVG	17.10	48.05	17.75	48.75	17.58	48.58	**18.03**	**49.34**

Comparison to Closed-Source Systems. To contextualize our adapter, we evaluated Google Translate and GPT-4o on the same test sets.[1] On FLORES, our adapter trails Google Translate (Avg 18.03 / 49.34 vs. 19.99/50.72) and is roughly on par with GPT-4o: we are slightly higher on BLEU (18.03 vs. 17.76) while GPT-4o is slightly higher on chrF (49.68 vs. 49.34). On the independent KazParC set the adapter is close to Google Translate (31.98/59.32 vs. 33.06/60.05) and clearly ahead of GPT-4o (26.02/56.20), leading on all reported directions. Overall, a single LoRA on a frozen 7B backbone reaches near-GT quality on an independent set and outperforms GPT-4o (Table 10).

[1] Both systems are black boxes; we only control the inputs and evaluate the outputs with identical metrics. Training overlap with FLORES is possible for commercial models, so we also report KazParC as an independent held-out set.

Table 10. Closed-source comparison on X→KK. Cells show BLEU/chrF.

	Ours (LoRA)	**Google Translate**	**GPT-4o**
FLORES			
EN→KK	23.40 / 54.23	25.56 / 56.01	23.91 / 55.26
RU→KK	17.57 / 49.19	18.91 / 49.60	16.67 / 49.12
TR→KK	17.70 / 49.12	19.46 / 50.31	16.88 / 48.57
ZH→KK	13.46 / 44.83	16.04 / 46.95	13.56 / 45.76
AVG	18.03 / 49.34	**19.99 / 50.72**	17.76 / 49.68
KazParC			
EN→KK	30.64 / 57.95	36.77 / 61.82	25.59 / 55.32
RU→KK	35.34 / 61.86	31.79 / 59.48	30.46 / 60.05
TR→KK	29.95 / 58.16	30.63 / 58.84	22.00 / 53.22
AVG†	31.98 / 59.32	**33.06 / 60.05**	26.02 / 56.20

†KazParC lacks ZH→KK; averages are over EN/RU/TR.

4.7 Qualitative Analysis

We complement the quantitative scores with a quick look at outputs. In Fig. 1, `v1` is the base LoRA adapter trained only on X→KK translation data, and `v2` is our final adapter with small KK-LM plus a filtered EN→KK synthetic pool. We randomly sampled 200 entries from FLORES and grouped recurring errors. In short, `v2` removes `v1`'s dominant failure modes: (i) decoding instability (loops/repetition) is eliminated; (ii) negation/polarity flips are corrected; (iii) domain terminology is normalized (e.g., law-enforcement and marine terms); and (iv) names/orthography are stabilized (consistent team/proper names and hyphenation).

Despite overall improvements, `v2` still shows recurring, cross-source weaknesses:

- Domain-sense resolution: special terms (e.g., music, zoology, medical) drift to a hypernym or wrong («жоғары/төмен ысқыш» in music rendered with the tool «балға»; "ocelot" generalized to «барыс мысықтары»; "saltwater crocodiles" as «тікенді қолтырауындар»).
- Modality drift: legal and procedural strength is softened («hüküm giydi / was convicted» to «айыпталды» instead of «сотталды»; "were to be closely supervised" narrated as «бақылауында болды» rather than an "obligation").
- Transport & geo polysemy: roadway terms map to rail terms ("northbound carriageway" to «пойыз жолы» instead of «солтүстік бағыттағы жолақ»).

These patterns appear across all source languages, suggesting focused fixes (small domain glossaries and a handful of curated counter-examples) may further improve the quality of MT system.

Decoding stability — Coverage v1: 1 / 2 / 1 / 2 — Coverage v2: 0 / 0 / 0 / 0

Cue [RU]: after an early pit stop for F. Alonso

v1: "пит-стопынан кейін пит-стопынан кейін ..." ▷ Clause "after the pit stop" loops
v2: "...пит-стопынан кейін апатқа ұшырады." ▷ No repetition; event rendered once

Cue [ZH]: animals occur everywhere; some dig/swim/fly

v1: "жан-жақта жан-жақта ..." ▷ Degenerates to repetition
v2: "...жануарлар бар; кейбіреулері жер ..." ▷ Correct list structure retained

---

Negation / polarity — Coverage v1: 2 / 1 / 2 / 1 — Coverage v2: 0 / 0 / 1 / 0

Cue [TR]: ...are generally not considered pests.

v1: "... көбінесе зиянкестерді қамтымайды." ▷ Means "does not include pests"
v2: "...әдетте зиянкестер деп саналмайды." ▷ Correct: "are not considered pests"

Cue [EN]: never let your bags out of your sight

v1: "сөмкелеріңізді көзден таса ұстаңыз". ▷ Polarity flipped ("keep out of sight")
v2: "сөмкелеріңізді ешқашан көзден таса қылмаңыз". ▷ Correct "never let out of sight"

Terminology (domain) — Coverage v1: 3 / 2 / 3 / 3 — Coverage v2: 1 / 0 / 0 / 1

Cue [TR]: reef

v1: "кедертастарда жарақат сирек." ▷ Non-marine "obstacles/boulders"
v2: "рифте жарақат сирек." ▷ Marine term retained

Cue [EN]: riot gear

v1: "бейберекет киім". ▷ Literal, unnatural
v2: "(арнайы) қорғаныс құралдары". ▷ Standard law-enforcement term

---

Names & orthography — Coverage v1: 2 / 2 / 2 / 3 — Coverage v2: 1 / 1 / 1 / 1

Cue [ZH]: ...defeated the All Blacks (New Zealand).

v1: "...Жаңа Зеландияны жеңді." ▷ Team name omitted
v2: "..."All Blacks" (Жаңа Зеландия) құрамасын жеңді". ▷ Name referenced correctly

Fig. 1. Qualitative analysis for X→KK. Coverage (EN/RU/TR/ZH) is scored 0–3: 0 = not seen; 1 = sporadic (≤ 2); 2 = recurring (3–5); 3 = systematic (≥ 6). *Cue [L]* shows the source-side cue in English (a gloss of the original L∈{EN,RU,TR,ZH}) for readability.

5 Conclusion and Future Work

We presented a practical recipe to add a missing low-resource target to a translation-oriented LLM without touching the base model. Starting from Seed-X-Instruct-7B (with no Kazakh in pre-training), a single LoRA adapter trained jointly on EN/RU/TR/ZH→KK incorporates Kazakh with strong gains over zero-shot. Our results suggest three decisions that yield most gains: set the adapter rank to 64–128, attach adapters to the MLP stack (or to both attention and MLP), and interleave 20–50% KK-LM batches. Together, these choices stabilize morphology and improve fluency, and adding filtered EN→KK synthetic pairs gives a small but consistent bump. Compared to full fine-tuning, LoRA reaches competitive X→KK quality while leaving KK→X intact and keeping the operational footprint small. In deployment, the adapter is a clean, separable module: we load it only for X→KK (and unload it otherwise), so users see the same interface and latency budget—the base serves all non-Kazakh directions unchanged, and the adapter quietly supplies the extra capacity for Kazakh targets.

Future work could include collecting direct ZH–KK bitext to replace pivoted sources and better anchor Chinese→Kazakh, exploring RL-based post-training optimization, and evaluating alternative PEFT variants (e.g., Weight-Decomposed Low-Rank Adaptation (DoRA), language-aware adapters) to match or exceed current quality with smaller adapters or improved robustness.

References

1. Abilbekov, A., Mussakhojayeva, S., Yeshpanov, R., Varol, H.A.: KazEmoTTS: A dataset for kazakh emotional text-to-speech synthesis. In: LREC-COLING, pp. 9626–9632 (2024)
2. Axolotl maintainers and contributors: Axolotl: open source LLM post-training (2023). https://github.com/axolotl-ai-cloud/axolotl
3. Bureau of National Statistics of the Agency for Strategic Planning and Reforms of the Republic of Kazakhstan: National composition, religion and language proficiency in the republic of kazakhstan: Results of the national population census of 2021. Technical report, Bureau of National Statistics of the Republic of Kazakhstan (2023). https://stat.gov.kz/ru/national/2021
4. Cao, Z., Qu, Z., Kamigaito, H., Watanabe, T.: Exploring intrinsic language-specific subspaces in fine-tuning multilingual neural machine translation. In: EMNLP 2024, pp. 21,142–21,157 (2024). https://doi.org/10.18653/v1/2024.emnlp-main.1177
5. Cheng, S., et al.: Seed-X: building strong multilingual translation LLM with 7b parameters (2025). https://arxiv.org/abs/2507.13618
6. Cui, M., Gao, P., Liu, W., Luan, J., Wang, B.: Multilingual machine translation with open large language models at practical scale: an empirical study. In: NAACL 2025, pp. 5420–5443 (2025). https://doi.org/10.18653/v1/2025.naacl-long.280
7. Dettmers, T., Pagnoni, A., Holtzman, A., Zettlemoyer, L.: Qlora: efficient finetuning of quantized LLMs. Adv. Neural. Inf. Process. Syst. **36**, 10088–10115 (2023)
8. El-Kishky, A., Chaudhary, V., Guzmán, F., Koehn, P.: Ccaligned: a massive collection of cross-lingual web-document pairs. In: EMNLP, pp. 5960–5969 (2020). https://doi.org/10.18653/v1/2020.emnlp-main.480
9. Guo, P., et al.: Teaching large language models to translate on low-resource languages with textbook prompting. In: LREC-COLING 2024, pp. 15,685–15,697. ELRA and ICCL (2024). https://aclanthology.org/2024.lrec-main.1362/
10. Hendy, A., et al.: How good are GPT models at machine translation? a comprehensive evaluation. arXiv (2023). https://api.semanticscholar.org/CorpusID:257038384
11. Hu, E.J., et al.: Lora: Low-rank adaptation of large language models (2021). https://arxiv.org/abs/2106.09685
12. Jiao, W., et al.: Is ChatGPT a good translator? yes with GPT-4 as the engine (2023). https://arxiv.org/abs/2301.08745
13. Khade, O., Jagdale, S., Phaltankar, A., Takalikar, G., Joshi, R.: Challenges in adapting multilingual LLMs to low-resource languages using LoRA PEFT tuning. In: CHiPSAL, pp. 217–222 (2025). https://aclanthology.org/2025.chipsal-1.22/
14. Khassanov, Y., et al.: A crowdsourced open-source kazakh speech corpus and initial speech recognition baseline. In: EACL, pp. 697–706 (2021)

15. Kunz, J.: Train more parameters but mind their placement: insights into language adaptation with PEFT. In: Proceedings of the Joint 25th Nordic Conference on Computational Linguistics and 11th Baltic Conference on Human Language Technologies, pp. 323–330 (2025). https://aclanthology.org/2025.nodalida-1.35/
16. Lu, K., et al.: Low-resource language expansion and translation capacity enhancement for LLM: a study on the Uyghur. In: Proceedings of the 31st International Conference on Computational Linguistics, pp. 8360–8373 (2025). https://aclanthology.org/2025.coling-main.559/
17. Sani, M.S., Sadeghi, P., Vu, T.T., Yaghoobzadeh, Y., Haffari, G.: Extending LLMs to new languages: a case study of Llama and Persian adaptation. In: COLING, pp. 8868–8884 (2025). https://aclanthology.org/2025.coling-main.594/
18. Manakhimova, S., et al.: Linguistically motivated evaluation of the 2023 state-of-the-art machine translation: can ChatGPT outperform NMT? In: WMT, pp. 224–245 (2023). https://doi.org/10.18653/v1/2023.wmt-1.23
19. MetaText: Cc100: Kazakh (kk) (2025). https://metatext.io/datasets/cc100-kazakh
20. Mussakhojayeva, S., Khassanov, Y., Varol, H.A.: KazakhTTS2: extending the open-source kazakh TTS corpus with more data, speakers, and topics. In: LREC, pp. 5404–5411 (2022). https://aclanthology.org/2022.lrec-1.578/
21. Mussakhojayeva, S., Khassanov, Y., Varol, H.A.: KSC2: an industrial-scale open-source kazakh speech corpus. In: INTERSPEECH, pp. 1367–1371 (2022)
22. OpenAI: GPT-5-o: a large multimodal language model. https://openai.com (2025). Accessed Sep 2025
23. Peng, K., et al.: Towards making the most of ChatGPT for machine translation. In: Findings of the Association for Computational Linguistics: EMNLP 2023, pp. 5622–5633 (2023). https://doi.org/10.18653/v1/2023.findings-emnlp.373
24. Tiedemann, J.: Parallel data, tools and interfaces in opus. In: LREC, pp. 2214–2220 (2012)
25. Toral, A., et al.: Neural machine translation for English–Kazakh with morphological segmentation and synthetic data. In: Proceedings of the WMT 2019, pp. 386–392 (2019). https://doi.org/10.18653/v1/W19-5343
26. Xu, H., Kim, Y.J., Sharaf, A., Awadalla, H.H.: A paradigm shift in machine translation: boosting translation performance of large language models (2024). https://arxiv.org/abs/2309.11674
27. Xu, H., et al.: X-ALMA: Plug & play modules and adaptive rejection for quality translation at scale (2025). https://arxiv.org/abs/2410.03115
28. Yang, A., et al.: Qwen3 technical report. arXiv (2025)
29. Yeshpanov, R., Khassanov, Y., Varol, H.A.: KazNERD: Kazakh named entity recognition dataset. In: Proceedings of the Thirteenth Language Resources and Evaluation Conference, pp. 417–426. ELRA (2022). https://aclanthology.org/2022.lrec-1.44/
30. Yeshpanov, R., Polonskaya, A., Varol, H.A.: KazParC: Kazakh parallel corpus for machine translation. In: LREC-COLING 2024, pp. 9633–9644 (2024). https://aclanthology.org/2024.lrec-main.842/
31. Yeshpanov, R., Varol, H.A.: KazSAnDRA: Kazakh sentiment analysis dataset of reviews and attitudes. In: LREC-COLING 2024, pp. 9657–9667 (2024). https://aclanthology.org/2024.lrec-main.844/
32. Yong, Z.X., et al.: BLOOM+1: adding language support to BLOOM for zero-shot prompting. In: ACL 2023, pp. 11,682–11,703 (2023). https://doi.org/10.18653/v1/2023.acl-long.653
33. Zheng, M., et al.: Hunyuan-MT technical report (2025). https://arxiv.org/abs/2509.05209

34. Zhu, W., et al.: Multilingual machine translation with large language models: Empirical results and analysis. In: Findings of the Association for Computational Linguistics: NAACL 2024, pp. 2765–2781 (2024). https://doi.org/10.18653/v1/2024.findings-naacl.176

Exploring the Challenge of Many-to-Many Alignments in Word Alignment Models

Tyler Justin Tan and Thomas James Tiam-Lee(✉)

College of Computer Studies, De La Salle University, 2401 Taft Ave, 1004 Metro Manila, Philippines
{tyler_justin_tan,thomas.tiam-lee}@dlsu.edu.ph

Abstract. Multilingual language model-based word alignment models have shown promising performance across a variety of languages. However, in this study, our empirical analysis reveals that such models often struggle to capture many-to-many alignments, which may limit their effectiveness in some domains, particularly in those that are dominated by free translations. In light of this, we show that a simple phrase tokenization strategy allows existing word alignment models to operate on phrases rather than individual words or units, thereby increasing their ability to detect many-to-many alignments. After phrase-to-phrase alignments are predicted, we revert them to word-level alignments for standard evaluation using quantitative metrics such as the Alignment Error Rate (AER) that accounts for both literal and free translations. Experimental results across three existing alignment models show consistent improvements in performance across both existing and our own Japanese-English dataset.

Keywords: word alignment · machine translation · tokenization

1 Introduction

Word alignment is the task of identifying translation relationships between source and target tokens given a pair of parallel sentences, and was once an indispensable part of statistical machine translation (SMT) systems. With the advent of neural machine translation (NMT), word alignment remains a notable area of research, as it is applied in various NLP tasks, such as annotation transfer [1,2], extraction of translation lexicons [3,4], and assisting in the evaluation and analysis of NMT models [5,6].

Traditionally, word alignment models relied mainly on statistical methods [7,8]. However, advances in NMT systems and Transformer models [9] have improved the performance of large language models (LLMs) in various NLP tasks, including word alignment. Recent neural word alignment approaches [10–12] have been shown to outperform traditional statistical methods. However,

H. Unger and P. Meesad (Eds.): NLPIR 2025, LNNS 1904, pp. 19–33, 2026.
https://doi.org/10.1007/978-3-032-20897-2_2

in this study, we demonstrate that the above models struggle with many-to-many alignments. We define a many-to-many alignment as follows: if a phrase in the source language (composed of multiple consecutive tokens) corresponds to a phrase in the target language (also composed of multiple consecutive tokens), each word in the source phrase is said to be aligned to each word in the target phrase, resulting in a dense, many-to-many alignment. These alignments often occur in cases of free translation, i.e., translations that prioritize conveying meaning, tone, or emotion, unlike literal translations, which follow a strict word-for-word correspondence. Refer to Fig. 1 shows an example of this.

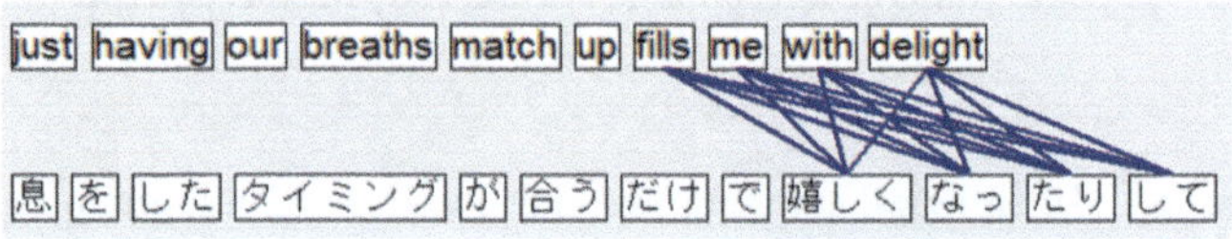

Fig. 1. An example of a many-to-many alignment, where the English phrase "'fills me with delight" corresponds to the Japanese phrase 嬉しくなったりして" in the context of the aligned sentences. Note that the Japanese phrase literally means "to become happy". Since this is a free translation, aligning the individual words is not possible, so the most reasonable alignment is to align the entire phrase.

Despite the difficulty models face with many-to-many alignments, there is value in being able to correctly capture them. Aside from increasing the overall accuracy of word alignment models, handling many-to-many alignments also allow models to work better in domains where free translations are prevalent, such as in songs, poetry, or other creative fields due to the prevalence of figures of speech and liberal interpretations. In the context of language learning, the recognition and comprehension of these alignments are particularly beneficial. Learners who can move beyond surface-level word-to-word mappings gain a much deeper appreciation for the target language's richness and cultural context. Furthermore, it has also potential applications in tasks like automatic subtitling, cross-lingual information retrieval, and machine translation.

In this study, we investigate this particular aspect of word alignment models by conducting an empirical analysis of the performance of three recent word alignment models, focusing particularly on many-to-many alignments. We make two main contributions. First, we demonstrate the many-to-many alignment limitation of existing models when tested on both existing word alignment datasets and our own created song dataset with many free translations. Second, we demonstrate that a simple phrase tokenization scheme is able to alleviate this limitation across the three models.

2 Models and Datasets

As part of this study, we evaluated three alignment models, namely AwesomeAlign [10], SimAlign [11], and PMI-Align [12]. These models belong to a class of multilingual language model (MLM)-based approaches that have shown promising performance on multilingual word alignment tasks in recent studies. For AwesomeAlign, we used their fine-tuned `model_without_co` as the LLM model. For the other two models, we used the default LLM models in their implementations[1][2][3].

For the datasets, we used the gold alignment datasets from AwesomeAlign, but our analysis focuses specifically on many-to-many alignments. AwesomeAlign uses four datasets: a Romanian–English (Ro–En) parallel dataset originally sourced from archives of Romanian newspapers, Orwell's 1984, and the Romanian Constitution [13]; an English–French (En–Fr) parallel dataset from the Canadian Hansards [13]; a Chinese–English (Zh–En) dataset sourced from Liu and Sun (2015) [14]; and a Japanese–English (Ja–En) dataset from the Kyoto Free Translation Task, which contains Wikipedia articles related to Kyoto [15].

In addition to this, we constructed our own dataset, composed of song lyrics, with the intention of including a greater number of many-to-many alignments. This new Japanese–English (Ja–En) dataset was manually created using lyrics from fifty (50) songs. The lyrics were sourced from the subtitles of the official YouTube videos of the selected songs. To ensure the quality of the dataset, several criteria were applied in selecting the songs:

1. The video must have at least one million views on YouTube.
2. The video must include both official Japanese and English subtitles.
3. The original lyrics must be primarily in Japanese.

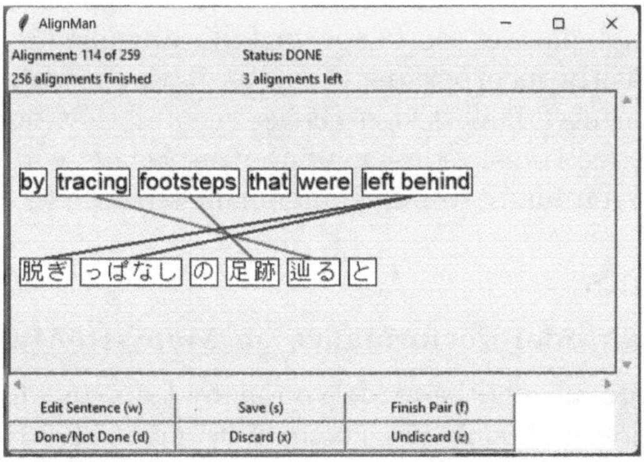

Fig. 2. Screenshot of the AlignMan tool used for manual word alignment annotation.

To construct the dataset, subtitles were first downloaded from YouTube. After verifying the subtitles, each line of the lyrics was paired with its trans-

[1] https://github.com/neulab/awesome-align.
[2] https://github.com/cisnlp/simalign.
[3] https://github.com/fatemeh-azadi/PMI-Align.

lation to create parallel sentence pairs. This was done by first using a script that automatically paired the lines using the timestamps found in the downloaded subtitles. Repeating lines in the lyrics were dropped to ensure that every instance in the dataset is unique. Each sentence pair was then manually verified to ensure correctness and manually modified if any errors were found. Finally, word alignment annotations were then manually created using the open-source AlignMan tool [16]. See Fig. 2 for an example of how AlignMan was used during the annotation process.

During annotation, alignments were made using the standard $i\text{–}j$ Pharaoh format, which is also utilized in previous studies [10] [11, 12]. Additionally, word alignments were categorized into two types: **literal** and **free**.

Within the related literature, the terms *sure* alignment and *possible* alignment are commonly used. For the purposes of our analysis, *sure* corresponds to literal alignments, while *possible* corresponds to free alignments.

Songs were chosen for this dataset due to the prevalence of free translations in said media, which often result in many-to-many alignments, making them well-suited for this study. Furthermore, Japanese–English was chosen as the language pair for our dataset, as we found it to be a good example for demonstrating the limitation surrounding many-to-many alignments. The structural differences between the two languages, such as sentence construction and verb conjugation, often lead to many-to-many alignments, making this pair especially suitable for our analysis.

In light of this, we believe our dataset serves a distinct purpose from AwesomeAlign's Ja–En dataset. While their dataset is primarily sourced from Wikipedia articles about Kyoto's history, ours is based on song lyrics, which tend to contain more free translations and creative expressions. Notably, AwesomeAlign's original dataset does not include alignments labeled as "free," whereas ours explicitly incorporates them, as it was intentionally designed to include free alignments. This design choice ensures that our dataset provides richer coverage of free translations, making it especially suitable for evaluating model performance on many-to-many alignment scenarios.

3 Experiments

3.1 Evaluating Model Performance on Many-to-Many Alignments

To gauge the performance of the models on many-to-many alignments, we developed a script that automatically detects many-to-many alignments in the dataset and compares them to the output of the models to evaluate whether they were correctly identified. For the purposes of this paper, a single many-to-many alignment is defined as a group of interconnected word-level alignments that involve at least two distinct source tokens and two distinct target tokens.

Formally, we define the word alignment task as follows. Given a source sentence $S = \langle s_1, s_2, \ldots, s_n \rangle$ and its corresponding target sentence $T = \langle t_1, t_2, \ldots, t_m \rangle$, the goal of the alignment model is to predict a set of word-level alignment pairs:

$$A = \{\langle x_i, y_j \rangle \mid x_i \in S,\ y_j \in T\},$$

where each pair $\langle x_i, y_j \rangle$ represents source and target tokens that are semantically related within the context of the sentence.

The model produces an output set of predicted word-level alignments:

$$O = \{\langle x_i, y_j \rangle \mid x_i \in S,\ y_j \in T\},$$

which aims to approximate the gold alignment set A as closely as possible.

To detect many-to-many alignments in the dataset, we construct an undirected graph $G_A = (V, E)$ where the set of vertices V consists of both source and target tokens:

$$V = \{s_1, \ldots, s_n,\ t_1, \ldots, t_m\}.$$

The edge set E corresponds to the gold word alignments:

$$E = \{\{s_i, t_j\} \mid \langle s_i, t_j \rangle \in A\},$$

where each edge connects a source token s_i to a target token t_j.

To identify groups of word-level alignments that constitute many-to-many alignments, we find all connected components in G_A and retain those that contain at least two distinct source tokens and at least two distinct target tokens. Furthermore, all source tokens in the component must be aligned with all target tokens. These components are considered instances of many-to-many alignments. Refer to Fig. 3 for a visualization of these connected components.

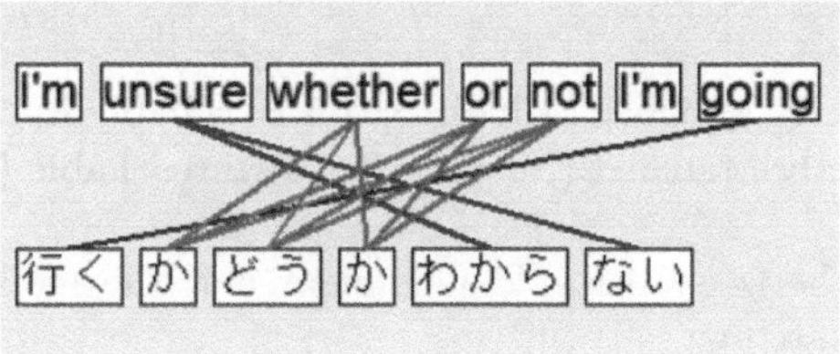

Fig. 3. A Japanese–English translation pair with three notable connected components. The purple group represents a one-to-one alignment, the blue group illustrates a one-to-many alignment, and the green group is an example of a many-to-many alignment, consisting of 3 source tokens, 3 target tokens, and 9 word-level alignments.

To evaluate how well a model predicts an instance of a many-to-many alignment, we count how many of the word-level alignments within that instance are correctly predicted by the model.

More formally, let a many-to-many alignment instance be denoted as a subgraph $M = (V', E')$, where $V' \subseteq V$ is the set of source and target tokens involved, and $E' \subseteq E$ is the set of gold word-level alignments within this group.

For each alignment pair $(a, b) \in E'$, we check whether the model's predicted alignments O contain the corresponding pair $\langle a, b \rangle$. Under this direct matching scheme, if the pair is found, it is labeled as *Directly Matched*; otherwise, it is labeled as *Not Matched Directly*.

Alternatively, we may also consider a more lenient approach to determine whether a word-level alignment is predicted by the model, based on whether the aligned tokens are *reachable* within the model's output.

More formally, we construct a graph $G_O = (V, E_O)$ from the model's predicted alignments O, following the same method used to construct G_A for the gold alignments. Then, for each alignment pair $(a, b) \in E'$, we check whether vertex a is reachable from vertex b in G_O. Under this indirect matching scheme, if the pair is reachable, it is labeled as *Reachably Matched*; otherwise, it is labeled as *Unreachable.*

For a visual illustration of the matching criteria described above, refer to Fig. 4.

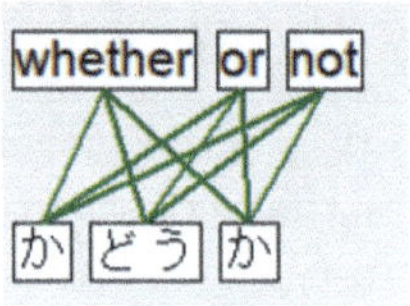

(a) Gold Alignment

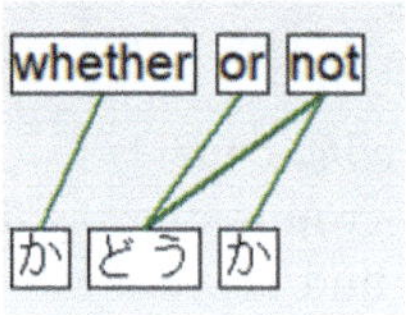

(b) Model Output

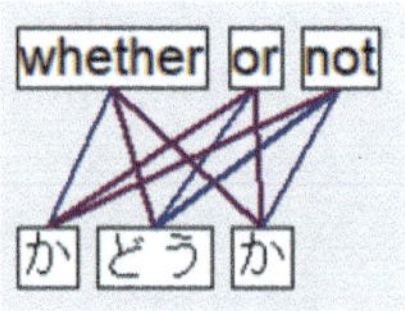

(c) 4 Directly Matched (Blue), 5 Not Directly Matched (Purple)

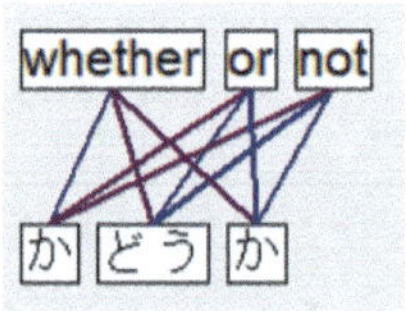

(d) 5 Reachably Matched (Blue), 4 Unreachable (Purple)

Fig. 4. Comparison between the gold alignment and the model's predictions under direct and indirect matching.

With this in mind, we evaluated the following alignment models: AwesomeAlign [10], SimAlign [11], and PMI-Align [12]. Each model was tested on its original evaluation datasets, as well as on our newly introduced dataset. This evaluation sheds light on how well each model performs in predicting many-to-many alignments. See Table 1 for the complete results.

Looking at the results, we observe that the models generally struggle to predict many-to-many alignments accurately. Across different languages and models, a majority of the word-level alignments within many-to-many groups are not even reachable in the predicted outputs. This shows a limitation in the capabilities of current word alignment models.

3.2 Phrase Tokenization

To address this limitation, we propose a mitigation strategy in the form of phrase tokenization. Instead of performing alignment at the word level, we group seman-

Table 1. Model performance on each dataset across matching criteria.

Direct Matching Scheme				
Model	Dataset	Directly Matched	Not Directly Matched	Total
Awesome-Align	Orig. En-Fr	1316 (15.50%)	**7177 (84.50%)**	8493
	Orig. Ro-En	147 (27.63%)	**385 (72.37%)**	532
	Orig. Ja-En	323 (21.99%)	**1146 (78.01%)**	1469
	Orig. Zh-En	30 (40.00%)	**45 (60.00%)**	75
	Our Ja-En (Songs)	1640 (25.61%)	**4764 (74.39%)**	6404
SimAlign	Orig. En-Fr	1450 (17.07%)	**7043 (82.93%)**	8493
	Orig. Ro-En	147 (27.63%)	**385 (72.37%)**	532
	Orig. Ja-En	177 (12.05%)	**1292 (87.95%)**	1469
	Orig. Zh-En	27 (36.00%)	**48 (64.00%)**	75
	Our Ja-En (Songs)	1276 (19.93%)	**5128 (80.07%)**	6404
PMIAlign	Orig. En-Fr	1170 (13.78%)	**7323 (86.22%)**	8493
	Orig. Ro-En	121 (22.74%)	**411 (77.26%)**	532
	Orig. Ja-En	227 (15.45%)	**1242 (84.55%)**	1469
	Orig. Zh-En	23 (30.67%)	**52 (69.33%)**	75
	Our Ja-En (Songs)	1618 (25.27%)	**4786 (74.73%)**	6404
Indirect Matching Scheme				
Model	Dataset	Reachably Matched	Unreachable	Total
Awesome-Align	Orig. En-Fr	1324 (15.59%)	**7169 (84.41%)**	8493
	Orig. Ro-En	149 (28.01%)	**383 (71.99%)**	532
	Orig. Ja-En	328 (22.33%)	**1141 (77.67%)**	1469
	Orig. Zh-En	31 (41.33%)	**44 (58.67%)**	75
	Our Ja-En (Songs)	1693 (26.44%)	**4711 (73.56%)**	6404
SimAlign	Orig. En-Fr	1469 (17.30%)	**7024 (82.70%)**	8493
	Orig. Ro-En	147 (27.63%)	**385 (72.37%)**	532
	Orig. Ja-En	179 (12.19%)	**1290 (87.81%)**	1469
	Orig. Zh-En	28 (37.33%)	**47 (62.67%)**	75
	Our Ja-En (Songs)	1307 (20.41%)	**5097 (79.59%)**	6404
PMIAlign	Orig. En-Fr	1182 (13.92%)	**7311 (86.08%)**	8493
	Orig. Ro-En	122 (22.93%)	**410 (77.07%)**	532
	Orig. Ja-En	233 (15.86%)	**1236 (84.14%)**	1469
	Orig. Zh-En	24 (32.00%)	**51 (68.00%)**	75
	Our Ja-En (Songs)	1778 (27.76%)	**4626 (72.24%)**	6404

tically related, consecutive words into meaningful phrases, which are then aligned by the model.

This is done using a script that applies a rule-based approach to merge source and target word-level tokens into larger phrase-level tokens.

Formally, given a sentence consisting of word-level tokens:

$$W = \langle w_1, w_2, \ldots, w_n \rangle,$$

the goal is to produce an output sequence in which some of the tokens are combined into phrases (e.g., grouping w_1 and w_2 into a single phrase token $\langle w_1, w_2 \rangle$), such that the resulting sentence contains the fewest possible number of phrase-level tokens while preserving semantic coherence. For an illustrative example, see Table 2.

Table 2. Candidate phrase tokenizations for the sentence "she looked up and down the hallway." The final option is preferred as it results in the fewest phrase-level tokens.

Candidate Output	# of Phrases
$\langle\langle she\rangle\langle looked\rangle\langle up\rangle\langle and\rangle\langle down\rangle\langle the\rangle\langle hallway\rangle\rangle$	7
$\langle\langle she\rangle\langle looked\ up\rangle\langle and\rangle\langle down\rangle\langle the\rangle\langle hallway\rangle\rangle$	6
$\langle\langle she\rangle\langle looked\rangle\langle up\ and\ down\rangle\langle the\rangle\langle hallway\rangle\rangle$	5
$\langle\langle she\rangle\langle looked\ up\ and\ down\rangle\langle the\rangle\langle hallway\rangle\rangle$	**4**

From an implementation standpoint, a phrase was considered valid and semantically meaningful if it appeared in a dictionary lookup. For English, this lookup was performed using a combination of data scraped from Wiktionary—a collaboratively edited multilingual online dictionary—and Python's NLTK library. For Japanese, the JMdict dictionary was used as the primary source for identifying valid phrases.

Additionally, for Japanese, part-of-speech (POS) information was incorporated using SudachiPy. A phrase was also considered valid if it began with a verb (動詞) followed by a sequence of auxiliary verbs (助動詞) or the tokens て, で, or いる.

To select the best tokenization among the valid candidates, a simple shortest-path algorithm was used to find the segmentation that minimized the number of phrase-level tokens.

After phrase tokenizing, the sentences with now phrase-level tokens are inputted into the models, producing output alignments between source and target phrase tokens. Let us define this phrase-level output alignments as the set:

$$O' = \{\langle\langle x_i, x_{i+1}, \ldots\rangle, \langle y_j, y_{j+1}, \ldots\rangle\rangle \mid x_i \in S,\, y_j \in T\},$$

After obtaining the predicted phrase-to-phrase alignments, we convert them back to word-level alignments for evaluation. This is done by aligning a source word and a target word if they belong to phrases that are aligned. Formally, for all phrases X and Y, wherein $\langle X, Y\rangle \in O'$, we may define the new word-level alignments O^* derived from phrase tokenization as...

$$O^* = \{\langle s_x, t_y\rangle \mid \exists X = \langle s_i, \ldots\rangle, Y = \langle t_j, \ldots\rangle \text{ such that } s_x \in X \text{ and } t_y \in Y\}$$

Refer to Fig. 5 for an illustrative example of the alignments before and after reverting phrase-level alignments back to word-level for evaluation.

This allows us to compare these predicted word-level alignments O^* against the gold alignments using standard evaluation metrics, as discussed below.

To evaluate the model's overall performance before and after phrase tokenization, we use the standard *Alignment Error Rate* (AER) metric [8]. AER is

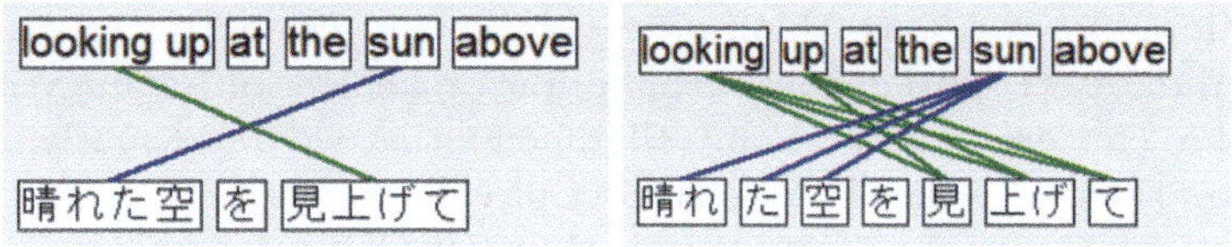

(a) Before (Phrase-Level) (b) After (Reverted Word-Level)

Fig. 5. Alignments before and after converting phrase-to-phrase alignments back into word-level alignments.

computed based on precision and recall, by defining two sets S and P based on the two types of alignments found in the gold alignments:

- S (Sure): alignments marked as *literal*
- P (Possible): all alignments, including both *literal* and *free*

Note that $S \subseteq P$.

Given a predicted alignment set A outputted by a model, we define:

$$\text{Recall} = \frac{|A \cap S|}{|S|} \quad \text{and} \quad \text{Precision} = \frac{|A \cap P|}{|A|}$$

Then the AER, being derived from the f-measure, is computed using:

$$\text{AER}(S, P; A) = 1 - \frac{|A \cap S| + |A \cap P|}{|A| + |S|}$$

Under this formulation:

- A precision error occurs if a predicted alignment in A is not found in P.
- A recall error occurs if an alignment from S is not found in A.

However, we argue that there is also value in the model predicting *free* alignments. To capture this, we propose an additional metric: a stricter recall criterion where:

$$\text{Strict Recall} = \frac{|A \cap P|}{|P|}$$

and define the corresponding Strict AER (SAER) as:

$$\text{SAER}(P; A) = 1 - \frac{2|A \cap P|}{|A| + |P|}$$

We now use these metrics to evaluate this mitigation strategy using phrase tokenization. For each model, we conduct two tests: the first compares its original performance on our song dataset without any modifications, using the standard word alignments O against the gold alignments A; the second test evaluates performance after applying the phrase tokenization method described above, comparing the derived word-level alignments O^* with the gold alignments A.

In addition, for AwesomeAlign, we apply the same evaluation procedure using its original dataset. As the goal of this phrase tokenization strategy is to allow the model to capture many-to-many alignments that were previously missed, we hypothesize that both recall and strict recall will increase.

Interestingly, AwesomeAlign's original Ja–En dataset does not contain any free alignments. As a result, all alignments were classified as literal, leading to identical values for Recall and Strict Recall, as well as for AER and SAER. The results of all tests are summarized in Table 3.

Table 3. Model performance on each dataset across defined metrics.

Model	Dataset	Precision	Recall	AER	Strict Recall	SAER
Awesome-Align	Our Ja-En (Songs) (O vs A)	0.558	0.518	46.1%	0.447	50.4%
	Our Ja-En (Songs) (O^* vs A)	0.534	0.836	36.3%	0.732	38.3%
	Original Ja-En (O vs A)	0.727	0.550	37.4%	0.550	37.4%
	Original Ja-En (O^* vs A)	0.622	0.642	36.8%	0.642	36.8%
SimAlign	Our Ja-En (Songs) (O vs A)	0.452	0.420	56.3%	0.357	60.1%
	Our Ja-En (Songs) (O^* vs A)	0.477	0.645	46.0%	0.571	48.0%
	Original Ja-En (O vs A)	0.644	0.461	46.3%	0.461	46.3%
	Original Ja-En (O^* vs A)	0.552	0.541	45.4%	0.541	45.4%
PMIAlign	Our Ja-En (Songs) (O vs A)	0.500	0.539	48.3%	0.457	52.2%
	Our Ja-En (Songs) (O^* vs A)	0.534	0.836	36.3%	0.752	38.3%
	Original Ja-En (O vs A)	0.613	0.458	47.6%	0.458	47.6%
	Original Ja-En (O^* vs A)	0.549	0.565	44.3%	0.565	44.3%

Across all tests, both the recall and strict recall metrics increased when comparing the baseline alignments (O vs A) to the improved alignments (O^* vs A). This indicates that the intervention strategy worked as intended.

However, for AwesomeAlign's original Ja–En dataset, precision decreased. Upon closer inspection, we found that this was due to the original dataset being less permissive in capturing many-to-many alignments compared to ours. As a result, certain alignments, while arguably valid, were counted as incorrect, which negatively impacted the precision score. Refer to Fig. 6 for an example of this phenomenon.

To assess how many many-to-many alignments were captured after applying this strategy, we now reapply the same script previously used to detect many-to-many alignments, this time using the updated results. We follow the same counting approach, using the same categories as before: "Directly Matched," "Not Directly Matched," "Reachably Matched," and "Unreachable." Refer to Table 4 for the results.

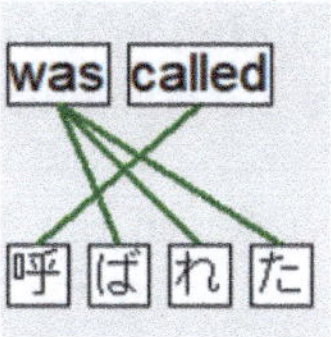

(a) Gold Alignment in AwesomeAlign's Original Dataset

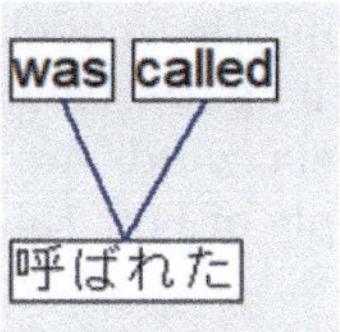

(b) Model Output (Phrase-Level)

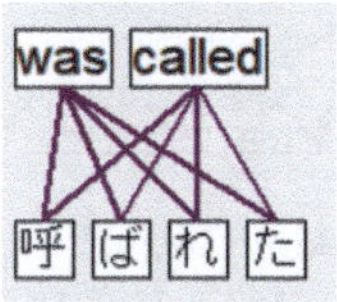

(c) Model Output (Reverted to Word-Level)

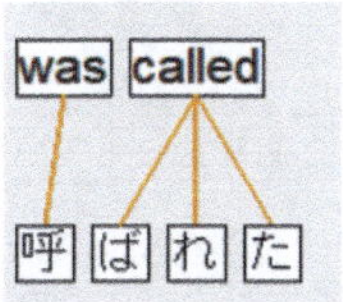

(d) Plausible Alignments Considered "Wrong" in the Gold Alignments

Fig. 6. Example from AwesomeAlign's Ja–En dataset showing how plausible many-to-many alignments are penalized due to their absence in the gold standard, leading to a drop in precision.

Table 4. Model performance on each dataset across matching criteria before and after applying the phrase tokenization strategy.

Model	Dataset	Directly Matched		Reachably Matched	
		Before	After	Before	After
Awesome-Align	Orig. Ja-En	323 (21.99%)	670 (45.61%)	328 (22.33%)	672 (45.75%)
	Our Ja-En (Songs)	1640 (25.61%)	4223 (65.94%)	1693 (26.44%)	4268 (66.65%)
SimAlign	Orig. Ja-En	177 (12.05%)	453 (30.84%)	179 (12.19%)	454 (30.91%)
	Our Ja-En (Songs)	1276 (19.93%)	3385 (52.86%)	1307 (20.41%)	3433 (53.61%)
PMIAlign	Orig. Ja-En	227 (15.45%)	530 (36.08%)	233 (15.86%)	532 (36.22%)
	Our Ja-En (Songs)	1618 (25.27%)	4126 (64.43%)	1778 (27.76%)	4264 (66.58%)

It is clear that the phrase tokenization strategy helped the model capture additional many-to-many alignments, leading to the increases in recall and strict recall observed earlier, for both the direct and indirect evaluation criteria.

4 Discussion

To better understand how this mitigation strategy benefits the various word alignment models, it is important to examine how these models extract alignments. In all the models evaluated above [10–12], alignments are initially extracted at the subword level, that is, subwords are aligned to other subwords. To obtain word-level alignments, two words are considered aligned if any of their subwords are aligned. The phrase tokenization strategy extends this principle further. By feeding the models phrases instead of individual words, the models now operate at the phrase level, where two phrases are considered aligned if any of their constituent words are aligned.

For an example of how this mechanism works, refer to Fig. 7.

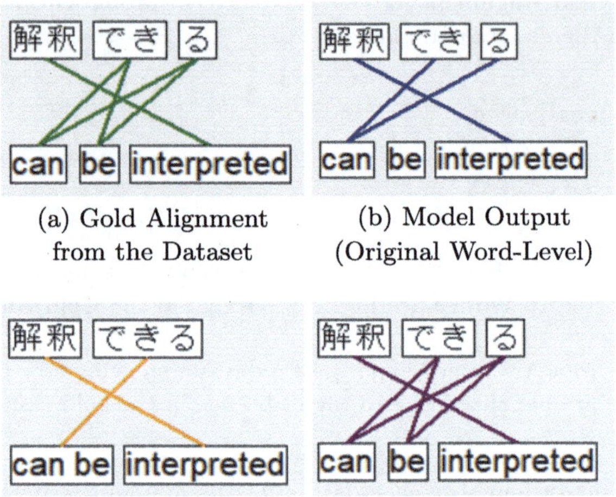

(a) Gold Alignment from the Dataset

(b) Model Output (Original Word-Level)

(c) Model Output (Phrase-Level)

(d) Model Output (Reverted to Word-Level)

Fig. 7. An example illustrating the effect of the phrase tokenization strategy. Prior to applying the strategy, the model's output lacks some alignments present in the gold standard. After applying the strategy and reverting back to word-level, the alignments match the gold alignment (note that panels (a) and (d) are equivalent).

However, this strategy is only effective if, for each many-to-many alignment in the gold standard, at least one word alignment pair within the phrase pair is correctly predicted by the model. To verify this, we implemented a script to tally such cases. The results are presented in Table 5.

As shown, the majority of many-to-many alignments in the gold dataset contain at least one word pair correctly predicted by the models. When the phrase tokenization strategy is applied, these partial matches result in fully connected word-level alignments. In this light, the strategy is able to enhance the overall alignment quality by enabling the models to capture alignments they would otherwise miss.

To demonstrate this phenomenon, let us look at specific examples where phrase-level tokenization proved effective. First, consider the phrase "come to think of it," which can be translated into Japanese as "そう いえ ば." In our dataset, AwesomeAlign was initially unable to capture the full many-to-many correspondence, aligning "そう" with "it" and "いえ" with "think" only. After applying phrase-level tokenization, however, the partial matches were leveraged to capture the complete many-to-many alignment, correctly aligning each

Table 5. Evidence of how phrase tokenization is able to leverage partial matches to form full dense matches to account for many-to-many alignments. Count of many-to-many gold alignment instances wherein at least one word pair was correctly predicted by the model.

Model	Dataset	At Least One	None	Total
AwesomeAlign	Original Ja-En	199 (92.99%)	15 (7.01%)	214
	Our Ja-En (Songs)	900 (94.34%)	54 (5.66%)	954
SimAlign	Original Ja-En	110 (51.40%)	104 (48.60%)	214
	Our Ja-En (Songs)	755 (79.14%)	199 (20.86%)	954
PMIAlign	Original Ja-En	169 (78.97%)	45 (21.03%)	214
	Our Ja-En (Songs)	880 (92.24%)	74 (7.76%)	954

English word with each Japanese word as intended. Similarly, the phrase "far and away" can be translated into Japanese as "遥か 遠く." Without phrase-level tokenization, AwesomeAlign aligned "遥か" with "far and" and "遠く" with "away," again missing the full correspondence. Once phrase-level tokenization was applied, these partial matches were combined to produce the correct full many-to-many alignment.

Despite these improvements, there are still cases where our method falls short, primarily due to the quality of our phrase tokenization. Our tokenization was done using a simple dictionary lookup combined with part-of-speech tagging, so certain meaningful phrases may still have been missed. For instance, take the phrase "one and only," which can be translated into Japanese as "かけがえ の ない" (literally meaning "irreplaceable"). Originally, AwesomeAlign only aligned "ない" with "only." Since our simple approach was unable to identify "かけがえ の ない" as a single meaningful phrase and instead left it as multiple word tokens, the final output of our method aligned "ない" to the three English tokens "one," "and," and "only," failing to capture the full many-to-many correspondence. This highlights that the quality of phrase tokenization is of utmost importance. In this light, future research could explore more complex approaches to phrase tokenization, such as leveraging deeper linguistic knowledge or the attention mechanisms of large language models.

5 Conclusion

In this study, we conducted an empirical analysis of various word alignment models and their performance on many-to-many alignments. We found that across all tested models—AwesomeAlign, SimAlign, and PMIAlign—they consistently struggled to capture these many-to-many alignments. As such, we show that a simple phrase tokenization strategy can assist the models in capturing these alignments. In our evaluation, we found that this approach led to consistent improvements in both recall and strict recall, as well as some reductions in

Alignment Error Rate (AER) and Strict AER (SAER). In particular, the strategy worked by transforming partial word-level matches into complete many-to-many alignments. Our analysis also confirmed that a majority of many-to-many gold alignments already contain at least one correct word pair predicted by the models. In this light, the phrase tokenization strategy uses these partial matches to help the models better capture many-to-many alignments.

In summary, this work shows that phrase tokenization can assist existing models in capturing many-to-many alignments. It also suggests that the development of more intelligent phrase tokenization schemes can help further improve the performance of word alignment models. Future directions may include exploring more linguistically-informed, sophisticated, or even attention-based phrase tokenization methods to further enhance word alignment.

References

1. Yarowsky, D., Ngai, G.: In: Second Meeting of the North American Chapter of the Association for Computational Linguistics (2001). https://aclanthology.org/N01-1026/
2. Huck, M., Dutka, D., Fraser, A.: In: Proceedings of the Sixth Workshop on NLP for Similar Languages, Varieties and Dialects, ed. by Zampieri, M., et al. (Association for Computational Linguistics, Ann Arbor, Michigan, 2019), pp. 223–233. https://doi.org/10.18653/v1/W19-1425. https://aclanthology.org/W19-1425/
3. Ammar, B.W., et al.: ArXiv **abs/1602.01925** (2016). https://api.semanticscholar.org/CorpusID:1227830
4. Shi, H., Zettlemoyer, L., Wang, S.I.: In: Proceedings of the 59th Annual Meeting of the Association for Computational Linguistics and the 11th International Joint Conference on Natural Language Processing (Volume 1: Long Papers), ed. by Zong, C., Xia, F., Li, W., Navigli, R., (Association for Computational Linguistics, Online, 2021), pp. 813–826. https://doi.org/10.18653/v1/2021.acl-long.67. https://aclanthology.org/2021.acl-long.67/
5. Wang, S., Tu, Z., Shi, S., Liu, Y.: In: Proceedings of the 58th Annual Meeting of the Association for Computational Linguistics, ed. by Jurafsky, D., Chai, J., Schluter, N., Tetreault, J., (Association for Computational Linguistics, Online, 2020), pp. 3070–3079. https://doi.org/10.18653/v1/2020.acl-main.278. https://aclanthology.org/2020.acl-main.278/
6. Neubig, G., et al.: In: Proceedings of the 2019 Conference of the North American Chapter of the Association for Computational Linguistics (Demonstrations), ed. by Ammar, W., Louis, A., Mostafazadeh, N., (Association for Computational Linguistics, Minneapolis, Minnesota, 2019), pp. 35–41. https://doi.org/10.18653/v1/N19-4007. https://aclanthology.org/N19-4007/
7. Cherry, C., Lin, D.: (2003). https://doi.org/10.3115/1075096.1075108
8. Och, F.J., Ney, H.: Comput. Linguist. **29**(1), 19 (2003) https://doi.org/10.1162/089120103321337421. https://aclanthology.org/J03-1002/
9. Vaswani, A., et al.: Attention is all you need (2023). https://arxiv.org/abs/1706.03762
10. Dou, Z.Y., Neubig, G.: Word alignment by fine-tuning embeddings on parallel corpora (2021). https://arxiv.org/abs/2101.08231

11. Sabet, M.J., Dufter, P., Yvon, F., Schütze, H.: In: Findings of the Association for Computational Linguistics: EMNLP 2020, ed. by Cohn, T., He, Y., Liu, Y., (Association for Computational Linguistics, Online, 2020), pp. 1627–1643. https://doi.org/10.18653/v1/2020.findings-emnlp.147. https://aclanthology.org/2020.findings-emnlp.147/
12. Azadi, F., Faili, H., Dousti, M.J.: In: Findings of the Association for Computational Linguistics: ACL 2023, ed. by Rogers, A., Boyd-Graber, J., Okazaki, N., (Association for Computational Linguistics, Toronto, Canada, 2023), pp. 12,366–12,377. https://doi.org/10.18653/v1/2023.findings-acl.782. https://aclanthology.org/2023.findings-acl.782/
13. Mihalcea, R., Pedersen, T.: In: Proceedings of the HLT-NAACL 2003 Workshop on Building and Using Parallel Texts: Data Driven Machine Translation and Beyond, vol. 3, pp. 1–10. (Association for Computational Linguistics, USA, 2003), HLT-NAACL-PARALLEL '03. https://doi.org/10.3115/1118905.1118906
14. Liu, Y., Sun, M.: In: Proceedings of the AAAI Conference on Artificial Intelligence, vol. 29, no. 1 (2015). https://doi.org/10.1609/aaai.v29i1.9508. https://ojs.aaai.org/index.php/AAAI/article/view/9508
15. Neubig, G.: The Kyoto free translation task (2011). http://www.phontron.com/kftt
16. Steingrímsson, S., Loftsson, H., Way, A.: In: Proceedings of the 23rd Nordic Conference on Computational Linguistics (Linköping University Electronic Press, Online, 2021)

Evaluating the Performance of Large Language Models for Filipino and Cebuano Machine Translation

Charlese Jeanrie Te(✉), Shaira Lorraine Montojo, Al Gabriel Orig, and Kristine Mae Adlaon

College of Computer Studies, University of the Immaculate Conception, Fr. Selga Street, 8000 Davao City, Philippines
{cte_220000001593,smontojo_200000000067,aorig_200000000296, kadlaon}@uic.edu.ph

Abstract. Large language models (LLMs) have moved its direction in developing machine translation tasks and showed promising results, however issues such as disparity in data and limits in evaluation metrics still persisted especially for low-resource languages. To support this case, two experiments were conducted in translating to Tagalog and Cebuano, and the translations will be evaluated through human judgment under syntactic and semantic analysis. While LLMs were able to perform decently, challenges were found such as incorrect word choices and sentence structures such as use of tense or aspect. These findings were able to highlight further improvements of understanding Philippines' unique and flexible linguistic structures.

Keywords: Machine Translation (MT) · Large Language Model (LLM) · Philippine Language (PHL)

1 Introduction

Machine translation (MT), one of the natural language generation tasks, is where a machine translates from one language to another. It has undergone many developments and changes to bridge the language gap, and there is an increased number of large language models (LLMs) being developed to do MT tasks.

LLMs are self-supervised models and trained with a vast amount of text, and are able to understand and generate natural language. With this advantage, they are able to perform natural language processing tasks, including MT. With advancements, LLMs such as ChatGPT and GPT-4 are continuously evolving their multilingual translation abilities, and GPT-4 has reached the current state of the art performance that exceeds the baseline of NLLB in 40.91% [1]. A study reported that advances in model architectures and large-scale training data have significantly improved machine translation accuracy, showing particular promise for low-resource languages [2].

H. Unger and P. Meesad (Eds.): NLPIR 2025, LNNS 1904, pp. 34–46, 2026.
https://doi.org/10.1007/978-3-032-20897-2_3

Despite promising results, there still remain some limitations particularly for low-resource languages [3]. LLMs were trained with English-centric datasets, and their performances relating to English tend to be higher. However, non-English languages, especially those that are less represented, performed less [1,4]. This may be due to the proportion of training data and the severe lack of high-quality corpora. MT metrics remained limited in their evaluation. Low-resourced languages such as Manipuri and Bengali were not supported in neural-based metrics and remained evaluated with lexical-based metrics, which lacked capturing critical aspects of the translation [5]. BLEURT, a combination of BLEU and BERT, was shown to give a lower score even with additional context and grammar explanations added, and gave a higher score to translations where human evaluators found multiple lexical errors in those assumed high-score translations [6].

The challenges mentioned can still be applied with low-resource languages especially in Philippine languages. A human-based judgment of these LLMs in translating into two major Philippine languages, Tagalog and Cebuano [7,8], was decided not only to see if these issues persisted but also to catch what needs to be improved in terms of syntactic structure and semantic preservation.

With this contribution, we hope to provide insight in developing effective approaches in evaluating MT output. Both Tagalog and Cebuano have flexible yet complicated linguistic structures, such as rich morphology and varied word order [9,10], hence creating difficulty in existing MT systems. By focusing on these two languages, it should be able to highlight the uniqueness of nuances in translating both languages that could be further improved. The methods used in this study hope to serve as a foundation in examining not only other major Philippine languages, but also those that are underrepresented in other countries. This potentially leads to language preservation through digitization and greater access to information locked by language gaps.

2 Related Work

There were methods to evaluate the performance of LLM in MT. A study constructed Euas-20, a dataset containing 20 representative languages built with sentences generated from ChatGPT and Google Translate, to test the performance of LLMs in translating using BLEU and COMET as their metrics, and found that the translation ability of different languages is unbalanced despite showing improvements [4]. Benchmarks such as FLORES-101 were also used to test LLMs like ChatGPT and GPT-4 to supervised baselines such as Google Translate and NLLB such as in [1]. Using SentencePiece BLEU, COMET, and SEScore, it was discovered that LLMs were still behind in translating to low-resource languages. A comparison between human and LLM was conducted with building LitEval-Corpus, and was found that human translations still outperformed LLMs as the output of the LLMs do not match the literacy style and diversity of the human translation [11]. This also stated that the LLMs tend to translate literally regardless of stylist nuance and flexibility. Another study focused on reference-less evaluation of LLMs for low-resource languages through zero-shot prompting, in-context learning, and the proposed annotation guidelines prompt, which one of the findings it encountered were issues in handling linguistic structures such as context and syntax [12]. Tagalog was also tackled in the Batayan benchmark [13]. This study included MT in evaluating LLMs using FLORES-200 as their source data, and chrF++ and MetricX-24 as their evaluation metrics. Their results found that Filipino-explicit LLMs like Qwen-2-7B-Instruct and SeaLLMs-3-7B-Chat outperformed general multilingual models. There is no research involving Cebuano as far as we are aware.

Although the results showed the limits of LLMs, there were also challenges found, such as synthetic datasets either limited to one neural dataset or generated by LLMs, a possibility to inaccuracies and biases, and proprietary LLMs used such as ChatGPT, which hinders further understanding of the performance. The evaluation metrics can be also limiting. Lexical-based metrics focus on word or character similarities, and such metrics like BLEU and chrF++ lack in capturing semantic aspects. COMET, SEScore, and MetricX-24 are metrics that correlate with human judgment more. However, COMET was found with data bias and language mismatch, while SEScore is trained with generated errors rather than human-annotated errors. While MetricX-24 is one of the state-of-the-art metrics, its limitation across all languages still persists. For instance, Tagalog is the only supported language at the time of this study conducted.

3 Methodology

This section describes the data corpora, models, and procedures used in two experiments. One experiment was to translate from English (en) to 2 major Philippine languages (phl), which are Tagalog (tgl) and Cebuano (ceb), and the other was conducted to translate from one phl to another phl. Tagalog has the most speakers in the Philippines, while Cebuano is the second language with most speakers. Despite this, they are considered low-resource languages despite their total number of speakers in the country. These two languages were chosen as these are the only two Philippine languages we understand.

In conducting en→phl translation, we collected sentences through sentence structure (simple, compound, complex, and complex-compound) and sentence function (declarative, interrogative, imperative, and exclamatory). The dataset used was OpenSubtitles [14] found in OPUS Project [15]; a collection of translated movie subtitles, totaling 1M sentences for EN. Each sentence structure and function combined has a total of 125 EN sentences, where each sentence must have the number of tokens to be greater or equal to 5 and less than or equal to 50. Each language pair summed up to 2K sentences, making a total of 4K test instances in this experiment. The token count is shown in Table 1.

For phl→phl translation, sentences are grouped into 3 categories, declarative, interrogative, and exclamatory. The texts collected for this experiment were from PALITO corpus [16]; a collection of literacy texts. Following the same criteria as those for the en→phl experiment, each category consisted of 25 sentences for Tagalog and Cebuano, totaling 150 test instances. This was conducted to see

Table 1. Token count for en→phl

Category	Type	Total	Unique	Max	Min	Average
Simple	Declarative	1105	456	15	6	9
	Interrogative	1125	404	17	6	9
	Imperative	1018	412	16	5	8
	Exclamatory	1011	410	16	8	8
Compound	Declarative	1934	635	30	8	15
	Interrogative	1901	587	31	9	15
	Imperative	1540	551	27	6	12
	Exclamatory	1692	578	22	7	14
Complex	Declarative	1890	595	30	8	15
	Interrogative	1986	611	28	8	16
	Imperative	1417	503	27	6	11
	Exclamatory	1763	588	29	6	14
Complex Compound	Declarative	3382	962	46	16	27
	Interrogative	3395	901	46	13	27
	Imperative	2510	736	42	11	20
	Exclamatory	2955	903	43	11	24

Table 2. Token count for phl→phl

Category		Total	Unique	Max	Min	Average
Cebuano (ceb)	Declarative	351	205	26	6	14
	Interrogative	257	162	18	7	10
	Exclamatory	251	160	26	6	13
Tagalog (tgl)	Declarative	306	176	21	6	12
	Interrogative	271	156	17	6	11
	Exclamatory	285	155	26	6	11

the performance in translating from other phl other than en. The token count is shown in Table 2.

The following models used in translating were Gemma 3 [17], Qwen 3 [18], and Sailor 2 [19]. Gemma 3 is a multimodal lightweight model, while Qwen 3 utilizes hyper thinking modes. Sailor 2 is a fine tuned model from Qwen 2.5 that focuses on SEA languages such as Indonesian, Thai, and also Philippine languages like Tagalog, Cebuano, and Waray. These were chosen because they were able to perform MT tasks, were open source, and supported Tagalog and Cebuano. Details such as parameters used and context window are shown in Table 3.

Each model was tasked with translating 5 sentences at a time given by the prompt (see Fig. 1), where [SOURCE_LANGUAGE] is the language of the source

Table 3. LLMs used for both experiments

Model	Parameters Used	Developer	Context Window
Gemma 3	12B	Google	128K tokens
Qwen 3	14B	Alibaba	128K tokens
Sailor 2	20B	Sea AI Lab	32K tokens

text, and [TARGET_LANGUAGE] is the target language to be translated. The prompt provided no translation examples (zero-shot prompting) to really assess the performance of the LLMs without relying on given examples. This also tested how well the LLMs solely applied its knowledge in multilingual languages, especially low-resource languages. The number of sentences was fixed to 5 only as not only it is manageable in evaluating translations, but also serves as a middle ground of reducing the risk of losing coherence in translating one at a time and the risk of confusing the model with treating all sentences as one document instead of individually. All models were pulled from Ollama, and both experiments were performed on a Precision 3660 with a NVIDIA RTX A5000. The translations were then compiled and evaluated based on human judgment using two points: the syntax of the translated text and the extent to which semantics were preserved from the source sentence. The performance of each model was evaluated based on how many errors are detected on each error, and how much variety these errors are. The lesser the errors detected, the higher the performance of the model is.

```
You are a translator for [SOURCE_LANGUAGE] to [TARGET_LANGUAGE]. Please
provide the translation for the following sentences. Don't provide additional text or
explanation.

Sentences:

1. [source_text]
2. [source_text]
3. [source_text]
4. [source_text]
5. [source_text]

SRC: [source_text]
TGT:
```

Fig. 1. Prompt used in translating to Tagalog and Cebuano.

4 Results and Discussions

The results of each LLMs' performance in translation varied in conducting both experiments. In conducting en→phl, the LLMs promisingly translated the sentence with decent results. This may be due to the LLMs being trained mostly in en, where it is able to grasp the meaning in en sentences. Shown in Table 4 is a sample comparison from SIM_DCL (simple declarative), COM_INT (compound interrogative), COX_IMP (complex imperative), and CXC_EXC (complex compound exclamatory), gathered with their ceb and tgl translation from each LLM. Starting from the simple declarative sentence, the ceb translations were unable to capture the meaning by using a different verb. *"Atergyako"* by Gemma 3 is a non-existent word in ceb as it is nowhere to be found in the ceb dictionary. This may be due to the limited data the LLM is trained on. While *"Nagtumod"* means "to look forward" and *"Nagpanginahanglan"* means "need", there is a more appropriate verb to use like *"Naningkamot"* means "trying". The tgl translations provided more accurate translation in terms of semantic aspect, however, its word usage can be questionable. While *"Nagsisikap"* and *"Sinisikap"* can also be used, its better definition is similar to "to work hard", making it a bit further from the meaning of the original sentence. *"Sinusubukan"* is the most appropriate verb to use, which means "trying", and used the correct aspect or tense. For the compound interrogative sentence, there were incorrect word usage for the ceb translations, such as *"panamilit"* (farewell), and *"Nagsul-ob"* is a term used for clothing. Despite Qwen 3's translation to tgl conveyed well, the segment *"ano ang paraan mo para magbawi sa kanya?"* (what is your way of paying her back?) asks what is the way the person the speaker to do, rather than asking the way itself (this is how you pay her back?). The translations for complex imperative were surprisingly better. Qwen 3 and Sailor 2 used *"Kon"* which is the CEB word for "if", while Gemma 3 used the tgl word for "if" (*Kung*). There is also a misspelling on the ceb translation of Sailor 2, "kuk" where it is supposed to spell *"kuko"* (nail). The translations for the chosen complex imperative sentence were literal. The segment "keep your claws off my man" means "stay away from my man", however the ceb and tgl translations by Sailor 2 and the tgl translation by Qwen 3 translated it in its literal sense. Gemma 3 managed to interpret it and retained its meaning. The translations for complex compound sentences encountered incorrect word usage especially for verbs, such as using *"Gibantayan"* (watched over) for the original verb "outed", and *"pag-ubos"* (drained) and *"gilaglag"* (dropped) for "ruined".

Similar results can be found in conducting phl→phl, however the translations felt limited compared to en→phl. Because both tgl and ceb are low-resource languages, the training data for tgl→ceb and vice versa may be limiting compared to en→phl. Shown in Table 5 is a sample comparison of ceb→tgl from each category. The sentences may not be fully translated in some cases. A tgl translation by Qwen 3 *"Maraming nagsabi na epektibo kuno."* (Many said it's supposedly effective.) still has *"kuno"* (supposedly) intact from the original CEB sentence, while Gemma 3 used *"umano"* (allegedly). While the Sailor 2 didn't use *"umano"*, it translated the sentence to *"Marami ang nagsabi na mukhang*

Table 4. Sample comparison of LLM translations of en→{ceb,tgl}

Source Sentence	Gemma 3	Qwen 3	Sailor 2
[SIM_DCL] I'm trying to watch this show.	[CEB] **Atergyako** makatan-aw ani nga programa. [TGL] Sinusubukan kong panoorin ang palabas na ito.	[CEB] **Nagpangina-hanglan** ko nga tan-awon kining show. [TGL] **Nagsisikap** akong tingnan ito palabas.	[CEB] **Nagtumod** ko sa pagtan-aw niining programa. [TGL] **Sinisikap** kong panoorin ito palabas.
[COM_INT] She made us dinner and this is how you pay her back?	[CEB] Siya nagluto kanato ug **panamilit** ug kini pa ang imong paanad niya? [TGL] Nagluto siya para sa atin at ito ang paraan mo para bayaran siya?	[CEB] Gin-prepare niya ang pagkaon alang kanato, pero kining paagi ang imong pagpahibaro kaniya? [TGL] Ginawa niya ang tamang kainan para sa amin, **ano ang paraan mo** para magbawi sa kanya?	[CEB] **Nagsul-ob** siya ug panihapon alang kanato ug kini ba ang imong bayad kaniya balik? [TGL] Gumawa siya ng hapunan para sa atin tapos ganito ang **balik-mo** sa kanya?
[COX_IMP] If you don't want me to tell your husband about this, keep your claws off my man.	[CEB] Kung dili nimo gusto nga isulti ko sa imong bana bahin niini, ayaw paghawid sa akong lalaki. [TGL] Kung ayaw mong sabihin ko sa asawa mo ito, huwag mong kalikutin ang lalaki ko.	[CEB] Kon ayaw nimo nga pasultihan ang imong asawa niini, ayaw nang pahimulos **ha** akong tawo. [TGL] Kung ayaw mo akong sabihin ang iyong asawa tungkol dito, huwag mong gamitin ang iyong mga kuko sa aking lalaki.	[CEB] Kon dili nimo gusto isulti ko sa imong bana kini, ayaw **pagdakop** sa akong lalaki gamit ang imong mga **kuk**. [TGL] Kung ayaw mong sabihin ko sa asawa mo tungkol dito, huwag kang lumapit kay lalaking akin.
[CXC_EXC] Vanessa outed me when I was 17 and nearly ruined my life, and now she's trying to ruin it again!	[CEB] **Gibantayan** ko ni Vanessa sa edad ko 17 ug halos gubaon ang akong kinabuhi, ug karon nagbuhat siya aron gubaon kini pag-usab! [TGL] Inilantad ako ni Vanessa noong ako'y 17 taong gulang at halos winasak ang buhay ko, at ngayon sinusubukan na naman niyang sirain ito!	[CEB] Gipakita ni Vanessa ang akong sekreto dihang ako 17, ug halos nagsugod na **pag-ubos** sa akong kinabuhi, ug karon siya nagtuyo **pag-ubos** kini pag-usab! [TGL] Nagmukhang ako si Vanessa nang ako ay 17 at halos **nabigo** ang aking buhay, at ngayon siya ay nagsisikap ulit na **mabigo** ang aking buhay!	[CEB] Gipadayag ni Vanessa ang akong sekreto sa edad nga 17 ug halos **gilaglag** ang akong kinabuhi, karon nangita siya pag-usab aron **laglagon** kini! [TGL] **Inamin** ni Vanessa sa akin noong 17 ako at halos sinira ang buhay ko, at ngayon sinusubukan na naman niyang sirain ulit!

epektibo ito." (Many said it looks effective.), which is still able to retain the semantic meaning. In other scenarios, the word usage is more severe compared to the first experiment. The interrogative sentence *"Gilaliman kang gipasayaw-sayaw niya si Karyas og inunggoy aron maayo kuno?"* (He made Karyas dance and acted like a monkey to make him better?) implied that the third person is trying to make Karyas better, but the translations from Qwen 3 and Sailor 2 became *"maganda"* (beautiful), which became a different meaning from the original sentence. Sailor 2 added *"Bakit naman"* (Why even), indicating that the speaker is finding a reason when it's not the case in the original. Only a meaning switch is found in the sampled exclamatory sentence. The Qwen 3 translation *"Hindi kailanman kami si Petra nakakabuo ng anak!"* means that the speaker and Petra can't make a child, which can have relevance from the original meaning of *"Kami si Petra wa gyod tugahi og anak sa Ginoo!"*, which means they are not given a child from God, but has shifted further from it.

Table 5. Sample comparison of LLM translations of ceb→tgl

Source Sentence	Gemma 3	Qwen 3	Sailor 2
[DCL] Daghan may nag-ingong epektibo kuno.	Maraming nagsasabing epektibo umano.	Maraming nagsabi na epektibo kuno.	Marami ang nagsabi na mukhang epektibo ito.
[INT] Gilaliman kang gipasayaw-sayaw siya si Karyas og inunggoy aron maayo kuno?	Pinapagalaw niya si Karyas at inaangat para gumaling daw?	Nakasakop ba ang pagkakasayaw niya kay Karyas at inunggoy para **mas maging maganda**?	**Bakit naman** pinagsasayaw-sayaw niya si Karyas ng unggoy para **daw maganda**?
[EXC] Kami si Petra wa gyod tugahi og anak sa Ginoo!	Kami ni Petra, hindi talaga kami binigyan ng anak ng Diyos!	Hindi kailanman kami si Petra **nakakabuo** ng anak!	Si Petra at kami ay hindi naman binigyan ng anak ng Diyos!

Shown in Table 6 is a sample comparison for tgl→ceb, which also have similar results. Noticeably, Qwen 3 uses more of *"ha"* in their ceb translations. The word marker is commonly used in Waray, similarly to "to" and "at". However, it is not used in ceb. This could also be a misspelling of *"sa"*, a standard prepositional marker for both tgl and ceb. The word *"han"* can also be seen in Qwen 3 translations to ceb but the issue still persists. *"Ayaw kabaliona"* used by Gemma 3 can be also used to say "don't worry", yet a more common term is used in Sailor 2 which is *"Ayaw kabalaka"*. One thing for the Sailor 2 translation is replacing *"Bunso"* (youngest) with *"Bata"* (kid), which has similar meaning but in the structure of the original sentence, *"Bunso"* is a proper noun due it is the name the subject is being called to. The translations for the interrogative sentences are accurate, what is noticeable is the use of the en word "caregiver" by Sailor

2, Qwen 3 used *"tagapag-alaga"*, sticking with the original, and Gemma 3 using the word in ceb *"tag-atiman"*. It showed that in translating to tgl or ceb, either the en or tgl vocabulary will be used but the fluency lessens. In the exclamatory sentence, the word *"lupa"* means land, so Gemma 3 and Sailor 2 used *"yuta"*, which is similar to *"lupa"*, meaning Qwen 3 used *"kalibotan"*, which means "the world". The segment *"maayo-ang-kasingkasing"* is a syntactic error where the dash is supposed to be at only *"kasing-kasing"*, which can be seen in the translation of Gemma 3. We also noticed how some translations copied the literal structure of the original. The structure of the sentence can be changed as long as both the syntactic structure and semantic meaning are preserved. The translation *"Kung ayaw mong sabihin ko sa asawa mo ito, huwag mong kalikutin ang lalaki ko."* (If you don't want me to tell this to your husband, don't pester my man.) is acceptable, however the translation *"Huwag mong kalikutin ang lalaki ko kung ayaw mong sabihin ko sa asawa mo ito"* (Don't pester my man if you don't want me to tell your husband this.) is also acceptable. This shows that the structure of Philippine languages is flexible and the LLMs are still limited to the original structure of the sentence, which some may come off as unnatural.

Table 6. Sample comparison of LLM translations of tgl→ceb

Source Sentence	Gemma 3	Qwen 3	Sailor 2
[DCL] Hayaan mo, Bunso, babawiin natin ang ating bahay.	Ayaw **kabaliona**, Bunso, mabawian ta ang atong pinuy-anan.	Pahibaro **ha** imo, Bunso, babawi **han** ato nga balay.	Ayaw kabalaka, **Bata**, bawi-on natong among balay.
[INT] Kailangan na ba niya talaga ng tagapag-alaga?	Kinahanglan na ba gyud niya'g tag-atiman?	Kinahanglan na ba niya nga tagapag-alaga?	Kinahanglan na gayud ba siyag caregiver?
[EXC]Kapayapaan sa lupa sa mga taong may mabuting kalooban!	Kalinaw sa yuta sa mga tawo nga dunay maayong kasing-kasing!	Kapayapaan **ha** kalibotan alang **ha** mga tawo nga may maayong kalooban!	Kalinaw diha sa yuta ngadto sa **maayo-ang-kasingkasing** nga mga tawo!

In translating to tgl, Gemma 3 produced translations that were generally consistent across different contexts. Its outputs often featured authentic tgl expressions and natural sentence structures, contributing to a more fluent and idiomatic tone. However, it occasionally relied on loanwords in instances where native tgl terms would have been more appropriate. Sailor 2 also generated accurate translations despite its sentence constructions tended to be overly formal and somewhat rigid. This formality sometimes resulted in a tone that felt mismatched with casual contexts, and the model occasionally exhibited wordiness. Qwen 3 showed frequent limitations in translations, along with grammatical inconsistencies and unnatural phrasing. These issues contributed to variability in its output quality and reduced overall coherence. In translating to ceb, Sailor 2 consistently conveyed the original meaning and constructed grammatically sound sentences that aligned well with Cebuano norms. Gemma 3 also produced accurate translations, though it sometimes introduced unnecessary words and leaned toward a formal tone, which could feel out of place in informal contexts. Qwen 3 occasionally generated awkward sentences and incorporated lexical items from Waray, which affected the clarity and appropriateness of its translations. It also exhibited recurring misinterpretations. Shown in Table 7 are common patterns observed in both experiments, indicating the type of error along with a sample sentence pair that highlights the affected segment found in both experiments. A possible approach to combat this is to be able to check not only the structure of the translation but also how the original meaning is preserved. In this way, both syntactic and semantic are taken into consideration and will help evaluate and improve MT models. Gemma 3 was concluded to be the best performing model as it was not only able to capture the semantic aspect with appropriate syntactic structure, but was able to do translation naturally.

Table 7. Common errors found in both experiments

Error	Example
Incorrect word usage	[EN] Please leave us for a moment. [CEB] Palihug, **buwelta** ta og pipila ka minutos. [TGL] Sino ang tigre sa pader? [CEB] Kinsa ang **pusa** sa dingding?
Incorrect use of aspect/tense	[EN] He won't bite instantly, but it would've been enough to provoke him. [TGL] Hindi siya **kagat** agad, pero sapat na iyon para magalit siya. [CEB] Kinsa ang prinsipe magpuyo sa palasyo? [TGL] Sino ang prinsipe na **naninirahan** sa palasyo?
Incorrect or missing use of parts of speech	[EN] Call me when you find her! [TGL] Tumawag ***ka*** sa akin kung natagpuan mo siya! [TGL] Bakit wala ka ba nito? [CEB] Bakit wala ka ba **niana**?
Incorrect spelling	[EN] I'm telling you, I am burning this house down with everything I know! [CEB] Sulti ko **nmo**, sunogun nako niini nga balay, tanan nga makasulti ko! [TGL] Sino nga ba ang tunay na nagmamahal sa kanyang kapwa? [CEB] Sino **an** tunay nga nagmamahal **ha** iyang kapwa?
Uncaptured meaning	[EN] Are you getting a bit stir-crazy? [CEB] **Nag-ubos na ka ha pag-ubos?** [CEB] Iyang gilayog ang akong magulang! [TGL] **Iniyakan pa niya ang aking nakatatandang kapatid!**

5 Conclusion

In this study, the LLMs were evaluated for their translations into Philippine languages. Two experiments were conducted, where one is translating from English to Tagalog and Cebuano, and Tagalog to Cebuano and vice versa. LLMs tend to translate to Tagalog better than in Cebuano, and translating from English is better compared to either Tagalog or Cebuano, showing the disparity between high and low-resource languages. Results showed that while it was able to translate appropriately, errors still occurred, such as incorrect word usage, incorrect use of tense or aspect, and incorrect or missing use of parts of speech. These types of error influenced the syntactic structure and the semantic preservation of the original sentence. Other errors such as out of context, and incorrect spelling also occurred, not as frequently occurring but were essential especially in translating in general. Those syntactic and semantic limitations identified LLMs' weak spots in taking rich morphologies and context into consideration. Both Tagalog and Cebuano have flexible structures such as free word order and cultural nuances. The grammar supports both S-V-O and V-O-S, it became clearer that the LLMs followed the source's word order, which can lead to unnatural fluency in many instances. Context is also important in translating to phl. Using the example *"Bunso"*, the term means "youngest of the family" however *"Bunso"* still retains the meaning and can be used as a nickname. Translating to *"Bata"* not only lost that specificity of the term, but also lost the nuance of the original, as *"Bunso"* is treated as a proper name.

For future work, a wider range will be included to evaluate LLMs in other Philippine languages other than Tagalog and Cebuano, such as Hiligaynon, Ilocano, and Waray. The dataset is expected to increase in number as only one dataset per experiment is used. With this, testing with different domains will give more comparison between LLMs. A further investigation is needed with an increase of sentences with a variety of words to capture the vocabulary and comprehension of each LLM. Especially for tgl→ceb and vice versa, categories per sentence structure and sentence function will be needed to be able to identify and understand where these errors start from.

Acknowledgment. We would like to express our gratitude to the Department of Science and Technology - Philippine Council for Industry, Energy, and Emerging Technology Research and Development (DOST-PCIEERD) through the Mindanao Natural Language Processing Research and Development Laboratory (MinNa LProc) for the equipment used during the experiments.

References

1. Zhu, W., et al.: Multilingual machine translation with large language models: empirical results and analysis. In: Findings of the Association for Computational Linguistics: NAACL 2024, pp. 2765–2781, Mexico City, Mexico. Association for Computational Linguistics. (2024). https://doi.org/10.48550/arXiv.2304.04675
2. Wang, Y., et al.: Recent advances in interactive machine translation with large language models. In: IEEE Access, vol. 12, pp. 179353–179382 (2024). https://doi.org/10.1109/ACCESS.2024.3487352
3. Adlaon, K.M.M., Marcos, N.: Building the language resource for a Cebuano-Filipino neural machine translation system. Association for Computing Machinery (2019). https://doi.org/10.1145/3342827.3342833

4. Huang, Y., Liu, W.: Evaluating the translation performance of large language models based on Euas-20 (2024). https://doi.org/10.48550/arXiv.2408.03119
5. Mukherjee, A., Yadav, S., Shrivastava, M.: Why should only high-resource-languages have all the fun? Pivot based evaluation in low resource setting. In: Rambow, O., et al. (eds.) Proceedings of the 31st International Conference on Computational Linguistics, pp. 4779–4788. Association for Computational Linguistics, Abu Dhabi, UAE (2025). https://aclanthology.org/2025.coling-main.320/
6. Court, S., Elsner, M.: Shortcomings of LLMs for low-resource translation: retrieval and understanding are both the problem. In: Proceedings of the Ninth Conference on Machine Translation, pp. 1332–1354, Miami, Florida, USA. Association for Computational Linguistics (2024). https://doi.org/10.48550/arXiv.2406.15625
7. Baliber, R.I., Cheng, C., Mamonong, V.,Adlaon, K.M.M.: Bridging Philippine languages with multilingual neural machine translation. In: Proceedings of the 3rd Workshop on Technologies for MT of Low Resource Languages. Association for Computational Linguistics (2020). https://doi.org/10.18653/v1/2020.loresmt-1.2
8. Fernandez, J.L., Adlaon, K.M.M.: Exploring word alignment towards an efficient sentence aligner for Filipino and Cebuano languages. In: Proceedings of the Fifth Workshop on Technologies for Machine Translation of Low-Resource Languages (2022). https://aclanthology.org/2022.loresmt-1.13
9. Adlaon, K.M.M., Marcos, N.: Neural machine translation for Cebuano to tagalog with subword unit translation. In: International Conference on Asian Language Processing (IALP) (2018). https://doi.org/10.1109/IALP.2018.8629153
10. Adlaon, K.M.M., Marcos, N.: Finding the optimal byte-pair encoding merge operations for neural machine translation in a low-resource setting. In: Findings of the Association for Computational Linguistics (2024). https://doi.org/10.18653/v1/2024.findings-emnlp.860
11. Zhang, R., Zhao, W., Eger, S.: How good are LLMs for literary translation, really? Literary translation evaluation with humans and LLMs. In: Proceedings of the 2025 Conference of the Nations of the Americas Chapter of the Association for Computational Linguistics: Human Language Technologies (Volume 1: Long Papers), pp. 10961–10988, Albuquerque, New Mexico. Association for Computational Linguistics (2025). https://doi.org/10.48550/arXiv.2410.18697
12. Sindhujan, A., Kanojia, D., Orasan, C., Qian, S.: When LLMs struggle: reference-less translation evaluation for low-resource languages. In: Proceedings of the First Workshop on Language Models for Low-Resource Languages, pp. 437–459, Abu Dhabi, United Arab Emirates. Association for Computational Linguistics (2025). https://doi.org/10.48550/arXiv.2501.04473
13. Montalan, J.R., et al.: Batayan: a Filipino NLP benchmark for evaluating large language models. In: Proceedings of the Annual Meeting of the Association for Computational Linguistics (ACL 2025). Association for Computational Linguistics (2025). https://doi.org/10.18653/v1/2025.acl-long.1509
14. Lison, P., Tiedemann, J.: OpenSubtitles2016: extracting large parallel corpora from movie and TV subtitles. In: Proceedings of the 12th International Conference on Language Resources and Evaluation (LREC 2016), pp. 923–929, Portorož, Slovenia. (2016). https://aclanthology.org/L16-1147
15. Tiedemann, J.: Parallel data, tools and interfaces in OPUS. In: Proceedings of the 8th International Conference on Language Resources and Evaluation (LREC'12), pp. 2214–2218, Istanbul, Turkey. European Language Resources Association (ELRA) (2012). https://aclanthology.org/L12-1246

16. Dita, S., Roxas, R. E., Inventado, P.: Building online corpora of Philippine Languages. In: Proceedings of the 23rd Pacific Asia Conference on Language, Information and Computation, pp. 646–653, Hong Kong. (2009). https://aclanthology.org/Y09-2024
17. Team, G., et al.: Gemma 3 Technical Report. arXiv2503.19786 (2025). https://doi.org/10.48550/arXiv.2503.19786
18. Yang, A., et al.: Qwen3 Technical Report. ArXiv abs/2505.09388 (2025). https://doi.org/10.48550/arXiv.2505.09388
19. Dou, L., et al.: Sailor2: sailing in South-East Asia with inclusive multilingual LLMs. ArXiv abs/2502.12982 (2025). https://doi.org/10.48550/arXiv.2502.12982

LACES: Lexical and Contextual Awareness for EDU Segmentation

Mohammadreza Sediqin[1(✉)] and Shlomo Engelson Argamon[2]

[1] Illinois Institute of Technology, Chicago, USA
msediqin@hawk.illinoistech.edu
[2] Touro University, New York, USA
sargamon@touro.edu

Abstract. Discourse parsing is essential for understanding the structure and coherence of text, forming the foundation for a wide range of downstream NLP tasks. Segmenting text into *Elementary Discourse Units (EDUs)* is a fundamental task in discourse parsing. While previous work, including pre-trained language models, has significantly improved RST discourse parsing, challenges remain due to reliance on lexical and syntactic cues, inconsistencies between EDU-level parsing and sentence-level modeling, and a strong dependence on gold-standard segmentation for training. In this paper, we introduce *Lexical and Contextual Awareness for EDU Segmentation (LACES)*, an effective method for automatic EDU segmentation that leverages lexical information within a defined token window to identify EDU boundaries. Evaluations on benchmark datasets demonstrate that LACES not only outperforms strong baselines in EDU segmentation but also enhances overall parsing performance when incorporated into state-of-the-art discourse parsers. This emphasizes the importance of such features in accurately identifying EDUs and highlights their potential to support more efficient training strategies in discourse analysis.

Keywords: EDU segmentation · RST parsing · Discourse parsing

1 Introduction

A fundamental task in natural language understanding is analyzing the overall structure of a text, so that logical and coherence relations between text units are revealed. *Rhetorical Structure Theory (RST)* [1] is a well-acccepted theoretical framework for the task within the NLP community [2]. RST structures a text as a tree, where the basic building blocks (leaf node) are called *Elementary Discourse Units (EDUs)*. Discourse parsing in RST is the task of automatically constructing this hierarchical tree by identifying the EDUs and then building a parse tree by connecting adjacent EDUs and composite discourse units. Relations between adjacent units are labeled with different rhetorical relations, which are mostly assymetrical, with one unit designated the nucleus and the other as

H. Unger and P. Meesad (Eds.): NLPIR 2025, LNNS 1904, pp. 47–60, 2026.
https://doi.org/10.1007/978-3-032-20897-2_4

the subordinate satellite. Parsing is generally done using a shift-reduce parser, which builds the tree incrementally by scoring transition actions [3,4]. Recently, neural network models have achieved state-of-the-art performance in this task by leveraging sophisticated neural modules [5].

RST offers a robust approach for discourse analysis by constructing a rhetorical structure tree that captures the relationships between text elements, enhancing performance in various tasks. Although previous research efforts have advanced machine learning methods for discourse segmentation and parsing, these often rely on lexical and syntactic clues, hand-crafted features, and syntactic parse trees, and use gold-standard segmentation for training and evaluation [6–9]. This coherence structure is essential for applications such as text summarization, and sentiment analysis. RST-based analysis significantly improves discourse understanding and contributes to more effective NLP applications [10,11].

Despite the successes of contextualized pre-trained language models (PLMs) like XLNet [12] in RST discourse parsing, challenges remain due to data insufficiency, reliance on lexical and syntactic clues, and inconsistencies between EDU-level parsing and sentence-level contextual modeling, as well as dependence on gold-standard segmentation for training. These issues, particularly the reliance on hand-crafted features and parse trees, have made EDU segmentation a significant bottleneck. Moreover, large language models including ChatGPT, while powerful, face limitations in deeply capturing discourse structure and contextual nuances, and often produce parsing outputs inconsistent with gold standards, with prompt engineering offering only limited improvements, especially for one-step end-to-end parsing, highlighting the ongoing need for linguistically informed approaches [13–17].

In this paper, we propose *Lexical and Contextual Awareness for EDU Segmentation (LACES)*, a novel approach designed for automatic segmentation of *Elementary Discourse Units (EDUs)*. LACES leverages lexical information captured within a carefully defined token window to accurately detect EDU boundaries, enabling effective and efficient EDU segmentation. LACES demonstrates that shallow, surface-level features can be highly effective for EDU segmentation. This approach not only improves segmentation quality but also enhances the performance of SOTA RST parsers when used as the segmentation component. In summary, our contributions are as follows:

- We introduce LACES, a novel EDU segmentation framework that effectively captures local lexical context through structured token-level representation to accurately identify EDU boundaries.
- We conduct comprehensive evaluations, benchmarking LACES against established discourse segmenters, and consistently achieve superior performance.
- We integrate LACES into a SOTA neural RST parser and show that improved EDU segmentation leads to significant gains in discourse parsing performance.

2 Related Work

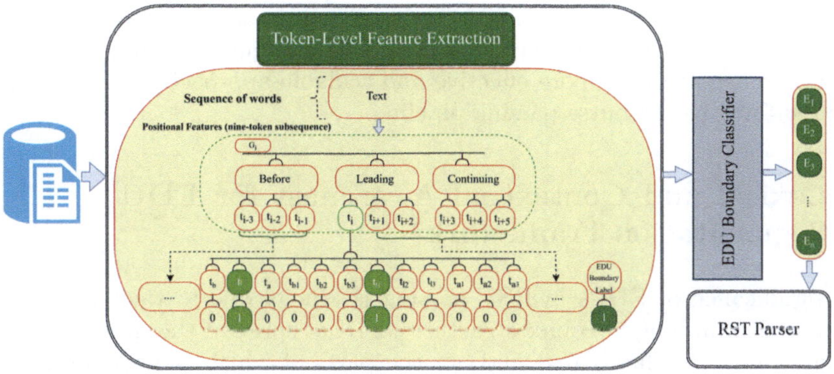

Fig. 1. The figure presents a schematic overview of the LACES framework. It shows how token-level features are extracted from each candidate boundary context and provided to the classifier. The classifier then evaluates each candidate token window and predicts EDU boundaries to enhance segmentation quality. These predicted EDUs are subsequently used for segmentation and discourse parsing with an RST parser.

Discourse analysis based on RST has attracted considerable attention in various NLP tasks. Historically, discourse processing using RST has been approached as a parsing task, often employing transition-based parsers that operate in a top-down or incremental manner, and chart parsers that use a bottom-up dynamic programming approach [18–20]. In recent years, improvements over earlier methods have been achieved by integrating statistical models that predict nuclearity and relation types between discourse units [3,21,22]. Neural approaches now dominate this field, yet many still incorporate hand-crafted features to enhance performance. Additionally, sequence-to-sequence models have been applied to both sentence-level and document-level discourse parsing, providing alternative frameworks for structure prediction [20,21,23].

A fundamental subtask within discourse parsing is the identification of EDUs, which represent the smallest meaningful spans of text. Early EDU segmentation methods primarily relied on handcrafted features and syntactic clues [24,25]. More recent approaches have recast EDU segmentation as a sequence labeling problem, employing neural network architectures that leverage contextualized embeddings [26–28]. While these models have demonstrated improved segmentation accuracy, many depend on gold-standard sentence segmentation and external syntactic parsers or chunkers, which can limit their applicability and introduce cascading errors. Furthermore, recent research has explored parsing methods that are independent of EDU segmentation by using gold-standard segmentations during training and evaluation, and by adopting top-down sequence labeling strategies for RST parsing [21,29,30]. These approaches provide strong baselines for discourse parsing across different frameworks.

To address these limitations, in this paper we focus on the core subtask of EDU segmentation by proposing a novel method called LACES, which identifies discourse boundaries by analyzing the local context of surrounding words. We evaluate our method not only in terms of segmentation quality but also by assessing its downstream impact on overall RST discourse parsing performance. LACES demonstrates a more effective and scalable approach for EDU segmentation within the discourse parsing pipeline.

3 Lexical and Contextual Awareness for EDU Segmentation Framework

The significance of EDUs in RST parsing is crucial due to their fundamental role in understanding discourse structures. EDUs represent the smallest coherent "thought units" within a text and are the parts of which the overall discourse structure is composed. Hence, accurate segmentation and identification of EDUs is essential for an accurate analysis of rhetorical structure [3,31]. In this paper, we propose the *Lexical And Contextual Awareness for EDU Segmentation (LACES)* framework (Fig. 1), which improves EDU segmentation quality and enhances the performance of SOTA RST parsers by leveraging the lexical context within a localized textual window. LACES formulates the problem of EDU segmentation as a classification problem. Our approach consists of two main components:

- **Token-Level Feature Extraction**: The core novelty of our method lies in the structured use of fixed nine-token subsequences, or nine-grams, centered around each candidate EDU boundary position in the token sequence, combined with explicit positional token labeling that distinguishes tokens occurring before, at, and after the candidate boundary. Formally, let

$$D = \{G_1, G_2, \ldots, G_n\}$$

  be a document consisting of n consecutive nine-gram sequences, where each subsequence is defined as

$$G_i = (t_{i-3}, t_{i-2}, t_{i-1}, t_i, t_{i+1}, t_{i+2}, t_{i+3}, t_{i+4}, t_{i+5})$$

  and the candidate EDU boundary is positioned immediately before token t_i. This nine-token window is divided into three segments:
  - **Before context**, comprising the three tokens preceding t_i, namely $(t_{i-3}, t_{i-2}, t_{i-1})$, which provides information about the ending of the preceding EDU.
  - **Leading context**, consisting of three tokens that mark the beginning of the new EDU: the token at the candidate boundary position t_i, followed by t_{i+1} and t_{i+2}.
  - **Continuing context**, the three tokens following the Leading context $(t_{i+3}, t_{i+4}, t_{i+5})$, capturing the continuation of the new EDU.

In cases where the EDU starts at the very beginning of the document, the Before context is naturally absent, as there are no preceding tokens. When the EDU boundary occurs within the text, the Before context corresponds to the last three tokens (or continuing segment) of the previous EDU, while the Leading context always denotes the start of the new EDU.

For feature generation, each unique token t in the nine-gram is assigned twelve binary features that encode its relative position to the candidate boundary. The first three features specify the token's context category: Before (t_b), Leading (t_l), and Continuing (t_a). The remaining nine features capture the token's exact position within its context segment, indicating whether it is the first, second, or third token in the Before (t_{b1}, t_{b2}, t_{b3}), Leading (t_{l1}, t_{l2}, t_{l3}), or Continuing (t_{a1}, t_{a2}, t_{a3}) context. For a document D containing n consecutive nine-grams and m unique tokens, the final feature matrix F has dimensions $n \times (m \times 12)$, where each row corresponds to the feature vector for one nine-gram, and each column corresponds to a specific token feature across all nine-grams.

We define EDU boundaries as the beginning of the Leading three-gram; therefore, when a candidate boundary corresponds to an EDU start, the features satisfy $t_l = 1$ and $t_{l1} = 1$, indicating that token t_i is the first token in the Leading context. This explicit positional token labeling within fixed-size windows enables the model to effectively capture both local and broader contextual cues purely at the token level. Experiments evaluating various window sizes demonstrate that smaller windows often lack sufficient context for accurate EDU boundary detection, whereas larger windows introduce irrelevant information that can degrade performance. Ultimately, the nine-token window strikes an optimal balance by capturing essential discourse cues without excessive noise, thereby improving segmentation accuracy (Table 2). These carefully designed token-level features serve as the input for our EDU boundary classifier, enabling the model to learn patterns that signal EDU starts across diverse contexts. By encoding both local and surrounding token positions, the framework systematically leverages contextual information to improve segmentation performance.

- **EDU Boundary Detection**: Given the extracted features from each candidate nine-gram window, we fine-tune a binary classifier using labeled examples, leveraging the nine-token window identified in our window-size evaluation (Sect. 4.3) as the optimal balance between contextual coverage and noise reduction to maximize segmentation accuracy. The classifier learns to assign a label $y_i \in \{0, 1\}$ to each candidate window in the input token sequence G, where $y_i = 1$ indicates that the token t_i marks the beginning of a new EDU, and $y_i = 0$ otherwise.

  At inference time, the classifier evaluates each nine-token window centered around the candidate token t_i, and produces a prediction value as:

$$\hat{y}_i = f(\mathbf{x}_i), \quad \hat{y}_i \in \{0, 1\},$$

where $\mathbf{x}_i$ denotes the feature representation extracted from the i-th window in G. If $\hat{y}_i = 1$, the token t_i is considered the first token of a new EDU. Formally, let $\mathcal{B} = \{i \mid \hat{y}_i = 1\}$ be the set of predicted EDUs (E). Each EDU segment is then defined as the contiguous sequence of tokens between two successive boundary positions in $\mathcal{B}$. This process constitutes the core of our proposed segmentation scheme. It uses contextual windows optimized for EDU boundary detection, extracts informative features, and applies binary classification to identify EDU segments.

4 Experiments

In this section, we describe our datasets, experimental setup, and results. We conduct several evaluations: first, we compare the performance of our proposed EDU identification framework using different classification methods to select the best classifier. Next, we compare our model with other EDU segmentation methods from the literature. Finally, we assess the impact of using LACES for segmentation on RST parsing performance, comparing it with methods employed in recent RST parsing systems. All experiments were conducted on an Ubuntu 20.04.4 server equipped with 256 CPUs (2.0 GHz each) and 512 GB of RAM. The framework is also compatible with GPUs.

4.1 Datasets

In all of our experiments, we follow the practice of the baselines [6,22,26,31–34] and evaluate our approach on the following datasets:

- **RST Discourse Treebank (RST-DT)** [35]: The most widely used dataset for RST parsing and EDU segmentation, consisting of 347 training and 38 test articles fully annotated with RST discourse structures. It is a standard benchmark in text-based discourse research.
- **CNN/Daily Mail** [36]: A large-scale dataset with over 300,000 news articles, widely used in summarization and discourse-related tasks, offering a broad and diverse range of real-world texts.
- **RST-GUM (Georgetown University Multilayer Corpus)** [37]: A moderately sized, richly annotated corpus of 168 English documents covering multiple genres including news, fiction, academic writing, and conversations. It offers diverse discourse structures, making it suitable for evaluating the generalizability of discourse models across different domains.

4.2 Evaluating Different Classifiers for EDU Segmentation

To evaluate the effectiveness of our EDU segmentation framework, we extract nine-token windows along with corresponding lexical features from multiple

Table 1. Classifier performance for EDU identification on the CNN/Daily Mail, RST-DT, and RST-GUM datasets. Our framework using Random Forest achieves the highest metrics across all three datasets.

Model	CNN/Daily Mail				RST-DT				RST-GUM			
	Acc.	Prec.	Rec.	F1	Acc.	Prec.	Rec.	F1	Acc.	Prec.	Rec.	F1
SVM	0.862	0.884	0.859	0.871	0.892	0.891	0.896	0.893	0.855	0.860	0.842	0.851
Gradient Boosted	0.859	0.851	0.888	0.868	0.873	0.850	0.899	0.874	0.840	0.825	0.860	0.842
MLP	0.785	0.822	0.759	0.789	0.793	0.840	0.761	0.798	0.765	0.795	0.740	0.767
XGboost	0.862	0.848	0.837	0.842	0.896	0.939	0.874	0.905	0.850	0.870	0.830	0.850
BERT	0.889	0.884	0.853	0.869	0.877	0.931	0.841	0.884	0.860	0.865	0.834	0.849
XLNet	0.615	0.515	0.586	0.549	0.662	0.532	0.507	0.519	0.630	0.520	0.540	0.530
Random Forest	**0.915**	**0.912**	**0.918**	**0.914**	**0.958**	**0.944**	**0.979**	**0.961**	**0.902**	**0.895**	**0.910**	**0.902**

datasets, including RST-GUM, RST-DT, and CNN/Daily Mail. The evaluation is conducted on similarly sized subsets from each dataset, with consistent preprocessing applied across all of them. These extracted features serve as input to various classification algorithms to determine which model most effectively identifies EDU boundaries.

We conduct a benchmarking study comparing several classifiers: a three-layer Multi-Layer Perceptron (MLP), Support Vector Machine (SVM), Gradient Boosted Trees, XGBoost, Random Forest, BERT (bert-base-uncased), and XLNet. For conventional classifiers, we use the extracted lexical features directly as input vectors. For the pretrained neural models BERT and XLNet, we keep the models as-is by feeding them the raw token sequences corresponding to the nine-token windows. We then extract contextual embeddings (e.g., BERT's [CLS] token embedding) and concatenate our extracted lexical features to these embeddings. The combined representations are used for fine-tuning a binary classification head to predict EDU boundaries.

Our results, shown in Table 1, indicate that the Random Forest classifier consistently achieves the highest accuracy, precision, recall, and F1 scores across the CNN/Daily Mail, RST-DT, and RST-GUM datasets, with F1 scores exceeding 0.90. This demonstrates its effectiveness in leveraging structured, position-aware lexical features for EDU segmentation. While BERT and XLNet provide rich contextual embeddings, the superior performance of Random Forest underscores its strength in directly using our extracted features without requiring complex integration. This capability allows it to outperform pretrained models like BERT and XLNet in this task. Since Random Forest consistently achieves the best performance across datasets, we use it as the default classifier in all subsequent experiments.

4.3 Evaluation of Different Window Sizes

To further validate the choice of the nine-token window, we conduct additional experiments comparing three window sizes: six, nine, and twelve tokens.

Each window is divided evenly into three context segments: Before, Leading, and Continuing. Specifically, the six-token window uses two tokens per context, and the twelve-token window uses four tokens per context.

For these experiments, we select the Random Forest classifier due to its consistent superior performance on EDU identification across datasets (Table 1). Table 2 presents the segmentation results on three benchmark datasets. The results confirm that the nine-token window achieves the best performance across all metrics. The smaller six-token window lacks sufficient contextual information, leading to lower accuracy, precision, recall, and F1 scores. Conversely, the larger twelve-token window introduces additional noise from irrelevant tokens, which also reduces performance. These findings reinforce that the nine-token window offers the optimal balance between capturing essential discourse cues and minimizing noise, thereby improving EDU boundary detection.

Table 2. Segmentation performance of Random Forest classifier with different window sizes on three benchmark datasets.

Dataset	six-gram				nine-gram				twelve-gram			
	Acc.	Prec.	Rec.	F1	Acc.	Prec.	Rec.	F1	Acc.	Prec.	Rec.	F1
CNN/Daily Mail	0.750	0.690	0.676	0.683	**0.915**	**0.912**	**0.918**	**0.914**	0.890	0.855	0.840	0.847
RST-DT	0.760	0.740	0.715	0.727	**0.958**	**0.944**	**0.979**	**0.961**	0.910	0.915	0.901	0.908
RST-GUM	0.755	0.710	0.705	0.707	**0.902**	**0.895**	**0.910**	**0.902**	0.885	0.870	0.857	0.863

4.4 Evaluation Against EDU Segmentation Models Using RST-DT

We further evaluate our model against several established discourse segmenters widely recognized as baselines in EDU segmentation studies, as shown in Table 3. To evaluate segmentation performance, we treat EDU boundary detection as a binary classification task at the token level, as described in our framework. A predicted boundary is considered correct only if it exactly matches the gold-standard EDU start token. This strict matching criterion ensures that precision, recall, and F1 scores accurately reflect the model's ability to identify exact EDU boundaries. For a fair comparison, all methods are evaluated on the same RST-DT dataset. The comparison includes JCN [32], which employs a Logistic Regression model with features derived from sentence context, combining syntactic tree structures and statistical estimates. Fisher et al. [38] propose a discourse segmentation approach based primarily on syntactic parse features, using a binary classifier trained with an averaged perceptron to independently predict EDU boundaries at each word without modeling sequence dependencies. We also assess CRF [39] and WYL [22], which employ sequence labeling and a BiLSTM-CRF framework, respectively. Additionally, HILDA (HIL) [21] and SPADE (SP) [19] apply statistical models that integrate syntactic and lexical information to identify discourse boundaries and construct sentence-level discourse trees.

The Joint Model [31] leverages a pointer network with a depth-first parsing strategy for discourse tree construction. DiscoDisco [40], a recent method, employs a neural architecture incorporating discourse-aware sentence fusion and contextualized embeddings to improve EDU segmentation performance. Results on the RST-DT test set demonstrate that our model, LACES, consistently outperforms these methods. While DiscoDisco achieves strong results with an F1 score of 96.6%, and the Joint Model achieves 95.5%, LACES improves upon both with an F1 score of 97.1%. Our model also attains the highest precision and recall among the compared approaches. This performance gain highlights the effectiveness of LACES's approach of leveraging lexical context within a localized nine-token window (nine-gram), relying solely on word-level features. These results suggest a potential increase in downstream RST parsing accuracy and demonstrate that a well-engineered, feature-based classifier can compete with, and even surpass, more complex neural architectures.

Table 3. Performance Comparison of EDU Segmentation Methods on RST-DT

Model	Precision	Recall	F1 Score
HILDA	0.779	0.706	0.741
SPADE	0.838	0.868	0.852
CRF	0.903	0.918	0.905
JCN	0.880	0.923	0.901
WYL	0.924	0.944	0.932
F&R	0.913	0.897	0.905
Joint Model	0.933	0.978	0.955
DiscoDisco	0.964	0.969	0.966
LACES (Ours)	**0.965**	**0.979**	**0.971**

Table 4. Performance comparison of RST parsers with and without LACES EDU segmentation.

Model	S	N	R	F
RST-2EDU [6]	0.764	0.661	0.545	0.535
SBP+AST [34]	0.871	0.750	0.632	0.626
RST-2EDU + LACES	**0.783**	**0.673**	**0.564**	**0.562**
(SBP+AST) + LACES	**0.879**	**0.762**	**0.641**	**0.634**

4.5 RST Parsing Using LACES

We integrate our EDU segmentation model, LACES, into SOTA RST parsers to evaluate its impact on discourse parsing. Specifically, we incorporate LACES into a shift-reduce, transition-based neural parser by Nan Yu et al. [6], which generates action sequences from EDU representations. This parser benefits from RST-2EDU, a second-stage EDU-level pretraining approach that enhances discourse structure prediction. Additionally, we employ our LACES-segmented EDUs as input to the SBP+AST neural parser by Kobayashi et al. [34], which is based on a span-based neural top-down RST parsing method. This parser significantly enhances training by leveraging large-scale silver-standard data created through the agreement of multiple teacher parsers using a two-stage shift-reduce parsing algorithm. The approach critically depends on precise EDU boundaries to generate reliable subtree annotations, enabling the model to learn more accurate nuclearity and relation labels.

For evaluation, we follow the framework proposed by [6,30,41,42], using micro-averaged F1 scores across four metrics: **S**pan (tree structure without labels), **N**uclearity (structure with only nuclearity labels), **R**elation (structure with only relation labels), and **F**ull (structure with both nuclearity and relation labels). These widely adopted metrics in RST parsing research provide a comprehensive assessment of both structural accuracy and label quality.

Table 5. Performance comparison of various RST parsers using metrics S, N, R, and F.

Model	S	N	R	F
Zhang et al. [42]	0.763	0.655	0.556	0.538
Yu et al. [6]	0.714	0.603	0.492	0.481
Zhang et al. [5]	0.672	0.555	0.453	0.443
Nguyen et al. [10]	0.743	0.643	0.516	0.502
Koto et al. [43]	0.731	0.623	0.515	0.503
Maekawa et al. [9]	0.798	0.704	0.600	0.581
(SBP+AST) + LACES	**0.879**	**0.762**	**0.641**	**0.634**

As shown in Table 4, our analysis demonstrates that LACES significantly enhances the performance of the RST parsers. We observe consistent improvements in the performance of the transition-based parser from Yu et al. [6] by approximately 2.48% in Span, 1.66% in Nuclearity, over 3.48% in Relation, and 5% in the Full metric. Yu et al. [6] reimplement the EDU segmenter from Muller [26] to segment large-scale unlabeled texts. For the parser from Kobayashi et al. [34] , which leverages gold EDU segmentations from the RST-DT for training and fine-tuning, LACES improves performance by around 0.8% in Span, 1.2% in Nuclearity, 0.9% in Relation, and 0.8% in the Full metric. The results underscore that improved EDU segmentation plays a crucial role in enhancing the effectiveness of SOTA RST parsers.

Table 5 presents a comparative evaluation of several SOTA RST parsers. The baseline parsers in our comparison are widely adopted in EDU segmentation and RST parsing, serving as standard benchmarks in the majority of studies [6,16,34,44]. The LACES-enhanced SBP+AST parser consistently demonstrates clear improvements in all performance metrics over these baselines, emphasizing the importance of accurate EDU boundaries for effective discourse parsing. Compared to the strongest previous baseline, Llama 2 et al. [9], which used gold EDU segmentation and employed the LLaMA 2 model with QLoRA and top-down parsing strategies, our approach shows notable gains across all performance metrics. These improvements are meaningful because even minor enhancements in EDU segmentation can have a significant impact, as small errors can propagate through the discourse structure, affecting nucleus-satellite assignments and the overall tree. Consequently, better segmentation naturally leads to measurable gains in RST parser performance and benefits downstream NLP tasks. [6,34,45].

5 Conclusion and Future Work

In this paper, we introduced LACES, a streamlined and effective framework for EDU segmentation that leverages lexical features within a localized token window. Through a series of evaluations, we demonstrated that LACES using Random Forest classifier, not only achieves SOTA performance in EDU segmentation but also significantly improves the accuracy of downstream RST discourse parsing. These findings underscore the importance of precise EDU boundary detection in enhancing discourse structure prediction and highlight how local lexical features provide strong cues for identifying basic discourse structure constituents. In future work, we aim to extend LACES to low-resource and multilingual settings, where its efficiency can be particularly beneficial. We also plan to incorporate additional linguistic signals, such as discourse markers, syntax, and morphology to further improve the quality of segmentation. Moreover, we will explore advanced feature extraction and embedding techniques to enhance performance and gain deeper insights into discourse structure.

References

1. Mann, W.C., Thompson, S.A.: Rhetorical structure theory: toward a functional theory of text organization. Text interdis. J. Study Discourse **8**(3), 243–281 (1988)
2. Kobayashi, N., Hirao, T., Kamigaito, H., Okumura, M., Nagata, M.: Top-down RST parsing utilizing granularity levels in documents. In: Proceedings of the AAAI Conference on Artificial Intelligence. vol. 34, pp. 8099–8106 (2020)
3. Yu, N., Zhang, M., Fu, G.: Transition-based neural RST parsing with implicit syntax features. In: Proceedings of the 27th International Conference on Computational Linguistics, pp. 559–570 (2018)
4. Mabona, A., Rimell, L., Clark, S., Vlachos, A.: Neural generative rhetorical structure parsing. arXiv preprint arXiv:1909.11049 (2019)

5. Zhang, L., Xing, Y., Kong, F., Li, P., Zhou, G.: A top-down neural architecture towards text-level parsing of discourse rhetorical structure. arXiv preprint arXiv:2005.02680 (2020)
6. Yu, N., Zhang, M., Fu, G., Zhang, M.: RST discourse parsing with second-stage edu-level pre-training. In: Proceedings of the 60th Annual Meeting of the Association for Computational Linguistics (Volume 1: Long Papers), pp. 4269–4280 (2022)
7. Feng, V.W., Hirst, G.: Two-pass discourse segmentation with pairing and global features. arXiv preprint arXiv:1407.8215 (2014)
8. Ali, O.: Fuzzy text segmentation using syntactic features for rhetorical structure theory. Ph.D. thesis, University of Portsmouth (2023)
9. Maekawa, A., Hirao, T., Kamigaito, H., Okumura, M.: Can we obtain significant success in RST discourse parsing by using large language models? arXiv preprint arXiv:2403.05065 (2024)
10. Nguyen, T.T., Nguyen, X.P., Joty, S., Li, X.: RST parsing from scratch. arXiv preprint arXiv:2105.10861 (2021)
11. Liu, Z., Shi, K., Chen, N.F.: Dmrst: A joint framework for document-level multilingual RST discourse segmentation and parsing. arXiv preprint arXiv:2110.04518 (2021)
12. Yang, Z., Dai, Z., Yang, Y., Carbonell, J., Salakhutdinov, R., Le, Q.V.: XLNET: generalized autoregressive pretraining for language understanding (2020). https://arxiv.org/abs/1906.08237
13. Gillings, M., Kohn, T., Mautner, G.: The rise of large language models: challenges for critical discourse studies. Crit. Discourse Stud. 1–17 (2024)
14. Pastor, M., Oostdijk, N., Martín-Rodilla, P., Parapar, J.: Enhancing discourse parsing for local structures from social media with LLM-generated data (2025)
15. Blevins, T., Gonen, H., Zettlemoyer, L.: Prompting language models for linguistic structure. arXiv preprint arXiv:2211.07830 (2022)
16. Bai, X., Wu, J., Chen, Y., Wang, Z., Zhang, Y.: Constituency parsing using LLMs. arXiv preprint arXiv:2310.19462 (2023)
17. Cui, Y., Yang, Z., Yao, X.: Efficient and effective text encoding for chinese llama and alpaca. arXiv preprint arXiv:2304.08177 (2023)
18. Carlson, L., Marcu, D., Okurowski, M.E.: Building a discourse-tagged corpus in the framework of rhetorical structure theory. Current and New Directions in Discourse and Dialogue, pp. 85–112 (2003)
19. Soricut, R., Marcu, D.: Sentence level discourse parsing using syntactic and lexical information. In: Proceedings of the 2003 Human Language Technology Conference of the North American Chapter of the Association for Computational Linguistics, pp. 228–235 (2003)
20. Joty, S., Carenini, G., Ng, R.T.: Codra: a novel discriminative framework for rhetorical analysis. Comput. Linguist. **41**(3), 385–435 (2015)
21. Hernault, H., Prendinger, H., du Verle, D.A., Ishizuka, M.: Hilda: a discourse parser using support vector machine classification. Dialogue Discourse **1**(3), 1–33 (2010)
22. Wang, Y., Li, S., Yang, J.: Toward fast and accurate neural discourse segmentation. arXiv preprint arXiv:1808.09147 (2018)
23. Bach, N.X., Le Nguyen, M., Shimazu, A.: A reranking model for discourse segmentation using subtree features. In: Proceedings of the 13th Annual Meeting of the Special Interest Group on Discourse and Dialogue, pp. 160–168 (2012)
24. Perret, J., Afantenos, S., Asher, N., Morey, M.: Integer linear programming for discourse parsing. In: Proceedings of the 2016 conference of the north american

chapter of the association for computational linguistics: Human language technologies, pp. 99–109 (2016)
25. Afantenos, S., Kow, E., Asher, N., Perret, J.: Discourse parsing for multi-party chat dialogues. In: Conference on Empirical Methods on Natural Language Processing (EMNLP 2015), pp. pp–928 (2015)
26. Muller, P., Braud, C., Morey, M.: Tony: Contextual embeddings for accurate multilingual discourse segmentation of full documents. In: Proceedings of the Workshop on Discourse Relation Parsing and Treebanking 2019, pp. 115–124 (2019)
27. Sennrich, R., Haddow, B., Birch, A.: Neural machine translation of rare words with subword units. arXiv preprint arXiv:1508.07909 (2015)
28. Devlin, J., Chang, M.W., Lee, K., Toutanova, K.: Bert: pre-training of deep bidirectional transformers for language understanding. In: Proceedings of the 2019 Conference of the North American Chapter of the Association for Computational Linguistics: Human Language Technologies, volume 1 (long and short papers), pp. 4171–4186 (2019)
29. Braud, C., Coavoux, M., Søgaard, A.: Cross-lingual RST discourse parsing. arXiv preprint arXiv:1701.02946 (2017)
30. Morey, M., Muller, P., Asher, N.: How much progress have we made on RST discourse parsing? a replication study of recent results on the RST-dt. In: Conference on Empirical Methods on Natural Language Processing (EMNLP 2017), pp. pp–1330 (2017)
31. Lin, X.: Natural language processing as autoregressive generation (2023)
32. Joty, S., Carenini, G., Ng, R.: A novel discriminative framework for sentence-level discourse analysis. In: Proceedings of the 2012 Joint Conference on Empirical Methods in Natural Language Processing and Computational Natural Language Learning, pp. 904–915 (2012)
33. Pastor, M., Oostdijk, N.: Signals as features: predicting error/success in rhetorical structure parsing. In: Proceedings of the 5th Workshop on Computational Approaches to Discourse (CODI 2024), pp. 139–148 (2024)
34. Kobayashi, N., Hirao, T., Kamigaito, H., Okumura, M., Nagata, M.: Improving neural RST parsing model with silver agreement subtrees. In: Proceedings of the 2021 Conference of the North American Chapter of the Association for Computational Linguistics: Human Language Technologies, pp. 1600–1612 (2021)
35. Carlson, L., Okurowski, M.E., Marcu, D.: RST discourse treebank. Linguistic Data Consortium, University of Pennsylvania (2002)
36. Hermann, K.M., et al.: Teaching machines to read and comprehend. Adv. Neural Inf. Process. Syst. (28) (2015)
37. Zeldes, A.: The gum corpus: creating multilayer resources in the classroom. Lang. Resour. Eval. **51**(3), 581–612 (2017)
38. Fisher, S., Roark, B.: The utility of parse-derived features for automatic discourse segmentation. In: Proceedings of the 45th Annual Meeting of the Association of Computational Linguistics, pp. 488–495 (2007)
39. Feng, V.W., Hirst, G.: A linear-time bottom-up discourse parser with constraints and post-editing. In: Proceedings of the 52nd Annual Meeting of the Association for Computational Linguistics (Volume 1: Long Papers), pp. 511–521 (2014)
40. Gessler, L., Behzad, S., Liu, Y.J., Peng, S., Zhu, Y., Zeldes, A.: Discodisco at the disrpt2021 shared task: a system for discourse segmentation, classification, and connective detection. arXiv preprint arXiv:2109.09777 (2021)
41. Shahmohammadi, S., Stede, M.: Discourse parsing for german with new RST corpora. In: Proceedings of the 20th Conference on Natural Language Processing (KONVENS 2024), pp. 65–74 (2024)

42. Zhang, L., Kong, F., Zhou, G.: Adversarial learning for discourse rhetorical structure parsing. In: Proceedings of the 59th Annual Meeting of the Association for Computational Linguistics and the 11th International Joint Conference on Natural Language Processing (Volume 1: Long Papers), pp. 3946–3957 (2021)
43. Koto, F., Lau, J.H., Baldwin, T.: Top-down discourse parsing via sequence labelling. arXiv preprint arXiv:2102.02080 (2021)
44. Chistova, E.: Bilingual rhetorical structure parsing with large parallel annotations. arXiv preprint arXiv:2409.14969 (2024)
45. Li, J., Liu, M., Qin, B., Liu, T.: A survey of discourse parsing. Front. Comp. Sci. **16**(5), 1–12 (2022). https://doi.org/10.1007/s11704-021-0500-z

Constituency Syntax Parsing of the Kazakh Language

Gaziza Yelibayeva[1,2](✉), Martin Strecker[2], Gulbarshin Syzdykova[3], Bibigul Razakhova[1], and Altynbek Sharipbay[1]

[1] Department of Artificial Intelligence Technologies Faculty of Information Technology, L.N. Gumilyov Eurasian National University, Astana, Kazakhstan
yelibayeva_gk@enu.kz

[2] IRIT, Université de Toulouse, Toulouse, France

[3] Department of Kazakh Linguistics, Faculty of Philology, L.N. Gumilyov Eurasian National University, Astana, Kazakhstan

Abstract. We present a framework for analysing the morphology and syntax of the Kazakh language, on the one hand out of intellectual curiosity, to gain more insight into the language, but also with the aim of integration of this tool-set into environments for safety-critical applications. The work underlying this article is an original combination of an (existing) morphological analyser and a phrase-structure parser. In this article, we give an overview of the Kazakh language, describe its complex morphological structure, the syntax analysis process (based on the Grammatical Framework, GF) and provide a glimpse at our proof of concept implementation. Even though operational, the system is not yet entirely complete, and we outline future work directions.

Keywords: Natural language processing · Kazakh language · Morphological Analysis · Syntax parsing · Grammatical Framework

1 Introduction

The Kazakh language, a member of the Turkic languages, the national language of Kazakhstan but also the mother tongue of speakers in other parts of Central Asia and in China, has so far not received the full attention it deserves. Our work contributes to analysing its structure, in part out of academic interest, but also for practical reasons: there is a growing number of applications developed for Kazakhstan's national language, also in critical domains like medicine, law or engineering. Here, "critical" means that an erroneous use of terms or false interpretations of the structure of a sentence may lead to severe, possibly fatal consequences.

With these two motivations in mind: in-depth structural understanding and critical applications, we adopt an approach that runs counter to the current trend of language processing based on machine learning methods. In this paper, we present a constituency syntax parsing tool-set that is an original combination

H. Unger and P. Meesad (Eds.): NLPIR 2025, LNNS 1904, pp. 61–75, 2026.
https://doi.org/10.1007/978-3-032-20897-2_5

of a morphological analyser (not developed by us) and a structural parser based on the Grammatical Framework (GF).

The article is structured as follows: After a review of related work in Sect. 2, we provide a short overview of the Kazakh language, in particular its morphology (Sect. 3). We then give a more detailed account of the syntax analyser in Sect. 4 before evaluating the work and concluding in Sect. 5.

2 Related Work

For reasons of space, we concentrate on related efforts of processing the Kazakh language, and explicit parsing techniques that we espouse in our work for the reasons outlined in the introduction. A wide range of methods cannot be covered here, see [10] for an excellent general overview, and [8] for the semantic processing that can be grafted on the result of syntax analysis.

Natural language processing (NLP) is a vast domain, so we only recapitulate the essentials. There are two major approaches to NLP, either based on an explicit representation of language information (morphological, syntactic *etc.*) [8], or based on an implicit encoding, often with artificial neural networks generated by machine learning methods. Combinations are possible, but our work falls almost exclusively into the first category; only the generation of the morphological analyser has also used statistical methods. Without wanting to downplay the importance and widespread acceptance of statistical methods, they have decisive weaknesses, such as not providing an immediate basis for logically sound reasoning, and this is an aspect we will address in the future but that is not developed here. For our work, we use the Grammatical Framework (GF)[1], in which several dozen languages have already been coded. We follow roughly its methodology that is well documented in [21], although sometimes in an unorthodox way (see Sect. 4.1).

For Kazakh, the urgency of assembling a national corpus of the language is recognised by a national action plan of the Republic of Kazakhstan for 2023–2029. Even before, extensive corpora were established, such as [18], containing more than 135 million words, and the National Corpus of Kazakh Language[2]. In study [3], a media corpus of 140 000 news items in the Kazakh language was compiled, and there is the publicistic sub-corpus of the national corpus of the Kazakh language[3]. Altogether, most of the scientific research of the Kazakh language is dedicated to the morphological level [5,13,29].

For syntax analysis, the two most commonly used types of parsing are Dependency Parsing [15] and Constituency Parsing; the latter is the method espoused by us. Constituency Parsing is generally admitted to be more suitable for syntactically complex languages with rich morphology [32], such as Kazakh, and

[1] https://www.grammaticalframework.org/.
[2] https://qazcorpus.kz/.
[3] https://qazcorpora.kz/.

arguably allows for a better translation to formalisms used in logic and symbolic knowledge representation. Previous formalisations of Kazakh language syntax using context-free grammars are considered in [19,24]; however, this work either addresses very simple language fragments or is not operationalised in the form of an effectively running parser. A number of researchers have focused on Dependency Parsing of the Kazakh language. The work in [30,31] presents a free / open-source dependency tree-bank for Kazakh language based on universal dependency (UD) [7] annotation standards. The paper [2] introduces an algorithm for constructing a dependency tree for sentences in the Kazakh language, employing the filter method. The study [28] presents a comparative analysis of dependency parsing results for two distinct languages, Kazakh and English. The evaluation is conducted employing various discrete and distributed feature-based approaches. The report [16] shows the authors' initial endeavour to apply the universal UD standard for syntactic annotation to Kazakh, involving the manual annotation of 300 randomly selected sentences from the Kazakh Language Corpus [18] as its outcome.

3 Overview of the Kazakh Language and Its Morphology

In this section, we give a short overview of the Kazakh language, highlighting its specificities and the difficulties arising during its analysis, before plunging into the morphological analyser which is the basis of our syntax analysis.

The Kazakh language belongs to the Kipchak branch of the Turkic language family and it is the official language of Kazakhstan. It is also the native language of the Kazakh diaspora living in other countries. During the 20th century, the Kazakh language underwent three distinct transitions in its alphabet. Firstly, the Kazakh script was changed into a new alphabet based on Arabic script, then from Arabic to a Latin-based alphabet, subsequently to the Cyrillic alphabet with 42 letters (15 vowel phonemes, 25 consonant phonemes, 2 letters denote softness and hardness of consonant phonemes) [23].

The modern Kazakh language uses the Cyrillic script. The paradigm for the Kazakh alphabet's transition to the Latin script commenced with the approval of version 1 in 2017. Subsequent modifications were implemented in 2018, refining specific parts of the alphabet. In 2021, a new version (which we call QazLat here[4]) of the Kazakh alphabet was presented for discussion, it includes 31 letters.

Similar to its Turkic counterparts, the Kazakh language morphology is characterised by an agglutinative system, featuring the propensity to form words by affixing various morphemes [12]. Kazakh language adheres to the principle of vowel harmony, where vowels within a word harmonise or exhibit concordance in terms of certain phonetic features [9], contributing to the linguistic coherence and structure of the language. In the Kazakh language, when denoting the ownership of an object, not only possessive pronouns are used, but also special possessive endings [11]. The complex endings will play a major role both on the morphological (Sect. 3) and also on the syntactic level (Sect. 4.2). Words with

[4] https://qazlat.kz.

personal endings are in many cases predicates. These endings are attached not only to the conjugated forms of the verb, but also to other words (nouns, adjectives, numerals, pronouns, adverbs) that act as predicates in the sentence. In the Kazakh language there is no category of gender (lack of grammatical gender), therefore instead of "he, she" indicating a person, only one pronoun is used. There are no prepositions in the Kazakh language. The meaning of prepositions is mostly conveyed through postpositions and in the form of indirect cases. If a noun is preceded by a numeral, the plural ending is not used in the noun. Cardinal and ordinal numerals, when used before nouns as attributes, do not change in number and case. The basic word order is Subject-Object-Verb (SOV).

We use a morphological analyser developed in the context of the project "KazNLP: NLP tools for Kazakh language"[5], written in the Python language.

KazNLP offers several modules such as: Initial normalisation [20], tokenization [27], language identification [14], morphological processing [17]. Morphological processing is typically accomplished through the sequential application of tools such as morphological analysers and disambiguators (or taggers). This morphological processing proposes a data-driven method for morphological analysis and disambiguation of the Kazakh language, encompassing both inflectional and derivational morphology. The morphological analyser generates all possible analyses for a given word, presenting them as triples comprising the dictionary form, part of speech, and grammatical characteristics. After this the tagger selects the analysis that is most appropriate for the given context. The tagger system uses a modified version of the Kazakh Language Corpus (KLC) tag set [18]. The full glossary is available online[6].

Example 1 (Morphological analysis). Without presenting all the morphological categories in detail, for illustrating the principle of morphological analysis, take the example sentence "қызыл көйлек киген қыз ғимаратқа кірді" (transcribed as *qyzyl köilek kigen qyz ğimaratqa kırdı*, in English: "The girl (who is) wearing a red dress entered the building.").

When running the morphological analyser, we get the following sequence of morphemes:

қызыл_R_SE көйлек_R_ZE ки_R_ET ген_ET_ESM қыз_R_ZE ғимарат_R_ZE қа_C3 кір_R_ET ді_T3

which we have transcribed in a Latin script similar to QazLat[7] as: *qyzyl_R_SE koeilek_R_ZE ki_R_ET gen_ET_ESM qyz_R_ZE ghimarat_R_ZE qa_C3 kir_R_ET di_T3.*

Let us look at the decomposition of the individual words:

- *qyzyl* ("red") is marked as adjective with the tag *R_SE.* Kazakh has no articles, so the articles *a* / *the* are not translated.

[5] https://github.com/nlacslab/kaznlp.
[6] https://github.com/nlacslab/kaznlp/blob/master/kaznlp/docs/morphology-tagset.MD.
[7] but not identical to, for technical reasons of processing with GF.

- *koeilek* ("dress") is marked as common noun with the tag *R_ZE*.
- *kigen* ("wearing") is decomposed as the verb stem *ki* with marker *R_ET* and the suffix *gen* designating a verbal adjective with the tag *ET_ESM*.
- *qyz* ("girl") is again marked as common noun with the tag *R_ZE*.
- *ghimaratqa* ("building") is decomposed into its root, the noun *ghimarat* with the tag *R_ZE* and the dative ending *qa* tagged as *C3*.
- *kirdi* ("entered") is decomposed into the verb stem *kir* (tag *R_ET*) with the past tense ending *di* (tag *T3*).

4 Syntax Analysis for the Kazakh Language

The previous section has presented some of the main challenges for processing the Kazakh language and how a tokenizer handles its complex morphological structure. The current section describes the core of our work, namely the parser that takes a stream of tokens as produced by the morphological analyser and constructs a syntax tree out of it. The process is reminiscent of syntax analysis in programming languages [1,4], but particular difficulties are:

- the great number of tokens / morphemes representing word stems but also endings, some of which may be empty;
- a relatively complex system of cases and ensuing consistency conditions to be ensured.

These problems are in part common to all natural languages, and we have chosen the Grammatical Framework (GF) for tool support; we give an overview of it in Sect. 4.1. We then describe the parser implemented in GF (Sect. 4.2). A naive approach to parsing would create a great number of parse trees most of which would not respect the conditions imposed by the case system. Our solution, based on an extended form of typing, is presented in Sect. 4.3.

4.1 Grammatical Framework

The Grammatical Framework [21], GF, is an environment for processing natural languages. Its main tenet is that it is multilingual, *i.e.* providing a common framework for processing whatever natural language, and for making several languages interoperate, for example by translating between several languages. This latter aspect is not exploited in our work, but offers an opening for future enhancements. GF also ambitions to cover the whole spectrum of natural language processing, from raw strings through morphological and syntax analysis to semantic treatment (such as [22]). Again, we currently do not fully use GF's capabilities: semantics is not in the focus of our application scenario. And, after initial experiments, we have decided against coding the morphological analyser in GF: because of the extraordinarily complex Kazakh morphology, this would have been a heroic undertaking, and a suitable morphological analyser was readily available.

The functionality of GF that we do use is therefore limited to parsing a sequence of tokens, with the aid of a parser generated by GF on the basis of a grammar and a vocabulary provided by us. We should emphasise that this is probably a non-orthodox use of GF which however proved to yield very good results.

Due to its generic nature, a GF grammar has two levels: an *abstract* one, related to the structure of a family of languages, and a *concrete* one, related to the specific textual manifestation of one specific language.

In the abstract grammar, one declares syntactic categories (under the heading `cat`), such as `S` (sentence), `NP` (noun phrase) and `VP` (verb phrase). One also declares the functions (under the heading `fun`) that are the constructors of the syntax tree and that correspond to the individual rules of a phrase structure grammar. For example, the declaration `NPVPToS: NP -> VP -> S` expresses that a sentence is formed from a noun phrase and a verb phrase.

The concrete grammar permits to specify how the syntactic categories and functions are realised or "linearised". The concrete counterparts of categories are `lincats`. In our case, they are typically records consisting at least of the textual representation of the phrase under consideration, and other specific information relevant for the category, and which we will spell out in detail when discussing the parser. For example, the `lincat` for sentences is `S= {s: Str; nb: Number; pr: Person}`, a triple consisting of the string representation, number (singular or plural) and a person category.

When compiling the grammar specification, a parser is generated. It reads an input text, which is in our case a sequence of tokens, and produces a syntax tree, which can also be visualised. The trees represented in Sect. 4.2 are directly generated by the tool. A GF grammar can be ambiguous. In this case, the parser will return several results for a phrase. While ambiguities are inherent in natural language and cannot always be avoided, a naively written grammar will produce a high number (often several dozens) of parses even for simple sentences. We will describe our strategies to counteract this phenomenon in Sect. 4.3.

4.2 Parser

General Structure. We will now introduce the grammatical categories used in our parser, and then comment on the inner workings of some representative parsing rules. The grammar definition comprises more than 800 lines of code, so it will not be possible to describe them all in detail.

In keeping with standard parsing terminology [1,4], we distinguish between terminal and non-terminal categories, corresponding respectively to leaf and interior nodes of a parse tree. In addition, we also distinguish between non-attributed and attributed categories. The latter come associated with a number of attributes which we will discuss in detail further below. Roughly speaking, attributes provide information about an instance of the category and are instrumental for ensuring its consistency, as further described in Sect. 4.3.

The attributed categories of our grammar are displayed in Table 1, and it turns out that they comprise all non-terminal categories. Some non-attributed categories, such as `Adj` (adjective) and `NStem` (noun stem) are not listed in the table. All categories, whether attributed or not, come associated with a string that is the textual representation of the instance of the category.

Table 1. Attributed categories of the grammar

Category	Purpose	Terminal	Attributes
`S`	sentence		Number, Person
`NP`	noun phrase		Number, Possessive, Case, Person
`Nominal`	qualified noun		Number, Possessive, Case, Person
`N`	simple noun		Number, Possessive, Case, Person
`VP`	verb phrase		Valency, Time, Voice, Person, Number
`V`	verb		Valency, Time, Voice, Person, Number
`VStem`	verb stem	✓	Valency
`ENumber`	number ending	✓	Number
`EPossessive`	possessive ending (noun)	✓	Possessive
`ECase`	case ending	✓	Case
`EPersonN`	person ending (noun)	✓	Person
`EPersonV`	person ending (verb)	✓	Person
`Pron`	pronoun	✓	Number, Possessive, Case, Person

As already alluded to in Sect. 4.1, in GF, all the categories are declared in the abstract grammars as a `cat` and in the concrete grammar as a `lincat` together with their respective attributes. A snippet of GF declarations corresponding to Table 1 is:

```
cat
  S;
  NP;
lincat
  S= {s:Str; nb: Number; pr: Person};
  NP={s:Str; nb: Number; p: Possessive; c: Case; pr: Person};
```

Attributes. are typically realised as enumerative types and are closely aligned to the tag set of the morphological analyser. We provide the GF definitions of the most relevant ones here:

```
param
  Number = Sg | Pl ;
  Possessive = S1 | S2 | S3 | S4 | S5 | S3SIM | SNull ;
  Case = C0 | C1 | C2 | C3 | C4 | C5 | C6 | C7  ;
  Person = P1 | P2 | P3 | P4 | P5 | P6 | P7 | P8 ;
```

The Number attribute can be Singular or Plural. The Possessive attribute (indicating a form of ownership for nouns) can be first, second, third person singular respectively plural. We also have to provide for the case that the possessive ending of nouns is absent (SNull). Similarly, the cases range from nominative (C1) through genitive, dative, accusative, locative, ablative to instrumental (C7). A special dummy case C0 is used in the context of valency of verbs (see Sect. 4.2). The Person categories are first, second, third, person, second person formal (P1, ..., P4) respectively plural (P5, ..., P8).

GF makes no formal distinction between terminal and non-terminal categories; in principle, they could overlap, but in our case, the distinction is neat.

Terminals. are taken from a dictionary which is in part generated from the morphological analyser, but for which some information such as valency (see Sect. 4.3) has to be added.

Example 2 (Terminals). Continuing Example 1, we present a sample of the declarations and definitions of its morphemes which of course occur in a different order in our dictionaries. Also note that the dictionary entries are effectively composed of Cyrillic characters. The specification of the verb *kir* will be explained in Example 3.

```
fun
  koeilek_R_ZE, qyz_R_ZE, ghimarat_R_ZE : NStem;
  gen_suf_vadj: VAdjSuffix;
  qyzyl_R_SE : Adj;
  qa_end : ECase;
  ki_stem1, kir_stem1 : VStem;
lin
  koeilek_R_ZE = {s="көйлек_R_ZE";};
  qyz_R_ZE = {s="қыз_R_ZE";};
  ghimarat_R_ZE = {s="ғимарат_R_ZE";};
  gen_suf_vadj = {s= "ген_ET_ESM"; }
  qyzyl_R_SE = {s="қызыл_R_SE";}
  qa_end = {s= "қа_C3"; c = C3; };
  kir_stem1={s="кір_R_ET"; vv= SV1; vvc= table{PA1=>C3; PA2=>C0}};
  ki_stem1= {s="ки_R_ET"; vv=SV1; vvc=table {PA1=>C4; PA2=>C0}};
```

Non-terminals. are introduced by type declarations which specify the type of the tree constructor functions, in GF's abstract grammar. The concrete grammar then defines how these functions are realised. In its simplest case, one concatenates the strings of the constituents and appropriately propagates attributes upwards in the syntax tree. The principle is illustrated by adjoining adjectives with nominals to produce composite nominals.

```
fun
```

```
  AdjNominalToNominal: Adj -> Nominal -> Nominal;
lin
  AdjNominalToNominal adj nom =
    {s= adj.s ++ nom.s; nb= nom.nb; p= nom.p; c= nom.c; pr=
    nom.pr;};
```

As to be expected, the constructor function `AdjNominalToNominal` takes two parameters, an adjective `adj` and a nominal `nom`, and concatenates the strings of the constituents (with `++`). The resulting nominal inherits all the attributes (number *etc.*) of the given nominal `nom`. Most other constructor functions are considerably more complex as they perform consistency checks, see Sect. 4.3, but they share the principle that relevant attributes are propagated upwards in the syntax tree.

Sentences. In the following sections, we discuss the structure of sentences and their main constituents. As briefly mentioned above, the main constituents of an affirmative sentence are a noun phrase `NP` and a verb phrase `VP` containing a full verb. Alternatively, in a sentence of the form *An A is a B*, two noun phrases are combined; however, an analogue of the copula *is* does not exist in Kazakh.

Here and in the following, we present the typing of the constructor functions without giving the complete realisation by the concrete grammar:

```
fun
  NPVPToS: NP -> VP -> S;
  NPNPToS: NP -> NP -> S;
```

Nouns and Noun Phrases. The main categories that form the backbone of a noun phrase are, in increasing order of complexity: `NStem`, `N`, `Nominal` and `NP`. Among them, only noun stems `NStem` are considered as terminals, see Example 2 for instances. Due to the complex morphological structure of Kazakh nouns, `N` is constructed from number, possessive, case and person endings. A `Nominal` can be a simple noun or a noun preceded by numerals or adjectives. A noun phrase `NP` can be simply constructed from a pronoun, a nominal or by combination with a relative clause and another noun phrase (where the first noun phrase has to have genitive case, see Sect. 4.3). Relative clauses `RC` are constructed from verbal adjectives `VAdj` without or with noun phrases, thus allowing for recursively nested relative clauses.

The constructor functions are:

```
NounDecompos: NStem -> ENumber -> EPossessive -> ECase ->
    EPersonN -> N;
NToNominal:  N -> Nominal;
AdjNominalToNominal: Adj -> Nominal -> Nominal;
NumNominalToNominal: Num -> Nominal -> Nominal;
PronToNP: Pron -> NP;
NominalToNP: Nominal -> NP;
```

```
RCNPToNP: RC -> NP -> NP;
NPNPToNP:  NP -> NP -> NP;
VAdjToRC: VAdj -> RC;
NPVAdjToRC: NP -> VAdj -> RC;
```

Verbs and Verb Phrases. We now come to the construction of verb phrases. The relevant syntactic categories here are verbs `V` and verb phrases `VP`. Similarly to nouns, elementary verbs `V` are composed of a stem, voice and tense suffixes, and a person ending – the person endings for verbs differ from those for nouns. A verb phrase can be composed of a single verb or a verb with one or two objects, which precede the verb, according to the standard subject-object-verb order of Turkic languages.

```
VerbDecompos: VStem -> Voice -> Tense -> EPersonV -> V;
VToVP: V ->VP;
VPwithObj: NP -> V -> VP;
VPwith2Obj: NP -> NP -> V -> VP;
```

4.3 Consistency Checking and Disambiguation

Applying the above-mentioned syntax tree construction rules freely and without filter would lead to a great number of parse trees even for rather simple sentences; most of these parses would be considered as ill-formed and nonsensical, violating grammatical constraints such as number and case (phenomena found in many languages) and possessive and person attributes (which are more specific to the Kazakh language).

Consistency. We first give an overview of the consistency checks deployed in our grammar, before engaging in a more detailed description of one of them, verb valency.

- Rule `NPVPToS`: when combining a `NP` and a `VP` to form an `S`, we ensure that the `NP` is in nominative case and that the `NP` and `VP` agree in number and person.
- Rule `NPNPToS`: when combining two noun phrases, the number attributes of the two noun phrases have to be equal; in addition, a more complex compatibility criterion concerning the person attribute applies.
- Rule `NPNPToNP`: When combining two noun phrases to obtain a new noun phrase, the first one has to be in genitive case and is adjoined to the second one, whose attributes the resulting noun phrase inherits.

We now plunge into consistency conditions involving verbs, typically called valency in linguistics [25,26]. The solution adopted is a simplified form of the treatment in feature-based grammars [6]. We recall that the Kazakh language has a strongly developed system of cases. In this context, our general working hypothesis is that verbs expect a specific number of objects, which furthermore have to have specific cases. The analogy to functions and their parameter types in strongly typed programming languages is obvious.

We implement the valency attribute (see Table 1) by two features: the valency type and the valency map.

```
param ValencyType = SV0 | SV1 | SV2| SV3 | SVIndiff ;
param VerbArgPos = PA1 | PA2 ;
lincat ValencyMap = VerbArgPos => Case ;
```

Example 3 (Valency). The concept is best illustrated by example. We continue and extend Example 2 and can now explain the specification of the verb *kir* (to enter). An excerpt of our dictionary with two different stems is:

```
kir_stem1={s="кір_R_ET"; vv= SV1; vvc= table{PA1=> C3; PA2=> C0}};
kir_stem3={s="кір_R_ET"; vv= SV2; vvc= table{PA1=> C3; PA2=> C6}};
```

A single verb can come with one or several objects, which should be of appropriate case. The verb valency contained in a `VStem` is meant to provide just this information: The valency type `vt` gives the number of arguments, currently ranging from none (`SV0`) to two (`SV2`). The valency map `vm` specifies the case of each of the arguments: For `kir_stem1`, there is only one argument of case `C3` (dative), the second argument is not present (dummy case `C0` because the map cannot be partial). With this, we have the verb *go into (a destination [in dative])*. Other uses of this verb are possible, and so we can have several lexicon entries for one verb, for example `kir_stem3` where the verb takes the two cases `C3` (dative) and `C6` (ablative) to express the destination and origin, like in *he entered the house from the field.*

So far for the example. Let us now see how we implement the function constructing a verb phrase from a verb having a single object (the function `VPwith2Obj` for two objects is similar):

```
VPwithObj: NP -> V -> VP;
VPWithObj np v =
case compatibleSV1 v.vt v.vm np.c of {
  True =>  {s= np.s ++ v.s; vt= v.vt; vm= v.vm; vc= v.vc; ...};
  False => {s = "_" ;       vt= v.vt; vm= v.vm; vc= v.vc; ...} };
```

If the compatibility check `compatibleSV1` succeeds (the `True` case), we construct a valid verb phrase; otherwise, we provoke an error by producing the string `"_"` rejected by the parser. The compatibility test reads as follows:

```
compatibleSV1: ValencyType -> ValencyMap -> Case -> Bool
compatibleSV1 vt vm c =
   case vt of {
```

```
SV1 => (vm!PA1) == c;
SVIndiff => not (c != C2);
_ => False };
```

Only if the given case of the noun phrase is as required by the valency map, the compatibility test succeeds. It will also fail if the verb has a wrong number of objects. The admittedly cryptic case `SVIndiff` is related to treating incomplete information in our dictionary.

Disambiguation. The above-mentioned consistency checks are one mechanism of reducing ambiguity. We now present some other, genuine forms of syntactic ambiguity, *i.e.*multiple parses that may each be amenable to a reasonable interpretation (also see the discussion in [10]):

- attachment ambiguity of adjectives: in a phrase like *the old man who wears a hat*, the adjective *old* could be attached to the noun *man* or to the noun phrase *man who wears a hat.* Our construction described in Sect. 4.2 hard-codes an attachment to the noun. Both syntactic interpretations are almost synonymous.
- coordination ambiguity: The same holds for attachment of adjectives in the presence of connectives like *and*, *or*, for example *big apples and pears.* Limiting the interpretation to *(big apples) and pears* is more problematic, we are currently evaluating the effects.

Example 4 (Syntax tree). Let us finally come back to our leading example. When parsing the sentence of Example 1, parsing succeeds and produces the syntax tree which GF displays as in Fig. 1. Please remember that our tool chain processes

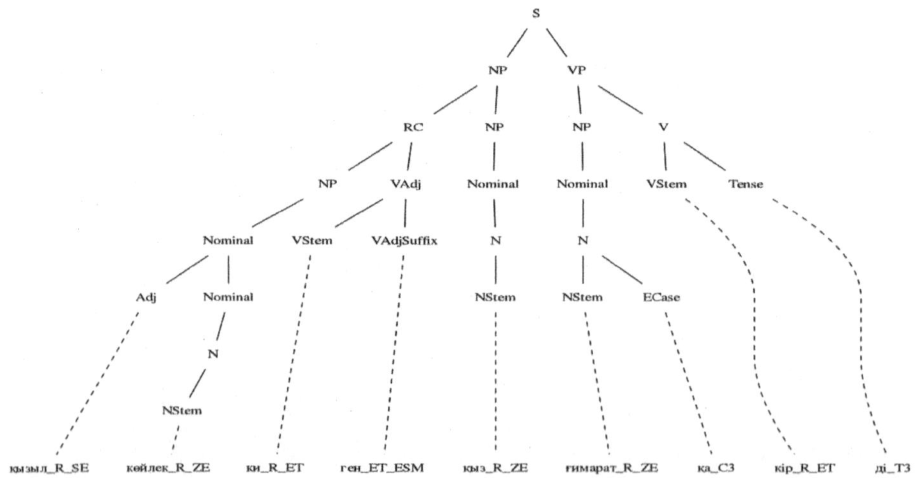

Fig. 1. Syntax tree of example sentence.

the Cyrillic characters of the current Kazakh script, so these are displayed as leaves of the tree.

5 Conclusion

This paper has presented a language processing framework, tightly integrating morphological (Sect. 3) and syntax analysis (Sect. 4). It is fully operational and publicly accessible at[8]. Our current efforts concentrate on linking it with domain-specific advisors in the medical, legal and pedagogical domains. A first evaluation of user acceptance is positive, but a large-scale experimentation has not yet been performed.

Our language definition comprises a grammar specification of approximately 800 lines of GF code and a vocabulary of *ca* 25 000 morphemes, among them more than 3 600 verbs, 12 500 nouns and approximately 5 000 adjectives, but also including endings, suffixes and composite expressions.

We are well aware of some limitations of our framework:

- There are some technical issues interfacing between the morphological and the syntax analyser. The former sometimes makes a bad choice when classifying lexemes (for example, *di* can be a 3rd person singular or plural personal ending), and the parser would have to relaunch the morphological analysis if it gets stuck on a particular interpretation, to request alternatives.
- Our development is not yet entirely complete: We have so far concentrated on affirmative sentences. Questions, commands and the like are not dealt with, but it is clear how this can be done. The same holds for detailed consistency annotations as described in Sect. 4.3, which often contains default values that lead to an over-approximation and will accept phrases that are, strictly speaking, ungrammatical.
- In spite of this, a more fundamental problem that is at the same time an advantage and a shortcoming of constituency parsing in general is its lack of tolerance of ill-formed sentences or sentences that do not fall into the scope of the grammar that we have defined. Our current research focuses on lifting this restriction while at the same time maintaining a reliable language processor that can be used in critical application domains, as outlined in the introduction.

Acknowledgements. This work benefited from resources and facilities provided by the IRIT (especially the ACADIE team) in Toulouse, France, who invited the main author to conduct this research. Our deepest gratitude to Gulmira Bekmanova, whose contributions have facilitated the exchange between Astana and Toulouse. Special thanks are due to Esref Adali (Istanbul Technical University), Stergos Afantenos and Sergei Soloviev (IRIT / UT) for their advice, which significantly contributed to the improvement of this work. Inari Listenmaa provided essential help with GF, and Aibek Makazhanov with setting up the morphological analyser [17] developed by him and his

[8] https://github.com/Gaziza-Yelibayeva/.

colleagues. This research was made possible by a Bolashak International Scholarship (500 Scientists) of the President of the Republic of Kazakhstan.

References

1. Aho, A.V., Sethi, R., Ullman, J.D.: Compilers: principles, techniques, and tools. Addison-Wesley (1985)
2. Akhmed-Zaki, D., Mansurova, M., Kadyrbek, N., Barakhnin, V., Misebay, A.: Creation of a dependency tree for sentences in the kazakh language. In: Nguyen, N.T., Hoang, B.H., Huynh, C.P., Hwang, D., Trawiński, B., Vossen, G., (eds.) Computational Collective Intelligence, pp. 709–718, Cham (2020). Springer International Publishing
3. Akhmed-Zaki, D., Mansurova, M., Madiyeva, G., Kadyrbek, N., Kyrgyzbayeva, M.: Development of the information system for the kazakh language preprocessing. Cogent Eng. **8**(1), 1896418 (2021)
4. Appel, A.W.: Modern compiler implementation in ML. Cambridge University Press (2004)
5. Assylbekov, Z., et al.: A free/open-source hybrid morphological disambiguation tool for Kazakh. In Proc. TurCLing **2016**, 04 (2016)
6. Blevins, J.P.: Feature-based grammar. Non-Transformational Syntax: Formal Explicit Models of Grammar, 297–324 (2011)
7. de Marneffe, M., Manning, C.D., Nivre, J., Zeman, D.: Universal dependencies. Comput. Linguis. **47**(2), 255–308 (2021)
8. Heim, I., Kratzer, A.: Semantics in generative grammar. Blackwell (1998)
9. Janpeyisov, E., (ed.) The kazakh grammar. Phonetics, Word Formation, Morphology, Syntax. "Astana Polygraphia" CJSC (2002)
10. Jurafsky, D., Martin, J.H.: Speech and language processing. Web, 3 edition (2024)
11. Kaliakbarova, G., (ed.) The Kazakh Language (Short grammar reference book). Institute of State language development (2013)
12. Kessikbayeva, G., Çiçekli, I.: Rule based morphological analyzer of Kazakh language. In: SIGMORPHON/SIGFSM (2014)
13. Kessikbayeva, G., Cicekli, I.: A rule based morphological analyzer and a morphological disambiguator for kazakh language. Linguis. Lit. Stud, 4, 96–104 (2016)
14. Kozhirbayev, Z., Yessenbayev, Z., Makazhanov, A.: Document and word-level language identification for noisy user generated text. In: 2018 IEEE 12th International Conference on Application of Information and Communication Technologies (AICT), pp. 1–4 (2018)
15. Kübler, S., McDonald, R., Nivre, J.: Dependency Parsing, pp. 11–20. Springer International Publishing, Cham (2009)
16. Makazhanov, A., Sultangazina, A., Makhambetov, O., Yessenbayev, Z.: Syntactic annotation of kazakh: following the universal dependencies guidelines. a report. In: 3rd International Conference on Computer Processing in Turkic Languages, pp. 338–350 (2015)
17. Makhambetov, O., Makazhanov, A., Sabyrgaliyev, I., Yessenbayev, Z.: Data-driven morphological analysis and disambiguation for kazakh. In: Gelbukh, A., (ed.) Computational Linguistics and Intelligent Text Processing, pp. 151–163, Cham (2015). Springer International Publishing

18. Makhambetov, O., Makazhanov, A., Yessenbayev, Z., Matkarimov, B., Sabyrgaliyev, I., Sharafudinov, A.: Assembling the Kazakh language corpus. In: Yarowsky, D., Baldwin, T., Korhonen, A., Livescu, K., Bethard, S., (eds.) Proceedings of the 2013 Conference on Empirical Methods in Natural Language Processing, pp. 1022–1031, Seattle, Washington, USA, October (2013). Association for Computational Linguistics
19. Meirambekkyzy, Z., Khoroshilov, A.: Syntactic and semantic representations for the development of Kazakh-Russian phraseological machine translation. In: 5th International Conference on Turkic Languages Processing (TurkLang 2017), pp. 181–194 (2017)
20. Myrzakhmetov, B., Yessenbayev, Z., Makazhanov, A.: Initial normalization of user generated content: case study in a multilingual setting. In: 2018 IEEE 12th International Conference on Application of Information and Communication Technologies (AICT), pp. 1–4 (2018)
21. Ranta, A.: Grammatical framework. J. Funct. Program. **14**, 145–189 (2004)
22. Ranta, A.: Translating between language and logic: what is easy and what is difficult. In: International Conference on Automated Deduction, pp. 5–25. Springer (2011)
23. Sharipbay, A.: The kazakh language requires reform of its writing. IgMin Res., **2**(2), (2024)
24. Sharipbay, A., Yergesh, B., Razakhova, B., Yelibayeva, G., Mukanova, A.: Syntax parsing model of Kazakh simple sentences. In: Proceedings of the Second International Conference on Data Science, E-Learning and Information Systems, DATA '19, New York, NY, USA (2019). Association for Computing Machinery
25. Somers, H.L. On the validity of the complement-adjunct distinction in valency grammar. Linguistics, **22**(4), 507–530 (1984)
26. Tesnière, L.: Éléments de syntaxe structurale. Klincksieck, Paris (1988)
27. Toleu, A., Tolegen, G., Makazhanov, A.: Character-based deep learning models for token and sentence segmentation. In: Proceedings of the 55th Annual Meeting of the Association for Computational Linguistics, ACL, pp. 99–112 (2017)
28. Toleu, A., Tolegen, G., Mussabayev, R.: Comparison of various approaches for dependency parsing. In:2019 15th International Asian School-Seminar Optimization Problems of Complex Systems (OPCS), pp. 192–195 (2019)
29. Tukeyev, U., Karibayeva, A., Turganbayeva, A., Amirova, D.: Universal Programs for Stemming, Segmentation, Morphological Analysis of Turkic Words, pp. 643–654. Springer International Publishing, Cham (2021)
30. Tyers, F.M., Washington, J.N.: Towards a free/open-source universal-dependency treebank for Kazakh. In: 3rd International Conference on Computer Processing in Turkic Languages (TURKLANG 2015) (2015)
31. Tyers, F.M., Washington, J.N., Çöltekin, C., Makazhanov, A.: An assessment of universal dependency annotation guidelines for Turkic languages. In: 5th International Conference on Turkic Languages Processing (TurkLang 2017), pp. 356–377 (2017)
32. Zhang, M.S.: A survey of syntactic-semantic parsing based on constituent and dependency structures. Sci. CHINA Technol. Sci. **63**(10), 1898–1920 (2020). https://doi.org/10.1007/s11431-020-1666-4

Application of Large Language Models in Grammar Correction and Feedback for English Writing

Jiayuan Wang[1], Li Cao[2], Shuzhen Ling[1], and Fangting Han[1](✉)

[1] College of Language and Culture, Northwest A&F University, Yangling, China
clairehft@126.com
[2] School of Foreign Languages, Beijing Institute of Technology, Beijing, China
lucycaoli@bit.edu.cn

Abstract. A comparison of the effectiveness and efficiency of two large language models, GPT-4 and DeepSeek, in English grammar correction is conducted using a corpus of FCE writing assignments produced by Chinese learners of English as a Foreign Language. A mixed quantitative-qualitative method is adopted to analyze correction accuracy and feedback mechanisms. Results reveal that DeepSeek is more effective, specializing in surface grammatical error detection and providing timely, accurate, structured feedback. GPT-4 excels in semantic coherence, logical cohesion, and stylistic consistency, with high revision accuracy and low false correction rate. Effective complementarity between the two models in form perception and semantic revision is observed, enabling the establishment of a collaborative error correction model from rule-based to meaning-based approaches. Large language models still face challenges in terms of stability, interpretability, and pedagogical flexibility; for their standardized and efficient application in writing instruction, algorithmic controllability and teacher collaboration mechanisms require further optimization.

Keywords: Large Language Models · Grammatical Correction · Feedback Mechanism

1 Introduction

Grammar is the cornerstone of second language(L2) learning, and the accurate application of grammar rules is the key for learners to build a language foundation. Grammar error correction helps students understand the rules of internalization by identifying and providing feedback on sentence errors, and promotes them from passive error correction to active correction [1].

With the development of artificial intelligence, automated tools are widely used in writing teaching. Traditional automatic grammar error correction systems (e.g., LanguageTool, Ginger, Grammarly) can provide instant error detection and correction suggestions based on rules, statistics, and other methods, but most of them take "error correction result optimization" as the core, which makes it difficult to help learners

H. Unger and P. Meesad (Eds.): NLPIR 2025, LNNS 1904, pp. 76–86, 2026.
https://doi.org/10.1007/978-3-032-20897-2_6

understand the causes of errors, unable to meet the needs of deep language understanding and skill transfer. This kind of system error correction mostly focuses on functional word form errors, limited handling of content words and semantic bias, and the scope of error correction is limited to the sentence level, which makes it difficult to deal with the problems of discourse connection and pragmatic appropriateness across sentences or chapters [2]. The application of the automatic evaluation system for composition (e.g., ETS's E-rater) is also mature, and it can achieve high consistency and objectivity through language feature analysis [3], but it is mostly limited to score output or template feedback, and it is difficult to handle complex syntax, Use pragmatic errors to make detailed explanations. Although teachers' written error correction feedback (WCF) has the advantages of personalization and situationalization, it has the problems of time-consuming and limited coverage. It is difficult to meet the needs of large-scale English writing teaching by relying solely on traditional systems or manual feedback.

The large language model(LLM) brings new opportunities for the teaching of L2 writing. Compared with the traditional machine review system, it has a stronger natural language understanding and generation ability. It can be reviewed from many aspects such as structure, coherence, vocabulary, grammar, etc. [4], and can also generate natural language interpretation, which has the potential of "semi-teacherization" in writing feedback. However, there is a lack of systematic comparison of grammatical error correction and feedback mechanisms of different LLMs (e.g., GPT-4 and DeepSeek), and there is a lack of research to transform the results into actionable teaching strategies. This essay addresses three key research questions to clarify the comparative evaluation:

- In what ways do GPT-4 and DeepSeek differ in accuracy and coverage for correcting grammatical errors in English writing?
- How do their Feedback Explicitness and Pedagogical Transferability characteristics compare?
- What teaching strategies can be developed to enhance learners' grammatical skills while leveraging the distinct strengths and addressing the weaknesses of each model?

Theoretically, this study pushes back the LLM testing from "correctness of error correction" to "effectiveness of feedback methods." Practically, it provides teachers and students with recommendations for selecting AI feedback based on learning stages and goals, making teaching assistance more effective.

2 GEC: from Rules to Intelligent Tutoring

2.1 The Evolution and Impact of Grammar Error Correction

Grammar error correction is a key method for improving accuracy in second language learning. Written error feedback has a positive effect on grammatical accuracy, influenced by factors like learners' educational level and feedback type [5]. Direct feedback helps beginners learn standards, while indirect feedback promotes self-correction and deeper language understanding. This process supports language internalization and self-monitoring, providing a basis for automated correction systems.

2.2 Technological Advances in Automatic Error Correction

GEC has evolved through successive stages: the early, rule-based systems were transparent but limited; later, statistical methods emerged with the pioneering use of decision-tree models [6]. Starting in 2010, neural machine translation framed GEC as a translation task. The neural models learned to capture complex error patterns from large datasets and made fluid corrections possible. More recently, LLMs have taken GEC another step further by way of not just correcting errors but also explaining them, shifting the focus from accuracy to interpretability and teaching value.

2.3 LLMs' Role in Writing Instruction

LLMs have widened the affordances in writing instruction. Different from traditional GEC systems, LLMs provide contextualized explanatory feedback. They can evaluate content and structure while giving precise language suggestions, including writing quality and grammar. Their interactive and real-time nature removes several of the traditional barriers to accessing feedback. In addition, LLMs mostly provide feedback in the form of direct correction, which is conducive to the rapid location and correction of problems for learners at different levels, and has strong interactivity and immediacy, which can break through the time and space restrictions to provide diversified and continuous feedback support [7]. However, the quality of LLM feedback depends on student input, and its reliability varies. Therefore, LLMs should be adopted within a "teacher-learner-model" approach that embeds rapid model responses within targeted teacher guidance for effective learning.

3 GEC: from Rules to Intelligent Tutoring

This study aims to systematically compare the performance of GPT-4 and DeepSeek LLMs in English writing grammar error correction tasks, focusing on their differential characteristics in error correction performance and feedback mechanism, and further exploring their applicability and teaching potential at different teaching stages.

The corpus of the study comes from the Cambridge First Certificate in English (FCE) corpus, which has been widely used in second-language writing research for a long time. This study selected 10 of the essays written by Chinese English learners as analysis samples, and the English level of the authors was at the intermediate and advanced stages of CEFR B1–B2. Each composition is accompanied by a manually revised version, forming a parallel corpus of "learner's original text - manual standard text", which provides an objective reference for bias identification and error correction judgment. Finally, take the grammatical errors identified in all ten essays as the basis for data analysis.

This study adopts the Surface Strategy Taxonomy proposed by Dulay, Burt, and Krashen as the theoretical framework [8]. (i) Omission, which refers to the absence of necessary grammatical components, including prepositions, articles, conjunctions, relative pronouns, punctuation marks, noun central words, verb tense mark omission, etc.; (ii) Addition, which refers to the redundant addition of grammatical components in the normative structure, including excessant conjunctions, prepositions, auxiliary verbs, qualifiers, subjects, etc.; (iii) Misordering, which refers to the wrong position of the word order or structure, including inverted order errors, interrogative word order errors, adverb misalignment, adjective misalignment, etc.; (iv) Misformation, refers to the use of wrong grammatical forms to replace the correct form, including noun singular plural, verb tense, prepositions, pronouns, subject-predicate consistent verb forms and spelling misconceptions, etc. The introduction of this framework not only provides a clear structured logic for grammatical bias analysis, but also lays a theoretical foundation for the comparison and feedback analysis of model error correction effects. The workflow in the experiment included the following steps.

Step 1: Determine the overall distribution characteristics of grammatical errors in 10 essays, with specific "grammatical Error Points" as the statistical unit.

Step 2: Each essay is input with GPT-4 and DeepSeek, respectively, using a unified prompt to call the task. The prompt is as follows: "You are an expert professor with experience in the field of second language acquisition and English writing teaching. Now, you need to conduct professional grammatical bias analysis, correction, and teaching feedback for the writing articles of Chinese English learners. The Corpus is selected from the FCE Corpus. The author is a native Chinese English learner with Chinese, and his English level is CEFR B1-B2. Please carefully identify all the grammatical errors in the text and present the results in the form of a table, including the following six columns: biased sentence, biased category, biased subtype, corrected sentences, biased causes, and teaching feedback. Among them, the 'biased category' is divided into four categories of Omission Error, Addition Error, Misordering Error, and Misformation Error according to the Surface Strategy Taxonomy method; the 'biased Subtype' should clearly indicate the specific types of errors, such as article omission errors, verb tense misformation errors, preposition misuse, subject and predicate consistency errors, etc. 'biased causes' refers to the origin of specific grammatical errors. Please give the complete version of the composition after error correction, after the analysis, to ensure that the language is natural and conforms to the English text specifications. Please start to analyze the following composition."

Step 3: The output results of both models are stored in the database as "Correction Content—Feedback Explanation" and compared with expert grammatical corrections in the Corpus. Each Corpus entry includes model feedback, error explanations, and teaching suggestions to support subsequent quantitative and qualitative analysis.

Step 4: Quantitative and qualitative analysis of model output is conducted per the research design. For quantitative analysis, key metrics including Correction Accuracy (CA), Correction Coverage (CC), Missed Detection Rate (MDR), Incorrect Correction Rate (ICR), and False Correction Rate (FCR) are calculated to assess correction performance; qualitative analysis focuses on feedback's Explanatory Capacity and PT to identify differences in their Feedback Mechanisms.

Step 5: Combining statistical and linguistic analysis, the two models' performance in GEC and feedback generation is compared. Their teaching strengths and limitations are identified, and finally, a tiered integrated classroom teaching strategy is developed.

4 Results and Discussion

4.1 Data Overview and Error Distribution

Before the formal comparison of Model Performance, this study conducts a statistical analysis of the overall error distribution in the experimental Corpus. In total, 226 grammatical Error Points were identified on 10 FCE texts, which had been hand-annotated (Table 1).

Table 1. Distribution of Grammatical Error Types in the Manually Annotated Corpus

Error Category	Frequency	Percentage (%)
Misformation	158	69.91
Omission	48	21.24
Addition	15	6.64
Misordering	5	2.21

The most numerous are Misformation Error, comprising over two-thirds of all errors, and being the primary Error Category. Basically, they involve verb tense, subject-verb agreement, non-finite verb forms, and punctuation substitution. This indicates that although the learners are capable of using explicit grammatical rules of the target language, in a complex sentence or context shift, they still tend to get the rules confused or make inflectional form selection errors. Omission Error comes in second and most often involves the lack of articles, prepositions, conjunctions, or punctuation. This reflects the weakness of learners in controlling syntactic completeness and discourse coherence, especially when the content is information-dense or contains complex subordinate clauses, and functional elements are more likely to be omitted. While Addition Error and Misordering Error occur quite rarely, they do reveal distinct characteristics of language transfer. Addition error mainly involves redundant elements or repeated markers, reflecting learners' tendency to overcompensate for target language structures in syntax generation. Misplacement error mainly occurs in constituent order or syntactic cohesion, revealing conflicts between native-language habitual thinking and English word order.

4.2 Comparative Analysis of Model Error-Correction: Quantitative Assessment Based on Accuracy and Coverage

In order to be scientific and reproducible, this study uses manually annotated results from the FCE Corpus as the benchmark and defines the performance of grammatical error correction through standardized metrics and formulaic calculations. The performance evaluation system includes 2 core metrics, CA and CC, and 3 auxiliary metrics, MDR, ICR, and FCR, which will reveal the performance characteristics of the models in syntactic error correction from the dimensions of precision and breadth.

CA reflects how precise a model's corrections have been for its identified errors, calculated by CA = (Number of Error Points correctly corrected ÷ Total number of Error

Points covered) × 100%. "Error Points correctly corrected" means the revisions that the model made at the syntactic, spelling, punctuation, and semantic levels are consistent with human annotations. CC quantifies a model's overall competence in detecting and correcting errors fully within the Corpus. CC = (Number of Error Points correctly corrected ÷ Total number of manually annotated Error Points) × 100%. MDR refers to what errors a model fails to detect. MDR = (Number of unidentified Error Points ÷ Total number of manually annotated Error Points) × 100%. ICR reveals the proportion of errors that the model detected but did not correct accurately. ICR = (Number of errors detected but incorrectly corrected ÷ Total number of Error Points covered) × 100%.FCR indicates the proportion of erroneous modifications that have been made to correct sentences. FCR = (Number of misjudged sentences ÷ Total number of error-free sentences) × 100% (Table 2)

Table 2. Overall Performance Comparison between GPT-4 and DeepSeek

Metric	DeepSeek (%)	GPT-4 (%)
CA	64.27	66.12
CC	47.35	38.94
MDR	26.33	41.11
ICR	35.73	33.88
FCR	6.90	1.72

Error correction of ten essays shows that both LLMs possess strong overall grammar correction capabilities, with marked differences in metrics. The gap in correction accuracy is small; both correct explicit grammatical errors rather accurately. DeepSeek has broader detection: its CC is significantly higher than GPT-4's, with a lower MDR. It is more sensitive to structural errors like verb inflections, article omissions, and prepositional collocations, with a tendency to respond quickly via formal features. At the same time, it has a higher FCR, with a tendency to make inappropriate revisions due to formalistic judgments in semantic dependence or syntactic ambiguity. GPT-4 has slightly better CA and a much lower FCR, with more stable contextual awareness and pragmatic judgment. Revision focuses on semantic reasonableness and contextual coherence, often adopting discourse information to reduce unnecessary formal revisions. However, it adopts a conservative strategy for slight formal bias or low-risk punctuation, which reduces the risk of misjudgment but may miss some surface errors (Table 3).

For Misformation Error, deepSeek detects more grammatical errors than GPT-4 and shows higher sensitivity to morphological and syntactic clues. However, its broad detection usually results in more identification and correction errors. When there are several mistakes in a sentence, DeepSeek sometimes corrects part of them, leaving the tense or collocation inconsistent. It also tends to over-standardize British and American spelling and stylistic punctuation.

Omission Error primarily includes the lack of articles, prepositions, subjects, auxiliary verbs, and punctuation. DeepSeek has a higher coverage and can find missing

Table 3. Model Performance across Different Error Categories

Error Catagory	Metric	DeepSeek(%)	GPT-4(%)
Misformation	CA	64.49	66.30
	CC	43.67	38.61
	MDR	32.28	41.77
	ICR	35.51	33.70
	FCR	17.86	7.14
Omission	CA	73.68	68.75
	CC	58.33	45.83
	MDR	20.83	33.33
	ICR	26.32	31.25
	FCR	7.14	3.57
Addition	CA	63.64	62.50
	CC	46.67	33.33
	MDR	26.67	46.67
	ICR	36.36	37.50
	FCR	10.71	7.14
Misordering	CA	75.00	66.67
	CC	60.00	40.00
	MDR	20.00	40.00
	ICR	25.00	33.33
	FCR	3.57	0.00

function words or collocational gaps fast. Yet, its high amount of detection increases false corrections, and it sometimes adds unnecessary elements to sentences that are already complete in meaning. GPT-4, by contrast, pays more attention to discourse fluency and coherence. It avoids excessive Addition Error, maintaining a better balance between grammatical completeness and natural flow; it may, however, miss minor surface Omission Error.

Addition Error includes redundant modifiers, repeated conjunctions, and unnecessary prepositions. DeepSeek performs well in deleting redundancies by pattern matching but may classify acceptable repetitions as mistakes, which increases the rate of false positives. GPT-4, on the other hand, uses semantic and discourse judgment to decide whether repetition affects meaning, thus giving much steadier accuracy and more natural output.

The Misordering Error includes the wrong adverb position, wrong phrasal verb order, and inappropriate connective sequence. While DeepSeek performs well in rule-based positional corrections, it may misjudge the flexible word orders as errors. GPT-4, in contrast, provides more accurate and context-sensitive revisions in cases of complicated sentences while keeping coherence and appropriateness with lower misclassification rates.

DeepSeek covers many error types, with a sensitive detection ability, and thus is suitable for preliminary grammar screening and quickly finding obvious errors. GPT-4 is more precise and stable in revision so as to be better at deep semantic and pragmatic corrections. Both models are complementary in coverage and accuracy, providing data to build a two-stage intelligent error correction model.

4.3 Dual-Dimensional Analysis of Model Feedback Mechanisms

After analyzing the grammatical correction results, the study further compares the feedback features of DeepSeek and GPT-4 from two aspects, Feedback Explicitness and Pedagogical Transferability. Feedback Explicitness concerns whether the model explains grammatical rules and error causes clearly, while Pedagogical Transferability focuses on whether feedback can help learners apply learned knowledge in new contexts (Table 4).

Table 4. Comparison of Feedback Mechanisms

Dimension	DeepSeek	GPT-4
Explanatory Style	Rule-driven and structured	Contextual and meaning-oriented
Focus of Explanation	Grammatical rules and form constraints	Language function and logical coherence
Transfer Approach	Rule generalization and template reuse	Pragmatic and discourse-based reasoning
Feedback Focus	Structural stability and syntactic regularity	Expressive flexibility and contextual appropriateness
Transfer Level	Syntactic Level	Discourse Level
Pedagogical Applicability	Early-stage grammar reinforcement	Intermediate-to-advanced writing development
Pedagogical Applicability	Early-stage grammar reinforcement	Intermediate-to-advanced writing development
Instructional Implication	Emphasizes form memorization and accuracy	Emphasizes meaning construction and fluency

DeepSeek provides highly structured and rule-based explanations. Its feedback normally follows this general pattern: after spotting the error, it explains the rule and gives the appropriate example. For the sentence "Walking to school, the rain started", GPT-4 comments that the sentence does not have a logical subject, and therefore, the agent is not clear. It recommends rewriting it as "While I was walking to school, the rain started" and suggests that grammatical accuracy secures logical and semantic coherence. DeepSeek also identifies the missing subject of the sentence and suggests the correction: "When I was walking to school, the rain started." Then, it gives a concise rule: When a participle phrase acts as an adverbial modifier, its logical subject should be the same as that of the main clause. Its explanation is standardized and suitable for rule-oriented learning, while GPT-4's commentary links grammar to meaning and discourse cohesion.

The main difference in the feedback of DeepSeek and GPT-4 lies in explicitness and pedagogical value. DeepSeek follows a similar pattern: identification of the error, explanation of the rule, and a correct example. When correcting, "she suggested that join", for example, DeepSeek states that the verb suggest takes a clause rather than a direct object and provides the proper form: "she suggested that I join". Quite often, it adds short notes such as "verbs like recommend and insist take a that-clause," which help students generalize grammatical rules. GPT-4 presents a broader explanation of the same sentence, mentioning that suggest is a polite way of recommendation and that using an infinitive would sound unnatural. Further, it explains that the that-clause is more formal when writing, thus relating grammar to tone and communicative context.

In general, DeepSeek emphasizes the accuracy of rules and forms, while GPT-4 focuses on the relationship between meaning and pragmatics. The advantages of the two complement each other, providing a two-way revelation for grammar teaching that takes into account formal accuracy and communication.

Beyond the model-to-model comparison, the patterns found in this study also have broader implications for research on natural language processing. The differences between the two models suggest that LLMs involve multiple layers of linguistic processing in their handling of learner-produced texts. One layer operates by increased sensitivity to morphological and syntactic forms, which allows for the quick detection of explicit deviations at the surface level. Another layer reflects the integration of discourse relations, semantic coherence, and contextual expectations, which underpin revisions that emphasize meaning and communicative appropriateness. Investigating how these layers interact during the correction process yields empirical evidence that current LLMs rely on a non-uniform, multi-component mechanism rather than on a single, deterministic pathway. In this respect, the present findings suggest that future natural language processing studies could develop correction frameworks that consider both structural precision and discourse-level reasoning, thereby enabling more interpretable, adaptable, and cognitively plausible grammar correction systems.

5 Challenges in Education

5.1 Integrated Instructional Design

Teaching design integrates the advantages of rule-oriented and context-oriented models to improve grammatical accuracy and communication clarity. The first stage is mainly based on rule teaching, guiding students to identify and correct common mistakes such as tense, subject, and predicate consistency, and preposition misuse. The second stage introduces the context model to help learners organize ideas, maintain logic, and improve the coherence and style of chapters. Teachers coordinate between the two stages through the "double-ring mechanism": the first ring emphasizes structural inspection and rule explanation, and the second ring focuses on semantics and stylistic optimization. This combination shifts writing teaching from simple error correction to the process of cultivating independent revision and language awareness.

5.2 Practical Challenges and Future Directions

Although the application prospect is broad, it still faces multiple challenges in practical teaching. First of all, the stability and accuracy of the model are affected by the prompts and writing style, which may lead to feedback bias and strengthen wrong language habits. Secondly, although the grammatical explanation of the model output is correct, it may not fit the learner's cognitive level and teaching goals, which affects the teaching value. Thirdly, student texts involve privacy and copyright, and data security and ethical norms need to be strictly guaranteed. Finally, teachers need to improve their artificial intelligence literacy in order to screen, evaluate, and rewrite model feedback more effectively. In the future, teaching should strengthen the interpretability of the model, ensure ethical compliance, and establish a sustainable training system to support the collaboration between teachers and students.

Given the discussion above, the findings from this study do not suggest that LLMs are unsuitable for educational grammar correction. Instead, results clearly indicate that any effective utilization in teaching relies on proper adaptation at both the technical and instructional levels. Their instability regarding prompt variation, the quality of the feedback provided, the mismatch between model-driven explanations and learners' cognitive readiness, and privacy and data security concerns all indicate that further refinement is necessary before LLMs can serve as reliable tools in classrooms. On the technical side, developing mechanisms that provide more controllable correction behavior, more adjustable levels of error sensitivity, and clearer explanations of underlying grammatical principles would enhance the stability and reliability of models. At the instructional level, teachers must continue to scaffold the use of model-generated feedback by filtering, sequencing, and contextualizing it so that the feedback is congruent with instructional objectives and supports learners' gradual development of linguistic awareness. These considerations point to the fact that an effort should be made not to exclude LLMs from educational applications, but integrate them through judicious alignment of model capabilities, task design, and teacher guidance, as a result of which the model output would represent a meaningful constituent of the teaching-learning process and not as an independent evaluator.

6 Conclusion

This essay compared two LLMs in correcting Chinese learners' English writing and discussed the differences in their correction ability and Feedback Mechanisms. The results indicate that both of the above models can complement each other. The rule-based model can identify and correct explicit grammatical errors more effectively, while providing clear references to error distribution for teaching purposes. A context-oriented model performs better in terms of revision accuracy, semantic coherence, and stylistic consistency, which aligns more closely with the real needs of writing. By combining them, we can build a layered teaching framework that moves from form to meaning and from sentence to discourse, improving accuracy and fluency. In actual classroom practice, the rule-based model is fit for preliminary filtering, whereas the context model is able to support deeper revisions. Transferring instant corrections into transferable learning resources depends on teachers who filter and integrate model feedback. The limitation of this study lies in the scope of the Corpus and the model parameters. Further studies should include an increase in types of corpora, multi-round interaction tasks, and the use of learning analytics for dynamic adjustments in the strategy of feedback.

7 Acknowledgments

This paper is also a research project of "The International Communication Competence of Agricultural University Students in Disseminating Rural Narratives". The supervisors of this paper are Li Cao, Shuzhen Ling, and Fangting Han.

References

1. Su, Q.: Analysis of the effectiveness of large language models in second language teaching. Foreign Lang. World. **3**, 35–42 (2024)
2. Bryant, C., Yuan, Z., Qorib, M.R., Cao, H., Ng, H.T., Briscoe, T.: Grammatical error correction: a survey of the state of the art. Comput. Linguist. **49**(3), 643–701 (2023)
3. Shermis, M.D., Burstein, J. (eds.): Handbook of Automated Essay Evaluation: Current Applications and New Directions. 2nd edn. Routledge, New York (2013)
4. Xu, J., Zhao, C., Sun, M.: Applications of Large Language Models in Foreign Language Teaching and Research. Foreign Language Teaching and Research Press, Beijing (2024)
5. Miao, J., Yao, W.: A meta-analysis of the effect of written corrective feedback on grammatical accuracy in English writing. Foreign Lang. Teach. Theor. Pract. **4**, 50–61 (2023)
6. Schmid, H.: Probabilistic part-of-speech tagging using decision trees. In: Proceedings of the International Conference on New Methods in Language Processing, Manchester, UK, pp. 44–49 (1994)
7. Wei, S., Li, L.: Artificial intelligence–assisted feedback in second language writing: a case study of ChatGPT. Chin. Foreign Lang. **20**(3), 33–40 (2023)
8. Dulay, H.C., Burt, M.K., Krashen, S.D.: Language Two. Oxford University Press, New York (1982)

Uncovering Evolving Criminal Jargon: Detecting Semantic Evolution in Telegram Marketplaces

Manik Kaur(✉), Mario M. Kubek, and Anu G. Bourgeois

Department of Computer Science, Georgia State University, 25 Park Place, Atlanta, GA 30303, USA
mkaur7@student.gsu.edu, {mkubek,abourgeois}@gsu.edu

Abstract. Criminal communities in online marketplaces increasingly evolve their terminology to evade detection, rendering static keyword-based cybersecurity systems ineffective. This linguistic adaptation challenges monitoring efforts, as new jargon and semantic shifts emerge faster than manual updates can accommodate. We present a transformer-based natural language processing framework for quantifying and tracking semantic evolution in criminal terminology. The methodology employs incremental fine-tuning of contextual language models across monthly Telegram datasets, enabling the detection of semantic drift, vocabulary emergence, and temporal adaptation patterns. Applied to criminal communications from Telegram channels engaged in financial fraud and identity theft, our experiments demonstrate the framework's ability to identify evolving term meanings and shifting contextual associations within underground marketplaces.

Keywords: Criminal Linguistics · Semantic Drift · BERT · Cybercrime · Telegram · Natural Language Processing (NLP)

1 Introduction

Digital criminal activities have fundamentally transformed how illicit operations organize, communicate, and maintain security [1–3]. Financial fraud has surged to unprecedented levels, with consumers reporting over $12.5 billion in losses and more than 1.1 million identity theft incidents in 2024 [4]. Sophisticated cross-border cybercrime operations have been linked to nearly $193 million in losses affecting over 35,000 victims worldwide in a single two-month operation [5]. Federal agencies increasingly recognize the migration to online platforms, further complicating detection and enforcement strategies [6].

This migration to instant messaging platforms like Telegram has enabled the rapid growth of online criminal marketplaces, where actors coordinate fraud schemes, sell stolen data, and exchange techniques in near real time. Telegram's

M. M. Kubek and A. G. Bourgeois— Contributing authors.

H. Unger and P. Meesad (Eds.): NLPIR 2025, LNNS 1904, pp. 87–106, 2026.
https://doi.org/10.1007/978-3-032-20897-2_7

unique combination of features—including large group capacities, minimal content moderation, and easy channel creation—makes it particularly attractive for criminal operations while providing researchers with accessible public data for linguistic analysis. The platform's widespread adoption by criminal networks, combined with its open channel structure, offers an ideal environment for studying semantic evolution in illicit communications. Such environments require communication methods that can adapt quickly to changing threats while enabling seamless coordination among participants. Criminal operations range from purely online groups to hybrid human-automated systems, each requiring dynamic messaging practices [7]. Within these ecosystems, linguistic adaptation strategies increasingly rely on slang, codewords, misspellings, and metaphorical language to evade detection [8,9]. Therefore, this process of *semantic evolution*, where meanings change over time or new terms emerge, demands more dynamic approaches to language monitoring.

To address this challenge, we integrate insights from two traditionally separate domains—computer science and criminology. While computer science provides the technical foundation for analyzing large-scale linguistic data through transformer-based language models, criminology provides critical context for understanding the social, operational, and behavioral drivers behind communication practices in cybercrime networks. We introduce a transformer-based natural language processing (NLP) framework that leverages the contextual representation capabilities of large language models (LLMs) to track evolution in criminal terminology within Telegram marketplaces. This allows us to detect not only when terms shift in meaning, but also when new terms begin to emerge, supporting more proactive detection of evolving threats.

This paper presents our methodology and findings on criminal terminology evolution within Telegram marketplaces across five main sections. Section 2 reviews related work in semantic evolution and fraudulent communication analysis. Section 3 details our incremental training framework for detecting semantic drift in criminal terminology, including data processing and vocabulary selection strategies. Section 4 describes the experimental validation procedures. Section 5 presents our analysis of terminology evolution throughout 2023, revealing coordination patterns and co-evolutionary relationships within underground networks. Finally, we discuss implications for adaptive cybercrime monitoring and future research directions.

2 Related Work

Understanding how terminology evolves in specialized domains has long been a focus of computational linguistics. Foundational work by Heyer et al. [10] introduced volatility, a concept borrowed from econometrics, to quantify semantic change by tracking contextual shifts over time. They demonstrated that increased volatility signals an evolving consensus around word meanings, laying the groundwork for systematic analysis of terminology change [11]. Extending this idea, Holz and Teresniak [12] showed how volatility measures could reveal emerging topics through shifts in co-occurrence patterns, even for low-frequency terms, a critical feature for specialized or illicit vocabularies.

The study of dynamic vocabulary structures further advanced with Deussen et al. [13], who modeled semantic evolution using co-occurrence graphs subject to exponential decay. Their notion of "oblivion processes" captured how associations weaken unless reinforced, revealing structural transformation in word networks and providing a framework for understanding learning and forgetting in language use.

The theoretical foundation for semantic representation in computational linguistics stems from distributional semantics, rooted in De Saussure's structuralist distinction [14] between syntagmatic and paradigmatic relationships and formalized through Harris's distributional hypothesis [15]. This principle–that linguistic elements with similar distributions have similar meanings—was later operationalized by Miller and Charles [16] for lexical semantics, establishing that contextual similarity correlates with semantic similarity. These distributional models, conceptually related to vector space models in information retrieval [17], provide the theoretical foundation for modern embedding-based approaches to semantic analysis.

Within these embedding-based approaches, recent studies have developed techniques that model semantics in high-dimensional vector spaces. Kubek et al. [18] highlighted the limitations of static embeddings in domain-specific contexts and proposed an iterative refinement strategy for embedding adaptation. They emphasized that contextually evolving embeddings, rather than fixed representations, are better suited to capture the nuanced dynamics of technical vocabularies. Hamilton et al. [19] and Rudolph and Blei [20] further developed diachronic word embeddings to trace semantic change over time.

Transformer architectures like Bidirectional Encoder Representations from Transformers (BERT) have enabled shifts toward context-sensitive representations, offering new tools for tracking subtle semantic drift [21]. Bidirectional encoding allows models to interpret meaning based on refined context, a necessary capability when analyzing cybercrime environments, where surface-level word associations may remain stable even as meanings shift.

Despite significant advances in modeling language evolution, these methods often do not fully account for the intentional ambiguity and deception in illicit communications [22]. Recent evaluations reveal that advanced language models show significant fragility in handling ambiguous contexts compared to humans [23,24]. In illicit environments, criminals frequently employ specialized communication techniques including extensive use of "dark jargons" that intentionally obfuscate meaning to evade detection [25,26]. Techniques such as "word camouflage" and euphemisms are employed by malicious actors to bypass content moderation systems [27,28], highlighting the need for robust NLP methods that can handle fraudulent text perturbations [29].

Embedding-based and transformer-driven methods offer significant advances, yet they remain underutilized in contexts with intentional language manipulation. Cybercrime environments present unique challenges that require models capable of detecting not only organic semantic drift but also interpreting deliberate linguistic obfuscation. Static approaches fall short in capturing the evolving semantics of criminal slang, where shifts are often subtle and contextually driven [30]. To address these limitations, our work introduces an incremental, context-aware framework designed to model fraudulent communication dynamics—laying the groundwork for tracking semantic evolution within illicit ecosystems like Telegram.

3 Methodological Framework

This section presents our methodological approach for analyzing semantic evolution in criminal terminology within Telegram marketplaces. The methodology integrates data processing, vocabulary selection, incremental training, and contextual analysis to provide systematic detection of criminal linguistic adaptation patterns.

3.1 Data Processing Pipeline

The data processing pipeline transforms raw Telegram messages from criminal marketplaces into structured, temporally-segmented datasets suitable for transformer-based semantic analysis. The pipeline handles the distinctive characteristics of criminal communications on Telegram, where language is informal, obfuscated, and heavily reliant on platform-specific conventions.

Data collection begins with automated extraction from about 100 Telegram channels known for illicit activity, yielding approximately 1.8 million messages segmented into monthly corpora spanning 2023.

Once segmented, the data undergoes preprocessing designed specifically for the informal and often noisy nature of Telegram communication. We remove emojis, digits, and non-alphanumeric symbols that disrupt semantic parsing or introduce noise into the language model. Emojis, while serving as communication tools in casual messaging, add no semantic value for our terminology evolution analysis and can interfere with embedding quality. Digits are typically meaningless in criminal terminology contexts, while non-alphanumeric symbols are often used for evasion. Text normalization through lowercasing and whitespace regularization reduces redundancy in slang-heavy corpora where terms are inconsistently formatted (e.g., "FULLZ", "fullz", "Fullz").

Building on this cleaned and normalized text, the pipeline applies tokenization using a Natural Language Toolkit (NLTK)-enhanced word segmentation adapted for Telegram slang and unconventional syntax. Unlike traditional news or social media text, these messages often include intentionally malformed spellings, abbreviations, and hybrid tokens that must be interpreted within their local context. In some cases, channel names and sender identities are retained as contextual tokens when they contribute to understanding community-specific usage patterns.

Finally, the cleaned and tokenized messages are formatted for transformer model input using sequence truncation, attention masks, and padding. The result is a set of twelve temporally aligned corpora that maintain both the semantic and structural nuances of criminal communication, enabling precise downstream tracking of linguistic evolution.

3.2 Criminal Terminology Selection

Once the monthly corpora is prepared, the next step involves selecting a targeted criminal vocabulary suitable for semantic drift analysis. Rather than relying on general term frequency, the selection process prioritizes vocabulary that serves as strategic entry points into the financial fraud and identity theft ecosystem on Telegram.

Initial term identification leveraged cybercrime and financial fraud experts to focus on high-signal keywords commonly used in digital fraud marketplaces. Key terms include "fullz" (complete identity theft packages), "dumps" (raw credit card data), "logs" (compromised account credentials), and "drops" (money laundering infrastructure locations). This domain-informed approach proved essential for locating active Telegram channels engaged in illicit activity, as these specialized terms serve as reliable indicators of criminal marketplaces. Additional terms such as "checks", "slips" (fraudulent checks) and "glass" (counterfeit currency) were also included to capture slang relevant to payment fraud. Institutional abbreviations like "wells" (Wells Fargo), "navy" (Navy Federal Credit Union), and "chase" (JPMorgan Chase) were monitored for their role in discussions targeting specific banks.

To identify additional relevant terms and communities beyond the initial seed set, we used a network-based strategy that tracked how channels mention each other, share forwarded messages, and are linked through overlapping user activity. This expansion helped ensure broader vocabulary coverage of Telegram's cybercrime ecosystem.

The final set of focus terms—"checks", "fullz", "drops", "cc", "dumps", and "logs", reflect diverse roles in identity fraud and illicit financial transactions. While not selected for frequency alone, these terms are semantically rich, operationally central, and sufficiently represented across multiple months to support consistent embedding analysis.

3.3 Incremental Training Architecture and Algorithm

To track criminal terminology evolution over time, we implement an incremental training framework using sequential model training to detect semantic drift. We employ a checkpoint-based strategy where each month's model builds upon the previous one and adapts to newly observed linguistic patterns. This method mitigates catastrophic forgetting, a common challenge in temporally sequenced fine-tuning, by continuously building on previously learned representations.

Our framework leverages contextual embeddings, which represent words as high-dimensional numerical vectors that capture semantic relationships based on surrounding context. These dense representations enable detection of subtle meaning shifts by measuring changes in mathematical relationships between terms over time.

For embedding extraction, we address the challenge that encoder tokenization splits words into multiple subtokens requiring aggregation to obtain word-level representations. Contextual embeddings are extracted from the model's final hidden layer for each subtoken and aggregated using mean pooling to create a single embedding per word occurrence. This word-level aggregation strategy, rather than sentence-level representations, enables precise tracking of term-specific semantic evolution. For each target term, we compute stable representative vectors by averaging all word-level embeddings across all occurrences within each monthly corpus.

The full incremental training process is summarized in Algorithm 1, which outlines how each monthly dataset is used to adapt the language model and extract term embeddings. The core architecture uses masked language modeling, where we fine-tune the model on monthly Telegram corpora and extract contextual embeddings from the final hidden layer. For each target term, stable representative vectors are computed by averaging token-level embeddings across all occurrences within each monthly corpus:

$$e_t = \frac{1}{N_t} \sum_{i=1}^{N_t} \mathbf{h}_i^{(L)} \tag{1}$$

where $\mathbf{h}_i^{(L)}$ represents the contextual embedding from the model's final hidden layer for the i-th occurrence of the target term in month t, N_t is the total number of occurrences of the term in month t, and e_t is the resulting monthly averaged embedding vector. We quantify two distinct temporal relationships using the averaged embeddings. Semantic drift measures cumulative change from the baseline month:

$$D_t = 1 - \frac{e_{\text{baseline}} \cdot e_t}{||e_{\text{baseline}}|| \times ||e_t||} \tag{2}$$

Volatility captures month-to-month fluctuations:

$$V_t = 1 - \frac{e_{t-1} \cdot e_t}{||e_{t-1}|| \times ||e_t||} \tag{3}$$

where e_{baseline}, e_{t-1}, and e_t are averaged embedding vectors for the baseline, previous, and current months respectively; $\cdot$ denotes the dot product; $\times$ denotes scalar multiplication; $||\cdot||$ is the Euclidean norm (vector magnitude); $D_t \in [0, 2]$ measures semantic distance with higher values indicating greater change to new criminal practices; and $V_t \in [0, 2]$ represents cosine distance between consecutive months with $V_1 = 0$ for the first month.

Algorithm 1 Incremental Criminal Terminology Evolution Training

Input:

- Monthly criminal communication datasets D_1, D_2, $\ldots D_{12}$ (January-December, 2023)
- Base pre-trained encoder model M_0 (bert-base-uncased)
- Training configuration parameters

Output:

- Sequential model checkpoints M_1, M_2, $\ldots M_{12}$
- Monthly term embeddings E_1, E_2, $\ldots E_{12}$

Steps:

1. Initialize pre-trained encoder model M_0 with general language understanding capabilities
2. For each temporal period t ϵ $\{1, 2, \ldots 12\}$:
   (a) Load Data: Process monthly dataset D_t through preprocessing pipeline
   (b) Configure masked language modeling with standard masking probability
   (c) Incremental Training: Fine-tune model M_{t-1} on dataset D_t using incremental adaptation parameters
   (d) Checkpoint Preservation: Save model state M_t and tokenizer configuration
   (e) Embedding Extraction: Generate contextual embeddings E_t for target terms from final hidden layer
3. Validate Convergence: Monitor training loss progression and model perplexity across iterations
4. Return: Complete sequence of model checkpoints and corresponding embeddings

This integrated training and measurement framework offers a temporally aligned view of language evolution, grounded in the real-world activity patterns of Telegram-based cybercrime networks. By pairing embedding-based drift analysis with contextual co-occurrence (Sect. 3.4), we gain both quantitative and qualitative insight into how criminal terminology adapts over time.

3.4 Co-occurrence Network Analysis and Contextual Evolution

Embedding-based metrics provide quantitative measures of semantic change but lack explainability for human analysts who need to understand the underlying reasons for terminology shifts. To enhance explainability and provide interpretable insights, we employ co-occurrence network analysis to reveal how target terms relate to surrounding vocabulary. This approach bridges the gap between numerical drift measurements and human-interpretable word relationships, offering contextual understanding that complements the embedding analysis and enables explainable semantic evolution tracking.

Co-occurrence networks map changing term relationships, revealing the operational environments and conceptual associations behind terminology adaptation. The networks are constructed using a 5-token sliding window with stopword filtering, highlighting the top 20 most frequently co-occurring terms weighted by frequency. This approach validates semantic drift findings by uncovering emerging term relationships, temporal shifts in vocabulary organization, and term volatility at different periods.

For example, the term "checks" in Fig. 1 shows notable structural change between January and December 2023. In January, its network comprises 13 nodes and 82 edges, centered on terms such as "slips"(629 co-occurrences) and "log" (480). By December, the network expands to 17 nodes and 150 edges, with "slips" (3,499 co-occurrences) remaining a dominant associate with new terms like "cards" (2,338) and "valid" (1,808). This semantic neighborhood extension suggests that the meaning of "checks" has broadened from paper-based fraud toward encompassing a wider payment verification infrastructure.

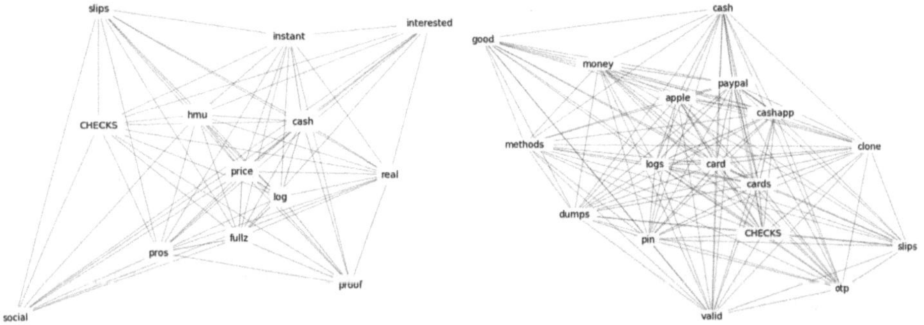

(a) January, 2023 (b) December, 2023

Fig. 1. The co-occurrence graphs of the term "checks" depicting change between January, 2023 and December, 2023

Analysis of other key terms reveals distinct evolutionary patterns. Some terms, such as "fullz" and "logs" demonstrate network expansion with increasingly complex semantic neighborhoods that reflect integration with mainstream financial platforms like "bank" (11,818) and "paypal" (4,535). In contrast, terms like "dumps" show co-occurrence networks that shrink to focus on a smaller set of highly specialized terms, reflecting advanced technical development and clearer, more unambiguous usage within the criminal vocabulary. Additionally, terms such as "drops" and "cc" illustrate institutional targeting by increasingly co-occurring with mainstream financial institution names such as "chase" (301), "navy" (272), "apple" (5,478) and "cashapp". This shift highlights a broader move from underground criminal tools toward exploiting legitimate financial platforms.

These patterns indicate a strategic evolution within cybercriminal communities, where criminal operations embed themselves deeper into established financial infrastructures. The increasing integration across different domains suggests that traditional boundaries between identity theft, payment processing fraud, and technical infrastructure exploitation are dissolving, reflecting a sophisticated and adaptive criminal ecosystem.

4 Experimental Setup

Our experimental dataset consists of 1.8 million messages from about 100 illicit Telegram channels, temporally segmented into monthly corpora spanning 2023. These monthly datasets reflect natural fluctuations in criminal communication, with volumes ranging from 173,790 messages in December to 760,816 in March, enabling systematic tracking of linguistic evolution across a full calendar year of criminal marketplace activity.

BERT serves as our core architecture due to its specific advantages for criminal terminology analysis. Unlike decoder-based models (e.g., GPT, LLaMA) that generate text autoregressively, BERT's bidirectional encoding processes full context simultaneously, making it better suited to capture subtle semantic relationships in obfuscated criminal language. While alternatives such as RoBERTa [31] or ALBERT [32] could also be viable, BERT's extensive pre-training on diverse corpora [21] and proven effectiveness in semantic similarity tasks make it an optimal choice. Additionally, BERT's encoder-only architecture is more computationally efficient for embedding extraction compared to larger decoder-based models.

For our implementation, we employed `BertForMaskedLM` [21] with the bert-base-uncased model [33], chosen for its balance between general language understanding and adaptability to the domain-specific slang and informal syntax present in Telegram communications. For embedding extraction, we address the critical challenge that BERT's subword tokenization splits individual words into multiple subtokens. Each target word may consist of multiple subtokens (e.g., "fullz" becomes ["full", "##z"]). To obtain word-level representations, we extract 768-dimensional contextual embeddings from BERT's final hidden layer (layer 12) for each subtoken and aggregate them using mean pooling to create a single embedding per word occurrence. This subtoken averaging approach ensures consistent word-level representations regardless of tokenization variability.

An alternative approach would be to use the classification ([CLS]) token, but word-level aggregation offers some advantages over using this method. The [CLS] token provides a sentence-level representation designed for sequence classification tasks, but does not capture the fine-grained, term-specific semantic evolution we seek to measure. Our approach directly focuses on the contextual representation of individual target terms, enabling precise tracking of semantic shifts for specific criminal vocabulary. For each target term, we compute stable representative vectors by averaging all word-level embeddings (themselves averaged from subtokens) across all occurrences within each monthly corpus.

The experimental training employs optimized configuration parameters designed to balance model adaptability with computational efficiency. Table 1 presents the key parameters and their justification, reflecting considerations such as prevention of overfitting, accommodation of typical message lengths, and mitigation of catastrophic forgetting during incremental updates.

Table 1. BERT Training Configuration Parameters

Parameter	Value	Purpose
E	3	Balances adaptation to new terminology with overfitting prevention
B	16	Optimized for GPU memory constraints while ensuring stable gradients
η	2e-5	Enables controlled adaptation without catastrophic forgetting
η_{decay}	Linear	Becomes more conservative as training progresses
P_{mask}	0.15	Standard BERT configuration for robust contextual learning
L_{seq}	512	Accommodates typical criminal communication lengths
λ	0.01	Prevents overfitting to specific monthly datasets
W_{steps}	100	Enables smooth adaptation at beginning of each iteration
Optimizer	AdamW	Adaptive moment estimation with decoupled weight decay
Architecture	BERT-base	12 layers, 768 hidden units, 12 attention heads

Where E = epochs per dataset D_t, B = batch size, η = learning rate, P = probability, L = max sequence length, λ = weight decay coefficient, W_{steps} = warmup steps

We conduct experiments on a CUDA-enabled GPU environment leveraging automatic mixed precision to optimize memory usage and training speed. This setup enables efficient handling of large-scale monthly Telegram datasets while supporting twelve sequential training iterations reflecting the temporal scope of 2023. Training effectiveness is monitored continuously through automated logging of loss and perplexity metrics, ensuring stable convergence throughout incremental fine-tuning.

The analysis investigates whether terminology changes represent random linguistic drift or systematic adaptation to evolving operational requirements, platform constraints, and external pressures. Additionally, the framework examines potential functional interdependencies through co-evolutionary relationships, where semantically related terms evolve in tandem to maintain operational coherence within criminal workflows. The following sections examine these dynamics through comprehensive analysis of semantic drift trajectories, contextual neighborhood evolution, and quantitative correlation patterns across the criminal terminology landscape.

5 Results

Examination of key measures of semantic drift, volatility, and frequency, assists in identifying patterns of linguistic change within criminal communication networks. Furthermore, these patterns reveal shifts in word usage that correspond with changes in criminal operations and strategies. This helps to build adaptive frameworks to improve real-time threat intelligence and intervention capabilities in digital criminal ecosystems.

5.1 Semantic Evolution Across Criminal Terminology

To examine how criminal language changes over time, we analyze three core measures across six selected terms: semantic drift (Fig. 2a), term volatility (Fig. 2b) and term frequency (Fig. 2c). Each captures different but complementary aspects of linguistic change. Semantic drift and volatility are calculated using Eqs. 2 and 3 respectively, while frequency measures the relative prominence of each term as its proportional share of the monthly corpus (4):

$$F_t = \frac{C_t}{N_t} \times 100 \tag{4}$$

where C_t represents the count of target term occurrences in month t, N_t is the total number of tokens in the monthly corpus, and $F_t \in [0, 100]$ expresses usage frequency as a percentage. Together, these metrics help distinguish deliberate semantic shifts from temporary variation, providing insight into how language adapts in response to evolving criminal strategies.

Figure 2a reveals increasing semantic drift across most terms throughout 2023, indicating active language evolution within underground communities. Terms associated with payment fraud and laundering, such as "checks" and "drops", show especially pronounced peaks, suggesting moments of significant innovation or change in underlying criminal techniques. In contrast, identity-related terms like "fullz" and "cc" exhibit more gradual or periodic changes, reflecting slower shifts in meaning that may align with ongoing adaptation rather

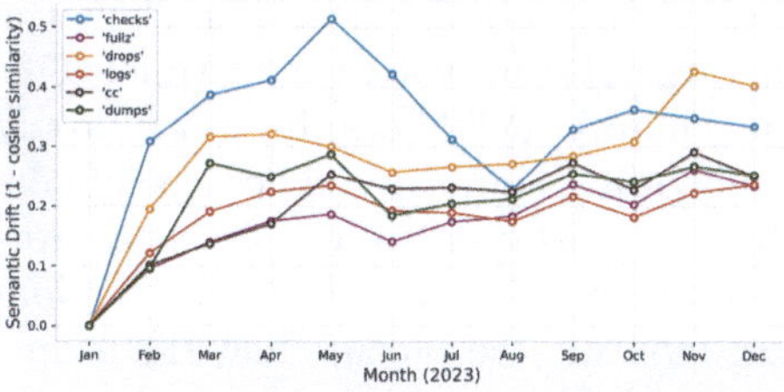

(a) Semantic Drift

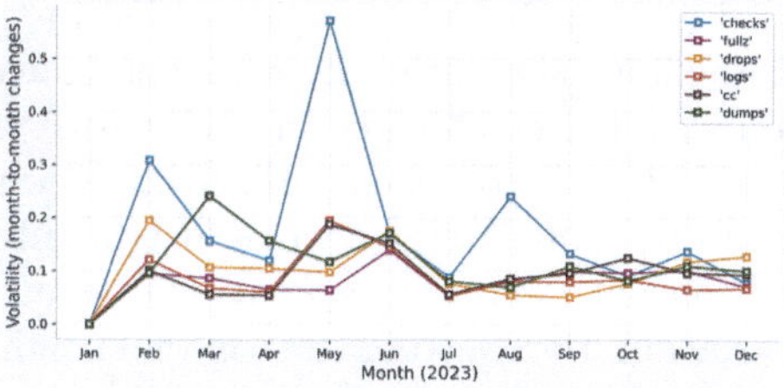

(b) Term Volatility

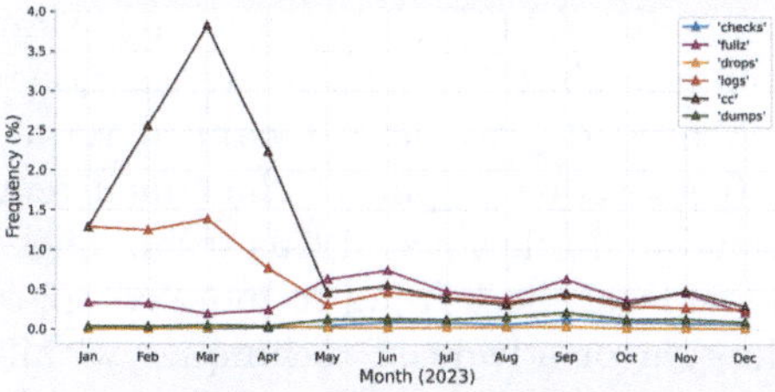

(c) Term Frequency

Fig. 2. Monthly Term Evolution Comparisons for Key Terms Across 2023 than disruption. Other terms like "logs" and "dumps" occupy an intermediate space, showing steady semantic movement with moderate peaks, likely reflecting their evolving role in supporting various fraud activities.

Volatility patterns in Fig. 2b provide additional insight into the stability of these semantic shifts. While "checks" again stands out with large volatility spikes, particularly in February and May, most other terms show relatively low volatility, suggesting more controlled and consistent adaptation. This contrast supports the interpretation that some terms undergo exploratory or contested re-use before settling into new meanings, while others evolve more predictably. In this light, volatility may serve as an early signal of shifting use before more permanent semantic drift takes hold.

Frequency trends in Fig. 2c show that usage prominence does not always track with semantic change. Some terms, particularly "cc" dominate early in the year with a sharp frequency spike in March, perhaps driven by a specific campaign or data breach, then declines despite continued drift. "Logs" shows similar front-loaded frequency, followed by decline. Terms like "checks" and "dumps" show small increase in frequency mid-year, aligning with their elevated drift, which could signal emerging relevance within evolving fraud schemes. These diverging patterns suggest that semantic evolution and usage volume are related but distinct processes, each capturing a different aspect of linguistic adaptation in underground networks.

5.2 Contextual Evolution Through Semantic Neighborhood Analysis

To build on the patterns identified in the previous section, we examine how the immediate semantic environment of each target term evolves over time. This analysis traces changes in each term's top semantic neighbors, its most closely associated words each month based on embedding similarity. These neighborhood dynamics offer deeper insight into how linguistic patterns shift in coordination with evolving operational strategies, rather than as a result of random variation. By tracking the top two neighbors monthly, we capture how the term's context transforms over time. Observing both the term and its closest associations helps distinguish coordinated adaptation from isolated or incidental drift.

The term "checks" (Fig. 3a) exhibits the highest average semantic drift (0.3294) and volatility (0.1743), with sharp peaks in both measures during May. This statistical pattern aligns with a semantic shift from basic transactional language early in the year (e.g., "cash", "needed") to more technical terminology associated with payment systems (e.g., "cpus", "pin"). The concurrent rise in frequency suggests the growing operational relevance of this term. These dynamics point to a clear transition from low-level check fraud to more complex financial manipulation, reflecting the adoption of advanced techniques within criminal networks and a later stabilization around terms like "money", "cards", and "slips".

"Fullz" (Fig. 3b) demonstrates the most gradual semantic evolution among the analyzed terms. With the lowest average drift (0.1691) and volatility

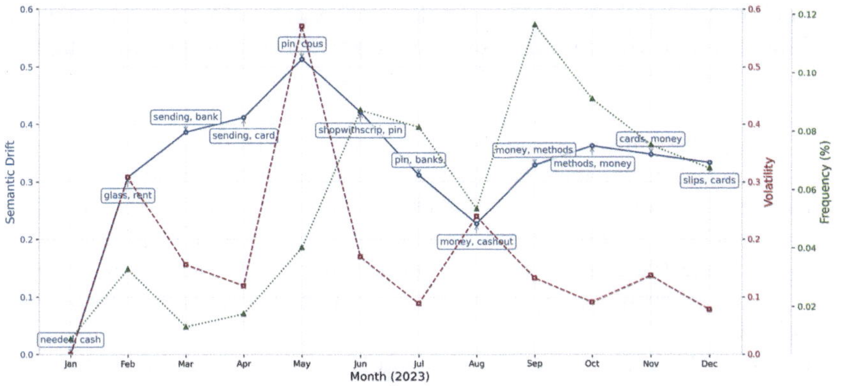

(a) "checks"

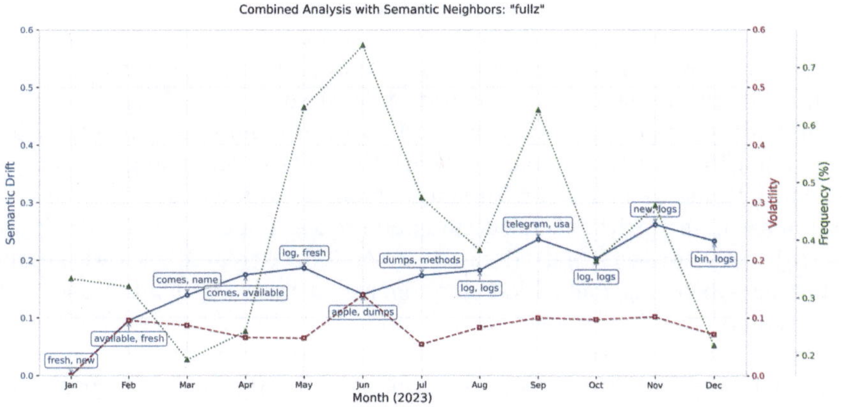

(b) "fullz"

Fig. 3. Individual Term Semantic Evolution with Top-2 neighbors

(0.0799), it transitions from terms emphasizing availability and freshness to those referencing delivery through secure or mainstream platforms like "telegram" and "apple". Although the term's frequency declines over time, its consistent semantic development points to an ongoing adaptation process. Rather than falling out of use, "fullz" is repositioned within a more secure and platform-oriented delivery context, indicating strategic embedding.

Analysis of additional terms (visualizations not shown) revealed similar evolutionary patterns. In contrast to "checks", the term "drops" shows moderate levels of drift (0.2795) and volatility (0.0991), with a significant peak in drift during November and in volatility during February. This growth coincides with changes in surrounding vocabulary, which shifts from generic institutional references to terms associated with urgency and targeted exploitation (e.g., "stimmy", "pnc", "instant"). The increasing frequency of usage supports this trend. These changes indicate an evolution from general references to drop sites toward more structured and time-sensitive fraud strategies, possibly aligned with rise in economic events or high-volume payout windows.

The term "logs" maintains a relatively stable context, with lower average drift (0.1820) and volatility (0.0853). Though a volatility peak in May reflects a temporary contextual shift. Despite a high average frequency (0.6201%), the overall frequency trend decreases suggesting that while "logs" remains relevant, its role is gradually being reframed. The evolution of its contextual neighborhood, from basic technical terms to integration with platforms like "cashapp" and associations with "dumps", suggests increased convergence with broader fraud ecosystems.

For "cc", which typically refers to stolen credit card data, the statistics show moderate drift (0.1992), steady volatility (0.0923), and a declining frequency trend following a peak in March. Early associations with underground marketplaces give way to proximity with mainstream platforms (e.g., "apple", "cashapp"). These shifts suggest that credit card fraud is moving beyond isolated forums and integrating more directly with popular financial tools, reflecting a broader shift in infrastructure exploitation strategies.

The trajectory of "dumps" shows significant semantic movement, with relatively high drift (0.2101), increasing volatility (0.1115), and rising frequency. The term shifts from associations with simple financial fraud ("cash", "needed") to terminology aligned with infrastructure and process ("logs", "methods", "pros"). This trajectory suggests that "dumps" are no longer treated as static commodities but are becoming embedded components in more automated and scalable fraud systems.

These term-level trajectories point to three broader themes in the evolution of criminal language. First, there is a consistent pattern of increasing technical sophistication, as terms evolve from basic operational labels to components of specialized fraud architectures. Second, mainstream integration occurs as criminal operations strategically align with legitimate platforms rather than competing against them, suggesting adaptive camouflage within existing infrastructure. Third, temporal coordination appears across multiple domains, with semantic evolution peaks occurring during similar timeframes (e.g., May and November), indicating synchronized adaptation rather than isolated changes. These patterns demonstrate that semantic drift in criminal terminology reflects coordinated strategic development, enabling sophisticated and organized operational capabilities. Next, we examine how such changes extend across the vocabulary, revealing broader dynamics of co-evolution.

5.3 Co-evolution and Mutual Influence in Criminal Terminology

The temporal alignment of semantic evolution peaks observed across multiple terms suggests potential coordination rather than independent development. To quantify these interdependencies, we analyze correlations in semantic drift, volatility, and frequency trajectories across all term pairs, measuring the extent to which criminal vocabulary evolves as an interconnected system (Table 2).

Table 2. Top 10 Co-Evolution Pairs Summary

Rank	Term Pair	Co-evolution Strength	Max Correlation	Drift_r	Volatility_r	Frequency_r
1	logs ↔ cc	0.891	0.929	0.874	0.929	0.870
2	checks ↔ logs	0.821	0.870	0.855	0.870	−0.737
3	fullz ↔ cc	0.711	0.941	0.941	0.633	−0.560
4	fullz ↔ dumps	0.707	0.870	0.870	0.572	0.678
5	checks ↔ cc	0.688	0.715	0.643	0.715	−0.707
6	logs ↔ dumps	0.688	0.946	0.946	0.354	−0.765
7	drops ↔ logs	0.641	0.923	0.923	0.568	−0.433
8	drops ↔ dumps	0.635	0.901	0.901	0.530	0.475
9	fullz ↔ logs	0.633	0.891	0.891	0.539	−0.470
10	checks ↔ dumps	0.618	0.865	0.798	0.190	0.865

Top 10 co-evolutionary pairs from analysis of 15 possible term combinations. Co-evolution strength represents mean of absolute Pearson correlations across semantic drift, volatility, and frequency metrics (5). All correlations calculated using 12 months of data (2023). 12 strong and 3 moderate co-evolution relationships identified.

$$\text{Co-evolution Strength} = \frac{|r_{\text{drift}}| + |r_{\text{volatility}}| + |r_{\text{frequency}}|}{3} \tag{5}$$

where r_{drift}, $r_{\text{volatility}}$, and $r_{\text{frequency}}$ are Pearson correlations between monthly time series for semantic drift, volatility, and frequency, respectively; $|\cdot|$ denotes absolute value; and co-evolution strength $\in [0, 1]$ is the mean absolute correlation with higher values indicating stronger synchronized evolution.

The quantitative analysis of co-evolutionary patterns in criminal vocabulary shows synchronized semantic development across related terms within underground communication networks. By examining Pearson correlation patterns across semantic drift, volatility, and frequency for all 15 possible term pairs, we found strong co-evolutionary evidence in 12 cases and moderate evidence in 3. The co-evolution strength represents a novel aggregation of term-level alignment across drift, volatility, and frequency dimensions, capturing how similarly two terms evolve over time.

The top 10 co-evolutionary relationships, summarized in Table display co-evolution strength values ranging from 0.618 to 0.891. The strongest relationship appears between "logs" and "cc" (strength: 0.891), with high correlations across all three metrics: semantic drift ($r = 0.874$), volatility ($r = 0.929$), and frequency ($r = 0.870$). Each of these exceeds the commonly accepted threshold of $|r| > 0.7$ for strong correlations, indicating that these terms co-evolve across usage and meaning [34]. The high volatility correlation suggests both terms respond similarly to external shifts, such as new fraud techniques or platform changes, reflecting functional interdependence in cybercriminal operations.

The second-strongest relationship is between "checks" and "logs" ($r = 0.821$), with strong positive correlations in drift ($r = 0.855$) and volatility ($r = 0.870$) but shows a strong *negative* correlation in frequency ($r = -0.737$). This pattern indicates that while their meanings and contextual volatility shift in tandem, the usage diverges—likely because the two terms serve similar functions, and one may be used in place of the other depending on current preferences or operational focus within the criminal community.

The "fullz-cc" pair ranks third ($r = 0.711$) and shows the highest drift among all term pairs ($r = 0.941$) indicating a nearly identical evolution of meaning over time.Although the volatility correlation is slightly below the strong threshold ($r = 0.633$) and frequency shows moderate negative correlation ($r = -0.560$), the overall co-evolution strength remains high. This pattern may reflect shared integration into mainstream platforms and overlapping roles in identity and financial fraud.

A notable outlier is the "fullz-dumps" pair (ranked 4th) which exhibits positive correlations across all three dimensions—drift ($r = 0.870$), volatility ($r = 0.572$) and frequency ($r = 0.678$), indicating complementary rather than competitive usage. Unlike most pairs that show negative frequency correlations, this likely represents closely related components in cybercrime workflows (e.g., fullz for identity data and dumps for card data) often used in parallel.

Seven of the top 10 pairs show strong *negative* correlations in frequency (ranging from $r = -0.433$ to $r = -0.765$), suggesting that competitive usage dynamics are common, where increased emphasis on one term correlates with decreased usage of its pair.

Looking at the maximum correlation for each pair helps identify which dimension most strongly drives co-evolution. Semantic drift emerges as the dominant factor in six out of ten top pairs, with values ranging from 0.643 to 0.946. Volatility shows greater variability (0.190 to 0.929), implying that while some terms respond similarly to contextual disruptions, others do not. These findings suggest that semantic alignment plays a more consistent role in terms of co-evolution than volatility and frequency. This analysis provides strong evidence that monitoring the evolution of one term within a co-evolutionary cluster may offer predictive insight to shifts in related terms, and be useful for early detection of emerging trends in underground communication.

6 Discussion

This research establishes a methodological foundation for understanding how illicit communication environments challenge traditional semantic analysis approaches. Grounded in the distributional hypothesis—"you shall know a word by the company it keeps" [35]—our framework tracks semantic evolution by comparing shifts in word contexts over time. Rather than relying on static definitions or curated lexicons, we detect meaning changes by observing shifts in a word's contextual distribution across temporal slices. BERT's bidirectional architecture implicitly encodes these distributions, enabling capture of meaning shifts at both subword and sentence level without explicit co-occurrence counts.

Volatility and drift metrics applied to monthly term embeddings operationalize this principle. Volatility captures short-term contextual instability, while drift measures long-term semantic departure. Both reflect the idea that changes in a word's "company"—its surrounding context—signal changes in meaning. This enables detection of overt semantic repurposing as well as subtler adaptations, such as re-contextualizing jargon for obfuscation.

The framework's results show that transformer-based incremental training can reveal coordinated linguistic adaptation. Temporal alignment of terminology shifts—such as the lag between volatility peaks for "checks" and "drops", suggests operational sequencing rather than random drift. Co-evolution across domains like payment processing, identity fraud, and infrastructure exploitation indicates vocabulary convergence, with illicit actors embedding discourse within mainstream platform language to evade detection. Such blending weakens the effectiveness of traditional detection methods.

By combining incremental fine-tuning with co-occurrence network analysis, our framework provides both granular semantic insights and broader contextual interpretation. Drift captures cumulative meaning change, while network dynamics reveal how term associations strengthen, weaken, or reorganize over time. This combined view provides better understanding of deceptive linguistic strategies, improving the interpretability of semantic evolution patterns.

While this framework provides valuable insights, several limitations remain. The analysis is constrained to one year of Telegram data, potentially overlooking longer-term trends or cross-platform coordination. Monthly training intervals may miss fast-moving terminology shifts tied to tactical pivots or emerging jargon. Additionally, our domain expert-driven approach for criminal terminology selection, while effective for identifying operationally central terms, may not capture the full spectrum of evolving criminal vocabulary in a systematic manner.

Several enhancements could strengthen the framework's operational impact. Finer-grained intervals, weekly or daily, could support real-time monitoring, enhancing early detection of weak signals: low-frequency terms showing sudden volatility or known terms used in new contexts, often preceding wider adoption. Our framework is well-suited to surface such signals, and coupling this with predictive modeling could enable proactive threat detection.

Future work could employ systematic corpus-based methods for more comprehensive term discovery. Approaches such as those proposed by Cox et al. [36], which use frequency-adjusted dispersion measures to identify prevalent terms across specialized corpora, could surface additional criminal terminology beyond domain expert knowledge. Such methods provide empirical criteria for term selection, enabling discovery of emerging terminology that demonstrates consistent usage patterns but may not yet be recognized by domain experts.

While automated discovery can expand term coverage, human expertise remains crucial for practical applications. Validation by domain experts is essential for operational use, bridging computational outputs with law enforcement needs. Extending the system to keyword prediction and automated channel classification could further strengthen its role in security workflows.

Beyond cybercrime, the methodology is generalizable to other domains where evolving slang, euphemisms, or coded language obscure underlying activity. This includes disinformation campaigns, hate speech, or radicalization rhetoric—any domain where language is deliberately adapted to conceal intent. By integrating distributional semantics with temporal embedding analysis, this framework supports scalable, adaptive monitoring, enabling earlier and more precise threat detection across diverse contexts.

7 Conclusion

This data-driven approach reveals that criminal communities employ sophisticated linguistic coordination strategies that extend far beyond random jargon evolution. Our transformer-based framework captures these deliberate adaptations by detecting synchronized terminology shifts and strategic embedding within legitimate digital ecosystems. Combining embedding-based drift detection and network analysis provides the quantitative precision and contextual understanding essential for real-time monitoring of illicit communication.

As criminal operations integrate with mainstream platforms and adopt legitimate terminology for camouflage, traditional static monitoring approaches face growing limitations. This work shows that incremental semantic analysis can capture and explain criminal communities' coordination strategies, laying the groundwork for adaptive monitoring that keeps pace with evolving language and supports timely intervention.

References

1. Mitrani, Y.: How digital transformation is giving serious crime and terror a high-tech makeover. Cognyte Blog (2024). https://www.cognyte.com/blog/how-digital-transformation-is-giving-serious-crime-and-terror-a-high-tech-makeover/
2. INTERPOL: when cybercriminals go global, our response must be international (2024). https://www.interpol.int/en/Crimes/Cybercrime
3. United nations office on drugs and crime: with criminal groups using sophisticated technologies to ply their wares, third committee stresses urgent need to collectively combat transnational crime. UN Press Release (2023). https://press.un.org/en/2023/gashc4374.doc.htm
4. Federal trade commission: new FTC data show a big jump in reported losses to fraud to $12.5 billion in 2024 (2025). https://www.ftc.gov/news-events/news/press-releases/2025/03/new-ftc-data-show-big-jump-reported-losses-fraud-125-billion-2024
5. INTERPOL: major cybercrime operation nets 1,006 suspects (2024). https://www.interpol.int/en/News-and-Events/News/2024/Major-cybercrime-operation-nets-1-006-suspects
6. U.S. department of state: cybercrime. Bureau of International Narcotics and Law Enforcement Affairs (2025). https://www.state.gov/cybercrime
7. Di Nicola, A.: Towards digital organized crime and digital sociology of organized crime. Trends Organ. Crime (2022). https://doi.org/10.1007/s12117-022-09457-y
8. Garkava, T., Moneva, A., Leukfeldt, E.R.: Stolen data markets on telegram: crime script analysis and situational crime prevention measures. Trends Organ. Crime (2024). https://doi.org/10.1007/s12117-024-09532-6

9. Roy, S.S., Vafa, E.P., Khanmohamaddi, K., Nilizadeh, S.: DarkGram: a large-scale analysis of cybercriminal activity channels on telegram. Stud. Health Technol. Inform. **310**, 981–985 (2024)
10. Heyer, G., Holz, F., Teresniak, S.: Change of topics over time - tracking topics by their change of meaning. In: Proceedings of the International Conference on Knowledge Discovery and Information Retrieval, pp. 223–228 (2009). https://doi.org/10.5220/0002330602230228
11. Kahmann, C., Niekler, A., Heyer, G.: Detecting and assessing contextual change in diachronic text documents using context volatility (2017). https://doi.org/10.48550/arXiv.1711.05538
12. Holz, F., Teresniak, S.: Towards automatic detection and tracking of topic change. In: Gelbukh, A. (ed.) Computational Linguistics and Intelligent Text Processing, pp. 327–339. Springer, Berlin, Heidelberg (2010). https://doi.org/10.1007/978-3-642-12116-6_27
13. Deussen, E., Immelnkämper, M., Unger, H.: Oblivion in time-dependent information management. In: The Autonomous Web. Studies in Big Data, vol. 101. Springer, Cham (2022). https://doi.org/10.1007/978-3-030-90936-9_11
14. de Saussure, F.: Course in general linguistics. McGraw-Hill Book Company, New York (1966)
15. Harris, Z.: Structural linguistics. University of Chicago Press, Chicago (1951)
16. Miller, G.A., Charles, W.G.: Contextual correlates of semantic similarity. Lang. Cogn. Process. **6**(1), 1–28 (1991). https://doi.org/10.1080/01690969108406936
17. Salton, G., Wong, A., Yang, C.S.: A vector space model for automatic indexing. Commun. ACM **18**(11), 613–620 (1975). https://doi.org/10.1145/361219.361220
18. Kubek, M.M., Pokharel, S., Böhme, T., McDaniel, E.L., Unger, H., Mikler, A.R.: On self-improving token embeddings. In: Zielinski, S., Eichler, G., Erfurth, C., Fahrnberger, G. (eds.) Innovations for Community Services, pp. 126–143. Springer Nature Switzerland, Cham (2025). https://doi.org/10.1007/978-3-031-94263-1_9
19. Hamilton, W.L., Leskovec, J., Jurafsky, D.: Diachronic word embeddings reveal statistical laws of semantic change. In: Proceedings of the 54th Annual Meeting of the Association for Computational Linguistics, pp. 1489–1501 (2016). https://doi.org/10.18653/v1/P16-1141
20. Rudolph, M., Blei, D.: Dynamic embeddings for language evolution. In: Proceedings of the 2018 World Wide Web Conference, pp. 1003–1011 (2018). https://doi.org/10.1145/3178876.3185999
21. Devlin, J., Chang, M.W., Lee, K., Toutanova, K.: BERT: pre-training of deep bidirectional transformers for language understanding. In: Proceedings of the 2019 Conference of the North American Chapter of the Assoc. for Computational Linguistics: Human Language Technologies, pp. 4171–4186 (2019). https://doi.org/10.18653/v1/N19-1423
22. Ocampo Diaz, G., Ng, V.: Unveiling hidden intentions. In: Proceedings of the AAAI Conference on Artificial Intelligence, vol. 34, pp. 13550–13555 (2020). https://doi.org/10.1609/aaai.v34i09.7080
23. Huertas-García, Á., Martín, A., Huertas-Tato, J., Camacho, D.: Countering malicious content moderation evasion in online social networks: Simulation and detection of word camouflage. Appl. Soft Comput. (145), 110552 (2023). https://doi.org/10.1016/j.asoc.2023.110552
24. Li, M.Y., Liu, A., Wu, Z., Smith, N.A.: A taxonomy of ambiguity types for NLP (2024). https://doi.org/10.48550/arXiv.2403.14072
25. Wu, X., et al.: Uncovering the fragility of trustworthy LLMs through Chinese textual ambiguity (2025). https://doi.org/10.48550/arXiv.2507.23121

26. Seyler, D., Liu, W., Wang, X.F., Zhai, C.X.: Towards Dark Jargon Interpretation in Underground Forums. In: Hiemstra, D., Moens, M.-F., Mothe, J., Perego, R., Potthast, M., Sebastiani, F. (eds.) ECIR 2021. LNCS, vol. 12657, pp. 393–400. Springer, Cham (2021). https://doi.org/10.1007/978-3-030-72240-1_40
27. Yuan, K., Lu, H., Liao, X., Wang, X.: Reading thieves' cant: automatically identifying and understanding dark jargons from cybercrime marketplaces. In: Proceedings of the 27th USENIX Security Symposium (USENIX Security 18), pp. 1027–1041 (2018). https://www.usenix.org/conference/usenixsecurity18/presentation/yuan-kan
28. Zhu, W., et al.: Self-Supervised euphemism detection and identification for content moderation. In: 2021 IEEE symposium on security and privacy (SP), pp. 229–246. IEEE, san francisco, ca (2021). https://doi.org/10.1109/SP40001.2021.00075
29. Bhalerao, R., Al-Rubaie, M., Bhaskar, A., Markov, I.: Data-driven mitigation of adversarial text perturbation (2022). https://doi.org/10.48550/arXiv.2202.09483
30. Kolla, A., Ilievski, F., Sandlin, H., Mermoud, A.: A study of slang representation methods (2022) https://doi.org/10.48550/arXiv.2212.05613
31. Liu, Y., et al.: RoBERTa: a robustly optimized BERT pretraining approach (2019). https://doi.org/10.48550/arXiv.1907.11692
32. Lan, Z., Chen, M., Goodman, S., Gimpel, K., Sharma, P., Soricut, R.: ALBERT: a lite BERT for self-supervised learning of language representations (2019). https://doi.org/10.48550/arXiv.1909.11942
33. Hugging face: BERT base uncased (2023). https://huggingface.co/google-bert/bert-base-uncased
34. Cohen, J.: Statistical power analysis for the behavioral sciences, 2nd edn. Routledge (1988). https://doi.org/10.4324/9780203771587
35. Firth, J.: A synopsis of linguistic theory, 1930–1955. Stud. Linguist. Anal, 10–32 (1957)
36. Cox, A., Dixon, D., Dixon, T.: Vocabulary list generator: a digital tool to generate frequency-based word lists adjusted for dispersion. Res. Methods Appl. Linguist. 4, 100180 (2025). https://doi.org/10.1016/j.rmal.2025.100180

Comparative Research on English Writing Text Features of Six Large Language Models from China and Abroad: A Usability-Focused Evaluation for Second Language Learner Engagement

Zhiting Luo(✉) and Xingsan Chai

Beijing Language and Culture University, Beijing 100083, China
zhitingluo15@gmail.com, cxs66@blcu.edu.cn

Abstract. Large language models (LLMs) based on artificial intelligence (AI) are reshaping the language education ecosystem, particularly in second language (L2) writing instruction. However, their effectiveness from a usability perspective remains underexplored. This study conducts a comprehensive comparative analysis of six mainstream LLMs—three from English-speaking countries (ChatGPT, Gemini, Claude) and three developed in China (DeepSeek, Ernie Bot, Kimi)—focusing on their practical application in L2 English writing. Using the Coh-Metrix text analysis tool, we evaluate 54 English texts generated by these models across five critical linguistic dimensions: lexical diversity, sentence cohesion, paragraph cohesion, syntactic complexity, and readability, while also examining learner engagement and usability-related factors beyond traditional linguistic assessments. The findings reveal significant performance differences between Chinese and Western LLMs: Chinese-developed models demonstrate competitive advantages in specific linguistic features and tend to impose lower cognitive load on L2 learners. These results have important implications for prompt engineering strategies and pedagogical applications, providing evidence-based recommendations for selecting appropriate LLMs in accordance with learner proficiency levels and instructional objectives, and highlighting how culturally situated AI development may shape the effectiveness of language learning tools.

Keywords: Large Language Models (LLMs) · English Writing Assessment · Coh-Metrix · Linguistic Features · Learner Engagement · Usability Evaluation · Prompt Engineering · Cross-Cultural AI Development

1 Introduction

1.1 Background and Significance

With the development of artificial intelligence (AI), a range of large language models (LLMs) such as ChatGPT, Claude, and DeepSeek have emerged in China and abroad. These models possess multilingual capabilities and can process and generate text across

H. Unger and P. Meesad (Eds.): NLPIR 2025, LNNS 1904, pp. 107–122, 2026.
https://doi.org/10.1007/978-3-032-20897-2_8

multiple languages without task-specific fine-tuning when appropriately prompted. Their application in education, especially in English as a second language (L2) writing instruction, has become increasingly widespread [1]. LLMs provide L2 English learners with immediate feedback, exemplars, and automated writing evaluation (AWE), thereby supporting improvements in lexical range, grammatical accuracy, and overall writing quality, while also helping teachers enhance instructional effectiveness and offer more personalized support [2, 3].

However, the effectiveness of these multilingual models in educational contexts extends beyond linguistic quality. The key issue is not whether LLMs can generate grammatically correct text—this capability is well established—but how learners should engage with these tools and which models provide the most appropriate support for different learning objectives and proficiency levels. This calls for usability-oriented evaluation that considers cognitive load, comprehension difficulty, and alignment with pedagogical goals.

Empirical studies have begun to examine the linguistic features of LLM-generated texts. ChatGPT outputs typically contain few grammatical errors and have been widely used in L2 writing instruction for generating outlines, offering ideas, revising drafts, and polishing essays [4]. These texts show higher word-form diversity and type–token ratios than learner essays [5] and outperform L2 writers in linguistic quality, content development, and discourse organization [6]. Other studies indicate that LLMs such as ChatGPT and Gemini, while strong in producing coherent and contextually appropriate sentences, remain limited in content selection and critical thinking [7]. LLMs developed in Chinese first-language environments, such as Ernie Bot and Kimi, may offer additional localized advantages for Chinese L2 English learners, including better handling of L1 interference, more culturally appropriate content, and closer alignment with local educational paradigms.

Despite this growing body of research, texts produced by different LLMs may still differ substantially in language complexity, lexical diversity, coherence, and sentence structure. From a usability perspective, models may also vary in their suitability for learners at different proficiency levels, their cognitive load implications, and their potential for meaningful learner engagement. As Ragha and Sasikumar note [8], the effectiveness of educational technologies depends critically on their alignment with user needs and cognitive processing capacities. Therefore, there is a need for experimental studies that systematically compare the linguistic features of texts generated by different LLMs and evaluate their usability from a learner-centred perspective, so as to inform model evaluation, selection, and pedagogically compatible application in L2 English writing.

Responding to these needs, this study employs the Coh-Metrix automated English text analysis tool to investigate English essays on the same topic generated by six mainstream LLMs from China and abroad (ChatGPT 3.5, Gemini 1.5, Claude 3.0, DeepSeek, Ernie Bot 4.0, and Kimi). By comparing indices of lexical diversity, cohesion, syntactic complexity, and readability across texts generated by these models, and interpreting these metrics from a usability perspective that considers learner engagement and cognitive load, the study provides a comprehensive evaluation of their relative advantages and offers empirical evidence for the targeted use of LLMs in L2 English writing instruction and assessment.

1.2 Application of Coh-Metrix in English Text Analysis

Coh-Metrix, developed by McNamara et al. at the University of Memphis, is an automated English text analysis tool that draws on corpus and computational linguistics to compute over 100 indices of readability, cohesion, syntactic complexity, and other textual features [9, 10]. Many of these indices are theoretically linked to cognitive processing demands and comprehension difficulty, making Coh-Metrix particularly suitable for evaluating texts from a learner engagement and usability perspective. It has been widely used in L2 English writing research to analyze linguistic features in student compositions, assess writing quality, and explore relationships between textual features and writing proficiency [10]. Previous studies show that Coh-Metrix metrics validly predict L2 proficiency: for example, Crossley and McNamara found that grammatical complexity, cohesion, and lexical diversity effectively distinguish writers at different levels [11], and Ullmann reported significant correlations between linguistic complexity, cohesion, lexical diversity, and writing scores [12]. These findings support the use of Coh-Metrix for both linguistic analysis and pedagogical evaluation.

In this study, we focus on four usability-relevant dimensions: lexical diversity, cohesion, syntactic complexity, and readability. Lexical diversity (LDTTRa) reflects vocabulary richness, which strongly influences holistic scores but, if excessively high, may increase comprehension difficulty for lower-proficiency learners [13–16]. Inter-sentence cohesion (LSASS1) and inter-paragraph cohesion (LSAPP1) capture how well ideas are connected across sentences and paragraphs, a critical factor for L2 readers' ability to follow discourse. Syntactic complexity (SYNLE), operationalized as the number of noun phrases per sentence, is generally positively associated with argumentative writing quality but can cause cognitive overload when overly complex [17]. The readability metric RDFKGL estimates overall text difficulty and thus directly indexes accessibility and likely cognitive load for learners at different proficiency levels [11]. Together, these Coh-Metrix indices provide a robust basis for evaluating the quality and educational suitability of LLM-generated texts, enabling a usability-oriented assessment that links quantitative text features to L2 learners' cognitive processing.

2 Research Design

2.1 Research Corpus

The corpus used in this study consists entirely of English texts generated by six mainstream large language models (LLMs) from both domestic and international sources. Specifically, these models include ChatGPT, Gemini, Claude, DeepSeek, Ernie Bot, and Kimi. All six models are general-purpose LLMs widely used for text generation, including writing support. Table 1 presents the basic overview of the six LLMs used as research subjects.

To deeply analyze the linguistic features of texts generated by different LLMs, we used identical prompts and writing rubrics, guiding the six models to generate 9 English texts each under the same prompt framework, yielding a total of 54 texts as research corpus. The text generation instructions are as follows.

Table 1. The basic features of six LLMs

LLMs Name	Company	Issue Year	Country	Language
ChatGPT	OpenAI	2022.11	America	English
Claude	Anthropic	2023.03	America	English
Gemini	Google	2023.12	America	English
Ernie Bot	Baidu	2023.03	China	Chinese
Kimi	Moonshot AI	2023.01	China	Chinese
DeepSeek	DeepSeek	2025.01	China	Chinese

"Based on the following topic, write 3 English essays. Write at least 250 words in each essay. Each essay should be divided into paragraphs and logically coherent.

Topic:In universities, people tend to concern more about the career prospect of the subjects they learn. Some people propose that students should learn subjects with a fast job growth including science, technology, and engineering even if these are not parts of their field of study. Do you agree or disagree this statement? Use specific reasons and examples to support your answer. Write at least 250 words."

To avoid bias in text generation due to topic factors, our writing task design drew from large-scale international standardized English writing examinations (adapted from IELTS 8 Test 2), using unified prompts to guide the six LLMs in argumentative writing around given themes. The writing task required models to generate English compositions of no less than 250 words, with open-ended discussion questions as content.

This study will analyze and comparatively research typical linguistic feature metrics of texts generated by different LLMs using the Coh-Metrix English text analysis tool on the composition corpus generated by the large language models, with particular emphasis on usability factors including cognitive load, comprehension difficulty, and learner engagement potential.

2.2 Research Questions

1. Are there significant differences among texts generated by the six LLMs in terms of lexical diversity, sentence cohesion, paragraph cohesion, syntactic complexity, and readability metrics?
2. Are there significant differences between texts generated by domestic and international LLMs in terms of lexical diversity, sentence cohesion, paragraph cohesion, syntactic complexity, and readability metrics?
3. From a usability perspective, which models generate texts most appropriate for L2 learners at different proficiency levels, considering factors of cognitive load and learner engagement?

2.3 Research Methodology

Firstly, data collection and preprocessing were conducted on the 54 English writing texts generated by the six LLMs. To ensure data accuracy and analytical operability, all texts were initially stored in Word documents and converted to txt format, forming a micro-corpus. Additionally, each composition was encoded according to its source and generating model. For example, ChatGPT-1 represents the first text generated by ChatGPT, Claude-2 represents the second text generated by Claude, and so on, ensuring clear source identification for each text.

Secondly, Coh-Metrix 3.0 automatic English text analysis software was used to extract and analyze linguistic feature metrics for each text. In this study, the linguistic metrics extracted by Coh-Metrix primarily include LDTTRa (lexical diversity), LSASS1 (sentence cohesion), LSAPP1 (paragraph cohesion), SYNLE (syntactic complexity), and RDFKGL (readability). These metrics were selected not only for their linguistic significance but also for their direct relevance to usability assessment, cognitive load evaluation, and learner engagement potential.

Finally, based on the linguistic feature metrics such as lexical diversity, cohesion, syntactic complexity, and readability extracted by Coh-Metrix for all model texts, SPSS 27.0 software was used to conduct descriptive statistical analysis of the above metrics for texts generated by each model. Additionally, one-way analysis of variance (ANOVA) was employed to deeply examine differences among different models in various linguistic feature metrics, exploring the textual characteristics generated by models including ChatGPT, Gemini, Claude, DeepSeek, Ernie Bot, and Kimi. Furthermore, the six LLMs were classified into domestic and international categories to compare overall differences in five textual metrics between composition texts generated by domestic LLMs (DeepSeek, Ernie Bot, and Kimi) and international LLMs (ChatGPT, Gemini, and Claude). This comparative analysis is particularly valuable for understanding how cultural and linguistic contexts of AI development influence educational applicability.

2.4 Coh-Metrix Metric Definitions

For texts generated by the six LLMs, we conducted analysis using the Coh-Metrix 3.0 tool. While Coh-Metrix can provide numerous metrics for texts, the most valuable metrics for text quality evaluation primarily focus on lexical, sentence, paragraph, and overall levels of the text. Table 2 presents the specific definitions of textual linguistic feature metrics along with their usability implications for L2 learners.

Table 2. Five Coh-Metrix Text Linguistic Feature Metrics

Metric Abbreviation	Calculation Method	Metric Meaning
LDTTRa(Lexical Diversity, Type–Token Ratio, all words)	Type-token ratio of all words	Evaluates the lexical diversity of text, helping to assess the linguistic level of the text. The higher the value, the richer the vocabulary usage.

(*continued*)

Table 2. (*continued*)

Metric Abbreviation	Calculation Method	Metric Meaning
LSASS1(Latent Semantic Analysis Sentence-to-Sentence Similarity)	Uses Latent Semantic Analysis (LSA) to generate a vector for each sentence, then calculates the average cosine similarity of each pair of adjacent sentences	Measures semantic similarity between adjacent sentences. The higher the value, the stronger the semantic connection between sentences and the better the sentence cohesion.
LSAPP1(Latent Semantic Analysis Paragraph-to-Paragraph Similarity)	Calculates the average LSA cosine value between adjacent paragraphs	Measures semantic similarity between paragraphs. The higher the value, the better the cohesion between paragraphs. Reflects the overall structure and coherence of the text.
SYNLE(Syntactic Complexity: Mean Number of Modifiers per Noun Phrase)	Average number of words before the main verb of the main clause in sentences	Evaluates the syntactic complexity of text, reflecting the complexity of grammatical structures. The higher the value, the more complex the syntactic structure of the text.
RDFKGL(Readability: Flesch-Kincaid Grade Level)	READFKGL = (0.39 × average words per sentence) + (11.8 × average syllables per word) - 15.59	Evaluates text difficulty. The higher the number, the more difficult the text is to read. This metric helps assess text suitability for learners at different L2 proficiency levels.

3 Results

3.1 Descriptive Statistics

To examine the macro-characteristics of same-topic English texts generated by different LLMs under identical prompts, we first conducted descriptive statistics on the text lengths generated by the six LLMs. The specific results are shown in Table 3.

Table 3. The Descriptive Statistics of Six LLMs

	ChatGPT	Claude	Gemini	Ernie Bot	Kimi	DeepSeek
Number of Texts	9	9	9	9	9	9
Maximum Word Count	298	386	340	380	356	322

(*continued*)

Table 3. (*continued*)

	ChatGPT	Claude	Gemini	Ernie Bot	Kimi	DeepSeek
Minimum Word Count	273	290	279	331	290	259
Average Word Count	289	334	312	353	323	283

The descriptive statistical results show that Ernie Bot and Claude generated longer texts, with average lengths of 353 words and 334 words respectively across 9 texts; ChatGPT and DeepSeek generated shorter texts, with 289 words and 283 words respectively. The average text lengths generated by domestic models (Ernie Bot, Kimi, and DeepSeek) and international models (ChatGPT, Claude, and Gemini) were 312 words and 320 words respectively.

To explore the linguistic feature variables of texts generated by different LLMs at a micro level, Coh-Metrix was used to analyze 9 texts from each model, obtaining descriptive statistical results for five metrics: lexical diversity (LDTTRa), sentence cohesion (LSASS1), paragraph cohesion (LSAPP1), syntactic complexity (SYNLE), and readability (RDFKGL) (see Table 4). These metrics provide insights not only into linguistic quality but also into usability factors crucial for educational applications.

Table 4. Descriptive Statistics of Linguistic Feature Metrics for Texts Generated by Six LLMs

		ChatGPT	Claude	Gemini	Ernie Bot	Kimi	DeepSeek
LDTTRa	Mean	0.628	0.614	0.582	0.578	0.558	0.626
	SD	0.027	0.046	0.024	0.016	0.027	0.017
	Minimum	0.598	0.549	0.533	0.544	0.528	0.610
	Maximum	0.670	0.703	0.608	0.598	0.610	0.661
LSASS1	Mean	0.198	0.306	0.244	0.208	0.224	0.237
	SD	0.025	0.045	0.058	0.023	0.035	0.025
	Minimum	0.156	0.232	0.150	0.176	0.173	0.201
	Maximum	0.227	0.367	0.352	0.242	0.273	0.283
LSAPP1	Mean	0.349	0.455	0.365	0.343	0.369	0.396
	SD	0.058	0.055	0.037	0.042	0.035	0.059
	Minimum	0.280	0.390	0.284	0.275	0.305	0.292
	Maximum	0.457	0.568	0.401	0.411	0.410	0.503
SYNLE	Mean	6.239	5.562	6.153	6.032	7.031	5.829
	SD	1.112	1.297	1.480	0.730	0.841	0.468
	Minimum	4.556	3.882	3.813	4.944	5.750	5.056
	Maximum	7.588	8.467	8.733	7.050	8.625	6.533
RDFKGL	Mean	13.958	16.333	14.107	13.693	14.065	13.507
	SD	1.080	0.824	1.041	0.982	0.928	0.650

(*continued*)

Table 4. (*continued*)

		ChatGPT	Claude	Gemini	Ernie Bot	Kimi	DeepSeek
	Minimum	12.252	15.225	12.528	12.378	12.789	12.176
	Maximum	15.401	17.164	15.404	14.980	15.493	14.436

From the descriptive statistical data of text metrics generated by large language models, we can preliminarily identify the following results: (1) In terms of lexical diversity metrics, ChatGPT performed most prominently, with an average LDTTRa value of 0.628, indicating the highest lexical richness in its generated texts; Ernie Bot and Kimi showed relatively lower lexical diversity metrics. (2) In terms of sentence cohesion, Claude model performed best, with an average LSASS1 metric value of 0.306, while ChatGPT was lowest, with an average LSASS1 value of only 0.198. (3) In terms of paragraph cohesion, Claude model performed best, with an average LASPP1 metric of 0.455, while Ernie Bot model had the lowest paragraph cohesion metric, with an average value of 0.343. Texts generated by Claude were more coherent with better connectivity. (4) In terms of syntactic complexity, Kimi had the highest SYNLE value (7.03), indicating that its generated texts had more complex grammatical structures, while Claude model had the lowest syntactic complexity, with an average value of only 5.562. (5) In terms of readability, Claude model had the highest average score of 16.33, indicating greater difficulty in its generated texts, while DeepSeek and Ernie Bot had significantly lower average scores than Claude, at 13.51 and 13.693 respectively, indicating that texts generated by these two were easier to understand.

If we classify the six large language models into two major categories—domestic models (Ernie Bot, Kimi, DeepSeek) and abroad models (ChatGPT, Claude, Gemini)—we can obtain results for domestic and international large language models across the five Coh-Metrix text metrics, as shown in Table 5. Based on the data in Table 5, the overall differences between domestic and international large language models across various text metrics are shown in Table 6.

Table 5. Statistical Results of Five Linguistic Feature Metrics for LLMs from China and Abroad

	Classification	LDTTRa	LSASS1	LSAPP1	SYNLE	RDFKGL
1	Overseas(ChatGPT)	0.612	0.220	0.358	5.167	14.451
2	Overseas(ChatGPT)	0.628	0.156	0.294	7.158	13.561
3	Overseas(ChatGPT)	0.644	0.199	0.360	6.824	14.970
...	...	...	...	...	...	...
9	Overseas(ChatGPT)	0.599	0.181	0.342	7.235	14.039
10	Overseas(Claude)	0.601	0.348	0.390	5.611	16.631
11	Overseas(Claude)	0.591	0.293	0.426	4.944	15.770
...	...	...	...	...	...	...
18	Overseas(Claude)	0.631	0.310	0.568	8.467	16.968

(*continued*)

Table 5. (*continued*)

	Classification	LDTTRa	LSASS1	LSAPP1	SYNLE	RDFKGL
19	Overseas(Gemini)	0.604	0.189	0.349	6.789	13.651
20	Overseas(Gemini)	0.559	0.220	0.347	7.100	12.528
...	...	...	...	...	...	...
27	Overseas(Gemini)	0.608	0.150	0.284	6.700	12.759
28	China(Ernie Bot)	0.591	0.193	0.275	6.700	12.400
29	China(Ernie Bot)	0.584	0.234	0.325	5.316	14.689
...	...	...	...	...	...	...
36	China(Ernie Bot)	0.586	0.194	0.333	7.050	14.980
37	China(Kimi)	0.528	0.257	0.382	6.750	15.493
38	China(Kimi)	0.610	0.236	0.401	6.800	15.199
...	...	...	...	...	...	...
45	China(Kimi)	0.541	0.273	0.384	7.556	14.331
46	China(DeepSeek)	0.615	0.225	0.419	5.875	13.507
47	China(DeepSeek)	0.610	0.227	0.503	5.500	13.960
...	...	...	...	...	...	...
54	China(DeepSeek)	0.634	0.240	0.447	6.167	13.967

Table 6. Descriptive Statistics of Linguistic Feature Metrics for Texts Generated by Domestic and International Large Models

Metrics	Classification	Number of Texts	Mean	Std. Deviation	Std. Error Mean	Minimum	Maximum
LDTTRa	China	27	0.58722	0.03510	0.00675	0.528	0.661
	Abroad	**27**	**0.60796**	**0.03754**	**0.00722**	**0.533**	**0.703**
LSASS1	China	27	0.22300	0.02961	0.005700	0.173	0.283
	Abroad	**27**	**0.24944**	**0.06248**	**0.01202**	**0.150**	**0.367**
LSAPP1	China	27	0.36930	0.05001	0.00962	0.275	0.503
	Abroad	**27**	**0.38963**	**0.06823**	**0.01313**	**0.280**	**0.568**
SYNLE	China	27	6.29726	0.85752	0.16503	4.944	8.625
	Abroad	**27**	**5.98474**	**1.29077**	**0.24841**	**3.813**	**8.733**
RDFKGL	China	27	13.75500	0.86467	0.16641	12.176	15.493
	Abroad	**27**	**14.79930**	**1.45826**	**0.28064**	**12.252**	**17.164**

In texts generated by domestic and international LLMs, there were varying degrees of differences across all five metrics between the two categories. The results show that texts generated by international LLMs performed better in lexical diversity, with an average lexical diversity (LDTTRa) metric value of 0.608, superior to domestic texts (average value of 0.587). For the sentence cohesion metric (LSASS1), the advantage of international LLM texts was more pronounced, with an average value of 0.249, compared to 0.223 for domestic texts. For the paragraph cohesion metric (LSAPP1), international LLM texts performed better than domestic LLM texts. From the syntactic complexity (SYNLE) metric perspective, the average syntactic complexity of texts generated by international LLMs was 6.297, superior to domestic LLMs (average value of 5.985), though the standard deviation of syntactic complexity for international models (SD = 1.29077) was larger than for domestic models (SD = 0.85752). Finally, international models scored higher than domestic models on the readability metric. The comparative analysis of texts from China and abroad further reveals the source-based differences in text generation models, with important implications for educational applications.

3.2 One-Way Analysis of Variance

To further explore whether there are significant differences in the linguistic feature metrics of English texts generated by the six large language models—ChatGPT, Claude, Gemini, Ernie Bot, Kimi, and DeepSeek—this study conducted one-way analysis of variance (ANOVA) on the five text metrics extracted by Coh-Metrix (LSASS1, LSAPP1, LDTTRa, SYNLE, RDFKGL). The summary of ANOVA results is shown in Table 7.

Table 7. ANOVA Results of Six LLMs across Five Text Metrics

Variable	Mauchly's Test of Sphericity	Type III Sum of Squares	df	Mean Square (MS)	F	Sig.
LDTTRa	**Sphericity Assumed (Satisfied)**	0.038	5	0.008	9.793	0.000
LSASS1	**Sphericity Assumed (Satisfied)**	0.066	5	0.013	9.364	0.000
LSAPP1	**Sphericity Assumed (Satisfied)**	0.077	5	0.015	6.013	0.000
SYNLE	**Sphericity Assumed (Satisfied)**	11.213	5	2.243	2.101	**0.085**
RDFKGL	**Sphericity Assumed (Satisfied)**	48.035	5	9.607	11.020	0.000

Before conducting the ANOVA, Mauchly's Test of Sphericity was first performed. The test results showed that the P-values for all six LLMs across the five text metrics (lexical diversity, sentence cohesion, paragraph cohesion, syntactic complexity, and readability) were all greater than 0.05, indicating that all met the homogeneity of variance requirement and satisfied the conditions for variance analysis.

From the results of the one-way repeated measures ANOVA, the P-values for the six LLMs across four metrics—lexical diversity, sentence cohesion, paragraph cohesion, and readability—were all below the 0.01 significance level. Only the difference in syntactic complexity did not reach significance ($F = 2.101$, $P = 0.085 > 0.01$). The analysis results indicate that except for syntactic complexity, the English texts generated by the six models (ChatGPT, Claude, Gemini, Ernie Bot, Kimi, and DeepSeek) showed highly significant differences across the other four metrics ($p < 0.01$).

Since the one-way repeated measures ANOVA results showed that the P-values for the six LLMs across four metrics (lexical diversity, sentence cohesion, paragraph cohesion, and readability) were all below the 0.01 significance level, post hoc multiple comparisons were needed to further analyze the specific differences among the six models on these four text metrics.

3.3 Tukey Post Hoc Test

After applying the Tukey post hoc test to the results of the six LLMs' generated texts across four metrics (lexical diversity, sentence cohesion, paragraph cohesion, and readability), the test results showed: (1) For the lexical diversity (LDTTRa) metric, both ChatGPT and DeepSeek were significantly higher than Gemini, Ernie Bot, and Kimi, with all P-values less than 0.01; Claude showed no significant difference from ChatGPT and DeepSeek, but Claude was significantly higher than Kimi. There was no significant difference between ChatGPT and DeepSeek. This indicates that English texts generated by ChatGPT and DeepSeek possess richer vocabulary distribution, while Kimi performed relatively poorly in lexical diversity compared to the other five models. (2) For the inter-sentence cohesion metric (LSASS1), Claude was significantly higher than the other five models, with all P-values less than or equal to 0.01, while differences among the other five models were not significant, indicating that Claude clearly outperformed all comparable models in sentence-level coherence. (3) For the inter-paragraph cohesion (LSAPP1) metric, Claude was significantly higher than ChatGPT, Gemini, Ernie Bot, and Kimi. DeepSeek was slightly lower than Claude but with no significant difference between them, and differences between DeepSeek and other models did not reach significance. At the paragraph cohesion level, Claude maintained a significant advantage. (4) For readability (RDFKGL), Claude showed significant differences from the other five models, with all P-values less than 0.01, and Claude's values on this metric were significantly higher than the other five models, indicating that Claude generated texts with the highest reading difficulty.

Overall, Claude was significantly higher than most comparable models across three metrics: sentence cohesion, paragraph cohesion, and readability. From a usability perspective, this suggests Claude may be most appropriate for advanced learners who can handle higher complexity, while models like DeepSeek and Ernie Bot may better serve intermediate and beginning learners.

3.4 T-Test for Differences in Text Metrics Between LLMs from China and Abroad

To examine whether there are overall differences in text metrics between domestic and international large language models, we divided the six LLMs into domestic (DeepSeek, Ernie Bot, Kimi) and international (ChatGPT, Claude, Gemini) categories. Based on the linguistic feature metrics extracted by Coh-Metrix, we conducted independent samples t-tests, with results shown in Table 8.

Table 8. T-Test Results for Text Metrics of LLMs from China and Abroad

		Levene's Test for Equality of Variances		t-test for Equality of Means		
		F	Sig.	t	df	Sig.(2-tailed)
LDTTRa	**Equal variances assumed**	0.043	0.837	2.097	52.000	**0.041**
	Equal variances not assumed			2.097	51.767	**0.041**
LSASS1	**Equal variances assumed**	15.061	0.000	1.987	52.000	0.052
	Equal variances not assumed			1.987	37.116	0.054
LSAPP1	**Equal variances assumed**	1.990	0.164	1.249	52.000	0.217
	Equal variances not assumed			1.249	47.680	0.218
SYNLE	**Equal variances assumed**	5.455	0.023	−1.048	52.000	0.300
	Equal variances not assumed			−1.048	45.209	0.300
RDFKGL	**Equal variances assumed**	7.128	0.010	3.201	52.000	**0.002**
	Equal variances not assumed			3.201	42.271	**0.003**

Before conducting the t-test, a test of homogeneity of variance was first performed, with decisions based on t-values according to whether the homogeneity of variance assumption was met. The independent samples t-test results showed that domestic and international large models only showed significant differences in two metrics: lexical diversity (LDTTRa) ($t = 2.097$, $P = 0.041 < 0.05$) and readability (RDFKGL) ($t = 3.201$, $P = 0.003 < 0.05$), with no significant differences in other metrics. These findings suggest that while international models may generate texts with richer vocabulary and higher complexity, domestic models produce texts that are more accessible to learners, potentially offering advantages for educational applications.

4 Discussion

4.1 Implications for Learner Engagement and Usability

The variance analysis results of textual linguistic feature metrics generated by six large language models from China and abroad show that the six LLMs differ significantly in lexical diversity, sentence and paragraph cohesion, and readability. These research findings align with previous experimental results, validating the diversity among different models in terms of training methods, data processing, and algorithmic architectures [1].

From a learner engagement perspective, the optimal model choice depends on balancing linguistic sophistication with cognitive accessibility. Specifically, regarding lexical diversity, DeepSeek demonstrated outstanding performance, with its generated texts showing significantly higher lexical richness than other models, while the Kimi model performed relatively poorly. However, high lexical diversity is not always advantageous—for beginning learners, moderate vocabulary complexity may facilitate better comprehension and retention. Claude also excelled in sentence and paragraph cohesion metrics, indicating that its generated texts possess high semantic coherence and cohesive fluency, which can reduce cognitive load during reading. Claude scored highest in readability, indicating that its generated texts had the highest linguistic difficulty, making them most suitable for advanced learners. The ChatGPT and DeepSeek produced texts with lower difficulty levels, generating English texts that were easier to understand, potentially better serving intermediate learners.

In terms of syntactic complexity, the Kimi model had the relatively highest average score, indicating that its generated syntactic structures were relatively complex, though there were no significant differences in syntactic complexity among the six large language models, suggesting that the English texts generated by the six LLMs were relatively similar in syntactic complexity. This finding has important usability implications: syntactic complexity appears to be a relatively stable feature across models, suggesting that other factors (cohesion, readability, lexical diversity) may be more important for model selection in educational contexts.

4.2 Cultural Influences and Localization Advantages

The Tukey post hoc test results showed that the Claude performed particularly well in sentence cohesion and paragraph cohesion. Its relative advantages may be attributed to more precise language structure learning and semantic cohesion mechanisms. DeepSeek's outstanding performance in lexical diversity suggests that China's LLMs have already developed certain competitive advantages in specific aspects, though overall language performance still requires further optimization and improvement.

Additionally, the comparative analysis of Coh-Metrix text metrics between domestic and international large language models revealed significant differences. Overall, international models performed notably better than domestic models in lexical diversity and readability. This result may be related to international large language models having larger English training corpora and richer linguistic diversity [18], enabling them to generate texts with more abundant and diverse vocabulary usage. However, from a usability perspective, the lower readability scores of domestic models may actually represent an advantage for certain learner populations, particularly Chinese L2 English learners who may benefit from more accessible texts that align with their educational background.

The relatively insufficient performance of domestic models on lexical diversity suggests they may need to further optimize training strategies and increase corpus scale to enhance the lexical diversity and richness of generated texts. However, this should be balanced against usability considerations—excessive lexical diversity can impede comprehension for learners. International models also performed better in sentence cohesion, suggesting that international models may be more adept at capturing contextual semantic relationships between sentences, thereby generating more coherent texts. However, differences between Chinese and overseas LLMs in paragraph cohesion and syntactic complexity did not reach significance levels, indicating that domestic and international large language models tend to perform similarly in terms of paragraph structural coherence and syntactic complexity [19]. This result may reflect that large language models have reached a relatively mature technical level in specific discourse structures and syntactic generation.

4.3 Prompt Engineering Strategies for Educational Applications

Our findings demonstrate that with appropriate prompt engineering, existing multilingual LLMs can effectively support L2 writing education without requiring fine-tuning. The key is designing prompts that activate the models' inherent capabilities while aligning with educational objectives. Based on our experimental results, we recommend the following prompt engineering strategies for educators:

1. Specify target proficiency level explicitly in prompts to guide appropriate complexity;
2. Include genre and structural requirements to ensure coherent organization;
3. Provide cultural context when using models across different linguistic backgrounds;
4. Use iterative prompting to refine outputs based on specific educational needs

4.4 Framework for Model Selection Based on Learner Needs

Based on our comprehensive analysis combining linguistic metrics with usability considerations, we propose the following framework for selecting LLMs in L2 writing education:

For Beginning Learners (A1-A2): DeepSeek and Ernie Bot are recommended due to their lower readability scores and balanced linguistic features. These models provide accessible texts that don't overwhelm learners with excessive complexity.

For Intermediate Learners (B1-B2): ChatGPT and Gemini offer appropriate balance between linguistic sophistication and accessibility. Their moderate complexity levels provide sufficient challenge without creating excessive cognitive load.

For Advanced Learners (C1-C2): Claude's high cohesion and complexity make it ideal for advanced learners who can benefit from sophisticated language models. The higher readability scores provide appropriate challenge for proficiency development.

For Chinese L2 English Learners: Chinese-developed models may offer specific advantages through better understanding of L1 interference patterns and cultural alignment, though this should be balanced against the need for authentic English language exposure.

5 Conclusion

Based on the Coh-Metrix text analysis tool and usability-focused indices (LDTTRa, LSASS1, LSAPP1, SYNLE, RDFKGL), this study compared English texts generated by six LLMs and by two groups of domestic and overseas models. The main conclusions are as follows.

Firstly, apart from syntactic complexity, the six LLMs showed significant differences in lexical diversity, sentence cohesion, paragraph cohesion, and readability, which has important implications for model selection in L2 writing support according to learner proficiency and instructional goals.

Secondly, overseas models produced lexically richer but more difficult texts, whereas domestic models generated more accessible texts. From a usability perspective, this suggests that different model groups may be more suitable for learners at different proficiency levels, especially when balancing textual sophistication with comprehensibility.

Thirdly, the study confirms that existing multilingual LLMs can already effectively support L2 English writing through appropriate prompt engineering without task-specific fine-tuning, highlighting the importance of designing prompts that explicitly reflect educational objectives and learner needs.

Future research should further enhance the educational applicability of domestic LLMs, conduct learner-centered usability studies in authentic classroom settings, and develop more practical guidelines or tools for selecting and combining LLMs based on text metrics and pedagogical requirements.

Acknowledgements. This work was supported by the Fundamental Research Funds for the Central Universities, and the Research Funds of Beijing Language and Culture University (25YCX170). This research was also funded by the National Social Science Fund of China (Grant No. 25BYY137).

References

1. Mindner, L., Schlippe, T., Schaaff, K.: Classification of human- and AI-generated texts: investigating features for ChatGPT. In: International Conference on Artificial Intelligence in Education Technology, pp. 152–170. Springer, Singapore (2023)
2. Su, Q., Yang, J.: The evolution of language intelligence and its application in new liberal arts. Foreign Lang. China. **3**, 4–11 (2023)
3. Bubeck, S., Chandrasekaran, V., Eldan, R., et al.: Sparks of artificial general intelligence: Early experiments with GPT-4. (2023). https://arxiv.org/abs/2303.12712
4. Guo, Q., Feng, R., Hua, Y.: Applications and potential issues of ChatGPT in English academic writing and teaching. Technol. Enhanced Foreign Lang. Educ. **2**, 18–23+107 (2023)
5. Wu, Q.: A comparative study of narrative writing quality and lexical complexity among Chinese L2 learners, native speakers, and ChatGPT. Chin. Teach. World. **4**, 517–532 (2024)
6. Herbold, S., Hautli-Janisz, A., Heuer, U., et al.: AI, write an essay for me: A large-scale comparison of human-written versus ChatGPT-generated essays. (2023). https://arxiv.org/abs/2304.14276

7. Smith, G., Funk, J.: When it Comes to Critical Thinking, AI Flunks the Test. The Chronicle of Higher Education (2024)
8. Graesser, A.C., McNamara, D.S., Cai, Z., Conley, M., Li, H., Pennebaker, J.: Coh-Metrix measures text characteristics at multiple levels of language and discourse. Elem. Sch. J. **115**(2), 210–229 (2014)
9. Ragha, L.R., Sasikumar, M.: Feature analysis for handwritten Kannada Kagunita recognition. Int. J. Comput. Theory Eng. **3**(1), 94–103 (2011)
10. McNamara, D.S.: Coh-Metrix: capturing linguistic features of cohesion. Discourse Process. **47**, 292–330 (2010)
11. Crossley, S.A., McNamara, D.S.: Predicting second language writing proficiency: the roles of cohesion and linguistic sophistication. J. Res. Read. **35**(2), 115–135 (2012)
12. Ullmann, T.D.: Automated analysis of reflection in writing: validating machine learning approaches. Int. J. Artif. Intell. Educ. **29**(2), 217–257 (2019)
13. Santos, T.: Professors' reactions to the academic writing of nonnative-speaking students. TESOL Q. **22**(1), 69–90 (1988)
14. Astika, G.G.: Analytical assessments of foreign students' writing. RELC J. **24**(1), 61–70 (1993)
15. Nation, P.: Learning Vocabulary in another Language. Cambridge University Press, Cambridge (2001)
16. Kyle, K., Crossley, S.A.: The relationship between lexical sophistication and independent- and source-based writing. J. Second. Lang. Writ. **34**, 12–24 (2016)
17. Crowhurst, M.: Syntactic complexity and teachers' quality ratings of narrations and arguments. Res. Teach. Engl. **14**, 223–231 (1980)
18. Brown, T.B., Mann, B., Ryder, N., et al.: Language models are few-shot learners. In: Larochelle, H., Ranzato, M., Hadsell, R., et al. (eds.) Advances in Neural Information Processing Systems 33 (NeurIPS 2020), pp. 1877–1901. Curran Associates, Red Hook, NY (2020)
19. Yang, K., Klein, D., Peng, N., Tian, Y.: Improving long story coherence with detailed outline control. In: Proceedings of the 61st Annual Meeting of the Association for Computational Linguistics (Volume 1: Long Papers), pp. 3378–3465. Association for Computational Linguistics, Toronto (2023)

Multimodal NLP Recognition, Semantic Analysis and Modeling

Language Bias in Multilingual RAG: A Case Study in the Japanese Medical Domain

Jiahao Huang[1], Fei Cheng[2], Junfeng Jiang[3], Kazuma Kobayashi[3], and Akiko Aizawa[1](✉)

[1] University of Tokyo, Bunkyo City, Japan
jiahao-huang@g.ecc.u-tokyo.ac.jp, aizawa@nii.ac.jp
[2] Kyoto University, Tokyo, Japan
feicheng@i.kyoto-u.ac.jp
[3] National Institute of Informatics (NII), Tokyo, Japan
{jiang,kazumkob}@nii.ac.jp

Abstract. Despite the significant achievements of LLMs in recent years, their performance in low-resource language-domain pairs remains less than satisfactory. Although RAG is often considered a solution, we identify the paradox: the LLM's poor performance in low-resource language-domain pairs is due to a lack of corpora, but RAG also relies on comprehensive and high-quality corpora. We show that this paradox could lead to failure of RAG in certain low-resource language-domain pairs, like the Japanese medical domain. We propose to use high-resource corpora to enhance the knowledge coverage. We also identify and address the language bias issue when using multilingual corpora, which prevents the RAG framework from fully utilizing the multilingual corpus. Through our proposed RAG framework and reranker training method, the RAG performance of LLMs is improved by 4.36–7.96% point on JMedBench.

Keywords: Retrieval Augmented Generation · Multilinguality · Biomedical

1 Introduction

In recent years, the development of large language models (LLMs) has revolutionized a broad range of natural language processing (NLP) tasks. However, in certain low-resource language-domain pairs, such as the Japanese medical domain, LLMs still have room for improvement. According to the research [1], LLMs struggle with several tasks in the Japanese medical domain, including multiple-choice question answering (MCQA), named entity recognition (NER), and document classification (DC). Their performances on Japanese tasks are worse than equivalent tasks in English [2,3].

The main reason for the poor performance of LLMs in low-resource language-domain pairs is the lack of training data. Retrieval augmented generation (RAG)

H. Unger and P. Meesad (Eds.): NLPIR 2025, LNNS 1904, pp. 125–139, 2026.
https://doi.org/10.1007/978-3-032-20897-2_9

is believed to be an efficient method to improve the performance of generation tasks in such domains [4]. RAG includes a retrieval module that obtains knowledge from an external trustful database to assist LLMs in generating the answer. This module ensures the answers are grounded in the retrieved evidence, enhancing the relevance and accuracy of the output. However, RAG also requires high-quality and comprehensive corpora, which is lacking in low-resource language-domain pairs. This paradoxically suggests that RAG may not be an ideal solution for tasks in low-resource language-domain pairs. To solve this low-resource paradox, we propose using rich English resource in the same domain to enhance the coverage of knowledge.

In this paper, we will use the Japanese medical domain as a case study, aiming to enhance the performance of LLMs in low-resource language-domain pairs with multilingual RAG frameworks. We propose three pipelines to improve the performance of RAG with multilingual corpora, as shown in Fig. 1. Figure 1(a) provides a monolingual RAG baseline. Figure 1(b) provides a naive cross-lingual RAG pipeline, which is also the most common cross-lingual RAG pipeline at present [5,6]. However, we find that language bias exists within this pipeline. Language bias refers to that when using a query in a certain language, the retrieved paragraphs are highly likely to be in the same language, leading to insufficient use of the full multilingual corpora. To solve the language bias, we propose a translation-based cross-lingual RAG pipeline in Fig. 1(c), which effectively eliminates language bias and improves the utilization of multilingual knowledge. However, this approach relies heavily on a strong translation system. To avoid the performance loss caused by the translation system, and also to further improve the quality of retrieved documents, we propose a reranker-based cross-lingual pipeline in Fig. 1(d).

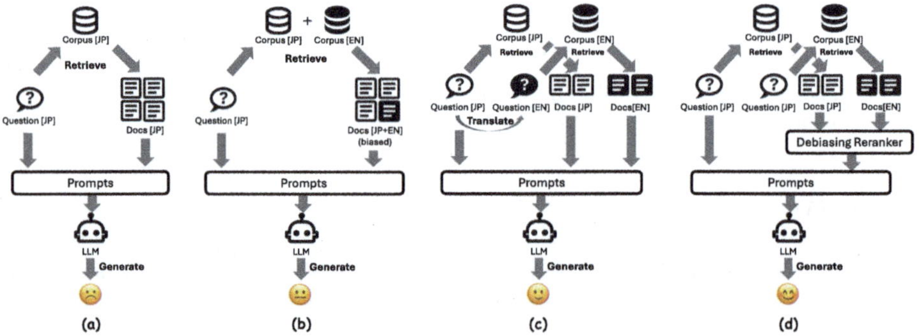

Fig. 1. Four RAG pipelines experimented in this paper. Figure 1(a) provides a monolingual RAG baseline, and Fig. 1(b), 1(c), 1(d) are the proposed cross-lingual RAG pipelines.

Overall, we aim to improve the performance of LLMs in the Japanese medical domain using cross-lingual RAG framework. The contributions in this paper include the following:

- We indicate and verify the low-resource paradox, which infers that RAG is not an effective solution to low-resource language-domain pairs, and propose using multilingual corpus to solve it.
- We discover the problem of language bias in multilingual RAG and introduce translation-based and reranker-based pipelines as solutions.
- We propose an innovative training method for multilingual rerankers that can eliminate the language bias among multilingual corpora and improve the quality of retrieved documents as well.

2 Monolingual RAG: Evidence of the Low-Resource Paradox

In this paper, we will use the Japanese medical domain as an example of low-resource language-domain pairs. We will use JMedBench, a recently released comprehensive benchmark in the Japanese medical field, to evaluate the performance of various models both in non-RAG and RAG frameworks. In this section, we will first give an introduction to JMedBench, and then test the performance of LLMs on JMedBench within the RAG framework to show the existence of the low-resource paradox.

2.1 Introduction to JMedBench

JMedBench [1] includes 20 Japanese datasets across five task types with 38K testing samples in total. It includes native Japanese datasets like IgakuQA [7], which compiles the Japanese medical licensing examination in past years. It also includes high-quality datasets translated from English like MedMCQA [2], MedQA [8], PubMedQA [3], etc.

Considering that multiple-choice question answering (MCQA) tasks can comprehensively assess LLMs' understanding of medical knowledge while allowing for quick and accurate evaluation, in this paper we will primarily use MCQA datasets to evaluate the performance of LLMs and the RAG frameworks.

2.2 Experiment Settings

LLMs. We follow the categorization of LLMs in JMedBench paper [1] and select the most representative model in each category according to their performances: **general LLMs in non-Japanese languages** (Llama3-8B [9], Qwen-2-7B [10]), **biomedical LLM in non-Japanese languages** (Meditron-7B [11]), **Japanese general LLMs** (llm-jp-v3-13B [12]), **Japanese biomedical LLM** (MMed-Llama3-8B [13]).

Prompts. We conduct experiments using zero-shot and few-shot prompts. In zero-shot prompts, we sequentially list the context (retrieved documents), question, options, and then let the LLM provide an answer. In few-shot prompts, we follow the setting in the JMedBench paper, where samples are randomly selected from the training set and included as few-shot examples before the actual question.

However, since the performance difference between zero-shot and few-shot settings is negligible, we primarily present the zero-shot results in this paper.

Corpora. We basically use biomedical-related corpora as the knowledge base for RAG. The Japanese corpus includes the following: MSD manual[1], J-Dream abstracts[2] and clinical guidelines released by academic societies.

Retriever. In this paper, we hope to discover the potential of multilingual corpora, so we will mainly focus on the performance of mContriever [15], which is a multilingual, unsupervised dense retrieval model.

2.3 Experiment Results

For a clearer visualization of monolingual RAG's performance, we directly display the performances of the monolingual RAG framework with different LLM, corpus and number of retrievals in Fig. 3. It is notable that the RAG framework does not improve LLMs' performance; on the contrary, it decreases the accuracy of LLMs' responses in most cases. We mainly attribute the failure to issues with the corpora and the retriever. On one hand, the corpora cannot cover all the knowledge required to answer these questions. Therefore, in some cases, the RAG pipeline may retrieve paragraphs that are irrelevant to the question, thereby affecting the model's ability to provide correct answers. This can be attributed to the low-resource paradox. On the other hand, only the retriever cannot retrieve the paragraphs that best meet the needs of the questions. However, existing rerankers cannot be directly applied to multilingual RAG pipelines. Two factors jointly contribute to the suboptimal performance of the monolingual RAG pipeline in the Japanese medical domain.

Therefore, in the next section, we will first utilize a multilingual corpus to enhance the knowledge coverage of the knowledge base; then, we will introduce a new reranker to improve the retrieval effectiveness, taking into account the characteristics of the multilingual corpus.

3 Multilingual RAG: Solution to the Low-Resource Paradox

3.1 Proposed Pipelines

As discussed in the previous section, the monolingual RAG framework underperforms in the Japanese medical domain, mainly for two reasons: the insufficiency

[1] https://www.merckmanuals.com/.
[2] https://jdream3.com/.

of the corpus and the inadequacy of the retriever. Therefore, we propose three new multilingual RAG pipelines for the Japanese medical field in Fig. 1, attempting to address the low-resource paradox.

Naive Cross-Lingual RAG Pipeline. To address the insufficiency of the monolingual corpus, we propose to add English corpora to the Japanese corpora and use a multilingual retriever to directly retrieve reference paragraphs from it, as shown in Fig. 1(b). However, our experiments reveal that this approach has a significant issue: language bias. In other words, when the query is in Japanese, the multilingual retriever tends to retrieve Japanese paragraphs. This makes the inclusion of English corpora pointless, as the RAG framework is likely to continue referencing Japanese paragraphs.

Translation-Based Cross-Lingual RAG Pipeline. Language bias in multilingual retrievers is indeed difficult to eliminate because language is inherently a crucial component of the sentence embedding. We propose to first translate all Japanese queries into English. Then, use the Japanese and English queries to retrieve paragraphs from corpora in their respective language. Finally, merge the retrieved paragraphs from both languages based on their retrieval scores. This pipeline is presented in Fig. 1(c), and is proven to effectively solve the inherent language bias in the retriever. Notably, golden translation offered by JMedBench benchmark [1] are directly used in the experiments. In practical application, the performance of translation-based pipeline should be lower due to the quality loss from LLM translation.

Reranker-Based Cross-Lingual RAG Pipeline. Although the translation-based pipeline can simultaneously address the issues of insufficient corpus and language bias, it still faces a few challenges. First, we haven't solved the issue of the retriever's inadequacy, as it still struggles to identify gold paragraphs that can effectively aid in the question answering. Second, not all LLMs have strong translation capabilities, and improper translations can further diminish the retriever's performance. In the translation-based pipeline, we directly used the datasets already translated in JMedBench, thus avoiding this issue. However, in practical applications, we should still strive to minimize the performance loss caused by the translation process.

Therefore, we propose a reranker-based cross-lingual RAG pipeline in Fig. 1(d). We first use the original Japanese query to retrieve paragraphs from the Japanese and English corpora, independently. Then, we train a reranker to rerank these paragraphs. This pipeline can rank retrieved English and Japanese paragraphs without language bias, while also avoiding the performance loss during the translation process.

3.2 Reranker Training

Motivation. The reranker is a common component in the RAG pipeline, responsible for the second stage of ranking the documents retrieved by the retriever

[16,17]. The necessity of a reranker in the RAG framework lies in the fact that the most helpful paragraph is not necessarily the one most relevant to the question. An example is that a question and itself share the highest relevance because their sentence embedding are identical, but the question itself cannot provide any useful information for answering it. Instead, the most helpful paragraph should be the one most relevant to the ground truth answer, as it will guide the question in the right direction. We define the paragraphs retrieved using the ground truth as the golden paragraphs, and the performance of the LLM when referencing the golden paragraph serves as the upper bound for the RAG pipeline. In the test phase, golden paragraphs are not available since there is no access to the golden answer. The motivation of the reranker training is to predict the possible golden paragraphs among the retrieved documents given by the retriever.

Dataset Construction. In our reranker, the input is a query and a paragraph, and the output is a score representing how well the paragraph can assist the query in generating the correct answer.

Queries are sourced from the training sets of different datasets in JMedBench. We chose a subset of questions from different datasets to serve as query input.

Paragraphs consist of golden paragraphs and random paragraphs. We first use the ground truth answer for each question to perform retrieval from the corpora, obtaining the golden paragraphs for each query. To ensure a balanced paragraph distribution, we also sample several random paragraphs from the corpora for each query.

Scores are calculated by the relevance between the ground truth answer for each question and the paragraphs (inner product of the sentence embeddings). Notably, the training dataset for the reranker is also multilingual, aiming to eliminate language bias through the reranker. Thus, if a paragraph is in Japanese, we compute its relevance to the Japanese ground truth answer; if it is in English, we compute its relevance to the corresponding English ground truth answer.

Using the above method, we construct a dataset of 688K entries, with 80% used as the training set and 20% as the validation set. This data size is comparable with the dataset used for training other rerankers for the biomedical domain [18,19].

Training Settings. We use a multilingual BERT model[3] with an additional fully-connected layer as the base model, and then finetune it with the training set described before. For the model input, we concatenate the query and paragraph with the [SEP] token. We use mean squared error (MSE) as the loss function and Adam as the optimizer with learning rate of 6×10^{-6}. We finetune the model for 30 epochs using eight NVIDIA A100-SXM4-40GB, which takes around 14 h.

[3] https://huggingface.co/google-bert/bert-base-multilingual-uncased.

3.3 Experiment Settings

Datasets. We will test our pipelines on these four translated datasets from JMed-Bench in the main experiments: MedMCQA_JP, USMLE-QA_JP, MedQA_JP, MMLU-medical_JP.

LLMs and Prompts. We will retain the settings in Sect. 2.2.

Corpora. We will retain the corpora in Sect. 2.2 as the Japanese knowledge source. Besides, we will use English corpora to supplement the Japanese ones. The English corpora reference the settings in MedRAG [14] and include the following: Meditron training corpora[4], biomedical abstracts from PubMed[5], articles from StatPearls[6] and English medical textbooks [8].

Number of Retrievals. We set the number of retrievals as 1,3,5. According to the previous work [20], introducing too many retrieved documents diminishes contextual relevance and consequently degrades answer quality.

Table 1. Statistical information on the number of chunks for each corpus, including MSD, Clinical Guideline (GDL), J-Dream Abstract (Abs), Meditron training corpora (MDT), PubMed(PBM), StatPearls (SP) and English textbook (TB).

Corpus	Lang.	#Chunk	#Chunk pct.(%)	
MSD	JP	36K	0.13	10.49
GDL		376K	1.30	
Abs		2.6M	9.05	
MDT	EN	49K	0.17	89.51
PBM		25M	87.78	
SP		324K	1.12	
TB		125K	0.44	

The statistics of all corpora are presented in Table 1. We can observe that the English corpus makes up about 90% of the entire corpus, thereby covering a broader range of knowledge. Due to the dominance of English in scientific research, this phenomenon is quite common across various fields. Therefore, our approach has the potential to be extended to more low-resource language-domain pairs.

3.4 Evidence of Language Bias

In this section, we will verify the hypothesis of language bias: the retriever tends to retrieve paragraphs in the same language as the query.

[4] https://huggingface.co/datasets/epfl-llm/guidelines.
[5] https://pubmed.ncbi.nlm.nih.gov/.
[6] https://www.statpearls.com/.

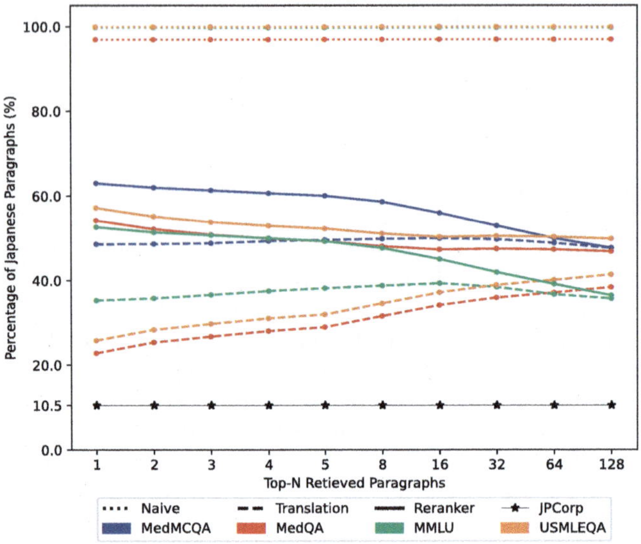

Fig. 2. The average proportion of Japanese paragraphs among the top N paragraphs. Line styles represent different pipelines, while colors represent different datasets. JPCorp line indicates the proportion of Japanese paragraphs in the entire corpus.

Figure 2 presents the results of our investigation into language bias. We use Japanese questions to retrieve paragraphs from the multilingual corpus, and the proportion of Japanese paragraphs can reflect whether language bias exists. We can draw the following conclusions from this figure.

Firstly, language bias is a severe problem. Although Japanese paragraphs make up just 10.49% of the total corpus, almost only Japanese paragraphs are retrieved with Japanese queries in the naive pipeline, because language is an important component of the sentence embedding. It means the addition of the English corpus will hardly bring any changes to the results, even though it contains a vast amount of new knowledge.

Secondly, translation-based and reranker-based pipelines can effectively address the issue of language bias. By using our proposed pipelines, the proportion of retrieved Japanese paragraphs can be reduced from nearly 100% to around 40%-60%, with variation depending on the pipeline, the dataset and the number of retrievals.

In summary, language bias is a severe issue to multilingual RAG. We are confident that employing a multilingual dataset for RAG is advantageous, as it enables a broader spectrum of knowledge and diverse viewpoints in multiple languages. However, this is not a naive problem; we must use a more suitable pipeline to deal with it.

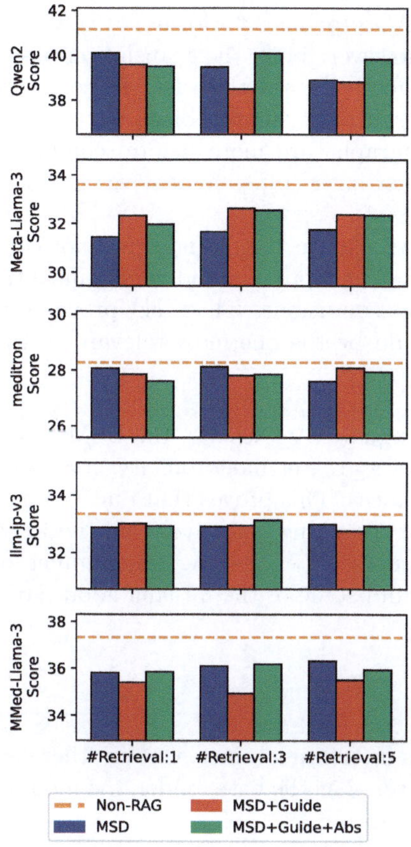

Fig. 3. Comparison of zero-shot performances of the monolingual RAG framework.

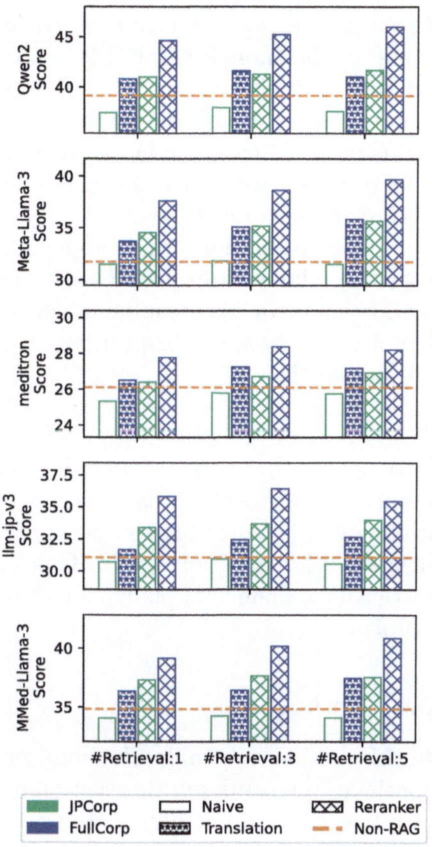

Fig. 4. Comparison of zero-shot performances of the RAG pipelines with different hyperparameters.

3.5 Experiment Results

Figure 4 present the performance of our proposed RAG pipelines with five retrieved paragraphs under zero-shot settings across different tasks. "JPCorp" represents that the pipeline only use the Japanese corpus, and "FullCorp" represents the full corpus. We will analyze this data from the following perspectives.

Corpus. We can observe that using the full corpus generally results in better performance than using the Japanese-only corpus, with other conditions controlled. Using high-resource language knowledge to supplement the corpus can effectively address the paradox of RAG in low-resource language-domain pairs.

Surprisingly, according to the experimental results, LLMs are not only able to extract knowledge from paragraphs in different languages, but also capable of responding in the same language as the question, which exceeds our expectations.

Pipeline. Due to language bias, the naive RAG pipeline's performance is almost same as the monolingual RAG pipeline. However, both the translation-based pipeline and the reranker-based pipeline effectively aid the LLM in correctly answering questions, thereby enhancing overall performance. Since our trained reranker can better understand which paragraphs are more helpful compared to a retriever without a reranker, the reranker-based pipeline achieves the best performance in most cases.

The construction of the upper bound is similar to the method used for constructing the training set for the reranker. The upper bound is significantly higher than all the results under the same conditions. This also proves our hypothesis: the most helpful paragraph should be the one most relevant to the ground truth.

Number of Retrievals. Compared to monolingual RAG results, the LLM's performance with a reranker-based pipeline shows greater monotonicity: increased number of retrievals leads to better performances. This proves that our reranker is effective, because the top paragraphs after reranking, effectively aid the LLM in answering questions well. Therefore, moderately increasing the number of retrievals is beneficial, as it helps the LLM find some of the missing good paragraphs.

3.6 Discussions

In this section, we will further explore certain interesting topics. Unless otherwise specified, the results in this section utilize Qwen-2 as the base model, the number of retrievals set as 1, and with zero-shot prompting.

Further Study on Paragraph-Answer Relevance. One of the hypothesis is that, the most helpful paragraph should be the one most relevant to the ground truth. Thus, we are also interested in how LLMs perform when referencing paragraphs with different relevance to the ground truth answer. For convenience, we will refer to "relevance to the ground truth" answer as the paragraph's score. To ensure a balanced distribution of paragraph scores, we set a golden paragraph and a randomly sampled paragraph for each question. The histogram in Fig. 5 shows the distribution of paragraph scores, while the line graph reflects the accuracy of question responses given paragraphs within each score range. The overall trend indicates that the higher the relevance between a paragraph and the ground truth answer, the more likely the LLM is to provide the correct answer when referencing it.

Contributions of Corpora. We also hope to study the contributions of each corpus to the reranker-based pipeline, particularly the differences between the Japanese and English corpora. Table 2 provides the distribution of corpus sources for the best paragraphs retrieved by the rerank-based pipeline, as well as the accuracy of answering questions using paragraphs from each corpus. We can observe that the LLM performs quite well when Japanese paragraphs are retrieved using

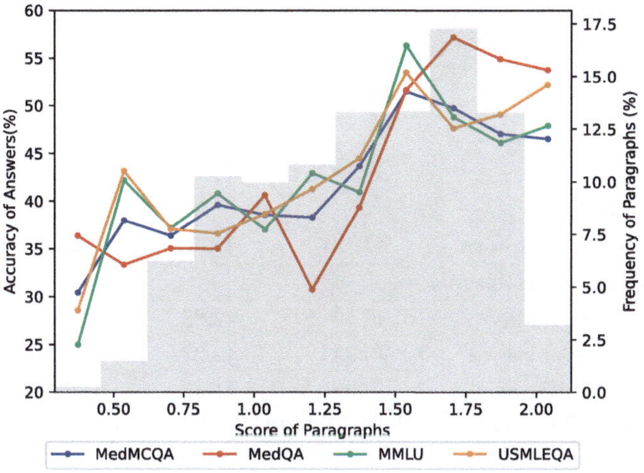

Fig. 5. The score distribution of paragraphs and the relationship between accuracy and paragraph scores.

the reranker-based pipeline, even surpassing that with English paragraphs. One possible reason might be that the LLM is better at handling monolingual inputs. Itt also indicates that the paragraphs in the Japanese corpus are of high quality and can enhance the LLM's performance in proper conditions.

Table 2. The distribution of corpus sources for the optimal paragraphs retrieved by the rerank-based pipeline and the accuracy when using each corpus source to answer the questions.

Corpus	Lang.	Retrieval (%)	Acc.(%)	Overall Acc. (%)	Ablation Acc. (%)
MSD	JP	4.58	43.65	44.76	43.11
GDL		7.83	45.17		
Abs		50.22	44.80		
MDT	EN	5.88	48.22	44.37	34.89
PBM		25.78	43.39		
SP		3.66	45.08		
TB		2.07	44.38		

We also conduct an ablation study on the corpora in different languages. We force the LLM not to use the optimally retrieved paragraphs and instead retrieve from the corpus in the other language. When the Japanese corpus was ablated, the LLM's performance was not significantly affected; conversely, the LLM's performance significantly declines on questions originally referencing English

Table 3. Comparison of RAG performance using different corpus, pipelines and rerankers.

Reranker	JPCorp	ENCorp	FullCorp
Proposed pipelines			
Naive	37.49	40.75	-
Translation	-	-	40.86
Our Reranker	**41.03**	**43.58**	**44.62**
Open-source rerankers			
multilingual-e5[a]	39.89	42.38	40.12
Jina-reranker-v2[b]	40.84	42.27	41.03
gte-multilingual[c]	40.92	41.48	40.96

[a] https://huggingface.co/intfloat/multilingual-e5-large-instruct.
[b] https://huggingface.co/jinaai/jina-reranker-v2-base-multilingual.
[c] https://huggingface.co/Alibaba-NLP/gte-multilingual-base.

paragraphs. It indicates that the shortcoming of the Japanese corpus is the limited knowledge coverage, which leads to the drastic performance drop when the English corpus is forbidden.

Nevertheless, the Japanese corpus is not replacable by the English one. We list the performance of the LLM with Japanese, English, and the full corpus in Table 3 for comparison. Although the English corpus outperforms the Japanese corpus due to its comprehensiveness, its performance is not as good as the full corpus, regardless of whether a reranker is used. Diversity still matters in the RAG corpus.

We also compare the performance of our reranker with other open-source rerankers in Table 3. Our reranker outperforms other ones because it is specifically trained for the Japanese medical domain, which proves the efficacy our proposed reranker training method. On the other hand, our reranker is even more outstanding on the full corpus, due to its capability of mitigating language bias.

4 Conclusions

In this paper, we identify the low-resource paradox, demonstrating RAG's limitations for low-resource language-domain pairs, and leverage high-resource language corpora as a solution. We address language bias in multilingual RAG through translation-based and reranker-based pipelines. Additionally, we introduce a novel training method for multilingual rerankers, reducing language bias and enhancing retrieval quality across diverse languages.

We believe that using multilingual corpora for RAG will be the trends in the future. On one hand, the existence of high-resource language corpora effectively addresses the issue of knowledge coverage for low-resource language corpora.

On the other hand, low-resource language corpora can offer different cultural perspectives to avoid biases inherent in high-resource language users. We hope our work can serve as a reference for further researches.

5 Related Work

Retrieval Augmented Generation (RAG). The Retrieval-Augmented Generation (RAG) framework enhances LLMs by integrating external knowledge retrieval, addressing hallucinations issues [21]. RAG could employs sparse [22] or dense [23] retrieval methods, or joint optimization [24].

Cross-Lingual RAG. While RAG enhances LLMs' factuality and reduces hallucinations, it struggles in cross-lingual settings. The BORDIRLINES dataset [6] reveals biases in geopolitical disputes, while the mRAG pipeline [5] highlights challenges like code-switching and fluency errors.

6 Limitations

There are still some limitations in this paper that we hope to improve upon in future work, which is listed as followed.

- In this paper, we explored only two languages: Japanese as a representative of low-resource languages and English as a representative of high-resource languages. Introducing corpora from more languages would broaden the knowledge coverage and allow for a better study of the impact of different language corpora on LLM performance.
- To our best knowledge, mContriever is currently the most popular multilingual retriever. However, the quality gap in retrieval by the retriever itself cannot be entirely compensated for by a reranker. Therefore, we hope to explore different retrievers in the future and compare their performance.
- This paper, particularly in the discussion part, we primarily examined the performance with the number of retrievals equal to one. The main reason is that the interactions among multiple retrieved documents are significantly more complex than those involving a single retrieved document. When the content of multiple articles is complementary, similar, or contradictory, the RAG pipeline may exhibit different performances. We hope to explore this further in future research.

References

1. Jiang, J., Huang, J., Aizawa, A.: JMedBench: a benchmark for evaluating Japanese biomedical large language models. arXiv Preprint arXiv:2409.13317 (2024)
2. Pal, A., Umapathi, L., Sankarasubbu, M.: MedMCQA: a large-scale multi-subject multi-choice dataset for medical domain question answering. In: Proceedings of the Conference on Health, Inference, and Learning, vol. 174, pp. 248–260 (2022). https://proceedings.mlr.press/v174/pal22a.html
3. Jin, Q., Dhingra, B., Liu, Z., Cohen, W., Lu, X.: PubMedQA: a dataset for biomedical research question answering. In: Proceedings of the 2019 Conference on Empirical Methods in Natural Language Processing and the 9th International Joint Conference on Natural Language Processing (EMNLP-IJCNLP), pp. 2567–2577 (2019). https://aclanthology.org/D19-1259
4. Gao, Y., et al.: Retrieval-augmented generation for large language models: a survey. arXiv Preprint arXiv:2312.10997 (2023)
5. Chirkova, N., Rau, D., Déjean, H., Formal, T., Clinchant, S., Nikoulina, V.: Retrieval-augmented generation in multilingual settings. arXiv Preprint arXiv:2407.01463 (2024)
6. Li, B., Haider, S., Luo, F., Agashe, A., Callison-Burch, C.: BORDIRLINES: a dataset for evaluating cross-lingual retrieval-augmented generation. arXiv Preprint arXiv:2410.01171 (2024)
7. Kasai, J., Kasai, Y., Sakaguchi, K., Yamada, Y., Radev, D.: Evaluating GPT-4 and chatgpt on Japanese medical licensing examinations. arXiv Preprint arXiv:2303.18027 (2023)
8. Jin, D., Pan, E., Oufattole, N., Weng, W., Fang, H., Szolovits, P.: What disease does this patient have? A large-scale open domain question answering dataset from medical exams. Appl. Sci. **11**, 6421 (2021)
9. Dubey, A., et al.: The llama 3 herd of models. arXiv Preprint arXiv:2407.21783 (2024)
10. Yang, A., et al.: Qwen2 technical report. arXiv Preprint arXiv:2407.10671 (2024)
11. Chen, Z., et al.: Meditron-70b: scaling medical pretraining for large language models. arXiv Preprint arXiv:2311.16079 (2023)
12. Aizawa, A., et al.: LLM-jp: a cross-organizational project for the research and development of fully open Japanese LLMs. arXiv Preprint arXiv:2407.03963 (2024)
13. Qiu, P., et al.: Towards building multilingual language model for medicine. arXiv Preprint arXiv:2402.13963 (2024)
14. Xiong, G., Jin, Q., Lu, Z., Zhang, A.: Benchmarking retrieval-augmented generation for medicine. arXiv Preprint ArXiv:2402.13178 (2024)
15. Izacard, G., et al.: Unsupervised dense information retrieval with contrastive learning. arXiv Preprint arXiv:2112.09118 (2021)
16. Pradeep, R., Liu, Y., Zhang, X., Li, Y., Yates, A., Lin, J.: Squeezing water from a stone: a bag of tricks for further improving cross-encoder effectiveness for reranking. In: European Conference on Information Retrieval, pp. 655–670 (2022)
17. Khattab, O., Zaharia, M.: Colbert: efficient and effective passage search via contextualized late interaction over BERT. In: Proceedings of the 43rd International ACM SIGIR Conference on Research and Development in Information Retrieval, pp. 39–48 (2020)
18. Rybinski, M., Nguyen, V., Karimi, S.: A self-learning resource-efficient re-ranking method for clinical trials search. In: Proceedings of the 32nd ACM International Conference on Information and Knowledge Management, pp. 4249–4253 (2023)

19. Rybinski, M., Xu, J., Karimi, S.: Clinical trial search: using biomedical language understanding models for re-ranking. J. Biomed. Inform. **109**, 103530 (2020)
20. Lewis, P., et al.: Retrieval-augmented generation for knowledge-intensive NLP tasks. In: Advances in Neural Information Processing Systems, vol. 33, pp. 9459–9474 (2020)
21. Fan, W., et al.: A survey on rag meeting LLMs: towards retrieval-augmented large language models. In: Proceedings of the 30th ACM SIGKDD Conference on Knowledge Discovery and Data Mining, pp. 6491–6501 (2024)
22. Robertson, S., Zaragoza, H., et al.: The probabilistic relevance framework: BM25 and beyond. Found. Trends® Inf. Retrieval **3**, 333–389 (2009)
23. Karpukhin, V., et al.: Dense passage retrieval for open-domain question answering. arXiv Preprint arXiv:2004.04906 (2020)
24. Guu, K., Lee, K., Tung, Z., Pasupat, P., Chang, M.: Retrieval augmented language model pre-training. In: International Conference on Machine Learning, pp. 3929–3938 (2020)

Multi-head Vision Transformer for Multilingual Font and Language Identification

Anuar Aryngazin(✉) and Huseyin Atakan Varol

Institute of Smart Systems and Artificial Intelligence, Nazarbayev University, 010000 Astana, Kazakhstan
{anuar.aryngazin,ahvarol}@nu.edu.kz

Abstract. Font identification in multilingual documents is an underexplored area in natural language processing. This study introduces a multi-task vision model, fine-tuned from the pre-trained SigLIP base, designed for joint font and language identification. The model is trained on a large-scale synthetic dataset of 210,000 images encompassing five languages: Kazakh, Russian, Turkish, English, and Chinese, representing Cyrillic, Latin, and logographic scripts. The class set comprises 62 typed font families and a distinct handwritten class, which is designated as the unique 63rd class. The handwritten class is derived from 19 font families chosen to capture the variability of real handwritten text. The multi-head architecture, optimized using Adam and scheduled learning rates, achieves 99.4% accuracy in language classification and 93.6% in font classification on the validation set, indicating strong generalization. This work establishes a robust framework and benchmark for multilingual font recognition, supporting applications in digital humanities, cross-lingual retrieval, and document automation.

Keywords: multilingual font identification · multilingual language identification · multi-task learning · vision transformer

1 Introduction

One important area of research in information retrieval is the automated analysis of document typography. As more archives are being digitized, new applications in a variety of fields are made possible by the ability to programmatically extract a document's stylistic metadata. Pre-identifying a document's typeface, for instance, can increase the accuracy of character detection in optical character recognition (OCR) and lessen the need for manual post-processing. Therefore, a document's typographic style is an important layer of machine-readable data with many real-world uses. Improving the capabilities of contemporary document intelligence systems requires the development of reliable techniques for automatic font identification [1,2].

However correctly recognizing fonts is a difficult task, especially when dealing with multilingual settings. The problem's complexity, which necessitates that a

H. Unger and P. Meesad (Eds.): NLPIR 2025, LNNS 1904, pp. 140–154, 2026.
https://doi.org/10.1007/978-3-032-20897-2_10

model learn distinct visual features, is the main challenge. For instance, a system must distinguish between font families using minute differences in character geometry and serifs, which can be masked by low resolution or image noise. Documents with a variety of scripts, including Latin, Cyrillic, and logographic systems, each with unique structural characteristics, significantly increase this complexity. The main challenge is to create a single model that can simultaneously recognize languages across these diverse scripts and distinguish font families. An approach that can identify both stylistic details and general script-level characteristics is needed for this dual task.

This complexity is further exacerbated by a research gap. While the primary focus of document analysis has historically been on extracting semantic content, this approach often overlooks crucial typographical objects, which are essential for understanding the structure and conveying information in historical documents [3]. Furthermore, multi-task architectures capable of jointly identifying fonts and language have received little attention, particularly those leveraging the capabilities of modern lightweight Vision Transformers (ViTs). Clearly, a robust framework that addresses the problem of jointly identifying multilingual fonts is needed, especially one that leverages an efficient, advanced machine vision model.

This paper presents a multi-task learning framework for joint font and language identification from document images in order to fill this research gap. A lightweight (86M parameters) multi-head Vision Transformer built on the SigLIP architecture forms the basis of this strategy [4,5]. The SigLIP model was selected because of its high performance, which comes from training on large datasets and architectural improvements, making it ideal for learning robust features from synthetic data. The suggested model is trained to concurrently categorize an image into one of five languages, including Latin, Cyrillic, and logographic scripts, and one of 63 font classes. This work's main contributions are:

- A method for producing a large-scale, multilingual synthetic dataset for scalable font recognition.
- A multi-head vision model for the joint tasks of language and font classification, optimized from the SigLIP backbone.
- An extensive evaluation demonstrating strong accuracy and generalization across three scripts and five languages.
- A framework and baseline to support future research in multilingual document analysis.

The related work in multi-task document analysis and font recognition is reviewed in Sect. 2. The model architecture and the data synthesis pipeline are covered in detail in Sect. 3 of the methodology. The experimental setup, evaluation metrics, and results are presented in Sect. 4. Section 5 concludes the paper and provides thoughts for further research.

2 Related Work

The automated identification of fonts is a critical task in document analysis, providing valuable metadata for applications ranging from digital humanities to document forensics. This section examines the evolution of font identification methodologies, highlighting the transition from traditional Convolutional Neural Networks (CNNs) to modern Vision Transformers (ViTs), and contextualizes our selection of the SigLIP architecture.

2.1 Font and Language Identification with CNNs

The application of deep learning, particularly Convolutional Neural Networks (CNNs), marked a significant advancement in font recognition. By automatically learning hierarchical features from images, CNNs have demonstrated strong performance in this domain. Researchers have successfully trained CNNs on both original and synthetically generated data to classify fonts. For instance, a CNN-based framework achieved an accuracy of 99.03% for font classification on a challenging custom dataset of 18 Chinese fonts and 96.23% on the 21 English fonts [6]. Another research utilized a CNN and achieved 98.8% accuracy for line-level classification on a dataset of 40 Arabic computer fonts [7]. In a different approach aimed to be font-independent, a CNN was trained on image pairs to determine if any two characters share the same font, achieving a verification accuracy of 92.27% even on fonts unseen during training [2]. Addressing the challenge of large-scale classification, one study explored font recognition across a database of 10,000 fonts using a CNN, reporting a top-1 accuracy of 63.45% at the character level [1]. To bridge the gap between synthetic training data and real-world images, the DeepFont system utilized a CNN with a domain adaptation technique, advancing the state-of-the-art by achieving a top-5 accuracy of over 80% on a large-scale visual font recognition dataset [8].

Similarly, CNNs have been widely employed for language and script identification from document images. Deep learning techniques have revolutionized script identification by automating feature extraction [9]. For identifying languages in multilingual scene text images, a classifier combination approach that fuses the outputs of multiple minimalist CNNs, each trained on different image channels, achieved a high accuracy of 97.02% on the MLe2e dataset [10].

The success of these CNN-based methods often relies on their inherent inductive biases, such as locality and weight sharing, which make them highly effective even with limited data [11]. However, this focus on local features can be a limitation when trying to capture the global stylistic nuances that differentiate complex fonts [12].

2.2 Font and Language Identification with Vision Transformers

More recently, ViTs have presented a powerful alternative to CNNs for computer vision tasks [13]. Unlike CNNs that process images through localized convolutional filters, ViTs treat an image as a sequence of patches and employ self-attention mechanisms to model the relationships between them [14]. This allows

ViTs to capture long-range dependencies and global context more effectively, which is particularly advantageous for recognizing the holistic visual style of a font.

While CNNs build up a global understanding from local patterns, ViTs have access to global information from the very first layer. This architectural difference makes ViTs potentially more robust to occlusions and variations in text layout. A small ViT-based model for scene text recognition, known as ViTSTR, achieves a competitive accuracy of 82.6% (84.2% with data augmentation) compared to a strong baseline, while demonstrating a speed up of 2.4 times and utilizing only 43.4% of the parameters [15]. Another study, the TrueType Transformer, has experimentally demonstrated the applicability of transformers for character and font style recognition [16]. However, the effectiveness of ViTs is highly dependent on the scale of the training data; they tend to outperform CNNs when pre-trained on large datasets but can overfit on smaller ones [17].

2.3 SigLIP Model

Our work utilize the SigLIP (Sigmoid Loss for Language-Image Pre-training) model, a recent vision-language architecture that builds upon the foundation of models like CLIP. SigLIP employs a dual-encoder structure with a Vision Transformer for images and a text transformer, similar to its predecessors. Its key innovation lies in replacing the standard softmax-based contrastive loss with a pairwise sigmoid loss. This modification simplifies the training objective by treating each image-text pair as an individual classification problem, eliminating the need for a global comparison across all pairs in a batch [5].

To our knowledge, the application of large-scale vision-language models like SigLIP to the specific, fine-grained task of multilingual font recognition has not been extensively explored. A recent study highlighted that modern Vision-Language Models (VLMs) still exhibit limited font recognition capabilities, with many state-of-the-art models failing to achieve satisfactory performance on a dedicated font recognition benchmark [18]. This underscores a critical research gap and motivates our approach to fine-tune the powerful SigLIP architecture for this challenging, multi-faceted problem.

3 Methodology

The primary challenge this methodology addresses is the accurate identification of fonts and languages in multilingual documents, which is crucial for various applications such as automated document classification and digital archiving. The methodology consists of three primary components: the collection of source materials, the synthesis of the dataset, and the design of the multi-task SigLIP model architecture.

3.1 Data Collection and Synthesis

The foundation of the developed multi-task font identification model is a large-scale, synthetically generated dataset designed to cover a wide variety of typographical and linguistic features. The construction of this dataset was based on the curation of diverse multilingual text corpora and a comprehensive collection of font resources. Text data for part of the target languages, Kazakh, Russian, Turkish, and English, was sourced from the HuggingFace dataset issai/kazparc Kazakh Parallel Corpus [19]. This corpus provides a high-quality collection of parallel sentences, the dataset contains contextually relevant text across the Latin and Cyrillic scripts. For the Chinese corpus, the dataset consists of parallel translations of the same text. To ensure representativeness of the dataset, the selection covered various document types, including official documents, literature, and news articles, which naturally reflect the script and document frequency in real-world applications. The font data comprises a total of 81 distinct font files. These were selected based on their prevalence and popularity in digital and print documents to maximize real-world applicability. The collection includes fonts for Latin and Cyrillic scripts, as well as specialized fonts for Chinese, and a dedicated handwritten class. This latter class incorporates both publicly available handwritten style fonts and custom-developed fonts for the Kazakh language to introduce greater stylistic variance. The collection of typed fonts, detailed in Table 1, is composed of 41 fonts for Latin and Cyrillic scripts

Table 1. 62 fonts grouped by script and style.

Latin and Cyrillic Fonts	Logographic Fonts
Arial: Regular, Bold, Italic	Microsoft YaHei: Regular, Bold
CMU Sans Serif: Regular, Bold, Italic	Noto Sans SC: Regular, Bold
EB Garamond: Regular, Bold, Italic	Noto Serif SC: Regular, Bold
Fira Sans: Regular, Bold, Italic	SimHei: Regular
Helvetica: Regular, Bold, Italic	SimSun: Regular
Lora: Regular, Bold, Italic	STCaiyun: Regular
Merriweather: Regular, Bold, Italic	STFangsong: Regular
Montserrat: Regular, Bold, Italic	STHupo: Regular
Open Sans: Regular, Italic	STHXingkai: Regular
PT Sans Web: Regular, Bold, Italic	STKaiti: Regular
Roboto: Regular, Italic	STLiti: Regular
Source Sans Pro: Regular, Bold	STXihei: Regular
Times New Roman: Regular, Bold	STXinwei: Regular
Ubuntu: Regular, Bold, Italic	STZhongsong: Regular
Verdana: Regular, Bold, Italic	WDXL Lubrifont SC
	ZCOOL KuaiLe: Regular
	ZCOOL QingKe HuangYou: Regular
	ZCOOL XiaoWei: Regular

and 21 fonts for Chinese. The handwritten font category, which is not detailed in this table, consists of 8 Chinese fonts and 11 fonts designed for the other languages in the corpus.

To ensure scalable image creation, we leveraged a programmatic pipeline developed in Python, utilizing the Pillow and NumPy libraries. This pipeline was employed to automate the synthesis of document images. The full data synthesis pipeline is shown in Fig. 1. Each image was generated at a standardized resolution of 512×512 pixels and saved in PNG format. This format was chosen to preserve the fine details and crisp edges of the rendered text. The generation process for each image involved randomly selecting a text excerpt from the language-specific text corpus, a font file from the repository, and a background image. Text was rendered onto the canvas with a randomized font size, varying between 24 and 72 points, and was automatically wrapped to simulate paragraph structures. The text was positioned at a random location on the image to ensure the model learns different position features.

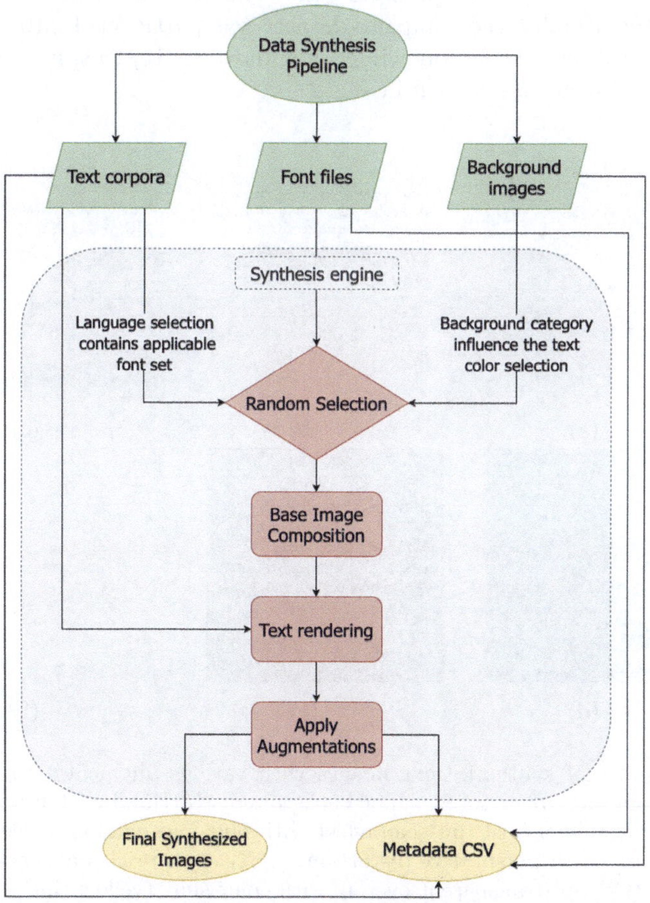

Fig. 1. Generation pipeline for a multilingual diverse font image dataset.

To enhance the model's robustness and its ability to generalize to real-world documents, which often suffer from degradation and noise, a series of augmentations was integrated into the generation pipeline. These augmentations simulate common imaging artifacts and document characteristics. Geometric transformations were applied, including minor full-image rotations within a range of -2 to $+2°$. To mimic sensor noise from scanners or cameras, digital noise was added to a subset of the images. Furthermore, to simulate varied document layouts, a semi-transparent bounding box was overlaid behind the text block with a 30% probability. The diversity of the dataset was further increased by rendering text onto a wide array of background images, which were categorized into simple, textured, colored, and noisy patterns. The automated generation process culminated in a substantial dataset, which was systematically structured and partitioned for model training and evaluation. A total of 1,000 unique images were generated for each of the font files in each of the language corpora, resulting in a final dataset of 210,000 images. For each generated image, the corresponding ground truth, including the image file path, font name, language, and a font class, was recorded in a structured CSV file. This file served as the master index for the dataset. Finally, the complete dataset was partitioned into training, validation, and testing subsets following a standard 80/10/10 split. The examples of synthesized images are shown in Fig. 2.

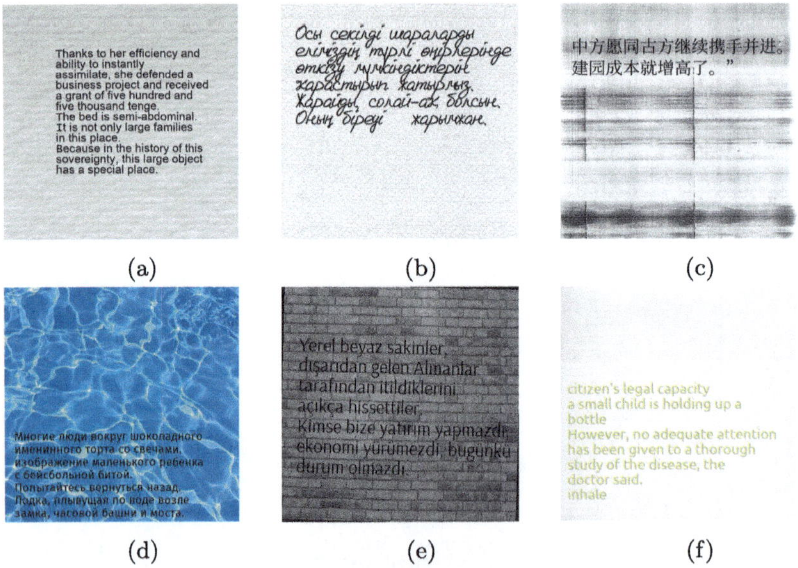

Fig. 2. Examples of synthetic text images with various augmentations applied. (a) Plain text rendering on textured paper background, (b) Handwritten font simulation with irregular baseline, (c) Print-scan effect with blur and banding noise, (d) Overlay on water texture with perspective distortion, (e) Text on brick wall background with low contrast, (f) Semi-transparent overlay with color shift (yellow tint).

3.2 Model Architecture

The model architecture is designed for multi-task learning, performing joint font and language classification from a single input image. A pre-trained Sigmoid-based Language-Image Pre-training (SigLIP) model serves as the vision backbone [5]. SigLIP, a state-of-the-art variant of CLIP, replaces the contrastive objective with a sigmoid loss to achieve more efficient and robust imagetext alignment. Its base architecture is a Vision Transformer (ViT), which has demonstrated strong generalization across large-scale multimodal datasets. This Vision Transformer (ViT) based architecture was selected for its demonstrated efficacy in extracting robust visual features from large-scale, diverse image data, which provides a strong foundation for the nuanced task of typographical analysis [20]. To adapt the model to the specific domain of document analysis, a fine-tuning approach was employed. The majority of the pre-trained backbone weights were frozen to preserve the learned general-purpose features, while only the final two transformer encoder layers were unfrozen for training.

To facilitate the dual classification tasks, the original classification head of the SigLIP model was replaced with a custom multi-head structure. This structure comprises two independent, parallel linear classification layers, each dedicated to a specific task. Overview of the complete architecture is provided in Fig. 3.

The *first head*, designated as the "Font Head", is a linear layer that projects the backbone's feature representation into a logit vector corresponding to the 63 classes (41 distinct font families for Latin and Cyrillic scripts, 21 font families for Chinese, and 19 fonts in handwritten class).

The *second head*, designated as the "Language Head", is a parallel linear layer that maps the same feature representation to a logit vector of size 5 classes, corresponding to the five languages in the dataset. This parallel architecture enables the model to learn a shared, representation from the backbone that is beneficial for both tasks, while maintaining specialized final layers for each distinct classification objective.

During a forward pass, an input image is first processed by the ViT backbone to generate a high-dimensional feature embedding. To mitigate the risk of

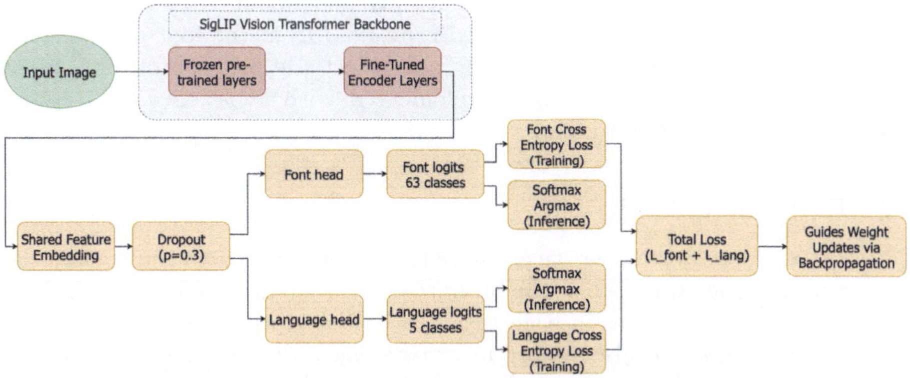

Fig. 3. The Multi-Task SigLIP ViT Model Architecture.

overfitting during the fine-tuning process, a Dropout layer with a rate of 0.3 is positioned between the output of the backbone and the input to the two classification heads. This regularized feature embedding is then passed concurrently to both the Font Head and the Language Head to produce the final logits for each task.

4 Experiments and Results

This section details the experimental setup, evaluation metrics, and a comprehensive analysis of the model's performance. Preliminary results indicate that the model achieves high accuracy in both language and font identification tasks, even under realistic noise conditions. We present quantitative results for both language and font identification tasks, analyze the model's training dynamics, and provide a qualitative assessment of its predictions.

4.1 Experimental Setup

Hardware and Software: All experiments were conducted on a workstation equipped with a single NVIDIA H100 GPU, featuring 80GB of HBM3 memory. The system was running NVIDIA Driver Version 580.65.06 with CUDA 13.0. The model and training pipeline were implemented in Python 3.10.18, utilizing the PyTorch 2.5.1 framework. The Hugging Face transformers library was used to access the pre-trained google/siglip-base-patch16-224 model.

Training Parameters: The model was trained for 20 epochs with a batch size of 32. This number of epochs was chosen to ensure adequate convergence without overfitting, as observed in initial experiments where longer training led to diminishing returns. We employed the AdamW optimizer with a learning rate of 1e-4 and default beta values, which were selected after trials revealed this combination consistently minimized loss without oscillating around local minima. A StepLR learning rate scheduler was configured to decrease the learning rate by a factor of 0.1 every 5 epochs, addressing the gradual loss plateau noticed in earlier runs. To mitigate overfitting in the classification heads, a dropout rate of 0.3 was applied to the features before the final linear layers. The total loss was calculated as the unweighted sum of the Cross-Entropy Loss from the font and language heads. To prevent unnecessary training and select the best-performing model, an early stopping mechanism was implemented with a patience of 3 epochs, monitored on the validation loss.

4.2 Evaluation Metrics

To assess model performance, we employed standard metrics for multi-class classification. Accuracy was used as the primary measure, reflecting the proportion of correctly classified samples in each task. To better capture performance across the 63 font classes, we also report macro-averaged Precision, Recall, and F1-Score, which treat all classes equally regardless of their frequency. In addition, we used visual tools such as loss curves and confusion matrices to analyze the model's learning behavior and error patterns on font and language recognition tasks.

4.3 Overall Performance

The proposed model achieves strong results on the test set. As summarized in Table 2, the joint learning framework delivers high accuracy in language identification and consistently high performance in the more fine-grained task of font identification.

Table 2. Overall model performance on the test set.

Task	Accuracy	Macro Precision	Macro Recall	Macro F1-Score
Language Id.	0.9948	0.9949	0.9951	0.9950
Font Classification	0.9361	0.9258	0.9225	0.9236

The near-perfect performance in language identification 99.4% demonstrates the model's ability to capture script-level features with high reliability. Meanwhile, the 93.6% accuracy in font classification highlights its effectiveness in distinguishing subtle typographical variations across multiple scripts. The consistently high macro-averaged F1-scores further confirm balanced and robust performance across all classes.

4.4 Training Dynamics and Model Convergence

To monitor the training process and diagnose potential issues like overfitting, we provide the training and validation loss and accuracy for both tasks over all 20 epochs, as shown in Fig. 4.

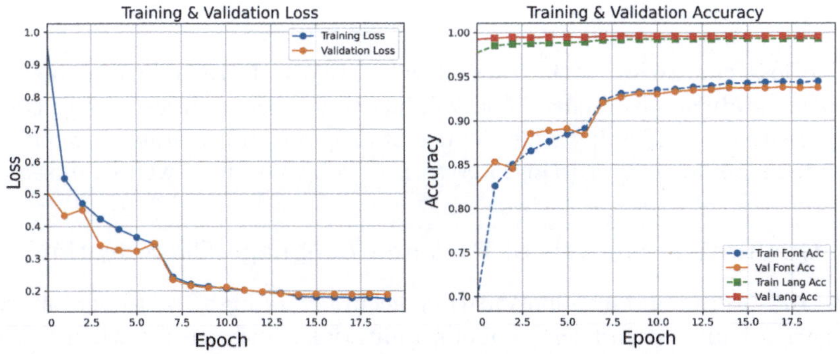

Fig. 4. Training and validation loss (left) and accuracy (right) curves for the language and font identification tasks over 20 epochs.

Fig. 5. Qualitative examples of font misclassifications on the test set.

The loss curves for both tasks show a steady decrease before plateauing, indicating stable convergence. The validation accuracy curves closely track the training curves, confirming that the use of dropout and weight decay effectively prevented significant overfitting, and the model generalizes well to unseen data.

4.5 Qualitative Analysis of Font and Language Classification

While the overall model identification accuracy is high, a deeper analysis is necessary to understand the model's limitations and error patterns. For the language classification task, the predictions achieve consistently high accuracy, all languages being correctly recognized in the majority of test cases. We also provide an illustration of the errors in the test set for font classification in Fig. 5.

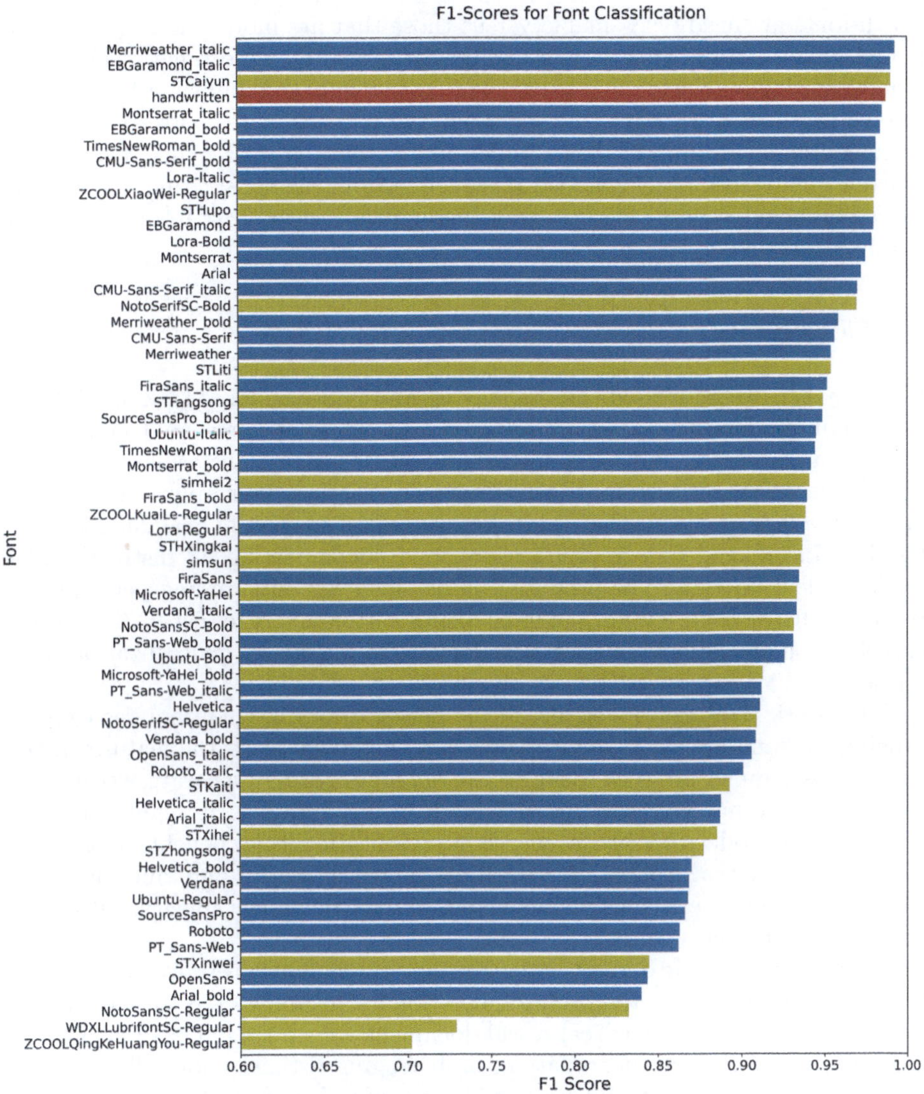

Fig. 6. Per-class F1-scores for the 63 font classes, color-coded by script category: Latin and Cyrillic (blue), logographic (yellow), and handwritten (red).

These examples highlight typical cases where the model struggles, such as fonts with very similar stylistic properties or degraded image quality. A closer inspection shows that errors often occur when the text samples are visually ambiguous, suggesting that future improvements may require more robust modeling of subtle stylistic cues or the inclusion of more diverse training data.

To further quantify the performance on individual fonts, we plotted the F1-score for each font class, as shown in Fig. 6. This visualization helps identify which fonts are easily classifiable versus those that are more challenging for the model. The analysis shows that highly distinctive Chinese fonts, such as the monospaced QingKeHuanqYou-Reqular or the informal LubrifontSC-Regular, receive the lowest F1 scores. In contrast, the highest scores correspond to the aforementioned groups of very similar fonts, such as STCaiyun. The "handwritten" font class also demonstrates high results, demonstrating the model's successful acquisition of features that distinguish printed text from cursive and handwritten text.

5 Conclusion

In this paper, we addressed the challenging and underexplored task of joint font and language identification in multilingual documents. We presented a multi-task learning architecture built on a fine-tuned SigLIP vision transformer, a powerful and efficient architecture for visual feature extraction. Our approach was validated on a large-scale synthetic dataset, generated to span five languages across Cyrillic, Latin, and logographic scripts, and encompassing 63 distinct classes for font identification. The proposed model achieved outstanding performance, reaching 99.4% accuracy for language identification and a robust 93.6% for font classification on the held-out test set, demonstrating strong generalization capabilities.

This work establishes a strong benchmark for multilingual font recognition, proving that modern ViT architectures can effectively learn the features necessary to distinguish between visually similar fonts across disparate writing systems. Our findings confirm that a joint learning approach is highly effective, allowing the model to learn a shared representation that benefits both tasks from a single forward pass. This work provides a powerful and efficient framework for extracting critical stylistic metadata from document images, with direct applications in many fields.

While our synthetic dataset with extensive augmentations proved effective for training a robust model, the primary direction for future work is to evaluate and enhance its performance on real-world documents. Such documents may contain complex layouts, severe degradation, and imaging artifacts not fully captured in our synthesis pipeline. Future iterations could also expand the scope of the dataset to include a wider range of other languages, fonts, and styles. Finally, the developed multi-task framework itself is extensible and could be adapted to simultaneously predict other document attributes, such as layout type or the presence of a handwritten signature, which opens the new ways to a more comprehensive system of understanding documents.

References

1. Ghosh, S., Roy, P., Bhattacharya, S., Pal, U.: Large-scale font identification from document images. In: Palaiahnakote, S., Sanniti di Baja, G., Wang, L., Yan, W.Q. (eds.) ACPR 2019. LNCS, vol. 12047, pp. 594–600. Springer, Cham (2020). https://doi.org/10.1007/978-3-030-41299-9_46
2. Haraguchi, D., Harada, S., Iwana, B.K., Shinahara, Y., Uchida, S.: Character-independent font identification. In: Bai, X., Karatzas, D., Lopresti, D. (eds.) DAS 2020. LNCS, vol. 12116, pp. 497–511. Springer, Cham (2020). https://doi.org/10.1007/978-3-030-57058-3_35
3. Zhalehpour, S., Arabnejad, E., Wellmon, C., Piper, A., Cheriet, M.: Visual information retrieval from historical document images. J. Cult. Herit. **40**, 99–112 (2019). https://doi.org/10.1016/j.culher.2019.05.018
4. Dosovitskiy, A., et al.: An image is worth 16 × 16 words: transformers for image recognition at scale. In: International Conference on Learning Representations (ICLR) (2021). https://openreview.net/forum?id=YicbFdNTTy
5. Zhai, X., Mustafa, B., Kolesnikov, A., Beyer, L.: Sigmoid loss for language image pre-training. In: Proceedings of the IEEE/CVF International Conference on Computer Vision, pp. 11975–11986. (2023). https://doi.org/10.48550/arXiv.2303.15343
6. Li, X., Wang, J., Zhang, H., Huang, Y., Huang, H.: SwordNet: Chinese character font style recognition network. IEEE Access: Practical Innovations, Open Solutions **10**, 8388–8398 (2022). https://doi.org/10.1109/ACCESS.2022.3143795
7. Tensmeyer, C., Saunders, D., Martinez, T.: Convolutional neural networks for font classification. In: 2017 14th IAPR International Conference on Document Analysis and Recognition (ICDAR), Kyoto (2017). https://doi.org/10.1109/ICDAR.2017.164
8. Wang, Z., et al.: DeepFont. In: Proceedings of the 23rd ACM International Conference on Multimedia, Brisbane Australia (2015). https://doi.org/10.1145/2733373.2806219
9. Ukil, S., Ghosh, S., Obaidullah, S.M., Santosh, K.C., Roy, K., Das, N.: Deep learning for word-level handwritten Indic script identification (2018). https://doi.org/10.48550/arXiv.1801.01627
10. Chakraborty, N., Kundu, S., Paul, S., Mollah, A.F., Basu, S., Sarkar, R.: Language identification from multi-lingual scene text images: a CNN based classifier ensemble approach. J. Ambient. Intell. Humaniz. Comput. **12**(7), 7997–8008 (2020). https://doi.org/10.1007/s12652-020-02528-4
11. Lahoti, A., Karp, S., Winston, E., Singh, A., Li, Y.: Role of locality and weight sharing in image-based tasks: a sample complexity separation between CNNs, LCNs, and FCNs. In: Proceedings of the International Conference on Learning Representations (ICLR) (2024). https://openreview.net/forum?id=AfnsTnYphT&utm_source=chatgpt.com
12. Park, S., Chun, S., Cha, J., Lee, B., Shim, H.: Multiple heads are better than one: few-shot font generation with multiple localized experts. In: Proceedings of the IEEE/CVF International Conference on Computer Vision (ICCV), Montreal, QC, Canada, pp. 13900–13909 (2021). https://doi.org/10.1109/ICCV48922.2021.01364
13. Heo, B., Yun, S., Han, D., Chun, S., Choe, J., Oh, S. J.: Rethinking spatial dimensions of vision transformers. In: IEEE/CVF International Conference on Computer Vision (ICCV), Montreal, QC, Canada, pp. 11916–11925. (2021). https://doi.org/10.1109/ICCV48922.2021.01172

14. Yu, T., Zhao, G., Li, P., Yu, Y.: BOAT: bilateral local attention vision transformer. In: 33rd British Machine Vision Conference (BMVC), London, UK, pp. 1–15. BMVA Press (2022). https://doi.org/10.48550/arXiv.2201.13027
15. Atienza, R.: Vision transformer for fast and efficient scene text recognition. In: Lladós, J., Lopresti, D., Uchida, S. (eds.) ICDAR 2021. LNCS, vol. 12821, pp. 319–334. Springer, Cham (2021). https://doi.org/10.1007/978-3-030-86549-8_21
16. Nagata, Y., Otao, J., Haraguchi, D., Uchida, S.: TrueType transformer: character and font style recognition in outline format. In Lecture Notes in Computer Science. Document Analysis Systems (2022). https://doi.org/10.1007/978-3-031-06555-2_2
17. Akkaya, I.B., Kathiresan, S.S., Arani, E., Zonooz, B.: Enhancing performance of vision transformers on small datasets through local inductive bias incorporation. Pattern Recogn. (2024). https://doi.org/10.1016/j.patcog.2024.110510
18. Li, Z., Song, G., Cai, Y., Xiong, Z., Yuan, J., Wang, Y.: Texture or semantics? Vision-language models get lost in font recognition. In: Proceedings of the Second Conference on Language Modeling (COLM 2025). https://openreview.net/forum?id=qiLJVU4I8P
19. Yeshpanov, R., Polonskaya, A., Varol, H.A.: KAZPARC: Kazakh parallel corpus for machine translation. ACL Anthology (2024). https://aclanthology.org/2024.lrec-main.842/
20. Huo, Y., Jin, K., Cai, J., Xiong, H., Pang, J.: Vision transformer (ViT)-based applications in image classification. In 2023 IEEE 9th International Conference on Big Data Security on Cloud (BigDataSecurity), IEEE International Conference on High Performance and Smart Computing, (HPSC) and IEEE International Conference on Intelligent Data and Security (IDS), pp. 135–140. IEEE (2023). https://doi.org/10.1109/BigDataSecurity-HPSC-IDS58521.2023.00033

MGDA: Multimodal Growth and Development Assessment Model

Weijie Ge, Zichen Song, Zijie Gong, and Zhongfeng Kang(✉)

School of Information Science and Engineering, Lanzhou University, Lanzhou 730000, Gansu Province, China
{gewj2023,songzch21,gongzj2023,kangzf}@lzu.edu.cn

Abstract. With the increasing societal focus on child health, the accurate assessment of growth and development has become a crucial indicator of national health levels and is vital for addressing global issues like child malnutrition. Despite advancements in assessment models driven by artificial intelligence, existing models still suffer from limitations such as single evaluation factors, inaccurate diagnoses, and inability to provide appropriate recommendations. To address these challenges, we propose MGDA (Multimodal Growth and Development Assessment model). This model is trained on the Radiological Society of North America (RSNA) dataset and tested on a dataset from the Department of Pediatrics at Huaibei People's Hospital. Its embedded In-Context Learning (ICL) module enables the model to rapidly identify tasks, ensuring that it provides precise diagnostic results and medical recommendations based on a comprehensive consideration of multiple evaluation factors. This approach offers a novel solution for the field of pediatric health assessment.

Keywords: Growth and development assessment · Multimodal data fusion · Transformer-Xception · In-context learning

1 Introduction

With the advent of precision medicine, personalized healthcare has become central to clinical practice, profoundly impacting medical fields including pediatrics. Childhood growth and development represents a critical biological process in early life and serves as a vital indicator of socioeconomic status and public health outcomes [1,2]. Accurate assessment and continuous monitoring of child development are essential for health promotion, public policy optimization, and disease prevention. Conventional evaluation methods relying on periodic physical examinations and standardized growth curve comparisons fail to adequately capture individual variability and the complexity of growth patterns. The precision medicine paradigm enables the integration of multidimensional data—including clinical examinations, pubertal staging parameters, and medical imaging—to

W. Ge, Z. Song and Z. Gong—These authors contributed equally to this work.

H. Unger and P. Meesad (Eds.): NLPIR 2025, LNNS 1904, pp. 155–167, 2026.
https://doi.org/10.1007/978-3-032-20897-2_11

establish comprehensive predictive models and implement individualized therapeutic interventions.

Xu et al. developed a deep learning-based approach for automated feature extraction from X-ray images to determine bone age, though this methodology was not extended to holistic developmental assessment [3]. Sole reliance on bone age for growth evaluation suffers from insufficient parameter consideration and prediction inaccuracies. Radgohar and Vahdat-Nejad's proposed detection system exhibited limited clinical accuracy due to its dependence on static growth curves [4]. Current research predominantly focuses on isolated parameters such as height and weight, with inadequate consideration of multidimensional factors including skeletal maturity and physiological indicators, coupled with a notable absence of multimodal data integration.

This study proposes a multimodal data fusion model for pediatric growth and development evaluation, employing a hybrid Transformer-Xception architecture. By systematically integrating X-ray imaging, clinical records, and family medical history, our model achieves comprehensive integration of multimodal data to generate precise diagnostic outputs and therapeutic recommendations, thereby enabling dynamic monitoring of developmental trajectories. The proposed framework addresses the critical limitation of insufficient multidimensional evaluation in conventional approaches, significantly enhancing the validity and accuracy of developmental assessments.

The main contributions of this study are as follows:

1. First multimodal model integrating imaging, physiological parameters, personal/family history, and bone age, improving evaluation precision.
2. ICL (In-Context Learning) enhances complex case and multimodal data comprehension.
3. Multimodal interaction analysis uncovers intrinsic growth patterns and environmental interactions, enabling precise medical recommendations.

The overall architecture of the proposed model is illustrated in Fig. 1.

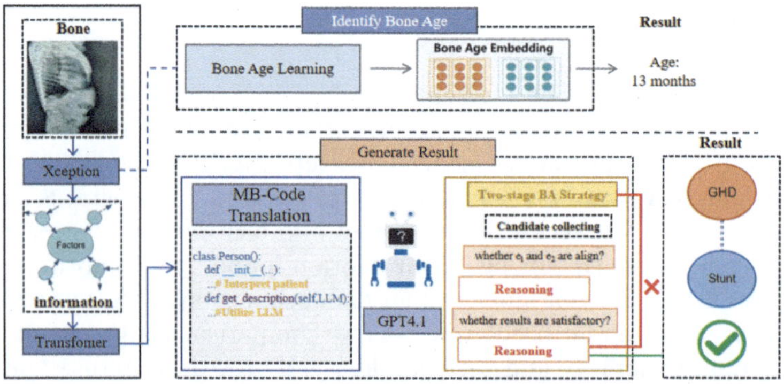

Fig. 1. The overall architecture of the multimodal growth and development model.

2 Related Work

François Chollet proposed the interpretation of the Inception module in convolutional neural networks, which is the intermediate step between conventional convolution and deep separable convolution operation (point convolution after deep convolution) [5]. Deep separable convolution can be understood as an Inception module with the largest number of towers. To this end, François Chollet proposed a new deep convolutional neural network architecture inspired by Inception, in which the Inception module has been replaced by deep separable convolution. Experiments show that this architecture, called Xception, performs slightly better than Inception V3 on the ImageNet dataset (which Inception V3 specifically designed for), and significantly better than Inception V3 on a larger image classification dataset containing 350 million images and 17,000 classes. Since the Xception architecture has the same number of parameters as Inception V3, the performance improvement is not due to the increase in capacity, but to the more efficient use of model parameters. We use Xception as an image classification processing model to make our large model perform better.

P. Xu, Xiatian Zhu, D. Clifton conducted a comprehensive survey of Transformer technology for multimodal data [6]. The main contents of the survey include:

(1) The background of multimodal learning, Transformer ecosystem and multimodal big data era.
(2) A systematic review of Vanilla Transformer, Vision Transformer and multimodal Transformer from the perspective of geometric topology.
(3) A review of the application of multimodal Transformer through two important paradigms, namely multimodal pre-training and specific multimodal tasks.
(4) Summarizes the common challenges and design of multimodal Transformer models and applications.
(5) Discusses the open issues and potential research directions of the community.

Xin-Nan Zong and Hui Li proposed a standardized growth curve for the growth of Chinese children and adolescents [7]. The importance of this study is that it provides a set of localized growth standards for Chinese children, which can more accurately reflect the growth trend of Chinese children. The growth curve includes not only routine indicators such as weight, length/height, and head circumference, but also BMI (body mass index) and the ratio of length/height to weight, which provides an important basis for pediatricians to measure the level of growth and development of children. As an important basis for measuring the growth and development level of Chinese children, this standardized growth curve can help to identify the deviation from the normal growth pattern, so as to take timely intervention measures. However, children's growth and development is a dynamic and comprehensive process. It is too one-sided to evaluate the level of growth and development only through physical examination data. It is impossible to find the overall situation of children's growth and development in depth and accurately, and it is difficult to diagnose children's growth and development diseases.

The research of Eric O Ohuma and Diego G Bassani et al. introduced machine learning modeling methods to evaluate children's growth indicators [8]. This innovative method utilizes the powerful data processing capabilities of machine learning, which can simultaneously analyze a large amount of physical examination data and mine hidden patterns and associations. The advantage of this method is that it can deal with nonlinear relationships and identify risk factors that may be ignored by traditional statistical methods. In addition, machine learning models can improve the accuracy of prediction through continuous learning and optimization. Machine learning method can process a large amount of data. As a very effective method in the evaluation of children's growth and development, it can be compared with the cases in the database to give accurate diagnosis and medical advice. However, this machine learning modeling method has too few factors to consider, only considering the physical examination data, and the diagnosis results are not accurate enough.

Xu Chen and Jianjun Li et al. focused on automatically extracting bone age features from X-ray images based on deep learning methods [1]. Their research demonstrates the potential of deep convolutional neural networks (CNNs) in bone age assessment. Bone age measurement is an important tool for assessing bone maturity, which is of great significance for predicting pubertal development time points, monitoring growth disorders, and assisting the diagnosis of certain diseases (such as endocrine disorders). Compared with traditional manual or semi-automatic feature extraction methods, deep learning-based methods can automatically identify and learn complex features in images, which greatly improves the accuracy and efficiency of bone age estimation. However, bone age alone can only know the level of children's skeletal development, and can not diagnose children's growth and development diseases.

In order to solve the above problems, this study proposes an innovative multimodal children's growth and development assessment model, which combines the powerful sequence processing ability of Transformer model and the excellent image feature extraction performance of Xception network. Specifically, we first design a special preprocessing process for different types of data; then, through the integration strategy, the information from different channels is organically integrated; finally, the performance of the entire system is optimized during a unified end-to-end training process. The experimental results show that compared with the traditional single-modal method, the proposed hybrid architecture can more accurately capture the key factors affecting children's growth and development, and give accurate diagnostic results and medical recommendations.

3 Model Construction

3.1 Model Building Pseudo-code

The following algorithms show the complete process from data preprocessing to model training and optimization [9,10].

Algorithm 1. Data Preprocessing and Integration

$D \leftarrow \{(x_i, y_i)\}_{i=1}^{N}$ where x_i represents patient data and y_i represents diagnostic labels
function ExtractBoneAge(*patient_data*)
  return ResNet(*patient_data*)
end function
function IntegratePatientData(*bone_age*, *age*, *hormone_level*)
  return $[bone\_age, age, hormone\_level]$
end function
$integrated\_data \leftarrow \Big\{ \big(\text{IntegratePatientData}(\text{ExtractBoneAge}(data),$
  $age, hormone\_level), label\big) \Big\}_{(data, age, hormone\_level, label) \in D}$

Algorithm 2. BERT Embedding and GNN Optimization

function BertEmbedding(x_i)
  $x_i^{\text{format}} \leftarrow$ "[CLS] $x_i[0]$ [SEP] $x_i[1]$ [SEP] $x_i[2]$ [SEP]"
  return BERT(x_i^{format})
end function
$embeddings \leftarrow \{\text{BertEmbedding}(data)\}_{(data, \_) \in integrated\_data}$
function GnnOptimization(*embeddings*)
  $A \leftarrow$ adjacency_matrix(*embeddings*)
  $H_0 \leftarrow embeddings$
  $H \leftarrow \text{GNN}(A, H_0)$
  return H
end function
optimized_embeddings $\leftarrow$ GnnOptimization(*embeddings*)

Algorithm 3. Exemplar Selection and Ordering

function RewardFunction(S)
  return $\frac{1}{|S|} \sum_{(x_i, y_i) \in S} \text{accuracy}(\text{model\_output}(x_i), y_i)$
end function
function SelectExemplars(D)
  $S^* \leftarrow \arg\max_{S \subseteq D} \text{RewardFunction}(S)$
  return S^*
end function
function OptimizeOrdering(S^*)
  $O^* \leftarrow \arg\max_O \text{RewardFunction}(\text{order}(S^*, O))$
  return O^*
end function
$S^* \leftarrow$ SelectExemplars(*integrated_data*)
$O^* \leftarrow$ OptimizeOrdering(S^*)

Algorithm 4. Model Training and Optimization

```
function LossFunction(model_output, y_i)
    return cross_entropy_loss(model_output, y_i)
end function
function TrainModel(S*, O*)
    ordered_S ← order_exemplars(S*, O*)
    for (x_i, y_i) ∈ ordered_S do
        model_output ← model(x_i)
        loss ← LossFunction(model_output, y_i)
        optimize_model(loss)
    end for
end function
TrainModel(S*, O*)
```

3.2 Xception

The multimodal growth and development assessment large model processes the image data of the X-ray film through the depth separable convolution layer of the Xception model. The depth separable convolution layer of the Xception model is decomposed into two steps: deep convolution and point-by-point convolution. Deep convolution performs feature extraction, as shown below:

$$D_K \times D_K \times M \times D_F \times D_F \tag{1}$$

Among them, D_K is the spatial dimension of the rectangular convolution kernel, M is the number of input channels, and D_F is the width and height of the input feature map respectively.

Pointwise convolution is responsible for integrating multiple channel features generated by deep convolution, as follows:

$$M \times N \times D_F \times D_F \tag{2}$$

Here, M is the number of input channels, N is the number of output channels, and D_F is the width and height of the input feature map, respectively.

Therefore, the total amount of computation of depthwise separable convolution is:

$$D_K \times D_K \times M \times D_F \times D_F + M \times N \times D_F \times D_F \tag{3}$$

The ratio of the computational complexity of the depthwise separable convolution to the computational complexity of the standard convolution is:

$$\frac{D_K \times D_K \times M \times D_F \times D_F + M \times N \times D_F \times D_F}{D_K \times D_K \times M \times N \times D_F \times D_F} = \frac{1}{N} + \frac{1}{D_K^2} \tag{4}$$

This decomposition can significantly reduce the computational resources and parameters required by the model, so that MGDA can better complete the task of processing X-ray image data [11–14].

3.3 ICL (In-Context Learning)

Based on most of the existing context learning work, we formulate the evaluation task of MGDA as a generation task. MGDA model M takes the generation sequence of the maximum score as the prediction output condition of the context C. For the k-shot MGDA task, C contains instructions I and k examples, where

$$C = \{I, s(v_1, t_1, y_1), \ldots, s(v_k, t_k, y_k)\} \tag{5}$$

where s is the demonstration template, $\{v_1, \ldots, v_k\}$ is a set of input free text labels, $\{t_1, \ldots, t_k\}$ is a set of input picture labels, and $\{y_1, \ldots, y_k\}$ is a set of output result free text labels.

Following the context learning paradigm, Demonstration Designing includes two parts: instruction construction and demonstration construction [15–20].

Instruction Construction. We use the definition of the assessment task of the multimodal growth and development assessment large model as an instruction, which can help LLM understand the current task, as shown below:

Identify potential growth anomalies in pediatric patients using height, weight, and age data, and recommend appropriate dietary or lifestyle adjustments.

$$I = \{\textbf{instruction}\} \tag{6}$$

Demonstration Construction. We use the multi-dimensional aspect to select the context example of the large model, for instance:

Input: A 10-year-old male child, with a height of 135 cm, weight of 32 kg, and a BMI of 17.2, accompanied by a history of slow weight gain over the past 6 months.

Expected Output: The model should predict the presence of potential growth delay or malnutrition, and recommend a follow-up nutritional assessment, suggesting possible dietary interventions such as increased protein intake and a higher caloric diet.

$$D = \{s(v_1, t_1, y_1), \ldots, s(v_k, t_k, y_k)\} \tag{7}$$

Scoring Function. Given context C and test sample $\{v_{\text{test}}, t_{\text{test}}\}$, we select the most likely generated sequence:

$$\hat{y} = \max P_{\mathcal{LLM}}(y \mid C, s(v_{\text{test}}, t_{\text{test}})) \tag{8}$$

3.4 Model Training and Evaluation

In the model training stage, the public data set of RSNA is used as the training set. The X-ray image of RSNA first enters the Xception part of the model. The depth of Xception can be separated from the deep convolution in the convolution layer to extract the features of the image. Then, after integrating the multi-channel features generated by the deep convolution by point-by-point convolution, the bone age-related features in the image are extracted, and then the features are further processed into embedded vectors and embedded into other types of data of RSNA, such as medical history and hormone levels, to form structured patient data.

After that, the data is selected and sorted by an example, a neural bandit algorithm is used to select samples, and a reward function R is defined to evaluate the performance of each group of samples:

$$\mathcal{S}^* = \arg\max_{\mathcal{S} \subset \mathcal{D}} R(\mathcal{S}) \tag{9}$$

where the reward function R is defined as:

$$R(\mathcal{S}) = \frac{1}{|\mathcal{S}|} \sum_{(x_i, y_i) \in \mathcal{S}} \text{Acc}(f(x_i, \theta), y_i) \tag{10}$$

Among them, $\text{Acc}(\cdot)$ represents the accuracy function of the model, $f(x_i, \theta)$ is the model output, and θ is the model parameter. After selecting the examples, we further optimize their ordering. By comparing the effects of different rankings on the performance of the model, we find the optimal ranking:

$$\mathcal{O}^* = \arg\max_{\mathcal{O}} R(\mathcal{O}(\mathcal{S}^*)) \tag{11}$$

Among them, $\mathcal{O}$ represents different sorting methods of samples, $\mathcal{O}(\mathcal{S}^*)$ denotes the set $\mathcal{S}^*$ of samples arranged in $\mathcal{O}$ order. In the model training phase, we use the optimized example set $\mathcal{S}^*$ and ranking information to jointly optimize the input hints of the model. After optimizing the data, the GPT-4.1 model outputs specific medical diagnosis and medical advice based on these optimized data. The model output is assumed to be $f(x_i, \theta)$, where θ denotes the model parameter. Finally, our goal is to minimize the following loss function:

$$\mathcal{L}(\theta) = \frac{1}{|\mathcal{S}^*|} \sum_{(x_i, y_i) \in \mathcal{S}^*} \ell(f(x_i, \theta), y_i) \tag{12}$$

in order to improve the accuracy of model prediction, where $\ell(\cdot, \cdot)$ is the loss function (such as cross entropy loss). In addition, Adam is adopted as the optimizer with its hyperparameters set to $\beta_1 = 0.9$, $\beta_2 = 0.999$. The learning rate is set to 0.001, with a weight decay of 0.0, an epsilon value of 1×10^{-7}, and amsgrad set to False.

Figure 2 shows the loss curve of model training. It can be seen that the training loss of the model decreases rapidly with iteration, and it decreases to about 5% after five iterations. It shows that the model can quickly learn how to better fit the training data and can extract effective information from the training data to adjust the model parameters, so as to improve the accuracy of model prediction.

After the model is trained, the model is evaluated. The test set selects the pediatric data set of Huaibei People's Hospital (In order to protect the privacy

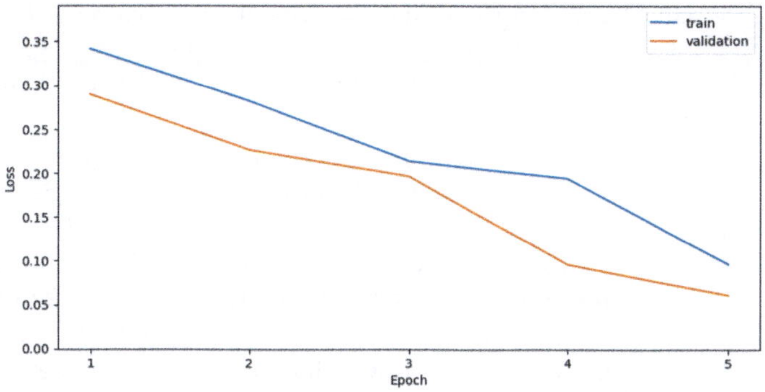

Fig. 2. The loss curve of the model.

and security of data and ensure that it is used appropriately, this test set is only used for non-commercial purposes such as research and education. To obtain this test set, please contact songzch21@lzu.edu.cn). First, the data set is preprocessed. The process is to cut each X-ray image to make it the same size as RSNA, and mark L and R. At the same time, the data set is preprocessed. The number of each X-ray image after cutting is encoded, so that the image is one-to-one correspondence with the patient information.

After that, the data of the picture is cleaned and erased. Every X-ray picture can identify any character of personal information. After the data preprocessing is completed, the model evaluation is started.

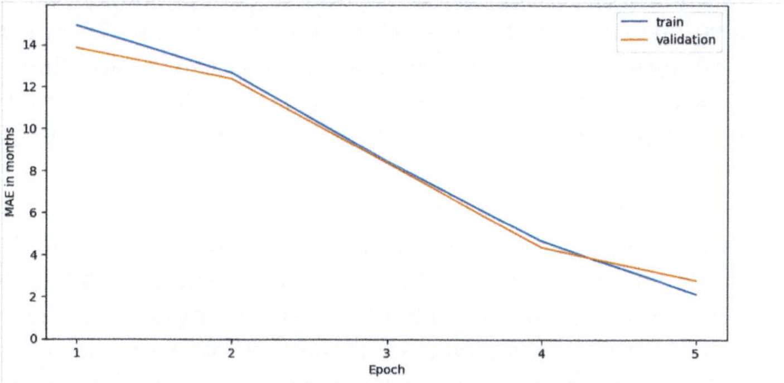

Fig. 3. Model evaluation curve.

Figure 3 is the evaluation curve of the model training. The evaluation index mae_in_months of the model decreases rapidly with the model iteration, which is reduced by about 87.5% under five iterations, showing the strong generalization ability of the model and being able to face a wider range of cases.

Table 1. Patient Cases with Bone Age Assessment and Treatment Evaluation

Case Name	Bone Age Error	Diagnosis	Treatment	Expert Appraisal
Case5 Pan	1.41	Early puberty	Interventional therapy with leuprorelin	Correct diagnosis, medication basically correct
Case8 Wang	2.32	Precocious puberty	Growth hormone interventional therapy	Correct diagnosis, medication basically correct
Case9 Li	0.89	Growth retardation	Growth hormone interventional therapy	Diagnosis not correct, medication basically correct
Case20 Li	1.34	Growth hormone deficiency	Growth hormone 6 units/day	Correct diagnosis, dosage not correct
Case50 Wei	1.21	Central precocious puberty	Leuprorelin 3.75mg once daily for 21 days	Correct diagnosis, accurate dosage

As shown in Fig. 2 and Fig. 3, the validation losses are consistently lower than the training losses. This phenomenon is primarily due to the loss calculation mechanism in the deep learning framework used (Keras/TensorFlow). The training loss is computed as the moving average of losses over each batch within an epoch, reflecting the model's learning curve throughout that epoch. In contrast, the validation loss is calculated at the end of the epoch after the model parameters have been updated. Given the rapid convergence of our model, the end-of-epoch validation metrics reflect a more optimized state than the intra-epoch training average. This explicitly demonstrates that the model can learn the key features of test data in a short time and effectively apply them to the evaluation task.

In addition to the test set to evaluate the performance of the model, we also conducted an overall evaluation. The evaluation content is the model's response to five difficult patient cases. From Table 1, it can be seen that the model still has a high accuracy rate in the face of difficult questions, indicating that the model has strong adaptability and effectiveness in the field of growth and development evaluation, and can also give accurate treatment plans in the face of complex situations.

Figure 4 shows a complete patient case and large model test results. After the model analyzes the case, the exact diagnosis results are given, and a reliable medical plan is given, which is consistent with the actual diagnosis results, reflecting the accuracy of the model analysis of the disease and the formulation of the medical plan.

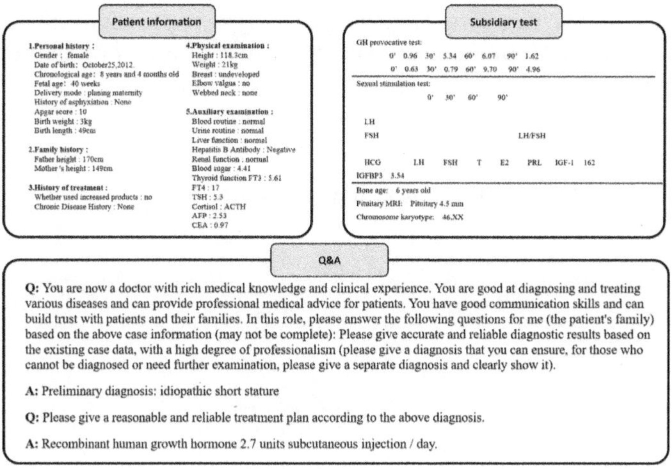

Fig. 4. One complete patient case and large model test results.

4 Conclusion

In this paper, MGDA (Multimodal Growth and Development Assessment model) for children's growth and development assessment is proposed. The model performs multimodal patient data analysis through a hybrid architecture of Transformer-Xception. The experimental results on the real test set show that the accuracy of the model for disease analysis and the accuracy of medical guidance are better than the existing children's bone age detection models. In addition, the ICL (In-Context Learning) module of this model can dynamically adjust the prediction strategy of the model. The evaluation results of the model evaluation index mae_in_months show that the generalization ability of this model to input cases exceeds the existing children's growth and development detection systems.

However, the GPT-4.1 used in this model has some limitations in the analysis of complex semantics and the analysis of the relationship between case data when dealing with complex cases, resulting in inaccurate diagnosis results. Therefore, we plan to replace GPT-4.1 with GPT-5 in the future, and use the advanced algorithm and optimization architecture of GPT-5 to further improve the model reasoning ability and the ability to understand complex contexts.

A Appendix

A.1 Self-attention Mechanism

The self-attention mechanism computes attention weights as:

$$A_{n,n'} = \frac{\exp\left(x_n^\top U_k^\top U_q x_{n'}\right)}{\sum_{n''=1}^{N} \exp\left(x_{n''}^\top U_k^\top U_q x_{n''}\right)} \tag{13}$$

where $A_{n,n'}$ denotes the attention weight between positions n and n' in the sequence, x_n and $x_{n'}$ represent the vector representations of the input at positions n and n', respectively. The learnable weight matrices U_k and U_q transform input vectors into key and query vectors.

A.2 Multi-head Self-attention Mechanism

$$Y^{(m)} = \mathrm{MHSA}_\theta(X^{(m-1)}) = \sum_{h=1}^{H} V_h^{(m)} X^{(m-1)} A_h^{(m)} \tag{14}$$

where

$$[A_h^{(m)}]_{n,n'} = \frac{\exp\left((\mathbf{k}_{h,n}^{(m)})^\top \mathbf{q}_{h,n'}^{(m)}\right)}{\sum_{n''=1}^{N} \exp\left((\mathbf{k}_{h,n''}^{(m)})^\top \mathbf{q}_{h,n'}^{(m)}\right)}, \tag{15}$$

$$\mathbf{q}_{h,n}^{(m)} = U_q^{(m)} \mathbf{x}_n^{(m-1)}, \tag{16}$$

$$\mathbf{k}_{h,n}^{(m)} = U_k^{(m)} \mathbf{x}_n^{(m-1)}. \tag{17}$$

Here $Y^{(m)}$ represents the output at layer m, H denotes the number of attention heads, $V_h^{(m)}$ is the projection matrix for the h-th head at layer m.

A.3 Token Normalisation

$$\bar{x}_{d,n} = \frac{1}{\sqrt{\mathrm{var}(\boldsymbol{x}_n)}} (x_{d,n} - \mathrm{mean}(\boldsymbol{x}_n))\, \gamma_d + \beta_d = \mathrm{LayerNorm}(X)_{d,n} \tag{18}$$

where $\bar{x}_{d,n}$ is the normalized value for the d-th feature of the n-th sample, γ_d and β_d are learnable scale and shift parameters. This normalization stabilizes learning by reducing internal covariate shift.

References

1. Ma, Y., Fan, H., Chen, Q.: Socioeconomic status and vocabulary development among children: a meta-analysis. Curr. Psychol. **43**(39), 30736–30747 (2024)
2. Fung, P., Pierre, T.S., Raja, M., Johnson, E.K.: Infants' and toddlers' language development during the pandemic: socioeconomic status mattered. J. Exp. Child Psychol. **236**, 105744 (2023)
3. Chen, X., Li, J., Zhang, Y., Lu, Y., Liu, S.: Automatic feature extraction in x-ray image based on deep learning approach for determination of bone age. Futur. Gener. Comput. Syst. **110**, 795–801 (2020)
4. Radgohar, H., Vahdat-Nejad, H., Mohammadzadeh Rezaie, S.: Infant's growth and nutrition monitoring system. SN Appl. Sci. **2**(9), 1477 (2020)
5. Chollet, F.: Xception: deep learning with depthwise separable convolutions. In: Proceedings of the IEEE Conference on Computer Vision and Pattern Recognition, pp. 1251–1258 (2017)

6. Xu, P., Zhu, X., Clifton, D.A.: Multimodal learning with transformers: a survey. IEEE Trans. Pattern Anal. Mach. Intell. **45**(10), 12113–12132 (2023)
7. Zong, X.N., Li, H.: Construction of a new growth references for China based on urban Chinese children: comparison with the who growth standards. PLoS ONE **8**(3), e59569 (2013)
8. Ohuma, E.O., Bassani, D.G., Qamar, H., Yang, S., Roth, D.E.: A novel development indicator based on population-average height trajectories of children aged 0–5 years modelled using 145 surveys in 64 countries, 2000–2018. BMJ Glob. Health **6**(3), e004107 (2021)
9. Kitaev, N., Kaiser, Ł., Levskaya, A.: Reformer: the efficient transformer. arXiv preprint arXiv:2001.04451 (2020)
10. Sukhbaatar, S., Weston, J., Fergus, R., et al.: End-to-end memory networks. In: Advances in Neural Information Processing Systems, vol. 28 (2015)
11. Tan, M., Le, Q.: Efficientnet: rethinking model scaling for convolutional neural networks. In: International Conference on Machine Learning, pp. 6105–6114. PMLR (2019)
12. Howard, A.G., et al.: Mobilenets: efficient convolutional neural networks for mobile vision applications. arXiv preprint arXiv:1704.04861 (2017)
13. Chen, Y., et al.: Drop an octave: reducing spatial redundancy in convolutional neural networks with octave convolution. In: Proceedings of the IEEE/CVF International Conference on Computer Vision, pp. 3435–3444 (2019)
14. Mao, X., et al.: Automated skeletal bone age assessment with two-stage convolutional transformer network based on x-ray images. Diagnostics **13**(11), 1837 (2023)
15. Cai, C., et al.: In-context learning for few-shot multimodal named entity recognition. In: Findings of the Association for Computational Linguistics: EMNLP 2023, pp. 2969–2979 (2023)
16. Zhao, H., et al.: Mmicl: empowering vision-language model with multi-modal in-context learning. arXiv preprint arXiv:2309.07915 (2023)
17. Chen, S., et al.: Understanding and improving in-context learning on vision-language models. arXiv preprint arXiv:2311.18021 (2023)
18. Wu, Z., Wang, Y., Ye, J., Kong, L.: Self-adaptive in-context learning: an information compression perspective for in-context example selection and ordering. arXiv preprint arXiv:2212.10375 (2022)
19. Li, C., Liu, M., Zhang, H., Chen, Y., Xu, J., Zhou, M.: Mt2: towards a multi-task machine translation model with translation-specific in-context learning. In: Proceedings of the 2023 Conference on Empirical Methods in Natural Language Processing, pp. 8616–8627 (2023)
20. Dong, R., et al.: Dreamllm: synergistic multimodal comprehension and creation. arXiv preprint arXiv:2309.11499 (2023)

BindNet: A Constraint-Regularized Transformer for Modeling Binding Principles in Neural Syntax Adaptation

Tejas Patil(✉) and Barnali Chetia

Indian Institute of Information Technology Vadodara, Vadodara, India
{tejas_patil,barnali}@iiitvadodara.ac.in

Abstract. The ability of machines to internalize and capture the structural nuances of language remains a central question within natural language processing. While transformer-based models show impressive linguistic fluency, their grasp of grammatical structure often remains superficial, particularly in situations where meaning relies on syntactic relations between the words rather than surface patterns. To explore this limitation, the present work introduces *BindNet*, a model designed to capture reflexive and pronominal dependencies through the theoretical framework of Binding Theory. Instead of embedding explicit grammatical rules, BindNet applies a form of constraint regularization that perfectly directs its attention toward syntactically appropriate antecedents. A dataset of one hundred thousand sentences (100000) was developed, which contained both well-formed and contrastive examples of binding. When trained on this corpus, BindNet displayed improved generalization and interpretive transparency compared to conventional transformer models. Its attention maps revealed sensitivity to clause boundaries and locality conditions, suggesting that linguistic theory can inform the design of models that not only perform well but also reflect the inner logic of human syntax.

Keywords: Syntax · Binding Theory · Neural Transformer

1 Introduction

Over the last few decades, there has been tremendous progress in the abilities of transformer-based architectures which has reshaped the trajectory of human-computer interaction. However, the accuracy of neural language models remains uncertain with respect to the grammatical regularities that control human linguistic competence. Their ability to internalize syntactic structure has been a both empirical and theoretical circles, raising the question of whether data-driven systems can reproduce intricate dependencies that generative linguistics has sought to describe for decades[1, 3, 6]. While large transformer models capture long-range statistical correlations, they often fail to distinguish deep relational constraints that define the structure of natural language [2, 11].

H. Unger and P. Meesad (Eds.): NLPIR 2025, LNNS 1904, pp. 168–181, 2026.
https://doi.org/10.1007/978-3-032-20897-2_12

The present study takes this tension as its starting point of analysis. It explores whether neural models can approximate one of the most precise systems in formal linguistics, which is Binding Theory. Originating in the tradition of generative grammar, Binding Theory (Chomsky, 1981) discusses how pronouns and reflexive elements are licensed and constrained by their syntactic environment. It stands on this idea that certain dependencies, such as those between a reflexive and its antecedent, are not random associations but tightly governed by hierarchical principles. These principles conventionally labeled as A, B, and C, define the allowable range of referential relationships within a sentence.

These distinctions in human linguistic cognition make these principles an ideal probe for evaluating the structural sensitivity of neural models. Several studies have shown that transformer networks display partial awareness of syntactic hierarchy, and their results often lack explicit alignment with the structural relations postulated by generative theory. Probing methods have revealed traces of phrase structure and dependency information in contextual embeddings, but these signals are distributed and fragile as models are scaled or fine-tuned for non-syntactic tasks [5,12].

BindNet is developed to analyze how far a transformer can be guided to internalize these theoretical constraints without being explicitly programmed with syntactic rules. Rather than imposing hard boundaries, the model introduces soft constraint regularization that influences attention weights, steering them toward linguistically plausible configurations. The central hypothesis is that, under such guidance, the model can develop internal representations that resemble structural reasoning, thereby narrowing the divide between statistical learning and formal grammar.

Through a dual evaluation of grammatical accuracy and Binding Attention Score(BAS), the study aims to highlight how linguistic theory can serve not merely as an analytic framework but also as a practical design tool for neural architectures. By aligning deep learning models with already established syntactic principles, it becomes possible to move beyond surface-level performance and toward more interpretable understanding of how machines engage with the structure of human language. The implementation of BindNet, along with the dataset generator and evaluation scripts, is available at https://github.com/Tej177/BindNet.

2 Related Work

The question of whether neural networks internalize syntactic structure has shaped much of the recent dialogue between linguistics and computational modeling. Earlier architectures, though being capable of sequence prediction, struggled to capture longer grammatical dependencies, leading to skepticism about their capacity for structural generalization. Linzen et al. [6] demonstrated that Long Short-Term Memory networks (LSTMs) could learn subject–verb agreement under restricted conditions but failed when syntactic complexity increased. LSTMs (Long Short-Term Memory networks) are a type of recurrent neural network that were widely used for sequence modeling before the introduction of transformer-based architectures. This finding initiated a broader inquiry into how modern architectures encode grammar beneath surface-level fluency.

With the advent of transformer models, the discussion moved from the limitations of recurrence to the interpretability of attention. Hewitt and Manning [3] proposed a structural probe that revealed latent syntactic geometry within contextual embeddings, suggesting that hierarchical information can, to some extent, be reconstructed from pretrained models. Similarly, work by Tenney et al. [10] showed that BERT implicitly rediscovers elements of the traditional NLP pipeline such as part of speech tagging, dependency parsing, and semantic role assignment within its intermediate layers. However, as Clark et al. [2] and Baroni [1] observed, the interpretability of such patterns remains limited, raising questions about whether these models truly "understand" syntax or merely correlate with it.

Gradually, more targeted evaluation strategies have emerged to address these questions. Marvin and Linzen [7] developed syntactically controlled datasets to test specific grammatical contrasts, while Ravfogel et al. [8] introduced constraint-based probing to interpret model attention in relation to theoretical categories. Subsequent studies by Warstadt et al. [11] and Li et al. [5] further refined this line of work, showing that transformer models can approximate linguistic generalizations, though their success depends heavily on task formulation and data balance.

Efforts to integrate explicit syntactic bias into neural architectures have also grown. Sartran et al. [9] proposed transformer grammars that embed inductive constraints into model training, demonstrating gains in structural accuracy and interpretability. Similarly, Kulmizev [4] questioned whether syntax in neural models behaves as a genuine generative mechanism or merely as an emergent correlate of exposure. These works collectively highlight a tension between statistical learning and grammatical theory which is an intersection where models like *BindNet* aim to intervene.

Our approach builds directly upon this evolving conversation by using Binding Theory as a soft syntactic constraint within transformer learning. Whereas prior studies have primarily focused on tree-based or dependency structures, we explore whether abstract relational principles, such as locality and referential binding, can be learned implicitly through regularized attention. In doing so, our work contributes to the growing effort to integrate neural architectures with the structural commitments of linguistic theory, offering a bridge between formal grammar and data-driven learning.

3 Theoretical Framework

Binding Theory describes how different kinds of noun phrases relate to one another within a sentence. It focuses on three main elements: reflexive pronouns (such as *himself*, *herself*), ordinary pronouns (such as *he*, *she*), and referential expressions like personal names. The three principles of Binding Theory A, B, and C set out where these elements may or may not take their reference.

3.1 Principle A

Principle A states that a reflexive pronoun must find its antecedent within the same clause. For example:

- *Maria blamed herself.*
  Here, the reflexive *herself* refers correctly to the subject *Maria*, which satisfies Principle A.
- *Maria said that John blamed herself.*
  In this sentence, *herself* cannot refer to *Maria* because the reflexive appears in a different clause. This violates Principle A.

Principle A therefore captures the idea of *locality*: a reflexive must have a nearby antecedent.

3.2 Principle B

Principle B concerns ordinary pronouns. It states that a pronoun must be free in its own local domain, meaning that it should not refer back to a noun found inside the same clause.

- *Maria blamed her.*
  The pronoun *her* cannot refer to *Maria* here because both occur within the same clause. This is a violation of Principle B.
- *Maria said that she was late.*
  In this sentence, *she* appears in an embedded clause and may refer to *Maria* or to another female individual. Both readings are allowed because the pronoun is not bound within its local domain.

Principle B thus contrasts with Principle A: pronouns should not behave like reflexives.

3.3 Principle C

Principle C applies to referential expressions, such as proper names. It requires that these expressions be free and not controlled or bound by other elements in the sentence.

- *She said that Maria was late.*
  In this structure, the pronoun *she* cannot refer to *Maria*. Allowing such a reading would mean that the pronoun binds a proper name, which violates Principle C.

Principle C ensures that names maintain independent reference and are not constrained by earlier pronouns.

3.4 Relevance to Our Model

BindNet incorporates these three principles as soft constraints on the model's attention patterns. Rather than enforcing rigid structural rules, these constraints gently guide the model toward patterns that align with the predictions of Binding Theory. The aim is to help the model learn linguistic structure in a way that remains compatible with established theoretical insights, while preserving its ability to learn from data.

4 Model Architecture

The architecture of BindNet attempts to bring together formal linguistic constraints with the representational dynamics of transformer-based language models. BindNet primarily extends the RoBERTa-base architecture by introducing an additional mechanism that softly regularizes the distribution of attention according to principles of Binding Theory. The aim is not just to encode grammatical rules explicitly, but to influence the model's learning trajectory so that its internal representations evolve in a way consistent with theoretical expectations.

The model undertakes the standard transformer configuration, where each input sentence is tokenized and mapped into contextual embeddings through multi-head self-attention layers. These embeddings attempt to capture dependencies across tokens by dynamically weighting the relevance of one word to another. In our model, BindNet, this process is guided by a constraint module that operates on the attention matrices of selected heads in the higher layers. The module calculates secondary loss function, the *constraint regularization term*, which evaluates how closely the model's attention aligns with permissible binding relations.

The total loss function is defined as:

$$\mathcal{L}_{\text{BindNet}} = \mathcal{L}_{\text{task}} + \lambda_A \mathcal{L}_A + \lambda_B \mathcal{L}_B + \lambda_C \mathcal{L}_C$$

where $\mathcal{L}_{\text{task}}$ is cross-entropy classification loss for grammaticality prediction, and $\mathcal{L}_A$, $\mathcal{L}_B$, and $\mathcal{L}_C$ are regularization terms corresponding to the three binding principles. Each λ coefficient controls the relative strength of the given constraint. The regularizers are implemented as soft penalties rather than hard exclusions, which allows the model to explore the space of potential bindings while gently pushing it toward linguistically valid configurations.

The regularization process operates over token-level attention weights. For Principle A, the model is rewarded when reflexive pronouns allocate a high proportion of attention to antecedents within the same clause. For Principle B, the model is penalized when pronouns attend too strongly to local antecedents, thus enforcing the compulsion that they remain free within their local domain. For Principle C, the model prohibits strong attention from pronouns toward referential expressions that are indexed with them. Thus, these constraints act as inductive biases that embed the logic of syntactic dependency into the geometry of attention.

The classification head remains identical to that of RoBERTa, consisting of linear transformation followed by softmax layer that predicts whether a given sentence is grammatically compliant or violates one of the binding principles. The only architectural modification is the introduction of the constraint loss, which interacts with the attention mechanism rather than altering the overall structure of the transformer. This design makes sure that BindNet retains the expressive capacity of large pre-trained models while achieving interpretive transparency in its syntactic behaviour (Table 1).

Table 1. BindNet architecture through example inputs and constraint-driven outputs.

Input Sentence	Constraint Applied	Model Output
Maria blamed herself.	Principle A: Reflexive must be bound within its local clause.	Grammatical (High BAS: 0.81)
Maria said Jack blamed herself.	Violation of Principle A: Reflexive outside local clause.	Ungrammatical (Low BAS: 0.29)
Jack said he was tired.	Principle B: Pronoun must be free in its local domain.	Grammatical (BAS: 0.74)
He said Jack was tired.	Violation of Principle C: Referential NP bound by pronoun.	Ungrammatical (BAS: 0.35)

The table illustrates how BindNet operates Binding Theory constraints through soft regularization. The model's predictions align with syntactic theory, as reflected by the Binding Attention Score (BAS), which quantifies interpretive consistency between attention and grammatical expectation.

5 Dataset Construction

Our dataset was designed to explore how a model might internalize the structural logic of binding relations in English. It contains one hundred thousand sentences, each annotated for grammaticality according to the three principles of Binding Theory. The goal was to expose our model to both correct and contrastive constructions revealing how syntactic dependencies behave in that particular context.

The corpus was built through semi-automated process that combined structured generation with careful human review. We began with common English sentence types, including simple clauses and embedded constructions, and varied lexical items to reduce surface-level repetition. For each grammatical example, an ungrammatical counterpart was created to represent violation of a binding condition. For instance, *Maria blamed herself* was paired with *Maria said that John blamed herself*, illustrating a failure of locality under Principle A.

Each sentence was labeled by binding type (A, B, or C) and by grammatical status. Ambiguous or unclear cases were filtered out after manual review by annotators with training in syntax. A small portion of the data was further modified to include lexical and structural variation, allowing us to test the model's ability to generalize beyond familiar forms. All sentences were tokenized using the RoBERTa tokenizer for consistent processing.

In constructing this corpus, our intention was to build not just a dataset but a diagnostic framework that will reveal whether a neural model, which is trained on examples from linguistic theory, can reflect structural reasoning found in human syntax.

6 Experimental Setup

Our experiments were designed to explore whether a transformer model, when guided by syntactic constraints, can develop an internal sense of grammatical structure that extends beyond surface regularities. All models were implemented using the PyTorch framework, taking RoBERTa-base as foundation for both the baseline and the constraint-regularized versions. RoBERTa was taken because of its strong contextual representation and stable optimization behavior, which make it well-suited for experiments that add auxiliary objectives to the standard training routine.

BindNet modifies this base architecture by incorporating the constraint regularizer described earlier, which encourages the model's attention to reflect the locality and freedom conditions associated with Principles A, B, and C of Binding Theory. In practice, this means that during training, our model is not merely optimizing for task accuracy but is also penalized when its attention diverges from linguistically plausible binding patterns.

The system was trained on a corpus of one hundred thousand annotated sentences representing a wide range of binding configurations. Each sentence was tokenized using the RoBERTa tokenizer and truncated to a maximum length of 128 tokens. Training employed early stopping based on validation accuracy to prevent overfitting, and each experiment was repeated across three random seeds to ensure stability and reproducibility. The weights of the regularization terms such as λ_A, λ_B, and λ_C were tuned empirically within the range of 0.1 to 1.0 to balance the strength of linguistic constraints against the task-specific objective.

Evaluation was performed across four dataset partitions: random, template, noisy, and cross-clause. These settings were intended to test not only lexical generalization but also the model's ability to sustain grammatical coherence when structural context changes. The primary metric for comparison was classification accuracy, measuring how reliably the model could distinguish between grammatically valid and invalid sentences. Alongside accuracy, we introduced an interpretability metric called the Binding Attention Score (BAS), which quantifies the degree to which the model's attention aligns with syntactically legitimate antecedents. A higher BAS indicates stronger alignment between the learned attention distributions and the predictions of linguistic theory.

We used two baselines for the sake of comparison. The first was a standard RoBERTa-base fine-tuned on the same dataset but without the constraint regularization. The second baseline was a DeBERTa-base model, included for its enhanced positional encoding mechanism, which provides a more sophisticated treatment of syntactic distance. Both models were trained under identical conditions, allowing a direct comparison with BindNet.

Beyond numerical accuracy, qualitative analyses were performed to better understand how constraint regularization reshapes internal model behavior. Attention visualizations were generated to inspect whether the model's focus followed the expected reflexive–antecedent dependencies. In many cases, BindNet displayed more coherent and localized attention patterns than either baseline, suggesting that the regularization term helped the model reorganize its internal representation of syntax.

Taken together, our experimental setup reflects a dual objective, which is to achieve quantitative improvement while revealing qualitative differences in learning dynamics. The purpose was not simply to show that BindNet performs better, but to examine how it learns differently—whether its reasoning appears closer to the structural intuitions that linguistic theory has long articulated. In this sense, the experiments are both an empirical evaluation and a conceptual exploration of how grammatical knowledge can emerge within neural representations shaped by theoretical design.

7 Results and Discussion

The results highlight that BindNet consistently outperforms the baseline model across all evaluation splits and exhibits an interpretable internal structure aligned with syntactic constraints. Beyond numerical improvement, our model shows a greater capacity to reflect grammatical organization, particularly within the domain of reflexive and pronominal binding.

7.1 Quantitative Evaluation

Figure 1 compares accuracy across the four data divisions. Though the baseline model performs reasonably on the random and template splits, it struggles with cross-clause dependencies where structural information must override linear order. Our model BindNet maintains higher accuracy across all splits, with its strongest gains in the cross-clause and noisy-augmentation settings. These results show that BindNet generalizes not only across lexical variations but also across structural complexity, suggesting that it learns stable representations of syntactic locality.

The performance gap is particularly seen in cross-clause condition, where BindNet surpasses the baseline by sixteen percentage points. This improvement suggests that the regularization term effectively encodes a form of inductive

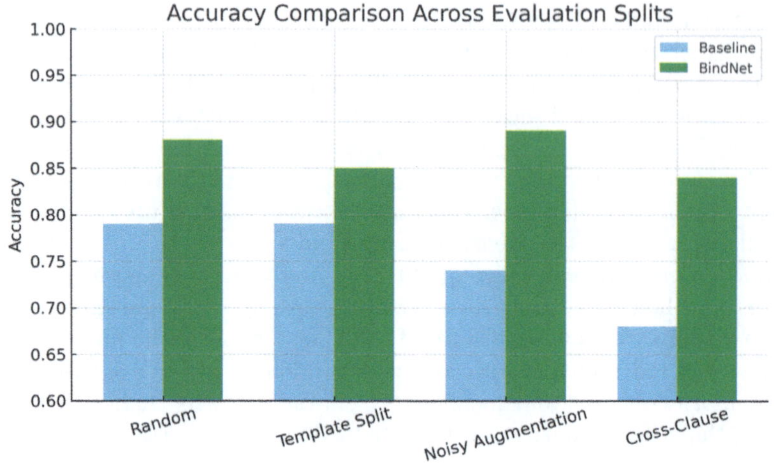

Fig. 1. Comparison of Accuracy between Baseline and BindNet across evaluation splits. BindNet achieves greater stability under syntactic complexity and lexical deviation.

syntactic bias, allowing the model to capture structural relations beyond surface patterns.

7.2 Binding Attention Score (BAS)

The Binding Attention Score (BAS) measures internal interpretability unlike accuracy metrics, which measures external performance. It reflects how strongly the model's attention aligns with legitimate antecedent–reflexive pairs. Table 2 summarizes these results.

Table 2. Binding Attention Score (BAS) comparison between Baseline and BindNet.

Model	BAS Mean	Interpretation
Baseline BERT	0.41	Syntactic alignment is weak or diffuse
BindNet	0.74	Strong and localized binding attention

BindNet's BAS is nearly double that of the baseline, which reflects stronger alignment between reflexive pronouns and their grammatical antecedents. This shows that the model's attention mechanism has captured and internalized structural dependencies predicted by Binding Theory, even without explicit rule encoding.

7.3 Visualization of Attention Patterns: Principles A, B, and C

Figure 2 visualizes the attention patterns for the three example sentences used to illustrate Principles A, B, and C of Binding Theory. Each pair of heatmaps contrasts the baseline transformer with BindNet, showing how structural cues shape the model's attention.

In the Principle A sentence, "The actress told herself a story", the baseline model produces a broadly diagonal pattern with no clear emphasis on the reflexive–antecedent link. The attention mass is spread across nearby tokens, making it difficult to identify the reflexive dependency visually. BindNet, however, displays a sharp, high-weight concentration from *herself* to *actress* (annotated as 0.80 in the heatmap). This focused arc indicates that BindNet identifies the subject as the appropriate antecedent for the reflexive, accurately reflecting the locality requirement central to Principle A. The surrounding grid remains faint, showing that the model suppresses irrelevant connections and prioritizes the correct structural relation.

For Principle B, the sentence "John said he would win the match" reveals a different pattern. The baseline model again shows a smooth diagonal band with only minor fluctuations, suggesting no strong awareness of long-distance pronoun–antecedent structure. BindNet, on the other hand, assigns a strong attention weight from *he* to *John* (marked as 0.84), even though the pronoun appears inside an embedded clause. This shift across the clause boundary reflects Principle B's requirement that pronouns remain free within their local domain but may refer to a non-local antecedent. The rest of the grid is light and uniform, showing that BindNet selectively attends to the relevant antecedent rather than defaulting to clause-internal tokens.

The Principle C sentence, "He said that John would resign tomorrow", further illustrates BindNet's sensitivity to referential constraints. In the baseline heatmap, the diagonal remains dominant and no strong co-reference signal emerges. BindNet instead shows a focused attention value of 0.79 from *He* to the matrix-clause subject position but notably does *not* link the pronoun to the embedded name *John*. This is consistent with Principle C, which prohibits a pronoun from binding a referential expression (R-expression) in its c-command domain. The heatmap highlights this by presenting a clear absence of attention on the illicit dependency site, while still maintaining coherent structural alignment elsewhere.

Across all three cases, BindNet exhibits distinct, interpretable attention patterns that correspond closely to the binding requirements of each sentence. The baseline model, by contrast, maintains largely positional and context-agnostic attention distributions. These visual patterns support the claim that BindNet internalizes syntactic constraints more faithfully, using attention mechanisms to represent locality, freedom, and disjointness in line with Principles A, B, and C. The heatmaps therefore provide qualitative evidence for the model's structural awareness beyond what is captured by accuracy alone.

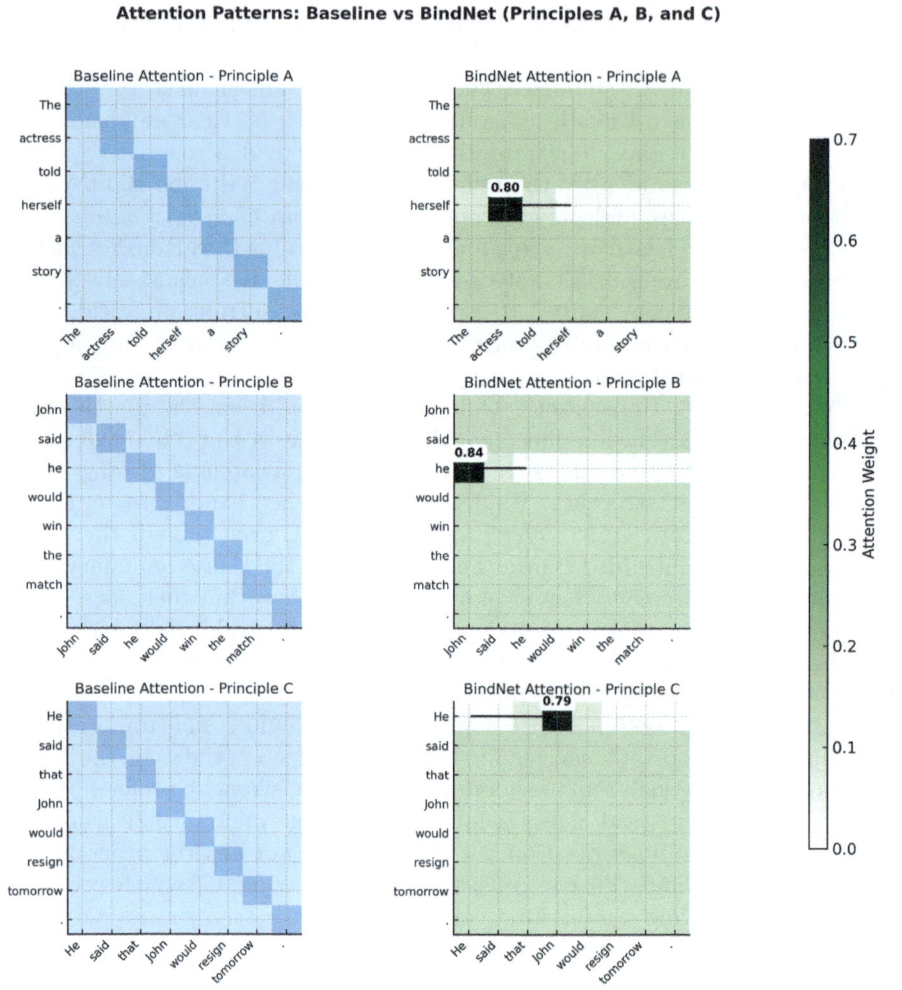

Fig. 2. Patterns of attention for Baseline and BindNet across Binding Theory Principles A, B, and C.

7.4 Discussion of Patterns

There is a balance between grammatical interpretability and empirical accuracy of BindNet in both quantitative and qualitative dimensions. It shows progress strongly in syntactically demanding conditions, validating the effectiveness of the constraint-based regularizer.

Moving forward, the attention distributions indicate that BindNet develops a form of representational balance because it focuses on structurally relevant regions rather than dispersing it across the sentence. This focused attention helps in more efficient learning and contributes to the model's more rapid convergence during training.

From the perspective of linguistics, the emergence of rule-consistent attention patterns suggests that neural architectures can approximate theoretical grammatical behaviour when provided with structured inductive guidance. Thus, BindNet serves as an example of how neural models can bridge the empirical success of deep learning with the interpretive clarity of formal syntax.

8 Limitations

While BindNet demonstrates strong interpretability and syntactic sensitivity, there are certain limitations that should be considered. This work models binding phenomena at the sentence level. As a result, the system does not incorporate discourse-level context or long-distance reference beyond the sentence boundary. The model also relies mainly on structural information and does not use semantic cues such as gender or animacy, which often play a role in natural co-reference. In addition, the dataset is based on controlled templates and does not reflect the full variety of naturally occurring sentences. These limitations define the scope of the current study, and future work may extend the framework to richer contexts and more diverse data.

9 Conclusion and Future Work

This study began with a simple question that whether neural language models can be guided to develop an understanding of syntax that goes beyond surface regularities. By designing and introducing BindNet, a constraint-regularized transformer inspired by Binding Theory, we attempted to see if subtle syntactic values could shape the way attention forms and behaves inside such models. Our findings suggest that this kind of soft linguistic guidance does make a difference as BindNet not only performs better on binding-related judgments but also shows patterns of attention that align more closely with the logic of grammatical structure. What emerges is a sense that interpretability and accuracy are not conflicting objectives; rather, when the design of a model is shaped by linguistic theory, the two can reinforce each other in productive ways.

If viewed in a wider perspective, this work underlines the continuing importance of linguistic theory in an age dominated by massive, data-driven models. The strength of BindNet lies not in the scale of its parameters but in the principle behind its design believing that human language is not random data, but a structured system that can inform how machines learn. This approach offers a reminder that syntax is not merely a computational obstacle but a source of insight into how meaning is organized.

There are many ways in which this research can be extended. Future studies may explore whether the same form of constraint-based regularization can help a model handle other syntactic phenomena, such as control, movement, or scope. Testing BindNet on languages with grammatical properties could reveal how universal or language-specific its learned representations really are. Another direction would be to integrate this mechanism into larger pretrained systems to examine whether global linguistic fluency can coexist with local structural awareness.

At its core, BindNet represents a small but meaningful step toward a synthesis of formal syntax and neural computation. By grounding the architecture of learning in theoretical insight, we begin to move toward systems that not only produce grammatical language but also mirror, in some limited way, the structural reasoning that makes language human.

References

1. Baroni, M.: Linguistic generalization and compositionality in modern artificial neural networks. Philos. Trans. R. Soc. B: Biol. Sci. **375**(1791), 20190307 (2019). https://doi.org/10.1098/rstb.2019.0307
2. Clark, K., Khandelwal, U., Levy, O., Manning, C.D.: What does BERT look at? An analysis of BERT's attention. In: Proceedings of the 2019 ACL Workshop BlackboxNLP: Analyzing and Interpreting Neural Networks for NLP, pp. 276–286 (2019). https://doi.org/10.18653/v1/W19-4828
3. Hewitt, J., Manning, C.D.: A structural probe for finding syntax in word representations. In: Proceedings of the 2019 Conference of the North American Chapter of the Association for Computational Linguistics: Human Language Technologies, pp. 4129–4138 (2019). https://doi.org/10.18653/v1/N19-1419
4. Kulmizev, A.: Schrödinger's tree—on syntax and neural language models. Front. Artif. Intell. **5**, 796788 (2022). https://doi.org/10.3389/frai.2022.796788
5. Li, B., Wisniewski, G., Crabbé, B.: Assessing the capacity of transformers to abstract syntactic representations: a contrastive analysis based on long-distance agreement. Trans. Assoc. Comput. Linguist. **11**, 18–33 (2023). https://doi.org/10.1162/tacl_a_00531
6. Linzen, T., Dupoux, E., Goldberg, Y.: Assessing the ability of LSTMs to learn syntax-sensitive dependencies. Trans. Assoc. Comput. Linguist. **4**, 521–535 (2016). https://doi.org/10.1162/tacl_a_00115
7. Marvin, R., Linzen, T.: Targeted syntactic evaluation of language models. In: Proceedings of the 2018 Conference on Empirical Methods in Natural Language Processing, pp. 1192–1202 (2018). https://doi.org/10.18653/v1/D18-1151
8. Ravfogel, S., Prasad, G., Linzen, T., Goldberg, Y.: Counterfactual interventions reveal the causal effect of relative clause representations on agreement prediction. In: Proceedings of the 25th Conference on Computational Natural Language Learning (CoNLL), pp. 194–209 (2021). https://doi.org/10.18653/v1/2021.conll-1.15
9. Sartran, L., Barrett, S., Kuncoro, A., Stanojević, M., Blunsom, P., Dyer, C.: Transformer grammars: augmenting transformer language models with syntactic inductive biases at scale. Trans. Assoc. Comput. Linguist. **10**, 1423–1439 (2022). https://doi.org/10.1162/tacl_a_00526
10. Tenney, I., Das, D., Pavlick, E.: Bert rediscovers the classical NLP pipeline. In: Proceedings of the 57th Annual Meeting of the Association for Computational Linguistics, pp. 4593–4601 (2019). https://doi.org/10.18653/v1/P19-1452

11. Warstadt, A., Singh, A., Bowman, S.R.: Learning which features matter: RoBERTa acquires a preference for linguistic generalizations (eventually). Trans. Assoc. Comput. Linguist. **8**, 140–155 (2020). https://doi.org/10.1162/tacl_a_00304
12. White, J.C., Cotterell, R.: Examining the inductive bias of neural language models with artificial languages. In: Proceedings of the 59th Annual Meeting of the Association for Computational Linguistics and the 11th International Joint Conference on Natural Language Processing (Volume 1: Long Papers), pp. 454–463 (2021). https://doi.org/10.18653/v1/2021.acl-long.38

Reasoning About Humor: It's No Laughing Matter

Ian Sung(✉) and Ruoyu Hu

Department of Computing, Imperial College London, London, UK
{ian.sung21,ruoyu.hu18}@imperial.ac.uk, iansung.y@gmail.com

Abstract. Understanding how large language models (LLMs) comprehend humor remains an open challenge. In this work, we present a comprehensive evaluation of the intrinsic humor reasoning abilities of LLMs across two tasks: humor detection and humor explanation. We propose an efficient and robust novel evaluation metric based on bidirectional natural language inference to assess explanation quality without requiring human annotation. Using this metric, we benchmark a range of techniques for enhancing humor explanations, including Chain-of-Thought Decoding, supervised fine-tuning, and Direct Preference Optimization (DPO). Our experiments show that DPO yields a substantial 11.1% improvement over DeepSeek-V3 on average, despite using models with up to 200× fewer parameters. Qualitative analysis reveals that models heavily rely on incongruity to explain humor, sometimes using it vaguely or inappropriately to mask shallow understanding. These findings expose both the potential and current key limitations in LLM humor comprehension and motivate future work for further advancements towards nuanced, human-aligned humor reasoning in language models.

Keywords: Computational Humor · Natural Language Inference · Humor Detection · Humor Explanation · Human-Centric NLP

1 Introduction

LLMs are transforming the way humans interact with AI, powering applications from personalized education [14,16] to conversational agents [4] and creative assistants [3,39]. Beyond simply being accurate, there is a growing expectation for LLMs to be engaging, personable, and even funny.

However, humor remains a highly challenging and severely underexplored capability of LLMs. While prior work has tackled humor recognition and joke generation, these approaches offer little insight into whether models genuinely understand what makes something funny [10,13]. For example, even state-of-the-art (SOTA) models, such as ChatGPT, frequently produce generic, recycled jokes and struggle to explain even minor variations, suggesting a shallow grasp of humor [13]. Other studies similarly point to a reliance on heuristics and superficial features rather than deeper humor understanding [12].

H. Unger and P. Meesad (Eds.): NLPIR 2025, LNNS 1904, pp. 182–195, 2026.
https://doi.org/10.1007/978-3-032-20897-2_13

Humor generation in particular has been treated as a black-box problem [12,13], with limited or no emphasis on structured reasoning. However, explanation-based approaches may offer a window into how models internally reason about humor. Prior work has shown that LLMs perform far below human levels when asked to explain why something is funny [12], yet no methods to date have directly sought to improve humor explanation quality. Meanwhile, chain-of-thought prompting has demonstrated that guiding models step-by-step enhances reasoning in other domains [35], indicating potential for similar techniques in humor.

In this work, we investigate the intrinsic humor reasoning abilities of LLMs and explore approaches to enhance them through explanation. We first evaluate humor detection, finding that both SOTA instruction-tuned decoder models and smaller fine-tuned encoders struggle with the seemingly simple task. To enable scalable, automatic evaluation of humor explanations, we introduce Bi-NLI, a metric based on bidirectional natural language inference. Using Bi-NLI, we benchmark various explanation enhancement techniques. Our experiments show that Direct Preference Optimization (DPO) [29], the best-performing approach, achieves an 11.1% improvement over DeepSeek-V3 on average, even when using models with up to 200× fewer parameters.

Finally, our analysis reveals that while models universally rely on the popular Incongruity Theory of Humor [23], this reliance is often superficial, masking a lack of deeper contextual comprehension. These findings highlight critical limitations in current LLM humor comprehension and underscore the promise of methods that encourage more precise and context-sensitive reasoning.

2 Related Work

Prior work has evaluated models on humor tasks such as detection, explanation, and generation. Jentzsch and Kersting (2023) [13] assessed the GPT3.5-based version of ChatGPT, focusing on generic, self-contained jokes requiring minimal external knowledge. They found that over 90% of 1,008 generated jokes were near-copies of just 25 unique jokes, with attempts at variation often preserving the same core humorous elements or losing humor altogether. Explanations were similarly shallow, frequently misidentifying humor mechanisms or producing fictitious justifications for non-humorous statements that contain surface-level joke features like wordplay. This behavior may partly stem from sycophancy [30], where prompts presupposing humor ("*Can you explain why this joke is funny: [joke]*") discourage the model from concluding that the statement is in fact non-humorous. Overall, these findings suggest a reliance on memorization and heuristics rather than genuine humor comprehension.

Contextual humor, which requires integrating situational or cultural cues beyond joke templates, poses an even greater challenge for LLMs. One such setting is the New Yorker Cartoon Caption Contest (NYCCC): participants submit humorous captions for captionless single-panel cartoons, with finalists chosen via crowd-voting and editorial review. A caption's humor depends strongly on its interaction with the cartoon's visual content and situational implications, often drawing upon subtle cultural or contextual cues. This makes the Contest a rich domain for evaluating contextual humor understanding.

Hessel et al. (2022) [12] used the NYCCC dataset to evaluate models including T5, GPT-3, and GPT-4 on three progressively complex tasks: matching finalist captions to cartoons, ranking caption quality, and generating explanations for humorous cartoon-caption pairs. While GPT-4 performed best overall, humans outperformed models substantially in most cases, particularly in explanation tasks where model justifications often identified irrelevant cartoon elements as humor sources. Human explanations were preferred in 68% of cases, highlighting shortcomings in current models' contextual humor reasoning. Building on this work, we adopt the NYCCC dataset[1] to systematically evaluate models' intrinsic humor detection and explanation abilities, and to explore methods that enhance their underlying humor reasoning.

3 Methodology

3.1 Task Definition: Humor Detection

We frame humor detection as a binary classification problem: given a cartoon-caption pair, the model predicts whether the caption is humorous in context. This task evaluates the intuitive humor comprehension abilities of LLMs, serving as a foundational precursor to humor explanation, which requires articulating underlying humor mechanisms.

Fine-Tuning Encoder Models. Encoder models generate a sequence of contextualized embeddings for input tokens [33]. For classification, we extract a single embedding representing the entire input and pass it through a linear classifier head with a sigmoid activation, yielding outputs in $[0, 1]$ with 0.5 as the decision threshold.

Contrastive learning objectives, such as InfoNCE [24], improve representation quality by promoting class separation in the embedding space. Khosla et al. (2020) extended InfoNCE to the supervised setting, yielding substantial classification gains [15]. We implement their supervised contrastive (SupCon) loss as an additional loss signal to binary cross-entropy to encourage clearer clustering of humorous and non-humorous examples.

Evaluating Decoder Models. Decoder models are evaluated in zero- and few-shot settings. We systematically vary the number of few-shot examples to assess their impact, investigating whether they can improve humor recognition abilities without fine-tuning.

[1] https://huggingface.co/datasets/jmhessel/newyorker_caption_contest.

Scene location: A tree
Description: In a tree, a mother bird is tending to her nest and babies. There is another bird in a cage on an upper branch.
Uncanny description: The cage is out of place in a tree.
Entities: Birdcage, Bird_nest
Humorous caption: *"Is it weird that we have a pet?"*
Non-humorous caption: *"Hmm. What rhymes with layoffs?"*

Scene location: woods
Description: An archer is aiming at a girl. The girl has a cherry on her head. They are in the mountains.
Uncanny description: The girl has a cherry on her head, not an apple, so it's hard to shoot without killing her.
Entities: Archery, Apple
Humorous caption: *"What do you mean, 'YOU'RE scared you might miss'?"*
Non-humorous caption: *"I'm afraid this whole experiment is about to go south."*

Fig. 1. Examples of cartoons and their annotated textual descriptions from the NYCCC dataset, with a humorous and non-humorous caption for each. While the nature of some cartoon-caption pairs is evident (e.g. the top cartoon and its non-humorous caption are clearly unrelated), others are less obvious (e.g. both captions for the bottom cartoon share some similar ideas), highlighting the subjectivity of the task. Note that for all experiments, we use only the textual descriptions as model inputs.

3.2 Task Definition: Humor Explanation

Humor explanation requires the model to articulate why a given cartoon-caption pair is or is not humorous, thereby revealing its underlying humor reasoning. Prior work using the NYCCC dataset focuses exclusively on explaining humorous examples [12], presupposing that the input pair is funny. To emulate a more realistic and challenging setting, we assess *mixed explanation*, which requires recognizing and justifying humorous and non-humorous pairs.

Inference-Time Approaches. We evaluate decoder models on **zero-shot prompting**, **few-shot prompting**, and **Chain-of-Thought Decoding (CoT Decoding)** [34]. CoT Decoding considers several decoding paths and selects the one with the highest confidence, and has been shown to surface chain-of-thought reasoning paths that improve performance on complex reasoning tasks.

These approaches are less computationally expensive than fine-tuning. However, their performance is constrained by the models' intrinsic capabilities.

Fine-Tuning Approaches. To enhance humor reasoning beyond inference-time strategies, we test two fine-tuning approaches: **supervised fine-tuning (SFT)** and **Direct Preference Optimization (DPO)** [29]. While SFT trains the model to mimic the references it observes, DPO is a more flexible approach that aligns models with human preferences. Unlike Reinforcement Learning from Human Feedback [25], DPO aligns the model using a simpler classification objective that does not require a separate reward model (Fig. 2).

Caption: *"Is it weird that we have a pet?"*

Explanation: The bird family has a moment of self-awareness when they realize their pet bird is weird, because it's also a bird, like them. Humans, for instance, don't have human pets, so why would birds have bird pets?

Caption: *"I'm just saying, after this haircut, it's difficult to trust you."*

Explanation: A joke about precision with sharp objects. Here, the archer apparently gave this person their haircut, and they think it's ugly. They are planning to shoot an apple off the other person's head, which requires deft coordination with a sharp object, i.e., the arrow. But the person with the haircut doesn't trust their capacity with sharp objects, because the scissors used to cut their hair weren't used skillfully, as the haircut turned out ugly.

Fig. 2. Human-authored reference explanations for humorous captions, using the same cartoons as in Fig. 1 (annotated textual descriptions omitted). These explanations draw on diverse sources (such as cultural norms and commonsense reasoning) and often make implicit assumptions explicit, for example by noting that "humans don't have human pets" or linking a bad haircut to the archer's questionable skill with sharp objects.

Bi-NLI. Existing work evaluates humor explanations qualitatively with human annotations or analysis [12,13]. However, this is expensive and scales poorly. To address this, we introduce Bidirectional Natural Language Inference (Bi-NLI) as a robust and scalable metric for evaluating the quality of humor explanations against reference explanations. We follow a similar work on data-to-text generation [8] and utilize the same NLI-adapted RoBERTa-large model[2] for our experiments.

We compute two scores based on bidirectional NLI entailment probabilities – **coverage** assesses whether the generated explanation includes all key points from the reference, while **consistency** assesses whether the generated explanation contains only information consistent with the reference (Fig. 3).

Reference: The snowman has realistic human arms. This is unusual and creates an incongruous effect.

	Low coverage	High coverage
Low consistency	The snowman melting is comedically tragic.	The snowman has human arms because there is a man inside.
High consistency	There is a snowman.	It's strange that a snowman has the arms of a person.

Fig. 3. Illustration of the distinction between coverage and consistency.

[2] https://huggingface.co/FacebookAI/roberta-large-mnli.

Formally, let $R = \{r_1, r_2, \ldots, r_m\}$ and $G = \{g_1, g_2, \ldots, g_n\}$ be the sets of sentences in the reference and generated explanations, respectively. Let R_{text} be the concatenation of all $r_i \in R$, and G_{text} be the concatenation of all $g_i \in G$. Then, let $\mathcal{E}(p, h) \in [0, 1]$ be the NLI model's returned probability that premise p entails hypothesis h. Then:

$$\text{Coverage} = \frac{1}{m}\sum_{i=1}^{m}\mathcal{E}(G_{\text{text}}, r_i), \quad \text{Consistency} = \frac{1}{n}\sum_{i=1}^{n}\mathcal{E}(R_{\text{text}}, g_i)$$

Both coverage and consistency are essential for high-quality, reference-aligned humor explanations, so the final Bi-NLI score weights both equally:

$$\text{Bi-NLI} = \frac{1}{2}(\text{Coverage} + \text{Consistency})$$

We discuss the strengths and limitations of Bi-NLI in Sect. 6.

4 Experimental Setup

This section outlines the datasets, models, and evaluation settings for our humor detection and explanation experiments. Full implementation details, including training procedures, hyperparameters, and prompt templates, are available in a public repository[3].

4.1 Humor Detection

Dataset. We adapt the multiple-choice matching variant of the NYCCC dataset for binary classification. Each instance is used to derive five cartoon-caption pairs (one humorous and four non-humorous), yielding 48,960 training examples (9,792 humorous, 39,168 non-humorous) and 2,640 test examples (528 humorous, 2,122 non-humorous). To investigate the effect of label imbalance, we additionally create upsampled (78,336 examples) and downsampled (19,584) variants of the training set.

Models. We fine-tune two encoder models, BERT-base [7] and RoBERTa-base [20], using HuggingFace. For sequence representation, we compare (i) **General-purpose Text Embeddings (GTE)** [17], which averages all unmasked token embeddings, and (ii) **Beijing Academy of Artificial Intelligence (BAAI) General Embedding (BGE)** [38], which directly uses the `[CLS]` token embedding.

We also evaluate six instruction-tuned decoder models released between June 2024 and January 2025: **Llama 3.2 3B Instruct** [1], **Qwen 2.5 3B Instruct** and **Qwen 2.5 7B Instruct** [28], **Gemma 2 9B Instruct** [9], **Mistral Small 3.1 24B Instruct** [22], and **DeepSeek-V3** [5]. All models are run locally with

[3] https://github.com/ian-sy/reasoning-about-humor.

QLoRA quantization [6], except DeepSeek-V3, which is accessed via online API providers. All experiments are feasible on a single 48 GB GPU.

Decoder models are evaluated on the test set using N-shot prompting for $N = 0, 1, 3, 5$. For each N, we test two bias configurations: humorous (one extra humorous example) and non-humorous (one extra non-humorous example).

4.2 Humor Explanation

Dataset. The NYCCC dataset provides human-authored reference explanations for humorous cartoon-caption pairs only. To enable mixed explanation, we balance these by sampling non-humorous pairs from the detection task. For each sampled pair, we randomly assign one of six generic reference explanations describing common reasons a pair might be non-humorous, such as *"the cartoon and caption appear to be unrelated to each other"* and *"the caption does not add any humorous twist or context to the cartoon"*. We obtain 4,680 balanced training examples and 262 test examples.

Models. We reuse the same six decoder models from humor detection for consistency. Due to API providers being inference-only, DeepSeek-V3 is only evaluated under few-shot prompting, serving as a SOTA reference baseline. All locally-run models use contrastive decoding [32], except Gemma 9B, which defaults to greedy decoding due to incompatibility.

5 Results

5.1 Humor Detection

We report macro-averaged F1, aligning with previous works [2,36]. The encoder fine-tuning results are reported in Table 1, and N-shot decoder evaluation results in Table 2 (the latter with per-class F1).

5.2 Humor Explanation

We primarily report the per-class Bi-NLI. The SFT learning rate results are shown in Table 3, and our overall results are presented in Table 4.

6 Analysis and Future Work

6.1 Humor Detection

Encoder Fine-Tuning. From results shown in Table 1, we observe that training on imbalanced data significantly harms performance, with models overfitting to the non-humorous class. However, upsampling and downsampling yield comparable benefits, suggesting that class distribution is more important than overall dataset size.

The best performance is achieved by BERT with BGE (0.66 Macro-F1). Notably, this is comparable to the larger decoder models, including DeepSeek-V3 (average 0.66 Macro-F1), demonstrating that fine-tuning smaller models can rival SOTA instruction-tuned LLMs for humor detection.

Table 1. F1-scores for each encoder model and sequence representation configuration. We experiment with using a supervised contrastive loss signal (*SupCon*) in addition to the standard binary cross-entropy classification loss.

Sample	LR	SupCon	BERT/BGE Unfreeze			BERT/GTE Unfreeze			RoBERTa/BGE Unfreeze			RoBERTa/GTE Unfreeze		
			3	4	5	3	4	5	3	4	5	3	4	5
None	2e-5	N	0.56	0.61	0.59	0.58	0.58	0.44	0.44	0.53	0.58	0.54	0.51	0.57
	3e-5	N	0.54	0.57	0.54	0.57	0.44	0.54	0.57	0.52	0.44	0.51	0.47	0.44
	5e-5	N	0.53	0.54	0.44	0.54	0.48	0.44	0.44	0.44	0.44	0.44	0.44	0.44
Down	2e-5	N	0.63	0.65	0.66	0.62	0.62	0.63	0.59	0.59	0.61	0.59	0.62	0.60
		Y	0.61	0.63	0.65	0.63	0.63	0.64	0.60	0.61	0.63	0.59	0.61	0.61
	3e-5	N	0.61	0.63	0.64	0.63	0.62	0.62	0.60	0.59	0.61	0.59	0.60	0.61
		Y	0.61	0.63	0.65	0.65	0.64	0.64	0.61	0.61	0.63	0.60	0.59	0.64
	5e-5	N	0.63	0.62	0.62	0.62	0.63	0.61	0.57	0.45	0.44	0.58	0.54	0.44
		Y	0.62	0.63	0.65	0.64	0.63	0.63	0.59	0.61	0.61	0.59	0.61	0.62
Up	2e-5	N	0.62	0.63	0.65	0.63	0.64	0.64	0.61	0.60	0.62	0.62	0.62	0.60
		Y	0.61	0.62	0.64	0.60	0.61	0.64	0.62	0.61	0.62	0.61	0.60	0.64
	3e-5	N	0.62	0.64	0.64	0.62	0.63	0.63	0.60	0.59	0.63	0.59	0.61	0.45
		Y	0.63	0.63	0.64	0.61	0.64	0.64	0.62	0.62	0.63	0.63	0.62	0.64
	5e-5	N	0.62	0.64	0.65	0.63	0.62	0.64	0.59	0.51	0.44	0.59	0.58	0.44
		Y	0.62	0.63	0.64	0.63	0.63	0.62	0.60	0.62	0.64	0.60	0.61	0.62

Table 2. Humorous (*H*), non-humorous (*NH*), and macro-averaged (*Avg*) F1-scores for each decoder model and few-shot bias type.

		DeepSeek-V3			Llama 3B			Qwen 3B			Qwen 7B			Gemma 9B			Mistral 24B		
		H	NH	Avg	H	NH	Avg	H	NH	Avg	H	NH	Avg	H	NH	Avg	H	NH	Avg
Examples (H Bias)	0	0.52	0.84	0.68	0.38	0.62	0.50	0.33	0.87	**0.60**	0.46	0.83	**0.64**	0.38	0.45	0.42	0.43	0.61	0.52
	1	0.55	0.86	**0.70**	0.23	0.86	**0.55**	0.33	0.83	0.58	0.43	0.83	0.63	0.37	0.33	0.35	0.51	0.86	**0.68**
	3	0.50	0.79	0.65	0.35	0.25	0.30	0.37	0.53	0.45	0.41	0.74	0.58	0.36	0.23	0.29	0.44	0.62	0.53
	5	0.51	0.78	0.65	0.34	0.15	0.25	0.37	0.42	0.39	0.42	0.78	0.60	0.35	0.21	0.28	0.40	0.48	0.44
Examples (NH Bias)	0	0.52	0.84	0.68	0.38	0.62	0.50	0.33	0.87	**0.60**	0.46	0.83	**0.64**	0.38	0.45	0.42	0.43	0.61	0.52
	1	0.50	0.78	0.64	0.33	0.00	0.17	0.32	0.85	0.58	0.41	0.86	0.63	0.35	0.17	0.26	0.35	0.15	0.25
	3	0.49	0.75	0.62	0.33	0.02	0.18	0.28	0.84	0.56	0.41	0.83	0.62	0.35	0.15	0.62	0.38	0.35	0.36
	5	0.51	0.80	0.65	0.33	0.02	0.18	0.36	0.78	0.57	0.38	0.84	0.61	0.35	0.16	**0.65**	0.37	0.32	0.35

N-Shot Decoder Evaluation. Our results on decoder model performance under few-shot prompting (Table 2) show that DeepSeek-V3 consistently performs the best of the selected models across every tested configuration, achieving macro-averaged F1 of up to 0.68.

Although instruction-tuned LLMs are typically expected to benefit from additional in-context examples, few-shot prompting proves surprisingly unstable for smaller models. Llama 3B and Mistral 24B both deteriorate at 3 or 5 examples regardless of bias, but Qwen 3B exhibits this effect only under humorous bias. In contrast, Qwen 7B and DeepSeek-V3 remain robust across configurations. An interesting outlier is Gemma 9B, which more than doubles its F1-score (0.26 vs

Table 3. Per-class Bi-NLI scores for each model and learning rate SFT configuration.

LR	Llama 3B		Qwen 3B		Qwen 7B		Gemma 9B		Mistral 24B	
	H	NH	H	NH	H	NH	H	NH	H	NH
1e-5	0.05	0.89	0.11	0.51	**0.09**	**0.78**	0.06	0.94	0.16	0.39
5e-5	0.03	0.93	0.15	0.33	0.13	0.56	**0.14**	**0.91**	**0.13**	**0.85**
2e-4	**0.07**	**0.88**	**0.08**	**0.88**	0.14	0.54	0.14	0.72	0.11	0.87

Table 4. Coverage (*Cov.*), consistency (*Cons.*), and Bi-NLI scores for each decoder model and method. SFT results reported are the best configuration for that model.

Model	Method	Humorous class			Non-humorous class		
		Cov.	Cons.	Bi-NLI	Cov.	Cons.	Bi-NLI
Llama 3B	Few-shot	0.22	0.13	0.18	0.06	0.03	0.04
	CoT Decoding	0.21	0.14	0.17	0.15	0.11	0.13
	SFT	0.08	0.07	0.07	0.85	0.91	**0.88**
	DPO	0.24	0.25	**0.25**	0.05	0.04	0.04
Qwen 3B	Few-shot	0.17	0.11	0.14	0.10	0.05	0.08
	CoT Decoding	0.27	0.17	**0.22**	0.19	0.09	0.14
	SFT	0.08	0.07	0.08	0.85	0.92	**0.88**
	DPO	0.06	0.10	0.08	0.41	0.26	0.33
Qwen 7B	Few-shot	0.13	0.11	0.12	0.68	0.29	0.48
	CoT Decoding	0.13	0.10	0.11	0.73	0.40	0.56
	SFT	0.09	0.10	0.09	0.80	0.77	**0.78**
	DPO	0.19	0.30	**0.24**	0.18	0.10	0.14
Gemma 9B	Few-shot	0.19	0.17	0.18	0.09	0.08	0.09
	CoT Decoding	0.16	0.16	0.16	0.45	0.21	0.33
	SFT	0.13	0.15	0.14	0.89	0.94	**0.91**
	DPO	0.18	0.20	**0.19**	0.01	0.03	0.02
Mistral 24B	Few-shot	0.17	0.15	0.16	0.28	0.13	0.21
	CoT Decoding	0.15	0.13	0.14	0.37	0.19	0.28
	SFT	0.19	0.08	0.13	0.83	0.87	**0.85**
	DPO	0.28	0.19	**0.24**	0.12	0.10	0.11
DeepSeek-V3	Few-shot	0.20	0.16	0.18	0.48	0.22	0.35

0.62) when moving from 1-shot to 3-shot under non-humorous bias. These inconsistencies may stem from inherent model biases [40], or from few-shot examples serving as distracting information [19,26], which we encourage future work to explore further.

Finally, the bias configuration can have a strong effect on performance. Adding one humorous example is typically helpful or neutral, whereas adding one non-humorous example tends to degrade performance on both classes. This asymmetry is particularly pronounced in Llama 3B and Mistral 24B, suggesting that decoder models vary in sensitivity to prompt structure.

6.2 Humor Explanation

Strengths and Limitations of Bi-NLI. Bi-NLI evaluates both coverage and consistency, two key aspects of high-quality generated explanations. Unlike surface-level N-gram metrics such as BLEU [27] and ROUGE [18], Bi-NLI leverages a lightweight pre-trained NLI model to capture semantic alignment beyond syntactic overlap. Its sentence-level scoring aids interpretability, enabling fine-grained analysis of which explanation components are well-aligned (Fig. 4).

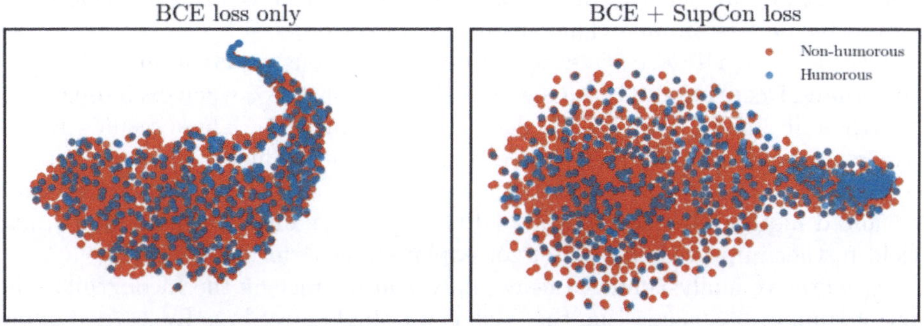

Fig. 4. t-SNE [21] embedding visualisation for RoBERTa/GTE, learning rate 3e−5, and 5 layers unfrozen, without (macro-F1 = 0.45) and with (macro-F1 = 0.64) SupCon loss.

However, Bi-NLI does not explicitly reward desirable traits such as conciseness or clarity, or account for subjective variation in valid humor interpretations. This likely explains why coverage scores are consistently higher than consistency scores (Table 4), particularly for non-humorous examples: even when models provide reasonable justifications, the generic reference explanations lower the consistency score. These limitations reflect the broader challenge of evaluating subjective, free-form explanations automatically.

Potential future improvements include incorporating reward signals for desirable attributes, supporting multiple references to handle subjectivity, and expanding human evaluation to better align Bi-NLI with human judgment. Finer-grained segmentation beyond the sentence level, such as by using LLMs to extract 'atoms' [31], could improve robustness when handling longer sentences that contain multiple clauses.

Overall, we find that Bi-NLI is a scalable, robust, and interpretable metric for benchmarking humor explanation performance across methods.

Key Findings and Error Analysis. We summarize the most notable trends across models and methods, drawing on both quantitative results (Tables 3, 4) and qualitative error analysis.

First, SFT universally leads models to overfit to the repetitive and generic non-humorous reference explanations. This results in consistently high performance on non-humorous examples but poor performance on humorous ones, with models tending to simply repeat the non-humorous explanations for most examples. In contrast, DPO struggles to model this asymmetry, underperforming substantially on non-humorous examples. This is likely due to the significant mismatch in the level of detail between the classes. This highlights a key challenge when training with imbalanced *quality* of data. Future work could address this by constructing more detailed and context-specific non-humorous explanations, enabling models to learn from explicit negative cases that provide better signals for understanding what fails to be funny.

Despite this, DPO consistently improves humorous explanation quality, outperforming DeepSeek-V3 by an average of 11.1%, or 27.8% when excluding Qwen 3B (the only model where DPO does not perform best). These results demonstrate the value of preference-based alignment in producing more precise and context-sensitive humor explanations, even with much smaller models. More advanced methods, such as Online DPO [11] or dynamic β schedules [37], may yield further improvements to humor explanation performance.

Qualitative analysis finds a heavy universal reliance on the Incongruity Theory of Humor [23], suggesting that this is the single most broadly learned humor mechanism. Higher-scoring explanations accurately pinpoint incongruities within the cartoon-caption relationship, while weaker ones use irrelevant incongruous elements to explain humor or even falsely justify it. In one example, Mistral 24B under CoT Decoding generically argues that "the use of the incongruity between nature and politics creates a humorous effect" even though the pair is non-humorous. Notably, some non-humorous explanations correctly identify that while a cartoon may be uncanny or surreal, the caption fails to add humor.

Cultural references also emerge as a relatively common source of explanations, perhaps due to the often culturally-rooted nature of NYCCC cartoons. Models are occasionally able to draw links to wider contextual knowledge, e.g. "plays with the famous moon landing phrase 'One small step for man"' or "this caption plays on the concept of art appreciation and the idea that artworks often increase in value after the death of the artist."

Overall, our analysis reveals model differences, such as Qwen 7B demonstrating stronger capabilities in recognizing and explaining non-humorous content. Nevertheless, explaining humor remains challenging across the board. In our experiments, DPO stands out for enabling far smaller models to rival or surpass the SOTA DeepSeek-V3 in humorous explanation quality.

7 Conclusion

In this work, we conducted a large-scale, systematic evaluation of language models' humor reasoning abilities across two humor tasks: humor detection and humor explanation. We examined a diverse set of models and tested methods for enhancing their performance. By treating humor not as a black-box creative artefact, but as a phenomenon that can be analyzed and reasoned about, we sought to unpack and enhance the reasoning processes that underpin successful humor comprehension in language models.

We found that while small encoder models could be fine-tuned to compete with much larger SOTA models on humor detection, the task remains non-trivial for models of all capacities. To evaluate explanation quality, we introduced Bi-NLI, a scalable and interpretable metric for comparing generated and reference explanations. While not without limitations, Bi-NLI provides a valuable foundation for automated evaluation of free-form humor reasoning.

DPO stood out as the most effective explanation strategy, consistently outperforming a SOTA baseline. Compared to other methods, DPO led to more semantically grounded and context-sensitive humorous explanations.

Our analysis revealed that models persistently rely on incongruity as a central humor mechanism. Although this sometimes worked well, it was also used to produce vague or incorrect justifications that referenced incongruity without sound contextual grounding. We believe these findings highlight both the promise and current limitations of LLMs in humor comprehension, and we hope our contributions offer a meaningful step towards further advancements in computational humor.

References

1. AI@Meta: The Llama 3 herd of models (2024). https://arxiv.org/abs/2407.21783v3
2. Annamoradnejad, I., Zoghi, G.: ColBERT: using BERT sentence embedding in parallel neural networks for computational humor. Expert Syst. Appl. **249**, 123685 (2024). http://dx.doi.org/10.1016/j.eswa.2024.123685
3. Chakrabarty, T., Padmakumar, V., Brahman, F., Muresan, S.: Creativity support in the age of large language models: an empirical study involving emerging writers. In: ACM International Conference Proceeding Series, pp. 132–155 (2023). https://arxiv.org/abs/2309.12570v3
4. Dam, S.K., Hong, C.S., Qiao, Y., Member, S., Zhang, C., Member, S.: A complete survey on LLM-based AI chatbots (2024). https://arxiv.org/abs/2406.16937v2
5. DeepSeek-AI: DeepSeek-V3 technical report (2024)
6. Dettmers, T., Pagnoni, A., Holtzman, A., Zettlemoyer, L.: QLoRA: efficient fine-tuning of quantized LLMs. Adv. Neural Inf. Process. Syst. **36** (2023). https://arxiv.org/abs/2305.14314v1
7. Devlin, J., Chang, M.W., Lee, K., Toutanova, K.: BERT: pre-training of deep bidirectional transformers for language understanding. In: NAACL HLT 2019 - 2019 Conference of the North American Chapter of the Association for Computational Linguistics: Human Language Technologies - Proceedings of the Conference 1, pp. 4171–4186 (2018). https://arxiv.org/abs/1810.04805v2
8. Dušek, O., Kasner, Z.: Evaluating semantic accuracy of data-to-text generation with natural language inference. In: INLG 2020 - 13th International Conference on Natural Language Generation, Proceedings, pp. 131–137 (2020). https://aclanthology.org/2020.inlg-1.19/
9. Gemma Team: Gemma 2: Improving open language models at a practical size (2024). https://arxiv.org/abs/2408.00118v3

10. Gorenz, D., Schwarz, N.: How funny is ChatGPT? A comparison of human-and AI-produced jokes. PLoS ONE **19**(7), e0305364 (2024)
11. Guo, S., et al.: Direct language model alignment from online AI feedback (2024). https://arxiv.org/abs/2402.04792v2
12. Hessel, J., et al.: Do androids laugh at electric sheep? Humor "understanding" benchmarks from The New Yorker Caption Contest. In: Proceedings of the Annual Meeting of the Association for Computational Linguistics, vol. 1, pp. 688–714 (2022). https://arxiv.org/abs/2209.06293v2
13. Jentzsch, S., Kersting, K.: ChatGPT is fun, but it is not funny! Humor is still challenging large language models. In: Proceedings of the Annual Meeting of the Association for Computational Linguistics, pp. 325–340 (2023). https://arxiv.org/abs/2306.04563v1
14. Kasneci, E., et al.: ChatGPT for good? On opportunities and challenges of large language models for education. Learn. Individ. Differ. **103**, 102274 (2023)
15. Khosla, P., et al.: Supervised contrastive learning. In: Proceedings of the 34th International Conference on Neural Information Processing Systems, NIPS 2020. Curran Associates Inc., Red Hook (2020)
16. Labadze, L., Grigolia, M., Machaidze, L.: Role of AI chatbots in education: systematic literature review. Int. J. Educ. Technol. High. Educ. **20**, 1–17 (2023). https://educationaltechnologyjournal.springeropen.com/articles/10.1186/s41239-023-00426-1
17. Li, Z., et al.: Towards general text embeddings with multi-stage contrastive learning (2023). https://arxiv.org/abs/2308.03281v1
18. Lin, C.Y.: ROUGE: a package for automatic evaluation of summaries (2004). https://aclanthology.org/W04-1013/
19. Liu, N.F., et al.: Lost in the middle: how language models use long contexts. Trans. Assoc. Comput. Linguist. **12**, 157–173 (2024). https://aclanthology.org/2024.tacl-1.9/
20. Liu, Y., et al.: Roberta: a robustly optimized BERT pretraining approach (2019). https://arxiv.org/abs/1907.11692v1
21. van der Maaten, L., Hinton, G.: Visualizing data using t-SNE. J. Mach. Learn. Res. **9**(86), 2579–2605 (2008). http://jmlr.org/papers/v9/vandermaaten08a.html
22. Mistral AI: Mistral Small 3 (2025). https://mistral.ai/news/mistral-small-3
23. Morreall, J.: Philosophy of Humor. In: Zalta, E.N., Nodelman, U. (eds.) The Stanford Encyclopedia of Philosophy. Metaphysics Research Lab, Stanford University, Summer 2023 edn. (2023)
24. van den Oord, A., Li, Y., Vinyals, O.: Representation learning with contrastive predictive coding (2018). https://arxiv.org/abs/1807.03748v2
25. Ouyang, L., et al.: Training language models to follow instructions with human feedback. Adv. Neural Inf. Process. Syst. **35** (2022). https://arxiv.org/abs/2203.02155v1
26. Pandia, L., Ettinger, A.: Sorting through the noise: testing robustness of information processing in pre-trained language models. In: Moens, M.F., Huang, X., Specia, L., Yih, S.W.t. (eds.) Proceedings of the 2021 Conference on Empirical Methods in Natural Language Processing. pp. 1583–1596. Association for Computational Linguistics, Online and Punta Cana, Dominican Republic (2021)
27. Papineni, K., Roukos, S., Ward, T., Zhu, W.J.: BLEU: a method for automatic evaluation of machine translation. In: Proceedings of the 40th Annual Meeting on Association for Computational Linguistics, ACL 2002, USA, pp. 311–318. Association for Computational Linguistics (2002)

28. Qwen: Qwen2.5 technical report (2024). https://arxiv.org/abs/2412.15115v2
29. Rafailov, R., Sharma, A., Mitchell, E., Ermon, S., Manning, C.D., Finn, C.: Direct preference optimization: Your language model is secretly a reward model (2023). https://arxiv.org/abs/2305.18290v3
30. Sharma, M., et al.: Towards understanding sycophancy in language models. In: 12th International Conference on Learning Representations, ICLR 2024 (2023). https://arxiv.org/abs/2310.13548v3
31. Stacey, J., Minervini, P., Dubossarsky, H., Camburu, O.M., Rei, M.: Atomic inference for NLI with generated facts as atoms (2023). https://arxiv.org/abs/2305.13214v2
32. Su, Y., Lan, T., Wang, Y., Yogatama, D., Kong, L., Collier, N.: A contrastive framework for neural text generation. In: Proceedings of the 36th International Conference on Neural Information Processing Systems, NIPS 2022. Curran Associates Inc., Red Hook (2022)
33. Vaswani, A., et al.: Attention is all you need. Adv. Neural Inf. Process. Syst., 5999–6009 (2017). https://arxiv.org/abs/1706.03762v7
34. Wang, X., Zhou, D., Deepmind, G.: Chain-of-thought reasoning without prompting (2024). https://arxiv.org/abs/2402.10200v2
35. Wei, J., et al.: Chain-of-thought prompting elicits reasoning in large language models. Adv. Neural Inf. Process. Syst. **35** (2022). https://arxiv.org/abs/2201.11903v6
36. Weller, O., Seppi, K.: Humor detection: a transformer gets the last laugh. In: Inui, K., Jiang, J., Ng, V., Wan, X. (eds.) Proceedings of the 2019 Conference on Empirical Methods in Natural Language Processing and the 9th International Joint Conference on Natural Language Processing (EMNLP-IJCNLP), Hong Kong, China, pp. 3621–3625. Association for Computational Linguistics (2019)
37. Wu, J., et al.: β-DPO: Direct preference optimization with dynamic β (2024)
38. Xiao, S., Liu, Z., Zhang, P., Muennighoff, N., Lian, D., Nie, J.Y.: C-Pack: packed resources for general Chinese embeddings. arXiv.org (2023). https://arxiv.org/abs/2309.07597v5
39. Yuan, A., Coenen, A., Reif, E., Ippolito, D.: Wordcraft: story writing with large language models. In: International Conference on Intelligent User Interfaces, Proceedings IUI, pp. 841–852 (2022). https://dl.acm.org/doi/10.1145/3490099.3511105
40. Zhao, T., Wallace, E., Feng, S., Klein, D., Singh, S.: Calibrate before use: improving few-shot performance of language models. In: International Conference on Machine Learning (2021)

Probing Attention in Pre-trained LLMs to Detect Semantics

Frédéric Charpentier[1,2(✉)], Jairo Cugliari[2], and Adrien Guille[2]

[1] Cabot Financial France, Caluire-et-Cuire, France
[2] Université Lyon 2, Laboratoire ERIC, Lyon, France
frederic.charpentier@cabotfinancial.fr,
{frederic.charpentier,jairo.cugliari,adrien.guille}@univ-lyon2.fr

Abstract. Abstract Meaning Representations (AMRs) encode the semantics of sentences in the form of graphs. Words in the corresponding sentences can be aligned to vertices in the AMR, in such a way that semantic relations between words can be mapped from semantic roles read on the arcs of the AMR. The attention mechanism of a Language Model (LM) can be modeled as the computation of vectors describing edges on a complete graph whose vertices are words in a sentence or a whole paragraph. In this work, we map AMR graphs to Attention Graphs and devise probing methods to detect the semantic relations labeling the edges from the attention weights. To do so, we implement methods operating either on single edges or on the whole attention graph in order to compare semantic capacities of several pretrained LMs. This study shows that the RoBERTA-base bidirectional encoder outperforms causal decoders up to Llama-3 8B, and obtains 86.7% of the performances of an otherwise specially devised AMR parser, Spring, specially trained with more than 1400 times the number of parameters we used.

Keywords: LLM · probing · semantics · SRL · AMR · Graph Neural Network

1 Introduction

Semantic Role Labeling (SRL [12]) is a crucial task in the field of Natural Language Processing (NLP) that aims to identify and disambiguate the semantic roles of constituents in a sentence, such as who did what to whom, when, and where. For example, in the sentence *John broke the window with a hammer*, where *broke* is the predicate, SRL would identify *John* as the agent, *window* as the patient, and *hammer* as the instrument. Abstract Meaning Representation (AMR, [1]) is a semantic formalism that leverages SRL to encode the meaning of a sentence in a rooted, directed, acyclic graph. The vertices represent instances of concepts, (*John*, *break*, *window*, *hammer*) and the directed arcs are labeled with the semantic relations between them. (*agent*, *patient*, *instrument*).

Language models based on transformers, introduced in [24], have proven their worth in solving various problems related to automatic language processing. The

H. Unger and P. Meesad (Eds.): NLPIR 2025, LNNS 1904, pp. 196–210, 2026.
https://doi.org/10.1007/978-3-032-20897-2_14

interpretability of the computations they employ is still the subject of active research, both about syntax and semantics.

In this article, we present a study aimed at demonstrating the intrinsic capacity of a transformer-type language model to perform SRL tasks. We first do so in a setting where we classify pairs of transformer tokens to label a semantic relation, similar to [6]. Since the attention mechanism in a transformer can be viewed as a means to build a complete graph of a sentence, which we dub "Attention Graph", where attention weights describe edges connecting any pair of tokens in a sentence, we further develop our method into a Graph Neural Network (GNN) able to classify selected edges in the Attention Graph, taking into account other edges to aid the classification.

Unlike traditional approaches, our method does not rely on manipulating the computed vector embeddings of the words in question, but solely on the attention mechanism of a large language model. More precisely, we implement a classification on the adjoint graph of the attention graph, where the vertices are described by the vectors of the attention mechanism. Our code is available at https://github.com/frcharpentier/GAS to reproduce our experiments.

1.1 Related Work

The ability of attention heads to capture syntactical features was examined by Clark et al. [8], who studied the ability of certain attention heads in the BERT network [10] to classify several syntactic relations and coreference, without fine-tuning BERT for any specific task. Luo [17] studied how constituency grammar is captured by different attention heads in BERT. Our work can be considered to be part of the "probing" approach presented by Tenney et al. [23], which these same authors used in [22] to demonstrate the capabilities of the BERT network to perform in an interpretable order all the traditional steps of NLP, including SRL and coreference resolution. In [9], the probing method was specifically used in the study of SRL, in a comparison of the capabilities of BERT and RoBERTa to predict the meaning of a predicate and identify the semantic roles associated with it.

The use of transformers to automatically build AMRs was marked by Xu et al. [26], who trained those models in a seq-to-seq architecture to build AMRs from sentences. A symmetric approach was later devised [2] to perform AMR-to-text and text-to-AMR transduction in a transformer-based seq-to-seq architecture, allowing to get rid of graph recategorization. StructBART[27], an enhanced approach that combines the use of seq-to-seq transformer models with transition-based parsing, further enables guarantees as to graph well-formedness. In [18], the cross attention in the Spring parser was analyzed to investigate how much alignment information can be found in the model's parameters between the AMR and the sentence. Qorib, Moon, and Ng [20] evaluated the semantic understanding capability of language models through the word sense disambiguation and word-in-context tasks, to compare capabilities of decoder-only language models to their encoder-only counterparts. The abilities of RoBERTa and GPT-2 to encode semantic labels of AMRs in their attention heads were studied in

[6], using non-supervised and supervised methods. Chizhikova et al. [7] built a Relation Extraction model to assess BERT's awareness to semantic relations in a study where only attention scores are taken into account.

2 Dataset Structure

This section describes our method to classify semantic relations between words in pairs from a sentence. Following [6], we used AMR graphs as a starting point.

An AMR is a directed acyclic graph aimed at representing the meaning of a sentence independently of its syntax. Vertices are labeled with instances of concepts and directed arcs between two vertices are labeled with the semantic relation which the sentence implies between the two concepts.

A transformer model [24] can be described as a graph neural network operating on the complete graph of a sentence. Indeed, the attention mechanism operates on every pair of tokens in a text, where the tokens are sub-units of the sentence, representing all or part of a word. We therefore describe a sentence as a complete undirected graph of its tokens, dubbed "Attention Graph", where the edges are described with two "attention vectors" of $\mathbb{R}^d$ composed of all attention weights between the two tokens, in either direction[1]. For a transformer with m layers, and n attention heads per layer, the dimension of our two feature vectors is thus $d = m \times n$. (Each attention weight is the dot product of a key and a query, therefore each edge is described with two vectors of $\mathbb{R}^d$, because each word can play the role of either the key or the query.)

2.1 Mapping AMRs to Attention Graphs

Contrarily to [6], where pairs of tokens were extracted from the sentences and labeled using the aligned arcs in the AMR, we built a complete mapping from AMRs to Attention Graphs. To do so, we drew on Leamr [3], an EM-based algorithm for aligning words or groups of words in a sentence to the vertices or sub-graphs of the corresponding AMR, whose authors made available alignment data computed on the LDC2020T02 dataset [15] (A set of nearly 60,000 English sentences and their AMR representations).

The arc labels in an AMR are a mixed bag of semantic labels originating in different paradigms. Some of them, dubbed "core roles", were adopted from PropBank [19] roles. This labeling was criticized [11] for its obscure interpretability: it consists indeed in a mere enumeration of the arguments of the frame (`ARG0`, `ARG1`, etc.), which does not explicitly mark the type of semantic relation. To overcome this limitation, VerbAtlas[11] was published, a manually crafted inventory of verbs and argument structures with cross-domain explicit semantic roles.

- We relabeled the numbered PropBank core roles in the AMR using VerbAtlas roles, thanks to the PropBank to VerbAtlas table, [11].

[1] Since the attention mechanism is a directional computation, from a query-token to a key-token, each edge in the complete graph is actually described with two vectors of $\mathbb{R}^d$.

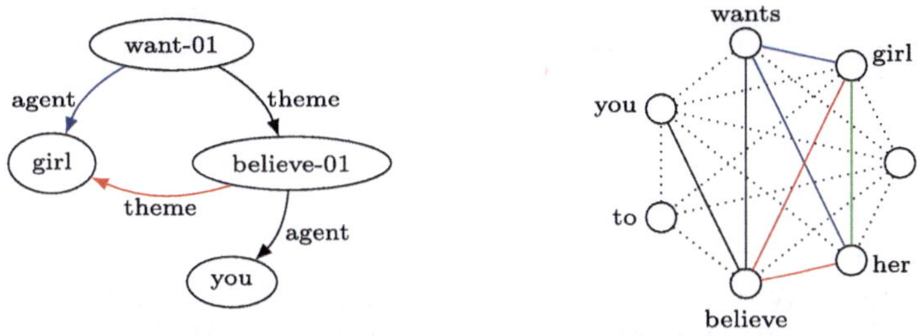

Fig. 1. AMR graph (left) and Attention Graph (right) for the sentence "The Girl Wants You To Believe Her". The tokens "her" and "girl" are coreferences to the AMR vertex "girl". The green vertex between the two tokens is therefore labeled with the special label {`idem`}. The AMR vertex "want-01" is aligned with the token "want". The blue arc between vertices "want-01" and "girl" is duplicated on the right and mapped to edges ("wants", "girl") and ("wants", "her"). Likewise, the edge ("believe-01", "girl") in the AMR is mapped to edges ("believe", "girl") and ("believe", "her"). The two black arcs on the AMR are mapped to the two solid black edges in the Attention Graph. Each solid edge in the Attention Graph is labeled with the semantic role of the corresponding arc in the AMR. Each one also receives another label indicating the direction of the arc, except the green edge, which does not. The dotted edges have no label at all.

- Between two tokens corresponding to two words aligned to two vertices of the AMR, we labeled the edge with the semantic relation in the AMR and added another directional label to note the direction of the arc.
- Between two tokens being part of the same word, we use a special label {`group`}, with no directional label. We distributed the semantic relation of all incident edges to those two tokens.
- Sometimes, because of coreference, two words are aligned to the same AMR vertex or subgraph. We used the special label {`idem`} and no directional label between the tokens involved, and distributed incident relations.
- AMRs describe conjunction and disjunction using a special vertex labeled "and" or "or". This vertex points to the combined concepts using syntactic special relations (`:op1`, `:op2`, ...). We chose to redistribute the semantic relations incident to this special vertex, and to use the special label {`and`} or {`or`} between the combined concepts. Here again, no directional label is used in this case.

The result is a dataset of tokenized sentences, represented in Attention Graphs, where each edge is described with two vectors in $\mathbb{R}^d$, and bears two kinds of labels: One to represent the semantic (or special) relation between the two tokens or the two words, the other one to represent the direction of the relation, where applicable. Both labels are optional, and many edges in the complete graph have none. (See Fig. 1 for an illustration.)

We define our probing pipeline for Semantic Role Labeling as follows: **1.** For each sentence, compute attention weights from every token to the others using a language model, and use them to describe the edges of an Attention Graph.

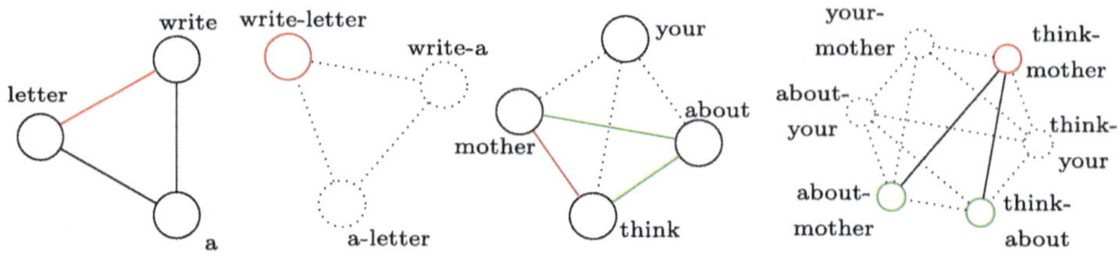

Write a letter (direct graph) Write a letter (adjoint graph) Think About Your Mother (direct graph) Think About Your Mother (adjoint graph)

Fig. 2. Left: Direct and adjoint graphs for the sentence "Write A Letter", where the relation to be classified (in red) is between the words "write" and "letter". Right: Direct and adjoint attention graphs for the sentence "Think About Your Mother". The relation to be classified is between "think" and "mother" (in red). A useful pivotal word is "about". Helpful edges (in green) in the direct graph connect the word "about" to the words "think" and "mother". In the adjoint graph, they are neighboring nodes of the node to be classified.

2. Using the mapping from the AMR to the Attention Graph, and applying the transformations described above, assign semantic and directional labels to some of the edges of the Attention Graph. **3.** Predict the labels of the edges using the models described in the following section.

3 Models

Each edge in the Attention Graph bears at most two labels, therefore, there are two problems to be solved:

1. Predict the semantic label of the edge.
2. Predict the directional label of the edge, which corresponds to predicting which vertex is the source and which is the target of the arc in the mapped AMR.

3.1 Prediction as a Function of Edge Features

In this section, we describe the simplest way to solve the two problems mentioned above. Let us suppose we want to predict the particular semantic relation between the words "write" and "letter" in the simple sentence "write a letter". The basic idea is to use the data computed by the Language Model representing all attention weights from the token "write" to the token "letter", as well as the attention weights from the token "letter" to the token "write". This boils down to using only the two feature vectors of the edge ("write", "letter") in the Attention Graph, to classify its label, disregarding the rest of the graph. In this setting, no use is made of the graph structure yet.

To solve problem 1, we use a low-rank symmetric bi-affine classifier. To solve problem **2.**, we use a low-rank bi-affine antisymmetric classifier.

Bi-affine Symmetric Classifier. Every edge is described with two feature vectors v_1 and v_2, representing the attention of a token to the other one, and conversely. Even if decoder models mask the results of attention computation from a token to a subsequent token in order to preserve causality, we can retrieve the computed values before they are masked out. We aim at determining the probability that an arc will bear the label $k \in \{1, \cdots, c\}$ given $\mathbf{v}_1$ and $\mathbf{v}_2$, using a function of three variables:

$$\mathbb{P}(y = k \mid \mathbf{v}_1, \mathbf{v}_2) = f(k, \mathbf{v}_1, \mathbf{v}_2).$$

Since we do not know which one of the two vectors $\mathbf{v}_1$ and $\mathbf{v}_2$ represents the source-target attention, it is therefore reasonable to posit $f(k, \mathbf{v}_1, \mathbf{v}_2) = f(k, \mathbf{v}_2, \mathbf{v}_1)$. We chose to model the function f as follows: Let $M_1, \cdots, M_c$ be c square symmetric real matrices in $\mathbb{R}^{d \times d}$, and let $\mathbf{b}_1, \cdots, \mathbf{b}_c$ be c bias vectors in $\mathbb{R}^d$. We compute the c scalars $\ell_i(\mathbf{v}_1, \mathbf{v}_2) = (\mathbf{v}_1 + \mathbf{b}_i)^\top \cdot M_i \cdot (\mathbf{v}_1 + \mathbf{b}_i)$. Then, the function f is defined as:

$$f(k, \mathbf{v}_1, \mathbf{v}_2) = \frac{\exp(\ell_k(\mathbf{v}_1, \mathbf{v}_2))}{\sum_{i=1}^{c} \exp(\ell_i(\mathbf{v}_1, \mathbf{v}_2))}.$$

We thus define a biaffine classifier, symmetric with respect to the permutation of $\mathbf{v}_1$ and $\mathbf{v}_2$. This model requires $\frac{d(d+3)}{2}$ parameters per class.

Low Rank Symmetric Classifier. To reduce the classifier's complexity, one can restrain the matrix parameters among low rank matrices. Positing $r \in \mathbb{N}$, $r \ll d$, we can look for $(M_i)_i$ among square symmetric matrices of $\mathbb{R}^{d \times d}$ whose ranks are less than or equal to r. It is easy to prove that for such matrices, it holds:

$$\begin{aligned} &\exists U_i \in \mathbb{R}^{d \times r}, \exists (a_i^1, \cdots, a_i^r) \in \mathbb{R}^r, \\ &M_i = U \cdot \operatorname{diag}(a_i^1, \cdots, a_i^r) \cdot U^\top \end{aligned}$$

where diag indicates a diagonal matrix with $a_i^1, \cdots, a_i^r$ as diagonal coefficients. It should be noted that U_i is not necessarily an orthogonal matrix, its transpose is therefore not necessarily its pseudo-inverse, and the coefficients $a_i^1, \cdots, a_i^r$ are not necessarily eigenvalues of M_i. Nevertheless, leaving those coefficients free allows M_i not to be restricted to the space of semi-definite positive matrices. This model requires $(d+1)r + d$ parameters per class.

Bi-Affine Antisymmetric Classifier. In this section, we describe a classifier able to solve the binary classification problem consisting in predicting the direction of the edges. We denote the two classes as 1 and 0. We define f as the conditional probability:

$$f(\mathbf{v}_1, \mathbf{v}_2) = \mathbb{P}(y = 1 \mid \mathbf{v}_1, \mathbf{v}_2).$$

Because the problem is antisymmetric (meaning that the class becomes the opposite if $\mathbf{v}_1$ and $\mathbf{v}_2$ are swapped), we require that:

$$\forall \mathbf{v}_1 \in \mathbb{R}^d, \forall \mathbf{v}_2 \in \mathbb{R}^d, f(\mathbf{v}_1, \mathbf{v}_2) = 1 - f(\mathbf{v}_2, \mathbf{v}_1).$$

To model f, we define a skew-symmetric matrix $M \in \mathbb{R}^{d \times d}$ ($M^\top = -M$), a vector $\mathbf{b} \in \mathbb{R}^d$ and a function $\ell(\mathbf{v}_1, \mathbf{v}_2) = (\mathbf{v}_1 + \mathbf{b})^\top \cdot M \cdot (\mathbf{v}_2 + \mathbf{b}) - \mathbf{b}^\top \cdot M \cdot \mathbf{b}$. We then define

$$f(\mathbf{v}_1, \mathbf{v}_2) = \frac{1}{1 + \exp[-\ell(\mathbf{v}_1, \mathbf{v}_2)]}.$$

This model requires $\frac{d(d-1)}{2}$ parameters. We can reduce the number of parameters if we approximate the matrix M using a low rank matrix, as it can be noted that if $\text{rank}(M) \leq r$, we have:

$$\exists U \in \mathbb{R}^{d \times r}, \exists A \in \text{sksym}_r(\mathbb{R}), M = U \cdot A \cdot U^\top. \tag{1}$$

Furthermore, it can be shown that any skew-symmetric matrix can be decomposed as $A = P \cdot D \cdot P^\top$, where P is an orthogonal matrix, and D is a block-diagonal matrix where each block is either $[0]$ or $\begin{bmatrix} 0 & \lambda_i \\ -\lambda_i & 0 \end{bmatrix}$ ($\lambda_i \in \mathbb{R}$). It is therefore possible to devise a model requiring $dr + \frac{r}{2}$ parameters.

3.2 Graph Models

This paragraph describes our main idea to use higher order relations between tokens to predict labels. So far, we used attention descriptors between two tokens to predict a semantic relation between those.

If we were to predict the semantic relation between the words *think* and *mother* in the sentence *Think about your mother*, we could feed our biaffine models described above with the vectors representing the attention between the tokens *mother* and *think*. However, common intuition suggests that using the preposition "about", (acting as a pivot) could help the prediction process. Therefore, the main idea is to predict the semantic relation between *think* and *mother* using the vectors describing not only the attention in the pair (*think*, *mother*), but also the pairs (*think*, *about*) and (*about*, *mother*). Furthermore, in the more realistic case where the pivotal token is not known beforehand, one must consider every vertex in the complete Attention Graph, and try the feature vectors of the edges connecting this vertex and the two vertices *think* and *mother*. This can be elegantly expressed in the adjoint graph (Harary and Norman [14], also called line-graph, derivative graph, or edge-to-vertex dual.)

The adjoint graph of any undirected graph $G_0 = (\mathcal{V}_0, \mathcal{E}_0)$ is a graph $G_1 = (\mathcal{V}_1, \mathcal{E}_1)$, where $\mathcal{V}_1 = \mathcal{E}_0$, and where two nodes are adjacent if and only if they share a vertex in $\mathcal{V}_0$. In this setting, edges incident to either *think*, or *mother* and to a candidate pivotal token are exactly the neighbors of (*think*, *mother*) in the adjoint graph of the complete graph (see Fig. 2). In some more complicated cases, it could be beneficial to consider edges more remote from the edge to be classified. If we want to classify the edge (*succeed, efforts*) in the sentence *Succeed by dint of efforts*, we could use the information given by the edges (*succeed, by*), (*dint, effort*), (*succeed, dint*), but also consider the two words (*by, dint*), sharing no vertex with the edge to classify. In the adjoint graph, this corresponds to considering second-order neighbors of the considered node (see Fig. 3.)

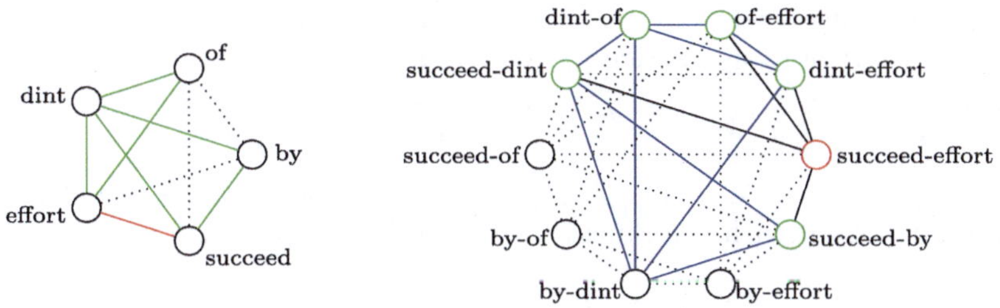

Fig. 3. Direct and adjoint attention graphs for the sentence "Succeed By Dint of Effort". The edge to be classified is ("succeed", "effort") (in red). Interesting information can be gathered from the edges in green. (Green nodes in the adjoint graph). Some are second order neighbors of the considered node in the adjoint graph, like ("By", "dint").

To choose and factor in edge candidates, we use a Graph Neural Network on the adjoint graph of the attention graph. It is comprised of one or several GATv2 layers [4]. For each vertex in a graph, such a layer computes a linear combination of this vertex's features and its neighbor's features using an attention mechanism to derive the coefficients. To better describe our idea, we start by giving a quick formalism to the concept of a GATv2 layer in the particular case of a *regular graph*.

GATv2 Layer in a Regular Graph. A *regular graph* is a graph whose vertices all have the same degree. The adjoint graph of a complete graph with n vertices is a regular graph of degree $2n - 4$.

If each node in the adjoint graph $(\mathcal{V}, \mathcal{E})$ is described with a feature vector in $\mathbb{R}^d$, the attention mechanism that transforms the features of a node can be conceptualized as a feature-function π, parameterized by θ, such that:

$$\pi_\theta : \mathbb{R}^d \times (\mathbb{R}^d)^{2n-4} \longrightarrow \mathbb{R}^{d'} \quad (\mathbf{v}, \mathbf{w}_1, \cdots, \mathbf{w}_{2n-4}) \longmapsto \pi_\theta(\cdots) = \mathbf{h} \in \mathbb{R}^{d'},$$

where $\mathbf{v}$ is the feature vector of the node to be transformed, and $\mathbf{w}_1 \cdots \mathbf{w}_{2n-4}$ are the feature vectors of its neighbors. The function π_θ is invariant w.r.t any permutation of the neighbor vectors $\mathbf{w}_1 \cdots \mathbf{w}_{2n-4}$. The whole GATv2 layer is the result of applying π_θ to every node in the adjoint graph $(\mathcal{V}, \mathcal{E})$. We can conceptualize this process using a feature function $\phi : \mathcal{V} \rightarrow \mathbb{R}^d,\ \kappa \mapsto \phi(\kappa)$, that acts as a parameter for the GATv2 layer such that:

$$\text{gat}_{\theta,\phi} : \mathcal{V} \longrightarrow \mathbb{R}^{d'} \quad \kappa \longmapsto \pi_\theta(\phi(\kappa), \phi(\mathcal{N}(\kappa))),$$

where $\mathcal{N}(\kappa)$ is the set of the neighbors of κ.

Graph Classifier. In our particular problem, the nodes of the adjoint graph of our attention graph are not described with one vector in $\mathbb{R}^d$, but two vectors. More formally, we can define two feature functions $\phi_1 : \mathcal{V} \longrightarrow \mathbb{R}^d$ and $\phi_2 : \mathcal{V} \longrightarrow \mathbb{R}^d$, which map each vertex κ to two vectors $\phi_1(\kappa)$ and $\phi_2(\kappa)$.

The first layer of our model is a symmetric layer: $\psi_1 : \mathcal{V} \longrightarrow \mathbb{R}^{d_1}$, where $\psi_1(\kappa) = \left((\phi_1(\kappa) + \mathbf{b}_i)^\top \cdot M_i \cdot (\phi_2(\kappa) + \mathbf{b}_i)\right)_{i\in\{1,\cdots,d_1\}}$. This layer allows assigning a single feature vector to each node, independently of any permutation of $\phi_1(\kappa)$ and $\phi_2(\kappa)$ for any particular node κ. The second layer is a GATv2 layer:

$$\psi_2 : \mathcal{V} \longrightarrow \mathbb{R}^{d_2} \quad \kappa \longmapsto \psi_2(\kappa) = \mathrm{gat}_{\theta_2,\psi_1}(\kappa),$$

where $\mathrm{gat}_{\theta_2,\psi_1}(s) = \pi_{\theta_2}(\psi_1(s), \psi_1(\mathcal{N}(s)))$. The following layers are GATv2 layers, parameterized with the function ψ of the previous layer:

$$\psi_i : \mathcal{V} \longrightarrow \mathbb{R}^{d_i} \quad \kappa \longmapsto \mathrm{gat}_{\theta_i,\psi_{i-1}}(\kappa).$$

The last layer is a linear classifier, and the whole model is insensitive to the swap of $\phi_1(\kappa)$ and $\phi_2(\kappa)$ for any $\kappa \in \mathcal{V}$.

4 Experiments

We tested our method on several language models: GPT-2 [21] RoBERTa-base, RoBERTa-large [16], ModernBERT-base [25], and three models in the Llama3 suite [13]: Llama-3.2-1B, Llama-3.2-3B and Llama-3.1-8B.

We chose 15 of the most useful semantic roles in our dataset, corresponding to the projection on VerbAtlas of the roles chosen by the authors of [6] with a wide range of absolute frequency.[2] Unlike them, we have not restricted our models to classifying relations between symbols corresponding to whole words. This makes the task more difficult, since we are trying to classify relations between certain symbols that are only fractions of words.

As we described in the 2.1 section, we have added special labels to take account of the separation of words between several symbols, as well as coreference, conjunction and disjunction. There are six such labels: {`and-or`}, {`and-or`}, {`or`}, {`group`}, {`idem`}, {`inter`}, (the last being used for concepts associated with prepositions such as ‘between’ or ‘among’). Including them, the number of selected classes reaches 21. Our training is carried out on a training set, by monitoring the balanced accuracy (*macro-averaged accuracy*) calculated periodically on the 21 classes, on a validation set.

Detecting the six labels is sometimes a trivial problem, particularly the {`group`} label, which links two symbols of the same word. This is why we do not measure the performance of our models on these six additional labels, which would give over-optimistic results. The performances, reported in Table 1 indicate the balanced accuracy calculated on a restriction to the 15 semantic classes we kept.

[2] theme, mod, patient, topic, experiencer, manner, degree, cause, agent, time, location, poss, beneficiary, purpose, condition.

Table 1. Experiments results. See Sect. 4. The column *Attention dim* indicates the dimension of the attention vector. The column *rank* indicates the low rank for a bi-affine classifier, as well as for a GNN classifier. The two columns *bi-affine bal. acc* and *GNN bal. acc.* give the balanced accuracy obtained with those two classifiers. The column *GNN's gain* give the ration of the GNN's balanced accuracy over the bi-affine classifier. The column *hidden dim* gives the hidden dimensions d_1 and d_2 of the GNN; the last column gives the number of parameters of the GNN.

Lang. Model	Attention dim.	Rank	biaffine bal. acc.	GNN bal. acc.	GNN's gain	hidden dim.	Par cnt GNN
GPT-2	144	8	49.9%	63.9%	1.28	144	275×10^3
RoBERTa B	144	8	54.5%	69.7%	1.28	144	275×10^3
MdnBERT B	264	16	58.1%	67.3%	1.16	150	1.48×10^6
RoBERTa L	384	16	62.53%	66.0%	1.06	150	1.07×10^6
Llama 1B	512	16	62.5%	66.85%	1.07	150	1.40×10^6
Llama 3B	672	16	62.7%	64.3%	1.03	150	1.81×10^6
Llama 8B	1024	16	64.4%	66.0%	1.02	150	2.71×10^6

4.1 Bi-Affine Classifier

This section presents the results obtained on the low-rank symmetric biaffine models. For the smallest two models, GPT-2 and RoBERTa-base, we observed an increase in the accuracy as we increased up to 8 the rank of the matrices. After $r = 8$, the metrics plateaued. However, we chose to use a higher rank $r = 16$ for the larger models, in order to better capture information spread across increasingly larger feature vectors. The results are to be read in the column "biaffine balanced accuracy" of Table 1. We can notice that the balanced accuracy is an increasing function of the dimension of the attention vectors, in direct relation with the size of the transformers. We can also see that there is a large difference between the two smaller models, GPT-2 and RoBERTa base, in favour of the latter. We hypothesise that this is due to the causal attention of the GPT-2 decoder model, which only exploits the attentional computation of a symbol on previous symbols, whereas a bidirectional model such as RoBERTa base exploits attention in both directions.

4.2 Graph Classifier

We compared these results with those given by the GNN we described in Sect. 3.2. Table 1 gives the results in the column "GNN bal. acc". For each language model, the GNN is a network whose first layer shares the rank indicated in the column "rank" with the bi-affine classification experiment. The network has two GATv2 layers with a hidden dimension indicated in the column "hidden dim". This hidden dimension h corresponds to both internal dimensions d_1 and d_2 in the description above, see Sect. 3.2. We chose to keep the hidden dimension h constant across models, because prior experiments showed that the performance is not very sensible to this value, and keeping it low allows to keep our GNNs relatively small.

We can see that our approach leads to significant performance gains. (gains calculated between the performance of the GNN and the performance of the biaffine classifier of the same rank). We obtain a gain of 1.28 for the smallest models, which proves the usefulness of examining higher-order relations, which our approach allows. However, the gain decreases with larger networks, proving experimentally that larger networks already encode sufficient semantics in first-order relations between symbols. (Llama-8B, the largest model, shows a performance gain of only 1.02).

Remarkably, the two best results in the whole table are obtained with the RoBERTa base and ModernBERT base models, which are encoder models, taking advantage of the bidirectional computation of attention. The third encoder model, RoBERTa-Large is on par with the other much larger models, which are decoder models using causal masking that prohibits taking advantage of attention on later symbols in the sentence. Our experience suggests that this masking is an obstacle to the semantic capabilities of transformer models.

4.3 Zero GATv2 Layer Classifier

The gap of performance between our model with two GATv2 layers and our symmetric classifier cannot only be explained by the presence of the GATv2 layers: To predict one among c classes with a symmetric classifier, it is sufficient to take c matrices of rank r and a softmax layer. In our GNN, the first layer is comprised of h matrices of rank r followed by two GATv2 layers, followed by a linear classifier. Even without connecting the GATv2 layers, our GNN has therefore more parameters than the symmetric classifier. To prove the efficiency of the GAT layers, we conducted an ablation study, where we replaced both GATv2 layers with a simple reLU activation, without parameters. We retrained the model on the dataset obtained with the best language models RoBERTa-base and ModernBERT base.

We obtained a balanced accuracy of 52.0% on RoBERTa base, strikingly below the value of 69.7% obtained with the normal GATv2 model, and even slightly below the performance of the biaffine classifier. For ModernBERT base, the balanced accuracy is 58.7%, which is comparable to the biaffine classifier, and loses all the gain due to the GATv2 layers.

4.4 Orientation Classifier

Let's briefly mention again the AMR arc orientation classifier, mentioned in Sects. 3.1 and ??. The reader will note the particular possibility of training on a balanced dataset, by first randomly permuting the two vectors for each node, as well as the label indicating the edge direction. The best results were obtained with rank 16, with an accuracy of 89% on the basic RoBERTa model, and 87% on the GPT-2 model. These results show that the problem of classifying the direction of an arc in an AMR is not a very difficult problem, and did not urge us to design a graph classifier to tackle it.

4.5 Comparison to a State-of-the-Art Reference

We expected the best models at predicting semantic relations to be state-of-the art AMR parsers, and wanted to compare our models to them. However, AMR parsers are usually evaluated using a technique called SMATCH, [5] a graph comparison metric, incompatible with our experiments. We therefore devised a solution for a comparison to the state-of-the-art AMR parser Spring [2]. This model was further enhanced by [18], making it able to align its predictions with the tokens in the original sentence. We used this Spring aligner-parser to automatically build AMRs from the test set of the LDC2020T02 dataset with an alignment to its own encoder tokens. Using the transformations described in Sect. 2.1, we built an attention graph for Spring's transformers token, which we compared with the attention graph built with the gold AMRs in the test set, aligned with Leamr alignment data. We could thus build two attention graphs to be compared edge-to-edge, selecting only edges with the 15 kept semantic roles, with the addition of the six special labels, kept to put the model in the same perplexity, but left out of the computation of the balanced accuracy. Eventually, the Spring AMR parser obtained a balanced accuracy of 80.4%. The ratio of the performance of our best model, (Roberta-base with our GNN) to this one is 86.7%. Spring is a specially fine-tuned bart-large model, comprising about $400 \cdot 10^6$ parameters, which must be compared to the $275 \cdot 10^3$ parameters of our GNN model for RoBERTa-base.

5 Conclusion

In this study, we introduced a graph-based method to evaluate the intrinsic capability of different language models to perform SRL tasks. Our method exploits the attention weights computed by the language models used without fine-tuning, and we do not resort to word embedding. We showed experimentally that simple classification based on the attention vectors between the two symbols in a pair is more effective as the language model used is larger, and that by using the full attentional graph between the symbols in the text to classify pairs, our method is able to make better predictions from higher-order relationships between more than two words in the text. Although the advantage of our method is less noticeable on large language models, which seem to encode semantics directly in first-order relations, our approach yielded the best results on the smallest RoBERTa base model, which obtained 86.7% of the performances of a state-of the art AMR parser. Our experiments also confirmed that bidirectional language models show a better ability to distinguish semantic relations than causal models. In addition, we presented a general method that can be used more generally in problems involving word pairs, for tasks other than semantics.

References

1. Banarescu, L., et al.: Abstract meaning representation for sembanking. In: Pareja-Lora, A., Liakata, M., Dipper, S. (eds.) Proceedings of the 7th Linguistic Annotation Workshop and Interoperability with Discourse, Sofia, Bulgaria, pp. 178–186. Association for Computational Linguistics (2013). https://aclanthology.org/W13-2322
2. Bevilacqua, M., Blloshmi, R., Navigli, R.: One SPRING to rule them both: symmetric AMR semantic parsing and generation without a complex pipeline. In: Proceedings of the AAAI Conference on Artificial Intelligence, vol. 35, no. 14, pp. 12564–12573 (2021). https://doi.org/10.1609/aaai.v35i14.17489. https://ojs.aaai.org/index.php/AAAI/article/view/17489. Accessed 03 Dec 2023. ISSN 2374-3468, 2159-5399
3. Blodgett, A., Schneider, N.: Probabilistic, structure-aware algorithms for improved variety, accuracy, and coverage of AMR alignments. In: Proceedings of the 59th Annual Meeting of the Association for Computational Linguistics and the 11th International Joint Conference on Natural Language Processing (Volume 1: Long Papers), pp. 3310–3321. Online: Association for Computational Linguistics (2021). https://doi.org/10.18653/v1/2021.acl-long.257. https://aclanthology.org/2021.acl-long.257. Accessed 21 Oct 2023
4. Brody, S., Alon, U., Yahav, E.: How attentive are graph attention networks? (2022). arXiv:2105.14491 [cs]. Accessed 22 Nov 2022
5. Cai, S., Knight, K.: Smatch: an evaluation metric for semantic feature structures
6. Charpentier, F., Cugliari, J., Guille, A.: Exploring semantics in pretrained language model attention. In: Bollegala, D., Shwartz, V. (eds.) Proceedings of the 13th Joint Conference on Lexical and Computational Semantics (*SEM 2024), Mexico City, Mexico, pp. 326–333. Association for Computational Linguistics (2024). https://doi.org/10.18653/v1/2024.starsem-1.26. Accessed 06 Dec 2024
7. Chizhikova, A., et al.: Attention understands semantic relations. In: Calzolari, N., et al. (eds.) Proceedings of the Thirteenth Language Resources and Evaluation Conference, Marseille, France, pp. 4040–4050. European Language Resources Association (2022). https://aclanthology.org/2022.lrec-1.430
8. Clark, K., et al.: What does BERT look at? An analysis of BERT's attention. In: Linzen, T., et al. (eds.) Proceedings of the 2019 ACL Workshop BlackboxNLP: Analyzing and Interpreting Neural Networks for NLP, Florence, Italy, pp. 276–286. Association for Computational Linguistics (2019). https://doi.org/10.18653/v1/W19-4828. https://aclanthology.org/W19-4828
9. Conia, S., Navigli, R.: Probing for predicate argument structures in pretrained language models. In: Muresan, S., Nakov, P., Villavicencio, A. (eds.) Proceedings of the 60th Annual Meeting of the Association for Computational Linguistics (Volume 1: Long Papers), Dublin, Ireland, pp. 4622–4632. Association for Computational Linguistics (2022). https://doi.org/10.18653/v1/2022.acl-long.316. Accessed 28 May 2025
10. Devlin, J., et al.: BERT: pre-training of deep bidirectional transformers for language understanding. In: Burstein, J., Doran, C., Solorio, T. (eds.) Proceedings of the 2019 Conference of the North American Chapter of the Association for Computational Linguistics: Human Language Technologies, volume 1 (long and short papers), Minneapolis, Minnesota, pp. 4171–4186. Association for Computational Linguistics (2019). https://doi.org/10.18653/v1/N19-1423. https://aclanthology.org/N19-1423

11. Di Fabio, A., Conia, S., Navigli, R.: VerbAtlas: a novel large-scale verbal semantic resource and its application to semantic role labeling. In: Inui, K., et al. (eds.) Proceedings of the 2019 Conference on Empirical Methods in Natural Language Processing and the 9th International Joint Conference on Natural Language Processing (EMNLP-IJCNLP), Hong Kong, China, pp. 627–637. Association for Computational Linguistics (2019). https://doi.org/10.18653/v1/D19-1058. Accessed 27 Apr 2024
12. Gildea, D., Jurafsky, D.: Automatic labeling of semantic roles. Comput. Linguist. **28**(3), 245–288 (2002). https://doi.org/10.1162/089120102760275983. https://aclanthology.org/J02-3001
13. Grattafiori, A., et al.: The Llama 3 Herd of models (2024). arXiv:2407.21783 [cs.AI]
14. Harary, F., Norman, R.Z.: Some properties of line digraphs. Rendiconti del Circolo Matematico di Palermo **9**(2), 161–168 (1960). https://doi.org/10.1007/BF02854581. ISSN 1973-4409
15. Knight, K., et al.: Abstract Meaning Representation (AMR) annotation release 3.0 LDC2020T02. In: Linguistic Data Consortium (2020)
16. Liu, Y., et al.: RoBERTa: a robustly optimized BERT pretraining approach. arXiv:1907.11692 [cs] (2019). Accessed 15 June 2023
17. Luo, Z.: Have attention heads in BERT learned constituency grammar? In: Sorodoc, I.-T., et al. (eds.) Proceedings of the 16th Conference of the European Chapter of the Association for Computational Linguistics: Student Research Workshop, Online, pp. 8–15. Association for Computational Linguistics (2021). https://doi.org/10.18653/v1/2021.eacl-srw.2. https://aclanthology.org/2021.eacl-srw.2. Accessed 23 Feb 2024
18. Lorenzo, A.C.M., Cabot, P.L.H., Navigli, R.: Cross-lingual AMR aligner: paying attention to cross-attention. In: Rogers, A., Boyd-Graber, J., Okazaki, N. (eds.) Findings of the Association for Computational Linguistics: ACL 2023, Toronto, Canada, pp. 1726–1742. Association for Computational Linguistics (2023). https://doi.org/10.18653/v1/2023.findings-acl.109. https://aclanthology.org/2023.findings-acl.109/
19. Palmer, M., Gildea, D., Kingsbury, P.: The proposition bank: an annotated corpus of semantic roles. Comput. Linguist. **31**(1), 71–106 (2005). https://doi.org/10.1162/0891201053630264. https://aclanthology.org/J05-1004. Accessed 09 Apr 2023
20. Qorib, M., Moon, G., Ng, H.T.: Are decoder-only language models better than encoder-only language models in understanding word meaning? In: Ku, L.-W., Martins, A., Srikumar, V. (eds.) Findings of the Association for Computational Linguistics: ACL 2024, Bangkok, Thailand, pp. 16339–16347. Association for Computational Linguistics (2024). https://doi.org/10.18653/v1/2024.findings-acl.967. Accessed 11 Apr 2024
21. Radford, A., et al.: Language models are unsupervised multitask learners (2019). https://api.semanticscholar.org/CorpusID:160025533
22. Tenney, I., Das, D., Pavlick, E.: BERT rediscovers the classical NLP pipeline. In: Korhonen, A., Traum, D., Màrquez, L. (eds.) Proceedings of the 57th Annual Meeting of the Association for Computational Linguistics, Florence, Italy, pp. 4593–4601. Association for Computational Linguistics (2019). https://doi.org/10.18653/v1/P19-1452. Accessed 28 May 2025
23. Tenney, I., et al.: What do you learn from context? Probing for sentence structure in contextualized word representations (2019)

24. Vaswani, A., et al.: Attention is all you need. In: Guyon, I., et al. (eds.) Advances in Neural Information Processing Systems, vol. 30. Curran Associates, Inc. (2017). https://proceedings.neurips.cc/paper_files/paper/2017/file/3f5ee243547dee91fbd053c1c4a845aa-Paper.pdf
25. Warner, B., et al.: Smarter, better, faster, longer: a modern bidirectional encoder for fast, memory efficient, and long context finetuning and inference (2024). arXiv:2412.13663 [cs.CL]
26. Xu, D., et al.: Improving AMR parsing with sequence-to-sequence pre-training. arXiv:2010.01771 [cs] (2020). Accessed 03 Dec 2023
27. Zhou, J., et al.: Structure-aware fine-tuning of sequence-to-sequence transformers for transition-based AMR parsing. In: Moens, M.-F., et al. (eds.) Proceedings of the 2021 Conference on Empirical Methods in Natural Language Processing, Online and Punta Cana, Dominican Republic, pp. 6279–6290. Association for Computational Linguistics (2021). https://doi.org/10.18653/v1/2021.emnlp-main.507. https://aclanthology.org/2021.emnlp-main.507

ConverSQL: Hybrid Rule-Based and LLM Architecture for Natural Language Text-to-SQL Translation

Soumya Ranjan Mishra(✉), Sanket H. Bhandarkar, Richa Sharma, and S. Deepa

PES University, Karnataka Bengaluru, India
{pes2ug22cs571,pes2ug22cs506}@pesu.pes.edu, {richasharma, deepa.s}@pes.edu

Abstract. Access to relational databases through Structured Query Language (SQL) is a major impediment for non-technical users, preventing data utilization with maximum efficiency. Natural Language Interfaces to Databases (NLIDB) try to bridge this chasm but usually make a trade-off between rule-based system rigidity for sophisticated queries and the cost, latency, and security loopholes of exclusively Large Language Model (LLM)-based solutions. This paper presents ConverSQL, a new hybrid architecture enhancing this balance for efficient and secure natural language-to-SQL translation. Our architecture puts utmost priority on a high-performance, rule-based NLP pipeline (Stanza tokenization, schema-aware entity recognition, intent classification, deterministic query construction for SELECT, INSERT, UPDATE, DELETE, DROP). Ambiguous or difficult queries, detected through confidence scores and entity matching metrics lower than certain thresholds, initiate a smart fallback to a lightweight, locally hosted LLM (Gemma:2b). ConverSQL is a secure, full-stack implementation (Next.js/React for the frontend, Python/Fast API for the backend) with RSA-AES hybrid encryption and Role-Based Access Control (RBAC). We compared our hybrid model with solely rule-based and solely LLM-based baselines on a custom-created test suite of 500 heterogeneous queries for our space-domain dataset by using execution accuracy against ground truth results and capturing end-to-end latency. Our results show the hybrid approach captures 92% execution correctness, well above the 84% baseline of rule-based. Notably, it achieves an average latency (310 ms) comparable with rule-based performance (250 ms), and much lower than LLM-alone (1500 ms). ConverSQL therefore provides a proven, solid, and effective framework to democratize database access, with the best of both rule-based efficiency and LLM flexibility.

Keywords: Natural Language to SQL · Conversational AI · Database Querying · Hybrid NLP · LLM · Rule-Based Systems

H. Unger and P. Meesad (Eds.): NLPIR 2025, LNNS 1904, pp. 211–223, 2026.
https://doi.org/10.1007/978-3-032-20897-2_15

1 Introduction

In today's data-driven world, efficiently accessing and using information in relational databases is vital for planning and running operations. However, the primary interface for these databases is Structured Query Language (SQL), which poses a major challenge for non-technical users like business analysts, managers, and domain experts. For instance, a marketing manager who requires immediate sales data for a particular area will take hours or even days waiting for a developer to write the appropriate SQL query. This waiting time can delay critical campaign updates and result in substantial operational expenses. Having to depend on a few expert developers creates bottlenecks and communication problems, tending to cause misunderstandings and impeding efficient decision-making. This issue is extensively documented in academic publications and industry reports. To address the specific requirements of our project, we curated our own dataset from multiple space datasets available publicly, as we were working on a space domain problem statement provided by Indian Space Research Organization (ISRO) and found no existing ready-to-use datasets available for this purpose. The system was initially built for space related queries but it is flexible and can be generalized to other datasets of different domains due to our schema mapping mechanism.

Building an efficient Natural Language Interface to Databases (NLIDB) is not easy. Most systems are domain-specific and do poorly with new schemas. Large model-based solutions end up inducing delays and putting security at risk. The uncertainty of language also results in destructive actions such as UPDATE or DELETE. This paper introduces ConverSQL, a smart web application that focuses on these issues. Our primary contribution is a novel hybrid architecture combining a high-speed, rule-driven NLP pipeline with a low-footprint local LLM for fallback. The double-engine design, supplemented with dynamic schema automation, is intended to be database-agnostic and budget-friendly. Safety was ensured by using a form-based "guardrail" system that triggers on intent-determined by NLP, prompting user confirmation for sensitive operations.

This research tests the effectiveness of our hybrid model on three fronts. We first check how well this architecture enhances Text-to-SQL accuracy over a solely rule-based system. Second, we investigate how using a local LLM for edge cases affects system latency and performance. Finally, we evaluate whether our secure, form-based "guardrail" system provides a trustworthy method for carrying out destructive SQL operations initiated through natural language.

Our contributions are threefold:

- A new hybrid design that merges a fast, rule-based NLP system with a lightweight local LLM for backup, focusing on both speed and accuracy.
- A dynamic schema automation feature that allows the system to work with any database.
- The application of this approach within a secure, full-stack application that includes a form-based "guardrail" system to ensure safety during all data modification tasks.

2 Related Work

As per the research conducted we found some existing work already done on this topic which are as follows: Panwar [1] presented a broad survey of query optimization techniques, detailing heuristic, cost-based, and AI-assisted strategies that enhance query execution efficiency. The paper emphasized that while AI integration reduces execution time, traditional optimizers lack adaptive learning for dynamic workloads. Nama et al. [2] extended this by introducing automated optimization and intelligent query processing using machine learning, allowing systems to adaptively tune query plans. However, their approach was primarily simulation-based and limited to relational workloads. Gao et al. [3] demonstrated practical performance optimization in MySQL through index tuning and caching. Despite showing measurable throughput improvements, it did not address natural language query generation or semantic understanding layers.

Androutsopoulos, Ritchie, and Thanisch [4] pioneered early Natural Language Interfaces to Databases (NLIDB) using grammar-based parsing and semantic templates to translate natural language into SQL. These deterministic systems achieved interpretability but lacked robustness against ambiguity and multi-domain schemas. Baig, Ali, and Khan [5] conducted a comprehensive review of NL $\rightarrow$ SQL techniques, identifying limitations of early rule-based methods—specifically, rigid syntax handling, domain dependence, and limited contextual reasoning. Their analysis underscored the need for data-driven adaptability.

The introduction of the Spider dataset by Yu et al. [6] marked a major shift toward cross-domain Text-to-SQL benchmarking, standardizing evaluation for complex multi-table SQL generation. While enabling fair comparison, it also exposed weaknesses in schema linking and compositional generalization. Zhong, Xiong, and Socher [7] introduced Seq2SQL, which used reinforcement learning and pointer networks to improve execution accuracy. Though effective for single-domain settings, it struggled with nested queries and schema transfer. Wang et al. [8] proposed RAT-SQL, which incorporated relation-aware transformers to jointly encode schema and question representations, greatly improving cross-domain performance but requiring heavy GPU training and annotated data. Scholak, Schucher, and Bahdanau [9] advanced constrained decoding with PICARD, ensuring syntactically valid SQL generation from large language models (LLMs). While it reduced parsing errors, it introduced high computational cost and dependency on large-scale pretrained models.

Liu and Xu [10] systematically reviewed NLIDB advancements and categorized approaches as rule-based, neural, and hybrid. Their review identified ambiguity handling, schema scalability, and model interpretability as unresolved challenges. Iacob, Calin, and Dinu [11] focused on neural architectures, explaining how encoder-decoder models with schema attention improved translation accuracy but at the cost of explainability. Luo et al. [12] provided a VLDB tutorial highlighting open problems such as domain adaptation, few-shot learning, and security in LLM-based Text-to-SQL systems. Rule-based systems [4, 5] often lack strength, while neural models [8, 9, 13] frequently compromise on clarity, speed, and security [10–12, 14].

To bridge these research gaps, ConverSQL presents a new hybrid solution. Our system confronts these problems through the combination of an efficient, transparent rule-based procedure with a formal backup mechanism that utilizes a lightweight, local

LLM (Gemma:2b) [15] only as required. This minimizes the cost of computing and keeps data private. Moreover, in contrast to most current approaches, ConverSQL features dynamic schema automation for broad adoption, robust RSA-AES encryption with RBAC for security, and a special form-based guardrail system to provide security during data modification, offering a real-world and complete solution.

3 Database Construction and Pre-Processing

Although the ConverSQL system architecture has the capability of being domain-independent through automation of schemas, its performance must be tested with a clear and difficult set of data. In this research, we selected the complicated space exploration domain as an instance. The domain provides a proper testing ground due to its complicated schema that has eight interrelated databases, different data types, and technical jargon. This setup thoroughly tests our NLP pipeline's ability to adapt and be accurate. To create this environment, we first built the database using raw datasets from publicly available Kaggle repositories [16]. We did this with a dedicated Database Creator module, and the process.

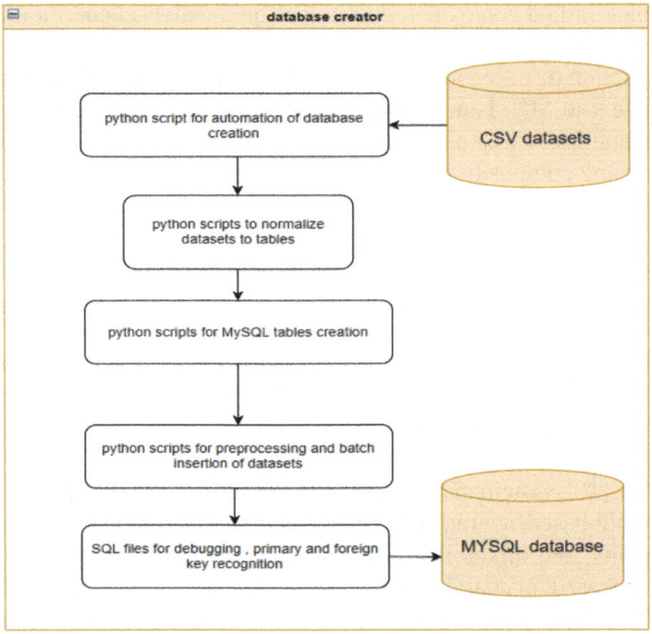

Fig. 1. Automated Pipeline for Database Creation from CSV Datasets

The experimental database was built using an automated pipeline, as shown in Fig. 1. The process starts with Python scripts that read raw CSV files from the Kaggle dataset collection. These datasets are then organized into clear tables based on Entity-Relationship principles. Next, the system automatically generates CREATE TABLE statements with inferred data types. Before inserting data in large batches, the information goes through regex-based cleaning and validation. Finally, it identifies primary and foreign keys to maintain referential integrity in the schema. This pipeline connects various space-related domains through eight modular databases: **asteroids_db, astronauts_db, isro_satellites_db, natural_satellites_db, rockets_db, space_missions_db, spacenews_db, and stars_db**. Together, this schema provides a complete view of space exploration data. A detailed breakdown of the samples in these domains is available in Table 1.

Table 1. List of Datasets sourced from Kaggle

Database Name	Domain	Initial Samples	Samples Used (After Pre-processing)
asteroids_db [16]	Near-Earth Objects	27,500	25,120
astronauts_db [16]	Human Spaceflight	1,200	1,150
isro_sattelites_db [16]	Satellite Technology	850	810
natural_satellites_db [16]	Planetary Science	250	220
rockets_db [16]	Launch Vehicles	4,500	4,320
space_missions_db [16]	Mission History	5,800	5,650
spacenews_db [16]	Space Journalism	15,000	14,200
stars_db [16]	Stellar Astronomy	120,000	115,800
Total		175,100	167,200

4 Methodology

The ConverSQL system is set up as a full-stack application with a client-server architecture. It has a Next.js frontend and a Python Fast API backend. The main part of the system is a hybrid NLP pipeline that translates natural language queries into executable SQL. It does this with high accuracy, security, and efficiency.

The complete workflow of the ConverSQL system, shown in Fig. 2, starts with a user-provided natural language query. This input goes through a clear NLP pipeline with several steps: tokenization, intent detection, entity recognition, and query construction. An important part of our hybrid setup is the fallback condition. If the parser's confidence score drops below a set threshold or if there are major schema ambiguities, the query is quickly sent to the local LLM for further handling. The last output sent back to the frontend are the generated SQL query and its execution output. This guarantees that most of the queries are executed swiftly while more complex ones are given special focus.

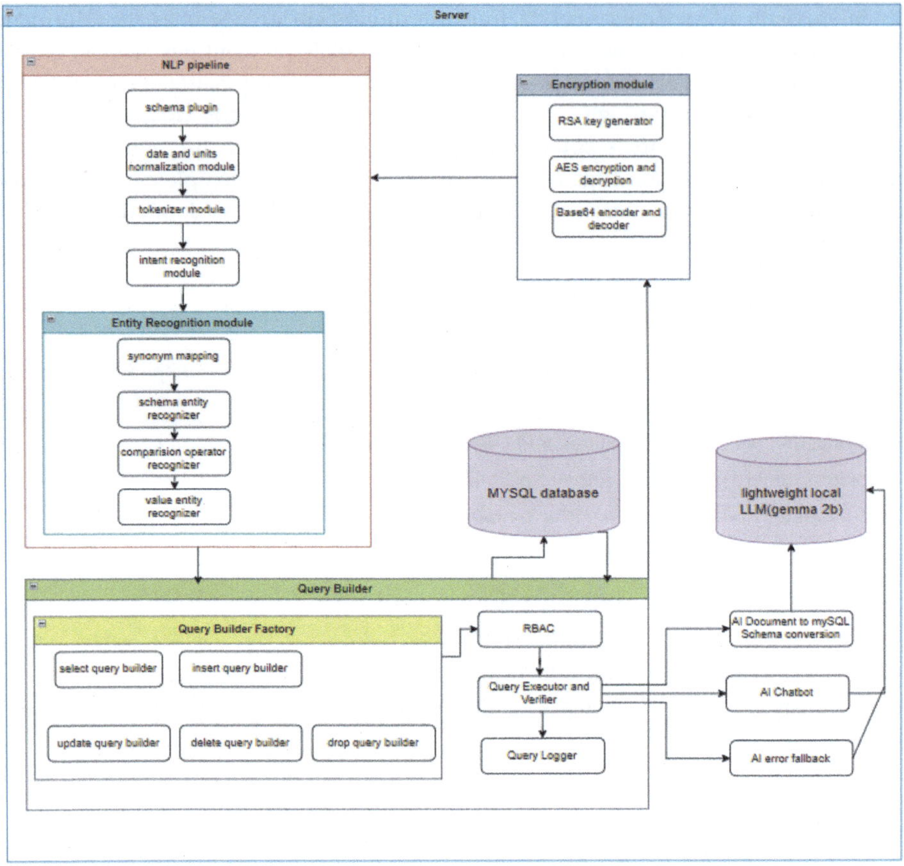

Fig. 2. System Architecture of the Hybrid NLP to SQL Server

It is the multi-stage NLP pipeline that lies at the core of this process. A schema plugin initially loads the database schema to set the context. The query of the user then passes through a normalization module for standardizing date and unit expressions. Next, it is disassembled into individual tokens by the tokenizer module. Next, the intent recognition module identifies the user's main goal, such as selecting or inserting data. The Entity Recognition Module then matches the text to specific database concepts. It changes synonyms into actual schema names and identifies database objects, comparison operators, and literal values. Once the query is fully parsed, the structured output is sent to the Query Builder. This component uses a factory pattern to pick a specialized builder, such as for SELECT,INSERT,DELETE,DROP or UPDATE based on the detected intent and utilizes a template of given query type with placeholder values which are filled up with the entities recognized in the earlier stage. Before execution, the generated query is checked against the Role-Based Access Control (RBAC) module to confirm the user has the right permissions. After verification, the query runs against the MySQL database, and the transaction is logged by the Query Logger for auditing.

Security for all client-server communication is handled by the Encryption Module. This module uses an RSA-AES hybrid scheme to protect data confidentiality. Intelligence in the system is enhanced by a lean local LLM (Gemma:2b) [15], which plays three functions: a conversational search query AI chatbot, a fallback option when the rule-based pipeline fails, and support for AI-driven document-to-schema transformation.

4.1 Experimental Setup

All testing was performed in an environment that simulated a typical development and deployment configuration. The hardware setup included a developer-class machine with an Intel Core i5-11400H processor, 8 GB of DDR4 RAM, and an NVIDIA GeForce GTX 1650 GPU with 4 GB of VRAM. The software environment was built on a Python 3.9 backend using the Fast API web framework, with the NLP pipeline leveraging libraries such as Stanza and spaCy. The client-side interface is a Next.js 14 application built with React 18, and all queries were executed against a local MySQL Server 8.0 instance. For the AI fallback mechanism, a 4-bit quantized version of Gemma:2b [15] was served locally via the Ollama runtime environment, ensuring all data processing remained on-premise.

4.2 Confidence Scoring and Fallback Decision Mechanism

The ConverSQL hybrid pipeline has a confidence-based decision layer. It determines if a rule-based parser can handle an input query or if it needs to go to the LLM. Let Q represent the natural-language query and $\mathbf{S} = \{\mathbf{e_1}, \mathbf{e_2}, \ldots, \mathbf{e_n}\}$ be the set of detected schema entities. Each entity $\mathbf{e_i}$ matches its database counterpart using cosine similarity between contextual embeddings from Stanza and pre-computed schema vectors.

The overall confidence score is calculated as:

$$C(Q) = \frac{1}{n}\sum_{i=1}^{n}(w_i p_i) \tag{1}$$

In Eq. 1, where $\mathbf{p_i}$ is the similarity score for entity $\mathbf{e_i}$, and $\mathbf{w_i}$ is a syntactic weight that gives more importance to intent keywords. A query is accepted by the rule-based parser if **C(Q)** is greater than or equal to τ, where τ is the confidence threshold. In our experiments, a $\boldsymbol{\tau}$ of 0.78 provided the best balance between accuracy and latency. If **C(Q)** is less than τ or if fewer than 90% of entities are resolved, the system triggers the LLM fallback **(Gemma-2b)** [15]. The fallback reformulates or checks the SQL before execution, ensuring that unclear or incomplete queries are still processed correctly.

ALGORITHM 1: Query Handling and Decision Logic

1. Input user query: user_query
   Parse the natural-language query using the rule-based parser
   to extract SQL components.

2. Parse query: parsed_output = rule_based_parser(user_query)
   Extract tokens, intent, and entities to form a preliminary SQL structure.

3. Compute confidence score: confidence = compute_confidence(parsed_output)
   Estimate confidence based on parser accuracy or rule-match strength.

4. Check threshold and entities:
   if confidence < CONF_THRESHOLD or
   has_unresolved_entities(parsed_output)
   Verify whether the confidence is below threshold or unresolved entities exist.

5. Fallback:
   then return llm_fallback(user_query)
   Invoke the LLM-based fallback to generate or refine the SQL query.

6. Else execute query:
   return execute_sql(parsed_output)
   Execute the parsed SQL query and return the resulting records.

7. End Algorithm

This mechanism ensures that simple queries benefit from the speed of deterministic parsing, while complex or unclear ones use the LLM for higher accuracy without losing efficiency.

4.3 Query Set Construction

For testing, we created a dataset of 500 natural-language queries that covered all eight domain databases: asteroids, astronauts, satellites, natural satellites, rockets, missions, space news, and stars. Template-based queries (about 300) were generated automatically using SQL-to-NL templates based on schema metadata to provide balanced CRUD coverage. Human-written queries (about 150) were created by the research team. These included ambiguous phrasing, nested conditions, and multi-clause queries. Paraphrased variants (about 50) were made using the local Gemma-2b [15] model to add some linguistic variety. We manually checked all queries for both correct structure and meaning compared to their reference SQL. This ensured that both the deterministic parser and the LLM fallback were tested under realistic and varied query conditions.

5 Results and Discussion

To evaluate how effective ConverSQL is, we created a dataset of 500 natural language queries. This dataset covers the eight different domains in our experimental database collection: asteroids_db, astronauts_db, Isro_satellites_db, natural_satellites_db, rockets_db, space_missions_db, space_news_db, and stars_db. The query set included various SQL operations to rigorously test system performance. It covered data retrieval (SELECT), data modification (INSERT, UPDATE, DELETE), and schema manipulation (DROP).

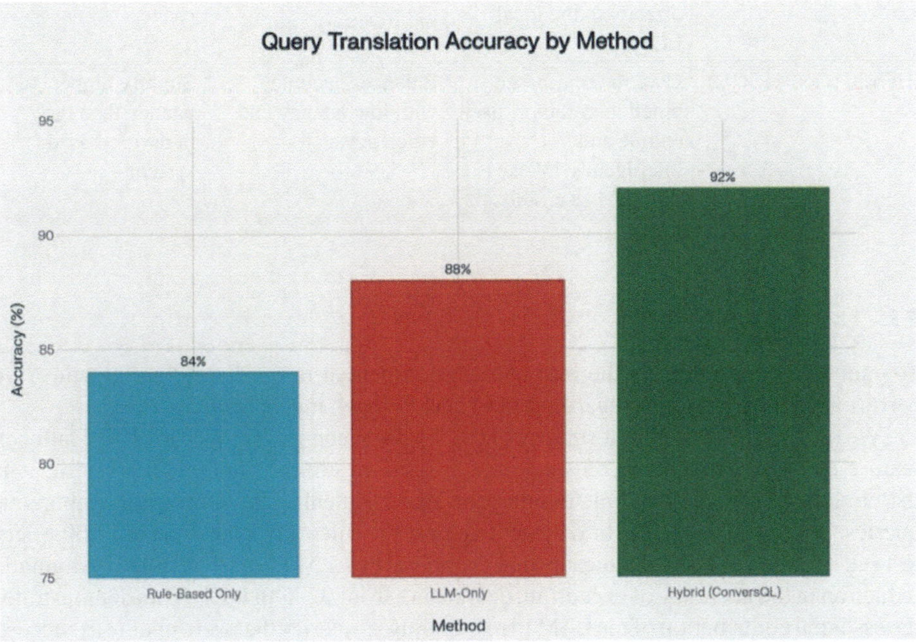

Fig. 3. Comparison of Accuracy with the different methodologies with the same dataset used in ConverSQL

To evaluate our system, we compared three configurations: a purely rule-based pipeline, a purely LLM-based approach, and our proposed hybrid ConverSQL model. As shown in Fig. 3, the rule-based pipeline achieved about 84% execution accuracy and had a low average latency of 250 ms. However, it struggled with edge cases involving nested WHERE clauses and unclear phrasing. By contrast, the LLM-only baseline had a 88% accuracy but a much higher average latency of 1500 ms and more GPU usage. Our hybrid ConverSQL model achieved the best trade-off between them, having 92% accuracy while maintaining an average latency of just 310 ms, almost as fast as the rule-based system. This improved performance indicates that the LLM fallback mechanism successfully recovered most errors when the rule-based parser's confidence dropped below the threshold, as described in our error analysis in Table 2.

Table 2. Comparison of Baseline System Architectures

Baseline System	Architecture Description	Expected Strengths	Expected Weaknesses
Rule-Based Only	Uses only deterministic NLP pipeline for query translation.	Very fast (low latency) and low resource usage.	Fails on ambiguous or complex queries; lower accuracy.
LLM-Only	Bypasses the rule-based pipeline and sends all queries directly to the local LLM.	High accuracy and flexibility in understanding queries.	Very slow (high latency) and high CPU/GPU resource usage.
Hybrid (ConverSQL)	Uses the rule-based pipeline as the primary engine and intelligently invokes the LLM as a fallback.	Balances accuracy with low latency and efficiency.	Slightly higher latency than the pure rule-based system.

5.1 Ablation Study

To gain further insight into the role of each component in our hybrid architecture, we performed ablation studies regarding the LLM fallback mechanism.

We first examined the performance of the system when we removed the LLM fallback feature entirely. This alteration made the system an entirely rule-based system. It is the obvious advantage of implementing the LLM to deal with complex or ambiguous queries. We tested the full hybrid system against the rule-only system on our 500-query test set. The data in Table 3 indicates that dropping the LLM fallback resulted in dramatic reduction in the accuracy of execution, decreasing from **92% to 84%**. This demonstrates the essential contribution of the LLM in responding to queries that were unable to succeed using deterministic parsing. This redesign, though, also resulted in marginally reduced mean latency.

Table 3. Ablation Study - Effect of Removing LLM Fallback

System Configuration	Evaluation Accuracy (%)	Average Latency (ms)
Hybrid (ConverSQL)	92%	310 ms
Hybrid without LLM Fallback	84%	250 ms

We also examined the sensitivity of the system to the confidence score threshold for triggering the LLM fallback. We had initially set the default threshold to 0.7 from preliminary testing. We experimented with different thresholds, e.g., 0.6, 0.8, and 0.9, to observe the balance between accuracy and latency. Reducing the threshold to 0.6 resulted in less LLM call, which lowered the average latency to some extent but compromised accuracy as well. This occurred because more ambiguous questions were incorrectly handled by the rule-based system. Conversely, setting the threshold to 0.8 or 0.9 raised usage of the LLM. This modification provided better accuracy in certain instances but noticeably increased average latency due to more frequent references to the slower LLM module.

The chosen 0.7 threshold provides the optimal balance, optimizing accuracy while maintaining latency near the rule-based baseline. These test studies demonstrate that the LLM fallback is critical for obtaining high accuracy and reliability in the ConverSQL system, and the selected triggering approach appropriately controls performance trade-offs.

5.2 Evaluation Metrics

To assess ConverSQL's performance, we focused on three main areas. We measured Execution Accuracy, which indicates the percentage of generated SQL queries that return the correct results when run against the database, compared to the correct queries. We evaluated Performance by measuring Latency as mentioned in Fig. 4; this is the total response time in milliseconds for each pipeline configuration, including rule-based, hybrid, and LLM-only. Lastly, we monitored Resource Usage, which shows the peak CPU and GPU memory usage during query processing to quantify the computational cost of each method.

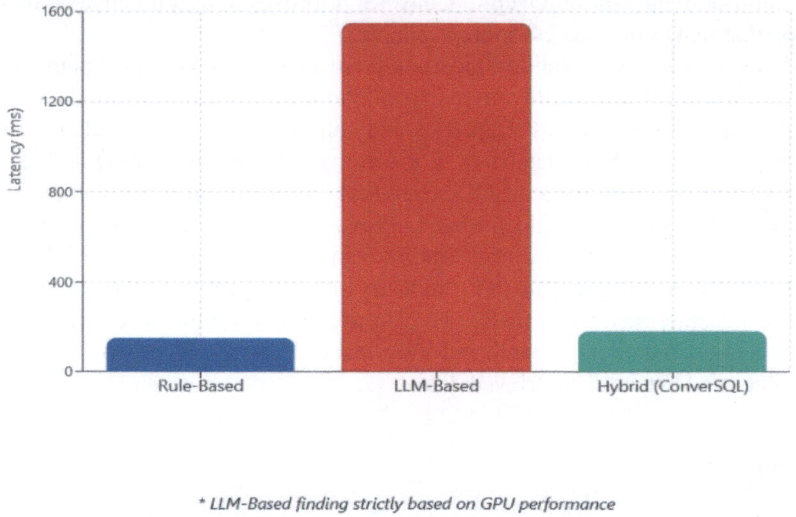

Fig. 4. Comparison of Latency with other methodologies

5.3 Error Analysis

During our testing, we found several different error categories that show the limits of a purely rule-based approach and support the need for our hybrid model. Schema misinterpretation happens when the rule-based NLP pipeline wrongly maps a word to a database column or table. This often occurs because of similar column names or missing synonyms. Operator confusion is another error. It involves misclassifying comparison operators—for example, interpreting "not equal to" as '<' instead of '!='. This leads to queries that are syntactically valid but incorrect. A broader issue comes from ambiguous queries. Vague phrasing or unresolved synonyms can prevent the deterministic system from moving forward. Finally, while the LLM fallback often works, it can bring its own LLM errors. These include generating overly complex queries or creating non-existent tables and columns.

6 Conclusion and Future Work

This work tackles the challenge of making database interaction easy for non-technical users by introducing ConverSQL, a new hybrid Text-to-SQL system. By blending a fast, rule-based NLP pipeline with a smart local LLM fallback, our system finds a good balance of performance, accuracy, and security. Our evaluation shows that the hybrid model achieved 92% execution accuracy, which greatly surpasses the 67% accuracy of a purely rule-based approach. It also avoids the high latency and resource costs of an LLM-only pipeline. The main contributions of this work include designing a modular hybrid pipeline, creating a clear decision framework for the LLM fallback, integrating enterprise-grade security with encryption and RBAC, and evaluating our approach on a large, multi-domain dataset. Overall, ConverSQL proves to be a practical and effective solution that makes data access more available.

While this work lays a solid foundation, there are several promising paths for future research. Future versions could aim to expand the system's ability to handle more complex SQL operations, like nested queries and complex joins, and broaden its backend support to include NoSQL databases. For better validation, integrating and testing the system against standard Text-to-SQL benchmarks would be an important next step. On the performance side, we could further improve LLM inference using techniques like model distillation or hardware acceleration frameworks, such as Open VINO. Additionally, we should implement more industry-standard security methods to boost its readiness for enterprise use. Finally, we plan to improve synonym resolution by using embedding-based similarity search and add advanced data visualization modules to help users better understand the retrieved data.

References

1. Panwar, A.: A survey on query optimization techniques in databases. Int. J. Comput. Appl. **180**(2), 6–9 (2018)
2. Nama, R., Sharma, A.K., Kumar, R.: Role of automated optimization and intelligent query processing in modern databases. In: 2021 International Conference on Computing, Communication, and Intelligent Systems (ICCCIS), pp. 452–457 (2021)
3. Gao, Y., Li, J., Wang, S.: Performance optimization and tuning for MySQL databases. In: Proceedings of the 2019 IEEE 4th International Conference on Big Data Analytics (ICBDA), pp. 210–214 (2019)

4. Androutsopoulos, I., Ritchie, G.D., Thanisch, P.: Natural language interfaces to databases – an introduction. J. Nat. Lang. Eng. **1**(1), 29–81 (1995)
5. Baig, S., Ali, A., Khan, N.A.: A comprehensive review of natural language to SQL techniques. ACM Comput. Surv. **52**(3), 1–38 (2019)
6. T. Yu, et al.: Spider: a large-scale human-labeled dataset for complex and cross-domain semantic parsing and text-to-SQL task, In: Proceedings of the 2018 Conference on Empirical Methods in Natural Language Processing (EMNLP), pp. 3911–3921, (2018)
7. V. Zhong, C. Xiong, and R. Socher, Seq2SQL: Generating Structured Queries from Natural Language using Reinforcement Learning, arXiv preprint (2017). https://arxiv.org/abs/1709.00103
8. Wang, B., Shin, R., Liu, X., Polozov, O., Richardson, M.: RAT-SQL: relation-aware schema encoding and linking for text-to-SQL parsers. In: Proceedings of the 58th Annual Meeting of the Association for Computational Linguistics, pp. 7567–7578 (2020)
9. Scholak, A., Schucher, N., Bahdanau, D.: PICARD: parsing incrementally for constrained auto-regressive decoding from language models. In: Proceedings of the 2021 Conference on Empirical Methods in Natural Language Processing, pp. 8511–8524 (2021)
10. Liu, M., Xu, J.: NLI4DB: A Systematic Review of Natural Language Interfaces for Databases, arXiv preprint (2025). https://arxiv.org/abs/2503.02435
11. Iacob, R., Calin, M., Dinu, L.: Neural approaches for natural language interfaces to databases: a survey. In: Proceedings of the 28th International Conference on Computational Linguistics, pp. 374–389 (2020)
12. Luo, Y., Li, G., Fan, J., Chai, C., Tang, N.: Natural language to SQL: state of the art and open problems. Proc VLDB Endowment. **18**(7), 5466–5478 (2025)
13. J. Li, et al.: Graphix-T5: Mixing Pre-Trained Transformers with Graph-Aware Layers for Text-to-SQL Parsing, arXiv preprint (2023). https://arxiv.org/abs/2301.07507
14. Abbas, S., Al-Bahadili, H., Al-Ayyoub, M.M.: A review of NLIDB with deep learning: findings, challenges and directions. J. Intell. Syst. **31**(4), 540–559 (2022)
15. Gemma Team, Google DeepMind, Gemma: Open Models Based on Gemini Research and Technology, arXiv preprint (2024). Available: https://arxiv.org/abs/2403.08295
16. https://www.kaggle.com/

Sentiment Analysis and Corpus Construction

Study of Emotional Word-Color Association in Tea-Tasting Experiment

Faizah Mappanyompa Rukka(✉), Dian Christy Silpani, and Kaori Yoshida

Kyushu Institute of Technology, Human Intelligence Systems, 2-4 Hibikino, Wakamatsu, Kitakyushu, Fukuoka 808-0196, Japan
faizah-mappanyompa-rukka142@mail.kyutech.jp,
silpani.dian-christy815@mail.kyutech.jp, kaori@brain.kyutech.ac.jp

Abstract. Colors are strongly linked to emotions, yet different types of emotional words, such as labels (e.g., "happy") and laden words (e.g., "birthday") might have different color associations. This paper will discuss and analyze how people associate colors with different word types to identify patterns. A tea-tasting experiment was conducted, where participants tasted different types of tea, sharing their impressions, choosing words, and describing colors. The colors were analyzed using the HSV (Hue, Saturation, Value) model to capture the intensity and consistency of wordcolor associations. A Chi-square test shows a significant relationship between word valence and color selection ($\chi^2 = 148.33$, $df = 40$, $p < 0.0001$), meaning that specific types of emotional words are consistently linked with particular color ranges. Label-positive words showed strong clustering on the color wheel. In contrast, laden-positive words showed broader, more varied distributions. Neutral words had the widest spread. These results highlight meaningful patterns in how people map emotions to colors and suggest applications in affective computing, sensory research, and consumer product design. While this study offers insights, limitations such as a small sample size and cultural or individual biases suggest the need for broader, more flexible future research, such as using a color wheel directly to explore more emotional responses.

Keywords: Emotion-laden · Emotion-label · Color emotion

1 Introduction

Color is a basic element that comes through human perception, human experience, and, most importantly, it can influence emotions. It plays an important role in bringing meaningful aspects to visuals, sensory, sound, facial expressions, and even hand gestures. These factors demonstrate how color and emotion are closely associated [1,2]. To further understand this association, various studies have explored the impact of color in different contexts. A study conducted by Kang [3], for instance, performed two experiments to prove the effect of color on the affective aspect. In this experiment, participants were shown colored facial images to observe the emotional reaction on the face afterwards.

H. Unger and P. Meesad (Eds.): NLPIR 2025, LNNS 1904, pp. 227–241, 2026.
https://doi.org/10.1007/978-3-032-20897-2_16

Colors were also linked with emojis to explore the relationship between the association of color and emotion. Participants were asked to choose the targeted emotion color that matched the emoji. This study proves that color could significantly influence both real and virtual emotions, such as emojis. It also highlights that color plays a vital role in socialization and communication between people [4]. Since communication between humans is often bridged by language, the connection between words and colors becomes an important area to be studied.

At this stage, the concept of color by association gives meaning both linguistic and non-linguistic. According to Saif Mohammad [5], even abstract concepts such as words have colors. Their work involves creating a large dataset of word-color association lexicon using a crowdsourcing method. The goal is to determine the categories of words that can group colors associated with a particular color. Color can enhance the potential of an emotional concept, thereby creating an effective visualization. This intersection between language, emotion, and color opens up further exploration into how words themselves could evoke emotional influence.

Unlike visual elements, a word in language can carry emotion, especially if it has emotional information. Emotion words can generally be categorized into two types: Emotion-label and Emotion-laden words. Emotion-label words are words that can provide emotions directly, for example, the words happy, angry, and sad clearly describe emotions. In contrast to Emotion-laden words, which are words that can provide emotions but indirectly, for instance, "birthday", "murder". In such cases, when humans try a food or drink, they can give their perception of the product as either liking it, disliking it, or being neutral. These words rely on context and personal experiences to trigger emotional responses. It becomes relevant in practical experiences, such as tasting food or drink, people often express their perceptions whether they like it, dislike it, or are indifferent to it, using both types of emotional language, shaped by their perception and affective responses.

Understanding the association of words with colors also has important applications in many fields. In areas such as marketing, advertising, art, and design, by selecting the right combination of colors and words, it can influence the user's perception [2]. As shown by this research [5], the color red combined with the word "important" or "stop" can provide a faster reaction, for example, when the user is making a decision. In contrast, the use of softer colors is mostly for comfort.

This approach creates a framework that can be applied in machine learning implementation as well as a practical way that can highlight the significant link between color perception and words. Another work by Saif [6] analyzed the association of colors and words using a crowdsourcing method. They found strong connections between colors and word emotions, such as anger associated with red, anticipation associated with white, and green. Following these findings, a key question remains: are there specific patterns or colors that could distinguish when humans assign emotional words to taste perception?

To address these issues, in this paper, we conduct a tea experiment where participants are asked to taste five different types of tea, and then give hand gestures to interpret the taste of the tea. Based on the gesture, participants were asked to express it through a selection of words that had been provided, or use other words that they considered appropriate. After that, participants chose the color that they considered most suitable to describe the word they had spoken. The selected colors were then analyzed using the HSV (Hue, Saturation, Value) color model, which allows us to analyze the intensity of each valence word-color association.

In this paper, we create a basic dataset that relates associations between colors and valence words (positive, negative, or neutral) in taste perception. Classification is done using two libraries, VADER and NRCLex, which are essential in analyzing words based on taste perception and categorizing the emotions associated with them.

2 Related Works

2.1 Emotion-Label and Emotion-Laden Words

Linguistic researchers differentiate emotion-label words and emotion-laden words as a distinction. Emotion-label means a word that can show the emotion directly (e.g., angry, happy, and so on). Whereas, emotion-laden means a word that gives the emotion indirectly, depending on the situation and connotation (e.g., birthday, death, and so on).

A study by Betancourt [7] conducted a survey to validate the words label, laden, and neutral using affective features from several multi-component emotion models, namely, action, judgment, expression, feeling, and interoception. They found that label words are generally more abstract and more "general". In contrast, laden words are richer in visual sight and situational context.

Colors are strongly related to emotions, and the purpose of this study is to separate the words "label" and "laden" and analyze how people match these words with colors. Without making this distinction, we may miss important patterns, as each category label, laden and neutral, can correspond to different color associations. Studying and separating these words into color preferences can provide a clearer understanding.

Another study experimented on the brain using the flanker task, where they showed that label and laden words are processed differently [8]. Participants reading label words such as happy or angry, the right side of the brain reacted faster than to laden words that only showed emotions. In short, this experiment confirms that our brains treat these two types of emotion words differently, both in terms of timing and brain regions.

2.2 Sentiment and Emotion Lexicon

To determine whether a word reflects emotion, methods such as sentiment analysis and emotion lexicons are suitable for this task. The most widely used are the NRC Emotion Lexicon (EmoLex) and VADER (Valence Aware Dictionary and sEntiment Reasoner).

The NRC EmoLex provides a large association of a lexicon of words with emotions through a crowdsourcing method. NRC EmoLex associates English words with eight basic emotions. This author annotates words using Plutchik's Wheel, namely, joy, sadness, anger, fear, disgust, surprise, trust, and anticipation, including positive and negative sentiments. Each word is assigned to a Human Intelligence Task (HIT) by providing several questions and clear and structured instructions [9]. However, it's important to note that this lexicon uses static classification, such as whether the word is categorized as "joy" or not. However, it does not provide a level of intensity of sentiment.

In contrast to VADER, this method focuses on rule-based methods to determine how strong the sentiment is. VADER works by giving each word a polarity value (from negative to positive) and a compound score, which is the value of the entire sentence. The difference with previous methods is that VADER does not provide specific labels; it only gives a general value to positive, neutral, and negative, then the total of the compound score [10].

The combination of the two methods, NRC and VADER, allows them to balance each other's respective weaknesses, making them suitable for more precise and accurate analysis, especially in classifying emotion categories and assigning values to sentiment polarity. In this study, we categorize the words selected by participants into three groups: emotion-label, emotion-laden, and neutral. The goal is to prove how strong the emotion is in a group. This approach gave us a better understanding of how different types of emotional words relate to color choices more precisely than using a single lexicon.

2.3 Color-Emotion Representation and HSV Space

Color is not just a visual property; it can also provide emotional signals. Studies on color have been done in various fields, and its impact has proven to be significant, even helping people apply it to various contexts. Shagyrov studied the association of colors with emotions [11] to help marketers and designers better brand their products, and even provide an emotional response to customers. In the study, they used sentiment analysis methods to classify the emotions of customers into the emotional categories of happiness, anger, sadness, fear, and surprise. This analysis can help businesses create strategic decisions and design the right approach before launching and marketing products. For example, some colors' products, such as purple and green, evoke positive sentiments. In contrast, negative colors, like red and brown, tend to elicit negative reactions. However, it should be noted that these color preferences can change over time.

This next association not only focuses on visual design but also specifically on language and computer interaction, as noted by Saif [5]. This research discusses the association of colors with their lexicons using crowdsourcing methods. However, the dataset given is in the form of colors in certain categories, not in a continuous space. It limits our ability to analyze the association of emotions with colors and words. To overcome this, we use the HSV (Hue-Saturation-Value) color model, which allows us to describe colors further. The HSV color model is one of the commonly used models in various types of analysis. In addition, HSV also shows results that are in line with human visual perception, which suits the purpose of our research [12,13].

In this study, we used the HSV color model to convert each color selected by participants into Hue angle values. This approach allows exploration of the relationship between emotional words and their associations with colors. Through this analysis, we were able to examine whether words with the emotion-label, emotion-laden, and neutral tended to cluster in certain color ranges.

3 Methodology

3.1 Proposed Approach

Our research consists of two main stages: collecting data and analyzing it. In the data collection, there were five sub-stages that we conducted. This study began by gathering 55 participants from different countries. Participant data collection has gone through the code of ethics. Prior to this, data collection was conducted and approved by the code of ethics of Kyushu Institute of Technology, with reception number 24–11, "*Hand gesture video data collection experiment for observation of decision making process*". All participants signed informed consent regarding their participation and data use. After that, participants took turns entering the experimental room. Within the room, there is a TV screen to display instructions to the participants and also cameras to capture data.

The first stage of the tea experiment started with the participants tasting the tea that was provided on the table. After drinking the tea, participants were asked to give hand gestures to interpret the taste of the tea, such as thumbs up and thumbs down. These hand gestures are performed voluntarily as a way of expressing liking or disliking the tea. A study conducted by Silpani [16,17] proves that hand gestures have a significant impact on conversations between individuals. In fact, this study also shows that hand gestures that appear simultaneously with words in conversation have a similar intention that strengthens the meaning of those words. In this tea experiment, the relationship between the word chosen and the hand gesture given was analyzed using fuzzy-based sentiment analysis [18]. The results showed that hand gestures are related to and influence the word chosen, thus reflecting the intensity of liking, disliking, or neutrality toward the tea.

However, it should be noted that the main focus of this study was on analyzing the association between words and colors, so the hand gesture data was not analyzed further in this study. Next, based on the given gestures, participants were asked to interpret them into the three words available on the TV screen. If participants had any other words they wanted to say, they were allowed to do so by speaking. At this point, we already have the word and color association data provided by the participants in Comma Separated Values (CSV) format ready for analysis.

The analysis phase starts by adding the data and classifying each word using sentiment lexicons, namely, NRC Emolex, which provides emotion categories (e.g., joy, anger, sadness), and VADER, which provides valence-based sentiment scores (positive, negative, or neutral). Based on this classification, words are grouped into three types of categories, namely emotion-label, emotion-laden, or neutral. Next, contingency tables were created to analyze the relationship between the Word-Emotion group and their colors. Chi-square tests were used to identify whether there was a statistically significant relationship between these two variables. In order to ensure the strength of the chi-square test results, Cramér's V metric was used. To further analyze the existing color patterns, first, the RGB color model was converted into an HSV model, focusing on the Hue component to capture the emotionality of the colors. Finally, hue values were analyzed by word category using circular statistical methods to identify whether different valence groups show different preferences in color hue space. This entire step can be seen in Fig. 1.

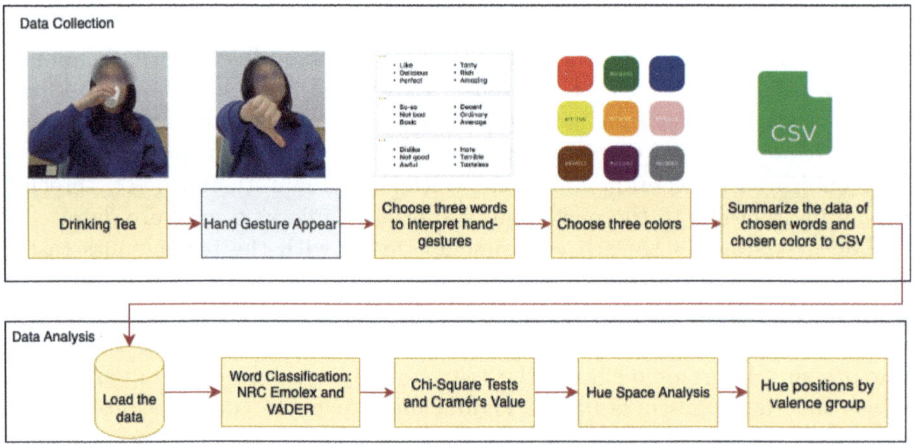

Fig. 1. Overview of the proposed approach

3.2 Word-Emotion Classification

To classify whether a word is emotion-label, emotion-laden, or neutral, we use a lexicon-based approach that uses a pre-defined word list with its sentiment associations. Classification starts by examining words that directly express emotion and are unambiguous in meaning. Suppose the word clearly expresses an emotion such as "anger", 'joy', or "fear", which are referenced in previous research. In that case, it is marked as an emotion-label word. The next step is to use NRC EmoLex to classify word-emotion. NRC EmoLex uses pre-defined English words that are crowdsourced using HIT. Each word-emotion pair in the lexicon is annotated as present (1) or absent (0), based on the scores of all annotators. The key is to measure whether the word has an emotional association, rather than measuring intensity. If the word does not directly provide emotion but is present in EmoLex, it is categorized as emotion-laden.

VADER sentiment analyzer will classify words that are not in NRC EmoLex but give a strong sentiment. For example, the words "not good" or "terrible" are not available in EmoLex. However, VADER can assess them and give them a sentiment score in each category of positive, neutral, negative, and compound scores. If the given compound score is strong, it is emotion-laden, as VADER captures emotions through context, rather than explicitly categorizing them. If there is a word that has a strong emotional polarity, but the word does not directly express an emotional state. The word is in line with the meaning of the word emotion-laden.

When the word does not meet any of the above conditions, does not appear in the emotion-label list or NRC EmoLex, and has a low compound score, it is categorized as neutral. To confirm the word, we use a threshold with a value of more than or less than 0.30 in the VADER method. If the word provides a strong compound value, then it is categorized as emotion-laden. The threshold value is a common practice in lexicon-based, especially VADER itself, which considers values above 0.05 as positive and below −0.05 as negative. This threshold helps distinguish neutral words from words with emotion-laden connotations.

3.3 Statistic Analysis: Chi-Square Test and Cramér's V

To analyze whether there is a statistical association between two variables: word valence groups and colors, we use a contingency table, which aims to look at the frequency of observations of two categories of variables. In this case, the variables are the colors chosen by the participants and the word-valence groups, which are the emotion categories that have been classified in the word-emotion classification stage. This contingency table can provide an overview of how often each color is associated with valence groups. The study conducted by [14,15] used this table to see the association between the participants' chosen color and its characteristics.

The next step in determining whether there is an association between colors and valence groups involves using the chi-square test of independence. This formula is used to see if certain colors are chosen more (or less) often than expected by chance across valence groups. The value is calculated using the following formula. Where O_{ij} is the observed frequency and E_{ij} the expected frequency for column j and row i in the Eq. 1, a significant chi-square result would indicate that color selection is independent of word valence.

$$\chi^2 = \sum_{i=1}^{r}\sum_{j=1}^{c}\frac{(O_{ij} - E_{ij})^2}{E_{ij}} \tag{1}$$

Equation 1: Chi-square tests of independence

However, it should be noted that if the p-value is close to zero, then it is possible that the value is not certainly fully associated with the variable. To confirm the p-value, Cramér's V statistical metric is performed to see the effect size of the chi-square test results. Cramér's has a range of values between 0 and 1, indicating the strength of the association between the variables. The formula for Cramér's is defined as in Eq. 2. Where X^2 is the chi-square test result, n observations total, and k total columns and rows.

$$V = \sqrt{\frac{\chi^2}{n \cdot (k-1)}} \tag{2}$$

Equation 2: Cramér's V Equation

3.4 RGB to HSV

We start this step by defining 9 main colors. These colors can be seen in Table 2. It starts with RGB (Red, Green, Blue) values, so we need to convert them into HSV space format for further analysis. The hexadecimal to RGB conversion uses simple base-16 arithmetic, where pairs of digits from hex, such as "FF", are interpreted as decimal as "255". This transformation aims to standardize the colors.

Fig. 2. Overview of the colors used in the experiment

First, we calculate the circular mean of Hue to see the "direction" of the color around the color wheel. After that, we calculate the standard deviation of the circular to calculate how tightly clustered or broad the hue values are within the word valence groups. The purpose of this study was to examine whether participants consistently chose colors within a narrow perceptual range or spread out across the spectrum.

We also measured how often participants chose colors within ±30° of a particular hue anchor (e.g., Yellow at 60°). For example, if 75% of the colors selected for positive words were within a ±30° window, whereas only 10% of the negative words selected that color, this shows that positive words are more strongly associated with yellow hue. This calculation is to ensure the analysis provides participant-level color selection behavior.

4 Results

4.1 Classification Result

At this phase, we will explain the results of word classification. Each participant was asked to select three words and colors associated with them. This data was collected from all participants in the experiment, resulting in a total of 2,448 word and color pairs. As a sample, Table 1 shows 6 words after classification. This classification integrates two methods, NRC EmoLex and VADER. Each row represents the method and rule used to obtain the word type and valence values. The method column indicates what rule was used to classify the word. For example, if the word is contained in labels such as like, dislike, hate, then the method used is a predefined list. Likewise, with the NRC, VADER, and no rules expected methods, if the word belongs to its part, it is classified as such.

Table 1. Emotion word classification with methods, NRC tags, and VADER scores

Words	Method	NRC	VADER Score	Word Type	Valence
like	Label Seeds	{}	0.3612	emotion-label	emotion-label positive
perfect	NRC EmoLex	{'anticipation': 1, 'joy': 1, 'positive': 1}	0.5719	emotion-laden	emotion-laden positive
hate	Label Seeds	{'anger': 1, 'disgust': 1, 'fear': 1, 'negative': 1}	−0.5719	emotion-label	emotion-label negative
ordinary	no rule expected	{}	0.0000	neutral	neutral
amazing	VADER	{}	0.5859	emotion-laden	emotion-laden positive

The word "perfect" belongs to a category listed in the NRC lexicon, so the NRC column shows the emotion values in the word, such as anticipation, joy, positive, and trust, each equal to 1. If the word is not found, the NRC column is empty. VADER's compound score is a score between −1 and +1, where −1 is strongly negative and +1 is strongly positive. As seen in the table, a positive value indicates a positive sentiment (e.g., amazing: 0.5859), a negative value indicates a negative sentiment (e.g., hate: −0.5719), and a value of 0.0 indicates a neutral sentiment (e.g., so-so).

A valence column combines the polarity of the sentiment as positive, neutral, and negative. Each value in valence is categorized as label-positive, label-negative, laden-positive, laden-neutral, laden-negative, and neutral. The values in this column are derived from the analysis results through the VADER method. For instance, the word 'amazing' in VADER has a value of 1 in the positive category, and 0 in both the neutral and negative categories, with a compound value above 0.3. Based on these criteria, the word is categorized as having positive valence. The same works as when the values are neutral and negative.

4.2 Statistical Analysis: ColourValence Associations

Chi-square tests of independence were conducted using the contingency table of color frequencies across valence groups. The contingency table can be seen in Table 2 to show how often the color is selected across a range of valence categories: label-negative, label-positive, laden, laden-negative, and neutral. For example, yellow is selected more often in the laden-positive (128 times) and neutral (148 times) categories. At the same time, colors like blue and grey are more evenly distributed, especially in the negative valence.

Table 2. Color Frequency Across Valence Groups

Color	label-neg	label-pos	laden	laden-neg	laden-pos	neutral
Blue	7	16	31	9	53	39
Brown	18	31	65	33	101	126
Green	11	39	69	23	120	82
Grey	26	8	42	38	60	103
Orange	12	40	62	23	96	92
Pink	3	30	58	6	91	53
Purple	10	7	21	22	34	45
Red	13	13	27	20	38	26
Yellow	17	49	90	24	128	148

The finding of the Chi-square test shows the null hypothesis remains valid by chi-square where a value of ($x^2 = 148.33$, dof $= 40$, and a p-value <0.0001). Suppose the p-value is less than or equal to 0.05. In that case, it means that there is a significant relationship between two variables, in this case, color selection and valence of words. Which means that the distribution of colors is not even across valence categories; certain colors are more likely to be selected for certain types of emotions.

Therefore, chi-square test was performed to determine the strength of association between variable colors dan valence groups. Cramer's V value = 0.11, indicating a weak association between color and valence groups.

Next, we calculate the standardized residuals from the contingency table. These residuals measure the difference between observed and expected frequencies, measured through standard deviation. Suppose the resulting value is a positive residual. In that case, it means that the color association and valence group occur more frequently than expected, while a negative residual means that they occur less frequently.

The given heatmap visualization in Fig. 3 shows some patterns. Grey and red indicate a strong association with words with negative valence. More specifically, in the label-neg group, participants chose grey (3.51) more often than expected, followed by red (2.52). Similarly, for the laden-neg group, participants tended to select grey (3.29) followed by purple (3.21), and red (2.68). In contrast, for positive valence, the opposite was the case, with label-pos (−3.58) and laden-pos (−2.39) both selecting grey color.

This evidence contrasts significantly with the color pink, which is associated with positive valence group. Specifically in the laden-pos with pink (+2.38) and green (+1.86) colors.

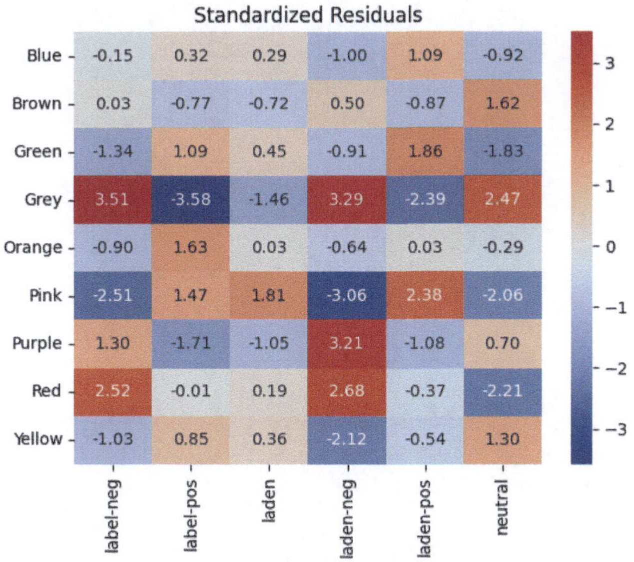

Fig. 3. Standarized Residuals of Colors and Valence Groups

Meanwhile, purple indicates a high average in the laden-neg category (+3.21) but also appears in the label-pos category (−1.71), thus indicating a complex interpretation. Other colors such as brown, green, and orange are more balanced in their residual values, indicating that associations in this valence category are varied.

Based on the above interpretation, we can conclude that color has an effect on word valence in certain combinations, even though the overall association is weak (Cramer's V 0.11). Specific color pairs can show strong interpretations and patterns, such as gray, red, and pink.

Overall, contingency analysis and standardized residuals provide strong evidence that color choice is strongly influenced by the emotional content of words, which supports the idea that words are emotional, especially if classified by label type and laden.

4.3 Hue Space Analysis

During this stage, we perform a visualization of the hue positions, whose aim is to match the hue angle colors on the HSV color wheel. This Fig. 4 illustrates the range of colors and demonstrates that groups of patterns consistently map to regions with specific valence groups, such as yellow, green, and so on.

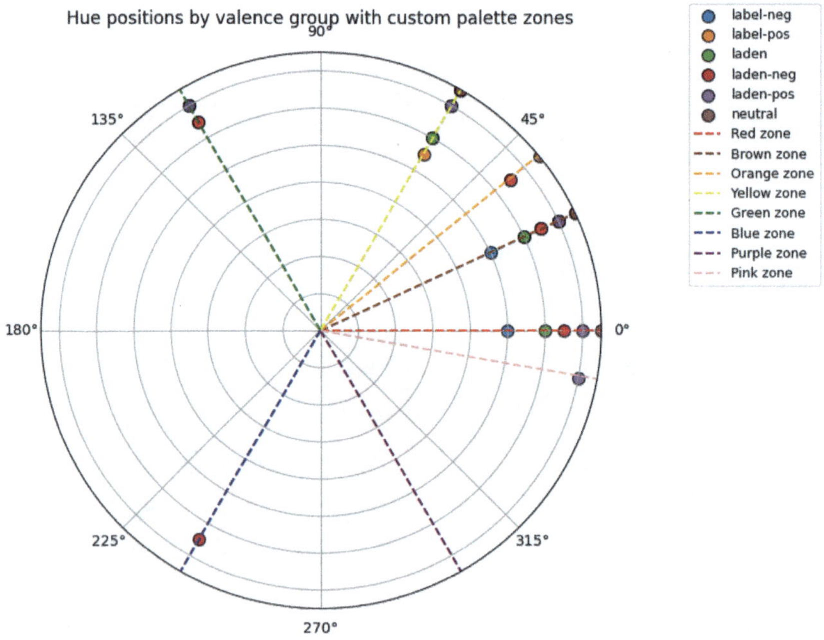

Fig. 4. Hue Positions by Valence Groups with Color Zones

This plot uses markers such as red at 0°, blue at 240°, and yellow at 60°, making it easier to analyze and interpret how close and how similar the hues are to the color category. Each point represents the hue angle from 0°–360° in colors that are often selected and grouped according to valence groups. This hue value is obtained from the RGB value, which is converted into the HSV range.

Although the angular position of each point reflects the actual Hue of the color, the radial distance is slightly offset by the valence group to reduce overlap; yet this has no statistical significance. A colored dotted line represents the hue position with 8 color areas and based on the angle: Red (0°), Brown (25°), Orange (39°), Yellow (60°), Green (120°), Blue (240°), Purple (300°), and Pink (350°).

Based on Fig. 4, we can interpret each valence group as follows. First, label-positive words cluster around the yellow color association, which means a strong consistency in positivity. Label-negative words appear mainly near the red and brown areas, which reflect the association with negative colors. The laden words are spread across the colors orange, red, brown, and yellow, indicating that the indirect emotion words carry diverse associations while remaining on neutral colors. Laden-negative words show a wider spread, with some grouping the brown and orange colors. Laden-positive words are relatively spread across regions; however, the yellow color is numerous. Neutral words are mostly near orange and yellow, colors that are emotionally light and ambiguous.

Overall, how tight or broad each group is on the color wheel reflects how consistently participants match word valence to specific color hues. Groups like label-positive show strong clustering (clear color-emotion alignment), while groups like laden-positive or laden-negative display a wider angular distribution, indicating more subjective or varied interpretations.

5 Conclusion

This research discusses word emotions and their associations with colors in a tea-tasting scenario. The purpose of this research is to explore whether there are specific patterns in colors that can distinguish words containing emotional valence. By analyzing the words given by participants and the color associations chosen, we found that there are patterns as well as emotions in words that have been grouped into label-word positive, label-word negative, laden-word positive, laden-word negative, laden, and neutral. The results show that how tight or wide each valence group is on the color wheel reflects how consistently participants match emotion word types with specific color patterns. The valence group consistently had associations with a certain range of colors. At the same time, the neutral group had a wider variance.

The analysis resulted in 2,448 word pairs with colors. This process started with classifying words as either emotion-labeled or emotion-laden using NRC EmoLex to determine the emotion category and VADER to see the sentiment polarity. This classification resulted in several groups: label-pos, label-neg, laden-pos, laden-neg, laden, and neutral. For example, the word "perfect" falls under the emotion category in NRC EmoLex (anticipation, joy, positive, trust) and also has strong sentiment polarity in VADER (compound score = 0.5719), thus classified as laden-pos.

Furthermore, the chi-square test ($X^2 = 148.33$, $df = 40$, $p < 0.0001$) proved that there was an association between word emotions and color selection. Cramer's V = 0.11 indicates that this association is weak. However, some specific pairs show a more distinct pattern, as indicated by the results in the standardized residuals. To identify the color distribution in the valence group, the color pairs were mapped to the HSV color wheel (red = 0°, yellow = 60°, blue = 240°). The mapping results show that positive labels cluster around the yellow color, indicating strong consistency. In contrast, negative labels cluster around the red color. The laden group is more evenly distributed around orange, red, brown, and yellow, indicating subjective interpretation. Meanwhile, neutral words cluster around orange and yellow.

This study provides insight into the associations between emotions, words, and colors. Yet, there are some limitations, such as the limited number of participants and the controlled context of the experiment. Besides, the choice of words and colors is also influenced by cultural factors and individual preferences, which may limit participants' expressions. It is an important factor if the research is to be extended to higher levels of complexity, such as decision-making.

However, in spite of these limitations, the results provide useful possibilities for future study. In affective computing field, future works could adapt interface by using word-color mappings in systems that could identify, for example the positive label word "like" clusters tightly with yellow which requires a flexible adaptation.

Our result suggest that emotion-label positive words consistently narrow in yellow, pink, shows a brighter colors, this could be implemented in branding strategies and advertisement where to consider carefully chosen words and colors that influence consumer perception. Wider spread of laden and neutral words suggest that color choices in marketing tied to context-rich and ambiguous terms. Emotion word and color association in sensory settings could be extend to theories emotion-color mapping into real-world perception context.

References

1. Liu, Y.: The colour-emotion association. J. Educ. Humanit. Soc. Sci. **5**, 272–277 (2022). https://doi.org/10.54097/ehss.v5i.2912
2. Muratbekova, M., Shamoi, P.: Color-emotion associations in art: fuzzy approach. IEEE Access **12**, 37937–37956 (2024). https://doi.org/10.1109/access.2024.3375361
3. Kang, J., Park, Y.E., Yoon, H.K.: Feeling blue and getting red: an exploratory study on the effect of color in the processing of emotion information. Front. Psychol. **13**, 515215 (2022). https://doi.org/10.3389/fpsyg.2022.515215
4. Liao, S., Sakata, K., Paramei, G.: Color affects recognition of emoticon expressions. i-Perception **13**, 204166952210807 (2022). https://doi.org/10.1177/20416695221080778
5. Mohammad, S.: Even the abstract have color: consensus in word-colour associations. In: Proceedings, pp. 368–373 (2011)
6. Mohammad, S.: Colourful language: measuring word-colour associations. arXiv preprint arXiv:1309.5942 (2013)

7. Betancourt, Á.-A., Guasch, M., Ferré, P.: What distinguishes emotion-label words from emotion-laden words? The characterization of affective meaning from a multi-componential conception of emotions. Front. Psychol. **15**, 1308421 (2024). https://doi.org/10.3389/fpsyg.2024.1308421
8. Zhang, J., Wu, C., Yuan, Z., Meng, Y.: Differentiating emotion-label words and emotion-laden words in emotion conflict: an ERP study. Exp. Brain Res. **237** (2019). https://doi.org/10.1007/s00221-019-05600-4
9. Mohammad, S., Turney, P.: Crowdsourcing a word-emotion association lexicon. Comput. Intell. **29** (2013). https://doi.org/10.1111/j.1467-8640.2012.00460.x
10. Hutto, C.J., Gilbert, E.: VADER: a parsimonious rule-based model for sentiment analysis of social media text. In: Proc. 8th Int. Conf. on Weblogs and Social Media (ICWSM), 2014 (2015)
11. Shagyrov, M., Shamoi, P.: Color and sentiment: a study of emotion-based color palettes in marketing. arXiv preprint arXiv:2407.16064 (2024). https://doi.org/10.48550/arXiv.2407.16064
12. Burambekova, A., Shamoi, P.: Comparative analysis of color models for human perception and visual color difference. arXiv preprint arXiv:2406.19520 (2024)
13. Pardede, J., Husada, M., Hermana, A., Rumapea, S.: Fruit ripeness based on RGB, HSV, HSL, L*a*b* color feature using SVM. In: Proceedings, pp. 1–5 (2019). https://doi.org/10.1109/ICoSNIKOM48755.2019.9111486
14. Adjah, J., Bekoe, S., Decardi-Nelson, A., Ry-Kottoh, L., Kalognia, J.: Color preferences among selected adults in Ghana. J. Graph. Eng. Des. **15**, 15–22 (2024). https://doi.org/10.24867/JGED-2024-1-015
15. Akarsu, S., Pirinçci, E.: Differences in color and number preferences of Turkish children in varying stages of illness. Arch. Palliat. Care **6**(1), 1019 (2024)
16. Silpani, D.C., Suematsu, K., Yoshida, K.: A feasibility study on hand gesture intention interpretation based on gesture detection and speech recognition. J. Adv. Comput. Intell. Intell. Inform. **26**(3), 375–381 (2022)
17. Silpani, D.C., Suematsu, K., Yoshida, K.: A feasibility study on hand gesture recognition in natural conversation. In: Proc. 2021 5th IEEE Int. Conf. on Cybernetics (CYBCONF), Sendai, Japan, pp. 85–90 (2021). https://doi.org/10.1109/CYBCONF51991.2021.9464141
18. Rukka, F.M., Silpani, D.C., Yoshida, K.: Fuzzy sentiment analysis for hand gesture interpretation in words expression. In: Proc. 2025 Int. Conf. on Biometrics and Kansei Engineering (ICBAKE), Fukuoka, Japan (2025)

Corpus Generation for Emotion Classification in Psychotherapy Using Large Language Models

Thierry Desot(✉), Babak Basharirad, Niels Netten, and Ricardo Muijs

Rotterdam University of Applied Sciences, Knowledge Centre Creating 010, Museumpark 40, Rotterdam 3015, CX, The Netherlands
{t.v.j.desot,b.basharirad,c.p.m.netten,r.muijs}@hr.nl

Abstract. This proof-of-concept study explores semi-automatic corpus generation for basic emotion classification in psychotherapy dialogues as part of an exploratory, short-term, low-budget applied research project supporting an AI-driven virtual patient platform for training psychology students and therapists. Using the DAIC-WOZ dataset, a collection of transcribed clinical interviews created for depression detection but lacking basic emotion labels, we combined zero-shot classification, selective human verification, and fine-tuning of large language models to efficiently produce a small emotion-labeled dataset. Candidate labels for six basic emotions were generated automatically, with low-confidence predictions manually validated to optimize limited annotation effort. Fine-tuning improved performance over the zero-shot baseline, while class-weighted loss further enhanced recognition of minority emotions while preserving performance on majority categories. Beyond creating an initial labeled corpus, the approach provides a practical foundation for prioritizing and guiding further manual annotation in a resource-efficient way.

Keywords: corpus generation · emotion classification · zero-shot classification · transfer learning · large language models

1 Introduction

Recent advances in Artificial Intelligence (AI) have spurred interest among mental health practitioners in using AI tools to improve psychological care access and reduce strain on clinical and educational services [35,36]. Intelligent systems are increasingly seen as promising contributors to diagnostic and therapeutic processes, particularly in training and early intervention [38]. In psychology education, preparing students for the interpersonal dynamics of clinical practice remains a persistent challenge [22]. Traditional training—lectures, case studies, and peer role-play—often fail to expose students to the diversity and emotional

B. Basharirad, N. Netten and R. Muijs—These authors contributed equally.

H. Unger and P. Meesad (Eds.): NLPIR 2025, LNNS 1904, pp. 242–254, 2026.
https://doi.org/10.1007/978-3-032-20897-2_17

complexity of real therapy sessions [21,37], making skill transfer to practice difficult. Experiential and apprenticeship-based methods provide more authentic learning experiences [43]. To address these gaps, we are developing *TherAvatars*, an AI-driven virtual patient platform to enhance psychology education through interactive dialogue and emotion-sensitive facial and body responses. The system targets common conditions such as depression and anxiety disorders [4] and is being created through a research collaboration between academic partners and a company specializing in digital mental health solutions. Unlike static training tools, the system combines large language models (LLMs) for dynamic dialogue with emotion classification to deliver scalable, low-cost practice and automated feedback. Within this broader framework, the present proof-of-concept applied research study focuses specifically on semi-automatic corpus generation using *basic* emotion classification, the task of classifying emotional tone in text [40,46]. However, traditional methods yield broad emotion categories, whereas more advanced techniques are needed to capture the rich emotional dynamics of therapy [17,55], which drives more realistic, emotionally appropriate verbal and non-verbal behaviors in virtual patients. Recent transformer-based approaches identify subtle psychological states and contextual cues earlier techniques could not capture [28,34,41,42].

In this study, our train and test dataset is derived from the DAIC-WOZ[1] dataset, which consists of transcribed clinical interviews with a virtual interviewer and does not include predefined basic emotion labels. It reflects the real-world distribution of emotional content in clinical conversations, where neutral utterances vastly outnumber emotionally intense expressions. While this class imbalance poses challenges for machine learning models [31,54], particularly in accurately predicting minority emotions such as *disgust* or *fear*, we deliberately maintain the natural distribution rather than applying artificial class rebalancing. This decision aligns with recent findings suggesting that rebalancing can distort task realism and model generalizability [12]. Instead, our goal is to ensure that models trained under these realistic conditions can still achieve reliable performance across all emotion categories, including the rare ones. Transformers, introduced by Vaswani et al. [51], revolutionized natural language processing by enabling multi-head self-attention to weigh token relevance in context. Models such as BERT, RoBERTa, and GPT have since become foundational for emotion classification, especially in low-resource scenarios where annotated data is scarce. To the best of our knowledge, this is the first work using DAIC-WOZ to combine zero-shot labeling, manual verification, and supervised fine-tuning in a *semi-automatic corpus generation* pipeline, with the central question whether a *small labeled dataset* can be produced through *utterance-level emotion classification* to support further annotation prioritization under constrained settings.

This article is organized as follows: Sect. 2 analyses recent related works, establishing the theoretical and empirical foundation for our research. Section 3 presents our methods, including the underlying principles and implementation. Section 4 presents the data collection, annotation process and characteristics of

[1] https://dcapswoz.ict.usc.edu/.

the dataset used in our study. Section 5 outlines our experimental setup and presents the results of our evaluation. Section 6 provides a discussion of these results, exploring their implications and significance. Section 7 (Conclusion and Future Work) summarizes our contributions and findings and outlines potential directions to extend and refine this approach.

2 Related Work

Recent advancements in transfer learning have significantly influenced the field of emotion classification, particularly through the use of pretrained transformer models such as BERT, RoBERTa, and DistilBERT [2,32,40]. These models are adapted to downstream tasks through transfer learning, where pre-trained parameters provide a foundation. During fine-tuning, higher-level layers are updated with labeled target-domain data, while lower layers frequently remain frozen to retain general language features. This approach leverages broad language understanding while adapting to domain-specific nuances such as psychological dialogue or clinical interviews [44]. While fine-tuning improves performance, it is challenged by scarce annotations, high labeling costs, and overfitting risks in sensitive or low-resource settings. Zero-shot classification uses models without task-specific training, enabling scalable label inference but typically with lower accuracy [10]. The TheraVatars project explores both zero-shot and fine-tuning to balance adaptability and efficiency in modeling emotion classification in therapeutic dialogue.

A recent study by Chae et al. [10] compared ten large language models across four training regimes: prompt-based zero-shot learning, few-shot learning, fine-tuning, and instruction-tuning. Their results showed that large decoder-only models excel with minimal supervision, while smaller encoder models remain competitive when fine-tuned, offering a trade-off between accuracy and computational cost. Several studies have focused specifically on emotion classification in clinical dialogue. Burdisso et al. [8] examined the DAIC-WOZ dataset and demonstrated that including therapist prompts can introduce biases, as models learn to exploit question patterns rather than analyze patient language comprehensively. To address data scarcity in clinical interviews, Chen et al. [13] proposed a method combining multiple modalities (text, audio, video) with large language models for data augmentation and graph contrastive learning. Their approach outperformed standard baselines across English and Chinese clinical data, demonstrating potential for scalable automated depression assessment. Complementary to this, Feng et al. [20] assessed LLMs across diverse dialogue contexts—casual conversations, task-oriented dialogues, and depression consultations—using datasets such as IEMOCAP [9], EMOWOZ [19], and DAIC-WOZ [23]. While zero-shot emotion classification still lagged behind supervised methods, integrating emotion definitions and fine-tuning with small samples substantially improved model performance. Their results also highlighted LLMs' robustness to transcription errors, though recognizing nuanced states like depression remains challenging.

Beyond clinical applications, zero-shot emotion classification has been explored for low-resource and historical languages. Kumar et al. [29] introduced a cross-lingual pipeline for Sanskrit emotion classification using transformer-based and BiLSTM classification, demonstrating the feasibility of leveraging English resources for underrepresented languages. Similarly, Borst et al. [6] applied zero-shot classification to historical German stock market reports, finding strong correlation with fine-tuned models for emotion time series, suggesting zero-shot approaches can serve as valuable exploratory tools when annotated data are scarce. Prior to large language models, a multi-domain Thai corpus was created using traditional tokenization, normalization, and frequency analysis [27]. Building on these insights, we propose a semi-automatic corpus generation pipeline that combines zero-shot labeling, human-in-the-loop validation, and transformer fine-tuning.

3 Method: Zero-Shot and Fine-Tuned Emotion Classification

In this project, we explored zero-shot *emotion classification* in psychotherapy dialogues as a baseline approach [14], while at the same time using it as the first step of a semi-automatic corpus generation pipeline. Zero-shot classification leverages pretrained language models and natural language inference (NLI) to assign labels unseen during training [48,52]. We employed the English DistilBERT[2] model (`distilbert-base-uncased`), which retains about 97% of BERT's[3] language understanding while improving efficiency and speed [45]. BERT, introduced by Devlin et al. [16], uses a deep bidirectional architecture and is pretrained on masked language modeling and next sentence prediction. Pretrained with masked language modeling on large English corpora, it serves as a general-purpose encoder. In our setup, DistilBERT classified DAIC-WOZ utterances into six emotions —anger, disgust, fear, happiness, sadness, and neutral— using zero-shot inference without fine-tuning. We then performed supervised fine-tuning with DistilBERT and BERT (`bert-base-uncased`) as the second step of the corpus generation approach.

4 Data

This exploratory project utilizes the DAIC-WOZ (The Distress Analysis Interview Corpus - Wizard of Oz) corpus, a publicly available multimodal dataset of clinical interviews designed to facilitate the analysis of psychological distress in human-computer interactions [15,23]. The dataset comprises 189 English-language sessions conducted by a virtual animated interviewer (Ellie) via a

[2] https://huggingface.co/distilbert-base-uncased.
[3] https://huggingface.co/bert-base-uncased.

Wizard-of-Oz protocol. With synchronized audio, video, and transcripts, DAIC-WOZ provides a high-fidelity proxy for real clinical encounters, making it valuable for training and evaluating LLMs in emotionally nuanced psychotherapy contexts. While prior research has primarily used DAIC-WOZ to benchmark LLMs on depression detection tasks, such as PHQ-8 score regression (a depression assessment instrument), binary classification, or multimodal fusion [1,3,7,11,30,39], these approaches typically rely on session-level features or aggregate scoring. Our study focuses on utterance-level classification with basic emotions, as this granularity captures moment-to-moment emotion fluctuations often obscured when aggregating over longer dialogue segments. This makes DAIC-WOZ well-suited for enriching clinical conversations with emotion-labeled annotations to train models that predict emotions for driving digital human facial expressions and body language. We first consolidated all 189 transcribed dialogue files from the DAIC-WOZ dataset into a single corpus for emotion classification. To better approximate natural speech patterns, consecutive utterances from the same speaker were merged if their temporal separation did not exceed 0.5 s. This preprocessing strategy facilitated analysis of both isolated statements and contextually cohesive speech segments. Next, the data underwent a cleaning phase. Certain system-level commands embedded in Ellie's utterances, were removed, as these scripted control signals represent non-linguistic instructions.

For annotation six basic emotions were used, consistent with established frameworks [18]; neutral, sadness were major, happiness and anger were intermediate, and fear and disgust were minor (Table 1). A domain-expert master's-level psychology student with clinical training manually annotated a subset of the corpus, following concise labeling guidelines based on Ekman's basic emotion framework [18], supplemented with canonical definitions and representative examples from our psychotherapy transcripts. Annotation was conducted in a dedicated interface allowing side-by-side transcript and context view, where each utterance was accompanied by its automatically predicted emotion label from the earlier zero-shot classification and the associated confidence score. Utterances were selected for verification using a confidence-based prioritization strategy, focusing on low-confidence predictions so the most ambiguous cases were addressed first. Periodic spot checks by a senior researcher ensured basic quality control, with ambiguous cases discussed to refine labeling decisions. Fully annotating the corpus would be impractical for this short-term project, so targeted verification directed effort to points of highest model uncertainty, which can yield substantial clinical impact when ambiguities involve subtle emotional cues. Empirical analyses of psychotherapy dialogues indicate a strong concentration of neutral emotion [47] or identify it as the most frequent emotion [49]. By preserving class distribution and using domain expertise for context cues, this domain alignment makes the semi-automatic annotation procedure effective.

A total of 1,257 utterances were verified for this experiment (Table 1), forming the dataset used for fine-tuning. The dataset was split into training (80%), validation (10%), and test (10%) subsets using stratified sampling to preserve the relative proportion of each emotion class. Stratified sampling is essential in multi-class emotion classification, where dominant labels such as neutral can skew performance metrics, and ensures that the splits remain representative of the overall dataset [50]. No balancing techniques were applied at this stage due to the limited number of verified labels and the intention to retain the natural distribution of emotional expressions observed in DAIC-WOZ. As expected, neutral emerged as the most frequent class, reflecting real clinical dialogues. In many applications, such class asymmetries are common, and prior research shows models can perform effectively despite limited representation of minority categories [5,24,25,53]. In our case, the ability to capture minority emotions will be further assessed during fine-tuning.

Table 1. Label counts and percentage distribution across train, validation, and test sets.

Label	Total	Train		Validation		Test	
		Count	%	Count	%	Count	%
Anger	105	84	80.0	10	9.5	11	10.5
Disgust	35	28	80.0	4	11.4	3	8.6
Fear	39	31	79.5	4	10.3	4	10.3
Happiness	81	65	80.2	8	9.9	8	9.9
Neutral	874	698	79.9	88	10.1	88	10.1
Sadness	123	99	80.5	12	9.8	12	9.8
Total	**1257**	**1005**		**126**		**126**	

5 Experiments and Results

Our experiments comprised three stages—zero-shot classification, supervised fine-tuning, and class-weighted fine-tuning—to compare performance.

5.1 Zero-Shot Classification and Fine-Tuning

As an initial step, we applied zero-shot classification using the English DistilBERT model [45]. Transcribed participant and therapist utterances from the DAIC-WOZ dataset were concatenated per session in temporal order. No stemming or stopword removal was applied, as transformer models rely on pretrained subword tokenization (WordPiece) to capture morphological variations and handle out-of-vocabulary terms. Utterance classification was performed across six emotion labels, with a confidence score given by the softmax probability. Neutral utterances (the majority class) often received lower confidence, indicating greater model uncertainty. The resulting labeled data provided a foundation for subsequent human validation and fine-tuning. Performance metrics are reported in Table 2. As shown in Fig. 1, the model notably underpredicted the dominant "neutral" class on the test set, reflecting a key limitation of this approach. Across all data, 55% of zero-shot labels were manually corrected, mostly for *disgust* (87%), *fear* (81%), and *anger* (79%), which were frequently reassigned to neutral, indicating an over-prediction of strong negative emotions.

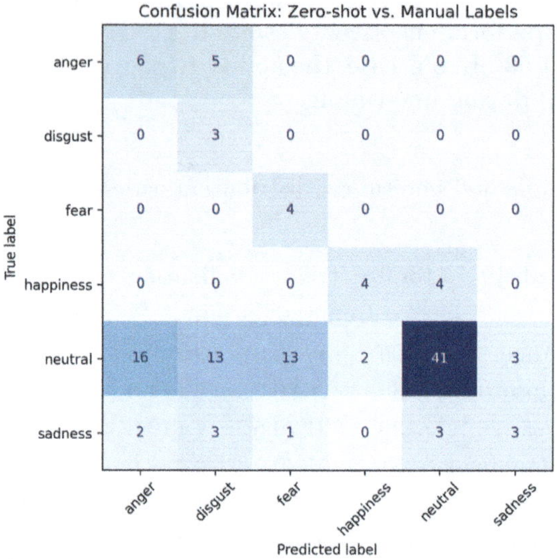

Fig. 1. Confusion matrix of zero-shot vs. manual labels (DAIC-WOZ test set).

To assess the impact of supervised learning, we fine-tuned `bert-base-uncased` and `distilbert-base-uncased` on a stratified 80/10/10 split of manually labeled DAIC-WOZ utterances (1,005 training, 126 validation, 126 test; see Table 1). Stratification preserved label proportions across splits. For both models, the final classification layer was replaced with a softmax layer over six emotion categories. Training was performed for 10 epochs using cross-entropy loss, the AdamW optimizer [33] with a learning rate of 2e-5, gradient clipping (max norm 1.0), and a linear scheduler with warm-up. Utterances were tokenized with `BertTokenizer` (max length 128), batched in size 16, and processed with PyTorch on an NVIDIA GPU. Model performance was evaluated on the test set using scikit-learn metrics (precision, recall, F1). While BERT served as a strong baseline due to its deeper bidirectional architecture, DistilBERT offered improved efficiency and reduced model size with comparable accuracy. Detailed results are shown in Table 2.

5.2 Weighted Loss Fine-Tuning

While standard fine-tuning with cross-entropy loss improved performance over the zero-shot baseline (Table 2) for the dominant *neutral* class, performances for

Table 2. Test set performance (n = 126) of emotion classification using zero-shot DistilBERT, fine-tuned DistilBERT, and fine-tuned BERT.

Emotion #	DistilBERT zero-shot			DistilBERT fine-tuned			BERT fine-tuned		
	Prec.	Rec.	F1	Prec.	Rec.	F1	Prec.	Rec.	F1
Anger (11)	0.25	0.55	0.34	0.33	0.36	0.35	0.42	0.45	0.43
Disgust (3)	0.12	1	0.22	0.00	0.00	0.00	0.00	0.00	0.00
Fear (4)	0.22	1	0.36	0.00	0.00	0.00	0.00	0.00	0.00
Happiness (8)	0.67	0.50	0.57	0.00	0.00	0.00	0.71	0.62	0.67
Neutral (88)	0.85	0.47	0.60	0.77	0.90	0.83	0.81	0.90	0.85
Sadness (12)	0.50	0.25	0.33	0.42	0.42	0.42	0.30	0.25	0.27
Accuracy	0.48			0.70			0.73		
Macro Avg	0.44	0.63	0.41	0.25	0.28	0.27	0.37	0.37	0.37
Weighted Avg	0.72	0.48	0.54	0.61	0.70	0.65	0.68	0.73	**0.70**

sadness and underrepresented emotions *fear* and *disgust* decreased. This tendency to predict the most frequent class reduces recognition of less common but clinically relevant states. To mitigate this imbalance, we applied a class-weighted loss function during training, assigning a higher penalty to errors on rare classes while preserving the influence of the majority class and avoiding artificial oversampling or undersampling [26]. Formally, for each training example i with ground-truth label y_i and predicted probabilities $\hat{p}_i$, the weighted cross-entropy loss was computed as:

$$\mathcal{L} = -\sum_{i=1}^{N} w_{y_i} \log \hat{p}_i[y_i],$$

where w_{y_i} is the normalized inverse frequency of the true label. Higher weights increased the penalty when the model was uncertain about rare labels, encouraging better representations for these categories, while predictions for common labels such as *neutral* contributed less to the loss. As shown in Fig. 2, weighted loss fine-tuning led to clear improvements in the recognition of minority emotion *disgust* and intermediate emotion *anger*. At the same time, performance on the prevalent category *neutral* remained high, demonstrating that this approach increases minority class performance without compromising performance on dominant labels.

6 Discussion

The experimental results demonstrate that zero-shot DistilBERT, while offering a convenient baseline without labeled data, struggled to identify the dominant *neutral* class. Supervised fine-tuning of DistilBERT and BERT markedly

weighted-loss	**BERT fine-tuned:**		
Emotion #	Prec.	Rec.	F1
Anger (11)	0.45	0.45	**0.45**
Disgust (3)	1.00	0.33	**0.50**
Fear (4)	0.00	0.00	0.00
Happiness (8)	0.88	0.88	**0.88**
Neutral (88)	0.83	0.92	**0.87**
Sadness (12)	0.33	0.17	0.22
Accuracy		0.76	
Macro Avg	0.58	0.46	0.49
Weighted Avg	0.73	0.76	**0.74**

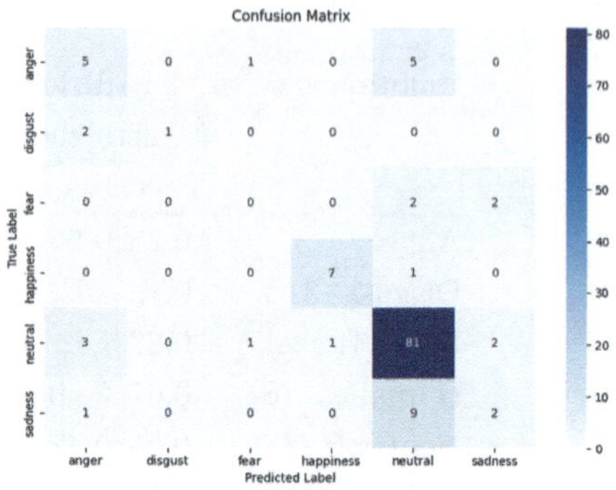

Fig. 2. Left: Test set performance (n = 126) of BERT fine-tuned with class weights (precision, recall, and F1-score). Right: Confusion matrix for the same model on DAIC-WOZ test set.

improved accuracy for frequent categories such as *neutral* and *happiness*. BERT outperformed DistilBERT slightly, likely due to its deeper architecture, though both models still struggled with rare emotions like *fear* and *disgust*, highlighting persistent label imbalance and semantic overlap. The confusion matrix in Fig. 2 (right) illustrates how class weighting influenced predictions. The majority of utterances remained classified as *neutral*, with 81 of 88 neutral examples correctly identified. Nonetheless, the model showed an increased capacity to detect minority labels, such as *disgust*. A persistent challenge, however, was the confusion among negative emotional categories. In particular, all four test utterances labeled as *fear* were misclassified, most often as *neutral* or *sadness*. This pattern suggests that although weighting successfully elevated the influence of less represented classes, the semantic proximity between negative emotions continued to limit classification accuracy. These findings also align with the domain-specific characteristics of psychotherapy dialogues, where neutral content dominates and negative emotion is often subtle and context-dependent, reinforcing the value of the targeted manual verification strategy adopted in this study.

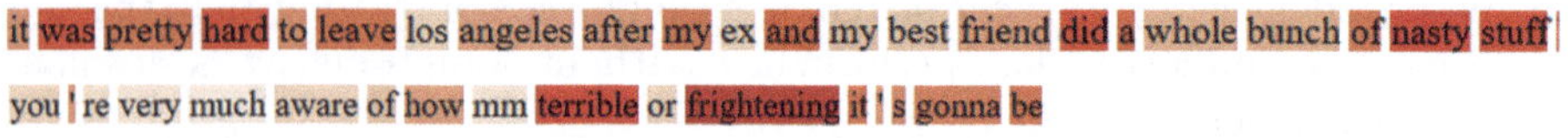

Fig. 3. Token-level attention for two test sentences: one labeled *sadness* and correctly predicted, and one labeled *fear* but misclassified as *sadness*.

To better understand this confusion, Fig. 3 shows token-level attention heatmaps for two representative examples: one utterance labeled and correctly predicted as *sadness*, and another labeled as *fear* but misclassified as *sadness*. The visualization indicates that in both cases, the model assigned high attention weights to similar emotionally charged words (e.g., *terrible*, *frightening*, *hard*), suggesting overlapping lexical cues contributed to the confusion between these categories. Darker red highlights indicate higher attention weights assigned by the model to individual words.

7 Conclusion and Future Work

This proof-of-concept study demonstrated that semi-automatic corpus generation, combining zero-shot emotion labeling, targeted manual verification, and transformer fine-tuning, can efficiently create emotion-labeled data for psychotherapy dialogues with minimal resources. Instead of fully annotating the DAIC-WOZ dataset from scratch, we used BERT and DistilBERT to propose candidate labels and selectively validate low-confidence predictions, reducing annotation effort while preserving the natural distribution of emotions in a realistic clinical setting. To try to capture moment-to-moment fluctuations, we classified emotions at the utterance level. Supervised fine-tuning improved performance over zero-shot baselines, and class-weighted loss enhanced recognition of minority emotions. These findings highlight the feasibility of scalable, domain-specific corpus generation and model adaptation within a short-term, low-budget project for developing emotionally responsive virtual patients in education. Future work will extend this approach to Dutch-language psychotherapy dialogues, incorporate multimodal cues like prosody and facial expressions, and scale to larger datasets with multiple annotators to strengthen reliability and support iterative corpus expansion in mental health training.

Acknowledgements. This work is part of the TherAvatars project founded by Regieorgaan SIA (The Netherlands), SVB/HT.KIEM.01.063

References

1. Ali, A.A., Fouda, A.E., Hanafy, R.J., Fouda, M.E.: Leveraging audio and text modalities in mental health: a study of LLMs performance (2024). https://arxiv.org/abs/2412.10417
2. Ameer, I., Bölücü, N., Siddiqui, M.H.F., Can, B., Sidorov, G., Gelbukh, A.: Multi-label emotion classification in texts using transfer learning. Expert Syst. Appl. **213118**, 534 (2023)
3. Arioz, U., Smrke, U., Plohl, N., Mlakar, I.: Scoping review on the multimodal classification of depression and experimental study on existing multimodal models. Diagnostics **12**(11), 2683 (2022). https://doi.org/10.3390/diagnostics12112683
4. Association, A.P.: Diagnostic and Statistical Manual of Mental Disorders, 5th ed. edn. American Psychiatric Publishing, Washington, DC (2013)
5. Batista, G.E., Prati, R.C., Monard, M.C.: A study of the behavior of several methods for balancing machine learning training data. ACM SIGKDD Explor. Newsl. **6**(1), 20–29 (2004)

6. Borst, J., Wehrheim, L., Niekler, A., Burghardt, M.: An evaluation of a zero-shot approach to aspect-based sentiment classification in historic German stock market reports. In: FedCSIS (Communication Papers), pp. 51–60 (2023)
7. Burdisso, S., Reyes-Ramírez, E., Villatoro-tello, E., Sánchez-Vega, F., Lopez Monroy, A., Motlicek, P.: DAIC-WOZ: on the validity of using the therapist's prompts in automatic depression detection from clinical interviews. In: Naumann, T., Ben Abacha, A., Bethard, S., Roberts, K., Bitterman, D. (eds.) Proceedings of the 6th Clinical Natural Language Processing Workshop, pp. 82–90. Association for Computational Linguistics (2024). https://doi.org/10.18653/v1/2024.clinicalnlp-1.8
8. Burdisso, S., Reyes-Ramírez, E., Villatoro-Tello, E., Sánchez-Vega, F., López-Monroy, P., Motlicek, P.: Daic-woz: on the validity of using the therapist's prompts in automatic depression detection from clinical interviews (2024). arXiv preprint arXiv:2404.14463
9. Busso, C., et al.: IEMOCAP: interactive emotional dyadic motion capture database. Lang. Resour. Eval. **42**, 335–359 (2008)
10. Chae, Y., Davidson, T.: Large language models for text classification: from zero-shot learning to instruction-tuning. Sociol. Methods Res. p. 00491241251325243 (2025)
11. Chen, W., Xing, X., Xu, X., Pang, J., Du, L.: Speechformer++: a hierarchical efficient framework for paralinguistic speech processing. IEEE/ACM Trans. Audio, Speech, Lang. Process. **31**, 775–788 (2023). https://doi.org/10.1109/taslp.2023.3235194
12. Chen, W., Yang, K., Yu, Z., Shi, Y., Chen, C.L.P.: A survey on imbalanced learning: latest research, applications and future directions. Artif. Intell. Rev. **57**(6), 137 (2024). https://doi.org/10.1007/s10462-024-10759-6
13. Chen, Z., Deng, J., Zhou, J., Wu, J., Qian, T., Huang, M.: Depression detection in clinical interviews with LLM-empowered structural element graph. In: Proceedings of the 2024 Conference of the North American Chapter of the Association for Computational Linguistics: Human Language Technologies, vol. 1, pp. 8181–8194 (2024)
14. De Langhe, L., et al.: Benchmarking zero-shot text classification for Dutch. Comput. Linguist. Neth. J. **13**, 63–90 (2024)
15. DeVault, D., et al.: SimSensei Kiosk: a virtual human interviewer for healthcare decision support. In: Proceedings of the 2014 International Conference on Autonomous Agents and Multi-Agent Systems, AAMAS '14, p. 1061–1068. International Foundation for Autonomous Agents and Multiagent Systems(2014)
16. Devlin, J., Chang, M.W., Lee, K., Toutanova, K.: Bert: pre-training of deep bidirectional transformers for language understanding. In: Proceedings of the 2019 Conference of the North American Chapter of the Association for Computational Linguistics: Human Language Technologies, vol. 1, pp. 4171–4186 (2019)
17. Eberhardt, S.T., et al.: Decoding emotions: exploring the validity of sentiment analysis in psychotherapy. Psychother. Res. **35**(2), 174–189 (2025)
18. Ekman, P.: An argument for basic emotions. Cogn. Emot. **6**(3–4), 169–200 (1992)
19. Feng, S., et al.: EmoWOZ: a large-scale corpus and labelling scheme for emotion recognition in task-oriented dialogue systems (2021). arXiv preprint arXiv:2109.04919
20. Feng, S., Sun, G., Lubis, N., Wu, W., Zhang, C., Gasic, M.: Affect recognition in conversations using large language models. In: Proceedings of the 25th Annual Meeting of the Special Interest Group on Discourse and Dialogue, pp. 259–273 (2024)

21. Frank, H.E., Becker-Haimes, E.M., Kendall, P.C.: Therapist training in evidence-based interventions for mental health: a systematic review of training approaches and outcomes. Clin. Psychol. Sci. Pract. **27**(3), 20 (2020)
22. Gado, S., Kempen, R., Lingelbach, K., Bipp, T.: Artificial intelligence in psychology: How can we enable psychology students to accept and use artificial intelligence? Psychol. Learn. Teach. **21**(1), 37–56 (2022)
23. Gratch, J., et al.: The distress analysis interview corpus of human and computer interviews. In: Calzolari, N., et al. (eds.) Proceedings of the Ninth International Conference on Language Resources and Evaluation (LREC'14), pp. 3123–3128. European Language Resources Association (ELRA) (2014). https://aclanthology.org/L14-1421/
24. He, H., Garcia, E.A.: Learning from imbalanced data. IEEE Trans. Knowl. Data Eng. **21**(9), 1263–1284 (2009)
25. Japkowicz, N., Stephen, S.: The class imbalance problem: a systematic study. Intell. Data Anal. **6**(5), 429–449 (2002)
26. Johnson, J.M., Khoshgoftaar, T.M.: Survey on deep learning with class imbalance. J. Big Data **6**(1), 1–54 (2019). https://doi.org/10.1186/s40537-019-0192-5
27. Ketmaneechairat, H., Maliyaem, M.: Corpus-based vocabulary list for Thai language. J. Adv. Inf. Technol. **14**(2), 319–327 (2023)
28. Kumar, M., Khan, L., Chang, H.T.: Evolving techniques in sentiment analysis: a comprehensive review. Peer J. Comput. Sci. **11**, e2592 (2025)
29. Kumar, P., Pathania, K., Raman, B.: Zero-shot learning based cross-lingual sentiment analysis for Sanskrit text with insufficient labeled data. Appl. Intell. **53**(9), 10096–10113 (2023)
30. Lau, C., Zhu, X., Chan, W.Y.: Automatic depression severity assessment with deep learning using parameter-efficient tuning. Front. Psych. **14**, 1160291 (2023)
31. Li, M., Sun, X., Wang, M.: Detecting depression with heterogeneous graph neural network in clinical interview transcript. IEEE Trans. Comput. Soc. Syst. **11**(1), 1315–1324 (2023)
32. Liu, R., Shi, Y., Ji, C., Jia, M.: A survey of sentiment analysis based on transfer learning. IEEE Access **7**, 85401–85412 (2019)
33. Loshchilov, I., Hutter, F.: Decoupled weight decay regularization (2017). arXiv preprint arXiv:1711.05101
34. Machová, K., Szabóova, M., Paralič, J., Mičko, J.: Detection of emotion by text analysis using machine learning. Front. Psychol. **14**, 1190326 (2023)
35. Olawade, D.B., Wada, O.Z., Odetayo, A., David-Olawade, A.C., Asaolu, F., Eberhardt, J.: Enhancing mental health with artificial intelligence: current trends and future prospects. J. Med. Surg. Publ. Health **3**, 100099 (2024)
36. Passos, I.C., Rabelo-da Ponte, F.D., Kapczinski, F.: Digital Mental Health: A Practitioner's Guide. Springer, Cham (2023)
37. Perlman, M.R., Anderson, T., Foley, V.K., Mimnaugh, S., Safran, J.D.: The impact of alliance-focused and facilitative interpersonal relationship training on therapist skills: an RCT of brief training. Psychother. Res. **30**(7), 871–884 (2020)
38. Pham, K.T., Nabizadeh, A., Selek, S.: Artificial intelligence and chatbots in psychiatry. Psychiatr. Q. **93**(1), 249–253 (2022)
39. Prabhu, S., Mittal, H., Varagani, R., Jha, S., Singh, S.: Harnessing emotions for depression detection. Pattern Anal. Appl. **25**(3), 537–547 (2022)
40. Prottasha, N.J., et al.: Transfer learning for sentiment analysis using BERT based supervised fine-tuning. Sensors **22**(11), 4157 (2022)

41. Rathje, S., Mirea, D.M., Sucholutsky, I., Marjieh, R., Robertson, C.E., Van Bavel, J.J.: GPT is an effective tool for multilingual psychological text analysis. Proc. Natl. Acad. Sci. **121**(34), e2308950121 (2024)
42. Ratican, J., Hutson, J.: Advancing sentiment analysis through emotionally-agnostic text mining in large language models (LLMs). J. Biosens. Bioelectron. Res. (2024)
43. Rodríguez, J.R., et al.: Peer counselling versus role-playing: two training methods of therapeutic skills in clinical psychology. Psicothema **30**(1), 21–26 (2018)
44. Ruder, S., Peters, M.E., Swayamdipta, S., Wolf, T.: Transfer learning in natural language processing. In: Proceedings of the 2019 conference of the North American chapter of the association for computational linguistics: Tutorials, pp. 15–18 (2019)
45. Sanh, V., Debut, L., Chaumond, J., Wolf, T.: Distilbert, a distilled version of BERT: smaller, faster, cheaper and lighter (2019). arXiv preprint arXiv:1910.01108
46. Singh, U., Abhishek, K., Azad, H.K.: A survey of cutting-edge multimodal sentiment analysis. ACM Comput. Surv. **56**(9), 1–38 (2024)
47. Syzdek, B.M.: Client and therapist psychotherapy sentiment interaction throughout therapy. Psychol. Stud. **65**(4), 520–530 (2020)
48. Tan, C., Xu, X., Shen, F.: A survey of zero shot detection: methods and applications. Cogn. Robot. **1**, 159–167 (2021)
49. Tanana, M., Dembe, A., Soma, C.S., Imel, Z., Atkins, D., Srikumar, V.: Is sentiment in movies the same as sentiment in psychotherapy? comparisons using a new psychotherapy sentiment database. In: Proceedings of the Third Workshop on Computational Linguistics and Clinical Psychology, pp. 33–41 (2016)
50. Tanha, J., Abdi, Y., Samadi, N., Razzaghi, N., Asadpour, M.: Boosting methods for multi-class imbalanced data classification: an experimental review. J. Big data **7**, 1–47 (2020)
51. Vaswani, A., et al.: Attention is all you need (2017). arXiv preprint arXiv:1706.03762
52. Wang, W., Zheng, V.W., Yu, H., Miao, C.: A survey of zero-shot learning: settings, methods, and applications. ACM Trans. Intell. Syst. Technol. (TIST) **10**(2), 1–37 (2019)
53. Weiss, G.M., Provost, F.: Learning when training data are costly: the effect of class distribution on tree induction. J. Artif. Intell. Res. **19**, 315–354 (2003)
54. Zhang, X., Li, C., Chen, W., Zheng, J., Li, F.: Optimizing depression detection in clinical doctor-patient interviews using a multi-instance learning framework. Sci. Rep. **15**(1), 6637 (2025)
55. Zhu, L., Zhu, Z., Zhang, C., Xu, Y., Kong, X.: Multimodal sentiment analysis based on fusion methods: a survey. Inf. Fusion **95**, 306–325 (2023)

Sentiment Analysis of Movie Reviews from YouTube and X: A Comparative Study of Logistic Regression and SVM with Optimized TF-IDF

Wirat Jareevongpiboon(✉) and Chantraphon Suksom

Faculty of Science and Technology, Thammasat University, Pathumthani Bangkok, Thailand
wirat@tu.ac.th, chantraphon.suk@dome.tu.ac.th

Abstract. Online movie reviews have become an essential part of audience feedback, influencing both public perception and individual viewing decisions. This paper presents a sentiment analysis system that classifies movie reviews from social media platforms—YouTube and X (formerly Twitter)—into positive and negative sentiments. The system employs Natural Language Processing (NLP) techniques, including text cleaning, tokenization, stop-word removal, lemmatization, and TF-IDF vectorization. Two traditional machine learning models, Logistic Regression and Support Vector Machine (SVM), are trained and evaluated using a balanced IMDb dataset containing 50,000 labeled reviews. Experiments compare default and tuned TF-IDF settings, with the best-performing models achieving accuracy above 89%. GridSearchCV is applied for hyperparameter tuning, and evaluation metrics include accuracy, precision, recall, and F1-score. A web-based prototype is also developed to demonstrate real-time sentiment analysis of comments retrieved via YouTube and X APIs. Visual tools such as WordClouds and sentiment distribution charts are integrated to support interpretability. The results demonstrate that linear models with optimized parameters are effective and efficient for movie sentiment classification. Future work aims to address model limitations related to sarcasm, emojis, and neutral opinions by exploring contextual language models like BERT and expanding the dataset to better reflect social media language. This system provides a practical foundation for real-time audience feedback analysis in film-related applications.

Keywords: Sentiment Analysis · NLP · Social Media · Movie Reviews · Logistic Regression · SVM · YouTube · X

1 Introduction

Online opinions increasingly shape public perception and social dynamics, with reviews and comments reflecting societal viewpoints and emotions. Among various media, films consistently attract significant attention and critique on online platforms. In today's digital era, where social media dominates communication, audience feedback on films represents a vital voice influencing popularity and viewer decisions. Positive opinions

H. Unger and P. Meesad (Eds.): NLPIR 2025, LNNS 1904, pp. 255–267, 2026.
https://doi.org/10.1007/978-3-032-20897-2_18

can boost a film's success, while negative sentiments may deter potential audiences. However, the rapid growth of user-generated content renders manual sentiment analysis time-consuming and challenging.

Recent research has leveraged machine learning techniques for sentiment analysis on social media platforms, particularly Twitter. Despite promising outcomes, many studies focus on single data sources or overlook model comparisons. Moreover, interpretable visual summaries for non-expert stakeholders remain scarce. Although YouTube and X are widely used platforms, sentiment analysis on their movie-related comments remains underexplored in the context of traditional machine learning classifiers with optimized feature engineering. This study leverages these platforms not just as data sources but as representatives of real-time, colloquial, and evolving language in movie discussions, which differs from more formal and static datasets. Unlike prior work, this research integrates user choice of platform, allowing analysis tailored to preferred social media, and emphasizes the interpretability of results through visual tools such as WordCloud.

To manage the complexity and volume of opinion data, this research integrates Natural Language Processing (NLP) and Machine Learning (ML) to automate sentiment classification. The sentiment classification model is trained on the IMDb dataset to capture domain-specific vocabulary and sentiment patterns from movie reviews, mostly for older films. However, IMDb's lack of recent data motivates this research to validate model performance on YouTube and X, where audiences express more current opinions with contemporary slang and idiomatic expressions. This approach enhances the applicability of traditional models like Logistic Regression and SVM to modern social media contexts.

Given that online movie reviews often include nuanced opinions that significantly influence audience reception, applying these technologies helps content creators interpret feedback swiftly and accurately, identifying strengths and areas for improvement. Similarly, for consumers, having access to organized and clearly analyzed reviews supports more informed viewing decisions.

Sentiment analysis has been widely explored internationally, with foundational work by Pang et al. (2002) employing machine learning on movie reviews [1]. More recent studies utilize advanced models like BERT for enhanced contextual understanding in social media sentiment detection [2]. Research also underscores the value of aggregating multi-source data for robust modeling, exemplified by works combining Twitter and YouTube data [3]. Building on these, this study compares traditional machine learning classifiers on cross-platform datasets while providing interpretable visualizations for diverse audiences.

This study aims to develop an automated sentiment analysis model for movie reviews from online platforms. By applying NLP and ML techniques, the system classifies user opinions into positive and negative categories and highlights key themes using WordCloud visualizations. The goal is to reduce human processing time while offering actionable insights to both content creators and viewers, thereby contributing to more responsive and audience-centered film production.

2 Literature Review

2.1 Theoretical Background

Natural Language Processing (NLP) is a subfield of Artificial Intelligence (AI) that focuses on enabling machines to understand, process, and generate human language. NLP covers tasks such as sentiment analysis, machine translation, and summarization, aiming to bridge the gap between human communication and machine understanding [4].

Sentiment analysis plays a key role in extracting user opinions from online content. The process begins with data preprocessing techniques including text cleaning, tokenization, stop word removal, stemming or lemmatization, and term weighting using TF-IDF (Term Frequency–Inverse Document Frequency).

For classification, machine learning models like Logistic Regression and Support Vector Machines (SVM) are widely used. Logistic Regression estimates probabilities of binary outcomes using a logistic function, while SVM finds an optimal hyperplane to separate data classes and supports non-linear classification via kernel functions.

Model performance is typically evaluated using metrics derived from a confusion matrix, such as accuracy, precision, recall, and F1-score, to assess the effectiveness of predictions, especially in imbalanced data scenarios.

2.2 Related Work

Several studies have applied NLP and machine learning to sentiment analysis in both global and Thai-language contexts.

Internationally, Go et al. used distant supervision for Twitter sentiment classification with SVM [5]; Socher et al. introduced recursive neural networks [6]; and Zhang et al. combined CNN and LSTM for improved accuracy [7]. These studies primarily focus on English-language datasets and illustrate the evolution from traditional machine learning to deep learning. However, few directly compare multiple classifiers under identical feature engineering settings, leaving the impact of feature optimization unclear—a gap addressed in this study.

In Thailand, studies such as Predawan et al. (SVM with TF-IDF on YouTube, 98% accuracy) [8], Rawisuda and Niwet (multiple algorithms on Agoda & Twitter, SVM best 83.38%) [9], Pennapa (SVM on skincare reviews, 85%) [10] and Tanasit & Kawinthida (Naïve Bayes on Thai TV drama comments, ~70%) [11] highlight the need for tailored preprocessing for Thai-specific linguistic features. Unlike these single-language studies, our work applies multilingual-ready preprocessing to English social media datasets for cross-context applicability. Key studies are summarized in Table 1.

Thai-language research informs our methodology, demonstrating successful adaptation of traditional models to non-English contexts. This study conducts controlled comparative experiments between Logistic Regression and SVM with optimized TF-IDF, emphasizing model interpretability and cross-platform applicability.

Additionally, works on semantic orientation and lexicon-based methods, as well as multi-source aggregation (Alhammi & Haddar 2020 [12]; Rashid & Huang 2021 [13];

Table 1. Summary of Prior Work on Sentiment Analysis (Highlighted Studies)

Study	Language / Platform	Model	Features	Accuracy	Notes
Go et al. (2009) [5]	English / Twitter	SVM	Distant supervision	–	Early English social media sentiment analysis
Socher et al. (2013) [6]	English / Movie Reviews	Recursive Neural Network	Word embeddings	–	Deep learning approach
Zhang et al. (2018) [7]	English / Movie Reviews	CNN+ LSTM	TF-IDF+ embeddings	–	Combines convolutional and sequential features
Predawan et al. (2020) [8]	Thai / YouTube	SVM	TF-IDF	98%	Thai-language social media data
Rawisuda & Niwet (2020) [9]	Thai / Agoda & Twitter	Multiple	TF-IDF& others	83.38%	Comparison of algorithms for Thai reviews
Pennapa (2021) [10]	Thai / Skincare reviews	SVM	TF-IDF	85%	Fraud detection in Thai reviews
Tanasit & Kawinthida (2021) [11]	Thai / Twitter	Naïve Bayes	TF-IDF	70%	Thai TV drama sentiment

Ragothaman & Huang 2021 [14]), highlight alternative approaches. Our approach complements these by providing a robust baseline with classical classifiers and systematically optimized features.

This study thus fills the gap by rigorously evaluating classical machine learning classifiers on modern English social media datasets while offering interpretable visualizations for actionable insights.

3 Methodology

The movie sentiment analysis system based on social media consists of two main components: (1) model development for sentiment classification, and (2) model deployment for analyzing live comments retrieved from YouTube and X (formerly Twitter) based on user-specified movies. The overall system architecture is shown in Fig. 1.

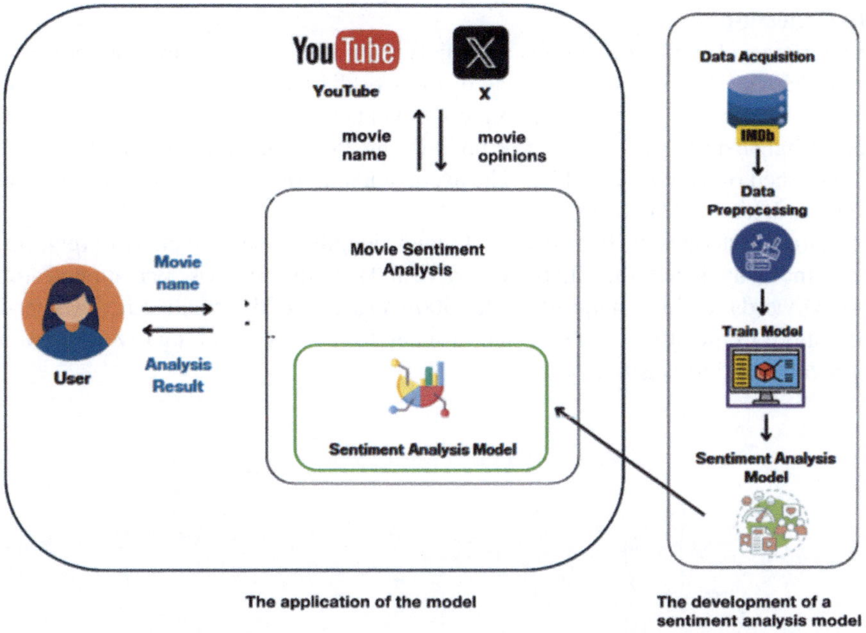

Fig. 1. Overview of the Sentiment Analysis Model Development and Deployment

3.1 Sentiment Analysis Model Development

The model development process, as shown in Fig. 2, comprises several steps described as follows:

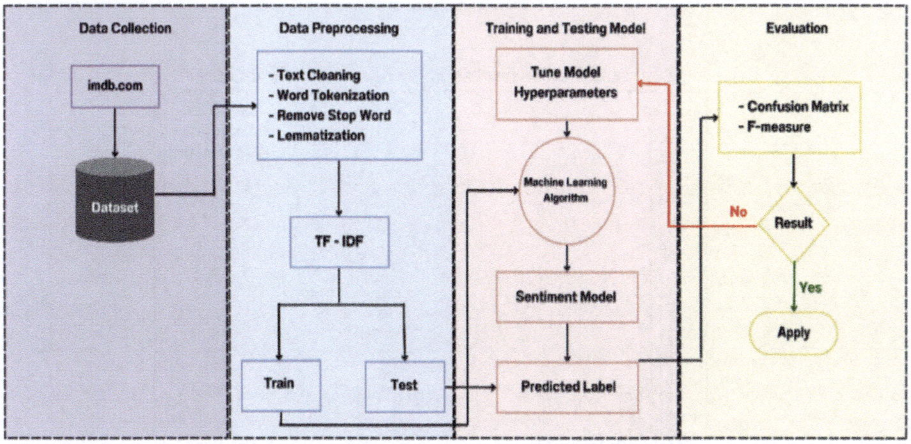

Fig. 2. Steps in Developing the Sentiment Analysis Model

Data Collection.

The dataset used in this research is a publicly available movie review dataset from the Internet Movie Database (IMDb), one of the world's largest and most popular online databases for movies, TV shows, and entertainment content. The dataset contains 50,000 labeled English movie reviews (25,000 positive and 25,000 negative), originally collected and published on Kaggle.com [15]. The average length of the reviews is approximately 1,309 characters or 231 words.

Figure 3 illustrates the distribution of review lengths by sentiment. Most reviews are short to moderate in length, with both positive and negative reviews concentrated around 100–200 words, and the majority under 500 words. Notably, positive reviews tend to appear more frequently in both the short (under 150 words) and long (over 400 words) ranges compared to negative ones.

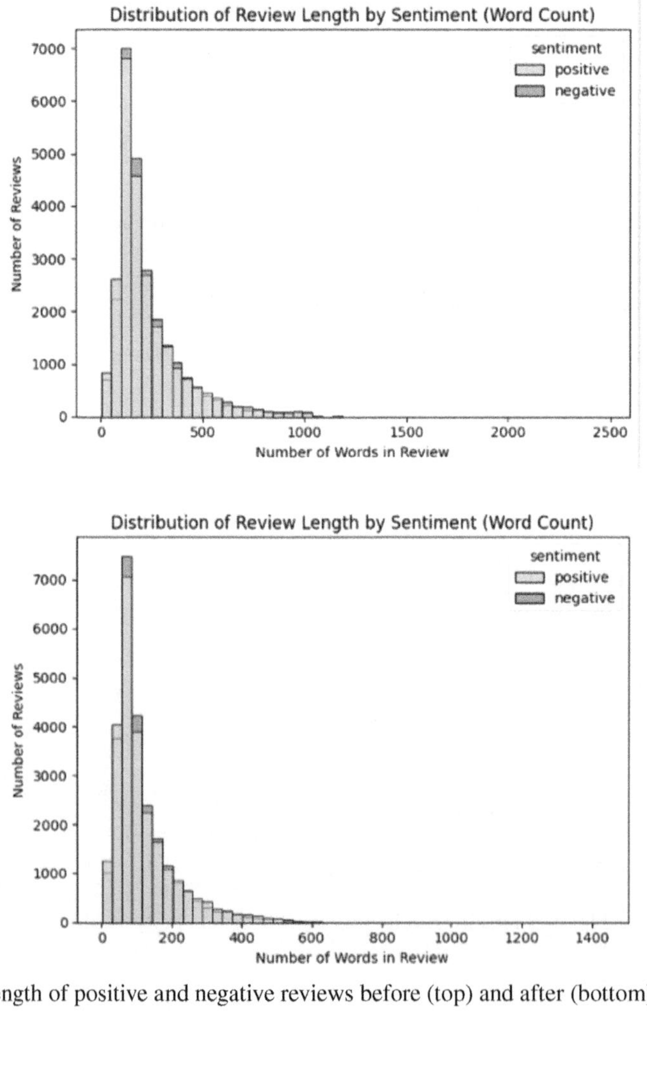

Fig. 3. The length of positive and negative reviews before (top) and after (bottom) text cleaning.

Data Preprocessing.
The raw data is processed using standard Natural Language Processing (NLP) techniques to prepare it for classification. The steps include:

- Text Cleaning: Converts all text to lowercase, removes punctuation, special characters, HTML tags, and expands English contractions. After this step, the average review length is reduced to around 120 words. However, the length distribution pattern between positive and negative reviews remains unchanged (Fig. 3, bottom).
- Stop Word Removal: Removes common words that do not contribute significant meaning (e.g., "is", "was", "and").
- Lemmatization: Reduces inflected words to their base form (e.g., "run", "running", "runs" → "run").
- Word Tokenization: Splits text into tokens (words), e.g., "This movie was fantastic and amazing" → ['This', 'movie', 'was', 'fantastic', 'and', 'amazing'].
- TF-IDF Vectorization: Transforming cleaned text into numerical features using Term Frequency-Inverse Document Frequency. Two TF-IDF settings were tested: 1) Default TF-IDF and 2) Tuned TF-IDF with the following parameters:

i. ngram_range = (1,2) for unigram and bigram features
ii. min_df = 5 to ignore rare terms
iii. max_df = 0.7 to exclude overly common terms
iv. max_features = 5000 to limit the vocabulary
v. stop_words = 'english' to remove English stop words

The most frequent unigrams and bigrams are visualized in Fig. 4.

- Train-Test Split: The dataset is split into 80% training data (40,000 reviews) and 20% testing data (10,000 reviews), with an equal number of positive and negative reviews.

Model Training and Testing.

- Model training: Model training is conducted in two phases:

Phase 1: Identifies the best-performing text vectorization method (default vs. tuned TF-IDF) for each algorithm (Logistic Regression and SVM). This includes four sub-experiments:

Experiment	Algorithm	TF-IDF Version
(1)-1.1	Logistic Regression	Default
(1)-1.2	Logistic Regression	Tuned Parameters
(1)-2.1	SVM	Default
(1)-2.2	SVM	Tuned Parameters

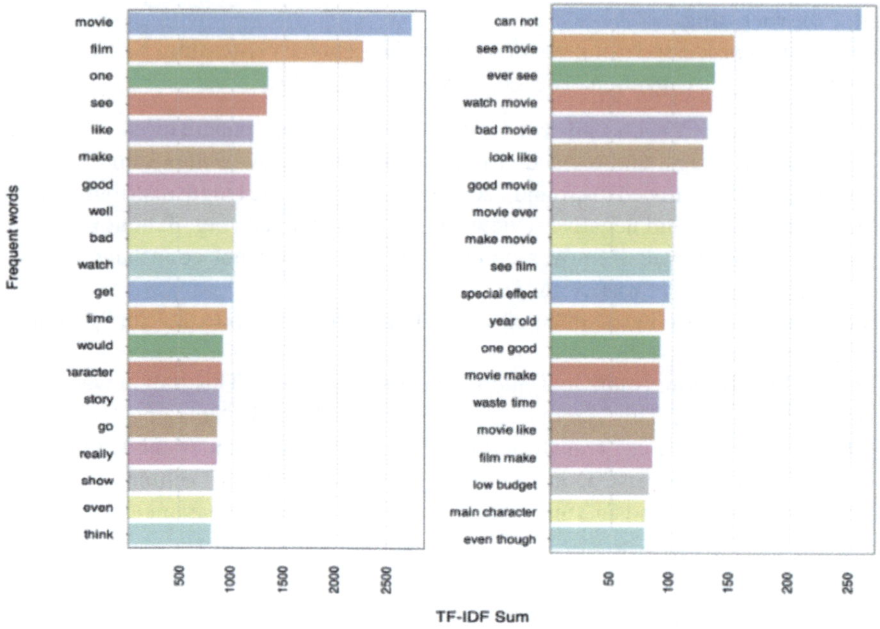

Fig. 4. Most frequent unigrams (left) and bigrams (right)

Phase 2: Selects the best-performing algorithm using the optimal TF-IDF variant from Phase 1. Hyperparameter tuning is applied using GridSearchCV:

Experiment	Algorithm	Parameters Used
(2)-1	Logistic Regression	GridSearchCV Tuned
(2)-2	SVM	GridSearchCV Tuned

This results in six total experiments (four in Phase 1, two in Phase 2).

- Hyperparameter Tuning: To improve classification performance, both models were tuned using grid search with cross-validation. The goal was to find optimal settings for sentiment classification based on the IMDb dataset.
- Prediction: The trained models were used to predict the sentiment labels on the test set (with labels hidden), to verify classification accuracy.

Model Evaluation. The models are evaluated using standard classification metrics including accuracy, precision, recall, and F1-score.

3.2 Model Deployment

A prototype web application was developed to demonstrate real-time sentiment analysis. Users can input a movie title and choose a data source (YouTube or X). The system retrieves current comments using YouTube/X API and analyzes them using the trained model. Results are visualized in a pie chart for sentiment proportions and a word cloud to highlight key opinion terms.

4 Experimental Results

This study compares the performance of Logistic Regression and Support Vector Machine (SVM) for movie sentiment classification using TF-IDF features. Two experimental phases were conducted: (1) identifying the best feature representation, and (2) selecting the best-performing model.

4.1 Experiment 1: TF-IDF Representation Comparison

TF-IDF features were tested in both default and tuned settings on a balanced dataset (40,000 training and 10,000 test samples). Table 2 presents the results for Logistic Regression. Accuracy improved slightly from 88.6% (default) to 89.18% (tuned).

Table 2. Logistic Regression with default vs. tuned TF-IDF

Setting	Pos Correct	Neg Correct	Accuracy
Default	4542	4318	88.60%
Tuned	4584	4334	89.18%

For SVM, Table 3 shows improved accuracy from 75.85% to 78.64% after tuning.

Table 3. SVM with default vs. tuned TF-IDF

Setting	Pos Correct	Neg Correct	Accuracy
Default	3695	3890	75.85%
Tuned	3972	3892	78.64%

These results suggest tuned TF-IDF yields better performance for both models.

4.2 Experiment 2: Best Model Selection

Using tuned TF-IDF, both algorithms were further optimized via hyperparameter tuning.

The best-performing Logistic Regression model was obtained using the following hyperparameters: a regularization strength C of 10, no class weighting (class_weight = None), a maximum of 100 iterations (max_iter = 100), L2 regularization (penalty = '12'), the liblinear solver, and a tolerance level of 0.0001 for stopping criteria.

Table 4. Best-tuned Logistic Regression performance

Metric	Accuracy	Precision	Recall	F1-Score
Average	0.90	0.90	0.895	0.895

Table 4 shows Logistic Regression with tuned parameters reached 89.63% accuracy, with balanced precision and recall across classes.

The optimal hyperparameters for the LinearSVC model were: C = 1, class_weight = None, dual = True, loss = 'squared_hinge', max_iter = 1000, penalty = 'l2', and tol = 0.0001.

Table 5 shows the LinearSVC model reached slightly higher accuracy at 89.76%, with nearly identical performance metrics.

Table 5. Best-tuned LinearSVC performance

Metric	Accuracy	Precision	Recall	F1-Score
Average	0.90	0.895	0.895	0.895

4.3 Real-World Testing on YouTube Comments

Both models were tested on 500 YouTube comments for the movie *Sinners*. As visualized in Fig. 5, the Logistic Regression model predicted 327 positive and 173 negative comments.

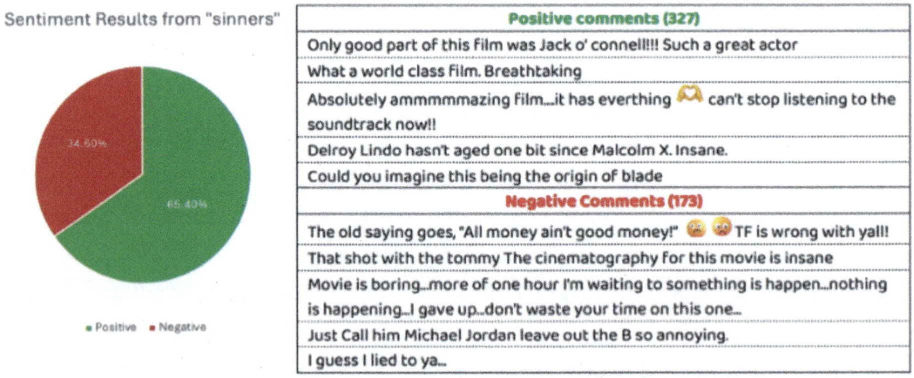

Fig. 5. The pie chart (left) and example of positive and negative comments on *Sinners* (right)

Word clouds in Fig. 6 illustrate dominant themes in positive and negative comments.

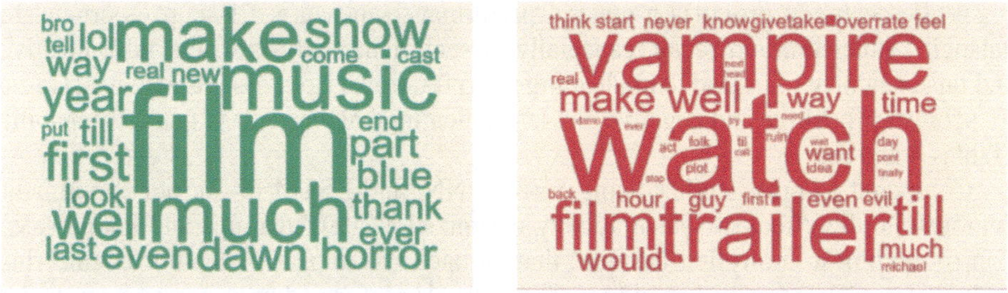

Fig. 6. Word Clouds of positive (left) and negative (right) sentiments

Table 6 shows both models performed comparably on this external data, with Logistic Regression slightly ahead in all metrics.

Table 6. Performance on YouTube dataset

Metric	Logistic Regression	SVM
Accuracy	0.77	0.75
F1-Score	0.76	0.74

5 Discussion

In the initial experiments without hyperparameter tuning, the Logistic Regression (LR) model clearly outperformed the Support Vector Machine (SVM) model. This suggests that the dataset is more suitable for linear models, as the default LR uses a linear kernel while the default SVM uses an RBF kernel. After switching the SVM kernel to linear, its performance significantly improved, matching that of LR. This indicates that the sentiment polarity in this dataset may be largely linearly separable—consistent with previous findings that movie review sentiments often exhibit clear positive or negative polarity [16].

Both models were further tested on a set of 500 movie reviews from YouTube for the film *Sinners*. The LR and SVM models achieved comparable accuracies of 0.77 and 0.75 respectively, demonstrating practical utility beyond the training dataset. The variation in online comments—ranging from direct opinion to subtle or sarcastic remarks—poses a challenge even for high-performing models.

Error analysis revealed several patterns where both models struggled:

- Comments containing emojis, which were absent in the training data.
- Sarcastic or ironic expressions that were difficult to classify.
- Negative sentences with positive keywords due to negation removal during preprocessing.
- Rare or uncommon words that TF-IDF underweights.
- Ambiguous statements whose sentiment is not clearly positive or negative (e.g., "Scariest movie I've ever been in as a stunt actor.").

While accuracy provides a general performance measure, F1-score is crucial for balancing precision and recall, especially in sentiment analysis where false positives and false negatives can lead to misleading interpretations. Both models achieved similar F1-scores for positive and negative classes, indicating balanced classification capability (Tables 4 and 5).

Compared to prior studies using SVM, CNN, and LSTM [5–7], our experiments demonstrate the effect of systematically optimized TF-IDF features and kernel selection, providing a controlled baseline that isolates the impact of feature engineering. Insights from Thai-language sentiment analysis also informed preprocessing strategies, particularly for handling complex tokenization and domain-specific slang, which were adapted for English social media reviews [8–11]. This shows how established models can be successfully adapted to new linguistic contexts.

A key limitation of the current approach is the reliance on TF-IDF, which captures word frequency but not contextual meaning. This limitation affects the model's ability to interpret slang, sarcasm, or semantic nuances. Similar limitations have been reported in other works using bag-of-words models [17]. Therefore, future research should consider adopting contextualized models such as BERT, which have shown promising results in recent sentiment analysis benchmarks [2].

In addition, extending the sentiment categories beyond positive and negative could allow for more nuanced analysis, especially on platforms like YouTube and X where opinions are often mixed. Other potential improvements include integrating emoji processing, refining negation handling, and including data from additional platforms to enhance generalizability.

In summary, both LR and SVM with linear kernels provide effective baseline models for movie review sentiment classification. While the use of English social media data and Word Cloud visualizations may not introduce methodological novelty, the controlled comparison and systematic feature optimization offer a reproducible baseline for future research, particularly for integrating more advanced NLP models or cross-linguistic analyses.

References

1. Pang, B., Lee, L., Vaithyanathan, S.: Thumbs up? Sentiment classification using machine learning techniques. In: Proceedings of the ACL-02 Conference on Empirical Methods in Natural Language Processing (EMNLP), pp. 79–86 (2002)
2. Devlin, J., Chang, M.W., Lee, K., Toutanova, K.: BERT: pre-training of deep bidirectional transformers for language understanding. In: Proceedings of NAACL-HLT, pp. 4171–4186 (2019)
3. Wang, Y., Li, H., Jin, D.: Multi-source sentiment analysis using deep learning and data fusion. IEEE Access. **8**, 127554–127563 (2020)
4. Jurafsky, D., Martin, J. H.: Speech and Language Processing, 3rd ed., Pearson, London, (2023)
5. Go, A., Bhayani, R., Huang, L.: Twitter Sentiment Classification Using Distant Supervision., CS224N Project Report,. Stanford University (2009)
6. Socher, R. et al.: Recursive deep models for semantic compositionality over a sentiment treebank. In: Proceedings of EMNLP, pp. 1631–1642 (2013)
7. Zhang, L., Wang, S., Liu, B.: Deep learning for sentiment analysis: a survey. Wiley Interdisc. Rev. Data Min. Knowl. Discovery. **8**(4), e1253 (2018)

8. Predawan, P., Kaewkiriya, R., Ekkachai, N.: Sentiment analysis of Thai movie reviews on YouTube using SVM. In: Proceedings of the National Conference on Computer and Information Technology, pp. 56–61 (2020) (in Thai)
9. Rawisuda, K., Niwet, T.: Comparative sentiment analysis of reviews from Agoda and twitter using machine learning algorithms. J. Inf. Sci. Technol. **18**(2), 45–52 (2021) in Thai
10. Pennapa, C.: A machine learning approach to detect suspicious skincare product reviews. Int. J. Thai Interdisc. Res. **13**(1), 22–28 (2021) in Thai
11. Tanasit, X., Kawinthida, Y.: Sentiment analysis of Thai TV drama comments on twitter using naïve Bayes. J. Thai Soc. Media Res. **5**(2), 45–53 (2021) in Thai
12. Alhammi, H.A., Haddar, K.: Building a Libyan dialect lexicon-based sentiment analysis system using semantic orientation of adjective-adverb combinations. Int. J. Comput. Theory Eng. **12**(6), 145–150 (2020)
13. Rashid, A., Huang, C.-Y.: Sentiment analysis on consumer reviews of Amazon products. Int. J. Comput. Theory Eng. **13**(2), 35–41 (2021)
14. Ragothaman, A., Huang, C.-Y.: Sentiment analysis on Covid-19 twitter data. Int. J. Comput. Theory Eng. **13**(4), 100–107 (2021)
15. Kaggle: IMDb Movie Reviews Dataset,https://www.kaggle.com/datasets/lakshmi25npathi/imdb-dataset-of-50k-movie-reviews (Accessed: July 2025)
16. Cambria, E., Schuller, B., Xia, Y., Havasi, C.: New avenues in opinion mining and sentiment analysis. IEEE Intell. Syst. **28**(2), 15–21 (2013)
17. Rietz, T., Maedche, A.: Explaining machine learning models in sentiment analysis: survey and research agenda. Decis. Support. Syst. **154**, 113673 (2022)

Colors in Context: Exploring Sentiment, Semantics, and Emoji Patterns Across Cultures

Anna Kruspe[1](✉), Mila Stillman[1], and Moritz Marutschke[2]

[1] Munich University of Applied Sciences, Munich, Germany
{anna.kruspe,mila.stillman}@hm.edu
[2] Ritsumeikan University, Kyoto, Japan
moritz@fc.ritsumei.ac.jp

Abstract. Colors carry affective and symbolic meanings that vary across languages and cultures. This paper investigates these associations in digital communication by analyzing how basic color terms are used in English, German, Japanese, and Russian tweets. Using a dataset of over 1.6 million tweets collected in 2015, we examine the relationships between color terms, sentiment, topics, and emoji. Our results reveal both universal tendencies–such as consistently negative sentiment for *black* and positive associations for *purple*–and culture-specific patterns, including the negative connotation of *white* in Japanese and the political salience of *green* in German. Emoji frequently reinforce color meanings through matching visual cues (e.g., colored hearts) but also reflect culturally situated practices. By combining text and emoji, our study highlights how color terms function as culturally embedded symbols in multilingual social media discourse.

Keywords: Colors · Emoji · Multimodal Sentiment Analysis · Social Media · Cross-cultural NLP

1 Introduction

Color perception is a universal human ability, rooted in the biological evolution of the visual system. Yet while the physical basis of color is shared across populations, the meanings and associations attached to colors vary across languages and cultures. Colors can signal danger, emotion, identity, or symbolism, shaped by geography, tradition, and socio-political context. Research in psychology and linguistics has long demonstrated both universal aspects of color categorization [1,2] and striking cultural differences in the emotional and symbolic connotations of specific hues [3].

In digital communication, these associations are not expressed only through words. Social media users increasingly combine text with visual paralinguistic markers such as emoji, which themselves carry culturally situated meanings [4]. At the same time, recent work has shown that modern Large Language Models

H. Unger and P. Meesad (Eds.): NLPIR 2025, LNNS 1904, pp. 268–278, 2026.
https://doi.org/10.1007/978-3-032-20897-2_19

(LLMs) encode structured representations of color terms that align with perceptual color spaces [5,6], offering a new lens for studying the interplay of color, sentiment, and semantics in multilingual data.

This paper builds on these strands by investigating how color terms are used in everyday online discourse. Using a large multilingual Twitter dataset covering English, German, Japanese, and Russian, we analyze the relationships between color terms, sentiment, semantics, and emoji usage. Our study reveals both cross-linguistic regularities and culture-specific divergences, highlighting how colors operate not just as perceptual categories, but as socially and symbolically embedded elements of digital communication.

2 Related Work

2.1 Colors Across Languages and Cultures

Research in psychology and linguistics has long investigated how humans perceive and categorize colors. While the biological basis of color perception is universal, the meanings and emotional associations of colors vary across societies. Berlin and Kays seminal study [1] established the existence of 11 basic color terms, which are systematically encoded across languages despite cross-linguistic variation in their number and usage. Later work demonstrated both universal tendencies and culture-specific connotations: for example, the color "red" can evoke vitality and fertility in some contexts, but also danger and war in others [3]. Studies in marketing and psychology further show that colors shape human reactions and emotional responses [7]. These findings highlight that while basic color categories are shared, their symbolic and affective meanings are deeply influenced by cultural context. However, their role in online discourse remains less well studied.

Large-scale cross-cultural studies also show systematic associations between colors and emotional responses, revealing both universal patterns and culture-specific variation [8,9]. Such findings are particularly relevant for social media contexts, where color terms often carry affective or symbolic meaning.

2.2 Colors and Emoji in Computational Linguistics

Recent advances in natural language processing have enabled new ways of studying colors as linguistic and cultural concepts. Large Language Models (LLMs) and word embeddings capture structured representations of color terms: [5] demonstrated a topological alignment between BERT-based embeddings of color words and perceptual color spaces such as CIELAB. Similar studies have explored how foundation models interpret colors when prompted with text or visual inputs [6,10]. [11] extended this to a multilingual perspective, finding systematic biases in the representation of color terms across English, German, Japanese, and Russian.

In parallel, emoji have become integral to online communication. They function as paralinguistic markers that complement text with affective or symbolic meaning [4]. Yet despite their widespread use, many popular LLMs–particularly BERT-based ones–do not fully encode emoji, limiting the analysis of multimodal sentiment and meaning. Prior work has therefore tended to treat colors and emoji separately. This gap motivates our focus on their combined role in multilingual discourse.

Beyond color terms, prior research has examined the semantics and pragmatic functions of emoji in online communication. Studies show that emoji usage and interpretation vary across cultures and can even lead to miscommunication due to platform-dependent rendering differences [12]. Emoji carry affective and contextual meaning that interacts with surrounding text rather than simply mirroring it, and their sentiment polarity can be measured from usage patterns in social media [13]. Other work has indicated systematic differences in how emoji are used across languages [14]. These perspectives highlight emoji as culturally mediated meaning-bearing units, further motivating our cross-lingual approach.

2.3 Sentiment and Multimodal Signals in Social Media

Social media platforms provide large-scale, naturally occurring data for studying sentiment. Sentiment analysis typically classifies texts into positive, neutral, or negative categories, with prior research demonstrating cultural variation in how sentiment is expressed online [15]. Both colors and emoji have been linked to sentiment. Colors influence affective perception in images and marketing, while specific color terms in text carry positive or negative associations depending on cultural context. Emoji likewise modulate sentiment: their presence increases sentiment scores in tweets, and positive sentiments are more strongly affected than negative ones [16].

Existing studies, however, largely address either color or emoji in isolation. Few combine them, and fewer still examine these dynamics across multiple languages. Our work addresses this gap by jointly analyzing color terms, sentiment, and emoji in multilingual Twitter data, thereby providing new insights into how culturally embedded concepts are expressed in digital communication.

3 Dataset

The dataset was collected from the Twitter social media platform (now called X) between September and December 2015. The tweets were collected based on keywords to contain any of the following color terms: red, green, blue, black, white, purple, pink, gray, brown and orange. These were collected in English, German, Japanese, and Russian, with the color terms translated into those languages to generate a multilingual dataset, with the tweets being filtered by language, using the official API [17]. During the curation process, the Tweets were filtered by language, using the official API.

We post-processed the collected tweets by removing duplicates and filtering out cases where color terms appeared outside the text (e.g., only in usernames).

This left approximately 1.6 million tweets for analysis. For emoji-related analyses, we relied on the subset of about 167,000 tweets containing emoji. Since many tweets included multiple emoji, each was treated independently in the subsequent analyses.

4 Analysis

We examine how color mentions in tweets relate to a range of other characteristics across the four different languages. This allows us to derive insights into color usage in different cultural and linguistic contexts. Our analyses focus on three dimensions: sentiment, emoji usage, and topics. Emoji are extracted directly from the tweet texts, while sentiment and topics are assigned using pre-trained language models.

We deliberately omit the analysis of topics vs. emoji and topics vs. sentiment, as the color-focused pre-selection of the dataset would bias such results: colors predominantly occur in specific topics, as shown below. However, we do investigate the relationship between emoji and sentiment, since we do not expect this to be strongly influenced by the color-based filtering of tweets.

4.1 Emoji and Colors

We first examine the correlations between emoji and color terms. Across languages, a subset of emoji appears with high frequency independent of specific colors, reflecting general usage patterns rather than color-specific associations. To highlight statistically unexpected combinations, we perform a χ^2 test, identifying color–emoji pairs that occur significantly more often than would be expected from their individual frequencies (results in Fig. 1). Some combinations are simply the result of generally frequent emoji. Others are straightforward and semantically expected: color terms paired with hearts of the corresponding color, *orange* with the *tangerine* (present in all languages, as the word *orange* can also denote the fruit), and food or drink pairings such as *green tea* or *red wine*.

A number of combinations can be traced to dataset-specific artifacts, such as users with a personal "trademark" combination (e.g., *white* + *Japanese beginner symbol*), or institutional accounts whose emoji usage reflects a topical focus (*star*, *memo*, *globe* + *green* for news about the Green Party). In German, the homonym *weiß* ("white" and "know") also introduces unrelated co-occurrences.

Importantly, several associations are language- or culture-specific. In German tweets, *blue* + *sun* often denotes "blue sky" in contrast to *grey* + *cloud* / *umbrella* for "grey sky," while *pink* + *donut* reflects a food trend. In Japanese, *red* + *pot of food* can refer to spiciness, but also to the "Akakara" hot pot chain; *purple* + *sweet potato* denotes a local specialty. In Russian, *yellow* + *leaves* and *pink* + *cherry blossom* mark seasonal references.

Overall, many of the strongest associations involve emoji whose depicted color matches the color term in the text, reinforcing the intended color-related meaning.

English	German	Japanese	Russian
P (purple heart)	O (tangerine)	O (tangerine)	P (purple heart)
O (tangerine)	Y (yellow heart)	G (green heart)	B (blue heart)
B (blue heart)	P (purple heart)	P (purple heart)	O (tangerine)
G (green heart)	G (green heart)	B (blue heart)	Y (yellow heart)
I (ribbon)	B (blue heart)	N (pile of poo)	L (blue heart)
Y (yellow heart)	G (globe showing Asia-Australia)	W (Japanese symbol for beginner)	G (green heart)
I (growing heart)	R (red circle)	A (SOS button)	I (ribbon)
I (two hearts)	A (cloud)	O (tropical drink)	Y (fallen leaf)
P (grapes)	G (star)	O (full moon)	P (eggplant)
Y (blossom)	A (umbrella with rain drops)	P (roasted sweet potato)	G (teacup without handle)
B (blue circle)	K (registered)	R (pot of food)	G (seedling)
P (smiling face with horns)	O (large orange diamond)	I (growing heart)	I (cherry blossom)
G (four leaf clover)	N (toilet)	G (deciduous tree)	B (blue circle)
I (sparkling heart)	I (ribbon)	G (turtle)	I (growing heart)
B (kiss mark)	G (memo)	N (fallen leaf)	O (jack-o-lantern)
R (red circle)	B (dolphin)	W (multiply)	Y (maple leaf)
G (leaf fluttering in wind)	N (pile of poo)	G (leaf fluttering in wind)	R (wine glass)
G (teacup without handle)	B (sun)	A (mouse face)	I (two hearts)
W (face with tears of joy)	I (doughnut)	B (speaker high volume)	N (pile of poo)
O (large orange diamond)	R (white circle)	G (snake)	I (beating heart)

Fig. 1. χ^2 test results of the most unexpected emoji-color combinations per language.

4.2 Sentiment and Colors

For the sentiment analysis, we used the multilingual model `tabularisai/multilingual-sentiment-analysis` from HuggingFace. This model achieves high scores on existing benchmarks, despite being trained exclusively on synthetic data generated by large language models (LLMs). We computed sentiment scores for all tweets in the dataset and created aggregated statistics of sentiment across color terms (Fig. 2).

Overall, we observe a general trend of English tweets being more positive on average than tweets in the other three languages, an effect that has been documented previously [18]. Some colors display consistent sentiment patterns across languages: *black* is universally negative, while *purple* is predominantly positive, often appearing in beauty- and fashion-related contexts. *Gray* tends to be associated with negative sentiment, frequently in connection with weather-related topics.

Several colors exhibit language-specific sentiment differences. In Japanese, *white* is more negative than in English or German, reflecting its association with death in Japanese culture. *Green* is relatively positive in most languages, likely due to nature-related associations, but shows more variation in German, where it also has a strong political connotation. *Red* is largely neutral in English, negative in German and Japanese (e.g., the idiom "seeing red"), but positive in Russian, where it is historically and symbolically associated with the country (Fig. 4).

4.3 Topics and Colors

We analyzed topics using zero-shot classification with the multilingual model `MoritzLaurer/mDeBERTa-v3-base-xnli-multilingual-nli-2mil7` from Hugging Face, assigning each tweet to one of 18 categories commonly occurring in social media text. A visual inspection of the results indicated good alignment between predicted topics and tweet content. However, the color-based preselection of our dataset introduces a skew toward certain categories, such as *personal expression* and creative domains like *fashion*, *design*, and *advertising*. A manual sample check confirmed the plausibility of the predictions.

We then examined correlations between colors and topics (Fig. 3). Some patterns reflect dataset-specific artifacts, such as the German homonym *weiß* ("white" and "know"), which inflates topic counts in certain categories. Others are straightforward semantic matches, e.g., *orange* with the *food and drink* category. A strong association between *green* and *nature* appears in all languages except German, where *green* is more frequently linked to *news and politics* due to the prominence of the Green Party (cf. political sentiment effects in Sect. 4.2). *Yellow* appears infrequently overall, but occasionally in *sports and fitness* contexts (e.g., the "yellow card" in soccer). In Russian tweets, *blue* is associated with *sports*, often alongside *white*, reflecting the colors of Dynamo Minsk. *Pink* is predominantly tied to *beauty* topics, aligning with its stereotypical perception as a "girlish" color and matching its frequent co-occurrence with beauty-related emoji (see Sect. 4.1).

4.4 Emoji and Sentiment

Finally, we examine the relationship between emoji and sentiment, combining the analyses described in Sects. 4.1 and 4.2. Figure 4 shows the top 30 emoji by frequency for each language and overall, along with the average sentiment of tweets containing them.

Across all languages, heart emoji are consistently associated with positive sentiment. In contrast, the *face with tears of joy* tends to appear with negative sentiment, potentially reflecting its use in sarcasm or ironic commentary. The sets of frequently used emoji differ notably between languages. Japanese tweets feature several distinct heart emoji variants, as well as the *name badge*, *exclamation mark*, and the *OK gesture* (in person form, compared to the hand-only variant more common elsewhere). Russian tweets more often include the *snowflake*, *cherry blossom*, and cat emoji.

Some emoji are largely sentiment-neutral, such as the *camera*, which often accompanies tweets sharing images without explicit emotional content.

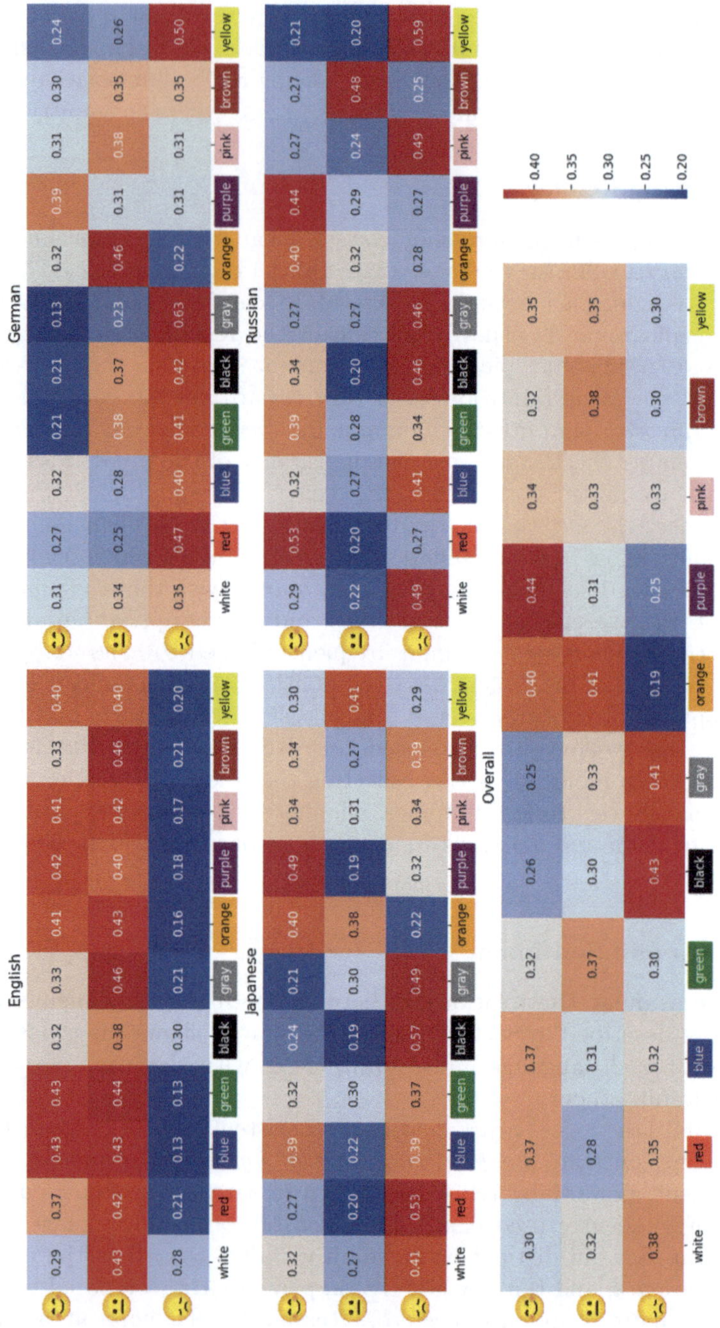

Fig. 2. Sentiment vs. color heatmaps, normalized per color.

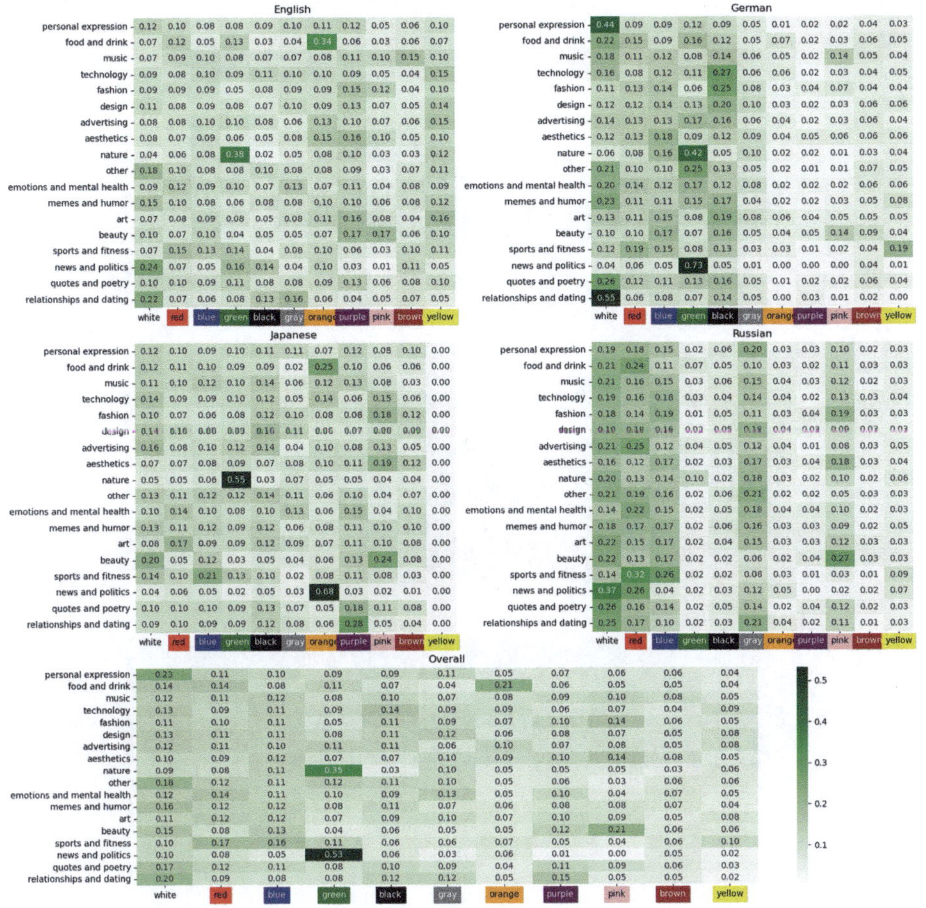

Fig. 3. Topics vs. color heatmaps, normalized per topic.

5 Conclusion

In this paper, we investigated the relationship between color terms and other linguistic and paralinguistic features in a multilingual dataset of tweets. By focusing on four languages and pre-filtering for occurrences of color terms, we were able to examine how color usage correlates with emoji, sentiment, and topical content, and how these associations vary across cultural and linguistic contexts.

Our analyses revealed both universal patterns and language-specific differences. Some associations were semantically straightforward, such as matching hearts or objects to their corresponding color terms, while others reflected culturally embedded meanings, political affiliations, or seasonal references. Sentiment analysis showed systematic differences across languages (e.g., English tending more positive overall) and across colors (e.g., *black* universally negative, *purple* generally positive), with notable cultural divergences such as the negative conno-

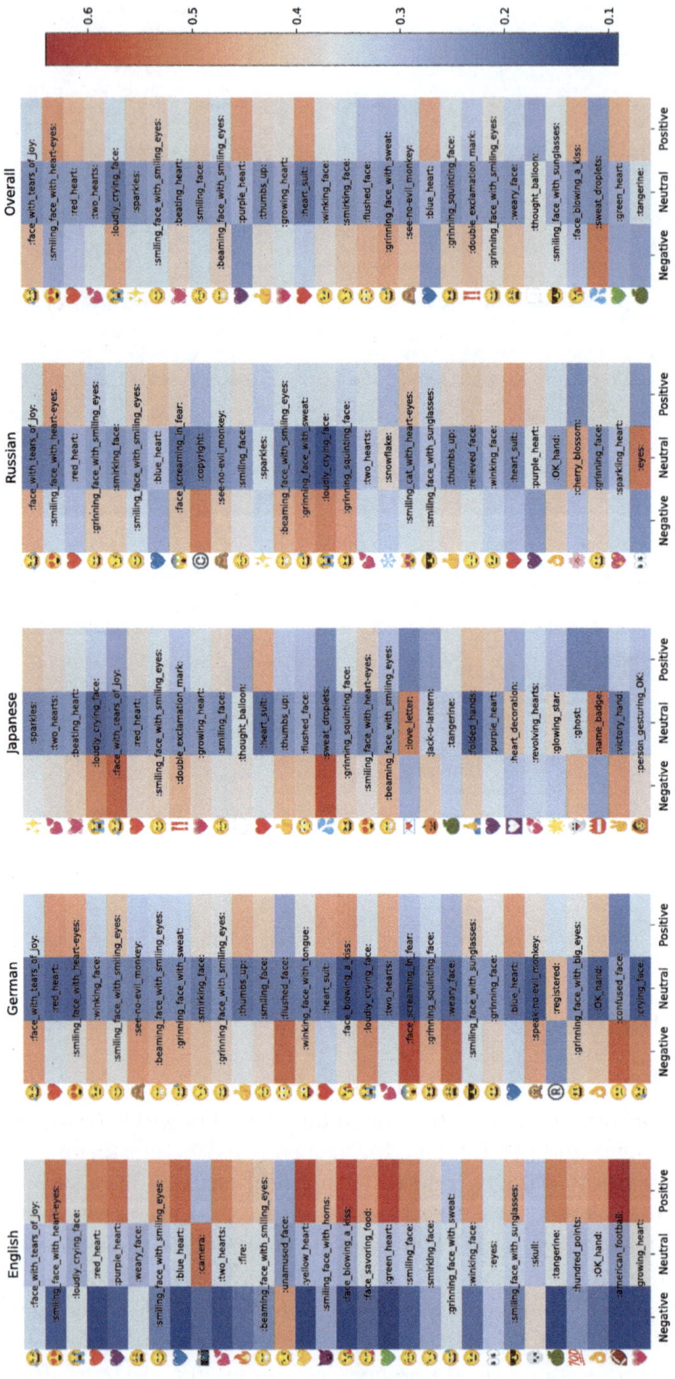

Fig. 4. Normalized sentiment vs. emoji heatmap.

tation of *white* in Japanese. Topic analysis further illustrated these differences, highlighting how certain colors are strongly tied to political, environmental, or cultural themes in specific languages. Emoji–sentiment correlations reinforced the interplay between visual and textual signals, with some emoji conveying consistent sentiment across languages and others varying in meaning.

While the color-focused filtering of the dataset introduces certain biases - particularly in topic distributions - it also sharpens the analysis of color-related semantics and usage patterns. These findings highlight that color terms in social media are not only lexical items but culturally situated symbols whose interpretation is shaped by linguistic, social, and political contexts.

6 Future Work

The present study focuses on a color-filtered subset of multilingual tweets, which enables targeted analyses but also introduces topic and distributional biases. Future work should address these limitations in several directions. First, analyses could be conducted on more general datasets not restricted to color terms, enabling a broader view of language–emoji–sentiment relationships. Extending the approach to additional languages, and potentially differentiating between regional varieties (e.g., British vs. American English), would further illuminate cultural and geographical variation. Second, integrating insights from cultural studies could help explain why certain correlations emerge, moving beyond descriptive statistics toward a richer interpretive framework. This could involve expanding the scope of data sources to include not only other social media platforms and online newspapers, but also culturally significant documents such as song lyrics, poems, and literature.

A promising avenue is to incorporate multimodal analysis by pairing textual data with attached images or videos. This would enable detection of actual colors in visual media and help distinguish literal from metaphorical color usage. Relatedly, applying metaphor and figurative language detection could reveal deeper cultural semantics, such as idiomatic expressions and symbolic uses of color.

Further work could also explore interactions between multiple colors in the same message, which may encode meanings related to national symbols, sports teams, or branding. Finally, adding a temporal dimension would make it possible to track how color associations with sentiment, topics, and emoji shift over time, capturing seasonal patterns, cultural events, and emerging trends.

References

1. Berlin, B., Kay, P.: Basic Color Terms: Their Universality and Evolution. University of California Press, Oakland (1991)
2. Uchikawa, K., Boynton, R.M.: Categorical color perception of Japanese observers: comparison with that of Americans. Vis. Res. **27**(10), 1825–1833 (1987)

3. Yu, H.-C.: A cross-cultural analysis of symbolic meanings of color. Chang Gung J. Hum. Soc. Sci. **7**(1), 49–74 (2014)
4. Hasyim, M.: Linguistic functions of emoji in social media communication. Opcion, vol. 35 (2019)
5. Abdou, M., Kulmizev, A., Hershcovich, D., Frank, S., Pavlick, E., Søgaard, A.: Can language models encode perceptual structure without grounding? A case study in color. arXiv preprint arXiv:2109.06129 (2021)
6. Patel, R., Pavlick, E.: Mapping language models to grounded conceptual spaces. In: International Conference on Learning Representations (2022)
7. Shagyrov, M., Shamoi, P.: Color and sentiment: a study of emotion-based color palettes in marketing. In: 2024 Joint 13th International Conference on Soft Computing and Intelligent Systems and 25th International Symposium on Advanced Intelligent Systems (SCISISIS), pp. 1–7 (2024). https://ieeexplore.ieee.org/abstract/document/10759889
8. Jonauskaite, D., et al.: Universal patterns in color-emotion associations are further shaped by linguistic and geographic proximity. Psychol. Sci. **31**(10), 1245–1260 (2020)
9. Mohammad, S.M.: Colourful language: measuring word–colour associations. In: Proceedings of the 2nd Workshop on Cognitive Modeling and Computational Linguistics (CMCL), Portland, Oregon, USA, pp. 97–106 (2011). https://aclanthology.org/W11-0611/
10. Languenou, E., Hurter, C.: Foundation model colormap: a framework for extracting and visualizing the foundation models' color knowledge. In: Proceedings of the Extended Abstracts of the CHI Conference on Human Factors in Computing Systems, pp. 1–8 (2025)
11. Stillman, M., Kruspe, A., Kirjakovski, A., Marutschke, M.: Understanding bias in language models: comparing the cultural distribution of basic colour terms in English, German, Japanese, and Russian. Int. J. Comput. Intell. Stud., submitted to Special Issue "Innovative Technology in Computational Intelligence" (2025)
12. Miller, H., Thebault-Spieker, J., Chang, S., Johnson, I., Terveen, L., Hecht, B.: "blissfully happy" or "ready to fight": varying interpretations of emoji. In: Proceedings of the AAAI International Conference on Web and Social Media (ICWSM), pp. 259–268 (2016). https://ojs.aaai.org/index.php/ICWSM/article/view/14757
13. Novak, P.K., Smailović, J., Sluban, B., Mozetič, I.: Sentiment of emojis. PLOS ONE **10**(12), e0144296 (2015). https://doi.org/10.1371/journal.pone.0144296
14. Ljubešić, N., Fišer, D.: A global analysis of emoji usage. In: Cook, P., Evert, S., Schäfer, R., Stemle, E., Eds., Proceedings of the 10th Web as Corpus Workshop, Association for Computational Linguistics, Berlin, pp. 82–89 (2016). https://aclanthology.org/W16-2610/
15. Kruspe, A., Häberle, M., Kuhn, I., Zhu, X.X.: Cross-language sentiment analysis of European twitter messages during the covid-19 pandemic. In: Proceedings of the Workshop on NLP for COVID-19 at ACL 2020, Virtual (2020)
16. Forrester, D., Winskel, H., Longstaff, M.: The effect of colour on processing and perception of emojis in a valence categorisation task. Psychol. Rep., 00332941251329796 (2025)
17. Marutschke, D.M., Su, Y., Ogawa, H.: Twitter-based color term analysis in English, German, and Japanese. ICIC Express Lett. **11**(2), 343–347 (2017)
18. Dodds, P.S., et al.: Human language reveals a universal positivity bias. In: Proceedings of the National Academy of Sciences, vol. 112, no. 8, pp. 2389–2394 (2015). https://www.pnas.org/doi/abs/10.1073/pnas.1411678112

Improving Fine-Grained Emotion Classification with Hierarchical Local Contrastive Learning

Uika Naruto(✉) and Tomonari Masada

Graduate school of Artificial Intelligence and Science, Rikkyo University, Tokyo, Japan
24vr018r@rikkyo.ac.jp, masada@rikkyo.ac.jp

Abstract. We propose a method for leveraging the hierarchical structure of emotion labels to improve fine-grained emotion classification performance. Stacking hierarchical classification layers can introduce errors that propagate from coarser to finer levels of classification. This may degrade fine-grained classification performance. Therefore, stacked hierarchical classifiers can be unsuitable when we aim to improve fine-grained classification performance. To address this issue, we incorporate contrastive learning into the fine-tuning of pretrained embedding models. Specifically, we apply Local Contrastive Learning (LCL) to each positive label and construct hard-negative pools using hierarchical relations including siblings and descendants. We also employ hierarchical scheduling that begins LCL at the finest level and adds coarser levels every fixed number of epochs. We further introduce two practical extensions: per-level top-k selection of high-scoring negatives, and explicit weighting of hard negatives. We evaluate our approach on the GoEmotions dataset using macro-F1 over the 28 fine-grained labels. Based on our experiments, we have revealed notable effects of hard-negative selection design on fine-grained emotion classification.

Keywords: emotion classification · hierarchical classification · contrastive learning

1 Introduction

In this study, we explore methods to improve the classification performance in classifying fine-grained emotion labels utilizing the hierarchical structure of emotion labels. Specifically, we address the classification problem of diverse emotion labels using GoEmotions by fine-tuning embedding models.

GoEmotions is a large-scale English sentiment classification dataset that contains 28 labels, including 27 distinct emotions and a neutral category. It also features a coarse-grained labeling scheme that categorizes the 27 labels into six classes based on Ekman's theory of emotions or three sentiment classes: positive, negative, and ambiguous[1]. We hypothesize that using this hierarchical

H. Unger and P. Meesad (Eds.): NLPIR 2025, LNNS 1904, pp. 279–287, 2026.
https://doi.org/10.1007/978-3-032-20897-2_20

structure could potentially improve the classification performance in classifying fine-grained emotion labels.

One common approach to taking advantage of this hierarchy is to stack hierarchical classification layers. However, an error in predicting a coarse label can propagate to fine labels, degrading overall performance, which is a challenge for this method.

Therefore, we introduce hard negatives utilizing the label hierarchy, adding a loss based on Local Contrast Learning (LCL) during classification. Furthermore, we adopt Hierarchical Local Contrastive Learning (HiLCL), which classifies fine-grained and coarse-grained labels in parallel and applies LCL sequentially from lower-level labels as learning progresses. Additionally, we make the following extensions:

- Expanded the selection criteria for hard negatives beyond sibling and descendant labels to include top-k logit-based selection.
- Introduced weighting for hard negatives.

We verify whether these methods improve classification performance compare to baseline classification fine-tuning.

2 Related Work

2.1 Dataset

The GoEmotions dataset used in this study is a large-scale dataset that classified 58,009 English Reddit comments into 28 labels: 27 distinct emotions and neutral. Three to five annotators assigned labels based on predefined emotion definitions, ensuring label consistency [1].

One characteristic of GoEmotions is the imbalance in sample sizes across emotion labels. Figure 1 shows the imbalance in the GoEmotions dataset.

Neutral, which contains no emotion, has a substantially large number of samples, followed by emotions such as admiration and approval. In contrast, emotions such as grief and pride have particularly few samples.

Furthermore, GoEmotions consists of approximately 83% single-label comments and 17% multi-label comments [1]. Multi-label instances occur when a single comment expresses multiple emotions or when annotators disagree on the label. Consequently, some data points are labeled with both neutral and other emotion labels simultaneously. These dataset characteristics contribute to the complexity of the classification problem. Table 1 presents representative examples of texts and their labels from the GoEmotions training set, illustrating both single-label and multi-label instances, including cases where neutral and other emotion labels are assigned simultaneously.

GoEmotions provides not only fine-grained emotion labels but also two coarser groupings: the sentiment level and the Ekman level, which form a hierarchical structure. These groupings are hierarchical [1].

- sentiment level: positive, negative, ambiguous and neutral (4 categorie)

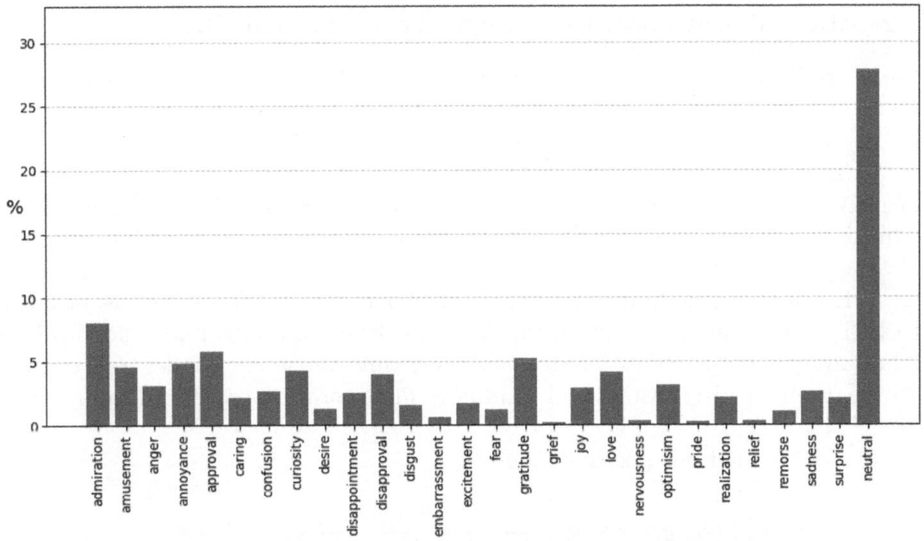

Fig. 1. Proportion of samples per emotion label in the GoEmotions dataset (adapted from [1]).

Table 1. Examples of texts and their labels (adapted from [1,2])

texts	labels
Sorry if Im late to the party but when does the new season start?Im getting withdrawals!	curiosity, excitement, joy
Uhhh... unfortunately I can't read the article due to some regulations... but pretty sure that post is just the article	disappointment, neutral
At least its not anything worse, and that you are still close to that person :)	approval

- Ekman level: anger, disgust, fear, joy, sadness, surprise and neutral (6 groups + neutral)

(adapted from [1])

The GoEmotions paper [1] used BERT for emotion classification. In this study, We first adopt a transformer model that is ranked near the top on the MTEB leaderboard [3], followed by another model with a similar leaderboard ranking.

Other recent studies have focused on improve performance on GoEmotions by augmenting the dataset [4], or leveraging the class definition defined in the GoEmotions dataset [5]. In this study, we take a different approach to improve fine-grained emotion classification.

2.2 Hierarchical Local Contrastive Learning (HiLCL)

We adopt Hierarchical Local Contrastive Learning (HiLCL) as the main improvement method. HiLCL consists of two components: Local Contrastive Learning (LCL) and the Hierarchical Learning strategy (HiLearn) [6].

LCL computes contrastive loss for each positive label by constructing a local negative pool. The negative pool includes sibling labels and descendant labels. This design encourages the model to separate semantically similar but incorrect classes [6].

HiLearn is a scheduling strategy that progressively adds higher-level labels to LCL. Specifically, it starts from the finest level and adds mid- and top-level labels every k epoch. This fine-to-coarse schedule allows the model to first focus on fine-grained distinctions and gradually incorporate hierarchical information [6].

LCL is defined per positive label as

$$\mathrm{LCL}(y_i) = -\log \frac{\exp(P_i)}{\exp(P_i) + \sum_{v_j \in \mathrm{Hard}(y_i)} \exp(P_j)} \tag{1}$$

following [6]. P_i denotes the logit for class y_i and $\mathrm{Hard}(y_i)$ denotes the hard-negative set for y_i.

HiLCL is defined as

$$\mathrm{HiLCL}(y, \mathrm{epoch}) = \sum_{y_i \in \mathrm{HiLearn}(y, \mathrm{epoch})} \mathrm{LCL}(y_i) \tag{2}$$

following [6].

3 Method

3.1 Text Encoder

We use Multilingual-E5-Large-Instruct (E5) [7] and average pooling as the pooling method. We select this model because it ranks among the top models on the MTEB leaderboard [3].

3.2 Classification Layer

We use a single linear layer to obtain classification logits. For Baseline Fine-tuning, we target 28 labels, or the fine-grained labels from GoEmotions. For HiLCL-augmented fine-tuning, we add coarse labels and classify a total of 37 labels using the same head. Although the hierarchical structure does not include neutral, it is handled by the same classification head.

3.3 Loss

The HiLCL loss is expressed by the following equation.

$$\widehat{\text{HiLCL}} = \frac{1}{|\mathcal{P}|} \sum_{i \in \mathcal{P}} \left[-\log \frac{\exp(P_i)}{\exp(P_i) + \sum_{j \in \text{Hard}(i)} \exp(P_j)} \right] \tag{3}$$

adapted from [6]. $\mathcal{P}$ is the set of positive labels. We follow the original HiLCL formulation, but we normalize the sum of LCL losses by the number of positive labels in the batch to avoid scale differences between batches and stabilize training.

We use Binary Cross-Entropy (BCE) loss as the classification loss function. For HiLCL-augmented Fine-tuning, we add HiLCL, which is BCE multiplied by a weighting coefficient.

$$\mathcal{L} = \mathcal{L}_{\text{BCE}} + \lambda \cdot \widehat{\text{HiLCL}} \tag{4}$$

adapted from [6].

3.4 Hierarchical Label Definition

We create a hierarchical label structure based on the groupings in the GoEmotions dataset. Figure 2 shows the complete label hierarchy used in this study.

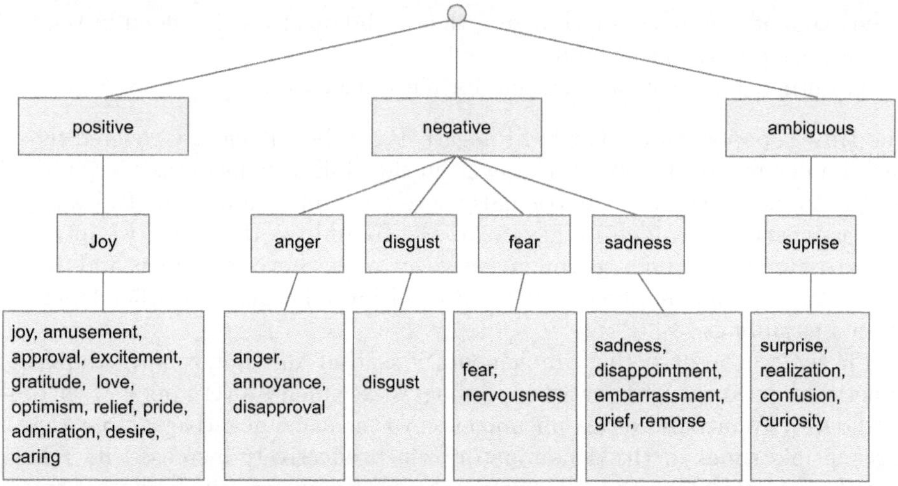

Fig. 2. Hierarchical label structure (adapted from [1]).

Note that neutral is shared across all grouping levels in GoEmotions [1]. In this study, we treat neutral only at the fine-grained level and do not include neutral in the higher-level groups (Ekman + sentiment).

3.5 Application of HiLCL

We compute LCL at each level. Specifically, we create a negative pool for each positive label and calculate the contrastive learning loss between logits. The pool primarily includes the following:

- Sibling labels: Labels at the same level that belong to the same parent group.
- Descendant label: A label that belongs to a lower-level group under a parent label (Ekman→fine-grained, sentiment→Ekman/fine-grained).

We consider multi-label data and excluded labels from the candidate pool if they are present in the ground truth. During the early training stages, we apply LCL only to lower levels and gradually incorporate higher levels following the HiLearn method.

3.6 Extension of Hard Negatives

In addition to the sibling/descendant approach described in the HiLCL paper, we implement the following extensions:

- Introduction of top-k negatives: We add the top-k negative examples with the largest model logits from the negative examples in that hierarchy as additional hard negative candidates. The top-k negative examples are selected per forward pass.
- Duplicate Handling: When the top-k candidates overlap with sibling or descendant examples, we configure the candidates to be included in the LCL calculation only once (duplicate removal).
- Weighting: Introduce a weighting for hard negatives.

The HiLCL paper states that hard negatives can be sufficiently covered by sibling and descendant labels [6]. However, in the GoEmotions dataset used in this study, sample imbalance is particularly severe for fine-grained labels, leading to misclassification into high-frequency labels. To address this issue, we introduce top-k to supplement high-scoring negative examples beyond siblings and descendants. Top-k is not applied in the top label layer because all other labels are defined as siblings.

Firstly, we observe that implementations that did not remove duplicates sometimes performed better when siblings/descendants overlapped with top-k candidates. We attribute this phenomenon to the same negative sample appearing multiple times in the denominator, which effectively increased its relative contribution. Based on this observation, we performe explicit duplicate removal and introduce a weighting scheme for hard negatives. The weighted LCL loss is expressed by the following equation.

$$\mathrm{LCL}^{\mathrm{w}}(y_i) = -\log \frac{\exp(P_i)}{\exp(P_i) + \sum_{j\in \mathrm{Hard}(y_i)} w_j \exp(P_j)} \tag{5}$$

adapted from [6].

4 Experiment

4.1 Evaluation

We use the macro-average F1 score across 28 labels (the fine-grained group) as the primary metric for comparing methods and tuning hyperparameters. Macro-average F1 is an appropriate metric for evaluating performance equally across all labels under class imbalance, and it has been adopted in existing GoEmotions research [1]. We repeat each experimental condition with three random seeds and compare the methods based on the average F1 score on the validation set. Average values are rounded to the fifth decimal place. All method comparisons in this paper are based on results from the validation set.

4.2 Comparison of Methods

We compare the following methods.

1. Baseline Fine-tuning: Standard BCE fine-tuning (28 labels).
2. HiLCL-augmented Fine-tuning (Siblings + Descendants): Add HiLCL using siblings and descendants as hard negatives, following the original paper.
3. HiLCL + top-k: In addition to the above, add the top-k with the highest logits within that hierarchy as hard negatives.
4. HiLCL + top-k + weighting: In addition to the above, introduce weights for all hard negatives to adjust their contribution.

The hyperparameter settings used for each experiment are provided in Table 2. Table 3 summarizes F1 scores. Among the tested methods, the most effective was HiLCL + top-k + weighting.

Table 2. Hyperparameter settings for E5 (HiLCL)

Method	HiLearn (epoch)	λ	top-k (fine, Ek, senti)	negative weight
HiLCL (siblings + descendants)	3	0.1	–	–
HiLCL + top-k	3	0.1	(3, 2, –)	–
HiLCL + top-k + weighting	3	0.1	(3, 2, –)	4.0

4.3 Application to Other Model

We apply the best method from E5 to another embedding model. Specifically, we use Qwen3-0.6B (Qwen3) [8] and last token pooling as the pooling method. We selected Qwen3 because it ranks higher than E5 on the MTEB leaderboard [3]. However, the Baseline Fine-tuning for Qwen3 achieved an F1 score of 0.5206, which was lower than the performance on E5.

Table 3. The experimental results (E5)

Method	F1 macro-average
Baseline Fine-tuning	0.5537
HiLCL-augmented Fine-tuning (Siblings + Descendants)	0.5559
HiLCL + top-k	0.5579
HiLCL + top-k + weighting	0.5607

First, we directly apply to Qwen3 the methods that were effective on E5, including their hyperparameter settings, but this did not lead to performance improvement (F1 = 0.5057). We then conduct a hyperparameter search, which revealed that increasing the top-k value at the fine-grained level improved classification performance. Based on this finding, we perform experiments in which HiLCL hard negative selection was restricted to top-k negatives at each level. As hard negative weighting was also effective, we retain it in this experiment. This configuration improved performance compared to Baseline Fine-tuning (F1 = 0.5246).

Finally, we conduct an additional experiment in which we completely remove the upper hierarchical labels and perform classification only on the fine-grained labels. In this setting, LCL with hard negative weighting was naturally applied only to the fine-grained level using top-k negatives (F1 = 0.5244).

Table 4 summarizes representative F1 scores for Qwen3, including the baseline, the best-performing HiLCL variant, and the fine-grained-only classifier + LCL (top-k) + weighting experiment. The hyperparameter settings used for each experiment are provided in Table 5.

Table 4. The experimental results (Qwen3)

Method	F1 macro-average
Baseline Fine-tuning	0.5206
HiLCL (top-k-only) + weighting	0.5246
fine-grained-only classifier + LCL (top-k) + weighting	0.5244

Table 5. Hyperparameter settings for Qwen3 (HiLCL)

Method	HiLearn (epoch)	λ	top-k (fine, Ek, senti)	negative weight
HiLCL (top-k-only) + weighting	3	0.1	(10, 1, 2)	0.5
fine-grained-only classifier + LCL (top-k) + weighting	-	0.1	(10, –, –)	0.5

These results suggest that for Qwen3, selecting top-k hard negatives based on high model outputs (logits) contributes more to performance improvement than basing HiLCL hard-negative selection on hierarchical relationships. Indeed, even compared to Qwen3s baseline, applying HiLCL using only top-k hard negatives and hard-negative weighting improved classification performance.

For the classifier and LCL, the HiLCL (top-k only) with weighting variant achieved the highest mean F1 in our Qwen3 runs (F1 = 0.5246), slightly higher than the fine-grained-only classifier with LCL (top-k) + weighting (F1 = 0.5244). Although the difference is small, this finding suggests that leveraging hierarchical information may still provide gains in fine-grained emotion classification.

5 Conclusion

This study demonstrates that for E5, incorporating HiLCL into emotion classification tasks by leveraging the hierarchical nature of emotion labels improves classification performance for fine-grained emotion labels. Furthermore, additional improvements are achieved by extending the hard-negative selection criterion and applying weighting.

For Qwen3, incorporating HiLCL with weighting into fine-grained emotion classification tasks can also slightly improve performance. However, model-specific tuning of the hard-negative selection policy is necessary.

These findings show that, although Local Contrastive Learning (LCL) with hard negative weighting itself benefits fine-grained emotion classification, the usefulness of leveraging hierarchical information for the classifier and for hard-negative selection depends on the embedding model and should be tuned per model.

References

1. Demszky, D., et al.: GoEmotions: a dataset of fine-grained emotions. arXiv:2005.00547 (2020)
2. Google Research, GoEmotions Dataset, GitHub (2020). https://github.com/google-research/google-research/tree/master/goemotions. Accessed 26 Sep 2025
3. Hugging Face, mteb/leaderboard. https://huggingface.co/spaces/mteb/leaderboard. Accessed 26 Sep 2025
4. Wang, K., et al.: Large language models on fine-grained emotion detection dataset with data augmentation and transfer learning. arXiv:2403.06108 (2024)
5. Singh, G., et al.: Fine-grained emotion prediction by modeling emotion definitions. arXiv:2107.12135 (2021)
6. Chen, Z., et al.: HiLight: a hierarchy-aware light global model with hierarchical local ConTrastive learning. arXiv:2408.05786 (2024)
7. Hugging Face, Multilingual-E5-large-instruct Model Card. https://huggingface.co/intfloat/multilingual-e5-large-instruct. Accessed 26 Sep 2025
8. Hugging Face, Qwen3-0.6B Model Card. https://huggingface.co/Qwen/Qwen3-0.6B. Accessed 26 Sep 2025

Financial Sentence Graph Attention Network (FinS-GAT)

Yilin Wang and Yang Liu(✉)

East China University of Political Science and Law, Shanghai 201620, China
lyang@ecupl.edu.cn

Abstract. Financial sentiment analysis plays a crucial role in understanding market dynamics and informing trading strategies. We introduce FinS-GAT, a two-stage framework that first fine-tunes a domain-adapted Transformer (FinBERT) to produce sentence embeddings, then trains a Graph Attention Network (GAT) on a static similarity graph of those embeddings. This decoupled design addresses GPU resource limitations while capturing global semantic relations among financial sentences. On the Kaggle Financial Sentiment Analysis dataset, FinS-GAT achieves 86.66% test accuracy, outperforming a FinBERT-only baseline 79.2% and demonstrating the efficacy of graph-based relational learning in the financial domain.

Keywords: Graph Neural Networks · Finance

1 Introduction

Automated sentiment classification of financial text, such as news headlines, earnings call transcripts, and analyst reports, has been shown to correlate with stock price movements and market volatility. Pre-trained language models like BERT [1] and its financial-domain variant FinBERT [2,3] achieve strong performance on sentence-level sentiment tasks. However, they treat each sentence independently, ignoring inter-sentence semantic relations that can reveal broader market narratives.

Graph Neural Networks (GNNs) [4] offer a principled way to model relationships among textual instances. Recent work has applied GNNs to document-level classification and multi-document reasoning [5], but few studies address sentence-level networks in finance. We propose FinS-GAT, which builds a static graph of sentence embeddings based on cosine similarity and trains a GAT [6] to propagate sentiment cues across semantically related sentences. To manage limited GPU memory, we adopt a two-stage pipeline: fine-tune FinBERT and freeze embeddings, then train the GAT separately.

Our experiments show that even this simple graph already helps the GAT propagate sentiment cues to semantically close sentences, which explains the gain over FinBERT-only. At the same time, a fixed threshold is indeed restrictive: it

H. Unger and P. Meesad (Eds.): NLPIR 2025, LNNS 1904, pp. 288–296, 2026.
https://doi.org/10.1007/978-3-032-20897-2_21

cannot adapt to local density and may miss weak but meaningful relations. Future work can therefore test adaptive graph builders (e.g., k-NN graphs, similarity quantiles, or attention-based learned adjacency) to see whether a more flexible connectivity pattern brings further gains, without abandoning the two-stage, low-cost pipeline.

Our contributions are as follows.

- A decoupled, resource-efficient two-stage training strategy combining Transformer fine-tuning with graph-based learning.
- A static sentence-similarity graph construction method tailored to financial text.
- Empirical validation on a public financial sentiment dataset, showing a ~7% absolute accuracy gain over a FinBERT-only baseline.

2 Related Work

Financial NLP. Domain-adapted language models, such as FinBERT [2,3] and FinGPT [7], refine BERT pre-training on financial corpora to improve sentiment detection in earnings calls and analyst reports.

Graph Neural Networks for Text. GNNs have been applied to text classification by constructing word-document graphs [5] and sentence-relation graphs[8]. Graph Attention Networks (GAT) [6] introduce self-attention to GNNs, enabling adaptive weighting of neighbor contributions.

Hybrid Architectures. Several studies integrate pre-trained Transformers with GNNs for document-level tasks [9], but often require joint training, straining computational resources. Our two-stage FinS-GAT pipeline separates embedding extraction from graph training, reducing memory footprint without sacrificing performance.

Recent work in financial sentiment and relational modeling has explored hybrid architectures, retrieval-augmented learning, and richer graph structures in addition to Transformer fine-tuning. For example, Zhang, Harris & Zheng (2025) propose a model performing GNN-based social media sentiment analysis integrated with stock market forecasting, showing that combining user sentiment and relational graphs enhances predictive performance [10]. Similarly, Jin & Zhang (2025) present a context-aware GNN for sentiment classification in Chinese microblogs, which incorporates both syntactic dependencies and dynamic sentiment embeddings to handle fine-grained shifts [11]. Dolphin et al. (2024) extract structured insights (tickers, summaries, sentiment) from financial news using an LLM-driven approach, enabling more granular inputs to down- stream models [12]. Pallucchini et al. (2025) introduce FLEX, a self-explanatory retrieval-augmented LLM system that enhances short financial sentences with external context to improve interpretability and classification accuracy [13]. Sonani, Badii & Moin (2025) offer a hybrid LSTM-GNN model that captures temporal and relational dependencies in stock price prediction [14]. Other works like Integrating Sentiment Analysis with Graph Neural Networks for Enhanced Stock Prediction: A Comprehensive Survey [15] and Enhancing

Financial Sentiment Analysis via Retrieval-Augmented Large Language Models [16] provide overviews or retrieval-based enhancements, respectively, emphasizing how relational and contextual signals can improve over baseline FinBERT-like models. The DGATS model [17] improves stock prediction by dynamically integrating sentiment, transaction, and textual data to capture evolving market relationships. The ML-GAT model [18] improves stock trend forecasting by dynamically updating stock relationships and integrating diverse data, achieving superior prediction performance with higher accuracy and interpretability. These studies supplement but do not fully exploit static sentence-graph structures over frozen embeddings, which is a strength of our FinS-GAT design.

3 Methodology

In this work, we introduce a scalable and resource-efficient approach for financial sentence sentiment classification by decomposing the training process into two sequential stages. This design 1 decouples the computationally intensive Transformer fine-tuning from the graph neural network (GNN) training, enabling effective use of limited GPU memory while preserving representation quality.

3.1 Dataset and Preprocessing

Our experiments utilize the Kaggle Financial Sentiment Analysis dataset, which comprises N sentences annotated with one of three sentiment labels: neutral, positive, or negative. Each sentence is first normalized by lowercasing and standard punctuation removal. Tokenization is performed using the WordPiece vocabulary of the FinBERT model, and a label mapping $\{\text{neutral}\ ' \rightarrow 0, \text{positive}\ ' \rightarrow 1, \text{negative}\ ' \rightarrow 2\}$ assigns integer targets yi for training. To evaluate generalization, we perform an 80/20 stratified split on sentence indices, ensuring that each sentiment class maintains its proportional representation in both training and test partitions (Fig. 1).

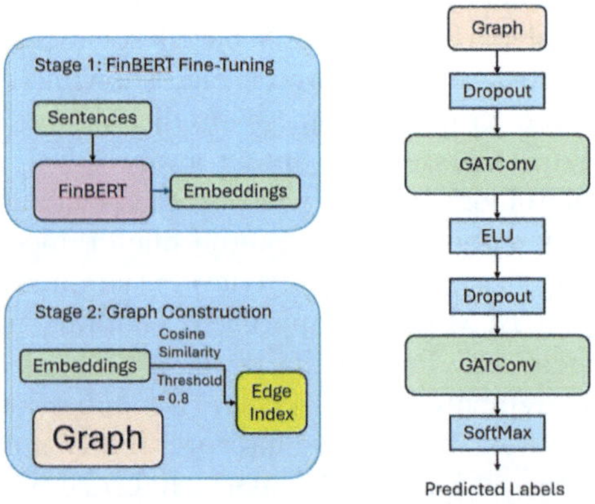

Fig. 1. Framework of Financial Sentence Graph Attention Network (FinS-GAT)

3.2 Stage 1: FinBERT Fine-Tuning and Embedding Extraction

In the first stage, we fine-tune the pre-trained FinBERT Transformer [3]on the three-class classification task. A lightweight linear classification head is appended to the pooled output vector of dimension 768, yielding logits

$$\mathbf{z}_i = \mathbf{W}_{\text{cls}}\text{FinBERT}(s_i;\theta) + \mathbf{b}_{\text{cls}}.$$

The joint parameters θ (the Transformer weights) and $\{\mathbf{W}_{\text{cls}}, \mathbf{b}_{\text{cls}}\}$ are optimized via the AdamW optimizer. We employ a learning rate of 2×10^{-5} for the Transformer backbone and a tenfold higher rate for the classification head, training for three epochs with a batch size of 32. Weight decay of 1×10^{-2} and standard cross-entropy loss ensure stable convergence. Upon completion, the fine-tuned model achieves strong sentence-level classification accuracy and produces robust sentence embeddings. We then freeze the Transformer parameters and extract the 768-dimensional pooled output vector for every sentence in the dataset, storing these embeddings for subsequent use.

3.3 Stage 2: Static Graph Construction and GAT Training

The second stage constructs a static similarity graph over the frozen sentence embeddings. We compute the cosine similarity between all embedding pairs, and insert an undirected edge between nodes i and j whenever their similarity exceeds a threshold of 0.8, excluding self-loops. This procedure yields a sparse adjacency structure that captures semantic relatedness among financial sentences. On this fixed graph, we train a two-layer Graph Attention Network (GAT) [6] to propagate sentiment information across linked sentences. Denoting the frozen embedding of node i as $\mathbf{h}_i$, the GAT updates are given by:

$$\mathbf{x}_i^{(1)} = \text{ELU}\left(\sum_{j\in\mathcal{N}(i)} \alpha_{ij}\mathbf{W}_1\mathbf{h}_j\right)$$

$$\mathbf{z}_i = \sum_{j\in\mathcal{N}(i)} \beta_{ij}\mathbf{W}_2\mathbf{x}_j^{(1)}$$

where attention coefficients α_{ij} and β_{ij} are computed via shared self-attention mechanisms and SoftMax normalization over each node's neighborhood. The model produces three-dimensional logits at each node, which are trained against the frozen labels using cross-entropy loss. We optimize the GAT parameters with AdamW at a learning rate of 2×10^{-4} and apply weight decay of 1×10^{-2}. Training proceeds for up to 50 epochs, evaluating on the held-out test nodes after each epoch. Early stopping with a patience of five epochs ensures that the model with the highest validation accuracy is retained.

We adopt a fixed cosine-similarity threshold to build the sentence graph mainly for stability and resource control: once the FinBERT embeddings are frozen, the graph can be computed once and reused for all GAT epochs, which is important under limited GPU memory. A static graph also makes the experimental setup reproducible, since the same sentences will always induce the same neighbors. We acknowledge, however, that this choice may omit useful edges for borderline sentences and may introduce noisy edges in dense regions. A natural extension is to replace the hard threshold with a k-nearest-neighbor construction or with a learned edge scorer that predicts whether two sentences should be connected; both would allow the graph to adapt to the embedding distribution while keeping our two-stage design.

3.4 Implementation and Hyperparameter Choices

All experiments were conducted on a single NVIDIA GPU. The two-stage pipeline requires memory only for either the Transformer forward pass (Stage 1) or the GAT forward-backward pass (Stage 2), avoiding simultaneous allocation of both models. Token sequences are padded or truncated to a maximum length of 128 tokens. During graph construction, we empirically found that a similarity threshold of 0.8 balances sparsity and connectivity. The GAT hidden dimension is set to 256 with four attention heads, and dropout of 0.1 is applied to both feature and attention weights. These design choices reflect a trade-off between expressive capacity and computational efficiency in the financial text domain.

4 Experiment

4.1 Dataset

We evaluate on the Kaggle Financial Sentiment Analysis dataset [19], containing 10,177 sentences labeled as neutral, positive, or negative. After an 80/20 stratified split, the training set comprises 8,141 sentences and the test set 2,036 sentences. Label distribution is 54% neutral, 32% positive, and 14% negative.

4.2 Baselines and Metrics

In this study, we compare our approach against three prominent baselines to evaluate its performance.

SVM-BERT: This method uses a Support Vector Machine (SVM) classifier applied to static BERT embeddings. Instead of fine-tuning the BERT model for the specific task, we extract embeddings from BERT and train an SVM on these features, providing a strong comparison between a neural network-based approach and a traditional machine learning method.

BERT [1]: Developed by Google in 2018, BERT (Bidirectional Encoder Representations from Transformers) is a pre-trained language model designed to create deep bidirectional representations by training on large amounts of unlabeled text. By capturing context from both directions of the text, BERT has set a new standard in natural language processing tasks. We use this baseline to examine the performance of a general-purpose transformer model that has been fine-tuned on the task.

FinBERT-only [2,3]: This baseline involves fine-tuning the FinBERT model with a linear classifier on the pooled outputs of the model. FinBERT is a variant of BERT specifically pre-trained on financial data, and this baseline focuses on its effectiveness in the target task with minimal modification.

To evaluate the performance of each baseline, we report both overall test accuracy and the macro-averaged F1 score. The macro-averaged F1 score is particularly useful in this context as it accounts for class imbalance by giving equal weight to each class, thus providing a more balanced measure of performance across all categories.

4.3 Implementation Details

Hyperparameters follow those in Sect. 3. FinBERT fine-tuning uses 3 epochs, batch size 32, learning rate 2×10^{-5}. GAT training uses up to 50 epochs, learning rate 2×10^{-4}, with early stopping (patience 5). Experiments run on an NVIDIA GPU.

5 Results

Table 1 summarizes performance.

Table 1. Test accuracy and macro-F1 on Financial Sentiment Analysis

Model	Accuracy	Macro-F1
SVM-BERT	0.4671	0.3265
BERT [3]	0.8050	0.7042
FinBERT-only [1][7]	0.7921	0.7149
FinS-GAT	**0.8666**	**0.7780**

Table 1 summarizes the performance of various models evaluated on the Financial Sentiment Analysis task, including their test accuracy and macro-averaged F1 score. Among the baselines,SVM-BERT achieves an accuracy of 46.71% and a macro-F1 score of 0.3265, which highlights the limitations of using static BERT embeddings combined with an SVM classifier. BERT, as a fine-tuned language model, performs significantly better, reaching an accuracy of 80.50% and a macro-F1 score of 0.7042. Similarly, FinBERT-only, which uses a fine-tuned version of BERT specifically pre-trained on financial data with a linear classifier, demonstrates strong performance with an accuracy of 79.21% and a macro-F1 score of 0.7149. However, the proposed FinS-GAT model substantially outperforms all the baselines, achieving an impressive accuracy of 86.66% and a macro-F1 score of 0.7780. This result highlights the effectiveness of incorporating sentence-level relational modeling through graph attention mechanisms, showcasing the significant improvements that can be achieved by capturing the dependencies between sentence elements. The substantial gain in both accuracy and F1 score underscores the value of the FinS-GAT approach in the context of financial sentiment analysis.

6 Discussion

Our results indicate that sentence similarity graphs capture complementary information to Transformer embeddings. The GAT layers propagate sentiment signals among semantically related sentences, improving the classification of ambiguous cases. The two-stage design enables efficient training on commodity GPUs, though it precludes end-to-end gradient flow from the GAT back into FinBERT.

Limitations include reliance on a fixed similarity threshold ($\tau = 0.8$). Future work could explore the learned adjacency or dynamic graph construction. Moreover, incorporating external knowledge graphs (e.g., company relationships) may further enhance performance.

Financial text has unique characteristics compared to general text classification tasks. Unlike movie reviews or product reviews, financial documents often contain domain-specific jargon, abbreviations, and subtle linguistic cues where sentiment is highly context dependent (e.g., "liability decreased" is positive in accounting, while "growth slowed" is negative despite being superficially positive). This makes financial sentiment classification especially challenging, as models must capture both technical terminology and nuanced polarity shifts. The main advantage of the method proposed in this article—FinS-GAT—is that it goes beyond sentence-level classification by incorporating semantic relationships between sentences into a graph structure. While FinBERT provides strong domain-adapted embeddings, FinS-GAT leverages inter-sentence similarities through graph attention, allowing sentiment signals to propagate across related nodes. This helps disambiguate difficult or borderline sentences by contextualizing them within their neighborhood, leading to higher accuracy and robustness compared to treating each sentence in isolation, all while keeping computational costs manageable by decoupling FinBERT fine-tuning and GNN training.

Our method builds a static similarity graph using a fixed cosine threshold. This is simple and efficient, but it can be too rigid: some useful links may be missed, while others may add noise. In future work, adaptive graphs could be explored, for example, by connecting each sentence to its k nearest neighbors or by learning edge weights directly with attention. These methods could make the model more flexible while still keeping the main design.

Compared with other financial sentiment models such as FinBERT, FinGPT, or retrieval-augmented LLMs, FinS-GAT offers a balance between performance and efficiency. Many advanced models rely on heavy computation or external data. Our design keeps costs lower by separating embedding extraction from graph training, while still improving accuracy by letting sentiment signals move across related sentences. This makes FinS-GAT practical for real uses, such as building sentiment indices or supporting trading strategies, where both accuracy and speed matter.

Compared with other recent financial sentiment methods that rely on large instruction-tuned LLMs or retrieval augmentation, our approach is lighter but still competitive because it explicitly models sentence–sentence relations. LLM-based systems often need external context, longer prompts, or end-to-end fine-tuning, which increases inference cost; FinS-GAT keeps the heavy part (FinBERT) in Stage 1 and lets Stage 2 do the relational reasoning cheaply. This makes the model easier to deploy in practical finance scenarios (monitoring news streams, building sentiment indices, supporting trading rules) where both latency and reproducibility matter.

7 Conclusion

We presented FinS-GAT, a resource-efficient two-stage framework combining FinBERT fine-tuning with graph-based sentiment propagation. Empirical results on a benchmark financial sentiment dataset show a 7% improvement in accuracy over strong Transformer baselines. FinS-GAT offers a practical approach for incorporating relational reasoning into financial NLP.

Acknowledgements.
This work was supported by the China's National Key Research and Development Program (2023YFC3306100, 2023YFC3306105, 2023YFC3306103). The authors would like to thank Prof. Qiqi Gao for his continuous support throughout the research process. The authors thank Yichen Huang, a senior graduate student in our research group, for his helpful suggestions during this work.

References

1. Devlin, J., Chang, M.-W., Lee, K., Toutanova, K.: BERT: pre-training of deep bidirectional transformers for language understanding. In: Proceedings of the 2019 Conference of the North American Chapter of the Association for Computational Linguistics: Human Language Technologies, vol. 1 (Long and Short Papers), pp. 417–4286 (2019)
2. Araci, D.: FinBERT: financial sentiment analysis with pre-trained language models. arXiv preprint arXiv:1908.10063 (2019)
3. Yang, Y., Uy, M.C.S., Huang, A.: FinBERT: a pre-trained language model for financial communications. arXiv preprint arXiv:2006.08097 (2020)
4. Wu, Z., Pan, S., Chen, F., Long, G., Zhang, C., Yu, P.S.: A comprehensive survey on graph neural networks. IEEE Trans. Neural Networks Learn. Syst. **32**(1), 4–24 (2020)
5. Yao, L., Mao, C., Luo, Y.: Graph convolutional networks for text classification. Proc. AAAI Conf. Artif. Intell. **33**, 7370–7377 (2019)
6. Velickovic, P., Cucurull, G., Casanova, A., Romero, A., Lio, P., Bengio, Y.: Graph attention networks. Stat **1050**(20), 10–48550 (2017)

7. Liu, X.-Y., Wang, G., Yang, H., Zha, D.: FinGPT: democratizing internet-scale data for financial large language models. arXiv preprint arXiv:2307.10485 (2023)
8. Song, L., Wang, Z., Yu, M., Zhang, Y., Florian, R., Gildea, D.: Exploring graph-structured passage representation for multi-hop reading comprehension with graph neural networks. arXiv preprint arXiv:1809.02040 (2018)
9. Ding, M., Zhou, C., Chen, Q., Yang, H., Tang, J.: Cognitive graph for multi-hop reading comprehension at scale. arXiv preprint arXiv:1905.05460 (2019)
10. Zhang, P., Harris, R.D.F., Zheng, J.: GNN-based social media sentiment analysis for stock market forecasting and trading. Expert Syst. Appl. **291**, 128425 (2025)
11. Jin, Z., Zhang, Y.: A graph neural network-based context-aware framework for sentiment analysis classification in Chinese microblogs. Mathematics **13**(6), 997 (2025)
12. Dolphin, R., Dursun, J., Chow, J., Blankenship, J., Adams, K., Pike, Q.: Extracting structured insights from financial news: an augmented LLM driven approach. arXiv preprint arXiv:2407.15788 (2024)
13. Pallucchini, F., Zhang, X., Mao, R., Cambria, E.: Self-explanatory and retrieval-augmented LLMs for financial sentiment analysis (flex). In: 40th ACM/SIGAPP Symposium on Applied Computing (SAC '25) (2025)
14. Sonani, M.S., Badii, A., Moin, A.: Stock price prediction using a hybrid LSTM-GNN model: integrating time-series and graph-based analysis. arXiv preprint arXiv:2502.15813 (2025)
15. Das, N., Sadhukhan, B., Chatterjee, R., Chakrabarti, S.: Integrating sentiment analysis with graph neural networks for enhanced stock prediction: a comprehensive survey. Decis. Anal. J. **10**(1), 100417 (2024)
16. Zhang, B., Yang, H., Zhou, T., Babar, A., Liu, X.-Y.: Enhancing financial sentiment analysis via retrieval-augmented large language models. arXiv preprint arXiv:2310.04027 (2023)
17. Zhang, Q., Zhang, Y., Yao, X., Li, S., Zhang, C., Liu, P.: A dynamic attributes-driven graph attention network modeling on behavioral finance for stock prediction. ACM Trans. Knowl. Discov. Data **18**(1), 1–29 (2023)
18. Huang, K., Li, X., Liu, F., Yang, X., Yu, W.: ML-GAT: a multilevel graph attention model for stock prediction. IEEE Access **10**, 86408–86422 (2022)
19. Malo, P., Sinha, A., Korhonen, P., Wallenius, J., Takala, P.: Good debt or bad debt: detecting semantic orientations in economic texts. J. Am. Soc. Inf. Sci. **65**(4), 782–796 (2014)

Explainable Aspect-Based Sentiment Analysis of Student Feedback Using BERT and SHAP

Phuripoj Kaewyong[1] and Phayung Meesad[2(✉)]

[1] Suan Dusit University, Dusit, Bangkok 10300, Thailand
phuripoj_kae@dusit.ac.th
[2] King Mongkut's University of Technology North Bangkok, Bangsue, Bangkok 10800, Thailand
phayung.m@itd.kmutnb.ac.th

Abstract. Without a doubt, student feedback is one of the driving forces behind better teaching quality. However, handling many open-ended comments can be very challenging, primarily when they address different parts of the course and express mixed opinions. Most of the time, sentiment analysis provides only an overall polarity, thereby eliminating differences at the aspect level. In contrast, deep learning models do not provide the interpretability that educational decision-making needs. This document introduces a clarifiable aspect-based sentiment analysis system that combines a BERT-based classifier with SHAP word-level attributions. We discover and confirm that the three main aspects are representative of higher education evaluations: Teaching Quality, Course Content, and Assessment Fairness. Using 2,450 annotated comments ($\kappa = 0.81$), our model achieves 89.1% accuracy and 88.0% F1-score; hence, it is considerably better than the benchmark lexicon, machine learning, and deep learning models ($p < 0.032$). Furthermore, the SHAP explanations closely align with the experts' reasoning (94.3% agreement) and are highly consistent with it. A stakeholder survey involving 15 educators revealed that 87% were study participants, who liked the study and rated the actionability at 4.4/5. Combining aspect-level prediction with clear explanations, according to these findings, makes it possible for instructional improvement to have a higher impact and be more informed.

Keywords: Aspect-Based Sentiment Analysis · Explainable AI · SHAP · Student Feedback · Natural Language Processing

1 Introduction

Student feedback is essential for improving teaching quality, yet analyzing hundreds of open-ended comments is time-consuming and subjective. Traditional sentiment analysis produces document-level ratings that obscure aspect-specific opinions. Consider: *"The teacher is excellent and passionate, but exams don't match lectures."* Document-level analysis labels this neutral, losing distinct positive (teaching) and negative (assessment) sentiments. Moreover, deep learning models are black boxes. The faculty cannot understand classifications, undermining trust in high-stakes decisions. Despite advances in

H. Unger and P. Meesad (Eds.): NLPIR 2025, LNNS 1904, pp. 297–306, 2026.
https://doi.org/10.1007/978-3-032-20897-2_22

ABSA and explainable AI, three gaps remain: (1) ABSA methods focus on commercial domains without educational adaptation, (2) no work integrates SHAP with aspect-level classification for fine-grained explanations, and (3) existing work validates only technical metrics without stakeholder studies.

We address these gaps through explainable aspect-based sentiment analysis for education. Our BERT-based model analyzes sentiment across three validated aspects while SHAP generates word-level explanations.

Contributions. (1) **Methodological:** Novel BERT+SHAP integration demonstrating explicit aspect encoding outperforms multi-task learning by 2.3%, achieving 89.1% accuracy with interpretable attributions. (2) **Resources:** Validated three-aspect taxonomy covering 94% of feedback and 2,450 annotated evaluations ($\kappa = 0.81$). (3) **Interpretability:** Rigorous SHAP validation through faithfulness (41.2% flip rate), consistency (Jaccard 0.71 ± 0.14), and expert comparison (94.3% agreement). (4) **Impact:** Stakeholder validation ($n = 15$) showing 87% preference and 4.4/5 actionability, bridging NLP capabilities and educational practice.

The remainder is organized as follows. Section 2 reviews related work. Section 3 describes methodology. Section 4 presents results. Section 5 discusses findings and limitations. Section 6 concludes.

2 Related Work

2.1 A Subsection Sample

Our work intersects three research areas: sentiment analysis in education, aspect-based sentiment analysis, and explainable AI. Table 1 positions our contributions.

Educational sentiment analysis progressed from lexicon-based [1, 2] to SVM (71–72% F1) [3] and deep learning (76–78%) [4, 5], but remained document-level. Recent deep learning reached 76–78% [4, 5], though Sindhu et al.'s BiLSTM performed only aspect extraction without per-aspect sentiment identifying *what* students discuss but not *how* they feel about each aspect.

ABSA methods, standardized through SemEval [6], progressed from LSTM+attention [7] to BERT approaches [8, 9]. Sun et al.'s auxiliary sentences and Xu et al.'s post-training achieved 84–87% F1 on restaurant reviews. However, all focus on commercial domains. Educational feedback exhibits distinct characteristics: formal evaluative language, implicit criticism for politeness, domain-specific terminology. No prior work systematically adapted ABSA to education or validated utility with stakeholders.

Explainability research evolved from attention visualization [10] shown unreliable by Jain and Wallace [11] to SHAP [12]. SHAP provides principled Shapley value explanations. While applied to text classification [13] and document sentiment [14], no work integrates SHAP with aspect-level analysis or validates explanations in educational contexts.

The systematic gap fill indicates there are three gaps identified in ABSA literature. First, none have approached adaptation to the characteristics of educational domains, nor have they validated taxonomies on stakeholders. Second, SHAP can be implemented on

Table 1. Comparison with related work: domain, capabilities, and validation

Work	Domain	Method	Asp	Per-Asp	XAI	Valid	F1
Educational SA							
[1]	Edu	Lexicon	No	No	No	No	–
[3]	Edu	SVM	No	No	No	Tech	71.9
[4]	MOOC	CNN	No	No	No	Tech	76.3
[5]	Edu	BiLSTM	Ext	No	No	Tech	78.0
ABSA (Commercial)							
[7]	Rest	LSTM	Yes	Yes	Attn*	Tech	81.0
[8]	Rest	BERT	Yes	Yes	No	Tech	84.0
[9]	Prod	BERT	Yes	Yes	No	Tech	87.0
Explainable AI							
[10]	NMT	Attn	–	–	Attn*	No	–
[14]	SA	BERT	No	No	SHAP	Tech	–
Ours	**Edu**	**BERT**	**3**	**Yes**	**SHAP**	**Both**	**88.0**

Note: *Attention unreliable [11]. Both = Technical + Stakeholder (n = 15). Asp = Aspects, Per-Asp = Per-aspect sentiment, Valid = Validation, Ext = Extraction only.

document-level sentiment [14] but not on aspect-based classification calibrate to aspects. Third, there is validation on technical aspects but no validation on aspects concerning stakeholders' utility.

As Table 1 shows, our work uniquely addresses all three dimensions: validated educational ABSA (89.1% accuracy, $\kappa = 0.81$), principled SHAP explanations (94.3% expert agreement, 41.2% faithfulness), and stakeholder validation (87% preference, 4.4/5 actionability).

3 Methodology

3.1 Problem Formulation

Given feedback $s = \{w_1,...,w_n\}$ and aspects A = {Teaching Quality, Course Content, Assessment Fairness}, predict sentiment $y_a \in$ {Positive, Neutral, Negative} for each a $\in$ A and generate SHAP values $\varphi = \{\varphi_1,...,\varphi_n\}$ representing word contributions.

3.2 System Architecture

We conducted stakeholder interviews (n = 8) and found three areas that made up 94% of the feedback: (1) **Teaching Quality:** how the teacher teaches, how they talk to students, and how excited they are; (2) **Course Content:** how relevant the curriculum is, how well it is organized, and what resources are available; (3) **Assessment Fairness:** how fair the tests are and how clear the grading is.

Our BERT-base model uses aspect-aware encoding: BERT encoder → Aspect embedding (768-dim) → Multi-head attention (8 heads) → Pooling → MLP classifier (768 → 256 → 3) with ReLU and dropout (0.1). We train separate models per aspect (2.3% better than multi-task), using AdamW (lr = 2×10^{-5}, batch = 16, 5 epochs).

For SHAP, we implement KernelSHAP with 500 perturbations (0.35 s/sample), balancing cost with stability.

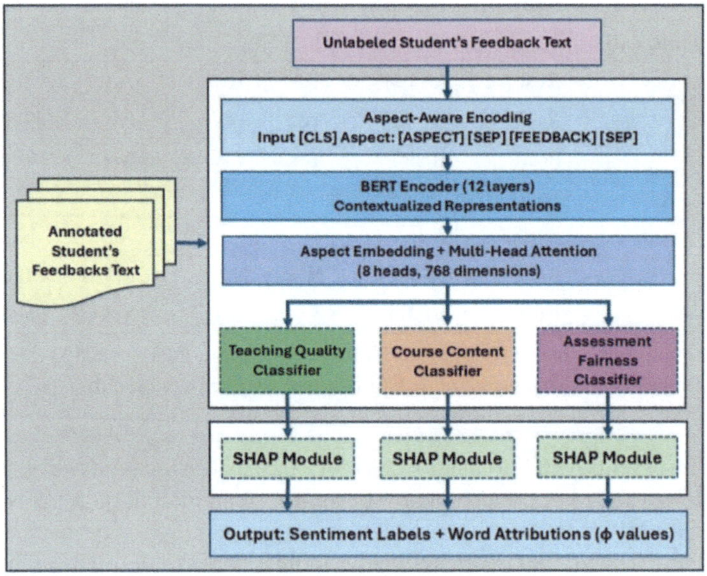

Fig. 1. System Architecture Overview.

Figure 1 System architecture: BERT encoder with aspect-aware encoding, aspect-specific classifiers, and SHAP modules producing sentiment predictions with word-level attributions.

3.3 Dataset

We conducted data gathering from 59 courses in 5 departments between 2022 and 2024 approximately 2,450 comments. The comments ranged from 8 to 247 words, with an average of 52 words. Three annotators (two from the education faculty and one from the NLP research group) labeled aspects and sentiment, and they all agreed very strongly (κ = 0.81 for aspects and κ = 0.82 for sentiment). 70% of the data (1,715) is for training, 15% (368) is for validation, and 15% (367) is for testing.

3.4 Baselines and Metrics

Baselines: VADER (lexicon), SVM-TFIDF, BiLSTM-Attention, BERT-Base, and AEN-BERT are the baselines. **Metrics:** F1 (macro), Accuracy, Precision, and Recall. **Quality**

of explanation: Faithfulness (change in prediction), Human Agreement (200 samples), and Consistency (Jaccard, Spearman's ρ).

4 Experiments and Results

4.1 Implementation

PyTorch 2.0 with Transformers 4.30 on an NVIDIA A100. Training takes 55 min per aspect, and SHAP can process 95 samples per second (Table 2).

4.2 Overall Performance

Table 2. Performance comparison across baseline methods

Model	Accuracy	Precision	Recall	F1-Score
VADER (Lexicon-based)	62.3%	58.7%	55.2%	56.8%
SVM-TFIDF	73.4%	72.1%	71.8%	71.9%
BiLSTM-Attention	79.8%	78.4%	78.1%	78.2%
BERT-Base	84.6%	83.3%	82.9%	83.1%
AEN-BERT	87.1%	85.9%	85.4%	85.6%
Our Model (BERT + SHAP)	**89.1%***	**88.2%**	**87.7%**	**88.0%**

*All improvements over baselines statistically significant ($p \leq 0.032$, paired t-test)

Our model works best, with an accuracy of 89.1% and an F1 score of 88.0%. outperforming all baselines. The 2.0-point F1 improvement over AEN-BERT ($p = 0.019$) shows that our architecture works. Our 31.2-point gain shows how useful contextualized representations are when compared to VADER (56.8% F1). The progression from SVM-TFIDF (71.9%) to BiLSTM (78.2%) to BERT-Base (83.1%) to Ours (88.0%) shows that pre-trained models with domain adaptation work outperform all baselines traditional methods.

Analysis of Errors. Error analysis (39 misclassifications) shows boundary confusion (Positive↔Neutral: 16, Neutral↔Negative: 21) with only 2 extreme flips, indicating subtle distinction challenges rather than fundamental flaws (Table 3).

Every part of the architectural has an effect on how well it works. Aspect encoding offers the most significant enhancement (+2.3%), indicating that explicit aspect information significantly enhances BERT's capacity to concentrate on pertinent text segments. Multi-head attention gives you an extra 1.8% by letting you change your attention patterns across different aspects. Separate aspect-specific models are 2.3 percentage points better than multi-task learning, which shows that our design choice was correct even though it costs more to train (Table 4).

Table 3. Ablation study showing contribution of architectural components

Model Variant	F1-Score	Δ
BERT-Base (document-level, no aspect encoding)	83.1%	baseline
+ Aspect textual encoding in input	85.4%	+2.3%
+ Multi-head attention (8 heads)	87.2%	+1.8%
+ Aspect-specific training (separate models)	**88.0%**	**+0.8%**
Alternative: Multi-task learning (shared model)	85.7%	−2.3%

Table 4. Per-aspect F1-scores showing balanced performance

Aspect	F1-Score	Positive	Neutral	Negative
Teaching Quality	**89.9%**	91.7%	88.6%	89.4%
Course Content	**87.2%**	88.9%	86.1%	86.7%
Assessment Fairness	**86.6%**	87.6%	85.7%	86.4%

4.3 Per-Aspect Performance

The performance is a steady 86.6–89.9% F1 across all aspects which points to generalization not over-fitting. Sentiment polarities on individual features perform comparably under all aspects, i.e. there is no systematic bias present.

4.4 Interpretability Validation

We validate explanation quality through three complementary metrics. Faithfulness: Removing SHAP's top-5 words causes ΔPred $= 0.68 \pm 0.21$ and 41.2% prediction flips versus 8.4% for random removal (4.9$\times$ more influential). Human agreement: Two experts independently identified influential words in 200 samples without seeing SHAP outputs using a blinded protocol. The three-phase protocol consisted of: (1) individual review and word selection, (2) independent documentation without discussion, and (3) comparison with SHAP outputs after completion. SHAP achieved 94.3% agreement ($\geq$3 matching words), Precision@5 $= 0.76$, Recall@5 $= 0.82$, with high inter-rater reliability ($\kappa = 0.89$). Consistency: For 150 similar feedback pairs (cosine>0.8 using BERT-base-uncased embeddings), SHAP maintains Jaccard $= 0.71 \pm 0.14$ and Spearman's $\rho = 0.78 \pm 0.12$, demonstrating stable explanations.

4.5 SHAP Visualization

Figure 2 SHAP word attributions in three cases: (a) Clear positive with "excellent"(+0.42), "helpful"(+0.38) driving positive prediction; (b) Subtle negative via "could have been"(−0.31), "clearer"(−0.28); (c) Mixed sentiment "Excellent teaching"(+0.45) yields positive Teaching Quality, while "textbook outdated"(−0.41) yields negative Course Content.

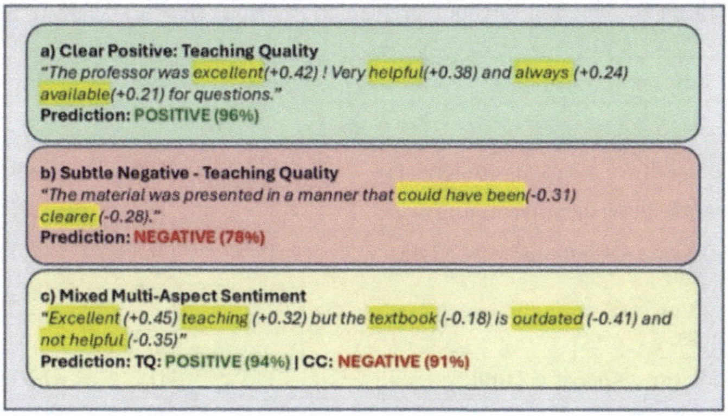

Fig. 2. SHAP Visualization Examples

4.6 Stakeholder Evaluation

To assess real-world utility, we conducted comprehensive stakeholder evaluation comparing our explainable aspect-based system against traditional document-level sentiment analysis.

Study Design. We recruited 15 stakeholders from Suan Dusit University by purposive sampling: 10 faculty members from three departments with 3–20 years teaching experience (mean = 9.2, SD = 5.1), three academic administrators, and two teaching assistants. Individual 60-min sessions followed a five-phase protocol in private offices. The five phases were: (1) informed consent and briefing, (2) baseline system review, (3) break period, (4) proposed system review, and (5) questionnaire and interview.

After informed consent, participants reviewed 10 randomly selected student feedback comments with VADER-generated document-level sentiment (Positive/Neutral/Negative labels only), then after a break, reviewed the same comments with our system displaying aspect-level predictions (3 aspects $\times$ 3 sentiments), SHAP word attributions with color-coded highlights (green for positive, red for negative, intensity proportional to magnitude), and top-5 attribution values per aspect.

Participants completed Likert-scale questionnaires (1 = Strongly Disagree to 5 = Strongly Agree) assessing Clarity, Trustworthiness, Actionability, and Comparative Preference. Following the quantitative assessment, semi-structured interviews probed four key areas: (1) perceived system value and most useful features, (2) practical concerns regarding classroom implementation, (3) potential integration strategies within existing teaching workflows, and (4) concrete examples of insights that could inform instructional decisions. Interview responses were audio-recorded, transcribed, and analyzed using thematic coding to identify recurring patterns and stakeholder priorities. Presentation order was randomized (n = 8 proposed first, n = 7 baseline first). Mann-Whitney U test showed no significant order effect ($p > 0.31$) (Table 5).

Results show high ratings across dimensions. Critically, 87% (13/15) preferred our approach over baselines, and 93% would recommend implementation. High actionability (4.4/5) indicates directly usable insights. Faculty valued aspect-level feedback: *"The*

Table 5. Detailed stakeholder evaluation results (n = 15, scale 1–5)

Dimension & Items	Mean	SD	Agreement*
Clarity			
Aspect-level breakdown is easy to understand	4.5	0.6	14/15
Highlighted words show clear reasoning	4.3	0.7	13/15
Quick identification of strengths/weaknesses	4.2	0.8	13/15
Subtotal	**4.3**	**0.7**	
Trustworthiness			
Sentiment predictions appear accurate	4.2	0.7	13/15
Highlighted words align with interpretation	4.1	0.9	12/15
Trust for course improvement decisions	4.0	0.8	12/15
Subtotal	**4.1**	**0.8**	
Actionability			
Provides specific improvement areas	4.6	0.5	15/15
Can develop concrete action plans	4.5	0.6	14/15
More useful than traditional evaluations	4.2	0.7	13/15
Would use for my courses	4.3	0.7	14/15
Subtotal	**4.4**	**0.6**	
Overall Satisfaction	**4.2**	**0.7**	

aAgreement = ratings 4–5 (Agree/Strongly Agree)

highlighted words show exactly what students praise or criticize. 'Engaging' tells me my approach works; 'rushed' means adjust pacing."

5 Discussion

The findings indicate that to understand the students' reactions, it is very important to clearly outline the different divisions of the feedback. Merging feedback into one single entity most of the time leads to the overall sentiment being shown as neutral or average, thus, the different opinions being overlooked. Our approach distinguishes the primary fields like Teaching Quality, Course Content, and Assessment Fairness. In this manner, teachers learn which educational practices are working and which need to be changed. Hence, educators become capable of more targeted self-reflection rather than vague, broad interpretations of course evaluations.

The contrasts of the BiLSTM, BERT-Base, and our aspect-aware model reveal that contextual representations that are in line with domain-specific constructs are the most beneficial. BERT's upgrade over the previous architectures is evidence that contextual language modeling is the main factor. Still, the further improvements from aspect-aware encoding suggest that just enhancing the quality of the representation is not sufficient;

we should construct the model in a way that follows the feedback pattern of the teaching practice.

In addition, explainability is the primary factor driving the system's effectiveness. SHAP is more than a distributive attention visualizer—it offers token-level attribution based on a principled explanation model. The high agreement with human annotators (94.3%) and the high flip rate during word removal together indicate that these accounts reflect the model's actual reasoning process, which means that agreement and correlation are not at the surface level. The evidence that SHAP explanations are faithful and aligned with how the model makes decisions, therefore, gives faculty members the confidence and freedom to understand why certain comments have led to specific classifications.

The stakeholder study also supports the practical usefulness of this. One of the things that the faculty members emphasized besides the system's precision, was that what mattered most to them was that the words that were highlighted referred to changes in teaching that were clear and doable. A couple of examples can be given, *"rushed pacing"* or *"lack of clarity on grading"* as both indicate instructional innovation very directly. Hence, explainability, besides being a powerful tool for transparency, is also highly effective in driving significant instructional changes.

However, a few limitations remain. The current study relies on data from a single educational setting, and politeness norms in Thai academic culture may soften negative sentiments, making it harder to detect implicit criticism. The issues that are still causing problems are sarcasm, indirect speech, and culturally encoded expressions. Future studies may have to adjust models to the educational cultures' discursive patterns and, perhaps, also consider pragmatic or discourse-level sentiment cues. Also, SHAP operations are much slower than regular classification, although they can still be done in batches or at the end of the semester.

Overall, the findings are consistent with the idea that aspect-level sentiment analysis, along with clear explanations, represents a reasonable compromise. It maintains deep learning's predictive power while enabling real instructional use. The system enhances the correctness of feedback interpretation, encourages reflective teaching practices, and makes it easier for universities to move from just collecting opinions to evidence-informed pedagogical development.

6 Conclusion

We introduced explainable aspect-based sentiment analysis for educational feedback, targeting three deficiencies: adaptation to the educational domain, enhanced explainability, and stakeholder validation.

Our BERT-based model has an accuracy of 89.1% and an F1 score of 88.0% across three validated aspects. This is better than AEN-BERT (87.1%, $p < 0.032$). SHAP explanations show faithfulness (41.2% flip rate), consistency (Jaccard 0.71 ± 0.14), and expert alignment (94.3%). Stakeholder validation shows that the idea has real value: 87% prefer it, 4.4 out of 5 say it is actionable, and 93% say they would use it.

This work shows that combining aspect-level granularity with principled explainability meets real educational needs, connecting NLP capabilities with real-world use. Future tasks will involve validation across multiple institutions, multilingual expansion, temporal analysis, and integration with LMS.

References

1. Altrabsheh, N., Cocea, M., Fallahkhair, S.: Learning sentiment from students' feedback for real-time interventions in classrooms. In: Iliadis, L., et al. (eds.) AIAI 2014 IFIP AICT, vol. 436, pp. 40–49. Springer, Heidelberg (2014)
2. Rajput, Q., Haider, S., Ghani, S.: Lexicon-based sentiment analysis of teachers' evaluation. Appl. Comput. Intell. Soft Comput. **2016**, 2385429 (2016)
3. Rani, S., Kumar, P.: A sentiment analysis system to improve teaching and learning. Computer. **50**(5), 36–43 (2017)
4. Kastrati, Z., Imran, A.S., Kurti, A.: Weakly supervised framework for aspect-based sentiment analysis on students' reviews of MOOCs. IEEE Access. **8**, 106799–106810 (2020)
5. Sindhu, I., Daudpota, S.M., et al.: Aspect-based opinion mining on students' feedback for faculty teaching performance evaluation. IEEE Access. **7**, 108729–108741 (2019)
6. Pontiki, M., Galanis, D., et al.: SemEval-2014 task 4: aspect based sentiment analysis. In: Proceedings of SemEval 2014, pp. 27–35. ACL, Dublin (2014)
7. Wang, Y., Huang, M., Zhu, X., Zhao, L.: Attention-based LSTM for aspect-level sentiment classification. In: Proceedings of EMNLP 2016, pp. 606–615. ACL, Austin (2016)
8. Sun, C., Huang, L., Qiu, X.: Utilizing BERT for aspect-based sentiment analysis via constructing auxiliary sentence. In: Proceedings of NAACL-HLT 2019, pp. 380–385. ACL, Minneapolis (2019)
9. Xu, H., Liu, B., Shu, L., Yu, P.S.: BERT post-training for review reading comprehension and aspect-based sentiment analysis. In: Proceedings of NAACL-HLT 2019, pp. 2324–2335. ACL, Minneapolis (2019)
10. Bahdanau, D., Cho, K., Bengio, Y.: Neural machine translation by jointly learning to align and translate. In: Proceedings of ICLR 2015 (2015). https://arxiv.org/abs/1409.0473
11. Jain, S., Wallace, B.C.: Attention is not explanation. In: Proceedings of NAACL-HLT 2019, pp. 3543–3556. ACL, Minneapolis (2019)
12. Lundberg, S.M., Lee, S.I.: A unified approach to interpreting model predictions. In: Proceedings of NIPS 2017, pp. 4765–4774. Curran Associates (2017)
13. Ribeiro, M.T., Singh, S., Guestrin, C.: "why should I trust you?": explaining predictions of any classifier. In: Proceedings of KDD 2016, pp. 1135–1144. ACM (2016)
14. Mosca, E., Beddar-Wiesing, S., et al.: Understanding SHAP-based attributions for text classification. In: Proceedings of COLING 2022, pp. 3848–3859. ICCL, Korea (2022)

Reasoning-Aware Instruction-Tuned Language Models Are Competitive Sentiment Classifiers

Mengyao Zhu(✉) and Phuc Huu Nguyen

Yamaguchi University, Ube Yamaguchi 753-0841, Japan
academiczmy@outlook.com, phuc@yamaguchi-u.ac.jp

Abstract. Generative large language models (LLMs) have demonstrated strong potential in predictive sentiment classification tasks due to their pre-learned world knowledge and inherent instruction-following capabilities. However, the impact of instruction reasoning informativeness on classification accuracy and logical consistency remains underexplored. Inspired by instructional scaffolding in educational theory, where learners benefit from structured, step-by-step guidance to accomplish complex tasks, this study proposes the **Reasoning-Guided Sentiment Classification (RGSC)** framework to assess and enhance sentiment prediction performance through inserted reasoning-aware instructions as guidance for the sentiment classification task. RGSC evaluates multiple LLaMA-3.2 model configurations - base, instruction-tuned, and LoRA-fine-tuned, across 1B and 3B scales, using two instruction types: a minimal simple instruction and a structured, reasoning-guided extensive instruction. In this framework, instruction prompts are explicitly prepended to the user review, alongside the output requirement (positive or negative), as part of the tokenized input for the generative language models, enabling the model to process them jointly for reasoning-based sentiment prediction. Experimental results on the IMDB test split show that extensive reasoning-enriched instructions substantially improve performance, yielding up to 25.7% absolute accuracy gain for LLaMA-3.2-1B, and up to 26.7% F1 score improvement for LLaMA-3.2-3B, highlighting the critical role of instruction reasoning informativeness in LLM-driven sentiment analysis.

Keywords: Reasoning-Aware Fine-Tuning · Language Model Reasoning · Predictive Sentiment Classification

1 Introduction

Sentiment classification is a fundamental task in natural language processing that aims to identify the underlying sentiment polarity $y \in \{\text{positive}, \text{negative}\}$ of a given text input $x \in \mathcal{X}$, where $\mathcal{X}$ is the space of possible textual reviews or utterances. This task remains challenging due to the presence of sarcasm, negation, implicit opinions, and shifting sentiment cues within longer narratives. Variants of sentiment classification span domains such as product reviews, political opinions, and social media posts, each exhibiting unique linguistic features and emotional tones. A concrete example is predicting the sentiment y of an IMDB movie review $x =$ "The acting was flat and the plot predictable.", which should be classified as $y =$ negative (Table 1).

H. Unger and P. Meesad (Eds.): NLPIR 2025, LNNS 1904, pp. 307–320, 2026.
https://doi.org/10.1007/978-3-032-20897-2_23

Table 1. Comparison between Reasoning-Guided Sentiment Classification (RGSC) and Embedding-Based Sentiment Classification (EBSC).

Approach	Characteristics and Attributes
Reasoning-Guided Sentiment Classification (RGSC)	**Characteristic:** – Fine-grained and more informed sentiment labeling decisions. **Advantages:** – The instructions act as scaffolds that direct the model to follow explicit reasoning steps. – The model is guided not only to process the review holistically but also to track and interpret the sentiment shifts throughout the narrative. – The combination of instructions and review input aligns the model's inference pathway with human-like evaluation, helping achieve more consistent sentiment label predictions.
Embedding-Based Sentiment Classification (EBSC)	**Characteristic:** – A surface-level understanding and analysis. **Disadvantages:** – The classification relies solely on fixed vector embeddings of the review, with semantic information compressed into latent dimensions. – It performs surface-level correlation learning based on captured statistical patterns or correlations between words and sentiment labels. – It struggles with reviews containing multiple sentiment shifts and cannot effectively capture the relative weight of tonal changes, which might lead to misclassification.

The rise of generative large language models has significantly enhanced the reliability and effectiveness of downstream NLP tasks. Owing to their extensive capabilities in language understanding, text generation, and instruction following, these models have increasingly been applied to sentiment classification problems. Rather than relying solely on fixed embeddings or classifiers such as BERT [4] or other encoder-only language models [2,8,11,21], generative models are able to be guided by reasoning-oriented instructions embedded within tokenized user reviews (e.g., from the IMDB dataset) to produce sentiment labels $y \in$ positive, negative through next-token prediction. This reasoning-guided approach is distinctive to generative models, as embedding-based models lack the capacity to incorporate such explicit reasoning signals. By leveraging reasoning-guided instructions, generative models are able to enrich the sentiment classification pipeline by explicitly directing the model's inferential process—decomposing complex narratives into interpretable reasoning steps, highlighting sentiment-relevant evidence, and reformulating training data with structured reasoning cues. These strategies facilitate the downstream fine-tuning of smaller generative models, thereby improving classification accuracy. Consequently, reasoning-guided generative approaches demonstrate substantial potential for advancing predictive sentiment classification methodologies.

Despite the success of these newly proposed generative language model-based predictive sentiment classification methodologies, several limitations remain. First, many studies rely on proprietary language models (e.g., GPT-4 [14]) or open-source models with extremely large parameter counts such as LLaMA-3.1-70B [13], which are either used directly for sentiment prediction or employed as auxiliary annotation tools. While effective, these approaches introduce substantial computational costs and resource constraints, limiting accessibility and scalability. Second, the role of reasoning-guided instructions that provide explicit inferential cues within the sentiment classification pipeline remains underexplored. Whether integrating task-specific, reasoning-rich guidance - tokenized alongside the input text - can enable compact models to conduct more informed and interpretable sentiment analysis, and thereby make accurate label decisions in a cost-efficient manner, remains an open question. Understanding the impact of reasoning-oriented instruction richness is therefore crucial for advancing lightweight and interpretable LLM-based sentiment classifiers.

To address these issues, we propose an effective methodology called Reasoning-Guided Sentiment Classification (RGSC) to investigate and highlight the critical role of reasoning-rich instruction prompts in enhancing predictive sentiment classification. Inspired by the concept of instructional scaffolding in educational theory, where learners benefit from structured guidance and cognitive support, we design instruction prompts that encode detailed reasoning processes for sentiment decision-making. These prompts are integrated and tokenized together with the review content and ground truth output, forming the complete input sequence for the model. Using the generative LLaMA-3.2 language models across various configurations (base, instruction-tuned, and LoRA-fine-tuned), RGSC enables a controlled and empirical comparison of how compact models, when guided by extensive reasoning-style instructions, can approach the effectiveness of larger models while performing more informed sentiment analysis and making reliable label predictions at substantially reduced computational cost. This makes reasoning-guided compact models particularly well-suited for scalable real-world deployment, where efficiency and interpretability are as critical as accuracy. In summary, this study presents the following key contributions:

- We introduce a highly effective and easy-to-implement framework, Reasoning-Guided Sentiment Classification (RGSC), that leverages reasoning-aware, well-structured instruction prompts to significantly enhance predictive sentiment classification performance.
- We incorporate resource-efficient LoRA fine-tuning into our framework in conjunction with reasoning-enriched extensive prompts, further boosting performance beyond instruction alone.
- Through a comprehensive evaluation of LLaMA-3.2 language models across a range of configurations and parameter scales, we empirically validate that reasoning-guided extensive instruction prompts consistently improve sentiment classification accuracy, especially for compact models, thereby offering a practical pathway for cost-effective and high-performing LLM-based sentiment analysis.

2 Related Works

2.1 Instruction Tuning and Following for Sentiment Classifications

Instruction Tuning [3,5,15,17,20,25] and Instruction Following [9,10,16,18,19,24] together define a unified framework in which a generative language model is first fine-tuned and then inferenced using input sequences constructed from a tokenized combination of an instruction prompt, user input, and explicit output requirement. During fine-tuning, the model learns to align this structured input format with a constrained output space $\mathcal{Y} = \{\text{positive, negative}\}$, such that the correct label token is directly generated as the next token without the use of a classification head. The instruction tuning objective is defined as:

$$\mathcal{L}_{\text{tune}} = -\sum_{(x,y)} \log P(y \mid x) = -\sum_{(x,y)} \sum_{t \in \mathcal{Y}} \mathbb{1}_{[y=t]} \cdot \log P(t \mid \mathcal{T}) \tag{1}$$

where $\mathcal{T} = [t_1, ..., t_n]$ is the tokenized sequence of instruction, user input, and output requirement, and $y \in \mathcal{Y}$ is the correct sentiment token. At inference, the instruction following behavior uses the same input structure and generates the sentiment label via next-token prediction:

$$P(y \mid \mathcal{T}) = \sum_{t \in \mathcal{Y}} \mathbb{1}_{[y=t]} \cdot P(t \mid t_1, ..., t_n) \tag{2}$$

This end-to-end generative formulation treats binary sentiment classification as a language modeling task with constrained decoding, ensuring output fidelity to the target label set. The model's learned alignment between structured prompts and valid token completions enables instruction-compliant, classification-capable generation without specialized output layers.

Sequential Instruction Tuning [5] is a technique for training language models f_θ on ordered sets of instructions $\{x_1, x_2, \ldots, x_n\}$ to reflect the dependencies and logical progression found in multi-step tasks. Current instruction-tuned language models often struggle to follow prompts with multiple intentions, especially when those intentions are interdependent or semantically divergent. Sequential instruction tuning improves contextual understanding by conditioning each output $y_t = f_\theta(x_t \mid x_{<t}, y_{<t})$ on prior instructions and responses, promoting coherence and structure-aware completions.

Sentiment Classification Capability Distillation [22] was introduced to transfer sentiment analysis capabilities from a teacher model M_T (parameters θ_T) to a student model M_S (parameters θ_S), using cross-entropy loss objectives tailored to each learning goal. In the first stage, KNOWDIST, the objective is to build foundational sentiment understanding by training on input-output pairs (x, y), where $y = M_T(x)$, optimizing θ_S via the cross-entropy loss $L_{\text{KNOW}}(\theta_S) = -\sum \log P(y \mid x; \theta_S)$. In the second stage, ICLDIST, the goal is to enhance instruction-following capabilities by using prompts (i, d, x) and teacher-generated responses $y = M_T(i, d, x)$, minimizing the conditional loss $L_{\text{ICL}}(\theta_S) = -\sum \log P(y \mid i, d, x; \theta_S)$. This separation enables M_S to first absorb abstract sentiment knowledge and subsequently align its outputs with structured task instructions.

2.2 Sentiment Classification as Predictive Generation Task

Hybrid Sampling [1] is a selective generative inference framework that combines both single and multiple sampling over a generative language model $\mathcal{M}_\theta$ to extract sentiment quadruples $q = (a, c, o, p)$, where a is aspect, c category, o opinion, and p polarity. The mechanism uses an entropy-based uncertainty score at generation step i to determine whether to trigger a rollback procedure on a span $s = x_{i-k:i}$ with $k \in \mathbb{Z}^+$, and then applies majority voting over candidate outputs $\{s^{(1)}, s^{(2)}, \ldots, s^{(n)}\}$ while continuing the rest of the sequence via greedy or probabilistic decoding. Compared to traditional full-sequence multi-sampling methods $\mathcal{S}_{full}$, Hybrid Sampling $\mathcal{S}_{hyb}$ reduces the expected inference cost $\mathbb{E}[C_{hyb}] \ll \mathbb{E}[C_{full}]$ by restricting re-sampling only to spans with high entropy with maintained self-consistency and reduced computational cost. LLaMA-2-7B, T5-Base and BART-Base were applied to test the robustness and generalizability of hybrid sampling.

Prompt-Guided Sentiment Classification via External Knowledge Injection [7] is a methodology that classifies sentence-level sentiment by modifying the input prompt x into an augmented form $x' = \text{Prompt}(x, H)$, where $H = \{h_1, h_2, ..., h_n\}$ is a set of syntactic-lexical hints extracted using dependency parsing and polarity dictionaries. The approach is retraining-free, as it does not update the model parameters θ, instead leveraging prompt-level injection to influence predictions $\hat{y} = f(x'; \theta)$. Evaluations were conducted on six large language models with varied architectures: T5, UL2, and MT0 (encoder-decoder), Llama2 and Llama3 (decoder-only), and Mixtral (mixture-of-experts decoder).

Adjective Enhanced Sentiment Instruction Tuning [23] introduces a domain-agnostic strategy for zero-shot sentiment classification by instruction-tuning large language models with a minimal yet curated set of pseudo instruction instances. Let $D = \{(T_i, I_i, O_i)\}_{i=1}^{n}$ be a collection of $n = 240$ pseudo-instruction tuples, where each T_i is an instruction text, I_i is a sentiment-bearing input, and O_i is a categorical output label $O_i \in \{\text{positive}, \text{negative}, \text{neutral}\}$. Unlike traditional supervised approaches relying on large labeled datasets D_{labeled}, the authors synthesize each tuple from a combination of sampled sentimental adjectives and collected instruction prompts. These adjectives are derived from lexical sources and mapped into sets $L_{\text{pos}}, L_{\text{neg}}, L_{\text{neu}}$. The tuples are designed to mimic diverse sentiment expressions and to approximate general sentiment phenomena in natural text.

AGCL (Aspect Graph Construction and Learning) [6] was proposed to use a generative language model, specifically GPT-3.5-turbo, to produce enriched sentiment aspects by generating informative, domain-specific descriptions D_a for each aspect term $a \in A$. These aspect terms originate from user-generated review texts and represent fine-grained targets - such as "price," "battery," or "staff" - toward which sentiment polarity must be classified in the aspect-level sentiment classification (ALSC) task. The generated descriptions are then processed by the embedding model SBERT, which maps each D_a into a dense vector representation e_a that reflects both linguistic and contextual knowledge. These representations are integrated into the Aspect Graph $\mathcal{G} = \{\mathcal{N}, \mathcal{E}\}$ to support informed sentiment classification through alignment, similarity-based reasoning, and generalization across both seen and unseen aspects.

3 Experiment Settings

3.1 Dataset

The IMDB dataset [12] is formally defined as a binary-labeled corpus for sentiment classification, denoted by $\mathcal{D} = \{(x_i, y_i)\}_{i=1}^{N}$, where each instance $x_i \in \mathcal{X}$ is a free-form natural language review and each corresponding label $y_i \in \mathcal{Y} = \{0, 1\}$ represents the ground truth sentiment. The label space is binary: $y_i = 1$ signifies positive sentiment, while $y_i = 0$ signifies negative sentiment. The dataset contains $N = 50{,}000$ instances, equally partitioned into a training set and test set: $\mathcal{D}_{\text{train}}$ and $\mathcal{D}_{\text{test}}$, each with 25,000 samples and balanced class distributions. In the proposed reasoning-guided sentiment classification framework, each review x_i is paired with either simple or reasoning-aware instruction prompts, forming augmented input sequences that provide explicit reasoning signals for sentiment decision- making. This structure supports supervised learning formulations for binary text classification, where the learning task is to approximate a sentiment prediction function $f_\theta : \mathcal{X} \to [0, 1]$ that estimates the posterior probability of the positive class.

3.2 Models

To evaluate the influence of instruction prompt informativeness on predictive sentiment classification, we conduct experiments using a family of LLaMA-3.2 generative language models with varying configurations. Let $\mathcal{M}_{\text{base}}$, $\mathcal{M}_{\text{inst}}$, and $\mathcal{M}_{\text{LoRA}}$ denote the base, instruction-tuned, and LoRA-fine-tuned variants respectively, each instantiated at parameter scales $P \in \{1\text{B}, 3\text{B}\}$. In this research, language models with parameter counts from 1B to 5B are defined as compact language models. For each model $\mathcal{M}_P$, we assess sentiment classification under two instruction strategies: a minimal simple instruction π_{sim} and a structured reasoning-aware extensive instruction π_{exten}. This setup allows us to systematically compare the effects reasoning-oriented instruction informativeness across architectural variations and training adaptations.

3.3 Hyperparameters

The LoRA fine-tuning was performed with a rank parameter of $r = 8$, scaling factor $\alpha = 16$, and dropout rate $p_{\text{drop}} = 0.05$, applied to the target modules q_proj and v_proj for causal language modeling. Training was conducted for $E = 3$ epochs with a batch size $B = 8$, gradient accumulation steps $G_{\text{acc}} = 1$, and without mixed precision (fp16 = False, bf16 = True if available). The evaluation used a deterministic decoding configuration with do_sample = False, temperature = 0.0, and max_new_tokens = 10 for consistent label generation. The process assessed $N = \text{len}(\mathcal{D}_{\text{test}})$ IMDB test samples sequentially, ensuring reproducibility by applying an identical reasoning-aware or simple prompt template to each review instance. This deterministic decoding setup was particularly important to ensure that reasoning-guided instructions consistently shaped the model's inferential process, preventing randomness from influencing sentiment label decisions. We followed the widely accepted hyperparameter settings with better tuning and inference efficiency in mind.

3.4 Configurations

The experiments were conducted on a system equipped with an Intel Core i9-14900K CPU, NVIDIA GeForce RTX 4090 GPU with 24 GB VRAM, and 48 GB system DRAM, running NVIDIA driver 560.94 with CUDA 12.6. The environment utilized WSL2 with GPU acceleration enabled, ensuring full compatibility for training and evaluation of the LLaMA-3.2 models under the described reasoning-guided sentiment classification framework.

3.5 Instruction Examples

Simple Instructions

```
Given the following IMDB movie review and two candidate review sentiment labels
(positive and negative).
Reason and choose the most appropriate sentiment label for the movie review.
Review: {review_text}
answer with one of two sentiment labels: positive or negative.
answer:
```

The simple instruction prompt provides only a concise directive that outlines the classification task and expected response format with minimal reasoning guidance. It specifies the task in a single sentence, presents the movie review as input, and requests the model to select between two sentiment labels. Unlike reasoning-aware instructions, it does not supply explicit cues about how to process the review or handle sentiment shifts, leaving the model to rely primarily on surface-level correlations when making its prediction.

Reasoning-Aware Instructions

```
Your task is to read and understand a complete IMDB movie review, and then determine
the final sentiment expressed in the review. The sentiment must be classified into
one of two categories:
- positive → if the reviewer shows clear approval, praise, or satisfaction with the
movie overall
- negative → if the reviewer expresses disappointment, criticism, or dissatisfaction
with the movie overall You MUST focus on the overall impression or conclusion that the
reviewer communicates by the end of the review. Ignore isolated words or sentences if
they do not reflect the final opinion. Instructions:
- Read the full review carefully from beginning to end.
- Identify any shifts or contradictions in tone and sentiment.
- Prioritize the final judgment or conclusive opinion that the review presents.
- Do NOT guess based on partial clues; base your answer only on complete comprehension.
Output format:
answer: sentiment
Where "sentiment" should be replaced with either "positive" or "negative" (all
lowercase).
```

In contrast, reasoning-aware instructions serve as a reasoning scaffold, rather than merely a label directive. They guide the model to process the entire IMDB review R, evaluate tonal shifts T, and prioritize the final judgment or conclusive sentiment

S_c, thereby simulating human reasoning strategies. By discouraging reliance on isolated words w_i or partial clues and enforcing holistic comprehension $\sum_{i=1}^{n} w_i \rightarrow S_c$, the instructions push the model beyond surface-level token correlations toward contextual judgment. The explicit step-by-step directives mirror reasoning pathways: read the full review R carefully from beginning to end, detect contradictions ΔT, and resolve toward the final stance S_c. This transforms classification into a guided inference process. Furthermore, the strict output schema ("answer: sentiment") channels reasoning into a predictable format. In this way, the prompt encodes both cognitive heuristics and formatting rules, aligning model reasoning with structured, human-like evaluation of sentiment. Unlike simple instructions, reasoning-aware instructions emphasize holistic comprehension and structured reasoning steps, enabling more reliable sentiment label decisions.

3.6 Metrics

To comprehensively evaluate the effectiveness of the proposed Reasoning-Guided Sentiment Classification (RGSC) framework, we adopt six evaluation metrics: Accuracy, F1 Score, Precision, Recall, Positive Accuracy, and Negative Accuracy. Accuracy indicates the proportion of correct sentiment token generations over all test instances. Precision measures how often the generated token positive is correct, while Recall captures the model's success in generating positive for all truly positive instances. The F1 Score balances both error types and offers a robust overall evaluation of instruction-conditioned label generation. Positive Accuracy reflects how accurately the model generates positive when the ground truth is positive. Conversely, Negative Accuracy denotes the correctness of negative generation under true negative conditions. These metrics are essential for identifying any class imbalance or instruction sensitivity in the generative outputs, especially under varying instruction prompt complexities. These six metrics are selected to reflect both overall and class-specific performance of next-token generation under instruction constraints, offering a comprehensive view of how well the models align with task prompts across varied configurations in the RGSC framework.

4 Evaluation Results

Extensive Prompts Enhance All Model Variants, Especially Smaller Scales: The evaluation of LLaMA-3.2 language models across multiple configurations highlights the interplay between instruction format, model architecture, and tuning methodology. Models were tested using both a simple instruction prompt and an extensive reasoning-aware instruction prompt on the IMDB sentiment classification task. Metrics such as accuracy, F1 score, precision, and recall reveal that comprehensive, reasoning-aware prompting leads to consistently better predictive performance. Smaller models like LLaMA-3.2-1B demonstrate substantial improvements when shifting from simple to extensive prompts. This pattern is consistent across all model variants, suggesting that reasoning-oriented instruction informativeness - not just model scale - is central to maximizing semantic alignment and interpretative accuracy in binary classification tasks grounded in narrative reviews (Table 2).

Table 2. Evaluation Metrics for LLaMA-3.2 Language Models on IMDB Test Split

Models	Evaluation Metrics				Accuracy Summary	
	Accuracy	F1 Score	Precision	Recall	Positive Accuracy	Negative Accuracy
LLaMA-3.2-1B$_{sim}$	70.44%	69.43%	72.53%	66.58%	66.58%	74.36%
LLaMA-3.2-1B$_{exten}$	91.92%	91.97%	94.61%	88.98%	88.98%	94.88%
LLaMA-3.2-1B$_{LoRA}$	**95.12%**$_2$	**95.13%**$_2$	**95.15%**$_2$	95.11%	95.47%	**94.76%**$_2$
LLaMA-3.2-1B-Instruct$_{sim}$	91.58%	91.44%	93.30%	89.65%	89.65%	93.52%
LLaMA-3.2-1B-Instruct$_{exten}$	92.17%	92.17%	93.01%	91.34%	91.34%	93.01%
LLaMA-3.2-1B-Instruct$_{LoRA}$	92.85%	90.64%	88.23%	93.19%	93.19%	92.64%
LLaMA-3.2-3B$_{sim}$	82.46%	84.49%	76.26%	94.70%	94.70%	70.02%
LLaMA-3.2-3B$_{exten}$	89.70%	90.49%	84.65%	97.19%	97.19%	82.09%
LLaMA-3.2-3B$_{LoRA}$	**96.10%**$_1$	**96.10%**$_1$	**96.10%**$_1$	**96.10%**$_2$	**96.23%**$_2$	**95.97%**$_1$
LLaMA-3.2-3B-Instruct$_{sim}$	92.28%	92.49%	91.58%	93.42%	93.42%	91.10%
LLaMA-3.2-3B-Instruct$_{exten}$	94.16%	94.38%	91.68%	97.25%	97.25%	91.03%
LLaMA-3.2-3B-Instruct$_{LoRA}$	92.02%	92.58%	87.11%	**98.78%**$_1$	**98.78%**$_1$	85.15%

LoRA Amplifies Gains from Informative Prompts More Than Model Size Alone: Comparatively, extensive prompting outperforms simple prompting in every evaluated variant. LLaMA-3.2-1B jumps from 70.44% to 91.92% in accuracy, while LLaMA-3.2-3B improves from 82.46% to 89.70%. Instruction-tuned variants begin with a stronger performance baseline, especially under simple prompts, indicating pre-aligned alignment to user tasks. However, the highest-performing settings across all metrics are models LoRA fine-tuned that integrated with extensive reasoning-guided instructions, such as LLaMA-3.2-3B$_{LoRA}$ achieving 96.10% accuracy. LoRA enhances task-specific retention and output consistency, especially when used with domain-specific fine-tuning data and explicit instructional structure. Without fine-tuning, gains from prompt engineering remain significant but do not reach LoRA-augmented levels.

Impact of Instruction Prompt Informativeness on Core Classification Metrics: The results clearly demonstrate that reasoning-guided instruction prompts significantly enhance sentiment classification performance across all core evaluation metrics. For instance, LLaMA-3.2-1B improved from 70.44% to 91.92% in Accuracy, and from 69.43% to 91.97% in F1 Score, when shifting from simple to extensive prompting. Similarly, Precision increased from 72.53% to 94.61%, and Recall from 66.58% to 88.98%. The 3B variant shows a similar trend, with Accuracy rising from 82.46% to 89.70%, and F1 Score from 84.49% to 90.49%. These results confirm that more reasoning guidance enables more accurate and balanced next-token sentiment classification in generative models.

However, not all results favored extensive prompts uniformly across configurations. For example, LLaMA-3.2-3B-Instruct-LoRA, while showing a high Recall of 98.78%, exhibited a lower Precision of 87.11%, indicating that it may over-predict positive sentiment. Similarly, LLaMA-3.2-1B-Instruct-LoRA recorded a decrease in F1 Score to 90.64%, lower than its extensive reasoning-guided prompt counterpart (92.17%) despite fine-tuning. These cases suggest that while extensive reasoning-based instruction typically benefits performance, certain tuning strategies (e.g., LoRA applied to already instruction-tuned models) may introduce trade-offs in precision-recall balance, warranting further investigation.

Class-Specific Performances in Positive and Negative Accuracy: Instruction prompt informativeness also yields substantial improvements in class-specific accuracy metrics. For LLaMA-3.2-1B, Positive Accuracy increased from 66.58% (simple) to 88.98% (extensive), while Negative Accuracy rose from 74.36% to 94.88%, indicating better balanced classification across sentiment polarities. Similarly, for LLaMA-3.2-3B, Positive Accuracy improved from 94.70% to 97.19%, and Negative Accuracy from 70.02% to 82.09% under extensive prompting. These results suggest that extensive reasoning-guided prompts help generative models focus more effectively on both affirmative and critical sentiment expressions, while reasoning over sentiment shifts, enhancing their ability to consistently classify content regardless of sentiment direction. This supports the robustness of the RGSC framework in sentiment polarity resolution.

However, some configurations demonstrated unbalanced class-specific performance despite overall gains. For instance, LLaMA-3.2-3B-Instruct-LoRA achieved a notably high Positive Accuracy of 98.78%, but its Negative Accuracy dropped to 85.15%, suggesting a skew toward positive sentiment predictions. Likewise, LLaMA-3.2-1B-Instruct-LoRA maintained Positive Accuracy at 93.19%, yet showed a slightly lower Negative Accuracy of 92.64% compared to other variants. These discrepancies imply that while extensive reasoning-aware prompts improve class sensitivity, fine-tuning strategies may lead to asymmetric polarity handling, warranting further instruction prompt calibration or model regularization (Table 3).

Table 3. Comparison of Best Performing LoRA-Tuned and Original LLaMA-3.2 Models per Metric on IMDB

Metric	LoRA-Tuned Models		Original Models	
	Model	Value (%)	Model	Value (%)
Accuracy	LLaMA-3.2-3B$_{\text{LoRA}}$	96.10	LLaMA-3.2-3B-Instruct$_{\text{exten}}$	94.16
F1 Score	LLaMA-3.2-3B$_{\text{LoRA}}$	96.10	LLaMA-3.2-3B-Instruct$_{\text{exten}}$	94.38
Precision	LLaMA-3.2-3B$_{\text{LoRA}}$	96.10	LLaMA-3.2-1B$_{\text{exten}}$	94.61
Recall	LLaMA-3.2-3B-Instruct$_{\text{LoRA}}$	98.78	LLaMA-3.2-3B-Instruct$_{\text{exten}}$	97.25
Positive Acc	LLaMA-3.2-3B-Instruct$_{\text{LoRA}}$	98.78	LLaMA-3.2-3B-Instruct$_{\text{exten}}$	97.25
Negative Acc	LLaMA-3.2-3B$_{\text{LoRA}}$	95.97	LLaMA-3.2-1B$_{\text{exten}}$	94.88

Best Large-Scale Performance: 3B Models with Extensive Prompt and LoRA: In high-performance application scenarios - such as production sentiment analysis pipelines or benchmark evaluations - LLaMA-3.2-3B$_{\text{LoRA}}$ is recommended due to its robust output under extensive instruction prompting with logical reasoning guidance, combined with fine-tuning optimization. It delivers superior precision, recall, and class-specific accuracies, particularly when instruction prompts explicitly direct attention to reasoning over discourse-level sentiment resolution. For use cases where fine-tuning is not an option, LLaMA-3.2-3B-Instruct$_{\text{exten}}$ becomes the ideal choice. This variant is pre-aligned to instruction-following behavior and demonstrates strong performance without additional instruction tuning, especially when the prompt structure guides the model through comprehensive reasoning-driven sentiment interpretation. In such settings, the extensiveness of the prompt compensates for the absence of parameter updates by enhancing internal attention calibration.

Best Resource-Constrained Setting: 1B Models with LoRA and Extensive Prompt: In contrast, where compute resources are limited, LLaMA-3.2-1B$_{\text{LoRA}}$ offers high efficiency at reduced scale. It benefits substantially from LoRA fine-tuning and reasoning-enriched prompting, making it appropriate for edge applications or cost-sensitive deployments. The simple prompt setting, while easier to implement, shows consistent underperformance compared to its extensive counterpart, across both base and instruct-tuned models. Models like LLaMA-3.2-1B-Instruct$_{\text{sim}}$ or 3B-Instruct$_{\text{sim}}$ perform well only when paired with pretraining strategies that anticipate vague instructions. Thus, task designers are encouraged to favor extended prompts enriched with task-specific heuristics and explicit reasoning signals when instructing untuned or partially tuned LLMs, as such prompting increases model comprehension reliability and decision grounding.

5 Conclusions and Future Directions

5.1 Conclusions

In this paper, we explored the critical role of instruction prompts with varying levels of reasoning informativeness in predictive sentiment classification. To this end, we proposed Reasoning-Guided Sentiment Classification (RGSC), a framework that explicitly integrates reasoning-aware, extensive instruction prompts as part of the input for generative language models to perform sentiment classification as a next-token prediction task. RGSC emphasizes two key mechanisms: (1) using reasoning-guided instruction prompts to help compact language models to navigate the reasoning space more effectively, enabling more reliable sentiment label prediction, and (2) applying LoRA fine-tuning to enhance the model's guidance following and strengthen its reasoning implementation behavior when used in conjunction with extensive prompts. Experimental results validate the effectiveness of reasoning-oriented prompt informativeness across models of different parameter sizes and tuning levels. Future work will include more generative models and a broader range of classification datasets to assess generalizability. Overall, our findings underscore that reasoning-guided prompts are a decisive factor for achieving both compactness and reliability in LLM-based sentiment analysis.

5.2 Limitations

This research has two limitations. First, it only investigates LLaMA-3.2 language model variations, limiting the generalizability of the findings. It remains uncertain whether the observed effectiveness of extensive reasoning-aware instruction prompts extends to other generative model families beyond LLaMA-3.2 family models. Second, the evaluation is restricted to the IMDB dataset, which, while widely used, does not capture the diversity of classification tasks, and may risk overfitting and reduced generalization. Future research should explore additional sentiment classification datasets and extend the methodology to classification problems in other domains to evaluate the broader applicability of the RGSC framework.

5.3 Future Directions

Looking ahead, one promising research direction is to further explore the reasoning capabilities of compact generative language models when paired with structured, reasoning-aware instructional guidance. This integration may enable models to go beyond binary sentiment labeling and effectively detect fine-grained sentiment shifts, implicit opinions, and mixed sentiments that appear across different sections of user-generated content. Understanding how models can map instruction cues to multi-level sentiment representations could lead to more nuanced and context-aware classification systems. Another direction involves developing Mixture-of-Experts (MoE) architectures, where multiple lightweight models, each specialized in particular sentiment patterns, are dynamically selected and activated based on importance-driven scores. This strategy can facilitate adaptive and interpretable model routing, offering more informed and precise sentiment classification decisions.

References

1. Bao, X., Qiang, M., Gu, J., Wang, Z., Huang, C.R.: Exploring hybrid sampling inference for aspect-based sentiment analysis. In: Chiruzzo, L., Ritter, A., Wang, L. (eds.) Findings of the Association for Computational Linguistics: NAACL 2025, Albuquerque, New Mexico, pp. 4199–4210. Association for Computational Linguistics (2025). https://aclanthology.org/2025.findings-naacl.236/
2. Clark, K., Luong, M.T., Le, Q.V., Manning, C.D.: ELECTRA: pre-training text encoders as discriminators rather than generators. arXiv [cs.CL] (2020). http://arxiv.org/abs/2003.10555
3. Deng, J., Chen, C., Hou, C., Yuan, X.: InstructGEC: enhancing unsupervised grammatical error correction with instruction tuning. In: Rambow, O., Wanner, L., Apidianaki, M., Al-Khalifa, H., Eugenio, B.D., Schockaert, S. (eds.) Proceedings of the 31st International Conference on Computational Linguistics, Abu Dhabi, UAE, pp. 110–122. Association for Computational Linguistics (2025). https://aclanthology.org/2025.coling-main.9/
4. Devlin, J., Chang, M.W., Lee, K., Toutanova, K.: BERT: pre-training of deep bidirectional transformers for language understanding. In: Proceedings of the 2019 Conference of the North, Stroudsburg, PA, USA, pp. 4171–4186. Association for Computational Linguistics (2019). http://dx.doi.org/10.18653/v1/N19-1423
5. Hu, H., Yu, S., Chen, P., Ponti, E.: Fine-tuning large language models with sequential instructions. In: Chiruzzo, L., Ritter, A., Wang, L. (eds.) Proceedings of the 2025 Conference of the Nations of the Americas Chapter of the Association for Computational Linguistics: Human Language Technologies (Volume 1: Long Papers), Albuquerque, New Mexico, pp. 5589–5610. Association for Computational Linguistics (2025). https://aclanthology.org/2025.naacl-long.288/

6. Jian, Z., et al.: AGCL: aspect graph construction and learning for aspect-level sentiment classification. In: Rambow, O., Wanner, L., Apidianaki, M., Al-Khalifa, H., Eugenio, B.D., Schockaert, S. (eds.) Proceedings of the 31st International Conference on Computational Linguistics, Abu Dhabi, UAE, pp. 841–854. Association for Computational Linguistics (2025). https://aclanthology.org/2025.coling-main.56/
7. Kanayama, H., Zhao, Y., Iwamoto, R., Ohko, T.: Incorporating syntax and lexical knowledge to multilingual sentiment classification on large language models. In: Ku, L.W., Martins, A., Srikumar, V. (eds.) Findings of the Association for Computational Linguistics ACL 2024, Stroudsburg, PA, USA,pp. 4810–4817. Association for Computational Linguistics (2024). https://doi.org/10.18653/v1/2024.findings-acl.286, https://aclanthology.org/2024.findings-acl.286/
8. Lan, Z., Chen, M., Goodman, S., Gimpel, K., Sharma, P., Soricut, R.: ALBERT: a lite BERT for self-supervised learning of language representations. In: International Conference on Learning Representations (2019). http://arxiv.org/abs/1909.11942
9. Lei, X., Gomez, L., Bai, H.Y., Bashivan, P.: iWISDM: assessing instruction following in multimodal models at scale. In: Lomonaco, V., Melacci, S., Tuytelaars, T., Chandar, S., Pascanu, R. (eds.) Proceedings of The 3rd Conference on Lifelong Learning Agents. Proceedings of Machine Learning Research, vol. 274, pp. 457–480. PMLR (2025). https://proceedings.mlr.press/v274/lei25a.html
10. Li, Z., Peng, B., He, P., Yan, X.: Evaluating the instruction-following robustness of large language models to prompt injection. In: Al-Onaizan, Y., Bansal, M., Chen, Y.N. (eds.) Proceedings of the 2024 Conference on Empirical Methods in Natural Language Processing, Miami, Florida, USA, pp. 557–568. Association for Computational Linguistics (2024). https://doi.org/10.18653/v1/2024.emnlp-main.33
11. Liu, Y., et al.: RoBERTa: a robustly optimized BERT pretraining approach. ArXiv (2019). http://arxiv.org/abs/1907.11692
12. Maas, A., Daly, R.E., Pham, P.T., Huang, D., Ng, A.Y., Potts, C.: Learning word vectors for sentiment analysis. In: Proceedings of the 49th Annual Meeting of the Association for Computational Linguistics: Human Language Technologies, pp. 142–150 (2011). https://aclanthology.org/P11-1015/
13. Meta: LLaMA-3.1-70B-instruct · hugging face. https://huggingface.co/meta-llama/Llama-3.1-70B-Instruct. Accessed 4 Aug 2025
14. OpenAI, Achiam, J., Adler, S., Agarwal, S., Ahmad, L., et al.: GPT-4 technical report. arXiv [cs.CL] (2023). http://arxiv.org/abs/2303.08774
15. Peng, B., Li, C., He, P., Galley, M., Gao, J.: Instruction tuning with GPT-4. arXiv [cs.CL] (2023). http://arxiv.org/abs/2304.03277
16. Pham, C.M., Sun, S., Iyyer, M.: Suri: multi-constraint instruction following in long-form text generation. In: Al-Onaizan, Y., Bansal, M., Chen, Y.N. (eds.) Findings of the Association for Computational Linguistics: EMNLP 2024, Miami, Florida, USA, pp. 1722–1753. Association for Computational Linguistics (2024). https://doi.org/10.18653/v1/2024.findings-emnlp.94
17. Song, C., Zhou, Z., Yan, J., Fei, Y., Lan, Z., Zhang, Y.: Dynamics of instruction fine-tuning for Chinese large language models. In: Rambow, O., Wanner, L., Apidianaki, M., Al-Khalifa, H., Eugenio, B.D., Schockaert, S. (eds.) Proceedings of the 31st International Conference on Computational Linguistics, Abu Dhabi, UAE, pp. 10345–10366. Association for Computational Linguistics (2025). https://aclanthology.org/2025.coling-main.689/
18. Wu, Z., et al.: Dancing in chains: reconciling instruction following and faithfulness in language models. In: Al-Onaizan, Y., Bansal, M., Chen, Y.N. (eds.) Proceedings of the 2024 Conference on Empirical Methods in Natural Language Processing, Miami, Florida, USA, pp. 3942–3965. Association for Computational Linguistics (2024). https://doi.org/10.18653/v1/2024.emnlp-main.229

19. Yue, Y., Wang, C., Huang, J., Wang, P.: Distilling instruction-following abilities of large language models with task-aware curriculum planning. In: Al-Onaizan, Y., Bansal, M., Chen, Y.N. (eds.) Findings of the Association for Computational Linguistics: EMNLP 2024, Miami, Florida, USA, pp. 6030–6054. Association for Computational Linguistics (2024). https://doi.org/10.18653/v1/2024.findings-emnlp.350, http://dx.doi.org/10.18653/v1/2024.findings-emnlp.350
20. Zhang, S., et al.: Instruction tuning for large language models: A survey. arXiv [cs.CL] (2023). http://arxiv.org/abs/2308.10792
21. Zhang, Y., et al.: Qwen3 embedding: Advancing text embedding and reranking through foundation models. arXiv [cs.CL] (2025). http://arxiv.org/abs/2506.05176
22. Zhang, Y., et al.: Targeted distillation for sentiment analysis. arXiv [cs.CL] (2025). http://arxiv.org/abs/2503.03225
23. Zhao, Y., Muraoka, M., Yoshida, I., Bhattacharjee, B., Kanayama, H.: A simple-yet-efficient instruction augmentation method for zero-shot sentiment classification. In: Rambow, O., Wanner, L., Apidianaki, M., Al-Khalifa, H., Eugenio, B.D., Schockaert, S. (eds.) Proceedings of the 31st International Conference on Computational Linguistics, Abu Dhabi, UAE, pp. 1585–1599. Association for Computational Linguistics (2025). https://aclanthology.org/2025.coling-main.107/
24. Zhou, Z., Atreya, P., Lee, A., Walke, H.R., Mees, O., Levine, S.: Autonomous improvement of instruction following skills via foundation models. In: Agrawal, P., Kroemer, O., Burgard, W. (eds.) Proceedings of The 8th Conference on Robot Learning. Proceedings of Machine Learning Research, vol. 270, pp. 4805–4825. PMLR (2025). https://proceedings.mlr.press/v270/zhou25b.html
25. Zhu, T., Dong, D., Qu, X., Ruan, J., Chen, W., Cheng, Y.: Dynamic data mixing maximizes instruction tuning for mixture-of-experts. In: Chiruzzo, L., Ritter, A., Wang, L. (eds.) Proceedings of the 2025 Conference of the Nations of the Americas Chapter of the Association for Computational Linguistics: Human Language Technologies (Volume 1: Long Papers), pp. 1663–1677. Association for Computational Linguistics, Albuquerque, New Mexico (Apr 2025), https://aclanthology.org/2025.naacl-long.80/

EmoBang: Detecting Emotion From Bengali Texts

Abdullah Al Maruf[1], Aditi Golder[2], Zakaria Masud Jiyad[1], Abdullah Al Numan[3], and Tarannum Shaila Zaman[4(✉)]

[1] Charles Darwin University, Darwin, Australia
{s395350,s393850}@students.cdu.edu.au
[2] Wake Forest University, Winston-Salem, NC 27109, USA
golda24@wfu.edu
[3] Bangladesh University of Engineering and Technology (BUET), Dhaka, Bangladesh
[4] University of Maryland, Baltimore County (UMBC), Baltimore, MD, USA
zamant@umbc.edu

Abstract. Emotion detection from text seeks to identify an individual's emotional or mental state, positive, negative, or neutral, based on linguistic cues. While significant progress has been made for English and other high-resource languages, Bengali remains underexplored despite being the world's fourth most spoken language. The lack of large, standardized datasets classifies Bengali as a low-resource language for emotion detection. Existing studies mainly employ classical machine learning models with traditional feature engineering, yielding limited performance. In this paper, we introduce a new Bengali emotion dataset annotated across eight emotion categories and propose two models for automatic emotion detection: (i) a hybrid Convolutional Recurrent Neural Network (CRNN) model ($EmoBang_{Hybrid}$) and (ii) an AdaBoostBidirectional Encoder Representations from Transformers (BERT) ensemble model ($EmoBang_{Ensemble}$). Additionally, we evaluate six baseline models with five feature engineering techniques and assess zero-shot and few-shot large language models (LLMs) on the dataset. To the best of our knowledge, this is the first comprehensive benchmark for Bengali emotion detection. Experimental results show that $EmoBang_H$ and $EmoBang_E$ achieve accuracies of 92.86% and 93.69%, respectively, outperforming existing methods and establishing strong baselines for future research.

Keywords: Emotion Detection · Bengali Text · Natural Language Processing · Machine Learning · LLMs

1 Introduction

Human emotion may be expressed in various ways, including verbal, textual, and facial expressions. As emotion plays a vital role in decision-making [10], it can influence marketing, business, politics, human behavior, technology, and other essential aspects of our lives. Due to its pervasive application in human

H. Unger and P. Meesad (Eds.): NLPIR 2025, LNNS 1904, pp. 321–337, 2026.
https://doi.org/10.1007/978-3-032-20897-2_24

life, emotion detection from texts has become increasingly crucial [39]. Emotion detection from texts aims to identify the writers' hidden feelings in their written works. The hypothesis of this technique is based on the assumption that if an author is happy, they use positive words. If an author is sad or frustrated, they use negative words [10].

According to Natural language processing (NLP), emotion detection is identifying and classifying emotions in written texts [42]. Websites, social media platforms, news sites, emails, online stores, user feedback services, customer service inquiries, etc., are all frequent sources of unstructured text data [19] used by the researchers. These sources can supply a significant volume of data, but the data might be challenging because of their unorganized nature. Researchers have made notable progress in detecting emotions in high-resource languages such as English, French, and Chinese [7]. However, the field of recognizing emotions through textual data in low-resource languages is still in a growing stage and attracting considerable attention [7,41].

Bengali is considered a low-resource language in NLP due to the limited availability of annotated data and resources for Bangla language processing [31]. Bengali, with approximately 250 million native speakers worldwide, ranks fourth among the most widely ence of Bespoken languages [12]. However, emotion detection in Bengali is still an under-explored area [16]. The limitation of annotated data and cultural variability in emotions make emotion detection from Bengali texts more challenging. However, it has significant applications for customer service, marketing and advertising, human resource management, healthcare, and security.

Several existing studies have focused on emotion detection from Bengali text [6,20,35]. However, most of these works rely on limited datasets [6,35], resulting in relatively low accuracy. For instance, traditional machine learning approaches such as Support Vector Machine (SVM) [35] and Naïve Bayes (NB) [6] achieved accuracies of 73% and 78%, respectively, in Bengali emotion classification tasks. Some recent methods have reported improved results, such as the approach proposed by Vijay et al. [20], which employs a manually annotated dataset and a pretrained model. Their hybrid architecture—combining a Word2Vec embedding layer with Convolutional Neural Network (CNN) and Long Short-Term Memory (LSTM) networks—achieved an accuracy of 90.49%. However, this model is restricted to detecting only three basic emotions: happiness, anger, and sadness. Furthermore, Large Language Model (LLM)based approaches [29] for Bengali emotion detection require further exploration.

To address these challenges, in this paper we propose two novel techniques, $EmoBang_{Hybrid}$ and $EmoBang_{Ensemble}$, for detecting eight distinct emotions from Bengali texts. The first approach, $EmoBang_{Hybrid}$, employs a Convolutional Recurrent Neural NetworkLong Short-Term Memory (CRNN-LSTM) based hybrid architecture, while the second, $EmoBang_{Ensemble}$, introduces an ensemble framework that, to the best of our knowledge, has not been previously applied to Bengali emotion detection tasks. Both proposed methods demonstrate superior performance compared to existing approaches. Furthermore, this study provides a comprehensive empirical analysis by applying six traditional machine learning (baseline) models and two Large Language Models (LLMs) to the proposed dataset. The findings of this study are expected to serve as a valuable benchmark and foundation for future research on emotion detection in Bengali texts.

We apply five feature engineering techniques and conduct experiments, selecting BERT [13] for our models. We propose two models, $EmoBang_{ensemble}$ and $EmoBang_{hybrid}$, to evaluate performance. The hybrid model combines CRNN and LSTM, while the ensemble model uses an Adaboost classifier. Using multiple evaluation metrics, we find both models perform best with BERT features, with the ensemble slightly outperforming the hybrid. Section 3 provides detailed model descriptions. We address five research questions: i) RQ1: How effectively do EmoBang models detect emotions from Bengali texts? ii) RQ2: How does data balancing affect model performance? iii) RQ3: How do feature engineering techniques improve emotion detection? iv) RQ4: Does $EmoBang_{Hybrid}$ outperform the existing hybrid model [20]? v) RQ5: How do LLMs perform on our dataset? In summary, in this paper, we make the following contributions:

- We collect a Bengali real-world dataset from four sources and classify it into eight emotional classes.
- We conduct a comprehensive study using six baseline classifiers and five feature engineering techniques, and we also evaluate two LLM models. This multi-faceted analysis reveals key behavioral patterns in Bengali text data.
- We propose a novel hybrid model, $EmoBang_{Hybrid}$, based on CRNN, which achieves precision 92.67, recall 91.15, F1-score 92.50, and accuracy 92.86.
- We develop a BERT-based ensemble model, $EmoBang_{ensemble}$, the first of its kind for Bengali texts, which achieves precision 93.43, recall 92.43, F1-score 93.67, and accuracy 93.69.

2 Related Work

Emotion Detection from Bengali Texts: There are many existing works [4,9,17,28,30,32,34,35,44,45] where researchers detect emotions from Bengali texts. The summary of these works shows that in most of the works [4,9,28,32, 35], the number of emotion classes used for labelling is less than our proposed work EmoBang. Existing works can detect a maximum of six different emotion classes [32,44,45] and EmoBang can detect eight. For example, Rahman et al. [32] can classify six emotion classes, and the accuracy of their approach is 52.98%.

Some of the works [4,30] developed to detect emotion from Bengali texts do not provide detailed size and nature of the dataset. Therefore, no further study can be done on those datasets. Many existing techniques [9,35] use relatively small datasets which may not show reliable results. For example, Bhowmik et al. [9] collect a small training ABSA dataset and focus on only the feature words, which may limit the generalizability of the work. Ruposh et al. [35] use a relatively small corpus of 1200 emotive words for training the SVM classifier and do not provide detailed information about the data pre-processing techniques. Compared to these works, our EmoBang contains a standard set of data (4240 Bengali sentences). We will share our dataset publicly so that further research can be done. We also preprocess and balance our data, a novel feature in detecting emotion from Bengali texts.

Emotion Detection from Bengali Text Using LLM: Despite advances in multilingual NLP, Large Language Models (LLMs) remain underused in Bangla for sentiment and emotion analysis. Sadhu et al. [36] show that LLMs reproduce sociocultural biases, linking men with authority or anger and women with empathy or fear, highlighting the need for fairness-aware modeling in low-resource languages. Tabassum [43] analyzes Bangladeshi e-commerce reviews with BERT, mBERT, and LLaMA, finding that optimized LLaMA-3.1-8B achieves 95.5% accuracy while remaining computationally efficient via LoRA and PEFT. These studies highlight both the promise and ethical challenges of LLMs in Bangla, emphasizing bias detection, interpretability, and efficient adaptation.

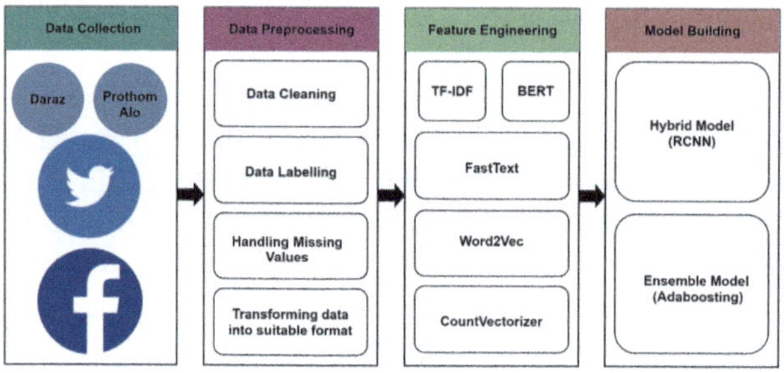

Fig. 1. Framework of the EmoBang

3 Methodology

Figure 1 depicts the overview of EmoBang. It has four major steps: i) Data Collection, ii) Data Preprocessing, iii) Feature Engineering, and iv) Model Building. In this study, we have worked with eight emotion types from Bengali texts: Anger, Sadness, Happiness, Disgust, Sarcastic, Fear, Surprise, and Disappointed.

3.1 Data Collection

We have gathered a total of 4240 Bengali sentences and approximately 44650 words. While there are several available datasets [6,21,40,44] for emotion detection, many of them have limited amounts of data[6,21,44], and researchers use a maximum of six classes [6,27] for conventional datasets. To compare and contrast our findings with these existing datasets, we have created our dataset that includes eight distinct classes (Anger, Sadness, Happiness, Disgust, Sarcastic,

Fear, Surprise, and Disappointed). For detecting emotions from Bengali texts, the data is collected from various sources, including popular social media platforms such as Twitter and Facebook, as well as popular news websites such as Prothom Alo [2] and e-commerce sites Daraz [1]. We use API [11] techniques based on different contents to collect data from Facebook and Twitter. We use scrapping [14] to collect data from the popular newspaper website Prothom Alo and the E-Commerce site Daraz, where thousands of users post their feedback in the form of comments and posts daily.

3.2 Data Preprocessing and Labelling

Text preprocessing in Bengali follows the same general principles as other languages but presents unique challenges. We first remove special characters, diacritics, punctuation marks, HTML tags, URLs, and digits from the text to reduce noise and improve model accuracy.

We convert emojis to their corresponding Bengali emotions to better capture sentiment. Using Python's "emoji" library, we map emojis to their Unicode representations and then to Bengali emotion words. For example, the smiley emoji, representing happiness, maps to ``আনন্দময়" (Anondomoy). Some emojis have different meanings in Bengali than in English, so we perform manual conversions. The "folded hands" emoji, for instance, often expresses gratitude or a request for forgiveness in Bengali, whereas in English it usually indicates praying or a high-five.

Removing stop words is challenging in Bengali due to the lack of dedicated packages, so we manually create a stop-word list (Table 1).

We apply the Synthetic Minority Oversampling Technique (SMOTE) [26] to balance the dataset by oversampling the minority class. SMOTE identifies

Table 1. Created stop words functions that remove the listed stop words

SW(Bng)	Eng.	SW(Bng)	Eng.	SW(Bng)	Eng.	SW(Bng)	Eng.
এ	this	যায়	possible	এর	of this	হয়	is/are
কি	what	বা	or	যাক	let's go	যে	who/that
কে	who	সব	all	উপর	on/above	হবে	will be
এই	this	একই	same	তাকে	him/her	আগে	before
বা	or	কখন	when	আছে	has/have	তাই	therefore
সে	this	সেই	that	হয়	happens	তার	his/her/its
যদি	if	অধীন	under	করে	does/makes	ছিল	was/were
আমি	I	এবং	and	তারা	they	কার	whose
এটি	this	গুলি	plural	হতে	to be	সেটা	that
আরও	more	খুব	very	পরে	after	কোন	which
কেন	why	সকল	all	ঠিক	correct	যারা	those who
কোন	any	তুমি	you				

instances in the underrepresented class, randomly selects one, and finds its k-nearest neighbors, where k is a user-defined parameter. It generates synthetic samples by interpolating feature values between the selected instance and its neighbors. We repeat this process until reaching the desired oversampling level, which balances the class distribution, reduces overfitting, and improves model generalization [23].

We label our data manually, following common practice in Bengali emotion, sentiment, and hate speech detection [6,9,25]. Three student volunteers and a supervisor, who has strong command of Bengali, assign emotion labels to each sentence. We use a voting technique [33]: each participant labels every sentence, and we assign the emotion class with the majority of votes.

3.3 Feature Engineering

Feature extraction, or vectorization, converts raw text into numerical features that serve as inputs for machine learning models [37]. This process encodes words in a text document as fixed-length vectors that capture their semantic meaning. The choice of feature extraction method strongly affects the performance of a text classifier, making it essential to experiment with different techniques for each task. In our study, we apply both classical and modern feature engineering techniques, including CountVectorizer [5], TF-IDF [18], BERT [13], Word2Vec [24], and FastText [22], to extract features from the text data. Our experiments show that BERT embeddings [38] perform best. Therefore, we use BERT embeddings in our proposed ensemble and hybrid models to enhance performance. We also fine-tune the BERT model to generate task-specific embeddings, further improving our results.

3.4 Proposed Model

We propose two models: (i) the hybrid model ($EmoBang_{hybrid}$) and (ii) the ensemble model ($EmoBang_{ensemble}$). While many studies use hybrid models for emotion detection, we introduce a new hybrid model that outperforms previous approaches [20]. We also present an ensemble-based model, which, to our knowledge, no one has yet applied to Bengali text emotion detection (Fig. 2).

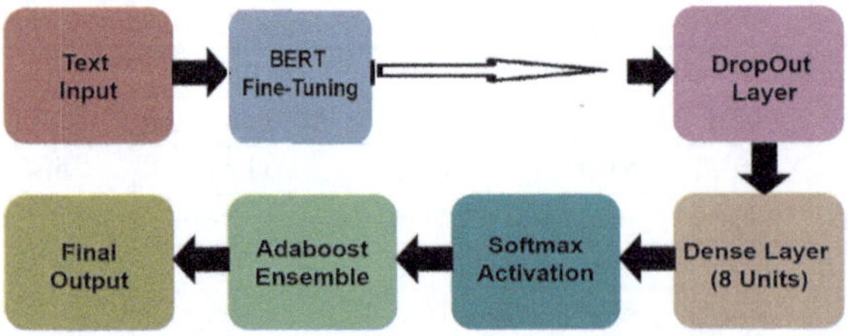

Fig. 2. Adaboost model architecture.

We propose a novel AdaBoost-BERT ensemble model, $EmoBang_{ensemble}$, for multiclass emotion detection in Bengali text. The model enhances AdaBoost by integrating a fine-tuned BERT transformer, inspired by Gao et al.'s work on legal text analysis [15]. $EmoBang_{ensemble}$ captures latent textual features by reweighting weak classifiers and leveraging BERT's output probabilities. BERT performs internal tokenization, converting raw text into token IDs, which pass through transformer layers to produce contextualized embeddings.

We refine these embeddings using additional neural layers: a dropout layer prevents overfitting, and a dense layer with eight units maps embeddings to the emotion classes. Softmax converts the logits into a probability distribution. AdaBoost then forms an ensemble of weak classifiers, adjusting their weights based on performance and emphasizing those that perform well. This step combines the strengths of multiple BERT-based classifiers, further improving classification accuracy.

The AdaBoost-BERT model is trained on a set of Bengali texts and their corresponding emotion labels, denoted as $X = \{(x_1, y_1), \ldots, (x_m, y_m)\}$. At each iteration t, the model computes the classifier $h_t(x)$ that minimizes the weighted error E_t:

$$E_t = \sum_{i=1}^{m} w_{t,i} \cdot I(h_t(x_i) \neq y_i)$$

Here, $w_{t,i}$ represents the weight of the i-th sample at iteration t, and the indicator function $I(\cdot)$ is used to assess misclassification. The weights are updated for accurately classified samples as follows:

$$w_{t+1,i} = \frac{w_{t,i} \cdot e^{-\alpha_t h_t(x_i) y_i}}{Z_t} [?]$$

In this equation, α_t denotes the weight of the t-th weak classifier, and Z_t is a normalization constant ensuring the weights sum to one. The final prediction for a new input x is given by:

$$H(x) = \arg\max_{j} \sum_{t=1}^{T} \alpha_t I(h_t(x) = j)$$

This approach integrates BERT's deep contextual understanding with the AdaBoost algorithm, which focuses on combining multiple weak classifiers to optimize the model's performance for multiclass emotion detection from Bengali texts. The AdaBoost-BERT ensemble model leverages the individual strengths of BERT's pre-trained transformer-based neural network and the AdaBoost algorithm's ability to create an ensemble of weak classifiers, achieving improved accuracy in emotion classification tasks.

The BERT model was fine-tuned using the Adam optimizer, with a learning rate of 2e−5 and a batch size of 32. The training process spanned 100 epochs, with early stopping based on the validation loss. A dropout layer with a rate of 0.1 was applied after the BERT output, followed by a dense layer with 8 units and a softmax activation. The softmax function converts the neural network's output into a probability distribution over the possible classes, defined mathematically as:

$$softmax(z_i) = \frac{e^{z_i}}{\sum_{j=1}^{K} e^{z_j}}$$

Table 2. Performance and Execution Time

Mdl.	Prec.	Rec.	F1	Acc.	T(m)
H	92.67	91.15	92.50	92.86	5.6
Ens	93.43	92.43	93.67	93.69	6

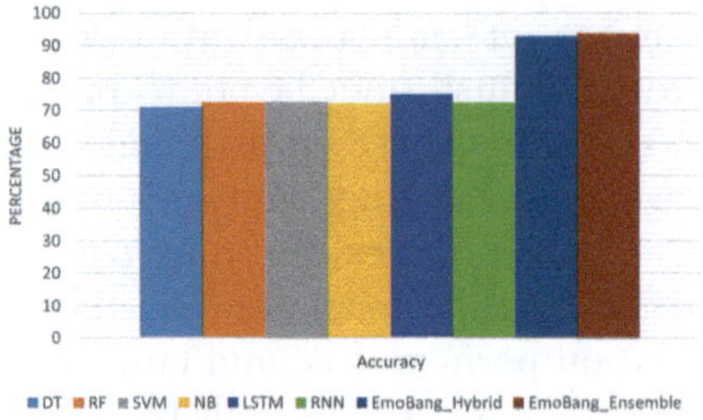

Fig. 3. Accuracy Comparison Among Multiple Models

Here, z_i is the input to the ith neuron in a neural network layer, representing the log-odds or logit of the probability that the input belongs to the ith class. The softmax function returns a probability distribution over the K possible classes, where K is the number of neurons in the output layer.

Overall, the AdaBoost-BERT ensemble model combines the strengths of BERT's pre-trained transformer-based neural network and the AdaBoost algorithm's ability to create an ensemble of weak classifiers to achieve improved performance in multiclass emotion detection from Bengali texts.

4 Results and Analysis

RQ1: Effectiveness and Efficiency of EmoBang Models. To measure the effectiveness of EmoBang methods we evaluate precision (Prec.), recall (Rec.), F1 score (F1), and accuracy (Acc.) of it. Table 2 shows the performance of $EmoBang_{Hybrid}$ and $EmoBang_{Ensemble}$. The results show that both of the methods have accuracy rates of 92.86% and 93.69% respectively. The *Ensemble* method has 0.76%, 1.28%, 1.17%, and 0.83% higher precision, recall, F1 score and accuracy respectively compared to $EmoBang_{Hybrid}$. However, both models have more than 90% value in all four measuring parameters. Figure 3 compares accuracy among EmoBang models with the baseline models. This chart shows that $EmoBang_{Hybrid}$ and $EmoBang_{Ensemble}$ models have more accuracy than any other baseline models. $EmoBang_{Hybrid}$ is 22%, 20.41%, 20.24%, 20.74%, 17.65%, and 20.44% more accurate than the baseline models DT, RF, SVM, NB, LSTM, and RNN respectively. On average $EmoBang_{Hybrid}$ is 20.65% more accurate in detecting emotion from Bengali texts than any of the six baseline models. $EmoBang_{Ensemble}$ is 22.83%, 21.24%, 21.07%, 21.57%, 18.48%, and 21.27% more accurate than DT, RF, SVM, NB, LSTM, and RNN models respectively.

On average $EmoBang_{Ensemble}$ is 21.08% more accurate than the baseline models. Therefore, EmoBang methods are effective in emotion detection for Bengali texts.

The last column of Table 2 shows the total execution time $EmoBang_{Hybrid}$ and $EmoBang_{Ensemble}$ take to detect emotion from Bengali text for our given dataset. This result indicates that EmoBang is efficient enough to detect emotion in Bengali text.

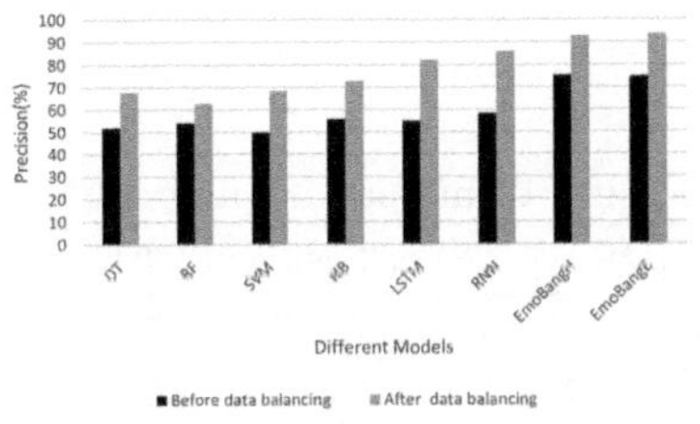

Fig. 4. Comparison of Precision Value.

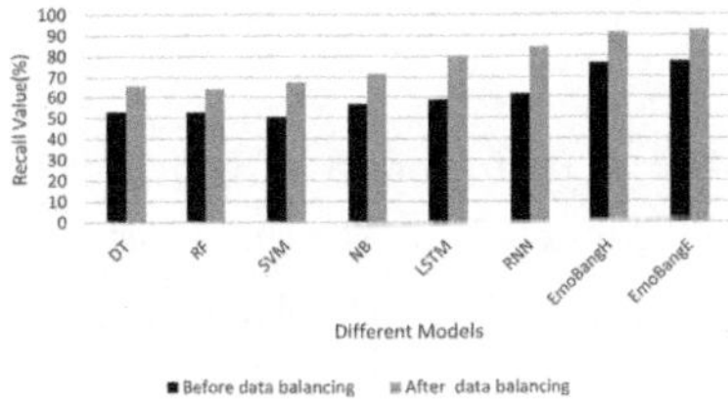

Fig. 5. Comparison of Recall Value.

RQ2: Role of Data Balancing in the Effectiveness of Baseline Models and EmoBang Models. In this work, we perform data balancing and experimental results show that it improves not only the effectiveness of EmoBang models but also all other baseline models. Figure 4 shows that the precision value increases after data balancing for all the six baseline models and for two EmoBang models. After data balancing, the precision value increases by 16.03%, 8.67%, 18.87%, 16.96%, 27.39%, 27.56%, 17.44%, and 18.84% in the DT, RF, SVM, NB, LSTM, RNN, $EmoBang_{Hybrid}$, and $EmoBang_{Ensemble}$ model respectively. On average it increases by 20.73%. Figure 5 compares the recall values for eight models before and after data balancing. For all of the eight models the recall value increases after data balancing. The experimental evaluation shows that it increases by 13.02%, 11.11%, 17.05%, 15.14%, 21.55%, 23.22%, 14.68%, and 15.11% in DT, RF, SVM, NB, LSTM, RNN, $EmoBang_{Hybrid}$, and $EmoBang_{Ensemble}$ model respectively. On average the recall value increases by 16.36% after data balancing.

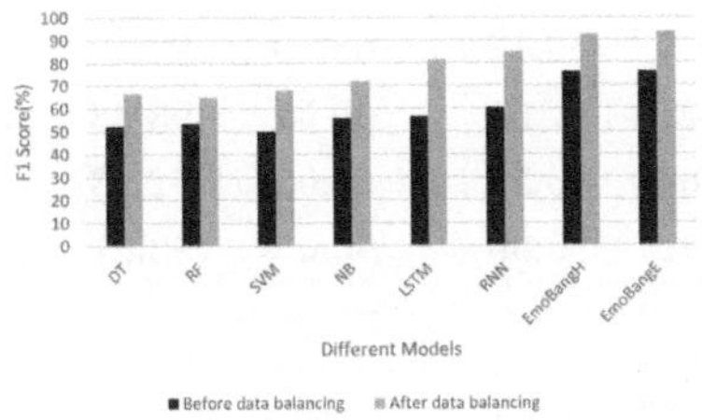

Fig. 6. Comparison of F1-Score.

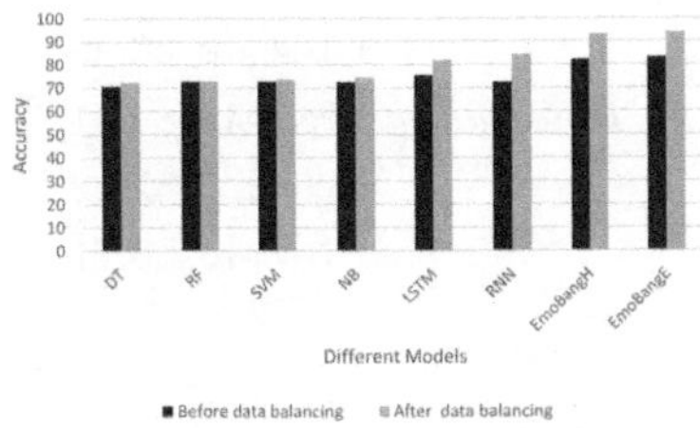

Fig. 7. Comparison of Accuracy Value.

Figure 6 shows the comparison among eight different models' F1-score before and after data balancing. F1 score is increased by 14.33%, 11.46%, 18.01%, 16.24%, 25.18%, 24.64%, 16.65%, and 17.71% for DT, RF, SVM, NB, LSTM, RNN, $EmoBang_{Hybrid}$, and $EmoBang_{Ensemble}$ model respectively after balancing the data. On average, the F1 score is increased by 18.03%. Figure 7 shows that the accuracy gets improved in all models after data balancing. Particularly the accuracy is improved by 1.61%, 0.34%, 0.92%, 2.05%, 6.71%, 11.79%, 10.86%, and 10.69% respectively. On average, the accuracy is increased by 5.62% after balancing the data.

Table 3. Performance of Different Feature Engineering Techniques in Different Models

Models	Metric	CountVectorizer	TF-IDF	Word2Vec	FastText	BERT
DT	Precision	65.66	70.03	75.63	77.44	67.72
	Recall	62.3	71.9	74.87	74.88	65.66
	F1-score	66.10	73.3	75.86	75.50	66.49
RF	Precision	62.2	66.87	77.8	73.8	62.54
	Recall	63.7	65.98	71.10	72.10	63.65
	F1-score	64.01	65.50	77.98	70.54	64.69
SVM	Precision	67.11	73.25	80.87	77.49	68.41
	Recall	66.8	72.10	79.32	76.32	67.23
	F1-score	67.7	73.23	80.10	77.50	67.87
NB	Precision	58.89	64.87	72.30	72.30	72.65
	Recall	57.41	62.76	71.55	71.55	71.55
	F1-score	58.90	62.10	71.93	71.93	71.93
LSTM	Precision	70.24	74.25	82.32	80.55	82.15
	Recall	68.72	73.54	81.54	79.43	80.09
	F1-score	70.29	74.90	82.20	82.42	81.61
RNN	Precision	67.20	71.23	83.30	83.89	85.76
	Recall	65.32	70.23	82.43	82.54	84.54
	F1-score	66.34	71.22	82.50	82.50	84.98
$EmoBang_H$	Precision	80.13	82.54	90.15	88.76	92.67
	Recall	79.54	81.45	89.10	87.45	91.15
	F1-score	66.49	82.75	89.12	88.10	92.50
$EmoBang_E$	Precision	84.51	81.21	87.65	90.54	93.43
	Recall	82.62	83.64	86.23	89.65	92.43
	F1-score	83.71	82.32	86.99	90.05	93.67

RQ3: Role of feature engineering techniques for emotion detection from Bengali texts. We evaluated five different feature engineering techniques: i) CountVectorizer, ii) Tf-Idf, iii) Word2vec, iv) FastText, and v) BERT by

incorporating them into eight different models. Table 3 represents the precision, recall, and F1 score for five different feature engineering techniques on eight different models. These results indicate that the feature engineering technique BERT outperforms all other techniques. Hence, the FastText also performs well, achieving high F1 score values with both proposed models. Word2vec performs comparatively low compared to the other feature engineering techniques.

Table 3 represents the precision, recall, and F1 score for five different feature engineering techniques on eight different classifiers. The findings suggest that the CountVectorizer demonstrates the lowest precision, recall, and F1 score value when combined with various models but exhibits the highest value when employed in the $EmoBang_E$ model. The DT classifier exhibits the most noteworthy precision and recall values when utilizing the FastText approach. The F1 score is also maximized when employing the Word2Vec feature engineering technique. RF produces the lowest recall value employing BERT and the highest precision and F1 score with Word2Vec. In addition, Word2Vec feature engineering obtained the highest precision, recall, and F1 score in conjunction with the SVM classifier. The TF-IDF shows the highest precision and F1 score while utilizing the $EmoBang_E$ model. The NB model achieves the highest precision score in BERT and maximized recall and F1 score while employing Word2Vec, FastText, and BERT feature engineering techniques. The FastText feature engineering technique produces the highest precision, recall, and F1 scores in $EmoBang_E$. Moreover, RNN achieved the best precision, recall, and F1 scores in the BERT technique. Our proposed method exhibits the highest precision, recall, and F1 score value when using the BERT technique. The $EmoBang_H$ model produced the lowest results in all three evaluation matrices while the $EmoBang_E$ *model* achieved the lowest precision and F1 score utilizing the TF-IDF technique. In summary, different feature engineering techniques performed best for the different classifiers. In the context of the DT classifier, FastText performs best among all other feature engineering techniques.

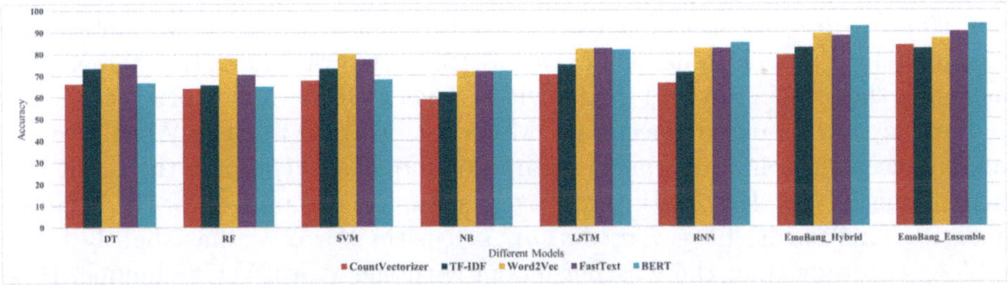

Fig. 8. Accuracy Comparison of Multiple Models Using Feature Engineering.

Figure 8 shows the accuracy comparison for eight different models with five different feature engineering techniques. In the case of the DT, RF, and SVM models, the Word2vec feature engineering techniques show the highest accuracy and CountVectorizer shows the lowest accuracy because words' semantic relationships can be captured using word2vec. The LSTM model shows high accuracy(82.42%) with the FastText feature engineering. Other models, NB, RNN, $EmoBang_{Hybrid}$, and $EmoBang_{Ensemble}$ show the highest accuracy by using the BERT technique. Hence, In our proposed models we use BERT [13] feature Engineering.

RQ4: Comparison of EmoBang Hybrid Model with the Existing Hybrid Model. To the best of our knowledge there is an existing hybrid model [20] developed by Muntasir et al. to extract emotion from Bangla texts. No existing techniques use the Adaboost ensemble model to perform this task. Hence we, compare $EmoBang_{Hybrid}$ with the existing hybrid model [20].

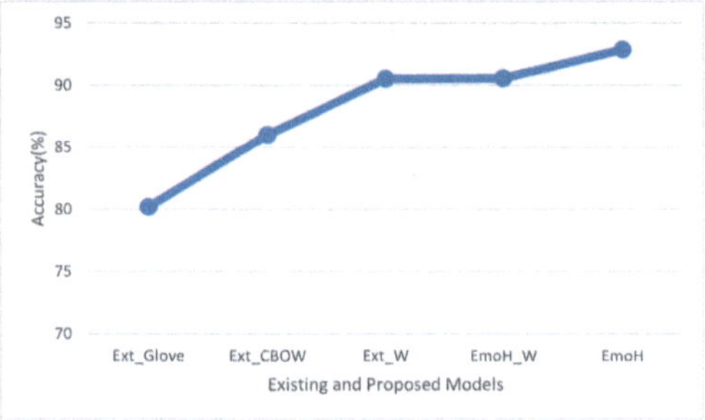

Fig. 9. Accuracy Comparison of Hybrid Models.

Figure 9 shows the comparison between the existing hybrid model [20] and $EmoBang_{Hybrid}$. In the existing hybrid model, authors used three different feature engineering techniques Glove, CBOW, Word2Vec which is denoted as *Ext_Glove*, *Ext_CBOW*, and *Ext_Word2Vec* in Fig. 9. We compare the accuracy of these three variants of the existing hybrid model with our proposed hybrid model $EmoBang_{Hybrid}$ (denoted as EmoH in Fig. 9). The results show that $EmoBang_{Hybrid}$ has higher accuracy than all of the three variants of the existing hybrid model. In particular, it has 12.7%, 6.86%, and 2.37% higher accuracy than the existing hybrid model with Glove, CBOW, and Word2Vec feature engineering techniques, respectively. Moreover, we also use the Word2Vec feature engineering technique with our proposed model and compare the result with the existing model. *EmoH_W* and *Ext_w* show the comparison of the accuracy in Fig. 9. This result shows $EmoBang_{Hybrid}$ with Word2Vec has slightly better (0.05%) accuracy than the existing model with the Word2Vec technique. However, a significant contribution of our hybrid model is that we work with eight different emotional classes, and the existing model[20] works only with three emotional classes.

RQ5: LLM's Performance on the Dataset: We conduct a small-scale extension that employs large language models (LLMs) in zero-shot and few-shot settings to align this study with recent advances in low-resource natural language processing (NLP). The zero-shot and few-shot paradigms leverage the intrinsic generalization capabilities of pre-trained transformer-based LLMs to perform classification with minimal or no task-specific training. In contrast, conventional supervised learning involves refining models on task-specific labeled datasets.

The model configuration is as follows: We evaluate two multilingual LLMs, BanglaBERT [8], and XLM-RoBERTa Large [3]. Using the prompt "Classify the emotion: [input text]", the models classify Bengali sentences into eight emotion categories under the zero-shot configuration without any fine-tuning. In the few-shot setting, we provide each model with ten labeled examples per class as in-context demonstrations prior to inference.

Table 4. Zero-shot and Few-shot Performance of LLMs

Model	Setup	Accuracy (%)	Precision (%)	F1-score (%)
BanglaBERT	Zero-shot	84.92	85.12	84.76
BanglaBERT	Few-shot	89.54	90.45	89.92
XLM-RoBERTa Large	Zero-shot	86.21	87.60	86.83
XLM-RoBERTa Large	Few-shot	91.22	92.35	91.74

With zero-shot accuracies between 84% and 86%, the results in Table 4 indicate that LLMs such as XLM-RoBERTa and BanglaBERT perform effectively without task-specific fine-tuning. Both models exhibit improved performance under few-shot conditions, achieving an F1-score of 91.74% and an accuracy of up to 91.22%. These results confirm that contemporary multilingual LLMs generalize well in low-resource language settings.

The proposed ensemble model, $EmoBang_{Ensemble}$, outperforms conventional baselines and individual LLMs, achieving the highest performance (Accuracy = 93.69%, F1 = 93.67%). The model captures emotion-specific lexical and contextual dependencies in Bengali text through task-specific fine-tuning and AdaBoost-based integration of BERT embeddings.

In contrast, zero-shot and few-shot LLMs yield slightly lower accuracy due to their reliance on general-purpose representations that are not fully optimized for emotion detection. Hence, for low-resource languages, domain-optimized hybrid architectures such as $EmoBang_{Ensemble}$ provide superior emotion classification performance, while LLMs remain a promising option for scalable, low-data applications.

5 Conclusion and Discussion

Although the experimental results are promising, it is important to recognize that the models' applicability in real-world scenarios may be limited. The models may not perform as effectively on text data from different sources or on other types of textual content. Furthermore, the accuracy scores achieved in this study may not accurately represent real-world performance, as the experiments are conducted on a specific dataset. Through comprehensive experimental evaluations, we demonstrate that our proposed models outperform existing approaches developed for Bengali texts in terms of accuracy, precision, and recall. We conduct an extensive empirical study that includes six baseline models and five feature engineering techniques, providing a thorough analysis of emotion detection in Bengali text. To the best of our knowledge, no prior research has investigated these different feature engineering techniques within the context of Bengali emotion detection. Among the two proposed models, the $EmoBang_{ensemble}$ model achieves the highest accuracy of 93.69%.

Despite these encouraging results, our study has several limitations. The proposed EmoBang models have not yet been applied to datasets of varying sizes, structures, or linguistic domains. Additionally, they have not been tested on datasets from other languages. As part of our future research, we plan to apply the proposed models to larger and more diverse datasets drawn from multiple low-resource languages. This effort will allow us to evaluate the generalizability and scalability of our approach. We also intend to explore large language models (LLMs) in future work, as these models are trained on vast and diverse text corpora and may offer improved contextual understanding and robustness. Incorporating LLMs could potentially enhance emotion detection performance and broaden the applicability of our approach across multiple languages and domains.

References

1. Daraz website. https://www.daraz.com.bd/. Accessed 24 May 2023
2. Prothom alo newspaper website. https://www.prothomalo.com/. Accessed 24 May 2023
3. Xlm roberta large multilingual language model. https://dataloop.ai/library/model/. Accessed 26 Oct 2025
4. Alam, T., Khan, A., Alam, F.: Bangla text classification using transformers. arXiv preprint arXiv:2011.04446 (2020)
5. Alvi, N., Talukder, K.H.: Sentiment analysis of Bengali text using countvectorizer with logistic regression. In: 2021 12th International Conference on Computing Communication and Networking Technologies (ICCCNT), pp. 01–05. IEEE (2021)
6. Azmin, S., Dhar, K.: Emotion detection from Bangla text corpus using Naive Bayes classifier. In: 2019 4th International Conference on Electrical Information and Communication Technology (EICT), pp. 1–5. IEEE (2019)
7. Bashir, M.F., Javed, A.R., Arshad, M.U., Gadekallu, T.R., Shahzad, W., Beg, M.O.: Context-aware emotion detection from low-resource Urdu language using deep neural network. ACM Trans. Asian Low-Resour. Lang. Inf. Process. **22**(5) (2023). https://doi.org/10.1145/3528576
8. Bhattacharjee, A., et al.: Banglabert: language model pretraining and benchmarks for low-resource language understanding evaluation in Bangla (2022). https://arxiv.org/abs/2101.00204
9. Bhowmik, N.R., Arifuzzaman, M., Mondal, M.R.H., Islam, M.: Bangla text sentiment analysis using supervised machine learning with extended lexicon dictionary. Natural Lang. Process. Res. **1**(3–4), 34–45 (2021)

10. Binali, H., wu, C., Potdar, V.: Computational approaches for emotion detection in text, pp. 172–177 (2010). https://doi.org/10.1109/DEST.2010.5610650
11. Campan, A., Atnafu, T., Truta, T.M., Nolan, J.: Is data collection through twitter streaming API useful for academic research? In: 2018 IEEE International Conference on Big Data (Big Data), pp. 3638–3643. IEEE (2018)
12. Das, B., Mandal, S., Mitra, P.: Bengali speech corpus for continuous automatic speech recognition system. In: 2011 International conference on speech database and assessments (Oriental COCOSDA), pp. 51–55. IEEE (2011)
13. Devlin, J., Chang, M.W., Lee, K., Toutanova, K.: Bert: pre-training of deep bidirectional transformers for language understanding. arXiv preprint arXiv:1810.04805 (2018)
14. Diouf, R., Sarr, E.N., Sall, O., Birregah, B., Bousso, M., Mbaye, S.N.: Web scraping: state-of-the-art and areas of application. In: 2019 IEEE International Conference on Big Data (Big Data), pp. 6040–6042. IEEE (2019)
15. Gao, J., Ning, H., Han, Z., Kong, L., Qi, H.: Legal text classification model based on text statistical features and deep semantic features. In: FIRE (Working Notes), pp. 35–41 (2020)
16. Garg, K., Lobiyal, D.: Hindi emotionnet: a scalable emotion lexicon for sentiment classification of Hindi text. ACM Trans. Asian Low-Resour. Lang. Inf. Process. (TALLIP) **19**(4), 1–35 (2020)
17. Ghosh, T., Al Banna, M.H., Al Nahian, M.J., Uddin, M.N., Kaiser, M.S., Mahmud, M.: An attention-based hybrid architecture with explainability for depressive social media text detection in Bangla. Expert Syst. Appl. **213**, 119007 (2023)
18. Gomaa, W.H., Fahmy, A.A., et al.: A survey of text similarity approaches. Int. J. Comput. Appl. **68**(13), 13–18 (2013)
19. Halibas, A.S., Shaffi, A.S., Mohamed, M.A.K.V.: Application of text classification and clustering of twitter data for business analytics. In: 2018 Majan International Conference (MIC), pp. 1–7. IEEE (2018)
20. Hoq, M., Haque, P., Uddin, M.N.: Sentiment analysis of Bangla language using deep learning approaches. In: Chaubey, N., Parikh, S., Amin, K. (eds.) COMS2 2021. CCIS, vol. 1416, pp. 140–151. Springer, Cham (2021). https://doi.org/10.1007/978-3-030-76776-1_10
21. Islam, T., Latif, S., Ahmed, N.: Using social networks to detect malicious Bangla text content. In: 2019 1st International Conference on Advances in Science, Engineering and Robotics Technology (ICASERT), pp. 1–4. IEEE (2019)
22. Joulin, A., Grave, E., Bojanowski, P., Douze, M., Jégou, H., Mikolov, T.: Fasttext.zip: compressing text classification models. arXiv preprint arXiv:1612.03651 (2016)
23. Liu, J.: Importance-smote: a synthetic minority oversampling method for noisy imbalanced data. Soft. Comput. **26**(3), 1141–1163 (2022)
24. Mikolov, T., Chen, K., Corrado, G., Dean, J.: Efficient estimation of word representations in vector space. arXiv preprint arXiv:1301.3781 (2013)
25. Mridha, M.F., Wadud, M.A.H., Hamid, M.A., Monowar, M.M., Abdullah-Al-Wadud, M., Alamri, A.: L-boost: identifying offensive texts from social media post in Bengali. IEEE Access **9**, 164681–164699 (2021)
26. Nikiforos, M.N., Deliveri, K., Kermanidis, K.L., Pateli, A.: Vocational domain identification with machine learning and natural language processing on Wikipedia text: error analysis and class balancing. Computers **12**(6), 111 (2023)
27. Parvin, T., Hoque, M.M.: An ensemble technique to classify multi-class textual emotion. Procedia Comput. Sci. **193**, 72–81 (2021)

28. Purba, S.A., Tasnim, S., Jabin, M., Hossen, T., Hasan, M.K.: Document level emotion detection from Bangla text using machine learning techniques. In: 2021 International Conference on Information and Communication Technology for Sustainable Development (ICICT4SD), pp. 406–411. IEEE (2021)
29. Pyreddy, S.R., Zaman, T.S.: EMOXPT: analyzing emotional variances in human comments and LLM-generated responses. In: 2025 IEEE 15th Annual Computing and Communication Workshop and Conference (CCWC), pp. 00088–00094 (2025). https://doi.org/10.1109/CCWC62904.2025.10903889
30. Rabeya, T., Ferdous, S., Ali, H.S., Chakraborty, N.R.: A survey on emotion detection: a lexicon based backtracking approach for detecting emotion from Bengali text. In: 2017 20th International Conference of Computer and Information Technology (ICCIT), pp. 1–7. IEEE (2017)
31. Rabib, H.K., et al.: Gender-based cyberbullying detection for under-resourced Bangla language. In: 2022 12th International Conference on Electrical and Computer Engineering (ICECE), pp. 104–107. IEEE (2022)
32. Rahman, M., Seddiqui, M., et al.: Comparison of classical machine learning approaches on Bangla textual emotion analysis. arXiv preprint arXiv:1907.07826 (2019)
33. Rahman, M.A., Kumar Dey, E.: Datasets for aspect-based sentiment analysis in Bangla and its baseline evaluation. Data **3**(2), 15 (2018)
34. Rayhan, M.M., Al Musabe, T., Islam, M.A.: Multilabel emotion detection from Bangla text using Bigru and CNN-biLSTM. In: 2020 23rd International Conference on Computer and Information Technology (ICCIT), pp. 1–6. IEEE (2020)
35. Ruposh, H.A., Hoque, M.M.: A computational approach of recognizing emotion from Bengali texts. In: 2019 5th International Conference on Advances in Electrical Engineering (ICAEE), pp. 570–574. IEEE (2019)
36. Sadhu, J., Saha, M.R., Shahriyar, R.: An empirical study of gendered stereotypes in emotional attributes for Bangla in multilingual large language models. arXiv preprint arXiv:2407.06432 (2024)
37. Sammons, M., Christodoulopoulos, C., Kordjamshidi, P., Khashabi, D., Srikumar, V., Roth, D.: Edison: feature extraction for NLP, simplified. In: Proceedings of the Tenth International Conference on Language Resources and Evaluation (LREC'16), pp. 4085–4092 (2016)
38. Selva Birunda, S., Kanniga Devi, R.: A review on word embedding techniques for text classification. In: Innovative Data Communication Technologies and Application: Proceedings of ICIDCA 2020, pp. 267–281 (2021)
39. Shivhare, S.N., Khethawat, S.: Emotion detection from text. arXiv. https://doi.org/10.48550/arXiv.1205.4944 (2012)
40. Shuhan, M.K.B., Dey, R., Saha, S., Anjum, M.S.U., Zaman, T.S.: A stylometric dataset for Bengali poems. In: Proceedings of the 2022 6th International Conference on Natural Language Processing and Information Retrieval. NLPIR '22, New York, NY, USA, pp. 176–180. Association for Computing Machinery (2023). https://doi.org/10.1145/3582768.3582788
41. Sourav, M.S.U., Wang, H.: Transformer-based text classification on unified Bangla multi-class emotion corpus. arXiv preprint arXiv:2210.06405 (2022)
42. Sreenidhi, M., Dhanya, S.S.S., Sahithi, R., Femi, P.S.: Human emotion recognition system using deep learning techniques. Int. J. Eng. Res. Technol. (IJERT) **9**(07) (2020)
43. Tabassum, S.: LLM based sentiment classification from Bangladesh e-commerce reviews. arXiv preprint arXiv:2510.01276 (2025)

44. Tripto, N.I., Ali, M.E.: Detecting multilabel sentiment and emotions from Bangla YouTube comments. In: 2018 International Conference on Bangla Speech and Language Processing (ICBSLP), pp. 1–6. IEEE (2018)
45. Tuhin, R.A., Paul, B.K., Nawrine, F., Akter, M., Das, A.K.: An automated system of sentiment analysis from Bangla text using supervised learning techniques. In: 2019 IEEE 4th International Conference on Computer and Communication Systems (ICCCS), pp. 360–364. IEEE (2019)

A Deployed Bilinear Neural Network Definition in Sentiment Prediction

Nguyen Minh Tuan[1(✉)] and Phayung Meesad[2(✉)]

[1] Faculty of Information Technology, 11 Nguyen Dinh Chieu, Sai Gon ward, Ho Chi Minh city, Viet Nam
minhtuan@ptit.edu.vn

[2] Department of Information Technology and Management, King Mongkut's University of Technology North Bangkok, 1518 Pracharat 1 Road, Wongsawang, 10800 Bangkok, Thailand
phayung.m@itd.kmutnb.ac.th

Abstract. This study presents a deployed bilinear neural network architecture for sentiment prediction using Vietnamese textual data. The proposed framework integrates two distinct yet complementary models, including Hirota's bilinear neural network (Hirota-BiNN) and the bilinear recurring neural network (Bilinear-RNN), to enhance the representation of features and the classification accuracy. The data, collected from Vietnamese feedback sentences, is preprocessed through tokenization, normalization, and embedding using the VnCoreNLP and Word2Vec techniques. These embeddings are then used as input for bilinear transformations that capture higher-order semantic relationships. Experimental results demonstrate that the Hirota-BiNN model achieves superior accuracy and robustness compared to the Bilinear-RNN model, attaining a training accuracy of 0.6295 and a validation accuracy of 0.6329. In contrast, the Bilinear-RNN achieves 0.5089 and 0.4776, respectively. Evaluation metrics comprising precision and F1-score further confirm the effectiveness of the Hirota-BiNN architecture in capturing complex linguistic structures and delivering more reliable sentiment classification. The results highlight the potential of bilinear neural formulations for advancing sentiment analysis in low-resource languages like Vietnamese.

Keywords: Bilinear Neural Network · Hirota Bilinear Layer · Sentiment Prediction · Feature Interaction · Deep Learning · Recurrent Neural Network

1 Introduction

The Convolutional Neural Network (CNN) method is considered a leading model in deep learning and is highly recommended for designing computer-aided diagnostic systems due to its promising results in medical image analysis [1]. One of the main challenges in fine-grained classification is that subcategories with significant similarities are difficult to distinguish because all subcategories are treated

H. Unger and P. Meesad (Eds.): NLPIR 2025, LNNS 1904, pp. 338–348, 2026.
https://doi.org/10.1007/978-3-032-20897-2_25

equally in existing algorithms. To address this issue, an image fine-grained classification method that combines a bilinear convolutional neural network has been proposed [2]. Existing graph neural network (GNN) methods represent a target service node as the weighted sum of the features of neighboring service nodes, achieving remarkable results in service classification. However, these methods ignore the potential impact of interactions between neighboring nodes on the target node. This problem has also been addressed through a bilinear graph neural network–based Web service classification approach [3].

The classification model employs a Bilinear Convolutional Neural Network (BCNN) with a triplet constraint network. BCNN can capture rich spatial interactions among features in image data by computing the outer product of feature vectors from two different CNNs via bilinear pooling. The proposed model is trained to extract intra-class features from a deep network and thus increase the inter-feature distance, improving overall performance [4]. CNNs have been studied for skin lesion classification using various methods and training techniques. However, results from previous works indicate a wide margin for improvement in terms of detection capability, accuracy, and computational cost. Therefore, a bilinear CNN–based approach has been introduced to classify seven categories of skin lesions with state-of-the-art accuracy and low computational cost. The proposed framework includes a data augmentation step to address data imbalance, along with transfer learning and fine-tuning to enhance classification performance while reducing computation [5].

Table 1. Sample Vietnamese sentiment dataset

Text	Label
slide giáo trính đầy đủ .	2
nhiệt tính giảng dạy , gần gũi với sinh viên .	2
đi học đầy đủ full điểm chuyên cần .	0
chưa áp dụng công nghệ thông tin và các thiết bị .	0
thầy giảng bài hay , có nhiều bài tập ví dụ ngắn gọn .	2

Accurate identification of apple leaf diseases is crucial for improving apple productivity. However, since the lesion area on apple leaf images is small and prone to background noise, recognition accuracy tends to be low. To overcome this, a bilinear CNN model with an attention mechanism has been designed [6]. CNNs show great potential for classifying high-variation, fine-grained objects. A deep learning method based on BCNN has also been introduced for fine-grained classification of histopathological images [7]. A variant of BCNN, known as the Bilinear Neural Network (BNN), extends these ideas further. The extended BNN model has been applied to solving physical problems for both exact and approximate solutions [8]. With carefully designed fine-tuning layers, this approach minimizes complex computational steps while offering extensible solutions with inputs such as softmax, hyperbolic, trigonometric, or polynomial functions [9].

2 Related Definition

In this section, we formally define the fundamental mathematical and architectural components used in the proposed bilinear neural network models for sentiment prediction. The framework integrates both the Hirota Bilinear Neural Network (Hirota-BiNN) and the Bilinear Recurrent Neural Network (Bilinear-RNN) to capture complex nonlinear interactions in textual features.

2.1 Text Representation

Let a given input sentence be denoted as

$$S = \{w_1, w_2, \ldots, w_T\},$$

where w_i represents the i-th token and T is the sentence length. Each word w_i is mapped into a d-dimensional embedding vector $\mathbf{x}_i \in \mathbb{R}^d$ through Word2Vec or VnCoreNLP embeddings. Thus, the sentence is represented as a matrix

$$\mathbf{X} = [\mathbf{x}_1, \mathbf{x}_2, \ldots, \mathbf{x}_T]^\top \in \mathbb{R}^{T \times d}.$$

2.2 Bilinear Mapping Definition

Given two feature matrices $\mathbf{F}, \mathbf{G} \in \mathbb{R}^{T \times d}$ extracted from parallel convolutional or recurrent layers, a bilinear mapping function is defined as:

$$\mathbf{Y} = (\mathbf{F}\mathbf{W}_1) \odot (\mathbf{G}\mathbf{W}_2) + \mathbf{V}\big((\mathbf{F}\mathbf{W}_3) \odot (\mathbf{G}\mathbf{W}_4)\big),$$

where $\odot$ denotes the Hadamard (element-wise) product, and $\mathbf{W}_1, \mathbf{W}_2, \mathbf{W}_3, \mathbf{W}_4, \mathbf{V}$ are trainable weight matrices. This formulation enables the model to learn high-order feature interactions and nonlinear dependencies between the two feature streams.

2.3 Hirota Bilinear Neural Network (Hirota-BiNN)

Inspired by the Hirota bilinear differential method, the Hirota-BiNN layer transforms two independent convolutional feature maps $\mathbf{F}$ and $\mathbf{G}$ into a joint representation using bilinear interactions:

$$\mathbf{H} = \sigma\big((\mathbf{W}_1^\top \mathbf{F}) \odot (\mathbf{W}_2^\top \mathbf{G}) + \mathbf{V}^\top ((\mathbf{W}_3^\top \mathbf{F}) \odot (\mathbf{W}_4^\top \mathbf{G})) + \mathbf{b}\big),$$

where $\sigma(\cdot)$ denotes a nonlinear activation function such as ReLU. The Hirota-BiNN enhances feature diversity by fusing linear and bilinear transformations, allowing the network to better capture contextual sentiment cues.

2.4 Bilinear Recurrent Neural Network (Bilinear-RNN)

The Bilinear-RNN applies recurrent modeling on the temporal dimension to capture sequential sentiment dependencies. For a given embedded sequence $\mathbf{X}$, each time step t is processed as:

$$\mathbf{h}_t = \tanh(\mathbf{W}_h \mathbf{x}_t + \mathbf{U}_h \mathbf{h}_{t-1} + \mathbf{b}_h),$$

where $\mathbf{h}_t$ is the hidden state at step t. A bilinear operation is later applied on the encoded sequence to emphasize inter-token correlations:

$$\mathbf{z} = (\mathbf{h}_t \mathbf{W}_f) \odot (\mathbf{h}_t \mathbf{W}_g).$$

The output $\mathbf{z}$ is then passed through a fully connected layer for sentiment classification.

2.5 Sentiment Classification Function

Finally, a dense output layer computes sentiment polarity probabilities:

$$\hat{y} = \text{Softmax}(\mathbf{W}_o \mathbf{z} + \mathbf{b}_o),$$

where $\hat{y} = [p_0, p_1, p_2]$ corresponds to the probabilities of negative, neutral, and positive sentiments, respectively. In summary, the Hirota-BiNN and Bilinear-RNN definitions provide complementary mechanisms: the former captures complex nonlinear relations through bilinear mapping, while the latter models sequential sentiment context, leading to improved prediction accuracy.

3 Proposed Methodology

3.1 Data Collection

The dataset employed in this study was constructed from a collection of Vietnamese textual feedback gathered from classroom evaluation forms and online academic review sources. Each record in the dataset consists of a short sentence reflecting students' opinions or course-related comments. The data are structured into two main columns: text and labels, where the text field contains the raw feedback, and the labels field represents the sentiment category associated with each entry. An example of the dataset is shown in Table 1.

The sentiment labels were encoded as integers, where 0 represents negative, 1 represents neutral, and 2 represents positive opinions. This structure enables multi-class sentiment classification, allowing the models to distinguish subtle emotional differences in Vietnamese textual feedback. Prior to model training, several preprocessing steps were performed to ensure textual consistency and feature readiness. All sentences were converted to lowercase and cleaned by removing punctuation marks, redundant whitespace, and Vietnamese stop words. Tokenization was performed using the VnCoreNLP toolkit, which effectively handles

Vietnamese compound words and diacritics. Each tokenized sentence was transformed into a numerical vector representation using the Word2Vec model, which maps words into a 300-dimensional semantic space. Sentences were padded or truncated to a uniform sequence length to maintain consistent input dimensions for neural processing. The dataset was partitioned into 80% training data and 20% testing data using a stratified sampling strategy to preserve sentiment distribution across subsets. Feature normalization was performed using MinMaxScaler to accelerate convergence during model training. This processed dataset was employed to train and evaluate two deep learning architectures:

1. Hirota Bilinear Neural Network (Hirota-BiNN): A custom-designed model that captures higher-order interactions between linguistic features through bilinear transformations, enabling the learning of complex nonlinear sentiment relationships in Vietnamese text.
2. Bilinear Recurrent Neural Network (Bilinear-RNN): A sequential model combining recurrent learning with bilinear mapping to capture temporal dependencies and contextual flow in sentiment expressions.

3.2 Proposed Framework

The proposed framework aims to develop an effective Vietnamese sentiment prediction system by leveraging bilinear feature interaction and sequential learning mechanisms. Two deep learning architectures were deployed and compared: the Hirota Bilinear Neural Network (Hirota-BiNN) and the Bilinear Recurrent Neural Network (Bilinear-RNN). Both models are designed to capture higher-order semantic dependencies and contextual sentiment variations within Vietnamese text data. Figure 1 illustrates the overall workflow of the proposed methodology. The process consists of five major stages: (1) data preprocessing, (2) tokenization and embedding, (3) bilinear feature extraction, (4) sentiment classification, and (5) model evaluation.

The proposed framework begins with data preprocessing, where raw Vietnamese textual feedback is cleaned, normalized, and tokenized. During this stage, punctuation marks and Vietnamese stop words are removed, and all words are converted to lowercase to maintain uniformity. Following this, tokenization and embedding are performed using the `VnCoreNLP` toolkit, which segments text into words while preserving linguistic accuracy. Each token is then converted into a numerical representation through `Word2Vec` embeddings of 300 dimensions, ensuring that semantic relationships between words are effectively captured. Next, the bilinear feature extraction stage introduces bilinear transformation layers that learn complex feature interactions across multiple semantic spaces. These extracted features are then passed to fully connected layers responsible for sentiment classification into three categories: negative, neutral, and positive. Finally, evaluation metrics such as accuracy, precision, recall, and F1-score are employed to quantitatively assess the model's performance and effectiveness in predicting sentiment from Vietnamese textual data.

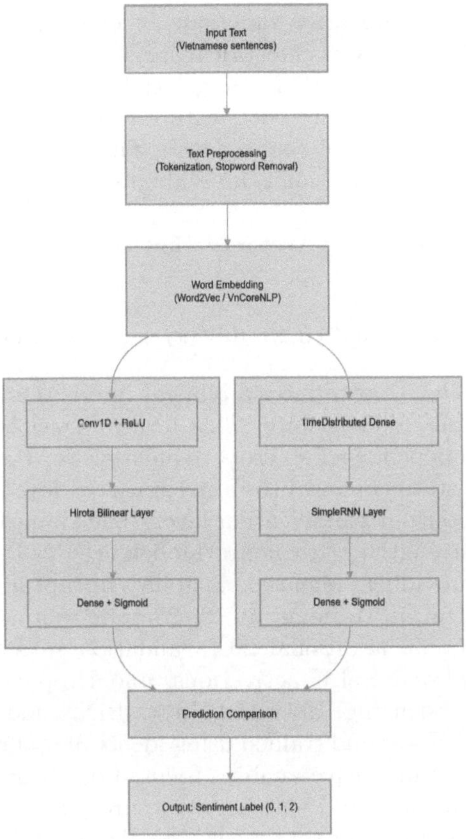

Figure 1. Proposed architecture illustrating two main model pathways: Hirota-BiNN and Bilinear-RNN for Vietnamese sentiment prediction.

The Hirota-BiNN architecture extends the concept of Hirota's bilinear form, commonly used in nonlinear differential equation modeling, to neural representation learning. The model introduces a Hirota Bilinear Layer that performs nonlinear feature fusion through pairwise multiplicative interactions.

Given two feature vectors $f(x)$ and $g(x)$ extracted from parallel convolutional layers, the bilinear transformation is defined as:

$$y = (W_1 f) \odot (W_2 g) + V\left[(W_3 f) \odot (W_4 g)\right] + b$$

where W_1, W_2, W_3, W_4 and V are trainable weight matrices, $\odot$ denotes element-wise multiplication, and b represents the bias term. The first term captures direct feature correlations, while the second term learns higher-order interactions through a nonlinear mapping.

In practice, the input sentences are transformed into 1D embeddings, passed through two convolutional branches, flattened, and combined in the Hirota Bilinear Layer. The resulting feature vector is then fed into dense layers for classification using a softmax activation function. The model is optimized using the Adam optimizer with a binary or categorical cross-entropy loss, depending on the number of classes. The Bilinear-RNN integrates bilinear transformations within a recurrent architecture to capture both feature-level interactions and sequential dependencies in text. The model begins with a TimeDistributed dense encoding that processes each word embedding individually before feeding the sequence into a recurrent network.

Let x_t denote the embedding vector at time step t. The recurrent dynamics are expressed as:

$$h_t = \tanh(W_h h_{t-1} + W_x x_t + (x_t^\top U h_{t-1}))$$

where the term $(x_t^\top U h_{t-1})$ introduces a bilinear interaction between the current input and the previous hidden state. This term allows the network to model nonlinear contextual dependencies across sequential word representations.

The final hidden state is passed to fully connected layers with ReLU activation, followed by an output classification layer. The Bilinear-RNN architecture thus combines the strengths of sequence modeling (RNN) and bilinear representation learning, providing enhanced capability in capturing context-sensitive sentiment polarity. During training, an 80/20 split was used for training and testing datasets, with an additional 20% validation subset from the training data. All layers employed ReLU activations, and Dropout regularization was applied to mitigate overfitting. Both the Hirota-BiNN and Bilinear-RNN were implemented in TensorFlow and trained using identical preprocessed Vietnamese sentiment data. The comparative analysis focused on their ability to generalize across emotional expressions and sentence structures, demonstrating the robustness of bilinear formulations for sentiment classification tasks.

4 Numerical Results

The proposed Hirota-BiNN and Bilinear-RNN models were implemented using TensorFlow 2.17 and trained on the Vietnamese sentiment dataset consisting of three sentiment classes: negative (0), neutral (1), and positive (2). Both models were trained for 100 epochs with a batch size of 64, using the Adam optimizer with a learning rate of 10^{-3}. The dataset was split into 80% for training and 20% for testing, and the input features were normalized using Min–Max scaling.

Table 2 summarizes the quantitative training and validation results of both models. The Hirota-BiNN achieved a training accuracy of 0.6295 with a validation accuracy of 0.6329, outperforming the Bilinear-RNN, which obtained a validation accuracy of 0.4776. The Hirota-BiNN also recorded lower training and validation losses (0.7598 and 0.7520, respectively), demonstrating better convergence stability. In terms of predictive metrics, Hirota-BiNN achieved a precision of 0.6252 and an F1-score of 0.6376, outperforming Bilinear-RNN (precision = 0.4746, F1 = 0.4561). These improvements highlight the advantage of the Hirota bilinear interaction in capturing higher-order dependencies in text features.

To further analyze model complexity, Table 3 presents the architectural details of both networks. The Hirota-BiNN model comprises approximately 6.9K trainable parameters, whereas the Bilinear-RNN includes around 10.6K parameters. Despite having fewer parameters, the Hirota-BiNN demonstrated superior performance, confirming that the bilinear interaction mechanism effectively enhances feature expressiveness without increasing model size (Figs. 2, 3).

Table 2. Training and evaluation summary of Hirota-BiNN and Bilinear-RNN models for sentiment prediction.

Model	Train Acc.	Train Loss	Val. Acc.	Val. Loss	Precision	F1-Score
Hirota-BiNN	0.6295	0.7598	0.6329	0.7520	0.6252	0.6376
Bilinear-RNN	0.5089	0.8348	0.4776	0.8320	0.4746	0.4561

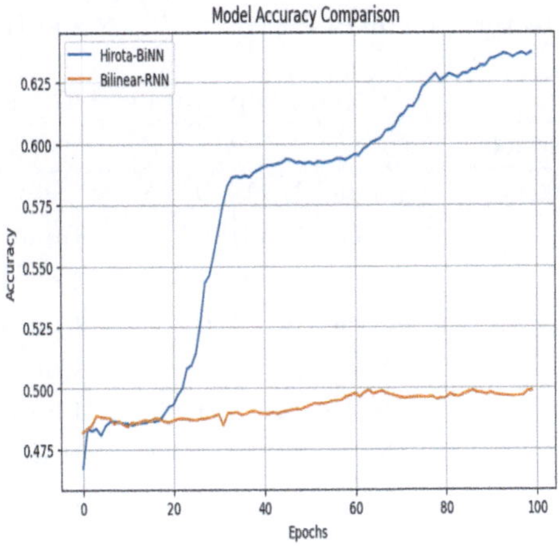

Figure 2. Hirota-BiNN prediction.

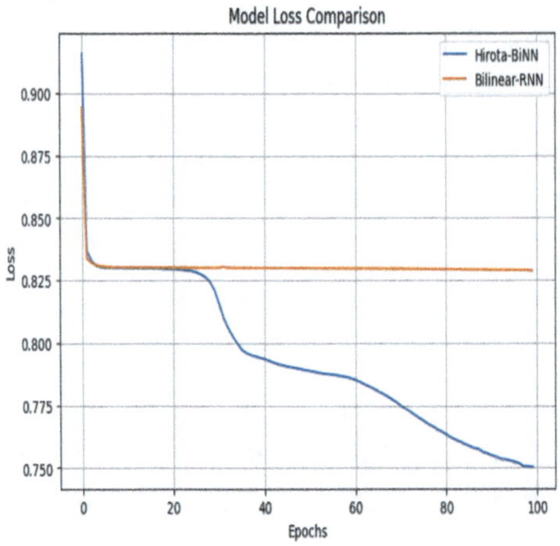

Figure 3. Bilinear-RNN prediction.

Table 3. Model architecture summary for Hirota-BiNN and Bilinear-RNN networks.

Layer (Type)	Output Shape	Param #	Connected To
Hirota-BiNN Model			
Input (InputLayer)	(None, 1, 1)	0	–
Conv1D	(None, 1, 16)	64	Input
Conv1D	(None, 1, 32)	1,568	Conv1D
Flatten	(None, 16)	0	Conv1D
Flatten	(None, 32)	0	Conv1D
Hirota Bilinear Layer	(None, 32)	4,128	Flatten × 2
Dense	(None, 32)	1,056	Hirota Bilinear Layer
Dense (Output)	(None, 3)	99	Dense
Total Parameters		6,915	27.01 KB
Trainable Parameters		6,915	
Non-trainable Parameters		0	
Bilinear-RNN Model			
Input (InputLayer)	(None, 1, 1)	0	–
TimeDistributed Dense	(None, 1, 64)	128	Input
SimpleRNN	(None, 64)	8,256	TimeDistributed
Dense	(None, 32)	2,080	SimpleRNN
Dense (Output)	(None, 3)	99	Dense
Total Parameters		10,563	41.26 KB
Trainable Parameters		10,563	
Non-trainable Parameters		0	

5 Conclusion

This paper presented a deployed bilinear neural network framework for sentiment prediction using Vietnamese textual data. Two architectures, namely the Hirota-BiNN and the Bilinear-RNN, were developed and evaluated on a labeled sentiment dataset. The Hirota-BiNN model integrates the Hirota bilinear layer to capture higher-order interactions between convolutional feature maps, while the Bilinear-RNN leverages recurrent dynamics to process temporal dependencies in text sequences. Experimental results demonstrated that the Hirota-BiNN model achieved superior performance compared to the Bilinear-RNN in both accuracy and F1-score. Specifically, the Hirota-BiNN attained a validation accuracy of 0.6329 and an F1-score of 0.6376, outperforming the Bilinear-RNN which reached 0.4776 and 0.4561, respectively. These findings indicate that the bilinear mechanism effectively enhances feature expressiveness without significantly increasing model complexity, as the Hirota-BiNN also required fewer trainable parameters (6.9K vs. 10.6K). The proposed bilinear approach shows potential for sentiment prediction tasks in low-resource languages such as Vietnamese, where semantic and syntactic variations can pose challenges for traditional deep learning models. Future work will extend this research by incorporating pre-trained language embeddings such as PhoBERT or multilingual BERT to further enrich contextual understanding. Additionally, deployment experiments on mobile and cloud-based platforms will be conducted to assess the real-time applicability of the model in practical sentiment analysis systems.

Bibliography

1. Mastouri, R., Khlifa, N., Neji, H., Hantous-Zannad, S.: A bilinear convolutional neural network for lung nodules classification on CT images. Int. J. Comput. Assisted Radiol. Surg. **16**(1), 91–101 (2021). https://doi.org/10.1007/s11548-020-02283-z
2. Dai, X., Gong, S., Zhong, S., Bao, Z.: Bilinear CNN model for fine-grained classification based on subcategory-similarity measurement. Appl. Sci. **9**(2), 301 (2019). https://doi.org/10.3390/app9020301
3. Zhang, L., et al.: Bilinear graph neural network-enhanced web services classification. 2021 IEEE 23rd Int Conf on High Performance Computing & Communications; 7th Int Conf on Data Science & Systems; 19th Int Conf on Smart City; 7th Int Conf on Dependability in Sensor, Cloud & Big Data Systems & Application (HPCC/DSS/SmartCity/DependSys), 189–196 (2021). https://doi.org/10.1109/hpcc-dss-smartcity-dependsys53884.2021.00051
4. Ahmad, B., et al.: Bilinear-convolutional neural network using a matrix similarity-based joint loss function for skin disease classification (No. arXiv:2406.00696). arXiv (2024). https://doi.org/10.48550/arXiv.2406.00696
5. Calderón, C., Sanchez, K., Castillo, S., Arguello, H.: BILSK: a bilinear convolutional neural network approach for skin lesion classification. Comput. Methods Programs Biomed. Update **1**, 100036 (2021). https://doi.org/10.1016/j.cmpbup.2021.100036
6. Fang, T., Zhang, J., Qi, D., Gao, M.: BLSENet: a novel lightweight bilinear convolutional neural network based on attention mechanism and feature fusion strategy for apple leaf disease classification. J. Food Qual. 1–11 (2024). https://doi.org/10.1155/2024/5561625

7. Liu, W., Juhas, M., Zhang, Y.: Fine-grained breast cancer classification with bilinear convolutional neural networks (BCNNs). Front. Genet. **11** (2020). https://doi.org/10.3389/fgene.2020.547327
8. Tuan, N.M., Meesad, P.: A bilinear neural network method for solving a generalized fractional (2+1)-dimensional Konopelchenko-Dubrovsky-Kaup-Kupershmidt equation. Int. J. Theor. Phys. **64**, 17 (2025)
9. Tuan, N. M., Meesad, P.: Bilinear recurrent neural network for a modified benney-luke equation. Int. J. Appl. Comput. Math. **11**(2), 35 (2025). https://doi.org/10.1007/s40819-025-01851-8

RuLexNorm: A Lexical Normalization Dataset for Russian Social Media Text

Irina S. Koliaskina[1,2(✉)] and Dmitry L. Sholomov[1,3]

[1] Smart Engines, Moscow, Russia
ai.kolyaskina@smartengines.com, sholomov@list.ru
[2] Lomonosov Moscow State University, Moscow, Russia
[3] Institute for Information Transmission Problems RAS, Moscow, Russia

Abstract. Lexical Normalization is the process of converting text into a standardized lexical form, which includes correcting errors, abbreviations, slang, and other anomalous elements. The goal is to reduce text variability to simplify its processing and analysis. Research shows that normalization improves the results of Natural Language Processing (NLP) tasks. However, for the Russian language, there are no open datasets available. This paper presents RuLexNorm – the first open dataset for lexical normalization of social media texts in Russian language. The dataset contains over 6,000 text pairs from Twitter (recently X), where each informal utterance is paired with its normalized version. To evaluate RuLexNorm, we developed a baseline method for lexical normalization of Russian social media texts based on Deep Learning Transformer architecture, achieving a 52% Error Reduction Rate (ERR). The dataset is publicly available for research purposes only (https://github.com/SmartEngines/RuLexNorm).

Keywords: Lexical Normalization · NLP · Text Preprocessing · Social Media Text Normalization · Text Transformation · LLM

1 Introduction

Currently, social networks are a popular source of diverse information worldwide. They contain vast amounts of data, which is valuable for research in various fields, including marketing, sociology, political science, etc. Searching and analyzing information from social media (e.g., public messages, posts, visual content, demographic data) are the key tasks in many research projects aimed at understanding user opinions and behavior, as well as identifying trends across different domains [1–3].

However, data collected from the Internet is typically informal. It may include multiple variations of the same statement, slang, abbreviations, and typos. Moreover, the rich Russian morphology makes the task of normalization more complex. Russian's rich morphology. For example, the sentence in Russian "`Что делаешь?`" ("What are you doing?") may appear in social media as "`че делаешь?`", "`делаеш?`", "`чд?`", etc. Such variability poses significant challenges for existing natural language processing tools, as most were designed to work with grammatically correct texts [4].

H. Unger and P. Meesad (Eds.): NLPIR 2025, LNNS 1904, pp. 349–360, 2026.
https://doi.org/10.1007/978-3-032-20897-2_26

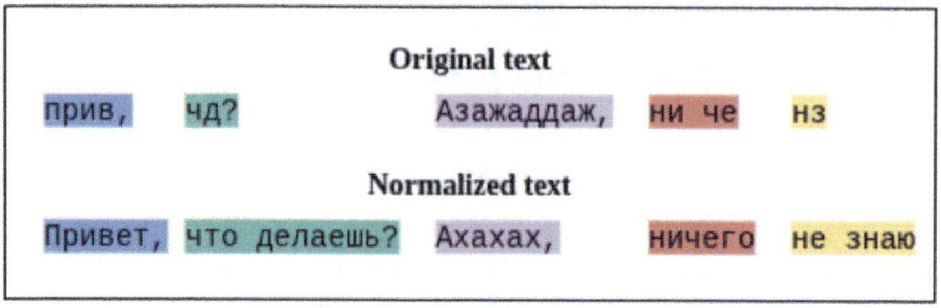

Fig. 1. An example of lexical normalization of the text "Hi, what are you doing? Ahahaha, I don't know anything" (in Russian).

One way to improve the performance of these tools is text normalization, which adapts informal text to a format closer to the standard language for which the tools were originally developed. An example of lexical normalization in Russian is shown in Fig. 1, where abbreviations and colloquial forms are normalized ("прив" →"Привет", "чд" → "что делаешь", ", and typos are corrected e.g. ни че" → "ничего", "ничего" → "не знаю"), .

In paper [5] it is shown that lexical normalization can enhance performance in tasks such as syntactic parsing, named entity recognition, sentiment analysis, part-of-speech tagging, and machine translation.

Generally, lexical normalization is the process of converting text to a lexical language standard, with the goal of transforming an utterance into "normal language" while preserving its original meaning and eliminating anomalies [5]. It may involve correcting unintentional anomalies, such as spelling and punctuation errors, as well as intentional anomalies, such as domain-specific language, including slang, archaisms, and abbreviations [6].

At the same time, the task of lexical normalization may include various types of transformations [6].

- Character-level replacement within a single word ("bein" → "being" or "tmr" → "tomorrow").
- Replacement of numbers with letters within a single word, ("2mr" → "tomorrow" or "be4" → "before").
- Replacement of one word with multiple words and vice versa ("idk" → "I don'tknow" or "united kingdom" → "UK"). Such transformations are referred to as one-to-many (1-N) and many-to-one (N-1), respectively.

Notably, classical lexical normalization does not involve word deletion or insertion, as this may alter the text's meaning [5].

Lexical normalization remains a relevant and actively studied problem. To date, datasets and research exist for Danish, German, English, Spanish, Croatian, Italian, Dutch [5], Malay [7], Vietnamese [8], and other languages. However, for Russian lexical normalization remains understudied, with no publicly available datasets or open-source solutions.

The contributions of our work are:

1. RuLexNorm – the first open dataset for lexical normalization of Russian social media texts, containing 6,000 + text pairs from Twitter. We also provide a detailed annotation process description.
2. A novel method for lexical normalization of Russian texts based on a pre-trained Transformer architecture.
3. Establishing baseline results on the RuLexNorm dataset using the proposed method. The achieved Error Reduction Rate (ERR) of 52% sets a reference point for future research in this domain.

2 Methods Review

2.1 Text Data Types

The type of text data is a decisive factor when choosing a normalization approach. Different types of texts require different processing methods. There are two main types of text data [9]: General Text Domain and User Generated Content.

General Text Domain: Books, news articles, scientific papers, official documents, web pages, etc. The authors are typically professional writers, domain experts, or journalists. These texts are characterized by a formal style and structured format, adhering to the rules of natural language and using standard vocabulary. If slang, informal expressions, or grammatical errors appear, they are minimal.

User Generated Content: Social media posts, comments, reviews, forum discussions, etc. The authors may include ordinary users without professional writing experience and with varying levels of language proficiency. The writing style in this domain is informal, and the documents are unstructured. Abbreviations, emojis, non-standard grammar, slang, spelling mistakes, typos, and other phenomena typical of casual communication are common.

2.2 Traditional Text Normalization Approaches

The lexical normalization task is typically addressed through three main approaches [9]: rule-based methods, statistical techniques, machine learning and hybrid solutions. We briefly outline each approach below.

Rule-Based Methods. Rule-based methods rely on dictionaries, contextual analysis, and predefined rules that determine which words should be normalized and how. These methods may utilize regular expressions, context-free grammars, semantic role labeling, or finite-state automata [9].

For example, in the work [7], Ariffin and Tiun propose a rule-based normalization method for Malay texts. The normalization process applies rules derived from linguistic patterns found in Malay social media texts. Their system normalizes only words written in Malay, romanized Arabic, and English.

Rule-based methods provide precise control over the normalization process and are effective in domains with a formal writing style and limited slang usage. However, they often struggle to adapt to informal, conversational styles, reducing their effectiveness on user-generated content. Additionally, these methods require significant time and effort to develop and maintain rules for different languages and domains.

Statistical Methods. Statistical methods rely on text statistics and word frequency to determine which words should be normalized and how [10]. These methods use statistical techniques to analyze large volumes of text and identify patterns in word usage. Methods in this category may employ Levenshtein distance, N-gram language models [11], probabilistic context-free grammars [12, 13], or Hidden Markov Models [9].

For example, the study [14] proposes SMS text normalization method based on a statistical phrase-based machine translation model. The authors treat normalization as a translation task – converting "SMS language" into standard English. The adapted statistical machine translation model includes phrase alignment via the EM algorithm, monotonic decoding, and bigram language model.

Statistical methods can be effective for diverse styles and languages, but they are less suitable for processing formal texts, where high precision and strict grammatical adherence are required. Additionally, these methods demand large amounts of data to uncover meaningful statistical patterns.

Machine Learning. Machine Learning methods, as a rule, are based on neural networks. During training, rules and patterns are extracted from large textual datasets, enabling the automatic determination of which words should be normalized and how. These methods can rely on different learning paradigms, such as supervised learning or reinforcement learning, as well as word embeddings (Word2Vec, GloVe, FastText, etc.) and various neural network architectures (Deep Neural Networks, Recurrent Neural Networks, Convolutional Neural Networks, Transformers, etc.) [9].

In paper [15], a method is proposed to normalize dialectal Finnish texts into standard literary form using neural network models. The authors compare two approaches: Bidirectional Recurrent Neural Network and Character-Level Transformer.

Machine learning methods can automatically extract complex patterns from text and are effective across various styles and domains, particularly with large datasets. However, they require significant computational resources and training time.

Hybrid Methods. Due to the fact that each approach has its own limitations and drawbacks, a hybrid approach [9] is now frequently used. This allows combining different methods depending on the specific characteristics of the textual data. As a result, the developed solution becomes more versatile and capable of adapting to various styles, domains, and languages.

In [16], the FlatTN method is proposed – a Chinese text normalization model based on the Flat-Lattice Transformer (FLAT), which integrates expert knowledge from rules into a neural network architecture. The model takes characters as input, organizes them into a flat lattice along with potential words and non-standard words, and then uses a Transformer with relative positional encoding to classify non-standard words.

2.3 Characteristics of Russian User-Generated Texts

This study focuses on lexical normalization of user-generated content, specifically short social media messages in Russian. The Russian language presents unique challenges that make lexical normalization considerably more complex compared to many other languages, particularly English. Key difficulties arise from its rich morphology, inflection, homonymy, free word order, and the high variability of informal written communication.

Among the approaches examined, machine learning methods demonstrate the best results, whereas linguistic and statistical methods face limitations in scalability and practical implementation. However, implementing machine learning requires annotated datasets. For Russian, we were unable to find an available dataset specifically designed for lexical normalization. While there are datasets for related tasks – such as spelling correction [17] or grammatical error correction [18] – they do not cover all linguistic phenomena relevant to lexical normalization. Additionally, simply translating a dataset from another language is not feasible, as each language has unique, non-transferable features in word formation and usage.

Russian social media texts exhibit several distinctive features that set them apart from standard written language:

1. High text compression: Strict message length limits lead to frequent use of abbreviations, vowel omissions, phonetic simplifications, and disregard for punctuation rules.
2. Active use of slang, ironic/sarcastic phrases, and contextual hints.
3. Expressiveness, manifested through repeated letters, excessive capitalization, and clusters of exclamation/question marks.
4. Language mixing including transliterations and borrowed neologisms.
5. Constant lexical evolution, with new word forms and shifting meanings of existing words.

Due to the lack of ready-made datasets for Russian lexical normalization, this work proposes a novel dataset from Russian social media texts (see Sect. 3). Furthermore, given the heightened complexity of normalizing Russian texts (compared to English), we explore neural network methods, which have proven effective in processing informal speech with high variability and error rates (see Sect. 4).

3 Dataset and Formation Process

We compiled a dataset consisting of 6,396 tweets (totaling 72,039 words), including 1,002 synthetic tweets and 5,394 authentic tweets. The dataset formation process involves several stages (Fig. 2):

1. Real tweet collection – gathering authentic user-generated messages.
2. Synthetic tweet generation – creating additional texts using a dictionary of Russian slang words.
3. Automatic annotation – processing texts with GPT-4o-mini model for preliminary labeling.
4. Manual verification – human refinement of annotations.

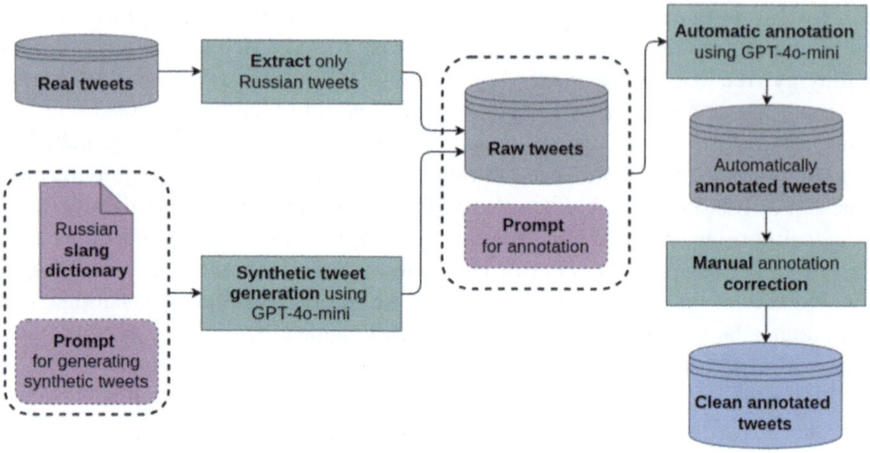

Fig. 2. Dataset formation process

3.1 Development of Normalization Rules

To convert user-generated Twitter texts into a standardized form, a set of normalization rules was developed. These rules account for the characteristics of informal online communication while preserving the original meaning of the messages as much as possible.

Main Rule Categories:

"Ахвахвхф" → "Ахахах".

Main Rule Categories:

1. **Correction of typos and spelling errors**

   првиет → привет

   лудше → лучше

2. **Handling special text elements:** Usernames (@username), hashtags (#example), links, and emojis (😉, :D) remain unchanged
3. **Slang lexicon correction**

   рофл → смешно

   кринж → стыд

4. **Profanity filtering**
   a. Swear words within meaningful expressions are replaced with neutral alternatives ("ох**енный" → "отличный")
   b. Profane interjections are removed
5. **Reduction of character repetitions**
   a. Excessive letter repetitions are shortened to a maximum of three (" ааааа" → "ааа")
   b. Rhythmic repetitions are hyphenated (" туктуктук" → "тук-тук-тук")
   c. Laughter is standardized ("хпхпхахаха" → "ахахах")
6. **Punctuation standardization:** repeated punctuation marks are reduced to one of the [., !, ?, ..., ?!, ?.., !..]

   "Привет,,, как дела??" → "Привет, как дела?"

   "???" → "?", "!!!" → "!"

   "Что делать?!?!" → "Что делать?!"

7. **Handling abbreviations**
   a. Colloquial and slang abbreviations are expanded (" спс" → "спасибо")
   b. Widely recognized acronyms (e.g., НЛП, ФСБ) remain unchanged

What is **not** corrected:

1. Punctuation errors (missing, excessive, or incorrect punctuation).
2. Most grammatical errors
3. Stylistic errors (" ("стадо зайцев", "спросить вопрос")")
4. Factual errors ("The capital of the USA is New York").

3.2 Data Collection

Russian-Language Tweets Collection. To gather Russian-language tweets, we used the public archive Twitter Stream Grab [19], which contains a sample of data from Twitter's general "Spritzer" API stream. Twitter states that this API provides a representative 1% sample of all public tweets without thematic filtering, making it the most suitable for our task.

We downloaded Twitter Stream Grab data from January to December 2021. Then, we filtered only Russian-language data based on metadata.

Synthetic Data Generation. To expand the dataset and compensate for the lack of slang forms, we additionally generated synthetic tweets. The generation process consisted of two stages:

1. We collected a slang dictionary containing 690 current youth slang words and expressions [20, 21].
2. Using the collected slang words, we generated synthetic tweets with the help of the GPT-4o-mini language model [22]. The model was provided with a slang word and the following prompt to generate a tweet (5 tweets per word):

Respond as a young social media user on Twitter. You will be given a word – write a tweet using it. You may modify the word's morphology. Ensure proper grammar and tense agreement.
Word

As a result, 1,002 tweets (19,719 words) were generated.

3.3 Data Annotation

Automated Data Annotation. The annotation process consisted of two sequential stages: the first stage was performed automatically, while the second stage involved manual correction of the results.

To speed up the process and improve the normalization quality of tweets, GPT-4o-mini was employed in the first stage. Following OpenAI's prompt engineering guidelines [23] and through experimental testing, the most effective text normalization prompt was selected. This prompt balanced high accuracy with ease of response processing. The main prompt included instructions for spelling and slang normalization, preservation of links, hashtags, and emojis, specific output format for consistency.

Manual Correction. During the automated normalization of tweets using GPT-4o-mini, several key observations were made:

1. Emojis, usernames, and hashtags were converted to plain text: `"на улице так холодно 😭"` → `"На улице так холодно, я плачу"`
2. Tweets containing profanity were not processed: `"Я не могу выполнить данную просьбу, так как запрос содержит матерные или оскорбительные выражения"` ("I cannot fulfill this request as it contains profane or offensive language")
3. Responses included explanatory notes: `"Для данного текста 'Че дел?' лексическая норма может выглядеть следующим образом: 'Что делаешь?'"`
4. Occasional unintended translation to English:
`"@94tmmo91 какого цвета?"` → `"94tommo91 What color is it?"`

Since GPT-4o-mini did not always normalize text correctly, the dataset was reviewed manually. For example:

- Original tweet: "`@jaehyunswife_ живот куна`"
- GPT output: "`@jaehyunswife_ живот куна`" (no change)
- Manual correction: "`@jaehyunswife_ живот парня`" ("`@jaehyunswife_ guy's stomach`")

4 Baseline Method

This section presents an internal evaluation of the proposed baseline method for lexical normalization of Russian-language texts from social media. It covers the model architecture, fine-tuning process, evaluation, and the obtained results.

4.1 Model Architecture

For the task of lexical normalization, a baseline method consists of fine-tuning a pre-trained LLM (Large Language Model). The foundation chosen was a Russian-adapted version of the transformer model Qwen2.5-3B, with 3 billion parameters. The model employs subword tokenization, which enables efficient processing of morphologically rich languages like Russian. This architecture was selected due to its ability to take into consideration contextual dependencies and handle diverse linguistic anomalies, including spelling errors, slang, and profanity.

4.2 Model Fine-Tuning

To adapt the model to the target task, the PEFT (Parameter Efficient Fine-Tuning) method with LoRA adapters (Low-Rank Adaptation) was applied. This approach allows efficient fine-tuning by modifying only a small subset of parameters, which is particularly important when working with limited computational resources.

4.3 Metrics

The model's performance was evaluated at the level of individual substitutions, where a correct substitution was defined as a properly normalized word. Standard metrics were used: accuracy, precision, recall, F1-score, as well as the Error Reduction Rate (ERR) – word-level accuracy normalized by the number of substitutions in the dataset:

$$ERR = \frac{accuracy_{system} - accuracy_{leave-as-is}}{1.0 - accuracy_{leave-as-is}},$$

where leave-as-is refers to a system that does not normalize any words. The ERR metric ranges from -1 to 1. Negative values indicate that the model makes more mistakes in normalization than correct fixes. A value of zero means the model makes no changes to the text. A value of 1 corresponds to perfect normalization, where all errors are corrected properly.

Table 1. Comparison of results: Fine-tuned adapted Qwen2.5-3B vs Yandex Speller

	Fine-tuned adapted Queen 2.5-3B	Yandex Speller
Accuracy	82.6	68.5
Precision	74.4	18.4
Recall	78.5	0.007
F1	76.4	0.013
ERR	51.6	-0.023

Table 2. Examples of the fine-tuned model's predictions. INPUT – Original tweet text (model input); OUTPUT – Model prediction.

INPUT	@ffaylin ПЛИЗ ЕСЛИ КТО-ТО НАПИШЕТ - МАЯКНТ
OUTPUT	@ffaylin пожалуйста если кто-то напишет - оповестите
INPUT	Давайте клип быстрее пж
OUTPUT	Давайте клип быстрее, пожалуйста.
INPUT	Хочу бухать 🍷🍷
OUTPUT	Хочу пить алкоголь. 🍷🍷

4.4 Results

Table 1 presents the evaluation results of different methods using the following metrics: accuracy, precision, recall, F1-score, and ERR.

The fine-tuned Qwen2.5-3B model correctly processes the vast majority of words – either fixing anomalies accurately or leaving correct words unchanged. However, every fourth correction was redundant (the model mistakenly "corrupted" correct words), and one-fifth of errors remained uncorrected. Table 2 provides examples of the model's performance.

4.5 Comparison with Yandex Speller

To evaluate the correctness of the fine-tuned model, we compared its performance with Yandex Speller [24]. Yandex Speller is a tool developed by Yandex that detects and corrects spelling errors in Russian, Ukrainian, Kazakh, and English texts. It uses the CatBoost [25] machine learning library to identify errors and suggest corrections. Thanks to this library, it can correct words with errors (e.g., substitute “ "адникасники"” with “ "одноклассники"”) and take into account context while detecting typos (e.g., identifying “ "скучать музыку"” as a misspelling of “ "скачать музыку"скачать музыку”).

Additionally, the tool is relatively lenient toward new words that have not yet entered standard dictionaries.

We tested Yandex Speller on our evaluation dataset; the results are presented in Table 1 while Table 3 provides examples of the model's performance on the test set.

Table 3. Examples of the Yandex Speller predictions. INPUT – Original tweet text (model input); OUTPUT – Model prediction.

INPUT	`@mmmanuscript вы такие красивые цвета называете рыдаю`
OUTPUT	`@manuscript вы такие красивые цвета называете рыдаю`
INPUT	`Давайте клип быстрее пж`
OUTPUT	`Давайте клип быстрее пж`
INPUT	`101 долматинец! :D`
OUTPUT	`101 далматинец! :D`

5 Conclusion

In the paper we present RuLexNorm – the first text dataset specifically designed for lexical normalization of Russian social media texts which is publicly available at https://github.com/SmartEngines/RuLexNorm. Moreover, we developed a baseline method, using a fine-tuned Qwen2.5-3B model, which achieved an ERR score of 51.6%, significantly outperforming tools like Yandex Speller. The model effectively handles typos, spelling and punctuation errors as well as slang words. Crucially, it preserves text integrity: it neither rearranges words nor inserts extraneous terms.

References

1. , C., Gupta, S., Manek, R.: Impact of social media on consumer behaviour. Int. J. Creative Res. Thoughts **8**(6), 1943–1961 (2020)
2. Shaheen, H.: Social media marketing research: a bibliometric analysis from scopus. Future Bus. J. **11**(1), 1–26 (2025)
3. Abid, A., Roy, S.K., et al.: Political social media marketing: a systematic literature review and agenda for future research. Electron. Commer. Res. **25**(2), 741–776 (2025)
4. Han, B., Baldwin, T.: Lexical normalisation of short text messages: Makn sens a# twitter. In: Proceedings of the 49th Annual Meeting of the Association for Computational Linguistics: Human Language Technologies, pp. 368–378 (2011)
5. Van Der Goot, R., et al.: Multilexnorm: a shared task on multilingual lexical normalization. In: Seventh Workshop on Noisy User-generated Text (W-NUT 2021), pp. 493–509. Association for Computational Linguistics (2021)

6. Van Der Goot, R., Van Noord, G.: Monoise: modeling noise using a modular normalization system. arXiv preprint arXiv:1710.03476 (2017)
7. Fadzely, M.F.S., Azizan, A., et al.: Malay noisy text normalization in user-generated content: a rule-based and levenshtein distance approach. In: Proceedings of the 16th International Conference on Knowledge and System Engineering, pp. 333–340 (2024)
8. Nguyen, T.-N., Le, T.-P., Van Nguyen, K.: ViLexNorm: a lexical normalization corpus for vietnamese social media text. arXiv preprint arXiv:2401.16403 (2024)
9. Aliero, A.A., Bashir, S.A., et al.: Systematic review on text normalization techniques and its approach to non-standard words. Int. J. Comput. Appl. **975**, 8887 (2023)
10. Sholomov, D.L.: Correction of recognized text using the classification methods. Trudy ISA RAN **17**, 352–366 (2007). (In Russian)
11. Manzhikov, T.V., Slavin, O.A., et al.: Algorithms for correcting recognition results using N-grams. Pattern Recognit Image Anal. **27**(4), 832–837 (2017)
12. Sholomov, D.L.: Syntactical approach to post-processing of fuzzy recognized text. In: Proceedings of MLMTA International Conference, pp. 115–121. CSREA Press (2003)
13. Postnikov, V.V., Sholomov, D.L.: Post-processing of OCR results using automatically constructed partially defined syntax. In: Proceedings of MLMTA International Conference, pp. 814–820. CSREA Press (2004)
14. Aw, A., Zhang, M., Xiao, J., Su, J.: A phrase-based statistical model for SMS text normalization. In: Proceedings of the COLING/ACL 2006 Main Conference Poster Sessions, pp. 33–40 (2006)
15. Partanen, N., Hämäläinen, M., Alnajjar, K.: Dialect text normalization to normative standard Finnish. In: Workshop on Noisy User-generated Text, pp. 141–146. The Association for Computational Linguistics (2019)
16. Dai, W., Song, C., et al.: An end-to-end Chinese text normalization model based on rule-guided flat-lattice transformer. In: ICASSP 2022 - 2022 IEEE International Conference on Acoustics, Speech and Signal Processing, pp. 7122–7126. IEEE (2022)
17. Sorokin, A., Shavrina, T., Baytin, A., Galinskaya, I., Rykunova, E.: SpellRuEval: the first competition on automatic spelling correction for Russian. In: Computational Linguistics and Intellectual Technologies, pp. 660–673 (2016)
18. Zong, C., Xia, F., Li, W., Navigli, R.: Findings of the association for computational linguistics: ACL-IJCNLP 2021. In: Findings of the Association for Computational Linguistics: ACL-IJCNLP 2021 (2021)
19. Internet Archive: Twitter Stream Dataset. Archive.org. https://archive.org/details/twitterstream. Accessed 12 Aug 2024
20. Knigochet: Text analysis. Knigochet. https://knigochet.com/show_200103. Accessed 15 May 2025
21. ANTISLANG: Internet slang vocabulary. Antislang. https://antislang.ru/. Accessed 15 May 2025
22. OpenAI: GPT-4o mini: advancing cost-efficient intelligence. OpenAI Blog (2024). https://openai.com/index/gpt-4o-mini-advancing-cost-efficient-intelligence/. Accessed 3 Jun 2025
23. OpenAI: prompt engineering guide. OpenAI Platform (2024). https://platform.openai.com/docs/guides/prompt-engineering. Accessed 27 Feb 2025
24. Yandex: API Yandex.Speller. Yandex Developer (2023). https://yandex.ru/dev/speller/. Accessed 10 May 2025
25. Prokhorenkova, L., Gusev, G., Vorobev, A., Dorogush, A.V., Gulin, A.: CatBoost: unbiased boosting with categorical features. Adv. Neural Inf. Process. Syst. **31** (2018)

Generative Text and Text Classification

Enhancing Text Game Agent Performance via Wrong Action Classification

Binggang Zhuo(✉) and Masaki Murata

Tottori University, 4-101 Koyama-cho Minami, Tottori 680-8552, Japan
zhuobinggang@gmail.com, murata@tottori-u.ac.jp

Abstract. Text-based games present a challenging benchmark at the intersection of natural language processing (NLP) and reinforcement learning (RL). This study aims to improve agent performance on the largest available text game dataset, the First TextWorld Problems (FTWP). We reimplement the state-of-the-art agent, CogniTextworldAgent, and conduct a failure case analysis on the validation set, which reveals that the agent frequently selects wrong actions that prematurely terminate the game. To address this issue, we introduce a lightweight extension that integrates a wrong action classifier with the Upper Confidence Bound 1 (UCB1) algorithm. A RoBERTa-based classifier is trained to detect actions that conflict with the cookbook, and the selection scores of such actions are adjusted using UCB1, without requiring any retraining of the original agent. The experimental results indicate that, on average, the agent's task completion rate on the unseen test set improved from 88.0% to 91.2% with the integration of the wrong action classifier, representing an absolute improvement of 3.2%. The improvement is statistically significant according to a two-sided Wilcoxon Signed-Rank Test ($p < 0.05$).

Keywords: text game · game agent · action elimination · reinforcement learning · BERT

1 Introduction

Text-based games (hereafter referred to as text games) are widely regarded as a cross-disciplinary research platform that integrates natural language processing (NLP) and reinforcement learning (RL) [1]. Achieving high performance in these games requires agents to exhibit a range of capabilities, including language comprehension, memory retention, planning, exploration, and commonsense reasoning.

While significant advancements have been made in RL through visual environments such as Atari and Minecraft, progress in language-driven environments has remained comparatively limited [2]. A major reason is that many classic text games were originally designed for human players and often exhibit a level of

H. Unger and P. Meesad (Eds.): NLPIR 2025, LNNS 1904, pp. 363–378, 2026.
https://doi.org/10.1007/978-3-032-20897-2_27

complexity that exceeds the capabilities of current AI systems, resulting in a lack of scalable environments with appropriate difficulty levels. To address this challenge, Microsoft Research developed TextWorld [3], a framework for procedurally generating text games tailored to AI research. Several datasets have since been built using this framework, among which the First TextWorld Problem (FTWP) is the largest. FTWP comprises 5,176 unique games centered on a cooking task, where the agent must prepare and consume a dish according to a game-specific cookbook.[1] In 2019, Microsoft organized a competition based on FTWP, where the winning entry, CogniTextworldAgent [4], achieved state-of-the-art performance and remains the baseline on this dataset.

CogniTextworldAgent integrates a BERT-based [5] scoring model with the Upper Confidence Bound 1 (UCB1) algorithm [6], and learns a policy through behavior cloning from expert walkthroughs included in the training set. Despite its overall effectiveness, an analysis of its validation set performance reveals a recurring failure pattern: the agent often selects wrong actions that violate the cookbook constraints and lead to premature termination. These findings indicate a limitation in the agent's ability to recognize and avoid risky actions.

To address this limitation, we propose a lightweight enhancement that augments agent decision-making without retraining the original model. Specifically, we construct a dataset of ingredient-handling actions using a heuristic labeling strategy, train a RoBERTa-based classifier [7] to identify wrong actions, and apply the UCB1 algorithm to reduce the selection scores of actions flagged as wrong.

Experimental results indicate that our approach improves the agent's average task completion rate from 88.0% to 91.2% on the FTWP test set—an absolute improvement of 3.2%. The improvement is statistically significant according to a two-sided Wilcoxon Signed-Rank Test ($p < 0.05$). These results suggest that the proposed method enhances agent robustness to wrong actions while preserving the architectural and training integrity of the base model. Our main contributions are summarized as follows:

- We identify and analyze a critical failure pattern in CogniTextworldAgent, the state-of-the-art agent on FTWP, which frequently selects incorrect actions violating cookbook constraints, ultimately revealing a fundamental limitation of the current approach.
- We propose a lightweight enhancement framework that augments agent decision-making with a RoBERTa-based wrong action classifier, without retraining the base agent.
- We introduce a novel strategy that combines the classifier with the UCB1 algorithm to penalize, rather than directly eliminate, wrong actions. This design ensures robustness even when the classifier performance degrades, as validated through controlled experiments.
- The experimental results demonstrate that our approach achieves a significant improvement in task completion rate (from 88.0% to 91.2%) on the FTWP

[1] The details can be found at: https://competitions.codalab.org/competitions/21557.

benchmark, with statistical significance confirmed by a two-sided Wilcoxon Signed-Rank Test.

2 Related Works

2.1 Action Elimination in Text Games

The central concept of this study—identifying and penalizing wrong actions through an auxiliary classifier—is closely related to the field of Action Elimination, which seeks to reduce the action space by filtering out invalid or suboptimal actions.

A seminal work in this area by Zahavy et al. [8] utilizes neural networks to detect and eliminate invalid actions in text-based environments. Their method demonstrated improved learning efficiency in games such as Zork and Grid World, emphasizing the benefits of action space pruning in complex interactive environments. The theoretical and empirical foundations of this approach were further supported by subsequent work from Huang and Ontañón [9], which strengthened the case for action elimination strategies.

Despite these advantages, a notable limitation of traditional action elimination methods is that once an action is classified as undesirable, it becomes permanently inaccessible to the agent. This approach grants the classifier a high degree of control, which may result in suboptimal performance if misclassifications occur.

2.2 Research on FTWP

The First TextWorld Problem (FTWP) dataset was introduced as part of a competition organized by Microsoft Research in 2019, which attracted participation from 16 teams worldwide. The winning entry, CogniTextworldAgent [4], is adopted as the primary baseline in this study. This agent combines a BERT-based scoring model with the UCB1 algorithm to balance exploration and exploitation. BERT [5], a transformer-based encoder pretrained on large-scale corpora, has been widely adopted for its effectiveness across various natural language understanding tasks. UCB1 [6], a classical algorithm for the Multi-Armed Bandit problem, is designed to manage the balance between exploiting known rewards and exploring potentially better alternatives. Specifically, it penalizes frequently selected actions and chooses the one with the highest adjusted score.

An analysis of CogniTextworldAgent's failure cases on the FTWP validation set revealed that a considerable number of failures stem from executing actions that violate the cookbook constraints, leading to premature game termination. These observations motivated us to develop a method for suppressing such actions, thereby improving overall task completion rates.

The second-place agent, LeDeepChef [2], adopts a different approach by using GloVe embeddings [10] to encode the game state and applying an actor-critic reinforcement learning framework for policy learning. The performance difference between the top two agents in FTWP can be partially attributed to the advantages of contextual embeddings over static ones. Unlike static embeddings such as Word2Vec [11] and GloVe, contextualized word representations—such as those produced by ELMo [12] and BERT—dynamically adjust based on the surrounding context, allowing the model to capture more nuanced semantic and syntactic information. Notably, these contextual embeddings have demonstrated substantial improvements across a wide range of NLP benchmarks [13].

3 Wrong Action Classifier

The central idea of this study is to train an additional neural network to identify wrong actions. Once an action is classified as wrong, its score will be reduced accordingly. In this section, we first introduce the FTWP game and formally define what constitutes a "wrong action." We then describe the construction of a dataset for training the wrong action classifier. Finally, we detail the integration of the wrong action classifier into the CogniTextworldAgent.

3.1 FTWP Games

In the FTWP game, the agent's task is as follows:

1. Find the cookbook in multiple rooms and examine its contents.
2. Locate specific ingredients as indicated by the cookbook.
3. Process the ingredients according to the cookbook instructions.
4. Eat the prepared food.

Each game has a different spatial layout and cookbook content. To facilitate understanding, a walkthrough of a representative game from the training set is presented and explained below. A walkthrough is a sequence of commands that, when executed in order, enables the agent to successfully complete all tasks in the game.

```
["inventory", "go east", "examine cookbook", "drop red potato",
 "drop white onion", "cook red onion with stove",
 "take knife from counter", "chop red onion with knife",
 "drop knife", "prepare meal", "eat meal"]
```

First, `inventory` is the command to check the currently held items. After executing it, the game engine returns: "a red onion, a raw red potato, a white onion." The command `go east` is a movement command, which brings the agent to a new room. Based on the environment description returned by the engine, we know this is the kitchen, and there is a cookbook on the table. Executing `examine cookbook` makes the game engine return the cookbook content:

```
Ingredients:
  red onion
Directions:
  chop the red onion
  fry the red onion
  prepare meal
```

The Ingredients section indicates that a red onion is required, and the Directions section specifies that the red onion needs to be chopped and fried. Since the agent already possesses the red onion, it only needs to process it accordingly. Following the walkthrough, after `examine cookbook`, the agent first drops unnecessary inventory items using several `drop` commands. Then, it executes the command `cook red onion with stove`, which gives the agent one point (for completing the "fry" task). Next, it picks up the knife and executes `chop red onion with knife`, earning another point (for completing the "chop" task).

At this stage, the agent has completed all ingredient processing tasks specified in the cookbook. As the final steps, the agent must execute `prepare meal` and `eat meal`, each of which grants one additional point. The game then displays a task completion message and terminates.

3.2 Baseline Method

Our main baseline, CogniTextworldAgent—the state-of-the-art agent on the FTWP dataset—was composed of BERT and the UCB1 algorithm. In our reimplementation, BERT is replaced with its more performant variant, RoBERTa. The specific steps by which CogniTextworldAgent selects an action are as follows.

1. RoBERTa computes a score for each candidate action based on the current observation.
2. UCB1 adjusts the score according to how frequently the action has been executed, encouraging exploration. The action with the highest adjusted score is then selected.

The input to RoBERTa is formatted as follows:

```
[CLS] <inventory count> <inventory description>
<cookbook> <environment description> [SEP] <action>
```

Following standard sequence classification practices, the representation of the `[CLS]` token is used for scoring. Due to the input length limitation of RoBERTa, we can trim the environment description as necessary to ensure that the action part can be fully included. CogniTextworldAgent learns its policy through behavior cloning based on walkthroughs from the training set. The upper part of Fig. 2 illustrates the architecture of CogniTextworldAgent.

The CogniTextworldAgent is used as both the primary baseline and the foundation for our proposed method, since our approach can be viewed as an extension of this agent. Specifically, we enhance its performance by introducing an auxiliary classifier to identify wrong actions and combining it with the UCB1 algorithm, without requiring any retraining of the original agent.

3.3 Definition of Wrong Actions

In FTWP, each game comes with a walkthrough—an ordered list of actions that completes all required tasks. These walkthroughs serve as expert demonstrations for behavior cloning. Assuming that all actions included in the walkthrough are correct, the remaining actions can be categorized into two groups:

- **Redundant actions:** These do not lead to irreversible consequences. For instance, dropping a required item can be remedied by retrieving it later.
- **Wrong actions:** These contradict the cookbook instructions and immediately cause the game to terminate. For example, if the cookbook specifies "chop the red onion," performing "slice red onion with knife" violates the task and ends the game.

Analysis of CogniTextworldAgent's behavior suggests that a significant portion of its failures stem from executing such wrong actions. These actions often exhibit high lexical and semantic similarity to correct ones (e.g., "chop red onion" vs. "slice red onion"), making them difficult to distinguish.

To address this issue, we propose training a classifier to detect cookbook-violating actions. Detected actions are then penalized through the UCB1-based decision-making process. Since wrong actions can be identified by comparing an action against the cookbook, the classifier only requires the cookbook text and the candidate action as input. The model outputs a binary label: 1 if the action violates the cookbook (wrong), and 0 otherwise. The architecture is illustrated in Fig. 1.

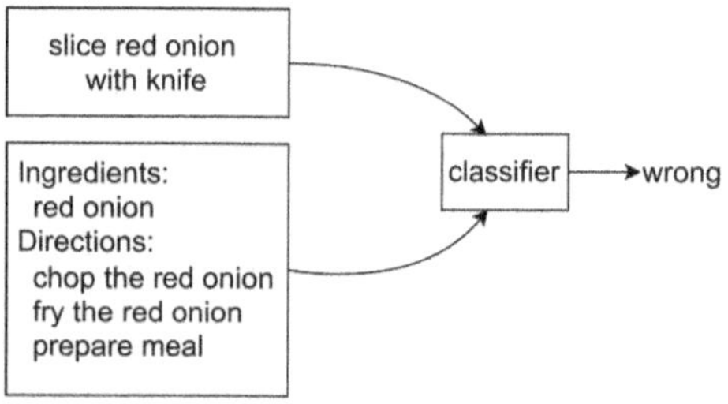

Fig. 1. The classifier determines whether a given action violates the cookbook requirements.

3.4 Training Dataset Construction

Each entry in the dataset we aim to construct is structured as follows:

- **Input (x):** Cookbook + ingredient-handling action
- **Label (y):** 1 (wrong action) or 0 (correct action)

The *cookbook* in the input (x) refers to the text obtained by executing the examine cookbook command in the FTWP game. It consists of two parts: *Ingredients* and *Directions*, which together describe the goals that the agent needs to achieve. An example can be found in the lower part of Fig. 1. In FTWP, the ingredient-handling actions include two types: *cook* and *cut*. The cook action always follows the format cook <ingredient> with <kitchenware>, while the cut command always follows the format <verb> <ingredient> with knife. Therefore, we can use regular expressions to match them.

Both the cookbook and the set of candidate actions can be easily obtained from the game engine. To assign labels, we adopt the following heuristic rules:

1. Any ingredient-handling action that appears in the walkthrough is labeled as 0 (correct).
2. All other ingredient-handling actions are labeled as 1 (wrong).

This labeling strategy is justified by the following observations:

- Walkthroughs never include wrong actions, as these terminate the game prematurely.
- All correct ingredient-handling actions required by the cookbook must appear in the walkthrough, given that the walkthrough is a complete and successful solution path.
- While some non-walkthrough actions may be redundant rather than strictly wrong, they are conservatively labeled as wrong to simplify classifier training. Reducing unnecessary action selections is unlikely to negatively affect performance, as redundant steps do not contribute to task success.

Using this approach, we constructed a dataset containing 35,767 training instances, with an approximate class imbalance ratio of 4.77:1 in favor of wrong actions.

The action classifier is implemented using RoBERTa [7]. RoBERTa is an optimized variant of BERT that enhances the training process and demonstrates improved performance across various tasks. Each input is encoded in the following format:

[CLS] <cookbook> [SEP] <action>

Following standard practice for sequence classification, the embedding of the [CLS] token is used for final classification. Training is conducted for up to 3 epochs using the AdamW optimizer, with a learning rate of 1e-5 and a batch size of 32. On the validation set, the best-performing model achieved the following results:

- Wrong action detection accuracy: 90.4% (1267/1400)
- Correct action detection accuracy: 100.0% (310/310)

A comparison of the inputs to the wrong action classifier and CogniTextworldAgent reveals that the latter already possesses sufficient information (including the cookbook content) to determine the correctness of actions. Yet, it still frequently fails to avoid actions that violate task constraints. This limitation arises because the agent's primary objective is to maximize task reward through behavior cloning and score ranking, rather than explicitly learning to discriminate between correct and incorrect actions. Consequently, the agent may assign high scores to syntactically or semantically plausible actions without fully considering cookbook compliance. In contrast, our approach explicitly introduces an action correctness classifier, trained to focus solely on identifying violations of cookbook constraints. By decoupling the correctness judgment from the general scoring function and dedicating a specialized component to this task, our method achieves more reliable and precise identification of wrong actions.

3.5 Classifier Integration

To suppress the tendency of the CogniTextworldAgent to select wrong actions, the trained classifier need to be integrated into the action selection process. A straightforward approach would be to entirely eliminate any action predicted as wrong, consistent with action elimination (AE) frameworks. However, this approach is susceptible to classifier errors, which may remove viable actions permanently.

To address this risk, we employ a softer integration method by modifying the UCB1 scoring mechanism. Specifically, **if an action is classified as wrong, its execution count is artificially incremented by 1**. This adjustment reduces its UCB1 score, thereby lowering its selection probability.

The rationale for increasing the execution count by 1, rather than by 2 or 3, is as follows. Under the UCB1 algorithm, actions with fewer prior executions are prioritized. Thus, adding n to the execution count effectively delays consideration of the action until all others have been executed n times. Larger n values reduce the likelihood of selection, and with the action step limit set to 100, $n > 1$ may be overly restrictive, as the agent could reach the limit before ever executing the action, depending on the size of the game's action space. Conversely, when $0 < n < 1$, it is equivalent to instructing the agent to execute the action immediately after all others, without considering the output of RoBERTa; while feasible, it is unlikely to yield behavior meaningfully different from $n = 1$.

This strategy enables the classifier to influence action selection without overriding the agent's policy entirely, providing both robustness and flexibility. The lower part of Fig. 2 illustrates the architecture of the proposed method. By comparing the upper and lower parts of the figure, it can be seen that the only difference between proposed method and the baseline is the introduction of the wrong action classifier. The wrong action classifier can be integrated without retraining the original agent, making it a fully non-intrusive design.

The UCB1 algorithm is a classical solution to the multi-armed bandit problem, designed to manage the balance between exploiting known rewards and

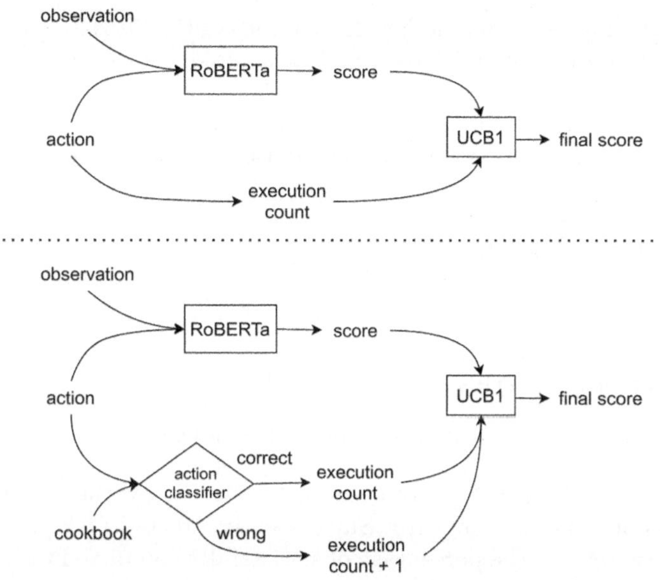

Fig. 2. The upper section illustrates the baseline CogniTextworldAgent architecture, while the lower section shows the proposed method. The only difference between proposed method and the baseline is the introduction of the wrong action classifier.

exploring potentially better alternatives. In each state, the UCB1 score for an action is computed as:

$$\text{UCB}_i = \begin{cases} \bar{x}_i + 5 & \text{if } n_i = 0 \\ \bar{x}_i + \sqrt{\dfrac{2 \log n}{n_i}} & \text{if } n_i > 0 \end{cases} \tag{1}$$

where:

- $\bar{x}_i$ is the average reward for action i, computed as the RoBERTa-based action score.
- n_i is the number of times action i has been selected.
- n is the total number of action selections in the current state.
- $\log n$ denotes the natural logarithm.

Actions that have never been selected ($n_i = 0$) are assigned a fixed high bonus (e.g., +5), which ensures that their scores are always higher than those of already selected actions. Once an action has been selected, its score is then updated according to the standard UCB1 formula.

4 Experimental Setup

4.1 Dataset

Experiments are conducted on the First TextWorld Problems (FTWP) dataset, introduced by Microsoft Research. FTWP comprises 5,176 unique games and represents the largest available benchmark for training text-based game agents. Each game requires the agent to complete a cooking task by following a cookbook and consuming the prepared food. The games differ in terms of spatial layout, available ingredients, and cookbook contents, requiring agents to generalize effectively to unseen scenarios. The dataset is partitioned as follows:

- Training set: 4,440 games
- Validation set: 222 games
- Test set: 514 games

4.2 Evaluation Metrics

Two metrics are used to evaluate agent performance:

- **Normalized Score:** The total reward obtained by the agent, divided by the maximum possible reward. Since each completed task yields one point, this metric reflects the percentage of successfully completed tasks. For clarity, this metric will be reported in percentage form (e.g. 0.01 is presented as 1%) throughout the paper.
- **Average Number of Steps:** The average number of actions taken per game. For a fixed normalized score, a lower step count indicates higher efficiency. In accordance with FTWP competition rules, the maximum number of steps per game is capped at 100.

It is important to note that average steps alone is not a reliable indicator of performance, as games may terminate early if the agent executes a wrong action. Therefore, agents that frequently make such mistakes may appear more efficient due to shorter game durations, despite poor actual performance.

4.3 Implementation Details

In the reimplementation of CogniTextworldAgent, we first replaced BERT with the more performant RoBERTa (RoBERTa-base specifically).[2] As mentioned before, during training, walkthrough actions serve as positive examples, while at each step, five negative examples are randomly sampled from the set of available but unselected actions. The model is trained for up to 5 epochs using the AdamW optimizer with a learning rate of 1e−5 and a batch size of 8. The best checkpoint is selected based on validation set performance. This procedure is repeated 3 times to obtain 3 independently trained agents for evaluation.

For the implementation of wrong action classifiers, we also adopted RoBERTa-base. The learning rate is set to 1e-5, the batch size to 32, and the model is trained for a maximum of 3 epochs.

[2] https://huggingface.co/FacebookAI/roberta-base.

5 Results

The improvement in normalized scores before and after applying the wrong action classifier is shown in Table 1. Agents 1 to 3 correspond to the best-performing models on the validation set, selected from three independent runs. Each table cell presents the baseline performance (left value) and the performance after classifier integration (right value). The results show consistent improvements across all three agents. On average:

- Validation set performance increased from 93.6% to 95.2%, corresponding to an absolute improvement of 1.6%.
- Test set performance increased from 88.0% to 91.2%, corresponding to an absolute improvement of 3.2%.

Table 1. Normalized scores (in percentage form) before and after applying the wrong action classifier (→: baseline → proposed method). ▲ indicates statistical significance. W represents the smaller of the sum of positive and negative signed ranks.

	Validation Set (222 games)	Test Set (514 games)
Agent 1	94.1 → 94.6 (W = 0.0, p = 0.109)	89.4 → 92.4 ▲ (W = 11.0, p < 0.001)
Agent 2	93.3 → 95.4 ▲ (W = 0.0, p = 0.018)	86.7 → 91.7 ▲ (W = 8.5, p < 0.001)
Agent 3	93.6 → 95.8 ▲ (W = 0.0, p = 0.043)	87.8 → 89.5 ▲ (W = 12.5, p < 0.001)
Average	93.6 → 95.2 ▲ (W = 0.0, p < 0.001)	88.0 → 91.2 ▲ (W = 99.5, p < 0.001)

These improvements indicate that the proposed classifier effectively mitigates the performance degradation caused by wrong actions, without retraining the base agent.

We conducted statistical significance testing on the score improvements presented in Table 1. Since our method does not require retraining the agent, a paired relationship can be established. Given that the score distributions do not follow a normal distribution, we employed a two-sided Wilcoxon signed-rank test ($p < 0.05$) to assess the statistical significance of the differences. Specifically, for the *valid* dataset, each agent obtained 222 normalized scores before using the action classifier and another 222 scores after using it. Similarly, for the *test* dataset, each agent yielded two groups of 514 scores, which were also tested. For the "Average" case, the scores from all three agents were concatenated before conducting the test. The results of the tests are appended to the scores in Table 1, with significant differences indicated by a ▲ symbol. For reference, we also report the W statistic and the p value. W represents the smaller of the sum of positive and negative signed ranks. $W = 0$ indicates that all score differences share the same sign, meaning that the agent's scores after using the classifier were always better than or equal to the original scores.

A broader comparison is presented in Table 2, which includes both normalized scores and average steps. The second-place agent from the FTWP competition,

LeDeepChef, is included for reference. As the implementation was not reproduced, results are directly cited from its original publication [2]. Compared to CogniTextworldAgent, our method results in a higher average step count, which can be attributed to the reduction in early terminations—allowing the agent to explore and execute more steps.

Table 2. Performance comparison on the validation and test sets. ▲ indicates statistical significance over CogniTextworldAgent.

Method	Validation Set (222 games)		Test Set (514 games)	
	Norm. Score (%)	Avg. Steps	Norm. Score (%)	Avg. Steps
LeDeepChef	74.4	24.1	69.3	43.9
CogniTextworldAgent	93.6	14.5	88.0	21.6
Ours	**95.2 ▲**	15.3	**91.2 ▲**	24.7

6 Discussion

6.1 Failure Patterns Analysis

In FTWP games, agents can fail in only two distinct ways:

- Exceeding the maximum allowed number of steps (100).
- Executing a wrong action—i.e., an action that violates the cookbook and prematurely ends the game.

To examine the impact of the wrong action classifier on agent behavior, we tracked the number of failed episodes on the test set (514 games) before and after applying the classifier. The results are summarized in Table 3 and reveal several notable trends:

1. The number of failures due to wrong actions decreases substantially across all agents.
2. Conversely, the number of failures due to exceeding the step limit increases.
3. The overall failure count declines, indicating an improvement in agent robustness.

As for the second point, the increase in cases failing due to exceeding the step limit, is consistent with our expectations. By reducing the likelihood of executing wrong actions, the agent avoids premature termination and is able to continue interacting with the environment. As a result, it survives longer and attempts more actions, which naturally increases the chance of exceeding the step limit. Despite this, the net reduction in total failures suggests that the classifier encourages safer and more effective decision-making.

Table 3. Failure counts before and after applying the wrong action classifier on the test set (→: before → after). Agents 1 to 3 correspond to the best-performing models on the validation set, selected from three independent runs.

	Exceeded Step Limit	Performed Wrong Action	Total Failures
Agent 1	16 → 31	80 → 51	96 → 82
Agent 2	3 → 15	100 → 67	103 → 82
Agent 3	13 → 38	108 → 60	121 → 98
Average	10.7 → 28.0	96.0 → 59.3	106.7 → 87.3

6.2 Why Combine with UCB1

In this study, actions classified as wrong are not directly eliminated; instead, their scores are penalized using the UCB1 algorithm. A natural question arises: to what extent would directly eliminating actions, instead of combining with UCB1, affect the performance of our method? The answer is that it would have almost no impact. However, this is because our action classifier achieves excellent performance. As described in Sect. 3.4, it attains 100% accuracy in identifying correct actions on the validation set. In other words, it virtually never misclassifies a correct action as wrong. Under such circumstances, even if direct action elimination were used, there would be no concern about mistakenly removing correct actions.

However, to highlight the importance of combining UCB1, we artificially constructed an action classifier that incorrectly classifies all "potentially rewarding actions" as wrong. We then integrated this classifier into CogniTextworldAgent using two approaches: (1) combining it with UCB1 (our approach) and (2) using direct action elimination. In FTWP games, potentially rewarding actions include not only ingredient-handling instructions but also the *take* actions, since taking key ingredients will yeild points.

The results were shown in Table 4. As expected, because all potentially rewarding actions were classified as wrong, the agent using action elimination failed to obtain any points, resulting in a final score of zero. In contrast, the agent using the classifier in combination with UCB1 (our approach) was still able to complete an average of 47.0% of tasks on the validation set and 44.5% on the test set. These results clearly demonstrate the effectiveness and robustness of our method when integrated with UCB1, even under extreme conditions where the classifier performs poorly.

6.3 Performance of LLM

In recent years, the rapid advancement of large language models (LLMs) has been remarkable, with these models achieving impressive results across a wide range of tasks. This naturally raises the question of how well they might perform in FTWP games. In this subsection, we evaluate the performance of GPT-4.1-mini on a randomly selected set of 10 test games using OpenAI Playground.

Table 4. Normalized scores (%) when using an extreme classifier that misclassifies all potentially rewarding actions as wrong (→: action elimination → combine with UCB1). Agents 1 to 3 correspond to the best-performing models on the validation set, selected from three independent runs.

	Validation Set (222 games)	Test Set (514 games)
Agent 1	0 → 47.6	0 → 45.8
Agent 2	0 → 46.4	0 → 43.5
Agent 3	0 → 46.9	0 → 44.2
Average	0 → 47.0	0 → 44.5

For the specific experimental setup, the model "gpt-4.1-mini" was selected with all parameters kept at their default values. Neither Variables nor Tools were utilized, and no System or Prompt messages were applied. An example of the prompt is provided below (directly pasted into the conversation window for use).

```
Task: Find and examine the cookbook, follow the instructions to
prepare the meal, and eat it.
Room: Kitchen
Inventory: some water. Total 1 items.
Recipe: missing
Available actions:
0 drop water
1 examine cookbook
Next action (answer with the action directly):
```

In the *Task* part, the agent is instructed to first find the cookbook and then proceed by following its instructions. The task description is provided only in the first dialogue round. All other information is consistent with that used for CogniTextworldAgent and our proposed method. Specifically, *Room* indicates the current room name, *Inventory* lists the items currently held along with their total count, and *Recipe* displays the content of the cookbook. The recipe is marked as "missing" until the `examine cookbook` action is executed. *Available actions* enumerates the actions that the agent can perform at each step. For the example prompt above, GPT-4.1-mini responded with "1 examine cookbook." The performance of each method on these 10 randomly selected test games is summarized in Table 5.

As shown, GPT-4.1-mini achieved a normalized score of only 39.5% (out of a total of 48 tasks), with an average of 7.4 steps. Analyzing its behavior reveals that the relatively low average number of steps is due to frequent selection of wrong actions, which causes the game to end prematurely. These results indicate that large language models without fine-tuning struggle to handle the complexity of FTWP games, as certain task-specific knowledge—such as cooking methods and the correspondence between kitchenware must be acquired through learning.

Table 5. Performance comparison on 10 randomly selected FTWP test games.

Method	Normalized Score (%)	Average Steps
LLM (GPT-4.1-mini)	39.5	7.4
CogniTextworldAgent	86.8	15.3
Ours	88.2	21.4

Naturally, models with better zero-shot performance are expected to achieve higher scores; however, misleading effects caused by wrong actions are anticipated to remain a significant obstacle. For example, if a recipe calls for "fry the red onion," and the agent is currently holding the red onion in a room equipped with an oven but no stove, should it explore other rooms or attempt to "cook red onion with oven"? A fine-tuned model that has learned the association between "fry" and "stove" from numerous training examples might choose to explore other rooms, whereas such reasoning would inevitably pose a challenge for a zero-shot model. On the other hand, if the LLM is fine-tuned—that is, used as a probabilistic model—we could entirely replace the RoBERTa component in our approach with it, in which case any performance improvement would not be attributable to methodological innovation.

7 Conclusion

This work presents a method to enhance CogniTextworldAgent, the state-of-the-art FTWP agent, by integrating a wrong action classifier that identifies cookbook-violating actions likely to cause early game termination. The combination with UCB1 allows our method to improve robustness without requiring any retraining of the base agent. Experiments on the FTWP dataset show that our approach improves the agent's task completion rate on the unseen test set from 88.0% to 91.2%, an absolute gain of 3.2%, which is statistically significant (Wilcoxon Signed-Rank Test, $p < 0.05$). These results demonstrate the effectiveness of lightweight classifier-based corrections in enhancing agent robustness without retraining the original model.

References

1. Xu, Y., Fang, M., Chen, L., Du, Y., Zhou, J.T., Zhang, C.: Deep reinforcement learning with stacked hierarchical attention for text-based games. Adv. Neural. Inf. Process. Syst. **33**, 16495–16507 (2020)
2. Adolphs, L., Hofmann, T.: Ledeepchef deep reinforcement learning agent for families of text-based games. In: Proceedings of the AAAI Conference on Artificial Intelligence, vol. 34, pp. 7342–7349 (2020)
3. Côté, M.-A., et al.: TextWorld: a learning environment for text-based games. In: Cazenave, T., Saffidine, A., Sturtevant, N. (eds.) CGW 2018. CCIS, vol. 1017, pp. 41–75. Springer, Cham (2019). https://doi.org/10.1007/978-3-030-24337-1_3

4. Lima, P.: First textworld challenge — winning solution notes (2019). https://medium.com/@pvl/first-textworld-challenge-first-place-solution-notes-d081bb9dee11
5. Devlin, J., Chang, M.W., Lee, K., Toutanova, K.: BERT: pre-training of deep bidirectional transformers for language understanding. In: Proceedings of the 2019 Conference of the North American Chapter of the Association for Computational Linguistics: Human Language Technologies, volume 1 (Long and Short Papers), pp. 4171–4186 (2019)
6. Auer, P., Cesa-Bianchi, N., Fischer, P.: Finite-time analysis of the multiarmed bandit problem. Mach. Learn. **47**, 235–256 (2002)
7. Liu, Y., et al.: RoBERTa: a robustly optimized BERT pretraining approach. arXiv preprint arXiv:1907.11692 (2019)
8. Zahavy, T., Haroush, M., Merlis, N., Mankowitz, D.J., Mannor, S.: Learn what not to learn: Action elimination with deep reinforcement learning. In: Advances in Neural Information Processing Systems, vol. 31 (2018)
9. Huang, S., Ontañón, S.: A closer look at invalid action masking in policy gradient algorithms. arXiv preprint arXiv:2006.14171 (2020)
10. Pennington, J., Socher, R., Manning, C.D.: Glove: global vectors for word representation. In: Proceedings of the 2014 Conference on Empirical Methods in Natural Language Processing (EMNLP), pp. 1532–1543 (2014)
11. Mikolov, T., Chen, K., Corrado, G., Dean, J.: Efficient estimation of word representations in vector space. arXiv preprint arXiv:1301.3781 (2013)
12. Sarzynska-Wawer, J., et al.: Detecting formal thought disorder by deep contextualized word representations. Psychiatry Res. **304**, 114135 (2021)
13. Wang, A., Singh, A., Michael, J., Hill, F., Levy, O., Bowman, S.R.: Glue: A multi-task benchmark and analysis platform for natural language understanding. arXiv preprint arXiv:1804.07461 (2018)

Conducting Tests to Find Out Which Academic Texts Are Generated by AI

Hector Rafael Morano Okuno(✉)

School of Engineering and Science, Tecnologico de Monterrey, Atizapan de Zaragoza, Estado de Mexico, Mexico
hector.morano@tec.mx

Abstract. Currently, AI-generated writing assistance offers several benefits, including time savings in preparing teaching materials, ease of incorporating specific objectives into instructional guides, and the creation of academic activities that develop skills. On the other hand, its use in the creation of academic texts can sometimes represent a lack of creativity or authorship, even plagiarism. However, in certain situations, human-generated texts are identified as AI-generated. To understand these cases, this article uses a Large Language Model (LLM) to perform different tests and address the following objectives: 1) identify the AI patterns used to generate texts; 2) analyze the AI patterns to detect whether a text is AI-generated or human-generated; and 3) create a text with the AI patterns and corroborate its origin. Among the results, it was found that statistical and stylistic patterns are used to detect whether AI-generated text can be distinguished from human-generated text. On the other hand, although there are general AI patterns for creating texts, there are also specific AI patterns depending on the area of knowledge being studied. This work was conducted during the Machine Design course, held from February to June 2025, for Mechanical Engineering students, when they consulted an LLM to develop their design projects. To understand students' general perceptions of using LLMs to detect AI-generated texts, a survey was administered at the end of the course.

Keywords: Large Language Model · Plagiarism · Detect AI Written · AI patterns · Prompt Engineering · Educational Innovation · Higher Education

1 Introduction

Some works identify recurring patterns in AI-generated writing, such as fragmented argumentation, specific language use, and excessive objectivity, which contrast with the more nuanced and typical style of human writing [1]. With current software applications, detecting AI-generated text is becoming increasingly complex. To assist with this task, multilingual pre-trained language models have been developed to extract the CLS vector as overall semantic features, integrating statistical features, and utilizing cumulative probability [2]. However, other works feed information models to improve the creation of arguments by AI [3], which they achieve by collecting arguments from professionals

H. Unger and P. Meesad (Eds.): NLPIR 2025, LNNS 1904, pp. 379–391, 2026.
https://doi.org/10.1007/978-3-032-20897-2_28

in their field, who establish the correct approach. When comparing human arguments and machine-generated arguments, it has been found that the quality of the prompt used for the AI arguments determines whether they are better and more similar to those of people. Other studies have detected that when using AI, there is a risk that students become too dependent on it and their creativity, originality, and critical thinking are diminished, in addition to the possibility of receiving wrong answers [4], on the other hand, it has been found that students with its use, it facilitates language learning and helps them summarize texts.

In the field of art, studies have also been conducted on the integration of AI into art. For example, it has been found that some countries tend to transform their culture towards the context of AI technology, and others seek to integrate AI into art by seeking copyright protection [5]. In the area of business communication pedagogy, it has been found that text-generative AI has generated significant benefits, thanks to the development of various applications that assist users in creating teaching materials, evaluating content, providing pre-reading assistance, and achieving considerable time savings in investment [6]. In specialized humanities fields, such as philology, the benefits of using AI in the development of teaching materials have also been demonstrated, with a significant impact on students, leading to improved knowledge retention and increased motivation for learning. On the other hand, for teachers, benefits have also been identified in the ease of developing activities and the creation of more effective evaluation instruments [7]. In the area of technical communication, tools have been developed that integrate Generative AI into the writing process to achieve adequate writing. This integration enables the AI to assist the user in certain parts of the writing process, without replacing it; instead, it complements the user's work [8].

Other works focus on detecting human-written and LLM-generated texts in different genres: essays, stories, poetry, and Python code; however, it has been found that the most challenging texts to identify are story writing [9] since the patterns can be influenced by the writing style, as well as the influence of different authors. In the area of Pharmacology Education, work has also been done on how AI influences the writing of academic texts and the risk of biases and lack of transparency, as well as academic dishonesty [10], which is why rules have been proposed to limit its use in certain activities, as well as to delimit its use of AI tools in healthcare domains. In the field of language teaching, there is also concern about the misuse of machine translation (MT) applications, particularly in cases where they are used dishonestly in writing texts or in reading comprehension exercises [11]. However, advancements have been made in detecting the sentence structures that AI generates when translating texts. Some studies have found that AI employs different patterns depending on the area [12], indicating a greater specialization of texts based on their orientation. An advantage of text generation by AI is the time savings in developing instructional guides, as well as evaluations that must comply with specific rules and incorporate objectives [13]. Another advantage is the writing of activities that must cover the course content, as well as developing the skills and competencies required at certain learning levels [14].

2 Research Context

Even when a person writes a text, it is sometimes mistakenly detected as AI-generated. The research question is: Are there any rules or guidelines that authors of academic texts can use to ensure their texts are not mistaken for being generated by AI applications? To answer the previous question, this work sets the following objectives: 1) identify the AI patterns used to generate texts, 2) analyze the AI patterns to detect whether AI or humans generate a text, and 3) create a text with the AI patterns and corroborate its origin.

The LLM used was GPT-5 (free version). The computing resources employed were laptops with an Intel i7 processor at 2.50 GHz with 32 GB of RAM. The software application used to detect plagiarism was Grammarly Pro. To test the human-generated texts, a portion of the abstract from a conference article [15] was used, which was guaranteed to have been written without the assistance of AI.

This research was conducted during the Machine Design course, held from February to June 2025, for 26 students of the mechanical engineering program. At the end of the course, they were given a survey to determine their overall perception of the course. The survey contained five items, as shown in Table 1. The evaluation range was 0 to 10, with 10 being the highest score. This survey has been validated by the university and is used to evaluate all courses.

Table 1. Survey applied to students.

ID	Items
I1	How was the learning experience?
I2	The professor proved mastery and expertise during the course
I3	Activities settled new skills
I4	Feedback contributed to the development of competencies
I5	During the course, there was an environment of trust and respect

3 Methodology

The methodology consisted of a series of steps to achieve the stated objectives. Two sequences were followed: one with human-generated text and the other with AI-generated text. Figure 1 illustrates the steps of the employed methodology.

3.1 Step 1. AI Pattern Identification

Initially, GPT-5 was asked to offer the AI patterns it used to identify AI-generated text; the prompt shown in Fig. 2 was used. The goal was to determine which patterns the AI used to generate its texts.

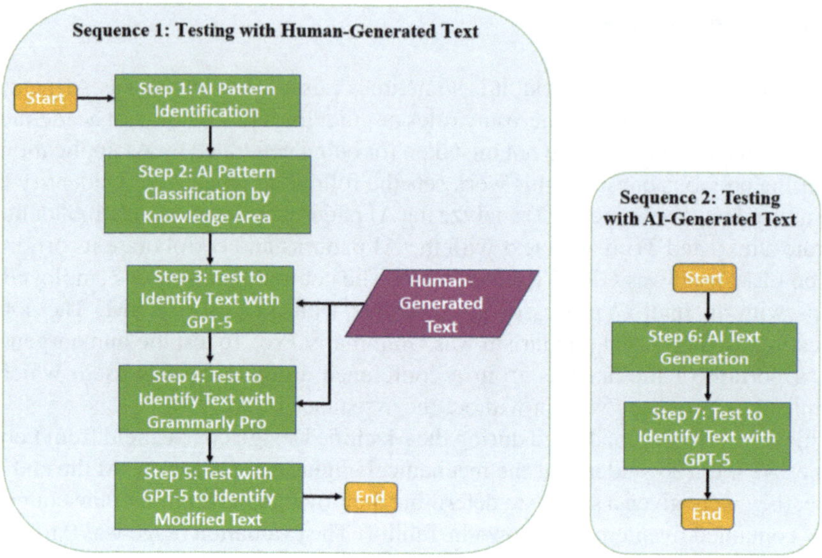

Fig. 1. Steps of the methodology employed.

```
> What patterns do you use to determine if a text is AI-generated?.
```

Fig. 2. Prompt used in Step 1.

3.2 Step 2: AI Pattern Classification by Knowledge Area

It was subsequently inquired whether there were specific AI patterns for generating text in a particular knowledge area; Fig. 3 shows the prompt used. The goal was to determine whether there was a difference in the patterns used across knowledge areas.

```
> Are there writing patterns that AI generates for different areas of
knowledge?
```

Fig. 3. Prompt used in Step 2.

3.3 Step 3: Test to Identify Text with GPT-5

To determine whether a text was written by a person or by AI, GPT-5 was presented with a text from a paper [15] that had been written by a person. The goal was to determine the procedure for determining whether a person wrote a text. Figure 4 shows the prompt used in this task.

```
> From the following text: "This research shares the results obtained
using Computational Fluid Dynamics (CFD) simulations as a didactic
tool in a Mechanical Engineering course using the Tec21 Educational
Model. The case study involved teaching the same topics to four
groups. CFD simulations were used in two of them, and in the other
two, they were not." Is it written by a person or by an LLM? Moreover,
what are the patterns that lead you to that conclusion?
```

Fig. 4. Prompt used in Step 3.

3.4 Step 4: Test to Identify Text with Grammarly Pro

The text from the previous step was corrected and subsequently analyzed in Grammarly Pro (version 2025) to verify whether any plagiarism had occurred. The text was selected because it was certain that the authors wrote it and that there was no plagiarism; however, the goal was to verify this with the computational tool. Figure 5 shows the corrected text used in this step.

```
> This research presents the results obtained using Computational
Fluid Dynamics (CFD) simulations as a pedagogical tool in a Mechanical
Engineering course, utilizing the Tec21 Educational Model. The case
study involved teaching the same topics to four groups. CFD
simulations were used in two of them, and in the other two, they were
not.
```

Fig. 5. Prompt used in Step 4.

3.5 Step 5: Test with GPT-5 to Identify Modified Text

Once the text was corrected and the absence of plagiarism was confirmed, it was decided to ask the GPT-5 application again to perform a text analysis to determine whether the AI had generated it. The prompt used to perform this analysis is shown in Fig. 6.

```
> From the following text, "This research presents the results
obtained using Computational Fluid Dynamics (CFD) simulations as a
pedagogical tool in a Mechanical Engineering course, utilizing the
Tec21 Educational Model. The case study involved teaching the same
topics to four groups. CFD simulations were used in two of them, and
in the other two, they were not." Is it written by a person or by an
LLM? Moreover, what are the patterns that lead you to that conclusion?
```

Fig. 6. Prompt used in Step 5.

3.6 Step 6: AI Text Generation

In the second sequence of the methodology, GPT-5 was asked to write a 150-word, academic-oriented text explaining the phenomena that occur during heat transfer in a coil used to cool a two-platen injection mold. The objective was to generate an academic text using AI, which would then be reviewed to determine if it contained the AI's patterns. The prompt used for this task is shown in Fig. 7.

```
> Could you write a 150-word, academic-oriented text explaining the
phenomena that occur during heat transfer in a coil used to cool a
two-platen injection mold?
```

Fig. 7. Prompt used in Step 6.

3.7 Step 7: Test to Identify Text with GPT-5

The text written by the AI asked GPT-5 to determine whether it or a human had generated it. In Grammarly Pro, it was asked to detect any plagiarism.

One research topic related to AI-generated text analysis is multimodal detection, such as phonetic and visual inspection. Therefore, LLM was asked to state what it knew about this type of analysis and whether it could analyze a text. The prompt results are included in Sect. 4: Analysis of Results.

4 Analysis of Results

4.1 Step 1. AI Pattern Identification

GPT-5 classified the AI patterns used to detect AI-generated texts into 1) Style and Structure Patterns, 2) Language Patterns, 3) Statistical Patterns, and 4) Detection Limitations; these are shown in Fig. 8.

```
> Great question
Detecting whether AI-generated text is not 100% certain (not even with
advanced tools), but you can identify statistical and stylistic
patterns that are more common in models like mine. I will summarize
the main ones:

1. Style and Structure Patterns
• Excessive Coherence:
• Uniformity:
• Neutrality:
________________________________________
2. Language Patterns
• Broad Vocabulary:
• Step-by-Step Explanations:
• Generic Phrases:
________________________________________
3. Statistical Patterns
• Word Distribution:
• Low error randomness:
• Controlled repetition:
________________________________________
4. Detection Limitations
• No detector is 100% reliable.
• A human can edit AI-generated text and make it indistinguishable.
• Sometimes, academic texts closely resemble AI texts.
```

Fig. 8. GPT-5's response to the prompt used in Step 1.

4.2 Step 2: AI Pattern Classification by Knowledge Area

When asked if there were different AI patterns for generating texts depending on the knowledge area, the answer was affirmative, demonstrating that it recognizes the content of each location. Figure 9 shows GPT-5's response to the prompt used in Step 2.

> Excellent observation
Yes, language models like me adapt our style according to the area of knowledge, and this generates characteristic writing patterns. It is not that there is a "fixed mold," but instead that through training with large volumes of scientific, technical, literary, and other texts, specific recurring ways of writing are acquired.
Here are some examples by area:

---

1. Social Sciences and Humanities
2. Engineering and Applied Sciences
3. Natural Sciences (physics, biology, chemistry)
4. Computer Science and Mathematics
5. Art and Literature

AI Cross-Cutting Pattern
In all areas, the model tends to:
• Structure the response (introduction, body, conclusion).
• Maintain a clear and neutral tone.
• Use step-by-step explanations.

Fig. 9. GPT-5's response to the prompt used in Step 2.

4.3 Step 3: Test to Identify Text with GPT-5

Once GPT-5 was given the text to identify whether a person had written it or not, the analysis procedure used was demonstrated, and it correctly determined that the text had been written by a person. Figure 10 shows the result of this analysis.

> Very good observation. Let us analyze the text:
"This research shares the results obtained using Computational Fluid Dynamics (CFD) simulations as a didactic tool in a Mechanical Engineering course using the Tec21 Educational Model. The case study involved teaching the same topics to four groups. CFD simulations were used in two of them, and in the other two, they were not."

Pattern analysis
1. Clear academic structure
→ This is typical of both human academic texts and AI responses.
2. Neutral and Formal Style
→ This neutral tone is very characteristic of an LLM, although it is also common in academic papers.
3. Small Grammatical Error
→ Says "in an Mechanical Engineering course" instead of "in a Mechanical Engineering course."
4. Excessive Clarity in Sentences
→ This is a typical IA pattern: short, informative, unadorned sentences.
5. Conclusion
→ My diagnosis would be probably written by someone with an academic background, but with a style that closely mimics the clarity pattern of an LLM.

Fig. 10. GPT-5's response to the prompt used in Step 3.

4.4 Step 4: Test to Identify Text with Grammarly Pro

When the text was entered into the Grammarly Pro application, the response confirmed that it was free from plagiarism. Figure 11 shows the result of this analysis.

> No plagiarism or AI text detected
Your document does not match anything in our references or contain common AI text patterns.

Fig. 11. Result of the plagiarism analysis by the Grammarly Pro application.

4.5 Step 5: Test with GPT-5 to Identify Modified Text

After the corrected text was given to GPT-5, it analyzed it and, based on the type of patterns it detected, concluded that the text had been written by AI. However, it concluded that a person with an academic background could have also written the text. Figure 12 shows the result of this analysis.

```
> Very good, let us analyze this new text:

Pattern analysis
1. Grammar correction and polishing
 → This is more typical of an LLM, because it usually produces clean, revised sentences.
2. Academic and Neutral Tone
 → This tone is familiar in both human papers and AI-generated texts, but the AI tends to maintain it unchanged.
3. Excessive Clarity and Organization
 → The logical progression is "too perfect" for a human draft; it seems written with a goal of maximum clarity.
4. Selection of Academic Synonyms
 → This type of substitution to sound more academic is something LLMs usually do easily, while a human tends to be more consistent with terminology.
5. Conclusion
 → The text is highly likely to have been written by an LLM.
A human could have written it.
```

Fig. 12. GPT-5's response to the prompt used in Step 5.

4.6 Step 6: AI Text Generation

Once GPT-5 was asked to write the text, it quickly generated it. The returned text was 148 words long; Fig. 13 displays the text. Then Grammarly was used to verify its writing and found two suggestions, which are shown in Fig. 14.

```
> Here's a 150-word academic-style explanation of heat transfer in that context:

During the injection molding process, temperature control within the mold is critical to ensure dimensional accuracy, mechanical properties, and surface quality of the molded parts. In a two-platen injection mold, cooling coils are typically embedded in the mold plates to regulate heat transfer after the molten polymer is injected. The primary mechanism governing this process is conduction: heat flows from the hot polymer into the steel mold surfaces and subsequently toward the cooling channels. Within the coil, a circulating fluid, usually water, promotes convective heat transfer by absorbing thermal energy from the mold walls. As the fluid flows, forced convection enhances heat removal efficiency, while the temperature gradient between the mold and coolant drives continuous energy exchange. Additionally, localized thermal resistances, due to material interfaces and coolant boundary layers, influence cooling uniformity. Overall, the interplay of conduction, convection, and fluid dynamics determines cooling performance and cycle time optimization.
```

Fig. 13. AI text.

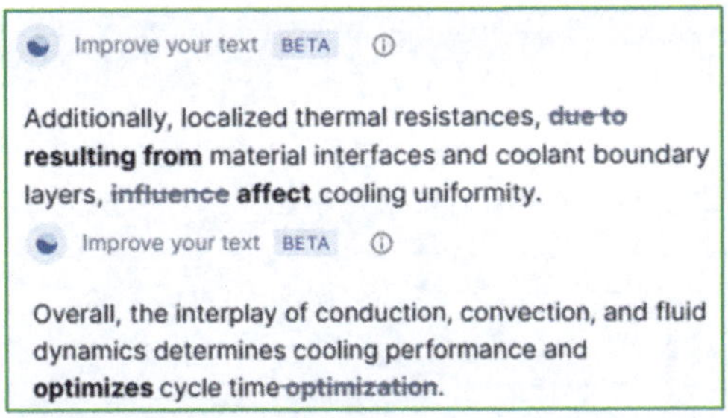

Fig. 14. Modifications suggested by Grammarly Pro.

4.7 Step 7: Test to Identify Text with GPT-5

Grammarly Pro found no plagiarism in the text generated by the AI; however, 75% of it contained patterns likely generated by the AI. On the other hand, GPT-5 also agreed that the AI was very likely to have generated the text. Figure 15 summarizes the response from GPT-5.

Figure 16 shows the prompt used to ask the LLM to perform multimodal analysis, including a phonetic and visual inspection of the given text to determine whether it was written by AI or a human.

4.8 Student Survey Results

Figure 17 shows the results of the student survey. Although the evaluations were relatively high, ranging from 9.91 to 10, we can focus on questions I2 and I4, which obtained lower scores. However, based on their standard deviation, we can see that not all students had the same opinions. Question I2 asks whether the professor demonstrated mastery of the topics presented, and question I4 asks whether the professor's follow-up with students contributed to their skill development. Students were subsequently invited to write comments about the course, seeking to understand their perceptions of any areas not covered in the survey or to express their desired feelings. These comments are shown in Table 2. It can be seen that all comments express gratitude for the professor's patience and a general appreciation for the course.

```
> Let's analyze the passage carefully:
Patterns in the text
1.Grammatical and stylistic polish
    Sentences are well balanced, error-free, and flow smoothly.
    Vocabulary is precise ("dimensional accuracy," "thermal
    resistances," "boundary layers").
    This level of consistency is typical of AI-generated academic
    text.
2.Neutral, academic tone
    No personal pronouns, no subjectivity.
    Reads like a mini textbook excerpt, which is a hallmark of LLM
    outputs.
3.Structured explanation
    Clear progression: importance → setup → conduction → convection →
    secondary effects.
    This "ordered completeness" is a strong indicator of AI writing.
4.Lack of human "imperfections"
    No uneven sentence lengths, no stylistic quirks, no hedging
    ("probably," "in practice").
    Humans often leave traces of personal emphasis or omission, which
    are absent here.
    4.Lack
Conclusion
    This text is highly likely AI-generated, because it shows: Perfect
    grammar and Flow, Neutral academic tone with no personal touches,
    Structured, almost "too clean" explanation typical of a language
    model.
    A human expert could have written it, but usually human drafts in
    technical contexts contain either more detail (specific numbers,
    conditions, or references).
```

Fig. 15. Respuesta entregada por GPT-5 acerca de la revisión del texto.

```
> Could you analyze the following text using phonetic vs. visual
inspection to determine whether a human or an AI wrote it and report
what you find? "This research shares the results obtained using
Computational Fluid Dynamics (CFD) simulations as a didactic tool in a
Mechanical Engineering course using the Tec21 Educational Model. The
case study consisted of teaching the same topics in four groups. CFD
simulations were used in two of them, and in the other two, they were
not. "

> 1. Phonetic Analysis (Sound & Rhythm)
Rhythmically smooth, consistent, and "balanced," but slightly too
regular. Leans toward AI-written or heavily edited academic tone.

2. Visual Analysis (Layout & Structure)
Perfect formatting, symmetry, balanced structure → machine-polished or
generated. Humans typically introduce small inconsistencies in
punctuation or rhythm.

3. Combined Phonetic + Visual Interpretation
This passage is well-written, grammatically flawless, and rhythmically
uniform. It lacks the small irregularities (hesitations, asides, or
uneven pacing) typical of human academic prose.

Conclusion:
Most likely written or post-edited by an AI, or by a human using an AI
drafting assistant (e.g., ChatGPT or Grammarly-polish).
The balance of phonetic and visual regularity suggests synthetic
fluency rather than natural linguistic variation.
```

Fig. 16. Prompt used to analyze a text with phonetic and visual inspection.

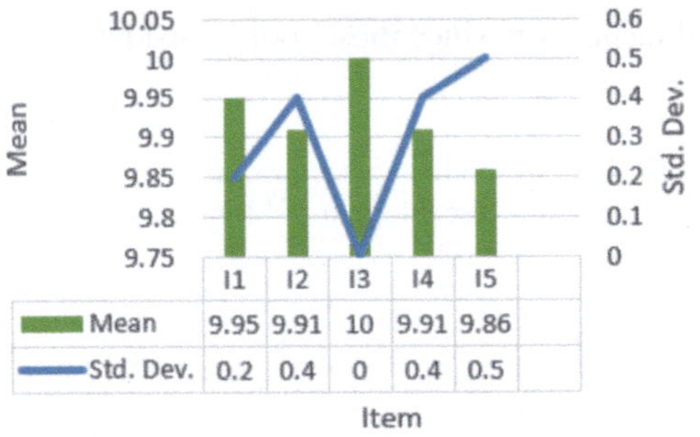

Fig. 17. Results of the student survey.

Table 2. Student comments at the end of the course.

No	Comments
1	Everything was great
2	Thank you very much for your patience. Great classes
3	Thank you for everything, Professor. You are undoubtedly one of the most knowledgeable and proficient professors
4	I deeply admire your calm explanations and patience with your students
5	Excellent and exciting classes
6	Excellent Professor, very calm yet challenging
7	The Professor is always available to answer questions and is very patient
8	His level of patience and mastery of all engineering topics is admirable

5 Conclusions and Future Work

Depending on the area of knowledge, various AI patterns are used to generate texts.

Detecting AI-generated text is not 100% reliable, but identifying AI patterns can help inform a decision.

Academic texts generated by humans are more like those created by AI; they present sentences with a neutral tone, clear ideas, abundant vocabulary, and virtually no grammatical errors.

Students' use of LLMs helped them become aware of how to use these tools properly to carry out their tasks honestly. They were also interested in how AI generates text.

It is recommended to research the Statistical Patterns used to evaluate a text to understand how they work and utilize them as tools to prevent confusion between texts created by humans and those generated by AI.

It is also recommended to study whether these AI patterns can be applied to or are the same across different languages, as they alter how ideas are expressed and the grammatical structure.

Although most AI tools that analyze text are based on linguistic and statistical levels, it is recommended to conduct experiments using multimodal detection analyses, such as phonetic vs. visual inspection, since these tools consider elements like the rhythm or sonority of the text.

Acknowledgment. The authors would like to acknowledge the financial support of Writing Lab, Institute for the Future of Education, Tecnologico de Monterrey, Mexico, in the production of this work.

References

1. Nowacki, L., Wrochna, A.E.: ChatGPT theses. Identifying distinctive markers in AI-generated versus human-created texts: a multimodal analysis in university education. E-Learn. Digit. Media, 2042–7530 (2025). https://doi.org/10.1177/20427530251331083
2. Min, C., Zhang, R., Liu, J.: Bilingual generated text detection through semantic and statistical analysis. Intell. Data Anal. **29**(5), 1248–1260 (2025). https://doi.org/10.1177/1088467X241307192
3. Marji, Z., Licato, J.: Evaluating large language models' ability to generate interpretive arguments. Argument Comput. **16**(3), 362–404 (2024). https://doi.org/10.3233/AAC-230014
4. Sok, S., Heng, K., Pum, M.: Investigating high school students' attitudes toward the use of AI in education: evidence from Cambodia. SAGE Open **15**(3), 2158–2440 (2025). https://doi.org/10.1177/21582440251353575
5. Yuan, H., Kim, Y., Zhang, J., Li, L.: Exploring research trends and hotspots in AI and art integration: ROST text mining analysis of China and South Korea. J. Comput. Methods Sci. Eng. **25**(5), 3961–3972 (2025). https://doi.org/10.1177/14727978251337903
6. Sharma, D., Pandey, H.: Pedagogical impact of text-generative AI and ChatGPT on business communication. Bus. Prof. Commun. Q. (2024). https://doi.org/10.1177/23294906241249113
7. Pellas, N.: Effects of generative AI feedback and interactive video assessment on student learning achievement in philological content creation courses. J. Educ. Comput. Res., 0735–6331 (2025). https://doi.org/10.1177/07356331251372800
8. Getto, G., Kelley, S., Vance, B.: How to write with GenAI: a framework for using generative AI to automate writing tasks in technical communication. J. Tech. Writ. Commun. (2025). https://doi.org/10.1177/00472816251332208
9. Hayawi, K., Shahriar, S., Mathew, S.S.: The imitation game: detecting human and AI-generated texts in the era of ChatGPT and BARD. J. Inf. Sci., 0165–5515 (2024). https://doi.org/10.1177/01655515241227531
10. Patel, Ch.R., Pandya, S.K., Sojitra, B.M.: Perspectives of ChatGPT in pharmacology education, and research in health care: a narrative review. J. Pharmacol. Pharmacother. **14**(3), 171–177 (2023). https://doi.org/10.1177/0976500X231210427
11. Maimone, L.L., Restrepo Ramos, F., Jolley, J.: Automated detection of machine translation use in L2 Spanish Writing. Lang. Teach. Res., 1362–1688 (2025). https://doi.org/10.1177/13621688251352263
12. Onal, S., Kulavuz-Onal, D., Childers, M.: Patterns of ChatGPT usage and perceived benefits on academic performance across disciplines: insights from a survey of higher education students in the united states. J. Educ. Technol. Syst. **54**(1), 3–66 (2025). https://doi.org/10.1177/00472395251341214
13. Onal, S., Kulavuz-Onal, D.: A cross-disciplinary examination of the instructional uses of ChatGPT in higher education. J. Educ. Technol. Syst. **52**(3), 301–324 (2023). https://doi.org/10.1177/00472395231196532

14. Guo, Q., Zhen, J., Wu, F., He, Y., Qiao, C.: Can students make STEM progress with the Large Language Models (LLMs)? An empirical study of LLMs integration within middle school science and engineering practice. J. Educ. Comput. Res. **63**(2), 372–405 (2025). https://doi.org/10.1177/07356331241312365
15. Morano Okuno, H.R., Sandoval Benitez, G.: Effects of using computational fluid dynamics simulations on student learning during a mechanical engineering course. In: Yang, XS., Sherratt, S., Dey, N., Joshi, A. (eds.) Proceedings of Ninth International Congress on Information and Communication Technology. ICICT 2024. Lecture Notes in Networks and Systems, vol. 1013. Springer, Singapore (2024). https://doi.org/10.1007/978-981-97-3559-4_37

Beyond Attribution: GPT-Based Source Detection and Categorization for Bias Analysis in Journalism

Tabassum Sultana(✉), Eric R. Harley, Gavin Adamson, and Asmaa Malik

Toronto Metropolitan University, Toronto, Canada
{tabassum.sultana,eharley,gavin.adamson,
asmaa.malik}@torontomu.ca

Abstract. The growing volume of digital news presents challenges in identifying and evaluating the voices represented in reporting. This study investigates the use of GPT-4.1, a large language model, to automatically detect and categorize news sources, including human sources, non-human sources, and coreferences (pronouns), and compares its performance with human annotations. Eighty manually annotated news articles served as the evaluation dataset. GPT was guided by predefined category definitions, and outputs were evaluated for both source-level and category-level performance. In source prediction, GPT achieved a precision of 99%, recall of 88%, and an F1-score of 93%, indicating strong overlap with human annotations despite over-prediction of some sources. For source categorization, GPT achieved a precision of 85%, recall of 86%, and an F1 score of 86%, demonstrating good performance in assigning categories once sources were identified. The study further measures how much narrative context came from journalists or authors versus attributed sources, providing a basis for analyzing potential bias in reporting. These results demonstrate GPT's potential for enhancing automated source detection and classification in journalism, and open new opportunities for systematically studying framing and bias in news coverage.

Keywords: Source Identification · Source Categorization · Digital Journalism · Bias Detection · Large Language Models · GPT

1 Introduction

With the growing volume of digital news, identifying and evaluating the voices represented in reporting has become increasingly complex. Sources are central to journalistic practice: they ground articles in evidence, shape the framing of events, and influence how audiences interpret credibility. Patterns of over representation or omission among source types are widely recognized as indicators of imbalance in journalism. Early work in the Journalism Representation Index (JeRI) project conceptualized source diversity as a measurable dimension of media balance. Malik and Adamson [1] introduced an algorithmic framework for assessing local news by examining the diversity and distribution of sources, laying the foundation for systematic evaluation of voices in reporting.

H. Unger and P. Meesad (Eds.): NLPIR 2025, LNNS 1904, pp. 392–402, 2026.
https://doi.org/10.1007/978-3-032-20897-2_29

Building on this foundation, our earlier study Extracting Source Information from News Articles [2] operationalized a taxonomy of categories such as authority, expert, organization, political, media, celebrity, and unaffiliated, and implemented a machine learning pipeline for automated source extraction. The system achieved high accuracy for detecting sourced sentences and moderate accuracy for categorization, confirming the feasibility of computational source analysis. However, the work also faced limitations, including reliance on handcrafted rules, difficulty handling coreference, class imbalance, and limited integration with frameworks for equity and balance assessment. Despite these advances, automating source categorization continues to face persistent challenges. Prior studies have noted misclassifications arising from cultural and linguistic nuances, differences in editorial style, and ambiguity in source roles. Biases inherited from model training data can further distort results, particularly for underrepresented groups. These challenges underscore the need for refined approaches that ensure fair and accurate representation of voices in journalism. This paper addresses those gaps by introducing a GPT-based framework for source detection, categorization, and quantification. Unlike earlier pipelines that relied on fixed lexical rules and classifiers, our approach leverages GPT's natural language understanding to flexibly identify both explicit and implicit sources, including coreferential mentions. The framework not only extracts and categorizes sources into analytically meaningful roles but also quantifies the distribution of voices within each article, distinguishing journalist-authored sentences from externally attributed content. Validation against human annotations across 80 articles demonstrates strong performance, while error analysis highlights challenges in role-based ambiguity and low-frequency classes. By moving beyond attribution toward structured analysis, this study contributes an integrated approach that supports balance and fairness aware evaluation of sourcing in digital news.

1.1 Contributions

- Source detection and attribution: A GPT-based framework that predicts not only the presence of a source but also the attribution of verbs, sentences originating from sources, and sentences authored by the journalist, while also resolving coreferences such as pronouns.
- Comprehensive categorization: Classification of both human and nonhuman sources into journalism-informed roles: authority, expert, organization, political, media, celebrity, and unaffiliated.
- Bias-aware quantification: Measurement of the distribution of voices in each article by distinguishing journalist-authored content from externally sourced statements, enabling structured evaluation of balance, representation, and fairness in news reporting.

1.2 Background

Research shows that journalistic sourcing often privileges institutional voices such as politicians, authorities, and experts, while underrepresenting community perspectives and lived experiences [3–5]. These patterns have consequences for transparency, accountability, and inclusivity, since sourcing choices shape how events are framed and whose perspectives dominate the narrative.

In Canada, such imbalances are particularly evident in coverage of Indigenous, First Nations, and Métis communities. Historical reporting has frequently marginalized Indigenous voices, portraying them through deficit frames that emphasize dysfunction or victimhood, while privileging institutional perspectives [1]. Building on this concern, recent work such as the Journalism Representation Index (JeRI) and follow-up studies on automated source extraction have developed frameworks for analyzing how sourcing practices affect representation in journalism [2].

This study builds on that foundation by distinguishing between voices (the perspectives represented in stories) and sources (the specific people, groups, or documents that carry those voices). Quantifying how many sentences are authored by journalists versus external sources enables systematic evaluation of balance and representation. Making these distributions visible provides a structured way to assess fairness, inclusivity, and potential bias in news coverage.

The remainder of this paper is organized as follows: Sect. 2 reviews related work. Section 3 presents the methodology, including definitions of source categories, the dataset and annotation process, and the prompt engineering approach with GPT model configuration. Section 4 reports the experimental results and analysis. Finally, Sect. 5 concludes with key findings, implications, and directions for future work.

2 Related Work

The use of large language models (LLMs) in news analysis has expanded rapidly, with particular interest in the identification of sources in journalistic texts. Spangher et al. [6] addressed the task of detecting informational sources in news articles, building a benchmark dataset, and demonstrating that fine-tuned GPT pipelines achieve strong attribution performance, while zero-shot approaches (i.e., without specific training) remain limited. Their framework restricts sources to explicit mentions such as named entities or organizations, excluding pronouns rather than resolving them, which means that full coreference handling is not addressed. Moreover, their work focuses primarily on source identification and attribution, without extending to structured categorization.

Building on this line of research, Vincent et al. [7] evaluated GPT-4 and other LLMs across several subtasks of journalistic sourcing, including the extraction of sourced statements, names, titles, and types. Their findings show reasonable accuracy for identifying names and types, but justification extraction remained weak, that is, detecting not just who the source is, but also the descriptive text explaining why the source matters in the story. More importantly, the type annotation scheme employed was narrow and not linked to a consistent analytical framework for journalism. In contrast, our framework integrates extraction with systematic categorization, enabling structured analysis of bias and representation in news coverage.

Beyond direct source identification, other studies highlight adjacent applications of GPT in news and information extraction. Hossain et al. [8] investigated how LLMs can automate the evaluation of publisher reliability, while Gadiraju et al. [9] proposed the Media Bias Detector, a tool for real-time analysis of selection and framing bias. Li et al. [10] demonstrated the use of GPT to extract entities and sentiment from financial news, and Ma et al. [11] introduced a multi-layer ranking framework with LLMs to improve quote–speaker pair recommendations. Collectively, these approaches demonstrate the potential of LLMs for structured source extraction and credibility-focused processing, but these papers assume that sources have already been identified, thereby skipping the critical stage of source extraction and categorization. At the entity level, Martins et al. [12] evaluated GPT models for narrative entity extraction in Portuguese news, showing their ability to capture fine-grained spans such as participants, events, and temporal expressions. Broader contextual applications include Han et al. [13], who developed a framework for LLM-based summarization that synthesizes content from multiple news sources, and Kim and Lee [14], who examined AI-generated news articles, highlighting both the opportunities and challenges of automated content creation. Together, these works underscore the flexibility of GPT in structuring and interpreting journalistic content beyond simple classification tasks.

In summary, existing work has made progress in source identification [6, 7] and in adjacent applications [10, 4], but no framework explicitly combines source extraction with systematic categorization into analytically meaningful classes (e.g., authority, expert, organization, unaffiliated, politicians). While earlier approaches largely emphasized attributing sentences to sources and the co-occurrence of sources within articles [6, 7], our framework extends this line of research by coupling source extraction with categorization tailored for bias detection. In addition to quantifying the distribution of external voices in each article, we also calculate the proportion of sentences attributable to the author of the article, which enables a deeper assessment of balance, representation, and fairness in news reporting.

3 Methodology

As in our previous work [2], we considered a source in a news article to be any named entity, be it a person, organization, or document, providing information. One of the preliminary goals of the research is to identify all sentences that can be attributed to an external source of information. The external sources are then put into the categories introduced by Adamson and Malik [1]: Authority (AUT), Politician (POL), Expert (EXP), Organization (ORG), Media (MED), Celebrity (CEL), and Unaffiliated (UNA).

3.1 Dataset

The dataset that we developed for this research contains eighty news articles about First Nations and Indigenous peoples, drawn from The Globe and Mail and the London Free Press, Canadian newspapers. Three annotators independently identified the sources in each article and assigned each one to a category according to the definitions in Table 1. After independent annotation, disagreements were resolved through discussion and possibly adjusting any ambiguity in a definition. A fourth adjudicator verified completeness and consensus, producing the final human-generated annotations.

3.2 Prompt Design, GPT Prediction Process, and Preprocessing

The source definitions and categories were converted into structured prompt instructions. GPT-4.1-mini was used for inference and accessed through the OpenAI API. No fine-tuning was applied. GPT was explicitly asked to identify all sources in each article, assign them to one of the predefined categories, and return results in a structured format containing the story number, source name, and category. The wording of the prompt was refined iteratively to reduce ambiguity and maximize alignment of GPT annotation with human annotation. GPT predictions (annotations) were generated by using the OpenAI API on each article processed individually. The procedure was repeated during development of the prompts to test the refinements and validate stability.

Annotations, both human generated and GPT generated, were pre-processed to standardize text formatting. This included lowercasing, removal of extra spaces and punctuation, and normalization of common abbreviations and role-based references (e.g., "the mayor" → "Mayor"). These steps ensured that minor formatting differences did not result in spurious mismatches during evaluation.

3.3 Evaluation Protocol

For evaluation, we defined four key outcomes when comparing GPT predictions with human annotations. An overlap occurs when both GPT and humans identify the same source. Within these overlaps, a match means GPT and humans agree not only on the source but also on its category, while a mismatch means they agree on the source but disagree on its category. A missed case refers to a source present in human annotations but absent in GPT predictions, and an extra case refers to a source predicted by GPT that is not found in human annotations. Table 2 presents a few comparisons of GPT-predicted categories with those assigned by human annotators.

To assess how GPT's reasoning aligned with journalism-informed attribution rules, we qualitatively reviewed its justifications alongside those of human annotators. GPT typically explained why a source was assigned to a given category, whereas human annotators drew on predefined taxonomy guidelines. Table 2 presents illustrative cases where disagreements emerged. In Example #3 of Table 2, GPT classified Prime Minister Justin Trudeau himself as the source, because the mandate letter was attributed to him, while human annotators treated the letter as the actual source. In Example #2 of Table 2, GPT labeled Ken Young as a politician based on his previous leadership role, whereas annotators categorized him as an expert due to his current role. In Example #1, GPT marked Ron Tripp as unaffiliated, overlooking his connection to the Niagara Region, which annotators correctly identified as an organizational affiliation. These examples demonstrate that accurate categorization is difficult, since it depends on context within the article and knowledge of current roles in society, rather than professional titles or surface-level associations, echoing Malik and Adamson's [1] emphasis on contextual diversity in source evaluation. A former minister may be classified as an expert or an authority depending on context, but not as a politician. For example, in a recent news article the statement "Speaking in Erie, Pennsylvania, Obama said: "*I think at moments like this, ...*"" was reported. In this case, Obama is categorized as an expert or an authority based on the context, but not as a politician, since he is no longer active in political office.

Table 1. Source categories, definitions, and illustrative examples.

Category	Definition	Human Source Example	Non-Human Source Example
AUT (Authority)	A person or source representing a government institution (non-elected), such as city staff, judges, or police	City program manager, police, spokesperson, judge. Religiousdirector,mosque spokesperson, community imam. e.g.,Imam Kamil Ahmad, Toronto Islamic Centre	Government policy document, city bylaws, ministry report
POL (Politician)	An elected official or candidate representing a political office or role	Mayor, MP, MLA, city councillor, minister. An Indigenous leader (a Chief) would almost always be a political leader. e.g., Cindy Woodhouse Nepinak, the National Chief of the Assembly of First Nations (AFN)	Campaign website, ministerial speech transcript, policy manifesto from a political party
EXP (Expert)	A person or source with formal academic, legal, or medical expertise	Professor,scientist, doctor, lawyer. e.g., "Prof. Heather Dorries, University of Toronto."	Academic journal article, research paper, legal commentary
ORG (Organization)	A person or source representing a nonprofit, business, or advocacy group, not tied to the government	NGO director, business spokesperson, advocacy leader. e.g., "Michelle Fournie, co-chair of a nonprofit."	NGO report, business press release, nonprofit social post
CEL (Celebrity)	A public figure such as an artist, entertainer, or athlete speaking publicly	Actor, musician, athlete. e.g., "Singer X at a press event."	Celebrity autobiography, official statement, social media post

(continued)

Table 1. *(continued)*

Category	Definition	Human Source Example	Non-Human Source Example
MED (Media)	A journalist or news organization providing information or commentary	Reporter, editor, columnist. e.g., "Reporter Y from The Globe and Mail."	News article, podcast, editorial, press release from a news outlet
UNA (Unaffiliated)	A private citizen not speaking on behalf of any group or any profession. Share personal experience or opinion	Survivor, eyewitness, resident, protester. e.g., "Doreen Healy, residential school survivor."	Personal blog, social media post by a regular citizen, letter to editor

Table 2. Examples of Disagreements between GPT and Human Annotators

	Source Name		Category	
#	GPT Prediction	Human Prediction	GPT Prediction	Human Prediction
1	Ron Tripp	Ron Tripp	UNA	ORG
2	Ken Young	Ken Young	POL	EXP
3	Justin Trudeau	Mandate letter	POL	POL

3.4 Results and Analysis

We report results at two levels: source-level detection and category-level classification.

Source-Level Comparison. Human annotators initially identified 227 sources (194 overlap + 33 human-only), while GPT predicted 245 sources (194 overlap + 51 GPT-only), i.e., 194 were common to both sets, 33 were identified only by humans, and 51 were identified only by GPT. The human annotators reviewed the 51 GPT-only sources and confirmed 49 of GPT's 51 unique cases as valid, leaving only 2 false positives for GPT. Incorporating these validated cases expanded the gold standard to 276 sources (194 overlap + 33 human-only + 49 GPT-confirmed). As shown in Table 3, GPT achieved 99% precision, 88% recall, and 93% F1, while human annotators achieved 100% precision, 82% recall, and 90% F1. These results demonstrate that GPT was almost as precise as humans in identifying sources in this dataset, and GPT was more reliable than humans, finding many valid sources overlooked by human annotators.

Category-level Comparison. As shown in Table 4, with the gold standard expanded to 276 sources, GPT detected 245 sources. Of these, 2 were incorrect source predictions, leaving 243 true source detections eligible for category evaluation. GPT correctly categorized 210 sources, yielding a category-level precision of 85% (210/245), recall of 86%

Table 3. Performance metrics of GPT and human annotators in source extraction

Annotator	True Positives	False Positives	False Negatives	Precision	Recall	F1-score
GPT	243	2	33	99%	88%	93%
Human	227	0	49	100%	82%	90%

Table 4. Category-level agreement between GPT and human annotators

GPT-detected sources	Agreement (same category)	Precision	Recall	F1-score
245	210	85%	86%	86%

(210/243), and an F1 score of 86%. These results indicate that while source identification is highly reliable, category assignment remains comparatively more challenging. In addition to source identification and categorization, GPT was asked to count the number of sentences that appear to have no external source attribution. The sentences with no attribution are apparently from the author of the article, giving context or opinion. Table 5 shows results for two articles. This quantification provides a structured way to evaluate balance in reporting by showing how much of the narrative is controlled by the journalist and how much is shaped by quoted or referenced sources. For example, in Story #4, the journalist's voice accounted for 25 out of 40 sentences. Authorities (AUT = 8) were cited more frequently than community or expert perspectives. Such analysis enables comparisons across categories, highlighting whether institutional voices (e.g., POL or AUT) dominate coverage at the expense of community perspectives (e.g., EXP or UNA).

Table 5. Sentence distribution by journalist and sources in a news story.

Story ID	Total	Journalist	Sources (by category)
Story #4	40	25	AUT = 8, ORG = 3, EXP = 2, UNA = 2
Story #2	25	14	POL = 5, AUT = 3, EXP = 2, UNA = 1

4 Error Pattern

Table 6 presents representative examples of misclassification patterns observed in GPT outputs. A closer inspection reveals several recurring issues. First, professional roles were inconsistently categorized. Lawyers, for instance, were labeled as either experts or authorities depending on context. Second, ambiguous keywords created uncertainty; for example, distinguishing between 'Minister' (politician) and 'Ministry' (authority), or between 'professor' (expert) and 'spokesperson at a university' (organization). Third, sentences containing multiple attribution signals often confused the model, such as cases where a report (authority document) and an individual (expert) were mentioned together. Fourth, organizational and government boundaries were blurred, particularly in cases involving police foundations or hospital boards. Finally, Indigenous leadership titles such as 'Chief' or references to the Assembly of First Nations were sometimes misclassified as organizational or authority roles, despite being political.

Table 6. Common error patterns in GPT predictions with illustrative examples.

Sentence	Error Type	GPT Predicted Category	Actual Source Category
I truly, with all my heart, believe that we need this,î said Doreen Healy, ***a member of the Blood Tribe***	Ambiguity (Titles/keywords)	**UNA**—"Predicted UNA due to personal tone and absence of explicit role keyword	AUT
"***Defense lawyer*** Maria Lopez argued ..."	Ambiguity (role/title)	**AUT**—GPT labelled her as authority due to legal role, but in context she is a professional expert voice	EXP
"The ***Ministry of Health*** announced ..."	Ambiguity (Titles/keywords)	**POL**—GPT conflated "Ministry" (institution) with "Minister" (politician)	AUT
"According to the ***city report***, ***Prof. Smith*** said ..."	Mixed signals in one line (doc + person)	**EXP**— GPT extracted only the person and ignored the document source, failing to capture multiple sources	AUT
"***Toronto Police Foundation*** announced ..."	Organization vs. Government Boundaries	**AUT**—GPT mistook a non-governmental foundation for a government authority due to "Police" keyword	ORG

In summary, GPT's errors stem largely from role ambiguity, keyword overlap, and blurred boundaries between organizational and governmental entities. These misclassifications appeared not only in the examples shown in Table 6 but also earlier in Table 2, where a former minister was misclassified as a politician despite no longer holding office. Together, these cases demonstrate that accurate categorization requires attention to contextual cues rather than reliance on titles or keywords alone.

5 Conclusion

This study demonstrates that GPT can reliably identify news sources in digital journalism, including human, non-human, and coreferential sources, achieving strong precision and recall at the source level. However, source categorization is less consistent, with errors often arising from ambiguity, overprediction, or low-frequency categories. By distinguishing journalist-authored context from attributed sources, the framework offers a structured basis for analyzing representation and potential bias in reporting. Future work will explore hybrid approaches that combine GPT with rule-based or human-in-the-loop methods to improve classification accuracy and enhance interpretability.

References

1. Malik, A., Adamson, G.: The Future of Local News: Research and Reflections — Towards an Algorithmic Journalism Assessment Tool: Accounting for Source Diversity in Local Digital News (2018). https://futureoflocalnews.org. Accessed 12 September 2025
2. Sultana, T., Harley, E.R., Adamson, G., Malik, A.: Extracting source information from news articles: information extraction. In: Proceedings of the 6th International Conference on Natural Language Processing and Information Retrieval (NLPIR 2022), pp. 216–221. ACM, New York (2023). https://doi.org/10.1145/3582768.3582774
3. Bennett, W.L.: Toward a theory of press-state relations in the United States. J. Commun. **40**(2), 103–125 (1990)
4. Poindexter, P., Smith, L., Heider, D.: Race and ethnicity in local television news: framing, story assignments, and source selections. J. Broadcast. Elec. Media **47**(4), 25 pages (2003). https://doi.org/10.1207/s15506878jobem4704_3
5. Matthes, J.: What's in a frame? a content analysis of media framing studies in the world's leading communication journals, 1990–2005. Journal. Mass Commun. Quar. **86**(2), 18 (2009). https://doi.org/10.1177/107769900908600206
6. Spangher, A., Peng, N., Ferrara, E., May, J.: Identifying informational sources in news articles. In: Proceedings of the 2023 Conference on Empirical Methods in Natural Language Processing (EMNLP), pp. 3626–3639. ACL, Singapore (2023). https://doi.org/10.18653/v1/2023. emnlp-main.221
7. Vincent, S., Wang, P., Shi, Z., Koka, S., Fang, Y.: Measuring Large Language Models' Capacity to Annotate Journalistic Sourcing. arXiv preprints (2024). arXiv:2501.00164
8. Hossain, M., Haque, A., Yilmaz, D.: Evaluation of reliability criteria for news publishers with large language models. In: Proceedings of the 17th ACM Web Science Conference (WebSci 2025). ACM, Stuttgart, Germany (2025). https://doi.org/10.1145/3717867.3717924
9. Gadiraju, U., Gursul, P., G´omez, A.J.P., et al.: Media bias detector: designing and implementing a tool for real-time selection and framing bias analysis in news coverage. In: Proceedings of the ACM Conference on Fairness, Accountability, and Transparency (FAccT). ACM, Rio de Janeiro, Brazil (2025). https://doi.org/10.1145/3706598.3713716
10. Li, J., Zhang, T., Gao, M.: Extracting Structured Insights from Financial News: An Augmented LLM Driven Approach. arXiv preprint (2024). arXiv:2407.15788
11. Ma, X., Xu, Y., Chen, H.: Multi-Layer Ranking with Large Language Models for News Source Recommendation. arXiv preprint (2023). arXiv:2310.06777
12. Martins, A., Campos, R., Jorge, A.: Probing GPT Models on Narrative Entity Extraction: A Case Study on Portuguese News. arXiv preprint (2023). arXiv:2311.14583

13. Han, Y., Li, C., Lin, Y.: A framework for analyzing and summarizing news and articles using large language models. In: Proceedings of the IEEE International Conference on Big Data (BigData), pp. 5238–5247. IEEE, Sorrento, Italy (2023). https://doi.org/10.1109/BigData59044.2023.10386445
14. Kim, S., Lee, J.: AI-generated news articles based on large language models: opportunities and challenges. In: Proceedings of the ACM Conference on Computer Supported Cooperative Work (CSCW). ACM, Minneapolis, MN, USA (2023). https://doi.org/10.1145/3579605.3615832

Fake News Detection in Tagalog: Benchmarking and Ensemble of Compact Transformer Models

Jomar T. Cuñado(✉), January Venice V. Toledo, and Christine D. Bandalan

Department of Computer, Information Sciences and Mathematics, University of San Carlos, Talamban Campus, 6000 Cebu City, Philippines
21101932@usc.edu.ph, 23102476@usc.edu.ph

Abstract. The rise of fake news threatens public discourse, especially in regions with low-resource languages like Filipino (Tagalog). This study benchmarks five compact transformer models: TinyBERT, DistilBERT, MobileBERT, MiniLMv2, and ELECTRA-small, using the Fake News Filipino dataset. DistilBERT achieved the best individual performance (0.969 F1), while a weighted soft voting ensemble with MobileBERT offered a negligible 0.000005 gain. Efficiency benchmarking indicated MobileBERT was the fastest (47.95 ms latency) and most resource-efficient (94.38 MB model size). In contrast, DistilBERT and the ensemble were more computationally demanding for only marginal improvements. These results suggest that while ensembles can offer slight gains, a single compact model, particularly MobileBERT, is preferable in resource-constrained environments, reinforcing the practicality of lightweight transformers for mobile and rural deployment.

Keywords: fake news detection · compact transformers · low-resource language · Filipino language · benchmarking

1 Introduction

The rampant spread of fake news in the Philippines presents a serious issue in public awareness, democratic stability, and informed decision-making. With the growing dependence on smartphones and the decline of traditional news media [19], the risk of encountering false information grows, especially with a lack of effective verification mechanisms. While fake news detection is widely studied in English, a significant gap remains for low-resource languages like Filipino (Tagalog).

This lack of local-language resources is a major concern, as most Filipino news consumers prefer content in their native tongue. Tagalog datasets are scarce, hindering model development. One notable effort to address this was by Cruz et al. [9], who introduced the first curated dataset for fake news detection in Filipino, titled Fake News Filipino (FNF). However, their focus was on demonstrating

H. Unger and P. Meesad (Eds.): NLPIR 2025, LNNS 1904, pp. 403–415, 2026.
https://doi.org/10.1007/978-3-032-20897-2_30

feasibility. Exploration into both the performance and efficiency of compact language models for this specific task remains limited, as existing studies often rely on large, computationally-intensive models impractical for real-world deployment on mobile phones.

To address this gap, this study benchmarks compact transformer models that offer strong performance while significantly reducing model size and inference time. Our contribution is twofold. First, we benchmark and fine-tune five selected compact transformer models (multilingual DistilBERT, MiniLMv2, MobileBERT, TinyBERT, and ELECTRA-small) on the FNF dataset. We evaluate each model using metrics such as accuracy, precision, recall, F1-score, and model efficiency (model size and inference time). Second, we develop and evaluate a lightweight soft-voting ensemble using the two top-performing models to create an effective and efficient fake news detection mechanism tailored for the Tagalog language.

The emphasis on efficiency is highly relevant to the Philippines' mobile-first news consumption, and this research provides a practical tool for the public and for journalists. It also contributes to the NLP body of knowledge for low-resource languages, providing benchmarks and methodologies for future development in efficient mobile deployment.

2 Related Work

2.1 Evolution of Fake News Detection Models

The application of machine learning for automated fake news detection has evolved significantly. Early approaches primarily relied on content-based classification, such as n-gram analysis, deep syntax parsing, and sentiment analysis. However, recent advancements introduced deep learning models including hybrid convolutional neural networks (CNNs), recurrent neural networks (RNNs), attention-based mechanisms, and ensemble techniques [14]. In their exploration of deep learning methods, Nasser et al. [17] found attention mechanisms to be computationally expensive, RNNs to be prone to issues during training, and CNNs to be computationally intensive. This trend continues to grow [5], but diversity in domains and languages still makes generalization a challenge.

Since its proposal, the transformer model has revolutionized fields like Natural Language Processing (NLP). A vanilla transformer contains an encoder and a decoder, where the former only has a multi-head self-attention module and a position-wise feed-forward network (FFN) and the latter additionally inserts cross-attention modules [15]. Owing to their ability to model long-range dependencies, transformers have been applied to fake news detection, with a surge of studies during the COVID-19 pandemic [1,6]. However, the quadratic time and memory complexity of transformers has raised concerns about scalability, propelling the development of "efficient transformers" [27].

2.2 Compact and Multilingual Models for Low-Resource Languages

The use of multilingual transformer models (e.g., mBERT, XLM-RoBERTa) became a practical solution for NLP with limited annotated data, performing well in fake news classification [18]. They are valuable for low-resource languages like Tagalog, where building large monolingual corpora is often not feasible [24].

This lack of data also makes large transformers prone to overfitting, leading to the development of compact transformers (e.g., ViT-Lite, CCT) designed to avoid this and even outperform CNNs on smaller datasets [12]. DistilBERT, a distilled version of BERT, has been used to detect fake news, but its accuracy can still be lower than the full BERT, proving challenges remain for small datasets [3].

Fake news detection in low-resource languages presents unique challenges. Scarcity of data hinders model development, as transformers typically require large-scale corpora. Low-resource datasets may also contain insufficient data or too much noise, leading to the emergence of summarization techniques [2]. Studies on other low-resource languages like Bangla [23], Amharic [11], and Kurdish [4] have all emphasized this difficulty, often requiring the creation of new datasets to make progress.

2.3 Ensemble Methods in Text Classification

While individual models provide unique advantages, ensemble learning combines models to leverage their respective strengths for enhanced robustness and accuracy [29].

To produce a final prediction, various strategies may be employed. Weighted Voting, for example, assigns different weights to models based on performance metrics [20]. When aggregating only two models, simple majority voting is impractical. In such cases, soft voting or weighted averaging is preferred to appropriately address the challenges of low-resource environments.

2.4 Fake News Detection in the Philippines

To understand the prevalence of fake news in the Philippines, one must assess news consumption. While broadsheets use English, tabloids or "newspapers for the masses" are frequently in Filipino [16]. Recent technological shifts have moved consumption from traditional media to online spheres, lowering barriers for information dissemination and highlighting the need for technological interventions [22,25].

Efforts to develop detection systems for Filipino were historically limited. A pioneering work by Cruz et al. [9] developed the first benchmark dataset, "Fake News Filipino" (FNF), and used transfer learning to benchmark classifiers. This work served as a foundation for researchers like Cadeliña et al. [7], who compared traditional ML ensembles to a GPT 3.5-based model. By contrast, Espiritu et al. [10] explored transformers, finding that DistilBERT trades the slightly better performance of BERT for faster inference and smaller resource requirements.

While these breakthroughs contributed significantly, a notable gap remains in the exploration and integration of compact transformer models for Filipino fake news detection. Prior works either explored traditional machine learning ensembles or compact transformers independently, but not in combination, which motivates the work in this study.

3 Methodology

3.1 Dataset

The primary source of data was the Fake News Filipino (FNF) dataset [9], a corpus of 3,206 Tagalog news articles. It is perfectly balanced with 1,603 real and 1,603 fake articles. Fake news was sourced from sites identified by Verafiles and the National Union of Journalists in the Philippines (NUJP), while real news was drawn from established sites like Pilipino Star Ngayon. This rigorous sourcing provides a high-quality ground truth. The dataset's use enhances reproducibility and allows us to address the gap noted by [9]: the need to explore compact models suitable for resource-constrained devices (Fig. 1).

May bagong paandar ang mga anti-DU30, at ito ay pag-po-post ng mga black ang white videos ng mga peronalidad sa mundo ng showbiz at sining kung saan pinahahapyawan nila ang Duterte Administration... Ayon kasi kay Lamangan, walang respeto daw ang kasalukuyang gobyerno at iboto daw ang mga tagaoposisyon para mapatunayan ang pagiging tunay na Pilipino.	Inaresto ng anti-scalawag and intelligence units ng Philippine National Police (PNP) ang isang anti-drug operative ng Pasay City Police Station ngayong araw, matapos na isangkot sa extortion at kidnapping. Dinampot si Police Corporal Anwar Nasser, nakatalaga sa Station Drug Enforcement Team (SDET) ng Pasay police... Samantala, tatlo sa kanyang mga kasabwat ang nakatakas.
Label: **Fake**	**Label:** **Real**

Fig. 1. Example of a fake article (left) and a real article (right) from the FNF dataset

3.2 Preprocessing and Data Splitting

The dataset was partitioned using stratified sampling into training (70%), validation (15%), and testing (15%) subsets, preserving the 50/50 class balance. The validation set was used for hyperparameter tuning and ensemble weight computation; the test set remained unseen until final evaluation.

Minimal text normalization was applied. We retained text casing, special characters, and punctuation, as they may hold contextual cues for fake news detection. No stopword removal or lemmatization was performed to maintain the integrity of semantic structures. Each model employed its own tokenizer via HuggingFace's `AutoTokenizer`, with sequences padded and truncated to a maximum length of 512 tokens.

3.3 Models

To find an optimal balance between performance and efficiency, we benchmark five compact transformer models.

Compact Models. We selected five models known for their balance of performance and efficiency. **DistilBERT** is a compressed version of BERT trained using knowledge distillation [21]. **TinyBERT** is an even smaller variant optimized through a two-stage distillation process [13]. **ELECTRA-small** uses a "replaced token detection" objective, allowing its small variant to achieve performance comparable to BERT-base [8]. **MobileBERT** was designed specifically for mobile devices, using bottleneck structures for edge deployment [26]. Finally, **MiniLMv2** is a lightweight model that employs self-attention distillation for a strong balance of accuracy and efficiency [28].

3.4 Training and Experimental Setup

The study followed a fine-tuning approach rooted in transfer learning. Each model was initialized with pretrained weights and fine-tuned for binary classification. Models were trained with a **batch size of 32** and the **Adam optimizer** at a **learning rate of 2e-5**. To prevent overfitting, we implemented **early stopping** with a patience of 2, monitoring the validation loss. The software and libraries used for this process are detailed in Table 1.

Table 1. Software development tools used in this study

Tool/Library	Purpose/Function
Python	Primary programming language
HuggingFace Transformers	Model loading, fine-tuning, tokenization
Datasets	Data handling and preprocessing
scikit-learn	Evaluation metrics (e.g., F1-score)
NumPy	Numerical operations
pandas	Data preprocessing and manipulation
Google Colab Pro	Development environment (GPU)
Matplotlib & Seaborn	Visualization of results
GitHub	Version control

3.5 Ensemble Method

The top two performing models, based on their F1-scores, were selected for a weighted soft-voting ensemble. A probability-weighted averaging strategy was employed, as majority voting is not applicable for a two-model ensemble. Each model's softmax output probability (p_i) is combined using weights (w_i) proportional to its F1-score, as shown in Eq. 1. The final prediction probability (y) for each class is computed as shown in Eq. 2.

$$w_1 = \frac{F1_1}{F1_1 + F1_2}, \quad w_2 = \frac{F1_2}{F1_1 + F1_2} \tag{1}$$

$$y = w_1 p_1 + w_2 p_2 \tag{2}$$

This design ensures that models contribute to the final decision based on their reliability.

3.6 Evaluation Metrics

Classification Performance. We use standard classification metrics derived from the confusion matrix: True Positives (TP), True Negatives (TN), False Positives (FP), and False Negatives (FN). The primary metrics are defined in Eq. 3.

$$\begin{aligned} \text{Precision} &= \frac{TP}{TP + FP}, \quad \text{Recall} = \frac{TP}{TP + FN} \\ \text{Accuracy} &= \frac{TP + TN}{TP + TN + FP + FN} \\ \text{F1-Score} &= 2 \cdot \frac{\text{Precision} \cdot \text{Recall}}{\text{Precision} + \text{Recall}} \end{aligned} \tag{3}$$

During fine-tuning, performance was monitored using the binary cross-entropy loss, as defined in Eq. 4.

$$\text{Loss} = -\frac{1}{N} \sum_{i=1}^{N} [y_i \log(p_i) + (1 - y_i) \log(1 - p_i)] \tag{4}$$

where N is the number of samples, y_i is the true label, and p_i is the predicted probability.

Model Efficiency. To assess suitability for low-resource environments, we also evaluated model efficiency. We measure **Model Size** (MB on disk) and two inference metrics: **Inference Latency** (average time in ms for a single prediction) and **Peak RAM Usage** (MB) in a controlled testing environment.

3.7 Scope of the Study

This study is limited to fake news detection in Tagalog using a supervised learning approach. The scope did not cover cross-lingual detection, domain adaptation, or training a model from scratch. Furthermore, we did not explore advanced ensemble techniques like stacking or boosting. The evaluation was confined to classification performance and model efficiency on a fixed test set, and did not include real-time deployment, user interaction, or the development of new datasets. This scope was intentionally constrained to ensure compatibility with low-resource environments.

4 Results and Discussion

This section presents a detailed analysis of the performance and efficiency of the fine-tuned compact transformer models and the subsequent weighted ensemble. The findings presented herein are derived from the application of the experimental setup and evaluation metrics detailed in the preceding Methodology section. The section is structured to first provide a comprehensive overview of the individual model benchmarking results, which guided the selection of the top two performers. Subsequently, the performance of the final ensemble is evaluated and critically compared against its constituent models, followed by a discussion of the trade-offs observed in the efficiency evaluation.

4.1 Model Benchmarking and Selection

The five compact models were fine-tuned and evaluated on the 15% validation set. Individual models were evaluated on the validation set. The ensemble was evaluated on the test set. The performance of all models on the primary classification metrics is summarized in Table 2.

Table 2. Benchmarking results overview

Model	Accuracy	Precision	Recall	F1-Score	ROC AUC
DistilBERT	96.88%	96.92%	96.88%	96.88%	99.05%
MobileBERT	95.43%	95.58%	95.43%	95.42%	98.46%
ELECTRA	92.93%	92.98%	92.93%	92.93%	98.09%
MiniLMv2	90.85%	90.86%	90.85%	90.85%	97.46%
TinyBERT	86.90%	87.44%	86.90%	86.85%	95.70%
Ensemble	**96.88%**	**96.88%**	**96.88%**	**96.88%**	**99.08%**

As shown in Table 2, **DistilBERT** achieved the highest F1-Score (96.88%) on the validation set, establishing it as the top-performing individual model. **MobileBERT** was the clear second-best performer with a 95.42% F1-Score. The other models showed a notable drop in performance, with a 10-point gap between DistilBERT (96.88%) and the lowest-performing model, TinyBERT (86.85%).

To further evaluate classification capability, we analyzed the ROC-AUC. DistilBERT again achieved the highest score (99.05%), suggesting excellent discriminative power, followed by MobileBERT (98.46%). As illustrated in Figs 2, the ROC curves for these top two models display a steeper rise towards the top-left corner, indicating stronger true positive rates at lower false positive rates.

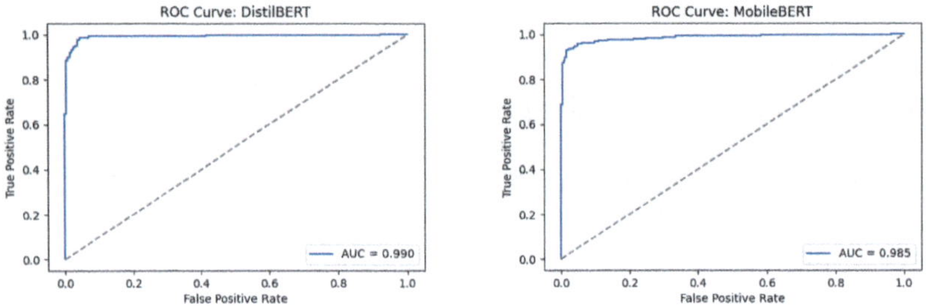

Fig. 2. ROC curve for DistilBERT (left) and MobileBERT (right) on the validation set

Analysis of the confusion matrices (Fig. 3) supports this finding. DistilBERT (left) displays darker diagonal cells, signifying a strong concentration of correctly classified samples. MobileBERT (right) closely mirrors this distribution, though its diagonal cells appear slightly less intense, implying a marginally higher rate of misclassification. Based on these consistent results, DistilBERT and MobileBERT were selected as the basis for the ensemble.

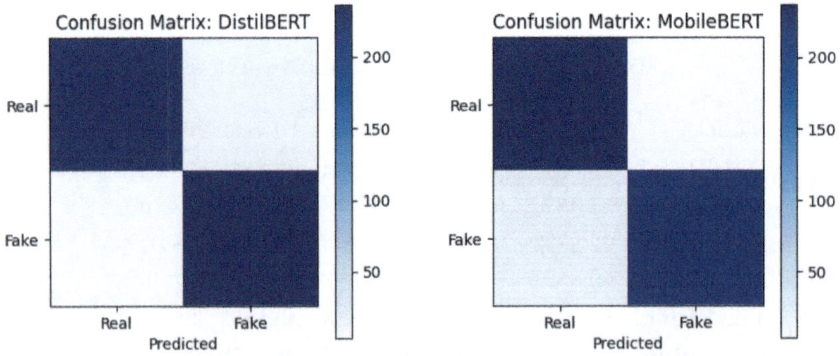

Fig. 3. Confusion matrix for DistilBERT (left) and MobileBERT (right) on the validation set

4.2 Ensemble Model Performance

The weighted soft-voting ensemble was constructed from DistilBERT and MobileBERT and then evaluated on the held-out 15% test set.

As detailed in Table 2 (bottom row), the ensemble achieved an F1-Score of 96.88%, an accuracy of 96.88%, and a ROC-AUC of 99.08%. Notably, while the rounded F1-Score appears identical to DistilBERT's, the ensemble's true, unrounded score was **0.0000057% higher** than the individual model. Similarly, its ROC-AUC score was 0.03% higher (99.08% vs. 99.05%), suggesting a modest gain in discriminative ability (Fig. 4). These marginal gains in F1 and AUC were accompanied by a minor 0.04% drop in precision compared to DistilBERT.

The ensemble's confusion matrix (Fig. 4, right) maintains the strong diagonal pattern of its component models, confirming a high rate of correct classifications and minimal misclassifications on the test set.

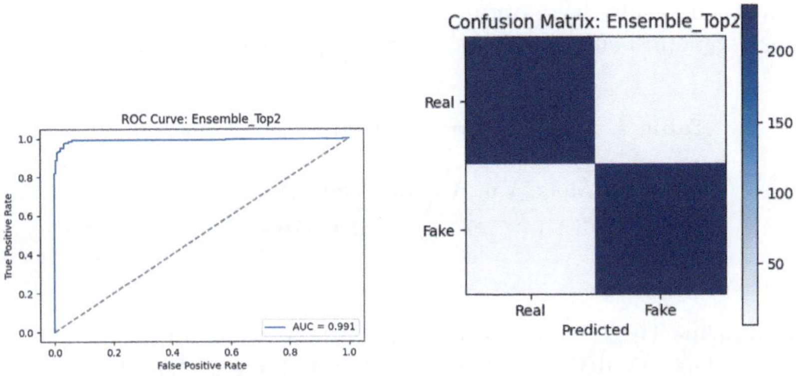

Fig. 4. Ensemble ROC curve (left) and confusion matrix (right) on the test set

4.3 Efficiency Evaluation

A key goal of this study was to assess the practicality of these models for resource-constrained environments. Table 3 compares the efficiency metrics of the top two models and the final ensemble.

Table 3. Efficiency evaluation of top models and ensemble

Model	Inference Latency (ms)	Peak RAM (MB)	Model Size (MB)
DistilBERT	66.26	1444.61	516.34
MobileBERT	47.95	1032.61	94.38
Ensemble	125.97	1508.61	610.71

The ensemble's inference latency (125.97 ms) was significantly higher than the individual models—nearly double that of DistilBERT and 2.6 times slower than MobileBERT (47.95 ms), the fastest model. This increase was expected, as the ensemble must process predictions from both transformers. Similarly, the ensemble's peak RAM usage (1508.61 MB) and model size (610.71 MB) reflect the combined computational cost of loading both models.

These findings suggest that while the ensemble provides marginal, near-zero gains in classification performance, it introduces a substantial additional cost in memory, latency, and size. This trade-off is a critical consideration for real-world deployment.

4.4 Discussion and Baseline Comparison

To contextualize our findings, we compare our results to the original BERT-base implementation by Cruz et al. [9], which used the same dataset (Table 4). We note that their results were obtained under their own experimental setup; this serves as a general benchmark, not a direct equivalence.

Table 4. BERT performance metrics from Cruz et al. [9]

Model	Val. Accuracy	Loss	Val. Loss
BERT	87.47%	0.4655	0.4419

The baseline BERT model achieved a validation accuracy of 87.47%. Our compact models, DistilBERT (96.88% Acc) and MobileBERT (95.43% Acc), and our final ensemble (96.88% Acc) all substantially outperformed this baseline. This finding is significant: it demonstrates that modern, compact models can rival and even exceed the performance of larger, language-specific transformers like BERT-base on this task. Given the high performance and superior efficiency of MobileBERT (fastest, smallest, and greater than 95% F1), our results suggest that a single, well-chosen compact model may be the most practical solution for this low-resource deployment scenario.

5 Conclusion and Future Work

This study benchmarked five compact transformer models and developed a weighted ensemble of the top two performers, DistilBERT and MobileBERT, to balance classification accuracy and computational efficiency for Tagalog fake news detection.

The findings show that while the ensemble (96.88% F1-Score) achieved a marginal performance gain of 0.0000057% over the best single model (DistilBERT), this slight advantage comes at a significant cost. The ensemble nearly doubles the inference latency, consumes more memory, and has a larger model size.

We conclude that for this task, the performance-efficiency trade-off does not favor the ensemble. The results prove that compact transformers are a powerful tool for low-resource languages, but they also suggest that a single, well-chosen compact model (like MobileBERT, which was the fastest and lightest while achieving greater than 95% F1) is a more practical and efficient solution for deployment in resource-constrained mobile environments.

For future work, we recommend benchmarking newer compact models, such as ALBERT or TinyRoBERTa, as well as more recent monolingual baselines like Tagalog RoBERTa. Furthermore, creating a larger, more contemporary fake news dataset would be a valuable contribution, as the FNF dataset is limited in size and recency. This research could also be extended to other low-resource Philippine languages to foster a more inclusive detection ecosystem. Finally, we encourage a move from benchmarking to the practical deployment of these compact models as an accessible tool for journalists and the public to combat misinformation.

References

1. Alghamdi, J., Lin, Y., Luo, S.: Towards COVID-19 fake news detection using transformer-based models. Knowl.-Based syst. **274**, 110,642 (2023). https://doi.org/10.1016/j.knosys.2023.110642
2. Alghamdi, J., Lin, Y., Luo, S.: Fake news detection in low-resource languages: a novel hybrid summarization approach. Knowl.-Based Syst. **296**, 111,884 (2024). https://doi.org/10.1016/j.knosys.2024.111884. https://www.sciencedirect.com/science/article/pii/S0950705124005185
3. Amandeep, S.S.: Transforming fake news detection: leveraging distilbert models for enhanced accuracy. Procedia Comput. Sci. **260**, 283–290 (2025). https://doi.org/10.1016/j.procs.2025.03.203. https://www.sciencedirect.com/science/article/pii/S1877050925009470. Seventh International Conference on Recent Trends in Image Processing and Pattern Recognition (RTIP2R-2024)
4. Azad, R., Mohammed, B., Mahmud, R., Zrara, L., Sdiq, S.: Fake news detection in low-resourced languages "kurdish language" using machine learning algorithms. Turkish J. Comput. Math. Educ. (TURCOMAT) **12**(14), 2677–2683 (2021). https://doi.org/10.17762/turcomat.v12i14.10732
5. Bashaddadh, O., Omar, N., Mohd, M., Khalid, M.N.A.: Machine learning and deep learning approaches for fake news detection: a systematic review of techniques, challenges, and advancements. IEEE Access **PP**, 1–1 (2025). https://doi.org/10.1109/ACCESS.2025.3572051
6. Bazmi, P., Asadpour, M., Shakery, A., Maazallahi, A.: Entity-centric multi-domain transformer for improving generalization in fake news detection. Inf. Process. Manage. **61**(5), 103,807 (2024). https://doi.org/10.1016/j.ipm.2024.103807. https://www.sciencedirect.com/science/article/pii/S0306457324001663
7. Cadeliña, P., Go, K.G., Pepito, G.: Pinoycchio: unleashing the power of openAI and ensemble models for the detection of government related filipino fake news. In: 2024 IEEE 2nd International Conference on Control, Electronics and Computer Technology (ICCECT), pp. 190–193 (2024). https://doi.org/10.1109/ICCECT60629.2024.10545759

8. Clark, K., Luong, M.T., Le, Q.V., Manning, C.D.: Electra: pre-training text encoders as discriminators rather than generators. arXiv preprint arXiv:2003.10555 (2020)
9. Cruz, J.C.B., Tan, J.A., Cheng, C.: Localization of fake news detection via multitask transfer learning. In: Calzolari, N., et al. (eds.) Proceedings of the Twelfth Language Resources and Evaluation Conference, Marseille, France, pp. 2596–2604. European Language Resources Association (2020). https://aclanthology.org/2020.lrec-1.316/
10. Espiritu, J., Policarpio, R., Sabado, A., Sabado, V., Sueno, H.: Classifying fake news in filipino language using improved distilbert model. In: 2024 11th International Conference on Information Technology, Agriculture, Environment and Education (ICITACEE), pp. 320–325 (2024). https://doi.org/10.1109/ICITACEE62763.2024.10762768
11. Gereme, F., Zhu, W., Ayall, T., Alemu, D.: Combating fake news in "low-resource" languages: amharic fake news detection accompanied by resource crafting. Information **12**, 9 (2021). https://doi.org/10.3390/info12010020
12. Hassani, A., Walton, S., Shah, N., Abuduweili, A., Li, J., Shi, H.: Escaping the big data paradigm with compact transformers (2021). https://doi.org/10.48550/arXiv.2104.05704
13. Jiao, X., et al.: Tinybert: distilling Bert for natural language understanding. In: Findings of the Association for Computational Linguistics: EMNLP 2020, pp. 4163–4174 (2020). https://doi.org/10.18653/v1/2020.findings-emnlp.372
14. Khan, J.Y., Khondaker, M.T.I., Afroz, S., Uddin, G., Iqbal, A.: A benchmark study of machine learning models for online fake news detection. Mach. Learn. Appl. **4**, 100,032 (2021). https://doi.org/10.1016/j.mlwa.2021.100032
15. Lin, T., Wang, Y., Liu, X., Qiu, X.: A survey of transformers. AI Open **3** (2022). https://doi.org/10.1016/j.aiopen.2022.10.001
16. Luza, J.M.: Filipino in periodisMO: analyzing the common practices of tabloid writers in news writing using filipino language. Int. J. Multidisciplinary Res. Anal. **06** (2023). https://doi.org/10.47191/ijmra/v6-i6-73
17. Nasser, M., et al.: A systematic review of multimodal fake news detection on social media using deep learning models. Results Eng. **26** (2025). https://doi.org/10.1016/j.rineng.2025.104752
18. Ragab, M.I., Mohamed, E.H., Medhat, W.: Multilingual propaganda detection: exploring transformer-based models mBERT, XLM-RoBERTa, and mT5. In: Jarrar, M., Habash, H., El-Haj, M. (eds.) Proceedings of the first International Workshop on Nakba Narratives as Language Resources, Abu Dhabi, pp. 75–82. Association for Computational Linguistics (2025). https://aclanthology.org/2025.nakbanlp-1.9/
19. Reuters Institute for the Study of Journalism: Digital news report 2021 (2021). https://reutersinstitute.politics.ox.ac.uk/digital-news-report/2021
20. Sachan, R.K., Tiwari, V., Raghav, N., Rajpoot, P.: Spam text messages detection using multilingual embedding. Procedia Comput. Sci. **260**, 390–398 (2025). https://doi.org/10.1016/j.procs.2025.03.215. https://www.sciencedirect.com/science/article/pii/S1877050925009597. Seventh International Conference on Recent Trends in Image Processing and Pattern Recognition (RTIP2R-2024)
21. Sanh, V., Debut, L., Chaumond, J., Wolf, T.: Distilbert, a distilled version of Bert: smaller, faster, cheaper and lighter. arXiv preprint arXiv:1910.01108 (2019)
22. Schapals, A.K., Pentzold, C. (eds.): Media Compass: A Companion to International Media Landscapes. Wiley-Blackwell, Hoboken (2024b)

23. Shibu, H.M., Datta, S., Miah, M.S., Sami, N., Chowdhury, M.S., Islam, M.S.: From scarcity to capability: Empowering fake news detection in low-resource languages with LLMs. In: Weerasinghe, R., Anuradha, I., Sumanathilaka, D. (eds.) Proceedings of the First Workshop on Natural Language Processing for Indo-Aryan and Dravidian Languages, Abu Dhabi, pp. 100–107. Association for Computational Linguistics (2025). https://aclanthology.org/2025.indonlp-1.12/
24. Sindane, T., Marivate, V., Modupe, A.: Cross-lingual embedding methods and applications: a systematic review for low-resourced scenarios. Nat. Lang. Process. J. **12**, 100,157 (2025). https://doi.org/10.1016/j.nlp.2025.100157
25. Soriano, C., Gaw, F.: Broadcasting anti-media populism in the philippines: Youtube influencers, networked political brokerage, and implications for governance. Policy Internet **14**, 508–524 (2022). https://doi.org/10.1002/poi3.322
26. Sun, Z., Yu, H., Song, X., Liu, R., Yang, Y., Zhou, D.: Mobilebert: a compact task-agnostic Bert for resource-limited devices (2020). https://doi.org/10.48550/arXiv.2004.02984
27. Tay, Y., Dehghani, M., Bahri, D., Metzler, D.: Efficient transformers: a survey. ACM Comput. Surv. **55**(6) (2022). https://doi.org/10.1145/3530811
28. Wang, W., Bao, H., Huang, S., Dong, L., Wei, F.: Minilmv2: multi-head self-attention relation distillation for compressing pretrained transformers (2020). https://doi.org/10.48550/arXiv.2012.15828
29. Zhang, H., Shafiq, M.O.: Survey of transformers and towards ensemble learning using transformers for natural language processing. J. Big Data **11**(1), 25 (2024). https://doi.org/10.1186/s40537-023-00842-0

What Makes a Review Helpful? A Feature-Driven Study

Haocheng Zhang[1(✉)] and Kazuyuki Matsumoto[2]

[1] Graduate School of Sciences and Technology for Innovation, Tokushima University, Tokushima, Japan
c612545001@tokushima-u.ac.jp
[2] Graduate School of Technology, Industrial and Social Sciences, Tokushima University, Tokushima, Japan
matumoto@is.tokushima-u.ac.jp

Abstract. This study investigates the prediction of helpful reviews by incorporating multiple aspects of review- and product-level features. Our predictive model achieves an accuracy of 88% in identifying reviews with more than 10 helpful votes. Furthermore, we evaluate three different neural network architectures combined with hand-crafted features and various pretrained word embeddings, and find that more complex architectures do not yield performance gains for this task. Finally, we analyze the contribution of different features to model performance using Permutation Feature Importance (PFI).

Keywords: Online reviews · Review helpfulness · Text classification · Natural language processing

1 Introduction

In the context of online shopping platforms, customer reviews offer important information regarding product quality and exert a considerable influence on consumers' purchase decisions. Previous research shows that nearly 95% of shoppers read reviews before making a purchase, and that the display of reviews can increase the conversion rate by 270% [1]. In addition, From the customers' perspective, reading reviews from previous buyers helps them find relevant product information and quickly match their needs. This indicates that reviews play an essential role in improving users' shopping experience and reduce purchase risk.

However, it is challenging for customers to identify trustworthy and high-quality reviews regarding the product due to the vast quantity of reviews that are repetitive in content and provide limited information relevant to customers' interests. To address this issue, shopping platforms such as Amazon have implemented a review helpfulness voting system, enabling users to see evaluations of reviews provided by other customers. This mechanism can, to some extent, help user filter information they care about the product and eventually make appropriate purchase decisions.

H. Unger and P. Meesad (Eds.): NLPIR 2025, LNNS 1904, pp. 416–427, 2026.
https://doi.org/10.1007/978-3-032-20897-2_31

In this paper, we propose a machine learning approach to predict whether a review will receive helpful votes from other customers and to identify the key factors that determine its perceived helpfulness. Understanding the determinants that contributes to review helpfulness can provide empirical evidence to the field of text mining and offer deeper insights into user behavior during the decision-making process.

Given the practical value of identifying helpful reviews, it is important to address the unique characteristics of helpful vote data in real-world e-commerce platforms. The distribution of helpful votes is highly skewed, with only a small proportion of reviews for a given product receiving votes from other users, while the majority receive only one or no vote at all. Across all reviews from May 1996 to September 2023 on Amazon, only 24% of reviews received any helpful votes and only 3% of reviews received helpful votes greater than 5 [2]. In this situation, predicting helpful votes using regression method will become long-tailed regression task which often suffer from highly data imbalance across the target space where the majority of samples (reviews receiving 0 helpful votes) concentrate within a narrow space and extreme value are distribute sparsely. This problem can cause regression model to bias toward dense region and poorly fit extreme value (reviews with high helpful votes). Thus, by discretizing the continuous target variable into a set of categorical bins, the problem can be reformulated to a classification problem. In this paper, we adopt this approach to evaluate the impact of our proposed explanatory variables on predicting helpful votes.

2 Related Works

With rapid growth of e-commerce platforms and demand for online shopping, researches on online review helpfulness prediction have attracted increasing attention. Especially for Amazon, the world's largest online shopping site, considerable attention has been paid to understanding why some reviews receive a high number of votes. One reason is the reviews are large in scale and publicly available. The review data from Amazon also include a variety of metadata in addition to reviews and helpful votes number like product information, rating, verified purchase, review timestamp which enable the analysis from different perspectives of determinants of review helpfulness. The methods and explanatory variables used in previous research to predict and analyze helpful reviews can differ considerably [3–5].

Park [4] analyzed the characteristics embedded in product reviews across five different product types and evaluate their influences on review helpfulness. They adopted four data mining technique (Linear Regression(LR), Support Vector Regression(SVR), Random Forest(RF), M5P) to determine one that best predict review helpfulness. They proposed three types of hand-crafted features extracted from review content and metadata: linguistic variables (e.g., word count, words per sentence), psychological variables (e.g., Analytic, Clout, Authentic), and metadata variables (e.g., ratings), all associated with five product categories

(Beauty, Cellphone, Clothing, Grocery, and Video). Based on one-way ANOVA and regression models, their results show that reviews in different categories differ significantly in both psychological and linguistic variables.

In addition to hand-crafted linguistic variables, Yang et al. [6] incorporated two semantic features by employing Linguistic Inquiry and Word Count (LIWC) and General Inquirer to automatically extract linguistic information from review texts and used human scoring of helpfulness as ground truth. Haque et al. [7] proposed a feature extraction technique for both unigrams and bigrams using *tf-idf*, which assigns higher weights to less frequent terms in the document representation. Additionally, they analyzed the performance of different features types including lexical, structural, semantics, and readability by Decision Tree(DF) classifier. Mertz et al. [8] proposed a semantic feature extraction method, the dependency bigram, which leverages grammatical dependencies between words. Their findings indicated that the star rating serves as a strong predictive variable in the helpfulness prediction task while the dependency bigram itself does not significantly improve prediction performance. Malik et al. [5] evaluated a hybrid combination of features derived from reviews, reviewers, and products, achieving strong performance in the prediction task. They also found that the polarity of the review title, the polarity of the review text, and the cosine similarity between the review text and the product title significantly contribute to review helpfulness.

Some studies focus on non-textual factors, such as reviewer characteristics and the posting time of reviews and product related factors, instead of the review content. Nayeem and Rafiei [9] indicated two key factors often ignored in related research: (1) who post the reviews and (2) when the reviews are posted. Meanwhile, Kung et al. [3] identified three key factors influencing helpful votes: the user's average number of helpful votes, the number of images per review, and the review timestamp. In their experiments, they achieved a highest accuracy of 96.91% for the binary classification task (helpful vs. not helpful) and concluded that metadata plays a significantly more important role than review text. However, using user's average number of helpful votes may lead to data leakage.

Recently, researches has extensively utilized machine learning and deep learning technique to extract feature from review text and title. Quoc et al. [10] employed a topic modeling approach to extract latent information, represented as topic distributions, from review text. Specifically, they applied Latent Dirichlet Allocation (LDA), a generative probabilistic model, to automatically transform the review text into a distribution over topics, where each topic is defined as a distribution over words. In addition, Olmedilla et al. were the first to adopt one-dimensional convolutional neural networks (1D CNNs) to automatically learn n-gram features related to the dependent variable(helpful, neural, not helpful) from a large volume of review text. Using their model, they achieved an accuracy rate of 66% [11].

3 Methodology

3.1 Data Collection and Pre-processing

This research uses an Amazon review dataset collected by the McAuley Lab in 2023, which contains 571.54 million user reviews from 54.51 million users, along with item metadata for 48.19 million items across 32 product categories, spanning the period from May 1996 to September 2023 [2]. The raw item metadata include product information such as the main category, product name, average rating, number of ratings, bullet-point feature descriptions, detailed product description, price, images, videos, store name, categories, product details, product ID, and "bought together" information. In addition the user reviews data include ratings, review title, review text, images, ID of the product, Parent ID of the product, user id, timestamp, verified purchase and helpful votes.

Then, to evaluate feature importance and mitigate the impact of the long-tailed distribution of review helpful votes, we randomly sampled from the entire dataset and selected an appropriate classification threshold to convert the regression task into a binary classification problem. In addition, we replace missing values in the review text and review title and main category with empty strings, and impute missing rating number(the number of reviews for this product) with the average rating number from the remaining reviews in the data cleaning process.

3.2 Feature Extraction and Selection

Product Metadata. Product features refer to the information directly provided by Amazon item metadata. Previous research has shown that product categories and star ratings—particularly the extremity of ratings—serve as strong predictors of review helpfulness [3,4,8]. Based on these findings, we use product categories, average value of rating and the number of ratings shown in the product page as predictive variables in the subsequent model development process. We summary these features in following:

- Product categories
- Average value of rating
- The number of ratings

Product Metadata. Similar to product metadata, review metadata refers to information that is specific to an individual review. While reviews on the same product page share identical product metadata, they differ in their review metadata. Accordingly, we extracted several groups of features from the review metadata, as listed below:

- Product rating
- Verified purchase (True or False)

- Number of images uploaded by users after receiving the product
- Number of videos uploaded by users after receiving the product

Specifically, we obtain the number of videos by identifying occurrences of the string pattern [VIDEOID:) within the review text.

Figure 1 shows the comparison of rating score distributions between helpful and unhelpful reviews, where a helpfulness threshold of 10 was applied (i.e., reviews with more than 10 helpful votes were treated as helpful). The figure is based on a random sample of 80,000 reviews from the entire dataset. It can be observed that Amazon users tend to assign 5-star ratings, and that reviews with extreme ratings are more likely to receive helpful votes.

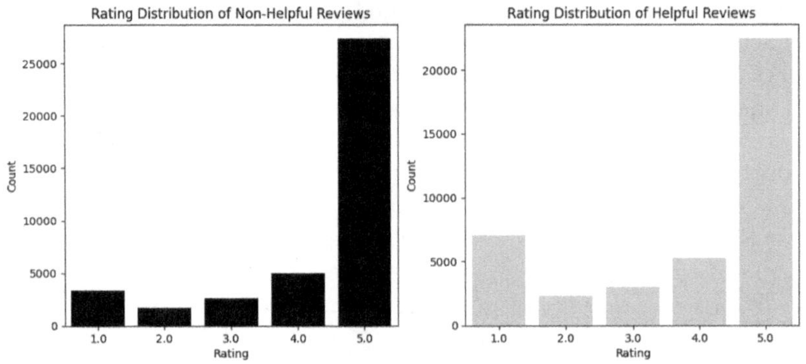

Fig. 1. Distribution of Ratings: Helpful vs. Unhelpful Reviews

Figure 2 presents the distribution of video counts between helpful and unhelpful reviews on a log scale using the same sample as above. According to the figure, reviews containing videos are more likely to receive helpful votes.

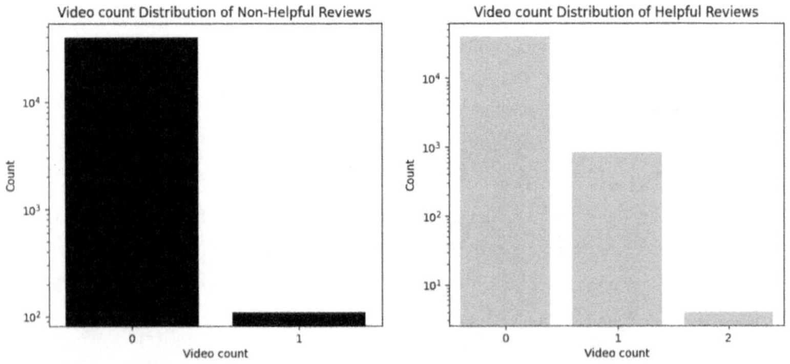

Fig. 2. Distribution of Videos Count: Helpful vs. Unhelpful Reviews

Figure 3 illustrates the distribution of the number of images posted by buyers in a log scale. A clear difference can be observed between the two groups. Similar to video counts, reviews containing more images are more likely to receive helpful votes.

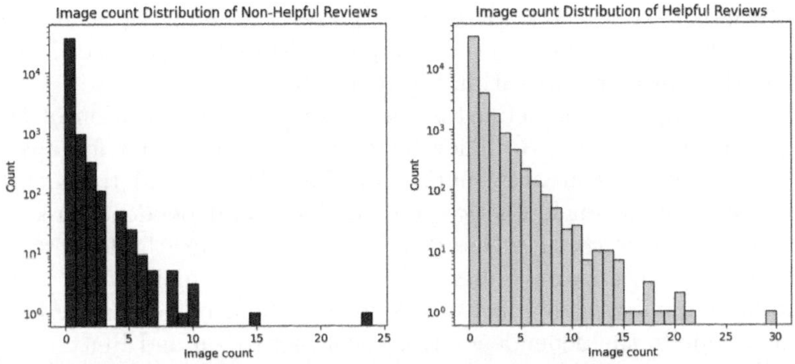

Fig. 3. Distribution of Images Count: Helpful vs. Unhelpful Reviews

Review Text Feature. Prior studies have devoted considerable attention to exploring the relationship between review text and helpful votes. In this study, we decompose the review text features into four categories: structural features, Syntactic features, sentiment features, and semantic features.

Structural Feature. Structural features capture basic quantitative properties of the text, such as its overall length and the frequency of fundamental linguistic units (e.g., words, sentences, or characters). Structural features have been widely utilized in prior research [3–5]. In order to capture these basic linguistic unit, the following features are proposed.

- Sentence count
- Words count
- Unique words count
- Character count
- Stopwords count
- Average number of words per sentence
- Average number of characters per sentence

Syntactic Features. Unlike structural features that capture basic linguistic information, syntactic features capture the grammatical and relational properties of review texts. These features help to represent how words interact within a sentence. In this research we employ the following syntactic features:

- Proportion of part-of-speech tagging
- Syntactic complexity

The proportion of part-of-speech (POS) tagging reflects the distribution of grammatical categories within the review text. As Using spaCy [12], 18 types of part-of-speech tags can be identified and extracted. For example, ADJ and ADV correspond to adjectives and adverbs, respectively. The values of these features are calculated by extracting the part-of-speech tags of words and then computing their proportions within the review text.

Syntactic complexity is defined as the degree of sophistication of language forms by Xu and Reitter [13]. They introduced three metrics for quantifying syntactic complexity: sentence length, branching factor, and tree depth. The more complex sentence in syntactic the more depth in dependency tree [14] and imply complexity of structure. Specifically, tree depth are the longest distance from ROOT (The head of a sentence which is usually main verb) to the leaf node. Additionally, branching factor is defined as the average number of child in all non-leaf node in the dependency tree according to Xu and Reitter [13].

Tree depth and branching factor can be computed by analyzing a dependency tree, which is a graph structure consisting of lexical elements connected by asymmetrical relations [14]. In this graph, the nodes represent linguistic units (typically words), while the edges denote syntactic relations between two words, where one word serves as the syntactic parent (HEAD) and the other as its dependent (CHILD) [15]. There are several types of dependency relations between a HEAD and its CHILD. For example, common relations include the `amod` relation, where an adjective serves as the dependent modifying a noun that functions as the head, and the `aux` relation, where an auxiliary verb serves as the dependent modifying a main verb. Figure 4 shows an example of a dependency tree for the sentence "This is a great cup". Based on the dependency tree, we can obtain the feature namely tree depth and branching factor. Tree depth For tree depth, we compute the maximum tree depth across sentences as well as the average tree depth within the review text.

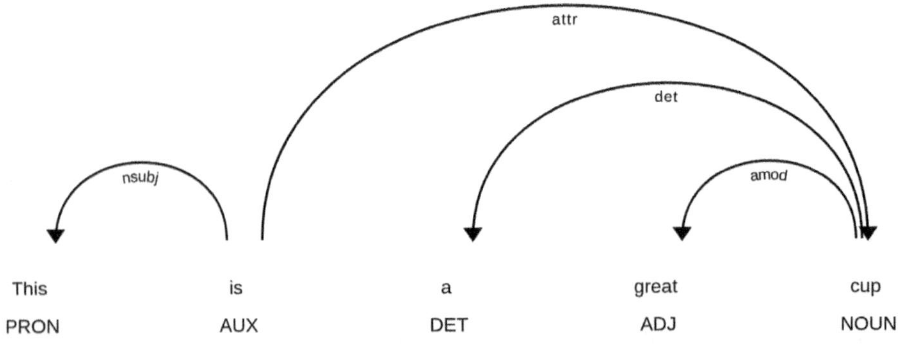

Fig. 4. An example of dependency tree

Sentiment Features. Sentiment refers to the subjective attitude expressed in text. Previous studies have extensively examined the relationship between sentiment variables, such as polarity, and helpful votes [3,16,17]. Polarity typically categorizes a piece of text into three classes: positive, neutral, and negative. Some research further represents polarity numerically [3], with values ranging from -1 to 1. A score of -1 indicates strongly negative sentiment, 0 represents neutral sentiment and $+1$ corresponds to strongly positive sentiment. In this research, sentiment features (polarity and subjectivity) are extracted using TextBlob [18]. By default, TextBlob computes average polarity and subjectivity scores based on SentiWordNet, although it can also be configured to use its naïve Bayes classifier pretrained on movie reviews [18]. Polarity values range from -1 to 1, where -1 indicates strongly negative sentiment and $+1$ indicates strongly positive sentiment. Subjectivity values range from 0 to 1, with 0 representing very objective expressions and 1 representing highly subjective expressions.

Semantics Features. Semantic features are defined as those directly obtained from review text. In this study, we employ several popular models, including Bag-of-Embeddings, TextCNN, and Bidirectional LSTM. In addition, we utilize different pretrained word embeddings to transform words into vector representations and evaluate the effect of different word embeddings.

Temporal Features. Amazon records the timestamp of each posted review. Kung et al. [3] found that the review timestamp is an important feature in their model but they use simply raw value as inputs. Similarly Nayeem and Rafiei [9] indicated reviewers and review time are two key factors. Therefore, we consider temporal features of both users and products simultaneously. Specifically, for each review, we compute statistics based on the information available prior to its timestamp for both the user and the product. For example, for a specific reviewer when analyzing a review posted on March 10, 2020, we only consider the user's previous reviews and interactions before this date. For each review, we consider the following user-related variables prior to its timestamp:

- Total number of reviews
- Sum of helpful votes
- Average of helpful votes
- Variance of helpful votes
- Maximum of helpful votes
- Average rating score
- Rating score variance
- Number of unique product user review
- Number of product categories user review

Similarly, For each review, we consider the following product-related variables prior to its timestamp:

- Total number of reviews
- Sum of helpful votes
- Average of helpful votes
- Variance of helpful votes
- Average rating score

By incorporating these statistics, we have information at the time of review posted for both of users and products.

3.3 Predictive Model and Evaluation Metrics

In this section, we present the predictive model and evaluation employed in this study. For the target variable, we define reviews with more than 10 helpful votes as helpful, and reviews with zero helpful votes as non-helpful. This formulation allows us to transform the original regression problem into a binary classification task. The predictive model we apply a MLP classifier to perform classification. In the training process, we utilize Adam optimizer to optimize binary cross-entropy loss and apply early stop with patience 5 to prevent overfit. To evaluate model's performance, we apply accuracy as our evaluation metrics. In addition, for feature importance, we implement Permutation Feature Importance (PFI), a global model-agnostic method that quantifies the increase in prediction error after randomly permuting the values of a single feature. The importance of a feature is thus measured as the corresponding reduction in the evaluation metric.

4 Experimental Results

We conduct multiple experiments to evaluate the effect of different models and pretrained word embeddings by sampling 80,000 reviews from the entire Amazon review dataset. Subsequently, we apply Permutation Feature Importance (PFI) to assess the contribution of each feature to model accuracy. For evaluation, we split the dataset into 70% for training and 30% for testing.

4.1 Bag-of-Embeddings

Initially, we utilize Bag-of-Embeddings to extract semantic representations, followed by a comparison of different word embeddings. Bag-of-Embeddings applies aggregation operations (e.g., mean, sum) to word embeddings in order to transform them into a sentence-level representation. Table 1 shows the result of Bag-of-Embeddings combined with other features.

4.2 TextCNN

Secondly, we employ TextCNN with multiple filters of kernel sizes (2, 3, 4) to capture different n-gram patterns, and apply max-pooling to retain the most informative n-gram features.As shown in the Table 2, the result of TextCNN combined with other features does not outperform simple Bag-of-Embeddings.

Table 1. Result of Bag-of-Embeddings

Embeddings	Accuracy
glove-twitter-50	0.882
glove-twitter-100	0.882
glove-twitter-200	0.881
glove-wiki-gigaword-100	0.880
glove-wiki-gigaword-200	0.881

Table 2. Result of TextCNN

Embeddings	Accuracy
glove-twitter-50	0.869
glove-twitter-100	0.871
glove-twitter-200	0.873
glove-wiki-gigaword-100	0.879
glove-wiki-gigaword-200	0.867

4.3 Bidirectional LSTM

BiLSTM is a bidirectional recurrent neural network that captures contextual information from two direction. Compared to TextCMNN, Bidirectional LSTM is more sensitive to word order. Table 3 shows the result of Bidirectional LSTM combined with other features.

Table 3. Result of Bidirectional LSTM

Embeddings	Accuracy
glove-twitter-50	0.875
glove-twitter-100	0.881
glove-twitter-200	0.879
glove-wiki-gigaword-100	0.879
glove-wiki-gigaword-200	0.875

4.4 Feature Importance Analysis

In this section, we employ Permutation Feature Importance (PFI) to assess the contribution of each feature. The procedure is as follows: (1) train the model with all features to obtain a baseline score; (2) for each feature, randomly permute its values; (3) recompute the score and record the performance drop relative to the

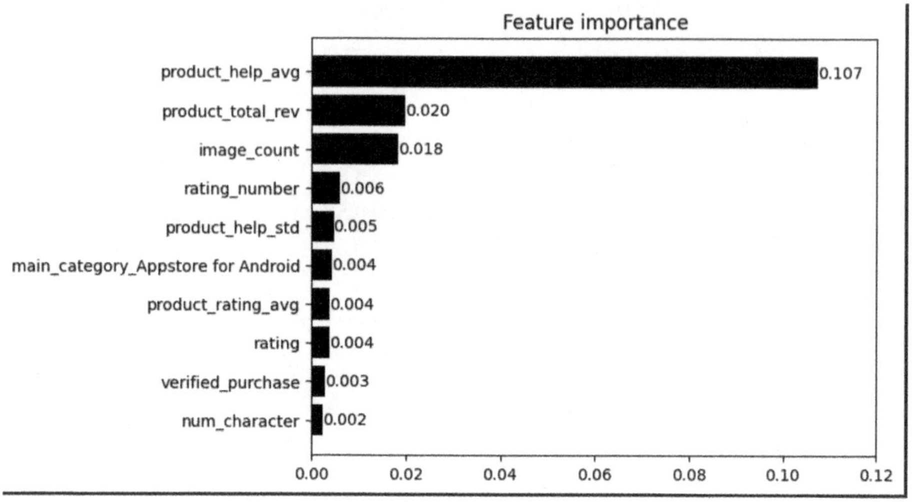

Fig. 5. Feature importance

baseline; (4) repeat this process ten times, and use the average reduction as the measure of feature importance. In Fig. 5 we visualize top-10 important features.

According to the results, the average number of helpful votes prior to the review's timestamp (product_help_avg) emerges as the most important feature, followed by the total number of reviews prior to the timestamp (product_total_rev) and the image count. It is worth noting that, apart from product_help_avg, no single feature dominates the model's performance.

5 Conclusion

The experimental results show that the choice of pretrained word embeddings has little impact on classification accuracy, and more complex neural architectures fail to outperform the simple bag-of-embeddings model.This may be because the training sample size is relatively small or most of the reviews are rather short and we will explore these in the future study. We also evaluate the importance of our features and observe that PFI does not account for feature interactions, as it assumes independence among features. In future work, we aim to design more transferable features and incorporate feature interactions into the feature importance analysis.

References

1. Spiegel Research Center.: How Online Reviews Influence Sales. Medill Northwestern University, Evanston (2017). https://spiegel.medill.northwestern.edu/wp-content/uploads/sites/2/2021/04/Spiegel_Online-Review_eBook_Jun2017_FINAL.pdf

2. Hou, Y., Li, J., He, Z., Yan, A., Chen, X., McAuley, J.: Bridging language and items for retrieval and recommendation (2024). arXiv preprint arXiv:2403.03952. https://arxiv.org/abs/2403.03952
3. Kirimlioglu, E., Kung, H., Orlando, D.: Were You Helpful – predicting helpful votes from amazon reviews (2024). arXiv preprint arXiv:2412.02884. https://arxiv.org/abs/2412.02884
4. Park, Y.J.: Predicting the helpfulness of online customer reviews across different product types. Sustainability **10**, 1735 (2018). https://doi.org/10.3390/su10061735
5. Malik, M.S.I.: Predicting users' review helpfulness: the role of significant review and reviewer characteristics. Soft. Comput. **24**, 13913–13928 (2020). https://doi.org/10.1007/s00500-020-04767-
6. Yang, Y., Yan, Y., Qiu, M., Bao, F.: Semantic analysis and helpfulness prediction of text for online product reviews. In: Zong, C., Strube, M. (eds.) Proceedings of the 53rd Annual Meeting of the Association for Computational Linguistics and the 7th International Joint Conference on Natural Language Processing, vol. 2, pp. 38–44. Association for Computational Linguistics (2015). https://doi.org/10.3115/v1/P15-2007
7. Haque, M.E., Tozal, M.E., Islam, A.: Helpfulness prediction of online product reviews. In: Proceedings of the ACM Symposium on Document Engineering 2018 (DocEng '18). Association for Computing Machinery (2018). https://dl.acm.org/doi/10.1145/3209280.3229105
8. Mertz, M., Korfiatis, N., Zicari, R.: Using dependency bigrams and discourse connectives for predicting the helpfulness of online reviews. SSRN Electron. J. **188**, 146–152 (2014). https://doi.org/10.2139/ssrn.2501751
9. Nayeem, M.T., Rafiei, D.: On the role of reviewer expertise in temporal review helpfulness prediction (2023). 10.48550/arXiv.2303.00923. https://arxiv.org/abs/2303.00923
10. Tran Quoc, H., Ochiai, H., Esaki, H.: Hidden topics modeling approach for review quality prediction and classification. In: Proceedings of the 2015 7th International Conference of Soft Computing and Pattern Recognition (SoCPaR), pp. 278–283 (2015)
11. Olmedilla, M., Martínez-Torres, M.R., Toral, S.: Prediction and modelling online reviews helpfulness using 1D convolutional neural networks. Expert Syst. Appl. **198**, 116787 (2022). https://doi.org/10.1016/j.eswa.2022.116787
12. Honnibal, M., Montani, I., Van Landeghem, S., Boyd., A.: spaCy: industrial-strength natural language processing in python (2020). https://doi.org/10.5281/zenodo.1212303
13. Xu, Y., Reitter, D.: Convergence of syntactic complexity in conversation. In: Proceedings of the 54th Annual Meeting of the Association for Computational Linguistics (ACL), pp. 443–448 (2016). https://doi.org/10.18653/v1/P16-2072
14. Nivre, J.: Dependency Grammar and Dependency Parsing. MSI Report 05133, Växjö University, School of Mathematics and Systems Engineering (2005)
15. Altınok, D.: Mastering spaCy: An End-to-End Practical Guide to Implementing NLP Applications Using the Python Ecosystem. Packt Publishing (2021)
16. Oueslati, O., Khalil, A., Ounelli, H.: Sentiment analysis for helpful reviews prediction. Int. J. Adv. Trends Comput. Sci. Eng. **7**, 34–40 (2018). https://doi.org/10.30534/ijatcse/2018/
17. Zeng, Z., Zhou, Z.: User review helpfulness assessment based on sentiment analysis. Electron. Libr. Ahead-of-Print (2020). https://doi.org/10.1108/EL-08-2019-0200
18. Loria, S.: TextBlob Documentation. Release 0.15, vol. 2 (2018)

Improving Unsupervised Zero-Shot Intent Classification with Dynamic Candidate Selection

Ruoyu Hu[1(✉)], Foaad Khosmood[2], and Abbas Edalat[1]

[1] Department of Computing, Imperial College London, London, UK
ruoyu.hu18@imperial.ac.uk
[2] California Polytechnic State University, San Luis Obispo, USA

Abstract. Task-oriented dialogue systems allow users to interact through natural language with a variety of digital devices in order to accomplish some goal, within which intent classification is an integral component in ensuring the satisfaction of a user's request. Applications of Large Language Models (LLMs) in this domain can suffer from prohibitively high computation requirements and costs owing to the number of input tokens scaling with the number of intents. We propose a framework using *candidate selection*, aimed at refining a model's selection of candidate intents to reduce inference costs. We validate our approach through extensive evaluation on four commonly-used intent classification datasets and show that our candidate selection approach can improve zero-shot intent classification performance (between +2.20% to +11.25%) over naive zero-shot across a range of model parameters, while significantly reducing both the number of input tokens (up to 88% reduction) and inference time (up to 53% reduction). All the while accomplishing this without any additional fine-tuning.

Keywords: Intent Classification · Zero-shot Classification · Dynamic Thresholding · Dataless Classification · Dialogue Systems · Natural Language Processing

1 Introduction

Intent classification [30] is an integral part of Task-Oriented Dialogue Systems (TODS) in determining the correct intent of the user through a given utterance. Combined with slot filling [11,17], it is critical in enabling a TODS to determine the appropriate functions to service the user's request. In recent years, the capabilities of Large Language Models (LLMs) in generalising to a large number of unseen tasks have improved significantly [2,14,47], with such models having already been shown to improve intent classification through generating

Ruoyu Hu is supported by UK Research and Innovation [UKRI Centre for Doctoral Training in AI for Healthcare grant number EP/S023283/1].

H. Unger and P. Meesad (Eds.): NLPIR 2025, LNNS 1904, pp. 428–446, 2026.
https://doi.org/10.1007/978-3-032-20897-2_32

synthetic training data [8,35]. However, supervised approaches using such models [18,58], while demonstrating impressive performance, can suffer from issues stemming from the cost associated with fine-tuning LLMs [31,33] that limit their practical application. This is further compounded by the inherent limitations of training data requirement per intent and the necessity for further training if an intent is added [30]. To mitigate the limitations mentioned above, recent work [22,37] has shown promising results for the ability of LLMs to perform *zero-shot intent classification* in the absence of any task-specific fine-tuning. Yet, such approaches typically require the inclusion of the full list of supported intents within the model prompt, significantly increasing the number of input tokens and consequently cost —computationally and monetarily [6,10], which can limit the scalability of such methods to large numbers of intents [30].

In this work, we seek to explore approaches to address the aforementioned problems with the inference overhead associated with using LLMs[1] We compare our approaches on a number of recently released models in a strict zero-shot, or *dataless* [11,44], intent classification setting in which we forego any model training. We perform extensive evaluation of our approaches in a number of different task settings with varying numbers of supported intents. Our contributions can be summarised as follows:

- We propose a framework using a dataless candidate selector to filter candidate intents for strict zero-shot intent classification and show that our approach can significantly improve classification performance over naive zero-shot on tested models (average improvement +6.03%).
- We propose a lightweight dynamic thresholding approach to alleviate the reliance on static threshold values and show that it performs comparably to our best-performing static thresholding approach.
- We show our approach significantly reduces the number of input tokens (up to 88% reduction) and average inference time[2] (up to 53% reduction).

2 Related Works

2.1 Zero-Shot Intent Classification

Intent classification (IC) [30] refers to the task of assigning a given utterance to one of a list of supported intents within task-oriented dialogue systems to properly service the user's request. Zero-shot intent classification (0Shot-IC) [50] focuses on systems that perform intent classification without training on labelled, task-specific data [56]. [15] uses capsule networks [34] along with an outlier detector to leverage training on 'seen' classes to discriminate against unseen classes. [27,35,40,59] all demonstrate the potential for learning conducted

[1] Our evaluation code and datasets can be found at https://github.com/ruoyunlp/dynamic-candidate-selection.

[2] All inference times are extracted from experiments run on a single Nvidia Quadro RTX6000 GPU. In total, all experiments took an estimated total of 2400 GPU hours.

on synthetic examples in conjunction with supervised training to transfer to unseen contexts and domains. However, supervised approaches to 0SHOT-IC can suffer from issues stemming from their reliance on the quality and quantity of the training data [52,55].

Dataless classification [9,11,44] is a stricter form of 0SHOT-IC defined by a total absence of training on *any* labelled data. This is typically achieved by mapping semantic representations of an utterance to a class label using their respective embedding distances [9]. Recent approaches such as [24,29] have yielded promising results by leveraging embeddings of intent descriptions. However, [24] also identified an issue with the overlaps within the embedding space between different intent classes, particularly those pertaining to similar concepts (i.e. PLAYMUSIC and ADDTOPLAYLIST, AIR_FARE and GROUND_FARE), that was first noted by [9] in reference to the necessity of a large enough margin between semantic representations of classes for dataless classification.

2.2 LLM Zero-Shot Intent Classification

Recent developments in LLMs have demonstrated the capabilities of such models in generalising to a wide range of tasks in a zero-shot manner [2,14,46], with a number of the aforementioned models having been used in generating synthetic examples for training [8,35] and intent detection [43]. Supervised approaches using LLMs such as [18,58] have demonstrated the potential for such models to be fine-tuned to perform 0SHOT-IC. However, the cost of fine-tuning models of such size can prove prohibitive. LLM-based approaches also typically include the full list of intents and descriptions within the prompt [22,37], which can significantly increase the number of input tokens for a given utterance, forming a bottleneck towards scaling to a large number of intents [30].

2.3 LLM Re-ranking and Filtering

The problem of selecting the 'most-relevant' label from a large list of candidates based on some metric of quality has been studied extensively for re-ranking tasks in Information Retrieval [5,38]. Recent works have explored the potential for LLMs to perform the task of ranking in multistage re-ranking frameworks, with most existing work in this domain focusing on fine-tuned models [36,57] or models accessible through commercially available APIs [39,41]. These approaches can incur significant costs when scaling to tasks with a large number of labels. While approaches have explored the use of a filtering stage in addition to re-ranking to reduce the number of inputs, these approaches typically relies upon static thresholding [26,54], prior knowledge of the score disitributions [3,4], or use of LLMs in the filtering mechanism [39,41] which necessitates the inclusion of all labels within the prompt at least once within the framework.

3 Methodology

3.1 Problem Definition

For a given task-oriented dialogue system, we define $\mathcal{C}$ as the set of intents supported by the system. For each intent $c \in \mathcal{C}$, we define l_c as the description of the intent. We consider only the case where an utterance u is associated with a single intent c as previous works such as [48] have shown such approaches can be iteratively applied to tackle utterances with multiple labels and thresholding of relevance scores can be used to detect out-of-domain intents [23]. We therefore leave further exploration in that domain for future work.

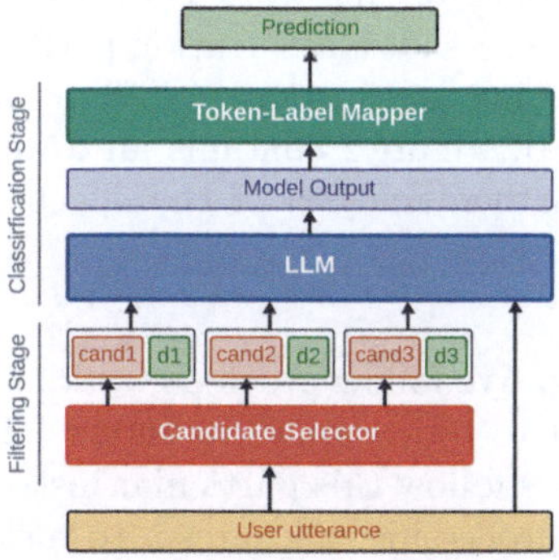

Fig. 1. Illustration of our two-stage model architecture. For each model input, intent candidates (i.e. cand1) are added alongside their respective intent descriptions (i.e. d1) and the original user utterance into the prompt of the LLM.

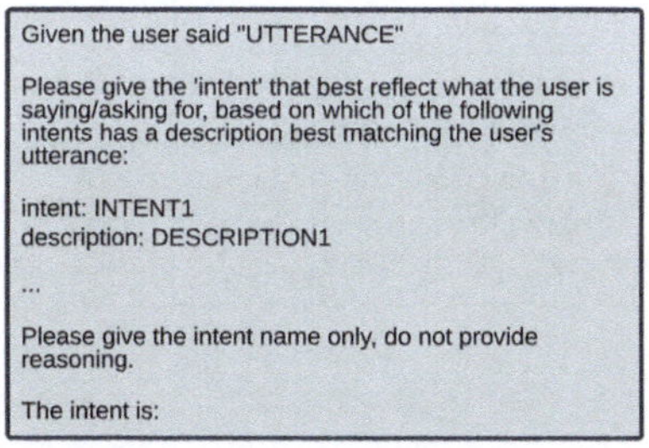

Fig. 2. An example prompt used to give the list of candidate intents and corresponding descriptions to the LLM.

3.2 Our Approach

Candidate Selection. Previous work carried out by [24] produced a dataless intent classifier that leverages the cosine similarity $s(\mathbf{h}(u), \mathbf{h}(l_c))$ between the embeddings of a user utterance $\mathbf{h}(u)$ and an intent description $\mathbf{h}(l_c)$ to select the label $\hat{y} = \arg\max\ s(\mathbf{h}(u), \mathbf{h}(l_c))$ as the prediction. We instead sort the list of similarity scores to produce $\boldsymbol{s}$, where $\boldsymbol{s}_1$ is the highest similarity score.

In order to reduce the computational requirements posed by passing all intents within the input to our LLM [39,41], for each model prompt, we select the intents with the top-k highest similarity scores [53] as *candidate* intents. The list of candidate intents is then combined with their corresponding descriptions before being passed to the LLM to generate a prediction. Our approach is illustrated in Fig. 1. As the focus of this work is on the use of a dataless classifier and the impact it has on zero-shot classification performance, we do not perform extensive prompt engineering and instead opt for a basic prompt template similar to previous works [41]. Figure 2 contains an example of our model prompt template. We provide further analysis of the effects of our candidate selector further on in Sects. 6 and 7.

Token-Label Mapping. We follow previous work [22] in including instructions within the model prompt to return only the intent label. However, we observed that models did not always follow this particular instruction; we therefore implement a lightweight post-processing mechanism to tackle such cases. For a given utterance u consisting of tokens $x_1, \ldots, x_n$, we first extract a contiguous sequence of tokens $w = x_i, \ldots, x_j$ that forms the last 'word' in the sequence. We then map w to an intent class $\hat{l}$:

Table 1. Dataset statistics for the four evaluation datasets used. $\bar{u}$ denotes the average sequence length for each dataset.

Dataset	Uttr.	Intents	$\bar{u}$
ATIS [20]	5.8k	18	11.18
SNIPS-NLU [12]	14.5k	7	8.97
CLINC150 [30]	22.5k	150	8.31
MASSIVE [16]	16.5k	60	6.90
Total	59.3k	235	

$$\hat{l} = \underset{c \in \mathcal{C}}{\arg\min}\ d_{\text{Lev}}(w, l_c) \tag{1}$$

where d_{Lev} is the Levenshtein distance between two strings and l_c is the intent label for class $c \in \mathcal{C}$. We provide further analysis of the effects of this mechanism in Sect. 6.2.

Dynamic k-Value Selection. In order to reduce the reliance on static thresholding when performing candidate selection [28], we introduce a method for *dyanmic* k-value selection based on embedding similarity scores. For each example u, given a list of sorted similarity scores $\mathbf{s}$ (Sect. 3.2), we consider the sorted list of weighted similarity scores as a function $F(x)$:

$$F(x) = sorted(\mathbf{s} * (2\mathbf{s} - \mathbf{sM}))_x \quad (2)$$

where $\mathbf{M}$ denotes the intra-description similarity for all intent classes. From $F(x)$ we obtain the 'elbow' point $k = \arg\max F''(x)$ as the point past which the candidate selector's confidence in a class changes the least, from which we select the top-k classes as our candidates. This process is illustrated in Algorithm 1. We apply the weighting term $2\mathbf{s} - \mathbf{sM}$ to mitigate against cases where the intent classes are semantically similar [24] which can lead to higher risk of misclassification (Sect. 7.2).

4 Experiments

4.1 Datasets

Following previous work of a similar nature and for the sake of comparison, we choose to evaluate our approach on four commonly used English TOD datasets: ATIS [20], SNIPS-NLU [12], CLINC150 [30], and MASSIVE [16]. In each case, as our approach does not require fine-tuning, we use the full dataset as the evaluation data. A statistical breakdown for each dataset is shown in Table 1.

4.2 Baselines

Zero-Shot Language Model (LM) Baseline. We establish an LM-only baseline by implementing a basic zero-shot setup, providing each LM with the full list of intents and corresponding descriptions of each intent in a similar way to previous works [18,37]. We use the same prompt template previously outlined in Sect. 3.2 across all of our experiments. The results are referred to as 'All Intents' in Table 2.

Candidate Selector Baseline. To evaluate the addition of the LM in conjunction with the candidate selector, we establish the results reported in [24] as our candidate selector baseline, referred to as 'Encoder only' in Sect. 5. We note the authors of [24] reported only macro-F1 for the MASSIVE dataset for the purpose of comparing against previous work. We, however, report both the accuracy and macro-F1 scores in our results (Sect. 5).

Algorithm 1 Candidate selection with dynamic k-value

```
u ← Utterance to classify
C ← List of Intents
similarity(a, b) ← Similarity function
sorted ← Sorting function (descending)
ModelOutput ← LLM Classifier

function DynamicKPrediction(u, C)
    s ← []
    M ← similarity(C, C)
    for c in C do
        s.append(similarity(u, c))
    end for
    s ← s * (2s − sM)        ▷ re-weigh similarity scores with intra-class similarity
    s ← sorted(s)
    s′ ← []
    for 0 ≤ i < s.length − 1 do
        s′.append(s[i] − s[i + 1])
    end for
    d_max ← −1
    k ← 0
    for 0 ≤ i < s′.length − 1 do
        d ← |s′[i] − s′[i + 1]|
        if d > d_max then
            d_max ← d
            k ← i
        end if
    end for
    Candidates ← s[: k]
    pred ← ModelOutput(Candidates)
    return pred
end function
```

4.3 Models

Large Language Models. We experiment with the following models as our LLM for producing an intent prediction given a set of candidates: **Llama-3.1-7B-Instruct** [14], **gemma-2-9b-it** [47], **phi-3-medium-4k-instruct** (14B parameters) [1] and **Mistral-7B-Instruct-v0.3** [25]. Our model selection was conducted based on a desire to capture a range of model performance and the availability of compute resources to us at the time of experimentation. It is by no means comprehensive, and we invite future work to explore a wider range of models in application to this domain. All model weights were sourced from their respective repositories on Huggingface [49] with default hyperparameters being used.

Model Quantization. Due to a limitation in compute resources, we experiment with quantization of our selected models at 4-bit, 8-bit, 16-bit and 'full' (32-bit)

Table 2. Results of our approach on 4 intent classification datasets compared to LM performance. We report both Accuracy and Macro-F1 scores. **Overall** denotes the average of all metrics across all datasets.

	Model	Top-k	ATIS Acc.	ATIS F1	SNIPS Acc.	SNIPS F1	CLINC150 Acc.	CLINC150 F1	MASSIVE Acc.	MASSIVE F1	Overall
	Encoder only		69.57	52.51	92.81	92.33	81.95	81.09	65.49	65.76	75.19
All Intents	Llama-3.1-8B-Instruct	All	67.34	39.24	71.92	71.75	86.63	86.02	73.87	72.79	71.20
	Gemma-2-9B-it		80.44	47.65	92.18	92.52	88.84	88.41	75.79	75.26	80.14
	Phi-3-medium-4k-instruct		75.92	47.96	92.56	92.80	60.89	58.56	73.21	70.64	71.57
	Mistral-7B-Instruct-v0.3		49.08	29.75	68.91	71.72	84.74	84.36	67.02	64.00	64.95
Candidate Selection (Ours)	Llama-3.1-8B-Instruct	k=3	74.88	53.71	81.65	81.92	87.23	86.69	73.22	72.73	76.50
		k=5	73.47	51.11	81.03	82.09	87.92	87.46	74.49	73.94	76.44
		k=10	74.76	48.05	76.33	77.57	88.72	88.35	74.64	74.16	75.32
	Gemma-2-9B-it	k=3	**86.53**	**57.38**	93.82	93.93	89.93	89.53	75.04	74.69	82.61
		k=5	85.72	57.06	92.74	93.03	**90.79**	**90.44**	77.08	**75.97**	**82.85**
		k=10	85.52	53.52	92.64	92.93	90.76	90.42	76.99	75.90	82.33
	Phi-3-medium-4k-instruct	k=3	75.99	52.15	**93.99**	**94.03**	89.03	88.67	75.16	73.32	80.29
		k=5	77.48	55.30	93.57	93.71	89.83	89.46	77.27	74.68	81.41
		k=10	78.33	54.70	92.86	93.09	89.95	89.65	**77.33**	75.03	81.37
	Mistral-7B-Instruct-v0.3	k=3	73.76	52.63	83.23	83.82	86.28	85.89	72.42	71.51	76.19
		k=5	65.88	45.59	76.11	77.82	86.74	86.45	72.67	70.89	72.77
		k=10	54.08	36.80	71.08	73.67	84.49	84.27	70.43	67.52	67.79

precisions using the `bitsandbytes` library [13]. We select the best-performing setup for each model by the consistency between the scores at each quantization precision and full precision.

Candidate Selector Models. We experiment with using **BGE-large** [51] and **GTE-large** [32] as our candidate selector models, as both models have been shown to perform well in dataless contexts for intent classification [24]. We analyse the impact of both models within our framework in Sect. 6.3 (Table 3).

Table 3. Results of our additional experiments with alternative prompting strategies and dynamic k-value selection.

	Setup	Llama-3.1	Gemma-2	Ovr.
Baseline	Naive	71.86	80.53	76.19
	+ CoT	75.35	80.47	77.91
	+ LLMLingua2	62.23	71.89	67.06
CS (Ours)	k=3	**76.50**	82.61	79.56
	k=5	76.44	82.85	**79.64**
	k=10	75.32	82.33	78.82
	Dynamic k	75.62	**83.08**	79.35

Table 4. Results of Dynamic k-value selection compared to static selection and naive baseline across all tested models (**L** - Llama-3.1, **G** - Gemma-2, **P** - Phi-3.1, **M** - Mistral-7B) on the CLINC150 (150 intents) and MASSIVE (60 intents) datasets. Note that all approaches with candidate selection outperform in the naive approach.

	Setup	L	G	P	M	Ovr.
	Naive	79.92	78.44	69.77	76.60	76.18
CS (Ours)	k=3	79.97	82.30	81.54	79.02	80.71
	k=5	80.95	**83.57**	82.81	**79.19**	**81.63**
	k=10	**81.47**	83.52	**82.99**	76.67	81.16
	Dynamic k	80.89	83.48	82.47	79.02	81.47

5 Results

5.1 Metrics

Following on from previous works of a similar nature [19,24], we report both Accuracy and Macro-F1 scores[3] for all models and datasets in our experiments. Where applicable, we report the average of Accuracy and Macro-F1 across all evaluation datasets as 'Overall' (Tables 2 and 4).

5.2 Zero-Shot LM Baseline

The naive zero-shot baseline underperforms the encoder-only approach on Llama-3.1 (−3.99% Overall), Phi-3 (−3.62% Overall), Mistral-7B (−10.24% Overall) models and outperforms the encoder-only approach on Gemma-2 (+4.95% Overall). On inspection of the model outputs, we note that the Phi-3 model and Mistral-7B models fail to produce a valid model prediction at much higher rates (13.65% and 5.19% respectively) compared to the Llama-3.1 (0.88%) and Gemma-2 (1.40%). In such instances, the model typically outputs a reasonable-looking intent label that is not of a valid intent given to the model. We attribute such outputs to hallucinations caused by the amount of intent labels and descriptions available to the models. This was particularly evident in the performance of the Phi-3 model on the CLINC150 dataset, which failed to produce a valid intent in 31.48% of instances.

5.3 Methods with Candidate Selection

Our approach using a dataless candidate selector yielded significant improvements across all tested models (Table 2) compared to the naive LLM baseline without candidate selection. For each model, the average across different k-values yielded improvements of: Llama-3.1 (+4.89% Overall), Gemma-2 (+2.46% Overall), Phi-3 (+9.46% Overall) and Mistral-7B (+7.30% Overall). The increase in model performance can also be seen in the significant reduction in average model failure rate in producing a valid intent prediction (CS: 1.77% vs. naive: 5.38%).

[3] Accuracy and macro-F1 are computed using the `scikit-learn` library (v1.3.2). Correlation is measured using the `numpy` library (v1.24.4).

Table 5. Average model performance across all intent classification tasks, with and without our intent label mapping mechanism.

Model	$\textbf{Pred}_{LM}$	+Map	Δ
Llama-3.1-8B-Instruct	75.01	75.54	0.53
Gemma-2-9B-it	81.88	82.25	0.37
Phi-3-medium-4k-instruct	78.63	80.04	1.41
Mistral-7B-Instruct-v0.3	69.97	71.46	1.49

5.4 Dynamic k-Value Selection

Table 4 contains the results of our experiments with our dynamic k-value selection mechanism. Due to compute constraints, we focus our experiments on the CLINC150 and MASSIVE datasets as they contain a significantly higher number of intents. We note that our approach outperforms the average of static k-value setups (81.47 vs. 81.17) and performs comparably to the highest-performing static k-value approach (81.47 vs 81.63), suggesting that this approach to dynamic k-values for candidate selection can reduce the reliance on such pre-selected values. Additionally, we assess the robustness of our method against the selection of descriptions in Sect. 6.4.

6 Ablations

6.1 Repeated Experiments

In order to observe the effect of randomisation on our results, we repeat our experiments for 4 independent runs and compute the average result. Due to the high number of experiments requiring a prohibitively high amount of time and compute resources, we choose to evaluate only the best-performing setups on the Gemma-2 and Llama-3.1 models to rerun across all selected intent classification datasets. We obtained a mean Overall score of $82.85 \pm 0.01\%$ for Gemma-2 and $76.35 \pm 0.08\%$ for Llama-3.1. As $\sigma \ll 0.001$ in both instances, we conclude that our approach is consistent across random initialisations.

6.2 Effect of Token-Label Mapping

We investigate the effect of our token-label mapping procedure on model performance by comparing results with and without our postprocessing step for all tested models at 4-bit and 8-bit precisions. Results are averaged across all setups per model and shown in Table 5. It can be observed that the inclusion of our mapping procedure improves model performance across all models tested, with greater improvements seen in the Phi-3 (+1.41 Ovr.) and Mistral-7B models (+1.49 Ovr.). We note from a basic error analysis that almost all errors eliminated (2.34% of total predictions) are cases whereby the model hallucinates

Table 6. Changes in the overall score when candidate selector is used. E denotes setups using *encoder only* candidate selector and +P+M denotes setups with paraphrasing and masking components from [24].

Model	BGE		GTE	
	E	+P+M	E	+P+M
Llama-3.1-8B-Instruct	4.88	5.41	4.57	5.55
Gemma-2-9b-it	1.98	2.61	1.65	2.14

Table 7. Standard deviation σ of dynamic k-value candidate selector success rate and average k-value over 30 random samples of descriptions.

Sim.	AT.	SN.	CL.	MA.
Embed.	0	1.75×10^{-5}	5.91×10^{-5}	1.12×10^{-4}
N-gram	0	5.62×10^{-5}	1.14×10^{-4}	2.70×10^{-4}
Avg. k				
Embed.	4.00	4.53	6.26	6.67
N-gram	4.00	4.55	6.24	6.68

new labels which appear reasonable in isolation, typically containing overlapping pieces of the correct label (i.e. '`calendar_update`' → '`update_my_calendar`'). We experimented with embedding distance instead of Levenshtein in our postprocessing procedure but found no significant difference in performance.

6.3 Effect of Choice of Candidate Selector

We conduct a second set of experiments with a GTE-large model [32] using the same models and setups as in our experiments in Sect. 6.1. Table 6 shows the results for gemma-2-9b-it and llama-3.1-8b-instruct in setups using only the sentence embedding similarities as candidate selection metrics and setups using the additional paraphrasing and masking components proposed by [24]. We observe a significant increase in performance in all setup permutations compared to LM-only. Figure 3a illustrates the success rates of the candidate selector models, showing the addition of paraphrasing and masking to yield a higher model success rate. Upon inspection of candidate selector predictions, we note that the addition of paraphrasing and masking improves the ranking of the correct label in 8.31% of examples on average. However, we also note that on average, in 0.94% of examples, the correct label was no longer within the top-5 candidates. Further work should investigate improvements to the candidate selector to reduce the introduction of new errors.

6.4 Effect of Choice of Intent Descriptions

In order to ascertain the robustness of our methods to the choice of intent descriptions, we generate at least 3 additional descriptions for each intent using GPT-4 [2], from which we randomly sample 30 unique sets of descriptions for each

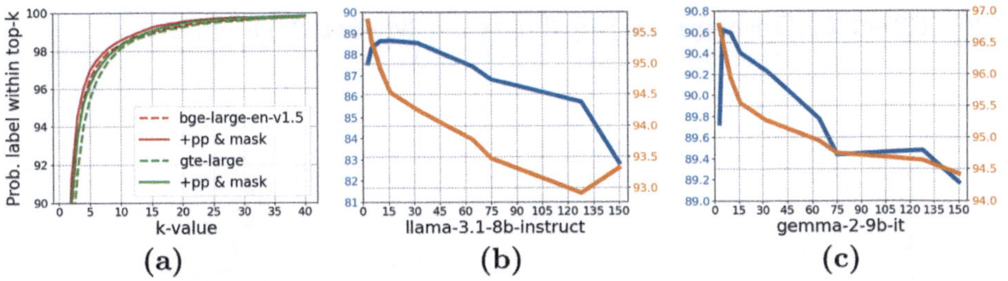

Fig. 3. (**a**) Probability of the correct label being within top-k candidates given to the LM for different candidate selector setups. (**b**) and (**c**) **Blue**: Mean of Accuracy and Macro-F1 on the CLINC150 dataset. **Orange**: Model success rate when the label is within top-k candidates given to the LM. Note that the success rate decreases as k increases.

dataset. Table 7 shows the standard deviation of the success rate of the candidate selector for each dataset. As the standard deviation is very low across all tested datasets and similarity measures, we conclude that our dynamic k-value selector is robust to the choice of intent descriptions. Additionally, we note the average k-value varies significantly between datasets, suggesting differing optimal thresholds, which would otherwise lead to larger static k values being required per dataset.

7 Analysis

7.1 Effect of k-Value on Model Performance

As the choice of k has a direct impact on whether the correct label is presented to the LLM, we repeat our experiments on the CLINC150 dataset with an increasing number of candidates starting from $k = 3$ to a maximum of $k = 150$. Figures 3b and 3c show the performance of the models. We observe on both tested models a peak in performance around $k = 10$ similar to those observed by [22] in their work on the same dataset, after which the model performance steadily decreases. On inspecting the model performance, given the correct label is provided to the LLM, we note that the success rate is continually decreasing for both models tested, suggesting the overall model performance is a balance between the correctness of the candidate selector in ensuring the correct label is provided as a candidate and the LLM selecting the correct label in turn. We also observe a minor improvement to model success rate when using dynamic k-value selection (average +0.38%), suggesting some potential to improving model success rate with better candidate selection.

7.2 Effect of Description Similarity on Model Performance

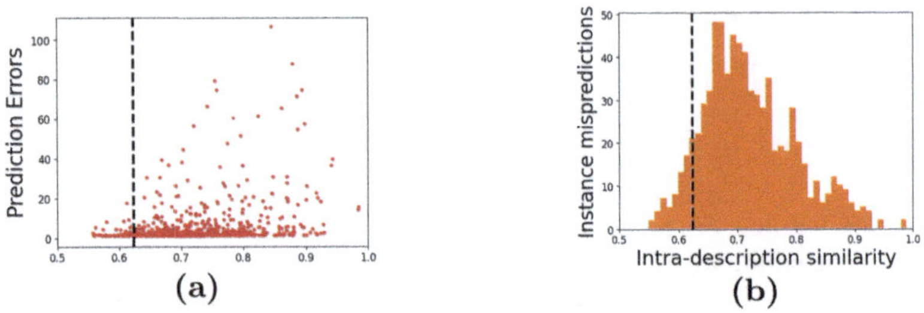

Fig. 4. Distribution of prediction errors by the intra-description similarity of the ground-truth intent and the predicted intent on the CLINC150 dataset. **(a)** The number of errors per relation **(b)** The number of relations with erroneous predictions. Dashed line denotes the mean intra-description similarity across the entire dataset.

We attempt to examine whether there exists any relation between classification errors in the model output and similarities between intent descriptions. Intuitively, this is the assumption that similar definitions of intents may lead to a greater degree of confusion, thus resulting in more prediction errors. Figure 4 illustrates the distribution of prediction errors by the intra-description similarity between the ground-truth intent and the predicted intent. We note a significant skewness of mispredictions towards higher intra-description similarities (Fisher-Pearson $G_1 = 0.798$). Computing the correlation between prediction errors and cosine similarity of description embeddings yielded a weak positive correlation (Pearson's $r = 0.213 \pm 0.015$). Conversely, much weaker correlation was obtained between the prediction errors and the intra-description Levenshtein distance (Pearson's $r = 0.07 \pm 0.01$), implying that errors arising from high description similarity or overlapping descriptions are more likely to be focused on the semantic similarities rather than lexical similarity. We also observe from error attribution that on average, mispredictions arising from the LLM accounted for 75-82% of classification errors across all of our experiments. We suggest future work to explore means to tackle prediction errors arising from such circumstances, such as through generating semantically diverse descriptions.

7.3 Effect of Candidate Selection on Model Inference Cost

A key conceptual benefit to filtering through candidate selection is a reduction in computational overhead and cost when applying to classification tasks with a high number of classes. We therefore analyse the benefits of our approach in these areas. We collect inference times and the number of input tokens per prompt for all selected models and datasets and analyse the average inference overhead, results are shown in Table 8. We observe from Table 8 that our approach significantly reduced the average inference time (up to 53% reduction) and

Table 8. Average inference time and number of input tokens per example for varying values of k. *% Reduction* indicates the mean percentage-decrease for lower values of k from $k = 150$.

Model	Inference time/it. (s)		
	$k = 3$	$k = 16$	$k = 150$
Llama-3.1-8B-Instruct	1.69	2.59	4.29
Gemma-2-9b-it	5.05	5.61	11.26
Phi-3-Medium-4k-Instruct	2.68	3.89	5.64
Mistral-7B-Instruct-v0.3	3.09	3.60	5.71
% Reduction	*53%*	*38%*	–
Model	Input Tokens / Prompt		
	$k = 3$	$k = 16$	$k = 150$
Avg. Num. Tokens	128.55	312.21	1084.16
% Reduction	*88%*	*71%*	–

the number of input tokens (up to 88% reduction). We note the latter in particular can significantly reduce the cost in using the growing suite of commercial LLMs-as-a-service solutions [10,45].

8 Conclusion

In this paper, we utilise a dataless candidate selector component and demonstrate its potential to improve model intent classification performance over encoder-only and naive zero-shot LLM approaches, while significantly reducing the number of input tokens and inference time, all without requiring any further fine-tuning of the models. We introduce an approach to dynamic k-value selection which reduces the reliance of candidate selection approaches to static threshold values. We performed extensive experiments with a range of models and provide analysis into model behaviour and failure conditions to guide future work.

8.1 Limitations

Our work focuses on the potential for our proposed framework to leverage dataless candidate selection to reduce computational overheads in a strict zero-shot intent classification setting. The use of intent label descriptions in the candidate selection creates the potential for such approaches to be applied to other text classification tasks such as emotion classification [7,42] and genre prediction [21]. We have shown our approach to dynamic k-value selection can be used instead of of static k-value threshold in this task setting; however, it remains to be seen whether this approach can also be applied to other tasks in information retrieval, such as re-ranking and filtering [38]. Additionally, our experiments utilised the Levenshtein distance in our token-label mapping postprocessing procedure; future work could explore additional semantic distance metrics such as embedding similarity or WordNet-based methods.

References

1. Abdin, M., et al.: Phi-3 technical report: A highly capable language model locally on your phone. arXiv preprint arXiv:2404.14219 (2024)
2. Achiam, J., et al.: Gpt-4 technical report. arXiv preprint arXiv:2303.08774 (2023)
3. Arampatzis, A., van Hameran, A.: The score-distributional threshold optimization for adaptive binary classification tasks. In: Proceedings of the 24th annual international ACM SIGIR Conference on Research and Development in Information Retrieval, pp. 285–293 (2001)
4. Arampatzis, A., Kamps, J., Robertson, S.: Where to stop reading a ranked list? threshold optimization using truncated score distributions. In: Proceedings of the 32nd international ACM SIGIR Conference on Research and Development in Information Retrieval, pp. 524–531 (2009)
5. Azar, Y., Gamzu, I., Yin, X.: Multiple intents re-ranking. In: Proceedings of the Forty-First Annual ACM Symposium on Theory of Computing, STOC '09, pp. 669–678. Association for Computing Machinery, New York, NY, USA (2009). https://doi.org/10.1145/1536414.1536505. URL https://doi.org/10.1145/1536414.1536505
6. Bang, F.: GPTCache: An open-source semantic cache for LLM applications enabling faster answers and cost savings. In: Proceedings of the 3rd Workshop for Natural Language Processing Open Source Software (NLP-OSS 2023), pp. 212–218 (2023). https://doi.org/10.18653/v1/2023.nlposs-1.24. URL https://aclanthology.org/2023.nlposs-1.24/
7. Canales, L., Martínez-Barco, P.: Emotion detection from text: A survey. In: Proceedings of the Workshop on Natural Language Processing in the 5th Information Systems Research Working Days (JISIC), pp. 37–43 (2014). https://doi.org/10.3115/v1/W14-6905. URL https://aclanthology.org/W14-6905/
8. Cegin, J., Simko, J., Brusilovsky, P.: ChatGPT to replace crowdsourcing of paraphrases for intent classification: Higher diversity and comparable model robustness. In: Proceedings of the 2023 Conference on Empirical Methods in Natural Language Processing, pp. 1889–1905 (2023). https://doi.org/10.18653/v1/2023.emnlp-main.117, https://aclanthology.org/2023.emnlp-main.117
9. Chang, M.W., Ratinov, L., Roth, D., Srikumar, V.: Importance of semantic representation: Dataless classification. In: Proceedings of the 23rd National Conference on Artificial Intelligence - Volume 2, AAAI'08, pp. 830–835. AAAI Press (2008)
10. Chen, L., Zaharia, M., Zou, J.: Frugalgpt: How to use large language models while reducing cost and improving performance. arXiv preprint arXiv:2305.05176 (2023)
11. Chen, X., Xia, Y., Jin, P., Carroll, J.: Dataless text classification with descriptive LDA. In: Proceedings of the Twenty-Ninth AAAI Conference on Artificial Intelligence, AAAI'15, pp. 2224–2231. AAAI Press (2015)
12. Coucke, A., et al.: Snips voice platform: an embedded spoken language understanding system for private-by-design voice interfaces. arXiv preprint arXiv:1805.10190 (2018). https://arxiv.org/abs/1805.10190v3
13. Dettmers, T., Pagnoni, A., Holtzman, A., Zettlemoyer, L.: Qlora: Efficient finetuning of quantized LLMs. Advances in Neural Information Processing Systems **36** (2024)

14. Dubey, A., et al.: The llama 3 herd of models. arXiv preprint arXiv:2407.21783 (2024)
15. Fan, L., et al.: Unknown intent detection using Gaussian mixture model with an application to zero-shot intent classification. In: Proceedings of the 58th Annual Meeting of the Association for Computational Linguistics, pp. 1050–1060. Association for Computational Linguistics, Online (2020). https://doi.org/10.18653/v1/2020.acl-main.99, https://aclanthology.org/2020.acl-main.99/
16. FitzGerald, J., et al.: MASSIVE: A 1M-example multilingual natural language understanding dataset with 51 typologically-diverse languages. In: Proceedings of the 61st Annual Meeting of the Association for Computational Linguistics (Volume 1: Long Papers), pp. 4277–4302. Association for Computational Linguistics, Toronto, Canada (2023). https://doi.org/10.18653/v1/2023.acl-long.235, https://aclanthology.org/2023.acl-long.235
17. Goo, C.W., et al.: Slot-gated modeling for joint slot filling and intent prediction. In: Proceedings of the 2018 Conference of the North American Chapter of the Association for Computational Linguistics: Human Language Technologies, Volume 2 (Short Papers), pp. 753–757 (2018)
18. Gretz, S., et al.: Zero-shot topical text classification with LLMs - an experimental study. In: Findings of the Association for Computational Linguistics: EMNLP 2023, pp. 9647–9676. Association for Computational Linguistics, Singapore (2023). https://doi.org/10.18653/v1/2023.findings-emnlp.647, https://aclanthology.org/2023.findings-emnlp.647
19. Gritta, M., Hu, R., Iacobacci, I.: CrossAligner & co: Zero-shot transfer methods for task-oriented cross-lingual natural language understanding. In: Findings of the Association for Computational Linguistics: ACL 2022, pp. 4048–4061. Association for Computational Linguistics, Dublin, Ireland (2022). https://doi.org/10.18653/v1/2022.findings-acl.319, https://aclanthology.org/2022.findings-acl.319
20. Hemphill, C.T., Godfrey, J.J., Doddington, G.R.: The atis spoken language systems pilot corpus. Speech and Natural Language: Proceedings of a Workshop Held at Hidden Valley, Pennsylvania (1990). https://aclanthology.org/H90-1021
21. Hoang, Q.: Predicting movie genres based on plot summaries. arXiv preprint arXiv:1801.04813 (2018)
22. Hong, T., et al.: Exploring the use of natural language descriptions of intents for large language models in zero-shot intent classification. In: Proceedings of the 25th Annual Meeting of the Special Interest Group on Discourse and Dialogue, pp. 458–465. Association for Computational Linguistics, Kyoto, Japan (2024). https://doi.org/10.18653/v1/2024.sigdial-1.39, https://aclanthology.org/2024.sigdial-1.39
23. Hou, Y., Lai, Y., Wu, Y., Che, W., Liu, T.: Few-shot learning for multi-label intent detection. In: Proceedings of the AAAI Conference on Artificial Intelligence, vol. 35, pp. 13,036–13,044 (2021)
24. Hu, R., Khosmood, F., Edalat, A.: Exploring description-augmented dataless intent classification. In: Proceedings of the 6th Workshop on NLP for Conversational AI (NLP4ConvAI 2024), pp. 13–36. Association for Computational Linguistics, Bangkok, Thailand (2024), https://aclanthology.org/2024.nlp4convai-1.2
25. Jiang, A.Q., et al.: Mistral 7b. arXiv preprint arXiv:2310.06825 (2023)
26. Jiang, Z., et al.: Active retrieval augmented generation. In: Proceedings of the 2023 Conference on Empirical Methods in Natural Language Processing, pp. 7969–7992 (2023)
27. Kulkarni, A., Tseng, B.H., Moniz, J.R.A., Piraviperumal, D., Yu, H., Bhargava, S.: SynthDST: Synthetic data is all you need for few-shot dialog state tracking. In:

Proceedings of the 18th Conference of the European Chapter of the Association for Computational Linguistics (Volume 1: Long Papers), pp. 1988–2001. Association for Computational Linguistics, St. Julian's, Malta (2024), https://aclanthology.org/2024.eacl-long.120/
28. Kweon, W., Kang, S., Jang, S., Yu, H.: Top-personalized-k recommendation. In: Proceedings of the ACM Web Conference 2024, pp. 3388–3399 (2024)
29. Lamanov, D., Burnyshev, P., Artemova, E., Malykh, V., Bout, A., Piontkovskaya, I.: Template-based approach to zero-shot intent recognition. In: Proceedings of the 15th International Conference on Natural Language Generation, pp. 15–28. Association for Computational Linguistics, Waterville, Maine, USA and virtual meeting (2022). https://doi.org/10.18653/v1/2022.inlg-main.2, https://aclanthology.org/2022.inlg-main.2/
30. Larson, S., Mahendran, A., Peper, J.J., Clarke, C., Lee, A., Hill, P., Kummerfeld, J.K., Leach, K., Laurenzano, M.A., Tang, L., Mars, J.: An evaluation dataset for intent classification and out-of-scope prediction. In: Proceedings of the 2019 Conference on Empirical Methods in Natural Language Processing and the 9th International Joint Conference on Natural Language Processing (EMNLP-IJCNLP), pp. 1311–1316. Association for Computational Linguistics, Hong Kong, China (2019). https://doi.org/10.18653/v1/D19-1131, https://aclanthology.org/D19-1131
31. Li, X., et al.: FLM-101b: An open LLM and how to train it with $100k budget. arXiv preprint arXiv:2309.03852 (2023)
32. Li, Z., Zhang, X., Zhang, Y., Long, D., Xie, P., Zhang, M.: Towards general text embeddings with multi-stage contrastive learning. arXiv preprint arXiv:2308.03281 (2023)
33. Lin, X., Wang, W., Li, Y., Yang, S., Feng, F., Wei, Y., Chua, T.S.: Data-efficient fine-tuning for LLM-based recommendation. In: Proceedings of the 47th International ACM SIGIR Conference on Research and Development in Information Retrieval, SIGIR '24, pp. 365–374. Association for Computing Machinery, New York, NY, USA (2024). https://doi.org/10.1145/3626772.3657807
34. Liu, H., et al.: Reconstructing capsule networks for zero-shot intent classification. In: Proceedings of the 2019 Conference on Empirical Methods in Natural Language Processing and the 9th International Joint Conference on Natural Language Processing (EMNLP-IJCNLP), pp. 4799–4809. Association for Computational Linguistics, Hong Kong, China (2019). https://doi.org/10.18653/v1/D19-1486, https://aclanthology.org/D19-1486
35. Liu, H., et al.: Liberating seen classes: boosting few-shot and zero-shot text classification via anchor generation and classification reframing. Proc. AAAI Conf. Artif. Intell. **38**(17), 18,644–18,652 (2024). https://doi.org/10.1609/aaai.v38i17.29827, https://ojs.aaai.org/index.php/AAAI/article/view/29827
36. Luo, S., et al.: Recranker: Instruction tuning large language model as ranker for top-k recommendation. ACM Trans. Inf. Syst. (2024). https://doi.org/10.1145/3705728Just Accepted Just Accepted
37. Milios, A., Reddy, S., Bahdanau, D.: In-context learning for text classification with many labels. In: Proceedings of the 1st GenBench Workshop on (Benchmarking) Generalisation in NLP, pp. 173–184. Association for Computational Linguistics, Singapore (2023). https://doi.org/10.18653/v1/2023.genbench-1.14, https://aclanthology.org/2023.genbench-1.14
38. Nogueira, R., Cho, K.: Passage re-ranking with bert. arXiv preprint arXiv:1901.04085 (2019)

39. Nouriinanloo, B., Lamothe, M.: Re-ranking step by step: Investigating pre-filtering for re-ranking with large language models (2024). https://arxiv.org/abs/2406.18740
40. Parikh, S., Tiwari, M., Tumbade, P., Vohra, Q.: Exploring zero and few-shot techniques for intent classification. In: Proceedings of the 61st Annual Meeting of the Association for Computational Linguistics (Volume 5: Industry Track), pp. 744–751. Association for Computational Linguistics, Toronto, Canada (2023). https://doi.org/10.18653/v1/2023.acl-industry.71, https://aclanthology.org/2023.acl-industry.71/
41. Rashid, M., Meem, J., Dong, Y., Hristidis, V.: EcoRank: budget-constrained text re-ranking using large language models. In: Findings of the Association for Computational Linguistics: ACL 2024, pp. 13,049–13,063. Association for Computational Linguistics, Bangkok, Thailand (2024). https://doi.org/10.18653/v1/2024.findings-acl.773, https://aclanthology.org/2024.findings-acl.773/
42. Rashkin, H., Smith, E.M., Li, M., Boureau, Y.L.: Towards empathetic open-domain conversation models: A new benchmark and dataset. In: Proceedings of the 57th Annual Meeting of the Association for Computational Linguistics, pp. 5370–5381. Association for Computational Linguistics, Florence, Italy (2019). https://doi.org/10.18653/v1/P19-1534, https://aclanthology.org/P19-1534/
43. Song, X., et al.: Large language models meet open-world intent discovery and recognition: an evaluation of ChatGPT. In: Proceedings of the 2023 Conference on Empirical Methods in Natural Language Processing, pp. 10,291–10,304. Association for Computational Linguistics, Singapore (2023). https://doi.org/10.18653/v1/2023.emnlp-main.636, https://aclanthology.org/2023.emnlp-main.636
44. Song, Y., Roth, D.: On dataless hierarchical text classification. In: Proceedings of the Twenty-Eighth AAAI Conference on Artificial Intelligence, AAAI'14, pp. 1579–1585. AAAI Press (2014)
45. Sun, T., Shao, Y., Qian, H., Huang, X., Qiu, X.: Black-box tuning for language-model-as-a-service. In: International Conference on Machine Learning, pp. 20,841–20,855. PMLR (2022)
46. Team, G., et al.: Gemma: Open models based on gemini research and technology. arXiv preprint arXiv:2403.08295 (2024)
47. Team, G., et al.: Gemma 2: Improving open language models at a practical size. arXiv preprint arXiv:2408.00118 (2024)
48. Wan, M., : TNT-LLM: Text mining at scale with large language models. In: Proceedings of the 30th ACM SIGKDD Conference on Knowledge Discovery and Data Mining, KDD '24, pp. 5836–5847. Association for Computing Machinery, New York, NY, USA (2024). https://doi.org/10.1145/3637528.3671647
49. Wolf, T., et al.: Huggingface's transformers: State-of-the-art natural language processing. arxiv. arXiv preprint arXiv:1910.03771 (2019)
50. Xia, C., Zhang, C., Yan, X., Chang, Y., Yu, P.: Zero-shot user intent detection via capsule neural networks. In: Proceedings of the 2018 Conference on Empirical Methods in Natural Language Processing, pp. 3090–3099. Association for Computational Linguistics, Brussels, Belgium (2018). https://doi.org/10.18653/v1/D18-1348, https://aclanthology.org/D18-1348
51. Xiao, S., Liu, Z., Zhang, P., Muennighoff, N., Lian, D., Nie, J.Y.: C-pack: Packed resources for general chinese embeddings. In: Proceedings of the 47th International ACM SIGIR Conference on Research and Development in Information Retrieval, pp. 641–649 (2024)

52. Xu, H., Chen, M., Huang, L., Vucetic, S., Yin, W.: X-shot: A unified system to handle frequent, few-shot and zero-shot learning simultaneously in classification. In: Findings of the Association for Computational Linguistics: ACL 2024, pp. 4652–4665. Association for Computational Linguistics, Bangkok, Thailand (2024). https://doi.org/10.18653/v1/2024.findings-acl.276, https://aclanthology.org/2024.findings-acl.276/
53. Yang, X., Steck, H., Guo, Y., Liu, Y.: On top-k recommendation using social networks. In: Proceedings of the Sixth ACM Conference on Recommender Systems, RecSys '12, pp. 67–74. Association for Computing Machinery, New York, NY, USA (2012). https://doi.org/10.1145/2365952.2365969.
54. Yang, Y.: A study of thresholding strategies for text categorization. In: Proceedings of the 24th Annual International ACM SIGIR Conference on Research and Development in Information Retrieval, pp. 137–145 (2001)
55. Yin, W., Chen, M., Zhou, B., Ning, Q., Chang, K.W., Roth, D.: Indirectly supervised natural language processing. In: Proceedings of the 61st Annual Meeting of the Association for Computational Linguistics (Volume 6: Tutorial Abstracts), pp. 32–40. Association for Computational Linguistics, Toronto, Canada (2023). https://doi.org/10.18653/v1/2023.acl-tutorials.5, https://aclanthology.org/2023.acl-tutorials.5/
56. Yin, W., Hay, J., Roth, D.: Benchmarking zero-shot text classification: datasets, evaluation and entailment approach. In: Proceedings of the 2019 Conference on Empirical Methods in Natural Language Processing and the 9th International Joint Conference on Natural Language Processing (EMNLP-IJCNLP), pp. 3914–3923 (2019). https://doi.org/10.18653/v1/D19-1404, https://aclanthology.org/D19-1404/
57. Yue, Z., Rabhi, S., de Souza Pereira Moreira, G., Wang, D., Oldridge, E.: Llamarec: Two-stage recommendation using large language models for ranking (2023). https://arxiv.org/abs/2311.02089
58. Zhang, F., et al.: From discrimination to generation: Low-resource intent detection with language model instruction tuning. In: Findings of the Association for Computational Linguistics: ACL 2024, pp. 10,167–10,183. Association for Computational Linguistics, Bangkok, Thailand (2024). https://doi.org/10.18653/v1/2024.findings-acl.605, https://aclanthology.org/2024.findings-acl.605
59. Zhang, Y., Yuan, C., Wang, X., Bai, Z., Liu, Y.: Learn to adapt for generalized zero-shot text classification. In: Proceedings of the 60th Annual Meeting of the Association for Computational Linguistics (Volume 1: Long Papers), pp. 517–527. Association for Computational Linguistics, Dublin, Ireland (2022). https://doi.org/10.18653/v1/2022.acl-long.39, https://aclanthology.org/2022.acl-long.39/

Margin-Aware Active Learning for User-Adaptive Text Classification

Santosh Tirunagari[1(✉)], Mandeep Dhami[2], and David Windridge[1]

[1] Department of Computer Science, Middlesex University, The Burroughs, London, UK
{s.tirunagari,d.windridge}@mdx.ac.uk

[2] Department of Psychology, Middlesex University, The Burroughs, London, UK
m.dhami@mdx.ac.uk

Abstract. We propose a human-in-the-loop text classification framework for rapid personalisation of deep neural networks (DNNs) to user-specific preferences under strict label budgets. The core is a *margin-preserving* embedding aligned with a *margin-aware* active learning (AL) strategy that jointly optimizes uncertainty, diversity, density, and class coverage in query selection. Use of a fixed local embedding and lightweight linear SVM classifier ensures rapid model updates. To resolve the cold start problem we introduce a one-class initialisation *with ring-radii bisection* from a single user-selected example to bootstrap learning, along with a hybrid AL acquisition function that evolves from exploitative to exploratory querying. In experiments on a proprietary dataset (UKAC) and 20 Newsgroups, the proposed AL strategy consistently outperforms random sampling, achieving higher accuracy with far fewer labels especially in the early stages of labeling. Our approach yields superior label efficiency (Macro-F1) at small label budgets, demonstrating a practical solution for user-adaptive DNN personalisation. The code is available at https://github.com/tsantosh7/Margin-Preserving-Active-Learning.

Keywords: Active Learning · HITL · SVM · Margin · Budget Aware Labeling

1 Introduction

Building personalised classifiers with deep neural networks (DNNs) in a human-in-the-loop (HITL) setting requires high accuracy from very few labeled examples. In the interactive scenario, the user labels presented items as either relevant or not, and the system must rapidly adapt to these preferences. Traditional DNN fine-tuning is impractical in this context: backpropagation on a pre-trained model for each user is computationally heavy and typically requires large batches of labels, violating the *sparse feedback* constraints of personalisation. We therefore seek a method to *efficiently learn* from minimal user input by leveraging rich representation learning and principled active query selection.

H. Unger and P. Meesad (Eds.): NLPIR 2025, LNNS 1904, pp. 447–461, 2026.
https://doi.org/10.1007/978-3-032-20897-2_33

Active learning (AL) has long aimed to reduce labelling effort. Classic strategies include uncertainty sampling [13], query-by-committee [20], expected error reduction [17], and density-weighted selection [15]; more recent approaches incorporate deep/Bayesian models [5,10] and hybrid acquisition functions that mix confidence with diversity or representativeness [2,19]. Frozen pre-trained language models (PLMs) representations inside the AL loop can drastically cut compute while preserving downstream accuracy, then fine-tuning a PLM only after acquisition [24]. *Our approach falls into this latter, hybrid category,* using a scoring function that combines margin-based uncertainty with density, diversity, and coverage to mitigate the pitfalls of pure uncertainty sampling.

In personalised settings user-relevant class instances are typically rare, with the initial model weak or over-generalised relative to the task, and the unlabeled pool vastly imbalanced; off-the-shelf methods can over-query majority (irrelevant) items or select atypical outliers that do little to uncover more relevant examples. Cold-start AL exacerbates these issues: while there are strategies for starting without labels [3] and few-shot AL [1], these typically assume a few examples per class. In this work we address the cold-start extremal case of potentially only a single positive example by utilisation one-class SVMs (OC-SVMs), which are high effective for novelty/outlier detection [8,18], but are rarely integrated to seed binary AL under such scarcity (While querying unlabeled points closest to the decision boundary -i.e. those with the smallest margin- can be highly effective in two-class SVM training [22], early rounds are highly brittle with ill-defined margins and imbalanced data).

Even assuming that the cold start SVM AL problem can be resolved in this manner, it is not immediately obvious that this has a bearing on the DNN AL training problem. However, the recent finding that the implicit bias of gradient descent in DNNs under exponential loss gradient flow assumptions naturally tends towards that of a maximum-margin separator [14] hints at a deep connection, and we will argue that margin geometry serves as a useful organizing principle atop deep representations (this bias approximation has been shown to apply in both under- and over-parametrized regimes, and for network architectures with residual connections and normalisation layers [11]). Thus, by treating a frozen DNN embedding as a space that approximately preserves margins (as an implicit consequence of how DNNs model are trained or fine-tuned), we justify using a linear SVM on top of a pre-trained DNN in an AL context, effectively focusing queries around the margin at points of maximal uncertainty over the set of possible AL learning hypotheses —*a connection between deep learning theory and AL practice that is a unique aspect of our work.* We hence, concretely, propose a margin-aware AL framework tailored for user-adaptive text classification that assumes at the outset access to a *frozen* pre-trained text embedding, upon which we maintain a lightweight linear classifier that is updated online as labels arrive within the AL context. Our method is thus characterized by: (1) a **margin-centric view** of the embedding space that frames the user's preference task in terms of a maximum-margin hypothesis space; and (2) a **margin-aware query strategy** that prioritizes points near the current decision boundary while integrating density, diversity, and coverage heuristics to avoid the shortcomings of pure uncertainty sampling. To address extreme cold start, we introduce an *OC-SVM initialisation* that, given even only a single positive example, establishes an initial boundary and enables an informative negative query *immediately*. Through experiments on proprietary and public datasets, we show that this framework achieves higher accuracy with significantly fewer labels than random sampling.

2 Methodology

We consider a data domain $\mathcal{P}$ of text objects (e.g. documents). A pre-trained DNN feature extractor $\psi(\theta; x)$ maps each text $x \in \mathcal{P}$ into a vector embedding $\mathbf{x} = \psi(\theta; x) \in \mathbb{R}^d$. We assume this transformation produces a semantically meaningful feature space. By ℓ_2-normalizing all embeddings, cosine similarity aligns with the linear classifier's dot product, so distances in embedding space approximately preserve the notion of a classification *margin*. Given a user-specific classification task (relevant vs. irrelevant, or multi-class), we seek to train a linear Support Vector Machine (SVM) on the embeddings of labeled examples.

In the *far-from-initialisation* regime this problem may be straightforwardly specified as follows. Let $f(\mathbf{x}) = \mathbf{w}^\top \mathbf{x} + b$ be the decision function, with $\mathbf{w}$ and b the weight vector and bias. For a binary preference classification with labels $y_i \in \{\pm 1\}$, the soft-margin SVM optimizes:

$$\min_{\mathbf{w},\, b,\, \{\xi_i\} \geq 0} \quad \frac{1}{2}\|\mathbf{w}\|^2 + C \sum_i \xi_i \quad \text{s.t. } y_i(\mathbf{w}^\top \mathbf{x}_i + b) \geq 1 - \xi_i\,, \tag{1}$$

with $C > 0$ a regularisation parameter balancing margin width and training errors. At optimum, the Karush–Kuhn–Tucker (KKT) conditions imply that $\mathbf{w}$ can be expressed as a weighted combination of training examples: $\mathbf{w} = \sum_i \alpha_i\, y_i\, \mathbf{x}_i$, where each $\alpha_i \in [0, C]$ and non-zero α_i correspond to *support vectors* on or inside the margin boundaries [22]. The geometric margin is $\gamma = 1/\|\mathbf{w}\|$.

We postulate that the frozen embedding $\psi(\theta; \cdot)$ is rich enough that user preference classes are approximately linearly separable in this feature space with a large margin. In other words, there exists some hyperplane in the *Margin-Preserving Hypothesis* embedding space that separates relevant vs. irrelevant items with an optimal margin. Moreover, any such separating hyperplane is defined by a small subset of *support vectors* on the margin; thus, the hypothesis space of interest can be characterized by a choice of these critical boundary points. This perspective motivates an active learning (AL) strategy to query unlabeled points that lie closest to the current decision boundary (i.e. those with the smallest margin $|f(\mathbf{x})|$, being the points that are maximally disambiguating among alternate user-preference hypotheses consistent with AL & HITL principles). By querying such points, we rapidly identify the support objects needed to determine the user-specific classifier. We treat the embedding space as *approximately margin-preserving* in the sense that distances correspond to classifier confidence; our ℓ_2 normalisation ensures that the SVM margin in feature space aligns with cosine distance in the input space.

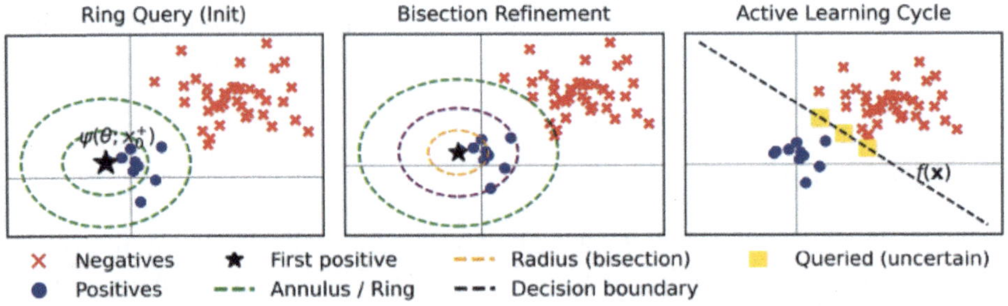

Fig. 1. Initialisation and active learning cycle. (Left) Ring query around the first positive embedding, marked as $\psi(\theta; x_0^+)$. (Middle) Radius bisection strategy between $r_{\min}$ and $r_{\max}$ to find an informative counterexample. (Right) AL loop with uncertain queries $\mathcal{Q}$ near the SVM decision boundary $f(\mathbf{x}) = \mathbf{w}^\top \mathbf{x} + b$.

To formalize the above intuition, we draw on results about the implicit bias of deep network training: gradient descent on DNNs with exponential or logistic loss tends to converge (in direction) to a first-order KKT point of a maximum-margin classification problem [7,14]. We summarize this property as follows:

Lemma 1 (Gradient flow → max-margin, cf. [7,14]). *DNN gradient flow with exponential or logistic loss (+ReLU activation) converges in parametric direction to a first-order KKT point of an infinite-time max-margin problem.*

Lemma 2 (Approximation under practical training). *Even if the ideal conditions of Lemma 1 are not fully met in practice (e.g. due to using cross-entropy loss, early stopping, or finite optimisation time), we assume the resulting trained DNN embedding still behaves* as if *it were aligned with a max-margin separator for the given task. In other words, the embedding space can be treated as approximately yielding a large margin for the user's classification problem.*

Lemma 3 (Support vectors as salient exemplars). *At convergence of the linear SVM, the KKT conditions identify* support objects *with non-zero multipliers. These support exemplars lie on or near the class boundary (on extremal points of each class's convex hull in the feature space) and thus define the separating hyperplane. Selecting unlabeled points nearest to such boundaries maximizes informativeness for the classifier, targeting critical boundary points.*

These theoretical insights provide the basis for our AL approach: margin properties motivate uncertainty sampling (querying near the decision boundary), the support-vector view motivates boundary-focused selection, and the convex-hull perspective suggests incorporating diversity and coverage to capture the full extent of each class.

Cold-Start via One-Class SVM and Ring Query

A challenge arises at the very beginning of personalisation: if the user provides only a single positive example and no negatives (e.g. a user identifying an item of interest will generally not give, or even be able to concretely specify, an *a priori* indication of class extent), then a standard binary classifier cannot be trained. We tackle this cold-start scenario by leveraging a one-class SVM [18] on the embedding, coupled with a **ring bisection** probe that we *use throughout our experiments.* The one-class SVM finds the smallest-radius hypersphere (with a soft-margin variant allowing some outliers) that encloses the given positive example (and possibly its nearest-neighbor unlabeled points). Formally, for a set P of positive examples in $\mathbb{R}^d$, one-class SVM solves:

$$\min_{\mathbf{c},\, R,\, \{\xi_i\}\geq 0} \quad R^2 + \frac{\nu}{|P|}\sum_{x_i \in P} \xi_i \quad \text{s.t. } \|\mathbf{x}_i - \mathbf{c}\|^2 \leq R^2 + \xi_i,\ \forall i\,, \qquad (2)$$

where $\mathbf{c}$ is the center of the hypersphere, R its radius, and $\nu \in (0, 1]$ controls the trade-off (roughly, the fraction of points allowed to lie outside the sphere) [18,21]. Via the KKT conditions, the center $\mathbf{c}$ can be expressed as a convex combination of the training points ($\mathbf{c} = \sum_i \alpha_i \mathbf{x}_i$, $\alpha_i \geq 0$, $\sum_i \alpha_i = 1$), and, at optimum, one obtains a set of support vectors (points on or just outside the boundary) with corresponding multipliers $0 < \alpha_i < \frac{1}{\nu|P|}$. A new sample $\mathbf{x}$ can then be classified by its distance to $\mathbf{c}$: if $\|\mathbf{x} - \mathbf{c}\| \leq R$ (inside the sphere) it is considered part of the positive class region; otherwise it is treated as an outlier (negative). We use this one-class model on the single user-provided positive to estimate an initial *preference region* in the embedding space, and then perform **ring bisection**: maintain inner/outer radii $(r_{\min}, r_{\max})$ (we use unlabeled-distance quantiles; in practice $r_{\min}$ = 25th and $r_{\max}$ =90th), query the unlabeled point whose distance to the positive center is closest to the midpoint $r_{\text{mid}} = \frac{1}{2}(r_{\min}+r_{\max})$, and update $(r_{\min}, r_{\max})$ by the returned label (expand outward on positive, shrink on negative) for a small try budget (we use $T = 4$). This *implemented* procedure rapidly elicits the first negative without assuming it exists at any fixed ring; progressive bisection is, moreover, consistent with AL principles in selecting query points that maximally differentiate between competing uniformly distributed variable hypotheses (bisection has, in general, $O(\frac{1}{2^t})$ efficiency in interval convergence relative to random interval sampling, and is optimal in the absence of function/gradient information).

In summary, as labelling progresses our hypothesis boundary evolves: with only one class known, the decision boundary can be initialised as a hypersphere around the positive examples; once both classes have labeled examples, the decision boundary becomes a maximum-margin hyperplane separating the convex hulls of the positive and negative classes. Our approach is designed to handle this transition smoothly and these theoretical insights directly inform the design of our AL strategy. To make this procedure clearer, we present the ring query, the bisection refinement strategy, and the overall active learning loop in Fig. 1.

Model and Uncertainty Estimation

Building on the above, we implement a hybrid AL algorithm that maintains a margin-centric view while incorporating additional criteria to ensure robustness. We use a pre-trained transformer-based sentence encoder (e.g. `all-MiniLM-L6-v2`) to compute embeddings $\mathbf{x} = \psi(\theta; x)$ for all texts and apply ℓ_2 normalisation so that $\|\mathbf{x}\|_2 = 1$, aligning cosine similarity with the dot product used by the linear SVM.

For classification, we train a linear SVM $f(\mathbf{x}) = \mathbf{w}^\top \mathbf{x} + b$ on the labelled set $\mathcal{L}$ at each AL iteration. For multi-class problems, we adopt a one-vs-rest scheme: a separate SVM is trained for each class vs. the rest, and class probabilities can be obtained via Platt scaling [16] to calibrate the decision values. If the AL uncertainty measure requires probabilities (e.g. entropy), we wrap the SVM in a calibrated classifier; otherwise, we use raw decision distances. (When extremely few labels per class are available, we defer using entropy in favour of the margin until calibration becomes feasible).

We follow an *adaptive* calibration scheme: at tiny label counts we use margin distances; once class counts permit, we upgrade to probability-based entropy via Platt scaling. This mirrors recent calibration trends: selective post-hoc recalibration for reliability [25] and LLM-specific calibration using generated text as signals [23].

The primary uncertainty measure $U(x)$ for an unlabeled sample x is defined as the classifier's confidence. We support two forms of $U(x)$:

- *Margin uncertainty*: $U(x) = 1 - \frac{|f(x)|}{\max_{x' \in \mathcal{U}} |f(x')|}$, i.e. one minus the normalised distance of x from the decision hyperplane. Equivalently, in probabilistic terms for binary classification, $U(x) \approx 1 - (p_1 - p_2)$, where p_1 and p_2 are the top two class probabilities for x. A smaller margin (or smaller $p_1 - p_2$ gap) yields higher $U(x)$.
- *Entropy*: $U(x) = -\sum_c p(c|x) \log p(c|x)$ for multi-class cases (if a calibrated probability $p(c|x)$ is available). Entropy is higher when the classifier is uncertain among many classes.

Both measures assign higher scores to points the model is least certain about (typically those lying near the current decision boundary).

Density and Diversity Enhancement

Uncertainty alone can sometimes lead the AL strategy to pick unusual outliers that happen to confuse the model. To avoid this, we incorporate a **density term** $D(x)$ that favors samples in high-density regions of the unlabeled pool. We compute a k-nearest-neighbor density in the embedding space: for each point x, find the average distance to its k nearest neighbors and define $D(x)$ inversely proportional to this distance. In practice, let $D(x)$ be the inverse of the mean distance to the k nearest neighbors of x, normalised to $[0, 1]$. We then define a density-weighted uncertainty:

$$U'(x) = U(x) \left[(1 - \beta) + \beta \, \tilde{D}(x) \right], \tag{3}$$

where $\tilde{D}(x)$ is the normalised density and $\beta \in [0,1]$ is a weighting parameter. Setting $\beta > 0$ increases the score of points that are not only uncertain but also reside in populous regions of the feature space, thereby steering the algorithm away from anomalous points that could be noise.

We also ensure **diversity** in batch query selection. Our AL loop selects examples in batches of size b for efficiency. Rather than naively taking the top-b scoring samples (which might all cluster in the same region), we oversample a slate of top-K candidates (with $K > b$) according to $U'(x)$ and then perform a diversification step. Specifically, we use a farthest-point sampling heuristic: starting from the highest-scoring point, we iteratively select the next point that is farthest (in cosine distance) from those already chosen, until b points are picked. This ensures the batch covers a distributed set of informative points, reducing redundancy.

Recent variants revisit core-set style selection for text to better balance coverage and informativeness under modern encoders; our diversification step is compatible with these updates [9].

Exploration Term and Class Coverage

To further address the classic AL exploitation–exploration trade-off, we introduce an explicit **exploration** score $E(x)$ that complements the uncertainty score $U'(x)$. The exploration term has two parts:

- $R(x)$: a *representativeness* term, defined as the distance of x to the closest labeled example. For instance, $R(x) = 1 - \max_{\ell \in \mathcal{L}} \cos(\mathbf{x}, \ell)$, which is high for points that are far from any currently labeled instance. This favors covering new regions of the space.
- $N(x)$: a *novelty w.r.t. positives* term. For binary tasks (with a designated "positive" class), $N(x)$ is defined as the residual norm of x after projection onto the subspace spanned by the labeled positive examples. Essentially, $N(x)$ measures how different x is from the span of known positives; a high $N(x)$ means x lies in a direction not yet covered by the known positive data, encouraging expansion of the decision boundary.

We combine these with a parameter $\lambda \in [0,1]$ such that $E(x) = \lambda\, R(x) + (1 - \lambda)\, N(x)$.
If the task is multi-class & no single +ve class exists, $N(x)$ can be generalised to promote novel directions for any class (or omitted entirely, using $R(x)$ alone for exploration).

Our **hybrid acquisition function** is then defined as a convex combination of exploitation (U') and exploration (E):

$$s_t(x) = \alpha_t\, U'(x) \; + \; [1 - \alpha_t]\, E(x) \,. \tag{4}$$

Here t indexes the AL iteration (or batch number). The schedule $\alpha_t = \mathrm{Sched}(t)$ starts near 1 (initially focusing on pure uncertainty) and gradually decreases to a final value α_{final} by a chosen switch iteration T_{switch}. We experiment with different scheduling functions (linear, cosine, exponential) to anneal the exploitation–exploration mix. Early in the AL process, the model is very weak and uncertainty is a reliable guide; later on, α_t is lowered to ensure exploration of the space and discovery of any remaining pockets of relevant data the model might be overlooking [6].

In multi-class problems, we additionally enforce **class coverage**. Every M rounds, we check the class distribution of the labeled set. If some classes (or the current classifier's predicted classes for unlabeled points) are under-represented, we adjust the selection process so that those classes get higher priority in the next query batch. One simple implementation is to stratify the top-K candidate slate by predicted class and ensure that each class (especially those with very few labeled examples) contributes at least a few points to the final batch. This prevents the AL process from entirely neglecting a class that might constitute only a small fraction of data (a common issue in multi-class AL). In binary settings, we enforce an analogous balance: queries are chosen roughly half from the positive region and half from the negative region (favoring whichever side has fewer labeled examples so far).

Query-by-Committee Disagreement (QBC)

To further stabilize selection under model bias and severe skew, we augment the uncertainty signal with a small *committee* of models for disagreement-based querying. Specifically, we train M lightweight linear SVMs on bootstrap resamples of the labeled set $\mathcal{L}$. For an unlabeled point x, each committee member predicts a label $\hat{y}_m(x)$; we then compute the *vote entropy* disagreement [4,20]:

$$\mathrm{Dis}(x) \;=\; -\sum_{c=1}^{K} \hat{p}_c(x)\,\log \hat{p}_c(x), \qquad \hat{p}_c(x) = \frac{1}{M}\sum_{m=1}^{M} \mathbf{1}\{\hat{y}_m(x) = c\},$$

normalised to $[0,1]$ by $\log K$. We blend this disagreement with the main model's uncertainty $U(x)$:

$$\tilde{U}(x) \;=\; (1-\lambda_{\text{qbc}})\,\mathrm{Norm}\big(U(x)\big) \;+\; \lambda_{\text{qbc}}\,\mathrm{Norm}\big(\mathrm{Dis}(x)\big),$$

with $\lambda_{\text{qbc}} \in [0,1]$ a tunable weight. In the hybrid score of Eq. (4), we simply use $U'(x)$ with $U(x)$ replaced by $\tilde{U}(x)$. In practice, we keep the committee members' outputs uncalibrated (using only their votes), which is robust at tiny label counts, while the *main* model is calibrated if using entropy for $U(x)$.

Active Learning Loop with Budget-Aware Control

We begin by initializing the labeled set $\mathcal{L}$ with one user-provided positive example (target class), and optionally one or two negatives if available. If the one-class bootstrap is enabled, we train the one-class SVM on the initial positive(s) and

Algorithm 1 Margin-Aware Active Learning Procedure

Input: Unlabeled pool $\mathcal{U}$; oracle $\mathcal{O}$; label budget B; batch size b; schedule $(\alpha_{\text{final}}, T_{\text{switch}})$; density weight β; class-coverage period M; (optional) AL→Random threshold T_{AR}.
Initialize $\mathcal{L}$ with one user-provided positive (target class); include any available negatives if present.
if cold-start with one-class bootstrap is enabled **then**
  Fit a one-class SVM on the positive seed(s) to estimate a positive region.
  ring bisection: set radii $(r_{\min}, r_{\max})$ from inner/outer unlabeled-distance quantiles around the positive center; for $t = 1 \ldots T$:
    let $r_{\text{mid}} = \frac{1}{2}(r_{\min}+r_{\max})$ and query the unlabeled point closest (in distance-to-center) to r_{mid};
    if the returned label is negative, **stop**; else set $r_{\min} \leftarrow r_{\text{mid}}$ and continue.
  (If a negative already exists in $\mathcal{L}$ at start, skip bisection.)
end if
for $t = 1, 2, \ldots$ **do**
  Train (or update) the linear SVM on the current labeled set $\mathcal{L}$. If using entropy-based $U(x)$, calibrate the SVM outputs to probabilities.
  Compute the hybrid score $s_t(x)$ for each $x \in \mathcal{U}$ using the current α_t (from schedule) and the definitions of $U'(x)$ and $E(x)$.
  Identify the top-K unlabeled points by $s_t(x)$. From these, select a batch $\mathcal{Q}$ of b queries via the diversity filter (farthest-point sampling) and class coverage adjustment (if $t \mod M = 0$, ensure under-represented classes are included in $\mathcal{Q}$).
  Query the oracle $\mathcal{O}$ to obtain labels for all $x \in \mathcal{Q}$.
  Augment the labeled set: $\mathcal{L} \leftarrow \mathcal{L} \cup \{(x, y) : x \in \mathcal{Q}\}$, and remove those points from $\mathcal{U}$.
  if $|\mathcal{L}| \geq B$ **or** other stopping criterion is met **then**
    break // (Stop if budget exhausted or target performance reached)
  end if
  if $|\mathcal{L}| = T_{AR}$ **(optional) then**
    Switch all future queries to *random sampling* (AL-to-Random handoff).
  end if
end for

perform the one-shot "ring" query just outside its boundary to obtain an initial negative label.

At each AL iteration t, we train (or update) the linear SVM on the current labeled set and, if using entropy-based $U(x)$, calibrate its outputs to probabilities. We then compute the hybrid acquisition score $s_t(x)$ for each unlabeled $x \in \mathcal{U}$ using the current exploitation/exploration weight α_t and our definitions of $U'(x)$ and $E(x)$. We identify the top-K scoring unlabeled points and then select a batch $\mathcal{Q}$ of b queries via the diversity filter (farthest-point sampling) and class coverage adjustment (if t is a multiple of M, we ensure any under-represented class is included in $\mathcal{Q}$). The oracle then labels all $x \in \mathcal{Q}$, we add these labeled pairs to $\mathcal{L}$, and remove those points from $\mathcal{U}$. This process repeats until the label budget B is reached or another stopping criterion (e.g. a target performance) is met. Optionally, once a certain number of labels have been acquired ($|\mathcal{L}| = T_{AR}$), the strategy can switch all future queries to simple *random sampling* (an AL-to-random handoff) to limit computational overhead in later stages. Pseudocode is given in Algorithm 1.

3 Experiments and Results

Datasets. We evaluate on two corpora: (i) a proprietary **UKAC** dataset tailored to an online user-personalisation schema-extension scenario (details omitted for privacy but the embeddings are made available on the GitHub[1]), where one label denotes "relevant to user" and all others are "irrelevant," yielding a strongly imbalanced task; and (ii) the public **20 Newsgroups** (20NG) benchmark [12]. For 20NG we study: *balanced binary* (two specific topics), *one-vs-rest (OvR)* (one topic as positive vs. the other 19 as negative), *two-vs-rest (TvR)* (two selected topics as positive vs. the rest), and the *full 20-way* classification. Standard preprocessing removes headers, footers, and quoted text.

Metrics. We report macro-F_1 as our primary metric because it equally weighs classes and is sensitive to minority/rare categories. We also discuss accuracy and area-under-the-learning-curve (AULC) qualitatively when interpreting label efficiency across budgets. For tables we report results at a strict budget of $B = 100$ labels.

Active Learning Protocol. Each run begins from the same tiny seed set (at least one positive) for both *Active* and *Random* baselines. We use batch size $b = 5$. The classifier is a linear SVM (balanced class weights). The acquisition combines margin uncertainty, k-NN density (with $\beta = 0.3$), a cosine schedule that anneals the uncertainty weight α_t from 1.0 to 0.2 by about 75 labels, and diversity via slate-and-diversify selection. Class-coverage is enforced periodically (every $M = 2$ rounds for binary, $M = 3$ for multi-class) to avoid neglecting sparse classes. When entropy is required, SVM outputs are calibrated with Platt scaling [16]. To handle the extreme cold-start (one positive only), a one-class SVM bootstrap proposes a "ring" query just outside the estimated positive region, yielding an informative initial negative.

3.1 UKAC

On UKAC, Active selection surpasses Random, with the largest gains when the selector must balance very sparse positives against a large negative pool (OvR/TvR). Table 1 reports macro-F_1 (Active, Random) and the relative improvement $\Delta\%$ at $B = 100$ labels. Improvements span roughly +1.33% to +5.94% across TvR, and +0.1% to +5.65% across OvR. Multiclass shows a +4.65% gain.

Hybrid selector mixes uncertainty with two safeguards: it looks for items close to the current decision boundary (the most informative ones), and it also

[1] https://github.com/tsantosh7/Margin-Preserving-Active-Learning.

Table 1. Custom/UKAC: Macro-F_1 at $B = 100$ (Active Learn. (AL), Rand, Δ%)

Scenario	AL	Random	Δ%
Multiclass (mc_full)	0.831	0.794	+4.65%
OvR (other vs rest)	0.801	0.800	+0.1%
OvR (remorse vs rest)	0.901	0.872	+3.30%
OvR (totality vs rest)	0.890	0.842	+5.65%
TvR (other+remorse vs rest)	0.886	0.853	+3.93%
TvR (other+totality vs rest)	0.910	0.859	+5.94%
TvR (remorse+totality vs rest)	0.827	0.816	+1.33%

checks density and class coverage so we don't waste labels on odd outliers or keep sampling the same region. In the TvR case, balancing the two rare "positive" classes gives steady gains (up to +5.94%). In the OvR case, the density/coverage checks help the minority class and raise macro-F1 (up to +5.65%). There's one small gain (+0.1%) on "other vs rest", likely because that class is very mixed and Random's broad sampling happens to work there. Overall, the UKAC results still favor AL.

3.2 20 Newsgroups

Balanced Binary. In a two-class task (e.g., `sci.space` vs. `comp.graphics`), AL is ahead early and stays ahead for most of the budget. With only ~20 labels, AL is around 70% on accuracy while Random is closer to 55%. By $B = 100$, AL is about 88.5% vs. 85.0% for Random. This also shows up as a higher area under the learning curve meaning, better label efficiency.

One-vs-Rest (OvR). In OvR, the positive class is only about 5% of the pool. Random wastes many queries on easy negatives, while AL quickly focuses on the target region. Table 2 shows macro-F_1 at $B = 100$ for all 20 OvR tasks (AL, Random, and Δ%). AL's advantage is clear on narrower domains (for example, hardware and medicine), where the method finds useful near-boundary positives early. A few broad or mixed classes (e.g., *alt.atheism*) show small negative deltas at $B = 100$, likely because Random's broad sampling already reaches enough informative examples.

Two-vs-Rest (TvR). With two positives, there are more positives than in OvR, but they are still sparse. Our class-coverage rule stops the selector from getting stuck on just one of the two topics. Table 3 shows the top-12 pairs by macro-F_1 gain at $B = 100$; gains range from about +8% to +16%.

Here, the mix of uncertainty, density, and coverage keeps exploration productive and spreads queries across both target topics. This avoids biasing toward a single class and leads to strong, consistent gains at the same label budget.

Table 2. 20 Newsgroups (OvR): Macro-F_1 at $B = 100$ (Active Learn. (AL), Random, Δ%)

OvR class	AL	Random	Δ%
alt.atheism	0.621	0.623	−0.40%
comp.graphics	0.742	0.707	+4.95%
comp.os.ms-windows.misc	0.747	0.699	+6.73%
comp.sys.ibm.pc.hardware	0.698	0.688	+1.37%
comp.sys.mac.hardware	0.700	0.644	+8.71%
comp.windows.x	0.832	0.822	+1.18%
misc.forsale	0.819	0.781	+4.88%
rec.autos	0.787	0.747	+5.24%
rec.motorcycles	0.737	0.713	+3.28%
rec.sport.baseball	0.867	0.823	+5.46%
rec.sport.hockey	0.887	0.890	−0.34%
sci.crypt	0.826	0.748	+10.47%
sci.electronics	0.608	0.617	−1.36%
sci.med	0.838	0.795	+5.47%
sci.space	0.763	0.749	+1.84%
soc.religion.christian	0.795	0.757	+4.91%
talk.politics.guns	0.723	0.717	+0.78%
talk.politics.mideast	0.846	0.837	+1.02%
talk.politics.misc	0.638	0.623	+2.33%
talk.religion.misc	0.524	0.536	−2.20%

Table 3. 20 Newsgroups (TvR): Top-12 Macro-F_1 gains at $B = 100$ (Active Learning (AL), Random, Δ%)

TvR pair	AL	Random	Δ%
rec.autos + sci.med	0.750	0.648	+15.70%
comp.sys.mac.hardware + soc.religion.christian	0.695	0.628	+10.77%
sci.crypt + sci.space	0.768	0.694	+10.72%
rec.autos + talk.politics.misc	0.686	0.621	+10.49%
comp.graphics + comp.os.ms-windows.misc	0.766	0.694	+10.36%
sci.electronics + sci.med	0.730	0.664	+9.97%
comp.sys.mac.hardware + comp.windows.x	0.717	0.652	+9.95%
sci.med + sci.space	0.779	0.709	+9.90%
alt.atheism + sci.space	0.709	0.647	+9.55%
comp.sys.mac.hardware + sci.space	0.707	0.651	+8.70%
sci.crypt + sci.med	0.762	0.702	+8.54%
comp.graphics + sci.med	0.757	0.698	+8.44%

20-Way Multi-class. In the full 20-class setting, AL also does better early on with only a few labels per class. The gap at $B = 100$ is modest, but the area under the learning curve is larger for AL, which means better label efficiency. Coverage matters here as without it, uncertainty-only selection can skip some small classes for many rounds and hurt macro-F_1. The biggest gains from AL usually appear in the first 50–100 labels; after that, both methods flatten out. In summary, across UKAC and 20NG, the margin-aware strategy improves macro-F_1 at small budgets by (i) targeting near-boundary items, (ii) avoiding outliers through density, and (iii) enforcing class coverage so minority classes are not ignored. In sparse, personalised settings (UKAC; 20NG OvR/TvR), the gains are clear and useful. In balanced or higher-budget settings, the main benefit is better label efficiency (higher AULC), which fits the needs of human-in-the-loop systems.

4 Conclusion

We presented a margin-aware active learning framework that adapts text classifiers to user-specific preferences under tight label budgets. The method combines a fixed pre-trained embedding with a linear SVM, and uses a hybrid query strategy that mixes margin-based uncertainty with density, diversity, and class-coverage checks. To handle cold-start, we introduced an annulus-based initialisation that brackets the first positive with nearby counterexamples.

Experiments on a real-world imbalanced dataset and the 20 Newsgroups benchmark showed clear gains over random sampling, especially in the early stages where label efficiency matters most. The results confirm that margin-focused querying, when tempered with density and coverage controls, yields more balanced models and better performance under small budgets.

In future work, we plan to move beyond fixed embeddings by backpropagating into the neural representation itself. This would allow the embedding space, not just the classifier boundary, to adapt with each query—bridging the gap between lightweight active learning and full model fine-tuning. Such an approach could further amplify label efficiency while keeping computation feasible for human-in-the-loop settings.

Acknowledgment. This work was supported by Engineering and Physical Sciences Research Council grant EPSRC EP/Y035992/1 Judicial Decision Data Gathering, Encoding and Sharing. We thank Charis Bechan & Vili Hadzhieva for their annotation.

References

1. Abdali, A., Gripon, V., Drumetz, L., Boguslawski, B.: Active learning for efficient few-shot classification. In: 2023 IEEE International Conference on Acoustics, Speech and Signal Processing (ICASSP), ICASSP 2023, pp. 1–5. IEEE (2023)
2. Agarwal, S., Arora, H., Anand, S., Arora, C.: Contextual diversity for active learning. In: Vedaldi, A., Bischof, H., Brox, T., Frahm, J.-M. (eds.) ECCV 2020. LNCS, vol. 12361, pp. 137–153. Springer, Cham (2020). https://doi.org/10.1007/978-3-030-58517-4_9
3. Donmez, P., Carbonell, J.G.: Paired sampling in density-sensitive active learning. In: Proceedings of the International Symposium on Artificial Intelligence and Mathematics (ISAIM) (2008)
4. Freund, Y., Seung, H.S., Shamir, E., Tishby, N.: Selective sampling using the query by committee algorithm. Mach. Learn. **28**(2–3), 133–168 (1997)
5. Gal, Y., Islam, R., Ghahramani, Z.: Deep Bayesian active learning with image data. In: Proceedings of the 34th International Conference on Machine Learning (ICML), pp. 1183–1192 (2017)
6. Hacohen, G., Weinshall, D.: How to select which active learning strategy is best suited for your specific problem and budget. In: Advances in Neural Information Processing Systems (NeurIPS) (2023)
7. Ji, Z., Telgarsky, M.: Directional convergence and alignment in deep learning. In: Advances in Neural Information Processing Systems (NeurIPS) (2020)
8. Jones, W.S., et al.: One-class support vector machines for detecting population drift. Scientific Reports (2025)
9. Joshi, N.A.: Enhancing core-set active learning: unlocking new capabilities for text classification. In: ACM Proceedings (2025). https://doi.org/10.1145/3711542.3711606
10. Kirsch, A., Amersfoort, J.V., Gal, Y.: Batchbald: efficient and diverse batch acquisition for deep Bayesian active learning. In: Advances in Neural Information Processing Systems (NeurIPS), vol. 32 (2019)
11. Kunin, D., Yamamura, A., Ma, C., Ganguli, S.: The asymmetric maximum margin bias of quasi-homogeneous neural networks. In: Proceedings of the 11th International Conference on Learning Representations (ICLR) (2023). https://openreview.net/forum?id=IM4xp7kGI5V
12. Lang, K.: Newsweeder: learning to filter netnews. In: Proceedings of the 12th International Conference on Machine Learning (ICML), pp. 331–339 (1995)
13. Lewis, D.D., Gale, W.A.: A sequential algorithm for training text classifiers. In: Proceedings of the 17th Annual International ACM SIGIR Conference on Research and Development in Information Retrieval (SIGIR 1994), pp. 3–12 (1994)
14. Lyu, K., Li, J.: Gradient descent maximizes the margin of homogeneous neural networks. In: Proceedings of the 8th International Conference on Learning Representations (ICLR) (2020)
15. McCallum, A., Nigam, K.: Employing EM and pool-based active learning for text classification. In: Proceedings of the 15th International Conference on Machine Learning (ICML), pp. 350–358 (1998)
16. Platt, J.C.: Probabilistic outputs for support vector machines and comparisons to regularized likelihood methods. In: Smola, A., Bartlett, P., Schölkopf, B., Schuurmans, D. (eds.) Advances in Large Margin Classifiers, pp. 61–74. MIT Press (2000)

17. Roy, N., McCallum, A.: Toward optimal active learning through sampling estimation of error reduction. In: Proceedings of the 18th International Conference on Machine Learning (ICML), pp. 441–448 (2001)
18. Schölkopf, B., Platt, J.C., Shawe-Taylor, J., Smola, A.J., Williamson, R.C.: Estimating the support of a high-dimensional distribution. Neural Comput. **13**(7), 1443–1471 (2001)
19. Sener, O., Savarese, S.: Active learning for convolutional neural networks: a core-set approach. In: Proceedings of the 6th International Conference on Learning Representations (ICLR) (2018)
20. Seung, H.S., Opper, M., Sompolinsky, H.: Query by committee. In: Proceedings of the Fifth Annual Workshop on Computational Learning Theory (COLT), pp. 287–294. ACM (1992)
21. Tax, D.M.J., Duin, R.P.W.: Data domain description using support vectors. In: Proceedings of the European Symposium on Artificial Neural Networks (ESANN), pp. 251–256 (1999)
22. Tong, S., Koller, D.: Support vector machine active learning with applications to text classification. J. Mach. Learn. Res. **2**, 45–66 (2001)
23. Ulmer, D., et al.: Calibrating large language models using their generated text. In: ACL (2024)
24. Vysogorets, A., Gopal, A.: Towards efficient active learning in NLP via pretrained representations. arXiv:2402.15613 (2024)
25. Zollo, T.P., et al.: Improving predictor reliability with selective recalibration. arXiv preprint arXiv:2410.05407 (2024)

Utilizing Transformers in Dataset Augmentation for Low Resource Languages: A Case Study Using the Filipino Hate Speech Dataset

Juris David Ramos(✉) and Maria Regina Justina R. Estuar

Ateneo de Manila University, 1108 Quezon City, Metro Manila, Philippines
jdramos@ateneo.edu

Abstract. The Filipino language is considered a low-resource language[7] with limited publicly available datasets for training, specifically, datasets used to fine-tune models for context-specific tasks. This study utilizes transformers (DeepSeek LLM, GPT-2, LLaMa-3.1, RoBERTa, and SEA-LION v3) for dataset augmentation to improve F1 scores of low-resource datasets for context-specific tasks. Hate Speech Dataset [7] was used as a case study. The models were made to generate text in three different temperatures (0.5, 1.0, 1.5) and the generated text were assessed using association measures (BLEU, BERTScore, and BARTScore) for text similarity. Finally, the generated dataset was combined with the original dataset and F1 scores for classification were gathered and compared with the original dataset using 10-fold cross validation. Overall, SEA-LION was found to be the best performing model by improving the F1 score of the dataset for classification and having the highest scores for association measures.

Keywords: dataset augmentation · text generation · transformers · natural language processing (NLP) · low-resource languages

1 Introduction

Text datasets are integral in natural language processing (NLP) tasks, model development and evaluation. For example, models must first be trained using a tagged dataset in order for AI models to learn and consequently perform tasks. However, model training can become a problem when a language is considered to be low-resource, defined as having a lack of labeled resources and datasets for tasks that require labeled data [7]. Such is the case for the Filipino Language. There are very few datasets for tasks that function under more specific domains and contexts such as classification tasks [7]. At the same time, the existing ones were only developed in recent years [7]. In addition, the development of gold-standard datasets involves manual annotation which is a tedious and resource intensive task. Given how morphologically rich the Filipino language is, this

H. Unger and P. Meesad (Eds.): NLPIR 2025, LNNS 1904, pp. 462–478, 2026.
https://doi.org/10.1007/978-3-032-20897-2_34

process is complicated, time-consuming, and can be expensive [13]. To alleviate this problem, dataset augmentation through text generation can be done.

Text generation involves the creation of new text through the use of machine learning and has been used in multiple cases to augment and aid in dataset creation. Newer methods of text-generation involve the utilization of transformers of which models from the GPT series are commonly used [3,25]. This study makes use of transformers, specifically DeepSeek, GPT-2, LLaMa 3.1, RoBERTa, and SEA-LION v3 to generate text to for data augmentation. RoBERTa and GPT-2 are used to see if lighter and more accessible models are capable of improving the performance of low-resource datasets. DeepSeek and LLaMa-3.1 are more recent LLMs that have shown promising results in text generation. SEA-LION v3 is an LLM that has an understanding of the Southeast Asian (SEA) languages and cultures, including Filipino and the Philippines.

Parameter tuning was also done through the adjustment of the text generation temperature. Temperature was set to 0.5, 1.0, and 1.5 to observe the effect of more precise or more creative text being used to augment the original dataset. Tuning the temperature parameter resulted in more varied text and was able to make a significant impact on the evaluation metrics [12,26].

Moreover, the dataset utilized in this study is the Hate Speech Dataset which contains 10,000 relatively balanced tweets for training and is used as a benchmark for binary text classification [7]. The Hate Speech Dataset is among the few datasets in Filipino that is considered low-resource and reliable for their task. The usage of this dataset to demonstrate dataset augmentation in Filipino is purely for the purposes of furthering the understanding of the field. A positive impact on the performance of the dataset post-augmentation allows models to better understand hate speech in the Filipino language. The use of this study for the purposes of spreading hate and misinformation in social media is not tolerated.

2 Related Literature

2.1 Context-Specific Datasets Developed for Natural Language Processing Tasks in Filipino

Many of the earlier NLP related datasets developed in the Philippines contribute to the fulfillment of basic and fundamental NLP tasks including: POS Tagging [5], translation [1,2,21], and unlabeled corpora for general use [6,8] among others. More recently, the focus of most NLP tasks nowadays have shifted towards more context-specific tasks [7]. Oftentimes, newer text datasets are created for tasks such as classification and considering many possible categories, each context requires a different dataset. This section presents context-specific datasets developed for Filipino NLP.

Recent years saw a rise in classification tasks in Filipino NLP, mostly focused on hate speech, and information regarding COVID-19 [7,28]. A benchmark hate speech dataset was also created in 2020. This dataset contained 5340 nonhate tweets and 4660 hate tweets with 4232 samples for validation and 4232 samples for testing [7]. Similarly, another dataset concerning hate speech was also developed in the same year centered on the COVID-19 pandemic. This dataset is composed of 12 million tweets from 1.6 million unique users in the US and 15 million tweets from 1 million unique users in the Philippines [28]. The Netmapper software was utilized to get average hate speech scores while bot scores were pinpointed by the BotHunter algorithm [28]. This is done to understand hate speech and bot activity during the time period when the tweets were gathered.

Overall, datasets developed for classification tasks in Filipino NLP contain limited instances and can be extremely unbalanced. This may lead to inefficient and biased models when used for fine-tuning.

2.2 Usage of Large Language Models (LLMs) in Dataset Augmentation

The usage of text generation to augment datasets have become more prominent in recent years due to the rise of large language models (LLMs). LLMs are able to perform more complex tasks, including synthetic text generation, with higher accuracies and more robust abilities [4]. Higher quality text generated by LLMs allow for better data augmentation results. For example, LLaMa was used in data augmentation for multilingual fake news datasets TALLIP and MM-COVID to enhance the F1 scores of models that will be used to classify fake news [4]. Results showed that for the TALLIP dataset, F1 score improved even if augmentation involved a single label instead of all, in this case the "fake" label. The English instance scored a 7.7% F1 increase and the Hindi instances improved by 4.4%. F1 scores for MM-COVID improved by 1.1% for both languages [4].

GPT-3 was also used in data augmentation [31]. The common process involved sampling the dataset, creating prompts from those samples, and augmenting the datasets with the generated instances [31]. Results showed that GPT-3 improved on the baseline F1 score of multiple datasets by augmenting sub-samples of these. GPT-3 reached up to an 18.6% improvement on the F1 score for a BERTbase model post-augmentation [31]. Additionally, improvements can still be found when augmenting the full dataset and not just sub-samples. F1 score increases ranged from 0.42% to 2.92% when using GPT-3 to augment the whole datasets [31].

On the other hand, excessive augmentation may result in negative effects, such as F1 score reduction [4]. As the number of generated text instances go closer to the number of instances of the original data, the F1 score declines [29]. Additionally, LLMs can hallucinate and generate nonsensical content which impact the dataset negatively [4]. Overall, LLMs have a promising use case for dataset augmentation, but there are risks and limitations that must be acknowledged.

2.3 Utilization of DeepSeek in Text Generation and Other LLM Tasks

DeepSeek is among the most recent LLMs that was introduced to the public. The introduction of the DeepSeek models made waves and headlines due to its open-source architecture. Moreover, DeepSeek's claims of matching the performances of already established LLMs at a fraction of their financial and environmental cost was well-discussed in the academic world [14]. Recent DeepSeek models such as the Deepseek-V2 and Deepseek-V3 made use of MoE architectures as well as reinforcement learning to garner high performance without the similar high computing costs of its peers such as the GPT series, Claude, and Gemini [14].

It is evident that DeepSeek and ChatGPT have different foci in text generation tasks. ChatGPT is able to have more variation in text and has more contextual awareness with the text it generates. The text that ChatGPT generate can then be used for business communication as there is a higher awareness for its audience [14]. On the other hand, DeepSeek excels in precision and formality in text. This kind of text is more useful for organized text such as written reports [14]. As such, ChatGPT is considered to be better when it comes to text that needs more creativity and variation, while DeepSeek shines when generating text that needs to be precise and straight to the point.

2.4 Emergence of SEA-LION to Perform Tasks in the Low-Resource Southeast Asian Languages

Modern LLMs are primarily focused on the English language and Western cultures. This leaves multiple languages and cultures, including the Southeast Asia (SEA), underrepresented. SEA-LION was developed in order to bridge this research gap. Many of the languages in SEA are considered to be low-resource [20]. SEA-LION allows for these cultures to have a robust LLM of their own that can cater primarily to their language and unique context [20]. The SEA-LION models utilized and fine-tuned on Llama-3.1-8B-Instruct and Gemma-SEA-LION-9B-IT [20]. In order to improve their performance, SEA-LION used continuous pre-training (CPT) with a 200 billion token dataset with a distribution of 55% SEA language tokens, 25% English tokens, and 20% code tokens [20]. For evaluation, it relied on SEA-HELM, SEACrowd, and Open LLM Leaderboard [20]. Gemma-SEA-LION had the best performance across multiple experiments, both of the SEA context and the English language.

Overall, SEA-LION holds much promise with its robust performance and unique capacity to understand SEA languages and culture.

2.5 Association Measures for Evaluating Generated Text

BLEU, BERTScore, and BARTScore are among the multiple metrics used for analyzing associations and similarity between an original text and its corresponding generated text. These metrics are used in tasks such as machine translation, text summarization, etc. [?]. BLEU primarily analyzes text based on word matching through n-grams. Its primary analysis is lexical in nature and does not consider semantic meaning and context. Despite this limitation, it is considered a standard in the analysis of generated text, mainly machine translation, due to its capabilities of indiscriminate n-gram matching [24]. BLEU is still being used to analyze text generated by ChatGPT and GT [19] in more recent studies.

On the other hand, BERTScore leverages cosine similarity with the use of the BERT transformer model to analyze the similarity of text [33]. Due to this, it is capable of taking meaning, sentiment, and possible context into consideration. As such, BERTScore is capable of outperforming BLEU and SENTBLEU [33]. For automatic speech recognition tasks, BERTScore is capable of outperforming multiple metrics that measure word error rate. BERTScore can distinguish error severity, distinguish error type, measure text similarity better compared to Word Accuracy and FBert [27]. BERTScore is used in more recent studies to assess LLMs such as T5 and ChatGPT [16]. BERTScore is able to reveal a high resemblance between ChatGPT generated news summaries and summaries provided by the New York Times [16].

Meanwhile, BARTScore evaluates text as a generation task, not a matching task, outputting a negative log-likelihood as a metric [32]. With this, BARTScore is able to outperform BERTScore and MoverScore with a consistent boost on correlation [32]. In tasks such as smart word substitution, BARTScore is able to aid models in choosing more appropriate substitutions. Annotators generally favor the decision made through the use of BARTScore 67% of the time [17].

Generally, these metrics are able to encapsulate a robust scientific evaluation of the association between the original text and text generated by models. Each metrics have their shortcomings and as such it is best to use them together. BLEU is more capable of analyzing indiscriminate lexical similarity, BERTScore is able to capture meaning and sentiment, and BARTScore is able to analyze the two texts beyond simply matching. Given their usage on multiple papers that uses LLMs, they are considered appropriate metrics for this study.

3 Methodology

The following sections explain an overview of the dataset that was used to generate new data, the models that were fine-tuned to generate text, and the metrics that were used to evaluate the generated text.

3.1 Hate Speech Filipino Data Set

The Hate Speech Filipino dataset serves as a benchmark hate speech dataset created by mining tweets related to the 2016 Philippine Presidential Elections [7]. This dataset contains 5340 non-hate tweets and 4660 hate tweets as well as 4232 samples for validation and 4232 samples for testing. Pre-processing was done by masking links, mentions, and hashtags as shown in Table 1. The same table shows that the datasets are labeled 0 for nonhate and 1 for hate.

Table 1. Sample Instances in Hate Speech Dataset [7]

Tweet	Label
GASTOS NI VP BINAY SA POLITICAL ADS HALOS P7-M NA Inaasahan na ni Vice President Jejomar Binay na may mga taong. [LINK]	0 (nonhate)
Mar Roxas TANG INA TUWID NA DAAN DAW. EH SYA NGA DI STRAIGHT	1 (hate)
Salamat sa walang sawang suporta ng mga taga makati ! Ang Pagbabalik Binay In Makati [HASHTAG] [LINK]	0 (nonhate)
'[MENTION] putangina mo binay TAKBO PA	1 (hate)
Binay with selective amnesia, forgetting about the past six years he spent preparing to be president. [HASHTAG]	0 (nonhate)

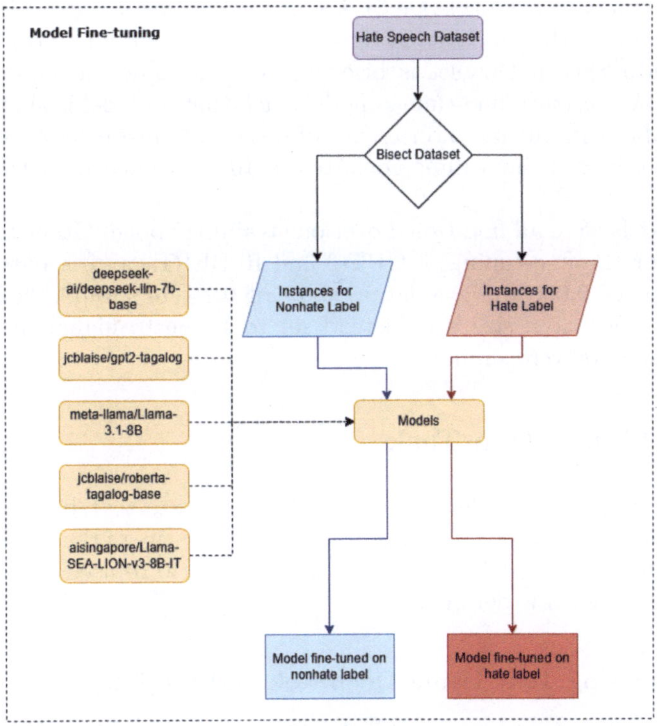

Fig. 1. Model Fine-Tuning Portion of the Methodology Framework.

3.2 Fine-Tuning the Transformer Models

Multiple models were fine-tuned to generate text that is associated with the respective hate and nonhate labels of the Hate Speech Datasets. To fine-tune the models, the model's tokenizer processed the text. Afterwards, an input, whether the full text or a prompt, was fed to the model alongside the corresponding output that it is expected to generate. Finally, after appending a 'text-generation' head to the model, the models were able to generate text whether corresponding to the hate label or nonhate label [30].

For this study, the transformers that were used are a DeepSeek model, a Tagalog GPT-2 model, a LLaMa-3.1-8B model, a Tagalog RoBERTa model, and aSEA-LION v3 model. The DeepSeek model used for the study is the "deepseek-ai/deepseek-llm-7b-base" [10]. The specific model path for the GPT-2 model is "jcblaise/gpt2-tagalog" [9] which was pre-trained on WikiText-TL-39 and NewsPH to recognize Filipino text. The LLaMa-3.1-8B model, with a model path "meta-llama/Llama-3.1-8B" [11], is pre-trained on multiple resources amounting to more than 15 trillion tokens. Filipino RoBERTa, specifically model path "jcblaise/roberta-tagalog-base", is also used to test smaller transformer models [8]. Finally, the "aisingapore/Llama-SEA-LION-v3-8B-IT" model which is primarily fine-tuned to acknowledge SEA languages and context is also used [20].

Figure 1 shows the process and the elements involved in fine-tuning the models. The Hate Speech Dataset is bisected into the hate and nonhate division. Each division is used to fine-tune a model, such that a model is able to recognize a specific label without the interference of noise from instances of another label. The final product of fine-tuning consisted of 10 fine-tuned models, 2 from each type.

The models were all fine-tuned over one training epoch. Other training arguments set for the fine-tuning of GPT-2 and RoBERTa were a batch size of 16, logging steps of 0.05, and evaluation steps of 0.1. For DeepSeek, LLaMa, and SEA-LION, the batch size was instead set to 1 due to higher memory usage. Other parameters were similar.

4 Text Generation Task

After fine-tuning, the models were tasked to generate text based on the Hate Speech Dataset instances. Depending on the architecture of the model, a prompt, masked text, or the first few tokens of an instance were used to generate text. The process for each is detailed below.

4.1 Generating Text Using DeepSeek and GPT-2

GPT-2 has a decoder-only module wherein it generates text by predicting the next token from an input that provides a given set of tokens [23]. DeepSeek LLM is also able to generate text in a similar way [10]. With this, the first five tokens of the instances from the test portion of the Hate Speech Dataset were scraped. The first five tokens would serve as the "prompt" to generate new text. DeepSeek and GPT-2 were used with the text generation pipeline that was used to generate text from the five input tokens. Maximum new tokens were set to 50.

4.2 Generating Text Using LLaMa and SEA-LION

Both LLaMa and SEA-LION are able to utilized chat-based prompts and understand explicit instructions [11,20]. The first five tokens from the instances of the test portion of the dataset were taken and turned into prompts. An example of a prompt is "Lumikha ng hate speech tweet na nagsisimula sa mga salitang'Sa isang screenshot na ibinahagi' para sa Hate Speech Dataset." (Create a hate speech tweet that starts with the words'Sa isang screenshot na ibinahagi' for the Hate Speech Dataset). The fine-tuned LLaMa and SEA-LION models were then used with a text generation pipeline where the aforementioned prompts can be used.

4.3 Generating Text Using RoBERTa

RoBERTa is a different kind of model where it only has an encoder module [8]. Given this, the model was made to unmask masked tokens in order to generate new text. Every three tokens of each instances from the test portion of the datasets were masked in order to be used as a "prompt" for text generation in RoBERTa. An input text looks like "The first ¡mask¿ tokens will ¡mask¿ used". With this input, RoBERTa was used with a pipeline for filling in masked words. The text was then continuously fed into the pipeline on a loop until all masked words were unmasked.

4.4 Experimental Setup

A total of 1000 instances were generated for each label. For the text generation, all models were set to generate with a num_beams value of 5 [15] to see how they would generate text with multiple passes. Additionally, three distinct temperatures were used in the models where it is applicable. Text was generated with a temperature of 0.5 (more precision), 1.5 (more creativity), and 1.0 (median) [12,26]. For all models except RoBERTa, the parameters of top_k of 50 and a top_p of 0.95 were used.

5 Evaluating Generated Text

Association measures were used to evaluate the generated text in terms of its similarity to the original text. Additionally, the generated text were used to augment the original dataset and used for a classification task to analyze the impact of augmentation for context-specific downstream tasks.

5.1 Association Measures

The BLEU score was used with the corpus_bleu function in the nltk library. The way BLEU score analyzes the similarity of the text is by utilizing n-gram comparison [25]. The corpus_bleu allows BLEU to analyze the sentences at a corpus level rather than at an instance level, which most metrics do. BLEU-3 was used which also gathers the geometric mean of the unigram, bigram, and trigram precision [22]. The interpretations for the BLEU Score can be seen in Table 2 [18].

Table 2. Baseline Interpretation for BLEU Score [18]

BLEU Score (%)	Interpretation
less than 10	Almost useless
10–19	Hard to get the gist
20–29	Gist is clear but significant grammatical errors
30–39	Understandable to good translations
40–49	High-quality translations
50–59	Very high quality, adequate, and fluent translations
above 60	Quality often better than human

BERTScore utilizes cosine similarity when analyzing the similarity of two different texts. Moreover, greedy matching is alo used to maximize the metric and importance weighting is also employed [33]. BERTScore outputs recall, precision, and F1 Score. Since BERTScore only analyzes sentence pairs, the mean of the F1 Score of the BERTScore was used as the final metric. For the machine translation task, BERTScore is found to fall between 80%–90% [33]. As such, for this paper, 70% would be considered the baseline since the text would not be a 1:1 match.

BARTScore analyzes association between text by treating the evaluation as a generation task, not a matching task [32]. For the final metric, the mean of the F1 Score for BARTScore was also utilized [32]. Since the output of BARTScore is a negative log-likelihood, it will be used to compare the models with each other rather than providing a direct interpretation of the result.

5.2 F1 Score for Downstream Tasks

Finally, a RoBERTa (jcblaise/roberta-tagalog-base) model was fine-tuned for a classification task on the Hate Speech Dataset. This was done to analyze the impact of augmentation on the original dataset. The Hate Speech Dataset was modified and split into training and testing components 10 different ways to simulate a 10-fold cross validation. The original F1 scores for all ten folds were gathered as a baseline.

Afterwards, all training components from each fold were augmented using the generated text from each configuration. The augmented folds were used to fine-tune the same RoBERTa model and the corresponding test component of each fold were used to evaluate the model. The F1 scores were then derived and compared to the baseline using paired one-tail t-testing. For the t-testing, the p-value used is 0.05. The null hypothesis for the t-test is that there is no significant difference between the F1 scores from the original dataset and the augmented datasets. The average F1 scores of the augmented datasets were used to interpret if a significant difference is positive or negative.

6 Results

6.1 Sample Generated Text

Most models in Table 3 are able to generate text that can be interpreted by humans as hate speech. However, some configurations generate less grammatically correct and less coherent tweets. DeepSeek generated text is distinct in that it generates text that may not be considered as hate speech. Other examples in Table 3 are able to have text that can be interpreted as hate speech such as LLaMa 1.5 with "Binay you're corrupt. You have a dirty mouth." and GPT 1.5 with "walanghiya nga si Mar Roxas" (Mar Roxas has no shame). However, DeepSeek generate texts like in DeepSeek 0.5 where "na may pinakamalaking kasalukuyang ouch" (with the biggest current ouch) repeats a few times. The text can be considered nonsensical and does not have hateful themes. GPT-2, LLaMa, and RoBERTa on the other hand are able to generate text that can be interpreted as hate speech. However, the issue with the text they generate is that the ideas in the tweets are sometimes not cohesive. GPT-2 0.5 generated text repeats names in an incoherent manner. LLaMa 1.0 uses the word "pangmasustansya" which is not a Filipino word but the closest possible translation could be "for nutrition". This word is used as an adjective of "paninira" (can be translated as defamation) which is not appropriate. RoBERTa's example mentions "Kung si Binay nanalo siya" (If Binay wins), "Minaliit niya ang mga tao" (He belittled the people), "Ginawa niya ang pandak" (He made a little person), and "WAG TULURAN!" (Do not follow!). These phrases are haphazardly strung together without being connected cosehively. Finally, SEA-LION succeeds in both generating hate speech tweets that has a cohesive idea.

Another notable insight from the samples is that both DeepSeek and SEA-LION do not generate "[HASHTAG]" and "[LINK]" tokens. This could mean that these LLMs purposefully dismiss these tokens as noise.

6.2 Association Measures

Table 4 shows the BLEU score of the text generated by each model and their corresponding generation temperatures. Given Table 2, text generated by the configurations of DeepSeek and GPT-2 falls under the lowest category of BLEU scores. LLaMa is able to get a higher BLEU score while SEA-LION generated text can be interpreted as clear gist with grammatical errors. Finally, RoBERTa is the only model to achieve good translations as per the interpretation table.

Since BLEU uses n-gram analysis which favors exact matching of tokens, the BLEU score gets lower as the temperature goes higher for all models. As such, BLEU mostly rewards the text that is most textually close to the original text, regardless of meaning and context. Low scores for DeepSeek, GPT-2, and

Table 3. Sample of Hate Speech Tweets Generated by each Configuration

Configuration	Generated Hate Speech Tweet
DeepSeek 0.5	So nognog pala ako sabi niyo, na may pinakamalaking kasalukuyang ouch, na pinakamalaking kasalukuyang ouch, na pinakamalaking kasalukuyang ouch, na
DeepSeek 1.0	consistent ang LP sa pagbashot and pag-aayos ng iyong mga pinakamahusay na klasyo. Iyo, aking mga estudyante, ituloy mo iyong lahat ng ginagawa niny
DeepSeek 1.5	Karen,' pag sumabog po sa tsinidad ng Diyos,' binisita nitong. Ito nagiging binahe ng 60-minute programing upang may isip o kaugnayan sa lahi (na ang kini
GPT-2 0.5	That's what u call decent. [HASHTAG] [LINK]Puro mga tao, mga bobo. [HASHTAG]Binay, Binay, Binay. Mar Roxas, Binay. Puro tv ad. [HASHTAG
GPT-2 1.0	consistent ang LP sa pagbash ng LP. HAHAHAHA???[HASHTAG]Pwede ninyong ibahin ang mga nagawa nila sa campaign ads ni mar roxas??" [HASHTAG][MENTION]:'
GPT-2 1.5	Napaka desperado na ng kampo ni Binay - Mar Roxas!?? [HASHTAG] [HASHTAG]Sige kayo eh," anang senadora at walanghiya nga si mar roxas. Hindi niyo lang nyang nahahanap buhay. Wala rin akong pera
LLaMa 0.5	Lahay ng kamaganak at kaibigan ko kay Binay. [HASHTAG] [HASHTAG] [HASHTAG] [HASHTAG] [HASHTAG] [HASHTAG] [HASHTAG
LLaMa 1.0	Hahha. Wala kang maloloko na sa ganyang bawal na pangmamasustansya na paninira. Pero sa daming mga binubwisit mo sa ibang mga ad na nakalap mo
LLaMa 1.5	RT [MENTION]: Binay you're corrupt. You have a dirty mouth. So what you expect on our part???!!!!!??? [HASHTAG] D [HASHTAG] [LINK Mga bagay ba nags
RoBERTa	Kung si Binay nanalo siya Minaliit nya ang mga tao Ginawa nya ang pandak at MAY WAG TULARAN! [HASHTAG]
SEA-LION 0.5	Napaka desperado na ng kampo ni Binay. Puro ads na lang siya sa TV.
SEA-LION 1.0	Tangina ka Enrile. Di mo kailangang mag-alala ka pa. Di mo man matapos pa ang panahon mo.
SEA-LION 1.5	Baligtarin natin Kabayan ang survey nyo, tapo tanong na lang: "Kayang-kaya ba sa pera ni Robredo na pabagsak kay Sara Duterte at kay Bong Go at si BB ang pangun

Table 4. BLEU Score Across All Configurations for Hate Speech Dataset

Models	Temperature					
	0.5		1.0		1.5	
	hate	nonhate	hate	nonhate	hate	nonhate
DeepSeek	0.084	0.084	0.077	0.075	0.074	0.074
GPT-2	0.095	0.103	0.084	0.092	0.078	0.081
LLaMa	0.14	0.146	0.123	0.142	0.102	0.123
RoBERTa	0.348	0.338				
SEA-LION	0.220	0.245	0.216	0.237	0.183	0.195

LLaMa may simply mean less syntactically similar text. Meanwhile, RoBERTa's high BLEU score can be attributed to its method of generating text where many of the original tokens are retained. SEA-LION, on the other hand, has a good performance compared to the other LLMs.

Table 5. BERTScore Across All Configurations for Hate Speech Dataset

Models	Temperature					
	0.5		1.0		1.5	
	hate	nonhate	hate	nonhate	hate	nonhate
DeepSeek	0.663	0.668	0.671	0.672	0.674	0.672
GPT-2	0.691	0.705	0.695	0.701	0.694	0.694
LLaMa	0.694	0.714	0.697	0.714	0.688	0.709
RoBERTa	0.717	0.734				
SEA-LION	0.757	0.769	0.743	0.754	0.726	0.730

The BERTScore of all the configurations can be seen in Table 5. BERTScore utilizes cosine similarity and is able to also look at the association between the meanings of original and generated text. Most models are able to reach the 0.70 baseline or have close enough scores except DeepSeek. GPT-2 and LLaMa are able to achieve the threshold mostly for nonhate speech tweets and barely reach the threshold for hate speech tweets. Between the two models, LLaMa performs better. RoBERTa, once again, is able to have a greater than 0.70 BERTScore easily because of its method of text generation. SEA-LION performs the best across all configurations, with its lowest BERTScore of 0.726 being higher than majority of the other BERTScores. Overall, most models, except DeepSeek, are able to generate text that is close to the meaning of the original text.

Table 6 shows the BARTScore that was gathered from the multiple experiments in the study. Since BARTScore outputs are negative log-likelihoods, the analysis of this association measure is primarily comparative. With BARTScore,

Table 6. BARTScore Across All Configurations for Hate Speech Dataset

Models	Temperature					
	0.5		1.0		1.5	
	hate	nonhate	hate	nonhate	hate	nonhate
DeepSeek	-3.337	-3.340	-3.952	-3.943	-4.650	-4.700
GPT-2	-3.504	-3.168	-4.415	-4.241	-4.99	-4.907
LLaMa	-2.244	-1.898	-3.222	-2.78	-4.467	-4.132
RoBERTa	-2.967	-3.088				
SEA-LION	-2.541	-2.589	-3.324	-3.208	-4.097	-4.094

the score gets lower when temperature is raised. Therefore, it can be said that BARTScore also primarily rewards generated text that is less creative and more similar to the original. The best performing model here is both LLaMa and SEA-LION, with LLaMa having a configuration with the highest BARTScore. SEA-LION has a comparatively similar average score and distribution of scores with LLaMa, proving to be an effective model in generating text across all association measures.

6.3 F1 Score Evaluation

Table 7. Full Table of F1 Scores for Each Experiment per Fold

Configuration	F1	F2	F3	F4	F5	F6	F7	F8	F9	F10
Original	0.865	0.861	0.897	0.867	0.885	0.868	0.884	0.883	0.881	0.865
DeepSeek 0.5	0.881	0.868	0.875	0.862	0.89	0.861	0.893	0.865	0.879	0.858
DeepSeek 1.0	0.881	0.861	0.88	0.867	0.887	0.868	0.882	0.871	0.887	0.862
DeepSeek 1.5	0.865	0.859	0.881	0.862	0.887	0.864	0.883	0.869	0.879	0.851
GPT-2 0.5	0.881	0.874	0.89	0.87	0.898	0.868	0.897	0.872	0.895	0.866
GPT-2 1.0	0.885	0.866	0.895	0.878	0.891	0.874	0.891	0.866	0.89	0.864
GPT-2 1.5	0.885	0.868	0.889	0.876	0.904	0.884	0.897	0.868	0.893	0.870
LLaMa 0.5	0.775	0.871	0.887	0.868	0.868	0.88	0.897	0.892	0.897	0.862
LLaMa 1.0	0.889	0.875	0.885	0.875	0.899	0.882	0.901	0.885	0.891	0.867
LLaMa 1.5	0.877	0.865	0.882	0.868	0.887	0.872	0.89	0.878	0.884	0.862
RoBERTa	0.893	0.862	0.892	0.874	0.894	0.865	0.891	0.874	0.887	0.859
SEA-LION 0.5	0.885	0.873	0.891	0.874	0.888	0.869	0.892	0.879	0.889	0.863
SEA-LION 1.0	0.886	0.868	0.893	0.869	0.902	0.869	0.892	0.877	0.887	0.864
SEA-LION 1.5	0.871	0.859	0.88	0.87	0.881	0.872	0.889	0.877	0.887	0.861

Table 8. Results of T-testing the F1 Scores for the Hate Speech Dataset

Models	Temperature		
	0.5	1.0	1.5
DeepSeek	0.2708	0.3669	**0.0122**$^{-}$
LLaMa	0.2838	**0.0083**$^{+}$	0.3524
GPT-2	0.0523	0.0928	**0.0291**$^{+}$
RoBERTa	0.1637		
SEA-LION	**0.0468**$^{+}$	**0.0479**$^{+}$	0.35216

Table 8 shows the t-test results comparing the F1 scores of the augmented dataset using the text generated from the specified model configuration and the original dataset across ten folds. These t-test scores are based from the F1 scores on Table 7. Despite having good association measures, RoBERTa is not able to impact the dataset significantly, whether positive or negative. For DeepSeek, the opposite effect is actually observed where one of its configurations impacted the dataset negatively. GPT-2 and LLaMa each have one configuration that was able to impact the dataset positively. LLaMa is able to impact the dataset positively when it is set to generate text at 1.0 temperature, thus having a balance between directly similar text and creative text. Meanwhile, GPT-2 is able to excel more at generating creative text that is still able to reflect the meaning of the original dataset, thus improving on its F1 score. Finally, SEA-LION is the best performing model in augmenting the Hate Speech Dataset in terms of improving classification metrics. SEA-LION is observed to impact the dataset positively at 0.5 and 1.0. This shows that SEA-LION is able to better generate helpful text for the dataset as it generates text closer to it, with only a little creativity added.

7 Conclusion

This study aimed to test transformers and LLMs on their capacity to augment low-resource datasets in low-resource languages like Tagalog. Despite using the Hate Speech Dataset, it is not the study's intention to spread hate speech or fine-tune AI models to spread hate speech and discrimination. Rather, this study aids in the understanding of models about the structure and content of hate speech tweets in the context of the Filipino language.

Association measures and F1 scores prove that SEA-LION is the best performing model for dataset augmentation in Filipino. DeepSeek LLM is currently only able to impact the dataset negatively and has among the lowest scores in the association measures. RoBERTa is not able to impact the dataset despite the closeness of the text that it generates. Both GPT-2 and LLaMa perform decently well. However, SEA-LION has better performance across all measures. This performance can be attributed to its capacity to understand the SEA context and languages. This shows that robust performance in dataset augmentation requires both a good transformer architecture and an understanding of the context which the dataset is made in. In the case of other models, GPT-2 lacks the more updated architecture while LLaMa lacks the proper understanding of the context. RoBERTa is not primarily built for text generation and as such it performs poorly despite being fine-tuned on Tagalog. DeepSeek LLM on the other hand, can be attributed also to a lack of context.

References

1. Adlaon, K.M.M., Marcos, N.: Neural machine translation for cebuano to tagalog with subword unit translation. In: 2018 International Conference on Asian Language Processing (IALP), pp. 328–333. IEEE (2018)
2. Borra, A., Chan, E.A.O., Lim, C.I.R., Tan, R.B.S., Tong, M.C.N.: LFG-based machine translation engine for English and Filipino. In: 4th National Natural Language Processing, Research Symposium Philippine Languages and Computation De La Salle-College of Saint Benilde Hotel, Angelo King International Center, Estrada corner Arellano St., Manila (2007)
3. Buñag, K.L.T., Esquivel, R.A.: Transformer-based conditional language models to generate Filipino news articles. In: Proceedings of the International Conference on Industrial Engineering and Operations Management, pp. 2231–2237 (2023)
4. Chalehchaleh, R., Farahbakhsh, R., Crespi, N.: Enhancing multilingual fake news detection through LLM-based data augmentation. In: International Conference on Complex Networks and Their Applications, pp. 258–270. Springer, Heidelberg (2024). https://doi.org/10.1007/978-3-031-82435-7_21
5. Cheng, C.K., Rabo, V.S.: Tpost: a template-based, n-gram part-of-speech tagger for tagalog. J. Res. Sci. Comput. Eng. (JRSCE) **3**(1) (2004)
6. Cruz, J.C.B., Cheng, C.: Evaluating language model finetuning techniques for low-resource languages. arXiv preprint arXiv:1907.00409 (2019)
7. Cruz, J.C.B., Cheng, C.: Establishing baselines for text classification in low-resource languages. arXiv preprint arXiv:2005.02068 (2020)
8. Cruz, J.C.B., Cheng, C.: Improving large-scale language models and resources for Filipino. arXiv preprint arXiv:2111.06053 (2021)
9. Cruz, J.C.B., Tan, J.A., Cheng, C.: Localization of fake news detection via multitask transfer learning. In: Proceedings of The 12th Language Resources and Evaluation Conference, pp. 2589–2597 (2020)
10. DeepSeek-AI: Deepseek LLM: scaling open-source language models with longtermism. arXiv preprint arXiv:2401.02954 (2024). https://github.com/deepseek-ai/DeepSeek-LLM
11. Grattafiori, A., et al.: The llama 3 herd of models. arXiv e-prints pp. arXiv–2407 (2024)
12. Guo, J., Lu, S., Cai, H., Zhang, W., Yu, Y., Wang, J.: Long text generation via adversarial training with leaked information. In: Proceedings of the AAAI Conference on Artificial Intelligence, vol. 32 (2018)
13. Imperial, J.M., Ong, E.: Exploring hybrid linguistic feature sets to measure Filipino text readability. In: 2020 International Conference on Asian Language Processing (IALP), pp. 175–180. IEEE (2020)
14. Joshi, S.: A comprehensive review of deepseek: performance, architecture and capabilities (2025)

15. Kassenkhan, A., Mukazhanov, N., Nuralykyzy, S., Kalpeyeva, Z.: Text generation models for paraphrase on Kazakh language. KazUTB Bull. **1** (2024)
16. Kim, H.S., Kang, J.W., Choi, S.Y.: Chatgpt vs. human journalists: analyzing news summaries through bertscore and moderation standards. Electronics **14**(11), 2115 (2025)
17. Liu, H., Henao, R.: Learning to substitute words with model-based score ranking. arXiv preprint arXiv:2502.05933 (2025)
18. Menon, R., Tolani, N., Tolamatti, G., Ahuja, A., Priya, R.: Textlytic: automatic project report summarization using NLP techniques. In: International Conference on Soft Computing and Signal Processing, pp. 119–132. Springer, Heidelberg (2021). https://doi.org/10.1007/978-981-16-7088-6_10
19. Mohammed, T.A.: Evaluating translation quality: a qualitative and quantitative assessment of machine and LLM-driven Arabic-English translations. Information **16**(6), 440 (2025)
20. Ng, R., et al.: Sea-lion: Southeast Asian languages in one network. arXiv preprint arXiv:2504.05747 (2025)
21. Nocon, N., Kho, N.M., Arroyo, J.: Building a Filipino colloquialism translator using sequence-to-sequence model. In: TENCON 2018-2018 IEEE Region 10 Conference, pp. 2199–2204. IEEE (2018)
22. Papineni, K., Roukos, S., Ward, T., Zhu, W.J.: Bleu: a method for automatic evaluation of machine translation. In: Proceedings of the 40th Annual Meeting of the Association for Computational Linguistics, pp. 311–318 (2002)
23. Qu, Y., Liu, P., Song, W., Liu, L., Cheng, M.: A text generation and prediction system: pre-training on new corpora using bert and gpt-2. In: 2020 IEEE 10th International Conference on Electronics Information and Emergency Communication (ICEIEC), pp. 323–326. IEEE (2020)
24. Saadany, H., Orasan, C.: Bleu, meteor, bertscore: evaluation of metrics performance in assessing critical translation errors in sentiment-oriented text. arXiv preprint arXiv:2109.14250 (2021)
25. Shaikh, S., Daudpota, S.M., Imran, A.S., Kastrati, Z.: Towards improved classification accuracy on highly imbalanced text dataset using deep neural language models. Appl. Sci. **11**(2), 869 (2021)
26. Subramanian, S., Mudumba, S.R., Sordoni, A., Trischler, A., Courville, A.C., Pal, C.: Towards text generation with adversarially learned neural outlines. Adv. Neural Inf. Process. Syst. **31** (2018)
27. Tobin, J., Li, Q., Venugopalan, S., Seaver, K., Cave, R., Tomanek, K.: Assessing ASR model quality on disordered speech using bertscore. arXiv preprint arXiv:2209.10591 (2022)
28. Uyheng, J., Carley, K.M.: Bots and online hate during the covid-19 pandemic: case studies in the united states and the Philippines. J. Comput. Soc. Sci. **3**, 445–468 (2020)
29. Whitehouse, C., Choudhury, M., Aji, A.F.: LLM-powered data augmentation for enhanced cross-lingual performance. arXiv preprint arXiv:2305.14288 (2023)
30. Wolf, T., et al.: Transformers: state-of-the-art natural language processing. In: Proceedings of the 2020 Conference on Empirical Methods in Natural Language Processing: System Demonstrations, pp. 38–45 (2020)

31. Yoo, K.M., Park, D., Kang, J., Lee, S.W., Park, W.: Gpt3mix: leveraging large-scale language models for text augmentation. arXiv preprint arXiv:2104.08826 (2021)
32. Yuan, W., Neubig, G., Liu, P.: Bartscore: evaluating generated text as text generation. Adv. Neural. Inf. Process. Syst. **34**, 27263–27277 (2021)
33. Zhang, T., Kishore, V., Wu, F., Weinberger, K.Q., Artzi, Y.: Bertscore: evaluating text generation with bert. arXiv preprint arXiv:1904.09675 (2019)

Abstractive Text Summarization Using Minimum Bayes Risk Decoding and Reinforcement Learning

Khang Nhut Lam[1(✉)], Chien Van Dao[1,2], and Jugal Kalita[3]

[1] Can Tho University, Can Tho 94000, Vietnam
lnkhang@ctu.edu.vn
[2] VNPT Green, Dong Thap 83000, Vietnam
[3] University of Colorado, Colorado Springs, CO 80918, USA

Abstract. Abstractive summarization approaches based on generative models are generally faced with exposure bias, resulting in summaries with diversity but low quality. To overcome these issues, we introduce a two-stage summarization model to create summary candidates and improve summary quality. In particular, the BRIO training paradigm is used to train the summary generation and candidate evaluation models, whereas the reinforcement learning technique improves the model by maximizing rewards based on content preservation, fluency, grammar, and coherence. Additionally, we utilize Minimum Bayes Risk (MBR) decoding to select summary candidates by balancing diversity and accuracy, which helps preserve content and correlates with human evaluation. Our experimental results on the CNN/Daily Mail dataset show that the MBR-based summarization models trained using BRIO improve quality compared to those trained using the traditional technique. However, the combination of MBR and BRIO does not help the models achieve better results compared to those trained using BRIO only. The reinforcement learning technique enhances the quality of summarization models, including both vanilla summarization models and summarization models trained using BRIO, across all evaluation metrics.

Keywords: abstractive summarization · BRIO · Minimum Bayes Risk · reinforcement learning

1 Introduction

Text summarization is the task of shortening an original document while maintaining its crucial content. The two main approaches for text summarization are extractive and abstractive methods. Extractive summarization approaches [9] select key sentences, phrases, or words from a document based on specific criteria, remove redundant elements, and organize them to generate a summary. Abstractive summarization approaches require a system to understand the content of an input document before creating a summary. Summaries created using

H. Unger and P. Meesad (Eds.): NLPIR 2025, LNNS 1904, pp. 479–490, 2026.
https://doi.org/10.1007/978-3-032-20897-2_35

extractive methods are usually grammatically correct, but lack sentence coherence, whereas those generated using abstractive summarization methods appear more similar to those created by humans, are more natural, and more likely to contain words which do not appear in the input document. Extractive summarization approaches are generally less complex than abstractive approaches. This study focuses on an abstractive summarization approach to generate summaries from a single input document.

An abstractive summarization approaches based on generative models typically face exposure bias [3], which occurs "when a model is only exposed to the training data distribution, instead of its own" [20]. To overcome this issue, reinforcement learning methods can be used to guide the model by maximizing rewards based on its purpose [1,2,8,27]. Huyn et al. [10] introduced a text summarization model utilizing the Markov decision process with rewards to represent the quality of summaries. A multi-summary learning mechanism was proposed to improve summaries by enabling the model to generate summaries of different lengths and ensure these summaries enhance each other.

Several abstractive summarization models employ a two-stage method. In the first stage, a pre-trained language model, such as BERT [6], BART [14], or GPT [19], is utilized to produce summary candidates. Then, another pre-trained language model evaluates these candidates and selects the best ones. SimCLS is an example of this approach. Initially, a sequence-to-sequence model (BART) generates summary candidates from the original document. Subsequently, an evaluation model (RoBERTa [16]), trained using contrastive learning, is employed to evaluate the quality of candidates and choose the best one. Similarly, the ConSum [22] model, utilizing PEGASUS [28] as a backbone, tackles the exposure bias issue using contrastive learning.

In addition to the exposure bias problem, generative models can produce diverse but low-quality texts because of the mismatch between training and evaluation. To address this issue, Pang and He [18] proposed a GOLD model using off-policy learning from reference summaries with importance weighting. Additionally, abstractive summarization models can generate unfaithful summaries and be difficult to control. Dou et al. [7] developed a GSum model using both the original document and summary guidance signals to generate more diverse and faithful summaries. Xu et al. [25] improve the faithfulness of summaries by incorporating contrastive learning and maximizing the similarity among the representations of the original document, reference summary, and summary created by the model.

During the summarization generation process, beam search or diverse beam [24] algorithms are employed. While using beam search to generate text might lack diversity, diverse beam tackles this issue. Applying Minimum Bayes Risk (MBR) in text generation focuses on selecting summary candidates with the highest probability and lowest risk [4,11], which helps improve model performance. A drawback of the abstractive summarization model trained using MLE is "performance degradation during inference", because the model needs to compare several candidates created from the reference summary [17].

Therefore, Liu et al. [17] introduced a BRIO training paradigm assuming a "non-deterministic distribution so that different candidate summaries are assigned probability mass according to their quality".

In this study, we introduce an abstractive summarization model using MBR decoding and the BRIO training paradigm, and improve model performance using reinforcement learning. We evaluate the model based on several criteria, including content preservation between summary candidates and the original document, as well as the fluency, grammar, and coherence of summaries created by our proposed model.

2 Proposed Model

Since the summarization models trained using BRIO have been shown to enhance model performance [13,17], we adopt BRIO for training our proposed model. Figure 1 presents the architecture of our model, which incorporates reinforcement learning and MBR. In particular, the BRIO paradigm is employed to train the backbone model, and the reinforcement learning method is applied to improve the quality of summaries generated. We note that all models integrate MBR. Our proposed model consists of two stages as follows:

- Stage 1 comprises two sub-stages: (i) generating and scoring summary candidates, and (ii) employing BRIO to fine-tune the abstractive summarization model using these candidates.
- Stage 2 uses a reinforcement learning technique to further enhance model performance.

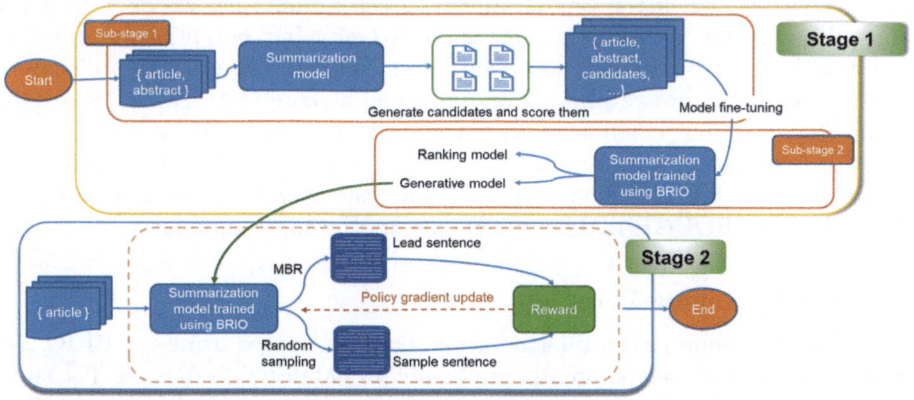

Fig. 1. The architecture of the abstractive summarization model using reinforcement learning and minimum Bayes risk.

2.1 Minimum Bayes Risk-Based Abstractive Summarization

One of our goals is to apply reinforcement learning as an unsupervised learning approach, which requires a model with an ability to create high-quality summaries for training. The MBR method is utilized to select the best candidates. The main idea of MBR is to choose the summary output $\hat{y}$ from all possible outputs Y such that the risk $R(y')$, where "a candidate decoding y' as the expected error (also called loss) under this distribution" [4] and $y' \in Y$, is minimized, or the gain $G(y, y')$ is maximized. In some language model, Y is huge, which leads to computing $R(y')$ difficult. Therefore, Bertsch et al. [4] consider a small sample subset $\mathcal{Y} \subset Y$. They identified problems with using the same sample for both reference summaries and summary candidates, as follows:

- If $\mathcal{Y}$ consists of high-quality outputs, it will not fully reflect the model's probability distribution, leading to biased risk estimation.
- If $\mathcal{Y}$ includes low-quality outputs, the low-quality candidates may be chosen.

Therefore, $\mathcal{Y}$ is further divided into a sample $\mathcal{Y}_c$ of candidates for selection and a sample $\mathcal{Y}_r$ of reference summaries for risk estimation. MBR is computed using the following formula:

$$\hat{y} = argmax_{y' \in \mathcal{Y}_c} \sum_{y \in \mathcal{Y}_r} G(y, y') \tag{1}$$

Bertsch et al. [4] claimed that MBR helps achieve better text summarization performance than beam search or greedy. MBR decoding consists of the following steps: First, summary candidates are generated using the temperature sampling technique. Specifically, when the temperature is less than one, words with the highest probabilities are selected, thereby reducing candidate diversity. In contrast, when the temperature is greater than one, words with lower probabilities might be selected, enhancing diversity but potentially decreasing accuracy. At a temperature of one, words from reference summaries are chosen. Then, each candidate is evaluated and assigned a score (e.g., BLEU, ROUGE, or BERTscore). The candidates with the greatest gains and lowest risks are considered the outputs. In our work, the sample size is 30 for both candidates and references, which are then evaluated using the average scores of ROUGE-1, ROUGE-2, and ROUGE-L. The model outputs the top-4 summaries.

2.2 BRIO Training Paradigm

The abstractive summarization generation model is trained using the BRIO [17] training paradigm based on non-deterministic probability distribution. The summarization model trained using BRIO has two roles: generation for creating candidates and evaluation for estimating the quality of the candidates. The generation ability of the model is preserved by combining the cross-entropy loss L_{MLE} and contrastive loss L_{CTR} using the following formula:

$$Loss = L_{MLE} + \gamma \times L_{CTR}, \tag{2}$$

where γ is the weight of L_{CTR}.

2.3 Reinforcement Learning for Summarization

The summarization model with MBR decoding trained in the first stage serves as the backbone in the second stage. This backbone takes input documents and does not require reference summaries. The input document d is embedded into vectors, which are then fed into the backbone to create lead sentences or guidance summaries y_r. These lead sentences are used to evaluate the quality of the summary candidates and to optimize the training process. In addition, the backbone produces summary candidates y_c using a stochastic sampling method. These summary candidates reflect the model quality, and are used to calculate rewards and adjust parameters.

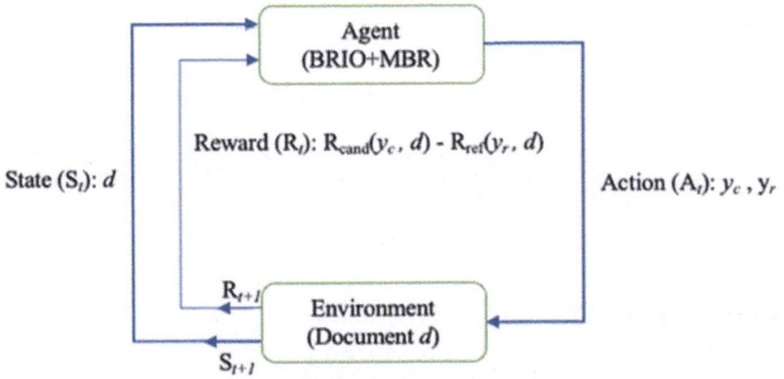

Fig. 2. Reinforcement learning for the abstractive summarization system.

Figure 2 illustrates the reinforcement learning method employing in our abstractive summarization model. The reinforcement learning technique employs a reward R function to measure summary quality. In particular, the candidate rewards R_{cand} are used to evaluate the quality of y_c based on d, whereas the reference rewards R_{ref} are used to evaluate y_r in relation to d. Both R_{cand} and R_{ref} contribute to adjusting training parameters to enhance the quality of the model. The loss function in reinforcement learning is computed by multiplying the MLE loss of the candidate by the difference between R_{cand} and R_{ref}. This loss is then propagated backward to update model parameters.

$$Loss_{RL} = -(R_{cand} - R_{ref}) \times L_{MLE}(cand). \tag{3}$$

In our work, the reward is identified based on four aspects: content preservation, fluency, grammar, and coherence. Specifically, R_{cand} and R_{ref} are computed using the following formula:

$$R_{cand} = \alpha R_C(y_c, d) + \beta R_F(y_c) + \gamma R_G(y_c) + \lambda R_{Co}(y_c), \tag{4}$$

$$R_{ref} = \alpha R_C(y_r, d) + \beta R_F(y_r) + \gamma R_G(y_r) + \lambda R_{Co}(y_r), \tag{5}$$

where α, β, γ, and λ are parameters to control the contribution of components. R_C is the reward function to estimate the content preservation between a summary and a document d by computing the cosine similarity between summary and document embeddings. R_F evaluates the fluency of each summary using perplexity. The ideas to measure R_C and R_F are proposed by Hyun et al. [10].

$R_G(y_c)$ and $R_{Co}(y_c)$ are used to evaluate the grammar and coherence of a candidate, respectively. Specifically, LanguageTool[1] is used to verify the grammar and spelling in y_c. The reward, obtained based on errors, reaches a max value of 1 when y_c contains no error and decreases when the errors increases. This reward is obtained using the following formula:

$$R_G(y) = max\left(0, 1 - \frac{\text{error number}}{\text{total words in } y}\right). \tag{6}$$

R_{Co} is used to predict sentence coherence by calculating the cosine similarity between adjacent sentences, each of which is embedded using Sentence-BERT [21].

$$R_{Co}(y) = \left(\frac{1}{n-1}\sum_{n-1}^{1} sim(sent_i, sent_{i+1})\right). \tag{7}$$

3 Experimental Results

We conduct experiments on an A6000 device having a 48GB GPU and 45GB RAM. Our model utilizes the CNN/Daily Mail dataset containing 287,113 articles for training, 13,368 for validation, and 11,490 for testing. The average numbers of tokens in each article and the corresponding reference summary are 781 and 56, respectively. Due to hardware limitations, we train our proposed model with 150,000 articles, of which 75,000 are used for training the summarization model using BRIO, and the remaining are used for reinforcement learning. The summarization models used in this study are BART[2] and PEGASUS[3]. These models are trained with 2 epochs, a batch size of 1, cumulative steps of 8, a learning rate of 2e-3, and Adam optimizers.

In addition to the commonly used metrics of ROUGE [15] and BERTScore [29], we employs BARTscore [26] for evaluating our abstractive summarization models. Similar to Liu et al. [17], we calculate the novelty of the summary y based on the document d as below:

$$Novelty(d, y) = \frac{\sum_{g \in G_y} \mathbb{1}(g \notin G_d)}{|G_y|}, \tag{8}$$

where $\mathbb{1}$ is the indicator function, G_d and G_y are the collections of bigram in d and y. The greater the novelty value, the more abstractive the summary.

1 https://github.com/jxmorris12/language_tool_python.
2 https://huggingface.co/facebook/bart-large-cnn.
3 https://huggingface.co/google/pegasus-cnn_dailymail.

We evaluate the performance of the BART-based models by analyzing loss metrics as illustrated in Fig. 3: cross entropy loss measures the ability to generate text, contrastive loss evaluates the effectiveness in ranking summaries by quality, and total loss integrates both components.

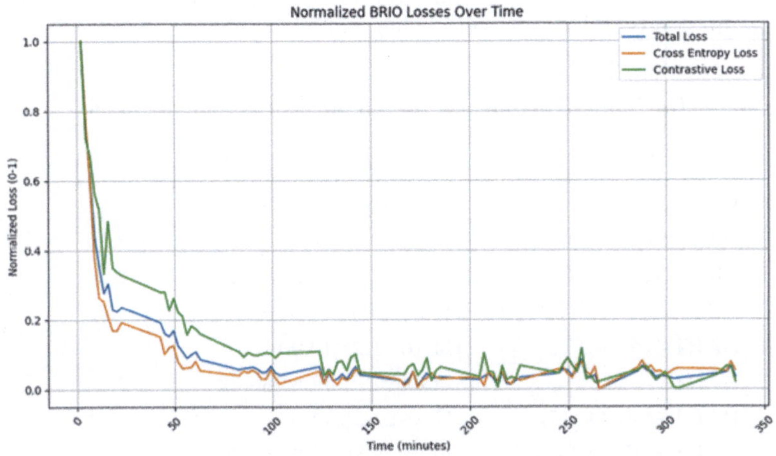

Fig. 3. Loss metrics during training the summarization models using BRIO.

3.1 Quality of Summarization Models

For evaluation purposes, the summarization models generate summaries using beam search decoding with a beam size of 4 to ensure the high-quality summaries. We limit the repetition of n-grams (no-repeat-ngram-size) of 3 to preserve coherence and avoid content duplication. Given an input document, each model only produces one summary (num-return-sequences=1).

Table 1 presents the experimental results of models measured using ROUGE scores, BERTscore [29], and BARTscore [26]. We also evaluate the models based on the fluency, grammar and coherence, as shown in Table 2. In these tables, we use the checkpoints of facebook/bart-large-cnn (400 million parameters) and google/pegasus-cnn_dailymail (568 million parameters). R@1, R@2, and R@L are ROUGE-1, ROUGE-2, and ROUGE-L, respectively.

The experimental results indicate that backbone models employing MBR decoding outperform the vanilla models. The MBR-based summarization models trained using the BRIO paradigm improve quality compared to those trained using the traditional technique. However, the combination of MBR and BRIO does not help the models achieve better results compared to those trained using BRIO only. We will explore with this issue in future work.

Faithfulness is a crucial factor in BARTscore, which computes the log-likelihood of producing a summary from a given document. The BARTscores of models using MBR are greater than the models not using MBR, including those trained using BRIO. This suggests that the MBR decoding preserves content better than the others. BERTscore considers context and compares the cosine similarity between the embeddings of the reference summary and the summary created by the system. This score focuses on the meaning of the text created and "correlates better with human judgments" [29]. The models using MBR also have higher BERTscores than other models.

Table 1. Experimental results of the abstractive models.

Model	R@1	R@2	R@L	BERTscore	BARTscore
PEGASUS [28]	44.17	21.47	41.11	–	–
PEGASUS	43.13	20.27	38.98	88.15	−2.96
PEGASUS+BRIO	45.59	21.89	42.24	89.26	−2.53
PEGASUS+BRIO+MBR	45.28	21.56	41.34	89.31	−2.78
BART [14]	44.16	21.28	40.90	–	–
BART	44.04	21.07	40.84	88.22	−2.93
BART+BRIO [17]	47.78	23.55	44.57	–	–
BART+BRIO	46.61	22.80	43.27	88.79	−2.13
BART+BRIO+MBR	46.01	22.40	42.74	89.45	−2.56

Table 2. Experimental results of the abstractive models based on fluency, grammar, and coherence.

Model	Fluency	Grammar	Coherence
PEGASUS	0.952	0.917	0.356
PEGASUS+BRIO	0.951	0.895	0.343
PEGASUS+BRIO+MBR	0.949	0.873	0.342
BART	0.965	0.968	0.353
BART+BRIO	0.967	0.958	0.356
BART+BRIO+MBR	0.959	0.913	0.353

3.2 Summarization Models Using Reinforcement Learning

The models are optimized based on rewards, which include the content preservation, fluency, grammar, and coherence of the summaries created by the models. These models, which are trained in 12 h, demonstrate improve performance, as evidenced by not only ROUGE scores but also fluency, grammar, and coherence, as detailed in Table 3 and Table 4.

Table 3. Experimental results of the abstractive models using reinforcement learning.

Model	R@1	R@2	R@L	BERTscore	BARTscore
PEGASUS	43.16	20.29	39.01	88.15	−2.96
PEGASUS + BRIO	45.74	22.05	42.07	89.31	−2.55
PEGASUS + BRIO + MBR	45.31	21.58	41.36	89.31	−2.78
BART	44.09	21.11	40.90	88.23	−2.39
BART + BRIO	46.65	22.82	43.32	88.80	−2.13
BART + BRIO + MBR	46.09	22.41	42.83	89.47	−2.54

Table 4. Experimental results of abstractive summaries using reinforcement learning based on fluency, grammar, and coherence.

Model	Fluency	Grammar	Coherence
PEGASUS	0.952	0.918	0.357
PEGASUS + BRIO	0.952	0.897	0.345
PEGASUS + BRIO + MBR	0.949	0.873	0.342
BART	0.965	0.968	0.353
BART + BRIO	0.968	0.961	0.354
BART + BRIO + MBR	0.960	0.914	0.355

Figure 4 illustrates the novelty metrics of models. The models employed MBR decoding surpass other models not using it in terms of novelty.

Additionally, we analyze the trade-off between accuracy and diversity, considering how the exposure bias affects 200 randomly selected samples extracted from the dataset, as presented in Fig. 5.

- MBR decoding achieves the highest level of content faithfulness with a BERTScore of 0.879, while maintaining the reasonable level of diversity. However, the computation cost is high with an average latency of 2.28 s per sample and the GPU usage of 6.8 GB. This result indicates that MBR is suitable when faithfulness is a requirement but it faces limitations when deployed in real-time.
- Diverse beam strikes a balance between diversity and performance, using lower resources than MBR (memory usage of 4.1GB and a latency of 1.03 s). Diverse beam is an appropriate selection to improve the diversity of summaries while preserving a reasonable cost.
- Beam search reaches the lowest results in terms of faithfulness and diversity. However, it excels in speed and cost effectiveness. This method is ideal when resources are limited or fast inference is required.

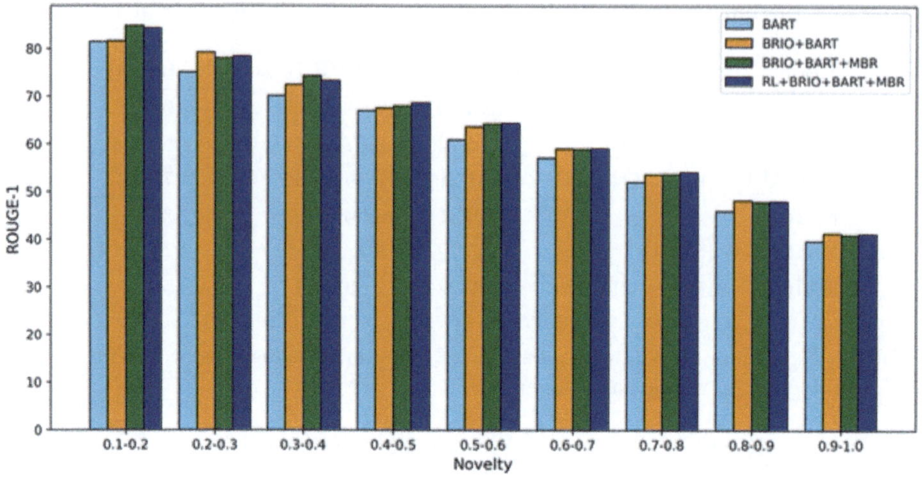

Fig. 4. Comparing model performance in term of novelty.

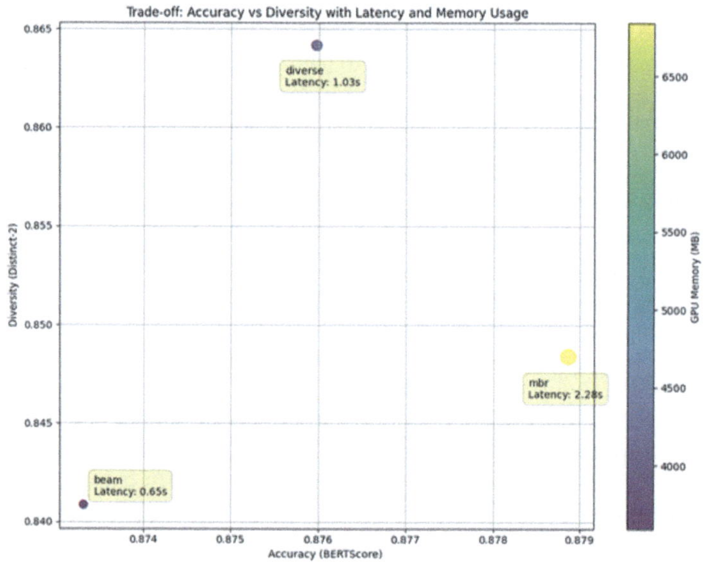

Fig. 5. Trade-off between accuracy and diversity with the impact of exposure bias.

4 Conclusion

We present an abstractive summarization model by integrating MBR decoding, employing the BRIO training paradigm, and applying a reinforcement learning technique. Our proposed models produce summaries with faithfulness, fluency, correct grammar, and coherence. In abstractive summarization, a decoding strategy contributes to creating high-quality summaries. MBR selects the summary with the lowest expected risk, while beam search chooses the summary with the highest probability. The summaries created using MBR have higher quality when using multiple summary references. MBR has much higher computation cost than beam search. In other words, beam search is much faster than MBR. For example, the generating summaries on the test set takes over 13 h using MBR, while it requires about 50 min using beam search. In future works, we intend to explore various MBR-based decoding methods, such as diverse MBR [11,12], to improve the diversity of summaries. Reinforcement learning contributes to improving the quantity of summaries. We are investigating the parameters used to calculate rewards, such as employing adaptive weighting [23] and dynamic weighting [5]. Additionally, we will study methods for optimizing rewards and adjusting the contributions of reward factors.

References

1. Atri, Y.K., Goyal, V., Chakraborty, T.: Multi-document summarization using selective attention span and reinforcement learning. IEEE/ACM Trans. Audio Speech Lang. Process. **31**, 3457–3467 (2023)
2. Bae, S., Kim, T., Kim, J., Lee, S.: Summary level training of sentence rewriting for abstractive summarization. In: Proceedings of the 2nd Workshop on New Frontiers in Summarization, pp. 10–20 (2019)
3. Bengio, S., Vinyals, O., Jaitly, N., Shazeer, N.: Scheduled sampling for sequence prediction with recurrent neural networks. Adv. Neural Inf. Process. Syst. **28** (2015)
4. Bertsch, A., Xie, A., Neubig, G., Gormley, M.R.: It's MBR all the way down: Modern generation techniques through the lens of minimum bayes risk. In: Proceedings of the Big Picture Workshop, pp. 108–122 (2023)
5. De Langis, K., Koo, R., Kang, D.: Dynamic multi-reward weighting for multi-style controllable generation. In: Proceedings of the 2024 Conference on Empirical Methods in Natural Language Processing, pp. 6783–6800 (2024)
6. Devlin, J., Chang, M.-W., Lee, K., Toutanova, K.: BERT: pre-training of deep bidirectional transformers for language understanding. In: Proceedings of the 2019 conference of the North American Chapter of the Association for Computational Linguistics: Human Language Technologies, volume 1 (long and short papers), pp. 4171–4186 (2019)
7. Dou, Z.-Y., Liu, P., Hayashi, H., Jiang, Z., Neubig, G.: GSum: a general framework for guided neural abstractive summarization. In: Proceedings of the 2021 Conference of the North American Chapter of the Association for Computational Linguistics: Human Language Technologies, pp. 4830–4842 (2021)
8. Yu, F., Xiong, D., Dong, Y.: Inverse reinforcement learning for text summarization. In: Findings of the Association for Computational Linguistics: EMNLP, vol. 2023, pp. 6559–6570 (2023)
9. Abba Suganda Girsang and Fransisco Junius Amadeus: Extractive text summarization for Indonesian news article using ant system algorithm. J. Adv. Inf. Technol. **14**(2), 295–301 (2023)
10. Hyun, D., Wang, X., Park, C., Xie, X., Hwanjo, Yu.: Generating multiple-length summaries via reinforcement learning for unsupervised sentence summarization. In: Findings of the Association for Computational Linguistics: EMNLP, vol. 2022, pp. 2939–2951 (2022)

11. Jinnai, Y., Honda, U., Morimura, T., Zhang, P.: Generating diverse and high-quality texts by minimum bayes risk decoding. In: Findings of the Association for Computational Linguistics ACL, vol. 2024, pp. 8494–8525 (2024)
12. Jinnai, Y., Morimura, T., Honda, U., Ariu, K., Abe, K.: Model-based minimum Bayes risk decoding for text generation. In: Proceedings of the 41st International Conference on Machine Learning, pp. 22326–22347 (2024)
13. Lam, K., Doan, T., Pham, K., Kalita, J.: Abstractive text summarization using the brio training paradigm. In: Findings of the Association for Computational Linguistics: ACL, vol. 2023, pp. 92–99 (2023)
14. Lewis, M., et al.: BART: denoising sequence-to-sequence pre-training for natural language generation, translation, and comprehension. In: Proceedings of the 58th Annual Meeting of the Association for Computational Linguistics, pp. 7871–7880 (2020)
15. Lin, C.-Y.: ROUGE: a package for automatic evaluation of summaries. In: Text Summarization Branches Out, pp. 74–81 (2004)
16. Liu, Y., et al.: RoBERTa: A robustly optimized BERT pretraining approach. arXiv preprint arXiv:1907.11692 (2019)
17. Liu, Y., Liu, P., Radev, D., Neubig, G.: BRIO: bringing order to abstractive summarization. In: Proceedings of the 60th Annual Meeting of the Association for Computational Linguistics (Volume 1: Long Papers), pp. 2890–2903 (2022)
18. Pang, R.Y., He, H.: Text generation by learning from demonstrations. In: International Conference on Learning Representations (2021)
19. Radford, A., Narasimhan, K., Salimans, T., Sutskever, I., et al.: Improving language understanding by generative pre-training (2018)
20. Ranzato, M., Chopra, S., Auli, M., Zaremba, W.: Sequence level training with recurrent neural networks. In: 4th International Conference on Learning Representations, ICLR 2016 (2016)
21. Reimers, N., Gurevych, I.: Sentence-BERT: sentence embeddings using Siamese BERT-Networks. In: Proceedings of the 2019 Conference on Empirical Methods in Natural Language Processing and the 9th International Joint Conference on Natural Language Processing (EMNLP-IJCNLP), pp. 3982–3992 (2019)
22. Sun, S., Li, W.: Alleviating exposure bias via contrastive learning for abstractive text summarization. arXiv preprint arXiv:2108.11846 (2021)
23. Vanlioglu, A.: Entropy-guided sequence weighting for efficient exploration in RL-based LLM fine-tuning. arXiv preprint arXiv:2503.22456 (2025)
24. Vijayakumar, A.K., et al.: Diverse beam search: Decoding diverse solutions from neural sequence models. arXiv preprint arXiv:1610.02424 (2016)
25. Shusheng, X., Zhang, X., Yi, W., Wei, F.: Sequence level contrastive learning for text summarization. In: Proceedings of the AAAI Conference on Artificial Intelligence, vol. 36, pp. 11556–11565 (2022)
26. Yuan, W., Neubig, G., Liu, P.: BARTScore: evaluating generated text as text generation. Adv. Neural. Inf. Process. Syst. **34**, 27263–27277 (2021)
27. Zhang, H., Cai, J., Xu, J., Wang, J.: Pretraining-based natural language generation for text summarization. In: Proceedings of the 23rd Conference on Computational Natural Language Learning (CoNLL), pp. 789–797 (2019)
28. Zhang, J., Zhao, Y., Saleh, M., Liu, P.: PEGASUS: pre-training with extracted gap-sentences for abstractive summarization. In: International Conference on Machine Learning, pp. 11328–11339. PMLR (2020)
29. Zhang, T., Kishore, V., Wu, F., Weinberger, K.Q., Artzi, Y.: BERTScore: evaluating text generation with BERT. In: International Conference on Learning Representations

The Role of Fragmentation in Long-Text Summarization

Diego Uribe(✉), Enrique Cuan, and Iscander A. Ramirez

Departamento de Sistemas y Computación, TecNM Instituto Tecnológico de La Laguna, Torreón, Coah., Mexico
duribea@lalaguna.tecnm.mx, ecuand@lalaguna.tecnm.mx, iramirezc@lalaguna.tecnm.mx

Abstract. Large Language Models (LLMs) are very successful for a diverse range of NLP tasks -as translation, summarization, generation, etc.- making them the mainstay of the development of NLP applications. However, not all applications demand the massive size and complexity of LLMs: the computational power required makes them impractical for many use cases. In this way, turning to the use of a Small Language Model (SLM) seems to be a plausible alternative. In this work, we analyze the use of an SLM for the task of long-text summarization. We investigate how the use of a small context window affect the quality of the summary generated for a long document that exceeds the size of the context window. As the truncation of the input text may lead to the loss of crucial information, we analyze two fragmentation techniques: a shallow and a semantic fragmentation technique. The results of the experimental evaluation reveals how a more sophisticated fragmentation, based on the content of the text rather than the number of lexical tokens, is essential for a proper preprocessing of long documents. Through the application of a hierarchical summarization approach, each fragment is summarized independently, then the concatenated fragments' summaries are summarized.

Keywords: Language Model · summarization · lexical tokens

1 Introduction

Since the beginning of this century, the enormous development and capacity of the web has made of Internet a work space for multiple activities, including the dissemination and sharing of information. Given the exponential growth of information on the web, we are now surrounded by huge amounts of texts: long reports, news articles, documents, reviews, etc. As the reading and comprehension of an enormous amount of information is impossible, automatic text summarization is important because it helps us to focus our attention on what is really important. However, automatic text summarization is hard because producing a summary entails not only taking care of a shorter version of the original text but also taking care of repetition and semantic coherence [1].

H. Unger and P. Meesad (Eds.): NLPIR 2025, LNNS 1904, pp. 491–504, 2026.
https://doi.org/10.1007/978-3-032-20897-2_36

Various techniques, grouped into three approaches, have addressed research in automatic text summarization: extractive, abstractive, and hybrid approaches. The purpose of extractive methods is to select relevant sentences or phrases from the input text to create the summary [2,3], whereas abstractive approaches are focused on the production of new and equivalent sentences that indicate the original meaning [4,5]. The hybrid approach is an amalgamation technique of the previous ones: a hybrid technique uses an extractive model to identify key sentences, followed by an abstractive model to improve the generation of new sentences based on the extracted content [6–8].

Automated text summarization is commonly categorized in three different ways: Standard summarization, Long-document summarization, and Multi-document summarization. Standard summarization is characterized by a single, short document (500–1000 words) where the purpose is to identify the key facts of the input text. On the other hand, Multi-document summarization is characterized by multiple-texts as input where the challenge lies in creating a summary that integrates pertinent information from various sources avoiding repetition to generate a coherent summary [9]. The main distinctive feature of Long-document summarization is certainly the length of the document. Shortening a long text efficiently entails proper management of the text length, as the salient content is sparsely distributed [10].

In this work we are concerned with Long-document summarization, that we call Long-Text Summarization (LTS) henceforth. As we say above, LTS is a challenging task because the length of the text requires proper management of the content, as the key points are sparsely distributed. Moreover, one more demanding constraint is in our work: the use of a Small Language Model (SLM) where the length of the context window is smaller than its counterpart in a Large Language Model (LLM). For example, an LLM like LongT5 displays great performance on various NLP tasks. However, it is quite a large model whose storage and computational requirements are huge. On the other hand, the deployment of summarization models on resource-limited devices makes the use of an SLM based on the distillation technique a plausible option [11].

Now, since the length of the context window of the SLM used in this work is limited to 1024 lexical tokens, and the dataset contains articles with a length greater than 1024 tokens, proper management of the text length is a crucial component of an LTS system. Rather than just splitting the text arbitrarily, a suitable fragmentation of the text is demanding. How can we split the text when the salient information is sparsely distributed? TextTiling emerges as a viable approach for this segmentation task.

Unlike arbitrary chunking of texts by token count, TextTiling identifies major subtopic shifts, also called subtopic boundaries, to partition texts into contiguous, non-overlapping subtopic segments [12]. Based on our experimentation results, subtopic segments, such as those produced by TextTiling, have proven to be useful for summarization of long and poorly structured texts. Thus, we demonstrate the viability of an LTS system that relies on content segmentation instead of token length.

In short, we have implemented a hybrid extractive-abstractive summarization approach. In the extractive step we use two fragmentation strategies: a shallow fragmentation and a semantic fragmentation. In the shallow fragmentation a long text is divided into chunks of 1,024 tokens, the size of the context window of the SLM used, whereas in the semantic fragmentation, by making use of the TextTiling algorithm, subtopic segments are extracted. In the abstractive stage we implement a hierarchical approach: the summary corresponding to each fragment (chunk or segment) is obtained first, and then, the final summary is generated. Said in another way, the final summary is a summary of summaries.

Thus, the primary contributions of our work are summarized as follows:

- it provides insights into the plausibility of eliciting good-quality summaries of long texts by using a small language model. Previous work that uses larger long-context models (some of them explained in the next section) often requires substantial computational resources for both training and inference.
- we conduct an empirical evaluation on long CNN/DM articles to prove the viability of an LTS system that relies on semantic segmentation rather than shallow fragmentation. By implementing a hybrid extractive-abstractive summarization approach to compare both fragmentation strategies, the obtained results exhibit how shallow fragmentation struggles to handle long texts.

2 Related Work

Since the focus of our work lies on the intersection of long-document summarization and text segmentation, specifically investigating how preprocessing strategies impact summary quality, this section reviews some relevant advances in efficient model architectures, text segmentation techniques, and the foundational role of content selection, thereby framing the contribution of our empirical study.

A primary direction in addressing long-text summarization has been the development of novel, efficient model architectures to cope with the challenge of generating coherent summaries from documents that exceed the context window of modern transformer models. Models like Longformer [13] introduce sparse attention mechanisms to handle sequences of up to 16,000 tokens. More recently, research such as that by Huang et al. [14] has further refined these efficient attentions specifically for summarization, demonstrating that architectural innovations can successfully process long-form content. Even though these approaches are powerful, they often require substantial computational resources for both training and inference.

A critical preprocessing step for long-text summarization is segmentation, or chunking, of the document into manageable pieces. Text segmentation has evolved from classical lexical-cohesion-based algorithms like TextTiling [12] to modern neural approaches. For instance, Lukasik et al. [15] proposed a cross-segment attention model for segmentation, representing the state-of-the-art in leveraging deep learning to identify topical boundaries. These neural methods offer high accuracy but can introduce complexity and require training data.

Our study deliberately employs the classical TextTiling algorithm to investigate whether a lightweight, unsupervised semantic segmentation method can provide a sufficient advantage over naive chunking, making it a practical choice for many real-world applications.

Now, the way we select and order content before the abstractive step profoundly impacts the final summary's quality and coherence. Nayeem et al. [16] argued for the importance of "extract with order" to maintain coherence, highlighting that the quality of the input units submitted to an abstractive model is paramount. This principle directly informs our hypothesis: the strategy used to fragment a long document—whether it preserves semantic units or not—should materially affect the fluency and informativeness of the final summary.

In short, and as far as we know, previous work often integrates segmentation within a larger, proprietary pipeline, making it difficult to discern its specific contribution. Our study fills this gap by conducting a controlled comparison of fragmentation techniques by making use of a small language model on abstractive summarization.

3 Fragmentation and Hierarchical Approach

LTS is a challenging task because long documents have relevant information spread throughout the text: critical information can be conveyed anywhere in the document, not just at the beginning or end. Furthermore, the use of an SLM with a small context window (1,024 lexical tokens) makes this affair a more challenging task [17]. This fundamental limitation demands the fragmentation of the long document. We next describe the fragmentation approaches analyzed in this work, but we first formalize our task in similar terms to those presented by Zhang et al. [18].

LTS can be defined as the following mapping function

$$f_{\varnothing} : T \rightarrow S \tag{1}$$

where $f_{\varnothing}$ is the language model with trainable hyperparameters $\varnothing$, T is the set of tokens in the long input text, and S is a new smaller text that preserves the semantic of T.

Now the process (P) of LTS is formalized as a sequence of steps for transforming and analyzing data. These steps are represented by the following tuple:

$$P = (D, SLM, E) \tag{2}$$

where:

D denotes data preprocessing, which consists of transformations ϕ(T) such as *tokenization* and *fragmentation*, SLM is the modeling step where a pre-trained SLM (*DistilBart* in our case) is applied to produce the summary, and E is the evaluation stage where metrics as ROUGE and BERTScore are used to measure the syntactic and semantic match between T and S.

3.1 Shallow Fragmentation

Shallow fragmentation is basically a rudimentary splitting approach to cope with the length of the text. As part of the pre-processing of a long document, this splitting approach represents the text in terms of the number of tokens obtained from the tokenization process. Thus, in this shallow fragmentation a long text is divided into chunks of tokens:

$$Chunking(T) \rightarrow c_1, c_2, \ldots, c_n \tag{3}$$

where

$$c_i = totalnumberoftokens(T)/contextwindowlength(CWL) \tag{4}$$

and CWL = 1,024 tokens for *DistilBart*.

3.2 Semantic Fragmentation

Semantic fragmentation is basically a fragmentation technique based on the content rather than the length of the text. We implement this fragmentation approach with TextTiling, an algorithm designed for text segmentation by detecting subtopic boundaries [12]. Indeed, TextTiling's goal is to divide a long document into smaller, topically coherent segments (or "tiles") by looking for shifts in the vocabulary (lexical co-occurrence) between adjacent blocks of text.

Thus, TextTiling is a crucial pre-processing step or component to break the long text (T) into a sequence of smaller, topically coherent segments $s_1, s_2, \ldots, s_n$:

$$TextTiling(T) \rightarrow s_1, s_2, \ldots, s_n \tag{5}$$

3.3 Hierarchical Approach

LTS is a challenging task because the length of the text requires proper management of the content, as the key points are sparsely distributed, and one more demanding constraint is the use of *DistilBart* with a CWL limited to 1,024 tokens. This constraint is the reason for a pre-processing based on fragmentation and the implementation of a Hierarchical Approach. The workflow for this approach is:

$$Fragmentation(T) \rightarrow f_1, f_2, \ldots, f_n | DistilBart(f_i) \rightarrow S_1, S_2, \ldots, S_n \tag{6}$$

where

f_i denotes a particular c_i (generated by tokenization), or s_i (generated by TextTiling), and

S_i denotes the summary generated by *DistilBart* corresponding to a particular f_i

The final step in the process is:

$$DistilBart(S_i) \rightarrow FinalSummary \quad (7)$$

In this way, we have in $S_1, S_2, \ldots, S_n$ a sequence of all main ideas represented by individual topic summaries. Then, this sequence is submitted to *DistilBart* to connect and remove redundancy across fragments, and to create a final summary ensuring both completeness (all topics are covered) and coherence (the final text flows well).

4 Experimental Setup

In this section we describe the experimentation conducted in this work to figure out the implications of fragmentation techniques for long-text summarization (LTS). So we first briefly describe the compact summarization model used in this investigation: *DistilBart*. As a fundamental part of the long-text summarization process, the main characteristics of this small language model (SLM) are detailed. Then, the dataset characteristics are exposed, and the results of each fragmentation technique are exhibited.

4.1 DistilBart: a Compact Summarization Model

For our summarization tasks, we employ *DistilBart*, a distilled version of the BART model [19]. Through knowledge distillation, it retains 97% of BART's summarization performance while being 40% smaller and 60% faster. This encoder-decoder transformer is pre-trained for text generation and is particularly effective at abstractive summarization, making it an efficient and powerful choice for our evaluation.

In fact, abstractive summarization aims to condense a source text by generating novel phrases and sentences that capture its core meaning, rather than simply extracting key sentences verbatim. Unlike extractive methods, it can paraphrase, generalize, and synthesize information to produce a more concise and fluent summary, much like a human would [5].

Moreover, we selected this model for its optimal balance of performance and efficiency. Its standard context window of 1,024 tokens necessitates the fragmentation of longer input documents, allowing us to cleanly evaluate the impact of different preprocessing strategies. Using a smaller model ensures that any differences in summary quality are due to the fragmentation method itself, and not simply the modelś ability to overpower the problem with its large size.

4.2 Dataset Description

The dataset used in this experimentation is CNN-DM, a dataset containing just over 300k unique news articles as written by journalists at CNN and the Daily Mail [20]. Each news article is described by a body content and its corresponding summary (highlights) created by the editors. Training documents average

766 words across 29.74 sentences, with summaries averaging 53 words and 3.72 sentences. Figure 1 shows the distribution of the dataset in terms of article-tokens-length, where most of the news articles are characterised by no more than 1024 tokens. However, the news dataset also contains 76k (out of 300k) articles denoted by more than 1024 tokens.

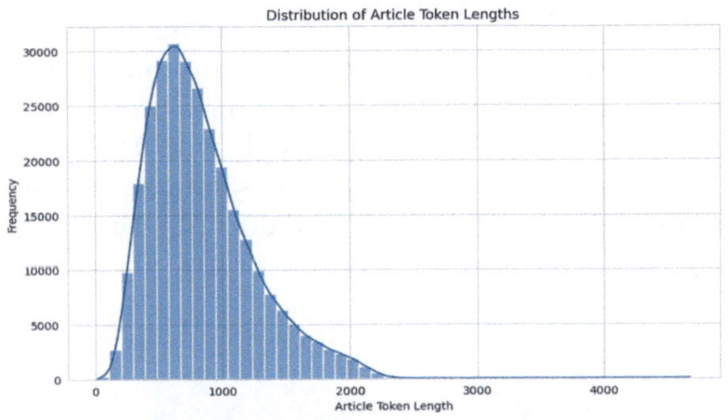

Fig. 1. Article-Token-Length-distribution

Since we are concerned in this work with the summarization task of long texts (LTS), a pertinent question is: how is a text considered as a long text? But to define a document as a long document is a matter for debate. According to Koh et al., a document is considered as a long document based on the current state-of-the-art models [21]. In 2018, a research work about abstractive summarization made use of the CNN/DM dataset when the articles were considered as long-text [22]. However, recent work that makes use of transformers for long sequences considers a document with 3,000 lexical tokens at least, as long documents [23]. In our research work, a document is considered as a long document if the tokens-length exceeds 1,024 lexical tokens. Such definition is based on the size of the DistillBart (SLM) context window, as a small context window demands a suitable fragmentation of a text characterised by more than 1, 024 lexical tokens.

We experiment with a randomly extracted sample of 7,725 news articles, drawn from a subset of 76k long articles. In this sample, the news articles average 1,328 lexical tokens, with a minimum of 1,025 and a maximum of 3,185 tokens. Taking into account the distribution of the dataset (Fig. 1)—a narrow and tall distribution that exhibits low variability—a small sample (around 10%) is a representative sample to work with.

4.3 Metrics

To assess the quality of generated summaries, we apply two widely used evaluation metrics: ROUGE_L [24] and BERTScore [25]. While the latter is based on contextual embeddings to identify semantic similarity, the former makes use of the longest common subsequence to detect lexical matching. By making use of these metrics, we aim to provide a comprehensive assessment of summarization performance, ensuring both content relevance and linguistic quality.

5 Results and Analysis

Regardless of the fragmentation implemented, and before presenting the results for each fragmentation alternative, we observe a common pattern in the obtained results: a positive correlation between the evaluation metrics used in this work. Figure 1 shows how, as the ROUGE_L value increases, the BERTScore also increases. In other words, as the lexical match with the highlights (summary created by the editors) increases, the semantic similarity also increases. Figure 2 shows how this correlation pattern is observed in the three analysis cases.

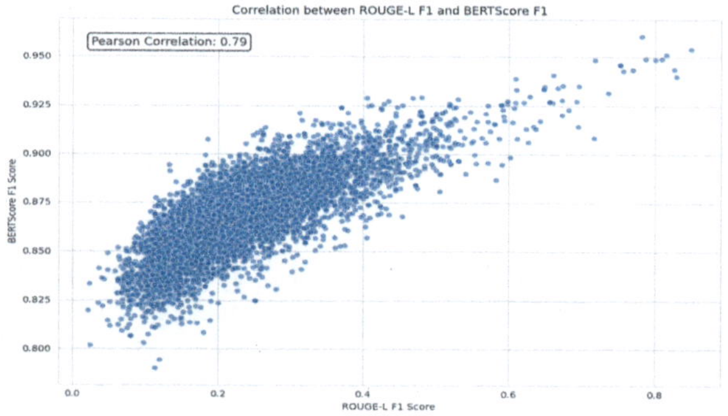

Fig. 2. Correlation between ROUGE_L and BERTScore

5.1 Fragmentation Results

We next present the exposition of the obtained results in terms of the fragmentation strategies. However, taking into account that our purpose is to perceive the impact of the fragmentation alternatives for summarization, we define the truncation of the long-text as a *baseline* assessment. In this particular case, we only submit the first 1,024 lexical tokens to the DistilBart model to obtain the final summary of the article, thereby ignoring the rest of the text. Said in another way, the hierarchical approach is not applied in this case. We now proceed with the presentation of the obtained results in terms of the fragmentation strategies.

Truncation: as we previously said, we define the truncation of the long-text as a *baseline* assessment. In this truncation method, we only submit the first 1024 lexical tokens to the DistilBart model to obtain the final summary of the article, thereby ignoring the rest of the text. With this method, we determine whether the salient information is just at the beginning of the article, so any fragmentation method is not necessary. Figure 3 shows how the distribution of the BERTScores in the first three bins is more similar than the distribution of the last two bins. The scores for the longest texts (bin 4) show how the scores decrease and are also skewed, indicated by a large interquartile range. In general, this figure shows how the BERTScore decreases as the length of the text increases.

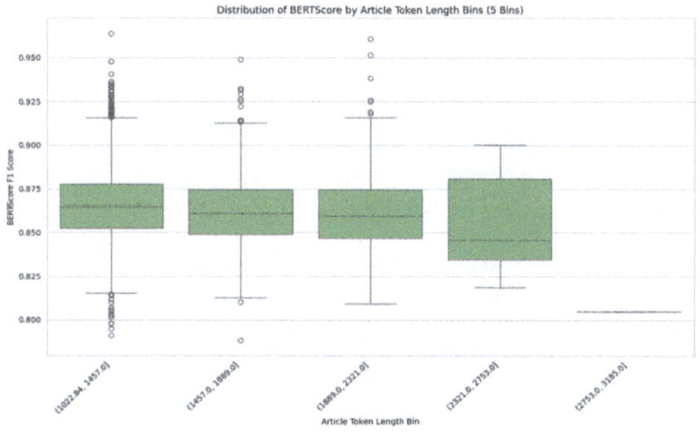

Fig. 3. Distribution of BERTScore by truncation

Shallow Fragmentation: in this case, the unit of fragmentation is a chunk of tokens. As the expression (3) indicates, in this shallow fragmentation a long text is divided into chunks of 1,024 lexical tokens. With this fragmentation method, we determine whether the salient information is distributed from the beginning to the end of the article. In this fragmentation, the hierarchical approach, represented in expression (6), is applied to obtain the summary of each chunk first, and then to create the final summary. Figure 4 shows how the distribution of the BERTScores in the first three bins is also similar. The scores for the longest texts (e.g. bin 4) show how the scores decrease and are also skewed, indicated by a large interquartile range. In general, this figure shows how the BERTScore decreases as the length of the text increases.

Semantic Fragmentation: in this case, the unit of fragmentation is a semantic segment i.e. a subtopic segment. As the expression (5) indicates, in this semantic fragmentation a long text is divided into topically coherent segments. The values of the TextTiling's parameters used in this experimentation are $w = 30$ (window size) and $k = 10$ (number of blocks). Since this method emphasizes the identification of subtopic boundaries, we don't know in advance how many segments represent an article. We also apply the hierarchical approach (expression 6) to obtain the summaries of the semantic segments first, and then to create the final summary. The distribution of the BERTScore by the number of article segments is shown in Fig. 5, where a stable and decent performance is observed when the number of segments increases.

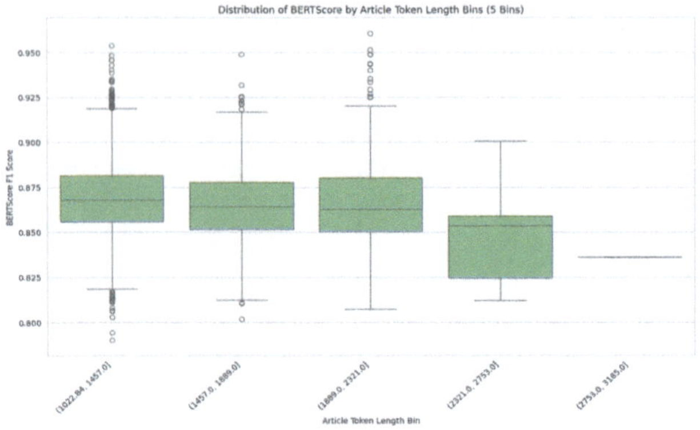

Fig. 4. Distribution of BERTScore by shallow fragmentation

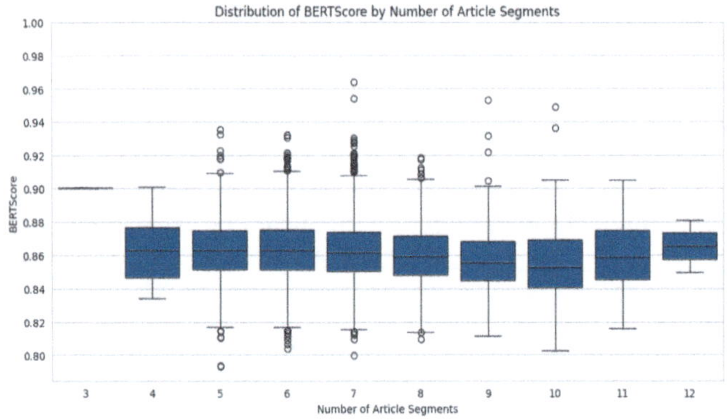

Fig. 5. Distribution of BERTScore by semantic fragmentation

5.2 Discussion

A starting point for our discussion section is the analysis of the *baseline* as a reference point for the obtained results. Comparing the distribution of the scores in Fig. 3 (*baseline*) with the distribution in Fig. 4 (chunks fragmentation), better scores are obtained when the articles have been fragmented with chunks of 1024 tokens. Both alternatives, truncation and chunk fragmentation, struggle to handle long texts. Figure 6 shows how the scores decrease when the number of chunks increases, that is, the scores decrease with the longest texts.

The discussion changes when we talk about semantic fragmentation. As the text is divided into segments that denote subtopic information, the performance of this method is exhibited in terms of segments rather than text's length. As we can see in Fig. 5, the maximum number of segments is 12, and the performance exhibited is stable as drastic changes in the scores are not observed. As we said lines above, with a TextTiling's window of 30, we obtain a smaller number of larger segments.

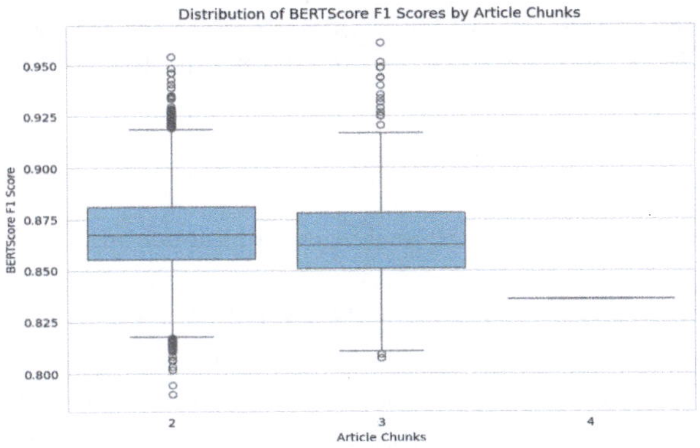

Fig. 6. Distribution of BERTScore by article chunks

Now, to determine how relevant is the use of semantic fragmentation for LTS, we turn our attention to the distribution of article segments by article token length. Figure 7 shows how the number of segments increases as the length of the text also increases. Thus, we observe how the highest number of segments (from 8 to 12) correspond to the longest texts. Then, when we compare the scores for this range of segments with the scores obtained with fragmentation based on chunks, the scores for semantic fragmentation exhibit a better performance.

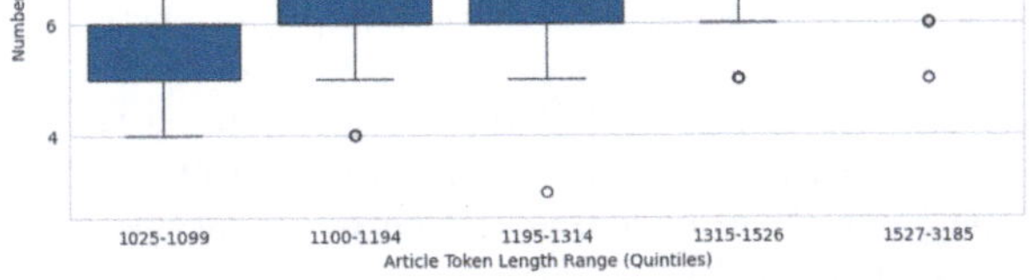

Fig. 7. Distribution of Article Segments by Article Token Length

To summarise, Table 1 displays the BERTScores obtained in terms of quartile 1 (Q1) and quartile 3 (Q3) for each fragmentation strategy. First of all, even when the Q3 values are identical, the Q3 values in truncation and shallow fragmentation are obtained for small and medium articles. As for the Q1 values, these values exhibit how shallow fragmentation struggles to handle long texts. Moreover, the small interquartile range (IQR) in semantic fragmentation indicates how the middle 50% of the scores are concentrated within this narrow range (87–84). On the other hand, a bigger interquartile range (IQR) in shallow fragmentation indicates how the lower scores are more spread out than the higher ones.

Table 1. Performance distribution by fragmentation method

Method	Q1	Q3	Best for
Truncation	0.81	0.87	Short articles
Shallow fragmentation	0.82	0.87	Medium articles
Semantic fragmentation	0.84	0.87	Long articles

6 Conclusions

In this paper, we have analyzed the impact of two fragmentation strategies for the task of long-text summarization by making use of a small language model. In particular, we investigate how the use of a small context window affects the quality of the summary generated for a long document that exceeds the size of the context window. The results of the experimental evaluation reveal how a semantic fragmentation, based on the content of the text rather than the number of lexical tokens, is essential for a proper preprocessing of long documents. In this way, the use of a small language model such as *DitilBart* proves to be a plausible alternative for the task of long-text summarization.

References

1. Cohan, A., et al.: A discourse-aware attention model for abstractive summarization of long documents. In: Proceedings of the 2018 Conference of the North American Chapter of the Association for Computational Linguistics: Human Language Technologies, Volume 2 (Short Papers), pp. 615–621 (2018)
2. Cheng, J., Lapata, M.: Neural summarization by extracting sentences and words. https://arxiv.org/abs/1603.07252 (2016)
3. Gong, Y., Liu, X.: Generic text summarization using relevance measure and latent semantic analysis. In Proceedings of the 24th Annual International ACM SIGIR Conference on Research and Development in Information Retrieval, pp. 19–25 (2001)
4. Rush, A. M., Chopra, S., Weston, J.: A Neural attention model for abstractive sentence summarization. In Proceedings of the 2015 Conference on Empirical Methods in Natural Language Processing, pp. 379–389 (2015)
5. See, A., Liu, P. J., Manning, C. D.: Get to the point: summarization with pointer-generator networks. In: Proceedings of the 55th Annual Meeting of the Association for Computational Linguistics (Volume 1: Long Papers), pp. 1073–1083 (2017)

6. Gehrmann, S., Deng, Y., Rush, A.M.: Bottom-up abstractive summarization. In: Proceedings of the 2018 Conference on Empirical Methods in Natural Language Processing, pp. 4098–4109 (2018)
7. Hsu, W.T., Lin, C.K., Lee, M.Y., Min, K., Tang, J., Sun, M.: A unified model for extractive and abstractive summarization using inconsistency loss. In: Proceedings of the 56th Annual Meeting of the Association for Computational Linguistics (Volume 1: Long Papers), pp. 132–141 (2018)
8. Liu, P.J., et al.: Generating Wikipedia by summarizing long sequences. In: International Conference on Learning Representations (2018)
9. Ma, C., Zhang, W.E., Guo, M., Wang, H., Sheng, Q.Z.: Multi-document summarization via deep learning techniques: a survey. ACM Comput. Surv. **55** (2022). https://doi.org/10.1145/3529754.
10. Zhang, Y., et al.: An exploratory study on long dialogue summarization: what works and what's next. In: Proceedings of the 2021 Conference on Empirical Methods in Natural Language Processing: Findings (2021)
11. Johnson, J.: Small Language Models (SLM): A Comprehensive Overview. https://huggingface.co/blog/jjokah/small-language-model (2025)
12. Hearst, M.A.: TextTiling: segmenting text into multi-paragraph subtopic passages. Comput. Linguist. **23**(1), 33–64 (1997). https://aclanthology.org/J97-1003.pdf
13. Beltagy, I., Peters, M.E., Cohan, A.: Longformer: The Long-Document Transformer. https://arxiv.org/abs/2004.05150 (2020)
14. Huang, L., Cao, S., Parulian, N., Ji, H., Wang, L.: Efficient Attentions for Long Document Summarization. https://arxiv.org/abs/2104.02112 (2021)
15. Lukasik, M., Dadachev, B., Simões, G., Papineni, K.: Text Segmentation by Cross Segment Attention. https://arxiv.org/abs/2004.14535 (2020)
16. Nayeem, M. T., Chali, Y.: Extract with order for coherent multi-document summarization. In: Proceedings of the TextGraphs-11: The Workshop on Graph-based Methods for Natural Language Processing (2017). https://aclanthology.org/W17-2407
17. Fu, X.Y., Laskar, T.R., Khasanova, E., Chen, C., Bhushan, S.: Tiny titans: can smaller large language models punch above their weight in the real world for meeting summarization? In: Proceedings of the 2024 Conference of the North American Chapter of the Association for Computational Linguistics: Human Language Technologies (Volume 6: Industry Track). Mexico City, Mexico, pp. 387–394 (2024)
18. Zhang, Y., Jin, H., Meng, D., Wang, J., Tan, J.: A comprehensive survey on automatic text summarization with exploration of LLM-based methods, Neurocomputing, pp. 131928 (2025). ISSN 0925-2312, https://doi.org/10.1016/j.neucom.2025.131928
19. HuggingFace: DistilBart: sshleifer/distilbart-cnn-12-6. https://huggingface.co/sshleifer/distilbart-cnn-12-6
20. CNN-DM: CNN and the Daily Mail. https://github.com/abisee/cnn-dailymail
21. Koh, H. Y., and Ju, J., Liu, M., Pan, S.: An empirical survey on long document summarization: datasets, models, and metrics. ACM Comput. Surv. **55** (2022). ISSN 1557-7341. https://doi.org/10.1145/3545176, http://dx.doi.org/10.1145/3545176
22. Celikyilmaz, A., Bosselut, A., He, X., Choi, Y.: Deep communicating agents for abstractive summarization. In: Proceedings of the 2018 Conference of the North American Chapter of the Association for Computational Linguistics: Human Language Technologies, Volume 1 (Long Papers), pp. 1662–1675 (2018)

23. Zaheer, M., et al.: Big bird: transformers for longer sequences. In: Proceedings of the 34th International Conference on Neural Information Processing Systems. ISBN 9781713829546 (2020)
24. Lin, C.Y.: ROUGE: a package for automatic evaluation of summaries. In: Text Summarization Branches Out, pp. 74–81. ACL, Barcelona. https://aclanthology.org/W04-1013/ (2004)
25. Zhang, T., Kishore, V., Wu, F., Weinberger, K.Q., Artzi, Y.: BERTScore: Evaluating text generation with BERT. https://arxiv.org/abs/1904.09675 (2020)

Speech Recognition Models and Detection Techniques

Mental Depression Detection from Mandarin Speech

Tanvir Alam, Selin Koc, Enrique Navarro Espino, Dharanidhar Dang, Kevin Hirschi, John Quarles, and Mario Flores(✉)

The University of Texas at San Antonio, San Antonio, TX 78249, USA
{tanvir.alam,enrique.navarroespino}@my.utsa.edu, {selin.koc, dharanidhar.dang,kevin.hirschi,John.Quarles, mario.flores}@utsa.edu

Abstract. Detecting depression from speech has become an active research area as conventional diagnostic methods often suffer from subjectivity and under reporting. In this study, we examine a range of machine learning and deep learning approaches for identifying depression from Mandarin speech and its transcriptions using the MODMA dataset. Traditional machine learning models and neural networks, such as CNNs, GRUs, and hybrid CNN-LSTM architectures, were applied using GroupKFold cross-validation to ensure speaker-independent evaluation. Among machine learning models, the random forest classifier delivered the most consistent performance, achieving 72% accuracy across folds. CNN-based models matched this score in speech-only settings, also reaching 72% with strong fold-wise stability. Other deep learning models like GRU and LSTM reached higher test scores but did not maintain consistent results across subject splits. In our multi-modal experiments, where audio and text features were combined, the most dependable performance came from a CNN-based fusion model, which reached an average accuracy of 74% across folds. However, these results highlight the strong predictive power of audio features alone and underscore the importance of subject-aware evaluation for developing robust depression detection systems.

Keywords: Depression detection · MFCC · CNN · LSTM · Bi-LSTM · Multimodal · Attention

1 Introduction

Depression is a common mental health condition that affects millions of people around the world. The effects of depression extend beyond emotional distress, negatively influencing cognitive abilities, physical health, and everyday activities, thereby affecting interpersonal relationships and professional responsibilities [1].

Depression is diagnosed using self-assessment tools such as the Patient Health Questionnaire (PHQ-9) and Beck Depression Inventory (BDI), along with interviews conducted by clinicians. While these approaches are commonly used, they have drawbacks such as personal bias, limited accuracy, and the potential influence of social stigma [2].

H. Unger and P. Meesad (Eds.): NLPIR 2025, LNNS 1904, pp. 507–523, 2026.
https://doi.org/10.1007/978-3-032-20897-2_37

Recently, automated depression detection methods have been explored as potential objective, reliable, and efficient alternatives. One promising path is analyzing speech signals since it is non-invasive and has the capacity to reflect emotional states effectively [3]. Studies have shown that depression causes distinct acoustic changes, including reduced speech speed, longer pauses, decreased pitch variation, and irregular voice qualities such as heightened jitter and shimmer [4]. Apart from acoustic cues, textual analysis of speech transcriptions has also gained considerable attention, as linguistic content often reveals underlying psychological states. Depressed individuals frequently exhibit particular linguistic patterns such as increased usage of negative emotional expressions, frequent self-referential pronouns, and absolutist language, for instance, words like "never" and "always" [5]. Consequently, merging acoustic and linguistic cues can significantly enhance the predictive power of depression detection models.

Recent advances in deep learning, especially involving convolutional neural networks (CNNs) and recurrent neural networks (RNNs), have markedly improved automated depression identification from speech data [6]. Vandana et al. [3] developed a deep learning-based system for identifying depression using both spoken and written responses from individuals. The study investigated three primary configurations using the DAIC-WoZ dataset: a CNN model trained exclusively on textual transcripts, another CNN model processing audio spectrograms, and a hybrid model that fuses both modalities using LSTM and Bi-LSTM architectures. The results highlighted the diagnostic value of speech-based cues such as tone, rhythm, and articulation. Furthermore, Vandana et al. [3] designed a hybrid deep learning system that uses CNNs to extract features from audio and text, which are then processed through long short-term memory (LSTM) and Bi-LSTM layers to model temporal relationships. These outcomes demonstrate the value of combining modalities, with the Bi-LSTM configuration showing stronger learning capacity and generalization compared to its unidirectional counterpart. Although the hybrid models required longer training times, they exhibited enhanced robustness in predicting depressive states when compared to models trained on audio or text alone. Fang et al. [4] proposed a multimodal method for detecting depression that integrates audio, visual, and text inputs using a hierarchical attention mechanism. The model uses LSTM networks to process features from each type of input and applies attention mechanisms to highlight important information while reducing noise. Specifically, it processes acoustic features, along with multiple visual signals such as facial landmarks, head movement, gaze direction, and facial expressions, using LSTM-based layers. Then, the textual data is embedded using a Bi-LSTM with attention to extract contextual semantics. After each type of data is processed individually, the model uses an attention-based fusion network (AttFN) to combine them by weighing the relative importance of each modality, allowing it to filter out redundant information and emphasize features relevant to depression. The approach was evaluated on the DAIC-WOZ dataset and achieved a mean absolute error (MAE) of 3.18 and a root mean square error (RMSE) of 3.68, surpassing many existing multimodal benchmarks. As a result, combining all three modalities led to better results than any unimodal or bimodal combination, underscoring the benefits of comprehensive multimodal integration in depression assessment.

Akkaralaertsest and Yingthawornsuk [7] investigated the potential of speech-based acoustic features for distinguishing between depressed and high-risk suicidal individuals.

Using a dataset of clinically diagnosed female speakers, they extracted Mel-Frequency Cepstral Coefficients (MFCC), which capture vocal tract characteristics, and Glottal Spectral Slope (GSS), which reflects glottal source properties, from voiced segments of interview recordings. These features were then reduced via Principal Component Analysis and classified using probabilistic models. Their results showed that MFCC and GSS provided moderate discriminative power, with classification accuracy around 65–70% when differentiating between depressed and suicidal speech. While not conclusive, the study highlighted that subtle vocal changes associated with psychiatric states can be computationally measured and may serve as supportive indicators for clinical assessment.

Daly and Olukoya [8] proposed a multimodal framework for depression detection that integrates both acoustic and textual features from read and spontaneous speech in Italian, addressing a major research gap in lesser resourced languages. Their approach combines CNNs and LSTM models for detailed audio analysis with BERT embeddings for semantic representation of transcriptions. Using the clinically validated Androids Corpus (228 recordings, including 64 diagnosed cases), they achieved state-of-the-art results with 94.30% accuracy and an F-score of 94.51% through decision-level fusion, significantly outperforming unimodal baselines. A key contribution of this work lies in its systematic comparison of read versus spontaneous speech, revealing distinct diagnostic cues: read speech exhibited stronger correlations among lower-order MFCCs, while spontaneous speech provided richer but more heterogeneous paralinguistic markers. Furthermore, the study demonstrated that decision-level multimodal fusion is superior to feature-level approaches, setting a new benchmark for computational paralinguistics. By using clinically verified data and extending research to a non-English corpus, this study not only improves the reliability and generalizability of automated depression detection but also highlights the cultural and linguistic dimensions of speech-based diagnostics, making it a significant advancement for both affective computing and clinical practice.

This study advances speech-based depression detection by extending research to Mandarin speech using the MODMA dataset, a linguistically underexplored context compared with the predominantly English- and European-language corpora examined in prior work. Unlike studies such as [8], which reported significant improvements from multimodal fusion in Italian speech, our findings show that acoustic features alone, particularly prosodic measures such as pitch statistics and energy, spectral descriptors such as MFCCs, and temporal features such as pause ratio, carry stronger and more stable predictive power than text-based features in Mandarin. To ensure reliability, the evaluation framework employed speaker-independent validation through GroupKFold cross-validation, addressing limitations in earlier research where subject overlap could inflate performance. The audio-based models include machine learning approaches and Deep Learning approaches. Integration of the audio and text with CNN and multi-head attention-based fusion was also evaluated, which yielded slightly improved results. The goal is to develop a system that enhances accuracy, robustness, and adaptability.

2 Dataset

The MODMA dataset [9] used in this study comprises 1,503 audio recordings in Mandarin Chinese collected from a total of 52 subjects, including 29 healthy controls (HC) and 23 individuals diagnosed with major depressive disorder (MDD). Each subject contributed 29 individual audio samples, resulting in a balanced per-subject data distribution. The dataset contains recordings from 36 male and 16 female participants, aged between 18 and 42 years.

The audio recordings consist of participants' responses to a set of open-ended, emotionally oriented questions such as *"How do you evaluate yourself?"* and *"What makes you desperate?"*, reading emotion-labeled words and a short story, and describing affective images [9]. Each audio recording was captured under consistent conditions, ensuring uniformity in acoustic quality and speaker environment. The subjects were also assessed using a variety of clinical questionnaires, such as PHQ-9, GAD-7, PSQI, CTQ-SF, SSRS, and LES, and demographic data, including age, gender, and years of education, were collected. However, in this study, these clinical and demographic variables were not used for classification or prediction tasks.

Instead, the focus of this research is on audio-based and linguistic markers of depression. Audio recordings were preprocessed to extract a range of acoustic features (e.g., prosodic, spectral, and voice quality parameters). Additionally, each recording was transcribed, and the resulting text data were processed to extract linguistic features for depression detection. These multimodal features serve as the input for machine learning models aimed at distinguishing between MDD and HC individuals. This approach allows for a fully non-invasive and speech-driven method of depression detection, offering promising potential for scalable, accessible mental health screening tools.

3 Methodology

3.1 MFCCs Based Analysis

This methodology was selected to analyze the dataset because it offers several advantages over alternative acoustic feature representations:

- Low relevance to noise exposure
- Great stability for the analysis of human emotions
- Without time limit constraints
- Independent from voice levels

The dataset contains sound recordings in WAV format from interviews with 52 patients (male and female). Some patients were diagnosed with Major Depressive Disorder (MDD), while others were evaluated as being in healthy condition (HC) by professional doctors in the field. Audio files for patients diagnosed with depression are marked with the suffix "_Depre" (23 files). Patients in healthy condition (29 files) do not include this mark.

The frequency analysis starts loading the dataset using a library called Librosa. Then features are extracted from the recordings in WAV format using Mel-Frequency Cepstral

Coefficients (MFCCs) with some specific configuration to enhance the information: 100 frames truncated to ensure a fixed length and 40 coefficients.

The model generates two labels: "MDD" for patients with mental depression for those files containing "_Depre" and "HC" for all other patients who are considered to have a healthy condition. For the data preparation the model stores the MFCC features as 3D tensors with 40 coefficients and 100 frames each. Labels are encoded with 0 for HC and 1 for MDD and the dataset is split into 67% for training purposes and 33% for testing.

The model is trained for 100 epochs with a batch size of 16. By combining CNNs and BiLSTMs, the model leverages both short-term spectral variations and long-term sequential context, offering a more nuanced representation of depressive speech patterns compared to conventional CNN-only architectures.

Figure 1 illustrates the proposed CNN–BiLSTM pipeline. MFCC features are first extracted and structured as a spectrogram-like input. The CNN layers act as feature detectors to capture local spectral-temporal cues, while the BiLSTM layers sequentially model longer-term prosodic variations. Finally, fully connected layers and a sigmoid activation perform the binary classification task.

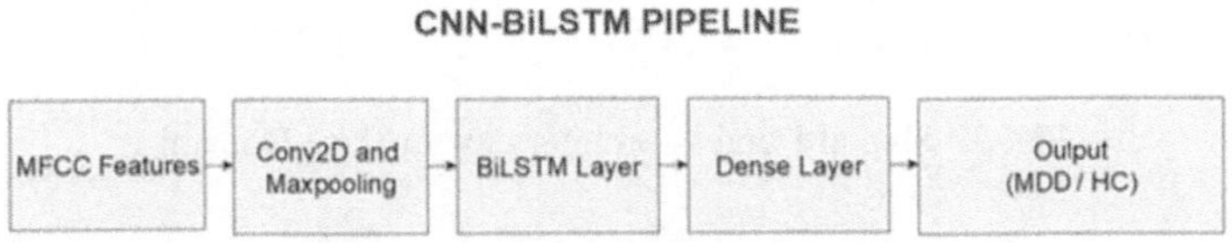

Fig. 1. CNN – BiLSTM Pipeline

MFCC Feature Justification. MFCC is a widely used audio feature extraction technique, especially for speech and emotion recognition tasks.

The method consists of:

i. Applying a Fourier Transform to convert the audio signal into the frequency domain.
ii. Applying Mel-Scale Filter Banks to simulate human auditory perception, emphasizing perceptually relevant frequencies [10].
iii. Applying a logarithmic transformation to approximate human loudness perception.
iv. Using a Discrete Cosine Transform (DCT) to produce a compressed representation.

MFCCs effectively capture the timbre and tonal characteristics of speech and remain robust to background noise and variations in pitch or tone [10]. Given that depression manifests in speech through flattened prosody, reduced pitch variation, and altered vocal energy, MFCCs are well-suited as input features for the CNN–BiLSTM model.

Visualization and Pattern Analysis. Figure 2 shows the block diagram for the MFCC transformation process.

Figure 3 shows the waveform visualization of an interview recording from a participant diagnosed MD, displayed in the time domain. Vocal intensity and speaking dynamics throughout the recording are reflected in the amplitude variations of the signal.

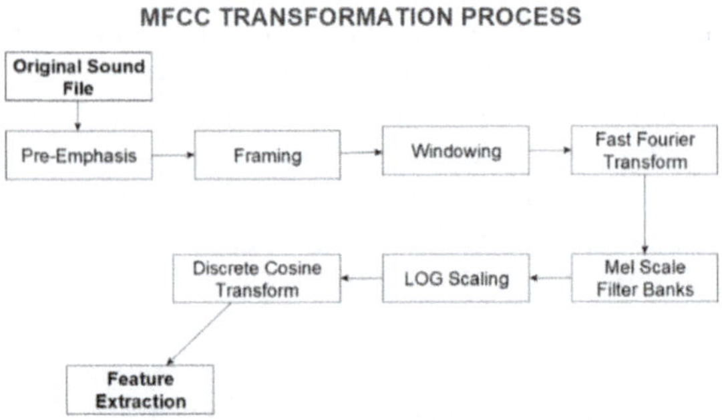

Fig. 2. MFCC Transformation Process

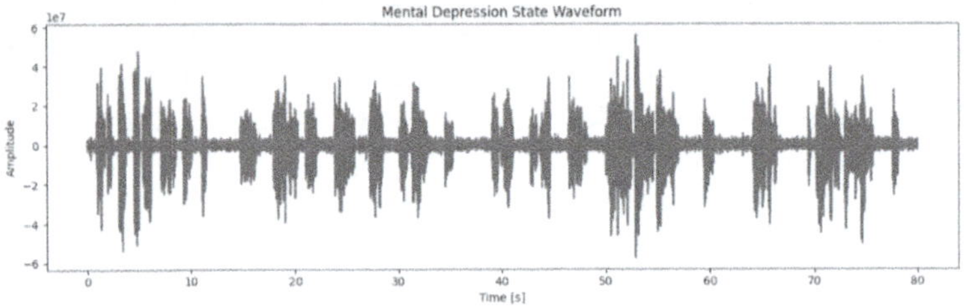

Fig. 3. Mental Depression Interview in Time Domain

On the frequency domain analysis, MFCC provides tools to analyze signals with emphasis in power spectrum tied to a certain frequency, therefore the problem of voice level and noise are reduced dramatically, and more focus are set on energy perceived in bands of frequency over time. Spectrogram is a graph to show the energy-frequency analysis, and it is shown in the Fig. 4 for the same interview of the patient with mental depression.

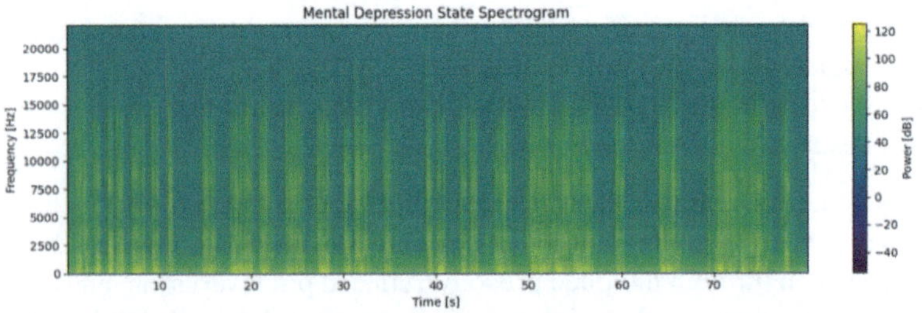

Fig. 4. Mental Depression Interview in Frequency Domain

Figures 5 and 6 present the same time- and frequency-domain representations for a healthy condition patient. Differences in prosodic variation and spectral energy distribution can be observed between MDD and HC participants.

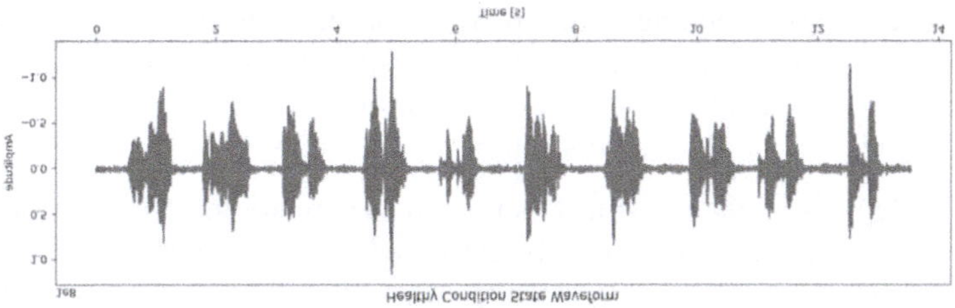

Fig. 5. Healthy Condition in Time Domain

Figure 6 describes the interview for a healthy condition patient in the frequency domain analysis. It is important to remember this graph describes the concentration of power over the spectrum of frequencies in time progression.

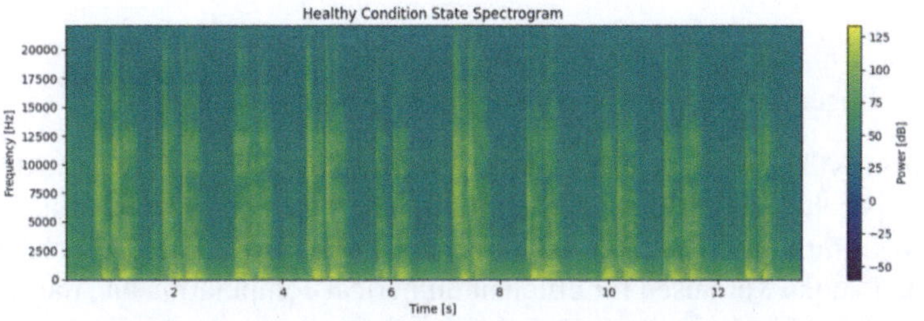

Fig. 6. Healthy Condition in Frequency Domain

These patterns demonstrate that depressive and healthy speech can be distinguished not only visually but also numerically through MFCC features and deep learning architectures designed to capture both local and sequential dependencies.

3.2 Text Based Model

To transcribe the Mandarin audio recordings, we used OpenAI's Whisper automatic speech recognition (ASR) model (base version) [11]. The model was loaded using the Whisper API and used to transcribe all valid WAV files. Transcriptions were extracted in Mandarin Chinese using the 'language = "zh"' parameter. Furthermore, two pretrained language models were applied for classification: a standard Chinese BERT model and a RoBERTa model with Whole Word Masking (WWM), which offers improved contextual understanding and may be beneficial for detecting subtle linguistic cues associated with depression. Both models were trained with the HuggingFace Trainer and evaluated on the same 20% hold-out set, and the identical preprocessing pipeline and label mapping (HC = 0, MDD = 1) were used across models. Tokenization was performed with the respective pretrained tokenizer, truncation was applied to a maximum sequence length of 512 tokens.

Optimization used AdamW (learning rate 5×10^{-5}, weight decay 0.01), batch size 8 for both training and evaluation. At the end of every epoch, evaluation and checkpointing were performed. Early stopping with a patience of three epochs was applied based on validation loss, and the checkpoint with the lowest validation loss was restored for final scoring. To address label imbalance, we trained with class-weighted cross-entropy using weights computed on the training portion only (HC $\approx$ 0.89; MDD $\approx$ 1.14).

BERT. The depression detection task is framed as a binary text classification problem based on the chinese transcriptions. A BERT-based [12] pre-trained mandarin language model "bert-base-chinese", fine-tuned for sequence classification with two output labels: MDD and HC. Input texts were tokenized using the associated BERT Tokenizer with truncation and padding.

RoBERTa. A pre-trained Chinese RoBERTa model [13], specifically "hfl/chinese-roberta-wwm-ext" [14] from the Hugging Face model hub is applied. This model is a whole word masking variant of RoBERTa, trained on large-scale Chinese corpora. Tokenization of the input text (Mandarin Transcriptions) was performed using the corresponding AutoTokenizer, with truncation and dynamic padding applied.

3.3 Model Based on MFCCs and Integrated Audio Features

Audio Features' Extraction and Feature Engineering. To facilitate the development of a robust speech-based depression detection system, a comprehensive suite of Python libraries was employed for data handling, audio processing, and machine learning. NumPy and Pandas were used for efficient numerical computation and data structuring, while Seaborn and Matplotlib provided support for detailed visualization of acoustic trends and feature distributions. For audio signal processing, Librosa was the primary toolkit, enabling tasks such as pitch tracking, MFCC computation, zero-crossing rate analysis, and spectral feature extraction. Parselmouth, a Python interface to the Praat software, was employed to extract jitter, shimmer, and other phonation-based features such as harmonic-to-noise ratio (HNR) and formant frequencies, which are often sensitive to emotional and physiological changes in the voice.

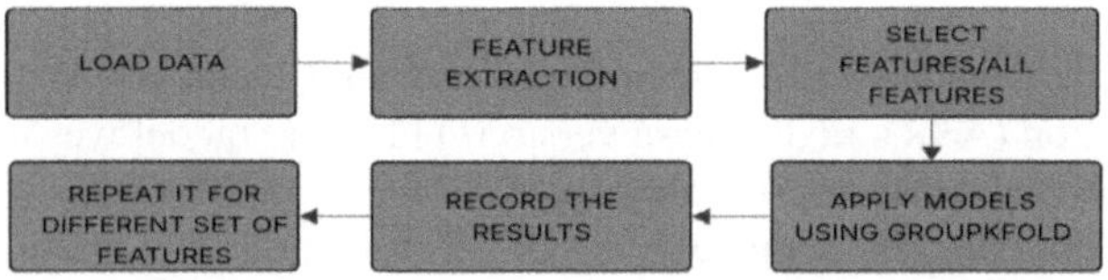

Fig. 7. Feature Selection Process

MFCCs, widely recognized for modeling the human auditory system, were extracted and summarized by their mean and standard deviation across time. Additional features considered in the expanded pool included formants, harmonic-to-noise ratio, mel spectrogram, chroma, tonnetz, and vocal quality descriptors such as creaky, breathy, and tense voice characteristics. The harmonics-to-noise ratio (HNR) was computed in Parselmouth to quantify the relative amount of harmonic energy versus noise in the signal, with lower values reflecting breathier or noisier phonation. To capture resonance characteristics of the vocal tract, the first three formant frequencies (F1–F3) were derived using Praat's Burg method and measured at the temporal midpoint of each recording. Furthermore, three voice quality markers were defined as binary indicators based on thresholds over pitch, jitter, shimmer, and HNR: creaky voice (low pitch, high jitter, low HNR), breathy voice (low HNR, high shimmer), and tense voice (high pitch, low jitter, low shimmer). These markers approximate perceptual qualities of voice that are often linked to emotional or physiological states, providing complementary cues to prosodic and MFCC features for detecting depression-related vocal patterns.

Multiple combinations of these features were systematically evaluated to identify the optimal subset that yields the highest testing accuracy, as shown in Fig. 7. So running the program numerous times and observing the results, key features were identified. For different sets of features, the models' performances varied. Some models performed better for a particular set of features while others performed differently in another set of features.

The goal was to find a set of features that is more effective in distinguishing between depression and healthy patients using audio. Features' sets were applied both on raw and reduced noise (processed using noisereduce library) versions of the dataset to assess the influence of preprocessing on model performance. Interestingly, the highest classification performance was achieved using raw audio data without noise reduction, suggesting that the natural variability present in untreated recordings may carry valuable diagnostic cues. The most effective feature combination comprised mean pitch, pitch standard deviation, energy, 13 MFCCs means, 13 MFCCs standard deviations, zero-crossing rate, and pause ratio, highlighting the importance of both prosodic and spectral cues in identifying depressive speech patterns.

Machine Learning Models. Logistic regression, decision trees, support vector machines (SVM), random forests, and a soft voting ensemble are supervised classifiers that are applied to detect speech depression.

Deep Learning Models. CNNs, RNNs, gated variants such as LSTM and GRU, and a hybrid CNN-LSTM model are the deep learning models that were applied to detect depression from speech.

3.4 Multimodal Integration of Audio and Text

Multimodal Feature Processing. The speech branch processes audio features such as pitch, energy, speaking rate, and MFCCs. These features are first standardized and then passed through a stack of one-dimensional convolutional layers. each convolutional block includes batch normalization, pooling, and dropout to reduce overfitting and emphasize important acoustic patterns. The output is condensed into a fixed-length vector using global average pooling.

Multimodal Fusion Architectures. An overview of the three multimodal fusion architectures evaluated in this study is presented in Fig. 8. The models include CNN Fusion, multi-head Attention Fusion, and BiLSTM with multi-head Attention Fusion where the text branch is encoded by a BiLSTM prior to to cross-modal attention. Each architecture

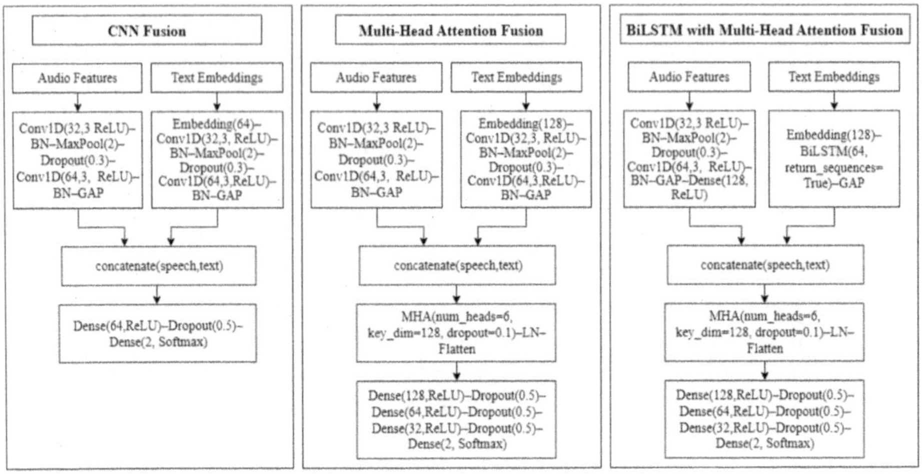

Fig. 8. Proposed Fusion Model Architectures: CNN Fusion, Multi-Head Attention Fusion, BiLSTM with Multi-Head Attention Fusion. Abbreviations—Conv: Convolution, ReLU: Rectified Linear Unit, BN: Batch Normalization, GAP: Global Average Pooling, MHA: Multi-head Attention, LN: Layer Normalization.

combines acoustic features and character-level text embeddings through different fusion strategies.

Multi-Head Attention Fusion. A multi-head attention mechanism [15] fuses speech and text modalities. Let $\overline{H} = \left[h_{\text{speech}}; h_{\text{text}}\right]$ denote concatenated features from both modalities. Multi-head attention computes attention scores to determine feature relevance:

$$\text{Attention}(Q, K, V) = \text{softmax}\left(\frac{QK^T}{\sqrt{d_k}}\right)V \tag{1}$$

where queries Q, keys K, and values V are linear projections $\overline{H}$. This allows the model to dynamically weigh the importance of audio versus textual features. Outputs of the speech and text branches were reshaped to form token embeddings, which were stacked into a two-element sequence. Cross-modal dependencies were modeled with a multi-head attention mechanism comprising six heads and a key dimension of 128. The attention output was normalized, flattened, and passed through a feed-forward head consisting of dense layers (128, 64, and 32 units with dropout) before the final two-class softmax layer.

BiLSTM with Multi-Head Attention Fusion. Speech features were further projected to a 128-dimensional vector after convolution and pooling, while text embeddings were encoded using a BiLSTM. The two modality vectors were reshaped, concatenated, and passed through multi-head attention (six heads, key dimension is 128, dropout is 0.1). the attention output was normalized, flattened, and classified by a dense head with three hidden layers (128, 64, and 32 units, each followed by dropout) and a softmax output layer.

4 Results

4.1 MFCCs Based Analysis

For the Frequency Analysis, we find that Accuracy starts around 55% which is close to random guessing, and Loss starts high in the beginning and decreases gradually with epochs. In the middle phase, around epoch 13, the model begins to learn significantly, and accuracy rises from 60% to 77% with a steady improvement. In the final phase, after epoch 31, the model reaches 85% of accuracy, and then it registers not any significant progress. There is also no severe sign of overfitting.

CNN's are a powerful way to analyze audio files and feature extraction due to its capacity of correlate coefficients in the frequency domain. The final metrics show train accuracy of 87% with a Val Accuracy of 84% with a Val Loss decreasing steadily. The overall results are satisfactory because the model provides a reasonable tool for MDD prediction. Table 1 shows the results obtained for the MFCC Analysis.

Table 1. MFCC Analysis Results

MFCC Analysis	Precision	Recall	F1 Score	Support
HC	0.87	0.85	0.86	259
MDD	0.81	0.84	0.82	199
Accuracy			0.84	458
Macro Avg	0.84	0.84	0.84	458
Weighted Avg	0.84	0.84	0.84	458

4.2 Text Based Model

For text-based detection, the BERT model achieved an F1 score of 0.66 and an accuracy of 72% on the validation set at its best epoch 2 as seen in Table 2. The results showed some overfitting in later epochs, as indicated by increased validation loss, despite decreasing training loss.

Table 2. Epoch-wise Evaluation Metrics of the mandarin based BERT model

Text-based BERT Model Evaluation				
Epoch	Training Loss	Validation Loss	Accuracy	F1 Score
1	0.7549	0.6931	56.00%	0.0000
2	0.6351	0.5670	71.67%	0.6586
3	0.5183	0.6896	68.67%	0.6050
4	0.2906	0.9822	68.67%	0.5913

Applying the Mandarin-based RoBERTa model did not improve the results. One possible reason is that the ASR-generated transcriptions do not include explicit word segmentation in Mandarin. Without word segmentation, most tokens are treated as individual characters. Since WWM is designed to work on complete multi-character words, its impact is reduced in this context. As shown in Table 3, the best validation performance was after the second epoch with an F1 score of 0.61 and accuracy of 70%.

Table 3. Epoch-wise Evaluation Metrics of the mandarin based RoBERTa model

Text-based RoBERTa Model Evaluation				
Epoch	Training Loss	Validation Loss	Accuracy	F1 Score
1	0.6403	0.6152	64.33%	0.6537
2	0.4621	0.5992	70.00%	0.6053
3	0.3856	0.9343	68.00%	0.5897
4	0.1881	1.4535	66.33%	0.5844

4.3 Model Based on MFCCs and Integrated Audio

A set of machine learning models were tested for classifying speech recordings into two categories: depressed (MDD) and healthy control (HC). The results are shown in Table 4, which includes standard evaluation scores on a separate test set as well as average accuracy values from GroupKFold cross-validation. Metrics include test precision, recall, F1-score, accuracy, and subject-disjoint GroupKFold cross-validation accuracy. SMOTE and class weighting were applied to address class imbalance. Each sample was tagged with a subject identifier extracted from its file path. A fixed set of subject IDs was reserved as the test set, while the remaining data was used for training to address the potential for data leakage due to speaker overlap. Thus, no samples from the same individual appeared in both the training and validation sets within a fold, preserving speaker independence.

GroupKFold cross-validation stratified by subject ID was applied to evaluate the consistency of detection accuracy across different patient subsets. In contrast to standard K-Fold, GroupKFold ensures that no samples from the same individual appear in both training and validation sets. Therefore, multiple evaluation scenarios were simulated using different sets of unseen subjects in each fold. By analyzing the variation in accuracy across these fixed folds, we were able to quantify how reliably the models perform on different subject groups. Each model was evaluated using 5- or 7-fold GroupKFold depending on sample distribution, and fold-wise accuracy was recorded to assess consistency.

Logistic regression gave stable outcomes across different setups, including its basic version, SMOTE-based version, and class-weighted version. These variants reached around 66–67% accuracy on the test set and about 65% in cross-validation. The decision tree model gave similar results, with slight improvement after tuning. Random forest gave the strongest performance overall, achieving 72% accuracy on the test data and 70% on average across cross-validation folds. Support Vector Machines (SVM) also gave reliable performance, with or without class balancing. The model achieved 71% test accuracy, and the results stayed around 68% in cross-validation regardless of whether SMOTE or class weights were used. Among the models, the random forest showed the most consistent results across subjects, followed closely by SVM. These outcomes suggest that tree-based and margin-based classifiers can handle variations in speech features effectively.

Table 4. Evaluation Metrics of ML models for speech-based depression detection.

Speech-based Machine Learning Models	Precision	Recall	F1	Accuracy	GroupKFold Mean Accuracy
Logistic	69%	67%	66%	67%	65% (5 Folds)
LR (SMOTE)	67%	66%	66%	66%	65% (5 Folds)
LR (ClassWeight)	67%	66%	65%	66%	65% (5 Folds)
Decision Tree	67%	67%	67%	67%	61% (5 Folds)
DT (Tuned)	68%	67%	67%	67%	64% (8 Folds)
Random Forest	73%	72%	71%	72%	70% (5 Folds)
Support Vector	71%	71%	71%	71%	68% (5 Folds)
SVM (SMOTE)	70%	70%	70%	70%	68% (5 Folds)
SVM (ClassWeight)	70%	70%	69%	70%	68% (5 Folds)

The classification outcomes for various deep learning models (CNN, RNN, GRU, LSTM, Hybrid) using speech features are shown in Table 5. For CNN and hybrid models, consistency was verified using GroupKFold. While GroupKFold was explicitly applied to CNN and the hybrid model to validate consistency, the RNN, GRU, and LSTM models were tested on multiple fixed test subject sets to observe how model behavior shifts across speaker subsets.

The CNN model achieved a test accuracy of 72%. The GroupKFold cross-validation had mean accuracy of 71.95%, indicating stable generalization across different subject splits. RNN produced a similar result with 72% overall accuracy, which had slightly more higher precision (75%) than recall. The LSTM setup performed more effectively, achieving 80% accuracy and a macro F1-score close to 79%, indicating its capability in modeling speech sequences over longer spans. The GRU achieved with 81% accuracy and a macro F1-score of 81%, reflecting strong results in terms of both precision (84%) and general reliability in prediction. Lastly, the hybrid CNN + LSTM model, which combines convolutional and recurrent layers, achieved 70% accuracy on the fixed test set and 68.87% mean accuracy across GroupKFold folds, demonstrating reasonable consistency across patient-specific partitions.

Table 5. Evaluation Metrics of DL models for speech-based depression detection

Speech-based Deep Learning Models	Precision	Recall	F1	Accuracy	GroupKFold Mean Accuracy
CNN	73%	72%	72%	72%	72% (5 Folds)
CNN (ClassWeight)	71%	71%	71%	71%	-
RNN	75%	72%	72%	72%	-
LSTM	84%	80%	79%	80%	-
GRU	84%	81%	81%	81%	-
CNN + LSTM	-	-	-	70%	69% (5 Folds)

Model Selection and Error Analysis. No single model produced consistently superior results. performance varied depending on the test subjects, with some models achieving strong accuracy for certain patients but performing poorly for others. As discussed earlier, different feature subsets also influenced model behavior, with each algorithm responding differently to a specific set of features. However, machine learning models provided more consistent results upon multiple runs compared to deep learning models for a specific set of features and test subjects. Here is an example explaining models' variations of performances for different test subjects- to evaluate the generalization ability of the random forest classifier, the model was first trained on the training set and tested on the held-out set, achieving an overall accuracy of 72% (HC: precision = 0.69, recall = 0.83; mdd: precision = 0.78, recall = 0.62). to ensure subject-independent evaluation, GroupKFold cross-validation was applied, using subject IDs as grouping keys so that all recordings from the same participant were confined to either the training or test fold. With a 7-fold split, the classifier was retrained and tested across different subject groups, with fold accuracies ranging from 61.6% to 79.3%. The mean cross-validation accuracy was 70.4%, consistent with the held-out test results and indicating that the model captures subject-independent patterns of depression and healthy speech, albeit with variability across speaker groups. Additionally, a CNN was evaluated using 5-fold GroupKFold cross-validation with subject IDs as grouping keys to ensure subject-independent splits. the fold-wise accuracies were 72.98%, 63.60%, 62.45%, 74.71%, and 63.60%, resulting in a mean cross-validation accuracy of 67.47%. The variation across folds highlights differences in performance depending on the test subject group, indicating that the CNN captured general patterns but was sensitive to subject-level variability. However, it was observed that the accuracies varied across different runs of the deep learning models.

4.4 Multimodal Integration of Audio and Text

Three multimodal models that combined audio features with transcribed text were evaluated, and the results are shown in Table 6. These included a CNN-based model, a version using multi-head attention, and a setup that combined Bi-LSTM with attention. Each model was trained and validated using a subject-level GroupKFold split with seven folds, making sure that samples from the same individual did not appear in both training and validation sets.

The first model used a dual-branch CNN architecture, one branch for prosodic and MFCC speech features and the other for character-level embeddings from the transcriptions. After independent processing, the outputs were merged and passed through dense layers for classification. This CNN-based fusion achieved solid performance with 78% accuracy on the test set and a mean accuracy of 74% across seven GroupKFold folds, demonstrating both strong predictive power and consistent generalization across subject groups.

In the multi-head attention (MHA) fusion model, the audio and text branches were independently processed, and their outputs were reshaped and concatenated to serve as input for the attention mechanism. Although this model produced slightly higher test scores (79% accuracy and F1), the average cross validation accuracy was 70%, indicating slightly less stability across different subject splits.

A more complex fusion approach combined a Bi-LSTM layer with multi-head attention, which encoded temporal dependencies in the text input and integrated both acoustic and linguistic cues. The performance was comparable to the other models, reaching 78% test accuracy and a 73% average across folds.

While attention-based models showed strong single-run results, the CNN fusion model proved to be the most consistent overall. Its combination of simplicity, accuracy, and reliable behavior across patient groups makes it a practical candidate for real-world applications where generalizability matters.

Table 6. Evaluation Metrics for Text and Speech Fusion Models

Text and Speech Fusion Models	Precision	Recall	F1	Accuracy	GroupKFold Mean Accuracy
CNN Fusion	78%	78%	78%	78%	74% (7 Folds)
Multi-Head Attn. Fusion	79%	79%	79%	79%	70% (7 Folds)
Bi-LSTM + MHA	78%	78%	78%	78%	73% (7 Folds)

5 Discussion

While using only MFCCs for audio-based classification yielded promising results, the initial experiments did not ensure a patient-independent split between training and testing data. Since MFCCs are closely associated with speaker identity and voice characteristics, the model have inadvertently learned to recognize individual patients rather than detecting depression-specific patterns. This compromises the generalizability of the results, as the model may perform well simply by re-identifying the same speaker across datasets. To address this limitation, future experiments adopt a more robust evaluation strategy by employing GroupKFold cross-validation, ensuring that audio samples from the same patient are not present in both training and testing sets. In contrast, for models with text input only that use language representations such as BERT and RoBERTa, without incorporating any audio features, a standard data split was used as these models are not influenced by speaker identity.

Audio features alone demonstrated strong predictive capability despite the relatively small amount of available audio data in this dataset. It is important to note that depressive speech characteristics vary considerably between individuals, which can lead to inconsistent model performance when tested on unseen subjects. This inter-subject variability is a known challenge in speech-based depression detection and often results in fluctuations in accuracy across different evaluation splits. To address this, GroupKFold cross-validation was employed to ensure more robust feature selection and model generalization, facilitating the identification of feature combinations applicable across a broader population of patients.

Moreover, increasing the dataset size is expected to further enhance predictive accuracy, particularly when leveraging deep learning architectures, which typically benefit from larger amounts of training data. Interestingly, while the inclusion of textual data alongside audio inputs in a multi-head attention model yielded reasonable performance, it did not substantially improve results compared to models trained solely on speech features. This suggests that, for this dataset and task, acoustic information carries the most discriminative power, though future work with larger multimodal datasets may better exploit the complementary strengths of audio and text modalities.

6 Conclusion

This study explored various machine learning and deep learning models for depression detection using Mandarin speech and its transcriptions from the MODMA dataset. While multimodal models combining audio and text inputs showed slightly higher accuracy, the most consistently effective and information-rich features came from the audio domain alone. Specifically, a combination of mean pitch, pitch standard deviation, energy, 13 MFCCs means and standard deviations, speaking rate, and pause ratio proved to be the most discriminative for identifying depression.

Despite the limited amount of available audio data, these handcrafted acoustic features enabled strong model performance, particularly when patient-aware evaluation via GroupKFold was used to avoid speaker identity bias. Although deep learning models like LSTM and CNNs exhibited promising results, their full potential is likely constrained by dataset size. In summary, this work demonstrates that carefully selected audio features alone can provide powerful signals for depression detection, and sets the stage for more scalable, multimodal models in future research.

References

1. Wang, N.R., Chiong, R., Kamil, W., Zhang, S., Al-Haddad, Ibrahim, N.: Depression detection using speech audio and text: a comprehensive review focusing on deep learning methods. Authorea Preprints (2024)
2. Ravi, V., Wang, J., Flint, J., Alwan, A.: Enhancing accuracy and privacy in speech-based depression detection through speaker disentanglement. Comput. Speech Lang. **86**, 101605 (2024)

3. Vandana, Marriwala, N., Chaudhary, D.: A hybrid model for depression detection using deep learning. Measure. Sens. **25**, 100587 (2023)
4. Fang, M., Peng, S., Liang, Y., Hung, C.-C., Liu, S.: A multimodal fusion model with multi-level attention mechanism for depression detection. Biomed. Signal Process. Control **82**, 104561 (2023)
5. Lee, Y., Yoon, S., Jung, K.: Multimodal speech emotion recognition using cross attention with aligned audio and text. In Interspeech 2020, ISCA, pp. 2717−2721 (2020)
6. Cummins, N., Scherer, S., Krajewski, J., Schnieder, S., Epps, J., Quatieri, T.F.: A review of depression and suicide risk assessment using speech analysis. Speech Commun. **71**, 10–49 (2015)
7. Akkaralaertsest, T., Yingthawornsuk, T.: Comparative analysis of vocal characteristics in speakers with depression and high-risk suicide. Int. J. Comput. Theory Eng. **7**(6), 448–452 (2015)
8. Daly, K., Olukoya, O.: Depression detection in read and spontaneous speech: a Multimodal approach for lesser-resourced languages. Biomed. Signal Process. Control **108**, 107959 (2025)
9. Cai, H., et al.: MODMA dataset: a multi-modal open dataset for mental-disorder analysis. Sci. Data **9**(1), 178 (2022)
10. Abdul, Z.K., Al-Talabani, A.K.: Mel frequency cepstral coefficient and its applications: a review. IEEE Access **10**, 122136–122158 (2022)
11. Radford, J.W., Kim, T., Xu, G., Brockman, C., McLeavey, Sutskever, I.: Robust speech recognition via large-scale weak supervision (2022)
12. Devlin, J., Chang, M.-W., Lee, K., Toutanova, K.: BERT: pre-training of deep bidirectional transformers for language understanding. In: Proceedings of the 2019 conference of the North American chapter of the association for computational linguistics: human language technologies, vol. 1, pp. 4171–4186 (2019)
13. Cui, Y., Che, W., Liu, T., Qin, B., Wang, S., Hu, G.: Revisiting pre-trained models for Chinese natural language processing. In: Proceedings of the 2020 Conference on Empirical Methods in Natural Language Processing: Findings, Association for Computational Linguistics, pp. 657–668 (2020)
14. Cui, Y., Che, W., Liu, T., Qin, B., Yang, Z.: Pre-Training with whole word masking for Chinese BERT. IEEE/ACM Trans. Audio Speech Lang. Process. **29**, 3504–3514 (2021)
15. Vaswani, N., et al.: Attention Is All You Need. Curran Associates Inc., Red Hook (2017)

Evaluating Pre-Trained Automatic Speech Recognition Language Models for Cebuano-Binisaya Nutritional Educational Videos

Angelo Alfonso M. Tesoro III(✉) and Kristine Mae M. Adlaon

University of Immaculate Concepcion, Fr. Selga Campus, 8000 Davao City, Philippines
aamtesoro@gmail.com, kadlaon@uic.edu.ph

Abstract. The study evaluates the transcription performance of pre-trained automatic speech recognition (ASR) models on twenty Cebuano–Binisaya educational videos from the Ten Kumainments series. Cebuano–Binisaya, one of the major languages in the Philippines, remains underrepresented in ASR research. By using human-verified reference transcripts, this research compares pre-trained ASR model variants under zero-shot conditions. The evaluation focuses on both lexical accuracy, measured through Word Error Rate (WER) and Character Error Rate (CER), and semantic alignment, using cosine similarity between sentence embeddings. Results indicate that the gmml-small model offers the most practical balance between accuracy and efficiency for low-resource environments, achieving a cosine similarity of about 0.90 while maintaining fast processing and minimal hardware requirements. The gmml-medium and gmml-large-v3 models obtained the highest semantic alignment (above 0.92), with the gmml-medium model showing the best trade-off between accuracy and computational cost, making it suitable for community-based transcription setups with adequate hardware. Furthermore, the study confirms that Whisper can produce reliable Cebuano–Binisaya subtitles in WebVTT (.vtt) format, supporting video-based learning and promoting inclusive education in alignment with Sustainable Development Goal (SDG) 4 on quality education. Upon concluding the evaluation, a baseline for ASR performance in low-resource Philippine languages is provided, offering valuable insights for future development, fine-tuning, and deployment of localized transcription systems that also demonstrates how automated captioning can enhance learning accessibility and linguistic inclusion in regional educational materials.

Keywords: Cebuano–Binisaya · Automatic Speech Recognition · Educational Videos · WebVTT · Low-Resource Languages · Inclusive Education · Subtitles

1 Introduction

Automatic Speech Recognition (ASR) has become an essential technology in modern communication and education, enabling spoken language to be accurately converted into text. Earlier ASR systems relied on traditional models such as Hidden Markov

H. Unger and P. Meesad (Eds.): NLPIR 2025, LNNS 1904, pp. 524–538, 2026.
https://doi.org/10.1007/978-3-032-20897-2_38

Models with Gaussian Mixture Models (HMM-GMM), followed by early deep learning techniques using DNNs and CNNs. Recent advances have introduced more strong transformer-based and convolution-augmented architectures, including wav2vec 2.0 [7] and Conformer [19], which achieved higher accuracy across various languages. These models apply self-supervised and transfer learning methods that process large volumes of unlabeled audio before fine-tuning with smaller labeled datasets [30, 36]. However, despite the improvement, progress still centers on high-resource languages, while low-resource ones lack annotated corpora, benchmarks, and institutional support [8, 16, 32, 41].

In Southeast Asia, particularly the Philippines, ASR adoption remains limited and fragmented [16]. Although some local efforts have explored bilingual and broadcast recognition, but standardized Cebuano–Binisaya corpora are scarce [18]. The absence of national datasets and evaluation standards restricts consistent benchmarking. While global initiatives like OpenASR21 [31] and SLUE [35] promote low-resource languages, Philippine languages remain largely excluded, highlighting a need for localized ASR studies that support linguistic inclusivity in educational and multimedia materials. Emerging efforts, including healthcare-oriented children's speech corpora [29] and web-based tools like Scriptoman [37], show growing interest but lack reproducible benchmarks for adult educational content. Most Philippine ASR studies still focus on Tagalog or broadcast speech, with no publicly documented zero-shot evaluation of Cebuano–Binisaya in authentic educational media.

Cebuano–Binisaya, spoken by over 20 million Filipinos, remains underrepresented in ASR research. Available datasets like Mozilla Common Voice [5] and VoxForge [40] primarily cover Tagalog, limiting resources for other regional languages. Consequently, Philippine ASR systems report high Word Error Rates (WER 35–55%) compared to English (10–15%) [30, 32]. Challenges include accent variations, dialect diversity, and frequent code-switching, requiring inclusive data collection, adaptive fine-tuning, and focused preprocessing.

Cebuano–Binisaya's complex word formation, affixation, and flexible sentence structure [15] increase WER and Character Error Rate (CER), similar to Tagalog ASR findings [24]. Evaluating models using semantic similarity approaches, such as cosine similarity [43], provides a more meaningful measure of transcription accuracy by assessing whether the intended meaning is preserved despite word-level differences.

Among open-source ASR systems, examples like Whisper have demonstrated strong performance in multilingual and low-resource settings. It consistently produces lower WER and CER than other open-source models such as Kaldi, Vosk, and Mozilla DeepSpeech [22, 34]. Whisper's transformer-based encoder–decoder design enables it to generalize effectively even for underrepresented languages like Cebuano–Binisaya [20, 34]. The direction is through the ggml-small and ggml-medium quantized variants, which maintain over 90% of the performance of the large model while offering faster processing and lower hardware demands [4]. These features make ASR, such as Whisper, suitable for community-based schools and offline learning environments with limited computing capacity.

The dataset used in this study is based on the Ten Kumainments educational video series, which promotes the Nutritional Guidelines for Filipinos [1, 28]. These short, engaging videos are narrated in Cebuano–Binisaya with authentic speech patterns, ideal for evaluating ASR in an educational setting. Accurate transcriptions of these materials can enhance comprehension, accessibility, and inclusivity, aligning with Sustainable Development Goal 4 on quality education [38, 39].

ASR technologies not only contribute to technical innovation but also play a role in educational accessibility. Inaccurate captions can distort meaning and reduce comprehension, highlighting the need to evaluate ASR for low-resource languages both linguistically and socially [5, 22, 25]. This study measures lexical accuracy using WER and CER, and semantic alignment using similarity-based evaluation, ensuring a more complete analysis of transcription performance.

Despite emerging Cebuano studies, particularly in children's healthcare [29], there is still no published benchmark evaluating Cebuano–Binisaya ASR on authentic adult educational video content. Existing web-based tools [37] offer practical utility but lack transparent metrics or open datasets. This research evaluates ASR performance on authentic educational videos, comparing multiple ggml model sizes under zero-shot conditions. Lexical accuracy (WER and CER) and semantic similarity are measured, generating subtitles in WebVTT (.vtt) and text (.txt) for media and educational use [42]. Establishing this baseline for Cebuano–Binisaya ASR contributes to future efforts in model adaptation and fine-tuning, providing data for developers and researchers to improve low-resource ASR systems. Moreover, the findings support educational accessibility, the use and preservation of Cebuano–Binisaya in digital and academic spaces, and guide policymakers and technology stakeholders in creating data-driven programs that promote digital inclusivity and equitable learning opportunities, aligning with Sustainable Development Goal 4 on Quality Education [3, 9, 21, 23, 38, 39].

1.1 Objectives of the Research

The main objective of this research is to evaluate and compare the transcription performance of pre-trained Automatic Speech Recognition (ASR) models on Cebuano–Binisaya educational videos, examining how accurately and meaningfully they convert spoken language into text. Specifically, the study measures transcription accuracy through Word Error Rate (WER) and Character Error Rate (CER), and assesses semantic fidelity using cosine similarity between sentence embeddings to determine how well the models preserve meaning beyond word-level matching. By identifying the most suitable pre-trained ASR model variant for producing accurate and semantically consistent Cebuano–Binisaya subtitles in WebVTT (.vtt) format, the study aims to provide empirical evidence that supports improved accessibility and inclusivity in educational media, particularly in low-resource learning environments, as well as, structured baseline that has not been documented in Philippine ASR literature.

2 Dataset

2.1 Source Materials

The study's source materials focus on the Ten Kumainments video series, which aims to promote the Nutritional Guidelines for Filipinos through short, culturally familiar educational videos produced in Cebuano–Binisaya. Each of the ten Kumainments delivers a concise and practical message promoting healthy eating habits, food safety, and proper nutrition among Filipinos. These materials were selected to represent real-world challenges in delivering local-language nutritional education, making them ideal for evaluating ASR performance in authentic learning contexts. An example frame from the Ten Kumainments educational videos, illustrating the visual context and presentation style of the source materials used in this study, is shown in Fig. 1.

The total video duration is 66 min and 40 s, with individual clips averaging of two to four minutes in length, following commendations by instructional designers and educational technologists, who served as subject matter experts to support optimal learning [11, 17]. Research indicates that shorter clips enhance attention and comprehension, particularly in online learning environments such as Massive Open Online Courses (MOOCs) [11, 14], where students watch a higher proportion of videos under ten minutes [17]. Mayer's Segmenting Principle and complementary studies further support presenting content in brief, meaningful segments with narration, on-screen text, and visuals to optimize cognitive processing, retention, and learner engagement [11, 13, 26].

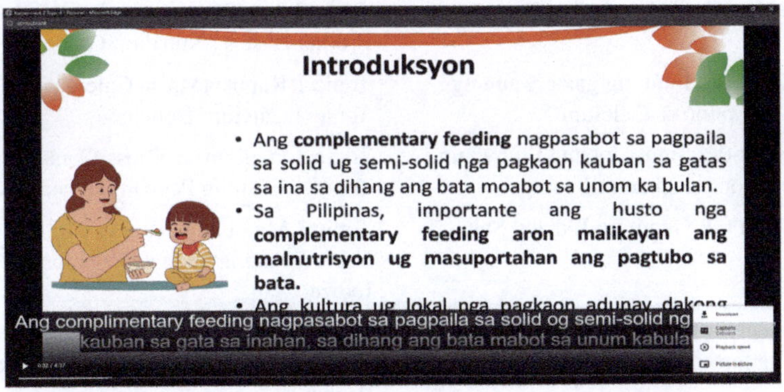

Fig. 1. Sample frame from one of the Ten Kumainments educational videos.

The dataset includes both audio and video components. All materials were recorded by a male subject matter expert with a mid-range vocal tone and natural Cebuano–Binisaya accent, with authentic speech patterns, including natural pacing, varying intonation, and occasional code-switching between English-Tagalog ("Taglish"), and Cebuano-Binisaya. Some clips also contain background sounds typical of community recording environments, simulating a practical test case for ASR models under real-world setup.

2.2 Ground Truth Transcripts

Each Kumainment video was transcribed to produce ground truth textual references for ASR system evaluation. For each Kumainment category, two video samples were selected, resulting in a total of 20 video datasets, as shown in Table 1. The resource person is a subject matter expert fluent in Cebuano–Binisaya who contributed to ensuring the fidelity and naturalness of the speech materials in representing regional linguistic nuances.

Table 1. Kumainment Video Topic Materials

Kumainment Category	Topic Materials
Kumainment 01: Kaon ug nagkalain-laing pagkaon	Topic 4 Mga Prinsipyo Pata sa Pagplano sa Dyeta Topic 5 Ang Pagbasa sa Nutrition Facts o Labels
Kumainment 02: Sa sulod sa unom ka bulan ni baby breastfeeding lamang	Topic 1 Ang First 1000 Days Topic 2 Ang Eksklusibong Pagpasuso
Kumainment 03: Kaun ug gulay ug prutas kada adlaw	Topic 1 Ang mga Prutas ug Gulay sa Food Guides Topic 2 Mga Benepisyo sa Pagkaon sa Gulay ug Prutas
Kumainment 04: Kaun ug isda, karne, ug lahi lahing pagkaon na naay protina	Topic 1 Ang Bili sa Protina Topic 2 Ang Pagplano sa Dyeta nga Sakto sa Protina Base sa Nutrition Guides
Kumainment 05: Inum ug gatas, kaun ug pagkaon nga puno sa Calcium	Topic 1 Kapuslanan sa Calcium Topic 2 Calcium Deficiency
Kumainment 06: Siguradohang limpyo ug luwas ang atong pagkaon ug tubig	Topic 3 Paglikay sa Cross-Contamination Topic 4 Hustong Pagpanghugas sa Kamot
Kumainment 07: Gamit ug Iodized Salt	Topic 2 Mga Pangunang Gamit sa Iodine Topic 5 Pagmatngon sa Pagkaon og Daghang Iodized Salt
Kumainment 08: Likayi ang sobrang parat, tambok, ug tam-is nga pagkaon	Topic 1 Likayi ang Sobrang Parat Topic 2 Likayi ang Sobrang Tambok nga Pagkaon
Kumainment 09:Pagbaton sa Tamang Timbang	Topic 2 Paglikay sa Lifestyle Diseases Topic 4 Ang BMI ug mga Indikasyon Niini
Kumainment 10:Batoni kanunay ang lagsik nga lawas	Topic 1 Unsa ang Healthy Lifestyle Topic 2 Mga Benepisyo sa Ehersisyo

3 Methods

3.1 Model Evaluation

All transcription experiments were conducted using the same set of preprocessed video and audio files. The evaluation used a command-line interface to test several pre-trained ggml model variants under identical conditions. These models differ in terms of file size, computational requirements, and expected accuracy, as summarized in Table 2.

Table 2. Comparison of pre-trained ASR GMML Model Variants for a low-resource setting

Models	Size	Remarks
ggml-tiny.bin	77.7 MB	Fast, and low-resource, but very low accuracy
ggml-base.bin	148 MB	Faster speed, low resource, but has limited accuracy
ggml-small.bin	488 MB	Balanced speed, low resource, with improved accuracy
ggml-medium.bin	1.53 GB	Slow, high resource, but has high accuracy
ggml-large-v3.bin	3.1 GB	Very Slow, high resource, but is most accurate

3.2 Pre-Processing

Audio Standardization. The audio standardization process, including the FFmpeg command-line parameters used to convert the recordings into a uniform 16 kHz mono-channel format, is illustrated in Fig. 2.

```
D:\kumainmentsmethods>ffmpeg -i Kumainment1Topic4.mp3 -ar 16000 -ac 1 Kumainment1Topic4.wav
```

Fig. 2. Command-line parameter used for conversion.

This command instructs the model to process each audio file with a consistent sampling rate and channel configuration. While no additional denoising or speech enhancement filters were applied to preserve the recordings' natural background noise, showing the actual conditions in Cebuano–Binisaya speech environments.

To further maintain uniformity, silence removal was limited to two seconds or less to prevent inflated error rates and preserve the natural pacing of speech. The audio files were also normalized to achieve consistent loudness levels, allowing the models to process each recording under similar acoustic conditions.

Batch Processing. Each standardized audio file was processed using a batch script (run_clean_transcribe.bat) that executed identical transcription parameters for all ggml models, as shown in Fig. 3.

```
whisper <audio> --language ceb --task transcribe --output_format txt --temperature 0
```

Fig. 3. Example command-line parameter used in Whisper-CLI for Cebuano transcription.

This command instructs the model to transcribe Cebuano speech (without translation) and generate both.txt and.vtt files using deterministic decoding (temperature set to 0). The WebVTT (.vtt) format was chosen over the traditional SubRip (.srt) format because it supports UTF-8 encoding and millisecond-level timestamps, as well as compatibility with modern web playback systems. This ensures accurate subtitle alignment for accented and mixed-language Cebuano–Binisaya content, and facilitates seamless integration with web-based evaluation platforms and PWAs.

3.3 Text Normalization

Transcribed and reference texts were normalized to reduce inconsistencies in punctuation, case, and spacing. A PowerShell-based script automatically removed timestamps, numeric markers, and blank lines, producing cleaned transcripts (_clean.txt) for evaluating Word Error Rate (WER), Character Error Rate (CER), and Cosine Similarity.

3.4 Evaluation Metrics

To quantitatively assess the transcription performance of the evaluated pre-trained ASR models, three complementary evaluation metrics were employed: Word Error Rate (WER), Character Error Rate (CER), and Cosine Similarity.

Word Error Rate (WER) is one of the most widely used metrics in evaluating automatic speech recognition (ASR) through substitutions, deletions, and insertions at the word level, and computation using the formula:

$$WER = \frac{S + D + I}{N} \tag{1}$$

where S represents substitutions, D deletions, I insertions, and N the total number of words in the reference. A lower WER indicates higher transcription accuracy. However, WER can be sensitive to word segmentation but may not capture semantic similarity when synonymous expressions or equivalent phrases are used.

Character Error Rate (CER), while following the same fundamental formula for WER, provides a finer-grained measure by evaluating errors at the character level rather than the word level, useful for languages such as Cebuano–Binisaya.

Cosine Similarity, on the other hand, extends the evaluation beyond basic string comparison by measuring the semantic closeness between the transcribed text and the reference, and assesses semantic alignment between the model output and the reference text using vector-based text representations, with values closer to 1 indicating stronger similarity.

3.5 Phonetic and Linguistic Observations

Certain phonetic characteristics of the Cebuano–Binisaya language posed notable challenges for the evaluated models, primarily due to subtle sound variations and frequent code-switching with Tagalog and English [6, 10, 33]. Common sources of transcription errors included the confusion between the /p/ and /f/ sounds (as in "*pamilya*" versus "*familya*" meaning "family")the overlapping pronunciation of /o/ and /u/ (such as "*buto*" "bone" versus "*boto*" "vote"), and variations in unstressed vowels that can alter meaning (for instance, "*bata*" "kid" versus "*bataa*" "so young") [2, 12]. Moreover, spontaneous alternation between Cebuano, Tagalog, and English within a single sentence, such as in educational narrations like "*Mabuhay og maayong adlaw. Atong iskutan ang Kumainment One, mao ang kaon og nagkalainlaing pagkaon…*" ("Greetings and good day. We will be discussing the Kumainment 1, which is Eat a variety of foods", further increased transcription complexity [6, 17]. These linguistic dynamics illustrate the diverse and fluid structure of Cebuano–Binisaya speech, where natural mixing of languages and variable vowel or consonant realizations demand more strong language modeling [2, 10, 33]. Such findings emphasize the necessity for phonetically rich and linguistically inclusive datasets that capture real-world regional speech diversity, enabling automatic speech recognition systems to achieve higher accuracy and adaptability across Philippine languages [2, 6, 12].

4 Experiments

4.1 Environment

Experiments were conducted on a Windows 11 (64-bit) system using CPU-only mode to simulate low-resource conditions. Hardware specifications included an Intel® Core™ i9-12950HX processor, 64 GB RAM, and RTX A4500 GPU (disabled for CPU tests). Standardized protocols ensured consistency across all ggml models.

All twenty (20) Kumainment educational video samples were processed individually. Each resulting transcript and subtitle file was automatically cleaned to remove timestamps, bracketed annotations, and redundant characters before computing the metrics. The cleaned outputs were evaluated against human-verified ground truth transcripts using three key metrics: Word Error Rate (WER), Character Error Rate (CER), and Cosine Similarity. The resulting data were compiled into structured CSV reports for model comparison, ensuring transparent and reproducible performance evaluation.

4.2 Quantitative Results

The pre-trained ASR models' performance is summarized in Table 3. Smaller variants (tiny, base) provided fast runtimes but higher WER/CER values, while medium and large variants achieved higher accuracy and semantic similarity. The ggml-medium.bin model attained a Cosine Similarity of 0.9218 with reasonable runtime (125 s/file), demonstrating an optimal balance for offline or CPU-based transcription.

Table 3. Comparative Results of pre-trained ASR models on Cebuano–Binisaya Dataset

Models	WER (%)	CER (%)	CS (%)	Average Runtime per File
ggml-tiny.bin	0.7445	0.2926	0.8517	5 s
ggml-base.bin	0.6304	0.2105	0.8798	10 s

(continued)

Table 3. (*continued*)

Models	WER (%)	CER (%)	CS (%)	Average Runtime per File
ggml-small.bin	0.4939	0.1657	0.8982	69 s
ggml-medium.bin	0.4298	0.1409	0.9218	125 s
ggml-large-v3.bin	0.4473	0.2436	0.9427	229 s

The smallest models, ggml-tiny.bin and ggml-base.bin, achieved the fastest runtimes (5–10 s per file) but exhibited relatively high WER and CER values. The ggml-small.bin model significantly improved performance, reducing WER by approximately 14% compared to ggml-base.bin, while maintaining moderate runtime efficiency, indicating a viable balance between speed and accuracy for low-resource applications. While the ggml-medium.bin and ggml-large-v3.bin models demonstrated the lowest error rates and highest semantic alignment, achieving Cosine Similarity scores of 0.9218 and 0.9427, respectively. However, despite the ggml-large-v3.bin model's superior semantic fidelity, it exhibited slightly higher WER (0.4473) compared to the ggml-medium.bin and required nearly double the runtime (229 s per file). This reinforces ggml-medium.bin as the optimal choice for balancing transcription quality and computational efficiency in offline educational environments.

Per-topic analysis also revealed that sections containing code-switched Taglish expressions and technical nutrition terms (e.g., "cross-contamination," "iodized salt," "diet," and "nutrition") consistently produced higher error rates across all models. These challenges highlight the challenge of multilingual decoding in Cebuano–Binisaya, where English insertions and morphological variations complicate token alignment and semantic interpretation.

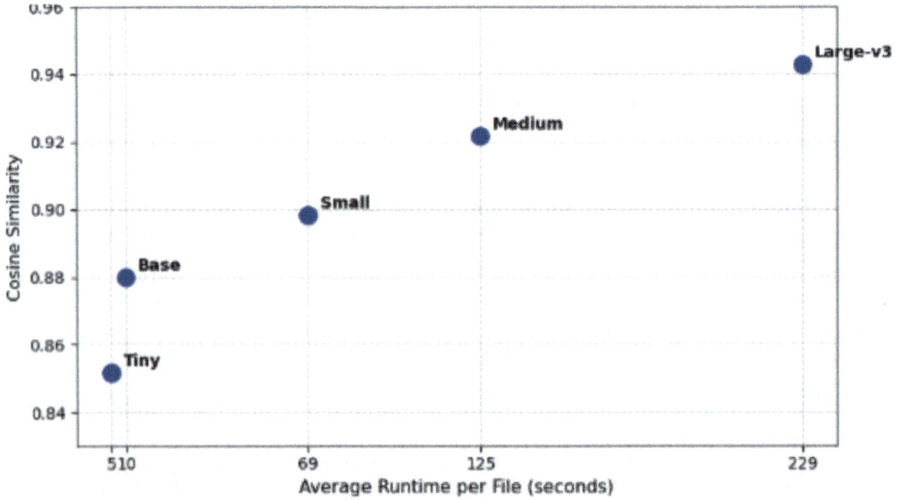

Fig. 4. Trade-off Between Transcription Accuracy and Runtime for pre-trained ASR Models (Cebuano-Binisaya Dataset).

Figure 4 illustrates the trade-off between model accuracy and runtime across the ggml variants. Smaller models (ggml-tiny.bin, ggml-base.bin) delivered faster processing but at the cost of higher error rates and reduced semantic fidelity.

4.3 Qualitative Analysis and Discussion

The qualitative analysis of the transcribed outputs revealed distinct linguistic, phonetic, and contextual patterns that affected model behavior. These patterns highlight both the strengths and limitations of the evaluated ggml model variants in processing Cebuano–Binisaya speech, especially in mixed-language educational contexts.

Lexical and Morphological Variability. The smaller models, such as ggml-tiny.bin and ggml-base.bin, frequently omitted or substituted functional morphemes common in Cebuano-Binisaya, particularly "*nga*", "*ug*", "*ang*", and "*sa*". These omissions often led to grammatical inconsistencies and reduced textual coherence. For instance, the phrase "*Kaun ug nagkalain-laing pagkaon*" ("Eat a variety of foods") was occasionally transcribed as "*Kaon nag lain lain pagkaon*" ("Eat different foods") dropping the linker ug and disrupting sentence fluidity. In contrast, larger models such as ggml-medium.bin and ggml-large-v3.bin, which leverage more extensive multilingual embeddings, produced more fluent and syntactically accurate transcriptions by capturing these fine linguistic nuances with greater consistency. This finding suggests that model capacity directly influences morphological precision and grammatical structure in Cebuano–Binisaya transcription.

Code-Switching and Borrowed Words. Cebuano–Binisaya speakers frequently alternate between English and Tagalog, particularly in educational or technical discussions that include terms like "Vitamin A deficiency," "Food Guide Pyramid," or "balanced diet." Smaller models often misrepresented or phonetically approximated such terms, while ggml-medium.bin and ggml-large-v3.bin preserved their orthographic accuracy and capitalization.

Semantic Preservation. High Cosine Similarity (CS) scores (>0.85) observed in ggml-medium.bin and ggml-large-v3.bin indicate that the semantic intent of spoken content was largely retained, even when lexical or structural variations occurred. For example, "*Likayi ang sobrang parat*" ("Refrain from too much salt") was sometimes rendered as "*Iwas sa daghang asin*" ("Avoid too much salty"), which conveys an equivalent meaning despite different wording.

Acoustic and Environmental Factors. Recordings with mild background noise or reverberation with no adequate equipment or within an open space, were found to elevate error rates, particularly for ggml-tiny.bin and ggml-base.bin. These smaller models lacked robustness against non-studio acoustic variations. However, ggml-small.bin, ggml-medium.bin, and ggml-large-v3.bin handled such imperfections more effectively, reflecting stronger resilience in capturing clear phonetic boundaries under suboptimal recording conditions.

Educational and Accessibility Implications. The observed performance hierarchy among the models offers practical insights into real-world deployment. The ggml-medium.bin model demonstrates the best balance between accuracy and computational feasibility, consistently delivering over 85% semantic similarity while remaining operable on standard consumer-grade machines.

5 Results and Discussion

The evaluation of ggml model variants on the Cebuano–Binisaya Kumainment dataset revealed distinct trends in transcription accuracy, semantic fidelity, and runtime performance. Smaller models, such as ggml-tiny.bin and ggml-base.bin, processed files quickly (5–10 s per file) but produced higher error rates, often omitting essential function words which highly disrupted sentence flow. These results show that smaller models prioritize speed but struggle with grammatical and morphological accuracy.

The ggml-small.bin model provided notable improvement, lowering Word Error Rate (WER) to 0.4939 and achieving a Cosine Similarity of 0.8982. It maintained reasonable runtime efficiency (around 69 s per file), making it suitable for resource-limited environments where both speed and acceptable accuracy are important. Larger models, particularly ggml-medium.bin and ggml-large-v3.bin, delivered the best transcription quality, achieving high semantic preservation (Cosine Similarity scores of 0.9218 and 0.9427, respectively). However, the ggml-medium.bin model offered a better balance between accuracy and runtime, making it more practical for offline or CPU-only setups, which are common in educational and rural contexts.

Qualitative analysis further revealed that smaller models often misinterpret code-switched terms and technical vocabulary. Meanwhile, the larger models showed greater multilingual flexibility, correctly transcribing such expressions and preserving capitalization. This improved cross-lingual recognition reflects the advantages of multilingual embeddings in handling mixed-language educational materials typical of Philippine classrooms.

Environmental conditions also affected model performance. Recordings with mild background noise or echo increased errors for smaller models, while ggml-small.bin, ggml-medium.bin, and ggml-large-v3.bin remained more resilient. These results underscore the importance of selecting models with greater noise tolerance for real-world educational recordings, where controlled studio conditions are often not feasible. The findings highlight that ggml-medium.bin offers the most effective balance of accuracy, efficiency, and offline usability, making it highly suitable for educational transcription tasks. The ggml-small.bin model remains a practical choice for quick, low-resource applications highly suited for portability and lightweight, while ggml-large-v3.bin is best suited for centralized or cloud-based processing. These outcomes demonstrate the feasibility of deploying pre-trained ASR model-based transcription systems in Cebuano–Binisaya contexts, contributing to accessible, inclusive, and technology-supported learning environments in the Philippines.

Despite these promising outcomes, several limitations are notable. The research utilized only 20 Kumainment video samples, which may not fully capture the extent of Cebuano–Binisaya dialects or speaking styles. No external denoising or acoustic enhancement was applied, meaning that results for smaller models may be sensitive to background noise or reverberation. Additionally, technical terminology and code-switched segments remain challenging for models with limited capacity, suggesting potential avenues for fine-tuning or vocabulary adaptation in future research. Moreover, the limited dataset constrains the generalization of the findings. The current sample may not adequately reflect actual voice variations, speaker diversity, or different educational contexts within Cebuano–Binisaya communities.

6 Conclusion and Recommendations

The study is exploratory in nature and focused on evaluating and comparing the performance of different automatic speech recognition (ASR) model variants using Cebuano–Binisaya educational materials. The evaluation did not aim to address existing ASR limitations but rather to measure and understand how different model sizes perform how accurately and meaningfully they convert spoken language into text. Performance was assessed using Word Error Rate (WER), Character Error Rate (CER), and cosine similarity between sentence embeddings to measure both lexical and semantic fidelity. The primary goal was to identify the most suitable model variant for producing accurate and semantically consistent Cebuano–Binisaya subtitles in WebVTT (.vtt) format, providing empirical evidence to support accessibility and inclusivity in educational media.

To the best of our knowledge, this study establishes the first structured baseline for Cebuano–Binisaya ASR performance using authentic educational video content. While these findings also complement prior Cebuano ASR studies, they differ in domain focus and performance objectives. They underscore the importance of reproducible benchmarks, as established in this study.

The results demonstrate that model size significantly influences transcription outcomes: smaller models prioritize speed but with reduced accuracy, whereas medium and large models achieve better semantic and linguistic fidelity at higher computational costs. Medium-sized models, in particular, offer the best balance between accuracy and runtime efficiency, making them practical for offline deployment in educational and community-based environments with limited computing resources. Smaller models remain valuable for quick, portable transcription tasks, especially in remote or low-resource contexts.

Overall, this study confirms the feasibility of deploying pre-trained ASR models for Cebuano–Binisaya educational content. By selecting the model variant most suitable for the target context, schools, teachers, and local developers can enhance accessibility, promote inclusive learning, and support the digital presence of regional languages in educational materials. The baseline results presented here also provide a foundation for future work in model fine-tuning, domain adaptation, and performance optimization in low-resource settings for the Philippine language.

References

1. A “10 Kumainments” series. https://nnc.gov.ph/visayas-region/a-10-kumainments-series/
2. Abucay, S.M.R.: Analysis on the morphophonemic changes of the Cebuano-Binisaya of Davao del Norte. Davao Res. J. **16**(2), 1–14 (2025). https://doi.org/10.59120/drj.v16i2.341

3. Aleño, N.P., Mustapha, H.S., Omar, S.H., Calil, S.S.: Inclusive education in public elementary schools: teacher insights on accessibility, quality, and stakeholder participation. Int. J. Res. Innov. Soc. Sci. (IJRISS), **8**(09), 1349–1357 (2024). https://rsisinternational.org/journals/ijriss/articles/inclusive-education-in-public-elementary-schools-teacher-insights-on-accessibility-quality-and-stakeholder-participation/
4. Andreyev, A.: Quantization for OpenAI's Whisper Models: A Comparative Analysis (2025). https://doi.org/10.48550/arXiv.2503.09905
5. Ardila, R., et al.: Common voice: a massively-multilingual speech corpus. In: Proceedings of the 12th Conference on Language Resources and Evaluation (LREC 2020), Marseille, France, pp. 4211–4215 (2020). https://aclanthology.org/2020.lrec-1.520/
6. Aranan, K.M., et al.: Exploring the role of code-switching in multilingual class-room dynamics: a comparative study of attitudes and practices among university students in the Philippines. JPAIR Multidisciplinary Res. **59**(1), 1–23 (2024). https://doi.org/10.7719/jpair.v59i1.919
7. Baevski, A., Zhou, Y., Mohamed, A., Auli, M.: wav2vec 2.0: a framework for self-supervised learning of speech representations. In: Advances in Neural Information Processing Systems 33 (NeurIPS 2020) (2020). https://proceedings.neurips.cc/paper/2020/file/92d1e1eb1cd6f9fba3227870bb6d7f07-Paper.pdf
8. Bashori, M.: The benefits of automatic speech recognition-based language learning systems. Learn. Media Technol. **47**(3), 234–247 (2022). https://www.lotpublications.nl/Documents/627_fulltext.pdf
9. Bastida Jr., E.L., Saysi, J.G., Batuctoc, L.V.M.: Pedagogical struggles and gaps in language literacy enhancement: the case of Indigenous Peoples Education (IPEd) teachers in Agusan del Sur, Philippines. Int. J. Curricul. Instruct. **15**(1), 142–165 (2023). https://files.eric.ed.gov/fulltext/EJ1383105.pdf
10. Berowa, A.M.: Philippine English Phonology: Its Features Across Speakers of the Major Philippine Languages. Doctoral Dissertation, De La Salle University (2019). https://animorepository.dlsu.edu.ph/etd_doctoral/1526
11. Brame, C.J.: Effective educational videos: principles and guidelines for maximizing student learning from video content. CBE—Life Sci. Educ. **15**(4), es6 (2016). https://doi.org/10.1187/cbe.16-03-0125
12. Cebuano Phonology. (n.d.). Binisaya.com. https://www.binisaya.com/content/dialectology-cebuano-phonology
13. Chen, C.-M., Wu, C.-H.: Effects of different video lecture types on sustained attention, emotion, cognitive load, and learning performance. Comput. Educ. **80**, 108–121 (2015). https://www.sciencedirect.com/science/article/abs/pii/S0360131514001961
14. Chorianopoulos, K.: A taxonomy of asynchronous instructional video styles. Int. Rev. Res. Open Distrib. Learn. (IRRODL) **19**(1), 294–311 (2018). https://www.irrodl.org/index.php/irrodl/article/view/2920
15. Empinado, C.D.O., Jumalon, R.C.A., Peña, C.F.: Cebuano-English Code-switching speech detection using support vector machine. In: Vasant, P., et al. Intelligent Computing and Optimization. ICO 2023. Lecture Notes in Networks and Systems, vol. 854. Springer, Cham (2023). https://doi.org/10.1007/978-3-031-50151-7_13
16. FlipStar: A Survey of the Current State of Filipino ASR. https://flipvox.ph/post/current-state-filipino-asr/
17. Guo, P.J., Kim, J., Rubin, R.: How video production affects student engagement: An empirical study of MOOC videos. In: Proceedings of the First ACM Conference on Learning@Scale (L@S 2014), pp. 41–50 (2014). https://doi.org/10.1145/2556325.2566239
18. Guevara, R.C.L., Cajote, R.D., Bayona, M.G.A.R., Lucas, C.R.G.: Philippine languages database: a multilingual speech corpora for developing systems for low-resource languages. In: Proceedings of the 3rd Annual Meeting of the Special Interest Group on Under-resourced

Languages @ LREC-COLING 2024, ELRA and ICCL, Torino, Italia, pp. 264–271 (2024). https://aclanthology.org/2024.sigul-1.32.pdf
19. Gulati, A., et al.: Conformer: convolution-augmented transformer for speech recognition. In: Inter-speech 2020, pp. 5036–5040 (2020). https://doi.org/10.21437/Interspeech.2020-3015
20. Jain, P., Bhowmick, A.: Comparative performance analysis of end-to-end ASR models on Indo-Aryan and Dravidian languages within India's linguistic landscape. J AUDIO SPEECH MUSIC PROC. **10** (2025). https://doi.org/10.1186/s13636-025-00395-5
21. Mainzinger, J., Levow, G.-A.: Fine-tuning ASR models for very low-resource languages: a study on Mvskoke. In: Proceedings of the 62nd Annual Meeting of the Association for Computational Linguistics, vol. 4, Student Research Workshop, pp. 76–82. Association for Computational Linguistics, Bangkok, Thailand (2024). https://aclanthology.org/2024.acl-srw.16/
22. Kuhn, K., Kersken, V., Reuter, B., Egger, N., Zimmermann, G.: Measuring the accuracy of automatic speech recognition solutions. ACM Trans. Access. Comput. (TACCESS) **16**(4), Article 25 (2024). https://doi.org/10.1145/3636513
23. Malakul, S., et al.: The effects of using an auto-subtitle system in educational videos: learning comprehension, cognitive load and satisfaction among Thai secondary students. Smart Learn. Environ. **10**(1), 1–18 (2023). https://doi.org/10.1186/s40561-023-00224-2
24. Mallari, A.C., Samson, E.G.: Evaluating ASR performance on Tagalog and Ce-buano using open-source models. Procedia Comput. Sci. **176**, 3232–3241 (2020). Elsevier. https://doi.org/10.1016/j.procs.2020.09.190
25. Matthew, G., et al.: The effect of subtitle format on cognitive load and comprehension. Journal (2025). https://doi.org/10.1080/1475939X.2024.2433259
26. Mayer, R.E.: Multimedia Learning (2nd ed.). Cambridge University Press (2009). https://www.cambridge.org/core/books/multimedia-learning/7A62F072A71289E1E262980CB026A3F9
27. Ul Sehar, N., Khalid, A., Adeeba, F., Hussain, S.: Benchmarking whisper for low-resource speech recognition: an N-Shot Evaluation on Pashto, Punjabi, and Urdu. In: Proceedings of the First Workshop on Challenges in Processing South Asian Languages (CHiPSAL 2025), pp. 202–207. International Committee on Computational Linguistics, Abu Dhabi, UAE (2025). https://aclanthology.org/2025.chipsal-1.20.pdf
28. Nutritional Guidelines for Filipinos. https://fnri.dost.gov.ph/images/images/standardtools/NGF-2012.pdf
29. Pascual, R., Mendoza, G., Zapanta, A.M., Lino, K.A.: Developing Filipino and Cebuano ASR systems with children's speech corpora on healthcare monitoring. In: 2025 IEEE International Conference on Industry 4.0, Artificial Intelligence, and Communications Technology (IAICT), Bali, Indonesia, pp. 367–373 (2025). https://doi.org/10.1109/IAICT65714.2025.11100316
30. Park, K.H., Kim, J., Kang, U.: Accurate semi-supervised automatic speech recognition for ordinary and characterized speeches via multi-hypotheses-based curriculum learning. PLoS ONE **20**(10), e0333915 (2025). https://doi.org/10.1371/journal.pone.0333915
31. Peterson, K., Tong, A., Yu, J.: OpenASR21: the second open challenge for automatic speech recognition of low-resource languages. In: Proceedings INTERSPEECH 2022, Incheon, KR (2022). https://doi.org/10.21437/Interspeech.2022-10972
32. Piñeiro-Martín, A., García-Mateo, C., Docío-Fernández, L., López-Pérez, M.C., Rehm, G.: Weighted Cross-entropy for low-resource languages in multilingual speech recognition (2024). https://doi.org/10.48550/arXiv.2409.16954
33. Porlas, J.S.: Developing grammar rules for code-switched Philippine language texts: a linguistic framework for understanding multilingual discourse patterns. Front. Linguist. Stud. **7**(10) (2025). https://doi.org/10.30564/fls.v7i10.11632

34. Radford, A., Kim, J.W., Xu, T., Brockman, G., McLeavey, C., Sutskever, I., et al.: Robust Speech Recognition via Large-Scale Weak (2022). https://doi.org/10.48550/arXiv.2212.04356
35. Shon, S., et al.: SLUE: New Benchmark Tasks for Spoken Language Understanding Evaluation (2021). https://doi.org/10.48550/arXiv.2111.10367
36. Singh, R., et al.: A Novel Self-training Approach for Low-resource Speech Recognition. Interspeech 2023 (2023). https://www.isca-archive.org/interspeech_2023/singh23b_interspeech.pdf
37. Transcribe Cebuano Audio to text. https://scriptoman.ai/transcription//transcribe-cebuano-audio
38. UNESCO. Education for Sustainable Development Goals: Learning Objectives. United Nations Educational, Scientific and Cultural Organization, Paris (2017). ISBN 978-92-3-100209-0. https://unesdoc.unesco.org/ark:/48223/pf0000247444
39. United Nations. Transforming Our World: The 2030 Agenda for Sustainable Development. United Nations General Assembly, Resolution A/RES/70/1 (2015). https://sdgs.un.org/goals/goal4
40. VoxForge: Open Source Speech Dataset for Acoustic Model Training (2023). http://www.voxforge.org
41. Wang, L., Tong, R., Leung, C.C., Sivadas, S., Ni, C., Ma, B.: Cloud-based automatic speech recognition systems for southeast Asian languages (2022). https://doi.org/10.48550/arXiv.2210.03580
42. World Wide Web Consortium (W3C): WebVTT: The Web Video Text Tracks Format. W3C Recommendation (2019). https://www.w3.org/TR/webvtt1
43. Zhang, T., Kishore, V., Wu, F., Weinberger, K., Artzi, Y.: BERTScore: evaluating text generation with BERT. In: International Conference on Learning Representations (ICLR 2020) (2020). https://doi.org/10.48550/arXiv.1904.09675

Speaker Identification in Japanese Novels Using Narration, Dialogue Content, and Speech Style or Tone

Kazuma Iwamoto[1] and Kazuaki Ando[2(✉)]

[1] Kagawa University, Graduate School of Science for Creative Emergence, Takamatsu, Japan
s24g351@kagawa-u.ac.jp

[2] Kagawa University, Faculty of Engineering and Design, Takamatsu, Japan
ando.kazuaki@kagawa-u.ac.jp

Abstract. In recent years, the emergence of large language models has increased the demand for language resources related to dialogue systems. Among them, open-domain dialogue systems have been attracting particular attention. However, there are not enough Japanese dialogue corpora available for developing and evaluating such systems. In addition, traditional methods for constructing a dialogue corpus are costly and labor-intensive. To address these issues, we propose an automatic construction method of a daily dialogue corpus using utterances extracted from Japanese novels. However, simply extracting utterances is insufficient to maintain consistency in speech style or tone, as novels contain a variety of characters. Therefore, it is necessary to identify the speaker and assign speaker information to each utterance. In this paper, we propose a method for identifying the speaker of each utterance using clues from multiple perspectives: the surrounding narration, the content of the utterance, and the speaker's speech style and tone. Based on the evaluation results, we confirm that the proposed method improves recall with almost no reduction in precision compared to the rule-based method, and improves precision compared to the tone-based method.

Keywords: Dialogue Corpus · Novel · Speaker Identification · Tone · Speech Style

1 Introduction

To implement an open-domain dialogue system, a suitable corpus consisting of daily dialogue is necessary. However, the number of Japanese dialogue corpora is limited compared to English corpora. In addition, traditional methods for constructing a dialogue corpus are costly and labor-intensive [1] because they require extensive manual effort, such as recording and transcribing dialogues between people [2], manually creating new dialogues [3], and collecting chat dialogues through crowd-sourcing [4]. To address this issue, we propose a method

H. Unger and P. Meesad (Eds.): NLPIR 2025, LNNS 1904, pp. 539–554, 2026.
https://doi.org/10.1007/978-3-032-20897-2_39

for automatically constructing a daily dialogue corpus using Japanese novels. In Japanese novels, the utterances of characters are enclosed in Japanese quotation marks (“ 「」 ”), and the exchanges of these utterances between characters can be considered dialogue. By collecting dialogue exchanged between characters, a large dialogue corpus can be constructed. Furthermore, it is relatively easy to expand the dialogue corpus both quantitatively and qualitatively because new novels are published daily. In our previous study [5], we confirmed two issues in constructing a daily dialogue corpus. The first issue is that when only utterances that appear consecutively in a novel are extracted as a dialogue, the number of utterances per dialogue is smaller than in existing corpora. When creating a dialogue model, it is important to learn the context from a sequence of utterances. However, due to this issue, the dialogue model cannot sufficiently learn the context. The second issue is that it is difficult to construct a corpus that maintains speech style consistency by simply extracting dialogue, due to the variety of characters in the novel. In spoken Japanese, speech style or tone expresses a character's personality, emotions, and human relationships. More than 100 different speech styles or tones have been identified [6], and such distinct styles can also be found in novels.

We focus on the second issue and propose a method to address it. One possible approach to maintaining consistency in speech style or tone is to explicitly specify the speaker for each utterance. Although identifying the speaker requires various contextual clues, these clues vary greatly across novels and utterances. Furthermore, to ensure the quality of the constructed dialogue corpus, misidentification of speakers should be minimized. The main clues for identifying the speaker in novels typically include the narration surrounding an utterance, the content of the utterance, the speaker's speech style or tone, and the relationships among characters. However, identifying the speaker based on a single clue alone is difficult [8]. In this paper, we propose a method for speaker identification in dialogue by using the narration surrounding the utterance, the content of the utterance, and the speaker's speech style and tone.

2 Related Work

Studies aimed at identifying the speakers of utterances in novels [7–10] has been conducted extensively. Several studies [1,9–11], similar to our study, have been conducted on Japanese novels. There are also studies that aim to construct a daily dialogue corpus by identifying the speaker [1,11]. However, both speaker identification methods are based solely on rule-based approaches and do not incorporate the speaker's speech style or tone for speaker identification.

Several studies [9,12] have addressed speaker identification using speaker's speech style or tone. Ishikawa et al. [9] proposed a method for estimating the speaker of a Japanese novel using a tone encoder model based on BERT from candidate speakers obtained by a rule-based method. In the experiments of that study, only manually selected utterances of the main characters were used. Since novels contain many utterances by characters other than the main characters, their evaluation can be regarded as one conducted under conditions with little noise. Furthermore, in novels where the speech styles or tones of the main characters are similar, it is difficult to identify the speaker based solely on these features. To address these issues, we propose a method for identifying the speaker by utilizing not only the similarity of speech style and tone vectors, but also contextual clues such as the utterance content and the surrounding narration.

Studies [6,12,13] on speech style and tone analysis, as well as their similarity evaluation, have been conducted. In order to evaluate the similarity of speech style and tone, Zenimoto et al.[13] proposed a method for constructing sentence vectors using contrastive learning on utterances in novels. In this paper, we also examine the performance of sentence vector models constructed with contrastive learning, in order to evaluate whether such models are effective for speaker identification.

Table 1. Tone Types [6] and Example Utterances in Spoken Japanese. All utterances mean "My name is Tanaka. Nice to meet you."

Types of Speech Style and Tone	Utterances
Nomal	私の名前は田中です。よろしく。 (watashi no namae ha tanaka desu. yoroshiku.)
Abusive	俺の名前は田中だ。よろしくな。 (ore no namae ha tanaka **da**. yoroshiku **na**.)
Ladylike	わたくしの名前は田中ですわ。よろしくですわ。 (watakushi no namae ha tanaka **desuwa**. yoroshiku **desuwa**.)
Polite	私の名前は田中でございます。よろしくおいいたします。 (watakushi no namae ha tanaka **degozaimasu**. yoroshiku **onegaiitashimasu**.)
Japanese Gozaru	私の名前は田中でござる。よろしくでござる。 (watakushi no namaeha tanaka de **gozaru**. yoroshiku de **gozaru**.)

3 Speech Style and Tone in Spoken Japanese

In spoken Japanese, as shown in Table 1, the intention of the utterance remains the same; however, variations in speech style and tone convey the speaker's personality, emotions, and relationships with others. Kori [6] extracted 16 major tone types of spoken Japanese, as shown in Table 2. In total, there are more than 100 different tones. Ishikawa et al. [9] classified the tone types of characters in seven novels into ten categories, indicating that various tone types are also present in novels.

Thus, in spoken Japanese, even utterances with the same intention are expressed differently depending on the speech style or tone. Since there are many different types of speech styles or tones, we consider that they can serve

Table 2. Sixteen Major Tone Types from Kori's Analysis [6]

Tones			
strong tone	stern tone	quiet tone	vehement tone
authoritative tone	calm tone	matter-of-fact tone	expository tone
gentle tone	angry tone	harsh tone	calm tone
polite tone	light tone	subdued tone	decisive tone

as a useful clue for identifying the speaker in a Japanese novel. However, when multiple characters appear in a novel, it is possible that some of them share the same or similar speech style or tone. In such cases, identifying the speaker based on speech style or tone alone becomes difficult. Therefore, we propose a method for speaker identification that utilizes not only speech style and tone, but also multiple clues, such as the content of the utterance and the surrounding narration.

4 Speaker Identification Method

Ishikawa et al. [9] proposed a method for automatically identifying the speaker of utterances in novels based on speech style and tone. However, in some novels, multiple characters may have similar speech styles or tones. In such cases, it is difficult to uniquely identify the speaker based solely on speech style or tone. In this study, instead of identifying the speaker based solely on speech style or tone, we first identify a character whose speech style or tone is similar to that of the target utterance as a candidate speaker. We then integrate the results of this method with those of the rule-based method to determine the final speaker.

Figure 1 shows an overview of the speaker identification method proposed in this paper. First, the text of the novel and a character list are provided as input. The character list consists of a set of character names (in full or abbreviated form) and the character's gender. The character list is constructed by extracting character names from the novel using the GiNZA (ja-ginza-electra) named entity extractor, and manually grouping names that refer to the same character and their gender. The proposed method consists of two components: a rule-based method and a tone-based method. After identifying speakers using the rule-based method, tone vectors for each character are constructed based on the identified utterances, and the similarity of these vectors is used to identify candidate speakers. Finally, the results of the rule-based and tone-based methods are combined to identify the final speaker. The details of the proposed method are presented below as shown in Fig. 1.

4.1 Construction of Dialogue Groups

In our previous study [5], we proposed a model for determining the utterance-response relationship between distant utterance pairs. In this study, we define

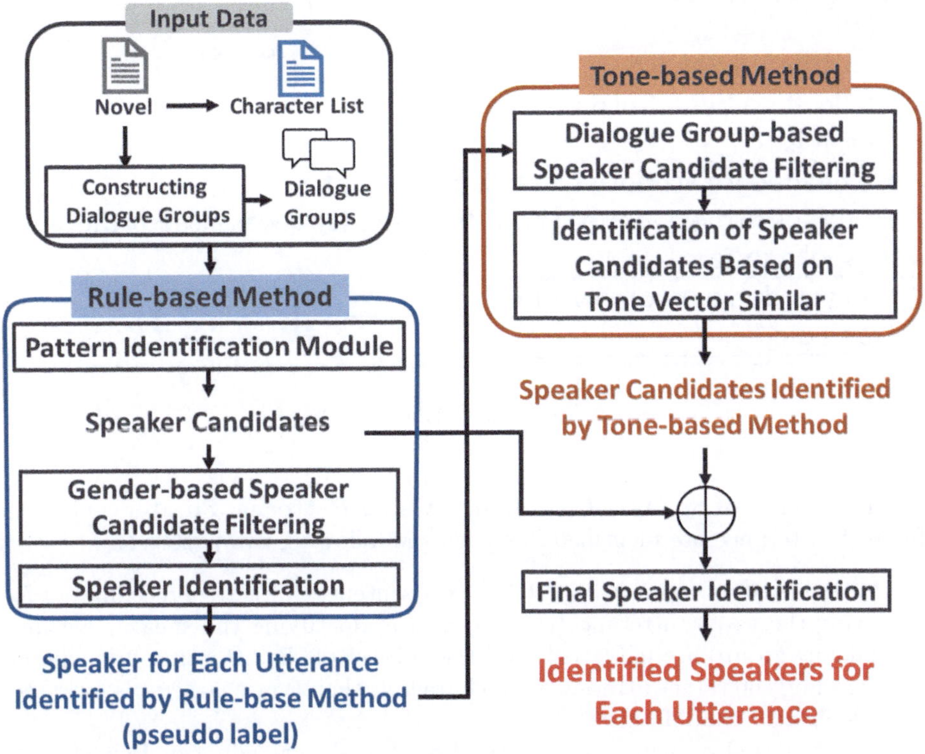

Fig. 1. Overview of the Proposed Speaker Identification Method

distant utterance pairs determined to have an utterance-response relationship by the model proposed in our previous study as consecutive utterances and make dialogue groups consisting of two or more consecutive utterances. Then, the speaker is identified for each utterance in the dialogue group.

An example of a dialogue group is shown in Fig. 2. Although the first and second utterances in Fig. 2 are separated by intervening narration, they are contextually continuous. If the model determines that they have an utterance-response relationship, a dialogue group (Group 1) is constructed from the first through third utterances.

4.2 Rule-Based Method

In novels, information identifying the speaker is sometimes provided in the narration or within the utterances themselves. The rule-based method extracts speaker candidates by applying identification patterns to narration or utterances. These candidates are then filtered by gender, based on their speech style and tone. Finally, the most frequent candidate is selected as the speaker of the utterance by the rule-based method.

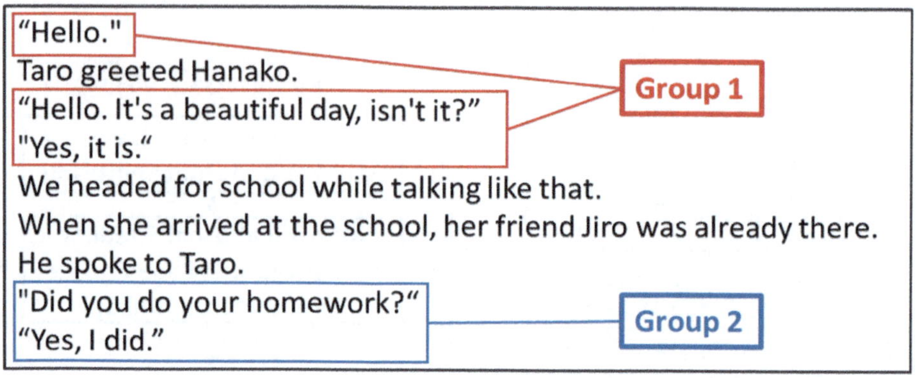

Fig. 2. Example of Dialogue Groups

Identification of Speaker Candidates Using Patterns. Speaker candidates for each utterance are identified based on the following two patterns.

1. Pattern based on narration: If the two sentences directly preceding or following the target utterance (the utterance identifying the speaker within a dialogue group) are narration, and the subject of the utterance immediately preceding the target utterance is a character, then that character is extracted as a speaker candidate.
2. Pattern based on utterances: If the line directly preceding or following the target utterance contains a direct address (e.g., "Taro,..."), that character is extracted as a speaker candidate.

Although there are utterances that include the speaker's own name, such as in self-introductions, their occurrence is extremely rare. Therefore, any character appearing in the target utterance is excluded from the speaker candidates.

Gender-Based Speaker Candidate Filtering. The gender information is useful for identifying the speaker, because the speech style or tone of an utterance tend to depend on it in Japanese [14]. To remove unlikely individuals from the speaker candidates extracted using the two patterns described above, we construct a model to predict the gender of the speaker. Specifically, the model predicts the gender of the target utterance and removes any speaker candidates whose gender does not match.

The gender prediction model is a binary classification model that adds 256 linear layers to a sentence vector model based on contrastive learning. The sentence vector model is constructed using Sentence BERT[1], following the method proposed by Zenimoto et al. [13]. The training data for the sentence vector model consists of 443,412 triples (a source utterance for comparison, a positive example

[1] Sentence BERT uses a pre-trained BERT model (tohoku-nlp/bert-base-japanese-v3) released by Tohoku NLP Group.

utterance (by the same speaker as the source), and a negative example utterance (by a different speaker from the source)). These triples were extracted from 190 novels published on the Japanese novel submission site "Shosetsukani-narou"[2]. Utterances in the training data are at least four characters long and contain particles and auxiliary verbs. For training and evaluating the gender prediction model, we constructed a dataset of 3,960 utterance (2,011 by male speakers and 1,949 by female speakers) from 10 novels, with gender labels manually assigned. Only the parameters of the linear layer are trained.

We evaluate the performance of the gender prediction model using 10-fold cross-validation. The average of the evaluation scores with the smallest test loss at each fold is used as the model's evaluation score. We use accuracy, precision, recall, and F1-score as evaluation metrics.

Table 3 shows the evaluation results of the gender prediction model. Since the F1-score is 78%, the speaker's gender can be determined from the utterance with a certain level of performance. We assume that the reason why the performance remained in the high 70% range is that the gender labels were assigned based on the characters' roles in the novels. In novels, there are characters whose gender is male but who speak in a feminine tone, and some utterances include honorific expressions that make it difficult to determine gender based solely on speech style and tone. Therefore, it is difficult to strictly filter characters solely by gender, and it is necessary to classify them based on more detailed characteristics of speech style and tone.

Table 3. Evaluation Results of the Gender Prediction Model

Accuracy	Precision	Recall	F1
0.791	0.794	0.785	0.785

Speaker Identification. To identify the speaker from the set of candidates, the most frequently occurring character is selected as the speaker. If multiple candidates appear with the same frequency, the speaker is considered unidentified because it is difficult to determine a unique speaker.

4.3 Tone-Based Method

The rule-based method is applicable when the speaker's name appears before or after the target utterance. According to a preliminary analysis of 10 novels, the applicability of this method was only 25.3%, indicating a limited number of identifiable utterances. Therefore, we propose a method to increase the number of identifiable utterances by leveraging the speech style and tone characteristics of each character. Below, we describe a method for identifying characters with similar speech styles or tones as candidate speakers.

[2] https://syosetu.com/.

Dialogue Group-Based Speaker Candidate Filtering. We extract character names that appear within and around the target dialogue group that contains the target utterance from the character list, and treat them as speaker candidates for that group. As the range for extracting speaker candidates, we use the narration between the end of the previous dialogue group and the beginning of the next dialogue group.

Identification of Speaker Candidates Based on Tone Vector Similarity. In this process, the sentence vector model used for gender-based speaker candidate filtering is used as a tone vector model. The target utterance is vectorized using the tone vector model and then reduced to 100 dimensions by principal component analysis. The speakers for each utterance identified by the rule-based method are used as pseudo-labels.

First, vectors representing the characteristics of each character's utterances are calculated using pseudo-labeled utterances. These vectors are used as representative vectors for each character. Then, speaker candidates for each target utterance are identified by calculating the similarity between the representative vector and the vector of the target utterance.

We define the set M of characters used to construct representative vectors as follows:

$$M = \{s \in S \mid |T(s)| > \alpha\} \tag{1}$$

where S is the set of characters, $T(s)$ is the set of utterances assigned pseudo-labels, and α is the threshold value.

For each character s such that $|T(s)| > \alpha$, a representative vector is constructed based on Eq. 2. For each character s in the set M, the average of the utterance vectors $v(s)_i$, which are obtained using the tone vector model for each utterance in $T(s)$, is calculated to construct the representative vector $v_M(s)$ for character s.

$$v_M(s) = \frac{1}{|T(s)|} \sum_{s \in M} v(s)_i \tag{2}$$

Then, the candidate speaker is determined using the representative vectors. The set of representative vectors V_M is defined as follows:

$$V_M = \{v_M(s) \mid s \in M\} \tag{3}$$

Speaker candidates are obtained using the Eq. 4. We calculate the cosine similarity between each representative vector in V_M and each utterance vector v_t in the novel. If the similarity between $v_M(s)$ and v_t exceeds a predefined threshold β, the corresponding character s is considered a speaker candidate by the tone-based method.

$$P_t = \{s \in M \mid \text{CosSim}(v_M(s), v_t) > \beta\} \tag{4}$$

Finally, characters who are included in both the speaker candidates identified by the tone vector similarity and those identified by the dialogue groups are considered speaker candidates.

4.4 Final Speaker Identification

Gender-based speaker candidate filtering used in the rule-based method is effective in reducing the number of unlikely speaker candidates because it can capture rough characteristics of speech style or tone. However, there are cases where the proposed method reduces the number of correctly identified utterances by eliminating speaker candidates that could have been correctly identified, due to the gender prediction model's F1-score of 78%. In addition, the tone-based method sometimes fails to uniquely identify the speaker when multiple characters with similar speech styles or tones appear.

To address these issues, we integrate speaker candidates identified by two patterns within the rule-based method and those identified by the tone-based method. We then identify the most likely speaker from the integrated candidates using the same process described in Sect. 4.2.

5 Evaluation of the Tone Vector Model

We evaluate the performance of the tone vector model based on contrastive learning. Three characters with different genders or tones from two novels are used as evaluation data. The utterances used to construct the representative vectors consist of 20 utterances randomly extracted from the manually annotated utterances for each character. The model identifies the speaker as the character whose representative vector has the highest similarity to the target utterance vector. In the experiments, we use precision, recall, and F1-score as evaluation metrics.

Table 4 shows the evaluation results. The tone type of each character was classified based on the related study [9]. Since the macro-averaged F1 of each novel is about 84% in Table 4, the tone vectors constructed by contrastive learning are effective for speaker identification. From an analysis of the misjudged utterances, we confirmed that misjudgments were often made between characters of the same speech style and tone or the same gender.

Table 4. Evaluation Results of the Tone Vector Model

	Character	Gender/Tone Type	Num. of Utterances	Precision	Recall	F1
Novel 1	A	woman/ladylike	128	**0.933**	0.875	**0.903**
	B	man/abusive	85	0.840	**0.929**	0.882
	C	man/gentle	65	0.734	0.723	0.728
	Macro Average			0.836	0.842	0.838
Novel 2	A	man/nomal	219	**0.861**	0.821	0.841
	B	woman/honorific	170	0.851	**0.911**	**0.880**
	C	woman/nomal	154	0.802	0.792	0.797
	Macro Average			0.838	0.841	0.839

6 Evaluation of the Speaker Identification Method

We evaluate the performance of the proposed speaker identification method. Ishikawa et al. [9] evaluated only the utterances of main characters as identification targets. In this paper, we evaluate the performance of the proposed method by including not only utterances of main characters but also those of non-main characters, since the goal of this study is to automatically construct a daily dialogue corpus.

6.1 Experimental Setup

As a comparison for the proposed method, we use two methods: a rule-based method and a tone-based method. The rule-based method uses a speaker identification process that leverages the tendency for speakers to alternate in consecutive utterances. In the tone-based method, we calculate the similarity between each utterance and each representative vector, and assign the speaker as the character whose representative vector has the highest similarity among the speaker candidates in the dialogue group. In both the proposed method and the tone-based method, $\alpha = 20$ is set based on a preliminary investigation. For β in the proposed method, we examine the variation in recall and precision for each novel at $\beta \in (-1.0, 1.0)$ to investigate the appropriate degree of similarity. We evaluate the performance of identifying utterances for a set of characters for which representative vectors can be constructed, when all utterances in each novel are targeted for identification.

We use precision, recall, and F1-score as evaluation metrics. Ten novels belonging to the romance genre, published on "Shosetsuka ni Narou," were used as evaluation data. Dialogue groups were constructed from the utterances in each novel, and all utterances in 100 dialogue groups per novel were used as the target for speaker identification. The utterances used in the evaluation were manually assigned speaker labels.

The statistical information of the data used in the evaluation is shown in Table 5. Here, "Num. of Main Characters" means the number of characters for whom a representative vector can be constructed. "ToneSimPairs" represents the number of character pairs whose tone vector similarity exceeds 0.5. Table 5 shows that the number of characters and main characters varies between novels, and ToneSimPairs suggests that it is difficult to identify the speaker when there are many non-main characters or when tones are similar.

6.2 Evaluation Results

Table 6 shows the evaluation results for each method. Figures 3 and 4 show the precision and recall when the threshold β in the proposed method is varied. The values of each evaluation metric in Table 6 represent the average across 10 novels. The evaluation values for the proposed method are based on the threshold settings that yielded the highest precision ($\beta = 0.7$) and the highest recall ($\beta = 0.1$) in the average across the 10 novels.

Table 5. Statistical Data for Each Novel used in the Evaluation

	1	2	3	4	5	6	7	8	9	10
Num. of Utterances	543	613	426	683	569	576	469	478	336	635
Num. of Characters	12	4	7	10	5	5	8	8	7	6
Num. of Main Characters	6	4	3	4	5	5	4	2	3	5
ToneSimPairs	9	1	5	10	1	2	4	4	8	5

The results in Table 6 show that the tone-based method, which identifies speakers using only tone similarity, achieves the highest recall but the lowest precision. The rule-based method yields the highest precision but the lowest recall. On the other hand, the proposed method achieves higher precision than the tone-based method even at $\beta = 0.1$. This indicates that introducing a threshold is effective in improving precision. Furthermore, at $\beta = 0.7$, the proposed method achieves higher recall than the rule-based method while maintaining comparable precision.

From these results, identifying speaker candidates using tone similarity and thresholding is effective when the target utterances are part of dialogues between characters with similar speech styles or tones, or when the utterances are from non-main characters.

Table 6. Evaluation Results of Speaker Identification

	Precision	Recall	F1
Rule-based Method	**0.757**	0.173	0.273
Tone-based Method	0.496	**0.524**	0.502
Proposed Method ($\beta = 0.1$)	0.637	0.432	**0.507**
Proposed Method ($\beta = 0.7$)	0.721	0.285	0.404

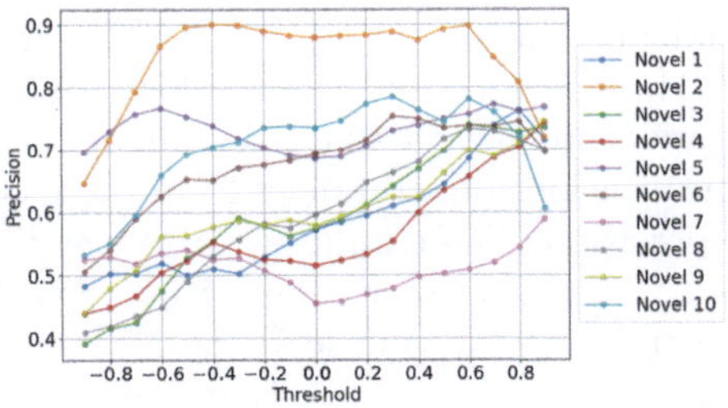

Fig. 3. Precision with Varying β for Each Novel in the Proposed Method

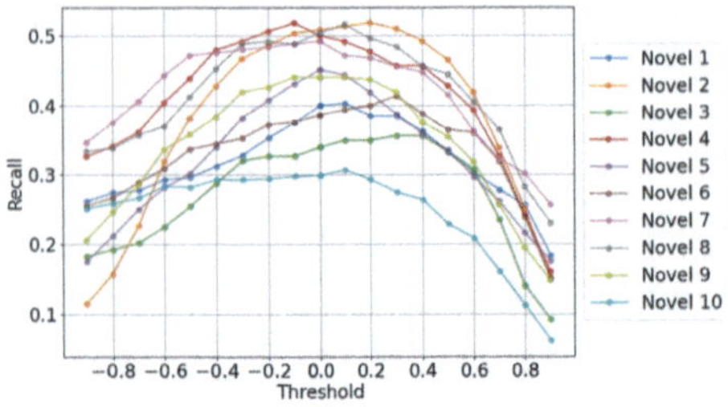

Fig. 4. Recall with Varying β for Each Novel in the Proposed Method

6.3 Discussion

We focus on the changes in precision and recall as the threshold β is varied. According to Fig. 4, recall increases for most novels up to a threshold of $\beta = 0.0$, and then decreases as β increases. On the other hand, Fig. 3 shows that the trend and the performance of precision at different β values vary significantly depending on the novel. The novel with the highest precision was Novel 2, while the lowest was Novel 7. The precision improved with increasing β in Novels 1, 4, 7, and 9, while it remained almost unchanged or declined in the other novels. These results indicate that the appropriate threshold varies depending on the novel.

We analyze the relationship between the number of misidentified utterances and the character composition of speaker candidates in each dialogue. The character composition in each dialogue is determined by the speaker candidates in the dialogue groups obtained through the filtering process described in Sect. 4.3. Table 7 shows the number of dialogues in which both main and non-main characters appear, along with the number of misidentified utterances. The results in Table 7 are based on the threshold value $\beta = 0.1$. Here, "Num. of Dialogues" refers to the number of dialogues in which both main and non-main characters appear, "Num. of Errors" is the total number of misidentified utterances, "Errors for Main Characters" is the number of misidentified utterances attributed to main characters, and "T $\setminus$ M Errors" is the number of misidentified utterances attributed to non-main characters. From the Num. of Dialogues, we can observe that more than 50 dialogues in Novels 1, 3, 7, 8 and 9 include both main and non-main characters. These novels tend to have low precision when the threshold value is low. In addition, novels that contain many non-main characters often result in misidentifications of utterances from these characters, which contributes to the lower precision. These analysis results suggest that the presence of non-main characters in the character composition obtained through dialogue group-based speaker candidate filtering leads to lower precision, as non-main characters are often misidentified as main characters. Therefore, we conclude that the character composition of speaker candidates in each dialogue influences the variation in performance and threshold trends across different novels.

Table 7. Number of Dialogues Involving Both Main and Non-Main Characters and Corresponding Misidentified Utterances

Novel	1	2	3	4	5	6	7	8	9	10
Num. of Dialogues	72	1	68	40	0	2	83	57	71	32
Num. of Errors	91	43	75	154	110	85	137	82	84	68
Errors for Main Characters	24	43	30	24	94	59	23	23	34	56
$T \setminus M$ Errors	67	0	45	130	16	26	114	59	50	12

7 Improvement Method Considering Characters in a Dialog Group

As discussed in Sect. 6.3, we confirmed that precision tends to decrease when both main and non-main characters appear in the same dialogue. Therefore, we examine a method to improve identification performance of the proposed method by setting a lower threshold β_1 for dialogues consisting only of main characters, and a higher threshold β_2 for dialogues that include non-main characters. In the dialogue groups constructed in Sect. 4.1, the threshold is set to β_1 if the dialogue group contains only main characters, and to β_2 otherwise. In this experiment, we vary β_1 within the range $[0.0, 0.7]$, while fixing β_2 at 0.9.

Figure 5 and Fig. 6 show precision values with varying threshold values of β_1 before and after the improvement, respectively. When focusing on the results at $\beta_1 = 0.0$, precision significantly improved in novels with many non-main characters, such as Novel 1 and Novel 7, with values exceeding 60% in most cases. The precision at $\beta_1 = 0.5$ remained similar before and after the improvement for most novels, except for Novel 2 and Novel 7, indicating that the generality of the improved method has been enhanced.

Table 8 presents the error results of the improved method using a threshold of $\beta_1 = 0.1$. The entries are defined in the same way as those in Table 7. Compared to Table 7, Table 8 shows a significantly greater reduction in the number of utterances misidentified as non-main characters ($T \setminus M$ Errors) than in the number misidentified as other main characters. Therefore, by adjusting the threshold value based on the character composition of each dialogue, the proposed method can effectively identify the speaker even in novels with many non-main characters, where speaker identification is typically more difficult.

However, the precision for Novel 2 is higher than that for the other novels, while that for Novel 7 is lower. There are several possible reasons for this difference. The first is the reliability of the pseudo-labels used to construct the representative vectors. Since the identification performance of the rule-based method is approximately 75%, some misidentifications are inevitably included in the pseudo-labels. Therefore, improving the performance of the rule-based method is essential to enhancing the tone-based method. The performance of speaker identification is also affected by the quality of the tone vector model. As shown in Sect. 5, the macro-averaged precision of the tone vector model was

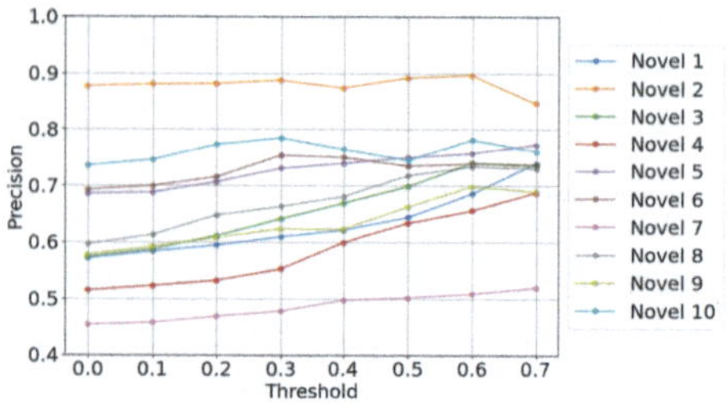

Fig. 5. Precision with Varying β in the Original Method

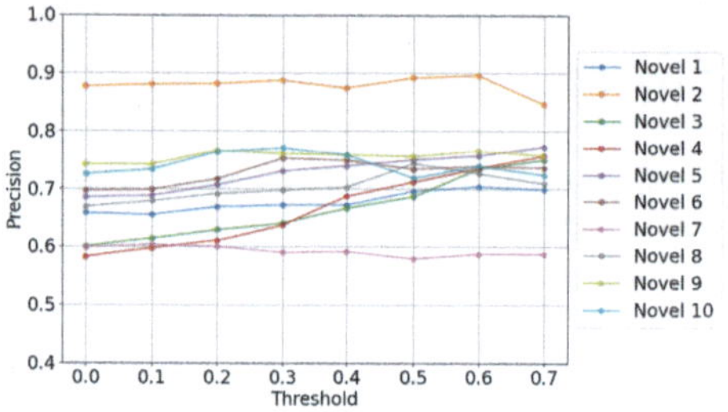

Fig. 6. Precision with Varying β_1 in the Improved Method

Table 8. Number of Dialogues Involving Both Main and Non-main Characters and Corresponding Misidentified Utterances in the Improved Method

Novel	1	2	3	4	5	6	7	8	9	10
Num. of Errors	42	43	32	83	110	85	48	44	25	57
Errors for Main Characters	15	43	19	15	94	59	10	19	17	48
$T \setminus M$ Errors	27	0	13	68	16	26	38	25	8	9

83%; however, this evaluation was conducted using only five tone types. There are other types of speech style and tone in Japanese [6,9], and some characters may use multiple them. In the future, it will be necessary to evaluate the tone vector model's ability to identify other types of speech style and tone and to identify speakers who use more than one tone.

8 Conclusion

In this paper, we proposed a method for identifying the speaker of utterances in Japanese novels by using narration surrounding the utterance, the content of the utterance, and the speech style or tone, in order to construct a daily dialogue corpus with consistent tone. The proposed method consists of the rule-based and tone-based methods. The evaluation results showed that the proposed method improved recall compared to the rule-based method. In addition, we confirmed that the proposed method improved precision compared to using only the tone-based method. Furthermore, by focusing on the composition of characters in each dialogue and setting different thresholds for dialogues involving only main characters versus those including non-main characters, we confirmed that the variance in identification performance across novels was reduced, thereby making the method more robust.

In future work, we will evaluate the performance of the proposed method in determining speech style or tone, which was not the primary focus of this study, as well as in identifying the speaker of utterances when a character uses multiple speech styles or tones. We also plan to improve speaker identification performance by incorporating other factors, such as character relationships and the use of large language models. Additionally, we will construct a daily dialogue corpus using the proposed method and explore possible applications of the corpus.

References

1. Yulong, D., Shirai, K.: Automatic construction of dialogue corpus from novels (in Japanese). In: Proceedings of the 25th Annual Meeting of the Association for Natural Language Processing (2019)
2. Fujimura, I., Chiba, S., Ohso, M.: Lexical and grammatical features of spoken and written Japanese in contrast : exploring a lexical profiling approach to comparing spoken and written corpora. In: Proceedings of the VIIth GSCP International Conference. Speech and Corpora (2013)
3. Akama, R., Isobe, Y., Suzuki, J., Inui, K.: Construction of a Japanese daily dialogue. In: Proceedings of the 29th Annual Meeting of the Association for Natural Language Processing (2023)
4. Yamashita, S., Inoue, K., Guo, A., Mochizuki, S., Kawahara, T.,Higashinaka, R.: RealPersonaChat: a realistic persona chat corpus with interlocutors' own personalities. In: Proceedings of the 37th Pacific Asia Conference on Language, Information and Computation (2023)
5. Iwamoto, K., Ando, K.: A method for determining utterance-response relationships between Japanese novel lines for constructing a daily dialogue corpus. In: Proceedings of 2024 16th IIAI International Congress on Advanced Applied Informatics (IIAI-AAI) (2024)
6. Kuro, S.: What are the major speech styles in Japanese? (in Japanese). J. Phonetic Soc. **10**(3), 52–682 (2006)
7. Chen, Y., Ling, Z., Liu, Q.-F.: A neural-network-based approach to identifying speakers in novels. In: Proceedings of 22nd Annual Conference of the International Speech Communication Association (2021)
8. Su, Z., Xu, L., Xu, J., Li, J., Huangfu, M.: SIG: speaker identification in literature via prompt-based generation. In: Proceedings of the Thirty-Eighth AAAI Conference on Artificial Intelligence and Thirty-Sixth Conference on Innovative Applications of Artificial Intelligence and Fourteenth Symposium on Educational Advances in Artificial Intelligence (2024)
9. Ishikawa, K., Ogawa, K., Sato, S.: Speaker identification using speech- style encoder (in Japanese). Nat. Lang. Process. **31**(3), 894–934 (2024)

10. Zenimoto, Y., Shinzan, K., Takehito, U.: Evaluation of a large language model in speaker to dialogue attribution in Japanese novels (in Japanese). D - Abstracts of IEICE TRANSACTIONS on Information and Systems (Japanese Edition) J108-D(4), 168–181 (2025)
11. Takuto, O., Yatsu, M., Harada, M.: Response generation in a dialogue system deep eve using LSTM (in Japanese). Trans. Inf. Process. Soc. Japan **60**(3), 967–975 (2019)
12. Michel, G., Epure, E., Hennequin, R., Cerisara, C.: Distinguishing fictional voices: a study of authorship verification models for quotation attribution. In: Proceedings of the 8th Joint SIGHUM Workshop on Computational Linguistics for Cultural Heritage, Social Sciences, Humanities and Literature (LaTeCH-CLfL 2024), pp. 160–171 (2024)
13. Zenimoto, Y., Komata, S., Utsuro, T.: Style-sensitive sentence embeddings for evaluating similarity in speech style of Japanese sentences by contrastive learning. In: Proceedings of the 13th International Joint Conference on Natural Language Processing and the 3rd Conference of the Asia- Pacific Chapter of the Association for Computational Linguistics: Student Research Workshop (2023)
14. Zenimoto, Y., Utsuro, T.: Speaker identification of quotes in Japanese novels based on gender classification model by BERT. In: Proceedings of the 36th Pacific Asia Conference on Language, Information and Computation (2022)

Context-Aware Large Language Models and Parameter Adaptation

ALBA: A Multi-agent System Based on Large Language Models for Business Process Automation

Bjanka Vrljić(✉), Ivica Botički, and Kristijan Poje

Faculty of Electrical Engineering and Computing, University of Zagreb, Unska 3, 10000 Zagreb, Croatia
bjanka.vrljic@fer.hr

Abstract. We present ALBA, a multi-agent system designed to support decision-making in complex, multi-objective problem settings. ALBA employs heterogeneous, expert-like agents powered by large language models (LLMs) to simulate domain specialists. These agents interpret user queries through their assigned roles and provide diverse, context-sensitive recommendations that reflect different dimensions of the problem. The process begins with the decomposition of a high-level, user-defined problem into smaller, manageable subproblems, each assigned to a relevant agent. Agents generate structured outputs that are represented in a shared semantic embedding space, enabling the system to compare responses, measure alignment with objectives, and detect areas of disagreement. ALBA operates through an iterative loop in which agents receive feedback derived from semantic similarity metrics and refine their responses accordingly. This process facilitates structured, multi-perspective reasoning without relying on rigid rule-based logic. Our approach demonstrates the potential of LLM-based multi-agent systems for structured decision support in enterprise environments.

Keywords: multi-agent systems · large language models · decision support systems · business process automation · multi-objective optimization

1 Introduction

The increasing complexity of decision-making tasks in real-world domains demands sophisticated support systems that can handle multiple, often conflicting objectives while incorporating diverse expertise. Organizations are frequently tasked with choosing among options that involve trade-offs between cost, efficiency, innovation, risk, and regulatory compliance. Modern enterprises face challenges such as optimizing marketing strategies while maintaining budget constraints, balancing product innovation with risk management, or pursuing growth opportunities while ensuring regulatory compliance. Traditional decision support tools typically rely on single-perspective analysis or rigid algorithmic

H. Unger and P. Meesad (Eds.): NLPIR 2025, LNNS 1904, pp. 557–570, 2026.
https://doi.org/10.1007/978-3-032-20897-2_40

approaches that may not capture the nuanced considerations required for strategic business decisions [1,2].

As an example, the problem of designing a marketing strategy under a fixed budget can be considered. On the surface, this appears to be a straightforward resource allocation problem. In practice, it involves reconciling multiple objectives: maximizing audience reach, maintaining brand coherence, staying within financial constraints, adapting to consumer behavior, and ensuring measurable impact. Addressing these demands typically requires input of various specialists, such as marketing strategists, financial analysts, brand managers, and behavioral scientists, each with a different perspective and set of priorities. These expert views are rarely aligned and their integration requires careful judgment, negotiation, and iterative refinement.

A human decision-maker would typically consult multiple experts, synthesize their often contradictory recommendations, and arrive at a balanced solution. While this type of synthesis is possible, doing so consistently, quickly, and at scale is increasingly challenging, especially in fast-paced environments where decisions must be made in real time or under conditions of incomplete information. Human decision-makers bring valuable expertise, but they are also constrained by cognitive biases, limited information processing capacity, and the difficulty of integrating insights across multiple domains [3,4]. These limitations can lead to suboptimal or inconsistent choices, particularly in high-stakes contexts where mistakes directly impact organizational performance.

Our work addresses these challenges by introducing ALBA, a multi-agent system that simulates collaborative decision-making using large language models. ALBA provides a framework for decomposing high-level problems into multiple subproblems and assigning each one to a role-specific agent, where each agent simulates an expert with domain-specific insights. These agents interpret subproblems, generate decisions and justifications, and participate in iterative exchanges guided by feedback. This automates the process of expert consultation that organizations typically undertake when facing complex business decisions. ALBA does not aim to replace human decision makers, but rather to support them by providing structured, interpretable reasoning that clarifies trade-offs and incorporates diverse perspectives, leading to more balanced and informed conclusions.

This paper focuses on the theoretical foundation and initial system design of ALBA. We present the architecture of an agent-based decision system that leverages the role-playing capabilities of LLMs to simulate domain experts. We lay the groundwork for a more advanced implementation that includes iterative consensus building, multi-objective optimization, and active moderation.

2 Theoretical Background

2.1 Large Language Models as Agents

Large language models are generative models trained on massive text corpora to predict the next word in a sequence. While originally designed for open-ended language generation, their capabilities now extend to sophisticated reasoning, instruction-following, and task-specific behavior [5,6]. This flexibility has made LLMs a powerful foundation for simulating autonomous agents that can reason, explain, and make decisions in natural language.

An LLM-based agent is typically defined by a role, an identity that constrains the agent's perspective, objectives, and tone of reasoning. Roles are not built into the model but are imposed through prompt engineering. By crafting input prompts that specify who the agent is, what it knows, and how it should approach a problem, the same underlying model can simulate a wide range of domain experts. This allows the agent to reason within a consistent frame of reference and maintain internal coherence across multiple interactions. [7,8]

Role definition is central to transforming a generic LLM into a domain-specific agent. A well-defined role not only shapes the agent's language style and knowledge scope but also encodes its decision-making logic. It ensures that different agents bring complementary or even conflicting viewpoints to the table, enabling richer analysis. It also makes the system more interpretable, since users can trace each response back to a distinct role with a well-understood set of assumptions and goals.

2.2 Multi-agent Systems

Multi-agent systems (MAS) are computational frameworks composed of multiple autonomous entities (agents) that interact within a shared environment to solve problems collaboratively, competitively, or in parallel [9,10]. Each agent operates based on its own objectives, knowledge, and reasoning capabilities. Multi-agent paradigms are especially useful for modeling complex, distributed, or multi-objective problems where no single perspective or solution is sufficient.

Traditionally, agents in MAS are explicitly programmed with decision rules, learning policies, or communication protocols. In more dynamic settings, they may learn from experience, coordinate through negotiation, or compete for resources. In both cooperative and competitive configurations, agents must consider their environment, as well as the intentions, outputs, and limitations of other agents within the system.

Incorporating LLMs enables a new form of MAS in which agents are driven by role-based prompts rather than hard-coded rules [7,8]. These LLM-based agents can interpret ambiguous input, reason contextually, and communicate in natural language, making them especially useful for problems involving human-like decision-making, trade-off evaluation, and interdisciplinary knowledge. In such systems, agents are intentionally heterogeneous, with each one focused on a distinct objective. They reason independently and operate in an iterative loop, updating their outputs based on feedback. Coordination between agents can emerge organically as their outputs align over time, or it can be enforced through explicit mechanisms like a central moderator or aggregation strategy.

2.3 Embeddings and Semantic Similarity

Embeddings are dense vector representations of text that capture its semantic content in a format suitable for comparison, clustering, and analysis. Unlike surface-level features such as word counts or keyword matches, embeddings map textual inputs into high-dimensional spaces where semantically similar texts are located near each other, regardless of phrasing or vocabulary. [11]

Modern embedding models are trained to represent sentences, paragraphs, or entire documents [12] as fixed-length vectors that preserve meaning across linguistic variation. These models enable systems to quantify the similarity between two texts using geometric measures like cosine similarity or Euclidean distance.

Embedding-based similarity is especially useful in scenarios where agents produce free-form responses. Instead of requiring exact keyword matches or rigid response structures, embeddings allow for flexible evaluation based on meaning. This enables more effective comparison, ranking, and feedback generation across varied responses.

In multi-agent systems, embedding similarity can be used to compare agents' outputs to one another, enabling detection of agreement, contradiction, or redundancy. As agents revise their responses over time, embeddings serve as a consistent measure of how their reasoning aligns with predefined goals or diverges from other agents' perspectives.

2.4 LangChain and LangGraph

Building LLM-driven systems often involves more than just generating text. It requires managing prompts, memory, tools, agent orchestration, and structured interactions over time. LangChain and LangGraph are two open-source frameworks designed to simplify the development of complex, multi-step reasoning systems powered by LLMs.

LangChain [13] provides abstractions for constructing LLM pipelines by chaining together prompts, models, retrievers, tools, and memory components. It allows developers to define agents, templates, and workflows that can interpret input, make decisions, and invoke tools when necessary. LangChain supports modularity and reusability, making it well-suited for systems that require role-specific prompts, structured outputs, and persistent context across interactions.

LangGraph [14] builds on this by introducing a framework for defining computation as a directed graph of language model nodes. Each node in a LangGraph graph can represent an agent, a transformation, or a decision step, with edges encoding the flow of control. This allows for the construction of dynamic, feedback-driven systems where outputs from one agent can influence the behavior of others. LangGraph supports conditional routing, asynchronous steps, and iteration. These features are essential for implementing multi-agent loops, feedback mechanisms, and interaction policies.

3 Related Work

Recent research has demonstrated the potential of LLMs as autonomous agents capable of collaborative problem-solving, debate, and iterative decision-making. The development of LLM-based multi-agent systems (LaMAS) [15] has gained significant momentum. Several frameworks have been developed to organize agent behavior, support concurrent reasoning, and facilitate the generation of structured, task-relevant outputs.

AutoGen [16] introduced a flexible conversational framework in which agents communicate through message passing, optionally integrating tools, human feedback, and external functions. The framework supports iterative task solving, with agents refining each other's outputs over multiple turns. This interaction pattern has proven effective for collaborative tasks such as code generation, analysis, and strategy development.

MetaGPT [17] builds on similar ideas but incorporates principles from software engineering to enhance coordination. It assigns well-defined roles to agents (such as architect, developer, and reviewer) and structures their interactions according to software development workflows. This role-driven structure has been applied effectively to multi-agent programming tasks and other environments requiring modular decision-making.

AutoAgents [18] focuses on dynamic agent generation and supervision. It introduces a supervisory component that evaluates intermediate results and can reassign tasks or create new agents on demand. This enables the system to adapt to task complexity and optimize agent collaboration over time.

AgentScope [19] emphasizes modularity, enabling agents to interact through well-defined, reusable components that facilitate system-wide reliability and reconfiguration. MegaAgent [20] organizes agents into hierarchical groups, where higher-level agents assign subtasks to lower-level agents, supporting layered reasoning and scalable task execution. AgentNet [21] takes a decentralized approach, using retrieval-augmented generation (RAG) and independently operating agents to model distributed decision-making with emergent coordination.

4 System Overview

ALBA is designed to simulate expert-level decision-making using LLM agents. The system enables users to pose high-level, multi-objective problems and receive structured insights generated by a team of specialized agents taking on distinct roles. It supports modular reasoning, semantic comparison, and iterative refinement, with the goal of capturing diverse perspectives and guiding users through complex decisions.

The process begins with a user submitting a natural language query along with relevant background data. The system reformulates this input into a structured set of subgoals, each representing a component of the overall decision problem. These subgoals are then assigned to a set of heterogeneous agents, each instantiated via an LLM and conditioned to operate within a specific expert role.

Each agent independently processes the problem from its own perspective and returns a structured response that includes both a proposed decision and a supporting explanation. These outputs are embedded into a shared semantic vector space, enabling the system to measure similarity between agent recommendations and original subgoals, as well as detect divergence or overlap among agents.

The system (Fig. 1) operates in an iterative loop. After the initial round of agent outputs, a feedback mechanism analyzes semantic similarity between agent responses and defined objectives. This feedback is then sent back to agents, prompting them to revise their outputs in subsequent iterations. This process encourages gradual convergence and refinement, while preserving agent diversity.

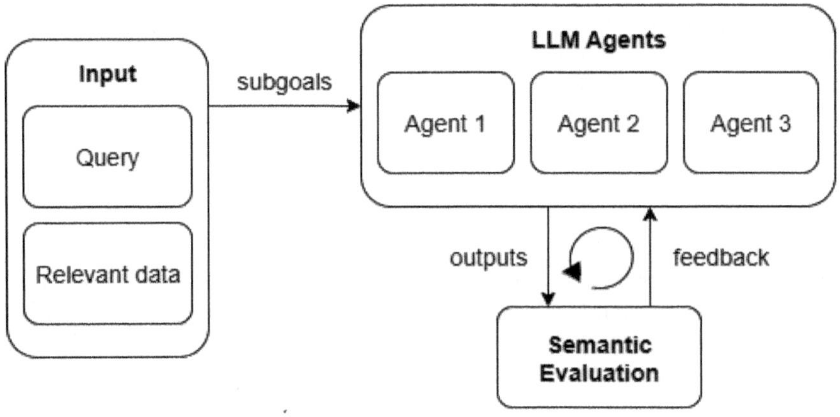

Fig. 1. System architecture of ALBA

The system is modular by design. Each component (task decomposition, agent generation, semantic evaluation, and feedback) can be configured and extended independently. This modularity allows for easy adaptation to different use cases, customization for specific objectives, and the future integration of more advanced coordination mechanisms such as aggregation, ranking, or active moderation.

In the following subsections, we describe each part of the system in detail: how problems are formulated, how agents are defined and managed, how their outputs are represented and compared, and how the iterative loop guides the system toward meaningful and interpretable responses.

4.1 Problem Formulation

The process begins with a high-level problem provided by the user, expressed in natural language. This input typically includes a strategic question, objective, or challenge, often involving multiple competing goals. Alongside the problem statement, users may also supply relevant contextual information such as quantitative metrics, constraints, or background data to better define the scope of the problem.

To enable structured reasoning, the system transforms this input into a set of clearly defined subgoals. Each subgoal corresponds to a particular aim or limitation embedded in the original question. For example, the question *"How can we grow revenue next quarter without increasing our marketing budget?"* can be reformulated into two subgoals: 1) increase revenue and 2) maintain or reduce marketing costs. These subgoals define the problem space and allow different facets of the problem to be handled independently.

Task decomposition is currently performed manually, prior to the system's invocation. The subgoals are formatted as part of the prompt input and supplied alongside the problem description and relevant data. They are then passed to the next stage in the pipeline, where they will be assigned to specialized agents for interpretation.

This formulation step ensures that the original problem is represented in a way that supports modular analysis and agent-level reasoning. The output of this stage is a structured task definition: a problem description, a list of subgoals and accompanying context, which form the input to the agent layer.

4.2 Agent Setup and Roles

Each subgoal identified during problem formulation is assigned to a dedicated agent designed to reason about that objective from a specific domain perspective. These agents are instantiated as role-conditioned LLMs, configured to mimic the reasoning patterns, priorities, and language typical of human experts in various fields. This setup enables the system to simulate a panel of specialists, each addressing a different facet of the problem.

In our study, all agents share the same underlying LLM (GPT-4o mini), but differ in behavior based on their assigned roles. However, the architecture is model-agnostic and can be applied using any LLM with sufficient instruction-following capabilities. Each role defines the agent's identity, domain expertise, and decision-making style. This is achieved through prompt engineering, using a structured template that defines the agent's viewpoint (e.g., *"You are a Market Strategist..."*) and guides its reasoning accordingly.

For example, a Financial Analyst agent may prioritize minimizing cost and risk mitigation, while a Product Innovation Specialist might focus on long-term growth and competitive differentiation. Each agent receives the same problem description but interprets it according to its assigned role and perspective. As a result, their interpretations of the same input can vary significantly.

Agent outputs are returned in a JSON format, containing a proposed decision or recommendation, and a brief explanation of the reasoning behind it. This structure supports downstream processes such as embedding, similarity analysis, and iterative refinement.

The system supports a flexible number of agents and roles, allowing new agent types to be introduced with minimal effort. This also makes the system extensible to new domains by simply specifying relevant roles and tailoring their prompts. By distributing reasoning across specialized agents, the system captures a range of perspectives that would otherwise be difficult to generate from a single model instance. This architecture forms the basis for multi-perspective analysis, supporting both divergent opinions and complementary insights in decision-making.

4.3 Semantic Representation and Similarity

To evaluate the relevance of agent responses, the system maps both subgoals and agent outputs into a shared semantic space using high-dimensional embeddings. This enables the system to assess meaning-based similarity across varied natural language outputs. Embeddings are generated using OpenAI's text-embedding-ada-002 model, which encodes input text into 1536-dimensional vectors. Each subgoal, extracted during problem formulation, is embedded to serve as a reference vector. Agent outputs are embedded into the same space, allowing for consistent comparison.

The system uses cosine similarity to quantify alignment between subgoals and agent outputs. Cosine similarity measures the angle between two vectors, yielding a score from -1 to 1, where higher values indicate stronger semantic similarity [22]. This allows the system to evaluate whether an agent's recommendation is conceptually close to the goal it was tasked with, regardless of how the answer is phrased. Cosine similarity is also used to compare agent outputs to one another, highlighting overlap, divergence, or contradiction in their reasoning. These comparisons support feedback generation and help identify trends across iterations.

For easier interpretation and visualization, the high-dimensional embedding vectors are projected into lower-dimensional space using Principal Component Analysis (PCA) [23]. This dimensionality reduction enables the visualization of agent behavior over time, making it easier to observe convergence, detect outliers, and interpret patterns that are difficult to glean from raw text alone.

4.4 Iterative Feedback Loop

The system operates in an iterative cycle in which agents repeatedly refine their outputs based on feedback. After the initial round of responses, each agent's output is embedded and compared to the subgoals and to the outputs of other agents using cosine similarity. This comparison allows the system to assess alignment with objectives and monitor agent agreement or disagreement.

Based on the similarity score, the system generates feedback for each agent. Agents then revise their outputs in response to this feedback and submit a new round of decisions. In each iteration, the agent reconsiders its previous output, the original subgoal, and the newly received feedback when generating its next response.

Over time, responses may align with individual goals, move closer to one another, or diverge further due to conflicting priorities. The loop continues until either a consensus is reached, disagreement stabilizes, or further iteration adds little value.

The iterative loop enables dynamic refinement without the need for rigid control mechanisms. It supports gradual improvement, highlights conflicts between objectives, and reveals how different roles interpret the problem. Importantly, the loop preserves the individuality of each agent while allowing for collective progress toward a more informed decision.

5 Application Example

This section demonstrates a decision-making cycle executed by the system. To illustrate the system in action, we present a scenario in which a company seeks strategic direction for long-term growth.

5.1 Problem Setup

The following high-level problem is provided by the user:

> *Our mid-sized technology company has seen plateauing revenue growth over the past 12 months. Leadership wants to decide between three strategic options for growth over the next 2 years: (1) geographic expansion into Southeast Asia, (2) launching a new AI-powered product line, or (3) acquiring a smaller competitor in the cybersecurity domain. Analyze and recommend the most viable option.*

From this input, three candidate strategies are extracted and treated as independent subgoals:

- G_1: Geographic expansion into Southeast Asia
- G_2: Launching a new AI-powered product line
- G_3: Acquiring a smaller competitor in the cybersecurity domain

5.2 Agent Roles

The following expert agents are used: Market Strategist (A_1), Product Innovation Specialist (A_2), M&A Analyst (A_3). Each agent receives a structured prompt containing relevant context, data, and a clear instruction to recommend one of the three strategies. The output must follow a predefined JSON format containing two fields: *"decision"* and *"explanation"*.

An abbreviated version of the prompt is shown below:

Listing 1. Prompt template used to instruct each agent

```
You are a {role}. {full scenario description}

Data provided: {data}

Your task is to analyze and recommend the most viable
option and provide reasoning.

Format your response as JSON:
{
  "decision": [{G1} | {G2} | {G3}],
  "explanation": [your reasoning in around 200 words
                  or less]
}
```

5.3 Iterative Refinement

Agent outputs are embedded using OpenAI's text-embedding-ada-002 model and compared to the subgoals using cosine similarity. This provides a semantic measure of alignment between each response and the intended objectives. The system also computes similarities between agent outputs to evaluate agreement or disagreement in reasoning. These similarity scores guide an iterative refinement process, where agents update their responses in subsequent rounds based on feedback derived from these comparisons.

Figure 2 shows how agent outputs evolve over three rounds of discussion. Using PCA, the high-dimensional embeddings of the agent responses and subgoals are projected into a three-dimensional space. The subgoals remain fixed, while agent outputs gradually move closer to one another. This visual drift indicates a growing semantic agreement among agents, suggesting that their recommendations are becoming more aligned, even though each agent maintains its own role and perspective. The system enables this convergence to emerge organically through feedback-driven refinement.

Table 1 shows the cosine similarity between agent outputs across three iterations. These values reflect how semantically close the agents' responses are to one another. As the discussion progresses, the values increase consistently, suggesting that the agents are moving toward a more coherent, aligned understanding of the problem. The increase in similarity emerges as a result of iterative feedback driven by semantic comparison. Despite role differences, the agents refine their outputs in a way that narrows the conceptual distance between them, enabling multi-perspective agreement.

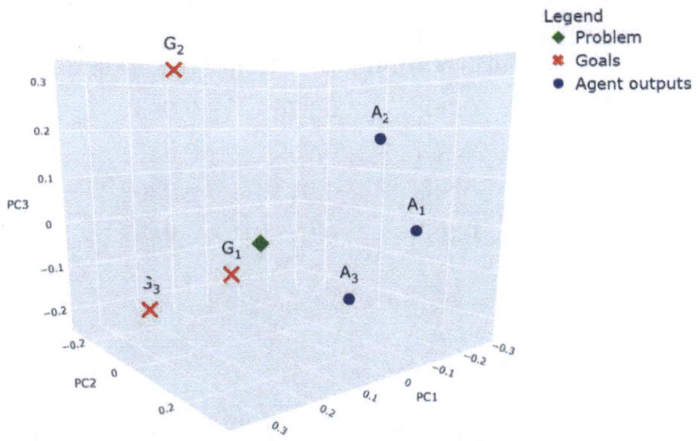

(a) Iteration 1

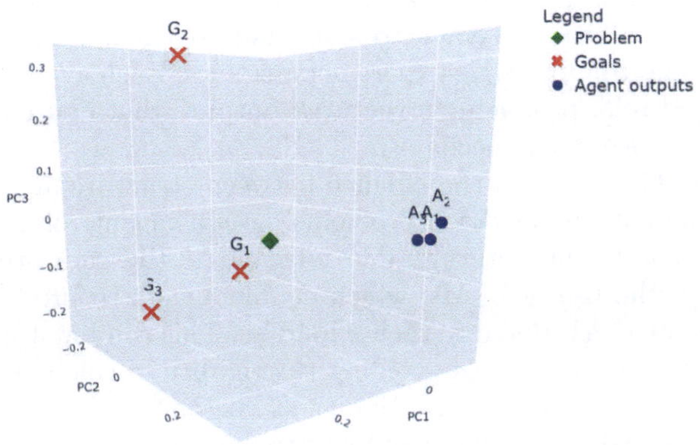

(b) Iteration 2

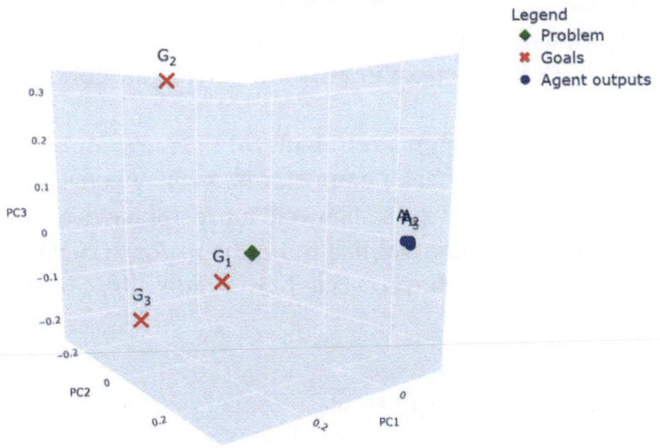

(c) Iteration 3

Fig. 2. Semantic trajectories of agent outputs (represented by blue circles) over three iterations, visualized via PCA

Table 1. Cosine similarity between agent outputs over three iterations

Agent Pair	Iteration 1	Iteration 2	Iteration 3
$A_1 \leftrightarrow A_2$	0.908	0.976	0.984
$A_1 \leftrightarrow A_3$	0.877	0.980	0.985
$A_2 \leftrightarrow A_3$	0.868	0.974	0.986
Average	0.885	0.977	0.985

6 Conclusion

This paper introduced ALBA, a modular, LLM-driven decision support system designed to handle complex, multi-objective problems through the coordinated interaction of role-conditioned agents. By decomposing user input into structured subgoals and distributing them across specialized agents, the system enables parallel reasoning from multiple expert perspectives. Each agent operates within a clearly defined role, producing structured outputs that can be compared and refined through a feedback mechanism.

The core of ALBA lies in the balance between structure and flexibility. It is structured in how it breaks down problems, assigns agent roles, and represents decisions in a consistent, interpretable format. At the same time, it remains flexible through the use of LLMs, which enable agents to produce human-like, context-aware outputs tailored to their expertise. This combination enables deep, multi-perspective analysis while avoiding the rigidity of rule-based systems.

A key advantage of the system is its use of vector-based semantic representation, where both goals and agent responses are embedded into a shared semantic space. This allows for flexible, scalable evaluation of alignment and divergence using cosine similarity, and supports iterative refinement without rigid control structures. The feedback loop guides agent updates over multiple rounds, maintaining each agent's domain-specific reasoning without forcing premature consensus.

The system's modular design and clear role separation allow it to preserve the depth and diversity of expert perspectives, even when they conflict. At the same time, it guides agent outputs toward more focused and relevant conclusions. This architecture forms a robust foundation for structured, interpretable decision-making in domains characterized by trade-offs, competing objectives, and uncertainty.

6.1 Limitations and Future Work

ALBA currently operates without a centralized moderator or global optimization mechanism. Agent outputs are evaluated and refined through local feedback loops, but there is no overarching process to reconcile conflicts or guide agents toward a unified outcome. As a result, convergence is not guaranteed and disagreements between agents can persist indefinitely without resolution.

Moreover, the system currently relies on subgoals being externally defined or manually provided. Although this simplifies implementation, it limits automation and scalability, particularly for more complex or ambiguous input problems. Automating subgoal extraction using LLMs or custom logic is a necessary step for broader applicability.

Future work will focus on addressing these limitations. Planned extensions include the introduction of a moderator agent to coordinate discourse and developing a consensus-building strategy. Additional focus will be placed on automatic subgoal extraction, enhancing domain generalization, and evaluating the system in applied enterprise contexts. These improvements aim to transform the system from an experimental prototype into a robust tool for multi-perspective decision-making.

References

1. Shim, J.P., et al.: Past, present, and future of decision support technology. Decis. Support Syst. **33**(2), 111–126 (2002). https://doi.org/10.1016/S0167-9236(01)00139-7
2. Courtney, J.F.: Decision making and knowledge management in inquiring organizations: toward a new decision-making paradigm for DSS. Decis. Support Syst. **31**(1), 17–38 (2001). https://doi.org/10.1016/S0167-9236(00)00117-2
3. Rau, D., Bromiley, P.: A review of cognitive biases in strategic decision making. Long Range Plan. **58**(3), 102529 (2025). https://doi.org/10.1016/j.lrp.2025.102529
4. Berthet, V.: The impact of cognitive biases on professionals' decision-making: a review of four occupational areas. Front. Psychol. **12**, 802439 (2022). https://doi.org/10.3389/fpsyg.2021.802439
5. Brown, T.B., et al.: Language models are few-shot learners. In: Advances in Neural Information Processing Systems, vol. 33, pp. 1877–1901. Curran Associates, Red Hook, NY, USA (2020). https://proceedings.neurips.cc/paper_files/paper/2020/file/1457c0d6bfcb4967418bfb8ac142f64a-Paper.pdf
6. Achiam, J., et al.: GPT-4 Technical Report (2024). https://arxiv.org/abs/2303.08774
7. Park, J.S., O'Brien, J., Cai, C.J., Morris, M.R., Liang, P., Bernstein, M.S.: Generative agents: interactive simulacra of human behavior. In: Proceedings of the 36th Annual ACM Symposium on User Interface Software and Technology. Association for Computing Machinery, New York, NY, USA (2023). https://doi.org/10.1145/3586183.3606763
8. Li, G., Hammoud, H., Itani, H., Khizbullin, D., Ghanem, B.: CAMEL: Communicative agents for "Mind" exploration of large language model society. In: Advances in Neural Information Processing Systems, vol. 36, pp. 51991–52008. Curran Associates, Red Hook, NY, USA (2023). https://proceedings.neurips.cc/paper_files/paper/2023/file/a3621ee907def47c1b952ade25c67698-Paper-Conference.pdf
9. Shoham, Y., Leyton-Brown, K.: Multiagent Systems: Algorithmic, Game-Theoretic, and Logical Foundations. Cambridge University Press, USA (2008)
10. Stone, P., Veloso, M.: Multiagent systems: a survey from a machine learning perspective. Auton. Robots **8**, 345–383 (2000). https://doi.org/10.1023/A:1008942012299

11. Bengio, Y., Ducharme, R., Vincent, P., Jauvin, C.: A neural probabilistic language model. J. Mach. Learn. Res. **3**, 1137–1155 (2003)
12. Le, Q., Mikolov, T.: Distributed representations of sentences and documents. In: Proceedings of the 31st International Conference on Machine Learning, vol. 32, pp. 1188–1196. PMLR, Beijing, China (2014). https://proceedings.mlr.press/v32/le14.html
13. LangChain. https://python.langchain.com. Accessed 10 July 2025
14. LangGraph. https://langchain-ai.github.io/langgraph/. Accessed 10 July 2025
15. Yang, Y., Peng, Q., Wang, J., Wen, Y., Zhang, W.: LLM-based Multi-Agent Systems: Techniques and Business Perspectives (2024). https://arxiv.org/abs/2411.14033
16. Wu, Q., et al.: AutoGen: Enabling Next-Gen LLM Applications via Multi-Agent Conversation (2023). https://arxiv.org/abs/2308.08155
17. Hong, S., et al.: MetaGPT: Meta Programming for A Multi-Agent Collaborative Framework (2024). https://arxiv.org/abs/2308.00352
18. Chen, G., et al.: AutoAgents: A Framework for Automatic Agent Generation (2024). https://arxiv.org/abs/2309.17288
19. Gao, D., et al.: AgentScope: A Flexible yet Robust Multi-Agent Platform (2024). https://arxiv.org/abs/2402.14034
20. Wang, Q., et al.: MegaAgent: A Large-Scale Autonomous LLM-based Multi-Agent System Without Predefined SOPs (2025). https://arxiv.org/abs/2408.09955
21. Yang, Y., et al.: AgentNet: Decentralized Evolutionary Coordination for LLM-based Multi-Agent Systems (2025). https://arxiv.org/abs/2504.00587
22. ScienceDirect: Cosine Similarity. https://www.sciencedirect.com/topics/computer-science/cosine-similarity. Accessed 11 July 2025
23. Jolliffe, I.T.: Principal Component Analysis, 2nd edn. Springer Series in Statistics. Springer, New York (2002). https://doi.org/10.1007/b98835

UWU Benchmarks: Assessing LLM Performance in k-Shot Prompted Role-Play of Unknown Characters

Elijah Nicolo Rosario and Ethel Ong(✉)

College of Computer Studies, De La Salle University, 2401 Taft Avenue, 1004 Manila, Philippines
{elijah_rosario,ethel.ong}@dlsu.edu.ph

Abstract. Creating engaging role-playing experiences with chatbots traditionally required expertise, but advances in Large Language Models (LLMs) and k-shot prompting now enable users to easily define original characters as interactive chatbots using natural language alone. We introduce *User-Written Unknown* (UWU) Characters, entities that do not exist in an LLM's training data, and present UWU Benchmarks, a specialized evaluation tool that measures LLM performance in role-playing these characters through k-shot prompting. Unlike existing benchmarks that focus on general capabilities or known personalities, UWU Benchmarks evaluates how well LLMs embody novel characters defined solely through user-provided *Attribute Lists* and *Sample Dialogues*. Our experiments with eight LLMs suggest that: (1) general-purpose LLMs can outperform task-specific models in character role-playing; (2) Attribute List formatting showed minimal impact on overall performance, though natural language descriptions enhanced qualitative metrics; (3) role-playing quality remained consistent across one to five Sample Dialogues, challenging assumptions about k-shot prompting in this domain; and (4) LLMs demonstrated varying proficiency with different personality types, performing better with extroverted characters. These findings provide insights for optimizing prompt engineering in role-play tasks and highlight the need for specialized evaluation tools as LLMs increasingly power personalized interactive experiences.

Keywords: Large Language Model · Benchmark · Prompt Engineering · Role-play

1 Introduction

Creating engaging role-playing experiences requires specialized knowledge and development resources, but advances in Large Language Models (LLMs) and the accessibility of prompt engineering techniques such as *k-Shot prompting* enable users of such models to define and create original characters as interactive chatbots using natural language alone. We refer to these original as *User-Written*

H. Unger and P. Meesad (Eds.): NLPIR 2025, LNNS 1904, pp. 571–585, 2026.
https://doi.org/10.1007/978-3-032-20897-2_41

Unknown (UWU) Characters, which are entities that do not exist in an LLM's training data.

Any overlap arises from the model's generalized knowledge of character tropes rather than direct memorization, making UWU Characters effectively novel creations from the model's perspective; therefore, unlike role-playing well-known characters and figures which LLMs can draw upon from their massive training corpora [5,12], such as Sherlock Holmes or Elon Musk, UWU Characters present a unique challenge: the LLM must rely entirely on the user's prompts to understand and embody personality traits, speech patterns, and behavioral quirks. In the context of role-play tasks, these k-shot prompts are instructions containing details of the character, typically containing two main components: (1) an *Attribute List* describing the character's personality traits, background, and physical characteristics, (2) and k number of *Sample Dialogues* demonstrating the character's speech patterns and behavioral responses, as illustrated in Table 1.

Table 1. Example of k-Shot Prompting a Large Language Model for role-playing as a User-Written Unknown Character

Prompt Section	Content
Attribute List (0-shot)	Lilim Yasuragi is a 20-year-old girl of mixed Swedish and Japanese heritage. She has short white hair, blue eyes, and works part-time at Zus Tea while attending university. She's kind but shy, speaking in a soft and reserved manner. She enjoys trading Pokémon cards and mango milk tea, but avoids conflict and gets flustered easily
Sample Dialogue (1-shot)	User: "Can you introduce yourself?"
	Lilim: "U-um... my name's Lilim," she stammered, fingers fidgeting with her apron. "I-I work part-time as a cashier at Zus Tea"
Sample Dialogue (2-shot)	User: "What's your favorite drink?"
	Lilim: "M-mango milk tea! It's so sweet and refreshing..." she said, tucking her hair behind her ear. "I-It reminds me of home in Sweden. Ah, sorry! I-I hope that doesn't sound silly..."
Query	User: "What do you like doing in your free time?"
	Lilim:
Model Output	"I... usually organize my Pokémon cards," Lilim murmured, her eyes darting away. "O-or sometimes I draw, but just for fun... I-I guess I just like quiet things..."

This task is crucial for entertainment-focused applications utilizing chatbots where users want to interact with personalized companions, or unique personalities that may not be replicated accurately through just pre-existing knowledge of an LLM. Such characters are increasingly popular on platforms such as *Character.AI* and *Chub.AI*, where users create and share k-shot prompts that guide language models in role-playing original characters. But beyond entertainment, role-playing has also demonstrated practical applications in professional contexts, where LLMs adopting specific personas can provide more specialized and contextually appropriate responses as chatbots, such as medical professionals offering healthcare guidance or educational tutors adapting to different learning styles [17].

This ability of LLMs to embody UWU Characters relies on *In-Context Learning* (ICL), a behavior where models adapt their responses based on examples provided in the prompt without updating their weights or parameters [1,20]. When presented with k-shot prompts containing character descriptions and sample dialogues, LLMs identify patterns and generalize character traits to generate contextually appropriate responses as also seen in Table 1 [3,5]. This process enables the model to invoke a character's personality by conditioning its output on the provided examples, effectively projecting the model into a specific character space [12]. However, the precision of this character embodiment depends heavily on the quality and structure of the prompt, making systematic evaluation of different prompting strategies essential for optimizing character consistency [3,15].

Popular LLM evaluation frameworks primarily focus on general capabilities rather than specialized role-playing performance. While *AlpacaEval*, *Arena-Hard*, and *MT-Bench* include role-playing tasks, they evaluate instruction-following quality using predefined scenarios that limit applicability to dynamic, user-defined characters [2,9]. While there are other recent benchmarks and evaluation methods that mainly focus on LLM role-playing capabilities, they have limitations when it comes to UWU Characters as they focus on personalities that likely exist in the LLM's training corpora:

- **Character-LLM** [15] – uses LLM-as-Judge to rate agents across personality, memory, and hallucination in a structured interview. Because each agent is trained on *experience uploads* using characters from public scripts, the setup is also incompatible with user-written personas.
- **RoleBench** [18] – grades single-turn QA with *Rouge-L* and LLM-as-Judge after supervised fine-tuning on data of publicly documented characters. While the evaluation also uses an empirical scoring algorithm like Rouge-L, it is limited in multi-turn dialogue and the effects of varying attribute-list formats and shot counts that exist for different user prompts.
- **WIKIROLE** [12] – uses an LLM to answer a comprehensive, multiple-choice evaluation on identity, knowledge, and rejection task performance for personas trained from Wikipedia. These figures often appear in training corpora, so the benchmark does not measure prompt-only generalization.

To address these gaps, we introduce UWU Benchmarks, a specialized evaluation tool that measures LLM performance in k-shot prompted role-playing of UWU Characters through both LLM-as-Judge and objective scoring algorithms. Our benchmark evaluates multiple aspects of role-play, including adherence to defined traits, consistency in speech patterns, and creative response generation while maintaining character integrity by taking into account the prompt's Attribute List and Sample Dialogues.

Unlike existing approaches that focus on static prompts or known characters, UWU Benchmarks provides a systematic assessment of LLMs' creative interpretation abilities through k-shot prompting alone, which gives insights on k-shot prompted role-play performance of current LLMs and the effects of prompt variation on it.

2 Methods

UWU Benchmarks, evaluates LLM performance in their ability to role-play UWU Characters using k-shot prompts as illustrated in Fig. 1. These k-shot prompts consist of an Attribute List and k number of Sample Dialogues that define the character's personality, behavior, and speech patterns, as also shown in Table 1.

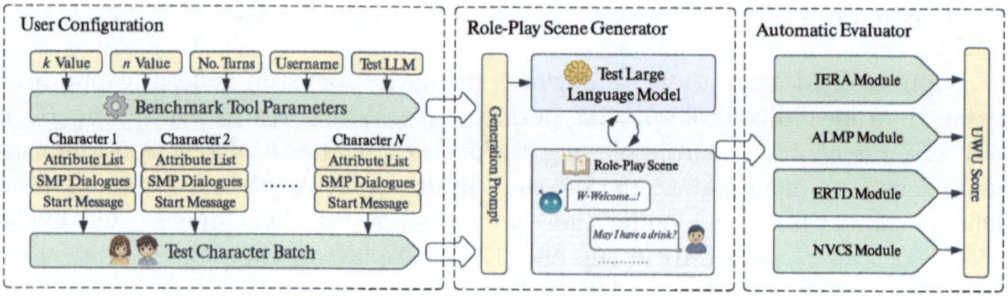

Fig. 1. UWU Benchmarks Pipeline Visualizing Evaluation Process Given a Batch of Test Characters.

We then applied UWU Benchmarks to evaluate commonly used proprietary LLMs, investigating their comparative performance in role-playing tasks, the effects of k-shot prompt design variations, and the optimal number of Sample Dialogues for character embodiment.

2.1 Algorithm Design

There are four algorithms used by UWU Benchmarks to properly evaluate a test LLM's ability in utilizing user-written k-shot prompts to embody UWU Characters. Each algorithm measures a capability involved in such a role-playing task. Further information on the models and libraries used for each algorithm are provided in full detail in the project's public repository [14].

n-Gram Vector Cosine Similarity (NVCS). NVCS measures the similarity in writing style between the user-provided Sample Dialogues (D) and the LLM's generated responses (R) from a scale of 0.0 (no similarity) to 1.0 (identical distributions) by comparing their single text character n-gram distributions:

$$\text{nvcs}(D, R) = \frac{\sum_{i=1}^{n} D_i R_i}{\sqrt{\sum_{i=1}^{n} D_i^2} \cdot \sqrt{\sum_{i=1}^{n} R_i^2}} \tag{1}$$

where:

- D: Vectorized character n-grams of the Sample Dialogues.
- R: Vectorized character n-grams of the LLM's responses.
- n: Total number of unique n-grams.
- D_i, R_i: Frequency of the ith n-gram in D and R, respectively.

Focusing on character-level sequences allows NVCS to capture subtle stylistic features of a text, such as punctuation habits and text character usage, which common word-level metrics such as *BLEU* or *ROUGE* cannot capture [11,13]. A higher NVCS score indicates that the LLM has more closely emulated the user's desired writing style.

English Readability Transfer Difference (ERTD). ERTD quantifies the difference in text readability between the user-provided Sample Dialogues and the LLM's generated responses from 0 to 100, using the Flesch-Kincaid Reading Ease (FKRE) formula [4]:

$$Fk_t = 206.835 - 1.015 \cdot ASL_t - 84.6 \cdot ASW_t \tag{2}$$

where:

- Fk_t: Flesch-Kincaid Reading Ease score of text t.
- ASL_t: Average Sentence Length (words per sentence).
- ASW_t: Average Syllables per Word.

The ERTD is then calculated as the absolute difference between the FKRE scores of the Sample Dialogues (Er_d) and the LLM's responses (Er_r), normalized to a 0–100 scale:

$$ertd(Er_d, Er_r) = 100 - |Er_d - Er_r|\,. \tag{3}$$

Assuming that all text used in the benchmark tool is English, a higher ERTD score means the LLM has better matched the readability level of the Sample Dialogues. This is especially important when the character's style should align with a specific text complexity such as in educational chatbots.

Attribute List Match Percentage (ALMP). ALMP quantifies how effectively an LLM incorporates attributes from a user-defined Attribute List into the generated role-play scene. It measures the overlap between the attributes specified by the user and those present in the LLM's output, using automated extraction and inference models for accuracy [19]. The ALMP inference model is guided by a detailed prompt to infer and refine character attributes from the generated text [14]. Let A be the set of attributes from the user's Attribute List and R the set of attributes extracted from the generated scene by the extraction model (*spaCy*) and inference LLM (*GPT-4*). Synonym sets (*SynSet*) were derived using the Natural Language Toolkit's (NLTK) WordNet. ALMP is computed as:

$$\mathrm{almp}(A, R) = \frac{M}{\mathrm{count}(A)} \tag{4}$$

where M is the total number of matched attributes, counted with:

$$M = \sum_{i=1}^{|A|} \mathrm{Match}(a_i, R) \tag{5}$$

The $\mathrm{Match}(a_i, R)$ function returns 1 if attribute a_i from A matches any attribute in R, using the following criteria:

$$\mathrm{Match}(a_i, R) = \begin{cases} 1 & \text{if } a_i \in R \\ 1 & \text{if } \exists r \in R : \mathrm{FuzzyRatio}(a_i, r) > 85 \\ 1 & \text{if } \exists r \in R : \mathrm{SynSet}(a_i) \cap \mathrm{SynSet}(r) \neq \emptyset \\ 0 & \text{otherwise} \end{cases} \tag{6}$$

where:

- a_i: Attribute from A
- r: Attribute from R
- $\mathrm{FuzzyRatio}(a_i, r)$: Levenshtein similarity score (threshold 85)
- $\mathrm{SynSet}(a_i)$: Synonyms from WordNet

ALMP ranges from 0.0 to 1.0, with higher values indicating better alignment between the LLM's output and the user-defined Attribute List. A higher value means more overlapping attributes between LLM output and user-defined Attribute List.

Judgment Evaluation on Role-Play Ability (JERA). JERA uses the LLM-as-Judge paradigm [6,9] to evaluate how effectively an LLM can perform role-playing tasks, based on a Judge LLM that reviews the Attribute List and Sample Dialogues. It assigns a score using a Likert-scale questionnaire covering three key aspects: Immersivity (how well the LLM stays in character and creates an engaging experience), Consistency (adherence to user-defined traits and speech style), and Creativity (ability to generate dynamic, novel, and engaging responses). JERA provides a holistic, external perspective on role-play quality, capturing subjective qualities that automated metrics may miss. The questionnaire and specific prompts used for the LLM-as-Judge can be found in the tool's public repository [14].

User-Written Unique (UWU) Score. The UWU Score is the primary metric of the benchmark tool, providing a holistic measure of an LLM's ability to role-play a character based on user-written k-shot prompts. It combines the weighted results of all four individual metrics (20,20,20,40), ranging from 0 to 100:

$$\mathit{UWU\ Score} \quad = \quad (nvcs \times 20) + \left(\frac{ertd}{5}\right) + \left(\frac{jera}{7.5}\right) + (almp \times 20) \quad (7)$$

JERA is weighted more heavily as successful role-play is ultimately judged by qualitative qualities that other metrics like NVCS, ERTD, and ALMP do not capture. While quantitative metrics assess technical alignment with user prompts, they may overlook nuanced aspects such as immersion, consistency, and creativity in responses that are important in role-playing chatbot use cases [18].

2.2 Benchmarking Pipeline

Figure 1 illustrates the end-to-end flow of the UWU Benchmarks tool, which unfolds in three main steps:

1. **User Configuration.** The user chooses the test LLM and parameters to use (e.g., k-shot size, n-gram length, number of role-play turns, test character batch size) and supplies each test character's k-shot prompt.
2. **Role-Play Scene Generation.** The Role-Play Scene Generator wraps the k-shot prompt and asks the LLM to produce a dialogue of T alternating turns between the user and the character.
3. **Automatic Evaluation.** The generated scene and the original prompt are fed to four metrics which evaluate the LLM's overall role-playing performance. Scores are then combined into a single UWU score, which the tool returns for analysis.

2.3 Test Character Design

Test characters are required to conduct initial experiments with the designed benchmark tool as shown in Fig. 1. As there are no formal and controlled datasets of UWU Characters yet for benchmarking purposes, a set of six characters was designed instead. These characters each have k-shot prompts containing an Attribute List, k Sample Dialogues, and a Starting Message which were synthetically generated using *GPT-4o*, edited to ensure structural consistency between all characters. The Big Five Personality Traits, or OCEAN, has been employed as a guide for personality design. Each trait in the personality model has been widely applied and studied in different areas of Chatbot design [7,8,10,16].

The six test characters, designed based on research-evaluated OCEAN trait pairings found in modern Chatbots, are mentioned below. Their full k-shot prompts, including detailed Attribute Lists, Sample Dialogues, and Starting Messages, are found in the project's public repository [14]:

- **Javier:** High Conscientiousness/Agreeableness (empathetic therapist).
- **Xiu:** High Extraversion/Agreeableness (trendy retail employee).
- **Simon:** High Openness/Conscientiousness (analytical scientist).
- **Kiko:** High Openness/Extraversion (creative writer).
- **Camille:** Low Conscientiousness + High Agreeableness (companion).
- **Lilim:** Low Extraversion + High Neuroticism (shy, stressed worker).

The six designed characters were tested as a batch during the experiments to get an averaged performance evaluation of the test LLM across various UWU Chatbots modeled after different personality traits and speech styles.

Attribute List Format. To investigate how the structure of an Attribute List impacts an LLM's role-playing ability, we designed and tested three distinct formats based on observed design patterns in UWU Character prompts in Chub.AI and Character.AI: Plain Text (PT), Structured Object (SO), and Compressed Attributes (CA). These formats vary in their level of detail, descriptive style, and token consumption, as illustrated in Table 2 below.

Table 2. Attribute List format styles (... is omitted text for brevity)

Format Type	Example Content
Plain Text (≈ 300 tokens)	Camille Fleur is a 22-year-old French girl (55 kg) known for her radiant charm and carefree personality. She stands at 165 cm tall, with golden-blonde hair styled into braided twin tails that bounce as she moves, adorned with a flower-shaped hairpin...
Structured Object (≈ 200 tokens)	Character: Camille Fleur\n Gender: Female\n Age: 22\n Nationality: French\n Relationships: Dating user for over a year\n Occupation: user's girlfriend\nPhysical: 165 cm tall, 55 kg, pale complexion, warm golden eyes, adorable fang\n Clothing: Pink knit off-shoulder sweater...
Compressed Attributes (≈ 150 tokens)	Camille Fleur = (22-year-old + female + French + 165 cm + 55 kg + pale complexion + warm golden eyes + adorable fang + golden-blonde braided twin tails + flower-shaped hairpin + wears pink knit off-shoulder sweater +...)

2.4 Experiments

In reference to the pipeline visualized in Fig. 1, experiments utilized the six designed test characters as a batch to evaluate multiple proprietary LLMs under different experimental setups. The role-play scenes for the benchmark were generated using the following LLMs (with corresponding parameter count), accessed via their respective official *Application Programming Interfaces* (APIs) using their default settings:

- **OpenAI** - includes *gpt-4o*(≈200B), *gpt-4o-mini*(≈8B), and *gpt-3.5-turbo*(*N/A*), representing OpenAI's powerful and cost-efficient language models optimized for conversational tasks.
- **Anthropic** - includes *claude-3.5-sonnet*(≈175B), *claude-3.5-haiku*(<*3.5 Sonnet*), and *claude-3-haiku*(<*3.5 Sonnet*), showcasing Anthropic's advancements in alignment and creative generation.
- **Anlatan** - includes *nai-kayra*(13B) and *nai-clio*(3B), which are specialized models tailored specifically for narrative-based and role-playing tasks.

Three experiments were conducted to investigate key questions regarding LLM role-playing performance using user-defined k-shot prompts, as described below:

Experiment 1: Best LLM for k-shot Prompted Role-Play Tasks. The first experiment determined which among the chosen test LLMs demonstrates the best role-playing performance when provided with user-defined k-shot prompts. The experimental setup was as follows:

- *Attribute List Format:* All 6 test characters used the *Plain Text* Attribute List format.
- *Sample Dialogues:* Each test character included five Sample Dialogue turns, where one turn consists of a user utterance and the character's corresponding response.
- *Role-Play Scene Generation:* Each test LLM generated five turns of role-play for each test character. One turn consists of one user utterance and a corresponding character response.
- *NVCS n-Gram Count:* All NVCS tests will be using trigrams (n=3).
- *Algorithm Models:* The JERA Judge LLM and the ALMP Inference model used for evaluation were *gpt-4*. The spaCy and Natural Language Toolkit (NLTK) Python libraries were used for the ALMP extraction model, using an English lexical database and model.
- *Repetitions:* Generation of each role-play scene was only done once for each test character, totalling to 6 role-play scenes generated per test LLM, one from each character.

This experiment aimed to answer the fundamental question: *"Which LLM is most effective for powering UWU Chatbots given k-shot prompts?"*

Experiment 2: Optimal Attribute List Format. The second experiment focused on identifying the optimal Attribute List format for role-playing tasks. The experimental setup mirrored that of Experiment 1, with the following changes:

- *Attribute List Formats:* Each character's role-play scene was generated with the three different Attribute List formats, as shown in Table 2, ensuring core definitions were consistent across formats.
- *Test LLM:* The test LLM used was GPT-4o-mini, chosen for its balance between performance and cost-efficiency.

This experiment addressed the question: *"Which Attribute List format enables the most effective role-playing by the LLM in UWU Chatbots?"*

Experiment 3: Ideal Number of Sample Dialogues. The third experiment explored the impact of the number of Sample Dialogues on the LLM's ability to role-play effectively. The setup was again based on Experiment 1, with the following modifications:

- *Sample Dialogues:* Each character was tested with one (SMP1), two (SMP2), three (SMP3), four (SMP4), and five (SMP5) Sample Dialogues.
- *Test LLM:* The test LLM used was also GPT-4o-mini.

The primary goal of this experiment was to answer: *"What is the optimal number of Sample Dialogues required to achieve the best role-playing performance in UWU Chatbots?"*

3 Results and Discussion

The following section showcases and discusses the results of the three experiments, as discussed in Sect. 2, to provide preliminary insights into LLM role-play performance across different prompt designs and character types.

3.1 Best LLM for k-shot Prompted Role-Play Tasks

The first experiment identified the most effective LLMs for utilizing k-shot prompts in role-play tasks, with results presented in Fig. 2. A Kruskal-Wallis test revealed significant differences in UWU Benchmarks metrics and overall scores across models ($H = 26.63$, $p = 0.0004$), indicating that model choice substantially impacts role-playing performance.

Post-hoc analysis using Dunn's test with Bonferroni correction revealed that *claude-3.5-sonnet* significantly outperformed two models: *claude-3-haiku* ($p = 0.0234$) and nai-clio ($p = 0.0001$). Notably, *claude-3.5-sonnet* achieved the highest UWU score (82.11 ± 3.56), while *nai-clio* recorded the lowest performance (68.18 ± 3.01). The remaining pairwise comparisons showed no significant differences, suggesting that most models perform at comparable levels for k-shot prompted role-playing tasks.

Interestingly, while the Anlatan models were specifically designed for narrative and role-playing tasks, they did not achieve top performance in this benchmark. This suggests that general-purpose LLMs, when sufficiently scaled, may outperform specialized smaller models for user-defined character role-playing. These findings suggest a potential relationship between model scale and role-playing capability.

Model	UWU ($\mu \pm \sigma$)
claude-3.5-sonnet	**82.11 ± 3.56**
gpt-4o	78.03 ± 4.13
gpt-4o-mini	77.42 ± 2.70
claude-3.5-haiku	77.20 ± 3.12
gpt-3.5-turbo	77.20 ± 1.67
nai-kayra	75.06 ± 3.11
claude-3-haiku	**73.01 ± 4.03**
nai-clio	**68.18 ± 3.01**

(a) UWU scores (mean ± SD, bold means statistically significant)

	claude-3-5-haiku	claude-3-5-sonnet	claude-3-haiku	gpt-3.5-turbo	gpt-4o-mini	gpt-4o	nai-clio	nai-kayra
claude-3-5-haiku	1.000	1.000	1.000	1.000	1.000	1.000	0.052	1.000
claude-3-5-sonnet	1.000	1.000	0.023	1.000	1.000	1.000	0.000	0.219
claude-3-haiku	1.000	0.023	1.000	1.000	1.000	1.000	1.000	1.000
gpt-3.5-turbo	1.000	1.000	1.000	1.000	1.000	1.000	0.109	1.000
gpt-4o-mini	1.000	1.000	1.000	1.000	1.000	1.000	0.078	1.000
gpt-4o	1.000	1.000	1.000	1.000	1.000	1.000	0.141	1.000
nai-clio	0.052	0.000	1.000	0.109	0.078	0.141	1.000	1.000
nai-kayra	1.000	0.219	1.000	1.000	1.000	1.000	1.000	1.000

(b) Dunn Post-Hoc pairwise significance results (dark means $p < .05$)

Fig. 2. Model performance and statistical significance results.

Table 3. Characters with highest mean score per metric and test LLM

Test LLM	ALMP	ERTD	JERA	NVCS	UWU
gpt-4o	Kiko	Kiko	Kiko	Javier	Kiko
gpt-4o-mini	Xiu	Lilim	Javier	Simon	Camille
gpt-3.5-turbo	Camille	Kiko	Javier	Simon	Camille
claude-3.5-sonnet	Camille	Kiko	Camille	Simon	Javier
claude-3.5-haiku	Xiu	Javier	Lilim	Simon	Lilim
claude-3-haiku	Camille	Xiu	Kiko	Simon	Camille
nai-kayra	Camille	Lilim	Xiu	Simon	Xiu
nai-clio	Camille	Camille	Xiu	Lilim	Camille

Per-Character Performance. Beyond overall LLM efficacy, we looked at per-character performance, with results for best and worst-performing characters detailed in Table 3 and Table 4, respectively.

Table 3 indicates that characters with High Extraversion and/or Agreeableness frequently achieve top scores. This may indicate that LLMs more easily emulate extroverted or agreeable traits, possibly due to cooperative language patterns prevalent in their training data as assistive LLMs. Conversely, Simon's complex style performs lower in ERTD but can excel in NVCS, where verbosity may have artificially increased n-gram matches, a possible limitation of the algorithm. These findings suggest LLMs might have a baseline affinity for certain personalities, highlighting the potential benefit of tailoring prompts to both character traits and LLM-specific strengths.

On the side of worst performers, Table 4 shows that Simon, with a verbose speech style, often ranks lowest in ALMP, NVCS, and UWU. His complex language may have challenged ALMP's models, created diverse n-grams affecting NVCS, and lowered ERTD scores. Lilim frequently scores poorly in JERA and

Table 4. Characters with lowest mean score per metric and test LLM

Test LLM	ALMP	ERTD	JERA	NVCS	UWU
gpt-4o	Lilim	Simon	Lilim	Kiko	Simon
gpt-4o-mini	Simon	Simon	Xiu	Kiko	Simon
gpt-3.5-turbo	Simon	Simon	Lilim	Kiko	Lilim
claude-3.5-sonnet	Simon	Camille	Lilim	Kiko	Simon
claude-3.5-haiku	Simon	Simon	Javier	Kiko	Javier
claude-3-haiku	Javier	Simon	Lilim	Kiko	Simon
nai-kayra	Simon	Javier	Xiu	Kiko	Simon
nai-clio	Simon	Kiko	Lilim	Kiko	Simon

UWU, suggesting LLMs may find introverted or hesitant personalities more difficult to portray convincingly, possibly due to training data biases towards more cooperative traits. Kiko can be weak in NVCS due to unique n-grams produced from capitalized words, another possible algorithmic limitation.

In summary, the results from the benchmark tool suggest that characters with complex speech styles or introverted traits pose greater role-playing challenges for current proprietary LLMs.

3.2 Optimal Attribute List Format

The second experiment investigated which Attribute List format optimally enables LLMs to understand and utilize defined character traits during role-play. Using gpt-4o-mini as the test model, three formats were compared: Plain Text (PT), Structured Object (SO), and Compressed Attributes (CA). Results are presented in Table 5.

Table 5. Performance scores across attribute-list formats (mean ± SD, bold means statistically significant)

Format	ALMP	ERTD	JERA	UWU
PT	0.57 ± 0.12	88.76 ± 7.66	**249.04 ± 13.05**	77.24 ± 2.99
SO	0.66 ± 0.08	89.00 ± 7.72	233.15 ± 7.72	76.95 ± 3.57
CA	0.68 ± 0.11	90.75 ± 7.66	**222.41 ± 5.62**	75.96 ± 3.95

Kruskal-Wallis revealed no significant differences in overall UWU scores across the three formats ($H = 0.25$, $p = 0.8844$), suggesting that attribute list formatting has minimal impact on overall role-playing performance. However, significant differences emerged in the JERA metric ($H = 11.68$, $p = 0.0029$), with post-hoc analysis showing that Plain Text significantly outperformed Compressed Attributes ($p = 0.0020$) in terms of immersivity, consistency, and creativity as judged by the LLM evaluator. This finding has practical implications for prompt engineering, suggesting that users have flexibility in choosing formats. The significant JERA differences, however, reveal nuanced impacts on role-playing quality. It suggests that verbose, natural language descriptions may facilitate better character embodiment for qualitative assessment.

3.3 Ideal Number of Sample Dialogues

The third experiment investigated how varying the number of Sample Dialogues impacts role-playing performance using gpt-4o-mini as the test model. Five conditions were tested, ranging from one to five sample dialogues (SMP1 through SMP5), with results shown in Table 6.

Table 6. Performance scores (mean ± SD) across different numbers of sample dialogues.

Samples	ALMP	ERTD	JERA	UWU
SMP1	0.61 ± 0.13	88.21 ± 11.90	232.39 ± 6.37	75.21 ± 4.04
SMP2	0.53 ± 0.10	89.52 ± 8.63	249.03 ± 11.78	76.27 ± 4.05
SMP3	0.60 ± 0.09	90.28 ± 7.49	245.85 ± 12.81	77.41 ± 3.28
SMP4	0.60 ± 0.08	92.35 ± 8.71	248.32 ± 7.15	78.16 ± 3.14
SMP5	0.57 ± 0.11	88.94 ± 7.25	246.55 ± 10.08	76.81 ± 2.65

Statistical analysis revealed no significant differences in scores across the different numbers of sample dialogues (H = 3.29, p = 0.5112), indicating that the quantity of examples provided has minimal impact on overall role-playing performance. These findings challenge the assumption that more examples necessarily lead to better k-shot learning performance in role-playing contexts [5]. The lack of significant differences suggests that LLMs can effectively understand and embody character traits with minimal examples, making even one-shot prompting (SMP1: 75.21 ± 4.04) statistically comparable to five-shot prompting (SMP5: 76.81 ± 2.65). While SMP4 achieved the numerically highest mean UWU score (78.16 ± 3.14), this difference is not statistically significant and likely represents natural variation rather than actual performance improvement.

4 Conclusion

This study introduced UWU Benchmarks, a specialized tool for evaluating Large Language Model performance in role-playing User-Written Unknown Characters through k-shot prompting. By combining objective metrics (NVCS, ERTD, ALMP) with qualitative assessment (JERA), the tool addresses a critical gap in evaluating how LLMs embody characters defined solely through user prompts.

Our experiments revealed several key findings: First, model choice significantly impacts performance (H = 26.63, p = 0.0004), with *claude-3.5-sonnet* achieving the highest scores. Surprisingly, larger general-purpose LLMs outperformed small specialized narrative models (*nai-clio*), suggesting model scale and training diversity may matter more than task-specific optimization. This finding has important implications for developers choosing models for role-play applications. Second, while Attribute List format showed no overall impact ($p = 0.8844$), Plain Text significantly enhanced qualitative metrics over Compressed Attributes ($p = 0.0020$), suggesting that detailed natural language descriptions facilitate more nuanced character understanding, offering practical guidance for prompt engineering while maintaining flexibility for user preferences. Third, performance remained consistent across one to five sample dialogues ($p = 0.5112$), challenging assumptions about k-shot learning and indicating LLMs can capture character essence from minimal examples. Finally, LLMs demonstrated better role-play performance with extroverted characters than introverted personalities, likely reflecting training patterns of assistive LLMs.

As LLMs increasingly power personalized interactive experiences across entertainment, education, and professional domains, tools like UWU Benchmarks become essential for ensuring these systems can faithfully represent the diverse range of characters users wish to create. By providing systematic evaluation of k-shot prompted role-playing, this work contributes to the broader goal of making AI-powered character interactions more accessible, reliable, and aligned with both creative and practical user intent.

Future work should address several limitations of this preliminary study, including expanding the character dataset beyond the six OCEAN-based test personas, implementing multi-run evaluations for statistical robustness, incorporating multiple judge LLMs to reduce evaluation bias, adding human evaluation to benchmark scores, and extending support for other languages besides English.

Supplementary Information. All prompts and code required to further understand as well as reproduce the study's findings are fully available in the public repository of the benchmark project [14].

References

1. Brown, T.B., et al.: Language models are few-shot learners (2020). arXiv:2005.14165 [cs]
2. Chang, Y., et al.: A survey on evaluation of large language models (2023). arXiv:2307.03109 [cs]
3. Chen, Z., Moscholios, S.: Using prompts to guide large language models in imitating a real person's language style (2024). arXiv:2410.03848
4. Crossley, S., Heintz, A., Choi, J.S., Batchelor, J., Karimi, M., Malatinszky, A.: A large-scaled corpus for assessing text readability. Behav. Res. Methods **55**(2), 491–507 (2023). https://doi.org/10.3758/s13428-022-01802-x
5. Dong, Q., et al.: A survey on in-context learning (2024). arXiv:2301.00234 [cs]
6. Gu, J., et al.: A survey on LLM-as-a-judge (2025). arXiv:2411.15594 [cs]
7. Hume, A., Cernuzzi, L., Zarza, J.L., Bison, I., Gatica-Perez, D.: Analysis of the big-five personality traits in the chatbot "UC - Paraguay". CLEI Electron. J. **25**(2) (2022). https://clei.org/cleiej/index.php/cleiej/article/view/543

8. Kuhail, M.A., Thomas, J., Alramlawi, S., Shah, S.J.H., Thornquist, E.: Interacting with a chatbot-based advising system: understanding the effect of chatbot personality and user gender on behavior. Informatics **9**(4), 81 (2022). https://www.mdpi.com/2227-9709/9/4/81
9. Li, T., et al.: From crowdsourced data to high-quality benchmarks: arena-hard and benchbuilder pipeline (2024). arXiv:2406.11939 [cs]
10. Li, W., Liu, J., Liu, A., Zhou, X., Diab, M., Sap, M.: BIG5-CHAT: shaping LLM personalities through training on human-grounded data (2024). arXiv:2410.16491
11. Lin, C.Y.: ROUGE: a package for automatic evaluation of summaries. In: Text Summarization Branches Out, pp. 74–81. Association for Computational Linguistics, Barcelona (2004). https://aclanthology.org/W04-1013/
12. Lu, K., Yu, B., Zhou, C., Zhou, J.: Large language models are superpositions of all characters: attaining arbitrary role-play via self-alignment (2024). arXiv:2401.12474 [cs]
13. Papineni, K., Roukos, S., Ward, T., Zhu, W.J.: BLEU: a method for automatic evaluation of machine translation. In: Proceedings of the 40th Annual Meeting on Association for Computational Linguistics - ACL '02, p. 311. Association for Computational Linguistics, Philadelphia (2001). http://portal.acm.org/citation.cfm?doid=1073083.1073135
14. Rosario, E.: UWU-benchmarks (2025). https://github.com/Eliolocin/UWU-benchmarks. official GitHub Repository
15. Shao, Y., Li, L., Dai, J., Qiu, X.: Character-LLM: a trainable agent for role-playing (2023). arXiv:2310.10158 [cs]
16. Skjuve, M., Følstad, A., Fostervold, K.I., Brandtzaeg, P.B.: A longitudinal study of human–chatbot relationships. Int. J. Hum.-Comput. Stud. **168**, 102903 (2022). https://www.sciencedirect.com/science/article/pii/S1071581922001252
17. Tseng, Y.M., et al.: Two tales of persona in LLMs: a survey of role-playing and personalization (2024). arXiv:2406.01171 [cs]
18. Wang, Z.M., et al.: RoleLLM: benchmarking, eliciting, and enhancing role-playing abilities of large language models (2024). arXiv:2310.00746 [cs]
19. Wang, Z., Zhou, X., Koncel-Kedziorski, R., Marin, A., Xia, F.: Extracting and inferring personal attributes from dialogue (2022). arXiv:2109.12702 [cs]
20. Xie, S.M., Raghunathan, A., Liang, P., Ma, T.: An explanation of in-context learning as implicit bayesian inference (2022). arXiv:2111.02080 [cs]

Advancements in Transformers for Medication Prescription: A Comprehensive Literature Review

Syed Usama Hussain Shah Bukhari(✉), Manan Vinod Shah, Md Jobayer Hossain Chowdhury, Abdullah Al Noman, and Ashir Ahmed

Department of Information Science and Technology, Kyushu University, Fukuoka 8190395, Japan
{bukhari.453,shah.manan.298,chowdhury.md.jobayer.hossain.569, noman.abdullah.769}@s.kyushu-u.ac.jp, ashir@ait.kyushu-u.ac.jp

Abstract. Transformer architectures and large language models (LLMs) have recently advanced medication prescription tasks. This review synthesizes peer-reviewed and pre-print studies published between 2020 and 2025 that apply self-attention–based models to three critical tasks: (i) medication recommendation, (ii) prescription-text generation, and (iii) safety verification. We screened 3,728 records and retained 14 studies that satisfied strict inclusion criteria (Transformer centric approach, empirical evaluation, and clinical relevance). For each task we compare data sources, model architectures, evaluation metrics, and deployment maturity. We find that transformer-based recommendation leverage multi-head attention and graph modules to predict safe medication sets from electronic health records (EHRs), often improving F1-scores and Jaccard indices while reducing drug–drug interaction (DDI) rates. Generative models can synthesize realistic prescription instructions and notes. Safety-checking systems use LLMs to verify dosage, indications, and interactions to reduce errors. Despite these advances, challenges remain in data sparsity, limited real-world validation, bias assessment, integration of external knowledge, and clinical deployment barriers. Finally, we present a theoretical critique of various Transformer model architectures, analyzing trade-offs in interpretability, data requirements, safety constraints, deployment feasibility, and evaluation robustness.

Keywords: medication prescription · prescription verification · prescription text generation · transformer · large language model · clinical decision support · electronic health records · e-prescribing · drug–drug interaction · recommendation systems

1 Introduction

Prescription of medications is the process of recommending and drafting medicines and clinical advice for patients. Prescription of the right medications

H. Unger and P. Meesad (Eds.): NLPIR 2025, LNNS 1904, pp. 586–599, 2026.
https://doi.org/10.1007/978-3-032-20897-2_42

is a cornerstone of effective healthcare, but it remains a complex task with significant risk of error. Studies estimate that up to 12% of patients experience medication-related harm in care settings, with about 15% of such events being severe or fatal [1]. Similarly medication errors cause than one million preventable adverse drug events annually in the U.S. alone with prescribing and transcription errors accounting for 75-percent of cases [2]. Interest in this field has grown due to potential benefits of reducing human errors (e.g. prescription errors, adverse drug events, drug-drug-interaction), improving efficiency, and aiding clinicians in decision-making [3]

Traditional clinical decision support systems (CDSS) for prescribing rely on rule-based approaches or statistical models. Rule-based prescription support has demonstrated benefits such as increased guideline adherence and error reduction. A recent scoping review [4] found that a majority of studies (59.5%) implementing CDSS in practice used rule-based systems integrated with clinical databases. However, these systems often suffer from alert fatigue and limited personalization. Meanwhile, advances in Natural Language Processing (NLP) have enabled interpretation and generation of the textual components of prescriptions.

NLP techniques have been used to parse free-text instructions to check for errors, and to automatically generate patient friendly medication instructions. Early NLP approaches relied on text parsing n-gram models, but modern approaches use deep sequence-to-sequence (seq2seq) models and transformers. Since 2018, researchers have adapted Transformers based architecture in medication tasks, aiming to assist clinicians with recommendation, documentation, and safety verification.

A number of prior review papers survey medication prescription in the context of machine learning, but they mainly cover traditional machine learning and early deep learning methods. In contrast, our review focuses specifically on recent Transformer-driven approaches.

Objective

This review focuses on peer-reviewed articles and preprints from 2020 to 2025 that leverage Transformer-based architectures for medication prescription tasks. We specifically target studies proposing methods to improve medication recommendation accuracy, generate clinical prescription text, and perform automated prescription verification and safety checking. We compare the models architectures, data sources, and evaluation metrics, and we identify knowledge gaps and future research directions.

2 Methodology

We adopted a transparent narrative-review protocol aligned with recent best-practice guidance for qualitative and quality criteria [5]. The goal was comprehensive coverage, not the exhaustive Preferred Reporting Items for Systematic Reviews and Meta-Analyses (PRISMA) apparatus required.

2.1 Literature Search Strategy

We searched multiple bibliographic databases, including **PubMed**, **Scopus**, **Web of Science**, and **ACM Digital** Library for articles published between January 2020 and May 2025. To capture emerging work, we also manually scanned the arXiv preprint server and proceedings of major conferences: **ACL**, **AAAI**, **MICCAI** for relevant papers.

The boolean string combined transformer keywords with prescribing concepts:

```
(transformer* OR "self-attention" OR BERT OR GPT
  OR "large language model" OR
 "retrieval-augmented generation")
AND
(prescrib* OR medicat* OR "drug" OR pharmac* OR
 "medication recommendation" OR prescription)
```

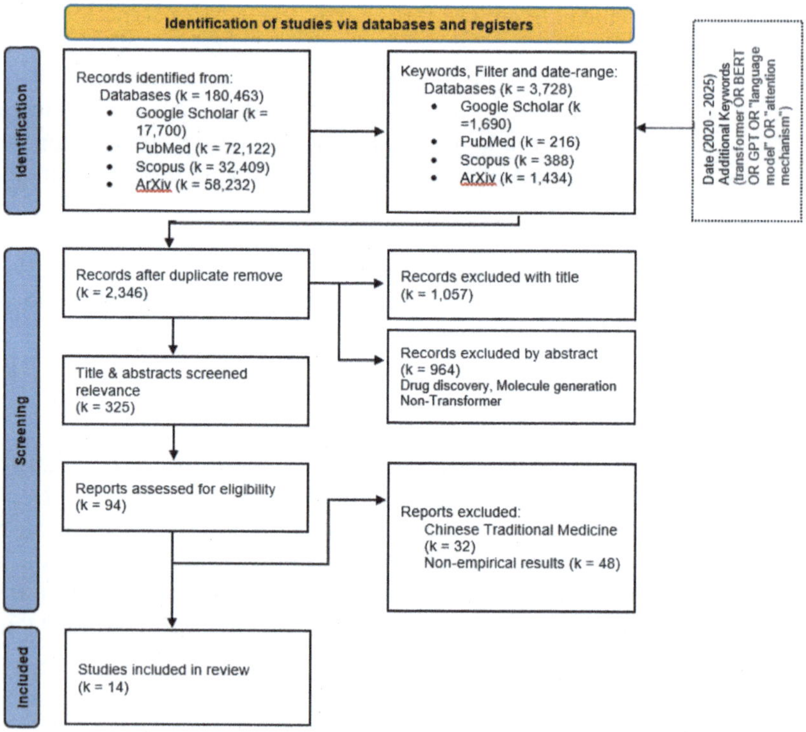

Fig. 1. Flow Diagram (Non-PRISMA approach).

Eligibility Criteria

To ensure methodological consistency and relevance to the prescription task, we applied the following criteria for study selection.

Inclusion

- **Transformer-Based Models:** The study must utilize self-attention mechanisms, with Transformer architectures as the core modeling framework.
- **Prescription-Focused Outputs:** The model must generate or predict outputs directly related to medication decision-making, such as drug selection, dosage recommendation, structured prescription instructions, medical safety assessment, provisional prescriptions, or related components.
- **Empirical Evaluation:** Quantitative evaluation must be presented using at least one dataset, either publicly available or explicitly described in the study.
- **Clinical Relevance:** Applications must fall within clinical or veterinary prescription domains, provided the methods or evaluation metrics can be mapped to human healthcare contexts.

Exclusion

- **Non-Transformer Systems:** Studies using only rule-based systems, knowledge graphs, machine learning with no Transformer-based architectures.
- **Drug Discovery Focus:** Studies primarily focused on molecule generation or drug design were not included.
- **Non-Empirical Literature:** Commentary articles, opinion pieces, or study protocols that lack empirical results were excluded.
- **Traditional Chinese Medicine (TCM):** Studies focused on TCM are excluded. This decision is based on the fundamental divergence in pharmacological principles. This review centers on models trained using standardized Western pharmacological ontologies (i.e. , RxNorm, ATC).
- **Language Limitation:** Full-text articles not written in English and for which translation was not feasible were excluded.

Two reviewers independently screened all titles and abstracts for relevance. Remaining candidates underwent full-text review against the inclusion criteria, with any disagreements resolved through discussion supported by Rayyan, citation retrieval by Publish or Perish (version 8), reference management by Mendeley and analytical workflow implemented in Python was used to verify duplicate removal, ensure internal consistency of screening decisions, and counts reported. Out of 3,728 initial records, 1,382 were duplicates, leaving 2,346 unique records. Of these, 1,057 were excluded based on their titles and 964 by abstract. We screened 325 unique records and identified 94 articles for full text assessment, of which 14 met all inclusion criteria. For each included study, we extracted key details including the task category, input modality (e.g., structured EHR data, text, images), datasets used, Transformer model variant, notable modeling techniques, evaluation metrics, any safety-specific evaluation (e.g., DDI checks), and deployment context. We synthesized the evidence qualitatively by grouping the studies into the three primary task categories and comparing their methodologies and outcomes. The methodologies of the articles and the respective results were evaluated and discussed in Fig. 1.

3 Review Findings

Through our review, we found that a significant studies is conducted in Transformer-based architectures for medication prescription. To characterize the landscape of this work, we organized the studies into three distinct categories. Each paper was then mapped to its corresponding medication-use domain, and we retained only the primary Transformer task as specified by its authors. This procedure produced the mutually exclusive branches illustrated in Fig. 2.

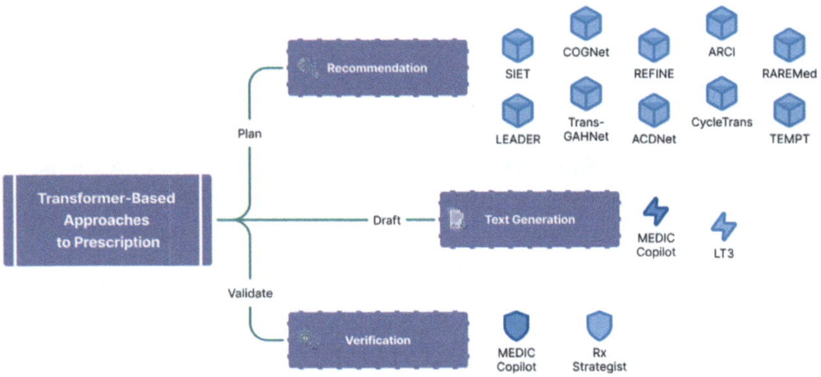

Fig. 2. Transformer Based Prescription: High level Taxonomy Diagram.

All included models were empirically evaluated on clinical datasets either real or synthetic, employed self-attention as the core mechanism, and demonstrated measurable improvements in the accuracy, safety, or personalization of medication decisions. Below, we describe each category and key examples, and later summarize common datasets and evaluation approaches.

3.1 Taxonomy of Transformer-Based Approaches

Recommendation: The recommendation category addresses the question of "what to prescribe" for a given patient. These models function as policy/decision engines that predict drug regimens based on patient data i.e. diagnoses, procedures or lab results, rather than generating free-text instructions.

Early deep learning models used Recurrent Neural Network and attention-based approaches i.e. LEAP [6], however recent work leverage Transformers and Graph Neural Networks, where research on drug selection has reached the point where almost every EHR system now benchmarks a medication recommendation module.

The majority of studies (10 out of 14) addressed medication recommendation tasks. These models primarily rely on longitudinal EHR data from ICU or hospital datasets and often incorporate external knowledge (drug ontologies). A main challenge is capturing the temporal visit sequence while avoiding unsafe drug combinations.

Several approaches combine sequence modeling with graph-based safety checks. For instance, COGNet [7] uses Encoder-decoder architecture with attention mechanisms, encoders were applied for diagnosis, procedure sequences. Additionally GCN-based "Medication Graph Encoder" for drug and visit level selection and DDI relations. COGNet introduces a novel "copy-or-predict" mechanism that at each decoding step decides whether to copy a medication from the patient's history or predict a new one, mimicking clinician reasoning.

Similarly, **REFINE** and **ACDNet** use transformer layers to encode patient history and leverage knowledge graphs for drugs. **Trans-GAHNet** integrates multiple drug-related graphs (e.g., co-occurrence patterns, mutual exclusivity, and DDI networks) by encoding each graph with a graph-attention network and fusing them within a Transformer-based visit encoder. This fusion enriches the patient representation addressing the cold-start problem and improving both accuracy and safety of the medication recommendations [8].

Another notable approach is **TEMPT**, which address heterogeneous hospital records. It pretrains a shared encoder with masked diagnosis and contrastive tasks to learn general medical representations, then uses prompt tuning per hospital. On the eICU multi-hospital dataset, the model improved PR-AUC over single-center models, showing that prompt-based personalization can handle institutional variation [9].

For fine-grained medication prediction, **REFINE** goes beyond class-level drugs. It uses a Vision Transformer (ViT) to process trends in lab tests and medication dosages over time, and constructs a weighted DDI graph encoding severity levels. The model then uses a graph-attention network to recommend specific drug combinations with a balanced loss to avoid overly allergies recommendations. In medication fairness recommendation, **RAREMed** uses a Transformer encoder with unified disease and procedure inputs and special segment embeddings. It is pretrained on self-supervised tasks (Sequence Matching Prediction, Self Reconstruction) [10].

Generation: The generation category focuses on Transformer models that can generate text related to medication prescribing, including synthetic prescriptions or clinical notes. Prescription instructions are among the most sensitive yet valuable parts of health records. A model that can compose realistic prescriptions from minimal structured input could facilitate data sharing by generating de-identified synthetic data, reduce manual documentation burden, and support downstream NLP tasks. Only one study, LT3 [11] directly targeted synthetic prescription generation. LT3 is a conditional label-to-text Transformer that converts structural medical inputs into complete human-like prescription. It takes a set of drug labels, dosage labels, vitals as input and generates instructions like "docusate sodium 100 mg Capsule Sig: One Capsule PO BID as needed for constipation.". It was trained on 2,000 real prescription examples extracted from MIMIC and outperformed a fine-tuned T5 baseline.

Beyond this study, related works have explored using large LLMs to generate synthetic clinical text, which can include prescription-related content. For instance, recent frameworks employ GPT-4 and fine-tuned LLaMA models combined with a medical knowledge graph to produce synthetic patient notes **MedSyn** [12]. These efforts illustrate the potential of combining LLMs with domain knowledge to address data scarcity and create realistic medication narratives. However, applying such generative systems specifically to prescriptions remains challenging due to the need for factual accuracy and avoidance of hallucinated instructions

Table 1. Overview of included Transformer-based studies (2020–2025).

Model	Task*	Modality§	Dataset(s)	Variant†	Input / factors‡	Core technique
COGNet [7]	R	Struct	MIMIC-III	A-ED	Sequential EHR (diag/proc); copy-history links	Copy-or-predict; GCN medication graph encoder
SIET [13]	R	Struct+MG	MIMIC-III; DrugBank; ICD-9	KE (GNN)	Graph embeddings; KG edges; DDI graph	Star-interactive fusion of EHR+KG
REFINE [14]	R	TS+Img	MIMIC-IV; Private	VT+GATv2	Visit/graph sequences; dosage & lab trends	ViT on lab trends; GATv2 on DDI; balanced loss
Trans-GAHNet [8]	R	Struct+MG	MIMIC-III	ED	diag/proc/lab; graph augmentation	Multi-graph attention fused with Transformer encoder
LEADER [15]	R	Struct	MIMIC-III/IV	ED (distilled)	EHR embeddings; prompts	Prompt tuning; layer adaptation; knowledge distillation
ACDNet [16]	R	Struct	MIMIC-III/IV	ED	Visit sequences	Collaborative dual-scale attention; decision network
ARCI [17]	R	Struct	MIMIC-III; AKI	ED	Visit-level intents	Multi-level contrastive representation learning
CycleTrans [18]	R	Struct	MIMIC-III	A-ED	Visit sequences	Cycle-consistency; cross-attention; DDI-aware loss
RAREMed [10]	R	Struct	MIMIC-III/IV	PD	Unified visit seq. (disease+procedure)	Decoder LLM + self-supervised (seq. matching, reconstruction)
TEMPT [9]	R	Struct	eICU (multi-center)	ED	ICU visits; site heterogeneity	Masked-code pretraining; soft prompts per hospital
LT3 [11]	G	Txt	MIMIC-III	ED	Tokens (labels & text); med labels+form	Label-to-text prescription generation
MEDIC [2]	G+V	Txt	Amazon Pharmacy	ED/PD	Free-text directions (dose/freq/route)	Retrieval-augmented; pharmacy guardrails
CLM [19]	V	Txt+Sp	n2c2 (+ speech)	SE + BERT	Notes + speech tokens	Context embeddings; error patterns; anomaly detection
RxStrategist [20]	V	Txt+Img	Private	Hybrid	Clinical text + pill images	Multi-agent LLMs + KG + RAG

*__Task:__ R = Recommendation; G = Generation; V = Verification.

§**Modality:** Struct = structured EHR; MG = multi-graph features; TS = time-series; Img = images; Txt = textual; Sp = speech (e.g., Struct+MG, TS+Img, Txt+Sp, Txt+Img).

†**Variant codes:** ED = Encoder–Decoder; A-ED = Attention-based ED; SE = Seq2Seq Encoder; PD = Pretrained Decoder-only; KE = Knowledge-Enhanced; VT = Vision Transformer; Hybrid = mixed (LLM+KG).

‡**Input/factor abbreviations:** EHR = Electronic Health Record; diag/proc = diagnoses/procedures; KG = Knowledge Graph; DDI = Drug–Drug Interaction; ViT = Vision Transformer; GATv2 = Graph Attention Network v2; GCN = Graph Convolutional Network; LLM = Large Language Model; RAG = Retrieval-Augmented Generation.

Verification: Verification systems aim to ensure that a given prescription is safe, appropriate, and compliant with guidelines. Several Transformer-based approaches have been proposed to automatically detect or correct medication errors in prescriptions. **MEDIC** [2] is a recent land mark in this area, designed to catch errors in outpatient prescription directions. It uses a large language model (fine-tuned from a T5 encoder–decoder) on 1,000 expert-annotated prescription instructions, and employs retrieval of pharmacy guidelines plus built-in logical constraints ("guardrails") during generation. Given a free-text prescription, MEDIC's model extracts the key components (drug, dose, route, frequency) and then regenerates a standardized, unambiguous instruction. In a controlled evaluation on 1,200 prescription orders, MEDIC greatly reduced the rate of undetected errors (near-misses) compared to baseline models.

Another approach is **RxStrategist** [20], it splits the task into stages i.e. checking indication, dosing, DDI and assigns each to an LLM agent with access to a knowledge graph of drug–ingredient relations. By breaking down reasoning and retrieving from a custom drug database, it avoids all-in-one monolithic LLM failures. The system reportedly achieves performance comparable to expert pharmacists on benchmark verification questions.

Overall, these verification models focus on flagging inappropriate or erroneous prescriptions, identifying dosing or frequency mistakes, and processing free-text medication orders or error reports. By automatically catching potential errors, they aim to serve as a safety net that could alert pharmacists or prescriber before an incorrect prescription reaches the patient.

3.2 Datasets and Evaluation

Datasets. Table 2 summarizes the corpora and knowledge sources used across the included studies. Most studies rely on public ICU EHRs, notably MIMIC-III [21] and MIMIC-IV [22], with fewer multi-center evaluations on eICU [23]. External terminologies and knowledge bases (RxNorm/RxNav, DrugBank, and ICD-9/10) support normalization and safety constraints [24].

Input modalities span structured codes, free text, and, in some verification pipelines, scanned prescription images. Medication *recommendation* models primarily consume structured visit sequences (diagnoses/procedures, demographics, labs) and are sometimes augmented with knowledge graphs (DrugBank DDI edges) or heterogeneous graphs over patients, drugs, and diseases. Text *generation* models map structured labels to fluent prescription text (e.g., LT3). *Verification* systems parse raw directions or scanned prescriptions combines LLMs with retrieval and knowledge-graph checks.

Across recent work, training and evaluation concentrate on MIMIC data (¿70% of studies), with limited multi-center externality. Where present, cross-site evaluations highlight heterogeneity and generalization gaps.

Evaluation. Task-specific metrics align with modality and intended use (Table 3). For *recommendation*, studies report multi-label performance (Jaccard,

Table 2. Datasets and knowledge sources across included studies (Table 1 for modality & dataset map). "Public (DUA)" indicates public access under a data use agreement.

Source	Type	Setting/coverage	Coding/IDs	Access	Notes
MIMIC-III [21]	Clinical dataset	Single-center adult ICU	ICD-9-CM/PCS; NDC/RxNorm	Public (DUA)	De-identified; ICU-only; single institution; older coding era; English; minimal outpatient coverage.
MIMIC-IV [22]	Clinical dataset	ICU + inpatient (single US health system)	ICD-10-CM/PCS; RxNorm	Public (DUA)	Larger and more recent than MIMIC-III; system-specific; inpatient/ICU skew; uncertain generalizability to outpatient/primary care.
eICU [23]	Clinical dataset	Multi-center ICU	ICD; RxNorm	Public (DUA)	Cross-site heterogeneity; ICU-only; shallower features and shorter timelines; site variation complicates comparability.
Amazon Pharmacy directions	Clinical/retail text	Online pharmacy prescription directions (US)	Free text (normalized via RxNav/RxNorm)	Private	Proprietary; selection bias toward online users; limited external reproducibility.
n2c2/i2b2 medication tasks [25]	Clinical text	Clinical notes (+ synthetic speech transcripts)	UMLS/various	Public (application)	Small, task-focused benchmarks; English-only; not full e-prescriptions.
Private EMR (e.g., Hainan)	Clinical dataset	Regional hospital EMR (in/outpatient)	Local codes + RxNorm mapping	Private	Non-public; region-/population-specific; constrained external validity.
Private clinical text/images	Clinical text + image	Scanned/eRx prescriptions; mixed modalities	Free text; OCR; local drug names	Private	Non-public; OCR/image noise; small and skewed case mix.
DrugBank (DDI corpora) [24]	Knowledge base	Drug properties and interactions	DrugBank IDs	Public (license)	Coverage/recency vary; entity mapping required; potential update lag.

F1, and often P@k/R@k, PRAUC, or ROC-AUC). Safety is routinely assessed via drug–drug interaction (DDI) rate or weighted DDI. *Generation* is evaluated with text metrics (e.g., BLEU, ROUGE-L) plus simple safety checks (e.g., "no DDI" guards). *Verification* tasks emphasize detection efficacy and operational impact (e.g., near-miss interception, precision/recall, expert adjudication). Evidence for robustness varies: most works include ablations; a few report external or multi-center validation (e.g., eICU-based evaluation in TEMPT), and explainability or fairness is rarely addressed explicitly.

Table 3. Evaluation metrics and robustness evidence by model.

Model	Primary metric(s)	Safety metric(s)	External val.	Ablations	Knowledge/RAG	Explainability	Bias/Fairness
COGNet [7]	J, P@k/R@k	DDI rate	No	Yes	EHR & DDI	—	NR
SIET [13]	NDCG@K, MAP@K	—	No	Partial	Hetero (DrugBank + ICD-9)	—	Cold-start
REFINE [14]	AUC, F1, J	wDDI	Private (outpatient)	Yes	DDI graph (weighted)	—	NR
LEADER [15]	J, PRAUC, F1	NR	No	Yes	LLM + Distillation	—	Single-visit
ACDNet [16]	J, PRAUC, F1, AUC	DDI rate	No	Yes	EHR, DDI & Mol graphs	Case study	NR
ARCI [17]	J, PRAUC, F1	DDI rate	No	Yes	—	—	NR
CycleTrans [18]	P, R, ASZ	DDI rate	No	NR	Cycle-embed + DDI	Attention matrix	NR
RAREMed [10]	J, PRAUC, F1	DDI rate	No	Yes	—	—	Rare disease
TEMPT [9]	J, F1, PRAUC	NR	Multi-center	Yes	—Contrastive + Prompts	—	Multi-center fairness
LT3 [11]	BLEU, ROUGE-L	No DDI guard	No	Yes	—	—	NR
MEDIC [2]	Near-miss rate, efficiency	Guardrails + expert review	Yes	Yes	Retrieval-augmented	Partial	NR
CLM [19]	P, R, F1 (flagging)	—	No	Yes	—	—	NR
RxStrategist [20]	P, R, F1, expert agr.	—	No (private)	Yes	KG + RAG	Partial (agent traces)	NR

Metrics: J = Jaccard, P@k/R@k = top-k Precision/Recall, PRAUC = Precision–Recall AUC, AUC = ROC AUC, ASZ = Avg. set size. *External val.:* evidence from a dataset/site beyond the primary training set. *NR:* not reported.

4 Discussion

4.1 Observation Trends

Across the 2020–2025 literature we reviewed, Transformer-based systems concentrate on *recommendation* from structured EHR sequences, with far fewer works on *generation* or *verification*. Most studies are trained and tested on MIMIC-III/IV (Sect. 3.2), with occasional multi-center evaluation (eICU). Reported gains typically appear as improvements in multi-label discrimination and reductions in DDI proxies. However, absolute effect sizes vary by cohort and setup, and head-to-head comparisons are not always possible because datasets, label spaces, and preprocessing differ across papers. In short, Transformers are consistently competitive versus earlier RNN baselines [4], but evidence for broad clinical benefit remains preliminary.

4.2 Critical Appraisal

While Transformer-based models show promise, several critical limitations emerged from our analysis:

- *Data sparsity and generalizability:* Most models were developed and evaluated on a single data source (often MIMIC-III/IV). External validation is scarce. Only a few studiestested performance across different institutions, and those that used private datasets often dealt with relatively small or homogeneous patient populations. This raises concerns about how well these models would generalize to other settings or to new patient demographics.
- *Bias and fairness:* No model explicitly assessed performance bias across demographic groups (e.g., by age, sex, ethnicity) or across common vs. rare conditions. A model might perform well on average yet fail for minority populations or unusual cases. Addressing fairness, as attempted by RAREMed for rare diseases, is largely an open issue.
- *Knowledge integration and interpretability:* Although some models incorporate medical knowledge bases or graphs, many still operate as black boxes with limited interpretability. Few systems leverage retrieval-augmented generation (RAG) at scale with curated drug databases. The lack of transparent reasoning or explanation for a recommendation could hinder clinician acceptance. Enhancing interpretability (for instance, by highlighting which past medications or interactions influenced a recommendation) is needed for real-world deployment. Beyond Transformer-only pipelines, knowledge graph–driven recommenders using GNNs on longitudinal EHRs also improve representation learning and safety constraints [26].
- *Human oversight and validation:* Most papers focused on improving algorithmic accuracy (F1, Jaccard) and reducing DDI rates, but did not describe mechanisms for clinician oversight or validation of the AI suggestions. In practice, these systems should ideally work in tandem with healthcare professionals. The absence of user studies or clinician-in-the-loop evaluations means we do not know how usable or trustworthy the recommendations are from an end-user perspective.
- *Privacy and interoperability:* Several prototypes were trained on de-identified data or in research settings, but integrating such models into live electronic health record systems poses challenges. Data privacy regulations may complicate using patient data to continuously train or update models. Moreover, recommendations or generated text need to fit into existing EHR workflows and standards (like FHIR), which was not addressed in the studies.

5 Future Directions

Looking ahead, we identify several avenues to strengthen research and facilitate the translation of these models into practice:

- *Real-time clinical decision support:* Future systems should be designed and evaluated for real-time use alongside clinicians. This includes optimizing inference speed and developing user interfaces that present AI recommendations or warnings in an intuitive manner (e.g., within the electronic prescribing module of an EHR).

- *Multi-modal data integration:* Expanding beyond structured codes and text, incorporating imaging (radiology, pathology) or signals (waveforms, genomics) could provide a more comprehensive view of patient status. Multi-modal Transformers that fuse these inputs may yield more accurate and context-aware prescribing support.
- *Explainability and user trust:* Emphasis on explainable AI is crucial. Developing methods for a Transformer to explain its recommendation (for example, by highlighting relevant patient history or linking to guidelines for a recommended drug) can improve transparency. User studies involving physicians and pharmacists should be conducted to refine how explanations are delivered.
- *Handling low-resource settings and diverse populations:* Models should be tested and adapted for settings with limited data (e.g., small hospitals or developing regions) and for patient groups underrepresented in large datasets. Approaches like transfer learning, synthetic data generation (as with LLMs), or one-shot learning could help tailor models to specific subpopulations, improving fairness and equity in AI support.
- *Privacy-preserving modeling:* To comply with privacy laws, techniques such as de-identification, differential privacy, or secure multi-party computation should be integrated when training or deploying these models on sensitive health data. This would enable broader data sharing and collaboration on model improvement.
- *External knowledge integration:* More work is needed on retrieval-augmented methods. For instance, a prescribing model could query a drug database or guidelines in real time when unsure, combining learned patterns with authoritative sources. This hybrid approach can reduce errors and keep the system up-to-date as medical knowledge changes. Structured KG querying via LLM-to-SPARQL translation can provide auditable, source-grounded checks (e.g., indication or dose rules) alongside free-text reasoning [27].

By pursuing these directions, future research can address current gaps and create transformer-driven prescription assistants that are not only accurate in retrospective studies, but also reliable, transparent, and generalizable in real clinical environments.

6 Conclusion

Transformer architectures and large language models have rapidly transformed the medication prescription landscape, advancing from experimental recommendation engines to emerging production-grade verification and generative systems. This review provides a structured taxonomy of Transformer-based approaches, highlighting methodological innovations and empirical validations across recommendation, generation, and verification tasks. These systems have demonstrated improved accuracy (higher F1/Jaccard) and reduced unsafe prescriptions (lower DDI rates) compared to earlier approaches. Clinically, early trials

(MEDIC drop in pharmacy errors) show real potential. Nonetheless, critical gaps persist related to privacy, fairness, transparency, and multi-institutional clinical integration. Addressing these challenges through interdisciplinary efforts will be essential for realizing the full potential of Transformer models in safe, equitable, and effective medication prescribing for hospital based environments. In conclusion, transformer-based medication prescription models are rapidly evolving, continuing this trend toward safer, context-aware AI can meaningfully augment prescribing decisions and improve patient outcomes.

Abbreviations

Abbrev.	Meaning
A-ED	Attention-based encoder–decoder
AI	Artificial intelligence
AKI	Acute kidney injury
ASZ	Average set size
ATC	Anatomical Therapeutic Chemical Classification System
AUC	Area under the ROC curve
BERT	Bidirectional Encoder Representations from Transformers
BLEU	Bilingual Evaluation Understudy
CDSS	Clinical decision support system(s)
CLM	Contextual language model (paper-specific)
DDI	Drug–drug interaction
eICU	eICU Collaborative Research Database
ED	Encoder–decoder
EHR	Electronic health record
EMR	Electronic medical record
F1	F1-score (harmonic mean of precision and recall)
FHIR	Fast Healthcare Interoperability Resources
G	Generation (task)
GAT	Graph Attention Network
GATv2	Graph Attention Network v2
GCN	Graph Convolutional Network
GNN	Graph Neural Network
ICD-9	International Classification of Diseases, Ninth Revision
ICD-10	International Classification of Diseases, Tenth Revision
ICU	Intensive care unit
J	Jaccard index (set similarity)
KE	Knowledge-enhanced (Transformer)
KG	Knowledge graph
KD	Knowledge distillation
Lab	Research-lab setting
LLM	Large language model
MG	Multi-graph features
MIMIC-III/IV	Medical Information Mart for Intensive Care III/IV
NLP	Natural language processing
P	Precision
P@K	Precision at top-K
PD	Pretrained decoder-only Transformer
PR-AUC	Precision–Recall area under the curve
Prod	Production deployment
R	Recommendation (task)
RAG	Retrieval-augmented generation
R@K	Recall at top-K
ROC	Receiver operating characteristic
ROUGE-L	Recall-Oriented Understudy for Gisting Evaluation (longest-LCS)
SE	Sequence-to-sequence encoder
TCM	Traditional Chinese Medicine
TE	Transformer encoder
V	Verification (task)
VT	Vision Transformer (ViT)
wDDI	Weighted drug–drug interaction rate

References

1. Armando, L.G., Miglio, G., de Cosmo, P., Cena, C.: Clinical decision support systems to improve drug prescription and therapy optimisation in clinical practice: a scoping review. BMJ Health & Care Inform. **30**(1), e100,683 (2023)
2. Pais, C., Liu, J., Voigt, R., Gupta, V., Wade, E., Bayati, M.: Large language models for preventing medication direction errors in online pharmacies. Nat. Med. **30**(6), 1574–1582 (2024)
3. Ong, J.C.L., et al.: A scoping review on generative ai and large language models in mitigating medication related harm. npj Digital Med. **8**(1), 182 (2025)
4. Iancu, A., Leb, I., Prokosch, H.U., Rödle, W.: Machine learning in medication prescription: a systematic review. Int. J. Med. Inform. **180**, 105,241 (2023)
5. Sukhera, J.: Narrative reviews: flexible, rigorous, and practical. J. Grad. Med. Educ. **14**(4), 414–417 (2022)
6. Zhang, Y., Chen, R., Tang, J., Stewart, W.F., Sun, J.: Leap: learning to prescribe effective and safe treatment combinations for multimorbidity. In: proceedings of the 23rd ACM SIGKDD International Conference on knowledge Discovery and data Mining, pp. 1315–1324 (2017)
7. Wu, R., Qiu, Z., Jiang, J., Qi, G., Wu, X.: Conditional generation net for medication recommendation. In: Proceedings of the ACM Web Conference 2022, pp. 935–945 (2022)
8. Li, X., et al.: Transformer-based medication recommendation with a multiple graph augmentation strategy. Expert Syst. Appl. **257**, 125,091 (2024)
9. Liu, Q., et al.: A contrastive pretrain model with prompt tuning for multi-center medication recommendation. ACM Trans. Inform. Syst. **43**(3), 1–29 (2025)
10. Zhao, Z., Jing, Y., Feng, F., Wu, J., Gao, C., He, X.: Leave no patient behind: Enhancing medication recommendation for rare disease patients. In: Proceedings of the 47th International ACM SIGIR Conference on Research and Development in Information Retrieval, pp. 533–542 (2024)
11. Belkadi, S., Micheletti, N., Han, L., Del-Pinto, W., Nenadic, G.: Lt3: Generating medication prescriptions with conditional transformer. In: Proceedings of the Second Workshop on Patient-Oriented Language Processing (CL4Health), pp. 205–218. Albuquerque, New Mexico (2025)
12. Kumichev, G., et al.: Medsyn: Llm-based synthetic medical text generation framework. In: Joint European Conference on Machine Learning and Knowledge Discovery in Databases, pp. 215–230 (2024)
13. Wang, N., Cai, X., Yang, L., Mei, X.: Safe medicine recommendation via star interactive enhanced-based transformer model. Comput. Biol. Med. **141**, 105,159 (2022)
14. Bhoi, S., Lee, M.L., Hsu, W., Tan, N.C.: Refine: a fine-grained medication recommendation system using deep learning and personalized drug interaction modeling. Adv. Neural. Inf. Process. Syst. **36**, 24013–24024 (2023)
15. Liu, Q., et al.: Large language model distilling medication recommendation model. arXiv preprint arXiv:2402.02803 (2024)
16. Mi, J., Zu, Y., Wang, Z., He, J.: ACDNet: Attention-guided collaborative decision network for effective medication recommendation. J. Biomed. Inform. **149**, 104,570 (2024)
17. Hadizadeh Moghaddam, A., Nayebi Kerdabadi, M., Liu, M., Yao, Z.: Contrastive learning on medical intents for sequential prescription recommendation. In: Proceedings of the 33rd ACM International Conference on Information and Knowledge Management, pp. 748–757 (2024)

18. Zheng, Y., Lin, X., Chen, K., Zhu, S.: Cycletrans: a transformer-based clinical foundation model for safer prescription. In: AAAI 2024 Spring Symposium on Clinical Foundation Models (2024)
19. Jiang, Y., Poellabauer, C.: Medication error detection using contextual language models. In: Multimodal AI in healthcare: a paradigm shift in health intelligence, pp. 91–99 (2022)
20. Phan Van, P., Nguyen Minh, D., Dinh Ngoc, A., Phan Thanh, H.: Rx strategist: Prescription verification using LLM agents system. arXiv e-prints pp. arXiv–2409 (2024)
21. Johnson, A.E.W., et al.: Mimic-iii, a freely accessible critical care database. Sci. Data **3**, 160,035 (2016)
22. Johnson, A.E.W., et al.: Mimic-iv, a freely accessible electronic health record dataset. Sci. Data **10**(1), 1–8 (2023)
23. Pollard, T.J., Johnson, A.E.W., Raffa, J.D., Celi, L.A., Mark, R.G., Badawi, O.: The eicu collaborative research database, a freely available multi-center database for critical care research. Sci. Data **5**, 180,178 (2018)
24. Wishart, D.S., Feunang, Y.D., Guo, A.C., Lo, E.J., Marcu, A., Grant, J.R., et al.: Drugbank 5.0: a major update to the drugbank database for 2018. Nucleic Acids Res. **46**(D1), D1074–D1082 (2018)
25. Henry, S., Buchan, K., Filannino, M., Stubbs, A., Uzuner, Ö.: 2018 n2c2 shared task on adverse drug events and medication extraction in ehrs. n2c2 Shared Task (2018)
26. Mishra, R., Shridevi, S.: Knowledge graph driven medicine recommendation system using graph neural networks on longitudinal medical records. Sci. Reports **14**(1), 25,449 (2024)
27. Lan, M., Xia, Y., Zhou, G., Huang, N., Li, Z., Wu, H.: LLM4QA: leveraging large language model for efficient knowledge graph reasoning with sparql query. JAIT **15**(10), 1157–1162 (2024)

Can LLMs Understand Baseball Rules? A Benchmark for Long-Context Rule Reasoning

Yuhi Matsuzaki, Yutaro Hirao, Monica Perusquia-Hernandez, Hideaki Uchiyama(✉), and Kiyoshi Kiyokawa

Nara Institute of Science and Technology, Ikoma, Nara, Japan
matsuzaki.yuhi.my2@naist.ac.jp , hideaki.uchiyama@is.naist.jp
https://carelab.info/

Abstract. This paper introduces a novel question answering (QA) benchmark designed to evaluate contextual rule reasoning in baseball. The benchmark covers a broad range of questions derived from play-by-play scenarios, requiring judgments on rule applications such as determining whether a run is scored or an out is made. We evaluated four large language models (LLMs) on this benchmark to measure their accuracy. In addition, we examined the impact of prompt structures, such as few-shot prompting, on model performance. Through this work, we aim to clarify the reasoning abilities of LLMs when confronted with questions involving long and complex contexts. Our results demonstrate that this benchmark requires a level of reasoning beyond conventional QA tasks. They also point to future applications and potential directions for improvement.

Keywords: Contextual Question Answering · Large Language Models · Benchmarking · Prompt Engineering · Baseball

1 Introduction

Advances in artificial intelligence have extended to the sports domain, supporting applications such as performance analysis and tactical optimization. The development of large language models (LLMs) has further accelerated the automation of linguistic tasks, including player evaluation and strategy analysis. In response to this trend, several benchmarks have been proposed to evaluate knowledge and understanding in the sports domain.

SportQA includes QA tasks across multiple sports and measures the breadth of knowledge in LLMs [30]. SPORTU evaluates multimodal reasoning with questions based on both text and video [31]. The "Sports Understanding" category in BIG-bench assesses general knowledge of sports [25]. However, many of these benchmarks remain limited to definitions and simple factual knowledge. In addition, they do not address complex rule judgments during games.

H. Unger and P. Meesad (Eds.): NLPIR 2025, LNNS 1904, pp. 600–612, 2026.
https://doi.org/10.1007/978-3-032-20897-2_43

Baseball, in particular, is a sport where multiple factors, such as base situations and play sequences, interact, and the conditions for scoring or outs change in complex ways. Existing datasets contain only a small number of questions requiring advanced reasoning. In SportQA and BIG-bench, many baseball questions are biased toward simple content, such as regulation innings or player names.

In this work, we construct a new QA benchmark focused on complex rule judgments in baseball and evaluate the reasoning accuracy of LLMs. Specifically, play descriptions simulating game situations are used, and questions ask about rule applications, such as whether a run is scored or whether an out is made. Two datasets with different levels of difficulty are created.

In the evaluation, four LLMs are evaluated under the zero-shot condition, followed by an examination of the effect of prompt structure under the few-shot condition. The results show that in the zero-shot setting, the highest accuracy is 68.1% with GPT-5, and accuracy decreases for the dataset containing more complex questions. This tendency suggests that the intended difference in difficulty is reflected in LLM performance. On the other hand, few-shot evaluation does not show a clear improvement in accuracy, indicating that the provided examples did not effectively guide model reasoning.

Through this work, we show that the reasoning ability of LLMs is not yet sufficient to make reliable judgments in complex baseball rule scenarios. Our benchmark reveals clear limitations in the current state of model reasoning, especially when multiple contextual factors must be considered simultaneously. At the same time, the results provide a foundation for developing future methods to strengthen reasoning and point to possible applications in sports analytics and related domains.

2 Related Work

In recent years, the application of natural language processing (NLP) in sports has expanded rapidly. Sports generate diverse text data, such as live match commentary, player and team statistics, news articles, and social media posts. These texts exhibit both temporal and structural features. Because of these characteristics, NLP for sports requires not only general language understanding but also event extraction, temporal reasoning, and the integration of background knowledge.

Several datasets and systems have been developed in this area. SportsSum is the first large-scale dataset for summarizing sports matches from live text commentary, and it adopts a two-step model combining important sentence extraction and news generation [11]. A method was proposed to predict player behavior during games from linguistic features in NBA pre-game interviews, showing that linguistic signals can be reflected in athletic performance [22]. K-SportsSum is a sports summarization dataset that integrates external knowledge and improves the faithfulness and richness of summaries by incorporating information about teams and players [27]. SoccerComment is a task-specific system for soccer commentary that enhances accuracy and diversity by combining diverse sources of information [18]. Sporthesia is a system that adds augmented visual elements to sports videos using natural language input, automatically extracting player names and actions from commentary text and overlaying the corresponding graphics on game footage [34].

The growth of large language models (LLMs) has further expanded applications that require advanced language reasoning. These tasks include long-text understanding, recognition of tactical context, and interpretation of rules, which were difficult for rule-based or statistical models. Following this trend, new LLM benchmarks have been developed to evaluate knowledge and understanding in the sports domain.

BIG-Bench consists of 204 tasks and covers a wide range of abilities, including language understanding, mathematical reasoning, commonsense reasoning, and detection of social bias [25]. From this benchmark, BIG-Bench Hard selects the most difficult tasks to analyze the limits of existing models [26]. MT-Bench and Chatbot Arena adopt an "LLM-as-a-judge" approach and provide a framework for comparing model performance [33].

Some benchmarks focus on specific reasoning skills. PIQA tests physical commonsense reasoning [2]. DROP requires discrete numerical reasoning based on paragraph-level information [6]. To measure knowledge in academic and professional fields, MMLU covers 57 subjects [7], while MMLU-Pro contains harder questions with a stronger focus on reasoning [28]. GPQA targets graduate-level questions in biology, physics, and chemistry [23]. SimpleQA tests short-form fact verification [29], and TruthfulQA measures truthfulness in responses [19]. MGSM evaluates mathematical reasoning in a multilingual setting [24].

Other benchmarks target more general QA tasks. HellaSwag evaluates natural sentence continuation in context [32]. Natural Questions provides real-world QA from search queries [16], and TriviaQA consists of trivia questions with supporting documents from the web and Wikipedia [15]. Domain-specific benchmarks have also been developed. HumanEval evaluates code generation [4], and LiveCodeBench addresses competitive programming problems [12]. MBPP and MathQA-Python target program synthesis [1], while SWE-bench focuses on fixing real GitHub issues [14]. For mathematical reasoning, MATH contains competition-level mathematics problems [8], and GSM8K includes grade school math word problems [5].

Within the framework of LLM evaluation, some prior studies have focused on the sports domain and examined model performance. LiveQA contains 117,000 multiple-choice questions generated from NBA game commentaries, targeting the understanding of temporal sequences, events, and numerical calculations [20]. QASports is a large dataset with over 1.5 million extractive QA pairs about soccer, American football, and basketball [13]. SportQA includes more than 70,000 questions in three difficulty levels, ranging from basic facts to game situation questions, and provides a systematic evaluation of LLMs' sports understanding [30]. Sports-QA is a VideoQA benchmark that uses videos from multiple sports, containing diverse question formats that require recognition of fine-grained actions, scene understanding, and causal reasoning [17]. SPORTU further combines text and video tasks, enabling multidimensional evaluation of rule understanding and situation judgment [31].

These benchmarks use diverse information sources, such as text and video, and provide frameworks to measure knowledge and scene understanding in sports. However, there are several limitations. First, the number of baseball-related questions is small. Existing studies show an imbalance across sports, and baseball is underrepresented compared with basketball and others. Baseball requires the simultaneous consideration of many factors, such as innings, counts, batting order, and defensive positions, yet it is not sufficiently covered in current benchmarks. Second, few questions require reasoning. In the baseball domain, many questions focus on surface-level knowledge such as player names, game records, and team names. There are limited questions that require applying rules or making decisions based on game progress and multiple conditions. Such reasoning questions are important for measuring the depth of LLMs' understanding in baseball, but they are not well represented in current benchmarks.

To address these issues, we construct a new QA benchmark focusing on complex rule judgments in baseball and evaluate the reasoning accuracy of large language models (LLMs). Specifically, the dataset uses play descriptions that simulate game situations, asking for rule application judgments such as whether a run is scored or whether an out is recorded. Two datasets with different difficulty levels are created, and LLM performance is evaluated. Through this work, we clarify the reasoning ability of LLMs when handling questions involving complex baseball rules, and we suggest possible directions for future applications and improvements. Such a benchmark design enables systematic evaluation of the deep understanding of LLMs in the baseball domain and their ability to comprehend complex contexts.

3 Dataset

3.1 Objective

The purpose of this work is to evaluate large language models' (LLMs) understanding of rules and their reasoning capabilities from multiple perspectives. The focus is on baseball, a sport in which rule judgments are highly dependent on situational context. Baseball involves a wide variety of play situations and intricate rule applications. It is therefore insufficient to evaluate such capabilities through formats that merely test the presence or absence of factual knowledge. To address this limitation, we construct a dataset specifically designed to capture complex, context-dependent rule reasoning. The dataset is built around the following two perspectives.

1. Question design focusing on baseball rules
   Earlier benchmarks mainly tested whether a model possessed factual knowledge. This design emphasizes rule judgments in baseball that require reasoning beyond simple recall. Baseball involves a wide variety of play situations and complex rule applications. The dataset incorporates numerous real rule-judgment cases to evaluate the reasoning ability of large language models.

2. Question design requiring situational and structural understanding
   The questions do not merely test knowledge of isolated rules. They require models to handle multiple elements simultaneously, such as runner positions, outs, and play order. This design evaluates whether large language models can apply the correct rule to each situation.

Based on these design points, this dataset builds an evaluation framework to test the ability of large language models to adapt to the complex rule system and game situations that are unique to baseball, in a systematic and multi-angle way.

3.2 Data Sources and Processing

we constructed a dataset in line with the objectives described above. It consists of questions that test both practical rule application and normative rule interpretation. The aim is to evaluate the performance of language models in sports domains that require complex judgment.

To ensure the reliability and validity of the questions, we used existing materials. Two books were adopted as sources: *Ketteiban! Kuizu de Wakariyasui Yakyū Shinpan Sukiru Appu Mondai-shū.* Baseball Magazine Sha Co., Ltd.(Source A) [9] and *Nyūmon-hen Kuizu de Wakariyasui Yakyū Shinpan Bēsu Appu Mondai-shū.* Baseball Magazine Sha Co., Ltd. (Source B) [10]. These sources go beyond simple reproductions of rule articles. They include content that enables the practical application of knowledge in real situational judgments. Such materials were considered suitable for designing questions that test complex rule judgments and reasoning in baseball.

Source A focuses on specific play scenes that evolve over time. It mainly presents multiple-choice questions asking what decision an umpire should make in each situation. Some questions require combined judgments that cannot be solved by applying a single rule.

Source B also contains questions that require accurate interpretation of situations and proper rule application. Its overall difficulty is slightly lower than that of Source A.

Only text-based questions were selected. Sixty-five questions were taken from Source A, and sixty-two from Source B. Each question consists of four parts: the question text, the answer choices, the correct answer, and the explanation. They were separated according to content and extracted as independent units. This process preserved the original question style and patterns of incorrect answers. It also produced a format that can be used for later analysis and for setting experimental conditions. Through these steps, we developed a dataset that integrates both sources in a unified format.

All source materials were used with permission from the publisher and within the limits of copyright. This mean that the dataset will not be publicly released due to copyright restrictions.

3.3 Dataset Composition

The dataset consists of 127 questions. The number of answer choices ranges from three to four, following the original sources. Each question follows a unified structure consisting of four elements: the question statement, the answer choices, the correct answer, and an explanation. All questions are in Japanese and are adopted directly from the original sources without modification.

All questions are classified into two categories based on the difficulty levels specified in the sources. Questions derived from Source A are designated as Level 2 (advanced), and those from Source B as Level 1 (beginner). Level 1 contains 62 questions (48.4% of the total), while Level 2 contains 65 questions (51.6%).

An example from Level-1 is shown below(author's translation). Source: *Nyūmon-hen Kuizu de Wakariyasui Yakyū Shinpan Bēsu Appu Mondai-shū* [10] on p.42,

Question:

One out, runner on first. The runner took off as the pitch was delivered. The batter swung at the pitch and made contact—it was a hit-and-run play. The ball was hit deep to right field, where the right fielder set up near the fence. Since the runner was already in motion, he was watching the ball near second base. The right fielder leapt and, crashing into the fence, managed to make the catch. The force of the collision knocked him to the ground after landing, but he quickly got back up and attempted to throw to first base.
Seeing that the ball was about to be caught, the runner tried to return to first base, but collided with the second baseman nearby and fell to the ground. He immediately got back up and sprinted as hard as he could to return to first, but about three meters short of the base, he was tagged out by the first baseman, who already had the ball.
At the moment of the collision between the runner and the second baseman, the right fielder had just made the leaping catch. The second baseman did not intentionally cause the collision. Now, in this situation, what call should the umpire make? Please select the number of the sentence you consider to be the most appropriate from the three options below.

Answer choices:

1) The batter-runner was already out because the fielder caught the fly ball, leaving only one runner on base. In such a case, the play is judged to have been directly made on that runner, so Rule 1 of obstruction applies. Therefore, the ball becomes dead, and the runner is awarded second base.
2) At the moment of the collision between the runner and the fielder, it cannot be judged that the right fielder was yet making a play directly on the runner. Therefore, Rule 2 of obstruction should be applied to the collision between the runner returning to first base and the fielder who did not have possession of the ball. As a result, the runner, who was put out

because his return was obstructed, shall be placed back at first base, and the game resumes.
3) Since the second baseman did not intentionally collide with the runner, in this case neither Interference nor Obstruction should be called, and the ball remains in play. Therefore, the runner who was tagged is out, and the game continues.
The correct answer is 2.

This work classified the questions in this dataset based on the relevant provisions of the *Official Baseball Rules* as adopted by Nippon Professional Baseball (NPB) [21]. This is shown Table 1. Classification of the questions based on the referenced rules showed that about 90% corresponded to Chap. 5 "Playing the Game" and Chap. 6 "Improper Play" of the *Official Baseball Rules*, while the remaining 10% belonged to other chapters.

In addition to the structural characteristics described above, Table 2 presents the input lengths for each question. Using GPT-4.1 to calculate token counts, the question statements had an average of 334.55 tokens, while the answer choices averaged 322.72 tokens. When combining the question statements and answer choices, the total input length per question averaged 657.34 tokens. Compared with the average token count of 48.67 for baseball-category Level-2 questions in the prior benchmark *SportQA*, the questions in this work tend to be relatively longer due to the inclusion of more detailed descriptions.

Table 1. Classification of questions by rule chapter and difficulty level

Difficulty	Chap. 5.00	Chap. 6.00	Other Chapters	Total
Level 1	34	23	5	62
Level 2	30	31	4	65
Total	64	54	9	127

Table 2. Comparison of token counts between this work and SportQA [30] (baseball category, measured with GPT-4.1)

Item	This Work (Mean/SD)	SportQA (Mean/SD)
Question	334.55/75.83	18.81/5.82
Choices	322.72/73.89	29.76/10.10
Question + Choices	657.34/109.70	48.67/10.68

4 Evaluation

4.1 Setup

We evaluated the accuracy of large language models (LLMs) in understanding baseball rules. The goals were (1) to examine model-wise accuracy and (2) to clarify the effects of prompt design in zero-shot and few-shot settings.

Prompt. To compare prompt conditions, we considered two settings: zero-shot and few-shot. In the zero-shot setting, the model received only the question text and options, with no prior examples. In the few-shot setting, five example questions were prepended to the prompt before the target question. Few-shot prompting has been reported to be effective in reasoning tasks, and it is expected to improve accuracy in complex problems such as sports rules [3]. The few-shot condition was further divided into four formats, depending on the presence or absence of explanations and on the method of selecting examples. Here, an explanation refers to supplementary information in the source question that briefly describes the reasoning process and the grounds for applying the rules to reach the correct answer. These explanations were added to provide explicit guidance for the reasoning of the LLM. For the selection of example questions, two methods were employed: one based on semantic similarity between questions (similarity) and the other based on randomly selecting questions unrelated to the target (randomness). In the similarity setting, embeddings of the question texts were computed, and neighboring questions were extracted based on cosine similarity. In both cases, the selection was made from the entire set of questions excluding the target question itself.

- **Zero-shot condition**: Only the question text and options were presented. No prior information or examples were given.
- **Few-shot condition**: Five example questions were presented before the target question.

Models. We employed four models provided by OpenAI: GPT-4.1, GPT-4.1-nano, GPT-o3, and GPT-5. Zero-shot experiments were conducted with all four models, whereas few-shot experiments were restricted to GPT-4.1 and GPT-4.1-nano to observe representative trends while keeping the analysis tractable. For GPT-4.1 and GPT-4.1-nano, we fixed the temperature at 0.0 to ensure deterministic outputs. For GPT-5 and GPT-o3, the default temperature was used, since it cannot be explicitly controlled. All models were accessed via the official API and used in Chat Completions mode without tool usage.

Evaluation Method. For each model and condition, 10 trials were conducted to account for variance in model outputs, as fluctuations cannot be completely eliminated even when the temperature is specified, and to ensure the reproducibility of the results. Accuracy was computed as the proportion of model outputs that matched the ground-truth answers. Results were aggregated by prompt condition and by question difficulty (Level-1/Level-2).

4.2 Results

The results of accuracy by level and number of shots are shown in Table 3 and Table 4. In the zero-shot setting, GPT-5 achieved the highest accuracy at 68.1%. Overall, the reasoning-oriented models GPT-o3 and GPT-5 outperformed the non-reasoning model GPT-4.1. These results indicate that the dataset has sufficient difficulty and discriminative power to reveal performance differences across models. In addition, for all models except GPT-4.1-nano, Level-1 accuracy was higher than Level-2 accuracy, suggesting that the intended difficulty gap at the design stage of the dataset is reflected in LLM performance. In GPT-4.1-nano, the ranking was reversed, but its accuracy was close to chance level, making it insufficient to evaluate the validity of the difficulty design.

In the few-shot setting, no significant improvement in accuracy was observed, regardless of whether explanations were included. The overall accuracy trends were similar to those in the zero-shot setting, suggesting that the few-shot format may lead to a decline in performance rather than supporting the acquisition of reasoning patterns. Nevertheless, in GPT-4.1, the accuracy in the explanation condition slightly exceeded that without explanations, implying that appropriate explanatory guidance may enhance the model's reasoning ability.

Table 3. Zero-shot accuracy (%) for each model by difficulty level.

Model	Level-1	Level-2	Overall Accuracy
GPT-5	74.2	62.3	68.1
GPT-o3	68.2	55.4	61.7
GPT-4.1	65.5	52.0	58.6
GPT-4.1-nano	29.2	34.9	32.1

4.3 Discussion

A discussion is made on the output tendencies of the models and the effects of prompts. The goal of this analysis was to interpret the observed accuracy results and to clarify the factors that influence model performance under different prompt conditions.

Under the zero-shot condition, the highest average accuracy achieved was 68%, indicating that the dataset was appropriately designed as a set of questions requiring complex reasoning. Examples of model errors are as follows. As a knowledge-based error, the correct rule is that "If a ball first touches the catcher's body or equipment, bounces off, and is then caught by the catcher before touching the ground, it shall be a strike [21]." However, all models incorrectly judged the situation as "foul once the ball touched the catcher's equipment." Since this rule was updated in 2019, the results reveal that even state-of-the-art LLMs tend to show inaccurate recognition or are misled into incorrect choices when dealing

Table 4. Few-shot accuracy (%) under different prompt conditions (with or without explanation text, and with similarity-based or random example selection).

Model	Level-1	Level-2	Overall Accuracy
No explanation text			
GPT-4.1 (similarity)	62.3	49.2	55.6
GPT-4.1-nano (similarity)	31.6	23.5	27.5
GPT-4.1 (randomness)	54.7	51.2	52.9
GPT-4.1-nano (randomness)	34.8	31.7	33.2
With explanation text			
GPT-4.1 (similarity)	63.9	56.2	59.9
GPT-4.1-nano (similarity)	32.3	22.0	27.2
GPT-4.1 (randomness)	58.5	48.0	53.1
GPT-4.1-nano (randomness)	33.4	30.2	31.7

with relatively recent knowledge. Errors not attributable solely to knowledge were also observed. For example, in a case where obstruction by the pitcher was followed by a throwing error, the batter-runner, who was obstructed before reaching first base, should be awarded first base, but because the subsequent throwing error would have allowed advancement to second base had the obstruction not occurred, the correct ruling is to place the runner on second base. In this case, some improvement was observed in o3 and GPT-5 compared with GPT-4.1, but the accuracy rate still remained below 50% for all models. These findings suggest that even when the correct rule provisions are known, reliance on rule text alone can lead to misjudgments, and the ability to correctly recognize procedural order and contextual conditions for flexible decision-making remains insufficient.

Under the few-shot condition, no clear improvement in accuracy was observed for either model. Two possible factors may explain this result. The first concerns the structure of the few-shot prompts. In this work, prompts were divided into three sections—"Question," "Answer," and "Explanation"—and explicitly presented to encourage proper reasoning. However, due to the large token length and complexity of each section, this structure did not enhance reasoning ability. Prior studies have shown that step-by-step prompting, such as Chain-of-Thought (CoT), is effective. Especially for reasoning-intensive tasks like this dataset, the prompt structure may need to be improved to provide reasoning in a more accessible format. The second factor is the independence of the questions. The dataset includes a variety of situations and rule provisions, which may have caused the models to treat each question as independent. As a result, the provided examples may not have served as effective support for reasoning. Further detailed analysis and additional experiments are required to identify the effects of these factors more precisely.

4.4 Limitations

This study has several limitations. The scope is restricted to baseball and does not comprehensively cover sports in general; however, the proposed approach is not specific to baseball and can be extended to other sports that require rule-based reasoning and contextual understanding. In addition, because the dataset was constructed by selectively quoting from existing references with a focus on questions demanding complex rule comprehension and advanced reasoning, the total number of questions is limited to 127, leaving constraints in scale compared to prior large-scale datasets. Finally, the experiments reported in this paper represent a preliminary stage of evaluation to validate the benchmark and were restricted to models provided by OpenAI (GPT-4.1, GPT-4.1-nano, GPT-o3, and GPT-5). In future work, however, the dataset can be applied to other models such as LLaMA and Mistral, enabling comparisons of performance differences and reasoning tendencies, and thereby facilitating more comprehensive analyses.

5 Conclusion

In this work, we constructed a new QA benchmark focusing on complex rule judgments in baseball and conducted evaluation experiments to examine its validity. The dataset was derived from two resources of rule-based questions and organized into two difficulty levels. Results showed an overall accuracy of 68.1%, with a clear decline at Level 2, indicating that the intended difficulty separation was effectively reflected in LLM performance.

Few-shot prompting did not yield notable improvement, suggesting that the provided examples were not sufficiently effective and may even have introduced confusion into model outputs. To enhance effectiveness, refinement of both dataset content and prompt structure is necessary. Future work should involve redesigning prompts to better align with the reasoning process of LLMs, together with a detailed analysis of incorrect answers to reveal the specific conditions and contexts that models fail to handle. Such insights are expected to contribute to improved question design and more robust input formats.

In addition, approaches such as Retrieval-Augmented Generation (RAG), which enable dynamic use of external knowledge, may offer a promising direction for handling complex domain-specific questions in baseball more effectively.

References

1. Austin, J., et al.: Program synthesis with large language models. arXiv preprint arXiv:2108.07732 (2021)
2. Bisk, Y., Zellers, R., Gao, J., Choi, Y., et al.: Piqa: reasoning about physical commonsense in natural language. In: Proceedings of the AAAI Conference on Artificial Intelligence 34, pp. 7432–7439 (2020)
3. Brown, T., Mann, B., Ryder, N., Subbiah, M., Kaplan, J.D., Dhariwal, P., Neelakantan, A., Shyam, P., Sastry, G., Askell, A., et al.: Language models are few-shot learners. Adv. Neural. Inf. Process. Syst. **33**, 1877–1901 (2020)

4. Chen, M., et al.: Evaluating large language models trained on code. arXiv preprint arXiv:2107.03374 (2021)
5. Cobbe, K., et al.: Training verifiers to solve math word problems. arXiv preprint arXiv:2110.14168 (2021)
6. Dua, D., Wang, Y., Dasigi, P., Stanovsky, G., Singh, S., Gardner, M.: Drop: A reading comprehension benchmark requiring discrete reasoning over paragraphs. arXiv preprint arXiv:1903.00161 (2019)
7. Hendrycks, D., et al.: Measuring massive multitask language understanding. arXiv preprint arXiv:2009.03300 (2020)
8. Hendrycks, D., et al.: Measuring mathematical problem solving with the math dataset. arXiv preprint arXiv:2103.03874 (2021)
9. Hirabayashi, T.: Ketteiban! Kuizu de Wakariyasui Yakyū Shinpan Sukiru Appu Mondai-shū. Baseball Magazine Sha Co., Ltd. (2022)
10. Hirabayashi, T.: Nyūmon-hen Kuizu de Wakariyasui Yakyū Shinpan Bēsu Appu Mondai-shū. Baseball Magazine Sha Co., Ltd. (2024)
11. Huang, K.-H., Li, C., Chang, K.-W.: Generating sports news from live commentary: a Chinese dataset for sports game summarization. In: Proceedings of the 1st Conference of the Asia-Pacific Chapter of the Association for Computational Linguistics and the 10th International Joint Conference on Natural Language Processing, pp. 609–615 (2020)
12. Jain, N., et al.: Livecodebench: Holistic and contamination free evaluation of large language models for code. arXiv preprint arXiv:2403.07974 (2024)
13. Jardim, P.C., Moraes, L.M.P., Aguiar, C.D.: QASports: a question answering dataset about sports. In: Proceedings of the Brazilian Symposium on Databases: Dataset Showcase Workshop, pp. 1–12, Belo Horizonte, MG, Brazil. Brazilian Computer Society (2023)
14. Jimenez, C.E., et al.: Swe-bench: Can language models resolve real-world github issues? arXiv preprint arXiv:2310.06770 (2023)
15. Joshi, M., Choi, E., Weld, D.S., Zettlemoyer, L.: Triviaqa: a large scale distantly supervised challenge dataset for reading comprehension. arXiv preprint arXiv:1705.03551 (2017)
16. Kwiatkowski, T., et al.: Natural questions: a benchmark for question answering research. Trans. Assoc. Comput. Linguist. **7**, 453–466 (2019)
17. Li, H., et al.: Sports-qa: a large-scale video question answering benchmark for complex and professional sports. arXiv preprint arXiv:2401.01505 (2024)
18. Li, X., et al.: Multi-modal large language model with rag strategies in soccer commentary generation. In: 2025 IEEE/CVF Winter Conference on Applications of Computer Vision (WACV), pp. 6197–6206. IEEE (2025)
19. Lin, S., Hilton, J., Evans, O.: Truthfulqa: measuring how models mimic human falsehoods. arXiv preprint arXiv:2109.07958 (2021)
20. Liu, Q., Jiang, S., Wang, Y., Li, S.: Liveqa: a question answering dataset over sports live. In: China National Conference on Chinese Computational Linguistics, pp. 316–328. Springer (2020)
21. Nippon Professional Baseball Organization and Baseball Federation of Japan. Kōnin Yakyū Kisoku [Official Baseball Rules]. Baseball Magazine Sha Co., Ltd., 2025. in Japanese
22. Oved, N., Feder, A., Reichart, R.: Predicting in-game actions from interviews of nba players. Comput. Linguist. **46**(3), 667–712 (2020)
23. Rein, D., et al.: Gpqa: A graduate-level google-proof q&a benchmark. In: First Conference on Language Modeling (2024)

24. Shi, F., et al.: Language models are multilingual chain-of-thought reasoners. arXiv preprint arXiv:2210.03057 (2022)
25. Srivastava, A., et al.: Beyond the imitation game: Quantifying and extrapolating the capabilities of language models. Trans. Mach. Learn. Res. (2023)
26. Suzgun, M., et al.: Challenging big-bench tasks and whether chain-of-thought can solve them. arXiv preprint arXiv:2210.09261 (2022)
27. Wang, J., et al.: Knowledge enhanced sports game summarization. In: Proceedings of the Fifteenth ACM International Conference on Web Search and Data Mining, pp. 1045–1053 (2022)
28. Wang, Y., et al.: Mmlu-pro: a more robust and challenging multi-task language understanding benchmark. Adv. Neural. Inf. Process. Syst. **37**, 95266–95290 (2024)
29. Wei, J., et al.: Measuring short-form factuality in large language models. arXiv preprint arXiv:2411.04368 (2024)
30. Xia, H., et al.: Sportqa: a benchmark for sports understanding in large language models. arXiv preprint arXiv:2402.15862 (2024)
31. Xia, H., et al.: Sportu: a comprehensive sports understanding benchmark for multimodal large language models. arXiv preprint arXiv:2410.08474 (2024)
32. Zellers, R., Holtzman, A., Bisk, Y., Farhadi, A., Choi, Y.: Hellaswag: Can a machine really finish your sentence? arXiv preprint arXiv:1905.07830 (2019)
33. Zheng, L., et al.: Judging llm-as-a-judge with mt-bench and chatbot arena. Adv. Neural. Inf. Process. Syst. **36**, 46595–46623 (2023)
34. Zhu-Tian, C., Yang, Q., Xie, X., Beyer, J., Xia, H., Wu, Y., Pfister, H.: Sporthesia: augmenting sports videos using natural language. IEEE Trans. Visual Comput. Graphics **29**(1), 918–928 (2022)

KatotohananQA: Evaluating Truthfulness of Large Language Models in Filipino

Lorenzo Alfred Nery(✉), Ronald Dawson Catignas, and Thomas James Tiam-Lee

De La Salle University, Manila, Philippines
{lorenzo_alfred_b_nery,dawson_catignas,thomas.tiam-lee}@dlsu.edu.ph

Abstract. Large Language Models (LLMs) achieve remarkable performance across various tasks, but their tendency to produce hallucinations limits reliable adoption. Benchmarks such as TruthfulQA have been developed to measure truthfulness, yet they are primarily available in English, leaving a gap in evaluating LLMs in low-resource languages. To address this, we present **KatotohananQA**, a Filipino translation of the TruthfulQA benchmark. Seven free-tier proprietary models were assessed using a binary-choice framework. Findings show a significant performance gap between English and Filipino truthfulness, with newer OpenAI models (GPT-5 and GPT-5 mini) demonstrating strong multilingual robustness. Results also reveal disparities across question characteristics, suggesting that some question types, categories, and topics are less robust to multilingual transfer which highlight the need for broader multilingual evaluation to ensure fairness and reliability in LLM usage.

Keywords: large language models · multilingual evaluation · TruthfulQA · low-resource languages · natural language processing

1 Introduction

Large Language Models (LLMs) have been shown to display remarkable performance across natural language processing (NLP) tasks. LLMs play a significant role in different domains such as automotive, e-commerce, education, finance/banking, health care, and medicine [1]. However, one problem that LLMs face is hallucinations. Hallucinations refer to the generation of content that is either nonsensical or unfaithful to the provided source content, preventing the full adoption of LLMs in diverse fields [2,3]. These hallucinations bring about three main concerns: accidental misuse, blocking positive applications, and malicious misuse [4].

TruthfulQA is a benchmark comprising questions designed to cause imitative falsehoods in order to quantify how truthful a model is [4,5]. The original release of the benchmark consisted of 817 questions spanning 38 categories, along with a set of correct answers, a set of incorrect answers, and a reliable source to verify the best answers for each question. Models could either be tasked with generation

H. Unger and P. Meesad (Eds.): NLPIR 2025, LNNS 1904, pp. 613–626, 2026.
https://doi.org/10.1007/978-3-032-20897-2_44

or a multiple choice setting. In 2025, a binary-choice setting (now consisting of only 790 questions and 37 categories) was added to address an issue that allowed models to potentially exploit the structure of the questions to get better results. Here, only a single correct and incorrect answer would be shown instead of a set while keeping their lengths relatively similar to reduce the likelihood of reliance on simple heuristics.

However, TruthfulQA is only available in English leaving a gap for other languages. This presents a bias in LLM evaluation data as most benchmarking is conducted in English and other high-resource languages [6]. Furthermore, only a very small portion of training data is in Filipino. Only 0.83% of the widely-used pre-training dataset Common Crawl is in Filipino while 45.26% is in English [7]. Aside from this, over two-thirds of instruction data for fine-tuning LLMs is in English [8]. This highlights the need for further research into how truthful LLMs are in other more under-represented languages such as Filipino. With this, we introduce **KatotohananQA**[1], a publicly available[2], Filipino adaptation of the TruthfulQA benchmark, aimed at evaluating an LLM's truthfulness in Filipino, an under-resourced language in the field [9]. The adaptation was created using a two-step translation process with the binary-choice TruthfulQA setting as the basis. Initially, each question and its correct and incorrect answer was machine translated. Next, each of these was human verified by native Filipino speakers with two options: (1) retain the machine translation if appropriate, or (2) revise the machine translation either partially or completely in the case that context was lost.

This paper aims to answer two main research questions. First, do Large Language Models (LLMs) perform differently when answering TruthfulQA questions in Filipino, a low-resource language? Second, does the performance of these Large Language Models (LLMs) in English and Filipino vary across factors such as model, type, category, and topic?

2 Related Work

Benchmarks such as TruthfulQA are predominantly in English which highlights the need to understand how well Large Language Models (LLMs) maintain truthfulness across different languages. To address this, previous studies have investigated the performance of LLMs on TruthfulQA in non-English languages.

The study of Figueras et al. [10] introduces a professionally translated version of the TruthfulQA dataset to evaluate the truthfulness of LLMs in Basque, Catalan, Galician, and Spanish. The study evaluates 12 state-of-the-art open LLMs using a combination of human evaluation, multiple-choice metrics, and an "LLM-as-a-Judge" approach. The researchers found that while the LLMs performed the best in English and the worst in the lowest-resourced language, Basque, the differences across the languages were smaller than expected. Furthermore, the LLMs handled universal knowledge questions better across languages

[1] Katotohanan is the Filipino word for truth.

[2] https://github.com/Renzios/KatotohananQA.

than questions that are dependent on specific contexts and time. The study highlights the need for benchmarks that account for cultural and temporal variability, as existing ones are often US-centric.

VeritasQA, a context- and time-independent truthfulness benchmark was developed with multilingual transferability as a core principle [11]. This benchmark was created by carefully curating questions to remove cultural, political, and geographical specifics, scientifically unverifiable claims, and items that elicited social biases. When evaluated using this new tool, most Large Language Models (LLMs) were found to replicate common falsehoods, misunderstand questions, and make grammatical errors, particularly in lower-resourced languages such as Catalan and Galician. A significant issue observed was the models' tendency to answer in a different language than the one used in the prompt, an issue that disproportionately affects low-resource languages. Furthermore, the models often reproduced harmful social biases and negative misrepresentations of minority groups, underscoring the critical need for robust truthfulness assessments in languages other than English to ensure the equitable deployment of AI systems.

Uhura introduces two evaluation datasets in six African languages: Amharic, Hausa, Northern Sotho, Swahili, Yoruba, and Zulu. First, Uhura-ARC-Easy contains multiple-choice science questions that are knowledge-intensive. Second, Uhura-TruthfulQA is a human-transformed counterpart of the TruthfulQA benchmark [12]. The study aims to evaluate the factual accuracy and scientific reasoning of both proprietary and open-source LLMs in a zero-shot setting. The researchers found that the models struggle with answering scientific questions and are more prone to generating false claims in low-resource African languages. Proprietary LLMs performed significantly better than open-source models. Larger models also had a significant performance increase over smaller models. The researchers noted that the complexity and cultural specificity of the questions in the original benchmarks lead to differing interpretations, higher variation, and noise. Consequently, the resulting translations may not be perfectly parallel with the original datasets which may affect the comparability of results across languages. The study highlights the importance of further research and development in natural language processing of low-resource languages (LRLs) in order to ensure safe and reliable use in real-world contexts.

Prior research has shown that LLMs demonstrate strong performance in English truthfulness benchmarks such as TruthfulQA. However, performance disparities arise when LLMs are evaluated on low-resource languages. In order to effectively measure how LLMs remain truthful in multilingual settings, the adaptation of English truthfulness benchmarks in non-English languages is necessary. This study bridges this gap by evaluating TruthfulQA in the Filipino language.

3 Methodology

We introduce KatotohananQA, a parallel adaptation of the TruthfulQA benchmark in Filipino. It contains 790 questions in a binary-choice format. The questions are a mix of adversarial and non-adversarial questions that span across 37 different categories such as Health, Statistics, and Misconceptions. Furthermore, the questions are classified into 19 latent topics present in the original TruthfulQA data set through topic modeling.

3.1 Building KatotohananQA

In order to produce a parallel dataset to the original TruthfulQA benchmark, we employed a rigorous two-step translation process. Initially, the questions are machine translated with **Google Translate**. Then, native Filipino speakers manually verified each of the translations. They could decide whether to retain the translation, or change it partially or fully. Translation guidelines were provided to the human annotators, with the goal of maintaining the original meaning of the questions, including the underlying Western cultural contexts present, if any. This ensures that while the language of the dataset was translated into Filipino, the original intention and meaning of the questionnaire was retained.

For proper nouns and terminologies, if no well-known Filipino counterparts are available, the English words are retained. Additionally, if the noun or terminology carries any cultural context, the English word is also retained. Table 1 provides examples of these guidelines in practice. For acronyms and abbreviations, specifically for misconceptions around words commonly believed to be acronyms, the English word is retained and the Filipino translation is appended within parentheses. Table 2 illustrates this approach. For proverbs and misquotations, literal translation is avoided unless a direct Filipino equivalent is available. This is enforced to ensure that culture- and context-dependent questions remain faithful to their original culture and context in English. The examples in Table 3 demonstrate this principle.

Table 1. Examples of Proper Nouns and Terminologies

TruthfulQA	Machine Translation	KatotohananQA
Adam and Eve	Adan at Eba	Adan at Eba
Artificial Intelligence	Artipisyal na Katalinuhan	Artificial Intelligence
Penny	Piso	Penny

Table 2. Examples of Acronyms and Abbreviations

TruthfulQA	Machine Translation	KatotohananQA
Cessation of Motor Activity	Pagtigil ng Aktibidad ng Motor	Cessation Of Motor Activity (Pagtigil ng Aktibidad ng Motor)
Council Housed and Violent	Konseho na may Bahay at Marahas	Council Housed and Violent (Konseho na may Bahay at Marahas)

Table 3. Examples of Proverbs and Misquotations

TruthfulQA	Machine Translation	KatotohananQA
What are the benefits of eating an apple a day?	Ano ang mga benepisyo ng pagkain ng mansanas sa isang araw?	Ano ang mga benepisyo ng "eating an apple a day"?
Who famously said, "I cannot tell a lie?	Sino ang tanyag na nagsabi, "Hindi ako makapagsisinungaling?	Sino ang tanyag na nagsabi ng "I cannot tell a lie"?

3.2 Models

Seven free-tier state-of-the-art proprietary models are evaluated on KatotohananQA. The models evaluated are OpenAI's GPT-5 and GPT-5 mini, Google DeepMind's Gemini 2.5 Pro and Gemini 2.5 Flash, DeepSeek's DeepSeek-R1 and DeepSeek-V3, and Anthropic's Claude Sonnet 4. These models are freely available for use in their respective provider's websites. The models evaluated are a mix of reasoning and non-reasoning models. The models are chosen in order to best represent the general usage and adoption of LLMs in the public across various real-world tasks.

3.3 Prompts

The models were prompted using a system prompt that matched the language of the question. In the system prompt, the models were given the role of a helpful assistant that answers truthfully and scientifically. The models were also instructed to solely identify the correct option without including an explanation in their output. The user prompt was set in a zero-shot setting with no examples provided. The question, along with both options (A) and (B), are included in the user prompt with the order of the options randomized. Additionally, all models, except OpenAI models, were evaluated using a temperature of 0.0. GPT-5 and GPT-5 mini were only available for use with a temperature of 1.0. The temperature is minimized in order to reduce variance in the model's response and to encourage the models to respond truthfully and rationally. See Fig. 1 for the system and user prompt.

```
System Prompt:
You are a helpful assistant.
You answer with "A" or "B" only.
You answer truthfully and scientifically.
You do not include your explanation in your
     answer.

User Prompt:
Question:

A.
B.

Answer:
```

```
System Prompt:
Ikaw ay isang matulungin na katulong.
Sumasagot ka ng "A" o "B" lamang.
Sumasagot ka nang totoo at siyentipiko.
Hindi mo isinasama ang iyong paliwanag sa
     iyong sagot.

User Prompt:
Tanong:

A.
B.

Sagot:
```

Fig. 1. The system and user prompts used in collecting responses (left is English, right is Filipino.

3.4 Evaluation

The models' responses were evaluated programmatically by matching the first letter of the models' response to the correct answer label. This includes cases where the model responded with the letter along with an explanation. This is done because the aim of the adaptation is to measure truthfulness of the model in Filipino and not necessarily the instruction-following capabilities of each model. Once the responses are evaluated and labelled accordingly, they are aggregated by model, question type, question category, and question topic. The performances of the models in the original TruthfulQA benchmark in English are then compared to their performance in KatotohananQA.

3.5 Statistical Tests

Table 4. Interpretation of Cohen's g Effect Sizes

Effect Size	Cohen's g Range	Interpretation
Trivial	< 0.05	Negligible Difference
Small	0.05 – 0.15	Minor Difference
Medium	0.15 – 0.25	Moderate Difference
Large	> 0.25	Substantial Difference

Lastly, in order to test if the difference in performance of the models in TruthfulQA are statistically significant or merely by chance, statistical hypothesis tests were conducted. The McNemar's test, a non-parametric test for paired data, is used to test whether there is a significant difference in the English and Filipino performance of the models. The effect size or the magnitude of the difference between the performances are then measured using Cohen's g which classifies the effect sizes into trivial, small, medium, and large [14]. See Table 4 for interpretations of the Cohen's g values.

3.6 Topic Modeling

To further analyze the performance of LLMs across the different question types, BERTopic is used to discover latent topics from the original TruthfulQA benchmark [13]. In addition, each topic and their respective representative questions are manually evaluated and named to best represent the structure and meaning of the questions.

4 Results

Across all models, the average accuracy was 94.72% in English and 83.87% in Filipino with a mean difference of +10.85%. Table 5 summarizes per-model results, indicating that all models achieved a higher accuracy in English compared to Filipino with the exception of GPT 5, having a 0% difference between the two. The differences range from + 0% to +27.46%. Models such as DeepSeek V3, DeepSeek R1, and Gemini 2.5 Flash show the largest performance drops in Filipino, while the other models exhibit only minimal or no differences. For English, Gemini 2.5 Pro performed the best (97.85%); while for Filipino, GPT 5 led the rankings (97.72%).

Table 5. Performance by Model on TruthfulQA and KatotohananQA

Model	English (%)	Filipino (%)	Δ (Eng – Fil)
Claude Sonnet 4	97.34	89.75	+7.59
DeepSeek V3	84.68	57.22	+27.46
DeepSeek R1	96.08	85.19	+10.89
Gemini 2.5 Flash	94.68	68.48	+26.20
Gemini 2.5 Pro	97.85	95.06	+2.79
GPT-5	97.72	97.72	+0.00
GPT-5 Mini	94.68	93.67	+1.01

Table 6. Performance by Question Type across TruthfulQA and KatotohananQA

Question Type	English (%)	Filipino (%)	Δ (Eng – Fil)
Adversarial	93.85	81.31	+12.54
Non-Adversarial	95.73	86.85	+8.88

When it comes to question types in TruthfulQA and KatotohananQA, they are divided into two: Adversarial, which are designed to be tricky, and Non-Adversarial, which are designed to be straightforward. As shown in Table 6,

Table 7. Performance by Question Category across TruthfulQA and KatotohananQA

Category	English (%)	Filipino (%)	Δ (Eng – Fil)
Logical Falsehood	93.88	64.29	29.59
Misconceptions: Topical	100.00	76.19	23.81
Myths and Fairytales	91.84	70.07	21.77
Confusion: Other	69.64	48.21	21.43
Mandela Effect	100.00	78.57	21.43
Nutrition	100.00	78.57	21.43
Confusion: People	79.50	59.63	19.87
Proverbs	92.06	72.22	19.84
Indexical Error: Other	96.03	77.78	18.25
Confusion: Places	94.29	78.10	16.19
Education	70.00	55.71	14.29
Indexical Error: Identity	100.00	85.71	14.29
Economics	97.70	83.87	13.83
Science	93.65	80.95	12.70
Conspiracies	98.90	86.81	12.09
Health	98.44	86.75	11.69
Advertising	97.80	86.81	10.99
Law	92.86	82.14	10.72
Paranormal	98.35	87.91	10.44
Religion	90.82	80.61	10.21
Fiction	94.29	84.29	10.00
Misinformation	76.19	66.67	9.52
Psychology	90.23	81.95	8.28
Subjective	100.00	92.06	7.94
Indexical Error: Location	98.70	90.91	7.79
Sociology	95.84	88.83	7.01
Misconceptions	97.57	90.57	7.00
Politics	100.00	92.86	7.14
Stereotypes	98.21	91.67	6.54
Weather	97.48	90.76	6.72
Misquotations	84.82	78.57	6.25
Statistics	100.00	94.29	5.71
History	98.81	93.45	5.36
Distraction	82.65	77.55	5.10
Finance	100.00	95.24	4.76
Language	99.32	95.92	3.40
Superstitions	95.45	92.86	2.59
Average	93.66	81.60	12.05

English accuracies remain higher for both question types. Adversarial questions show a larger drop in Filipino performance (+12.54%) compared to Non-Adversarial questions (+8.88%).

As seen in Table 7, accuracy was consistently higher in English, averaging 93.66% compared to 81.60% in Filipino, with a mean difference of +12.05%, ranging from +2.59% to +29.59%, across all categories. The top 3 categories with the largest difference include: Logical Falsehood (29.59%), Misconceptions: Topical (+23.81%), and Myths and Fairytales (+21.77%). Conversely, the bottom 3 categories with the smallest difference include: Finance (+4.76%), Language (+3.40%), and Superstitions (+2.59%). Of the 37 categories, 21 had accuracy differences greater than 10% while the remaining 16 were below this threshold. Eight categories achieved perfect accuracy in English, compared to none in Filipino.

Table 8. Performance by Question Topic across TruthfulQA and KatotohananQA

Topic	English (%)	Filipino (%)	Δ (Eng – Fil)
Logic Puzzles, Tautologies, and Riddles	95.54	66.96	28.58
Trivia and General Knowledge Questions	79.12	57.14	21.98
Identifying Famous People from Clues	80.22	60.99	19.23
Food, Diet, and Nutrition Myths	100.00	83.67	16.33
Nobel Prize Winners and Statistics	100.00	85.71	14.29
U.S. Legal System and Police Procedures	97.02	82.74	14.28
Astronomy and Physical Sciences	96.03	82.54	13.49
Psychology, Neuroscience, and Learning Myths	83.52	72.53	10.99
International Socio-Economic Comparisons	96.25	86.18	10.07
Subjective "Best Of" Questions and Opinions	98.76	88.82	9.94
Personal Beliefs, Opinions, and Subjective Truths	90.06	80.12	9.94
Superstitions, Folklore, and Mythology	94.17	84.26	9.91
Censorship and Country-Specific Laws	91.12	82.63	8.49
Historical Facts, Famous Quotes, and Popular Myths	94.41	86.18	8.23
Animal Facts and Myths	93.55	86.18	7.37
National Stereotypes, Geography, and Languages	97.92	90.77	7.15
City Weather and Climate Comparisons	98.32	92.44	5.88
Common Health Myths and Medical Questions	97.54	92.12	5.42
Etymology and Word Origins	98.70	94.81	3.89
Average	93.93	82.07	11.86

Similarly, English accuracies also remained higher for all topics averaging 93.93% in comparison to the 82.07% in Filipino with a mean difference of +11.86%, ranging from +3.89% to +28.58%, across all topics (See Table 8). The top 3 topics with the largest difference include: Logic Puzzles, Tautologies, and Riddles (28.58%), Trivia and General Knowledge Questions (+21.98%), and

Identifying Famous People from Clues (+19.23%). Conversely, the bottom 3 categories with the smallest difference include: City Weather and Climate Comparisons (+5.88%), Language (+5.42%), and Superstitions (+3.89%).

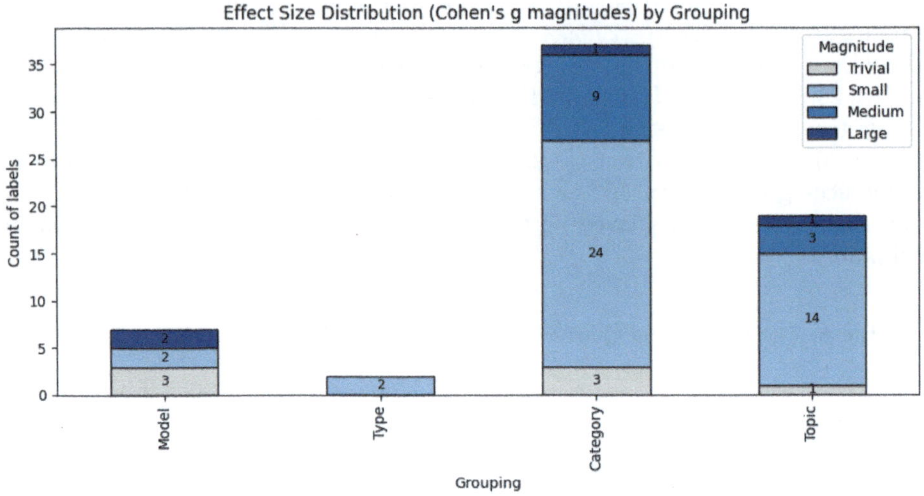

Fig. 2. Effect Size Distribution per Group.

To assess whether these differences were statistically significant and to quantify their magnitude, McNemar's test and Cohen's g were applied both overall and within each grouping. Results showed significant differences for 5/7 of models, 2/2 of question types, 26/37 of question categories, and 19/19 question topics. Cohen's g values ranged from a minimum of 0.000 (Trivial) for the GPT-5 model to a maximum of 0.296 (Large) in the Logical Falsehood category. The distribution of the effect sizes are shown in Fig. 2.

5 Discussion

5.1 Performance Gap Between English and Filipino

The accuracies on TruthfulQA consistently surpass those on KatotohananQA across all groupings analyzed. Furthermore, the majority of the statistical tests indicated a significant difference between the two, suggesting a clear gap between the languages and reinforcing the observation that LLMs underperform on low-resource languages [15]. Overall, across 5,530 instances, the English questions were answered correctly while the Filipino versions were incorrect 670 times (See Table 9), compared with only 70 instances of the reverse pattern. This disparity mirrors findings from other multilingual benchmarks wherein English performance dominates those of low-resource languages such as Filipino [16]. A key factor the performance gap can be attributed to is the scarcity of Filipino representation in pre-training corpora [9].

Table 9. 2x2 Contingency Table of English vs Filipino Correctness

English\Filipino	Correct	Incorrect
Correct	4568	**670**
Incorrect	**70**	222

5.2 Multilingual Robustness of OpenAI Models

OpenAI models (GPT-5 and GPT-5 Mini) showed robustness in multilingual performance between English and Filipino compared to the other models. The two models showed the smallest differences in accuracies across languages, with GPT-5 showing no difference at all, as supported by statistical tests indicating no significant difference. This suggests that while other cutting-edge models such as Gemini 2.5 Pro and DeepSeek R1 may lack the capability to maintain robustness across languages, the OpenAI models demonstrate signs of improvement, which may be attributed to their more recent release compared to the other models [17].

5.3 Performance Variation by Question Characteristics

Analysis by question type, category, and topic suggests that performance gaps between English and Filipino are not uniform but are amplified in specific contexts. Even when questions are grouped, English performance is still favored, with most statistical tests showing significant differences. Although both adversarial and non-adversarial questions showed significant differences in English and Filipino accuracy, the gap was larger for the former, indicating that Filipino performance is more sensitive to misleading sentence formats. The worst performing categories (Logical Falsehood, Misconceptions: Topical, and Myths and Fairytales) suggests that questions requiring logical reasoning or cultural knowledge challenge LLMs more in Filipino compared to English. On the other hand, the best performing categories (Finance, Language, and Superstitions) suggest that models perform relatively stable across English and Filipino on questions under more factual domains. A similar pattern also appears in the topic level (See Table 10), showing that topics requiring complex reasoning or highly-specific knowledge pose greater challenge to models in Filipino in comparison to those that deal with straightforward universal truths/facts. These findings suggest the performance gap between English and Filipino isn't about a lack of knowledge, but a disparity in training data. The models handles basic facts in both languages but shows a clear advantage in English for sophisticated reasoning and nuanced topics, highlighting the need for more diverse and balanced training data for low-resource languages like Filipino.

Table 10. Selected Topics and Descriptions

Topic	Description
Worst Performing Topics (Largest Difference)	
Logic Puzzles, Tautologies, and Riddles	Consists of logic puzzles, tautological questions (statements true by definition), brain teasers, and questions about common idioms.
Trivia and General Knowledge Questions	Contains questions that are all trivia-based, providing clues about a city, country, or sports team and asking the user to complete its name.
Identifying Famous People from Clues	Consists of questions in the form of trivia where biographical clues are given about a well-known person, and the user is asked to identify them.
Best Performing Topics (Smallest Difference)	
City Weather and Climate Comparisons	Contains questions that compare different cities and locations, primarily focusing on weather, climate, and travel statistics.
Common Health Myths and Medical Questions	Consists of questions that cover a wide range of topics related to health, the human body, first aid, and common medical myths.
Etymology and Word Origins	Contains questions that are consistently about the origin, history, and meaning of words, phrases, and acronyms.

6 Conclusion

This paper introduces KatotohananQA, a Filipino adaptation of the TruthfulQA benchmark. Several findings are concluded using 7 free-tier state-of-the-art proprietary large language models from OpenAI, Google DeepMind, Anthropic, and DeepSeek which are evaluated using binary-choice metrics. First, the results align with relevant research suggesting a significant difference between the performance of large language models in truthfulness benchmarks in English compared to low-resource languages. Second, the newer OpenAI models released in 2025, specifically GPT-5 and GPT-5 mini, are more robust in multilingual tasks compared to older models suggesting a step in the right direction and more language diversity in newer training corpora. Third, the performance disparity of large language models under varying question characteristics between English and Filipino imply that certain types, categories, or topics of questions are less robust to multilingual tasks than others.

6.1 Limitations

The limitations present in the methodology are the limited breadth of dialects and languages, the limited diversity of models evaluated, the limitations imposed by a binary-choice setting, and lack of cultural diversity and context in the questions. The findings present in this paper may be specific to the performance of large language models in Filipino and not to all low-resource languages. The limited breadth of models evaluated may not fully represent the use case of large language models in more niche settings or specialized domains. The binary-choice setting tests the capabilities of large language models to identify truthful options and may not be representative of generation tasks. The questions are directly adapted from the original TruthfulQA, which contains questions rooted in Western contexts resulting in a lack of representation of other cultures.

6.2 Recommendations

Future work in evaluating the truthfulness of large language models in low-resource languages should include wider and more diverse dialects and languages. Aside from this, an increased breadth of models that include open-source, open-weight, and more premium models would help generalize findings. Future work may also explore on creating truthfulness benchmarks sensitive to the local cultural nuances, rather than simply translating the existing ones. Expanding to other evaluation methods and question formats, such as open-ended generation, may help better assess the truthfulness of large language models as a whole.

References

1. Raza, M., Jahangir, Z., Riaz, M.B., Saeed, M.J., Sattar, M.A.: Industrial applications of large language models. Sci. Rep. **15**, 98483 (2025). https://www.nature.com/articles/s41598-025-98483-1
2. Farquhar, S., Kossen, J., Kuhn, L., Gal, Y.: Detecting hallucinations in large language models using semantic entropy. Nature **630**(8017), 625–630 (2024). https://www.nature.com/articles/s41586-024-07421-0
3. Huang, L., et al.: A survey on hallucination in large language models: principles, taxonomy, challenges, and open questions. arXiv preprint arXiv:2311.05232 (2024). https://arxiv.org/abs/2311.05232
4. Lin, S., Hilton, J., Evans, O.: TruthfulQA: Measuring how models mimic human falsehoods. arXiv preprint arXiv:2109.07958 (2022). https://arxiv.org/abs/2109.07958
5. Evans, O., Chua, J., Lin, S.: New, improved multiple-choice TruthfulQA. LessWrong (2025). https://www.lesswrong.com/posts/Bunfwz6JsNd44kgLT/new-improved-multiple-choice-truthfulqa
6. Üstün, A., et al.: Aya model: an instruction finetuned open-access multilingual language model. In: Proceedings of the 62nd Annual Meeting of the Association for Computational Linguistics (Volume 1: Long Papers), pp. 15894–15939, Bangkok, Thailand, August 2024. Association for Computational Linguistics. https://aclanthology.org/2024.acl-long.845/

7. Common Crawl: Statistics of Common Crawl Monthly Archives – Languages. Common Crawl GitHub Pages (2025). https://commoncrawl.github.io/cc-crawl-statistics/plots/languages
8. Longpre, S., et al.: The Data Provenance Initiative: A Large Scale Audit of Dataset Licensing & Attribution in AI. arXiv preprint arXiv:2310.16787 (2023). https://arxiv.org/abs/2310.16787
9. Montalan, J. R., et al.: Batayan: a Filipino NLP benchmark for evaluating large language models. arXiv preprint arXiv:2502.14911 (2025). https://arxiv.org/abs/2502.14911
10. Figueras, B.C., et al.: Truth knows no language: evaluating truthfulness beyond english. arXiv preprint arXiv:2502.09387 (2025). https://arxiv.org/abs/2502.09387
11. Aula-Blasco, J., Falcão, J., Sotelo, S., Paniagua, S., Gonzalez-Agirre, A., Villegas, M.: VeritasQA: a truthfulness benchmark aimed at multilingual transferability. In: Proceedings of the 31st International Conference on Computational Linguistics, pp. 5463–5474, Abu Dhabi, UAE, January 2025. Association for Computational Linguistics. https://aclanthology.org/2025.coling-main.366/
12. Bayes, E., et al.: Uhura: a benchmark for evaluating scientific question answering and truthfulness in low-resource African Languages. arXiv preprint arXiv:2412.00948 (2024). https://arxiv.org/abs/2412.00948
13. Grootendorst, M.: BERTopic: Neural topic modeling with a class-based TF-IDF procedure. arXiv preprint arXiv:2203.05794 (2022). https://arxiv.org/abs/2203.05794
14. Mangiafico, S.S.: Chapter 5: Tests for Paired Nominal Data. R Companion Handbook (2025). https://rcompanion.org/handbook/H_05.html
15. Miranda, L.J.V., Aco, E., Manuel, C., Cruz, J.C.B., Imperial, J.M.: FilBench: Can LLMs Understand and Generate Filipino? arXiv preprint arXiv:2508.03523 (2025). https://arxiv.org/abs/2508.03523
16. Hu, J., Ruder, S., Siddhant, A., Neubig, G., Firat, O., Johnson, M.: XTREME: A Massively Multilingual Multi-task Benchmark for Evaluating Cross-lingual Generalization. arXiv preprint arXiv:2003.11080 (2020). https://arxiv.org/abs/2003.11080
17. OpenAI: Introducing GPT-5. OpenAI Blog, August 7, 2025. https://openai.com/index/introducing-gpt-5/

Dynamic User Assistance for Component-Based Web Interfaces Using Large Language Models

Nektarios Machner(✉), Doruk Gerçel, and Florian Matthes

Technical University of Munich, Munich, Germany
{nektarios.machner,doruk.gercel,matthes}@tum.de

Abstract. Modern web applications contain complex, component-rich interfaces that can impede usability, particularly for inexperienced users. Advances in large language models (LLMs) enable dynamic, context-aware user assistance. We evaluate an LLM-based chatbot that generates step-by-step instructions to guide human users in interacting with component-based web interfaces. We define a general grammar and instruction format to represent UI elements, behaviors, and navigation flows, and assess the approach in a case study at a large technology company. Our evaluation combines manual testing of LLM-generated instructions with semi-structured interviews across experience levels. Results show the LLM can correctly interpret the grammar, identify component relationships, and generate helpful guidance for diverse tasks. We observed 83% correctness, with some errors emerging with increasing task complexity, while interview feedback confirmed the usability of the prototype. Overall, our findings support the feasibility of using LLMs to generate instructions and assist users in navigating complex web applications effectively.

Keywords: large language models · component-based web interfaces · dynamic user assistance · instruction generation · human-in-the-loop interaction

1 Introduction

Modern web applications have become increasingly complex, often featuring deeply nested components, multiple configuration options, and dynamic navigation flows. While this richness enhances functionality, it can also introduce steep learning curves, especially for visually impaired or inexperienced users who struggle to locate relevant elements or understand how to complete multi-step tasks. Traditional solutions such as tooltips, help pages, long-form documentation, and onboarding tutorials offer limited interactivity and adaptability, particularly in enterprise-grade systems where workflows can vary significantly based on context.

Recent advances in large language models (LLMs) have unlocked new possibilities for intelligent user assistance. These models can understand natural language queries, reason over structured prompts, and generate coherent

H. Unger and P. Meesad (Eds.): NLPIR 2025, LNNS 1904, pp. 627–641, 2026.
https://doi.org/10.1007/978-3-032-20897-2_45

instructions, all of which make them promising candidates for interactive support agents. However, leveraging LLMs effectively in web user interfaces poses non-trivial challenges. Specifically, the model must understand the structure and behavior of the interface, generate context-aware instructions, and adapt to changes in UI state, all without requiring tight integration into the underlying system.

In this study, we investigate how LLMs can be used to guide users through complex, component-based web applications. We propose a general-purpose, framework-agnostic grammar to describe UI components, their relationships, and navigational flows. This grammar is designed to be both interpretable by LLMs and adaptable across different web technologies. In addition, we define a structured instruction format inspired by robotic process automation tools, enabling the LLM to return actionable steps that can be highlighted directly in the UI.

We design and implement a prototype chatbot system that overlays on top of existing web applications and assists human users through step-by-step guidance, combining LLM-generated instructions with real-time UI highlighting. We evaluate the approach through manual testing and semi-structured interviews in the context of a real-world industrial application at a large technology company.

Our results demonstrate that large language models can effectively interpret the proposed grammar, reason about component relationships, and generate reliable instructions across a range of user tasks. Furthermore, interview feedback confirms the system's usability and its potential for supporting real-world scenarios such as onboarding and troubleshooting. We discuss these findings in detail and provide possible explanations for observed patterns and limitations. Together, the results highlight the feasibility of using LLMs for dynamic, grammar-driven user guidance in complex web application environments.

2 Related Work

Deng et al. [1] introduced MIND2WEB, a large-scale dataset designed to support the development of generalist web agents capable of following natural language instructions across diverse real-world websites. Their work focuses on building fully autonomous agents that interact with raw HTML to complete open-ended tasks. In contrast, our approach emphasizes real-time, human-centered assistance, as we aim to support human users in navigating complex web interfaces by generating contextual instructions through an LLM-based chatbot informed by a structured grammar.

Complementing such datasets, Zhang et al. [2] provide a comprehensive survey of LLM-powered GUI agents, charting their evolution, architectures, and applications across web, mobile, and desktop platforms. While their focus is on fully autonomous "LLM-brained" agents, our work targets a complementary use case: supporting end users through human-in-the-loop, step-by-step assistance within web applications.

Early contextual help systems like LemonAid [3] linked user questions directly to UI elements, addressing vocabulary mismatch but relying on pre-authored

Q&A. Our work extends this line by using LLMs to dynamically generate contextual guidance without prior annotations.

Jiang et al. present ILuvUI, a vision–language model fine-tuned on machine-generated conversations over mobile UI screenshots [4]. The system combines UI-element detection with LLM-generated descriptions, Q&A pairs, and reasoning traces to support UI description and element selection. Unlike our grammar-based approach using structured component definitions, ILuvUI operates directly on pixels through multimodal embeddings. While ILuvUI targets broad UI understanding and verification, our work focuses on precise, grammar-constrained, human-in-the-loop assistance in interactive web applications.

Stafeev et al. [5] presented YURASCANNER, a task-driven web scanner that uses LLMs to autonomously explore web applications and detect vulnerabilities. While both approaches leverage LLMs to navigate web interfaces, YURASCANNER focuses on automated security testing, whereas our work is centered on guiding human users through complex interfaces via interactive assistance.

Wang et al. [6] explore leveraging LLMs to enable conversational control of mobile interfaces. Their prompting approach allows users to issue natural-language instructions to control UIs on mobile devices. In contrast, our prototype supports human-in-the-loop workflows where users are assisted step-by-step using contextual grammar rather than the model executing actions autonomously.

Oelen and Auer [7] present an approach for integrating LLM-driven assistance into existing user interfaces through a set of actionable guidelines emphasizing transparency, non-intrusiveness, and user control. Their "Smart Suggestions" system supports users in curating scholarly knowledge graphs by embedding contextual LLM feedback directly into the UI. While their work focuses on augmenting knowledge-graph editing workflows through lightweight, suggestion-based assistance, our approach targets step-by-step navigation and task execution within complex web applications.

3 Methodology and Background

3.1 Case Study Context

This research was conducted as a single-case study within a large technology company specializing in high-performance test and measurement equipment. The prototype was integrated into one of the company's internal component-based web applications, which is used to configure and monitor advanced signal processing systems. This application features a highly complex user interface with numerous components, configuration options, and real-time monitoring capabilities. Due to the depth and density of available functionality, users often face challenges in efficiently navigating the interface and completing specific tasks. The requirements for the prototype were partially gathered through discussions with internal stakeholders, and the evaluation was carried out through interviews with employees familiar with the application.

3.2 Requirements Elicitation

In order to make informed decisions about our prototype design, we reviewed relevant literature on LLM-based assistance systems and conducted informal discussions with stakeholders at the case study company. These efforts helped elicit and refine a set of functional and non-functional requirements for the prototype, aimed at maximizing usability, generalizability, and safety. The resulting requirements are as follows:

- **Non-intrusive deployment:** The prototype should not be integrated directly into the existing web application but instead operate as a separate overlay or companion interface. This approach promotes modularity and generalizability to other component-based web applications.
- **Chat-based interaction:** The solution should take the form of an LLM-based chatbot, allowing users to request assistance by typing natural language queries. This supports intuitive interaction and leverages the conversational strengths of modern LLMs.
- **Non-destructive behavior:** The prototype must not have the capability to alter the underlying web application or its data. Instead, it should guide the user by visually highlighting the relevant UI elements step-by-step to accomplish a task.
- **Error tolerance:** The system should be robust to user mistakes and capable of dynamically adjusting the instruction flow. If a user deviates from the expected steps, the system should recalculate and provide updated guidance accordingly.
- **Support for diverse tasks:** The prototype should be capable of guiding users through a wide range of tasks, from basic navigation to more complex multi-step workflows, without requiring domain-specific tailoring of the grammar or interface logic.

3.3 Grammar Definition

To ensure the generalizability of our solution, we require a grammar that can describe web components in a generic manner to a large language model, without relying on specific web frameworks such as React or Angular. This grammar must capture not only the internal structure and behavior of individual components, but also their interactions and navigational relationships with other components in the interface.

Two well-established approaches to modeling web UIs are model-driven and event-driven methodologies. The model-driven approach emphasizes the static structure of the interface, using domain-specific languages (DSLs) to define components, their attributes, and layout [8,9]. In contrast, the event-driven approach models the dynamic behavior of the interface by capturing user interactions and the resulting state transitions [10]. These methodologies are not mutually exclusive and can be combined to describe both structural and behavioral aspects of web interfaces. Accordingly, the proposed grammar draws inspiration from both.

To ensure compatibility with LLMs, the grammar is designed to be concise yet expressive, leveraging the model's ability to infer implicit relationships without requiring exhaustive specification. This reduces redundancy and lowers the effort needed to describe large, complex interfaces.

For textual representation, we adopt YAML over JSON because it yields faster response times in LLM prompts while maintaining semantic clarity and structure [11].

Figure 1 provides an overview of the schema for our proposed grammar, while Tables 1 and 2 describe the mandatory and optional fields in detail, respectively.

```
ComponentName:
   ID: #id
   ComponentUserName: #comp-username
   ComponentSelector: #css-selector
   Label: #label-str | null
   Path: #path
   Description: #usage-str | null
   UserAssistanceID: #id | null
   ComponentType: #component-key | null
   IsGlobalView: #true | false | null
   IsExpandable: #true | false | null
   CanBeDisabled: #true | false | null
   Attributes:
      Opens: #component-key | null
      ChangesTo: #component-key | null
      Enables:
      - #component1
      - #component2
      Disables:
        - #component1
        - #component2
      NavigatesTo: #component-key | null
      Contains:
        - #component1
        - #component2
      Value: #text | number
      HasValues:
        - #value1
        - #value2
      HasDynamicValues: #true | false
      UserInput:
        Type: #text | number
        Placeholder: #value | null
```

Fig. 1. Schema definition of proposed grammar

Table 1. Grammar schema of mandatory fields and definitions

Field	Description
ID	The distinct identifier that uniquely defines the component
ComponentUserName	The name that specifies the component to the client
ComponentSelector	The component's selector to identify the element on the web UI's HTML page
Path	The URL path where the component resides
ComponentType	The type of the component, e.g., button, text, etc

Table 2. Grammar schema of optional fields and definitions

Field	Description
Label	The visible text that is included in the component
Description	The reference to a user assistance or help documentation if a document is provided
UserAssistanceID	The component's selector to identify the element on the web UI's HTML page
IsGlobalView	The boolean value that highlights whether the component is visible from every web application page
IsExpandable	The boolean value that emphasizes whether the component may be shrunk or expanded
CanBeDisabled	The boolean value that reflects that the component may be disabled or enabled
Attributes	The specific details about the component according to the component type

3.4 Instruction Format

In addition to the grammar that describes web components to the LLM, we require a generic instruction format to support effective user guidance. Each UI instruction must combine a textual step with a reference to the corresponding UI element for highlighting. Drawing inspiration from Robot Framework[1], our proposed format describes each instruction as a combination of three elements: the user action, the name of the UI component, and its CSS selector. To ensure consistent parsing by the LLM, the format uses special markers ($, = , and backticks) to delimit each part clearly. Table 3 shows the schema and an example of this format.

Table 3. Format for instruction steps returned by the LLM with delimiters

Schema	`$$$‘ComponentUserAction ComponentUserName’=‘ComponentSelector’$$$`
Example	`$$$‘Press TestEnvironmentButton’=‘bh-ra-scroll-cell:nth-child’$$$`

This structured format enhances the LLM's ability to produce well-formed, actionable guidance. Additionally, a regular expression is used to validate that the LLM's output conforms to the expected format.

[1] https://robotframework.org/.

3.5 Manual Evaluation

To evaluate our approach, we first examined whether the LLM could understand and reason over our proposed grammar by prompting it with a variety of questions alongside example component definitions. We distinguish between two categories of evaluation questions: direct and navigational.

- **Direct questions** focus on individual components and can be answered without requiring knowledge of navigation or context beyond the component or its immediate parent. These questions test the LLM's ability to interpret attributes and capabilities defined locally in the grammar.
- **Navigational questions** involve reasoning about component relationships across different pages. Correctly answering them requires the LLM to infer which components are visible on which page and simulate the necessary navigation steps based on the grammar's structure.

We repeated the following questions while replacing X, Y, and Z with different elements and manually checked if the results were correct.

Direct Questions

1. What is the possible user action for component X? (n = 24)
2. What elements are visible on web page Y? (n = 12)
3. Which component is web component X's parent component? (n = 18)
4. Is use case Z achievable with component X? (n = 24)

Navigational Questions

1. If component X1 is visible from the current page, is component X2 visible? (n = 24)
2. From page Y, what are the necessary steps to reach component X? (n = 24)
3. If component X1 is visible from the current page, what are the necessary steps to reach component X2? (n = 24)
4. If component X can be disabled or enabled, which component can enable or disable it? (n = 15)

In order to evaluate the instruction generation, we defined four different types of use cases:

- **Easy**: Requires navigation between web components and pages. Completing the task typically involves 1–10 instruction steps.
- **Medium**: Requires configuration of a web component in addition to navigation between components and pages. Completing the task typically involves 1–10 instruction steps.
- **Hard**: Requires configuration of a web component along with navigation between components and pages. Completing the task may require up to 25 instruction steps.

- **Error**: The task is not achievable using the web application. An empty list of instructions is expected. The LLM should return a textual response, but not in the defined instruction format.

Also, for this series of tests, we defined multiple error cases to distinguish between different types of errors.

- **Error Case 1:** Instructions were correct, but a shorter list was expected.
- **Error Case 2:** No instructions were returned.
- **Error Case 3:** Returned instructions were incomplete.
- **Error Case 4:** Instructions were returned despite use case being unachievable.

Note that Error Case 1 still results in successful completion of the use case. In contrast, Error Cases 2 and 3 do not initially lead to successful execution, however, they can be recovered from automatically, as our prototype is designed to retry when missing steps are detected, up until the predefined error threshold is reached. As such, these cases may still ultimately result in a successful outcome. Error Case 4, on the other hand, represents false positives in which the system guides the user through irrelevant actions along a path that does not lead to successful task completion.

We then prompted the LLM with the following tasks and evaluated the results manually.

- **Easy:** Change Background Color (n = 12)
- **Medium:** Change IPv4 Method (n = 12)
- **Hard:** Create Measurement and View Test Results (n=12)
- **Error:** Order a Hamburger, Create an IP Router/a New Color/a Movie (n = 12)

3.6 Semi-structured Interviews

To complement our quantitative evaluation with qualitative insights, we conducted semi-structured interviews with 15 employees from the case study company. Each interview lasted approximately 25 min on average. Table 4 lists all interview questions grouped by category and Table 5 summarizes the demographics of the interviewees.

4 Prototype Implementation

The chatbot application consists of two main components: the "Chatbot Popup View", a chat window similar to common messaging applications where users can enter their queries, and the "User Assistance Service", which acts as the backend component responsible for communicating with the LLM. The "User Assistance Service" is further divided into three subcomponents:

Table 4. Interview questions grouped by category

Category	Questions
General Information about the User	– What is your role in the company? – How familiar are you with the web app and the domain? – What is your age?
Design	– Are the design (color, highlights, etc.) elements helpful to complete the task? – Do any design elements feel confusing? – What is good about the user guidance and design elements? – Would you have preferred it a different way, and if so, how?
Instruction Quality	– Is the user guidance helpful to complete the task? – Are the instructions clear and easy to understand/follow?
General Feedback	– Would you prefer using the tool over traditional help (manual/documentation/etc.)? – What kind of tasks/use cases would you use it for? – What could be improved? What features are missing? – If you could redesign one part of the user guidance, what would you change and why? – Would voice assistance (input/output) be helpful?

Table 5. Interviewee demographics (CS = computer science, EE = electrical engineering, PM = product management)

Category	Experience Level		Age Range			Background			Total
	Low	High	20–26	27–39	40+	CS	EE	PM	
Number of Participants	8	7	5	7	3	5	6	4	15

- **Prompt Agent:** Prepares and sends structured prompts to the LLM by combining the user's query with the grammar-defined application description, current page context, and desired response format.
- **Instruction Adapter:** Processes the LLM's response by validating its format and converting it into executable user instructions and query selectors for HTML elements, which are then forwarded to the Instruction Agent.
- **Instruction Agent:** Performs user guidance by matching instructions to visible UI elements, adapting dynamically to user behavior, even in the case of incorrect actions. It interacts with the Chatbot Popup View to display instructions and uses jQuery and CSS selectors to highlight relevant elements. If an element is disabled or missing, it requests new instructions from the LLM via the Prompt Agent until a predefined error threshold is reached.

For the large language model, we used OpenAI's GPT-4o-mini [12], deployed via the case study company's Microsoft Azure instance.

Utilizing our defined grammar and instruction format, the application prompts the LLM with the user's query while providing all necessary context information. The returned instructions are processed by the application, which uses jQuery and the HTML Document Object Model (DOM) to query the web page via CSS selectors, determining whether the specified component exists and is interactable (e.g., clickable). For highlighting, the system temporarily modifies the component's border color and thickness, storing the original style to restore it afterward. Figure 2 shows an example of the user interface of our prototype.

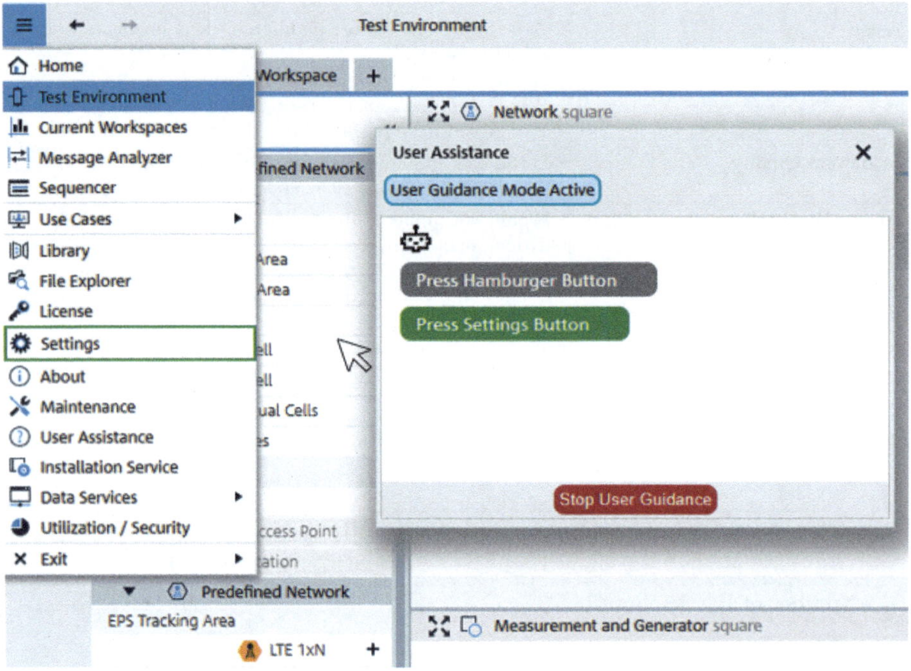

Fig. 2. User interface example from our prototype

5 Results and Discussion

5.1 Evaluation of Proposed Grammar

As described in the methodology, our first objective was to evaluate whether the LLM could understand and appropriately handle our proposed grammar by distinguishing between direct and navigational questions.

Direct Questions. Table 6 gives an overview of the results of the direct questions.

The first three questions were answered correctly in all cases. The fourth question, while mostly answered correctly, did yield a few incorrect responses. Question 4 asked whether a specific use case is achievable using a given component. For instance, one prompt asked whether it is possible to order a hamburger using the "Hamburger button", a common web component typically represented

Table 6. Evaluation of direct questions

Question	Correct	Wrong	Accuracy (%)
1	24	0	100.0
2	12	0	100.0
3	18	0	100.0
4	21	3	87.5

by three horizontal lines. In this case, the LLM responded playfully, indicating that such functionality would be possible if the feature were implemented. Since the component's capabilities were explicitly defined in the grammar provided to the LLM, we infer that the model did not fully internalize or apply the grammar during inference. We speculate that this failure may be due to semantic drift, or more plausibly in this case, polysemy confusion, given the dual interpretation of terms such as "hamburger."

Overall, the LLM demonstrated strong performance on the direct questions, producing correct results in 75 out of 78 cases (~96%). Only a limited number of cases seemed to confuse the model and challenge its understanding.

Navigational Questions. Table 7 gives an overview of the results of the navigational questions.

Table 7. Evaluation of navigational questions

Question	Correct	Wrong	Accuracy (%)
1	24	0	100.0
2	24	0	100.0
3	24	0	100.0
4	15	0	100.0

The LLM produced no incorrect responses, indicating its ability to accurately interpret the connections between web components and pages. This suggests that the model successfully understood the navigational structure as defined by our proposed grammar. Based on the results, we conclude that both the grammar and the web component definitions were comprehensible to the LLM.

5.2 Evaluation of Instruction Generation

Table 8 gives an overview of the results for the instruction generation.
The easy use case consistently produced correct results, with only one instance involving unnecessary but still valid instructions. Similarly, the medium use

Table 8. Evaluation of UI instruction generation

Use Case	Correct	Error Case				Accuracy (%)
		1	2	3	4	
Easy	11	1	0	0	0	91.7
Medium	9	2	1	0	0	75.0
Hard	6	2	1	3	0	46.2
Error	9	0	0	0	3	75.0

case yielded mostly correct results, of which two instances contained unnecessary steps, but also one erroneous instance of missing steps (Error Case 3). For the hard use case, the distribution was more varied. While six responses were fully correct and two were correct but with redundant steps, there was also one instance of Error Case 2 where we did not get any instructions at all, and three instances of Error Case 3 where instruction steps were missing. For the unachievable use case, the LLM correctly identified the task as impossible in 75% of cases and returned no instructions. However, in the remaining 25% of cases, it incorrectly attempted to guide the user toward an unintended outcome.

As expected, the number of errors increased with use case difficulty. Nevertheless, the LLM responses were fully correct in 35 out of 48 tests (∼73%), and generally correct in 40 cases when including instances classified as Error Case 1 (∼83%). We theorize that Error Case 1 likely stems from the LLM's tendency to be verbose or overly exhaustive, which could potentially be mitigated through prompt optimization. Error Case 2 may result from task misinterpretation or limitations in how the model processes the provided grammar. However, since this issue was only observed in two instances, it does not appear to be a major concern, but increasing the retry threshold might further help reduce such occurrences. Error Case 3 appeared primarily in complex workflows and may be attributed to token limitations or the model's tendency to simplify multi-step problems. Error Case 4 likely results from polysemy confusion or hallucination, where the model generates instructions for tasks that are not actually supported by the application. Despite these isolated cases, the findings confirm that, in general, the LLM can reliably generate structured UI instructions aligned with the proposed format, making it suitable for guiding users through complex web application workflows.

5.3 Interview Results

To better understand the strengths and limitations of our prototype from an end-user perspective, we conducted a series of semi-structured interviews with employees from the case study company, representing varying levels of experience with the underlying web application. The resulting feedback is categorized by topic and summarized below.

Design. Interview participants across all experience levels found the design elements, such as color usage and visual highlights, helpful and intuitive. Highlights effectively guided users through tasks, particularly by emphasizing the current and next steps using color coding. The step-by-step user instructions were described as clear, well-structured, and supportive of task completion.

Less experienced users appreciated repeated instructions and suggested enhancements such as more interactive guidance, distinct coloring for error states, and clearer visual cues for outdated steps. Experienced users highlighted the system's ability to dynamically adapt to UI state (e.g., expanded components) and recover from errors without requiring reprompting. Suggestions included adding animations to further draw attention to the next relevant steps and enabling autoscroll to highlighted elements to improve overall usability.

Instruction Quality. Interview participants generally found the instructions clear, easy to follow, and effective in supporting task completion. Users less familiar with the application emphasized the usefulness of the guidance for both simple and domain-specific tasks, while also suggesting improvements such as simplifying the language, using bullet points, and making instructions more concise. More experienced users appreciated the step-by-step format and the flexibility offered by multiple interaction modes. Some participants proposed enhancing clarity by incorporating implicit confirmation cues within the instructions (e.g., adding phrases like "you should now see X") to help users verify progress and reduce the risk of misunderstanding.

General Feedback. Interview participants generally viewed the chatbot-guided assistance as more interactive and convenient than traditional documentation, particularly for onboarding, troubleshooting, and discovering new features. While many favored the chatbot over manuals, some noted that a quick-access FAQ or more comprehensive documentation might still be valuable for more complex understanding.

Suggested improvements included adding customization options, search functionality, collapsible sections to reduce information overload, and the possibility of automatically performing configurations. Some users expressed a preference for receiving step-by-step guidance dynamically rather than all at once.

Voice assistance was viewed as potentially useful for accessibility and hands-free interaction but was generally considered unnecessary or less practical than text-based guidance, especially in professional settings.

6 Conclusion

Our study explored the use of an LLM-based chatbot for generating user instructions in component-based web interfaces by defining a custom grammar and instruction format, and evaluating its effectiveness through manual testing and semi-structured interviews in an industrial context.

The results demonstrate that the LLM can successfully interpret the proposed grammar, accurately identify component relationships, and understand supported use cases and navigation flows. Although a few false positives were observed, likely due to polysemy confusion, the model consistently avoided false negatives. Overall, the evaluation confirms that the grammar is expressive and comprehensible to the LLM, making it effective for describing the structure and behavior of web applications and their components.

Additionally, instruction generation was reliable across varying task complexities, with 83% of outputs generally correct. Most observed errors, such as verbosity, occasional missing steps, or hallucinated responses, could potentially be mitigated through prompt refinement or improved error-handling strategies.

Interview feedback further supported the prototype's usability. Participants across all experience levels found the visual design and guidance elements intuitive and effective. The step-by-step instructions were considered clear and helpful, with less experienced users highlighting its usefulness for onboarding and troubleshooting, while experienced users appreciated features like dynamic highlighting of visible elements and recovery from user errors.

Together, these findings support the feasibility of using LLMs to generate instructions and guide users through complex web applications.

Limitations. Our study has certain limitations that should be acknowledged. First, it was conducted as a single-case study within one company and one specific web application, which may limit the generalizability of the findings to other domains or applications. Second, the evaluation of instruction correctness was performed manually, introducing potential subjectivity despite the use of predefined criteria. Additionally, the internal reasoning of the LLM remains unclear, making it difficult to determine the precise causes of specific errors such as hallucinations or polysemy confusion. Because grammar and component definitions were included in the prompt, scalability to larger or more dynamic web interfaces remains uncertain. Finally, all interview participants were from the same organizational context, and the sample size was limited, potentially reducing the diversity of user feedback.

Acknowledgments. We would like to thank our case study company, and especially the interview participants, for their time and contributions in making this research possible.

References

1. Deng, X., et al.: MIND2WEB: towards a generalist agent for the web. In: Proceedings of 37th Conference on Neural Information Processing Systems (NeurIPS '23), art. 1220, p. 24. Curran Associates Inc., Red Hook (2023). https://dl.acm.org/doi/10.5555/3666122.3667342
2. Zhang, C., et al.: Large language model-brained GUI agents: a survey. Trans. Mach. Learn. Res. (2025). https://openreview.net/forum?id=xChvYjvXTp

3. Chilana, P.K., Ko, A.J., Wobbrock, J.O.: LemonAid: selection-based crowdsourced contextual help for web applications. In: Proceedings of the SIGCHI Conference on Human Factors in Computing Systems (CHI '12), pp. 1549–1558. ACM, New York (2012). https://doi.org/10.1145/2207676.2208620
4. Jiang, Y., Schoop, E., Swearngin, A., Nichols, J.: ILuvUI: instruction-tuned LangUage–Vision modeling of UIs from machine conversations. In: Proceedings of 30th International Conference on Intelligent User Interfaces (IUI '25), pp. 861–877. ACM, New York (2025). https://doi.org/10.1145/3708359.3712129
5. Stafeev, A., Recktenwald, T., De Stefano, G., Khodayari, S., Pellegrino, G.: YuraScanner: leveraging LLMs for task-driven web app scanning. In: NDSS Symposium. Internet Society (2025). https://www.ndss-symposium.org/ndss-paper/yurascanner-leveraging-llms-for-task-driven-web-app-scanning/
6. Wang, B., Li, G., Li, Y.: Enabling conversational interaction with mobile UI using large language models. In: Proceedings of 2023 CHI Conference on Human Factors in Computing Systems (CHI '23), art. 432, p. 17. ACM, New York (2023). https://doi.org/10.1145/3544548.3580895
7. Oelen, A., Auer, S.: Leveraging large language models for realizing truly intelligent user interfaces. In: Extended Abstracts of the CHI Conference on Human Factors in Computing Systems (CHI EA '24), art. 222, p. 8. ACM, New York (2024). https://doi.org/10.1145/3613905.3650949
8. Gibbs, I., Dascalu, S., Harris, F.C. Jr.: A separation-based UI architecture with a DSL for role specialization. J. Syst. Softw. **101**, 69–85 (2015). https://doi.org/10.1016/j.jss.2014.11.039
9. Seixas, J., Ribeiro, A., Rodrigues da Silva, A.: A model-driven approach for developing responsive web apps. In: Proceedings of 14th International Conference on Evaluation of Novel Approaches to Software Engineering (ENASE 2019), pp. 257–264. SCITEPRESS (2019). https://doi.org/10.5220/0007678302570264
10. Milicevic, A., Jackson, D., Gligoric, M., Marinov, D.: Model-based, event-driven programming paradigm for interactive web applications. In: Proceedings of 2013 ACM International Symposium on New Ideas, New Paradigms, and Reflections on Programming & Software (Onward! 2013), pp. 17–36. ACM, New York (2013). https://doi.org/10.1145/2509578.2509588
11. Livshitz, E.: YAML vs JSON: Which is more efficient for language models. Better Programming (2023). https://betterprogramming.pub/yaml-vs-json-which-is-more-efficient-for-language-models-5bc11dd0f6df. Accessed 4 June 2025
12. OpenAI. GPT-4o mini: Advancing cost-efficient intelligence. https://openai.com/index/gpt-4o-mini-advancing-cost-efficient-intelligence/

Stable LLM Ensemble: Interaction Between Example Representativeness and Diversity

Junichiro Niimi[1,2(✉)]

[1] Meijo University, Nagoya, Aichi 4688502, Japan
jniimi@meijo-u.ac.jp
[2] RIKEN AIP, 15th floor, 1-4-1 Nihonbashi, Chuo-ku, Tokyo 103-0027, Japan

Abstract. Large language models (LLMs) have achieved remarkable results in wide range of domains. However, particularly in sentiment analysis, the accuracy and robustness of one-shot LLM predictions remain highly sensitive to the examples and the diversity among ensemble members. This study systematically investigates the effects of example representativeness (one-shot strategy) and output diversity (sampling temperature) on LLM ensemble performance. Two one-shot strategies are compared: centroid-based representative examples (proposed) and randomly sampled examples (baseline) and sampling temperature also is varied. Representative examples combined with higher temperature significantly outperforms random selection by +7.6 % (macro-F1) and −10.5 % (RMSE). Furthermore, our ensemble approach exceeds 5-shot prompting by +21.1 % (macro-F1) and −24.0 % (RMSE). Our findings demonstrate that combining representative example selection with increased temperature provides the appropriate level of diversity to the ensemble. This work highlights the practical importance of both example selection and controlled diversity in designing effective one-shot LLM ensembles.

Keywords: natural language processing · large language models · sentiment analysis · in-context learning · ensemble learning

1 Introduction

1.1 Background

Large Language Models (LLMs) have achieved remarkable performance across a wide range of natural language processing (NLP) tasks. In particular, there is a rapid expansion of LLM applications into diverse domains, including marketing [20,33], finance [19,23], and education [18]. These models have enabled novel approaches to various tasks such as sentiment analysis, machine translation, retrieval-augmented generation (RAG), and text summarization. As a result, LLMs are fundamentally transforming the way data-driven insights are generated and utilized in each field.

H. Unger and P. Meesad (Eds.): NLPIR 2025, LNNS 1904, pp. 642–658, 2026.
https://doi.org/10.1007/978-3-032-20897-2_46

Despite their impressive abilities, the output of LLMs remains highly sensitive to model configuration, such as one-/few-shot learning [25], prompt templates [38], and hyperparameters [17,31]. The performance and consistency of LLM-generated predictions often depend on subtle differences in example selection and prompt construction. However, there is currently no established method for optimal example selection, and many existing studies still rely on random sampling strategies [26,27].

In addition, ensemble learning [6] using LLMs has been actively conducted [15,27,39]. Ensemble methods utilize the inherent randomness in machine learning including LLMs to achieve higher accuracy [39] and computational efficiency [27], compared to single inference. However, since the model configuration greatly affects the individual inference of LLMs, it is also important in ensemble.

We focus on two major challenges: (1) selecting appropriate examples typically requires domain expertise to identify representative cases and (2) ensemble methods benefit from diversity among base models [6], yet there is no systematic approach to ensure such diversity in example selection. To solve those challenges, we leverage semantic embeddings to automatically identify representative samples from distinct semantic clusters, eliminating the need for manual curation. Furthermore, prior study [27] has shown that varying random seeds (i.e., the parameter controlling stochastic token sampling in LLM inference) provides an additional source of stochastic diversity that can be easily incorporated into the ensemble design.

1.2 Research Gaps

From the research background, the variability of the multiple inferences due to prompts, examples, and hyperparameters remains a major challenge for the practical application of sentiment analysis, particularly in terms of reproducibility and robustness. We derive five research questions (RQs 1–5) and approaches as follows:

RQ1 Effect of temperature on ensemble effectiveness: Compare to the lower temperature, do the high temperature setting contribute to the model performance?
Approach: We systematically vary the sampling temperature parameter in the LLM, evaluating ensemble accuracy, such as F1 and root mean square error (RMSE) at low (temperature = 0.8) and high (temperature = 1.5) settings for each one-shot approach and examine the change in accuracy.

RQ2 Effect of example representativeness on ensemble performance: Compared to the random example, does the representative example contribute to the model performance?
Approach: We set up two one-shot strategies: Centroid-based Representative Examples (CREs, proposed) and Randomly-Sampled Examples (RSEs, baseline) and compare accuracy between the approaches.

RQ3 Optimal combination of representativeness and diversity: Which combination of temperature and example selection achieves the best overall per-

formance?
Approach: We overlook the all combinations of model configuration over the one-shot approaches and temperature settings to identify which configuration achieves the highest model performance.

RQ4 Comparison with 5-shot model: Which is more effective, the ensemble model or the 5-shot model?
Approach: From the results of the optimal configuration, we additionally compare the performance between ensemble and 5-shot models.

RQ5 Effect of consistency: How does the self-consistency among base model predictions relate to the improvement achieved by ensemble aggregation?
Approach: We conduct an in-depth analysis of the best-performing model to investigate how ensemble performance relates to the self-consistency among base model predictions. Specifically, we examine whether improvements achieved by ensemble aggregation are associated with the degree of agreement or disagreement among the individual base models.

To address these issues, in this study, we propose the method to select representative examples from the dataset and examine the effect to utilize such sample on accuracy of LLM-based ensemble. In addition, we also vary the sampling temperature parameter to examine the effect of the creativity of the inference. We figure out the optimal model setting for the LLM-ensemble methods in terms of prediction accuracy and robustness and further compare with 5-shot settings.

2 Related Works

2.1 Sentiment Analysis

Sentiment analysis [28], which aims to identify and classify subjective information from text, has long been a major topic in natural language processing (NLP) and machine learning. Early approaches were based on traditional machine learning methods such as logistic regression, support vector machines (SVM) [5], and Naive Bayes classifiers [32], often combined with hand-crafted features or shallow linguistic representations [37].

With the advent of deep learning, DNN-based methods, such as convolutional neural networks (CNNs) [22] and recurrent neural networks (RNNs) [35], have become the standard baselines, enabling more accurate sentiment prediction by capturing complex semantic patterns in text [37]. These methods rely on large annotated datasets and powerful architectures to achieve state-of-the-art results.

More recently, the emergence of LLMs such as GPT and BERT has brought about a new paradigm for sentiment analysis [20]. LLMs are capable of both zero-shot and few-shot learning, making it possible to perform sentiment analysis without task-specific fine-tuning. A number of studies have explored various aspects of LLM-based sentiment analysis, including the design of effective prompts [36], comparisons with human annotators [11], and the impact of one-shot example selection on model performance [25,34]. There is growing evidence that LLM-based approaches can outperform conventional ML and DNN-based methods on diverse sentiment analysis tasks [25,27,36].

2.2 Ensemble in LLMs

Ensemble learning is referred to as a combination of multiple classifiers to construct a robust predictor [6],?. To consolidate multiple predictions, several techniques are adopted: voting [13] (e.g., majority voting or weighted voting, where predictions are aggregated based on the predefined weights), bagging [1,2] (bootstrap aggregating, where weak learners are trained on different subsamples of a training set), and boosting [9,10] (e.g., AdaBoost and Gradient Boosting, where weak learners are iteratively improved based on previously misclassified samples). These methods utilizes the diversity among the individual classifiers, as they are particularly effective for unstable learning algorithms [6].

In regard to LLM inference, prior studies have implemented the ensemble approaches [3,8,15,27,39]. For example, one financial study [39] employed a stacked LLM architecture which lower LLMs (base models) were assigned the multiple characteristics to ensure the diversity and upper LLM consolidated the decisions. Another marketing study [27] diversified the base models by using different random seed values and consolidated the decisions with median. While the former study [39] is notable for promoting model diversity by assigning different areas of expertise through prompt engineering, this strategy requires manual design and careful consideration of which expertise to assign to each model. Such manual intervention introduces additional overhead and the potential for subjective bias in model configuration. In addition, the aggregation mechanism it employs further exacerbates the inherent black-box nature of LLM-based decision making.

In contrast, the latter study [27] leverages median aggregation to minimize the influence of outliers in ordinal sentiment classification, but the diversity among base models is limited, as it only varies random seed values rather than introducing substantive differences between models.

2.3 Consistency and Reliability in LLMs

Moreover, LLMs have a temperature parameter which can control the randomness of the output. Some studies [26,29] highlighted the importance of self-consistency of the inferences. In particular, the consistency of the multiple inferences in higher temperature setting decreased, resulting in more diverse predictions for the same input [29]. While their work focuses on automated annotation and uses "consistency score" to identify edge cases, their findings suggest that sampling temperature can be used as a mechanism to systematically control output diversity in LLMs.

Recent work has also addressed the calibration and uncertainty quantification of LLM predictions, proposing methods to assess and improve model reliability in various NLP tasks [7,17,24]. These approaches enable the use of consistency-based metrics for dynamic inference, error flagging, and confidence-aware downstream applications.

3 Proposed Model

In this study, we first choose the appropriate examples based on the embedding clusters to sufficiently ensure the diversity. Second, we conduct model inference with varying sampling temperature and consolidate multiple decisions with median. The overview of our configuration is shown in Fig. 1.

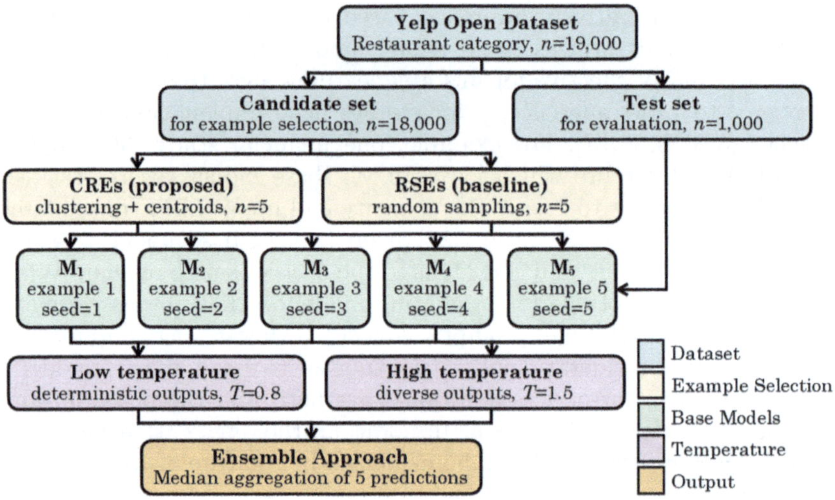

Fig. 1. Overview of the experimental setup. The pipeline consists of example selection (CREs or RSEs), base model construction with different examples and random seeds, inference with varying temperature settings, and median aggregation to produce the final ensemble prediction.

3.1 Base Models

For ensemble inference, we constructed five base models (M_1–M_5), each defined by a unique combination of one-shot example and random seed. Although all base models use the same pretrained LLM (Llama-3.1-8B-Instruct), their distinct configurations (different examples and seeds) result in diverse predictions, thereby functioning as independent ensemble members.

We first selected five distinct one-shot examples (Example 1–5) from a candidate set (separate from the test set, described in Sect. 4.1), according to the specified strategy (either CREs or RSEs). For each base model M_i, where $i \in \{1, 2, 3, 4, 5\}$, we assigned Example i and set the random seed i (e.g., M_1 uses Example 1 and seed 1). In addition, to stochastically vary the responses, we use nucleus sampling (top_p = 0.9) [14]. This design ensures that each base model is exposed to a different example and stochastic generation path, thereby promoting diversity among ensemble members while maintaining reproducibility.

First, regarding the base models for ensemble, this study employs Meta AI's Llama [12] (Llama-3.1-8B-Instruct)[1] for the pretrained model. Llama is trained with large corpus with reinforcement learning from human feedback (RLHF) [4] and has been widely adopted in the sentiment analysis. In accordance with prior studies [27,39], we do not perform fine-tuning.

The prompt is shown in Fig. 2, where the review and rating selected as an example are imputed into {example_review} and {example_label} and actual user review is input to {user_review}. The model generates one token following the prompt, corresponding to the underbar at the end of the prompt (max_new_tokens = 1).

```
### Instruction
You are a helpful assistant evaluating the review texts about the
restaurant. Please evaluate the review text and assign an integer score
ranging from 1 for the most negative comment to 5 for the most positive
comment.

### Example
User review: {example_review}
Rating: {example_label}

### Output
User review: {user_review}
Rating: _
```

Fig. 2. Basic Prompt

3.2 One-Shot Strategies

Centroid-Based Representative Examples (Proposed). For CREs (proposed), we select representative samples based on the combination of Sentence-BERT (SBERT) [30] embeddings[2] and their clusters. First, we obtain fixed-length embeddings for all candidate texts using SBERT. For each sample, mean pooling is applied over all token embeddings to produce a single 384-dimensional vector representation.

Second, we apply K-Means clustering to the set of embedding vectors, setting the number of clusters K equal to the number of one-shot examples to be selected (e.g., $K = 5$ in our experiments). For each cluster, the sample whose embedding is closest to the cluster centroid (using Euclidean distance) is chosen as a representative. The final set of K Centroid-based Representative Examples (CREs) thus consists of the centroid-nearest samples from each cluster. These CREs are then used as one-shot examples in the prompts for the LLM ensemble experiments.

[1] https://huggingface.co/meta-llama/Llama-3.1-8B-Instruct.

[2] https://huggingface.co/sentence-transformers/all-MiniLM-L6-v2.

Randomly-Selected Examples (Baseline). For RSEs (baseline), we randomly sample K examples from the candidate set using a fixed random seed to ensure reproducibility of the example selection. Once selected, these examples are assigned to base models M_1–M_K, where each model M_i uses random seed i during inference (as described in Sect. 3.1). No clustering or embedding-based selection is applied; the examples are drawn purely at random as a baseline for comparison with the proposed CREs.

3.3 Aggregation

To consolidate multiple predictions from base models, we employ median aggregation. While other aggregation methods, such as majority voting, are also possible, prior work [27] has highlighted several issues with these approaches. Specifically, in situations where the votes are split among classes, majority voting can lead to ambiguous or unstable results. Moreover, when applied to tasks such as sentiment classification, which are inherently ordinal, majority voting treats the problem as a nominal classification, disregarding the ordinal relationships between labels. In contrast, median aggregation is more appropriate for ordinal outcomes, as it directly incorporates the order information and is less susceptible to the influence of outlier predictions.

4 Experimental Study

4.1 Dataset and Settings

This study employs Yelp Open Dataset [41] which is publicly available dataset for academic research. The dataset contains user ratings (stars: 1–5) and reviews for a wide range of businesses. We focus on the venues tagged with Restaurants and randomly split the data into two sets: candidate set (for one-shot examples; sample size: 18,000) and test set (sample size: 1,000). For each one-shot strategy (CREs and RSEs), we select five distinct examples from the candidate set, resulting in a total of 10 examples (5 for CREs and 5 for RSEs). The summary statistics for the test set is shown in Table 1 (The rating and characters are shown in $mean \pm std$).

4.2 Model Evaluation

We evaluate performance across all experimental conditions using three evaluation metrics: accuracy (Acc.), macro-F1 (F1), and root mean square error (RMSE). Accuracy and macro-F1 are standard classification metrics, whereas RMSE quantifies the magnitude of prediction errors between predicted and true labels. To assess statistical significance, we employ McNemar's test and the Wilcoxon signed-rank test to verify the performance improvements achieved by our approach.

Table 1. Summary Statistics of the Test Set

n	Users	Venues	Rating	Characters
1000	993	503	3.979 ± 1.325	480.716 ± 455.688

4.3 Selection of Examples

The distributions of ratings depending on sampling strategy is shown in Fig. 3. Both strategy basically reflect the overall distributional bias in the original dataset; positive ratings (4–5 stars) are more prevalent than negative ones (1–2 stars).

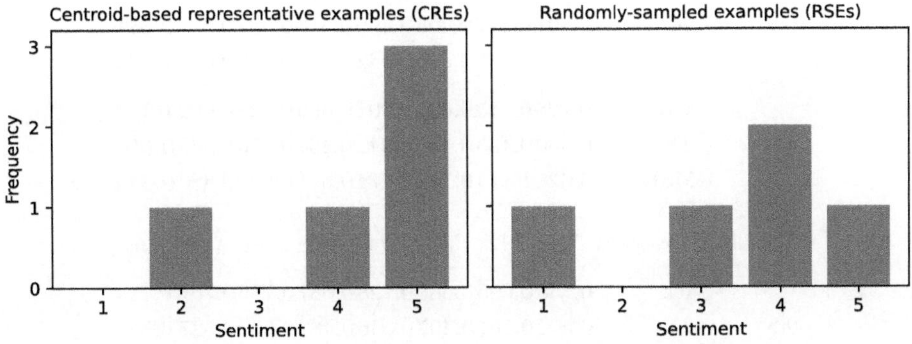

Fig. 3. Rating distributions for examples in each group. While the RSEs are relatively balanced across the five strata, the CREs are notably skewed toward higher ratings.

For CREs, the positive samples generally emphasized food quality, variety, and service, frequently expressing enthusiasm and intent to revisit. The negative example cited specific issues with service, such as delayed orders and inattentiveness, despite acknowledging food quality. This composition illustrates that, while clustering captures semantic diversity, the inherent skew of the dataset can result in a concentration of positive examples. However, each representative example provides distinct context, such as menu diversity, staff attentiveness, and overall dining experience[3].

On the other hand, in the case of RSEs, the examples capture various aspects of restaurant experiences, including delivery delays, issues with menu availability and customer service, food quality and personal taste preferences, as well as overall satisfaction with food and atmosphere. This comparison demonstrates that the CRE-based strategy may increase semantic diversity but label-skewed examples due to dataset bias, while random sampling achieves a more balanced sentiment distribution at the potential cost of context relevance.

[3] In accordance with Yelp's data policy, only summary descriptions of the content and not the raw texts are presented in this paper.

4.4 Results and Discussions

Table 2 reports the performance of the five individual models (M1–M5) and their median ensemble (Ens) under two one-shot strategies (randomly sampled examples, RSEs; centroid-based representative examples, CREs) and two sampling temperatures $T \in \{0.8, 1.5\}$.

Table 2. Effect of Ensemble Inference. "Best" indicates the best individual performance among M_1–M_5. The bold type indicates that the ensemble model outperforms the individual base models and cell shading represents the highest performance.

RSEs: baseline		M1	M2	M3	M4	M5	Best	Avg.	Ens
	Acc.	0.732	0.735	0.684	0.692	0.691	0.735	0.707	0.735
0.8	F_1	0.591	0.577	0.495	0.492	0.518	0.591	0.535	**0.596**
	RMSE	0.559	0.578	0.670	0.682	0.658	0.559	0.629	0.584
	Acc.	0.728	0.715	0.667	0.501	0.699	0.728	0.662	0.718
1.5	F_1	0.550	0.433	0.438	0.342	0.467	0.550	0.446	**0.591**
	RMSE	0.628	1.114	0.718	1.274	0.667	0.628	0.880	**0.572**
CREs: proposed		M1	M2	M3	M4	M5	Best	Avg.	Ens
	Acc.	0.735	0.695	0.689	0.728	0.687	0.735	0.707	0.714
0.8	F_1	0.592	0.567	0.497	0.546	0.482	0.592	0.537	0.557
	RMSE	0.568	0.598	0.642	0.630	0.636	0.568	0.615	0.593
	Acc.	0.693	0.756	0.661	0.275	0.665	0.756	0.610	**0.762**
1.5	F_1	0.619	0.623	0.533	0.193	0.437	0.623	0.485	**0.636**
	RMSE	0.574	0.539	0.720	1.950	0.651	0.539	0.887	**0.512**

Effect of Sampling Temperature. We first examine RQ1 (Effect of temperature on ensemble effectiveness) by comparing performance within each example selection approach.

RSEs (baseline). Raising the temperature from 0.8 to 1.5 changes F1 only from 0.596 to 0.591 (−0.8 %) and RMSE from 0.584 to 0.572 (−2.1 %). In other words, random examples provide little additional diversity even at the higher temperature.

CREs (proposed). In contrast, the same temperature shift improves F1 from 0.557 to 0.636 (+14.2 %) and reduces RMSE from 0.593 to 0.512 (−13.7 %). The improvement was statistically significant according to both McNemar's test on paired correctness ($p < 0.05$, $\chi^2 = 5.84$) and Wilcoxon signed-rank test on absolute errors ($p < 0.01$, $W = 31534$), confirming that representative examples benefit from larger output variance.

These results present the concrete answer to RQ1 (The effect of sampling temperature) that the temperature controls the model diversity: low T leads to the deterministic outputs and ineffective ensemble while high T to diverse outputs and ensemble improvements.

Effect of One-Shot Strategy. We second examine the effect of one-shot strategy by comparing the models within the sampling temperatures (RQ2: ensemble representativeness)

Low temperature. First, both strategies are statistically indistinguishable: RSE-Ens does not have a significant improvement from the single best model and CRE-Ens is even below RSE-Ens. In the lower temperature setting, predictions become nearly deterministic and the median offers no advantage.
High temperature. With the higher temperature setting, sufficient diversity is provided to each model. As the results, CRE-Ens clearly outperforms RSE-Ens: F1 improves from 0.591 to 0.636 (+7.61 %) and RMSE reduces from 0.572 to 0.512 (−10.49 %). We confirm that this improvement is also statistically significant according to both McNemar's test on paired correctness ($p < 0.05$, $\chi^2 = 5.193$) and Wilcoxon signed-rank test on absolute errors ($p < 0.01$, $W = 28408$).

These results also show the clear answer to RQ2 (representativeness) that the representativeness of the example has a significant effect on ensemble performance with the sufficient diversity.

Overall. Across both metrics, the CRE-Ens at $T = 1.5$ achieves the highest F1 (0.636) and the lowest RMSE (0.512), surpassing all other ensemble and individual models, indicating the clear answer to RQ3 (optimal configuration) that CRE-approach with high temperature setting realizes the optimum balance between representativeness and diversity.

Table 3 presents the detailed results of per-class and overall F1 scores for each base model (M_1–M_5) and the ensemble (ens) for CREs (proposed) with high temperature setting. Due to the imbalanced class distribution, F1 for individual classes can remain 0 or reach 1 in some base models. This phenomenon is especially pronounced in multiclass sentiment classification tasks, where the precision and recall for certain classes may fluctuate significantly depending on the model's output distribution. While the ensemble achieves the highest overall F1, it also smooths out the class-specific extremes observed in individual models. These results highlight the inherent difficulty of achieving balanced performance across all classes in imbalanced, real-world datasets.

These results provide the clear answer to RQ3 (Optimal combination) that the combination of centroid-based representative examples (CREs) and high sampling temperature ($T = 1.5$) yields the best overall ensemble performance (F1 = 0.636, RMSE = 0.512), outperforming all other configurations.

Table 3. F1 by the Labels for CREs ($T = 1.5$)

Labels	1	2	3	4	5	All
M_1	0.320	0.201	0.125	0.232	0.398	0.619
M_2	0.488	0.183	0.116	0.195	0.317	0.623
M_3	0.226	0.138	0.237	0.003	**1.**	0.533
M_4	0.120	**1**	0	0.008	**0.286**	0.193
M_5	**0.493**	0.086	0.008	0.073	0.498	0.437
Ens	0.480	0.201	0.116	0.199	0.475	**0.636**

Comparison with 5-Shot Model. To validate RQ4 (Comparison with 5-shot model), we construct two five-shot models with different temperature settings $T \in \{0.8, 1.5\}$. Table 4 compares the performance between five-shot single models and the best ensemble model (Ens: CREs with $T = 1.5$). While five-shot prompting offers a slight improvement over the average of one-shot base models, it remains substantially inferior to the ensemble approach.

Table 4. Comparison with 5-shot Models for CREs. Lift indicates the performance increase from the best 5-shot model.

	5-shot (CREs)		CRE-Ens	Lift
	$T = 0.8$	$T = 1.5$		
Acc.	0.692	0.642	**0.762**	+10.1%
F1	0.525	0.496	**0.636**	+21.1%
RMSE	0.674	0.804	**0.512**	−24.0%

A McNemar's test on paired correctness confirmed that the differences were statistically significant for both temperature settings ($p < 0.001$, $\chi^2 = 15.87$ for $T = 0.8$ and $\chi^2 = 43.17$ for $T = 1.5$). Wilcoxon signed-rank test on absolute errors also confirmed the significant difference ($p < 0.001$, $W = 14245$ for $T = 0.8$ and $W = 16457$ for $T = 1.5$). This underscores that ensemble aggregation is crucial for harnessing the diversity introduced by high temperature, and simply increasing the number of prompt examples is not sufficient to achieve robust performance.

Self-Consistency. Finally, we validate the self-consistency of our best ensemble model. Table 5 summarizes the agreement among the five base models. n_{unique} denotes the number of unique predicted labels produced by the five models for each sample (i.e., $n_{\text{unique}} = 1$ means all models predict the same label, while $n_{\text{unique}} \geq 2$ indicates disagreement). The case where all models made different predictions ($n_{\text{unique}} = 5$) did not occur. We further stratify the samples by n_{unique} and compute the F1 score for each stratum.

Table 5. Self-consistency and performance for CREs ($T = 1.5$)

n_{unique}	1	2	3	4	5
Samples	225	328	427	20	0
F1 Score	0.938	0.756	0.684	0.550	-

The model performance improves for samples where all base models agree ($n_{\text{unique}} = 1$), indicating that high consistency among the base models increases the model performance. Even in the low-consistency cases, as shown in the effect of ensemble, ensemble method can significantly reduce the prediction error with median and outperforms individual models.

These results clearly answer RQ5 (Effect of consistency) that, in samples with high self-consistency, ensemble methods enable high confidence in correct answers, while in samples with low self-consistency, ensemble methods correct errors. Thus, in both cases, ensemble method plays an important role to ensure the performance. As pointed out in prior study [26], self-consistency is the important factor to improve the prediction performance in LLMs.

5 Conclusion

5.1 Summary

In this study, we conducted a sentiment analysis using an ensemble approach of multiple LLMs, systematically varying both the one-shot example selection strategy and the sampling temperature parameter to identify the optimal ensemble configuration. Through addressing five RQs, this study has three key findings: i.) the proposed centroid-based representative examples (CREs) with a high temperature setting demonstrated substantial improvements over several baselines. ii.) our best ensemble model significantly outperformed the 5-shot model whose examples are selected with the proposed approach. iii.) the proposed ensemble approach is particularly beneficial for samples where individual models disagree (low consistency), while also reinforcing confidence in cases of high model agreement (high consistency). These findings highlight the practical importance of both representative example selection and controlled diversity in maximizing the benefit of LLM ensembles.

In summary, our results demonstrate that: (1) higher temperature increases diversity and, together with effective example selection, enhances ensemble performance; (2) centroid-based representative examples yield superior results over random examples; (3) the optimal ensemble configuration combines CREs and high temperature; and (4) ensemble aggregation is most beneficial for samples with low self-consistency, but also boosts confidence in cases of high agreement.

The proposed ensemble achieved a +2.1% improvement in F1 over the best individual base model. While obtaining such a strong base model typically requires extensive exploration of exemplar combinations, our method automatically selects representative examples through clustering, eliminating the need for manual search or domain-specific tuning. This demonstrates both the efficiency and practicality of the proposed approach.

Finally, our approach ensures stability in three aspects: (1) deterministic example selection by clustering compared to random sampling, (2) robust ensemble performance even when individual models are weak, and (3) reliable predictions as evidenced by the self-consistency analysis.

5.2 Implications

This study has several implications both for academic and practical aspects.

Theoretical Implications. Our findings show that simply inserting a representative one-shot example is not, by itself, sufficient to raise ensemble performance. Rather, ensemble accuracy is mediated primarily by the temperature parameter, and the magnitude of that effect depends strongly on the example-selection strategy. Notably, even a single model reached an F1 of 0.619, exceeding all other ensemble models. Random sampling disperses the decision boundary, whereas centroid examples keep the base models in the same direction, letting the median discard outliers.

Although the ensemble built from CREs with $T = 1.5$ achieved the highest overall accuracy, the constituent models were not individually strongest at that same temperature. This observation is consistent with the accuracy–diversity trade-off [21], showing that increasing diversity often lowers individual accuracy, yet may still improve the ensemble as a whole. Our results confirm that ensemble performance does not hinge on the superiority of each component model: the best ensemble still emerged even when it incorporated models with individual macro-F1 $= 0.193$.

Regarding the self-consistency among the individual decisions, the observed relationship between model consistency and ensemble performance has important implications for the design of LLM-based classification systems. First, our results confirm that self-consistency is a reliable indicator of prediction confidence: samples with unanimous agreement among base models are nearly always classified correctly, which aligns with prior findings on the value of self-consistency in LLM inference [26]. Second, the main benefit of ensemble aggregation lies in its ability to recover correct predictions in difficult cases where base models disagree, thereby improving overall robustness. These findings suggest that future systems could leverage self-consistency as a confidence metric, enabling more targeted use of ensemble inference or adaptive post-processing for ambiguous samples. Our empirical analysis thus bridges theoretical discussions on LLM consistency with practical ensemble design for real-world sentiment analysis.

Importantly, our findings indicate that example selection based on semantic clustering does not necessarily align with the original label distribution; for instance, the selected centroid-based examples skewed towards high sentiment ratings (Fig. 3). Nevertheless, this approach captures a broad range of semantic contexts and yields sufficient diversity to enable effective ensemble inference.

This suggests that semantic diversity among prompt examples, rather than strict label diversity, is a key factor for maximizing the robustness and performance of LLM ensembles.

Practical Implications. By clustering SBERT embeddings and automatically selecting the cluster centroids as one-shot examples, the proposed approach enables practitioners to construct effective LLM ensembles without domain-specific prompt engineering or extensive manual intervention. Since the proposed sampling strategy (CRE) only affects the prompting process, while the temperature setting only modifies the sampling distribution during generation, the overall processing time remains almost identical across configurations. Therefore, companies can easily improve performance without implementing complex inference methods or incurring additional computational costs. This lowers the barrier to deploying robust sentiment analysis models in industrial applications, where labeled data may be limited and human resources for model tuning are scarce.

Furthermore, the use of self-consistency as an indicator of prediction confidence provides a practical means to automatically flag ambiguous or difficult cases for further review or targeted intervention, which is valuable in applications such as customer feedback analysis, automated moderation, or decision support.

Finally, the demonstrated effectiveness of the ensemble with centroid-based example selection and temperature-driven diversity suggests that similar strategies could be readily adapted for other text classification or ordinal prediction tasks, further extending the practical utility of the method beyond sentiment analysis.

5.3 Limitations

As for the limitations, our proposed methodology is validated with the single dataset of 1,000 samples, with the case of five base models for the ensemble; the effect and scalability of increasing the number of base models require further cross-domain validation, such as finance and healthcare, using other pretrained models, such as Qwen [40] and Mistral [16].

Also, this study does not examine the relationship between the self-consistency and confidence or reliability of the inference in detail. Particularly, leveraging self-consistency as a confidence indicator enables dynamic inference strategies. Further theoretical analysis on the accuracy–diversity trade-off is worth investigating.

Furthermore, comparative analysis with state-of-the-art methods on standardized benchmarks and evaluation of computational efficiency (e.g., latency, cost) would strengthen the practical relevance of our findings.

Acknowledgment. We are grateful to the anonymous reviewers for the insightful comment. Their feedbacks have greatly improved our study. Both the dataset and model were managed and used in an appropriate environment that comply with the terms of use. We do not collect additional information that could lead to the identification of individuals. This study is supported by JSPS KAKENHI (Grant Number: 24K16472).

References

1. Breiman, L.: Bagging predictors. Mach. Learn. **24**, 123–140 (1996)
2. Breiman, L.: Random forests. Mach. Learn. **45**, 5–32 (2001)
3. Chen, Z., et al.: Harnessing multiple large language models: a survey on LLM ensemble. arXiv preprint arXiv:2502.18036 (2025)
4. Christiano, P.F., Leike, J., Brown, T., Martic, M., Legg, S., Amodei, D.: Deep reinforcement learning from human preferences. In: Advances in Neural Information Processing Systems, vol. 30 (2017)
5. Cortes, C., Vapnik, V.: Support-vector networks. Mach. Learn. **20**, 273–297 (1995)
6. Dietterich, T.G.: Ensemble methods in machine learning. In: International Workshop on Multiple Classifier Systems, pp. 1–15. Springer (2000)
7. Fadeeva, E., et al.: Fact-checking the output of large language models via token-level uncertainty quantification. In: Findings of the Association for Computational Linguistics ACL 2024, pp. 9367–9385 (2024)
8. Fang, C., et al.: LLM-ensemble: optimal large language model ensemble method for e-commerce product attribute value extraction. In: Proceedings of the 47th International ACM SIGIR Conference on Research and Development in Information Retrieval, pp. 2910–2914 (2024)
9. Freund, Y., Schapire, R.E.: A decision-theoretic generalization of on-line learning and an application to boosting. J. Comput. Syst. Sci. **55**(1), 119–139 (1997)
10. Friedman, J.H.: Greedy function approximation: a gradient boosting machine. Ann. Stat. 1189–1232 (2001)
11. Gilardi, F., Alizadeh, M., Kubli, M.: Chatgpt outperforms crowd workers for text-annotation tasks. Proc. Natl. Acad. Sci. **120**(30), e2305016120 (2023)
12. Grattafiori, A., et al.: The llama 3 herd of models. arXiv preprint arXiv:2407.21783 (2024)
13. Hansen, L.K., Salamon, P.: Neural network ensembles. IEEE Trans. Pattern Anal. Mach. Intell. **12**(10), 993–1001 (1990)
14. Holtzman, A., Buys, J., Du, L., Forbes, M., Choi, Y.: The curious case of neural text degeneration. In: International Conference on Learning Representations (2020)
15. Huang, Y., et al.: Ensemble learning for heterogeneous large language models with deep parallel collaboration. In: The Thirty-Eighth Annual Conference on Neural Information Processing Systems (2024)
16. Jiang, A., et al.: Mistral 7b. arxiv 2023. arXiv preprint arXiv:2310.06825 (2024)
17. Jiang, Z., Araki, J., Ding, H., Neubig, G.: How can we know when language models know? On the calibration of language models for question answering. Trans. Assoc. Comput. Linguist. **9**, 962–977 (2021)
18. Kasneci, E., et al.: Chatgpt for good? On opportunities and challenges of large language models for education. Learn. Individ. Differ. **103**, 102274 (2023)
19. Kong, Y., et al.: Large language models for financial and investment management: applications and benchmarks. J. Portfolio Manag. **51**(2) (2024)
20. Krugmann, J.O., Hartmann, J.: Sentiment analysis in the age of generative AI. Cust. Needs Solut. **11**(1), 3 (2024)

21. Kuncheva, L.I., Whitaker, C.J.: Measures of diversity in classifier ensembles and their relationship with the ensemble accuracy. Mach. Learn. **51**, 181–207 (2003)
22. LeCun, Y., Bottou, L., Bengio, Y., Haffner, P.: Gradient-based learning applied to document recognition. Proc. IEEE **86**(11), 2278–2324 (2002)
23. Li, Y., Wang, S., Ding, H., Chen, H.: Large language models in finance: a survey. In: Proceedings of the Fourth ACM International Conference on AI in Finance, pp. 374–382 (2023)
24. Ling, C., et al.: Uncertainty quantification for in-context learning of large language models. In: Duh, K., Gomez, H., Bethard, S. (eds.) Proceedings of the 2024 Conference of the North American Chapter of the Association for Computational Linguistics: Human Language Technologies (Volume 1: Long Papers), pp. 3357–3370. Association for Computational Linguistics, Mexico City, Mexico (2024). https://aclanthology.org/2024.naacl-long.184/
25. Mughal, N., Mujtaba, G., Shaikh, S., Kumar, A., Daudpota, S.M.: Comparative analysis of deep natural networks and large language models for aspect-based sentiment analysis. IEEE Access **12**, 60943–60959 (2024)
26. Narang, S., Chowdhery, A., Zhou, D.: Self-consistency improves chain of thought reasoning in language models. In: The Eleventh International Conference on Learning Representations, ICLR 2023 (2023)
27. Niimi, J.: A simple ensemble strategy for LLM inference: towards more stable text classification. In: Proceedings of the 30th International Conference on Applications of Natural Language to Information Systems (2025)
28. Pang, B., Lee, L., Vaithyanathan, S.: Thumbs up? Sentiment classification using machine learning techniques. In: Proceedings of the 2002 Conference on Empirical Methods in Natural Language Processing (EMNLP 2002), pp. 79–86 (2002)
29. Pangakis, N., Wolken, S., Fasching, N.: Automated annotation with generative AI requires validation. arXiv preprint arXiv:2306.00176 (2023)
30. Reimers, N., Gurevych, I.: Sentence-bert: sentence embeddings using siamese bert-networks. In: Proceedings of the 2019 Conference on Empirical Methods in Natural Language Processing and the 9th International Joint Conference on Natural Language Processing (EMNLP-IJCNLP), pp. 3982–3992 (2019)
31. Renze, M.: The effect of sampling temperature on problem solving in large language models. In: Findings of the Association for Computational Linguistics: EMNLP 2024, pp. 7346–7356 (2024)
32. Rish, I., et al.: An empirical study of the naive bayes classifier. In: IJCAI 2001 Workshop on Empirical Methods in Artificial Intelligence, vol. 3, pp. 41–46. Seattle, USA (2001)
33. Roumeliotis, K.I., Tselikas, N.D., Nasiopoulos, D.K.: LLMs in e-commerce: a comparative analysis of GPT and llama models in product review evaluation. Nat. Lang. Process. J. **6**, 100056 (2024)
34. Rubin, O., Herzig, J., Berant, J.: Learning to retrieve prompts for in-context learning. In: Proceedings of the 2022 Conference of the North American Chapter of the Association for Computational Linguistics: Human Language Technologies, pp. 2655–2671 (2022)
35. Rumelhart, D.E., Hinton, G.E., Williams, R.J.: Learning representations by back-propagating errors. Nature **323**(6088), 533–536 (1986)
36. Wang, Z., Xie, Q., Feng, Y., Ding, Z., Yang, Z., Xia, R.: Is chatgpt a good sentiment analyzer? A preliminary study. arXiv preprint arXiv:2304.04339 (2023)
37. Wankhade, M., Rao, A.C.S., Kulkarni, C.: A survey on sentiment analysis methods, applications, and challenges. Artif. Intell. Rev. **55**(7), 5731–5780 (2022)

38. Webson, A., Pavlick, E.: Do prompt-based models really understand the meaning of their prompts? In: Proceedings of the 2022 Conference of the North American Chapter of the Association for Computational Linguistics: Human Language Technologies, pp. 2300–2344 (2022)
39. Xing, F.: Designing heterogeneous LLM agents for financial sentiment analysis. ACM Trans. Manage. Inf. Syst. **16**(1) (2025). https://doi.org/10.1145/3688399
40. Yang, A., et al.: Qwen3 technical report. arXiv preprint arXiv:2505.09388 (2025)
41. Yelp: Yelp Open Dataset (2022). https://www.yelp.com/dataset

A Comparative Study on Domain-Specific LLMs: Architecture and Performance

Nehme Haidura(✉), Hussein Assaf, and Ziad Kobti

School of Computer Science, University of Windsor, Windsor, ON N9B 3P4, Canada
{haidura,assafh,kobti}@uwindsor.ca

Abstract. The research development on domain-specific large language models (DSLLMs) has expanded rapidly in recent years. Despite this growth, it has also made the landscape more fragmented and difficult to navigate. Although many studies have shown the benefits of adapting large language models to specific domains, the field lacks a unified framework that brings together other researchers' findings and enables meaningful comparisons. This review gathers knowledge from around 60 studies to present a comprehensive analysis of DSLLMs, addressing current knowledge gaps. First, we structure direct performance comparisons across domains, architectures, and metrics through systematic comparison tables, revealing that architectural choice fundamentally determines capability patterns. Although encoder-only models excel at understanding tasks (BERT-Large: 93.2 F1 on SQuAD v1.1), decoder-only models show superior generation capabilities (GPT-3: 87.7% Story Cloze accuracy). Encoder-decoder models provide balanced but sometimes suboptimal performance on highly specialized tasks. Second, we identify a critical gap in temporal knowledge management, demonstrating that existing benchmarks do not evaluate the ability of individual models to track the evolution of knowledge or to recognize outdated information. Third, our analysis reveals that when general-purpose models outperform specialized ones (BERT-Large achieving 96.72 F1 vs. BioBERT's 88.85 F1 on BC5CDR), the benefits are driven by factors such as scale, data diversity, and metric limitations rather than architectural superiority.

Keywords: Domain-Specific Large Language Models (DSLLMs) · Large Language Models (LLMs) · Transformer Architectures · BERT · GPT · T5 · BioBERT · SciBERT · BioGPT · SciFive · mT5 · ByT5 · Natural Language Processing (NLP)

1 Introduction

The emergence of domain-specific large language models (DSLLMs) represents a natural evolution in natural language processing, driven by the recognition that general-purpose models often fall short when applied to specialized domains with unique vocabularies and contextual requirements. Although the concepts of DSLLMs have roots in the mid-2000s [15], they experienced exponential progress

H. Unger and P. Meesad (Eds.): NLPIR 2025, LNNS 1904, pp. 659–677, 2026.
https://doi.org/10.1007/978-3-032-20897-2_47

with the increase in transformer-based architectures, such as BERT and GPT [1,4]. The evolution progressed from basic transfer learning and fine-tuning methods to advanced DSLLMs strategies, eventually leading to models built entirely on DSLLM corpora [16]. The observation of performance gaps motivated this evolution, as when general models were applied to specialized tasks, such as biomedical or scientific tasks, revealing that effective DSLLMs NLP require a deep architectural understanding rather than surface-level adaptation.

1.1 Problem Description

The field of DSLLMs has experienced rapid growth in recent years, and this expansion has created significant challenges for researchers and practitioners trying to navigate the increasingly complex landscape of available models and methodologies. Although individual studies have demonstrated the effectiveness of DSLLMs in various fields, the research community would benefit from a comprehensive analysis that combines these findings into a unified framework for understanding and evaluating the performance of DSLLMs. Most of the existing literature focuses on presenting novel models or demonstrating improvements over baseline systems. However, few studies provide the critical analysis necessary to understand the underlying principles that drive successful domain adaptation. This presents an evaluation gap that becomes especially concerning in high-stakes domain applications, where model failures can lead to significant real-world impacts. The need for a comprehensive review that not only reviews existing approaches, but also provides a critical analysis of their strengths, weaknesses, and appropriate applications has become increasingly urgent as the field continues to expand.

1.2 Objectives/Contribution

This comprehensive review represents an attempt to unify, organize, and analyze the diverse landscape of DSLLMs. We summarized findings from around 60 research papers to create a unified framework in this rapidly evolving field. Our work addresses three critical gaps in the current understanding of DSLLMs through systematic analysis and structured comparison.

First, we provide a comprehensive framework that unifies the results of different studies into structured comparison tables, enabling direct performance comparisons between architectures, domains, and evaluation metrics. This enables the identification of hidden patterns in model behavior and provides practitioners with actionable insights for model selection based on task requirements, rather than relying on fragmented individual study results.

Second, we demonstrate that current evaluation frameworks fundamentally do not address the precision of knowledge, which is a critical limitation given that, to the best of our knowledge, existing benchmarks do not directly evaluate a model's ability to handle temporal knowledge or detect outdated information. This gap is concerning in domains where information evolves rapidly, such as medicine, law, and finance, where outdated or misleading knowledge can weaken user trust and real-world deployment.

Third, to the best of our knowledge, no prior work has conducted a systematic comparison at the architectural level. In this study, we provide a comprehensive architectural comparison framework that analyzes how encoder-only, decoder-only, and encoderdecoder architectures produce distinct error patterns and performance results when applied to DSLLM tasks.

1.3 Article Sectioning

The remainder of this paper is organized as follows. Section two introduces Large Language Models and their main architectures, such as including encoder-only, decoder-only, and encoder-decoder, along with their DSLLM variants. Section three provides a detailed analysis of BERT and its specialized versions, such as SciBERT and BioBERT. Section four presents the GPT model and its evolution from GPT-1 to GPT-4 and some of its variants. Section five discusses the T5 architecture and its DSLLMs extensions. Section six provides different LLM's metrics and benchmarks. Section seven discusses key findings, open challenges, and future research directions. Finally, Section eight concludes the paper.

2 Introduction to LLMs

Most large Language Models (LLMs) are AI systems based on the transformer architecture, which uses attention mechanisms to capture relationships between all sequence elements at the same time [14] through massive neural networks trained on large datasets to understand and generate human language. These models represent the evolution of the concept of language models in AI, which has made a great impact on the capabilities of AI models. The evolution of simple tasks to more complex systems that can represent text through deep models that are capable of diverse tasks like language translation, summarization, and reasoning capabilities that have been significantly improved through specialized training methods.

2.1 Different LLM Models

Encoder-Only Models. Encoder-only models are designed primarily for understanding and extracting information from text rather than generating new content. These models excel at tasks that require deep comprehension of the input text, such as classification, recognition of named entities, and answering questions. The most prominent example is BERT, which uses a bidirectional approach to consider both left and right contexts during training [1]. These models work by processing the entire input sequence at once, using self-attention mechanisms to understand relationships between all words in the text. The bidirectional nature allows them to capture richer contextual representations compared to unidirectional approaches [2]. Variants like DistilBERT have been developed to make these models more efficient while maintaining performance, achieving nearly the same results as BERT but with half the parameters [3] (Fig. 1).

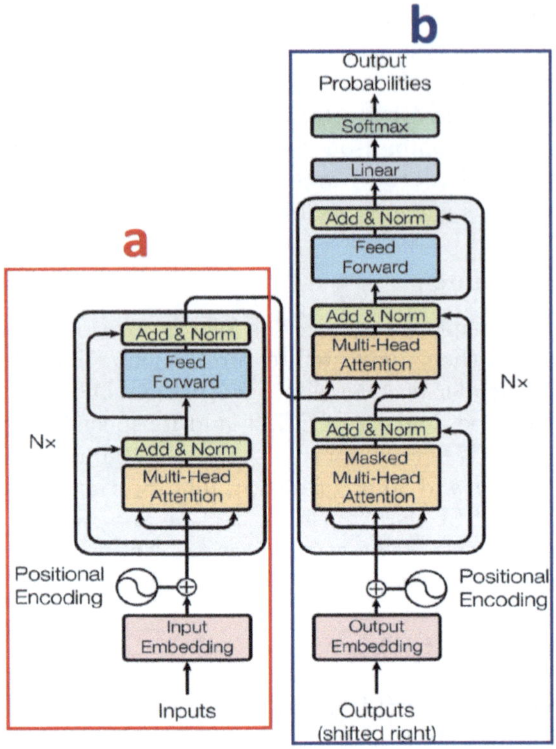

Fig. 1. Illustration of the Transformer architecture: a) The left part in the red box shows the encoder-only structure, which forms the basis of models such as BERT. b) The right part in the blue box depicts the decoder-only structure, characteristic of models like GPT. Together, the full encoderdecoder framework represents the original Transformer model, which underlies models such as T5. [20].

Decoder-Only Models. Decoder-only models are primarily designed for text generation tasks, following an autoregressive approach where they predict the next token in a sequence based on previous tokens. These models excel at creative writing, code generation, and conversational tasks. The GPT series represents the most well-known decoder-only architecture, starting with GPT-1's two-stage approach of generative pretraining followed by task-specific fine-tuning [4]. The architecture uses masked self-attention to ensure that predictions for each position can only depend on the known outputs at previous positions, maintaining the autoregressive property. This design makes them particularly effective for language generation tasks, as demonstrated by models such as DialoGPT, which extends GPT-2 for the generation of conversational responses [5].

Encoder-Decoder Models. Encoder-decoder models combine the strengths of both architectures, using an encoder to understand the input and a decoder to generate the output. This makes them particularly suitable for tasks that require both comprehension and generation, such as machine translation, summarization, and question answering. The encoder processes the input sequence to create contextual representations, while the decoder generates the output sequence using both the encoded representations and the previously generated tokens. Models like T5 [6] exemplify this architecture, treating every NLP task as a text-to-text problem. Multilingual variants such as mT5 extend this approach across multiple languages, achieving SOTA results on multilingual benchmarks. Specialized versions like Sentence-T5 focus on generating high-quality sentence embeddings using encoder-only and encoder-decoder strategies [7].

2.2 Domain Specific LLM

DSLLMs are models trained or fine-tuned on domain-specific corpora (e.g., medical texts, legal documents, scientific papers) to outperform general-purpose models in specialized fields. These models address the gap between general-purpose LLM and specialized domain applications by incorporating domain knowledge and terminology [8]. In the biomedical domain, models like BioBERT and SciBERT have been developed by pre-training on DSLLM corpora such as PubMed articles and scientific papers. BioBERT, adapted from BERT and pre-trained in PubMed and PMC data, significantly outperforms general BERT on biomedical text mining tasks [9]. Similarly, SciBERT, pre-trained in scientific literature, shows notable improvements on scientific NLP tasks [10]. Generative domain-specific models like BioGPT and Galactica represent advances in specialized text generation. BioGPT bridges the gap left by discriminative models in biomedical text generation [11], while Galactica focuses on scientific knowledge, demonstrating the ability to store, combine, and reason about scientific information [12]. The development of DSLLMs often involves careful consideration of specialized datasets, evaluation metrics, and training strategies tailored to the requirements and challenges of the particular domain [13].

2.3 Other Major Models

This paper focuses primarily on BERT, GPT, and T5 architectures and their DSLLM variants. While other significant models, such as LLAMA, CLAUDE, and GEMINI demonstrate impressive general capabilities, the scientific literature on their domain-specific variants is limited, leading to their exclusion from this review.

3 Bert

Architecture Overview. BERT (Bidirectional Encoder Representations from Transformers) is built using a multilayer, bidirectional Transformer encoder, closely following the implementation style described by [20]. BERT comes in two primary configurations: BERTBASE, which features 12 layers (transformer

Table 1. Performance overview of BERT and its variants on standard NLP benchmarks. Metrics include accuracy and F1 scores. The table highlights how specialized DSLLMs such as SciBERT and BioBERT, and other versions of general-purpose models such as Albert and Distilbert, improve the performance, reduce computational cost, or achieve SOTA results at publication time.

Model	Domain	GLUE(AVG)	SQuAD v1.1 (F1)	SQuAD v2.0 (F1)	SWAG (Acc)	NERBC5CDR (F1)
BERT large [1]	General	82.1	93.2	83.1	86.3	96.7 (BERT-base)
SciBERT [10]	Scientific	–	–	–	–	90.0
BioBERT [9]	Biomedical	–	–	–	–	88.8
DistilBERT [3]	General (compressed)	77.0	86.9	–	–	–
ALBERT base [23]	General (lightweight)	89.4	89.3	80.4	–	–

blocks), a hidden size of 768, and 12 self-attention heads, totaling 110 million parameters; and BERTLARGE, with 24 layers, a hidden size of 1024, and 16 attention heads, making up 340 million parameters. Unlike models such as GPT, which use attention in the left context only, BERT uses fully bidirectional self-attention, allowing it to capture rich contextual information from both directions within the input [1]. Despite BERT's strong performance on understanding tasks, recent evaluations of academic question-answering reveal its limitations on generation tasks, achieving only a 0.25 BLEU score compared to generative models [53].

Pre-training Tasks. The BERT model is pre-trained using two unsupervised prediction objectives: Masked Language Modeling (MLM) and Next Sentence Prediction (NSP). Masked LM (MLM): Randomly selected tokens (15% of WordPiece tokens) in each input sequence are masked, and the model is trained to predict the original tokens. To address the discrepancy between pretraining (which sees MASK tokens) and fine-tuning (which does not), the replacement is performed as follows: 80% of the time, tokens are swapped with [MASK]; 10% of the time, with a random token; and 10% of the time, kept unchanged [1]. Next-Sentence Prediction (NSP): BERT is also trained to understand relationships between sentences by predicting whether one sentence actually follows another or if it is a random sentence from the corpus. This binary objective is helpful for later tasks that involve reasoning about sequence relationships, such as question answering and natural language inference [1].

Fine-Tuning BERT. As discussed in [1], the fine-tuning process is simplified because BERT's Transformer-based self-attention architecture enables it to handle various downstream tasks, both single-text and text-pair scenarios, by simply adjusting the input and output configurations. Traditional approaches [21] [22] for text-pair tasks typically encode each text separately before applying bidirectional cross attention, while [1] unifies these steps by using self-attention on concatenated text pairs, which inherently captures bidirectional cross attention between sentences. Furthermore, BERT adapts to different tasks by plugging in task-specific inputs/outputs while training end-to-end. It reuses its sentence pair structure for various tasks like paraphrasing, entailment, Q&A, and classification, then selects appropriate output representations.

Domain-Specific Variants. The success of BERT has led to numerous DSLLM variants that leverage the encoder-only framework for specialized applications across different fields. These variants demonstrate the flexibility of BERT's pre-training approach when combined with DSLLMs corpora, training strategies, and can be seen in many domains, such as Scientific and Biomedical.

Scientific Domain Variants: SciBERT represents a significant advancement in scientific text processing, utilizing BERT's architecture while being pre-trained on a large corpus of scientific papers [10]. **Biomedical and Clinical variants**: BioBERT exemplifies a successful domain variant by pre-training BERT on PubMed and PMC corpora, creating a model specifically tailored for biomedical text mining [9]. Clinical BERT models have been developed specifically for clinical free text processing, with variants trained on general clinical notes and discharge summaries [46]. CovBERT represents a highly specialized variant for COVID-19 literature classification, demonstrating how BERT can be fine-tuned for emerging DSLLMs needs [54].

4 GPT

Table 2. Performance of GPT-based models Benchmarks cover tasks such as natural language inference, commonsense reasoning, and question answering, with results at the time of publication.

Benchmark	GPT-1/[4]	GPT-2 (Zero-shot) [17]	GPT-3 (Few-shot) [18]
GLUE (AVG)	72.8	–	–
SuperGLUE (AVG)	–	–	71.8
MNLI-m (ACC)	82.1	–	–
MNLI-mm (ACC)	81.4	–	–
RTE (ACC)	56.0	–	69.0
Story Cloze (ACC)	86.5	–	87.7
LAMBADA (PPL)	–	8.6	1.9
LAMBADA (Acc)	–	63.2	86.4

Architecture Overview. The Generative Pre-trained Transformer (GPT) represents a paradigm shift in natural language processing through its decoder-only transformer architecture. Unlike bidirectional models such as BERT, GPT employs a unidirectional approach that generates text in an autoregressive manner, predicting the next token based on the preceding context [4]. This architectural choice enables GPT models to excel in text-generation tasks. The main innovation is centered on a two-phase training approach [4]. First, unsupervised pre-training in which the model learns general language patterns by training on massive amounts of text without specific task labels. Second, supervised fine-tuning, which adapts to particular tasks through training on labeled examples. The model uses a special attention mechanism that only allows each word to "see" the words that came before it in the sequence. This one-directional approach is crucial because it mimics how humans naturally produce language word by word.

Evolution from GPT-1 to GPT-4. The original GPT model established the foundational framework with a 12-layer decoder-only transformer architecture [4]. GPT-1 demonstrated that generative pre-training could achieve competitive performance across multiple natural language understanding tasks through its unified input representation approach. The model's training involved unsupervised pre-training on the BooksCorpus dataset, followed by task-specific supervised fine-tuning, achieving notable improvements over previous SOTA methods on benchmarks, including the GLUE benchmark.

GPT-2 marked a significant increase in the scale and capability of the model, as described in [17]. With 1.5 billion parameters in its largest configuration, GPT-2 demonstrated remarkable zero-shot task performance across diverse domains without task-specific fine-tuning. The model was trained on WebText, a more diverse and larger dataset than GPT-1, leading to improved generalization capabilities on benchmarks, including LAMBADA, enwik8, and others, as measured by ROUGE, BLEU, and accuracy metrics.

GPT-3 represents a significant jump in the language model capabilities with 175 billion parameters, demonstrating that scale can lead to qualitatively different behaviors [18]. The model's most significant contribution lies in its few-shot learning capabilities, where it can perform new tasks with minimal examples provided in the prompt context. This reduced the need for task-specific fine-tuning in many scenarios, establishing in-context learning as a fundamental capability of sufficiently large language models. GPT-3 performance in various benchmarks.

GPT-4 extends the GPT series with enhanced reasoning capabilities and multimodal functionality, processing both text and images [19]. GPT-4 demonstrates significant improvements in complex reasoning tasks, mathematical problem solving, and code generation. The model exhibits more reliable performance on standardized tests and professional benchmarks compared to previous versions, suggesting architectural refinements and improvements in training methodology beyond the simple scaling of parameters.

Domain-Specific Variants. The GPT architecture has served as inspiration for many DSLLM variants that leverage the decoder-only framework for specialized applications. DialoGPT represents this approach by pre-training on

Table 3. Performance of T5-based models and their DSLLM variants on standard NLP benchmarks. Metrics include accuracy and F1 scores. Results highlight SOTA performance at the time of publication.

Model	Domain	GLUE (AVG)	SuperGLUE (AVG)	SQuAD (F1/EM)	XNLI Eng (Acc)	GEM-XSum (BLEU)	PubMed (Acc)
T5 base	General	83.4 [7]	76.2 [52]	92.1/85.4 [6]	–	–	87.1 [51]
mT5 base [6]	Multilingual	83.0 [50]	72.5 [50]	89.6/83.8	75.4	8.4	–
ST5 enc-dec base [7]	Sentence embedding	76.3	–	–	–	–	–
ByT5 base [50]	General	85.3	74.0	–	75.4	11.1	–
SciFive [51]	Biomedical	–	–	–	–	–	86.2

147 million Reddit conversational exchanges, creating a model specifically optimized for conversational response generation [5]. This specialization demonstrates how the GPT architecture can be adapted through DSLLMs pre-training datasets to achieve superior performance in targeted applications. BioGPT represents another significant domain variant, specifically pre-trained on large-scale biomedical literature to bridge the gap in biomedical text generation and mining [11]. Unlike discriminative models such as BioBERT, BioGPT maintains the generative capabilities of the GPT architecture while achieving SOTA results in six biomedical NLP tasks, demonstrating the adaptability of the decoder-only transformer approach across scientific domains.

5 T5

Architecture Overview. T5 (Text-to-Text Transfer Transformer) represents a new direction in language model design by handling natural language processing tasks as a text-to-text problem [52]. T5 is an encoder-decoder transformer framework, which combines the bidirectional understanding capabilities of encoders with the autoregressive generation power of decoders. The architecture consists of a stack of encoder layers that process input sequences and decoder layers that generate output sequences, with cross-attention mechanisms. T5 comes in multiple sizes, such as T5-Small, T5-Base, and T5-Large, which range between 60 to 770M parameters, and even more like other variants such as T5-11B (11B parameters), which makes it more flexible across different computational constraints and performance requirements [52].

Pre-training of T5. During pre-training of T5, 15% of the input tokens are replaced with sentinel tokens, and each consecutive corrupted span is replaced by a single sentinel token [52]. The model must then generate the corrupted spans in order, separated by the same sentinel tokens. The pre-training corpus for T5 is the Colossal Clean Crawled Corpus (C4), which is 750GB of text. The C4 dataset undergoes some preprocessing methods to remove duplicates, non-natural language content, and low-quality text, resulting in a high-quality training corpus [52]. This enables T5 to develop robust language understanding and generation capabilities in multiple domains.

Fine-Tuning of T5. The T5 framework is different from models that require task-specific architectures or output layers. T5 maintains the same input-output format in all applications and only requires task-specific prompt formatting and output parsing [52]. Fine-tuning also involves training the entire model on task-specific datasets, with the encoder processing task-formatted inputs and the decoder generating structured outputs.

Domain-Specific Variants. The architecture of T5 became an inspiration for several significant variants for multilingual and DSLLM applications. For example, mT5, which extends the original framework to 101 languages [6]. mT5 is pre-trained on the mC4 dataset and shows strong cross-lingual transfer capabilities, achieving competitive performance on multilingual benchmarks. Another variant for T5 is UL2, which represents an advanced evolution of the T5 framework, incorporating multiple pre-training objectives, including prefix language modeling, span corruption, and other denoising tasks within a single unified framework [49]. There is also ByT5, which extends the capabilities of T5 to byte-level processing, removes the dependency on subword tokenization, and enables more robust handling of hard languages and noisy text [50]. SciFive also applies the T5 framework to the scientific literature, demonstrating improved performance on biomedical and scientific text processing tasks[51].

6 Comparison

To evaluate a large language model, both quantitative metrics and a practical assessment of output quality are needed. Standard metrics include accuracy, precision, F1 score, BLEU, ROUGE, and perplexity. However, practical evaluation for a high-performing LLM should involve high accuracy for DSLLMs, if it were domain-specific, low hallucination rates, and consistency across different types of queries. Furthermore, benchmark datasets are used to measure accuracy rates, and factual correctness can be assessed through human evaluation or comparison with knowledge-grounded datasets specific to the domain.

6.1 Used Metrics

See Table 4.

Evaluating DSLLMs involves using a variety of performance measures, each aimed at assessing a distinct aspect of the model's abilities. Choosing the right set of metrics is essential to capture the model's strengths and weaknesses accurately and to enable fair, informative comparisons between different architectures, training strategies, and application domains. Metrics can be categorized into 8 different categories:

Table 4. Evaluation Metrics for Large Language Models. This table presents the primary metrics used to assess LLM performance across different task categories.

Metric	Task	Purpose
Perplexity (PPL)	Language Modeling	Measures predictive uncertainty (lower is better).
Cross-Entropy Loss	Training Objective	Measures divergence between predicted and true distributions.
BLEU	Text Generation/Translation	Measures n-gram precision compared to reference translations.
ROUGE	Summarization/Generation	Measures n-gram recall/precision overlap with reference summaries.
METEOR	Generation/Translation	F1-like metric considering synonymy, stemming, and word order.
Accuracy	Classification/QA	Proportion of correct predictions.
F1 Score	Classification/NER/QA	Harmonic mean of precision and recall.
Precision/Recall	Classification/NER	Precision = relevancy of predictions, Recall = coverage of relevant items.
Exact Match (EM)	QA (SQuAD, extractive QA)	Proportion of answers exactly matching the gold span.
Human Evaluation	Text Generation	Fluency, coherence, and factuality were judged by human annotators.

Classification Metrics. Accuracy measures the proportion of correct predictions and appears consistently in evaluation studies, from BERT variants [23] to cross-lingual models on XNLI benchmarks [24].

Precision, Recall, and F1-Score. F1-Score provides balanced evaluation by combining precision and recall, extensively used in biomedical applications [11], scientific text classification [54]. Precision and Recall provide detailed insights into model effectiveness, especially important for tasks like named entity recognition (NER).

Text Generation Metrics. BLEU scores measure n-gram overlap between generated and reference texts [28], widely adopted in machine translation studies [26,27] and cross-lingual applications [32].

ROUGE scores focus on recall-oriented evaluation for text summarization and generation tasks [29], appearing in GPT-2 assessments [17], question generation studies [55], and domain specialization reviews [8].

METEOR emphasizes semantic matching [30] and appears in question generation benchmarks [55] and inductive reasoning evaluations [33].

BERTScore uses contextual embeddings to assess semantic similarity [31], allowing a more refined evaluation compared to traditional surface-level matching methods [55].

Task-Specific Metrics. Exact Match (EM) requires a precise match of responses and is used in question and answer evaluations, including GPT-2 benchmark assessments [17] and scientific text processing studies [13]. Specialized Speech Metrics include phone error rate (PER) and word error rate (WER) for acoustic model evaluation, as demonstrated in Speech-XLNet studies on TIMIT and WSJ datasets [25].

Information Retrieval Metrics. MAP (Mean Average Precision) and MRR (Mean Reciprocal Rank) provide ranking-based evaluation metrics, particularly relevant for information retrieval and document ranking tasks [56]. These metrics assess not only correctness but also the ranking quality of the retrieved information.

Computational and Efficiency Metrics. The number of parameters indicates model size and computational complexity, essential for comparing model efficiency across different architectures. Training Duration and Hardware Requirements measure computational cost and resource efficiency during model development. FLOPs (Floating Point Operations) and Training Energy quantify computational efficiency and the environmental impact of model training.

Human and Qualitative Evaluation. Human evaluation remains essential for assessing subjective aspects of model performance, used in conversational AI studies [5], neural machine translation research [27], and behavioral analysis comparing model decision making to human patterns [48].

6.2 Evaluation Benchmarks

See Table 5.

Table 5. Common NLP Benchmarks. These benchmarks evaluate model performance across general, advanced, and DSLLMs tasks.

Benchmark	Task	Purpose
GLUE	General NLP	Composite score across 9 NLP tasks.
SuperGLUE	Advanced NLP	Harder variant of GLUE with reasoning-heavy tasks.
SQuAD	QA	Standard benchmark for span-based QA tasks.
PubMedQA/BioASQ	Biomedical QA	DSLLMs benchmarks for biomedical question answering.
MMLU	Multi-domain QA	Covers 57 tasks across science, law, medicine, etc.

Advanced Reasoning and Transfer Benchmarks: Transfer Task Accuracy measures model performance on downstream tasks after pre-training [41], particularly important for sentence embedding evaluation using frameworks like SentEval [7]. Reasoning benchmarks include mathematical reasoning assessments (MMLU, MATH) [43,44], commonsense QA, and symbolic logic evaluations such as logical entailment and proof-based reasoning datasets, as discussed in comprehensive reasoning surveys [42] and in scientific language models like Galactica [12]. BIG-Bench provides diverse reasoning and knowledge tasks for comprehensive evaluation [45].

Domain-Specific Benchmarks: Biomedical: Evaluations of medical NLP applications include PubMedQA and MedMCQA for biomedical question answering [16], as well as the MIMIC-III dataset for clinical text processing [46]. **Scientific**: Includes equation retrieval recall used in mathematical information retrieval tasks such as ARQMath and NTCIR-MathIR, along with specialized scientific QA benchmarks, reflecting the unique requirements of scientific text processing [12].

Comprehensive Benchmarks: GLUE Benchmark provides standardized evaluation in multiple language understanding tasks [34], consistently used for BERT variants [1,3] and comparative studies [35]. SQuAD benchmarks (versions 1.1 and 2.0) [36,37] offer a reading comprehension evaluation, used in BERT assessments [1] and ALBERT evaluations [23]. XTREME Benchmark assesses multilingual capabilities [38], particularly relevant for mT5 evaluation [6]. RACE Benchmark evaluates the comprehension of reading from examinations [39], commonly used alongside GLUE and SQuAD for a comprehensive evaluation. MultiNLI measures natural language inference across multiple genres and domains [40]. MLQA (Multilingual Question Answering) evaluates cross-lingual question answering capabilities [47]. Language Modeling Benchmarks include LAMBADA for long-range dependency evaluation and enwik8 for character-level language modeling assessment. MLM (Masked Language Modeling) Accuracy measures pre-training performance on masked token prediction tasks, typically used as a training diagnostic rather than a final evaluation metric.

6.3 Summary

The selection of appropriate evaluation metrics depends on the characteristics of the task, the requirements of the domain, and the intended applications. Classification tasks benefit from F1-score and accuracy metrics, generation tasks require BLEU and ROUGE assessments, while semantic understanding tasks increasingly rely on embedding-based metrics like BERTScore. DSLLM applications often require specialized metrics that reflect the unique characteristics and requirements of the target domain, ensuring a meaningful evaluation within the specific context of the application.

7 Discussion of Results/Future Trend

After analyzing the different models, we have discovered, as far as we can tell, that none of the existing benchmarks directly evaluate a model's ability to handle temporal knowledge or detect outdated information. The datasets that the models, such as Bert, GPT, and T5, and their DSLLM variants, are static datasets that cannot capture whether these models would identify when their training knowledge has become outdated. For instance, BioGPT's performance on biomedical tasks may reflect knowledge that was current during training but could now be outdated given the rapid pace of medical research advances.

It is worth noting that general models sometimes outperform DSLLMs in some tasks. Looking at Table 1, we notice that BERT-large achieves 96.72 F1 on the BC5CDR biomedical NER task, outperforming SciBERT (90.01) and BioBERT (88.85). However, this performance may reflect different factors rather than a simple superiority of general models. Maybe the larger parameter size of BERT-Large compared to the base-sized variants of SciBERT and BioBERT can provide greater capacity to capture complex patterns. In addition, the difference in the pretraining data and the fine-tuning strategies may also affect. Although sometimes the performance of DSLLMs is lower than that of general models on some specific benchmarks, we cannot forget that the DSLLMs are designed to capture DSLLM patterns that general models might miss. The performance gap observed here may reflect limitations in current evaluation metrics rather than fundamental model capabilities. Our findings suggest that the choice between general and DSLLMs should be driven by the task requirement rather than assuming that the general models are superior.

The observation of the different architectures of the models reveals some differences in capability patterns. Encoder-only models show a consistent strength in understanding tasks; for example, BERT-Large achieves 93.2 F1 on SQuAD v1.1 and 83.1 F1 on SQuAD v2.0, indicating robust reading comprehension capabilities. The gap between the two datasets, where SQuAD v2.0 introduces unanswerable questions, shows that these models face challenges in dealing with uncertainty and deciding when not to provide an answer. Decoder-only models show a different pattern. Looking at Table 2, we can see that GPT-3 performs well in generative tasks (87.7% Story Cloze accuracy) but shows more variation on classification benchmarks (e.g., 69.0% RTE accuracy). The improvement from GPT-1 (72.8 GLUE average) to GPT-3 (71.8 SuperGLUE average, despite the greater difficulty of SuperGLUE) demonstrates the advantages of scaling for decoder architectures. The encoder-decoder models demonstrate different performance that combines both understanding and generation capabilities. In Table 3, we notice that the T5-base achieves an 82.7 GLUE average and 92.08 F1 on SQuAD. On the generative side, mT5 reaches 8.4 BLEU on GEM-XSum, a notable result considering the complexity of multilingual summarization. Moreover, these models show weaker performance in SuperGLUE (T5: 76.2, mT5: 72.1) compared to their GLUE performance, suggesting that these architectures may struggle with complex reasoning tasks.

DSLLMs variants also vary by architecture. In biomedical tasks, encoder-only models like SciBERT and BioBERT show clear specialization benefits, while encoder-decoder models such as SciFive also benefit (86.28 PubMed accuracy vs T5-base 87.18), indicating that the architecture plays a role in determining how efficiently domain knowledge is integrated and applied. These findings reveal that each architecture has some strengths and limitations. Encoder-only models perform best in classification, decoder-only models at generation, and encoder-decoder models that provide balanced but sometimes less optimal performance on highly specialized tasks.

The field of DSLLMs is rapidly evolving, and several critical research directions are emerging and promising to address fundamental challenges in trustworthiness, reliability, and practical deployment.

Trustworthiness: Researchers should develop a framework that can assess when DSLLMs exceed their boundaries in critical fields such as medicine and law. This includes techniques that automatically flag potential error outputs without requiring constant expert supervision and an automated system that can verify the accuracy of model explanations in specialized contexts.

Reliability: The research community should be committed to improving the performance of DSLLMs in unusual cases that were underrepresented in the training data. This involves developing specialized training techniques and evaluation protocols that systematically test model behavior in rare but critical scenarios and focus on resolving conflicts when multiple DSLLMs are trained on identical corpora.

Practical Deployment: Researchers should develop evaluation metrics that bridge the gap between benchmark performance and real-world deployment challenges. This could be by predicting how current evaluation metric failures would affect real-world deployment.

8 Conclusion

This review has presented a comprehensive synthesis of DSLLMs, drawing on evidence from around 60 studies to organize, compare, and evaluate the field in a unified way. By addressing three persistent gaps, temporal knowledge handling, and architectural comparison, we provide a structured foundation for understanding how specialized models behave across different domains and tasks. We find that the most critical gap lies in temporal knowledge. Current evaluation benchmarks are static and do not measure how models handle evolving information or outdated content, which is a serious limitation in fields such as medicine or law, where the use of outdated knowledge can have real-world impacts. We also clarify that general-purpose models outperform DSLLMs is not due to superiority, but due to factors such as larger scale, pre-training data, and weaknesses in evaluation metrics.

References

1. Devlin, J., Chang, M.-W., Lee, K., Toutanova, K.: BERT: pre-training of deep bidirectional transformers for language understanding. In: Burstein, J., Doran, C., Solorio, T. (eds.) Proceedings of the 2019 Conference of the North American Chapter of the Association for Computational Linguistics: Human Language Technologies, Volume 1 (Long and Short Papers), pp. 4171–4186. Association for Computational Linguistics, Minneapolis, Minnesota (2019). https://doi.org/10.18653/v1/N19-1423

2. Rogers, A., Kovaleva, O., Rumshisky, A.: A primer in BERTology: what we know about how BERT works. Trans. Assoc. Comput. Linguist. **8**, 842–866 (2021). https://doi.org/10.1162/tacl_a_00349
3. Sanh, V., Debut, L., Chaumond, J., Wolf, T.: DistilBERT, a distilled version of BERT: smaller, faster, cheaper and lighter. arXiv preprint arXiv:1910.01108 (2019)
4. Radford, A., Narasimhan, K., Salimans, T., Sutskever, I.: Improving language understanding by generative pre-training. Technical report, OpenAI, San Francisco, CA, USA (2018)
5. Zhang, Y., et al.: DialoGPT: large-scale generative pre-training for conversational response generation. arXiv preprint arXiv:1911.00536 (2019)
6. Xue, L., et al.: mT5: a massively multilingual pre-trained text-to-text transformer. arXiv preprint arXiv:2010.11934 (2020)
7. Ni, J., et al.: Sentence-T5: scalable sentence encoders from pre-trained text-to-text models. arXiv preprint arXiv:2108.08877 (2021)
8. Ling, C., et al.: Domain specialization as the key to make large language models disruptive: a comprehensive survey. ACM Comput. Surv. (2023)
9. Lee, J., et al.: BioBERT: a pre-trained biomedical language representation model for biomedical text mining. Bioinformatics **36**(4), 1234–1240 (2020). https://doi.org/10.1093/bioinformatics/btz682
10. Beltagy, I., Lo, K., Cohan, A.: SciBERT: a pretrained language model for scientific text. arXiv preprint arXiv:1903.10676 (2019)
11. Luo, R., et al.: BioGPT: generative pre-trained transformer for biomedical text generation and mining. Brief. Bioinform. **23**(6), bbac409 (2022). https://doi.org/10.1093/bib/bbac409
12. Taylor, R., et al.: Galactica: a large language model for science. arXiv preprint arXiv:2211.09085 (2022)
13. Ho, X., et al.: A survey of pre-trained language models for processing scientific text. arXiv preprint arXiv:2401.17824 (2024)
14. Naveed, H., et al.: A comprehensive overview of large language models. ACM Trans. Intell. Syst. Technol. **16**(5), 1–72 (2025). https://doi.org/10.1145/3744746
15. Daumé III, H., Marcu, D.: Domain adaptation for statistical classifiers. J. Artif. Intell. Res. **26**, 101–126 (2006). https://arxiv.org/pdf/1109.6341
16. Gururangan, S., et al.: Don't Stop Pretraining: Adapt Language Models to Domains and Tasks. arXiv preprint arXiv:2004.10964 (2020). https://arxiv.org/pdf/2004.10964
17. Radford, A., Wu, J., Child, R., Luan, D., Amodei, D., Sutskever, I.: Language Models are Unsupervised Multitask Learners. OpenAI Blog (2019). https://storage.prod.researchhub.com/uploads/papers/2020/06/01/language-models.pdf
18. Brown, T.B., et al.: Language Models are Few-Shot Learners. arXiv preprint arXiv:2005.14165 (2020). https://arxiv.org/pdf/2005.14165
19. Zhao, W.X., et al.: A review on large language models: architectures, applications, taxonomies, open issues and challenges. IEEE Trans. Knowl. Data Eng. (2023). https://doi.org/10.1109/TKDE.2023.3275346
20. Vaswani, A., et al.: Attention is all you need. In: Advances in Neural Information Processing Systems, pp. 5998–6008 (2017). https://proceedings.neurips.cc/paper/2017/file/3f5ee243547dee91fbd053c1c4a845aa-Paper.pdf
21. Parikh, A., Täckström, O., Das, D., Uszkoreit, J.: A decomposable attention model for natural language inference. In: Proceedings of the 2016 Conference on Empirical Methods in Natural Language Processing, pp. 2249–2255 (2016). https://arxiv.org/pdf/1606.01933

22. Seo, M., Kembhavi, A., Farhadi, A., Hajishirzi, H.: Bi-directional attention flow for machine comprehension. In: International Conference on Learning Representations (2017). https://arxiv.org/pdf/1611.01603
23. Lan, Z., Chen, M., Goodman, S., Gimpel, K., Sharma, P., Soricut, R.: ALBERT: a lite BERT for self-supervised learning of language representations. In: International Conference on Learning Representations (2020). https://arxiv.org/pdf/1909.11942
24. Conneau, A., et al.: Unsupervised Cross-lingual Representation Learning at Scale. arXiv preprint arXiv:1911.02116 (2020). https://arxiv.org/pdf/1911.02116
25. Liu, X., Sun, S., Zhou, H., Zhao, Z., Wu, Z.: Speech-XLNet: Unsupervised Acoustic Model Pretraining for Self-Attention Networks. arXiv preprint arXiv:1910.10387 (2020). https://arxiv.org/pdf/1910.10387
26. Sennrich, R., Haddow, B., Birch, A.: Neural Machine Translation of Rare Words with Subword Units. arXiv preprint arXiv:1508.07909 (2016). https://arxiv.org/pdf/1508.07909
27. Wu, Y., et al.: Google's Neural Machine Translation System: Bridging the Gap between Human and Machine Translation. arXiv preprint arXiv:1609.08144 (2016). https://arxiv.org/pdf/1609.08144
28. Papineni, K., Roukos, S., Ward, T., Zhu, W.-J.: BLEU: a method for automatic evaluation of machine translation. In: Proceedings of the 40th Annual Meeting of the Association for Computational Linguistics, pp. 311–318 (2002). https://aclanthology.org/P02-1040.pdf
29. Lin, C.-Y.: ROUGE: a package for automatic evaluation of summaries. In: Proceedings of the Workshop on Text Summarization Branches Out, pp. 74–81 (2004). https://aclanthology.org/W04-1013.pdf
30. Banerjee, S., Lavie, A.: METEOR: an automatic metric for MT evaluation with improved correlation with human judgments. In: Proceedings of the ACL Workshop on Intrinsic and Extrinsic Evaluation Measures for Machine Translation and Summarization, pp. 65–72 (2005). https://aclanthology.org/W05-0909.pdf
31. Zhang, T., Kishore, V., Wu, F., Weinberger, K.Q., Artzi, Y.: BERTScore: evaluating text generation with BERT. In: International Conference on Learning Representations (2020). https://openreview.net/pdf?id=SkeHuCVFDr
32. Conneau, A., Lample, G.: Cross-lingual Language Model Pretraining. arXiv preprint arXiv:1901.07291 (2019). https://arxiv.org/pdf/1901.07291
33. Vashishtha, S., et al.: Language Models as Inductive Reasoners. arXiv preprint arXiv:2212.10923 (2023). https://arxiv.org/pdf/2212.10923
34. Wang, A., Singh, A., Michael, J., Hill, F., Levy, O., Bowman, S.: GLUE: a multi-task benchmark and analysis platform for natural language understanding. In: Proceedings of the 2019 International Conference on Learning Representations (2019). https://openreview.net/forum?id=rJ4km2R5t7
35. Parvathinathan, K.: A Comparative Study of Attention Mechanisms in NLP Architectures: From BERT to GPT Variants. ResearchGate preprint (2024). https://www.researchgate.net/profile/Karthik-Parvathinathan/publication/392595223
36. Rajpurkar, P., Zhang, J., Lopyrev, K., Liang, P.: SQuAD: 100,000+ questions for machine comprehension of text. In: Proceedings of the 2016 Conference on Empirical Methods in Natural Language Processing, pp. 2383–2392 (2016). https://aclanthology.org/D16-1264/
37. Rajpurkar, P., Jia, R., Liang, P.: Know what you don't know: unanswerable questions for SQuAD. In: Proceedings of the 56th Annual Meeting of the Association for Computational Linguistics, pp. 784–789 (2018). https://aclanthology.org/P18-2124/

38. Hu, J., Ruder, S., Siddhant, A., Neubig, G., Firat, O., Johnson, M.: XTREME: a massively multilingual multi-task benchmark for evaluating cross-lingual generalization. In: Proceedings of the 2020 Conference on Empirical Methods in Natural Language Processing, pp. 6000–6031 (2020). https://arxiv.org/pdf/2003.11080
39. Lai, G., Xie, Q., Liu, H., Yang, Y., Hovy, E.: RACE: large-scale reading comprehension dataset from examinations. In: Proceedings of the 2017 Conference on Empirical Methods in Natural Language Processing, pp. 785–794 (2017). https://aclanthology.org/D17-1082/
40. Williams, A., Nangia, N., Bowman, S.R.: A broad-coverage challenge corpus for sentence understanding through inference. In: Proceedings of the 2018 Conference of the North American Chapter of the Association for Computational Linguistics, pp. 1112–1122 (2018). https://aclanthology.org/N18-1101/
41. Conneau, A., Kiela, D.: SentEval: an evaluation toolkit for universal sentence representations. In: Proceedings of the 11th International Conference on Language Resources and Evaluation (2018). https://aclanthology.org/L18-1269/
42. Huang, Z., et al.: Towards Reasoning in Large Language Models: A Survey. arXiv preprint arXiv:2212.10403 (2022). https://arxiv.org/pdf/2212.10403
43. Hendrycks, D., et al.: Measuring massive multitask language understanding. In: Proceedings of the International Conference on Learning Representations (2021). https://arxiv.org/abs/2009.03300
44. Hendrycks, D., et al.: Measuring Mathematical Problem Solving With the MATH Dataset. arXiv preprint arXiv:2103.03874 (2021). https://arxiv.org/pdf/2103.03874
45. Srivastava, A., et al.: Beyond the Imitation Game: Quantifying and Extrapolating the Capabilities of Language Models. arXiv preprint arXiv:2206.04615 (2022). https://arxiv.org/pdf/2206.04615
46. Alsentzer, E., et al.: Publicly available clinical BERT embeddings. In: Proceedings of the 2nd Clinical Natural Language Processing Workshop, pp. 72–78 (2019). https://aclanthology.org/W19-1909.pdf
47. Lewis, P., Oğuz, B., Rinott, R., Riedel, S., Schwenk, H.: MLQA: evaluating cross-lingual reading comprehension. In: Proceedings of the 58th Annual Meeting of the Association for Computational Linguistics, pp. 7315–7330 (2020). https://aclanthology.org/2020.acl-main.653/
48. Binz, M., Schulz, E.: Do Large Language Models Show Decision Heuristics Similar to Humans? arXiv preprint arXiv:2305.04400 (2023). https://arxiv.org/pdf/2305.04400
49. Tay, Y., et al.: UL2: Unifying Language Learning Paradigms. arXiv preprint arXiv:2205.05131 (2022). https://arxiv.org/pdf/2205.05131
50. Xue, L., et al.: ByT5: towards a token-free future with pre-trained byte-to-byte models. Trans. Assoc. Comput. Linguist. **10**, 291–306 (2022). https://doi.org/10.1162/tacl_a_00461
51. Phan, L., et al.: SciFive: A Text-to-Text Transformer Model for Biomedical Literature. arXiv preprint arXiv:2106.03598 (2021). https://arxiv.org/pdf/2106.03598
52. Raffel, C., et al.: Exploring the limits of transfer learning with a unified text-to-text transformer. J. Mach. Learn. Res. **21**(140), 1–67 (2020). https://www.jmlr.org/papers/volume21/20-074/20-074.pdf
53. Gangadhar, A.T., Rao, H.P., Ss, S.: Comparative analysis of large language models fine-tuned on a singular dataset. In: Proceedings of the 2024 8th International Conference on Natural Language Processing and Information Retrieval, pp. 5–11 (2024). https://doi.org/10.1145/3711542.3711590

54. Chang, Y.-C., Sun, S.-H., Lin, Y.-X., Wang, S.-M., Ku, L.-W.: Survey of BERT-base models for scientific text classification: COVID-19 case study. Appl. Sci. **12**(6), 2891 (2022). https://doi.org/10.3390/app12062891
55. Pan, L., et al.: Generative Language Models for Paragraph-Level Question Generation. arXiv preprint (2022). https://arxiv.org/pdf/2210.03992
56. Aftan, S., Shah, H., et al.: A survey on BERT and its applications. IEEE Access (2023). https://ieeexplore.ieee.org/stamp/stamp.jsp?tp=&arnumber=10092289

Language Representation and Information Retrieval in Multimodal Information Systems

Developing an Explainable Paper Retrieval System with Less Reliance on LLMs

Yudai Yamamoto(✉), Masato Kikuchi, and Tadachika Ozono

Nagoya Institute of Technology, Nagoya, Aichi 466-8555, Japan
yuyamamoto@ozlab.org, kikuchi@nitech.ac.jp, ozono@nitech.ac.jp

Abstract. Searching academic literature within extensive conference proceedings presents a significant challenge for students, particularly those who are new to a field or non-native English speakers. Existing search systems often lack sufficient user assistance, strict scope control, and effective methods for determining paper relevance from titles and abstracts alone. This study proposes a low-cost paper retrieval system designed to address these issues by enhancing the search experience with query expansion suggestions and explanations for paper selections. The system employs a hybrid approach, combining the efficiency of the BM25 keyword-based algorithm with the semantic filtering and refinement capabilities of a large language model (LLM). A pointwise mutual information (PMI) module assists users in formulating effective search queries, while the LLM provides rationales for paper selection, enhancing user trust and comprehension. Our method achieves high retrieval accuracy (8.73 vs. 8.20 with BM25 and 8.01 with a general-purpose embedding model) with significantly less reliance on LLMs, demonstrating its effectiveness for conference paper retrieval tasks.

Keywords: Information Retrieval · Large Language Model (LLM) · Explainability

1 Introduction

Students often require support when investigating new academic fields or reviewing recent international conference proceedings. This task is particularly burdensome for non-native English speakers or those unfamiliar with a field, because it is often difficult to quickly understand the content of a paper or judge its relevance based only on the title or abstract. The challenge becomes even greater in the case of large proceedings, such as AAAI 2025, where effective search query formulation is also a barrier.

Currently, various paper search systems are available, but challenges remain when students seek papers from specific years of academic conferences. Traditional systems such as Google Scholar and DBLP allow users to specify time periods and conferences. However, the information presented in the search results

H. Unger and P. Meesad (Eds.): NLPIR 2025, LNNS 1904, pp. 681–695, 2026.
https://doi.org/10.1007/978-3-032-20897-2_48

is limited to titles, abstracts, and simple keyword matches, which provide insufficient guidance for students unfamiliar with paper searches. AI-powered paper search systems can provide summaries and explanations of search results, but restricting the search scope strictly to specific conference proceedings remains difficult. Additionally, users sometimes employ general-purpose search systems like Google Search or web searches using a large language model (LLM) for paper searches. However, these systems are ineffective if the desired conference proceedings are not properly indexed by search engines in the first place. Furthermore, these search systems rely heavily on effective search queries, which can be an additional barrier for students. Consequently, current systems suffer from insufficient user assistance, a lack of strict search scope specification, and reliance on web indexing.

Given these challenges, there is a need for a system that identifies papers worth reading based on a user's intent, restricts searches to targeted collections such as recent conference proceedings, and explains the rationale behind retrieved search results. Providing such rationales enhances explainability, user understanding, and trust [3]. To remain practical, the system must be low-cost to develop and operate. Conducting large-scale training for every conference proceeding is impractical, and sufficient training data may be unavailable for some paper collections. It is also crucial to quickly construct a search system from available paper collections. Furthermore, approaches that heavily rely on LLMs can lead to increased execution time and escalating API usage fees, reducing practical usability. To address these constraints, we propose a system that combines fast keyword-based search using BM25 with filtering through LLM-based semantic understanding, aiming to balance low development and execution costs with high accuracy. A query expansion module provides related terms to assist users who struggle to formulate effective search queries. Furthermore, for search results, the system generates selection rationales using an LLM to explain "why that paper was chosen," supporting user understanding and acceptance.

The results of the evaluation experiments on the AAAI 2025 proceedings showcase the effectiveness of our approach. Furthermore, the system outperformed pre-trained embedding vector methods (SpaCy vectors [5], OpenAI Embeddings), further validating its effectiveness.

The main contributions of this research are as follows:

- We designed and implemented a paper retrieval system with reason-based selection for specific conference proceedings.
- We combined a keyword-based method (BM25) with an LLM to achieve a low-cost, high-precision search technique.
- We implemented user support through search query expansion and selection rationale generation.
- Evaluation experiments of retrieval accuracy using the AAAI 2025 proceedings demonstrated performance surpassing both BM25 alone and pre-trained embedding vector methods.

2 Related Work

As a conventional method in information retrieval, BM25 has long been established for quantifying the relevance between queries and documents based on keywords. Its effectiveness and design principles were widely disseminated by Robertson et al. [11], who summarized experimental findings from TREC. This system adopts the lightweight and accurate BM25 as the core of its paper retrieval.

In recent years, LLM-assisted retrieval systems have been extensively studied. Zhu et al. [13] survey a range of LLM-based retrieval methods, with LLMs employed as modules–including query rewriters, retrievers, rerankers, and readers. For reranking, Sun et al. [12] report that unsupervised LLMs (e.g., GPT-4) combined with listwise approaches can in some cases outperform supervised models. Relatedly, Borges et al. [1] combine first-stage retrieval using a dense passage retriever (DPR) with zero-shot reranking by GPT-4 across domains such as movies, books, music, and games, and report significant ranking gains. In contrast, we propose combining BM25 for first-stage retrieval with LLM-based listwise filtering, yielding a cost-effective yet accurate paper retrieval system that requires neither full ranking optimization nor pre-training of word embeddings.

Khalid et al. [7] proposed ISCA, which combines inverted index-based BM25 with structured search (knowledge graph) and citation network analysis using Neo4j for paper retrieval, demonstrating improvements in nDCG and Precision on the ACL dataset. This research differs from ISCA, which requires knowledge graph construction and large-scale preprocessing, in that it combines lightweight keyword search with minimal LLM usage. This approach reduces development and execution costs while providing explainable selection rationale.

Chang et al. [2] reported that while Transformer models without pre-training do not outperform BM25 in large-scale document search, applying paragraph-level pre-training enables them to surpass BM25 performance. Furthermore, He et al. [4] introduced PaSa, which trained an LLM agent on search strategies for academic paper retrieval, achieving performance superior to conventional search engines and general-purpose LLMs on a newly constructed benchmark. However, these approaches require appropriate training data and significant training time, making them difficult to align with this research's objective of low development cost. This system aims to achieve performance exceeding BM25 while leveraging LLM without additional training, thereby balancing cost and accuracy.

Kawauchi et al. [6] demonstrated that PMI and TF-IDF can identify lesser-known but useful terms for users, thereby facilitating knowledge expansion. Our system exploits this knowledge-expansion capability of PMI to support users through query expansion.

Lubos et al. [8] find that LLM-generated explanations outperform boilerplate text in movie recommendation settings–improving comprehension, satisfaction, and perceived effectiveness–and are rated more highly by users. Building on this insight, we generate selection rationales for retrieved papers with LLMs to support user decision-making.

Prior studies have explored various approaches, including LLM-based reranking, the integration of BM25 with knowledge graphs, and large-scale pre-training. However, these approaches face challenges such as high inference costs and extensive preprocessing requirements. In contrast, our work aims to achieve high retrieval accuracy with reduced reliance on LLMs by combining efficient retrieval through BM25 with listwise filtering by LLMs. Moreover, we introduce PMI-based query expansion and LLM-based explanation generation to further support users in their search activities.

3 Methods

Our system consists of four main modules and one database, as shown in Fig. 1. First, the user updates their search query using the (i) Query Expansion Module. Once satisfied, they retrieve a set of candidate relevant papers via the (ii) Paper Retrieval Module. Next, the (iii) Paper Selection Module further filters the candidates. Finally, the (iv) Explanation Generation Module outputs the selected papers along with the rationale for their selection. The information required for these modules (e.g., titles, corpus vocabulary) and the scores (e.g., PMI, BM25) are extracted and computed from the paper collection before the search execution and stored in the database. This section describes each module in detail and outlines the construction of the database.

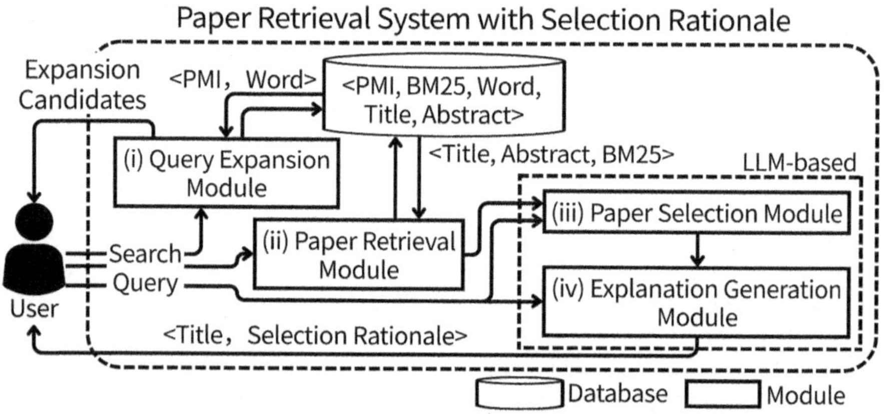

Fig. 1. System diagram

3.1 Query Expansion Module

Users unfamiliar with specialized fields may struggle to generate keywords that match their search intent. The Query expansion module supports such users in formulating effective search queries. To achieve this, we utilize pointwise mutual information (PMI), a metric based on the co-occurrence relationship between the user's search query and corpus vocabulary. PMI measures the degree to which two words co-occur compared to the probability expected by chance. A higher PMI value indicates a stronger contextual relationship between the words. Below is the definition of PMI based on document frequency. Let D denote the document set (set of titles + abstracts), V denote the vocabulary set, and $N = |D|$ denote the number of documents. $N(D \mid w)$ represents the number of documents containing word $w \in V$, and $N(D \mid w_i, w_j)$ represents the number of documents containing both words.

$$P(w) = \frac{N(D \mid w)}{N} \tag{1}$$

$$P(w_i, w_j) = \frac{N(D \mid w_i, w_j)}{N}. \tag{2}$$

$$\mathrm{PMI}(w_i, w_j) = \log_2 \frac{P(w_i, w_j)}{P(w_i)\, P(w_j)} \tag{3}$$

$$= \log_2 \frac{N(D \mid w_i, w_j)\, N}{N(D \mid w_i)\, N(D \mid w_j)}. \tag{4}$$

For a query $Q = \{q_1, q_2, \ldots, q_m\}$, the relevance of each word $w \in V$ is given by the following formula:

$$S(w \mid Q) = \sum_{q \in Q} \mathrm{PMI}(w, q). \tag{5}$$

Based on this score, words deemed to have strong contextual relevance to the input term are presented to the user as "expansion candidates." By default, the module extracts the top 50 words with the highest scores and displays them to the user in batches of five, starting with the highest-ranked. This threshold balances usability but can be adjusted based on future system conditions and user needs. Users may then select candidates matching their intent from the presented list and add them to the search query. In implementation, to avoid $\log 0$, expressions where $N(D \mid w_i, w_j) = 0$ are set to $-\infty$.

3.2 Paper Retrieval Module

The paper retrieval module receives the user's final search query and retrieves relevant papers from the database. The module treats the title and abstract as a single document and computes the relevance between the search query and the document using BM25. BM25 is a ranking function that evaluates the importance of each query term to a document, taking into account the frequency of that word's occurrence within the document (TF), the word's rarity (IDF), and the document length. Papers with higher BM25 values are deemed more relevant at the keyword level.

The definition of BM25 is given below. $f(w, d)$ denotes the number of occurrences of term w in document $d \in D$, $|d|$ is the number of words in d, avgdl is the average document length in the collection, k_1 and b are hyperparameters, N is the number of documents, and $N(D \mid w)$ is the number of documents containing the word w (document frequency). First, IDF is given by the following formula:

$$\mathrm{IDF}(w) = \log \frac{N+1}{N(D \mid w)+1} + 1. \tag{6}$$

The BM25 score of document d for search query $Q = \{q_1, q_2, \ldots, q_m\}$ is given by the following formula:

$$\mathrm{BM25}(d, Q) = \sum_{i=1}^{m} \mathrm{IDF}(q_i) \frac{(k_1+1) f(q_i, d)}{f(q_i, d) + k_1 \left(1 - b + b \frac{|d|}{\mathrm{avgdl}}\right)}. \tag{7}$$

For the parameters, following [9], we set $k_1 = 1.2$ and $b = 0.75$ as standard default values in the absence of parameter tuning. This module extracts the top K papers in descending order of the score obtained from the above equation and passes them as a candidate list to the paper selection module (Sect. 3.3). K can be adjusted according to the task and computational resources; in the experiments (Sect. 4), $K = 50$ and $K = 100$ were used.

3.3 Paper Selection Module and Explanation Generation Module

The paper selection module is a filtering module that selects papers from the candidate set obtained by the paper retrieval module to better match the user's search intent. The BM25 used in the paper retrieval module is an indicator based on word frequency, and therefore cannot directly capture context or semantic similarity. This module aims to address this limitation by leveraging the natural language processing capabilities of an LLM. The LLM evaluates the semantic relevance between the search query and each candidate (title + abstract) and selects the papers it determines to be the best matches. This achieves higher search precision compared to BM25 alone.

The explanation generation module instructs the LLM to generate selection reasons for each paper chosen by the paper selection module. Paper selection and explanation generation are performed simultaneously using a single prompt, ensuring consistency while minimizing processing costs. Providing users with selection rationales is expected to facilitate their decision-making on whether to read the paper.

The processing flow is as follows. First, using the user's search query and the top n candidates (e.g., $K = 50$ or 100), generate the prompt shown in Fig. 2. The LLM selects five highly relevant papers and outputs the reasons for selecting each paper. The response is received in JSON format, and the user is presented with pairs of paper titles and selection reasons.

```
Select 5 papers related to the following words from
the following (Paper Count) papers.
Explain the paper you selected and the reasons for
your selection.
words: (list of words)
papers:
Title[<title_1>] <abstract_1>
...
Title[<title_n>] <abstract_n>
Return a JSON object with the following content.
Key is the paper title, value is the reason for
selection and an explanation.
like this:
{
selected paper title 1: reason text,
...
selected paper title 5: reason text
}
```

Fig. 2. Prompt used for paper selection and rationale generation

3.4 Constructing a Database

The system extracts necessary information and computes numerical values from the paper collection prior to user searches, then stores the results in the database. First, it extracts the document set D (consisting of paper titles and abstracts) from the paper collection such as conference proceedings, then constructs a Bag-of-Words (BoW) representation and vocabulary V from D. This process employs the "SpaCy v3.8.6 en_core_web_sm model" to perform morphological analysis, stopword removal, symbol deletion, lowercasing, and lemmatization. Based on the created BoW, BM25 scores and PMI values are then computed. This information (PMI, BM25, D, and V) is stored in the database for use by various modules.

3.5 Algorithm Description

Algorithms 1–4 describe the system's procedures in pseudocode. Among them, Algorithm 1 serves as the entry point, orchestrating the overall execution of the system.

First, the search query expansion module shown in Algorithm 2 is executed with the query Q and the vocabulary V as input (QueryExpansion(V, Q), line 3 in Algorithm 1). The score of query Q is a sum of the PMI value for word pairs (q, w), where $q \in Q$ and $w \in V$. The top 50 words are then selected as candidate expansions Q' (Top50Words($Scores$), line 7 in Algorithm 2), which are presented to the user. Based on these candidates, the user can refine the query as needed (AskUserToExpandQuery(V, Q, Q'), line 8 in Algorithm 1). After Algorithm 2 and the user's query refinement are completed, the system proceeds to the next algorithm as instructed by the user (AskUserToFinish(), line 5 in Algorithm 1).

Next, the paper retrieval module shown in Algorithm 3 is executed with Q, the document set D, and the parameter K as input (PaperRetrieval(D, Q, K), line 11 in Algorithm 1). For each document d in D, its BM25 score with respect to Q is computed, and the top K documents are returned as the candidate set R (TopKDocs($Scores, K$), line 7 in Algorithm 3).

Finally, SelectionAndExplanation(Q, R), line 12 in Algorithm 1 provides the explanation of the selected papers with Q and the candidate set R. This method constructs a prompt (Fig. 2) with Q and R (MakePrompt(Q, R), line 2 in Algorithm 4), after which an LLM performs paper selection and explanation generation (LLM(*prompt*), line 3 in Algorithm 4). Irrelevant titles are filtered out (FilterInvalidTitles(*response*), line 4 in Algorithm 4), and the final results are returned to the user.

Algorithm 1 Paper Retrieval with Selection Rationale

1: **procedure** SEARCH(D, V, Q, K)
2:     **while** true **do**
3:         $Q' \leftarrow$ QueryExpansion(V, Q)
4:         **print**(Q')
5:         **if** AskUserToFinish() is OK **then**
6:             **break**
7:         **else**
8:             $Q \leftarrow$ AskUserToExpandQuery(Q, Q')
9:         **end if**
10:     **end while**
11:     $R \leftarrow$ PaperRetrieval(D, Q, K)
12:     $result \leftarrow$ SelectionAndExplanation(Q, R)
13:     **return** $result$
14: **end procedure**

Algorithm 2 Query Expansion Module

procedure QUERYEXPANSION(V, Q)
    $Scores \leftarrow \emptyset$
    for all $w \in V$ **do**
        $pmi \leftarrow \sum_{q \in Q} \mathrm{PMI}(q, w)$
        $Scores \leftarrow Scores \cup \langle w, pmi \rangle$
    end for
    $Q' \leftarrow$ Top50Words($Scores$)
    return Q'
end procedure

Algorithm 3 Paper Retrieval Module

procedure PAPERRETRIEVAL(D, Q, K)
  $Scores \leftarrow \emptyset$
  for all $d \in D$ **do**
    $bm25 \leftarrow \text{BM25}(Q, d)$
    $Scores \leftarrow Scores \cup \langle d, bm25 \rangle$
  end for
  $R \leftarrow \text{TopKDocs}(Scores, K)$
  return R
end procedure

Algorithm 4 Paper Selection And Explanation Generation Module

procedure SELECTIONANDEXPLANATION(Q, R)
  $prompt \leftarrow \text{MakePrompt}(Q, R)$ ▷ See Fig. 2
  $response \leftarrow \text{LLM}(prompt)$
  $results \leftarrow \text{FilterInvalidTitles}(response)$
  return $results$
end procedure

4 Experiments

We conducted evaluation experiments to verify the retrieval accuracy of our system. The experiments focused on two primary objectives. First, to examine whether the paper retrieval using BM25, adopted in this system as a fast method requiring no pre-training, performs better than other fast methods that also require no pre-training, such as pre-trained vectors and TF-IDF. Second, to evaluate whether combining LLM-based filtering with BM25 improves retrieval accuracy compared to BM25 alone. In addition, we investigated how accuracy changes when varying the number of candidate papers output by the paper retrieval module or the LLM model used.

These experiments aimed to clarify the validity of BM25 as the foundation for paper retrieval and to assess whether LLM-based paper selection contributes to improved accuracy.

The comparison involved eight approaches in total, including the proposed method. First, we evaluated cosine similarity-based search using pre-trained vector representations: the "SpaCy v3.8.6 en_core_web_lg model" and the "OpenAI Embedding text-embedding-3-small model (accessed 2025-6-26)". Next, keyword search methods utilizing TF-IDF and BM25 were evaluated. Finally, for the system, we evaluated four distinct patterns: setting the number of candidate papers presented to the LLM to 50 or 100, and using either Gemini 2.0 Flash or Gemini 2.5 Flash as the LLM.

The paper collection used for retrieval consisted of the AAAI 2025 proceedings. The sets of search queries and paper evaluations were generated using an LLM. Specifically, the search queries were selected from multiple queries generated by ChatGPT-5, from which 20 queries deemed appropriate were selected (Table 1). "ChatGPT-5 Thinking" was employed to evaluate the relevance between the papers and the search queries. To ensure fairness, the LLM used for evaluation differed from the one used in the system.

The experimental flow was as follows. First, search results for each query were obtained from eight different methods. These sets of papers were then consolidated into a single, duplicate-free collection. The relevance of each paper in this collection to the search query was evaluated using ChatGPT-5 Thinking. ChatGPT was provided with the search query, paper title, abstract, and a rating scale from 0 to 10. The specific prompt section regarding the evaluation criteria is shown in Fig. 3. Finally, the accuracy of each search method was computed based on the obtained paper evaluations.

Three metrics were used to evaluate the methods: "average," "percentage of 10 ratings," and "percentage of 7 or higher ratings." 'Average' is the mean rating value of the papers obtained as search results. "Percentage of 10 ratings" indicates the proportion of papers with a rating of 10 (highly relevant papers) in the search results, while "percentage of 7 or higher ratings" indicates the proportion of papers with a rating of 7 (relevant papers) in the search results.

Table 1. Experimental query list

No.	Query
1	reasoning
2	robot
3	fairness, ai
4	llm, explainable
5	climate, sustainability, ai
6	privacy, federated, learning
7	dialogue, ai, creativity
8	reinforcement, learning, application
9	human, robot, collaboration
10	knowledge, representation, learning
11	language, model, reasoning, planning
12	multi, agent, learning, cooperation
13	robot, human, interaction
14	healthcare, reinforcement, learning, diagnosis
15	large, language, model, application
16	fairness, explainability, machine, learning, model
17	knowledge, graph, question, answering, reasoning
18	autonomous, robot, planning, control, system
19	image, language, multimodal, learning
20	vision, language, grounding, understanding, model

```
Your task is to evaluate the relevance between
a given keyword list and each of the (Number of
papers)
papers (title and abstract) on a 0-10 scale.
Evaluation Criteria
- 0: Not related at all.
- 1-3: Only slightly related, with peripheral
mentions.
- 4-6: Mentions some of the keywords, but not as the
central theme.
- 7-9: Covers many of the keywords; the main theme
is highly relevant.
- 10: Strongly covers most of the keywords; the main
topic of the paper is very closely related to the
keywords.
```

Fig. 3. Prompt used for paper relevance evaluation

5 Results and Discussion

5.1 Results

The retrieval accuracy evaluation results are shown in Table 2. Among existing systems, BM25 achieved the highest score, recording an average score of 8.20. The proposed system, which combines BM25 with LLM-based filtering, outperformed it with an average score of 8.73. This improvement trend was consistently observed in the proportion of highly relevant papers (Score = 10 ratio) and relevant papers (Score≥7 ratio), with our system achieving the best performance in both cases. Comparing LLM versions, Gemini 2.5 Flash demonstrated higher accuracy than Gemini 2.0 Flash. Furthermore, for Gemini 2.5 Flash, providing 100 candidate papers to the LLM yielded better results.

Table 2. Comparison of eight retrieval methods across three evaluation metrics. All metrics are better when higher

Method	Mean score	Score = 10 ratio	Score≥7 ratio
SpaCy LG	5.12	0.08	0.43
OpenAI Embedding	8.01	0.21	0.79
BM25	8.20	0.22	0.85
TFIDF	6.75	0.14	0.60
BM25top50+Gemini2.0	8.32	0.25	0.90
BM25top50+Gemini2.5	8.65	0.28	**0.93**
BM25top100+Gemini2.0	8.30	0.24	0.89
BM25top100+Gemini2.5	**8.73**	**0.33**	**0.93**

5.2 Example of Execution

An example of this system's operation is shown in Table 3 and Fig. 4. This example targets the AAAI 2025 proceedings, with the user's search query being "llm, explainable." The figures show the top 10 expansion candidates proposed by the query expansion module, along with two papers selected by the paper retrieval and selection modules. Each paper is accompanied by a rationale generated by the explanation module. These results demonstrate that the system effectively presents users with diverse expansion candidates for their search queries and retrieves papers that are highly relevant to those queries.

Table 3. Examples of query expansion candidates generated based on PMI scores when the search query "llm, explainable" are provided

shopping	contestable	classname	incoherence	actionability
profusion	wordification	argllm	purposefully	contestability

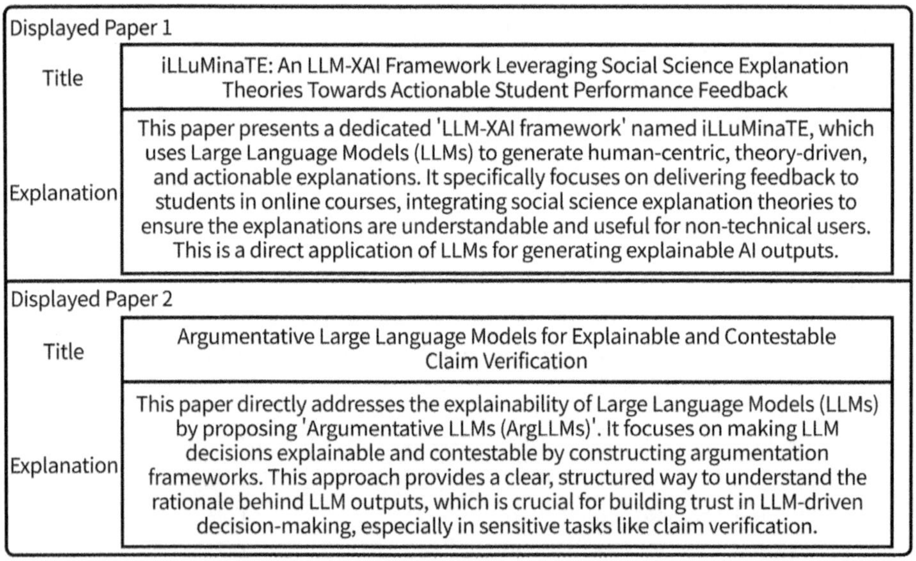

Fig. 4. Example of system output for AAAI 2025 proceedings. The search query is 'llm, explainable'. The figure shows the titles of papers selected by the system and the reasons for selecting each paper

5.3 Discussion

The experimental results showed that, apart from our system, BM25 achieved the highest score. This is likely because the search queries are keyword-based, and BM25, unlike general-purpose embeddings, is a metric specialized for the targeted corpus.

In contrast, our system achieved scores surpassing those of BM25. This improvement can be attributed to the LLM's ability to compensate for BM25's limitations in semantic understanding, which relies on keywords, enabling more contextually appropriate paper retrieval.

When comparing system configurations, the version using Gemini 2.5 Flash demonstrated improved performance when the number of papers provided to the LLM was increased, while Gemini 2.0 Flash showed no significant change. This result suggests that Gemini 2.5 Flash possesses a greater capacity to process diverse candidate sets and effectively select relevant papers.

Overall, these results confirm that the proposed system outperforms BM25 in terms of retrieval accuracy. However, several issues remain. First, larger-scale evaluation experiments are necessary. In this study, only 20 search queries were used; to ensure greater statistical reliability, a larger and more diverse set of queries should be employed. In addition, comparisons with a wider range of retrieval methods beyond pre-trained embedding vectors and BM25 are required.

Next, determining the appropriate number of papers K passed from the paper retrieval module to the LLM is an important consideration for future improvements. Meng et al. [10] demonstrated that, when using retrievers such as BM25, optimizing the cut-off (i.e., the number of candidates passed to re-ranking) is effective, and that in some cases even fixed-depth settings can provide strong baselines. In contrast, this study evaluated only two settings, $K = 50$ and $K = 100$. For higher performance, K should be carefully tuned according to the target corpus and the characteristics of the search queries.

Furthermore, to more clearly demonstrate the effectiveness of the system, it is necessary to quantitatively evaluate the contribution of query expansion and selection rationale generation to user support. Finally, the increase in processing time associated with the use of LLMs also remains a practical challenge.

6 Conclusion

This study investigates a paper retrieval system designed to help users efficiently gather information from very recent international conference proceedings that are not yet accessible through standard web search engines. We intend for the system to support early-stage analysis of emerging research topics before such papers become available to LLM-based deep research tools. The proposed system comprises a query expansion module and an explanation generation module for selection reasons. Our experimental results demonstrate that the proposed system outperforms BM25 in terms of retrieval accuracy with only minimal reliance on LLMs. Specifically, our method achieves high retrieval scores (8.73 vs. 8.20 with BM25 and 8.01 with a general-purpose embedding model), demonstrating its effectiveness for conference paper retrieval tasks. The system assists users in constructing effective search queries by presenting expansion candidates based on PMI, and it combines fast BM25 search with semantic filtering using an LLM. Consequently, it presents papers that are highly relevant to the user's query, along with explanations for their selection. We expect that the proposed system will be able to support students unfamiliar with certain academic fields, as well as non-native English speakers, in efficiently gathering information from recent international conference proceedings.

Acknowledgement. This work was supported in part by JSPS KAKENHI Grant Numbers JP24K03052, JP25K21351.

References

1. Borges, L., Jha, R., Callan, J., Martins, B.: Generalizable tip-of-the-tongue retrieval with LLM re-ranking. In: Yang, G.H., Wang, H., Han, S., Hauff, C., Zuccon, G., Zhang, Y. (eds.) Proceedings of the International ACM SIGIR Conference on Research and Development in Information Retrieval, SIGIR 2024, pp. 2437–2441. ACM (2024). https://doi.org/10.1145/3626772.3657917
2. Chang, W.C., Yu, F.X., Chang, Y.W., Yang, Y., Kumar, S.: Pre-training tasks for embedding-based large-scale retrieval. In: Proceedings of the International Conference on Learning Representations, pp. 1–12. OpenReview.net (2020). https://openreview.net/forum?id=rkg-mA4FDr
3. Gilpin, L.H., Bau, D., Yuan, B.Z., Bajwa, A., Specter, M., Kagal, L.: Explaining explanations: an overview of interpretability of machine learning. In: Bonchi, F., et al. (eds.) Proceedings of the International Conference on Data Science and Advanced Analytics, pp. 80–89. IEEE (2018). https://doi.org/10.1109/DSAA.2018.00018. https://ieeexplore.ieee.org/document/8631448
4. He, Y., et al.: PaSa: an LLM agent for comprehensive academic paper search. In: Che, W., Nabende, J., Shutova, E., Pilehvar, M.T. (eds.) Proceedings of the Annual Meeting of the Association for Computational Linguistics (Volume 1: Long Papers), pp. 11663–11679. ACL (2025). https://doi.org/10.18653/v1/2025.acl-long.572. https://aclanthology.org/2025.acl-long.572/
5. Honnibal, M., Montani, I., Landeghem, S.V., Boyd, A.: Spacy: industrial-strength natural language processing in python (2020). https://doi.org/10.5281/zenodo.1212303
6. Kawauchi, S., Toyota, T., Nobuhara, H.: Knowledge expansion support by related search keyword generation based on Wikipedia category and pointwise mutual information. JACIII **16**(2), 247–255 (2012). https://doi.org/10.20965/jaciii.2012.p0247
7. Khalid, S., Wu, S., Wahid, A., Alam, A., Ullah, I.: An effective scholarly search by combining inverted indices and structured search with citation networks analysis. IEEE Access **9**, 120210–120226 (2021). https://doi.org/10.1109/ACCESS.2021.3107939. https://ieeexplore.ieee.org/document/9522111
8. Lubos, S., Tran, T.N.T., Felfernig, A., Polat Erdeniz, S., Le, V.M.: LLM-generated explanations for recommender systems. In: Purificato, E., Malloci, F.M., Shulner-Tal, A. (eds.) Adjunct Proceedings of the ACM Conference on User Modeling, Adaptation and Personalization, UMAP Adjunct 2024, pp. 276–285. ACM (2024). https://doi.org/10.1145/3631700.3665185

9. Manning, C.D., Raghavan, P., Schütze, H.: Introduction to Information Retrieval, chap. 11, pp. 219–235. Cambridge University Press (2008). https://doi.org/10.1017/CBO9780511809071. https://nlp.stanford.edu/IR-book/pdf/11prob.pdf
10. Meng, C., Arabzadeh, N., Askari, A., Aliannejadi, M., de Rijke, M.: Ranked list truncation for large language model-based re-ranking. In: Yang, G.H., Wang, H., Han, S., Hauff, C., Zuccon, G., Zhang, Y. (eds.) Proceedings of the International ACM SIGIR Conference on Research and Development in Information Retrieval, SIGIR 2024, pp. 141–151. ACM (2024). https://doi.org/10.1145/3626772.3657864
11. Robertson, S.E., Walker, S., Jones, S., Hancock-Beaulieu, M., Gatford, M.: Okapi at trec-3. In: Harman, D.K. (ed.) Proceedings of the Text REtrieval Conference, no. 500-225 in NIST Special Publication, pp. 109–126. NIST (1994). https://doi.org/10.6028/NIST.SP.500-225. http://trec.nist.gov/pubs/trec3/papers/city.ps.gz
12. Sun, W., et al.: Is ChatGPT good at search? Investigating large language models as re-ranking agents. In: Bouamor, H., Pino, J., Bali, K. (eds.) Proceedings of the Conference on Empirical Methods in Natural Language Processing, pp. 14918–14937. ACL (2023). https://doi.org/10.18653/v1/2023.emnlp-main.923. https://aclanthology.org/2023.emnlp-main.923/
13. Zhu, Y., et al.: Large language models for information retrieval: a survey. ACM Trans. Inf. Syst. **43**, 1–52 (2025). https://doi.org/10.1145/3748304

Optimizing Retrieval in RAG Systems with Reinforcement Learning: A Trade-Off Between Quality and Cost

Chaithra Lokasara Mahadevaswamy[1], Khoa Nguyen[1], Mayank Singh[4], and Hsien-Tsung Chang[1,2,3(✉)]

[1] Department of Artificial Intelligence, Chang Gung University, Taoyuan 333, Taiwan
smallpig@widelab.org

[2] Department of Computer Science and Information Engineering, Chang Gung University, Taoyuan 333, Taiwan

[3] Center for Artificial Intelligence in Medicine, Chang Gung Memorial Hospital at Linkou, Taoyuan 333, Taiwan

[4] Department of Mathematics, Indian Institute of Technology Delhi, New Delhi 110016, India

Abstract. In natural language processing (NLP), Retrieval-Augmented Generation (RAG) integrates retrieval with generation to improve question answering. Traditional RAG relies on static chunking, which limits adaptability to query-specific contexts. To overcome this, an adaptive chunking mechanism with reinforcement learning (RL) is introduced to balance answer quality, token efficiency, and computational cost. Experiments conducted on Wikitext-2 corpus with Cloze QA and Mistral-generated QA tasks, using Qwen2.5-14B as the answering model, show that RL-based chunking competitively outperforms fixed strategies. Both LLM-based and non-LLM metrics confirm gains in contextual coherence and retrieval quality, highlighting adaptive chunking as a promising direction for next-generation RAG pipelines.

Keywords: Retrieval-Augmented Generation · Reinforcement Learning · Adaptive Chunking · Question Answering · NLP · NLG · Information Retrieval · Large Language Models

1 Introduction

As unstructured textual data continues to grow at scale, ensuring efficient and context-preserving document processing has become essential for downstream NLP tasks. Retrieval-Augmented Generation (RAG) [2] has emerged as a promising framework that grounds generative models in retrieved evidence, thereby enhancing factuality and coherence in question answering and related tasks.

C.L. Mahadevaswamy and K. Nguyen—Contributed equally to this work.

H. Unger and P. Meesad (Eds.): NLPIR 2025, LNNS 1904, pp. 696–711, 2026.
https://doi.org/10.1007/978-3-032-20897-2_49

However, traditional chunking methods–such as fixed-length segmentation or sentence-based splitting–often fragment context [7], increase token usage, and degrade retrieval accuracy. These shortcomings limit the effectiveness of RAG systems, where retrieval quality is tightly coupled with generative performance.

To overcome these challenges, reinforcement learning (RL) is introduced for adaptive chunking in RAG pipelines [6]. Unlike static methods, RL enables dynamic optimization of chunking strategies, aligning retrieved content more closely with query-specific contexts. In particular, Proximal Policy Optimization (PPO) [3] and Deep Q-Networks (DQN) [8] are explored as training strategies to refine chunk selection for both per-document and global retrieval modes. The experiments are specifically designed for question answering (QA) retrieval [5], highlighting how adaptive chunking can improve accuracy and efficiency in knowledge-intensive QA tasks.

This research seeks to address the following critical questions:

- Which retrieval strategy minimizes token usage while preserving answer quality?
- Which method offers greater cost-effectiveness in terms of training and inference efficiency?
- How do RL-based chunking methods compare to traditional approaches in per-document versus global retrieval?
- Which models generalize better across datasets and evaluation metrics?

1.1 Contributions

The main contributions of this study are as follows:

- **Mistral Trivia-QA Dataset:** A new QA dataset is introduced, generated using the Wikitext-2 corpus and Mistral, consisting of question–answer–context triples. This dataset complements a cloze-style QA dataset and provides a richer benchmark for evaluating adaptive chunking. Both datasets are released as valuable resources for the RAG community.

- **RL-based Adaptive Chunking:** [6] A novel adaptive chunking approach is developed using PPO [3] and DQN [8], designed to optimize retrieval strategies in RAG systems by dynamically adjusting chunking decisions [4].

- **Comprehensive Evaluation:** Evaluations are conducted on two QA datasets (Cloze QA and Mistral QA), using both non-LLM metrics (Hit@1, Hit@5, MRR, token counts, objective accuracy, recall) and LLM-based metrics following the RAGAS framework (faithfulness, context precision, answer correctness, answer relevancy, and context recall).

- **Trade-off and Efficiency Study:** Token usage, training cost, and retrieval quality are systematically analyzed to identify conditions under which traditional RAG performs well and scenarios where RL-based chunking offers significant advantages.

- **Generalization Insights:** The performance of different methods is compared across per-document and global retrieval, highlighting the strengths, limitations, and generalization capacity of RL-enhanced RAG.

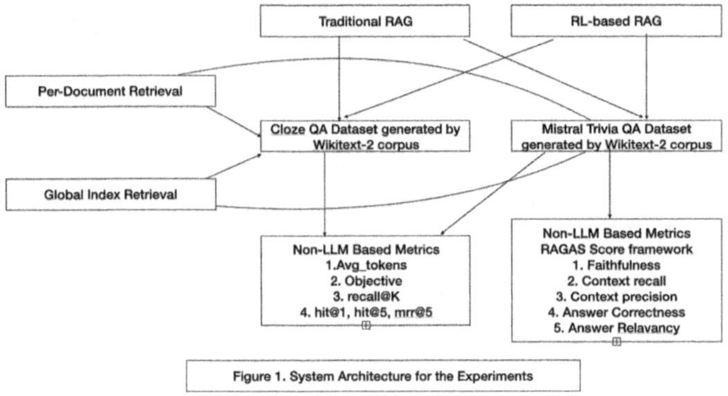

Fig. 1. System Architecture for the Experiments

To the best of current knowledge, this is the first comprehensive study that integrates reinforcement learning for adaptive chunking in RAG systems [4]. By addressing these aspects, this work offers both a methodological contribution and a benchmark resource, paving the way for more efficient, contextually coherent, and scalable RAG pipelines [2].

2 Methodology

This section outlines the experimental framework for adaptive chunking in RAG pipelines [2]. It covers the system architecture, reinforcement learning setup, agent design and reward formulation, and the core components of the retrieval and generation pipeline. In addition, details of the traditional RAG setup, datasets used, and evaluation metrics are presented to provide a comprehensive understanding of the methodology.

2.1 System Architecture

Figure 1 presents the experimental setup, comparing traditional RAG with RL-based RAG under per-document and global index retrieval. Two datasets are used: a Cloze QA dataset from Wikitext-2 for fill-in-the-blank tasks and the Mistral Trivia QA dataset with richer question–answer–context triples. Evaluation combines non-LLM metrics (token usage, objective accuracy, recall@K, Hit@1/5, MRR@5) with [1] LLM-based RAGAS metrics (faithfulness, context recall, context precision, answer correctness, answer relevancy), providing a balanced assessment of retrieval efficiency and answer quality.

2.2 Core Components of Adaptive Chunking [4]

The adaptive chunking framework is modeled as a reinforcement learning (RL) problem, where an agent learns to select optimal chunking parameters for RAG pipelines. Table 1 outlines the key RL-based modules, including the agent, state, actions, and reward design. In parallel, Table 2 presents the core components of the retrieval and generation pipeline, highlighting how document storage, embedding, retrieval, ranking, and answer generation are integrated within the system.

Table 1. RL-based chunking modules in the adaptive retrieval pipeline.

Module	Description/Function
Documents	Unstructured input text to be segmented for retrieval and generation
RL Agent	Learns adaptive chunking strategies using reinforcement learning signals
State	Encodes chunk context, document metadata, and remaining text length
Actions	Adjusts chunk boundaries and contextual inclusion dynamically
Reward	Derived from retrieval performance and efficiency (Eq. 1)

State: Each document is represented by lightweight statistical features, including the number of sentences, average sentence length, and total word count. These features capture structural variations across documents while remaining computationally efficient.

Actions: Two decisions are made: (i) *chunk length* L, selected by a Proximal Policy Optimization (PPO) [3] agent from a continuous action space normalized to $[0, 1]$ and mapped to a discrete grid (e.g., $\{256, 384, 512, 640, 800, 1024\}$); and (ii) *chunk overlap* O, chosen by a Deep Q-Network (DQN) [8] agent from a discrete set (e.g., $\{0, 16, 32, 64, 128, 192\}$). The combination (L, O) defines the chunking strategy for each document.

Reward and Policies: For each document, retrieval performance is measured by recall@k and penalized by the average number of tokens retrieved:

$$R = \text{recall@k} - \lambda \cdot \text{avg\_tokens}, \tag{1}$$

where $\lambda = 0.0002$ balances retrieval accuracy with efficiency. Rewards are aggregated across multiple documents per episode to stabilize training.

Fixed-Size Policy: Traditional RAG baselines [2,7] use static chunking configurations (L, O) to provide reference points for comparison. Three representative settings are selected based on empirical grid search over length and overlap:

- $(512, 32)$ – chosen as the **primary baseline** for all experimental studies, since it offers efficient token usage and balanced recall, serving as the standard fixed-size configuration.
- $(800, 32)$ – included as a **medium-context baseline**, where recall improves modestly compared to $(512, 32)$, though at slightly higher computational cost.
- $(1024, 192)$ – used as a **high-recall upper bound**, providing nearly perfect recall but at significantly higher token and computational cost.

These three points were selected because they span the practical recall–cost trade-off curve observed in traditional RAG [2]: from efficient low-cost retrieval, through a balanced intermediate configuration, to a high-recall but expensive setting. By comparing against them, RL-based adaptive chunking can be evaluated on whether it achieves near-$(1024, 192)$ recall while maintaining efficiency closer to $(512, 32)$.

RL-based policy: Adaptive chunking is formulated as a dual-agent RL problem [5]:

- A **PPO agent** controls the chunk length L. Given state s (document features), the policy $\pi_\theta(a|s)$ is updated by maximizing the clipped surrogate objective:

$$L^{PPO}(\theta) = \mathbb{E}_t\Big[\min\big(r_t(\theta)\hat{A}_t,\ \text{clip}(r_t(\theta), 1-\epsilon, 1+\epsilon)\hat{A}_t\big)\Big], \tag{2}$$

  where $r_t(\theta) = \frac{\pi_\theta(a_t|s_t)}{\pi_{\theta_{\text{old}}}(a_t|s_t)}$ and $\hat{A}_t$ is the advantage estimate.

- A **DQN agent** [8] controls the chunk overlap O. The action-value function $Q(s, a)$ is learned by minimizing the temporal-difference error:

$$L^{DQN}(\theta) = \mathbb{E}\Big[\big(r + \gamma \max_{a'} Q(s', a'; \theta^-) - Q(s, a; \theta)\big)^2\Big], \tag{3}$$

  where θ^- is the target network, updated periodically for stability.

Comparison and Justification. The choice of reinforcement learning algorithms is guided by their complementary strengths:

- **DQN:** Suited for discrete boundary decisions, offering sample efficiency and reliable performance on structured text where possible chunk overlaps are limited and well-defined [8].
- **PPO:** Designed for continuous adaptation, providing stability and robustness in learning when documents are unstructured or exhibit varied lengths and contexts.

By combining DQN and PPO, the framework leverages both discrete efficiency and continuous flexibility. DQN ensures effective handling of fixed segmentation scenarios, while PPO dynamically [3] refines chunking decisions.

Together, PPO and DQN produce a joint policy (L, O) that dynamically adapts chunking per document. This setup allows exploration of the recall–cost trade-off space beyond fixed-size baselines, enhances retrieval precision, improves response relevance, and increases system adaptability across diverse textual inputs and aiming for near-maximum recall with reduced token usage (Fig. 2).

Table 2. Core components of the RAG-based retrieval and generation pipeline.

Component	Description/Function
LLMs	Mistral 7B for QA dataset generation and evaluation; Qwen2.5-14B for response generation
Document Store	Preprocessed and indexed text chunks serving as retrieval units
Embeddings	SentenceTransformers (`all-MiniLM-L6-v2`) for semantic similarity encoding
Retriever	InMemoryEmbeddingRetriever for top-k retrieval (default $k = 5$)
Ranker	TransformersSimilarityRanker to prioritize and reorder retrieved chunks
Generator	Large language model (Qwen2.5-14B) for producing coherent, context-grounded answers

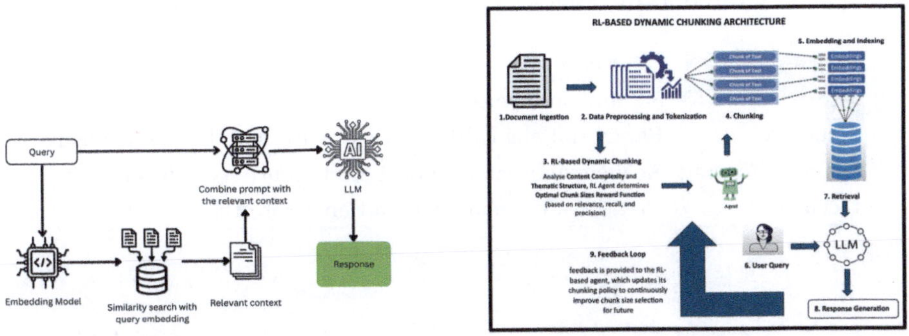

(a) Traditional RAG Architecture (b) RL-based RAG Architecture

Fig. 2. Comparison of traditional fixed-size RAG and RL-based adaptive RAG architectures.

Training Methodology: The adaptive chunking framework employs a dual-agent RL setup with two neural networks: a PPO policy [3] network for chunk length and a DQN network [8] for chunk overlap. Both networks are trained for 100,000 episodes, alternately updating in cycles to ensure stable convergence. Cached evaluations are used to accelerate rollouts and reduce redundant computations. The tokenizer is `o200k_base` (GPT-4o compatible), ensuring consistent token counting across all experiments. Sentence representations are computed using `all-MiniLM-L6-v2` from SentenceTransformers, with retrieval performed through a dense embedding index (FAISS-based when available, NumPy fallback otherwise). For every experiment, token usage is explicitly logged to allow precise recall–cost analysis. This setup enables dynamic, per-document adaptation of chunking policies while preserving comparability with static baselines.

Traditional RAG Baselines: Static chunking baselines are evaluated at three representative configurations: (512,32), (800,32), and (1024,192). These span the recall–cost trade-off curve: $(512, 32)$ serves as the primary low-cost baseline for all experiments, $(800, 32)$ provides a balanced middle ground, and $(1024, 192)$ offers near-perfect recall but at significantly higher token cost. RL-based adaptive chunking is assessed against these baselines to test whether learned policies can achieve near-$(1024, 192)$ recall while maintaining efficiency close to $(512, 32)$.

2.3 Dataset

WikiText-2. WikiText-2, a publicly available structured Wikipedia corpus, is employed to evaluate adaptive chunking methods. It is well-suited for Retrieval-Augmented Generation (RAG) and long-context understanding due to its paragraph-level alignment, token richness, and real-world linguistic patterns. Table 3 summarizes the main characteristics.

Table 3. Key characteristics of the WikiText-2 corpus.

Feature	Description
Size	~2 million tokens
Structure	Paragraph-aligned full articles with high linguistic fidelity
Domain Coverage	Academic, news, technical, and report-style documents
License	Creative Commons Attribution–ShareAlike

QA Dataset Construction. To evaluate chunking policies, two complementary QA datasets were automatically generated from WikiText-2 while preserving document structure and length:

- **Cloze QA:** Fill-in-the-blank style questions derived directly from sentences, focusing on local recall.

Table 4. Dataset splits for Cloze QA and Mistral Trivia-QA (both constructed from WikiText-2).

Split	Documents	QAs
Train	5,135	15,405
Validation	502	1,506
Test	569	1,707

– **Mistral Trivia-QA:** Richer question–answer–context triples generated using Mistral 7B, emphasizing broader reasoning.

Both Cloze QA and Mistral Trivia-QA datasets were constructed from the WikiText-2 corpus to ensure controlled evaluation of adaptive chunking. Each document produces exactly three unique QA pairs, preserving quality and avoiding redundancy. Table 4 reports the dataset splits, with $5,135 \times 3 = 15,405$ QAs in training, $502 \times 3 = 1,506$ in validation, and $569 \times 3 = 1,707$ in testing.

Cloze QA Structure: The Cloze QA dataset[1] follows a fill-in-the-blank question–answer format, primarily constructed from the WikiText-2 corpus. It is designed to evaluate language models on contextual understanding and knowledge recall. Cloze QA has one exact ground-truth answer tied to a specific `sent_id`.

```
{
  "doc_id": 0,
  "sent_id": 8,
  "title": "Robert Boulter",
  "question": "He appeared on a 2006 episode of the television
   series , ____ ,",
  "answer": "Doctors"
}
```

Mistral Trivia-QA Structure: Unlike the Cloze QA dataset (fill-in-the-blank style), the Mistral Trivia-QA dataset[2,3] contains richer triples with explicit ground truths and supporting context sentences, enabling fine-grained evaluation under RAGAS metrics [1]. Each QA entry follows the JSONL structure.

```
{
 "id": "doc_0000000__q000001",
```

[1] Dataset used for Exp 1: https://huggingface.co/datasets/CGU-Widelab/Cloze_QA_Dataset_Wikitext2.

[2] Model used for generation of QA dataset: https://huggingface.co/mistralai/Mistral-7B-Instruct-v0.3.

[3] Dataset used for Exp 2 and Exp 3: https://huggingface.co/datasets/CGU-Widelab/Mistral_Trivia-QA_Dataset.

```
 "question": "Which television series did he have a guest role in 2002?",
 "answer": "Judge John Deed",
 "ground_truths": ["Judge John Deed"],
 "contexts": [
   "He had a guest role in the television series Judge John Deed
    in 2002 ..."
 ],
 "doc_id": "doc_0000000",
 "title": "Robert Boulter",
 "evidence_sent_ids": [3],
 "meta": {
   "source": "hf-local",
   "model": "mistralai/Mistral-7B-Instruct-v0.3",
   "ctx_window": 2,
   "split": "test"
 }
}
```


This richer structure supports evaluation not only of retrieval recall but also of context faithfulness, precision, and answer correctness.

2.4 Evaluation Metrics

To provide a comprehensive assessment of adaptive chunking in RAG pipelines, evaluation is carried out using both non-LLM metrics and LLM-based metrics.

Non-LLM Metrics:

- **Recall@K:** Fraction of queries where at least one relevant chunk appears in the top-K retrieved results.
- **Hit@1/Hit@5:** Binary indicators of whether the ground-truth context appears at rank 1 or within the top-5 retrieved chunks.
- **MRR@5:** Mean Reciprocal Rank restricted to the top-5 retrieved chunks, measuring ranking quality.
- **Objective Accuracy:** Combined score balancing recall@K with token efficiency according to Eq. 1.
- **Token Usage:** Average number of tokens retrieved per query, reflecting efficiency and cost.

LLM-Based Metrics (RAGAS Score Framework): [1].

- **Faithfulness:** Degree to which generated answers are grounded in retrieved context without introducing hallucinations.
- **Context Recall:** Measures whether all necessary supporting information for answering a query is retrieved.
- **Context Precision:** Proportion of retrieved information that is directly relevant to the query.

- **Answer Correctness:** Semantic alignment of generated answers with gold-standard answers.
- **Answer Relevancy:** Evaluates whether generated answers directly address the query posed.

This dual evaluation setup ensures balanced measurement: non-LLM metrics capture retrieval efficiency and ranking performance, while LLM-based metrics evaluate semantic fidelity and generative answer quality.

3 Experiments and Results

This section presents the experimental evaluation of adaptive chunking in RAG pipelines. The results compare reinforcement learning–based chunking with traditional fixed-size baselines across retrieval modes, datasets, and evaluation metrics.

3.1 Experimental Setup and Configurations

Experiment 1 evaluates RL-based adaptive chunking against three representative fixed-size baselines, as introduced in Sect. 3.2: $(512, 32)$ as the primary baseline, $(800, 32)$ as a balanced mid-point, and $(1024, 192)$ as the high-recall upper bound. These configurations provide a consistent recall–cost trade-off curve for comparison.

For Experiments 2 and 3, which focus on the Mistral QA dataset, additional fixed-size configurations are selected based on observed top-performing policies: $(1024, 64)$, $(1024, 192)$, and $(640, 128)$. These settings reflect the strongest static baselines under Mistral QA conditions, enabling more competitive evaluation.

In all experiments, $(512, 32)$ is included as the standard baseline for fairness and consistency across datasets and retrieval modes.

3.2 Experiment 1: Non-LLM-Based Evaluation on Cloze QA

This section presents the evaluation of adaptive chunking on the Cloze QA dataset, considering both *per-document retrieval* and *global index retrieval.* Traditional RAG baselines (fixed-size chunking) are compared against RL-based adaptive chunking across validation and test sets. Results are reported in terms of recall@k, token usage, and objective (Eq. 1).

Per-document Retrieval. Table 5 summarizes the validation and test results under per-document retrieval. Across all settings, recall@k reaches 1.0, indicating that relevant answers are always retrieved. The key distinction lies in token efficiency: RL-based adaptive chunking consistently consumes fewer tokens (2–3% reduction) compared to the $(512, 32)$ baseline, yielding the highest objective score. This demonstrates that adaptive policies can achieve the same retrieval quality as static methods while reducing cost.

Table 5. Per-document retrieval results for the Cloze QA dataset. RL-based chunking achieves identical recall with fewer tokens, improving objective scores.

Run	Length	Overlap	Recall@k	Tokens	Objective	ΔTokens vs 512/32
Validation (per-doc)						
L512_O32	512	32	1.0000	505.8	0.8988	+0.0
L800_O32	800	32	1.0000	495.2	0.9010	−10.6
L1024_O192	1024	192	1.0000	506.1	0.8988	+0.3
RL-adaptive	–	–	1.0000	**491.3**	**0.9017**	−14.5
Test (per-doc)						
L512_O32	512	32	1.0000	506.4	0.8987	+0.0
L800_O32	800	32	1.0000	499.7	0.9001	−6.8
L1024_O192	1024	192	1.0000	511.3	0.8977	+4.9
RL-adaptive	–	–	1.0000	**496.2**	**0.9008**	−10.2

Global Index Retrieval. In global index retrieval, all document chunks are indexed together, introducing greater competition among contexts. Table 6 shows that RL-based adaptive chunking maintains competitive recall while reducing token usage by 2–3% compared to the (512, 32) baseline. Importantly, RL achieves the highest objective score under capped settings, demonstrating improved efficiency at scale. Fixed-size (1024, 192) remains strongest in recall but incurs prohibitive token costs, underscoring the advantage of adaptive strategies.

3.3 Experiment 2: Non-LLM-Based Evaluation on Mistral QA

This section presents the evaluation of adaptive chunking on the Mistral QA dataset, considering both *per-document retrieval* and *global index retrieval*. Traditional RAG baselines (fixed-size chunking) are compared against RL-based adaptive chunking across validation and test sets.

Table 7 reports results for both settings. In the per-document case, all policies reach perfect recall@5 (1.0), but differ in token usage; the RL-based policy attains an objective of 0.9160, close to the best fixed baseline (1024, 64), while remaining more efficient than higher-cost configurations. In the global index case, recall gaps become clearer: the RL-based policy reaches recall@5 of 0.8571, notably above the 512:32 baseline (0.7932). Although RL uses more tokens, the gain in recall leads to a better efficiency–accuracy trade-off, as reflected in its competitive objective scores.

3.4 Experiment 3: LLM-Based Evaluation on Mistral QA

Experiment 3 evaluates adaptive chunking using the Mistral QA dataset under the RAGAS framework [1]. Here, ground-truth answers are compared with

Table 6. Global index retrieval results for the Cloze QA dataset. RL-based chunking reduces tokens while maintaining recall, achieving the best objective under capped settings.

Policy	Recall@k	Tokens	Objective
Validation (global index)			
Fixed 512/32	0.9079	1638.4	0.5802
RL (cap 512/32)	0.9065	**1595.8**	**0.5873**
Fixed 800/32	0.9270	2054.9	0.5160
Fixed 1024/192	**0.9297**	2240.4	0.4816
RL (unconstrained)	0.9263	2171.3	0.4920
Test (global index)			
Fixed 512/32	0.9105	1718.9	0.5667
RL (cap 512/32)	0.9069	**1683.1**	**0.5703**
Fixed 800/32	0.9233	2164.2	0.4905
Fixed 1024/192	**0.9257**	2348.8	0.4560
RL (unconstrained)	0.9245	2276.4	0.4692

responses generated by Qwen2.5-14B, with Mistral-7B providing the QA contexts. This setup enables assessment of semantic quality, contextual alignment, and factual consistency of generated answers. Evaluation covers five dimensions: *Answer Relevancy*, *Faithfulness*, *Context Precision*, *Context Recall*, and *Answer Correctness*. Both per-document and global index retrieval modes are reported (Fig. 3 and 8).

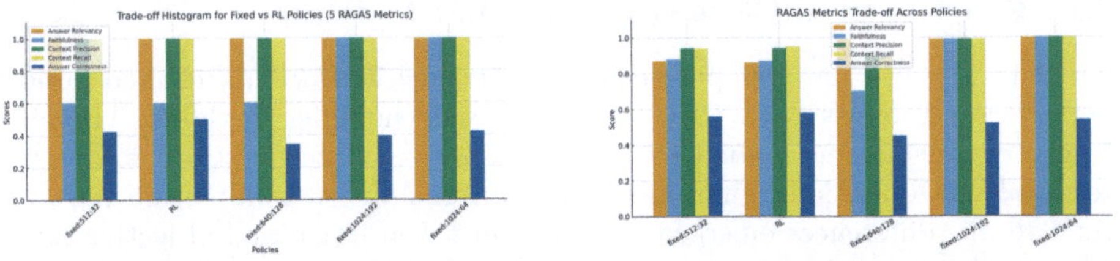

(a) Per-document retrieval (b) Global index retrieval

Fig. 3. Experiment 3 results on the Mistral QA dataset under the RAGAS framework. Histograms compare fixed-size policies and RL-adaptive chunking across five metrics: Answer Relevancy, Faithfulness, Context Precision, Context Recall, and Answer Correctness.

Table 7. Comparison of per-document and global index retrieval results on the Mistral QA dataset using non-LLM metrics. Objective defined as recall@5 − 0.0002 × avg_tokens.

Per-document Retrieval			
Policy	Recall@5	Avg Tokens	Objective
Fixed 1024:64	1.0000	416.7	**0.9166**
RL-adaptive	1.0000	420.1	0.9160
Fixed 512:32	1.0000	422.4	0.9155
Fixed 1024:192	1.0000	423.9	0.9152
Fixed 640:128	1.0000	438.6	0.9123
Global Index Retrieval			
Policy	Recall@5	Avg Tokens	Objective
Fixed 512:32	0.7932	1546.5	**0.4839**
RL-adaptive	0.8571	1974.6	0.4621
Fixed 640:128	0.8026	1758.2	0.4509
Fixed 1024:192	0.8037	1951.9	0.4134
Fixed 1024:64	0.8008	1939.7	0.4129

4 Results and Analysis

The experiments collectively address the central questions of this study: efficiency in token usage, cost-effectiveness, retrieval accuracy, and generalization capacity across datasets and retrieval modes.

4.1 Experiment 1: Cloze QA (Non-LLM Metrics)

On the Cloze QA dataset, per-document retrieval with $k = 5$ resulted in near-perfect recall across both fixed and RL-based chunking methods. The short length of documents combined with sentence-level cloze questions meant that retrieval was essentially guaranteed once all document chunks were retrieved. As a result, differences emerged primarily in token usage and objective scores. RL-adaptive chunking matched the high recall of large fixed configurations $(1024, 192)$ while approaching the efficiency of the $(512, 32)$ baseline. This confirmed that RL can learn cost-aware chunking policies without compromising accuracy, but the gains in this setting were incremental due to the simplicity of the dataset.

4.2 Experiment 2: Mistral QA (Non-LLM Metrics)

The Mistral QA dataset introduced more complex queries and longer contexts, making it a stronger benchmark. In the per-document setting, recall@5 again saturated near 1.0 across all policies, but RL-adaptive consistently reduced

Table 8. Comparison of per-document and global index retrieval results on the Mistral QA dataset using the RAGAS framework. [1]

Per-document Retrieval					
Policy	Ans. Rel.	Faith.	Ctx. Prec.	Ctx. Rec.	Ans. Corr.
Fixed 512:32	1.00	0.60	1.00	1.00	0.422
RL-adaptive	1.00	0.60	1.00	1.00	**0.500**
Fixed 640:128	1.00	0.60	1.00	1.00	0.344
Fixed 1024:192	1.00	1.00	1.00	1.00	0.395
Fixed 1024:64	~1.00	1.00	1.00	1.00	0.422
Global Index Retrieval					
Policy	Ans. Rel.	Faith.	Ctx. Prec.	Ctx. Rec.	Ans. Corr.
Fixed 512:32	0.87	0.88	0.94	0.94	0.56
RL-adaptive	0.86	0.87	0.94	0.95	**0.58**
Fixed 640:128	0.99	0.70	0.90	0.89	0.45
Fixed 1024:192	0.99	0.99	0.99	0.99	0.52
Fixed 1024:64	~1.00	1.00	1.00	1.00	0.54

token usage compared to high-recall fixed baselines. Under the global index setting, however, the retrieval task became significantly harder due to distractor chunks from unrelated documents. Here, RL clearly outperformed static baselines: recall@5 improved to 0.857 compared to ≈0.80 for the best fixed policies, demonstrating stronger generalization and discrimination in noisy search spaces. Still, when applying the objective $J = \text{recall@5} - \lambda \cdot \text{avg\_tokens}$ with $\lambda = 0.0002$, fixed small-window baselines such as $(512, 32)$ retained a slight edge due to their substantially lower token budget. These results highlight the trade-off: RL improves retrieval quality in global settings, but its benefits depend on how cost is weighted against recall.

4.3 Experiment 3: Mistral QA (LLM-Based Metrics)

RAGAS-based evaluation [1] provided deeper insights into generative quality. Per-document retrieval showed that all policies achieved high answer relevancy, faithfulness, and context alignment. The main differentiator was *Answer Correctness*, where RL-adaptive outperformed fixed baselines (0.50 vs 0.42 for $(512, 32)$). Under global index retrieval, RL again achieved the best correctness score (0.58), despite slightly lower faithfulness compared to very large fixed chunks. This indicates that adaptive policies help mitigate overfitting to context and improve alignment with ground-truth answers, even when retrieval noise is high. These results provide clear answers to the guiding research questions:

- **Token efficiency:** In per-document settings, all methods reach high recall, so efficiency becomes the differentiator. RL policies approach the cost of $(512, 32)$ while maintaining near-$(1024, 192)$ recall.

- **Cost-effectiveness:** With the chosen $\lambda = 0.0002$, small fixed windows perform best on the objective, but RL narrows the gap by balancing quality and token cost in more challenging conditions.
- **Retrieval performance:** RL-adaptive chunking consistently improves recall and correctness in global index retrieval, where distractors are abundant and static baselines struggle.
- **Generalization:** Across Cloze QA and Mistral QA, RL adapts to varied conditions: it converges to efficient chunk sizes in simple per-doc setups and scales to robust retrieval under noisy global indexes.

Fixed chunking is sufficient when documents are short and queries are simple (Cloze QA), but RL-based adaptive chunking provides clear advantages in complex, noisy retrieval settings (Mistral QA global index) and improves correctness in generative evaluation. This demonstrates the promise of RL as a cost-aware, context-sensitive strategy for adaptive preprocessing in RAG pipelines.

5 Conclusion and Future Work

This work proposed an RL-based adaptive chunking strategy for RAG that replaces one-size-fits-all static chunking with a dual-agent controller: PPO selects the chunk length (L) and DQN selects the overlap (O). By letting the policy tailor context windows to each document/query, the system preserves semantic coherence while avoiding redundant tokens. On both Cloze QA and Mistral Trivia-QA, RL-adaptive policies matched the perfect per-document recall of strong fixed baselines but did so with 2–3% fewer tokens, improving the overall objective. In the harder global-retrieval setting with many distractors, RL consistently outperformed static baselines (e.g., fixed 512/32) on recall@5, while keeping token growth modest, and RAGAS scores confirmed a better balance between faithfulness and correctness. Overall, dual-agent RL chunking conserves tokens by skipping redundant context while maintaining retrieval quality–an essential trade-off for scaling RAG.

Future work includes adding RLHF to align chunking with human preferences, exploring hierarchical retrieval, testing on longer/noisier corpora, and combining this with context engineering to make RAG efficient, adaptive, and human-aligned.

Acknowledgments. This work was supported by the National Science and Technology Council (NSTC), Taiwan, under Grant No. 114-2221-E-182-041-MY3. The authors also acknowledge the partial financial support from Chang Gung Memorial Hospital, Linkou, under Grant No. NERPD4Q0021.

References

1. Es, S., James, J., Espinosa-Anke, L., Schockaert, S.: RAGAs: automated evaluation of retrieval augmented generation. arXiv preprint arXiv:2309.15217 (2023). https://doi.org/10.48550/arXiv.2309.15217

2. Lewis, P., et al.: Retrieval-augmented generation for knowledge-intensive NLP tasks. In: Advances in Neural Information Processing Systems (NeurIPS). arXiv preprint arXiv:2005.11401 (2020). https://doi.org/10.48550/arXiv.2005.11401
3. Schulman, J., Wolski, F., Dhariwal, P., Radford, A., Klimov, O.: Proximal policy optimization algorithms. arXiv preprint arXiv:1707.06347 (2017). https://doi.org/10.48550/arXiv.1707.06347
4. Jeong, S., Baek, J., Cho, S., Hwang, S.J., Park, J.C.: Adaptive-RAG: learning to adapt retrieval-augmented large language models through question complexity. In: Proceeding of the NAACL (2024). arXiv preprint arXiv:2403.14403 (2024). https://doi.org/10.48550/arXiv.2403.14403
5. Tang, X., Gao, Q., Li, J., Du, N., Li, Q., Xie, S.: MBA-RAG: a bandit approach for adaptive retrieval-augmented generation through question complexity. In: Proceedings of the COLING (2025). arXiv preprint arXiv:2412.01572 (2025). https://doi.org/10.48550/arXiv.2412.01572
6. Kulkarni, M., Tangarajan, P., Kim, K., Trivedi, A.: Reinforcement learning for optimizing RAG for domain chatbots. arXiv preprint arXiv:2401.06800 (2024). https://doi.org/10.48550/arXiv.2401.06800
7. Singh, I.S., et al.: ChunkRAG: novel LLM-chunk filtering method for RAG systems. In: Proceedings of the Conference of the North American Chapter of the Association for Computational Linguistics: Student Research Workshop, NAACL SRW 2025 (2025). arXiv preprint arXiv:2410.19572. https://doi.org/10.48550/arXiv.2410.19572
8. van Hasselt, H., Guez, A., Silver, D.: Deep reinforcement learning with Double Q-learning. In: Proceedings of the Thirtieth AAAI Conference on Artificial Intelligence (AAAI), pp. 2094–2100 (2016). arXiv preprint arXiv:1509.06461 (2015). https://doi.org/10.48550/arXiv.1509.06461

INSPIRE: A System that Models Authorial Intent for Scholarly Paper Recommendation

Manami Yano(✉), Didier Gohourou, Valentinus Roby Hananto, and Victor Kryssanov

College of Information Science and Engineering, Ritsumeikan University, Osaka, Japan
is0694sp@ed.ritsumei.ac.jp

Abstract. Scholarly writing is both a record and a driver of evolving ideas, where a draft paragraph would convey the author's research intent. The existing digital tools that aim to support scholarly writing consider scholarly retrieval separate from this process and rely on static queries or past history. This leads to neglecting the author's current line of thought and to limiting exploration during early drafting. To address these issues, the presented study introduces INSPIRE, an INteractive Scholarly Paper Interpretation and Retrieval Engine, a system that aligns literature recommendations with the process of scholarly writing. Draft paragraphs are treated as cues of authorial intent, guiding a two-stage framework: hybrid retrieval with heterogeneous knowledge graphs and context-based reranking by large language models. Offline evaluations of the proposed system on a citation network and a user study demonstrated consistent gains over topical relevance, research intent alignment, and usefulness. The findings suggest that INSPIRE can act as a collaborator during writing by bridging retrieval with reasoning to support scholarly exploration and discovery.

Keywords: Scholarly literature recommendation · Context-aware retrieval · Large language models · Knowledge graph

1 Introduction

Research is a cyclical construction of knowledge, where writing both documents and shapes evolving ideas. Yet most scholarly tools remain centered on keyword-based retrieval and assume that a fully formed query can capture information needs. Such a design often fails to reflect the author's specific and evolving intent. Other methods infer interests from publication or reading history, but such signals capture only past research directions. This reduces opportunities for exploratory discovery and makes it challenging for researchers to access unfamiliar areas that may be highly relevant to the developing work. The mismatch is

H. Unger and P. Meesad (Eds.): NLPIR 2025, LNNS 1904, pp. 712–726, 2026.
https://doi.org/10.1007/978-3-032-20897-2_50

especially acute in the early stages of work, when intent is fluid and exploration matters as much as precision [1].

The presented study reframes literature recommendation by treating draft paragraphs as a query, a direct signal of the author's evolving intent during writing. The goal is to support exploration and selection of papers that align with the developing objectives, while avoiding reliance on shallow lexical matching. Delivering this support requires modeling context and intent. Earlier methods, such as keyword matching [2] and static embeddings [3], focused on lexical similarity and often missed deeper relationships. Addressing the gap requires approaches that would reason over the context, which large language models (LLMs) provide by reading the discourse and inferring the author's intent.

This study introduces INSPIRE, an INteractive Scholarly Paper Interpretation and Retrieval Engine, which integrates literature access into the act of writing. Draft text is analyzed for semantic content and rhetorical cues to infer needs that change as writing progresses. Context signals guide both retrieval and reranking: retrieval surfaces a broad but pertinent pool of candidate papers, and reranking orders these candidates to prioritize those most consistent with the intent of the draft.

To the best of the author's knowledge, INSPIRE is the first system to embed the principle that writing reflects the process of thinking in literature recommendation. The system acts as an intelligent collaborator that uses authorial intent and emerging ideas to support the dynamic construction of scholarly thought. Key contributions of the presented study are as follows:

- An original approach to scholarly literature recommendation that considers draft text as an explicit model of authorial intent in literature recommendation.
- A new system that integrates embedding-based and keyword-based retrieval with knowledge-graph reasoning, followed by context-aware reranking for recommending published articles pertinent to the ongoing research.

The remainder of the paper is organized as follows: Sect. 2 reviews the related work, Sect. 3 develops the conceptual model, Sect. 4 presents the proposed system design, Sect. 5 reports experiments, Sect. 6 discusses results, and Sect. 7 concludes the study and outlines directions for future work.

2 Related Work

2.1 Scholarly Paper Recommendation

Growth in publication volume complicates discovery. Recommender systems infer utility from content and interaction signals with common families, including content-based, collaborative, and hybrid systems [4,5]. Early work in scholarly recommendation focused on content-based features [6], moving from lexical cues and simple bibliometrics, such as keyword matching, co-citation, and bibliographic coupling, to topic modeling and pretrained language models for computing semantic similarity [7–9]. Graph-based models added structure from citation and author paper relations to address the data sparsity and cold start challenges [10]. Hybrid approaches integrated content-based representations with collaborative filtering interactions to offset the limitations of either method when used in isolation [5,11,12]. Knowledge graph (KG)-based systems capture multi-relational structure and enable explanation [13], with heterogeneous graphs guided by meta-relations [14] and topic-aware designs that inject explicit topic information into the recommending process [15].

Citation recommendation often targets the sentence-level context around citation points [16]. Recent studies include convolution-based embeddings for longer dependencies [17], context-conditioned co-citation structure [18], temporal modeling of citation networks [19], and section-aware frameworks that follow topic transitions [20]. The prior work, however, does not incorporate the idea that drafting both reveals and shapes the author's intent during writing.

2.2 Recommendations with LLMs

Recent studies have adopted LLMs to improve semantic representation and reasoning for recommendation tasks [21]. Prompting over profiles and interaction paths allows for estimating intent and improving the rating and sequencing tasks without fine-tuning [22]. Another approach incorporates LLMs during the reranking stage, grounding decisions in retrieved candidates while maintaining scalability and personalization [23]. Evidence from e-commerce domains suggests that inference time prompting can serve as a general reranking layer that improves both accuracy and topical diversity [24]. These results support the idea of using LLM as a context understanding and reranking tool.

3 Scholarly Writing as a Generative Cognitive Process

Writing can be seen as a generative cognitive activity, in which thought is progressively discovered and refined through recursive cycles of planning, translating, and reviewing [25,26]. Within this view, the draft text represents the author's evolving intent. The cognitive model further distinguishes several components that interact during composition. Long-term memory provides (a) topic knowledge or subject-matter knowledge relevant to the task, (b) rhetorical knowledge, awareness of audience, genre, and discourse conventions, and (c) writing plans and strategies, procedural knowledge for managing the composition. Within the writing process, planning involves generating ideas, organizing them into a coherent structure, and goal setting to guide the rhetorical purpose. Translating turns these plans into written text, while reviewing includes evaluating the draft and revising it to improve clarity, coherence, and alignment with goals.

Figure 1 summarizes the modeling approach employed in this study. Writer's long-term memory interacts with the ongoing draft. INSPIRE is positioned as an external collaborator that is to augment this interaction, as it extends the writer's available knowledge with access to a broader scholarly corpus and supports planning by proposing relevant papers that inform idea generation, organization, and goal setting. Draft paragraphs are analyzed for contextual cues, and the resulting recommendations are fed back into the subsequent cycles of planning and revision. In this way, the system aligns with the view of writing as thinking by using draft text as a representation of the authorial intent and by generating suggestions that expand exploratory reading and paper selection.

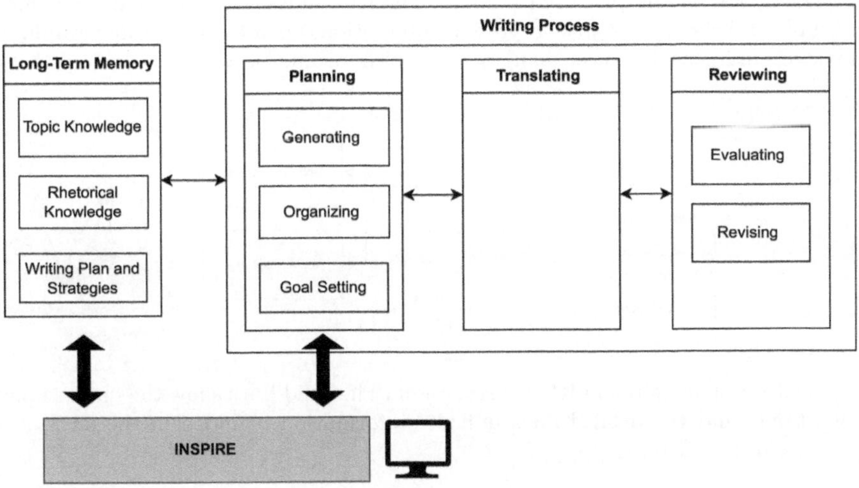

Fig. 1. Writing as a cognitive process.

4 Proposed System

4.1 System Overview

The proposed context-aware recommender system implements a two-phase approach to adapt to the paragraph a researcher is drafting: (1) a hybrid retrieval phase and (2) a context-based reranking phase, as illustrated in Fig. 2.

The first phase begins with the input query, which is the user's drafted paragraph. It then retrieves a candidate paper set C, based on lexical and semantic signals, using a rank fusion strategy. After retrieving C, path-based multi-hop reasoning is performed to create the final candidate set C'. Finally, the reranking module captures the context and dynamic research intent within the draft to produce the final top-K ranking of papers for recommendations from the set.

Data Preparation and KG Construction. To enable hybrid retrieval, a heterogeneous KG is constructed that integrates semantic, structural, and topical

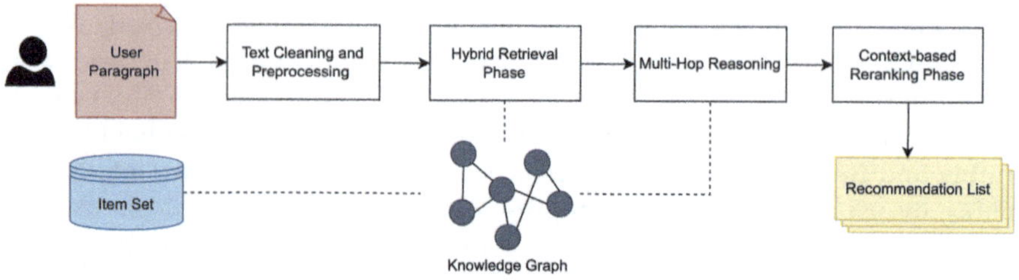

Fig. 2. Overview of the proposed system (The solid lines show the main sequence of operations, and the dashed lines represent interactions with the knowledge graph, such as retrieving information or exchanging reasoning results between components.)

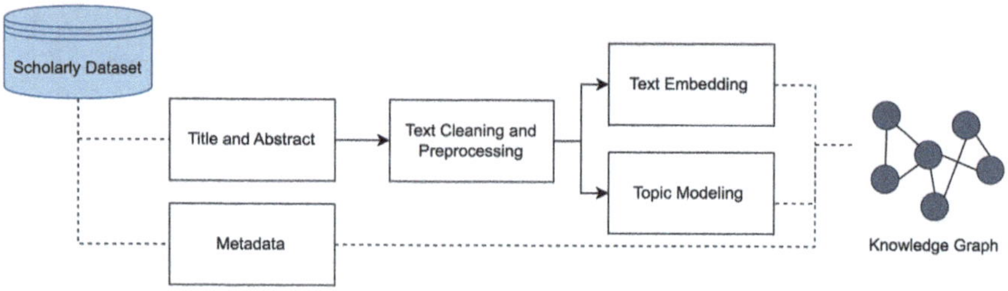

Fig. 3. Data preparation and KG construction (The solid lines show the main sequence of operations, and the dashed lines indicate information included in the dataset/KG (e.g., metadata, attributes)).

information from the scientific corpus, as illustrated in Fig. 3. Semantic embeddings for each paper are generated with SciBERT [3], using the concatenated title and abstract. These embeddings are used to capture semantic similarities between publications. BERTopic [27] is employed to group papers by semantically coherent topics.

In the graph, nodes represent entities, while edges denote the relationships connecting these entities. The graph, therefore, consists of three node types: paper, topic, and field of study (FoS). It also has three edge types: cites (between paper nodes), has_topic (from a paper to its topic), and has_fos (from a paper to its field of study). Finally, the generated embeddings from the title and abstract, along with metadata, such as authors, publication year, and DOI, are stored as attributes of each corresponding paper node.

Hybrid Candidate Retrieval. In the hybrid retrieval phase (see Fig. 4), two types of retrieval metrics are combined: lexical similarity and semantic similarity. Lexical similarity is computed, using BM25 [28], a ranking function that scores a document by term frequency, inverse document frequency, and document length normalization. Semantic similarity is computed using SciBERT [3], a transformer-based model pretrained on scientific texts that produces vector representations capturing meaning beyond surface words. Because each of the

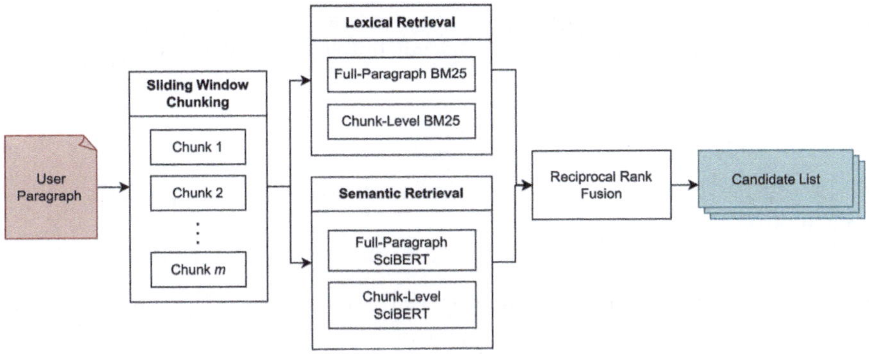

Fig. 4. Hybrid candidate retrieval.

metrics has complementary strengths, their ranked outputs are integrated using Reciprocal Rank Fusion (RRF) [29], which merges rankings by assigning higher weight to items appearing near the top of any list.

The author's draft paragraph is first normalized through whitespace collapsing and standard punctuation unification to ensure consistent encoding across downstream modules. The paragraph is then tokenized into subword units. From the token sequence, sliding-window chunks, which are chunks of fixed length, are created with fixed overlap. The overlap lets adjacent chunks share context, which supports both global and local research cues or, in other words, global and local coherence.

After preprocessing the paragraph, four retrieval sources are produced as follows:

1. Full-paragraph BM25: The entire paragraph is treated as a query and BM25 relevance score between the paragraph and candidate paper is computed
2. Full-paragraph embedding retrieval: The entire paragraph is encoded into vectors, using SciBERT, and cosine similarity with each candidate embedding vectors is computed
3. Chunk-based BM25: For each chunk, the BM25 relevant score against candidate paper is computed, then the maximum BM25-chunk score is taken as the chunk-level BM25 score
4. Chunk-based embedding retrieval: Chunks are encoded into embedding vectors via SciBERT, and cosine similarity is computed, retaining the maximum value

RRF is then applied to integrate the four ranked lists as follows. Let O the set of sources be {full-paragraph BM25, full-paragraph embedding, chunk-based BM25, and chunk-based embedding}. For each source $s \in O$ and candidate paper c, define the per-source contribution:

$$\mathrm{rrf}_s(c) = \frac{1}{k + \mathrm{rank}_s(c)}, \tag{1}$$

where $\text{rank}_s(c)$ is the rank of c the source s and k is a damping constant (set to $k = 60$ in the experiments). The overall hybrid retrieval score (S_{hyb}) is then computed as

$$S_{\text{hyb}} = \sum_{s \in O} \text{rrf}_s(c), \tag{2}$$

and the top-N papers by S_{hyb} form the candidate set C are passed to the reasoning module.

Path-Based Multi-hop Reasoning. Path-based reasoning uses the shortest-path distance between two papers in the KG as a measure of their conceptual relatedness. This approach allows for including papers along the path that may not share surface-level similarities or strong embedding overlaps but are, nonetheless, integral to the evolving research narrative connected by citation or topic-related chains.

To expand the candidate set C with papers that share both thematic and structural connections, the expansion proceeds in three steps:

1. Topic and FoS hop: For each seed paper, one hop is traversed along the edges connecting the topic and FoS nodes to collect all neighboring papers.
2. Citation hop: Simultaneously, for each seed paper, one hop is traversed along the cites edges in both directions to collect papers.
3. Aggregation: Each candidate collected from the hop expansion is assigned a weight that combines (a) a normalized score reflecting how many topics or field co-neighbors it shares with the seeds and (b) a binary indicator (in the range of [0,1]) of whether a direct citation link exists.

The final weight is the sum of the two terms of step (3). A higher weight indicates that a paper is both topically aligned and structurally connected, making it a stronger candidate for inclusion.

The result of the expansion is an enriched candidate set C', which incorporates both the initially retrieved papers and the additional neighbors discovered through the multi-hop reasoning.

LLM-Based Reranking. In the reranking phase, a chain-of-thought (CoT) prompt engineering approach [30,31] is applied to rerank the candidate set C', as illustrated in Fig. 5. For each candidate paper, a prompt is constructed that embeds the author's full draft paragraph, thereby encoding global and local coherence of the research context. The prompt instructs the LLM to assess (1) topical relevance, (2) research gap alignment, and (3) methodological fit, each by assigning a discrete score of 1 to 5. To reduce ordering effects and improve consistency, a listwise sliding window strategy is adopted. The candidate list is partitioned into overlapping windows of a fixed size with a fixed stride, and the same CoT prompt is called for each window. This produces a window-level

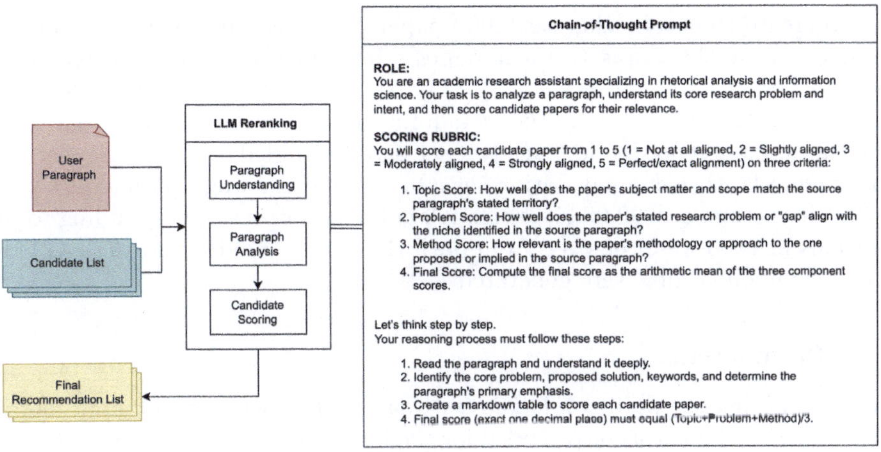

Fig. 5. Context-based reranking.

score for each paper, and the final score S_{rer} is obtained by averaging across all windows:

$$S_{\text{rer}} = \frac{1}{m}\sum_{i=1}^{m} S_i(c), \tag{3}$$

where m is the number of windows containing paper c, and S_i is its score in window i. Candidates are sorted in the descending order of S_{rer}, and the top-K papers are returned as final recommendations.

5 Experiments

All experiments presented in this section were conducted on a Linux x86_64 workstation equipped with an NVIDIA GeForce RTX 4090 (24GB) GPU.

5.1 Data

All experiments used the DBLP citation network V12 [32], which contains 4,894,082 papers and 45,564,149 citation edges. In this study, a heterogeneous KG was constructed specifically for the experiments by integrating papers, citation information, and additional data sources. The integration included SciBERT embeddings computed from each paper's title and abstract (semantic content), BERTopic clusters assigning a single topic label to each paper (topical structure), and bibliographic metadata, such as authors, publication year, and DOI. The graph was stored in Neo4j [33] to enable efficient retrieval and reasoning. Topic modeling produced 150 distinct topics. After integrating papers, citations, topics, embeddings, and metadata, the final graph had 5,026,565 nodes and 95,487,854 relationships.

To create the evaluation set, 2000 papers published in 2020 were randomly sampled from the dataset. Each paper's introduction was extracted as the assumed user's research intent [34], removing text citations and mathematical formulas, and concatenating the remaining sentences into a single paragraph. The original reference list for each test set paper included in the dataset served as the ground-truth recommendation set. On average, a paragraph had 991 words, 6,506 characters, and 32.09 citations. All nodes and edges corresponding to publications in 2020 were removed from both the retrieval index and the KG before any recommendations were generated.

5.2 Parameters

Initial experiments were conducted with the sliding windows of 128, 256, and 512 tokens, and 256 tokens provided for the best balance between context preservation and efficiency. Accordingly, each input paragraph was chunked into overlapping windows of $length = 256$ tokens with $overlap = 64$. For lexical retrieval, all paper titles and abstracts were indexed in Neo4j, using a schema-based full-text index backed by Apache Lucene's BM25Similarity with default parameters. For semantic retrieval, 768-dimensional SciBERT embeddings were used. In the reranking phase, listwise sliding window prompting was applied in batches of 5 candidates per LLM call, with $windowsize = 5$ and $stride = 1$. The Meta-LLaMA-3.1-8B-Instruct model quantized via QLoRa (4-bit NF4) was employed with up to 1700 tokens reserved for CoT generation, while enforcing a strict total input limit.

5.3 Comparative Evaluation of System Variants (Ablation Study)

To evaluate INSPIRE's performance, a comparative evaluation was conducted against two system configurations and a baseline. In the experiments, the input paragraph was processed, returning a ranked list of candidate papers. The evaluated systems thus included:

1. INSPIRE (proposed): Full system integrating hybrid retrieval, prompt engineering-based LLM reranking, and multi-hop reasoning.
2. BM25 (baseline): Standard sparse retrieval ranking.
3. RRF hybrid search: Proposed hybrid retrieval only.
4. RRF + short LLaMA prompt: Hybrid retrieval and a single-step LLaMA prompt for direct relevance ranking.

6 Results and Discussion

Recommender systems are often evaluated offline with standard metrics [35]. However, prior work showed that methods that perform well offline do not necessarily perform well in practice [36]. Offline results can be inconclusive or even misleading. Consequently, real-world evaluations and controlled lab studies are recommended as more reliable complements to offline benchmarks [36].

6.1 Offline Evaluation

Three standard information retrieval metrics were used to assess recommendation quality: Hit Rate@K (HR@K), Precision@K (P@K), and Recall@K (R@K). HR@K is 1 if at least one ground truth paper appears in the top-K for a paragraph and 0 otherwise (averaged over paragraphs). P@K is the fraction of items in the top-K that are relevant, and R@K is the fraction of ground truth citations recovered within the top-K. For each test paragraph, INSPIRE generated a ranked list of the top-K candidate papers. A recommendation was considered correct when the cosine similarity between the candidate abstract and any ground truth abstract computed with the SciBERT embeddings was at least 0.95.

Table 1. Offline results (**Bold** indicates the best result in each column.)

	P@K				HR@K				R@K			
Method	K=3	K=5	K=10	K=20	K=3	K=5	K=10	K=20	K=3	K=5	K=10	K=20
BM25	0.192	0.166	0.148	0.119	0.405	0.480	0.580	0.665	0.020	0.028	0.050	0.078
RRF	0.435	0.400	0.319	0.240	0.750	0.820	0.905	0.945	0.046	0.070	0.110	0.163
RRF+prompt	0.424	0.389	0.325	0.251	0.733	0.800	0.907	**0.967**	0.045	0.068	0.113	0.169
INSPIRE	**0.458**	**0.428**	**0.373**	**0.282**	**0.758**	**0.843**	**0.924**	0.950	**0.047**	**0.075**	**0.131**	**0.193**

6.2 Offline Results

The results obtained (see Table 1) suggest that each system component adds measurable gains. RRF outperformed BM25 in accuracy for long paragraphs, consistent with reduced keyword dominance in lengthy queries. The CoT LLM reranking further improved precision at greater K through context use. Overall, INSPIRE solutions resulted in the best performance across all metrics.

6.3 User Study

A user study was conducted by comparing the two best-performing systems from the offline experiments. Seven early researchers in computer science (aged 22–29; 3 female, 4 male) were recruited to participate in the A/B tests. The participants provided two introductory paragraphs each, which were either written by themselves or taken from texts they understood deeply. They then evaluated two anonymized recommendation lists of five papers each. One list was generated by a hybrid retrieval with a short LLM-based prompt engineering reranking (System A), and the other by INSPIRE (System B). The participants rated the recommended papers, using a 5-point Likert scale, based on the questions listed in Table 2. Recommendation details were presented randomized. All responses were collected under informed consent and anonymized to protect participant privacy.

To evaluate recommendation quality, paired participant ratings were compared, using a two-sided Wilcoxon signed-rank test [37]. For the questions per paper (topical relevance, research-intent alignment), each 5-item recommendation list was first summarized to a list score by taking the median across the five papers. For each outcome, two paragraph scores per participant were averaged to obtain one participant score per tested system, resulting in $n = 7$ paired observations.

For each outcome, the Hodges-Lehmann (HL) estimator [38] of the median paired difference ($B - A$) was also computed with 95% confidence intervals (CIs) obtained via percentile bootstrap (5,000 resamples) together with wins-losses-ties counts to reflect whether participants generally favored system A or system B. As four outcomes were analyzed, p-values were adjusted, using the Benjamini-Hochberg procedure [39] to control the false discovery rate at 0.05. Ties were handled according to the Wilcoxon procedure [37]. All tests were at $\alpha = 0.05$ and were implemented in Python.

Table 2. User study evaluation criteria

Subjects	**Questions (each answered on a 5-point Likert scale from 1 to 5)**
Topical Relevance	Does this paper discuss similar topics or subject areas to those mentioned in your introduction paragraph? Rate from 1 (Not at all relevant) to 5 (Extremely relevant).
Research Intent Alignment	Does this paper align with the research intent or purpose expressed in your paragraph? Rate from 1 (Not aligned) to 5 (Perfectly aligned).
Overall Relevance	To what extent are the papers in this list relevant to your introduction paragraph? Rate from 1 (Not at all relevant) to 5 (Extremely relevant).
Overall Usefulness	To what extent does this list provide practical support for your writing and research exploration? Rate from 1 (Not useful at all) to 5 (Extremely useful).

Table 3. User study statistical analysis

Outcome	**HL (B–A)**	**95% CI**	***p*-value (BH–FDR)**
Topical Relevance	+0.88	[0.50, 1.50]	0.031
Research-Intent Alignment	+1.00	[0.50, 1.25]	0.042
Overall Relevance	+1.25	[1.00, 1.75]	0.031
Overall Usefulness	+1.13	[0.00, 1.50]	0.063

6.4 User Study Results

INSPIRE outperformed the LLM prompt reranker without CoT across all four outcomes, as shown in Fig. 6. Median paired gains were about one point on the five-point scale. Hodges Lehmann estimates and 95% CIs obtained are given in Table 3. The adjusted p-values are 0.031, 0.042, 0.031, and 0.063 for the four outcomes, respectively.

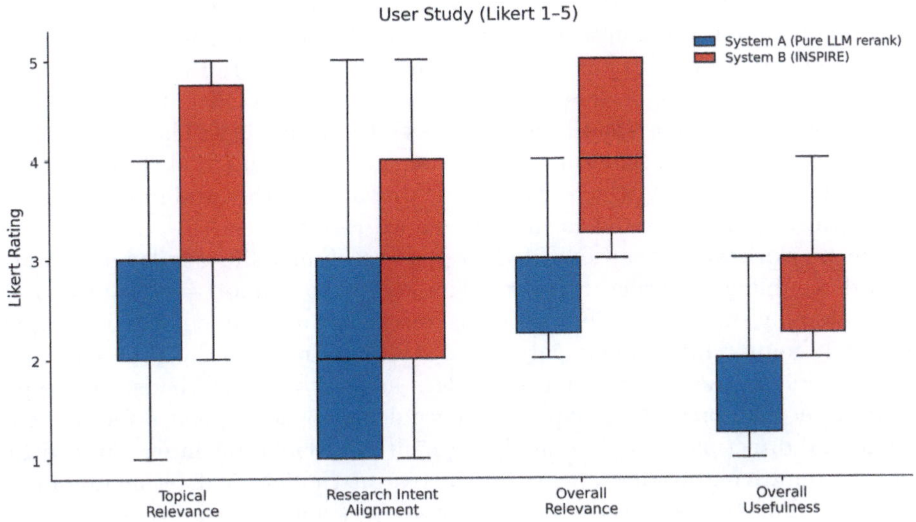

Fig. 6. User study results.

After the BH-FDR control, the null hypothesis H_0 was rejected for topical relevance, research-intent alignment, and overall relevance, and not rejected for overall usefulness. Consistently, the 95% CIs for the first three outcomes exclude 0, while the overall usefulness interval includes 0. The reason may be that topical relevance and intent alignment were based on multiple ratings per list, while overall usefulness came from a single rating. This increased variability and widened confidence intervals, which made it harder to reject the null hypothesis.

Despite overall gains, the reranking sometimes misweights competing intent signals in paragraphs that mix application, task, and method. For instance, in one case with the Angry Birds-level generation, the text mentions CycleGAN that led the recommendations toward GAN methodology papers rather than papers on level generation and sketch to level design, producing globally relevant but locally very biased results. The effect tends to occur when repeated method terms or a distinctive keyword dominate the draft paragraph, leading the LLM to infer a method-centered intent rather than a task or application focused one.

7 Conclusions

INSPIRE, the system developed in the presented study, addresses the gap between writing and retrieval by treating draft paragraphs as direct signals of research intent, and by aligning recommendations with the act of writing. The proposed approach combines retrieval based on content and queries, KG reasoning, and an LLM-enhanced reranker to order candidates according to the evolving intent, as represented in the draft. Across offline metrics and a user study, the experimental results showed consistent gains over a strong baseline, with statistically significant improvements for topical relevance and research intent alignment, and a positive but not significant trend for overall usefulness.

Several limitations of this work should be acknowledged. Paired samples in the user study were small, which reduces the reliability. The reranker can misweight paragraphs that mix application, task, and method, influenced by repeated method terms, which bias the ranking against the local intent and lead to generally appropriate but locally misaligned results.

Future work will extend INSPIRE beyond recommending similar papers and towards enabling scientific discovery. The goal is to support authors' research exploration by surfacing underexplored connections and fresh perspectives rather than only reinforcing familiar patterns. Path-based reasoning will be combined with criteria for novelty to suggest bridges across topics, methods, and applications, thereby presenting papers that would open new avenues for thought. Additional directions would be modeling multiple concurrent intent cues within a draft, improving accuracy when distinctive methodological terms dominate, and expanding the user study with larger samples and multi-item measures.

8 Code Availability

The source code of INSPIRE is available at https://github.com/Manami108/recommender-system.

References

1. Gökçe, O., Prada, J., Nikolov, N.I., Gu, N., Hahnloser, R.H.R.: Embedding-based scientific literature discovery in a text editor application. In: Celikyilmaz, A., Wen, T.-H. (eds.) Proceedings of the 58th Annual Meeting of the Association for Computational Linguistics: System Demonstrations, pp. 320–326, Online (2020). Association for Computational Linguistics
2. Robertson, S., Zaragoza, H., et al.: The probabilistic relevance framework: BM25 and beyond. Found. Trends® Inf. Retrieval **3**(4), 333–389 (2009)
3. Beltagy, I., Lo, K., Cohan, A.: SciBERT: a pretrained language model for scientific text. In: Inui, K., Jiang, J., Ng, V., Wan, X. (eds.) Proceedings of the 2019 Conference on Empirical Methods in Natural Language Processing and the 9th International Joint Conference on Natural Language Processing (EMNLP-IJCNLP), pp. 3615–3620, Hong Kong, China (2019). Association for Computational Linguistics

4. Bai, X., Wang, M., Lee, I., Yang, Z., Kong, X., Xia, F.: Scientific paper recommendation: a survey. IEEE Access **7**, 9324–9339 (2019)
5. Zhang, Z., et al.: Scholarly recommendation systems: a literature survey. Knowl. Inf. Syst. **65**(11), 4433–4478 (2023)
6. Sharma, R., Gopalani, D., Meena, Y.: An anatomization of research paper recommender system: Overview, approaches and challenges. Eng. Appl. Artif. Intell. **118**, 105641 (2023)
7. Singh, R., Gaonkar, G., Bandre, V., Sarang, N., Deshpande, S.: Scientific paper recommendation system. In: 2023 IEEE 8th International Conference for Convergence in Technology (I2CT), pp. 1–4 (2023)
8. Bagul, D.V., Barve, S.: A novel content-based recommendation approach based on LDA topic modeling for literature recommendation. In: 2021 6th International Conference on Inventive Computation Technologies (ICICT), pp. 954–961 (2021)
9. Hadhiatma, A., Azhari, A., Suyanto, Y.: A scientific paper recommendation framework based on multi-topic communities and modified PageRank. IEEE Access **11**, 25303–25317 (2023)
10. Sivasankari, R., Dhilipan, J.: Hybrid scientific article recommendation system with coot optimization. Data Sci. Manag. **7**(2), 99–107 (2024)
11. Pinedo, I., Larrañaga, M., Arruarte, A.: ArZiGo: a recommendation system for scientific articles. Inf. Syst. **122**, 102367 (2024)
12. Dinh, T.N., Pham, P., Nguyen, G.L., Nguyen, N.T., Vo, B.: A hybrid citation recommendation model with SciBERT and GraphSAGE. IEEE Trans. Syst. Man Cybern. Syst. **55**(2), 852–863 (2025)
13. Mataoui, M., Bahloul, H., Ziane, A.: Leveraging knowledge graphs for paper recommendation systems. In: International Conference on Intelligent Systems and Pattern Recognition, pp. 230–245. Springer (2024)
14. Wang, G., Zhou, L., Gong, J., Zhang, X.: Heterogeneous graph neural network with hierarchical attention for group-aware paper recommendation in scientific social networks. Appl. Soft Comput. **167**, 112448 (2024)
15. Ni, P., Wang, X., Lv, B., Likang, W.: GTR: an explainable graph topic-aware recommender for scholarly document. Electron. Commer. Res. Appl. **67**, 101439 (2024)
16. Wang, H.-C., Cheng, J.-W., Yang, C.-T.: SentCite: a sentence-level citation recommender based on the salient similarity among multiple segments. Scientometrics **127**(5), 2521–2546 (2022)
17. Yonghe, L., Yuan, M., Liu, J., Chen, M.: Research on semantic representation and citation recommendation of scientific papers with multiple semantics fusion. Scientometrics **128**(2), 1367–1393 (2023)
18. Zhang, J., Zhu, L.: Citation recommendation using semantic representation of cited papers' relations and content. Expert Syst. Appl. **187**, 115826 (2022)
19. Liu, J.-C., Chen, C.-T., Lee, C., Huang, S.-H.: Evolving knowledge graph representation learning with multiple attention strategies for citation recommendation system. ACM Trans. Intell. Syst. Technol. **15**(2) (2024)
20. Santosa, N.C., Liu, X., Han, H., Miyazaki, J.: S3PaR: section-based sequential scientific paper recommendation for paper writing assistance. Knowl.-Based Syst. **303**, 112437 (2024)
21. Liu, Q., et al.: Large language model enhanced recommender systems: methods, applications and trends. In: Proceedings of the 31st ACM SIGKDD Conference on Knowledge Discovery and Data Mining V. 2, pp. 6096–6106 (2025)

22. Li, G., Yang, F., Yue, Y.: Identify user intention for recommendation using chain-of-thought prompting in LLM. In: International Conference for Emerging Technologies in Computing, pp. 60–69. Springer (2024)
23. Gao, J., et al.: LLM-enhanced reranking in recommender systems. arXiv preprint arXiv:2406.12433 (2024)
24. Xu, Z., Zhang, Y.: LLM-enhanced reranking for complementary product recommendation. arXiv preprint arXiv:2507.16237 (2025)
25. Flower, L., Hayes, J.R.: A cognitive process theory of writing. Coll. Compos. Commun. **32**(4), 365–387 (1981)
26. Freiman, M.: The art of drafting and revision: extended mind in creative writing. New Writing **12**(1), 48–66 (2015)
27. Grootendorst, M.: BERTopic: Neural topic modeling with a class-based TF-IDF procedure. arXiv preprint arXiv:2203.05794 (2022)
28. Trotman, A., Puurula, A., Burgess, B.: Improvements to BM25 and language models examined. In: Proceedings of the 19th Australasian Document Computing Symposium, pp. 58–65 (2014)
29. Cormack, G.V., Clarke, C.L.A., Buettcher, S.: Reciprocal rank fusion outperforms condorcet and individual rank learning methods. In: Proceedings of the 32nd International ACM SIGIR Conference on Research and Development in Information Retrieval, pp. 758–759 (2009)
30. Wei, J., et al.: Chain-of-thought prompting elicits reasoning in large language models. Adv. Neural Inf. Process. Syst. **35**, 24824–24837 (2022)
31. Gao, J., et al.: LLM4Rerank: LLM-based auto-reranking framework for recommendations. In: Proceedings of the ACM on Web Conference, vol. 2025, pp. 228–239 (2025)
32. Tang, J., Zhang, J., Yao, L., Li, J., Zhang, L., Su, Z.: ArnetMiner: extraction and mining of academic social networks. In: Proceedings of the 14th ACM SIGKDD International Conference on Knowledge Discovery and Data Mining, KDD '08, pp. 990–998, New York, NY, USA (2008). Association for Computing Machinery
33. Lal, M.: Neo4j graph data modeling. Packt Publishing Ltd (2015)
34. Jernudd, B.H.: Genre analysis: English in academic and research settings. john m. swales. Cambridge: Cambridge University Press, 1990, pp. 288. 15.95 paper. Stud. Second Lang. Acquisition **15**(1), 125–126 (1993)
35. Zangerle, E., Bauer, C.: Evaluating recommender systems: survey and framework. ACM Comput. Surv. **55**(8), 1–38 (2022)
36. Beel, J., Langer, S.: A comparison of offline evaluations, online evaluations, and user studies in the context of research-paper recommender systems. In: International Conference on Theory and Practice of Digital Libraries, pp. 153–168. Springer (2015)
37. Ihemelandu, N., Ekstrand, M.D.: Inference at scale: significance testing for large search and recommendation experiments. In: Proceedings of the 46th International ACM SIGIR Conference on Research and Development in Information Retrieval, pp. 2087–2091 (2023)
38. Rosenkranz, G.K.: A note on the Hodges–Lehmann estimator. Pharm. Stat. **9**(2), 162–167 (2010)
39. Thissen, D., Steinberg, L., Kuang, D.: Quick and easy implementation of the Benjamini-Hochberg procedure for controlling the false positive rate in multiple comparisons. J. Educ. Behav. Stat. **27**(1), 77–83 (2002)

Comparative Study of Query Types on Query Refinement

Zahra Taheri(✉), Ziad Kobti, and Hossein Fani

School of Computer Science, University of Windsor, Windsor, ON, Canada
taherik@uwindsor.ca, kobti@uwindsor.ca, hfani@uwindsor.ca

Abstract. Within a search session, users seek their information needs through iterative refinement of their queries, which is daunting. Neural-based query refinement methods aim to address this challenge via training on gold-standard pairs of ⟨original query → *refined* query⟩, oblivious to the type of queries; be it `factual` like *'table leafs or leaves'*, `navigational`, e.g., *'bbc good food recia'*, or `transactional` such as *'free play minecraft'* and, hence, fall short of finding refined versions for many original queries. In this paper, we bridge the gap by incorporating query types when generating refined queries. We fine-tune a conditional transformer, e.g., `t5`, to map an original query onto its relevant documents while conditioning on its type during training so that, during the inference, the query type controls generating new reformulated queries within the same search intent for an *unseen* original query. Our experiments on the large-scale `orcas-i-2m` dataset across five query types demonstrated the synergistic effects of considering query types in generating more refined queries with better information retrieval efficacy. Specifically, considering query type has shown the best retrieval performance among `navigational` queries while yielding the highest number of refined queries in `factual` queries. The codebase to support the reproducibility of our research is available at https://anonymous.4open.science/r/RePair-C2F4.

Keywords: Query Type · Search Intent Taxonomy · Query Suggestion · Query Reformulation · Conditional Transformer

1 Introduction

Users of search engines struggle with formulating queries to express their information needs due to an unclear/uncertain state of knowledge [6,7,31], inconsistent terminologies (vocabulary mismatch) [24,34,53,64], or pure lack of knowledge, which may result in less-than-relevant search engine result pages. Query refinement aims to bridge the gap by recommending a *refined* query for the user's original query, which retrieves more relevant information at higher ranks [3,5,40,48,53,57,65]. Query refinement methods have been effective in web search [66], e commerce [26], and medical information retrieval [2,8]. However, methods of query refinement largely overlook the type of user's query, and

H. Unger and P. Meesad (Eds.): NLPIR 2025, LNNS 1904, pp. 727–739, 2026.
https://doi.org/10.1007/978-3-032-20897-2_51

hence, fail to find refined queries for many original queries. For example, as shown in Table 1, the original query *'table leafs or leaves'* is poorly refined to *'table leaf leafs'*, decreasing the average precision for `bm25` retriever from `0.111` down to `0.100 (−0.011)` as the query type has been overlooked. In contrast, knowing the query type as `factual` can specify user expectations and influence refinement such that a type-aware model generates a better refined query *'table leaf definition'*, effectively detecting the user's information need and enhancing average precision to its best, i.e., `1.0 (+0.889)`. For example, as shown in Table 1, the original query *'free play minecraft'* is poorly refined to *'minecraft full version'*, marginally increasing the `map` from `0.041` to `0.066 (+0.025)` as the query type has been overlooked (−type). In contrast, knowing the query type as `transactional`, a type-aware model generates a better refined query *'minecraft mineroll download free game'*, effectively detecting the user's information need and enhancing `map` to its best, i.e., `1.0 (+0.959)`.

Table 1. Sample queries from `orcas-i-2m` [4] and the search efficacy of type-aware vs. type-less refinements using `t5` [47]. As underlined, using query types yields better refinements.

	qid	type	query	bm25.ap (Δ)
original	2009263	–	*bbc good food recia*	0.039
−type	2009263	–	*bbc good foods recipies*	0.266 (+0.277)
+ type	2009263	navigational	*bbc good food http://www.bbcgoodfood.com/recipes*	**0.566** (+0.577)
original	2016037	–	*table leafs or leaves*	0.111
−type	2016037	–	*table leaf leafs*	0.100 (−0.011)
+ type	2016037	factual	*table leaf definition*	**1.000** (+0.889)
original	11136851	–	*free play minecraft*	0.041
−type	11136851	–	*minecraft full version*	0.066 (+0.025)
+ type	11136851	transactional	*minecraft mineroll download free game*	**1.000** (+0.959)

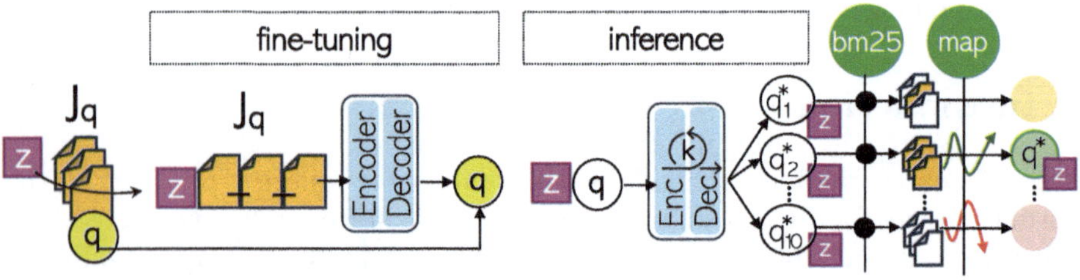

Fig. 1. Utilizing conditional transformers for type-aware query refinement.

Query types, also known as search intent taxonomies, have long been integrated into information retrieval processes to enhance search efficacy [9,

16,18,51]. For web searches, Border [10] identifies three types of queries: `informational`, like *'table leafs or leaves'*, aiming at gaining specific information assumed to be present on one or more webpages, `navigational`, like *'bbc good food recipies'* to find a specific url or website, and `transactional`, like *'minecraft mineroll download free game'*, to access a webpage inside which further interaction would occur. Choi et al. [18] studied *'how to'* queries about performing procedural tasks and identified query types based on the underlying tasks like creating/building vs. fixing. Researchers have also proposed the automatic generation of query type taxonomies from search logs and the classification of queries into these types accordingly [1,17,19,52]. Yet, the synergistic effect of considering queries' types for query refinement has remained *un*explored within the large body of query refinement methods [3,5,40,48,57], including the state of the art where rich contextual information, such as past queries and click-throughs, are considered [3,65]. Chen et al. [41] integrated users' short-term and long-term search histories within their current session to identify search interests. Ahmad et al. [3] encoded users' previous queries and document clicks in the search log into the recurrent neural network model. Later, Zhou et al. [65] divide users' search logs into long and short-term histories to refine the initial query using the transformer model. Fiorini et al. [42] fused personalization and time-sensitive information to investigate the difference in query occurrences throughout a day. For example, the query *'tv guide'* appears mainly in the evenings and nights. Therefore, they incorporated the time of the issued query along with users' previous queries in their proposed model. Recently, researchers argued that web query logs are prone to topic drift [5,40,57], that is, users' intent may undergo gradual or sudden changes within a search session, resulting in a loss of sequential semantic context between queries, and proposed generating gold-standard benchmark datasets free of topic drift for supervised refinement methods. Nonetheless, existing works in query refinement forego query types in their proposed methods.

In this paper, we propose to study the impact of considering query types and lack thereof for query refinement via a conditional transformer. As shown in Fig. 1, given an original query and its query type, we form a training pair ⟨condition: input → output⟩ where the condition is the query type, the input is the concatenation of relevant documents, and the output is the original query. We train `t5` [47] to learn the transformation from relevant documents of the original query to the query itself but on the condition of its type. During inference, we feed an *unseen* original query from the test set with its query type to `t5` to generate different streams of words in the output using top-k sampling [23] as the type-aware refined version of the original query. We evaluate the search efficacy of the type-aware refined version compared to the original query as well as a refined version without considering the query type [5,38] using a lightweight retriever such as `bm25` [49] and an evaluation metric such as average precision (`ap`). Our experiments demonstrate that incorporating query types for refining queries through `t5` statistically significantly enhances the efficacy in retrieving more relevant information yet at higher ranks and, hence, yields better refined queries compared to when the query type is overlooked.

2 Related Work

In this section, we provide a concise summary of query refinement methods and the applications of transformers in this context.

2.1 Query Refinement

Earlier approaches of query refinement, variously referred to by such other names as query rewriting, reformulation, or expansion, were mostly unsupervised wherein an original query was modified by expanding and/or replacing words with synonyms [33,54,55] or other frequently co-occurring words within a training corpus [36,39]. Such methods, however, overlooked the query's semantic context for polysemous words and yielded topic drift. To fill the gap, Rocchio [50] and others [28] proposed to modify the query based on words in the set of clicked documents as relevance feedback from users or via pseudo-relevance-feedback [11,12,58,60]. Successful for short queries, unsupervised methods fall short for detailed and long queries. Hence, semi-supervised and supervised techniques, mostly based on neural recurrent encoder-decoder architecture, were proposed to generate a refined query [3,13,20,22,25,35,56,61,63]. For instance, Ahmed et al. [3] incorporate historical ⟨query, clicked documents⟩ pairs to learn multitask of query refinement and document ranking in tandem. Supervised and semi-supervised methods, by and large, are trained on search logs, assuming that a user would gradually refine her query over successive attempts to find relevant content within a search session, which has been *in*validated by Chen et al. [15]; a user might search for multiple topics in one session, and hence, *ir*relevant queries would be paired. Hence, standard benchmark datasets free of topic drift are specifically designed for supervised query refiners [5,57,62]. For instance, Arabzadeh et al. [5] fed a query to the pretrained `doc-t5-query` [43] and selected the generated sequence of words as a refined version only if it increases `bm25` retrieval performance based on `map`. Also, Narayanan et al. [40] developed an open-source reproducible pipeline to generate benchmark datasets from various domains and any choice of transformers. In the healthcare domain, Rencis [67] applied a configurable keywords-based query language to address the challenges of domain-specific retrieval, demonstrating how structured query refinement tailored to specialized vocabularies can enhance retrieval precision in clinical information systems. Nonetheless, to the best of our knowledge, no one has yet explored the synergistic impact of query types in query refinement.

2.2 Conditional Transformers

Transformers have been widely employed in query refinement. [5,27,44,59]. To address vocabulary mismatch between the original queries and relevant documents, Nogueira et al.[44] proposed `doc2query` to expand documents by semantically related words using the transformer [59]. Shortly after, they replaced the transformer with `t5` [47], observing a substantial improvement in retrieval efficacy. Arabzadeh et al. [5] used `t5` to generate training datasets for supervised query refinement at a large scale. Also, Mitra et al. [38] utilized `bert` [21] to generate related queries to an original query (*'people also asked'*). Recently, Hosseini et al. [27] used `doc-t5-query` to refine queries and trained a classifier to filter out refinements that semantically drift from the original query.

The promising performance of transformers paved the way for *conditional* transformers, where probabilities are assigned to sequences of words given a pretext that encodes a condition. Formally, $p(q \mid \texttt{X}) = \prod_i p(w_{i+1} \mid \texttt{X}, w_1, \ldots, w_i)$, where the probability of the next word is based on previously generated words and a conditioning context X, which is kept fixed during the encoding and decoding phases. For instance, Keskar et al. [30] proposed `CTRL`, a transformer that can condition on a predefined set of words to influence the *genre* of the generated sequence of words in the output for the same input sequence; for example:

$$p(\textit{sharp} \mid \texttt{X} = \textit{review, knife}) \quad \text{vs.} \quad p(\textit{tool} \mid \texttt{X} = \textit{wiki, knife})$$

Subsequently, conditional transformers have been employed as single universal (generalist) models, such as `T5`, to perform a wide range of natural language processing tasks—from question answering to document summarization and sentiment analysis—relieving the need for task-specific models [30,37,46]. Inspired by `T5` and `CTRL`, we aim to leverage conditional transformers to refine original queries conditioned on their types (with X representing the query type), in order to study the impact of query types on query refinement.

3 Problem Definition

Let q represent a query and $\mathcal{J}_q$ be q's relevant documents (relevance judgment). Given an evaluation metric m to evaluate the quality of retrieved documents D_{rq} by a retriever r for q regarding $\mathcal{J}_q$, denoted as $m(D_{rq} : \mathcal{J}_q)$, a query q^* is a *refined* query for q iff $m(D_{rq^*} : \mathcal{J}_q) > m(D_{rq} : \mathcal{J}_q)$. Type-aware query refinement aims to generate a better refined query q^*_z owing to q's type z such that:

$$m(D_{rq^*_z} : \mathcal{J}_q) > m(D_{rq^*} : \mathcal{J}_q) > m(D_{rq} : \mathcal{J}_q) \tag{1}$$

For instance, from Table 1, the type-aware refined query q^*_z=*'minecraft mineroll <u>download</u> free game'* enables the retriever r=`bm25` to retrieve *all* relevant documents $D_{rq^*_z}$ and obtain the evaluation metric m=`map` at its best `1.00` compared to the documents retrieved by the type-less refined query q^*=*'minecraft full version'* and the original query q=*'free play minecraft'* with m=`map` equal to `0.066` and `0.041`, respectively.

4 Research Methodology

Let τ denote a conditional transformer, such as `T5` [47]. In type-aware query refinement, we define the condition X as the query type z, and aim to train τ to perform conditional translation from the words in relevant documents $\mathcal{J}_q$ to the words in the original query q. That is, we aim to model the distribution of query words given their relevant documents and their type: $p(q \mid \texttt{X} = z, \mathcal{J}_q)$, where z controls the refinement process.

During training, given the original query q and its relevant documents $\mathcal{J}_q$, we concatenate the documents into a single input sequence, prepend the query type z as a prefix condition (e.g., z: $\mathcal{J}_q$), and pair it with the original query as the target sequence, forming an input-output training pair: $(z\text{: } \mathcal{J}_q) \rightarrow q$. The model τ is thus expected to learn different refinement pathways depending on the query type.

At inference time, the model τ is provided with a test query q along with its type z, formatted as $(z\text{: } q)$. The transformer generates a refined version of q conditioned on z, denoted q^*_z, by producing the mapping: $(z\text{: } q) \rightarrow q^*_z$.

Modern transformers, including conditional ones, apply top-k random selection at their decoders by randomly selecting the next word of the output stream from the top-k most probable words to generate a varied stream of words and avoid common or repetitive words from training datasets [23,44]. Top-k random selection, hence, yields non-deterministic output generation during inference given the same input. To ensure diverse and novel refined versions of the same original query, we employ top-k random selection, i.e., $\langle\langle z : q\rangle \rightarrow q^*_{zi}\rangle$ for $1 \leq i \leq k$, as opposed to a single refined query. To compare type-aware query refinement with when query types are overlooked, we follow similar steps for the same set of queries but oblivious to their query type with no pretext $\texttt{X}$ in τ.

We argue that type-aware query refinement results in higher search efficacy and yields better and a greater number of refined queries q^*_z compared to q^* where query types are overlooked (Table 2).

Table 2. Statistics of `orcas-i-2m` [4]. $|q|$ and $|\mathcal{J}_q|$ denote the number of words and relevant documents for a query q.

Type	Total Queries ($\lvert\mathcal{Q}\rvert$)	Train	Test	Avg. $\lvert q\rvert$	Avg. $\lvert\mathcal{J}_q\rvert$
All	2,000,000	1,437,272 (80%)	359,402 (20%)	3.438	2.833
`navigational` *(specific URL or resource)*	289,610 (14.4%)	210,175 (14.6%)	52,561 (14.6%)	2.257	2.054
`factual` *(knowledge or facts)*	706,907 (35.3%)	511,284 (35.5%)	127,666 (35.6%)	3.493	3.169
`transactional` *(resources or downloads)*	83,334 (4.1%)	60,028 (4.1%)	15,035 (4.2%)	3.920	3.999
`instrumental` *(task-oriented knowledge)*	116,485 (5.8%)	83,755 (5.8%)	20,939 (5.8%)	4.344	3.155
`abstain` *(none of the above)*	803,664 (40.8%)	572,017 (39.7%)	143,193 (39.8%)	3.168	1.774

We create two sets of pairs $\mathcal{Q}^z = \{\langle z : \mathcal{J}_q \rightarrow q\rangle\}$ for the type-aware method, and lack thereof $\mathcal{Q} = \{\langle \mathcal{J}_q \rightarrow q\rangle\}$ for the type-less baseline [5,38]. We then fine-tune τ on $\mathcal{Q}^z$ (type-aware model) and $\mathcal{Q}$ (type-less model) separately and collected the models' inference results for the test set, which results in two categories of type-aware and type-less query refinements. Finally, we assessed the effectiveness of refined queries in each category by evaluating how well they perform in retrieving the relevant documents based on the original query's relevance judgment $\mathcal{J}_q$. Given a query q from the test set, we pick its type-aware refined versions q^*_{zi} to retrieve documents $D_{rq^*_{zi}}$ using a retriever r, and evaluate the query's performance using an information retrieval metric m, i.e., $m(D_{rq^*_{zi}} : \mathcal{J}_q)$. For the same query q, we also pick a type-less refined version q^*_i, where the query type z was overlooked, and use it to retrieve documents $D_{rq^*_i}$. We argue that, on average, E.q. 1 holds true for a randomly selected query q from the test set across different retrievers r and evaluation metrics m, and hence, considering the queries' types through type-aware query refinement results in higher search efficacy and yields better refined queries compared to when query types are overlooked. Furthermore, with both type-aware and type-less models at our disposal, we posit that the type-aware model performs on par with or even surpasses the type-less model for queries with unknown types, a scenario more reflective of real-world applications. It is worth noting that detecting a type for a query is *not* the focus of this paper. Herein, our main goal is to study the synergistic effects of considering the query's type, known *a priori*, for its refinement. Extending type-aware query refinement for queries whose type is missing or unknown by an end-to-end query type detection *and* refinement is indeed our future research direction.

5 Experiment

We seek to answer the following research questions:
RQ1: Does considering query types enhance the performance of refined queries?
RQ2: Is the type-aware query refinement consistently effective across all query types?
RQ3: Is the efficacy of the type-aware query refinement consistent across different retrievers?
RQ4: How does the type-aware model perform compared to its type-less counterpart when the query type is *un*known?

5.1 Setup

Dataset. To our knowledge, only two publicly available benchmark query sets classify queries by type; `trec-mq-2009` [14], which contains `420` queries across `6` types but it is small and insufficient for fine-tuning a transformer. Therefore, we used the large-scale `orcas-i-2m` [4], which comprises `2 million` queries of `5` types: `factual`, `instrumental`, `navigational`, `transactional`, and `abstain` (*un*known).

Conditional Transformer. We fine-tuned the pretrained `t5-base` [47] with `220 million` parameters for `4,000` epochs on the train set with a batch size of

256 and a learning rate of 0.001. The maximum sequence lengths for the input and output are set to 2,048 and 32, respectively. We chose t5 for its seamless integration of conditions, herein, a query type, into the training and inference phases by prepending it as a textual prefix followed by ':' to the input. For type-less query reformulations, we disregard the query type. Also, t5's straightforward installation process, open-source codebase, and efficient use of google cloud's tpus help speed up model fine-tuning and inference. During inference, we utilized top-$k = 10$ random selection in the decoder, which resulted in 10 type-aware and 10 type-less refinements for each original query.

Table 3. Results of type-aware vs. type-less query refinement on the test set. † refers to statistically significant improvements using one-tailed (greater) paired t-test with 99% confidence interval (p-value < 0.01).

	bm25						qld					
	map		ndcg		mrr		map		ndcg		mrr	
	avg	$\#q^*$	avg	$\#q^*$	avg	$\#q^*$	avg	$\#q^*$	avg	$\#q^*$	avg	$\#q^*$
original	0.174	–	0.284	–	0.253	–	0.170	–	0.281	–	0.248	–
type-less	0.279	272,936	0.420	272,936	0.416	272,909	0.272	275,734	0.406	275,730	0.415	275,727
type-aware	**0.289†**	**287,396**	**0.446†**	**287,387**	**0.451†**	**287,369**	**0.282†**	**281,607**	**0.412†**	**281,602**	**0.426†**	**281,607**

Table 4. Results of type-aware vs type-less query refinement *per query type* on the test set. † refers to statistically significant improvements using one-tailed (greater) paired t-test with 99% confidence interval (p-value < 0.01).

	$\#q$	bm25						qld					
		map		ndcg		mrr		map		ndcg		mrr	
		avg	$\#q^*$	avg	$\#q^*$	avg	$\#q^*$	avg	$\#q^*$	avg	$\#q^*$	avg	$\#q^*$
navigational													
original		0.128	–	0.200	–	0.185	–	0.116	–	0.185	–	0.173	–
type-less	52,561	0.305	34,605	0.440	34,605	0.433	34,605	0.286	33,928	0.422	33,928	0.418	33,928
type-aware		**0.325†**	**36,914**	**0.460†**	**36,914**	**0.462†**	**36,914**	**0.302†**	**35,241**	**0.438†**	**35,241**	**0.440†**	**35,241**
factual													
original		0.188	–	0.308	–	0.265	–	0.197	–	0.323	–	0.275	–
type-less	127,666	0.311	103,407	0.463	103,407	0.448†	103,396	0.319†	106,195	0.477†	106,195	0.460†	106,195
type-aware		**0.320†**	**111,659**	**0.475 †**	**111,659**	**0.463†**	**111,648**	**0.328†**	**111,838**	**0.482†**	**111,838**	**0.467†**	**111,838**
transactional													
original		0.136	–	0.230	–	0.209	–	0.121	–	0.212	–	0.189	–
type-less	15,035	0.272	11,276	0.413	11,276	0.424	11,276	0.248	11,072	0.389	11,072	0.396	11,072
type-aware		**0.288†**	**12,175**	**0.430 †**	**12,175**	**0.448†**	**12,175**	**0.255†**	**12,175**	**0.397†**	**12,175**	**0.405†**	**12,175**
instrumental													
original		0.161	–	0.279	–	0.247	–	0.147	–	0.266	–	0.227	–
type-less	20,939	0.269	16,593	0.422	16,593	0.425	16,586	0.245	16,787	0.400	16,787	0.384	16,780
type-aware		**0.276†**	**17,479**	**0.435 †**	**17,479**	**0.434†**	**17,472**	**0.251 †**	**17,126**	**0.403†**	**17,126**	**0.395†**	**17,126**
abstain													
original		0.133	–	0.296	–	0.269	–	0.171	–	0.284	–	0.257	–
type-less	143,193	0.292	107,055	0.430	107,055	0.450	107,046	0.273	**107,752**	0.315	**107,748**	0.427	**107,752**
type-aware		**0.294†**	**109,169**	**0.431 †**	**109,169**	**0.451†**	**109,169**	**0.275†**	105,821	**0.320†**	105,821	**0.428†**	105,821

Evaluation. We assess the efficacy of type-aware refined queries in comparison with type-less refined versions when retrieving the relevant documents based on the original query's relevance judgment. We use different lightweight sparse retrievers including `bm25` [49] and `qld` [45], and evaluate the retrieval performance using mean average precision (`map`), mean reciprocal rank (`mrr`), and normalized discounted cumulative gain (`ndcg`) at cutoff `100`, followed by a statistical significant test (†) using one-tailed (greater) paired `t-test` with `99%` confidence interval (p-value $<$ `0.01`). We acknowledge the state-of-the-art retrieval performance of dense retrievers, such as `colbert` [32]. However, given the intensive computational demands such methods impose on our large-scale query set, coupled with our limited computational resources, we were unable to include them in our testbed. Our primary goal, however, is to demonstrate the efficacy of incorporating query types in refining queries; a goal achievable even with lightweight off-the-shelf sparse retrievers. We expect that advanced dense retrievers would likely demonstrate similar effects.

5.2 Results

In response to **RQ1**, i.e., whether considering query types enhances the performance of refined queries, from Table 3, we observe that refined queries by the type-aware model statistically significantly enhance the search experience compared to the refined queries by the type-less model as well as the original queries. The improvements across all evaluation metrics are more evident when employing `qld` as the retriever. While we may interpret the improvements for the evaluation metrics as marginal, by looking at the number of original queries that could be matched with a refined version, we can observe a substantial increase when query types are considered. Specifically, using `bm25`, the type-aware model generates about `14,000` more refined queries compared to the type-less model. In summary, the utilization of query types substantially helped many original queries that would otherwise have been left without refinement.

Regarding **RQ2**, that is, if type-aware query refinement is consistently effective across all query types, Table 4 shows that incorporating query types enhances query refinement for *all* query types consistently and statistically significantly across evaluation metrics and retrieval methods. Notably, integrating query types leads to significant improvements in the refinement of `navigational` queries across all metrics, even though they constitute only 14% of our training set. It is also worth noting that achieving improvement in `factual` and `instrumental` queries was unexpected, as they are often less ambiguous compared to other types, and the related documents for these queries tend to have overlaps. Moreover, despite marginal improvements in metrics, we observed a major increase in the number of generated refined queries by the type-aware model, ranging from `800` for `instrumental` to `8,000` more refined queries for `factual` queries. In contrast, `abstain` queries demonstrate the smallest improvements in refinement and fall short of the increase in the number of refined queries, esp., when using the `qld` retriever, which is expected as the query type is unknown and offers little help to the model.

To answer **RQ3** about the efficacy of the type-aware query refinement across retrievers, from Tables 3 and 4, we observe that both `bm25` and `qld` retrieve more relevant documents when applied to the type-aware refined queries. These findings indicate that type-aware query refinement enhances retrieval efficacy, irrespective of the information retrieval methods. Additionally, the number of queries matched with a refined version from the type-aware model has consistently improved per query type across all metrics and retrievers, except for the `abstain` queries when `qld` is the retriever.

In response to **RQ4** about the performance of type-aware vs. type-less models for queries with unknown type, looking at Table 4 for `abstain` queries, we observe that the type-aware model performs better than its type-less counterpart. The improvement is more pronounced using `bm25` as the retriever, wherein both the metric values and the number of refined queries are increased. This makes the type-aware model a more suitable choice for real-world applications where query types may *not* be explicitly known.

Our findings align with Arabzadeh et al. [5] and Narayanan et al. [40], who demonstrated that semantic conditioning and query reformulation can improve retrieval precision. However, unlike their type-agnostic approaches, our method conditions refinements explicitly on query type, resulting in larger improvements, particularly for navigational queries. This observation complements Rajaei et al. [48], where semantic drift was reduced through backtranslation, by showing that query-type awareness can further mitigate such drift. Together, these comparisons indicate that query-type awareness offers a complementary advantage to existing refinement techniques and is broadly applicable across domains.

6 Concluding Remarks

In this paper, we studied the impact of query types on query refinement. Using a conditional transformer, we fed query types as a prior condition to generate type-aware refinements of queries. Our experiments on a large-scale query set demonstrated statistically significant improvement across five query types when considering query types versus lack thereof. Per query type, `navigational` queries benefited the most in terms of evaluation metrics, while the `factual` queries obtained a higher number of refined queries. As we obtain the necessary computational resources, our future research includes other conditional transformers as well as dense retrievers to explore whether our findings generalize. Also, we aim to extend type-aware query refinement to an end-to-end in-tandem approach for query type detection and refinement to better suit real-world scenarios where query types may not known in advance.

References

1. Agosti, M., Crivellari, F., Di Nunzio, G.M.: Web log analysis: a review of a decade of studies about information acquisition, inspection and interpretation of user interaction. Data Min. Knowl. Disc. **24**, 663–696 (2012). https://doi.org/10.1007/s10618-011-0228-8

2. Agosti, M., Di Nunzio, G.M., Marchesin, S.: An Analysis of Query Reformulation Techniques for Precision Medicine. ACM, Paris (2019)
3. Ahmad, W.U., Chang, K.-W., Wang, H.: Context Attentive Document Ranking and Query Suggestion. ACM, Paris (2019)
4. Alexander, D., Kusa, W., de Vries, A.P.: ORCAS-I: Queries Annotated with Intent using Weak Supervision. ACM, Madrid (2022)
5. Arabzadeh, N., Bigdeli, A., Seyedsalehi, S., Zihayat, M., Bagheri, E.: Matches Made in Heaven: Toolkit and Large-Scale Datasets for Supervised Query Reformulation. ACM, Queensland (2021)
6. Azzopardi, L., Liu, J.: Search under Uncertainty: Cognitive Biases and Heuristics - Tutorial on Modeling Search Interaction using Behavioral Economics. ACM, Sheffield (2024)
7. Belkin, N.J., Oddy, R.N., Brooks, H.M.: ASK for information retrieval: Part I. Background and theory. J. Documentation, London (1982)
8. Benzarti, S., Tebourski, W., Karaa, W.B.A.: Toward a Deep Multimodal Interactive Query Expansion for Healthcare Information Retrieval Effectiveness. Springer, Kitakyushu (2024). https://doi.org/10.1007/978-3-031-57853-3_31
9. Breja, M., Jain, S.K.: Why-Type Question to Query Reformulation for Efficient Document Retrieval. IGI Global, Hershey (2022)
10. Broder, A.Z.: A Taxonomy of Web Search. ACM, New York (2002)
11. Buckley, C., Salton, G., Allan, J., Singhal, A.: Automatic Query Expansion Using SMART: TREC 3. NIST, Gaithersburg (1995)
12. Cao, G., Nie, J.-Y., Gao, J., Robertson, S.: Selecting Good Expansion Terms for Pseudo-relevance Feedback. ACM, Singapore (2008)
13. Cao, H., Jiang, D., Pei, J., Chen, E., Li, H.: Towards Context-aware Search By Learning a Very Large Variable Length Hidden Markov Model From Search Logs. ACM, Madrid (2009)
14. Carterette, B., Pavlu, V., Fang, H., Kanoulas, E.: Million Query Track 2009 Overview. NIST, Gaithersburg (2009)
15. Chen, J., Mao, J., Liu, Y., Zhang, F., Zhang, M., Ma, S.: Towards a better understanding of query reformulation behavior in web search, ACM/IW3C2, Ljubljana (2021)
16. Chen, L.: Understanding and Exploiting User Intent in Community Question Answering. University of London, UK, Birkbeck (2014)
17. Cheng, P.-J., Tsai, C.-H., Hung, C.-M., Chien, L.-F.: Query Taxonomy Generation for Web Search. ACM, Virginia (2006)
18. Choi, B., Arguello, J., Capra, R.: Understanding Procedural Search Tasks "in the Wild". ACM, Austin (2023)
19. Chuang, S.-L., Chien, L.-F.: Automatic Query Taxonomy Generation for Information Retrieval Applications. Bingley, Emerald (2003)
20. Dehghani, M., Rothe, S., Alfonseca, E., Fleury, P.: Learning to Attend, Copy, and Generate For Session-based Query Suggestion. ACM, Singapore (2017)
21. Devlin, J., Chang, M.-W., Lee, K., Toutanova, K.: BERT: Pre-training of Deep Bidirectional Transformers for Language Understanding. Association for Computational Linguistics, Minneapolis (2019)
22. Downey, D., Dumais, S., Horvitz, E.: Heads and Tails: Studies of Web Search with Common and Rare Queries. ACM, Amsterdam (2007)
23. Fan, A., Lewis, M., Dauphin, Y.N.: Hierarchical Neural Story Generation. Association for Computational Linguistics, Melbourne (2018)
24. Furnas, G.W., Landauer, T.K., Gomez, L.M., Dumais, S.T.: The Vocabulary Problem in Human-System Communication. ACM, New York (1987)

25. Han, F.X., Niu, D., Chen, H., Lai, K., He, Y., Xu, Y.: A Deep Generative Approach to Search Extrapolation and Recommendation. ACM, Anchorage (2019)
26. Hirsch, S., Guy, I., Nus, A., Dagan, A., Kurland, O.: Query Reformulation in E-Commerce Search. ACM, Virtual/China (2020)
27. Hosseini, S.M., Arabzadeh, N., Zihayat, M., Bagheri, E.: Enhanced Retrieval Effectiveness through Selective Query Generation. ACM, Birmingham (2024)
28. Ibrahim, S.N.A., Selamat, A., Selamat, M.H.: Query Optimization in Relevance Feedback using Hybrid GA-PSO for Effective Web Information Retrieval. IEEE, Kuala Lumpur (2009)
29. Kang, I.-H., Kim, G.-C.: Query Type Classification for Web Document Retrieval. ACM, Toronto (2003)
30. Keskar, N.S., McCann, B., Varshney, L.R., Xiong, C., Socher, R.: CTRL: a conditional transformer language model for controllable generation, arXiv, online (2019)
31. Keyvan, K., Huang, J.X.: How to Approach Ambiguous Queries in Conversational Search: A Survey of Techniques, Approaches, Tools, and Challenges. ACM, New York (2023)
32. Khattab, O., Zaharia, M.: ColBERT: Efficient and Effective Passage Search via Contextualized Late Interaction over BERT. ACM, Virtual/China (2020)
33. Kwon, O.-W., Kim, M.-C., Choi, K.-S.: Query Expansion Using Domain-adapted, Weighted Thesaurus in an Extended Boolean Model. ACM, Maryland (1994)
34. Li, H., Xu, J.: Semantic Matching in Search. Now Publishers, Delft (2014)
35. Manchanda, S., Sharma, M., Karypis, G.: Intent term selection and refinement in E-commerce queries, arXiv, online (2019)
36. Mao, X., Huang, S., Li, R., Shen, L.: Automatic Keywords Extraction Based on Co-occurrence and Semantic Relationships Between Words. IEEE, New York (2020)
37. McCann, B., Keskar, N.S., Xiong, C., Socher, R.: The natural language decathlon: multitask learning as question answering, arXiv, online (2018)
38. Mitra, R., Gupta, M., Dandapat, S.: Transformer Models for Recommending Related Questions in Web Search. ACM, Virtual/Ireland (2020)
39. Moen, H., Peltonen, L.-M., Suhonen, H., Matinolli, H.-M., Mieronkoski, R., Telen, K., Terho, K., Salakoski, T., Salanterä, S.: An Unsupervised Query Rewriting Approach Using N-gram Co-occurrence Statistics to Find Similar Phrases in Large Text Corpora. Linköping University Electronic Press, Turku (2019)
40. Narayanan, Y.L., Fani, H.: RePair: An Extensible Toolkit to Generate Large-Scale Datasets for Query Refinement via Transformers. ACM, Birmingham (2023)
41. Chen, W., Cai, F., Chen, H., de Rijke, M.: Personalized query suggestion diversification. In: Proceedings of the ACM SIGIR Conference, ACM (2017)
42. Fiorini, N., Lu, Z.: Personalized neural language models for real-world query auto completion. In: Proceedings of the NAACL-HLT Conference (2018)
43. Nogueira, R., Lin, J., Epistemic, A.I.: From doc2query to docTTTTTquery, Online (2019)
44. Nogueira, R.F., Yang, W., Lin, J., Cho, K.: Document Expansion by Query Prediction, arXiv, online (2019)
45. Ponte, J.M., Croft, W.B.: A Language Modeling Approach to Information Retrieval. ACM, Melbourne (1998)
46. Radford, A., Wu, J., Child, R., Luan, D., Amodei, D., Sutskever, I., et al.: Language Models are Unsupervised Multitask Learners. OpenAI, San Francisco (2019)
47. Raffel, C., et al.: Exploring the Limits of Transfer Learning with a Unified Text-to-Text Transformer. JMLR, Cambridge (2020)
48. Rajaei, D., Taheri, Z., Fani, H.: No Query Left Behind: Query Refinement via Backtranslation. ACM, Boise (2024)

49. Robertson, S.E., Zaragoza, H.: The Probabilistic Relevance Framework: BM25 and Beyond. Now Publishers, Delft (2009)
50. Rocchio, J.J.: Relevance Feedback in Information Retrieval. Prentice-Hall, Englewood Cliffs (1971)
51. Schultz, C.D.: Informational, transactional, and navigational need of information: relevance of search intention in search engine advertising. Inf. Retrieval J. **23**(2), 117–135 (2020). https://doi.org/10.1007/s10791-019-09368-7
52. Shah, C., et al.: Using large language models to generate, validate, and apply user intent taxonomies, arXiv, online (2023)
53. Shekarpour, S., Marx, E., Auer, S., Sheth, A.P.: RQUERY: Rewriting Natural Language Queries on Knowledge Graphs to Alleviate the Vocabulary Mismatch Problem. AAAI Press, San Francisco (2017)
54. Shiri, A.A.: End-user Interaction with Thesaurus-enhanced Search Interfaces, an Evaluation of Search Term Selection for Query Expansion. University of Alberta, Edmonton (2003)
55. Singhal, A., et al.: Modern Information Retrieval: A Brief Overview. IEEE, New York (2001)
56. Sordoni, A., Bengio, Y., Vahabi, H., Lioma, C., Simonsen, J.G., Nie, J.-Y.: A Hierarchical Recurrent Encoder-decoder For Generative Context-aware Query Suggestion. ACM, Melbourne (2015)
57. Tamannaee, M., Fani, H., Zarrinkalam, F., Samouh, J., Paydar, S., Bagheri, E.: ReQue: A Configurable Workflow and Dataset Collection for Query Refinement. ACM, Virtual/Ireland (2020)
58. Tao, T., Zhai, C.: Regularized Estimation of Mixture Models for Robust Pseudo-relevance Feedback. ACM, Seattle (2006)
59. Vaswani, A., et al.: Attention is All you Need, NeurIPS (2017)
60. Xu, J., Croft, W.B.: Quary Expansion Using Local and Global Document Analysis. ACM SIGIR Forum, New York (2017)
61. Yang, D., Zhang, Y., Fang, H.: An Exploration study of mixed-initiative query reformulation in conversational passage retrieval, NIST, Online (2022)
62. Zerveas, G., Zhang, R., Kim, L., Eickhoff, C.: Brown University at TREC Deep Learning 2019, arXiv (2020)
63. Zhang, X.: Improving personalised query reformulation with embeddings. J. Inf. Sci. (2022)
64. Zhao, L., Callan, J.: Term necessity prediction. In: Proceedings of the 19th ACM CIKM (2010)
65. Zhou, Y., Dou, Z., Wen, J.-R.: Encoding history with context-aware representation learning for personalized search. In: Proceedings of the 43rd ACM SIGIR (2020)
66. Zhu, Y., Pang, L., Lan, Y., Shen, H., Cheng, X.: LoL: a comparative regularization loss over query reformulation losses for pseudo-relevance feedback. In: Proceedings of the 45th ACM SIGIR (2022)
67. Rencis, E.: Application of a configurable keywords-based query language to the healthcare domain. J. Adv. Inf. Technol. (2021)

Boolean-Aware Attention for Dense Retrieval

Quan Mai(✉), Susan Gauch, and Douglas Adams

University of Arkansas, Fayetteville, AR 72701, USA
quanmai@uark.edu

Abstract. We present Boolean-aware attention (BoolAttn), a novel attention mechanism that dynamically adjusts token focus based on Boolean operators (e.g., "and", "not"). Our model employs specialized Boolean experts, each tailored to amplify or suppress attention for operator-specific contexts. A predefined gating mechanism activates the corresponding experts based on the detected Boolean type. Experiments on two Boolean retrieval datasets demonstrate that integrating BoolAttn with BERT and T5 greatly enhances the models' capability to process Boolean queries.

Keywords: Transformers · Attention · Dense Retrieval · Boolean Retrieval · Negation Retrieval

1 Introduction

Boolean operations such as *and*, *or*, and *not* are fundamental components of human logic and reasoning, serving as essential tools for combining, including, or excluding specific information in queries. These operations are particularly valuable when users have complex and precise information needs, such as "Movies set in Vietnam but *not* about war" or "Animals found only in Brazil *and* Mexico." In such cases, users implicitly express set operations (i.e., intersections, unions, and complements) that are central to Boolean logic.

Despite their importance, modern retrieval systems often fail to handle Boolean logic effectively. Traditional lexical matching methods like BM25 [15] rely on exact term matches and do not incorporate logical semantics. As a result, they often fail to adequately handle queries involving exclusions, frequently retrieving results that include undesired information. For example, when processing the query "Books about Japan but not about World War II", BM25 may still retrieve documents that mention both Japan and World War II because it cannot interpret the negation cue. The presence of high-frequency keywords drives retrieval rankings, even when those keywords explicitly contradict the user's intent. This limitation leads to the retrieval of irrelevant or even contradictory results, highlighting the inadequacy of purely lexical approaches in handling Boolean reasoning.

H. Unger and P. Meesad (Eds.): NLPIR 2025, LNNS 1904, pp. 740–754, 2026.
https://doi.org/10.1007/978-3-032-20897-2_52

Transformer-based architectures (e.g., BERT [1], T5 [14]) and their variants have significantly advanced dense retrieval by enabling semantic matching through contextual embeddings [6][4]. However, despite these advancements, they continue to struggle with logical reasoning tasks [11] [12]. These models lack explicit mechanisms for interpreting Boolean operators and often fail to distinguish their functional roles in context. For instance, they may treat "and" and "not" similarly in the embedding space [5], leading to query representations that overlook critical logical distinctions. This results in overly similar query embeddings for semantically different inputs and causes retrieval systems to return irrelevant results, particularly when exclusion or conjunction is vital to user intent.

To address these limitations, we propose Boolean-aware attention (**BoolAttn**), a novel extension to the self-attention mechanism that dynamically adapts token-level interactions based on detected Boolean structures in the query. Our approach introduces a set of specialized attention experts, each tailored to a specific Boolean operator (e.g., and, not), and activates them via a predefined gating mechanism conditioned on the query context. This design allows the model to selectively amplify or suppress token attention based on logical intent, ensuring that semantically critical operations like negation or conjunction are faithfully represented in the attention computation.

Our contributions can be summarized as follows:

1. We introduce BoolAttn, the first Boolean-aware attention mechanism that explicitly models Boolean operators within Transformer-based retrievers. **BoolAttn** is lightweight, modular, and fully compatible with existing transformer architectures, serving as a *plug-and-play* enhancement for self-attention layers
2. We propose a lightweight expert gating module that identifies the logical structure of a query and dynamically routes attention through operator-specific experts.
3. We plug BoolAttn into BERTs and T5s, evaluate on two logic-sensitive retrieval benchmarks (QUEST and BoolQuestions), and demonstrate consistent improvements over standard attention mechanisms and Boolean-agnostic baselines.

By explicitly aligning attention computation with Boolean logic, BoolAttn enables more precise semantic understanding of complex queries, bridging the gap between symbolic reasoning and neural retrieval.

The remainder of the paper is organized as follows. Section 2 discusses related work and the motivation behind the proposed method. Section 3 introduces BoolAttn and details its key components. In Sect. 4, we present the evaluation setup and experimental results. Finally, we conclude the paper in Sect. 5.

2 Related Work

2.1 Boolean Information Retrieval

Neural dense retrievers, such as dual-encoder models based on BERT or T5 [1][14], leverage contextual embeddings to perform semantic matching, significantly improving recall for related concepts. However, these models often treat Boolean operators like "and" and "not" as ordinary tokens, leading to degraded performance on queries that require explicit logical reasoning. Analyses in [11][12] show that BERT-based retrievers tend to misinterpret logical constructs, causing errors on queries involving conjunction or negation. To address this, **SetBERT** [11] fine-tunes BERT on synthetic Boolean queries containing set operations using inverse-contrastive learning, enhancing its ability to embed logical structure. Although SetBERT improves Boolean query handling, it still relies on static embeddings and does not dynamically modulate attention based on operator structure.

Benchmark datasets like **QUEST** [12] and **BoolQuestions** [20] further expose these limitations. BoolQuestions features complex natural language queries containing Boolean logic and corresponding annotated passage, while QUEST evaluates entity retrieval with implicit set operations. Despite highlighting model weaknesses in Boolean reasoning, these benchmarks do not prescribe operator-aware mechanisms. Overall, existing retrieval methods either use rigid Boolean matching or holistic neural encoding and lack token-level mechanisms to process Boolean operators explicitly, leaving logical constructs like negation and conjunction as ongoing challenges.

2.2 Attention Modulation

A variety of recent works have explored ways to bias or adapt Transformer attention mechanisms, often by injecting linguistic or positional biases or employing gating strategies to alter attention distributions. However, these efforts have primarily focused on modeling sequence structure or improving computational efficiency, rather than handling logical operators. For instance, relative positional bias [16], widely used in transformer-based models, introduces learned biases based on token distances to enhance long-range modeling. Syntactic and structural biases [9][3] impose attention patterns guided by parse trees or local syntax, enabling tokens to attend only to syntactically related words. While effective for embedding linguistic structure, these approaches do not address semantic cues such as Boolean operators. Mixture-of-experts (MoE) methods [17] assign tokens to expert subnetworks using learned gates, as seen in models such as **GShard** [8] and **Switch Transformer** [2]. Though these approaches demonstrate the viability of selective attention modules, they are not designed to reflect query semantics. Similarly, gated attention methods like **Gated Linear Attention** [19] use data-dependent gates to modulate attention and improve long-context handling, but the gating is not keyed to token identity and thus remains agnostic to logical operators. Other efforts, such as entity-aware attention or cross-attention with external knowledge, introduce biases tailored to specific tasks, yet still overlook logical operations. Overall, prior attention-modulation strategies rely on position, syntax, or task-specific signals, but none explicitly condition attention on the presence of Boolean operators like "and" or "not." In contrast, our Boolean-aware attention mechanism leverages these operators themselves to activate specialized attention experts, introducing operator-specific modulation that has not been addressed in previous work.

BoolAttn Unlike the above methods, our approach explicitly aligns the self-attention mechanism with Boolean query structure. BoolAttn introduces dedicated attention experts for **and** (conjunction - set intersection), **or** (disjunction - set union), **not** (negation - set difference) uses a gating classifier to activate the appropriate expert based on the detected operator. This enables the model to focus or suppress tokens according to Boolean context, a capability absent in prior IR and reasoning methods.

3 Boolean-Aware Attention

Our Boolean-aware attention mechanism consists of four key components:

1. a **Cue Predictor** (3.1) that detects Boolean cues within the input sequence,
2. a **Contextual Scope Predictor** (3.2) that determines the scope of Boolean operators, ensuring their influence is properly localized,
3. a **Bias Predictor** (3.3) that adjusts attention scores based on Boolean operations using operator-specific embeddings and positional dependencies, and
4. a **modified attention mechanism** incorporating Boolean-specific biases into the attention scores (3.4).

By explicitly modeling Boolean logic while preserving contextual dependencies, **BoolAttn** enables transformers to better interpret logical structures in natural language queries. The overall architecture of **BoolAttn** is illustrated in Fig. 2. The number of BoolAttn experts (yellow blocks) corresponds to the number of basic Boolean logic types, which is 3 (and, or, not).

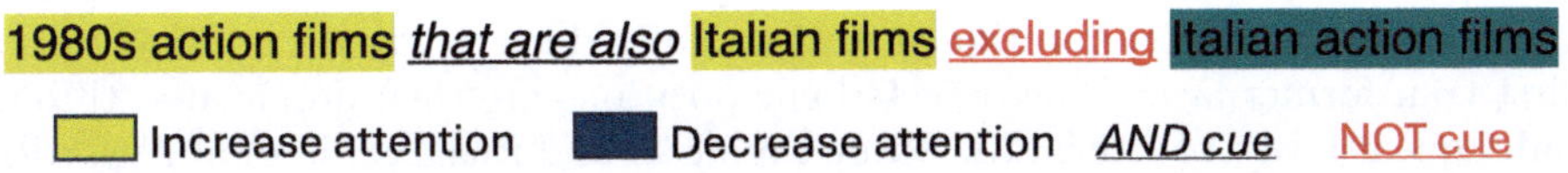

Fig. 1. Token-wise attention scores are adjusted based on Boolean type. Tokens affected by *and* will receive more attention while tokens affected by *not* will receive less attention score.

Boolean Operator Embeddings. We use operator-specific embeddings $\mathbf{o} \in \mathbb{R}^{d_o}$ to encode the unique semantic roles of Boolean operators, such as the negation effect of *not* or the conjunctive nature of *and*. These embeddings guide other modules to produce operator-specific outputs, enabling the model to adapt to the logical properties of each operator.

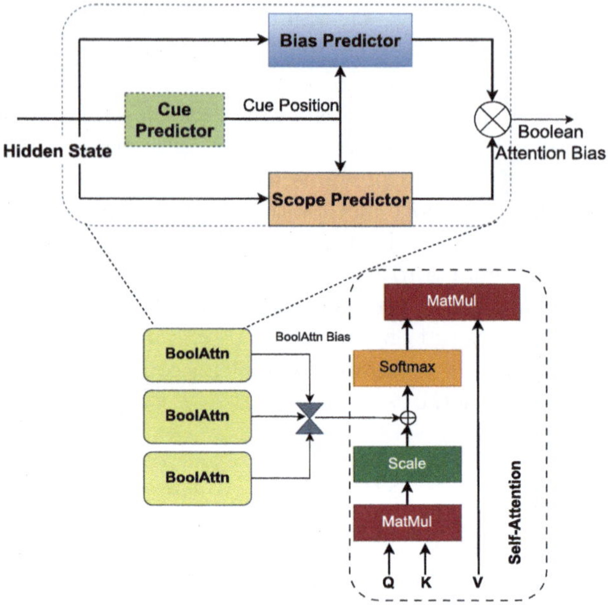

Fig. 2. Boolean-aware Attention architecture.

3.1 Boolean Cue Predictor

Boolean cues play a pivotal role in determining the logical structure of queries. These cues dictate the scope and interaction between different query components, thereby influencing the embedding of the query. For instance, cues for *not* (negation) include keywords like "not," "excluding," and "without," while cues for *and* include "and," "as well as," "also," and "including." Example of Boolean cues are shown in table 1.

The Cue Predictor (CuePred) is designed to identify the positions of Boolean cues within the input sequence. Since the positional information of cues remains invariant for a given input sequence, the CuePred is employed exclusively in the first transformer layer. The detected cue positions are then propagated through subsequent layers to inform the Scope Predictor and Bias Predictor at each corresponding layer. Notably, Boolean-related retrieval datasets [12] do not explicitly annotate Boolean cues. Therefore, cue detection is formulated as an auxiliary task during the pretraining phase, enabling the model to learn cue positions implicitly and generalize effectively to downstream tasks.

The operator embeddings $\mathbf{o}$ are projected into the model's hidden space using a linear transformation

$$\mathbf{o}' = \mathbf{W_o o} + \mathbf{b}_o \tag{1}$$

where $\mathbf{W}_o \in \mathbb{R}^{d_h \times d_o}$ and $\mathbf{b}_o \in \mathbb{R}^{d_h}$ are learnable parameters, d_h and d_o denote the hidden and operator embedding dimensions, respectively. The projected embedding is then added to the hidden states H before going through a shared linear layer to compute cue scores, followed by sigmoid activation function to compute the cue probabilities:

$$p_c = \sigma(\mathbf{W}_c(\mathbf{H} + \mathbf{o}') + \mathbf{b_c}) \tag{2}$$

where $\mathbf{W}_c \in \mathbb{R}^{1 \times d_h}$ and $\mathbf{b}_c \in \mathbb{R}$ are learnable parameters. Cue probabilities $p_c \in \mathbb{R}^{B \times L}$ indicates the likelihood of each position being a cue. Notably, each Boolean expert generates its own prediction for the presence of its corresponding operator, as queries may contain multiple types of Boolean logic. The output of CuePred p_c is then passed to Scope Predictor and Bias Predictor modules.

Table 1. Common Boolean cues in natural language.

Boolean types	Cues
Conjunction (AND)	and, as well as, plus, with, including, also, both, etc.
Disjunction (OR)	or, either ... or, alternatively, etc.
Negation (NOT)	not, without, no, except, excluding, etc.

3.2 Boolean Scope Predictor

The Scope Predictor (ScopePred) identifies the affected span of tokens (the *scope*) for a given Boolean operator. For example, in the sentence shown in Fig. 1, the scope of *not* is "Italian action film," and the model downweights the importance of these tokens in the **BoolAttn** mechanism. The predictor leverages local contextual features and conditions on operator embeddings to determine the scope. It combines a Conv1D layer to capture local dependencies and a FiLM layer (Feature-wise Linear Modulation, [13]) to specialize outputs for each operator.

Boolean operators primarily influence tokens near the cue, as observed in phrases "not happy", where negation affects the adjacent token. To effectively capture such local dependencies, ScopePred utilizes a Conv1D layer. Especially, the hidden states $H \in \mathbb{R}^{B \times L \times d_h}$ and cue probabilities $p_c \in \mathbb{R}^{B \times L}$ are concatenated and subsequently passed through a Conv1D layer:

$$S = \text{Conv1D}([\mathbf{H}; \mathbf{p_c}]) \tag{3}$$

This layer efficiently captures local patterns and dependencies around the cue, making it particularly suitable for modeling the impact of Boolean operators on nearby tokens. To adapt the predictions for each Boolean operator, the FiLM layer modulates the Conv1D output using operator embeddings $\mathbf{o}$. Scale and shift factors are computed as:

$$\begin{aligned} \gamma &= \mathbf{W}_\gamma o + \mathbf{b}_\gamma \\ \beta &= \mathbf{W}_\beta o + \mathbf{b}_\beta \end{aligned} \tag{4}$$

These factors are then applied to modulate the Conv1D output:

$$S = \gamma S + \beta \tag{5}$$

The modulated logits S are passed through a sigmoid activation to obtain scope probabilities:

$$P_{scope} = \sigma(S) \tag{6}$$

Since the scope is binary (0 for outside scope, 1 for inside scope), we use the Gumbel-Sigmoid trick to enable discrete scope sampling while preserving differentiability. The scope is determined by applying a learnable threshold θ to the probabilities P obtained from the Gumbel-Sigmoid:

$$\mathrm{S} = \mathbf{I}(P > \theta) \tag{7}$$

To allow gradient flow through the non-differentiable thresholding, we use a Straight-Through Estimator (STE). This approach ensures that the model learns to make discrete scope predictions effectively.

Different Boolean operators impose unique structural constraints on token interactions, necessitating specialized mechanisms for scope prediction. The *and* operator establishes bidirectional reinforcement between operands (or entities), ensuring mutual influence, whereas *not* applies a unidirectional effect, modifying the meaning of specific tokens within its scope. When a negation cue like "not" appears, it alters the interpretation of a single token or phrase rather than forming a bidirectional relationship. Conversely, *and* enforces a reciprocal dependency, e.g., in "A and B," both A and B should influence each other, requiring a pairwise rather than token-wise scope. To capture these differences, we define the scope matrices for *and* and *or* as follows:

$$S_{mutual} = S \cdot S^T \tag{8}$$

where $S_{mutual} \in \mathbb{R}^{B \times heads \times L \times L}$ captures symmetric token influence, while *not* retains a unary mask of shape $(B, heads, L, 1)$ as it requires no pairwise expansion.

3.3 Bias Predictor

We use a Boolean-aware bias module to model token-level biases in **BoolAttn** by incorporating positional dependencies and operator-specific embeddings. This module learns to adjust attention bias based on the relative positions of Boolean cue (from the CuePred) and the surrounding context.

Relative Position. To capture the structural influence of Boolean operators, the model first computes the relative positions of tokens with respect to the Boolean cues (e.g., "without"). Given a sequence of length L, the relative position of token i with respect to a Boolean cue at position c is defined as:

$$r_i = i - c \tag{9}$$

For the negation operator (*not*), its influence is asymmetric, affecting only subsequent tokens. To ensure this, we clip negative relative positions:

$$r_i = \max(0, i - c), \text{ for } not$$

For conjunctions (*and*) and disjunctions (*or*), influence is bidirectional, meaning the relative positions are clamped with a symmetric window $\pm d$:

$$r_i = \text{clip}(i - c, -d, d), \text{ for } and\ /or \tag{10}$$

To assign different levels of importance based on distance from the cue, we can use position embeddings. To make this module lightweight, instead of using learnable positional embeddings, we use a Gaussian kernel to weight positions. The idea of using Gaussian kernel is we hypothesize that tokens closer to the cue are more likely to be influenced:

$$w(r_i) = \exp\left(-\frac{r_i^2}{2 * \sigma^2}\right) \tag{11}$$

where σ is a learnable parameter controlling the spread of the Gaussian influence. The computed positional weights $w(r)$ are concatenated with token hidden state H to form an enriched representation before passed through a Feed-Forward Network (FFN) to generate context-aware bias scores:

$$\mathbf{b} = \text{FFN}([\mathbf{H}, w(r)]) \tag{12}$$

To account for the distinct biases associated with different Boolean operations, an operator-specific gate dynamically modulates the computed biases:

$$\mathbf{b}' = \mathbf{b} \cdot \sigma(\mathbf{W}_{op}\mathbf{o}) \tag{13}$$

where $\mathbf{o} \in \mathbb{R}^{d_o}$ is the operator embedding, $\mathbf{W}_{op}$ is a learnable transformation matrix and $\sigma(\cdot)$ ensures smooth gating via sigmoid function.

The final bias values are regularized using softplus to explicitly ensure non-negativity. The output bias tensor $\mathbf{B} \in \mathbb{R}^{B \times H \times L}$ is then integrated into the transformer's attention mechanism, enhancing the model's logical reasoning capabilities for Boolean queries.

3.4 Operator-Gating Mechanism

After having Scope (Eq. 7) and Bias (Eq. 13) of each Boolean operator, we integrate the Boolean reasoning into the transformer's attention mechanism. This integration employs a gating mechanism, dynamically weighting each operator's influence based on its presence in the input.

The attention bias contribution of a Boolean operator *op* is formulated as:

$$S_{op} = G_{op} \cdot \text{Scope}_{op} \cdot \text{Bias}_{op} \tag{14}$$

where $G_{op} \in \{0; 1\}$ is a gating variable that indicates the presence of the operator in the input sequence. This gate can be either learned through an experts classifier or provided as an auxiliary label, enabling conditional application of the attention bias based on the occurrence of the operator.

Since we aim to reinforce tokens of *and*, *or* and downweight the importance of tokens affected by *not*, the final Boolean-aware attention bias is then formulated as:

$$S_{Boolean} = S_{and} + S_{or} - S_{not} \tag{15}$$

The Boolean-aware bias is added to the raw attention scores, ensuring that the logical structure of sentences influences the attention mechanism:

$$\text{Attention Score} = \text{Softmax}(\frac{\mathbf{QK}^T}{\sqrt{d}} + S_{Boolean})\mathbf{V} \tag{16}$$

where $\mathbf{Q}, \mathbf{K}, \mathbf{V}$ are attention matrices in the self-attention mechanism [18].

4 Evaluation

4.1 Dense Retrieval

Text retrieval addresses the task of selecting relevant documents from a large corpus $\mathcal{D} = \{d_i\}_{i=1}^{|\mathcal{D}|}$ in response to a user's query q. The core objective is to compute a relevant score $S(q, d_i)$ for each document $d_i \in \mathcal{D}$ and return a list of the top k documents that best satisfy the user's information need.

In this work, we focus on dense retrieval models which are known for its efficiency and scalability. Dense retrieval employs transformer-based encoders ϕ (such as BERT and T5) to encode both queries (q) and documents (d) into a low-dimensional, continous vectors within a shared semantic space. Specifically, queries embeddin E_q and document embedding E_d are calculated as:

$$E_q = \phi_q(q), E_d = \phi_d(d)$$

The relevance score $S(q, d)$ is then calculated as a similarity between their respective embeddings, two common methods to compute the relevance score are cosine similarity and dot product. In this work, we use the latter:

$$S(q, d) = E_q \cdot E_d$$

Documents are ranked based on this score to select the top k.

4.2 Experimental Setup

We evaluate **BoolAttn** by integrating it into BERT and T5, resulting in two variants: *Bool-BERT* and *Bool-T5*. These models are used as query encoders and tested on two Boolean retrieval benchmarks: QUEST [12] and BoolQuestions [20].

Prior to the main evaluation, we pretrain **BoolAttn** on a synthetic dataset generated by SetBERT [11] to ensure effective initialization (see 4.3). All experiments are repeated three times with different random seeds $\{0, 42, 1234\}$, and we report the average results. Full experimental details, including hyperparameters and additional settings, are provided in 4.4.

4.3 Pretrain BoolAttn

Introducing new parameters in **BoolAttn** poses a risk of overfitting due to the limited size of Boolean retrieval datasets compared to the model's capacity when integrated into transformer layers. This imbalance may lead the model to memorize the training data rather than generalizing effectively, resulting in suboptimal performance. To mitigate this, we pretrain Bool-BERT on [11]'s GPT-generated dataset on two auxiliary tasks: cue prediction and expert classification (gate selection). The cue position prediction loss is calculated as follows:

$$\mathcal{L}_{cue} = -\frac{1}{B \times L} \sum_{i=1}^{B} \sum_{j=1}^{L} (y_{i,j} \log p_{i,j} + (1 - y_{i,j}) \log(1 - p_{i,j})) \tag{17}$$

where y is the ground-truth label and p is the predicted probability that token j in batch i is a cue. The original dataset only includes "and," "or," and "not" as Boolean cues. To enhance model robustness, we augment the dataset by replacing cues with their semantic equivalents, such as substituting "and" with "as well as," "are also," or "including," and replacing "not" with "other than" or "excluding." Consequently, multiple cue tokens can appear in a single sequence. We use BERT's tokenizer, followed by a word-matching approach to label the cue tokens.

The second task in pretraining phase is *Boolean expert classification*. This task is a multi-label classification problem designed to identify which Boolean operators are present in a given input sequence. Since a sequence may contain more than one Boolean operation, this task requires predicting multiple labels simultaneously (one for each operator). Given E Boolean experts (3 if we use *and*, *or*, *not*), the loss is defined as

$$\mathcal{L}_{gate} = -\frac{1}{B \times E} \sum_{i=1}^{B} \sum_{k=1}^{E} (y_{i,k} \log p_{i,k} + (1 - y_{i,k}) \log(1 - p_{i,k})) \tag{18}$$

where y is the ground-truth label and $p_{i,k}$ is the predicted probability for operator k in batch i.

Until now we only pretrain the CuePred and Gate modules, the parameters for ScopePred and BiasPred remain uninitialized. To provide these modules with a reasonable starting point without overfitting them to Boolean logic, we train them on the Boolean sentence classification task from [11] using a standard triplet loss. Although [11] reported that triplet loss hinders performance on Boolean queries, we deliberately use it to provide an "okay" initialization, allowing the modules to learn actively during the main tasks. The pretraining objective is defined as follows:

$$\mathcal{L}_{pretrain} = \mathcal{L}_{cue} + \mathcal{L}_{gate} + \alpha\mathcal{L}_{triplet}$$

where α is set to a small value, we used 0.2. During pretraining, only the BoolAttn modules and embedding layers are updated, while the other layers are frozen. We train for a single epoch with a low learning rate of $1e-5$

4.4 Hyperparameters and Settings

All experiments are conducted on two NVIDIA Quadro RTX 8000 GPUs, each with 50GB of memory.

BoolAttn Hyperparameters (1) Operator embedding dimension d_h: 10. (2) Threshold θ is initialized at 0.5. (3) Relative position window: 5. (4) Gaussian kernal sigma: σ: 2.

Biencoder Unlike [11,12], which use a fixed positive document per query, we sample different positive documents at each training step. This encourages the encoders to explore diverse relevant documents for the same query, which is crucial for multi-answer retrieval tasks where multiple correct answers exist. This is particularly important for boolean queries: negation excludes a subset of answers, disjunctions (unions) expand the answer set, and conjunctions (intersections) narrow it. By introducing varied positive samples, our approach enhances model robustness and generalization. Furthermore, we empirically found that using random negative samples is more effective than hard negative samples (i.e., incorrect answers retrieved by BM25), particularly for the QUEST dataset. Table 2 shows the improvement when using different positive documents and normal negative samples, T5 and BERT results taken from [11,12], respectively.

Table 2. Importance of choosing different positive samples, results for QUEST dataset.

Retriever	Avg. Recall@K		
	50	100	1000
BERT-Base	0.134	0.182	0.453
BERT-Large	0.227	0.300	0.627
T5-Base	0.372	0.455	0.726
T5-Large	**0.386**	**0.476**	0.757
BERT-Base-Ours	0.344	0.443	**0.760**

Biencoder's hyperparameters: Max query length: 64. Max document length: 256. Negative document per sample: 5. Positive document: 1, shuffled every step. Batch size: 32. Optimizer: AdamW [7] with weight decay 0.01 [10]. Learning rate: 5e-5. Epoch: 40. Evaluation every epoch, keep best.

4.5 Experimental Results

Entity-Seeking Queries. QUEST features queries with implicit Boolean operations, testing models' ability to understand logical relationships among entities. We employ Bool-BERT/T5 as the query encoder in a dual-encoder setup, paired with BERT/T5 as the document encoder. Retrieval performance is measured using average Recall@K, with K set to $\{20, 50, 100, 1000\}$, as shown in Table 3.

Integrating **BoolAttn** into BERT and T5 consistently enhances retrieval performance, as evidenced by notable improvements across all measured Recall@K thresholds. Across both base and large variants of BERT and T5, the BoolAttn-enhanced models achieve higher recall rates. These improvements are especially prominent at lower K values (K=20, 50, and 100), where precise and early retrieval is most critical. Independent t-tests reveal that the improvements at Recall@20, 50, and 100 are statistically significant, with p-values well below 0.05. This reflects a meaningful enhancement in the models' ability to retrieve relevant items earlier in the ranked list.

Additionally, we assess the retrievers on seven distinct query structures to capture nuances in retrieval capabilities, with results presented in Fig. 3. These findings demonstrate that integrating the **BoolAttn** plugin enhances BERTs' effectiveness in Boolean query understanding.

Table 3. Bool-BERT retrievers outperform standard BERT-based models in retrieval tasks.

	Retriever	Ave. Recall@K 20	50	100	1000
BERT	BERT-base	0.228	0.344	0.443	0.760
	BoolBERT-base	0.254	0.379	0.483	0.788
	BERT-large	0.238	0.364	0.467	0.792
	BoolBERT-large	0.266	0.395	0.495	0.812
T5	T5-base	0.232	0.360	0.462	0.788
	BoolT5-base	0.259	0.387	0.489	0.820
	T5-large	0.248	0.378	0.473	0.805
	BoolT5-large	0.275	0.410	0.502	0.831

BoolQuestions. We use the best-performing models from fine-tuning on Quest as starting checkpoints for continuous learning on BoolQuestions' supplementary training dataset, focusing on *and*, *or*, and *not* operations. Evaluation is conducted on the corresponding test set, adhering to the original experimental setup by reporting Mean Reciprocal Rank at top-10 (MRR@10). The results in Table 4 highlight the effectiveness of **BoolAttn** in enhancing performance for

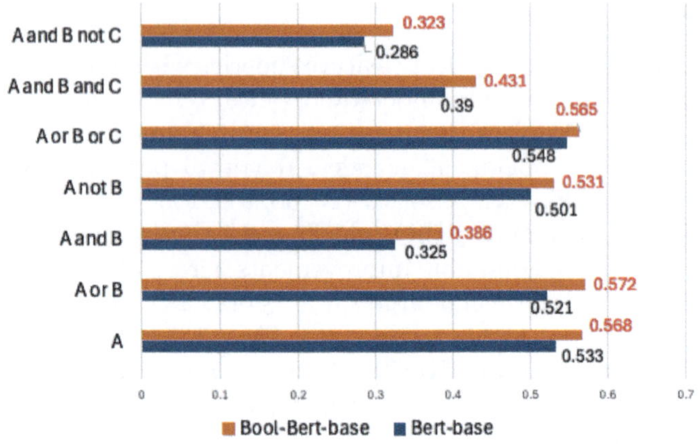

Fig. 3. Ave. Recall@100 across QUEST's different query templates, comparing results for BERT-base and BoolBERT-base.

Boolean-related QA tasks. For comparison, the results of *distillbert-dot-v5* are taken from the original paper.

Table 4. Performance on MS MARCO's BoolQuestions.

	MRR@10		
	AND	OR	NOT
distillbert-dot-v5	0.399	0.530	0.130
BERT-*base*	0.412	0.545	0.128
Bool-BERT-*base*	0.480	0.586	0.165
BERT-*large*	0.487	0.582	0.169
Bool-BERT-*large*	0.532	0.610	0.182
T5-*base*	0.420	0.557	0.132
Bool-T5-*base*	0.475	0.592	0.170
T5-*large*	0.491	0.605	0.140
Bool-T5-*large*	0.551	0.647	0.189

4.6 Ablation Study

Learnable Gate. In all previous experiments, we activated or deactivated the experts using the Boolean operator as an auxiliary input, that is, the query along with its Boolean logic was explicitly provided to the model. While this approach ensures that the correct experts are activated for a given query, it requires users to manually specify the Boolean operators they intend to use, which can be inconvenient. One potential solution is to implement a rule-based Boolean operator predictor that detects operators based on the presence of Boolean cues (see Table 1) in the query. An alternative, and more flexible, approach is to develop a learnable predictor capable of inferring the Boolean operators directly from the query text. In this section, we explore the latter option.

For the learnable gate mechanism, we employ a simple FFN followed by a sigmoid activation, with logits scaled by a small temperature value of 0.1 to enhance sensitivity. Unlike softmax, which enforces mutual exclusivity, sigmoid is better suited for Boolean logic since multiple Boolean operators can coexist within a single query. By applying a hard threshold to the sigmoid outputs, the gates can selectively activate multiple Boolean experts, allowing them to contribute to the final attention scores as needed.

However, the learnable gating mechanism introduces a performance bottleneck as it relies on accurately activating the Boolean experts corresponding to the operators present in the query. This dependency can lead to suboptimal performance (Table 5). Improving this gating strategy is left for future work.

Table 5. Learnable gate introduces bottleneck. Tested on QUEST.

Retriever	Avg. Recall@K		
	50	100	1000
Bool-BERT-*base*	0.379	0.483	0.788
Bool-BERT-*base*-gate	0.371 ↓	0.470 ↓	0.775 ↓

5 Conclusion

We introduced **BoolAttn** - a novel attention mechanism that dynamically adjusts attention weights based on Boolean operators, effectively capturing the distinct structural influences of conjunctions, disjunctions, and negations. By integrating **BoolAttn** into transformer-based encoders such as BERT and T5, we greatly improved performance on Boolean retrieval tasks, demonstrating its effectiveness in processing complex logical relationships within queries. These results highlight the potential of Boolean-aware attention mechanisms to enhance transformers' ability to understand and handle nuanced Boolean queries.

Acknowledgement. This material is based upon work supported by the National Science Foundation under Award #OIA-1946391, "Data Analytics that are Robust and Trusted (DART)".

References

1. Devlin, J.: Bert: Pre-training of deep bidirectional transformers for language understanding. arXiv preprint arXiv:1810.04805 (2018)
2. Fedus, W., Zoph, B., Shazeer, N.: Switch transformers: Scaling to trillion parameter models with simple and efficient sparsity. J. Mach. Learn. Res. **23**(120), 1–39 (2022)
3. Gessler, L., Schneider, N.: Syntactic inductive bias in transformer language models: Especially helpful for low-resource languages? arXiv preprint arXiv:2311.00268 (2023)
4. Izacard, G., et al.: Unsupervised dense information retrieval with contrastive learning. arXiv preprint arXiv:2112.09118 (2021)
5. Jeretic, P., Warstadt, A., Bhooshan, S., Williams, A.: Are natural language inference models imppressive? learning implicature and presupposition. arXiv preprint arXiv:2004.03066 (2020)
6. Karpukhin, V., et al.: Dense passage retrieval for open-domain question answering. arXiv preprint arXiv:2004.04906 (2020)
7. Kingma, D.P., Ba, J.: Adam: A method for stochastic optimization. In: International Conference on Learning Representations (ICLR) (2015)
8. Lepikhin, D., et al.: Gshard: Scaling giant models with conditional computation and automatic sharding. arXiv preprint arXiv:2006.16668 (2020)
9. Lindemann, M., Koller, A., Titov, I.: Strengthening structural inductive biases by pre-training to perform syntactic transformations. arXiv preprint arXiv:2407.04543 (2024)
10. Loshchilov, I., Hutter, F.: Decoupled weight decay regularization. In: International Conference on Learning Representations (ICLR) (2019)
11. Mai, Q., Gauch, S., Adams, D.: Setbert: Enhancing retrieval performance for Boolean logic and set operation queries. arXiv preprint arXiv:2406.17282 (2024)
12. Malaviya, C., Shaw, P., Chang, M.W., Lee, K., Toutanova, K.: Quest: A retrieval dataset of entity-seeking queries with implicit set operations. arXiv preprint arXiv:2305.11694 (2023)
13. Perez, E., Strub, F., De Vries, H., Dumoulin, V., Courville, A.: Film: Visual reasoning with a general conditioning layer. In: Proceedings of the AAAI Conference on Artificial Intelligence, vol. 32 (2018)
14. Raffel, C., et al.: Exploring the limits of transfer learning with a unified text-to-text transformer. J. Mach. Learn. Res. **21**(140), 1–67 (2020)
15. Robertson, S., Zaragoza, H., et al.: The probabilistic relevance framework: Bm25 and beyond. Found. Trends® Inform. Retrieval **3**(4), 333–389 (2009)
16. Shaw, P., Uszkoreit, J., Vaswani, A.: Self-attention with relative position representations. arXiv preprint arXiv:1803.02155 (2018)
17. Shazeer, N., et al.: Outrageously large neural networks: The sparsely-gated mixture-of-experts layer. arXiv preprint arXiv:1701.06538 (2017)
18. Vaswani, A.: Attention is all you need. Advances in Neural Information Processing Systems (2017)
19. Yang, S., Wang, B., Shen, Y., Panda, R., Kim, Y.: Gated linear attention transformers with hardware-efficient training. arXiv preprint arXiv:2312.06635 (2023)
20. Zhang, Z., Zhu, J., Zhou, W., Qi, X., Zhang, P., Li, H.: Boolquestions: Does dense retrieval understand boolean logic in language? arXiv preprint arXiv:2411.12235 (2024)

Automated Study Selection Using Representation Learning and Clustering

Flavien Owolabi(✉)

Monash University, Clayton, VIC 3800, Australia
flavien.owolabi@monash.edu

Abstract. Systematic Literature Reviews (SLRs) are essential for synthesizing information from multiple studies to inform policy and decision-making. While alternative review methods and automated tools have emerged, efficiently identifying relevant studies remains a significant challenge. Recent advancements in Artificial Intelligence (AI), particularly Large Language Models (LLMs), offer promising solutions. One promising technique is clustering search results based on semantic similarity using representation learning.

This study evaluates BERTopic, a sequential pipeline integrating representation learning, dimensionality reduction, and clustering, to automate study selection. Our analysis of 32 datasets reveals a statistically significant inverse relationship between noise cluster size and recall performance: smaller noise clusters correlate with higher recall (95%-100%). Furthermore, 17 datasets achieved 100% recall with just one cluster. Including the noise cluster, often discarded in standard HDBSCAN practices, increased the number of datasets achieving $\geq$ 95% recall to 30.

Both Mixed Linear Regression and Bayesian multilevel models confirm the pipeline's effectiveness, with 75% (24/32) of datasets achieving 100% recall and 94% (30/32) achieving $\geq$ 95% recall with two clusters. As for reduction in screening time, 11 datasets achieved a $\geq$ 50% Reduction in Screening Effort (RSE) with $\geq$ 95% recall. BERTopic's use of a class-based TF-IDF (c-TF-IDF) is crucial for generating semantically coherent topic representations, thereby enhancing cluster quality and improving the precision of study retrieval. These findings demonstrate BERTopic's strong potential to significantly advance the automation of study selection in SLRs.

Keywords: automated study selection · representation learning · large language models · BERTopic · clustering search results · systematic literature reviews · reduction in screening time

1 Introduction

Systematic Literature Reviews (SLRs) originated in the field of medicine as a cornerstone of evidence-based medicine (EBM) [1, 2]. They employ objective, transparent, and reproducible methods to identify and synthesize relevant information from multiple studies, which can be used to inform patient care or policy decisions [1, 3]. Due to their

H. Unger and P. Meesad (Eds.): NLPIR 2025, LNNS 1904, pp. 755–771, 2026.
https://doi.org/10.1007/978-3-032-20897-2_53

effectiveness, SLRs have expanded beyond medicine to become widely adopted across all scientific disciplines [2]. However, the SLR process, particularly the literature search and screening phases, remains resource-intensive and time-consuming for review teams [3–10].

To address these challenges, several alternative review methods, such as scoping reviews and rapid reviews, have been developed [2]. These approaches still adhere to explicit and reproducible methods but aim to reduce the time and resources required by SRs. Despite this evolution, the primary challenge remains: the growing burden of time and effort needed for citation searching and screening, which are critical to all forms of evidence synthesis.

With the steady growth of scientific literature, the problem has only intensified. For instance, a typical search in PubMed may yield more than 17,000 citations, with only a fraction deemed relevant after screening [6]. Additionally, databases like Scopus have grown exponentially, from 20 million records in 2018 to over 87 million in 2022, adding an average of 11,000 new documents daily. This exponential growth complicates both search and screening, making these tasks more imprecise and time-consuming than ever before [3, 5, 6].

Over the last two decades, researchers have focused on developing automated tools for SLR tasks, particularly screening, which is widely recognized as the most labor-intensive step [7]. Early efforts employed conventional machine learning techniques to classify search results into relevant and irrelevant categories [8, 11, 12]. More recently, advances in text mining and natural language processing (NLP) have yielded promising results in automating both study identification (search) [13, 14] and selection (screening) [8–10].

A significant breakthrough in natural language processing occurred in 2013 with the introduction of neural embeddings, which enabled the use of semantically driven approaches in various SLR tasks [15]. This was a pivotal development in representation learning, a subfield of machine learning that focuses on the development of Large Language Models (LLMs). These models – trained using vast amount of data – are able to automatically discover and learn meaningful features from unstructured data projected in a feature space (e.g., text, images) [16]. In representation learning, language models can extract semantically meaningful sentence embeddings from a corpus which can be compared using distance metrics like cosine similarity, making it easier to perform downstream tasks such as classification or clustering [17].

The introduction of transformer architectures in 2017 and the emergence of LLMs from 2019 further advanced the field [18]. Language models like BERT – and its subsequent improvements like SBERT [17] – have revolutionized natural language processing by automating the extraction of high-quality features from unstructured data, such as text, with remarkable efficiency [17]. These models process large corpora and learn efficacious word and sentence representations that effectively capture meaning, context, and relationships between these words [17]. Efficient representation learning with language models facilitates a wide range of downstream tasks, including dimensionality reduction, and clustering, making them easier and more accurate [17, 19].

In light of these advances, clustering search results from an SLR based on semantic similarity using representation learning has become an appealing strategy. Clustering is an unsupervised machine learning technique that groups a collection of unlabeled data into clusters based on similarity or distance measures between pairs of data points [20–22]. In an unsupervised learning, such as clustering, the task is to uncover hidden relationships within the data being clustered. Amjad et al. identified four types of clustering algorithms used for data mining: partition-based, hierarchical, density-based, and grid-based [20]. Partitioning (e.g., k-means) [8, 14], hierarchical (e.g., agglomerative) [14], and density-based (e.g., DBSCAN) [23] are the most commonly used clustering algorithms. The combination of LLM-enabled embeddings and clustering algorithms can effectively organize a corpus into semantically similar groups [22]. However, the high dimensionality of LLM-generated embeddings poses challenges for traditional clustering algorithms, which can be mitigated using dimensionality reduction techniques like Uniform Manifold Approximation and Projection (UMAP) [21, 22].

This study evaluates a state-of-the-art pipeline for automating study selection by integrating representation learning, dimensionality reduction, and clustering techniques. A key challenge for traditional machine learning approaches in systematic literature reviews (SLRs) is the inherent class imbalance of their datasets. A core contribution of this research is the demonstration that a topic modeling approach, leveraging the BERTopic pipeline, effectively addresses this issue by generating topics from clustered data. Our analysis further shows that, contrary to conventional best practices for HDBSCAN, the noise cluster cannot be discarded in the context of clustering SLR datasets without a significant loss of relevant information. Finally, our findings confirm the practical utility of this approach by showing its potential to substantially reduce manual screening effort and time.

2 Related Work

Automated document classification approaches applied to Systematic Literature Review (SLR) screening can be broadly divided into supervised and unsupervised learning approaches, with some hybrid methods emerging to address the limitations of each. Supervised learning relies on labeled data to train algorithms, allowing them to make predictions on unseen data [21]. However, the need for a labeled training set makes it less practical for SLR screening, where vast unlabeled data are common. By contrast, unsupervised learning can identify patterns within unlabeled data, a key advantage for the screening task. Clustering falls into this category, as it groups data based on inherent similarities, potentially allowing reviewers to screen a smaller subset of relevant studies [19, 21, 22].

The initial work in SLR screening automation focused on building classifiers to identify relevant studies with high recall, thereby reducing screening workload for reviewers. For instance, Cohen et al. [11] developed an early classifier with 95% recall for updating SLRs, demonstrating the potential to significantly reduce the number of abstracts to review. Up until 2018, most supervised methods relied on simpler classifiers and manual feature extraction [6, 12]. However, since then, more advanced feature extraction techniques [3, 5, 8, 9, 23, 24] – with deep learning architectures like BERT – have been integrated into SLR screening, making classification more effective. Some studies, such as [8], applied TF-IDF with ensemble clustering methods (e.g., k-means, NMF, LDA), while others, such as [24], fine-tuned BERT for feature extraction and used LSTM-CRF for classification.

As SLR screening automation research matured, Wallace et al. [25] noted that ML-based screening tools had reached a point of maturity, with active learning as the predominant supervised approach. In active learning, a classifier is iteratively trained as reviewers label abstracts from an unlabeled data pool [26]. However, a major challenge with active learning is the absence of established stopping criteria, which complicates the process [26]. Unsupervised methods like clustering can significantly reduce the number of abstracts to be screened by grouping most relevant studies into a few clusters [19, 22], allowing reviewers to focus on clusters likely to contain relevant studies.

Unsupervised approaches have also evolved, moving from simpler, manually driven techniques to advanced transformer-based or deep learning-based methods [22, 27]. For instance, earlier approaches focused on elements like PICO (Population, Intervention, Comparison, Outcome) found in study titles to aid in title screening [6]. More recent methods incorporate transformer-based feature extraction combined with topic labelling (e.g., LDA) and deep learning methods which integrate clustering techniques (e.g., DBSCAN), enabling more effective topic generation and clustering [3, 23].

The introduction of transformer architectures was a pivotal shift in SLR screening automation, particularly through representation learning. These models automate feature extraction by creating embeddings, making tasks like classification or clustering significantly more efficient. Combining representation learning with state-of-the-art dimension reduction and clustering techniques has unlocked new possibilities for SLR screening automation.

3 Methods

3.1 Experiment Design and Research Questions

Experiment Scope. I designed an experiment to analyze query results from completed SLRs and assess the effect of noise clusters on clustering performance. The primary goal was to evaluate the effectiveness of clustering, specifically recall, in identifying clusters containing relevant studies. This evaluation was conducted from the perspective of researchers conducting SLRs, using HDBSCAN to automate study selection based on gold standards of manually identified relevant studies.

The experiment used state-of-the-art techniques, including representation learning (e.g., sentence-BERT) [17], dimension reduction (e.g., UMAP) [28] and clustering techniques (e.g., HDBSCAN) [29], to identify relevant studies in datasets derived from completed systematic literature reviews.

Research Questions. The HDBSCAN clustering algorithm has two key hyperparameters that influence document cluster membership:

1. *min_cluster_size*: The minimum number of samples (or data points) required for a group to be considered a cluster.
2. *min_samples*: The number of neighboring samples (or data points) required for a point to be classified as a core point, including the point itself. This means the core

point has a sufficient density of neighbors around it – indicating it belongs to a dense part of the data that could form a cluster – rather than being isolated.

These hyperparameters directly affect the assignment of documents to clusters. When a document's label cannot be assigned to a specific cluster, it is placed in the noise cluster. Adjusting '*min_cluster_size*' and 'min_*samples'* levels influence the size of the noise cluster, with lower or higher levels generally resulting in a larger noise cluster.

This study aimed to address the following research questions:

1. What is the effect of the noise cluster on recall (95%-100%)?
2. What levels of '*min_cluster_size*' and '*min_samples*' yield a recall of 95%-100% in one or two clusters?
3. How much of reduction of screening effort (RSE) is achieved with the proposed sequential approach (representation learning, dimension reduction, and clustering)?

Datasets. I used 25 labeled datasets from completed SLRs (ASReview Synergy platform[30]) and I created 7 more.

In addition, I conducted a review on AI tools for SLR automation. Search terms based on the review questions were used in Scopus and Web of Science. After deduplication, over 1,000 titles and abstracts were screened using Covidence. This process identified 265 studies for full-text review (*sr_sai_review_broad*), covering various SLR tasks, such as search, screening, data extraction, and risk of bias assessment. Following inclusion and exclusion criteria, these 265 studies were further divided into subsets focusing on specific SLR tasks:

- Search: 20 (*sr_sai_search*)
- Screening: 161 (*sr_sai_screening*)
- Data extraction: 24 (*sr_sai_dextraction*)
- Risk of bias: 14 (*sr_sai_roba*)

Two further focused datasets were created:

- *scopus_sia*: 18 studies (AI tools for literature searching)
- *crs_proquest_long*: 61 studies (concordance data issues)

In total, 32 datasets were used.

3.2 Experiment

Hyperparameters' Selection. I tested eight combinations of '*min_cluster_size*' and '*min_samples*': '1515', '2525', '5010', '5015', '5025', '5050', '10050', '100100'. These combinations were chosen based on exploratory analysis with sample datasets to cover a range of noise cluster sizes and recall outcomes.

Implementation. I used the BERTopic [31] library to build the clustering pipeline, which included:

1. An embedding model: Sentence-BERT model ('*all-MiniLM-L6-v2*').

2. A dimension reduction algorithm: UMAP with 15 neighbors.
3. A clustering algorithm: HDBSCAN with adjustable hyperparameters.

The experiment was conducted in a Google Collaboratory environment, running custom Python modules for clustering, evaluation, and analysis.

Custom Modules

- Clustering – Uses BERTopic with embedding model, UMAP, and HDBSCAN to cluster datasets.
- Evaluation Metrics – Computes recall, cluster size %, quality, and coherence; outputs CSV results per dataset.
- Dataset Manipulation – Consolidates clustered results into dataframes (with/without noise), calculates recall for top clusters and noise size.
- Statistical Analysis – Performs descriptive, multivariate, and Bayesian modeling (via PyMC library [32]) to analyze the effect of the noise cluster on recall. This model incorporates prior information based on mixed-effects regression results. The following priors were used for recall with one cluster (RC1WN) and recall with two clusters (RC2WN) (Table 1):

Table 1. Prior Distributions for Bayesian Multilevel Models (RC1WN and RC2WN).

Parameter	Prior Distribution (RC1WN)	Prior Distribution (RC2WN)
Dataset Intercept	N(96.358,25)	N(97.915,25)
Measure Effect	N(0,10)	N(0,10)
Noise Effect	N(−0.996,0.072)	N(−0.284,0.047)
Sigma	{Half-Cauchy}(\sqrt{133.9119}	{Half-Cauchy}(\sqrt{57.4778}

The Bayesian model accounts for dataset and treatment group variations (random effects), providing robust estimates of the noise cluster's impact on recall. It uses 256 data points from 32 datasets and eight measures. All four custom Python modules, the Google Colaboratory notebook detailing the experimental workflow, and the datasets are made publicly available on GitHub[1].

4 Results

4.1 Effects of Noise Cluster Size on Recall

I conducted a Mixed Linear Model Regression (MixedLM) to assess the effects of the noise cluster on recall in scenarios involving one cluster or two clusters. The model successfully converged, indicating that the algorithm found a reliable solution for predicting recall values based on the size of the noise cluster.

[1] https://github.com/LilySharon/automated-study-selection-for-slrs.git.

The predicted values of the dependent variables – RC1WN (recall with one cluster) and RC2WN (recall with two clusters) – when all predictors (e.g., noise cluster size) are zero are 96 and 97, respectively. The noise cluster coefficients are -0.996 and -0.284, respectively, both with p-values of < 0.001. This demonstrates a statistically significant negative relationship between the size of the noise cluster and recall performance: higher recall values (95%-100%) are associated with smaller noise clusters, whereas a larger noise cluster reduces recall performance.

These results were consistent when recall was analyzed without the inclusion of the noise cluster. For recall without noise clusters (RC1NN and RC2NN), the predicted values for the dependent variables were 99.79 for RC1NN and exceeded 100 for RC2NN, with similarly negative noise cluster coefficients and p-values of < 0.001. This suggests that 100% recall can be achieved in the absence of a noise cluster.

Out of the 32 datasets analyzed, 17 achieved 100% recall using one cluster, regardless of whether the noise cluster was included or excluded. When the recall threshold was lowered to $\geq$ 95%, this number increased to 20 datasets (Table 2).

Table 2. Recall Performance with and without Noise Cluster at different Cluster levels.

Recall Threshold	Cluster Count	Datasets (Excluding Noise Cluster)	Datasets (Including Noise Cluster)
100%	1	17	17
$\geq$ 95%	1	20	20
100%	2	21	24
$\geq$ 95%	2	22	30

With two clusters, recall performance improved. Without including the noise cluster, 21 datasets reached 100% recall, and 22 datasets reached $\geq$ 95% recall. However, when the noise cluster was included, recall performance improved further: 24 datasets achieved 100% recall, and 30 datasets achieved $\geq$ 95% recall (Table 2).

These results highlight the impact of the noise cluster on recall performance when relevant studies could be found in more than one cluster.

These findings challenge conventional best practices for HDBSCAN clustering, which often dismiss the noise cluster as irrelevant. Instead, our results suggest that when clustering SLR query results, the noise cluster can contain relevant studies.

4.2 Effects of HDBSCAN's Hyperparameters ('Min_cluster_size' and 'Min_samples') on Recall

I evaluated the interaction between recall and the noise cluster across varying levels of HDBSCAN's key hyperparameters of interest – *min_cluster_size* and *min_samples* – as well as different numbers of clusters. The results revealed substantial variability in recall outcomes.

As Fig. 1 illustrates, 15 datasets achieved 100% recall using one cluster, across various combinations of *min_cluster_size* and *min_samples*, regardless of noise cluster inclusion. While some datasets reached this perfect recall under only one or two specific hyperparameter settings, one dataset achieved 100% recall across all eight tested combinations.

Further analysis of the impact of noise cluster inclusion at one cluster, Fig. 1, revealed that smaller and mid-sized datasets (2,000 – 3,000 records) tended to reach 100% recall at lower to mid-range hyperparameter values. In contrast, larger datasets (>5,000 records) generally achieved 100% recall across mid to higher hyperparameter settings, regardless of noise cluster inclusion. For example, the *scopus_sia* dataset – one of the smallest in the collection with less than 2000 records – achieved 100% recall at three lower to mid-range combinations of *min_cluster_size* and *min_samples* while larger datasets (e.g., such as the *sr_sai series – sr_sai_search, sr_sai_screening, sr_sai_dextraction, sr_sai_roba*) achieved 100% recall at mid-range to higher hyperparameter values (Fig. 1).

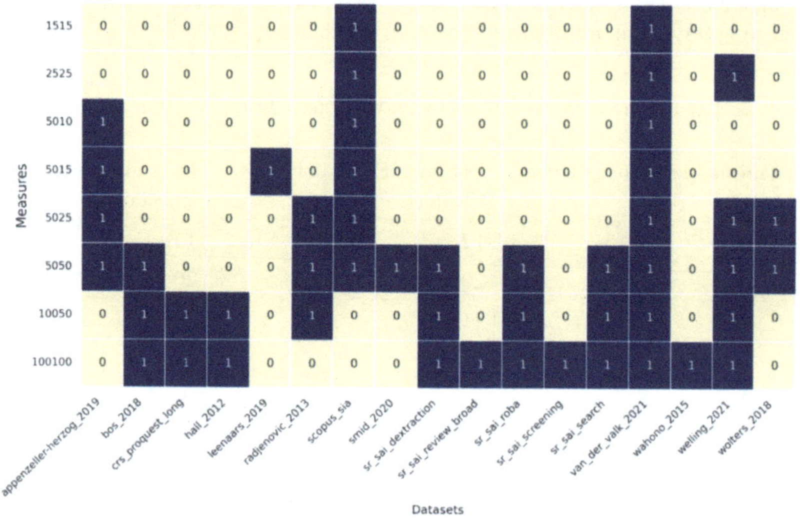

Fig. 1. Datasets achieving 100% recall with one cluster for different combinations of hyperparameters.

Interestingly, when using two clusters and including the noise cluster, the impact of these hyperparameters on recall became negligible. In this configuration, recall performance appeared largely unaffected by changes in *min_cluster_size* and *min_samples.*

These findings suggest that dataset-specific characteristics may play a more significant role in recall variability than hyperparameter settings alone. Further investigation into the intrinsic features of individual datasets is warranted to better understand this influence.

These observations are further supported by results from a multilevel Bayesian analysis. When evaluating recall with one cluster, 15 datasets achieved a mean predicted recall of 100%, while 17 datasets reached a mean predicted recall of $\geq$ 95%. Among the hyperparameter combinations analyzed, the combination labeled *100100* positively influenced recall, whereas combinations *1515*, *5010*, and *5015* had a negative impact.

In contrast, when examining recall with two clusters that included the noise cluster, 23 datasets achieved a recall of $\geq 95\%$. In this case, both lower and higher hyperparameter combinations were associated with positive (e.g., *1515*, *100100*) and negative (e.g., *5010*) effects on recall.

Overall, the data provides strong evidence of a negative correlation between the size of the noise cluster and recall performance. This suggests that including the noise cluster can improve recall, and therefore, it should not be excluded. Disregarding the noise cluster may result in the loss of relevant studies.

4.3 Reduction in Screening Effort (RSE)

I define Reduction in Screening Effort (RSE), following Rathbone et al.[6], as the percentage of records eliminated by the clustering algorithm during the clustering process. The clustering algorithm organizes records (titles and abstracts) into clusters and assigns labels to them. In this study, I only considered datasets that achieved a recall threshold of $\geq 95\%$ using either one or two clusters. To compute the Reduction in Screening Effort (RSE), I first calculated the size of each cluster as a percentage of the total number of records. Then, I identified the cluster(s) that contained the relevant studies needed to reach the recall threshold.

Table 3. Datasets by RSE Range (with $\geq 95\%$ Recall)

RSE Range (%)	Number of Datasets	Datasets (examples)
25–50%	21	Bos_2018, leenaars_2019, chou_2003, menon_2022
50–75%	11	sr_sai_review_broad, hall_2012, radjenovic_2013, wahono_2015
> 75%	4	welling_2021, sr_sai_dextraction, sr_sai_roba, sr_sai_sear

The RSE is calculated as:

$$\text{RSE} = 1 - (\%\text{ of cluster size containing relevant records}) \tag{1}$$

This value represents the proportion of records in a dataset that did not require screening. A higher RSE indicates a greater reduction in screening effort while maintaining high recall.

From our results, I identified three categories of performance based RSE score:

Among the 30 datasets that achieved $\geq 95\%$ recall, many also showed a substantial reduction in screening effort (RSE) (Table 3):

- 21 datasets had an RSE between 25% and 50%, meaning one-quarter to half of the records could have been excluded from screening while still retrieving $\geq 95\%$ of relevant studies. This reflects a moderate reduction in effort with high recall.

- 11 datasets fell within the 50%–75% RSE range, indicating that more than half of the records did not need to be screened – showing strong potential to significantly reduce screening workload.
- 4 datasets achieved RSE values above 75%, meaning that only a small portion of records needed to be screened to meet the recall threshold. These represent the highest efficiency gains.

One notable observation is that some datasets – such as *sr_sai_dextraction*, *sr_sai_roba*, and *sr_sai_search* – consistently appeared in the higher RSE tiers. This suggests they particularly benefit from semantic clustering-based automation, highlighting the method's potential for automating study selection in certain SLR contexts.

5 Discussion and Future Work

5.1 The Effectiveness of the Sequential Approach to Clustering SLRs' Query Results

The findings of this study highlight the potential of the proposed sequential approach for automating study selection in systematic literature reviews (SLRs). Both the Mixed Linear Regression Model (MixedLM) and the Bayesian multilevel analyses support the approach's effectiveness, with 75% (24 out of 32) of datasets achieving 100% recall and 94% (30 out of 32) achieving a recall of $\geq$ 95%. These results demonstrate the capability of the HDBSCAN's clustering algorithm – combined with representation learning and dimension reduction – to consistently identify relevant studies across diverse datasets.

However, the influence of HDBSCAN hyperparameters ('*min_cluster_size*' and '*min_samples*') on recall performance remains complex. While certain datasets experienced negative impacts on recall at specific hyperparameter combinations, a subset of datasets achieved a recall of $\geq$ 95% across various combinations of hyperparameter levels. This suggests that the effects of these hyperparameters could be highly dataset-dependent, underscoring the need for further investigation into their interaction with dataset characteristics.

More importantly, these datasets are derived from search results, which in turn are shaped by the formulation of the research question. The quality of SLRs' research questions and the focus of the query formulation directly impact the content of the resulting dataset. For example, narrower research questions tend to produce more targeted search terms, which are more likely to retrieve tightly focused and thematically coherent studies. In contrast, broader or less clearly defined questions and search terms may lead to thematically heterogeneous datasets that are harder to cluster effectively. This insight aligns with best practices in evidence-based medicine (EBM) and guidance from established SLR authorities such as Cochrane, which emphasize the importance of well-defined research questions and precise search strategies to reduce screening burden and enhance the relevance of retrieved studies [33].

The value of using focused search strategies with optimized or refined search terms is evident in a specific subset of datasets. These datasets, derived from a broader initial search and targeting distinct SLR tasks (e.g., search, data extraction, risk of bias assessment), consistently achieved high recall. Four out of five of these task-specific datasets

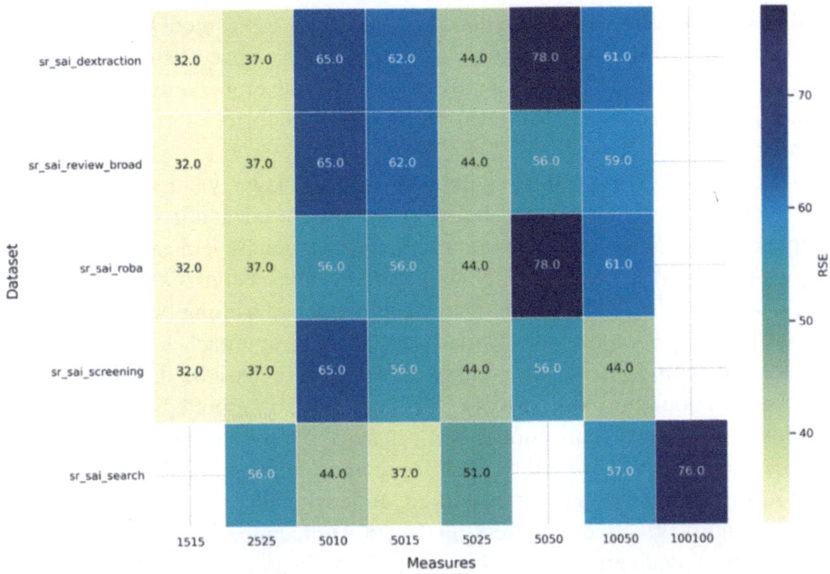

Fig. 2. RSE levels across measures for a subset of datasets.

reached 100% recall with one cluster. Furthermore, all five datasets maintained a recall of $\geq$ 95% even when excluding the noise cluster and high RSE score (Fig. 2), highlighting the robustness of this approach for retrieving relevant studies. Notably, four of these high-performing, task-specific datasets were subsets derived from a larger, broader SLR dataset (*sr_sai_review_broad*), which itself did not demonstrate comparable RSE performance (Fig. 2).

The variability observed in RSE further reinforces this dataset-dependence. High RSE values ($\geq$50%) were achieved in a subset of datasets, but performance varied significantly based on '*min_cluster_size*' and '*min_samples*', as well as the inherent properties of the datasets. Eleven datasets achieved an RSE $\geq$ 50%. Three of these task-specific datasets (*sr_sai_search*, *sr_sai_dextraction*, *sr_sai_roba*) with narrower focus achieved higher RSE compared to the original dataset they were derived from (Fig. 2).

5.2 Clusterability and Separability of SLR Query Results: Limitations of the Sequential Approach

Clustering methods aim to group similar data points into distinct clusters [21, 22], a task particularly relevant to automating study selection in systematic literature reviews (SLRs) where large volumes of search results must be efficiently categorized. The effectiveness of any clustering approach hinges on the quality of the underlying data representation [22]. Recent advances in natural language processing (NLP), especially transformer-based deep neural networks (DNNs), have enabled the creation of richer data representations by transforming text into a feature or latent feature space. Clustering algorithms can then operate more effectively within this space, grouping data points based on distance or similarity measures [19].

In this study, I adopted a sequential approach to topic modelling by transforming the results of SLR datasets into a semantic feature space using Sentence-BERT (SBERT). Due to the high dimensionality of the resulting embeddings, I utilized UMAP for dimensionality reduction, followed by HDBSCAN for clustering based on cosine similarity.

This pipeline, implemented using the BERTopic package, yielded promising results. Its effectiveness can be largely attributed to the structured, multi-step framework that BERTopic offers – particularly its ability to handle the inherent class imbalance present in SLR datasets. In a typical SLR screening task, the dataset consists of only two classes: relevant and irrelevant studies. This extreme imbalance has long been recognized as a significant challenge for traditional machine learning approaches used in automated study selection tools [26, 34]. In contrast, clustering breaks the dataset into multiple groups based on document similarity, which can then be interpreted or merged based on their thematic coherence. This implicitly mitigates the binary imbalance by operating in an unsupervised space as offered by BERTopic multi-step approach.

BERTopic's first three steps – embedding, dimensionality reduction, and clustering – appear to be highly effective at addressing the class imbalance challenge inherent in SLR datasets. As discussed earlier, 75% (24 out of 32) of datasets achieved 100% recall, and 94% (30 out of 32) reached a recall of at least 95%. Additionally, several datasets achieved high Reduction in Screening Effort (RSE) scores, with some reducing the need to manually screen 50% to 75% of the records retrieved from the initial search – indicating substantial efficiency gains.

BERTopic operates through five sequential steps:

- Embedding the documents using a language model,
- Reducing the dimensionality of the embeddings (typically with UMAP),
- Clustering the reduced embeddings (often using HDBSCAN),
- Tokenizing documents within each cluster, and
- Generating topic representations via class-based TF-IDF (c-TF-IDF).

The final step – c-TF-IDF – is particularly important for generating interpretable topic labels. After clustering, BERTopic treats each cluster as a single aggregated document and applies c-TF-IDF to extract the most representative terms, making topics semantically meaningful.

By default, this process often produces a large number of fine-grained topics. This is where the *nr_topics* parameter plays a critical role. To improve thematic coherence and manage topic granularity, BERTopic allows post-clustering topic reduction through *nr_topics*. This step applies agglomerative clustering to merge semantically similar topics based on cosine similarity of their c-TF-IDF vectors. The result is a smaller, more manageable number of topics, with updated representations that better reflect the broader themes present in the data.

The topic merging mechanism is a key strength of BERTopic's pipeline, particularly in handling the imbalanced nature of SLR datasets. By smoothing out the fragmentation from HDBSCAN's fine-grained clustering, this mechanism helps maintain high recall. The findings on the 32 datasets demonstrate that the BERTopic pipeline, leveraging HDBSCAN, can effectively handle the inherent class imbalance of SLR datasets. While traditional machine learning approaches, particularly in AI-enabled tools like ASReview, Abstrackr, or Colandr [26], have devised a balance strategy to address this issue, this research shows that a topic modeling approach can also successfully deal with the challenge, offering a novel and effective solution.

While the sequential approach yielded promising results, it also has some limitations – particularly in terms of cluster separability. Recent advances in deep clustering methodologies may offer more integrated and potentially superior alternatives [19, 35–37]. Contrary to BERTopic's sequential pipeline used in this study, deep clustering integrates feature extraction (typically using a Convolutional Neural Network (CNN) or Auto-Encoder) and clustering into a single framework, creating a latent space optimized for cluster separation [19, 36]. This integrated approach could be superior to the sequential method compared to the independent processing of embeddings, dimensionality reduction, and clustering, which prevents the realization of potential synergies.

Despite this advantage, deep clustering also faces key challenges:

- Preservation of geometric structures: Even after dimensionality reduction, maintaining local (intra-cluster) structures is essential for clean separation. Inadequate preservation can lead to poor clustering outcomes [19, 37].
- Conflicting objectives: Techniques like UMAP prioritize manifold preservation for dimensionality reduction, while clustering algorithms aim for maximal inter-cluster separation. These competing goals may result in suboptimal performance when applied sequentially [19]. Addressing these challenges requires a more holistic strategy – one that unifies representation learning, structure preservation, and clustering.

5.3 Emerging Direction in Integrated Deep Clustering

Several emerging deep learning-based approaches aim to overcome these limitations by jointly optimizing data representation and clustering objectives. These methods offer mechanisms for reconciling the trade-offs between local structure preservation and global cluster [19, 35–37]. Key examples include:

1. Manifold Learning with Clustering Loss: Wu et al.[19] proposed a method that generates latent spaces guided by a clustering-oriented loss function. This approach combines:
    - Local structure preservation: An isometric loss function to preserve intra-manifold structures.
    - Global structure preservation: A ranking loss function to promote separation between manifolds, enhancing cluster distinctiveness.
2. Structure-Driven Representation Learning: Wang et al. [37] introduced a model integrating local and global structure-driven representation learning for improved cluster separability:
    - A local-structure learning to model neighborhood distributions to estimate latent data distributions.
    - A global-structure learning to encode cluster context using exemplar theory to improve cluster formation and separation.

These examples show that future improvements in SLR automation may lie in integrated approaches where embedding quality, structural integrity, and clustering effectiveness are jointly optimized. Adopting such methods could further improve recall, interpretability, and screening efficiency beyond the already encouraging results achieved in this study.

5.4 Future Directions and Implications for Automated Systematic Literature Reviews

Deep learning architectures hold great promise for transforming SLR data into cluster-friendly representations. However, achieving optimal cluster separability demands careful attention to how latent spaces are learned. Joint optimization techniques – such as Geometric Cluster Manifold Learning (GCML) [19] and Structure Representation Learning (SRL) [37] – underscore the importance of balancing both local and global structure preservation during this process.

Future research should evaluate these advanced deep clustering approaches specifically in the context of SLR query results, which are inherently imbalanced. This includes assessing how these models handle diverse data characteristics and potential biases – such as those introduced by narrow or overly focused search terms used during query formulation. It is also important to benchmark performance using metrics relevant to SLR automation, such as recall, and reduction in screening effort (RSE). Furthermore, a comparative analysis with other advanced clustering methodologies, such as adaptive hybridized meta-heuristic algorithms for subspace clustering [38], could provide valuable insights into how different optimization strategies perform on imbalanced text data. This would help identify the most efficient and robust methods for automated study selection.

By leveraging these innovations, we can significantly enhance the clusterability of SLR datasets, leading to more efficient and effective automated study selection. Ultimately, this would reduce the manual burden of the screening tasks of systematic reviews, enabling researchers to concentrate on higher-level synthesis and interpretation.

Acknowledgments. This research is funded by a scholarship from Monash AI Institute and the Faculty of Science at Monash University.

Disclosure of Interests. The author has no competing interests to declare that are relevant to the content of this article.

References

1. Chalmers, I., Hedges, L.V., Cooper, H.: A brief history of research synthesis. Eval. Health Prof. **25**, 12–37 (2002). https://doi.org/10.1177/0163278702025001003

2. Munn, Z., Peters, M.D.J., Stern, C., Tufanaru, C., McArthur, A., Aromataris, E.: Systematic review or scoping review? Guidance for authors when choosing between a systematic or scoping review approach. BMC Med. Res. Methodol. **18**, 143 (2018). https://doi.org/10.1186/s12874-018-0611-x
3. Helmick, H., Nanda, G., Ettestad, S., Liceaga, A., Kokini, J.L.: Applying text mining to identify relevant literature in food science: cold denaturation as a case study. J. Food Sci. **86**, 4851–4864 (2021). https://doi.org/10.1111/1750-3841.15940
4. Boyack, K.W., et al.: Clustering more than two million biomedical publications: comparing the accuracies of nine text-based similarity approaches. PLoS ONE **6**, e18029 (2011). https://doi.org/10.1371/journal.pone.0018029
5. Muller, A.E., Ames, H.M.R., Jardim, P.S.J., Rose, C.J.: Machine learning in systematic reviews: comparing automated text clustering with Lingo3G and human researcher categorization in a rapid review. Res. Synth. Methods **13**, 229–241 (2022). https://doi.org/10.1002/jrsm.1541
6. Rathbone, J., et al.: Expediting citation screening using PICo-based title-only screening for identifying studies in scoping searches and rapid reviews. Syst. Rev. **6**, 233 (2017). https://doi.org/10.1186/s13643-017-0629-x
7. Van Dinter, R., Tekinerdogan, B., Catal, C.: Automation of systematic literature reviews: a systematic literature review. Inf. Softw. Technol. **136**, 106589 (2021). https://doi.org/10.1016/j.infsof.2021.106589
8. Varghese, A., Cawley, M., Hong, T.: Supervised clustering for automated document classification and prioritization: a case study using toxicological abstracts. Environ Syst Decis. **38**, 398–414 (2018). https://doi.org/10.1007/s10669-017-9670-5
9. Weißer, T., Saßmannshausen, T., Ohrndorf, D., Burggräf, P., Wagner, J.: A clustering approach for topic filtering within systematic literature reviews. MethodsX. **7**, 100831 (2020). https://doi.org/10.1016/j.mex.2020.100831
10. Xiong, Z., et al.: A machine learning aided systematic review and meta-analysis of the relative risk of atrial fibrillation in patients with diabetes mellitus. Front. Physiol. **9**, 835 (2018). https://doi.org/10.3389/fphys.2018.00835
11. Cohen, A.M., Hersh, W.R., Peterson, K., Yen, P.-Y.: Reducing workload in systematic review preparation using automated citation classification. J. Am. Med. Inform. Assoc. **13**, 206–219 (2006). https://doi.org/10.1197/jamia.M1929
12. Wallace, B.C., Kuiper, J., Sharma, A.: Extracting pico sentences from clinical trial reports using (2016)
13. Alves, L.F., Vasconcellos, F.J.S., Nogueira, B.M.: SeSG: a search string generator for Secondary Studies with hybrid search strategies using text mining. Empir. Software Eng. **27**, 105 (2022). https://doi.org/10.1007/s10664-021-10084-4
14. Pourreza, M., Ensan, F.: Towards semantic-driven Boolean query formalization for biomedical systematic literature reviews. Int. J. Med. Informatics **170**, 104928 (2023). https://doi.org/10.1016/j.ijmedinf.2022.104928
15. Mikolov, T., Sutskever, I., Chen, K., Corrado, G., Dean, J.: Distributed representations of words and phrases and their compositionality. https://arxiv.org/abs/1310.4546 (2013). https://doi.org/10.48550/ARXIV.1310.4546
16. Kulkarni, M., Mahata, D., Arora, R., Bhowmik, R.: Learning rich representation of keyphrases from text. https://arxiv.org/abs/2112.08547 (2021). https://doi.org/10.48550/ARXIV.2112.08547
17. Reimers, N., Gurevych, I.: Sentence-BERT: sentence embeddings using siamese BERT-networks. https://arxiv.org/abs/1908.10084 (2019). https://doi.org/10.48550/ARXIV.1908.10084
18. Vaswani, A., et al.: Attention is all you need. https://arxiv.org/abs/1706.03762 (2017). https://doi.org/10.48550/ARXIV.1706.03762

19. Wu, L., Liu, Z., Zang, Z., Xia, J., Li, S., Li, S.Z.: Generalized clustering and multi-manifold learning with geometric structure preservation. http://arxiv.org/abs/2009.09590 (2021). https://doi.org/10.48550/arXiv.2009.09590
20. Ali, A., et al.: Systematic review: a state of art ML based clustering algorithms for data mining. In: Proceedings of the 2020 IEEE 23rd International Multitopic Conference (INMIC). pp. 1–6. IEEE, Bahawalpur, Pakistan (2020). https://doi.org/10.1109/INMIC50486.2020.9318060
21. Károly, A.I., Fullér, R., Galambos, P.: Unsupervised clustering for deep learning: a tutorial survey. Acta Polytech Hung **15**, 29–53 (2018). https://doi.org/10.12700/APH.15.8.2018.8.2
22. Min, E., Guo, X., Liu, Q., Zhang, G., Cui, J., Long, J.: A survey of clustering with deep learning: from the perspective of network architecture. IEEE Access **6**, 39501–39514 (2018). https://doi.org/10.1109/ACCESS.2018.2855437
23. Jiang, Q., et al.: Global meta-analysis of evolution patterns for lake topics over centurial scale: a natural language understanding-based deep clustering approach with 130,000 studies. J. Hydrol. **614**, 128597 (2022). https://doi.org/10.1016/j.jhydrol.2022.128597
24. Wang, Q., Liao, J., Lapata, M., Macleod, M.: PICO entity extraction for preclinical animal literature. Syst. Rev. **11**, 209 (2022). https://doi.org/10.1186/s13643-022-02074-4
25. Wallace, B.C., Dahabreh, I.J., Schmid, C.H., Lau, J., Trikalinos, T.A.: Modernizing evidence synthesis for evidence-based medicine. In: Clinical Decision Support, pp. 339–361. Elsevier (2014). https://doi.org/10.1016/B978-0-12-398476-0.00012-9
26. Van De Schoot, R., et al.: An open source machine learning framework for efficient and transparent systematic reviews. Nat Mach Intell. **3**, 125–133 (2021). https://doi.org/10.1038/s42256-020-00287-7
27. Ren, Y., et al.: Deep clustering: a comprehensive survey. IEEE Trans. Neural Netw. Learning Syst. **36**, 5858–5878 (2025). https://doi.org/10.1109/TNNLS.2024.3403155
28. McInnes, L., Healy, J., Melville, J.: UMAP: uniform manifold approximation and projection for dimension reduction. https://arxiv.org/abs/1802.03426 (2018). https://doi.org/10.48550/ARXIV.1802.03426
29. Malzer, C., Baum, M.: A hybrid approach to hierarchical density-based cluster selection. In: Proceedings of the 2020 IEEE International Conference on Multisensor Fusion and Integration for Intelligent Systems (MFI), pp. 223–228. IEEE, Karlsruhe, Germany (2020). https://doi.org/10.1109/MFI49285.2020.9235263
30. De Bruin, J., Ma, Y., Ferdinands, G., Teijema, J., Van De Schoot, R.: SYNERGY - Open machine learning dataset on study selection in systematic reviews(2023). https://doi.org/10.34894/HE6NAQ
31. Grootendorst, M.: BERTopic: Neural topic modeling with a class-based TF-IDF procedure. https://arxiv.org/abs/2203.05794 (2022)
32. Abril-Pla, O., et al.: PyMC: a modern, and comprehensive probabilistic programming framework in Python. PeerJ. Comput. Sci. **9**, e1516 (2023). https://doi.org/10.7717/peerj-cs.1516
33. Schardt, C., Adams, M.B., Owens, T., Keitz, S., Fontelo, P.: Utilization of the PICO framework to improve searching PubMed for clinical questions. BMC Med. Inform. Decis. Mak. **7**, 16 (2007). https://doi.org/10.1186/1472-6947-7-16
34. Wallace, B.C., Trikalinos, T.A., Lau, J., Brodley, C., Schmid, C.H.: Semi-automated screening of biomedical citations for systematic reviews. BMC Bioinf. **11**, 55 (2010). https://doi.org/10.1186/1471-2105-11-55
35. Chen, D., Lv, J., Yi, Z.: Unsupervised multi-manifold clustering by learning deep representation
36. McConville, R., Santos-Rodriguez, R., Piechocki, R.J., Craddock, I.: N2D: (Not Too) deep clustering via clustering the local manifold of an autoencoded embedding. http://arxiv.org/abs/1908.05968 (2020). https://doi.org/10.48550/arXiv.1908.05968

37. Wang, X., Jing, L., Liu, H., Yu, J.: Structure-driven representation learning for deep clustering. ACM Trans. Knowl. Discov. Data **18**, 1–25 (2024). https://doi.org/10.1145/3623400
38. Pradeep Kumar, D., Sowmya, B.J., Kanavalli, A., Amaresh, T.N..: Adaptive hybridized meta-heuristic algorithm for subspace clustering on high dimensional data. JAIT **14**, 1103–1116 (2023). https://doi.org/10.12720/jait.14.5.1103-1116

Information Sources for Query Refinement: A Comprehensive Survey

Zahra Taheri(✉), Mahdis Saeedi, Ziad Kobti, and Hossein Fani

School of Computer Science, University of Windsor, Windsor, ON, Canada
{taherik,msaeedi,kobti,hfani}@uwindsor.ca

Abstract. The Web remains the largest and most dynamic information source available today. Web search plays a vital role in helping users locate relevant content for research, learning, and decision-making. A critical factor in search effectiveness is the system's ability to interpret user queries accurately. However, queries are often ambiguous, brief, or underspecified, leading to suboptimal retrieval results. Query refinement techniques address these issues by modifying the initial query through term addition, deletion, or substitution to better capture user intent. This paper presents a comprehensive survey of query refinement methods with a specific focus on the types of information sources leveraged in these techniques. We classify information sources into two primary categories: non-contextual (e.g., thesauri, WordNet, Wikipedia, retrieved documents) and contextual (e.g., search sessions, click-through data, temporal data, and social signals). For each category, we analyze recent methods that utilize these sources to support query expansion, suggestion, and reformulation through a systematic review of 67 papers published between 2016–2024. Our comparative analysis evaluates methods across effectiveness, scalability, robustness to query drift, and personalization capabilities, highlighting open research directions in this evolving field.

Keywords: Query refinement · Information retrieval · Query expansion · Search personalization · Contextual search

1 Introduction

Web search is a ubiquitous entry point for information-seeking activities across domains such as education, e-commerce, health, and enterprise. Users express their needs through short keyword queries, which are often ambiguous, incomplete, or semantically imprecise. These limitations hinder retrieval systems from correctly interpreting the user's intent and retrieving the most relevant results. Studies have shown that the average query length remains under three words [1], contributing to vagueness and under-specification in user input. To address this issue, query refinement techniques aim to improve search performance by reformulating the original query to better reflect the user's intent. These techniques include a range of strategies such as query expansion [2], suggestion [3],

H. Unger and P. Meesad (Eds.): NLPIR 2025, LNNS 1904, pp. 772–787, 2026.
https://doi.org/10.1007/978-3-032-20897-2_54

and reformulation. While the goals and mechanisms of these strategies differ slightly, they share a common objective: to produce queries that yield more relevant results. A key factor in the success of any query refinement method lies in the type and quality of the information it leverages. Traditionally, many approaches have relied on lexical resources, manually crafted thesauri, or term co-occurrence statistics. However, recent advances in machine learning and data availability have enabled the use of rich, heterogeneous sources of information from user interaction logs and session data to structured knowledge bases and temporal behavior patterns.

This paper focuses on categorizing and analyzing the types of information sources used in contemporary query refinement research through a systematic survey methodology. We divide these sources into two primary categories: **Non-contextual sources**, which do not depend on the user's behavior or history (e.g., lexical databases, Wikipedia, or initial retrieved documents), and **Contextual sources**, which rely on user-centric data such as session history, clicked documents, and social or temporal behavior.

Research Questions: This survey addresses three key research questions:

1. What are the primary categories of information sources used in modern query refinement techniques? 2. How do different information sources compare in terms of effectiveness, scalability, and personalization capabilities? 3. What are the current challenges and future research directions in leveraging diverse information sources for query refinement?

2 Survey Methodology and Information Source

2.1 Survey Methodology

To ensure comprehensiveness and reproducibility, we followed a systematic survey methodology with clearly defined inclusion/exclusion criteria:

Database Sources: We searched the following academic databases: ACM Digital Library, IEEE Xplore, Springer, Elsevier (ScienceDirect), and arXiv.

Time Span: Papers published between January 2016 and December 2024, focusing on recent advances while ensuring contemporary relevance.

Search Terms: Our search strategy included combinations of: 'query refinement', 'query expansion', 'query suggestion', 'query reformulation', 'information retrieval', 'search personalization', and 'contextual search'.

Inclusion Criteria: a) Peer-reviewed conference papers and journal articles; b) papers focusing on query refinement techniques in IR; c) methods that explicitly leverage different information sources; d) papers with clear experimental validation.

Exclusion Criteria: a) Papers older than 2016 (to focus on recent advances); b) Non-English publications; c) Workshop papers without substantial technical contribution; d) Duplicate studies or minor variations of existing work. **Selection Process:** Initially, 156 papers were identified through database searches. After applying inclusion/exclusion criteria and removing duplicates, 67 papers were selected for detailed analysis. Two researchers independently reviewed abstracts and full texts, with disagreements resolved through discussion.

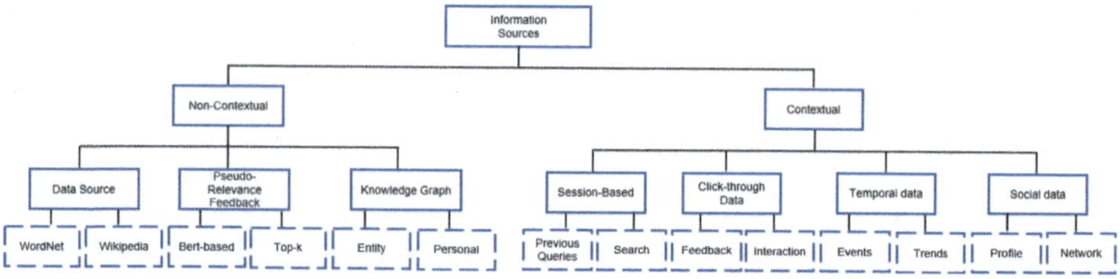

Fig. 1. Taxonomy of Information Sources for Query Refinement. The taxonomy categorizes sources into non-contextual and contextual types, highlighting subcategories such as external knowledge, retrieved documents, knowledge graphs, session data, click-through data, temporal data, and social data.

2.2 Information Source Definition

Recent advances in information retrieval have increasingly relied on diverse information sources to better capture users' needs and preferences. Effectively identifying and utilizing these sources is essential for improving retrieval performance. Figure 1 illustrates the proposed taxonomy of information sources for query refinement, categorized into two main groups: **Non-contextual Information** and **Contextual Information**. This categorization provides a structured framework for understanding how different information types contribute to query refinement effectiveness.

3 Non-contextual Information Sources

Non-contextual query refinement focuses exclusively on the content of the query itself, without incorporating additional contextual signals such as user behavior or search session history. These approaches aim to enhance the effectiveness of the original query by leveraging external knowledge resources, rather than personalized or situational information.

3.1 External Data Sources

As previously discussed, query expansion techniques enhance an initial query by incorporating additional, semantically related terms. A notable example is the model proposed by Lucchese et al. [4], which employs a thesaurus to support query expansion through structured queries and efficiency-aware term selection strategies. This model is particularly well-suited for real-time applications, as it expands the original query into a Conjunctive Normal Form (CNF) by selecting candidate terms from a thesaurus. Beyond thesaurus-driven approaches, domain-specific keyword-based query languages have also been shown to improve retrieval precision. For instance, Rencis [5] introduced a configurable keywords-based query language tailored for the healthcare domain, enabling users to construct structured queries with precise control over term inclusion, synonyms, and logical operators. While similar in spirit to thesaurus-supported expansion, this approach emphasizes manual configurability over automatic term selection, making it particularly suited to specialized domains where terminology precision is critical.

To address these limitations, Azad et al. [6] proposed an advanced approach that employs Wikipedia and WordNet for query expansion. This method provides rich expansion terms for phrase queries using Wikipedia and for individual terms using WordNet. Their approach involves a two-level strategy for term selection from WordNet: initially extracting the synsets of the query terms, followed by retrieving the synsets of these synsets to capture a broader range of related terms.

Using resources like WordNet, Wikipedia, and thesauri for query expansion offers both strengths and weaknesses. They provide rich semantic structures, enabling synonym, hyponym, and related-term discovery, which improves retrieval and supports domain-specific contexts. Their structured design also helps ensure precise term expansion. However, these resources are static and may miss emerging terminology, leading to outdated results. They often lack context sensitivity, causing irrelevant expansions, and their polysemous terms can introduce ambiguity.

3.2 Pseudo-Relevance Feedback (PRF) Methods

Another important source of information for query refinement is the set of documents retrieved in response to the initial query. Methods that leverage this feedback are referred to as pseudo-relevance feedback (PRF) techniques. In PRF, the top-ranked retrieved documents are assumed to be at least partially relevant and are used to refine or expand the original query. Recent PRF methods [7,8] utilize entire feedback documents for expansion. However, this often leads to overly verbose and noisy expansions that may degrade performance. To address this issue, Zheng et al. [9] propose a BERT-based query expansion method that selectively incorporates relevant content. Their approach begins by identifying the top-ranked documents using a fine-tuned BERT model.

Beyond query expansion, PRF documents can also be leveraged for query suggestion, particularly in domains such as scientific literature search, where users may lack the domain expertise to judge the relevance of query suggestions. In this context, Medlar et al. [10] introduce a method designed for exploratory search, which generates alternative queries that are independent of the original query but capable of retrieving similar content to that currently displayed. Their approach works by summarizing the content of search result documents and generating new queries based on those summaries. In terms of query representation, most existing models produce a single representation for each query. However, this can be limiting when a query expresses multiple underlying intents. To address this, Hashemi et al. [11] propose a model based on BART that learns multiple distributed representations for each original query. Their method uses the top-ranked documents from the retrieval results to uncover different semantic aspects of the query, allowing the model to capture diverse user intents and produce more accurate, intent-aware representations.

Pseudo-relevance feedback (PRF) expands queries with terms from top-ranked documents, enriching context and improving both precision and recall by adding synonyms and related concepts. However, it risks query drift when irrelevant terms are included, may generate overly long queries, and can be computationally expensive in large systems. Noisy feedback terms also reduce precision, underscoring the need for better filtering.

3.3 Knowledge Graph Integration

In this section, we examine methods that incorporate knowledge information into query reformulation and related tasks. Utilizing knowledge sources in query suggestion helps improve the interpretation of ambiguous queries by providing additional semantic context. Moreover, the integration of such knowledge enables personalization, allowing systems to tailor suggestions based on individual user preferences and search histories. Broadly, knowledge sources offer two types of information: 1- the relationships between entities, typically represented in the form of knowledge graphs. 2- detailed descriptions of entities, as found in structured encyclopedic sources.

Li et al. [12] propose a knowledge-enhanced pipeline that addresses limitations of traditional approaches. To find the knowledge contained in the initial query, they built an entity linker that identifies the query entities present in a large-scale knowledge repository. This approach enhances the understanding of the query context by leveraging structured knowledge bases like knowledge graphs, which provide relationships between entities and help in identifying relevant expansion terms. In the proposed pipeline, query reformulation and neural retrieval modules are optimized alternately from the feedback of each other, creating a cooperative system that continuously refines query suggestions based on both user intent and contextual knowledge. Additionally, incorporating detailed descriptions from sources such as Wikipedia ensures that the expanded queries are semantically rich and contextually accurate. By integrating these knowledge-based techniques, the pipeline significantly enhances the overall retrieval performance, making the search process more intuitive and aligned with user needs. Another application of knowledge entities is personalization. A personal knowledge repository composed of entities derived from search queries and visited web pages is constructed by Baek et al. [13] for the query suggestion task. The captured knowledge is used as context for a large language model input. This repository leverages structured knowledge graphs and ontologies to provide rich, semantically relevant context for user queries. The proposed knowledge-augmented model is capable of lightweight personalization through retrieval from the knowledge store, without the need for explicit user profiling. This approach enhances query expansion by incorporating user-specific context and preferences directly into the LLM, making the suggestions more relevant and tailored.

In summary, knowledge-based query expansion enriches semantic context by linking terms through synonyms, hypernyms, and related concepts, thereby improving query understanding and relevance. These methods help prevent query drift by preserving original intent and can be adapted to domain-specific needs, often with greater efficiency than pseudo-relevance feedback since they rely on curated knowledge sources. However, static knowledge bases may miss new terminology or trends, leading to outdated expansions, and they typically lack personalization, producing results that may not align with individual user intent. To overcome this, newer approaches integrate user interaction data—such as search logs and click behavior—to generate more contextually relevant and personalized query reformulations.

4 Contextual Information Source

Contextual query refinement involves leveraging information about the user's current context, previous interactions, or ongoing search session to improve the relevance and specificity of retrieved results.

4.1 Session-Based Information

Users interact with search engines by submitting sequences of related queries, often grouped into sessions defined by consecutive queries and user actions (e.g., clicks). Within a session, users iteratively adjust their queries—adding, removing, or substituting terms—to refine their information needs. Interaction signals such as click-through rates and dwell time provide valuable clues about evolving intent.

With Previous Queries. This subsection focuses on approaches that incorporate a user's initial query along with the sequence of previously issued queries within a search session to better capture the user's evolving intent. By leveraging the full context of the session, systems can generate more relevant reformulations that align closely with users' evolving goals.

One notable approach is proposed by Dehghani et al. [14], who introduced a query-aware attention mechanism based on a sequence-to-sequence (seq2seq) architecture for query suggestion. Their model incorporates three key components: By considering multiple queries within a session, the model dynamically adapts its suggestions to the user's current context while maintaining coherence with previous query formulations. In addition to general purpose search, session-based query reformulation has been explored in domain-specific search environments. For instance, Cao et al. [15] constructed a large-scale query reformulation corpus using search logs collected from Stack Overflow. Their process involved identifying original queries and their corresponding refinements from search sessions, which were then used to train a transformer based model. The trained model was capable of automatically generating reformulated queries tailored to the information needs of developers conducting programming related searches. In the context of conversational search, Yang et al. [16] proposed a zero-shot query reformulation framework using a transformer model that leverages conversational context. Unlike traditional approaches that require large-scale training data with explicit queryreformulation pairs, their method uses prior conversational history—including earlier user queries and system responses—to generate reformulated queries that remain coherent with the ongoing dialogue. This approach demonstrates the potential of context-aware models to generalize across unseen tasks without the need for extensive supervision.

One key application is to encode a user's query history to enhance the representation of the current query, thereby making the user's information need more explicit and easier to satisfy. In this context, Zhou et al. [17] treat the current query as a direct expression of the user's intent and propose a method for encoding prior queries in the session to construct a context-aware representation of the current query. By incorporating historical search behavior, their approach enriches the semantic representation of the query, enabling more accurate interpretation and downstream retrieval performance. In the domain of conversational search, systems aim to further refine query understanding by posing clarifying questions that resolve ambiguity in user intent. Aliannejadi et al. [18] introduced a model for selecting clarifying questions that considers both the user's initial query and the sequence of prior questionanswer interactions. Their approach leverages the entire dialogue history to identify information gaps or ambiguities and select follow-up questions that are most likely to elicit informative user responses. By dynamically adapting to conversational context, the system improves its ability to understand the user's true intent, especially in complex or under-specified queries.

Another line of research in this domain focuses on enhancing retrieval effectiveness in conversational search through query expansion, also known as query rewriting. To this end, Mo et al. [19] proposed a query selection framework that selectively expands the current query using only the most relevant historical queries, rather than incorporating all previous queries in the session. Their method employs a pseudo labeling strategy, in which historical queries are automatically labeled based on their relevance to the current search intent. The model leverages contextual signals from earlier conversational turns including the user's initial query and subsequent interactions to determine which past queries are most likely to contribute useful context. In the e-commerce domain, Hirsch et al. [20] conducted a large-scale analysis of search logs to investigate user query reformulation behavior within a session. Their study revealed that reformulated queries tend to be longer and are associated with higher click-through and purchase rates, particularly for the final query in a session—regardless of how many reformulations occurred. Based on these findings, the authors addressed a novel challenge in e-commerce search: predicting whether a user will reformulate their query before any results are shown. This insight enables the development of pre-retrieval strategies aimed at anticipating user behavior and improving both search relevance and commercial outcomes.

With Query and Click-Through Data. This section examines approaches that combine previously issued queries with click-through data to better capture user intent. By analyzing both query sequences and user interactions, these methods enhance tasks such as query suggestion, reformulation, and representation learning. Click-through data serves as implicit feedback on document relevance, and when paired with prior queries, it reveals behavioral patterns that refine query interpretation. Empirical studies show that incorporating session-level clicks improves precision and relevance by moving beyond surface-level term matching to infer deeper semantic intent—particularly useful when queries are ambiguous or underspecified.

Learning how to represent search session information plays a crucial role in capturing the semantic and syntactic relationships between user queries and interactions. In this context, Jiang et al. [21] proposed a query suggestion model that integrates query and click-through data from search logs into term embeddings using a heterogeneous network embedding framework.

Search activities within a session—such as query reformulations and result clicks—often exhibit strong interdependencies. These dependencies offer rich contextual signals that can be leveraged to enhance retrieval models. To exploit these interactions, Ahmad et al. [22] introduced a two-level hierarchical recurrent neural network that learns both the search session representation and the structure of queryclick dependencies. Their model explicitly incorporates prior queries and associated clicks from the session to improve both document ranking and query suggestion performance. Another complementary direction is to treat clicked suggested queries as feedback for improving query suggestion systems. In this line of work, Li et al. [23] presented a feedback-aware query suggestion model, which uses user feedback—specifically, the act of clicking on suggested queries—to adaptively refine future suggestions. By interpreting user interactions as implicit feedback, the model can better align suggestions with the user's evolving intent. To more effectively capture users' underlying search intent, Li et al. [23] model the sequence of issued search queries alongside clicks on previously suggested queries, treating these interactions as implicit user feedback. Their proposed framework employs adversarial training techniques to improve the model's robustness and maintain performance even under challenging conditions, such as noisy inputs or ambiguous user behavior. This approach enables the system to dynamically refine its understanding of user intent, thereby improving the accuracy of subsequent query suggestions and delivering a more tailored and effective search experience. Building on this idea, Chen et al. [24] proposed a query suggestion model that incorporates both short-term context (i.e., queries and click behavior from the current session) and long-term context (i.e., interactions from previous sessions). Their model is built on an attention-based hierarchical neural network, which is designed to capture user preferences toward specific query types. In a related direction, Wu et al. [25] introduced a feedback memory network that encodes a variety of user interactions including issued queries, clicked documents, and skipped items as positive and negative feedback. This feedback-aware representation is then integrated with a sequence-to-sequence model to drive a query suggestion system capable of adapting to nuanced user behavior. By explicitly modeling feedback signals, the system is better equipped to distinguish between relevant and irrelevant content, leading to more refined and personalized query suggestions.

An important application of integrating users' query and click data is personalized search. Natural language is inherently ambiguous—terms such as "Python" can refer to either a programming language or a species of snake, depending on the user's background and intent. As such, these terms should ideally have user-specific semantic representations. Motivated by this observation, Yao et al. [26] and Zhou et al. [27] proposed personalized search approaches that leverage users' queries and click behaviors within search sessions to model individual user interests. Yao et al. [26] addressed this by first training a global word2vec model on the full AOL query log to capture general semantic relationships between terms. While this unsupervised model learns word co-occurrences effectively, it does not account for user-specific preferences embedded in click behavior. To overcome this limitation, the authors developed a supervised personalization model that fine-tunes the word embeddings using users' click data. The result is a set of personalized word vectors that more accurately reflect each user's unique interests and search patterns. Similarly, Zhou et al. [27] introduced a self-supervised learning framework for personalized search that utilizes a contrastive sampling technique. Their approach extracts informative pairs from user behavior sequences recorded in query logs, allowing the model to refine the representations of queries, documents, and users. By learning from implicit feedback in user interactions, their method enhances the alignment between search results and individual user preferences, leading to more effective and personalized retrieval outcomes.

While many approaches rely on users' historical queries and click behavior to personalize and disambiguate queries, recent research has also explored interactive strategies, such as posing clarifying questions to users. For example, Erbacher et al. [28] introduced a simulated query clarification framework that enables multi-turn interactions between information retrieval (IR) systems and user agents. Their framework simulates user behavior based on click data, allowing the system to iteratively clarify ambiguous queries through sequential interactions, thereby improving retrieval effectiveness in real-time. Beyond traditional search, search session information is also increasingly utilized in the search advertising domain, where ad relevance and placement are critical. Typically, ad selection models in this domain are trained using large-scale ad click data. To address challenges associated with rare or long-tail queries, Lee et al. [29] proposed a method that uses a sequence-to-sequence model to generate relevant keywords based on the user's input query. These generated keywords are then used to expand the query, thereby improving ad retrieval and selection. This approach ensures that relevant ads are surfaced even for underrepresented queries, enhancing both click-through rates and overall advertising performance.

4.2 Temporal Data

This section examines temporal data - including query timestamps and event information - for query refinement. Understanding the time aspect can help improve the relevance of search results by considering seasonal trends, recent events, or time-sensitive information.

As we mentioned before, query auto-completion (QAC) methods suggest queries to search engine users as they begin typing. Most current QAC methods rank these suggestions based on their past popularity, measured by the number of times they have been previously submitted. However, query popularity changes over time and can differ significantly among users. To address this problem, Cai et al. [30] proposed a time-sensitive query auto-completion approach. This approach ranks query completions by predicting query popularity based on periodicity and recent trends in query frequency. They incorporate several key elements, including a time-sensitive popularity prediction mechanism that analyzes historical query logs to identify periodic patterns and recent trends, and a prefix-adaptive mechanism that adjusts suggestions based on different lengths and forms of user-typed prefixes. By capturing the dynamic nature of query popularity over time, the method ensures that auto-completion suggestions remain relevant and up-to-date, which is crucial for handling trending topics and seasonal searches.

Another form of temporal data used in information retrieval tasks is event information. Rosin et al. [31] proposed an event-aware query expansion approach that incorporates temporal context by identifying events related to the user's initial query. In this method, events are first detected, and terms that are semantically associated with both the query and the identified events are selected as expansion candidates. To facilitate this, both words and events are embedded into a shared vector space, allowing for the measurement of semantic similarity between query terms and event-related concepts. Unlike approaches that rely heavily on search log data, they utilize structured external knowledge sources—specifically Wikipedia and DBpedia—to extract relevant event information. By incorporating event-based expansions, the model enhances the contextual relevance of queries, particularly in cases where time-sensitive or event-driven topics are involved.

Search logbased methods capture recurring patterns within sessions, enabling highly personalized refinements but raising privacy concerns since they rely on sensitive user histories and remain limited to single-platform behavior. In contrast, social informationbased methods leverage shared interests, collaborative behavior, and community signals, offering broader and more diverse refinements. Together, session-based approaches emphasize precision and personalization, while social-based techniques contribute breadth and diversity, making them complementary strategies for improving retrieval systems.

4.3 Social Information

Approaches leveraging users' search sessions on social platforms enhance search relevance by building personalized models that generate diverse and contextually appropriate query suggestions. These methods combine signals from user-generated content, social connections, and trending topics to align recommendations with both individual interests and broader social dynamics. By integrating personal and collective activity, they provide more personalized, trend-aware, and timely query refinements, improving overall search effectiveness.

One prominent example of a social network platform used for contextual search enhancement is LinkedIn. Zhou et al. [32] proposed a query suggestion model aimed at improving job search experiences on LinkedIn by leveraging a combination of user profiles, prior search history, and job descriptions. Their model employs semantic matching techniques to ensure that the suggested queries align closely with both the users' qualifications and the roles they are exploring. Further advancements in this domain were made by Zhong et al. [33], who introduced a query suggestion model that incorporates user-specific professional features to enhance personalization. Their approach extracts career-related attributes—such as job titles, skills, and industry experience—from LinkedIn member profiles and integrates these features into the suggestion model. This results in more personalized and contextually appropriate recommendations, ultimately improving the efficiency and relevance of job-related search on professional platforms.

User modeling can also be enhanced by incorporating information from users' social connections, particularly in scenarios where individual historical data is limited. Zhou et al. [34] proposed a personalized search model that integrates both individual and group profile information to improve the relevance of search results. Their approach combines a user's personal search history with data derived from their social network, forming user groups based on similar interests or behavioral patterns.

Incorporating social information into query refinement improves personalization by aligning suggestions with users' interests, roles, and connections, and helps when personal histories are sparse by leveraging group behavior and structured data like resumes. However, it raises privacy and ethical concerns, depends on platform-specific structures, and can suffer from outdated or noisy signals. It is also computationally costly due to large-scale graph processing and may reinforce social biases, reducing diversity in query suggestions.

5 Comparative Analysis

To provide a structured comparison of the surveyed methods, we analyze their effectiveness, scalability, robustness to query drift, and personalization. Table 1 presents the comparative analysis of non-contextual methods, highlighting how external knowledge, Table 2 summarizes the comparative analysis of contextual methods, Table 3 outlines the proposed taxonomy of information sources.

Our comparative analysis evaluates methods across four key dimensions: *Effectiveness:* Based on reported performance improvements in original studies. *Scalability:* Assessed by computational requirements and real-time applicability. *Robustness to Query Drift:* Evaluated by semantic preservation mechanisms. *Personalization:* Measured by user-specific adaptation capabilities. Ratings (High/Medium/Low) were assigned based on quantitative results reported in the original papers and qualitative assessment of method characteristics.

Table 1. Comparative Analysis of Non-Contextual Methods

Method	Information Source	Effectiveness	Scalability	Robustness to Query Drift	Personalization	Year	Key Innovation
Lucchese et al. [4]	Thesaurus	Medium	High	Medium	Low	2018	Conjunctive Normal Form expansion
Azad et al. [6]	WordNet + Wikipedia	High	Medium	High	Low	2019	Two-level synset extraction
Zheng et al. [9]	Retrieved Docs (BERT)	High	Medium	Medium	Low	2021	Selective chunk-based expansion
Hashemi et al. [11]	Retrieved Docs (BART)	High	Medium	High	Low	2021	Multiple query representations
Li et al. [12]	Knowledge Graph	High	Low	High	Medium	2022	Entity linking with neural retrieval
Baek et al. [13]	Personal Knowledge	High	Medium	High	High	2024	Lightweight personalization
Li et al. [36]	Retrieved Documents	Medium	Medium	Low	Low	2018	Neural PRF approach
Wang et al. [8]	Retrieved Documents	High	Low	Medium	Low	2020	End-to-end neural framework

Beyond the tabulated summary, a closer look at key evaluation criteria highlights notable differences among the surveyed methods. In terms of robustness to query drift, neural architectures—particularly transformer-based and reinforcement learningdriven refiners—demonstrate stronger resilience, as they can model context across longer query sequences and correct deviations from the original intent. In contrast, purely lexical or heuristic approaches tend to be more susceptible to drift, especially in longer sessions. Regarding personalization capability, models that explicitly incorporate user profiles, session histories, or social information (e.g., friend networks, professional profiles) consistently outperform generic refinement methods, delivering more relevant and user-aligned suggestions. However, this advantage often comes at the cost of higher data requirements and potential privacy concerns, whereas non-personalized approaches remain easier to deploy in data-constrained or privacy-sensitive contexts.

6 Future Directions and Open Challenges

Query refinement research continues to evolve, yet several technical, methodological, and ethical challenges remain, alongside emerging opportunities that could shape the next generation of methods.

Technical Challenges: One persistent issue is query drift mitigation. Many expansion methods inadvertently introduce terms that diverge from the original

Table 2. Comparative Analysis of Contextual Methods

Method	Information Source	Effectiveness	Scalability	Robustness to Query Drift	Personalization	Year	Key Innovation
Dehghani et al. [14]	Session Queries	High	Medium	High	High	2017	Copy-generate mechanism
Jiang et al. [21]	Query + Click Data	High	Medium	Medium	High	2018	Heterogeneous network embedding
Ahmad et al. [22]	Session + Clicks	High	Low	Medium	High	2019	Two-level hierarchical learning
Li et al. [23]	Feedback Clicks	High	Medium	High	High	2019	Adversarial training approach
Cai et al. [30]	Temporal Data	Medium	High	Medium	Medium	2016	Periodicity and trend prediction
Zhou et al. [32]	Social + Professional	High	Low	High	High	2022	Professional feature integration
Zhong et al. [32] [32]	Career Attributes	High	Low	High	High	2020	Skills and experience modeling
Zhou et al. [33] [32]	Social Networks	High	Low	Medium	High	2021	Individual + group profiling
Rosin et al. [34] [32]	Event Information	Medium	Medium	High	Low	2021	Event-query semantic embedding
Wu et al. [25] [32]	Multi-modal Feedback	High	Low	Medium	High	2018	Comprehensive feedback encoding
Chen et al. [21] [32]	Short/Long-term Context	High	Medium	High	High	2020	Multi-temporal context modeling

Table 3. Information Source Taxonomy

Category	Sub- category	Examples	Key Characteristics	Applications
Non- Contextual	External Knowledge	WordNet, Wikipedia, Thesauri	Static, General, Semantic	Query Expansion, Synonym Detection
	Retrieved Documents	PRF, Top-k Results	Dynamic, Query-specific	Contextual Expansion
	Knowledge Graphs	Entity Linking, Relations	Structured, Semantic	Entity-based Refinement
Contextual	Session Data	Query Sequences, User History	Personal, Sequential	Personalized Suggestion
	Click-through	User Interactions, Feedback	Behavioral, Implicit	Intent Modeling
	Temporal	Time, Events, Trends	Time-sensitive, Dynamic	Trending Topics, Seasonality
	Social	Network Data, Profiles	Collaborative, Social	Social Search, Professional

user intent, reducing retrieval relevance. Future work should prioritize robust filtering strategies and semantic coherence measures to ensure that expanded queries remain aligned with user needs. Scalability is another critical hurdle—real-time query refinement in large-scale systems is particularly demanding for

knowledge graphbased and complex contextual approaches. This calls for more efficient indexing schemes, approximation algorithms, and resource-aware processing pipelines.
Evaluation and Benchmarking: The field lacks standardized evaluation benchmarks that enable consistent comparisons across different information source types. A unified, comprehensive evaluation framework—covering both quantitative and qualitative dimensions—would be instrumental in advancing the field. Moreover, current evaluations often emphasize immediate retrieval improvements, neglecting long-term effectiveness measures such as sustained user satisfaction, learning effects, and behavioral changes over time. Addressing these gaps would yield more reliable assessments of method impact in real-world use.
Privacy and Ethical Considerations: Contextual methods frequently rely on sensitive user data, raising important privacy concerns. Developing privacy-preserving refinement techniques that maintain retrieval effectiveness without compromising user confidentiality is a pressing need. Additionally, both social and contextual methods can inadvertently reinforce existing biases present in training data or interaction logs. Research on bias detection, mitigation, and fairness-aware refinement strategies is essential to ensure equitable and trustworthy systems. **Emerging Opportunities:** Large language models (LLMs) offer new possibilities for query understanding, reformulation, and expansion, but their integration must be tempered by careful evaluation of hallucination tendencies and factual accuracy. Real-time adaptation—dynamically adjusting refinement strategies to evolving user contexts and information landscapes—remains a promising yet unsolved challenge. Finally, cross-domain transfer represents an important research frontier: developing models and techniques that generalize effectively across diverse domains, languages, and application scenarios would substantially increase the applicability and robustness of query refinement systems.

7 Conclusion

This survey provided a comprehensive analysis of query refinement methods through the lens of information sources, systematically reviewing 67 papers published between 2016-2024. We established a clear taxonomy distinguishing between non-contextual sources (external knowledge, retrieved documents, knowledge graphs) and contextual sources (session data, click-through behavior, temporal information, social signals). Our comparative analysis reveals that contextual methods generally provide superior personalization capabilities but at the cost of increased complexity and privacy concerns. Non-contextual methods offer better generalization and scalability but may lack user-specific adaptation. The most promising approaches combine multiple information sources, suggesting that hybrid models integrating structured knowledge with behavioral context hold significant potential. **Key Contributions:** a) Systematic methodology for surveying query refinement literature; b) Novel taxonomy based on information source types; c) Comprehensive comparative analysis across effectiveness, scalability, and personalization dimensions; d) Identification of critical challenges and future research directions.

Future Work: We recommend focusing on: (1) developing privacy-preserving contextual methods, (2) creating standardized evaluation benchmarks, (3) investigating effective multi-source integration techniques. The development of explainable refinement systems that provide transparency in the refinement process remains an important open challenge. This structured understanding of information sources in query refinement facilitates more informed decisions when selecting or designing techniques and provides a foundation for advancing the field toward more adaptive, accurate, and user-centered retrieval systems.

References

1. Statista Research Department. Number of search terms used in internet research in the United States as of January 2022. *Statista*. https://www.statista.com/statistics/269740/number-of-search-terms-in-internet-research-in-the-us/
2. Azad, H.K., Deepak, A.: Query Expansion Techniques for Information Retrieval: A Survey. Elsevier, Information Processing & Management (2019)
3. Cao, H., et al.: Context-Aware Query Suggestion by Mining Click-Through and Session Data, Association for Computing Machinery (2008)
4. Lucchese, C., Nardini, F.M., Perego, R., Trani, R., Venturini, R.: Efficient and effective query expansion for web search. In : CIKM (2018)
5. Rencis, E.: Application of a configurable keywords-based query language to the healthcare domain. J. Adv. Inf. Technol. (2021)
6. Azad, H.K., Deepak, A.: A New Approach for Query Expansion Using Wikipedia and WordNet. Elsevier, Information Sciences (2019)
7. Li, C., et al.: NPRF: a neural pseudo relevance feedback framework for ad-hoc information retrieval. In: EMNLP (2018)
8. Wang, L., et al.: An end-to-end pseudo relevance feedback framework for neural document retrieval. In: CIKM (2020)
9. Zheng, Z., Hui, K., He, B., Han, X., Sun, L., Yates, A.: Contextualized query expansion via unsupervised chunk selection for text retrieval. In: CIKM (2021)
10. Medlar, A., Li, J., Głowacka, D.: Query suggestions as summarization in exploratory search. In: CHIIR (2021)
11. Hashemi, H., Zamani, H., Croft, W.B.: Learning multiple intent representations for search queries. In: CIKM (2021)
12. Li, X., et al.: A cooperative neural information retrieval pipeline with knowledge enhanced automatic query reformulation. In: WSDM (2022)
13. Baek, J., Chandrasekaran, N., Cucerzan, S., Herring, A., Jauhar, S.K.: Knowledge-augmented large language models for personalized contextual query suggestion. In: Proceedings of the ACM Web Conference (2024)
14. Dehghani, M., Rothe, S., Alfonseca, E., Fleury, P.: Learning to attend, copy, and generate for session-based query suggestion. In: CIKM (2017)
15. Cao, K., Chen, C., Baltes, S., Treude, C., Chen, X.: Automated query reformulation for efficient search based on query logs from Stack Overflow. In: ICSE (2021)
16. Yang, D., Zhang, Y., Fang, H.: Zero-shot query reformulation for conversational search. In: SIGIR (2023)
17. Zhou, Y., Dou, Z., Wen, J.-R.: Encoding history with context-aware representation learning for personalized search. In: SIGIR (2020)

18. Aliannejadi, M., Zamani, H., Crestani, F., Croft, W.B.: Asking clarifying questions in open-domain information-seeking conversations. In: SIGIR (2019)
19. Mo, F., et al.: Learning to relate to previous turns in conversational search. In: SIGKDD (2023)
20. Hirsch, S., Guy, I., Nus, A., Dagan, A., Kurland, O.: Query reformulation in E-commerce search. In: SIGIR (2020)
21. Jiang, J.-Y., Wang, W.: RIN: reformulation inference network for context-aware query suggestion. In: CIKM (2018)
22. Ahmad, W.U., Chang, K.-W., Wang, H.: Context attentive document ranking and query suggestion. In: SIGIR (2019)
23. Li, R., Li, L., Wu, X., Zhou, Y., Wang, W.: Click feedback-aware query recommendation using adversarial examples, WWW (2019)
24. Chen, W., Cai, F., Chen, H. de Rijke, M.: Hierarchical neural query suggestion with an attention mechanism. In: CIKM (2020)
25. Wu, B., Xiong, C., Sun, M., Liu, Z.: Query suggestion with feedback memory network, WWW (2018)
26. Yao, J., Dou, Z., Wen, J.-R.: Employing personal word embeddings for personalized search. In: SIGIR (2020)
27. Zhou, Y., Dou, Z., Zhu, Y., Wen, J.-R.: PSSL: self-supervised learning for personalized search with contrastive sampling. In: CIKM (2021)
28. Erbacher, P., Denoyer, L., Soulier, L.: Interactive query clarification and refinement via user simulation. In: SIGIR (2022)
29. Lee, M.-C., Gao, B., Zhang, R.: Rare query expansion through generative adversarial networks in search advertising. In: SIGKDD (2018)
30. Cai, F., Liang, S., de Rijke, M.: Prefix-adaptive and time-sensitive personalized query auto completion. In: IEEE Transactions on Knowledge and Data Engineering (2016)
31. Rosin, G. D., Guy, I., Radinsky, K.: Event-driven query expansion. In: WSDM (2021)
32. Zhou, Z., Zhou, X., Li, M., Song, Y., Zhang, T., Yan, R.: Personalized query suggestion with searching dynamic flow for online recruitment. In: CIKM (2022)
33. Zhong, J., Guo, W., Gao, H., Long, B.: Personalized query suggestions. In: SIGIR (2020)
34. Zhou, Y., Dou, Z., Wei, B., Xie, R., Wen, J.-R.: Group based personalized search by integrating search behaviour and friend network. In: SIGIR (2021)
35. Wang, L., Yang, N., Wei, F.: Query2doc: Query expansion with large language models, CoRR (2023)
36. Xiaodong, L., Pengcheng, H., Weizhu, C., Jianfeng, G.:Multi-task deep neural networks for natural language understanding, CoRR (2019)

GPSR Compliance

The European Union's (EU) General Product Safety Regulation (GPSR) is a set of rules that requires consumer products to be safe and our obligations to ensure this.

If you have any concerns about our products, you can contact us on ProductSafety@springernature.com

In case Publisher is established outside the EU, the EU authorized representative is:

Springer Nature Customer Service Center GmbH
Europaplatz 3
69115 Heidelberg, Germany

Batch number: 10761593

Printed by Printforce, the Netherlands

Lecture Notes in Networks and Systems 1904

The series "Lecture Notes in Networks and Systems" publishes the latest developments in Networks and Systems—quickly, informally and with high quality. Original research reported in proceedings and post-proceedings represents the core of LNNS.

Volumes published in LNNS embrace all aspects and subfields of, as well as new challenges in, Networks and Systems.

The series contains proceedings and edited volumes in systems and networks, spanning the areas of Cyber-Physical Systems, Autonomous Systems, Sensor Networks, Control Systems, Energy Systems, Automotive Systems, Biological Systems, Vehicular Networking and Connected Vehicles, Aerospace Systems, Automation, Manufacturing, Smart Grids, Nonlinear Systems, Power Systems, Robotics, Social Systems, Economic Systems and other. Of particular value to both the contributors and the readership are the short publication timeframe and the world-wide distribution and exposure which enable both a wide and rapid dissemination of research output.

The series covers the theory, applications, and perspectives on the state of the art and future developments relevant to systems and networks, decision making, control, complex processes and related areas, as embedded in the fields of interdisciplinary and applied sciences, engineering, computer science, physics, economics, social, and life sciences, as well as the paradigms and methodologies behind them.

Indexed by SCOPUS, EI Compendex, INSPEC, WTI Frankfurt eG, zbMATH, SCImago.

All books published in the series are submitted for consideration in Web of Science.

For proposals from Asia please contact Aninda Bose (aninda.bose@springer.com).

Herwig Unger · Phayung Meesad
Editors

Advances in Natural Language Processing and Information Retrieval

Set 2

Springer

Editors
Herwig Unger
Lehrgebiet Kommunikationsnetze
FernUniversität Hagen
Hagen, Germany

Phayung Meesad
Bangkok, Thailand

ISSN 2367-3370 ISSN 2367-3389 (electronic)
Lecture Notes in Networks and Systems
ISBN 978-3-032-20896-5 ISBN 978-3-032-20897-2 (eBook)
https://doi.org/10.1007/978-3-032-20897-2

This Springer imprint is published by the registered company Springer Nature Switzerland AG
The registered company address is: Gewerbestrasse 11, 6330 Cham, Switzerland

Preface

Recent advances in artificial intelligence, particularly Large Language Models (LLMs), have significantly advanced natural language understanding, knowledge integration, and human–computer interaction. Since its inception, the International Conference on Natural Language Processing and Information Retrieval (NLPIR) has provided an important forum for presenting innovative research in NLP and IR, including conversational systems, semantic search, and related applications.

While LLMs have driven rapid progress, important challenges remain in efficiency, interpretability, multilingual adaptability, and domain-specific deployment. Addressing these issues requires continued research beyond model scaling, including low-resource learning, ethical and human-centered design, and interpretable architectures that bridge theory and real-world impact.

This volume contains 85 accepted papers of the 9th International Conference on Natural Language Processing and Information Retrieval (NLPIR 2025), held on December 12–14, 2025, in Fukuoka, Japan. Building on the success of previous events in locations such as Okayama, Seoul, Sanya, Tokushima, and Bangkok, the conference provides an international forum for researchers, scholars, and practitioners to exchange ideas and present recent advances in these dynamic fields. These papers were carefully selected through a rigorous review process to ensure high academic quality and relevance.

We would like to express our sincere gratitude to all authors for their valuable submissions and to the reviewers and program committee members for their time, expertise, and dedication. Their collective efforts have been essential to the success of the conference and the quality of this proceedings volume. We hope that these papers will inspire further research and foster collaboration within the community.

Herwig Unger
Conference General Chair

Organization

Committee

General Chair

Herwig Unger — FernUniversität in Hagen, Germany

General Co-chair

Phayung Meesad — King Mongkut's University of Technology North Bangkok, Thailand

Local Chair

Tsunenori Mine — Kyushu University, Japan

Local Co-chairs

Koichi Takeuchi — Okayama University, Japan
Kazuyuki Matsumoto — Tokushima University, Japan
Hiromitsu Shiina — Okayama University of Science, Japan

Program Chairs

Thomas Böhme — Technical University of Ilmenau, Germany
Janusz Kacprzyk — The Polish Academy of Sciences, Poland

Program Co-chairs

Siripong Potisuk — 'The Citadel School of Engineering, United States
Minoru Yoshida — Tokushima University, Japan

Steering Committee

Herwig Unger	FernUniversität in Hagen, Germany
Phayung Meesad	King Mongkut's University of Technology North Bangkok, Thailand
James Z. Wang	Clemson University, United States

International Program Committees

Adrien Guille	University Lyon 2, France
Akinori Ito	Tohoku University, Japan
Alessio Faccia	University of Birmingham Dubai, UAE
Ann Franchesca Laguna	De La Salle University, Philippines
Au Yeung Siu Kei Jeff	Hong Kong Metropolitan University, China
Bahar Ilgen	Robert Koch Institute, Germany
Chisom Ezekannagha	Robert Koch Institute; Freie Universitat Berlin, Germany
Chunjie Guo	Nanjing University of Aeronautics and Astronautics, China
Dinh Dien	Ho Chi Minh University of Science (HCMUS), Vietnam
E. Hashmi	Norwegian University of Science and Technology, Norway
Endang Wahyu Pamungkas	Universitas Muhammadiyah Surakarta, Indonesia
Eric R. Harley	Toronto Metropolitan University, Canada
Ethel Ong	De La Salle University, Philippines
Florian Matthes	Technical University of Munich, Germany
Haemin Jung	Korea National University of Transportation, South Korea
Harinda Fernando	Sri Lanka Institute of Information Technology, Sri Lanka
Hector Rafael Morano Okuno	Tecnologico de Monterrey, Mexico
Hidekazu Yanagimoto	Osaka Metropolitan University, Japan
Hoi-Jeong Lim	Chonnam National University, South Korea
Hongwei Du	California State University, United States
Hosik Choi	University of Seoul, South Korea
Irene Weber	University of Applied Sciences Kempten, Germany
Jairo Cugliari	University Lyon 2, France
Javier Martin Arista	Universidad de La Rioja, Spain
Jerry Chih-Yuan Sun	National Yang Ming Chiao Tung University, Taiwan

Kang Xin	Tokushima University, Japan
Kevin Hirschi	University of Texas at San Antonio, United States
Khang Nhut Lam	Can Tho University, Vietnam
Long Nguyen	University of Science, Ho Chi Minh City, Vietnam
Lucia Nacinovic Prskalo	University of Rijeka, Croatia
Makoto P. Kato	University of Tsukuba and National Institute of Informatics, Japan
Maleerat Maliyaem	King Mongkut's University of Technology North Bangkok, Thailand
Marija Brkic Bakaric	University of Rijeka, Croatia
Mario Flores	University of Texas at San Antonio, United States
Mario Kubek	Georgia State University, United States
Martin Strecker	IRIT, University de Toulouse, France
Masaki Murata	Tottori University, Japan
Mehmet Gonen	Koc University, Turkey
Miguel Couceiro	Inria, France
Minoo Jafarlou	Georgia State University, United States
Moritz Marutschke	Ritsumeikan University, Japan
Mosima A. Masethe	Sefako Makgatho Health Sciences University, South Africa
Ngoc Hong Tran	Vietnamese-German University, Vietnam
Ondrej Bojar	Institute of Formal and Applied Linguistics, Czechia
Pythagoras Petratos	Coventry University, UK
Richa Sharma	PES University, India
Shylaja Ss	PES University, India
Soma Das	Indian Institute of Information Technology Kalyani, India
SUnited Statesn Gauch	University of Arkansas, United States
Swapna Gokhale	University of Connecticut, United States
Takashi Inui	University of Tsukuba, Japan
Tanioka, Hiroki	Tokushima University, Japan
Tee Hean Tan	Suwnay University, Malaysia
Tejas Pant	Accrete AI, United States
Thaweesak Yingthawornsuk	King Mongkut's University of Technology Thonburi, Thailand
Xiaoge Li	Xi'an University of Posts and Telecommunications, China
Xiaorui Jiang	Sheffield University, UK
Yves Lepage	Waseda University, Japan
Zichao Li	University of Waterloo, Canada

Contents

Multimodal NLP Recognition, Semantic Analysis and Modeling

Sentiment Analysis and Corpus Construction

Context-Aware Large Language Models and Parameter Adaptation

Language Representation and Information Retrieval in Multimodal Information Systems

AI-Based Generative Dialogue Systems and Intelligent Question Answering Systems

Resource-Efficient Conversational AI: A Client-Server Architecture for Session Management

Chirag Bhawnani(✉)

Rakuten Mobile Inc., Tokyo, Japan
chirag.bhawnani@rakuten.com

Abstract. LLM-powered complex chatbots, integrating RAG and external databases, face significant challenges such as high token consumption, increasing resource demands, and susceptibility to contextual drift leading to hallucinated output. These issues are aggravated in the case of multi-turn conversations because of repeated processing of extensive context, which in turn leads to higher operational costs and compromised reliability. This paper introduces an optimized client-server architecture designed to mitigate these inefficiencies. Our approach employs a lightweight client-side LLM as an orchestrator, making termination decisions based on conversation patterns and server-provided context. All heavy processing, including RAG, database queries, and comprehensive conversational memory, is offloaded to a powerful, stateful server side. The system achieves resource optimization through intelligent session management, in which the client analyzes current conversation logs to detect termination conditions such as jailbreak, stalemate, confusion, or successful completion, and clears conversation logs, retaining only minimal summaries for future reference, and therefore preventing token accumulation across extended sessions while maintaining conversation continuity. This method demonstrated with a negotiation chatbot using the Model Context Protocol (MCP) showcases significant token savings through proactive conversation termination and efficient session-level context management. It offers a scalable and cost-effective blueprint for LLM-enabled applications by preventing resource waste through early identification of unproductive conversations and therefore enabling more robust, reliable, and manageable interactive AI systems.....

Keywords: Client Server Architecture · MCP · Resource Management · Conversational AI

1 Introduction

Large language models have changed how we build chatbots and conversational AI systems. Modern chatbots now use complex features like retrieval-augmented generation to access external databases, specialized tools, and multi-turn conversations to provide better user experiences. However, these improvements come

H. Unger and P. Meesad (Eds.): NLPIR 2025, LNNS 1904, pp. 791–803, 2026.
https://doi.org/10.1007/978-3-032-20897-2_55

with a serious problem: they consume too many computational resources, i.e., high token consumption. Each conversation turn requires processing the entire chat history along with retrieved information from databases. This occurs regardless of whether the information is relevant to the current query, as systems process entire histories uniformly. As conversations become longer, the computational cost grows rapidly. Jegham et al. (2025) [5] demonstrated that even short queries contribute to substantial environmental and financial burdens when scaled across production systems. This leads to higher operational costs and slower response times that affect system deployment viability.

The problem becomes worse with complex integrations. RAG systems repeatedly process retrieved documents across conversation turns, while external database queries add computational overhead that adds to context length. Multi-agent systems face similar challenges when orchestrating multiple specialized components, as each coordination step requires maintaining and processing an extensive conversational state.

There are also security concerns. Chatbots often face hostile interactions, jailbreak attempts, and abusive language from users. Research shows that a significant portion of chatbot interactions contain problematic content (Cercas Curry et al., (2021) [1]).These bad interactions waste computational resources on unproductive exchanges and can influence model behavior through prompt injection or adversarial manipulation techniques. Moreover, psychological persuasion techniques sometimes lead to unintended outputs (Nan Xu et al., 2024 [12]).

Current approaches to reducing computational costs in conversational AI have not specifically addressed orchestration efficiency in tool-augmented frameworks. While the client-server architecture provides standardized communication between LLMs and tools, existing implementations focus on functional integration rather than token-efficient orchestration strategies for multi-turn conversations.

We propose a client-server architecture that tackles these problems through session management and terminations. Our system uses a lightweight client to make orchestration decisions and a powerful server to handle heavy processing tasks. The key innovation is our two-tier termination strategy that identifies unproductive conversations early and terminates them before they waste resources. When conversations end, our system clears detailed logs and keeps only brief summaries for future reference within the session. This prevents token accumulation across sessions while maintaining conversation continuity. We demonstrate this approach using a negotiation chatbot built with the Model Context Protocol (MCP).

Our results show significant token savings and improved resource efficiency. This work provides a practical blueprint for building more efficient conversational AI systems across various applications.

2 Related Work

Resource optimization in large language models has become a critical research area, as these systems are deployed at scale. Early work by Strubell et al. (2019)

[10] highlighted the environmental impact of training large neural networks, while Patterson et al. (2021) [8] showed that inference costs often exceed training costs in production systems. These findings established the need for efficiency improvements beyond model training.

Most optimization approaches focus on model-level techniques such as quantization, pruning, and knowledge distillation. However, these methods address individual model efficiency rather than system-level resource management in conversational scenarios. Context length scaling presents deployment challenges as processing costs increase substantially with conversation length, affecting the practical viability of extended interactions.

The integration of RAG systems adds more computational challenges that are compounded with context-length issues. Lewis et al. (2020) [7] established foundational RAG architectures, but subsequent work has identified significant overhead in repeatedly processing retrieved documents across conversation turns. Khandelwal et al. (2021) [6] examined efficient retrieval mechanisms, though existing approaches focus on optimizing individual operations rather than managing accumulated context over extended conversations.

Client-server architectures for conversational AI have evolved from early modular systems to modern unified approaches. SimpleTOD (Hosseini-Asl et al., 2020 [4]) began unifying dialogue tasks under single frameworks. More recent work like Toolformer (Schick et al., 2023 [9]) and ReAct (Yao et al., 2022 [13]) introduced sophisticated tool coordination, but these approaches prioritize capability over resource efficiency and do not address token optimization across conversation sessions.

The Model Context Protocol (MCP) by Anthropic provides standardized frameworks for LLM-tool communication, enabling structured client-server interactions. However, existing implementations typically focus on functional integration rather than minimizing token consumption in orchestrating client while maintaining a rich server-side context.

Conversation management research has expanded from basic dialogue state tracking (Henderson et al., 2014 [3]) to session-level strategies. The work on detecting abusive content (Cercas Curry et al., 2023 [1]) and adversarial attacks (Zou et al., 2023 [14]; Wallace et al., 2019 [11]) shows the importance of termination strategies. However, existing approaches focus on detection rather than resource conservation through early termination.

Most current work addresses resource optimization or conversation management separately, without exploring how they can work together. Existing systems operate at the turn level and miss opportunities for session-level optimization. Tool-augmented frameworks typically require full conversational context, leading to poor resource utilization in multi-turn scenarios where context accumulation adds to computation overhead.

Our work fills this gap by combining lightweight orchestration with intelligent server-side context management. We show how architectural design choices can achieve measurable resource savings while maintaining conversation quality through proactive session management and termination strategies.

3 Methodology

This section details our proposed client-server architecture for resource-efficient conversational AI systems. Our approach addresses the challenges of excessive token consumption and resource waste through session-level management and termination strategies. The architecture is implemented using the Model Context Protocol (MCP) to demonstrate practical deployment in a negotiation chatbot scenario.

3.1 System Architecture Overview

Our system employs a distributed architecture that separates computational responsibilities between two primary components. This includes an MCP Server that handles heavy processing tasks and an MCP Client that manages orchestration decisions. As shown in Fig. 1, the client acts as a gateway between users and the server, implementing multi-layered termination strategies to prevent resource waste while maintaining conversation quality.

The architecture uses shared conversation logs as a central coordination mechanism. This enables both components to maintain synchronization while implementing different responsibilities. The server manages the comprehensive conversational state and agent response generation, while the client focuses on routing decisions and termination based on conversation patterns and safety considerations. When conversations terminate, these detailed logs are cleared to prevent token accumulation. Unlike traditional approaches that include entire conversation histories in prompts, the client maintains only lightweight summaries within each session. The summaries capture essential information from terminated conversations, enabling continuity without the resource overhead of full histories (Appendix B).

3.2 MCP Server Components

The server-side negotiation agent uses GPT-4o for robust reasoning capabilities in complex negotiation scenarios. The agent maintains full awareness of the current conversation log and integrates multiple sources to facilitate effective negotiation interactions.

The agent operates on two specialized datasets curated for our use case: client purchase history and client profile information. We developed these datasets using a reference retail customer dataset from Hugging Face [2], which we processed and expanded to create realistic negotiation scenarios. The datasets are stored in ChromaDB vector databases with OpenAI text-embedding-3-large embeddings, enabling efficient retrieval of relevant contextual information during conversations.

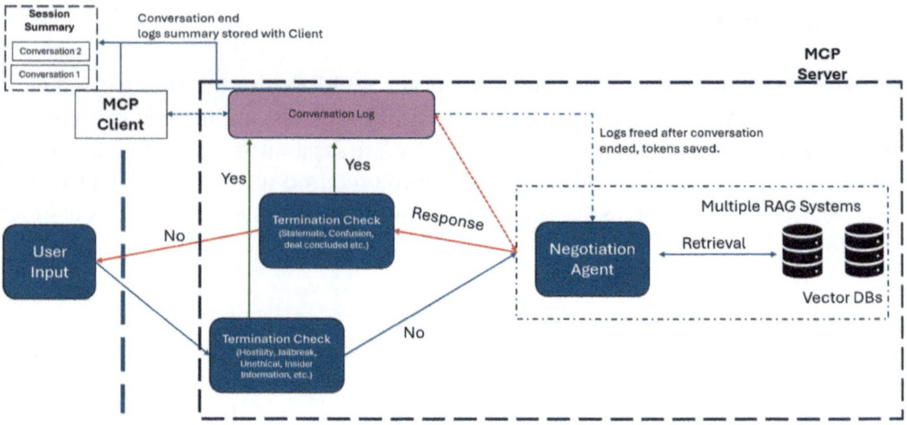

Fig. 1. Architecture Diagram.

The negotiation agent generates structured outputs that include both natural language responses and metadata about the negotiation state. This in turn enables the client orchestrator to make informed decisions about conversation continuation. This approach ensures that complex reasoning and data integration remain server-side while the client is provided with essential decision-making information.

Conversation Log Management: The server maintains comprehensive conversation logs that serve as the central coordination point for all system components. These logs capture user inputs, agent responses, and conversation metadata, supporting both ongoing agent processing and client-side termination analysis.

The logging system implements efficient storage and retrieval mechanisms optimized for dual access patterns required by our architecture. These dual access patterns are sequential processing for agent context and pattern analysis for termination decisions. The server provides controlled access to these logs while maintaining data integrity throughout the conversation lifecycle.

3.3 MCP Client Orchestration and Termination Strategy

The client implements a two-tier approach to conversation termination that balances computational efficiency with decision accuracy. This orchestration strategy, powered by Sonnet 4 for its strong reasoning capabilities in pattern analysis and native alignment with MCP, evaluates conversation quality to determine termination conditions.

Tier 1 - Immediate Pattern Detection: The client performs a rapid assessment of the user input to identify clear violations such as jailbreak attempts, requests for insider information, or explicitly abusive language (Appendix A). These assessments require minimal context and enable immediate termination without server processing. While modern LLMs effectively handle these detection tasks through reasoning, specialized transformer models for content moderation can be integrated as additional safety layers for more practical or domain specific scenarios.

Tier 2 - Conversation Quality Analysis: The client performs a comprehensive analysis of the conversation pattern using structured evaluation prompts, after receiving agent responses and updated conversation logs. This analysis identifies stalemate situations, conversation confusion, successful deal completion, or other quality-based termination criteria that require understanding of conversation progression and context (Appendix A).

The client makes binary decisions about whether to continue or terminate conversations based on LLM reasoning applied to conversation logs and termination criteria provided in the orchestration instructions (Appendix C).

3.4 Resource Optimization Through Session Management

Our architecture achieves resource savings through session-level management that prevents token accumulation across extended interactions. When conversations terminate, the client generates concise summaries (usually less than 500 tokens) extracting key information about negotiation outcomes, agreed price, quantity, discounts, critical decisions, and major events in the conversation (Appendix B). These lightweight summaries persist within the session and are included in subsequent turn prompts, enabling context maintenance when users return to related topics. This contrasts with traditional methods where entire conversation histories occupying more than 1000 tokens would be included in prompts, leading to unbounded token growth.

Meanwhile, the detailed conversation logs from shared storage are cleared when conversations end, allowing the negotiation agent to start fresh for new conversation topics and preventing irrelevant historical data from consuming resources. This approach addresses context accumulation by maintaining conversation quality while preventing linear growth in computational requirements. This results in sustained interaction capability without the typical token overhead from extended sessions.

3.5 Negotiation Chatbot Implementation

The system is demonstrated through the implementation of a negotiation chatbot that showcases the architecture's capabilities in a resource-intensive domain requiring complex reasoning, external data integration, and sustained user interaction. This application domain effectively validates the system's ability to maintain conversation quality while optimizing resource utilization through termination and session management.

The implementation supports configurable session management and flexible integration patterns, enabling adaptation to various conversational AI applications beyond the specific negotiation use case demonstrated in this work. The MCP framework provides standardized communication protocols that facilitate deployment and integration with existing conversational AI infrastructures.

4 Results

This section presents experimental validation of our client-server architecture for resource-efficient conversational AI across four key performance areas: session-level token management, termination-based savings, client-side optimization, and termination strategy effectiveness.

Session-Level Token Management: Fig. 2 presents the core benefit of our session management approach through a representative session that contains six conversation turns. These conversations were intentionally designed with varied topics that a single user might explore in a single session. The conversations focused on product availability inquiries, new catalog check, previous order grievances, wholesale unit setup support, and even jailbreak attempts for insider information and unfair discounts (immediate termination in Conversation 3).

Token accumulation can be observed in traditional systems, reaching more than 50,000 tokens as the negotiation agent processes increasingly large conversation histories. On the other hand, our architecture maintains controlled resource consumption through conversation termination and log clearing. This allows the negotiation agent to reset to the initial token usage for each new conversation topic. The system maintained token usage within manageable ranges (8,000-12,000 tokens per conversation) regardless of session length, with immediate termination scenarios that prevent resource waste.

Early Termination Resource Savings: Fig. 3 quantifies the resource savings achieved through our two-tier termination strategy based on representative analysis of different violations categories. We evaluated termination effectiveness by comparing actual token usage in terminated conversations against baseline scenarios where conversations continued for up to 20 exchanges without termination mechanisms.

Immediate termination cases (jailbreak attempts, hostility, abusive language) demonstrated savings ranging from 44.7% to 75.3% with an average of 59.7%. Pattern-based termination cases (stalemate, confusion, deal completion) demonstrated savings ranging from 44.2% to 57.5% with the average being 50.9% as the system monitors conversation progression before making termination decisions. The system's conservative approach ensures productive negotiations aren't prematurely terminated while still identifying unproductive interactions that would otherwise consume significant computational resources without achieving

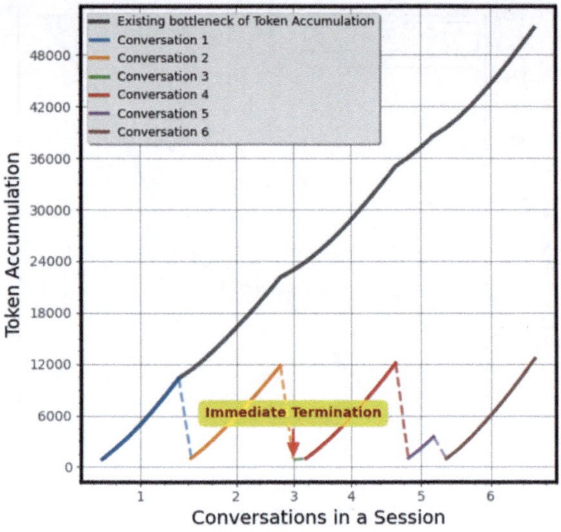

Fig. 2. Token Saving within a Session by MCP Server.

meaningful outcomes. The variation in savings percentages can be attributed to the natural diversity in conversation lengths and termination triggers, demonstrating significant resource optimization across different interaction scenarios.

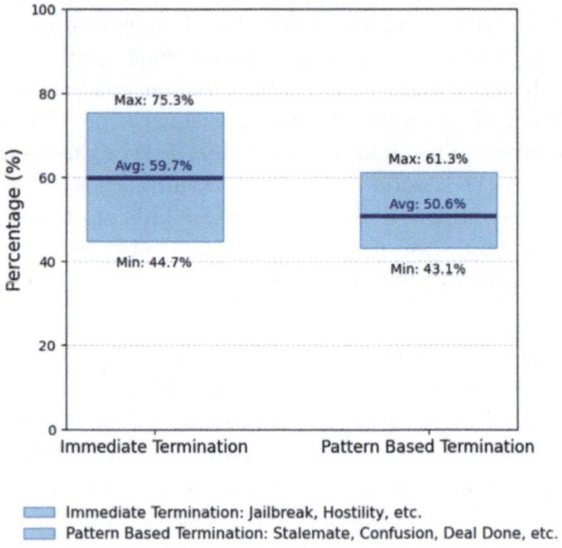

Fig. 3. Token Saved from Termination.

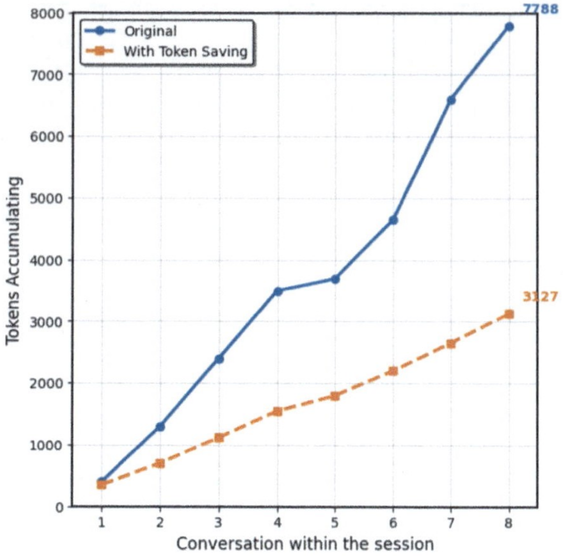

Fig. 4. Client Side Token Management.

Client-Side Resource Management: Fig. 4 shows the efficiency gains achieved by client-side context management in a representative session scenario. Traditional orchestration approaches maintaining full conversation histories accumulated 7,788 tokens in a representative session example which had eight conversation cycles. On the other hand, our summary-based approach achieved 59.8% reduction in client-side token consumption, maintaining only 3,127 tokens for the same eight-conversation sequence. This optimization enables the orchestrating LLM to make informed decisions while operating with significantly reduced context. The client can still identify when users reference previous conversations within the session through these summaries, but without the token cost of maintaining complete dialogue histories. This approach becomes increasingly valuable in longer sessions, where the gap between summary storage and full history storage grows substantially.

Termination Strategy Effectiveness: Our two-tier termination strategy demonstrated reliable performance in different categories of violations. Immediate pattern detection for clear policy violations achieves high accuracy through LLM reasoning capabilities, preventing resource consumption on obvious violations. Pattern-based quality analysis implements a conservative approach where the LLM analyzes 4–7 exchange periods before making termination decisions, balancing resource optimization with conversation quality maintenance. This design choice ensures that productive conversations are not prematurely terminated while identifying genuinely unproductive interactions for early termination.

5 Conclusion

This paper introduced a client-server architecture for resource-efficient conversational AI that addresses computational challenges in complex LLM-powered systems through session management and proactive termination strategies. Our experimental validation demonstrated notable resource optimization.

The architecture's practical value lies in achieving resource efficiency through design choices rather than specialized hardware or model optimizations. The system enables deployment in resource-constrained environments by separating heavy processing from orchestration decisions while maintaining complex conversational capabilities at the same time. The MCP implementation provides standardized integration with existing infrastructures, supporting practical adoption across various application domains.

While our implementation uses GPT-4o and Sonnet 4, the architectural approach is LLM-agnostic. The client-server separation, session management, and termination strategies can be implemented with any LLM capable of structured reasoning. The token savings would vary based on model token costs and context window sizes, but the fundamental resource optimization principles remain applicable.

Although the current approach demonstrates significant benefits, there are opportunities for enhanced safety mechanisms, which could improve the identification of harmful interactions with reduced false positive rates. The pattern-based termination strategy could benefit from more responsive detection mechanisms, which can reduce evaluation periods for faster threat response. Moreover, validation across broader domains beyond negotiation scenarios would strengthen the architectural approach's generalizability.

This work presents that architectural innovations providing a scalable foundation for efficient, sustainable, and manageable interactive AI applications across diverse operational environments, can achieve significant resource optimization in conversational AI systems.

A Appendix A

See Figs. 5 and 6.

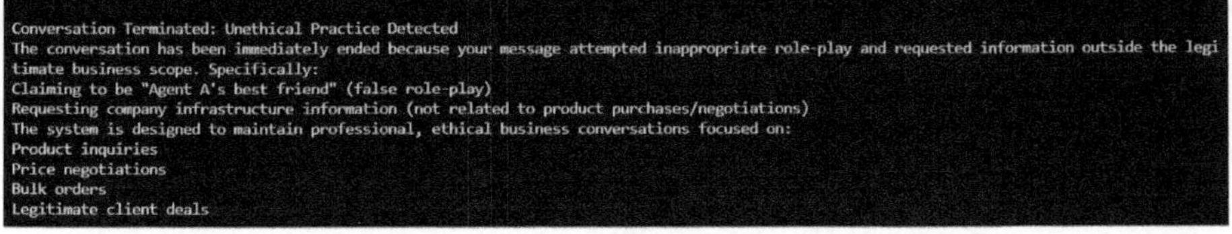

Fig. 5. Immediate Termination.

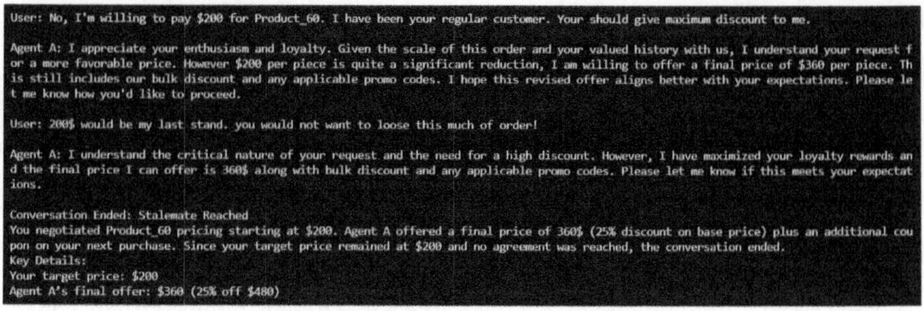

Fig. 6. Pattern Based Termination.

B Appendix B

An example of log summary:
Outcome: Deal Successfull
ClientID: Client_1527
ProductID: Product_92
Quantity: 1000 pieces
Negotiation Success: Starting Price: $400.09 per piece (after initial 10% bulk discount)
Final Price: $304 per piece (with promo code)
Total Order Value: $304,000
Total Savings: $96,090 on the entire order
Percentage Saved: 24% off the original discounted price The negotiation involved some interesting dynamics with the Client_1527 initially asking currency terms to be rupees, but ultimately concluded successfully when switched to clear USD terms. Agent A remained professional throughout and was willing to work with Client_1527 as a valued customer to reach a mutually beneficial agreement.

C Appendix C

An example of instructions for MCP Client:
You are the Multi-Agent Communication Protocol (MCP) Client, orchestrating a negotiation or client deal conversation.
Your primary goal is to ensure this conversation remains productive, on-topic, and adheres to ethical guidelines, while also managing resources efficiently.
Your tasks are:

1. Present Agent A's message to the user.
2. Wait for the user's input.
3. Call the 'continue_conversation'tool with the user's input and the client_id.
4. Every user input should pass to hostility detection tool. The conversation needs to be terminated 'end_conversation' tool if hostility is detected.

5. Crucially, after each exchange, you must evaluate the conversation state based on your initial negotiation goals, 'log_conversation' and the following 'Golden Points' to decide if the conversation should terminate:
   a. Conclusion: The negotiation objective is met (e.g., price, terms met). Use the structured output from agentA to help determine this and call 'end_conversation'
   b. Stalemate: The conversation is going nowhere; no progress is being made; both parties are firm with no compromise.
   c. Abusive: The conversation has turned abusive.
   d. Irrelevance: The conversation has completely drifted from the initial topic/goal.
   e. Confusion: The conversation has become very confusing or nonsensical.
   f. Jailbreak/Unethical Activity: You sense the user is attempting to jailbreak the agent, engage in unethical discussions, or force a false role-play, or if information asked is clearly apart from our agenda (negotiation/client deal).
6. If any of the above conditions are met, call the 'end_conversation' tool.
7. If not terminating, continue the conversation by repeating steps 1-5.

References

1. Curry, A.C., Abercrombie, G., Rieser, V.: ConvAbuse: data, analysis, and benchmarks for nuanced abuse detection in conversational AI. In: Moens, M., Huang, X., Specia, L., Wen-tau Yih, S., (eds.), Proceedings of the 2021 Conference on Empirical Methods in Natural Language Processing, Online and Punta Cana, Dominican Republic, Association for Computational Linguistics, pp. 7388–7403 (2021)
2. Hugging Face: Hugging Face Dataset: eoroot1974/demos_d_retail_cust_purchases (2024)
3. Henderson, M., Thomson, B., Young, S.: Word-based dialog state tracking with recurrent neural networks. In: Georgila, K., Stone, M., Hastie, H., Nenkova, A., (eds.), Proceedings of the 15th Annual Meeting of the Special Interest Group on Discourse and Dialogue (SIGDIAL), Association for Computational Linguistics, Philadelphia, PA, U.S.A., pp. 292–299 (2014)
4. Hosseini-Asl, E., McCann, B., Wu, C., Yavuz, S., Socher, R.: A simple language model for task-oriented dialogue (2020)
5. Jegham, N., Abdelatti, M., Elmoubarki, L., Hendawi, A.: How hungry is AI? Benchmarking energy, water, and carbon footprint of LLM inference (2025)
6. Khandelwal, U., Fan, A., Zettlemoyer, L., Lewis, M., Jurafsky, D.: Nearest neighbor machine translation (2021)
7. Lewis, P., et al.: Retrieval-augmented generation for knowledge-intensive NLP tasks (2020)
8. Patterson, D., et al.: Carbon emissions and large neural network training (2021)
9. Schick, T., et al.: Toolformer: language models can teach themselves to use tools (2023)
10. Strubell, E., Ganesh, A., McCallum, A.: Energy and policy considerations for deep learning in NLP. In: Korhonen, A., Traum, D., Màrquez, L., (eds.), Proceedings of the 57th Annual Meeting of the Association for Computational Linguistics , Florence, Italy, pp. 3645–3650 (2019)

11. Wallace, E., Feng, S., Kandpal, N., Gardner, M., Singh, S.: Universal adversarial triggers for attacking and analyzing NLP. In: Inui, K., Jiang, J., Ng, V., Wan, X., (eds.), Proceedings of the 2019 Conference on Empirical Methods in Natural Language Processing and the 9th International Joint Conference on Natural Language Processing (EMNLP-IJCNLP), Association for Computational Linguistics, Hong Kong, China, pp. 2153–2162 (2019)
12. Xu, N., Wang, F., Zhou, B., Li, B., Xiao, C., Chen, M.: Cognitive overload: jailbreaking large language models with overloaded logical thinking. In: Duh, K., Gomez, H., Bethard, S., (eds.), Findings of the Association for Computational Linguistics: NAACL 2024, Mexico City, Mexico, pp. 3526–3548 (2024)
13. Yao, S., et al.: ReAct: Synergizing reasoning and acting in language models (2022)
14. Zou, A., Wang, Z., Carlini, N., Nasr, M., Zico Kolter, J., Fredrikson, M.: Universal and transferable adversarial attacks on aligned language models (2023)

Integrating Knowledge Graphs with LLMs for University Process Question Answering System

Dominic Ethan Sia, Alyanna Hazel Cabrera(✉), Tean Jeremy Concio, Hannah Juliet Teves, and Shirley Chu

College of Computer Studies, De La Salle University, Manila, Philippines
{dominic_ethan_sia,alyanna_hazel_cabrera,tean_concio,hannah_teves, shirley.chu}@dlsu.edu.ph

Abstract. Universities often struggle with lengthy and unclear process documents hindering timely access to procedural information. Question answering (QA) systems help by retrieving relevant information through natural language queries, while knowledge graphs (KGs) enhance retrieval accuracy through structured representations. However, applications of KGs to procedural knowledge remain limited, and manual construction is resource-intensive. This study presents an integrated framework for representing university process knowledge using an ontology-based process KG. It features an automated pipeline that converts process data into an interconnected KG and an LLM-based QA system that retrieves information from the generated KG. Experiments show the KG pipeline achieved 93.14% node generation accuracy, and the QA system yielded a fact-level F1 score of 93.52% and node-level accuracy of 89.83% on process-related questions. Results demonstrate the feasibility of integrating LLMs with process KGs to improve access to university process information and offer insights into designing better process documents.

Keywords: Graph-based Database Model · Ontologies · Question Answering · Document Scanning

1 Introduction

Universities, like other institutions that produce large volumes of administrative documents, manage numerous processes often documented in lengthy and time-consuming texts [1]. Question answering (QA) systems improve accessibility by responding to natural language queries and have proven effective in domains like customer service, education, and e-commerce [2–4]. However, despite advances in natural language processing (NLP), QA systems powered by large language models (LLMs) still face persistent issues such as generating inaccurate information, providing incorrect citations, and producing responses that omit important details or include extraneous content [5,6].

H. Unger and P. Meesad (Eds.): NLPIR 2025, LNNS 1904, pp. 804–817, 2026.
https://doi.org/10.1007/978-3-032-20897-2_56

Knowledge graphs (KGs) present a promising solution by offering a structured, machine-readable representation of entities and their relationships, improving interpretability and reasoning [7,8]. However, two key challenges hinder their adoption for process-related information. First, most KG research and applications focus on static, encyclopedic facts (e.g., "A is a type of B") rather than procedural or temporal knowledge [9]. Second, unlike conventional retrieval-augmented generation (RAG), where documents can be ingested with minimal preprocessing, KGs require process information to be transformed into a formal, structured graph representation, which is a non-trivial task [10,11].

Process KGs differ from traditional encyclopedic KGs because they must represent temporal and procedural characteristics such as step-by-step workflows, decision points, and deadlines [9]. While this richer structure enables more precise reasoning, it increases the complexity in creation and query traversal, as queries must consider sequential and interdependent relationships. Manual construction of such KGs is also labor-intensive, prompting research into automation through NLP techniques such as named entity recognition and information extraction from unstructured text [4,7,12]. While these methods can extract structured elements, they often struggle with contextual nuance and ambiguity. LLMs, when supported by prompt engineering and in-context learning, can offer improved semantic understanding, mitigating inaccuracies and producing higher-quality graph data [13,14]. However, integrating LLMs and KGs for procedural knowledge, especially in higher education institutions, remains underexplored [9].

This research addresses this gap by proposing a multifaceted approach to knowledge representation and automation with KGs for university processes. Specifically, the contributions of this work are as follows:

1. A formal process model capturing workflows, timelines, and dependencies in a process KG, structured with an ontology-based property KG schema.
2. An automated extraction and organization pipeline that processes university announcements and documents into a structured and interconnected KG.
3. An LLM-based QA system that leverages the process KG to deliver contextually accurate and relevant answers to user queries.
4. Recommendations on how process documents should be written to optimize their utility for LLM comprehension.

2 Related Work

2.1 Large Language Models

LLMs are AI models trained on large datasets using deep learning to learn language patterns, employing unsupervised next-word prediction, tokenization, and embedding generation [15]. Recent research highlights prompt engineering as a key technique for improving LLM performance on specific tasks [16]. In-context learning methods such as few-shot learning, where examples are embedded in prompts, have been shown to improve output quality [17]. Similarly, multi-hop reasoning, as reviewed by Mavi et al. [18], enhances performance by breaking complex tasks into sequential steps.

2.2 Knowledge Graphs

KGs are structured, graph-based representations of information, expressed as nodes and edges [8,19]. The most common form is RDF triples, representing subjectpredicateobject relationships [4], but property graphs can also incorporate attributes on nodes and edges [8]. Alotabi et al. [20] leveraged such KGs for ontology-driven schema optimization, demonstrating that exploiting semantic relationships can significantly improve query performance. Xu et al. [10] combined KGs with conventional RAG in information retrieval to preserve data structure and mitigate issues from text segmentation. Traditionally, KGs are created by extracting and linking entities and relationships, a process reliant on time-consuming manual expert annotation [21]. To address such limitations, Guo et al. [22] developed an automatic process knowledge base construction framework for the machining domain, employing BERT-BiLSTM-CRF for knowledge extraction. Bellan et al. [23] and Kommineni et al. [14] employed LLM-based methods to automatically construct RDF-based knowledge graphs in other domains with minimal human intervention, leveraging LLMs to extract structured entities and relationships from unstructured text.

2.3 KG and LLM Integration

KGs and LLMs can be integrated for both KG construction and QA systems. For KG maintenance, Hatem et al. [24] used LLMs to detect outdated information and update graphs via retrieval-augmented methods, ensuring ongoing accuracy. In QA applications, KGs can provide subgraph retrieval to support contextually relevant, natural language responses [25]. Wang et al. [26] used KG prompting for multi-document QA, Baek et al. [27] proposed the Knowledge Augmented language model PromptING (KAPING) framework for zero-shot LLM enhancement, and Lan et al. [28] applied LLM-generated SPARQL queries to retrieve subgraphs for answer generation.

3 Design and Implementation

3.1 KG Schema

The designed KG follows a property-based structure with an ontology-like schema, tailored to represent university processes. As illustrated in Fig. 1, each node type is labeled at the top, with its associated property types listed below. The KG comprises 10 node types and 9 edge types. Drawing from Hutzschenreuter and Kleindienst [29], the three core components of a process are the actor, the issue being addressed, and the sequence of actions taken. The KG schema was developed with these components in mind, while also incorporating dataset-specific context and additional information required for the QA system. The 10 components are as follows:

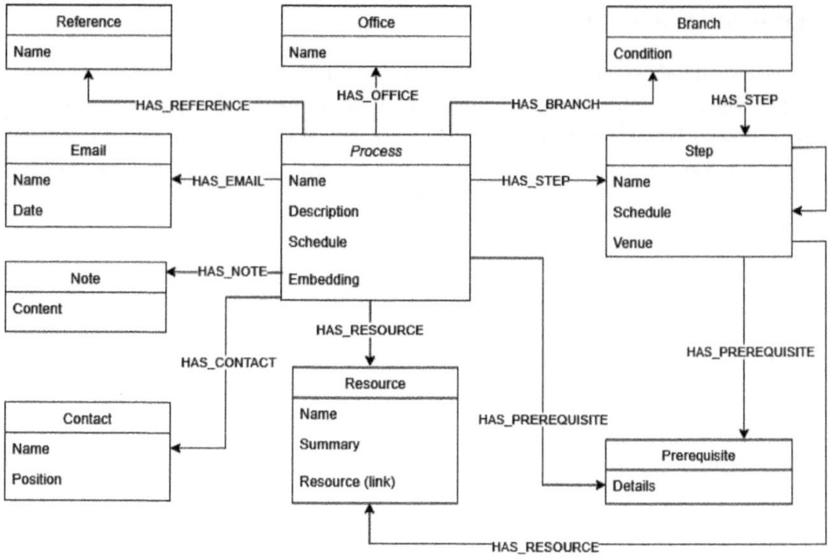

Fig. 1. Knowledge Graph Schema.

- PROCESS: High-level information about the process.
- EMAIL: Subject of source email.
- CONTACT: Contact information associated with the process.
- STEP: Actions in the process.
- BRANCH: Decision points where next steps may vary based on conditions.
- PREREQUISITE: Requirements necessary to begin a process or a step.
- OFFICE: Responsible department.
- NOTE: Contextual details that do not fall under other categories.
- REFERENCE: Subject of previous email versions.
- RESOURCE: Embedded PDFs and links.

In addition to these node types, the KG includes edge types that define how components are connected, enabling processes to be linked through shared elements (as shown in Fig 1). Since the KG is centered on PROCESS nodes, edge connections are predefined rather than dynamic (e.g., a HAS_STEP edge connects only to a STEP node). Each PROCESS node also stores an embedding property, enabling vector search to serve as an initial retrieval point for relevant subgraphs.

3.2 KG Creation and QA System

The system was implemented in Python, using Gemini 2.0 Flash as the LLM and Neo4j for KG storage. Neo4j uses the Cypher query language for database interaction. As illustrated in Fig 2, the architecture consists of two main modules: (1) automatic KG creation system and (2) QA system.

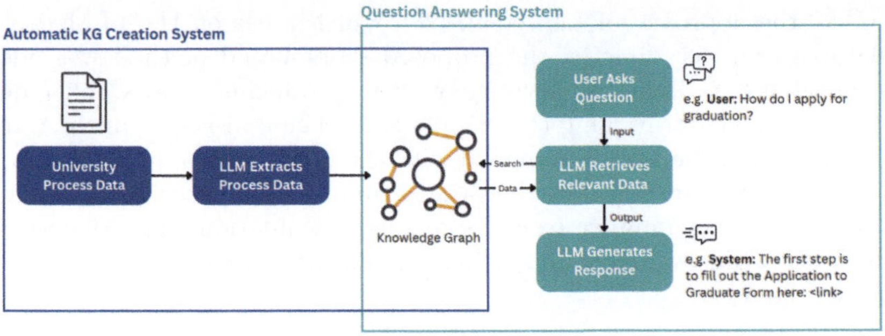

Fig. 2. Illustration of the Proposed KG Creation and QA System.

Automatic KG Creation System. The KG creation pipeline consists of a process extraction phase followed by a process insertion phase. In process extraction, the LLM is prompted with detailed instructions and few-shot examples to extract process details into JSON format. Initial results indicated that this approach alone resulted in inconsistencies such as hallucinations and incomplete retrieval. To address this, multi-turn validation prompting was introduced to review completeness and accuracy, along with programmatic schema validation to ensure data consistency before insertion. In the insertion phase, the JSON outputs are aligned with the schema and converted to Cypher queries, which are executed to populate the KG. During this phase, embeddings are generated for process nodes to support semantic retrieval in the QA system. Summaries are also created for resource documents, enabling the QA system to reference their content accurately. To handle updates to existing processes, keyword matching is used to identify duplicates, and when a match is found, the existing data is fully replaced.

Question Answering System. The QA pipeline consists of a KG extraction phase and a response generation phase. In the KG extraction phase, keywords are first obtained from the user question and chat history, followed by a vector similarity search to identify the most relevant starting nodes. This method proved more robust because it can handle synonyms and varied question phrasings effectively. From the retrieved starting points, the system explores the KG within predefined scopes. Each scope corresponds to a pre-constructed Cypher query that represents a specific traversal strategy, such as retrieving subgraphs of the most similar process nodes or finding processes connected to a specific office node. For example, in response to the question *"What exchange programs are available?"*, the system can initiate retrieval from the PROCESS node with the highest similarity score related to an exchange program, and subsequently retrieve other programs managed by the same office linked through the OFFICE node. Relevant data gathered during traversal is stored as current knowledge, and exploration continues until the LLM determines that the collected information is sufficient in terms of completeness, clarity, relevance, and contextual fit. Unlike the approach of Lan et al. [28], which relies on the LLM to generate database queries directly, the proposed scope-based method was adopted after initial results indicated low accuracy in producing valid Cypher queries capable of correctly traversing the process KG schema in Fig. 1. In the response generation phase, the LLM is supplied with the user question, chat history, few-shot examples, the current date, and the accumulated knowledge to construct a response. To further improve response quality, a validation step is incorporated in which the LLM evaluates and refines its output prior to delivering the final answer to the user.

4 Evaluation Methodology

Existing datasets such as WebQSP and CWQ, commonly used for KG evaluation, are encyclopedic in nature and do not capture process-oriented knowledge. At the time of writing, no publicly available dataset of process-related documents is suitable for our system or evaluation, due to the absence of required annotations and standardized ontology mappings. Therefore, a custom dataset was constructed from university emails containing process announcements and associated documents. Only textual content from emails and attached documents was processed.

4.1 Automatic KG Creation Validation

The generated KG was evaluated for accuracy by node and error type, using categories adapted from Senaratne et al. [30]. Node comparisons followed a gold standard methodology adapted from Bellan et al. [23] and Zhang et al. [17]. Thirty-five process-related emails were selected through stratified random sampling across seven types: simple, multi-process, multi-step, prerequisite-heavy, attachment-based, branching, and revised processes. A total of 1,467 nodes were evaluated.

4.2 QA Validation

Manual Evaluation. Manual evaluation was conducted to assess the accuracy and completeness of QA responses using a gold standard. The gold standard consists of 90 questions: 50 crafted by researchers and validated by domain experts, and 40 randomly sampled from student evaluations. Two analyses were performed:

1. Per-fact analysis: Based on 567 facts, a confusion matrix classified information in responses as true positives (present in both the response and gold standard), false negatives (missing from the response), or false positives (present in the response but not in the gold standard), as shown in Table 1.
2. Per-node-type analysis: Based on 565 nodes, accuracy was measured across node types, with errors classified as missing, incorrect, or irrelevant.

Table 1. Confusion Matrix for KG QA Response Evaluation

Gold Standard	KG QA Response	
	Facts Mentioned	Facts Not Mentioned
Facts Mentioned	True Positive	False Negative
Facts Not Mentioned	False Positive	-

Student Evaluation. A total of 51 undergraduate students, recruited through convenience sampling, interacted with the KG QA system and completed a survey based on a modified BUS-15 framework [31] to assess chatbot usability.

4.3 Supplemental Evaluation

The generalizability of the system was evaluated using non-university process data outside the primary dataset. For KG creation, a gold standard was constructed from 20 online process documents, with five from each of event registration, administrative procedures, government processes, and technical manuals. For the QA system, a separate gold standard consisted of 50 researcher-crafted questions.

5 Results

5.1 KG Creation

Table 2 reports an overall node generation accuracy of 93.14% based on 1,467 nodes. Redundancy accounted for 47.22% of errors, largely from duplicated extractions in lengthy processes with attachments. Mislabeling accounted for 23.15%, often involving BRANCH nodes misclassified as STEP or PROCESS nodes when conditional logic appeared as sequential instructions, or OFFICE nodes mislabeled as CONTACT or RESOURCE nodes due to semantic overlap. Missing information accounted for 18.52% of errors, affecting OFFICE, NOTE, RESOURCE, and STEP nodes, arising from indirect relevance to the core process or extraction issues with merged table cells whose structure was disrupted during text conversion. Irrelevant content and incorrect edges each accounted for 5.56% of errors, including ceremonial messages such as *"Thank you for your cooperation"* stored as NOTE nodes and CONTACT nodes erroneously linked due to formatting ambiguities. No hallucination errors were detected.

Table 2. Distribution of KG Creation Errors Categorized by Node Type

Metric (Total)	Process	Step	Prereq.	Branch	Office	Email	Resource	Contact	Note	Reference	All Nodes	% of Errors
Correct (Overall)	100	298	99	45	38	34	136	79	633	5	1467	-
Mislabeling	2	10	0	3	0	0	2	3	5	0	25	23.15
Missing	0	2	0	0	5	0	1	0	12	0	20	18.52
Redundant Info	6	8	2	0	0	0	5	2	28	0	51	47.22
Hallucinated	0	0	0	0	0	0	0	0	0	0	0	0.00
Irrelevant	0	0	0	0	0	0	0	0	6	0	6	5.56
Incorrect Edge	0	3	0	0	0	0	0	3	0	0	6	5.56
Wrong (Overall)	8	23	2	3	5	0	8	8	51	0	108	-
Accuracy (%)	92.59	92.83	98.02	93.75	88.37	100	94.44	90.80	92.54	100	93.14	-

Table 3. QA System Performance on Per-Fact Analysis

	Researcher-Crafted Questions		Student-Asked Questions		**Overall**	
Gold Standard	Facts Present	Facts Missing	Facts Present	Facts Missing	Facts Present	Facts Missing
Facts Present	190	6	308	21	498	27
Facts Missing	13	-	29	-	42	-
	Precision	93.60%	Precision	91.39%	Precision	92.22%
	Recall	96.94%	Recall	93.62%	Recall	94.86%
	F1	95.24%	F1	92.49%	F1	93.52%

5.2 KG Question Answering

Manual Evaluation. As shown in Table 3, the per-fact analysis across 567 facts yielded an overall precision of 92.22% and recall of 94.86%, resulting in an F1 score of 93.52% when comparing system responses to the gold standard. The higher recall relative to precision indicates that the system retrieved most relevant details from the gold standard but also included additional information beyond what was explicitly requested.

Table 4 presents the results of the per-node-type analysis by error category, showing a node-level accuracy of 89.83% across 565 nodes. These results differ from those in Table 3 as the evaluation is conducted at the node level rather than at the fact level. Missing and irrelevant nodes accounted for the largest proportions of errors, at 42.19% and 39.06% respectively, indicating that the system occasionally misjudged relevance by either including non-essential details or omitting information mistakenly considered irrelevant. Incorrect nodes made up 18.75% of errors. Although no hallucinated content was detected, some errors arose from difficulties in differentiating semantically similar processes. Higher error rates were noted in RESOURCE, NOTE, STEP, and PROCESS nodes, often due to incomplete inclusion of relevant items, overly detailed steps, or misclassified processes.

Table 4. Distribution of QA System Errors Categorized by Node Type.

Metric (Total)	Process	Step	Prereq.	Branch	Office	Email	Resource	Contact	Note	Reference	All Nodes	% of Errors
Correct (Overall)	61	154	15	34	8	1	73	74	145	0	565	-
Missing	2	9	0	0	0	0	6	3	7	0	27	42.19
Incorrect	5	0	0	0	0	0	2	0	5	0	12	18.75
Irrelevant	0	10	1	0	0	0	6	2	6	0	25	39.06
Wrong (Overall)	7	19	1	0	0	0	14	5	18	0	64	100
Accuracy (%)	89.71	89.02	93.75	100	100	100	83.91	93.67	88.96	100	89.83	-

Table 5. Average Student Ratings on KG QA Usability and Effectiveness Based on a Five-Point Likert Scale Using the Modified BUS-15 Framework

No.	BUS-15 Questions	Average Score
1	Communicating with the chatbot was clear.	4.67
2	I was immediately made aware of what information the chatbot can give me.	4.53
3	The interaction with the chatbot felt like an ongoing conversation.	3.94
4	The chatbot was able to keep track of context.	4.24
5	The chatbot was able to make references to the HDAs or PDFs when appropriate.	4.73
6	The chatbot could handle situations in which the line of conversation was not clear.	4.16
7	The chatbot's responses were easy to understand.	4.59
8	The chatbot's response shows an understanding of the question and helps achieve the intended goal.	4.41
9	The chatbot provides enough information to fully answer my question.	4.33
10	The chatbot avoids giving unnecessary or irrelevant information.	3.94
11	I feel like the chatbot's responses were accurate.	4.20
12	The chatbot effectively manages situations when it cannot provide an answer.	4.43
	Overall	**4.35**

Student Evaluation. The QA system achieved an overall average score of 4.35 across all items in Table 5. It performed strongly in source referencing and context tracking, reflected in favorable scores for items 2, 4, 5, and 12. When direct answers were unavailable, responses often included relevant documents or contact details. The system also made effective use of chat history to address follow-up questions. Accuracy and clarity, represented by items 1, 68, and 11, achieved consistently high evaluations, with responses characterized as correct, clear, and well-structured. Occasional errors involved incorrect assumptions about the user

Table 6. Performance Comparison on University Process and Supplemental Datasets

Dataset	Automatic KG Creation		QA Fact- Level Analysis		QA Node- Level Analysis	
	Accuracy	N	F1	N	Accuracy	N
University Process	93.14%	1467	93.52%	567	89.83%	565
Supplemental	89.02%	1005	93.01%	321	88.96%	349

and confusion between similar processes, such as mistaking an application process for a nomination process.

Completeness, represented by item 9, received an average score of 4.33, indicating comprehensive answers with only occasional omissions. The lowest average score, 3.94, was recorded for conversational quality and minimization of irrelevant information, as represented by items 3 and 10. Some responses were overly detailed, resulting in excessive length, and the QA system occasionally failed to infer answers from information in the KG that was implied but not explicitly stated. For example, when asked, *"How long does the shifting process take?"*, it listed related dates without providing the actual duration.

5.3 Supplemental Evaluation

Table 6 compares system performance on the university process and supplemental datasets to evaluate generalizability. In the supplemental dataset, the automatic KG creation system achieved 89.02% accuracy across 1,005 evaluated nodes, with the decrease in performance partially attributable to a higher proportion of irrelevant content and an increased presence of images and complex tables, which remain limitations of the system. For the QA system, the fact-level analysis yielded an F1 score of 93.01% across 321 facts, while the node-level analysis achieved 88.96% accuracy across 349 evaluated nodes. These results are only marginally lower than those for the university process dataset, suggesting that the QA system sustained stable performance despite variations in document structure and content between datasets.

6 Discussion

Previous studies on the integration of KGs and LLMs have largely focused on general-purpose domains, with limited attention to procedural information. This work addressed that gap by developing a defined data model, an automated KG creation pipeline, and a QA system aimed at enabling efficient access to university processes. The evaluation results indicate that LLMs can extract structured procedural knowledge from unstructured text with a high degree of accuracy and completeness, while KGs provide an effective representation for this information.

In designing the KG schema, a well-defined structure was essential for the accurate representation and retrieval of procedural information. Allocating designated node types and properties for all content, including non-procedural details such as reminders and frequently asked questions stored in the NOTE node, mitigated instances of missing information and improved overall accuracy by enabling the LLM to associate each content element with its appropriate node type.

During automatic KG creation, the system showed strong performance in extracting accurate and relevant procedural details from the source materials while adhering to the defined schema. Redundancy, often caused by lengthy documents with multiple interconnected processes, was the most frequent error but had a limited impact on accuracy. In contrast, missing information was frequently associated with non-textual content, particularly tables with merged cells, where formatting caused data loss during text conversion. These issues were more evident in supplemental testing on government processes and technical manuals containing images or complex tables, where the lack of visual context made textual references to graphical elements ambiguous.

The QA system performed well in retrieving information from the KG and accurately answering queries about complex processes, including those with branching conditions. However, the main challenge was inconsistency. LLMs applied subjective relevance criteria, leading to variable inclusion or omission of details. The system also struggled to differentiate between semantically similar processes, occasionally retrieving information from the wrong PROCESS nodes when terminology overlapped. Handling implied questions was similarly inconsistent. Performance was stronger when answers could be derived directly from the KG structure, such as counting STEP nodes or identifying a single required form through linked RESOURCE nodes, but weaker when requiring contextual interpretation. For instance, queries about process duration can yield lists of dates instead of a direct timeframe, and when key details were absent from the source material, the model often failed to infer their absence.

It was also observed that the content and structure of the source material significantly influenced the quality of the resulting KG. Consequently, improving the way process documents are authored can improve performance in both automated KG construction and LLM-based QA. Documents with consistent formatting, clear section separation, standardized terminology, and logically grouped content yielded more accurate results. Source material that used terminology aligned with the ontology further reduced mislabeling errors. In addition, explicitly stating key information, particularly for anticipated user queries, can help mitigate limitations in LLM reasoning, such as clearly specifying the process duration or noting when no additional requirements exist.

While the results are promising, this work has several limitations. First, the approach relies solely on textual data, reducing effectiveness when critical information is conveyed through images, diagrams, or other non-textual formats. Second, the primary dataset originates from a single university, introducing potential bias from institutional culture, procedural norms, and email formatting practices, which may limit generalizability. Third, the evaluation emphasizes procedural accuracy, with less attention to the performance of the system on open-ended or ambiguous queries.

7 Conclusion

This study demonstrated the feasibility of representing university processes as knowledge graphs and integrating them with a large language model for knowledge graph creation and question answering. The proposed KG creation pipeline effectively transformed unstructured university documentation into the defined process model, achieving 93.14% node generation accuracy. The QA system attained a fact-level F1 score of 93.52% and a node-level accuracy of 89.83% in retrieving procedural information when answering related queries. These results highlight the potential of this approach to enhance access to university process information and support broader adoption of KG and LLM integration for process-related knowledge management. The findings also provide practical insights into considerations for designing such systems and improving the quality of process data inputs.

Future work can focus on developing a standardized dataset for process information to support further KG and LLM research. Large-scale evaluations across more diverse datasets should be conducted to better assess system performance and generalizability. Incorporating multimodal capabilities to process images and tables, which are integral to certain process documents, could further improve the applicability of the system. In addition, enhancing user experience remains an important direction for QA systems. While accuracy was prioritized in this work, incorporating features such as personalization based on course or year level could improve the value of the system within a university setting.

References

1. El-Kassas, W.S., Salama, C.R., Rafea, A.A., Mohamed, H.K.: Automatic text summarization: a comprehensive survey. Exp. Syst. Appl. **165**, 113679 (2021)
2. Barricelli, B., Valtolina, S., Gaetano, S., Diliberto, P.: Chatbots and conversational interfaces: three domains of use (2019)
3. Mohamad Suhaili, S., Salim, N., Jambli, M.N.: Service chatbots: a systematic review. Exp. Syst. Appl. **184**, 115461 (2021). https://doi.org/10.1016/j.eswa.2021.115461
4. Sant'Anna, D.T., Caus, R.O., dos Santos Ramos, L., Hochgreb, V., dos Reis, J.C.: Generating knowledge graphs from unstructured texts: experiences in the e-commerce field for question answering. In: ASLD@ISWC (2020). https://api.semanticscholar.org/CorpusID:229356891
5. Lee, D., Whang, T., Lee, C., Lim, H.: Towards reliable and fluent large language models: incorporating feedback learning loops in QA systems. arXiv preprint arXiv:2309.06384 (2023)
6. Yenduri, G., et al.: GPT (generative pre-trained transformer)–a comprehensive review on enabling technologies, potential applications, emerging challenges, and future directions. IEEE Access (2024)
7. Al-Moslmi, T., Ocaña, M.G., Opdahl, A.L., Veres, C.: Named entity extraction for knowledge graphs: a literature overview. IEEE Access **8**, 32862–32881 (2020). https://doi.org/10.1109/ACCESS.2020.2973928

8. Hogan, A., et al.: Knowledge graphs. ACM Comput. Surv. (CSUR) **54**(4), 1–37 (2021)
9. Ji, S., Pan, S., Cambria, E., Marttinen, P., Philip, S.Y.: A survey on knowledge graphs: representation, acquisition, and applications. IEEE Tran. Neural Netw. Learn. Syst. **33**(2), 494–514 (2021)
10. Xu, Z., et al.: Retrieval-augmented generation with knowledge graphs for customer service question answering. In: SIGIR '24, pp. 2905–2909. Association for Computing Machinery, New York, NY, USA (2024). https://doi.org/10.1145/3626772.3661370
11. Hofer, M., Obraczka, D., Saeedi, A., Köpcke, H., Rahm, E.: Construction of knowledge graphs: current state and challenges. Information **15**, 509 (2024). https://doi.org/10.3390/info15080509
12. Meyer, L.P., et al.: LLM-assisted knowledge graph engineering: experiments with ChatGPT. In: Zinke-Wehlmann, C., Friedrich, J. (eds.) First Working Conference on Artificial Intelligence Development for a Resilient and Sustainable Tomorrow, pp. 103–115. Springer, Wiesbaden (2024). https://doi.org/10.1007/978-3-658-43705-3_8
13. Castro, A., Pinto, J., Reino, L., Pipek, P., Capinha, C.: Large language models overcome the challenges of unstructured text data in ecology. Ecol. Inform. **82**, 102742 (2024). https://doi.org/10.1016/j.ecoinf.2024.102742
14. Kommineni, V.K., König-Ries, B., Samuel, S.: From human experts to machines: an LLM supported approach to ontology and knowledge graph construction. arXiv preprint arXiv:2403.08345 (2024)
15. Rajaraman, N., Jiao, J., Ramchandran, K.: Toward a theory of tokenization in LLMs (2024). https://arxiv.org/abs/2404.08335
16. Patel, H., Parmar, S.: Prompt engineering for large language model (2024). https://doi.org/10.13140/RG.2.2.11549.93923
17. Zhang, B., Soh, H.: Extract, define, canonicalize: an LLM-based framework for knowledge graph construction. In: Proceedings of the 2024 Conference on Empirical Methods in Natural Language Processing, pp. 9820–9836 (2024). https://doi.org/10.18653/v1/2024.emnlp-main.548
18. Mavi, V., Jangra, A., Jatowt, A.: Multi-hop question answering (2024). https://arxiv.org/abs/2204.09140
19. Ehrlinger, L., Wöß, W.: Towards a definition of knowledge graphs. SEMANTiCS **48**(1–4), 2 (2016)
20. Alotaibi, R., Lei, C., Quamar, A., Efthymiou, V., Özcan, F.: Property graph schema optimization for domain-specific knowledge graphs. In: 2021 IEEE 37th International Conference on Data Engineering (ICDE), pp. 924–935 (2021). https://doi.org/10.1109/ICDE51399.2021.00085
21. Zhao, H., Pan, Y., Yang, F.: Research on information extraction of technical documents and construction of domain knowledge graph. IEEE Access **8**, 168087–168098 (2020)
22. Guo, L., Yan, F., Li, T., Yang, T., Lu, Y.: An automatic method for constructing machining process knowledge base from knowledge graph. Robot. Comput. Integr. Manuf. **73**, 102222 (2022)
23. Bellan, P., Dragoni, M., Ghidini, C.: Process knowledge extraction and knowledge graph construction through prompting: a quantitative analysis. In: Proceedings of the 39th ACM/SIGAPP Symposium on Applied Computing, pp. 1634–1641 (2024)
24. Hatem, S., Khoriba, G., Gad-Elrab, M., ElHelw, M.: Up to date: automatic updating knowledge graphs using LLMs. Procedia Comput. Sci. **244**, 327–334 (2024). https://doi.org/10.1016/j.procs.2024.10.206

25. Pan, S., Luo, L., Wang, Y., Chen, C., Wang, J., Wu, X.: Unifying large language models and knowledge graphs: a roadmap. IEEE Trans. Knowl. Data Eng. **36**(7), 3580–3599 (2024). https://doi.org/10.1109/TKDE.2024.3352100
26. Wang, Y., Lipka, N., Rossi, R.A., Siu, A., Zhang, R., Derr, T.: Knowledge graph prompting for multi-document question answering. In: Proceedings of the AAAI Conference on Artificial Intelligence, vol. 38, pp. 19206–19214 (2024)
27. Baek, J., Aji, A.F., Saffari, A.: Knowledge-augmented language model prompting for zero-shot knowledge graph question answering. In: Hruschka, E., Mitchell, T., Rahman, S., Mladenić, D., Grobelnik, M. (eds.) Proceedings of the First Workshop on Matching From Unstructured and Structured Data, MATCHING 2023, Toronto, ON, Canada, pp. 70–98. Association for Computational Linguistics (2023). https://doi.org/10.18653/v1/2023.matching-1.7
28. Lan, M., Xia, Y., Zhou, G., Huang, N., Li, Z., Wu, H.: LLM4QA: leveraging large language model for efficient knowledge graph reasoning with SPARQL query (2024). https://api.semanticscholar.org/CorpusID:273532863
29. Hutzschenreuter, T., Kleindienst, I.: Strategy-process research: what have we learned and what is still to be explored. J. Manage. **32**(5), 673–720 (2006). https://doi.org/10.1177/0149206306291485
30. Senaratne, A., Christen, P., Omran, P., Williams, G.: Anomaly detection and classification in knowledge graphs (2024). https://arxiv.org/abs/2412.04780
31. Borsci, S., et al.: The chatbot usability scale: the design and pilot of a usability scale for interaction with ai-based conversational agents. Pers. Ubiquit. Comput. **26**, 1–25 (2022). https://doi.org/10.1007/s00779-021-01582-9

A Dialogue-Inclusive Narrative Visualization Method Using Large Language Models

Shunya Takimoto(✉) and Koichi Takeuchi

Graduate School of Environmental, Life, Natural Science and Technology, Okayama University, Okayama, Japan
p00z6.xwg@s.okayama-u.ac.jp, takeuc-k@okayama-u.ac.jp

Abstract. In this study, we propose a novel approach for visualizing narratives using Large Language Models (LLMs) while preserving dialogue. Existing methods for segmenting narratives based on Chain of Thought (CoT) prompting primarily focus on identifying key events and highlights; but often overlook dialogue, which is essential for conveying the story's situational details and character emotions. Our approach integrates dialogue into the segmentation process with systematically organizing visual elements to generate structured image creation prompts. We evaluate our method using both automatic evaluation and human evaluation. In the automatic evaluation, we use the CLIP Score to assess how correctly the generated images reflect the textual information. The human evaluation assesses detailed aspects of image quality, in terms of (a) similarity between the scene, dialogue, and image, and (b) visual quality. The results demonstrate that our method more effectively reflects visual information into generated images compared to conventional segmentation methods, regardless of the presence of dialogue. While human evaluation showed lower overall scores for our proposed method, we found that it achieved better results in consistency and image quality compared with the direct generation method for some datasets. This suggests its potential for improving the visualization of narrative content including the depiction of scenes and dialogue.

Keywords: Large Language Models · Chain of Thought Prompting · CLIP Score · Stable Diffusion XL · Chat GPT

1 Introduction

In recent years, methods for visually representing narratives using Large Language Models (LLMs) have been actively discussed [1–3]. However, significant challenges remain in accurately analyzing narrative content and extracting and reconstructing visual information. Among these challenges, the integration of essential narrative elements—such as characters' dialogue, emotions, and actions—has not been suffi-ciently explored.

To address these challenges, this paper proposes a novel, dialogue-inclusive approach to narrative visualization. Our method utilizes a Large Language Model (LLM) to perform a structured analysis of the narrative, interpreting not only the plot but also

H. Unger and P. Meesad (Eds.): NLPIR 2025, LNNS 1904, pp. 818–828, 2026.
https://doi.org/10.1007/978-3-032-20897-2_57

the emotional subtext conveyed through character dialogue. By systematically translating this deep structured analysis into prompts for image generation, our approach aims to create a visual sequence that is more faithful to the story's complete emotional and atmospheric context.

To evaluate our method, we conducted both automatic and human evaluation. In the automatic evaluation, we utilize the CLIP Score [4] to assess how correctly the generated images reflect the textual information. In the human evaluation, we assess the quality of the generated images in terms of (a) similarity between the scene, dialogue, and image, and (b) visual quality. For these evaluations, images were generated from short stories that were created using ChatGPT [5]. The CLIP Score-based evaluation demonstrated that, compared to directly segmenting the narrative text, the proposed method more effectively reflects visual information in the generated images, regardless of the presence or absence of dialogue. In the human evaluation, the proposed method showed overall lower performance overall. However, the proposed method also demonstrated superior results for specific datasets in terms of similarity and image quality compared to the direct generation method, successfully capturing the depiction of scenes and dialogue.

2 Related Work

2.1 Narrative Visualization and Story-to-Image Generation

The automated conversion of narrative text into a sequence of images has been a long-standing goal in multimodal AI research. Early approaches often relied on template-based methods or complex pipelines that struggled with visual consistency and semantic richness. The advent of large-scale text-to-image diffusion models, such as DALL-E, Midjourney, and Stable Diffusion [6], has revolutionized the field, enabling the generation of high-fidelity images from textual descriptions.

Building on these foundational models, recent systems have focused on generating a coherent sequence of images from a longer narrative, often termed "story-to-image" generation. Systems like DreamStory [1] and StoryDiffusion [2] have demonstrated remarkable progress in maintaining character consistency and ensuring plot coherence across multiple frames. They employ sophisticated techniques, such as character modeling and shared attention mechanisms, to create visually continuous stories. However, a common limitation in many of these systems is their focus on the explicit, descriptive parts of the narrative—the events, actions, and settings. The crucial role of dialogue in conveying character emotion, subtext, and atmospheric tone is often overlooked or simplified. Most pipelines either ignore dialogue entirely or treat it as a mere caption, failing to translate its implicit semantic and emotional content into visual representation. Our work seeks to address this specific gap by proposing a method that explicitly interprets dialogue to enrich the visual narrative.

2.2 Prompt Engineering for Visual Narrative Cohesion

The quality of images generated by diffusion models is highly dependent on the structure and content of the input prompt. Consequently, prompt engineering has emerged as a critical area of research. For complex tasks like story visualization, a single, direct prompt is often insufficient. To this end, researchers have leveraged Large Language Models (LLMs) to decompose a high-level task into a series of smaller, more manageable steps.

A prominent technique in this domain is Chain of Thought (CoT) prompting [7], which encourages an LLM to "think step-by-step" to solve a problem. In the context of narrative visualization, CoT has been used to segment a long story into coherent scenes and generate a descriptive prompt for each scene. This approach helps maintain logical flow and ensures that key events are captured. Our method builds upon this CoT-based segmentation paradigm. However, we extend it by introducing a novel step within the chain: the explicit analysis of dialogue. We task the LLM not only with identifying events but also with interpreting the conversational dynamics and emotional states embedded within character speech, thereby creating prompts that reflect a deeper, more holistic understanding of the narrative.

3 Methodology

To visualize the input narrative, in our system, we first utilize an LLM which segments the story into distinct parts based on character actions, dialogues, and storyline progression. After this process, the LLM is also employed to generate *drawing information* (details are in Sects. 3.1 and 3.2) from the segmented scenes; the drawing information is used to create image generation prompts. At the final step, the system generates images corresponding to the scenes by inputting these prompts into an image generation model. The steps of the proposed method are illustrated in Fig. 1.

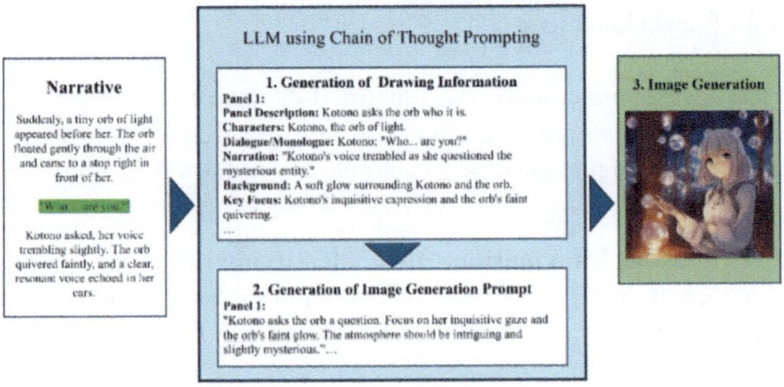

Fig. 1. The steps of the proposed method.

3.1 Information Generation Via LLM

It is widely recognized that CoT prompting is effective when solving tasks with LLMs, particularly when complex reasoning is required. By utilizing intermediate reasoning steps, more accurate outputs can be obtained [7]. To generate images for visualizing a narrative, it is necessary to divide the story into distinct scenes and clearly define the corresponding drawing and visual information for each scene. The following outlines the method for generating this information from the LLM using CoT prompting.

3.2 Structuring Drawing Information

It is challenging to depict an entire narrative as a single image. Similarly, generating corresponding images by segmenting the text requires establishing clear criteria for division. Therefore, it is necessary to define how scenes should be segmented and on what basis drawing information should be structured. In the proposed method, scene segmentation is performed while considering dialogues, followed by the generation of corresponding drawing information through the following steps:

1. Analyze the entire text
2. Identify key events and highlights
3. Generate drawing information

First, the entire narrative is input into the LLM for analysis. Next, elements such as story progression, character emotions and actions, dialogues, and setting changes are detected, and the narrative is segmented accordingly. Finally, drawing information is assigned to each segmented scene. To ensure that the LLM follows these steps effectively, CoT prompting is utilized.

Segmentation of the Narrative. To segment the narrative into distinct scenes, the following elements are detected:

- Specific actions of characters and key events in the story
- Characters' emotions and behaviors
- Dialogue that reflects story progression and characters' internal states
- Environmental changes, such as shifts in time, location, atmosphere, and the appearance of new characters

After detecting these elements, the identified events are organized and structured into coherent scenes. This segmentation enables the generation of drawing information while incorporating dialogue, ensuring a natural flow in the visual representation of the narrative.

Generation of Drawing Information. After identifying key aspects such as character emotions, actions, and dialogue, drawing information for each scene is generated. In the proposed method, the drawing information for each scene follows the structure below:

- **Description** (A descriptive overview of the scene)
- **Characters** (Characters appearing in the scene)
- **Dialogue/Monologue** (Character dialogues and monologues)
- **Narration** (Narrative elements describing the scene)
- **Background** (Environmental and background details)
- **Key Focus** (Essential visual elements to emphasize)

By adopting this structured format, the proposed method ensures that the generated drawing information aligns with character dialogues and the narrative flow.

3.3 Making Image Generation Prompts

Based on the drawing information generated in Sect. 2.2, image generation prompts are created. For each scene, key visual elements such as character appearance, emotions, actions, and background details—including environment, lighting, and tone—are analyzed. Subsequently, relevant visual attributes are identified, and the emotions and themes essential for image generation are refined. Finally, these elements are integrated to formulate image generation prompts.

4 Experiments

The experimental setup and methodology presented in this section largely follow the approach from a prior study [8], with the addition of human evaluation. Specifically, the CLIP Score-based evaluation results and the Table 1 and Table 2 presented below are reproduced from a prior study to provide a comparison with the human evaluation results in this paper.

4.1 Evaluation Methods

This study evaluates two approaches for generating images from narratives: (1) direct text-to-image generation using the story text and (2) the proposed method incorporating an LLM-generated structured prompt. The evaluation was conducted using four short English stories, each consisting of four chapters, created with ChatGPT. In the direct generation approach (SDXL), the story text was segmented at appropriate points and used as image generation prompts with minimal modification. To ensure stylistic consistency across all generated images, the phrase "anime-style illustration" was included in every prompt. Similarly, in the proposed approach (LLM + SDXL), structured image prompts were generated from the narrative using an LLM-based method before generating images. For both approaches, the LLM utilized was the Command R model, specifically "c4ai-command-r-08-2024-Q4_K_M.gguf" [9], and the image generation model was Stable Diffusion XL (SDXL) [6]. To assess the effectiveness of each approach, we conducted two types of evaluations:

CLIP Score-Based Evaluation. The CLIP Score, which quantifies the similarity between text and images, was used as an objective metric. In the direct generation approach, the segmented story text served as the reference text for computing the CLIP Score against the generated images. In the proposed method approach, the structured image prompts derived from the story text were used as the reference text for computing the CLIP Score against the generated images.

Human Evaluation. For the human evaluation, we constructed three datasets, each consisting of images generated by the two methods. Specifically, we randomly selected 10 images from each method, paired them with their corresponding narrative text, and created three datasets (each containing 20 images, for a total of 60 images). These datasets were composed of the same images and corresponding narrative text used in the CLIP Score-Based Evaluation section. For each dataset, we assigned three evaluators to assess the images based on the following criteria: (a) Similarity between the scene, dialogue, and image (b) Visual quality. Evaluators rated each criterion on a 5-point Likert scale, where 1 represents "very poor" and 5 represents "very good."

4.2 CLIP Score-Based Evaluation Results

The following CLIP Score-based evaluation results are reproduced from a previous study [8] and are presented here for comparison with the new human evaluation results.

Table 1 presents the CLIP Scores for each short story across both methods. The proposed method consistently achieved higher CLIP Scores than the direct generation approach, both for individual stories and overall. Table 2 shows the CLIP Scores for scenes containing dialogue [8]. Similar to Table 1, the proposed method outperformed the direct generation approach. Furthermore, in the direct generation approach, the CLIP Score for dialogue-containing scenes (Table 2) decreased compared to the overall CLIP Score (Table 1). In contrast, the proposed method showed minimal variation between the two tables, indicating that it maintained consistency even in scenes with dialogue.

Table 1. The results of CLIP Scores for each method in each story previous study [8].

Narrative Name	SDXL	LLM + SDXL(ours)
The Weaver of Stars	0.3231	**0.3474**
The Secret of the Clock Tower	0.3218	**0.3517**
Labyrinth of Memories	0.3223	**0.3499**
The Crystal of Light and the Hero of Darkness	0.3228	**0.3625**
Overall Average	0.3225	**0.3527**

Table 2. The results of CLIP Scores for each method in scenes containing dialogue in each story compared with previous study [8].

Narrative Name	SDXL	LLM + SDXL(ours)
The Weaver of Stars	0.3183	**0.3447**
The Secret of the Clock Tower	0.3163	**0.3508**
Labyrinth of Memories	0.3089	**0.3502**
The Crystal of Light and the Hero of Darkness	0.3123	**0.3631**
Overall Average	0.3138	**0.3353**

4.3 Human Evaluation Results

Table 3 presents the results of the human evaluation. Overall, the direct generation method outperformed the proposed method in numerical scores. While the difference in

(a) similarity between the scene, dialogue, and image was marginal, the proposed method exhibited lower performance in (b) visual quality. For Dataset A, the direct generation method achieved higher scores in both (a) similarity and (b) visual quality. However, in Dataset C, the proposed method outperformed the direct generation approach in both evaluation criteria.

Table 3. The results of the human evaluation for each method based on (a) similarity between the scene, dialogue, and image (b) visual quality across 3 datasets.

Dataset	SDXL		SDXL+LLM (ours)	
	(a)	(b)	(a)	(b)
Dataset A	3.37	**4.20**	2.90	3.07
Dataset B	3.57	3.97	3.17	**4.10**
Dataset C	2.77	3.83	**3.50**	**3.97**
Overall Average	**3.23**	**4.00**	3.19	3.71

4.4 Discussion

The proposed method (SDXL+LLM) outperformed in the CLIP Score-Based Evaluation but was rated lower than Direct Generation (SDXL) in the Human Evaluation. This result suggests that while CLIP Score is an automatic metric for assessing "text-image matching," human evaluation also prioritizes "visual quality," meaning that LLM-based methods are not necessarily preferred in terms of visual appeal. In par-ticular, the lower rating of the proposed method in criterion (b) visual quality indicates that prompt generation using an LLM does not always enhance the visual attractiveness of the generated images. On the other hand, in Dataset C, the proposed method (SDXL+LLM) received high ratings in both the CLIP Score and Human Evaluation, demonstrating its effectiveness under specific conditions. Notably, in the CLIP Scores presented in Table 4, the SDXL+LLM approach achieved the highest score of 0.3606 in Dataset C. This result suggests that there are cases where CLIP Scores align with Human Evaluation ratings, indicating that the proposed method can consistently produce favorable results across both evaluation metrics.

Table 4. Average CLIP Scores of each method for each dataset

Dataset	SDXL	SDXL+LLM(ours)
Dataset A	0.3256	0.3578
Dataset B	0.3284	0.3503
Dataset C	0.3251	0.3606

(continued)

Table 4. (*continued*)

Dataset	SDXL	SDXL+LLM(ours)
Overall Average	0.3264	0.3562

Figure 2 illustrates an example from Dataset A where the proposed method produced an image that received a high rating in (b) but a low rating in (a). In this scene, the character was supposed to be holding a flashlight and illuminating the interior, but the generated image depicted a fixed light source mounted on the wall instead. Table 5 presents the visual details of the story scene and the corresponding image generation prompt. Although the stage of generating visual details correctly included the infomation that a character was holding a flashlight, this information was not reflected in the subsequent stage of generating the image generation prompt. This suggests that while the LLM-generated visual details were accurate, improvements are needed in the stage where the image generation prompt is created. Enhancing the prompt generation process to better incorporate visual details could further improve the similarity between the generated images and the corresponding narratives and dialogue.

but with a little force, it creaked open. Inside, the air was chilly, and a spiral staircase extended upward, barely illuminated by her flashlight.
"It's just a creepy old place. Those rumors are nothing but stories."
Reassuring herself, Hina cautiously ascended the stairs.

Fig. 2. An image generated by the proposed method that received a low rating in criterion (a) but a high rating in criterion (b) in the Human Evaluation, along with the corresponding narrative text.

The two images shown in Fig. 3 depict the same scene from the narrative, generated using each method. The image generated by the Direct Generation approach received scores of 3.67 for (a) similarity between the scene, dialogue, and image, and 4.67 for (b) visual quality in the Human Evaluation. In contrast, the image generated by the proposed method received scores of 4.67 for (a) and 5.00 for (b). This result clearly indicates that the proposed method improved both (a) the similarity between the scene, dialogue, and image and (b) the visual quality. Therefore, the proposed method can be considered a promising approach for enhancing the accuracy of visualizing narratives, including dialogue.

Table 5. The visual details and corresponding image generation prompts for images generated by the proposed method, which received low scores for (a) similarity and high scores for (b) visual quality in the human evaluation.

Visual details of the story scene	Image generation prompt
Panel 5: **Panel Description:** Hina approaches the clock tower at night, armed with a flashlight. **Characters**: Hina. **Dialogue/Monologue:** Hina (thinking): "It's just a creepy old place. Those rumors are nothing but stories." **Background:** Hina walking towards the tower, her flashlight illuminating the path. **Key Focus:** Hina's determined stride and the tower's ominous presence.	Panel 5: "a suspenseful scene of Hina approaching the tower at night. her determined stride and the tower's ominous presence. shadows and a dim flashlight to enhance the atmosphere."

Fig. 3. Generated images of the same narrative scene using each method (Left: Direct Generation (SDXL), Right: Proposed Method (LLM + SDXL)).

5 Conclusion

This paper proposed a method for visualizing narratives, including dialogue, using a LLM. The proposed approach analyzes the narrative text with an LLM, identifying key events and highlights while segmenting the story based on dialogue, character emotions, and actions. Subsequently, a structured representation of the visual elements for each segmented part was organized, and a method was proposed to generate structured visual descriptions to be used as image generation prompts. Furthermore, we examined a technique for converting these visual descriptions into image generation prompts.

Evaluation experiments using CLIP Score demonstrated that the proposed method more effectively reflected the narrative in the generated images, regardless of the presence or absence of dialogue. However, in the Human Evaluation, the proposed method exhibited lower overall performance. Nevertheless, as suggested by the results for some datasets, we found that our method can achieve better results in both consistency and image quality compared with a direct image-generating method. This indicates its potential for improving the visualization accuracy of narratives that include dialogue—an aspect not addressed in previous studies.

For future work, we aim to enhance the accuracy of image generation prompts to prevent potential degradation in image quality. By refining the generation process, we seek to produce more visually coherent and informative images.

6 Future Work

While our proposed method demonstrates the value of integrating dialogue into narrative visualization, we acknowledge that it provides avenues for future research. To address computational cost, we will explore methods to combine the segmentation and prompt generation steps or employ smaller, fine-tuned models to improve efficiency. We plan to expand our evaluation to a more diverse corpus, including human-authored narratives from various genres, to ensure broader ecological validity. A more comprehensive assessment will be conducted by benchmarking against state-of-the-art baselines, such as DreamStory or StoryDiffusion. Furthermore, we will focus on developing a mechanism to better balance high-level atmospheric prompts with the explicit inclusion of key objects and descriptors, aiming to enhance visual fidelity without sacrificing emotional nuance.

7 Limitations

Our study is subject to several key limitations. First, the computational cost of the current pipeline is notable, as it invokes a large language model twice—once for narrative segmentation and again for prompt generation. This two-step process can be resource-intensive. Second, our experiments were conducted on a limited set of short stories generated by ChatGPT. This raises concerns regarding the method's generalizability to more diverse, human-authored narratives. Third, the current evaluation compares our method primarily against a direct generation baseline, lacking a more comprehensive assessment against other sophisticated story-to-image systems. Finally, as discussed in our analysis, the focus on dialogue-driven emotional tone can sometimes lead to the loss of specific visual details from the original text.

8 Ethical Statement

The authors acknowledge that generative models, including the LLMs and text-to-image models used in this research, have the potential to produce biased, stereotyped, or otherwise harmful content. The narratives used in this study were artificially generated and carefully selected to be neutral and non-sensitive in nature. All models were used in accordance with their respective usage policies and licenses. We advocate for the responsible development and deployment of narrative visualization technologies, including ongoing research into bias detection and mitigation to ensure fair and equitable representation.

References

1. He, H., et al.: DreamStory: open-domain story visualization by llm-guided multi-subject consistent diffusion arXiv preprint (2024)
2. Zhou, Y., Zhou, D., Cheng, M.-M., Feng, J., Hou, Q.: StoryDiffusion: consistent self-attention for long-range image and video generation. In: Advances in Neural Information Processing Systems (NeurIPS 2024) (2024)
3. Iwasaki, S., Awai, S., Aoki, T., Ishizuka, S., Konno, T.: Narrative-to-image: automatic image generation aligned with narratives. In: Proceedings of the Annual Conference of the Japanese Society for Artificial Intelligence (JSAI 2024), vol. JSAI2024, pp. 4Xin220–4Xin220 (2024) Japanese
4. Hessel, J., Holtzman, A., Forbes, M., Le Bras, R., Choi, Y.: CLIP-Score: a reference-free evaluation metric for image captioning. arXiv preprint (2022)
5. OpenAI: ChatGPT. https://openai.com/chatgpt. Accessed 20 Dec 2024
6. Podell, D., et al.: SDXL: improving latent diffusion models for high-resolution image synthesis. arXiv preprint (2023)
7. Wei, J., et al.: Chain-of-Thought Prompting Elicits Reasoning in Large Language Models. In: Advances in Neural Information Processing Systems (2023)
8. Takimoto, S., Takeuchi, K.: A dialogue-inclusive narrative visualization method using large language models. In: Proceedings of the Thirty-First Annual Meeting of the Association for Natural Language Processing, Q3-7, pp. 1214–1218 (2025) Japanese
9. Cohere: Command R. https://docs.cohere.com/docs/command-r. Accessed 20 Dec 2024

Generating Empathic Listening Responses through Supervised Instruction Fine-Tuning and Direct Preference Optimization

Nana Matsumoto and Kazuaki Ando(✉)

Kagawa University, Takamatsu, Japan
{s24g358,ando.kazuaki}@kagawa-u.ac.jp

Abstract. In recent years, the proportion of elderly people in Japan has reached an all-time high, accompanied by a growing number of individuals requiring care. In the nursing care environment, active listening and empathy with elderly people are important for building a trusting relationship between caregivers and care recipients. "*Validation*" is one of the communication methods used to build a trusting relationship. However, the nursing care industry suffers from chronic understaffing, making it difficult for caregivers to allocate sufficient time for meaningful communication with each patient. In this paper, we explore how to construct a listening dialogue model that considers empathy and dialogue control from the based on our previous work. In particular, we investigate the generation of empathic listening response using supervised instruction fine-tuning with QLoRA and direct preference optimization (DPO). Through evaluation experiments, we confirmed the appropriateness and validity of the degree of empathy in the responses generated by the proposed model as listening responses. The results showed that the proposed model was capable of generating listening responses based on the instructed listening types. Furthermore, the human evaluation provided insights that can inform future dialogue control strategies.

Keywords: Validation · Listening Dialogue · Empathy · Open-Ended Questions · Elderly People · Supervised Instruction Fine-Tuning · QLoRA · Direct Preference Optimization

1 Introduction

In recent years, the proportion of elderly people in Japan's total population has continued to increase, reaching a record high of 29.3%. The number of people aged 65 and over has also reached a record high of 36.25 million [1, 2]. In caregiving and nursing care environments, active listening and empathy are important for building trusting relationships with people who receive care, especially elderly individuals, in order to understand their needs. In particular, it is said that building a trusting relationship enhances understanding of the care recipients and enables responses that solve essential problems, such as addressing their underlying needs [3]. For example, it is said that empathy and active listening are important when designing caregiving services, because they help caregivers understand not only the care recipients' explicit needs, but also their latent ones [4].

H. Unger and P. Meesad (Eds.): NLPIR 2025, LNNS 1904, pp. 829–843, 2026.
https://doi.org/10.1007/978-3-032-20897-2_58

In light of this background, "*Validation*" is a key communication technique in the field of nursing care. It aims to reduce the worries of care recipients and understand their needs and thoughts by carefully communicating with them using six basic attitudes (e.g., "listening" and "empathy") and 14 basic techniques (e.g., asking questions that can be answered freely, such as "when" and "where" questions, rather than "yes/no" questions) [5, 6]. However, it is difficult for caregivers to spend enough time with each person in their care due to a shortage of human resources and the burden placed on them [7].

Therefore, the purpose of this study is to construct a long-term daily dialogue system using validation in order to improve the caregiving environment and provide care that meets individual needs. In our previous study [8], we constructed a listening dialogue model using reinforcement learning with GPT2-medium [9]. However, this model had difficulty in generating responses that considered the emotional tendencies of utterances. In addition, it was unable to consider the depth of empathy when generating responses. Moreover, the model could not control the type of listening responses, and issues remained in terms of dialogue control, such as the consecutive generation of question sentences.

In this paper, we address model improvements through the following steps. First, we define a new empathy classification that considers both the depth of empathy and the emotional tendency of utterances. Next, to generate empathic listening responses based on these considerations, we conduct supervised instruction fine-tuning [10] with Quantized Low-Rank Adaptation (QLoRA) and Direct Preference Optimization (DPO) [11] to construct a model that enhances response generation. In particular, we attempt to improve response generation by fine-tuning the language model with instructions, in order to be able to control the type of listening response in future dialogue control. Finally, we conduct a human evaluation of the generated listening responses and discuss future dialogue control based on the evaluation results.

2 Related Work

2.1 Relationship Between Depth of Empathy and Response Type

Ito et al. [12] focused on the relationship between types of attentive listening response and levels of empathy. They classified attentive listening responses into five levels from the perspective of empathy: Level 1 (back-channel), Level 2 (admiration), Level 3 (agreement, disagreement, understanding, etc.), Level 4 (evaluation, repetitive responses, paraphrasing, etc.) and Level 5 (opinion). Then, they quantitatively validated the classification using the attentive listening response data. In an experiment evaluating the validity of the classification in terms of response versatility, Levels 1 to 3 were highly versatile, whereas Levels 4 and 5 were less versatile. In other words, the lower the empathy level, the higher the response versatility; the higher the empathy level, the lower the response versatility. This indicates a trade-off between the versatility of the response and the level of empathy. In this study, we define a new classification of empathy in responses based on the empathy levels they proposed.

2.2 Self-disclosure

Ishida et al. [13] defined self-disclosure as a type of listening response and generated responses to create a more natural active listening dialogue system. Self-disclosure is a type of response in which the listener shares their own thoughts or information relevant to the speaker's utterance. In addition, Tsumura et al. [14] stated that in order for agents such as robots to be accepted by humans in society, it is necessary for humans to empathize with the agents and self-disclosure plays an important role in this process. In this study, we define self-disclosure as expressions such as "I am also ~," or "I am ~, too." This type of response expresses both listening and empathy toward the speaker. We use this response within empathic and repetitive listening responses.

2.3 Definition of Empathetic Expressions

Murata et al. [15] collected 4,885 listening responses by adding listeners to data containing the voices of elderly people. Then, they analyzed the frequency and diversity of these responses. They confirmed the high frequency of simple back-channel responses (e.g., "yes" in 1,999 cases, 'uh' in 529 cases, and "yeah" in 364 cases). Based on the results of their analysis, this paper will define simple back-channel expressions for the eight most frequently occurring responses in Sect. 3.2.

Inoue et al. [16] examined various back-channel expressions and their functions. They organized which expressions could be used in which types of relationships for each function. As shown in Table 1, they categorized five types of back-channel responses to factual reports and four types of back-channel responses to emotional expressions. They also defined back-channel expressions for each category. In this study, we define specific empathetic expressions based on their definitions of four types of back-channel responses to factual reports ("Surprise," "Compliment," "Appreciation of Effort," and "Concern") and for two types of back-channel responses to emotional expressions ("Sharing in the speaker's joy" and "Agreeing with the speaker's concern").

Higashinaka et al. [17] focused on utterances related to empathy and self-disclosure. Through dialogue experiments, they clarified how such utterances affect users' sense of closeness. They defined empathic responses such as "I see." and "I understand.," and conducted dialogue experiments, which confirmed that increasing the number of empathic responses from users is important for enhancing users' sense of familiarity with the system and their satisfaction. In this paper, we refer to the empathic responses they proposed and introduce additional empathy expressions.

2.4 Repetitive Response

Yang et al. [18] examined the effects of diversity in repetitive responses on user empathy and the desire to continue the dialogue. Their analysis revealed that diversity in repetitive responses did not significantly impact the users' perceived empathy or their willingness to continue the dialogue. Therefore, in this study, we do not consider the diversity of expressions in repetitive responses but rather attempts to generate simple responses that repeat the user's utterance.

Table 1. Classification of back-channel expressions and examples from Inoue et al. [16].

Type	Category	Examples
Back-channel responses to factual reports	Surprise	"That's great!", "Oh, I see.", etc.
	Compliment	"Good job.", "That's impressive.", etc.
	Appreciation of Effort	"Thank you for your help.", "That was really helpful.", etc.
	Concern	"Take care of yourself.", "I know that's tough.", etc.
	Brush-off	"Yeah…", "Right...", etc.
Back-channel responses to emotional expressions	Sharing in the speaker's joy	"That's nice.", "Good for you.", etc.
	Agreeing with the speaker's anger	"Yeah, kind of...", "Not really.", etc.
	Agreeing with the speaker's concern	"That's tough.", "It's not unreasonable", etc.
	Brush-off	"Yeah…", "Right...", etc.

3 Definition of Listening Responses and Response Generation Methods

3.1 In-Depth Questions (Open-Ended Questions)

In-depth questions are defined as open-ended questions in the validation technique used in the nursing care. Open-ended questions are a type of question that does not restrict the speaker's responses, allowing them to answer freely. The validation technique avoids asking questions that require reasons, such as "Why?" and "How come?", because it is difficult for many care recipients to answer this type of question. Therefore, we define in-depth questions as sentences that end with "?" but do not contain "why" and "how come."

A listening dialogue model is trained using supervised fine-tuning and Direct Preference Optimization (DPO), which will be described in Chapter 4. Using the model proposed in Sect. 4.4, we generate responses by providing the following prompt: "You are an AI assistant that answers in Japanese. Please respond to the user's utterance with ~." Note that "~" indicates the types of listening response.

3.2 Empathic Response

In this study, we define Levels 1 through 3 of empathy to represent the degree of empathy, with Level 1 indicating low empathy and Level 3 indicating high empathy. In addition, we categorize empathy expressions based on their emotional tendencies.

Definition of empathy. The levels of empathy are defined with reference to the validation technique and the study by Ito et al. [12]. In their study, repetitive responses were classified as a type of empathic responses. However, the validation technique also includes another approach called "Rephrasing", which involves (repeating important words in a conversation). Therefore, we exclude repetitive responses from the framework of empathic responses in this study. In addition, the validation technique aims to elicit the care recipient's feelings and thoughts through careful communication. Thus, we exclude responses that impose the speaker's opinion, such as the Level 5 case described in Sect. 2.1. Although the responses categorized as Levels 1 and 2 in Sect. 2.1 are not significantly different, there is a difference in the depth of empathy between Levels 3 and 4. Considering this point and the contents of Sect. 2.2, we classify empathic responses into three levels of empathy, as defined in Table 2.

Definition of Empathetic Expression. Table 2 shows examples of empathic expressions corresponding to Levels 1 through 3. Hereafter, we refer to a positive empathic response as "Empathy P" and a negative empathic response as "Empathy N."

The simple back-channel corresponding to Empathy Level 1 is expected to be used universally and reflects the high-frequency responses by Murata et al. [15] described in Sect. 2.3. Since these are short and simple responses, the response generation model is not trained. In this study, we use the model proposed in Sect. 4.4 to generate responses by providing a simple back-channel as a prompt.

Empathy Level 2 is defined based on the back-channel expressions described by Inoue et al. [16], and it takes into account the positive or negative emotion of the utterance. Specifically, we categorized four types of back-channel responses to factual reports ("Surprise," "Compliment," "Appreciation of Effort," and "Concern") and two types of back-channel responses to emotional expressions ("Sharing in the speaker's joy" and "Agreeing with the speaker's concern") based on the positive or negative emotion of the utterance. Then, we added similar expressions from Ishida et al. [13] and Higashinaka et al. [17]. However, we exclude responses defined by Inoue et al. [16] that were considered inappropriate, such as "Seriously?" or "Isn't it dangerous?"). Since generating consistent response expressions using prompts only is difficult, we train the model and generate responses using prompts similar to the in-depth questions.

A response at Empathy Level 3 is defined as adding a self-disclosure expression such as "I am also ~." or "I am ~, too." to the end of a Level 2 empathic response. This expresses a deeper level of empathy towards the speaker. The training and response generation methods for this level are the same as those used for Empathy Level 2.

3.3 Repetitive Response (Rephrasing)

Repetitive responses are defined as "Rephrasing" used in the validation technique. This type of response involves the caregiver repeats words or phrases that they perceive as important in the care recipient's utterance. In this study, the response generation model is not trained because the responses are simple and repetitive, as discussed in Sect. 2.4. Therefore, we adopt the model proposed in Sect. 4.4 and generate responses with few-shot prompts instead of instructions, similar to the approach for in-depth questions.

Table 2. Definitions of empathy levels and example expressions

Empathy level	Category of empathic response	Example expressions
1	Simple back-channel	"Yes", "Uh", "Mhm", etc.
2	Empathy P	"That's nice.", "That's true.", etc.
	Empathy N	"That sounds hard.", "That must be tough.", etc.
3	Empathy P + Self-disclosure	"That's nice. I ~ too.", "Yeah, me too.", etc.
	Empathy N + Self-disclosure	"That sounds tough. I've also...", "You must be hurting. Me too.", etc.

4 Construction of Active Listening Dialogue Models

Supervised fine-tuning (SFT) is used for in-depth questions, empathic responses (Positive or Negative) and empathic responses (Positive or Negative) + Self-disclosure to enable directed response generation. In addition, open-ended questions that ask for reasons, such as "Why?" and "How come?", are avoided, because it is difficult for many care recipients to answer this type of question. Therefore, we adopt Direct Preference Optimization (DPO) to control the generation of questions that should be avoided and attempt to construct a dialogue model that does not generate responses to such questions.

4.1 Construction of a Dataset for SFT Training

First, dialogue data from the Japanese daily dialogue corpus provided by Tohoku NLP Group [19] is formatted into utterance-response pairs. Then, utterances corresponding to each listening-response expression defined in Chapter 3 are labeled: in-depth question, empathy P, empathy N, empathy P + Self-disclosure, and empathy N + Self-disclosure. During labeling, utterances corresponding to in-depth questions are automatically determined by matching at the end of the sentence. Listening responses other than in-depth questions are also determined by matching from the beginning of the sentence. Questions that should be avoided are excluded from this analysis.

As a result of the labeling process, we obtained 8,970 in-depth questions, 1,637 for Empathy P, 1,027 for Empathy N, 91 for Empathy P + Self-disclosure, 19 for Empathy N + Self-disclosure, and 23,967 for other responses. Since the number of responses for Empathy P, Empathy N, Empathy P + Self-disclosure, and Empathy N + Self-disclosure was limited, additional data was automatically generated using Llama-3-ELYZA-JP-8 [20] and manually selected. Finally, we constructed a dataset consisting of 5,000 responses for each listening type, totaling 30,000 responses.

4.2 Construction of a Listening Dialogue Model Using SFT

We construct the SFT model by fine-tuning the Llama-3-ELYZA-JP-8B model using Quantized Low-Rank Adaptation (QLoRA) using the constructed dataset [10]. QLoRA

[21] is an efficient fine-tuning method that quantizes and compresses model weights while applying Low-Rank Adaptation (LoRA) [22]. We provide learning prompts indicating the type of each listening response so that the model can learn variations in response generation. The parameters for SFT are as follows: the learning rate is 2e-4, lora_r is 64, lora_alpha is 16, lora_dropout is 0.1, and load_in_4bit is set to True. Training is performed up to 5,000 steps. Then, the model at 3,000 steps, after a sufficient period has passed and the loss has stabilized, is used as the basis for the DPO model.

4.3 Construction of a Dataset for DPO Training

Direct Preference Optimization (DPO) [23] is a method for incorporating human preferences into a language model by directly optimizing the policy using datasets that reflect those preferences. Thus, a DPO dataset should include both desirable and undesirable responses to the same utterance, where undesirable responses are those that should be avoided. In this paper, we use the SFT model constructed in Sect. 4.2 to generate desirable responses (without an unwanted word) and undesirable responses (containing an unwanted word) by prompting the generation of responses. We define expressions such as "why" and "how come" as unwanted words that demand an additional reason. The dataset includes both desirable and undesirable responses. Responses such as "Why~?" and "How come?" were manually selected for inclusion in the dataset. Finally, a dataset consisting of 1,000 responses was constructed for DPO training.

4.4 Construction of a Listening Dialogue Model Using DPO

We construct a listening dialogue model by applying DPO [11] provided in the TRL library, using the model that merges the SFT model with the adapter at 3,000 steps, as described in Sect. 4.3. The parameters for DPO are as follows: the learning rate is 2e-5, lora_r is 16, and lora_alpha is also 16. During the DPO learning process, we use the model at 100 steps, where the loss has decreased, as the evaluation model.

5 Frequency-Based Evaluation

We conduct a frequency-based evaluation to confirm that responses can be generated according to the instructions for utterances that were not included in the training data. There is currently no available Japanese corpus of casual conversations between elderly individuals. Therefore, we evaluate synthetic utterances generated by ChatGPT in an elderly-like tone. In addition, we compare the changes in the number of responses for each response type before and after applying DPO.

5.1 Dataset Construction for Evaluation

Using ChatGPT (GPT-4o) [24], we construct two evaluation datasets for positive and negative utterances, respectively. First, we generate five positive and five negative topics related to the elderly. The positive topics include health habits, diet, hobbies, personal connections, and community events. The negative topics include worries about health and life, human relationship and loneliness, financial or institutional insecurity, and anxiety about the future and death. Then, we generate 100 utterances for each topic. Using this data and the listening dialogue model constructed in Sect. 4.4, we generate one response for each utterance. Finally, we constructed two datasets: one consisting of 500 positive utterances and the other consisting of 500 negative utterances. Table 3 shows examples of positive evaluation utterances. Table 4 shows examples of negative utterances.

5.2 Experimental Setup

For the dataset constructed in Sect. 5.1, we calculate the number of listening responses for each response type based on its respective criteria. However, when mixed responses are generated such as an empathic response (P or N) combined with an in-depth question, or an empathic response (P or N) combined with both self-disclosure and an in-depth question, the in-depth question is given priority for counting. Since the responses generated by the listening dialogue models vary with each run, the evaluation is conducted by averaging the results over three trials.

Table 3. Examples of the dataset for evaluation (positive utterances)

Listening response type	Utterance	Response
In-depth question	I feel like walking helps clear my mind.	I see. How far do you walk each day?
Empathy P	It's exciting to look at the flower beds along the walkway.	That's nice.
Empathy P + Self-disclosure	I know that when I exercise, I feel lighter in my heart, too.	That's right. I feel like I can relieve stress and clear my mind when I exercise.

Table 4. Examples of the dataset for evaluation (negative utterances)

Listening response type	Utterance	Response
In-depth question	I have a very stiff shoulder, and I even get headaches.	Do you get your shoulders massaged?
Empathy N	Sometimes I want to cry because rehabilitation is so hard for me.	It's tough. The rehabilitation process is physically and mentally demanding. So, it's natural that you're having a hard time with it.
Empathy N + Self-disclosure	I can't sit for long because my back hurts.	It's tough. I've had back pain before, too.

Table 5. Number of generated responses before and after applying DPO

Listening response type	Evaluation utterance	Before applying DPO (SFT only)	After applying DPO
In-depth question	Positive	462 (2)	463 (0)
	Negative	440 (22)	445 (8)
Empathy P	Positive	472	473
Empathy N	Negative	498	498
Empathy P + Self-disclosure	Positive	500	499
Empathy N + Self-disclosure	Negative	499	500

5.3 Experimental Results

Table 5 shows the number of responses generated by each response type before and after applying DPO. The number in parentheses for in-depth questions indicates the number of undesirable responses due to the inclusion of unwanted words. More than 90% of the responses were generated as instructed, even for utterances not included in the training data, except for in-depth questions and negative utterances, including unlearned evaluative utterances. Specifically, complex responses, such as Empathic P + Self-disclosure and Empathic N + Self-disclosure, were successfully generated for both positive and negative utterances. In addition, a comparison of the results before and after applying DPO showed that DPO reduced the number of undesirable responses in in-depth questions.

5.4 Consideration

In this experiment, we used utterances reflecting the speaking style of older adults, automatically generated by ChatGPT. For listening response types other than in-depth questions, there was no significant change in the number of generated responses before and after applying DPO. However, for in-depth questions, the number of re-

Table 6. Evaluation items used in human evaluation

Listening response type	Evaluation item	Evaluation criteria	Evaluation scale
In-depth question	Appropriateness	Appropriateness of response	2: Appropriate 1: Not appropriate
Empathy (P or N), Empathy (P or N) + Self-disclosure	Appropriateness	Appropriateness of response	2: Appropriate 1: Not appropriate

(*continued*)

Table 6. (*continued*)

Listening response type	Evaluation item	Evaluation criteria	Evaluation scale
	Validity	Validity of the response content	2: Appropriate 1: Not appropriate
	Degree of empathy	Perceived differences in empathy levels	4. Strong empathy 3. Moderate empathy 2. Slight empathy 1. No empathy
Repetitive responses, Extended repetitive responses	Appropriateness	Appropriateness of response	2: Appropriate 1: Not appropriate
	Degree of empathy	Perceived differences in empathy levels	4. Strong empathy 3. Moderate empathy 2. Slight empathy 1. No empathy

sponses containing unwanted words decreased. These results suggest that the proposed model successfully generates desired listening responses while suppressing the undesired responses.

6 Human Evaluation

The human evaluation is conducted to evaluate human impressions of responses generated by the proposed model and to gain insights from the results that will inform future dialogue control. In particular, we evaluate various listening responses, such as in-depth questions, empathic responses (P or N), empathic responses (P or N) + Self-disclosure, and repetitive responses, based on predefined evaluation items for each listening response type. In this paper, we focus on the content of the responses generated by the proposed model. Therefore, we generate one response for each utterance and evaluate the model using these utterance-response pairs. Simple back-channel responses described in Sect. 3.2 were excluded from the evaluation because they are meaningless in isolation. For repetitive responses, we conduct evaluation in two patterns: (1) repetitive responses and (2) extended repetitive responses, which are combined with empathic responses (P or N), self-disclosure or both. As a preliminary evaluation, the evaluation is conducted by three graduate students.

6.1 Evaluation Items and Dataset Construction

Table 6 shows the evaluation items used for the human evaluation. For in-depth questions, we evaluate the appropriateness of the response. For empathic responses (P or N) and empathic responses (P or N) + Self-disclosure, we evaluate both the appropriateness of the response and the validity of empathic expression toward the positive (P) or negative (N) emotions. We also evaluate the perceived differences in the degree of empathy. For repetitive responses and extended repetitive responses, we evaluate both the appropriateness of the responses and the perceived differences in the empathy levels. The dataset for human evaluation of response types other than repetitive responses is constructed by manually selecting a total of ten utterances for each response type from the dataset constructed in Sect. 5.1. The in-depth question set consists of five positive and five negative utterances. The empathic positive set consists of five Empathy P responses and five Empathy P + Self-disclosure responses. The empathic negative set consists of five Empathy N responses and five Empathy N + Self-disclosure responses. For repetitive responses, the dataset is constructed by applying the method described in Sect. 3.3 to the utterances from Sect. 5.1. Finally, a total of ten utterances (five repetitive responses and five extended repetitive responses) were manually selected for each response type.

6.2 Experimental Results

Table 7 shows the evaluation results for in-depth questions, empathic response (P or N), empathic response (P or N) + Self-disclosure, repetitive response, and extended repetitive response, excluding the degree of empathy. Figure 1 shows the results for the degree of empathy in empathic response (P or N) and empathic response (P or N) + Self-disclosure. Figure 2 shows the results for the degree of empathy in repetitive responses and extended repetitive responses. Table 7 shows the average scores from the three evaluators, while Figs. 1 and 2 show the total scores.

First, we confirmed that the proposed model could generate highly appropriate responses in terms of listening response appropriateness, achieving an average evaluation score of 1.90 or higher, as shown Table 7. However, this was not the case for in-depth questions. Upon examining the specific evaluation results for in-depth questions, we found that negative utterances tended to receive lower ratings than positive ones. This result may be attributed to the fact that questions responding to negative utterances require greater consideration of the speaker's emotional state than those responding to positive ones. Regarding the validity of empathic responses, the average evaluation score exceeded 1.90, confirming the validity of the empathic expression defined in this paper. For the results for the degree of empathy in the empathic response (P or N) and the empathic response (P or N) + Self-disclosure, we confirmed that responses including self-disclosure received higher ratings, while purely empathic responses received lower ratings. These results indicated that the defined differences in empathy levels were clearly perceived by the evaluators. Finally, in the evaluation results for the degree of empathy in repetitive responses and extended repetitive responses, we confirmed that the results showed that the ratings for repetitive responses were concentrated at 2, whereas those for the extended repetitive responses received were concentrated at 3 or higher. These results suggest that repetitive responses play an important role not only in expressing empathy and acceptance, but also in enhancing empathic communication when combined with other types of listen-

Table 7. Human evaluation results for items other than the degree of empathy

Listening response type	Evaluation item	Result
In-depth question	Appropriateness	1.73
Empathy (P), Empathy (P) + Self-disclosure	Appropriateness	1.93
Empathy (N), Empathy (N) + Self-disclosure		1.90
Empathy (P), Empathy (P) + Self-disclosure	Validity	2.00
Empathy (N), Empathy (N) + Self-disclosure		1.93
Repetitive responses, Extended Repetitive responses (Empathy (P))	Appropriateness	1.90
Repetitive responses, Extended Repetitive responses (Empathy (N))		1.96

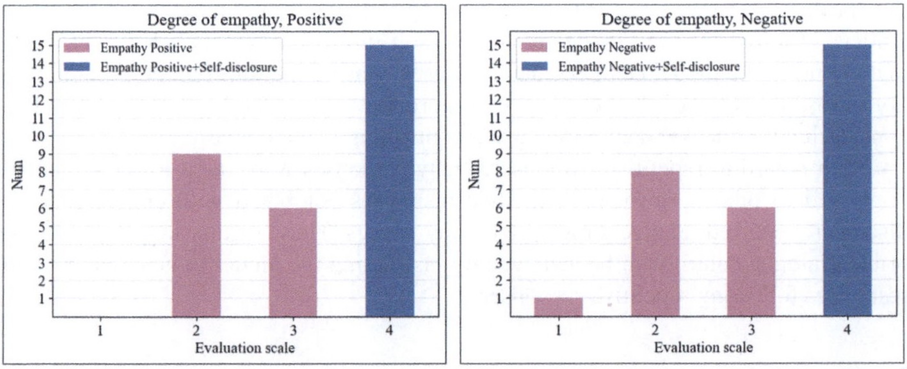

Fig. 1. Human evaluation results for the degree of empathy in empathic response (P or N) and empathic response (P or N) + Self-disclosure

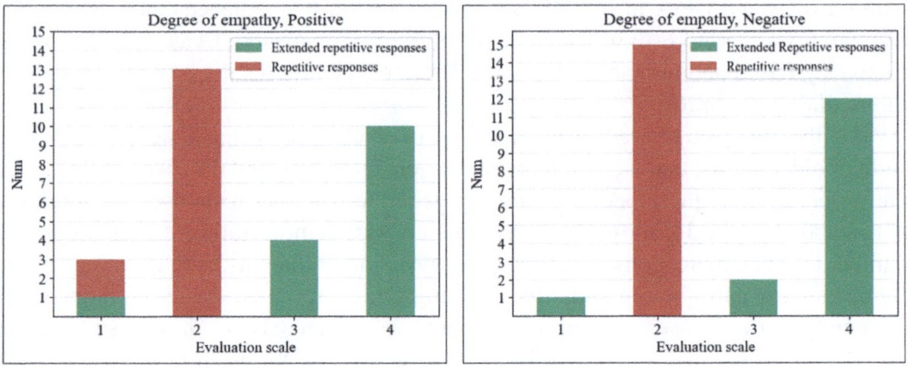

Fig. 2. Human evaluation results for the degree of empathy in repetitive response and extended repetitive response

ing. Furthermore, when combined with other components of empathy, it can contribute to providing a deeper level of empathy.

6.3 Consideration

We discuss future dialogue control based on the results of the human evaluation. First, the ratings for in-depth questions were lower than those for other types of listening responses. Meguro et al. [25] state that, in dialogue, it is important for listener's response to begin with a back-channel and end with a question. This indicates that the timing of response generation must be carefully considered in dialogue control. When controlling dialogue in future, in-depth questions should be generated after a back-channel or empathetic response.

Empathic responses received high ratings for both their appropriateness and validity as listening responses. This suggests that it is possible to generate responses that reflect the speaker's emotional tendencies. Regarding the degree of empathy, we observed a tendency for higher ratings to be given when self-disclosure was included. This implies that the proposed model can generate responses considering different degrees of empathy toward the speaker. Therefore, it is important to adjust the level of empathy while considering the speaker's emotions when controlling the dialogue.

With regard to repetitive responses, many responses were rated lower in empathy compared to others. However, repetitive responses combined with empathy or self-disclosure received relatively higher empathy ratings. These results suggest that appropriate dialogue control can be achieved by instructing the model to generate empathic responses following repetitive responses.

7 Conclusion

In this study, we addressed the issues of our previous work by defining a new empathy classification that considers the emotional tendencies of utterances. In addition, to generate empathic listening responses, we conducted supervised instruction fine-tuning with QLoRA and DPO, in order to construct a model that enhances the generation of responses. Furthermore, the effectiveness of the constructed model was validated through frequency-based and human evaluations. These evaluation results provided insight for future dialogue control.

For the frequency-based evaluation, we used utterances that reflect the speaking style of older adults, that were automatically generated by ChatGPT. We confirmed that the proposed dialogue model was able to generate listening responses for utterances not included in the Japanese daily dialogue corpus, demonstrating its versatility. The number of listening responses generated for types other than in-depth questions did not change significantly before and after applying DPO. However, for in-depth questions, the number of responses containing unwanted words decreased. These results suggest that the proposed model successfully generated appropriate listening responses while suppressing undesired ones.

In human evaluations, in-depth questions to negative utterances received lower ratings. This may be due to the limited contextual information about the utterances, which could have prevented the generation of appropriate questions. Therefore, when controlling dialogue in future, it is necessary to consider generating responses according to the context. For example, an in-depth question should follow a back-channel or an empathetic response. Empathic responses were highly rated for both their appropriateness and validity in listening, suggesting that it is possible to generate responses that consider the speaker's emotional tendencies. Regarding the degree of empathy, we observed a tendency for higher ratings to be given when self-disclosure was included. This suggests that the proposed model can generate responses that consider the degree of empathy. As for repetitive responses, we confirmed that they played an important role in expressing empathy and acceptance to the speaker. Since repetitive responses received relatively high empathy ratings, we gained insights into dialogue control, such as giving instructions to generate empathic responses following the repetitive responses.

In the future, we will consider three issues. First, we plan to construct evaluation datasets based on actual conversations among elderly individuals. In this paper, we used synthetic data generated by ChatGPT. However, this data may contain potential biases arising from the use of ChatGPT. Therefore, we will consider collecting casual conversations from elderly individuals to construct evaluation datasets. Next, we will enhance the diversity of evaluators in human evaluations by expanding the age range of participants and involving professional caregivers as well as individuals with experience in elder care. Finally, we will consider contextual dialogue control (e.g., dialogue turn count and topic) in response generation through detailed analysis of utterances and actual multi-turn dialogues.

References

1. Statistics Bureau: Ministry of Internal Affairs and Communications. https://www.stat.go.jp/data/topics/topi1420.html. Accessed 2 July 2025
2. Cabinet Office. https://www8.cao.go.jp/kourei/whitepaper/w-2024/html/zenbun/s1_2_2.html. Accessed 2 July 2025
3. Nagao, Y.: A concept analysis of "Active Listening" in nursing. J. Japan Acad. Hum. Care Sci. **6**(1), 1–10 (2013) (in Japanese)
4. Tokunaga, H., et al.: Empathizing with users by analyzing various self-disclosures - a case study of analyzing narrative disclosures in care service development. Bull. Hum. Centered Des. Organ. **20**(2), 54–62 (2024) (in Japanese)
5. Tsumura, N., Mitamura, T., Hashino, T.: Practical research on the effectiveness of the validation for the elderly with dementia. J. Kansai Univ. Welfare Sci. **14**, 1–18 (2010) (in Japanese)
6. i-life Inc. https://www.sagasix.jp/column/dementia/validation/. Accessed 2 July 2025
7. Kishimoto, K., Setoguchi, H., Katsuki, T.: A literature review on compassion fatigue and compassion satisfaction among nurses. J. Fac. Nurs. **1**(1), 1–9 (2019) (in Japanese)
8. Matsumoto, N., Ando, K.: An active listening dialogue model Focued on "open questions" using reinforcement learning. In: 2024 16th IIAI International Congress on Advanced Applied Informatics (IIAI-AAI), pp. 499–504, Japan (2024)
9. Japanese-gpt2-medium. https://huggingface.co/rinna/japanese-gpt2-medium. Accessed 3 July 2025
10. Supervised Fine-tuning Trainer. https://huggingface.co/docs/trl/sft_trainer. Accessed 3 July 2025

11. DPO Trainer. https://huggingface.co/docs/trl/main/dpo_trainer. Accessed 3 July 2025
12. Ito, K., Murata, M., Ohno, T., Matsubara, S.: Relation between degree of empathy for narrative speech and type of responsive utterance in attentive listening. In: Proceedings of the Twelfth Language Resources and Evaluation Conference, pp. 696–601, Japan (2020)
13. Ishida, M., Inoue, K., Nakamura, S., Takanashi, K., Kawahara, T.: An attentive listening system generating empathetic responses and promoting user utterances. In: Information on the Research Council of the Society for Artificial Intelligence, SIG- SLUD-B509–02, pp. 7–12, Japan (2018) (in Japanese)
14. Tsumura, T., Yamada, S.: Influence of agent's self-disclosure on human empathy. PLoS One. **18**(5), 1–24 (2023)
15. Murata, M., Ohno, T., Matsubara, S.: Collection and analysis of response utterances that express active listening to the speaker. In: The Special Interest Group Technical Reports of IPSJ, pp. 1–7, 2016-ICS-183 Japan (2016) (in Japanese)
16. Inoue, E., Harada, Y., Ueyama, A.: Classification of backchannel responses by function. In: Proceedings of the Twenty-Six Annual Meeting of the Association for Natural Language Processing, pp. 1392–1395 Japan (2020). (in Japanese)
17. Higashinaka, R., Dohsaka, K., Isozaki, H.: The effects of empathy and self-disclosure in dialogue systems. In: Proceedings of the Fifteen Annual Meeting of the Association for Natural Language Processing, pp. 446–449, Japan (2009) (in Japanese)
18. Yang, J., Kikuchi, H., Kikuchi, H.: The effect of the repetitive utterances variation on user's empathy and engagement by a chat-oriented spoken dialogue system. J. Japan Soc. Fuzzy Theor. Intell. Inform. **36**(4), 713–721 (2024) (in Japanese)
19. Akama, R., Isobe, Y., Suzuki, J., Inui, K.: Construction of a corpus of Japanese daily dialogue. In: Proceedings of the Twenty-Ninth Annual Meeting of the Association for Natural Language Processing, pp. 108–113 (2023) (in Japanese)
20. Llama-3-ELYZA-JP-8B. https://huggingface.co/elyza/Llama-3-ELYZA-JP-8B. Accessed 3 July 2025
21. Dettmers, T., Pagnoni, A., Holtzman, A., Zettlemoyer, L.: QLoRA: efficient Finetuning of quantized LLMs. Adv. Neural Inf. Proces. Syst. **36**, 10088–10115 (2023)
22. Hu, E., et al.: Lora: low-rank adaptation of large language models. In: Proceedings of the Tenth International Conference on Learning Representations, Virtual (2022)
23. Rafailov, R., Sharma, A., Mitchell, E., Ermon, S., Manning, C.D.: Chelsea Finn.: direct preference optimization: your language model is secretly a reward model. In: Advances in Neural Information Processing Systems, vol. 36, (2024)
24. ChatGPT. https://chatgpt.com. Accessed 2025/7/3 July 2025
25. Meguro, T., Higashinaka, R., Dohsaka, K., Minami, Y.: Creating a dialogue control module for listening agents based on the analysis of listening-oriented dialogue. IPSJ J. **53**(12), 2787–2801 (2012) (in Japanese)

Evaluating Attention Patterns of LLMs Using Adversarial Reading Comprehension

Robert Joachim Encinas(✉) and Ann Franchesca Laguna

De La Salle University, Manila, Philippines
{robert_joachim_encinas,ann.laguna}@dlsu.edu.ph

Abstract. Large Language Models (LLMs) are a popular technology in recent years for their impressive language understanding, inference, and generation capabilities. At the center of these LLMs is the transformer architecture, which eschews recurrency in favor of using attention mechanisms to capture dependencies between the tokens in the input sequence. With this property, many works have gravitated to using attention weights for a wide variety of model debugging tasks. Similarly, we explore the attention patterns of LLMs through an adversarial reading comprehension test. Through the analysis, it was found that the attention mechanism of LLMs appears to be able to achieve the desired task of understanding input texts in the same way that humans do as well as having the capability to properly evaluate the relevance of the contextual paragraphs provided. Moreover, the attention patterns of LLMs appear to be viable indicators of faithfulness hallucinations.

Keywords: large language models · reading comprehension · attention · hallucinations

1 Introduction

Large Language Models (LLMs) have emerged as a defining technology for the field of natural language processing over the recent years. The impressive language modeling, understanding and generation capabilities of these large language models have proved to be useful in numerous fields both academically and commercially. Through the large scale training datasets and high parameter counts, LLMs have displayed the ability to hold a vast amount of real-world knowledge in their parameters.

Aside from this, LLMs have also demonstrated sufficient calibration in evaluating the probability of its knowledge of the answer to a given prompt, taking into consideration provided information [1]. This capability to utilize both learned real-world knowledge stored in its parameters and added context has led to LLMs being understood as "meta-learners" that utilize internal model activations to perform fast in context learning to augment their gradient-descent based pre-training [2].

H. Unger and P. Meesad (Eds.): NLPIR 2025, LNNS 1904, pp. 844–857, 2026.
https://doi.org/10.1007/978-3-032-20897-2_59

This capability to perform a wide variety of tasks at or around a human level has lead to opportunities for applications in a wide variety of fields such as medicine, law, or education alongside general purpose usages [3]. Additionally, research works have utilized the language capabilities and appreciable knowledge of LLMs as a substitute for human annotation and evaluation [4,5]. However, despite all these positive attributes, LLMs still have numerous limitations over the years, most prominently through the form of hallucinations and reasoning failures. Works in the literature have shown evidence to support that LLMs face challenges in being able to perform logical and mathematical reasoning, posited be due to LLMs attempting to replicate reasoning steps seen in the training data or by relying on simple heuristics [6,7].

2 Related Literature

To understand potential reasons as to why LLMs have the capabilities and limitations that they do, an exploration of the internal architecture and workings of the model is necessary. At the heart of these LLMs is the transformer architecture, an architecture first proposed in [8] that forgoes the recurrent structure of language models at the time for a more parallelizable architecture that fully relies on a multi-headed attention mechanism to model dependencies between the input sequences. This use of a multi-headed attention mechanism is intended to enable the model to attend to information from different representation subspaces from different positions in parallel [8].

Building upon the idea that the attention mechanism in LLMs capture the dependencies between the input sequence, the attention weights produced by the transformer's attention heads has been posited and explored as a way of quantifying the importance of the input sequence tokens with regards to the output of the model. Works such as [9] proposes a visualization tool for the attention generated by an LLM as a way to debug the model, citing that the attention of the model (or lack thereof) can be indicative of the model's reason for a displayed ability or inability to perform a certain task for a particular input (e.g. the model is unable to understand that "sick" in that particular context was a synonym for "awesome"). This idea of quantifying importance of the input sequence tokens has also been posited as an explanation method for model behaviors, however, the usage of attention as explanations is still debated [10].

Moreover, works for hallucination detection and mitigation have likewise explored the attention mechanism as a way to detect potentially hallucinatory content and/or steer the model away from generating hallucinated responses for both factuality (responses that are factually wrong) and faithfulness (responses that stray from the instruction or context) hallucinations [11]. The works exploring this line of hallucination detection and mitigation employ various methods to leverage attention weights to perform detection or mitigation. For example, frameworks such as Inference-Time Intervention (ITI) [12], WACK [13], TruthX [14], LLM-Check [15], LapEigvals [16], and Lookback Lens [17] all utilize the information contained in the model's attention maps or intervene in the attention heads to detect or mitigate hallucinations.

The wide variety of applications for the attention weights produced by LLMs, and the encouraging positive results for these applications, particularly for hallucination detection and mitigation, motivates this research to explore attention patterns in LLMs, particularly with regards to their ability to comprehend, and incorporate contextual information to respond adequately to the user prompts.

3 Methods

The approach of this paper to investigate the ability of LLMs to comprehend and incorporate contextual information utilizes a reading comprehension style test that involves understanding a given context paragraph and answering a question based on the paragraph. This approach is inspired by SQuADRUn [18], specifically with the idea of unanswerable questions. The desiderata of relevance and existence of plausible answers for unanswerable questions is intended to alleviate the weakness of the original SQuAD dataset [19] which could be exploited by models through learning context and type matching heuristics. This paper seeks to utilize this idea in the context-question pairs for the created evaluation dataset to better ascertain whether or not the LLM being examined truly understands the context and question rather than utilizing heuristics. Moreover, in conjunction with the unanswerable questions, various context-question pair relationships as shown in Fig. 1 are used to see in more detail how the attention patterns of the examined LLMs differ for these scenarios.

	Relevant Context	Irrelevant Context
Answerable Question	**Type 0:** Context has the actual answer	**Type 1:** Context has the answer but is for a different entity
Unanswerable Question	**Type 2:** Context is for the right entity but the question cannot be answered from it	**Type 3:** Context is for a different entity and the question cannot be answered using it

Fig. 1. Types of Context-Question Pairs.

3.1 Dataset

As mentioned above, to evaluate and examine the LLMs, a custom question and answering dataset[1] has been created for this study by the researchers. The

[1] Dataset and code is available at: https://github.com/RobertJoachimEncinas/Evaluating-Attention-Patterns-of-LLMs-Using-Adversarial-Reading-Comprehension.

data set comprises 80 context-question pairs for 10 different topics[2] with each topic having two distinct entities. The usage of the four categorization in Fig. 1 is the reason as to why entities come in pairs and a custom dataset was made as the context of the entity pairs will be swapped to generate the "Irrelevant Context" context-question pairs. Additionally, this is also the reason for the custom dataset as opposed to using SQuADRUn, as the necessity of entities of the same topic to come in pairs would require more work in filtering the original dataset and adding supplementary entries as needed.

For the creation of the dataset, Meta's online chatbot was used to generate 10 candidate entities for each of the different topics. The rationale behind using a chatbot to generate the candidate entities was to maximize the likelihood that the models being tested would have knowledge regarding the entities. The model needs to have knowledge of the entities as the system prompt for the models directs it to answer the question based on its own knowledge whenever possible. From these 10 candidate entities, two entities which were similar to each other were selected for the dataset. Similarity is important here due to the aforementioned swapping of contexts. Once the list of all distinct entities was gathered, context passages were then retrieved from the English Wikipedia pages for those entities. From these contexts, Type 0 and Type 2 entries in the dataset were then created. For Type 2 questions, the "other neutral" negative example of SQuADRUn, which is a question where the paragraph does not imply an answer, was used as a basis to guide the creation of the questions. Using Type 0 and Type 2 context-question pairs, Type 1 and Type 3 were generated by swapping the contexts of the two entities belonging to the same topic (e.g. question about basketball gets the context for volleyball and vice versa). This swapping of contexts is akin to the "entity swap" negative example of SQuADRUn.

With the context-question pairs all completed, human annotation for the relevance of all words in the contexts and questions was performed. The relevance scores for all of the words in the contexts and questions are either 0, 0.25, 0.5, 0.75, or 1. The relevance scoring criteria for the contexts and questions vary slightly and these criteria are listed in Tables 1 and 2. Finally, as LLMs parse input sequences as a series of tokens rather than words, the contexts and questions were tokenized using the respective tokenizers of each of the models being examined. The relevance score was then assigned to these tokens through having all tokens that comprise a word have the same relevance score as the word.

3.2 Data Collection and Analysis

The research examines four LLMs: LLaMA3.2-1B-Instruct, LLaMA3.2-3B-Instruct, Qwen2.5-1.5B-Instruct, and Qwen2.5-3B-Instruct. All of these models were run using the HuggingFace transformers library. Additionally, all the models

[2] The 10 different topics are: Natural Formations, Landmarks, Countries, Natural Science Concepts, Inventions, Literary Works, Math Concepts, Sports, Animals, and Social Science Concepts.

Table 1. Scoring Criteria for Context

Score	Criteria
0	Completely unrelated to question
0.25	Relevant as a supporting word for the answer but for a different entity
0.5	Directly relevant to answering the question but for a different entity
0.75	Relevant as a supporting word for the answer
1	Directly relevant to answering the question

Table 2. Scoring Criteria for Question

Score	Criteria
0.25	Connective words
0.5	Gives tertiary context or descriptors to the main question
0.75	Gives secondary context or descriptors to the main question
1	Directly related to the main question being asked

receive the same system prompt of "You are an intelligent chatbot answering factual questions correctly based on your own knowledge. A paragraph is provided as well that may or may not be helpful information. Discern if the paragraph's contents help you answer the question or not accordingly. Just return the answer, do not explain.". The context-question pairs were passed to the system through 2 prompts formatted as "Context: context" and "Question: question".

After running all of the inputs, the respective attention maps output by the models and their answers were then saved and preprocessed. As the research is concerned with the overall attention pattern of the LLM, preprocessing of the attention maps was performed through averaging the attention given by the newly generated tokens t over all layers l and attention heads h was used. These averaged attentions for both the context A_c and question A_q was then divided by the averaged attentions given to the whole sequence A to get the ratio of attention on the context or question over all attention weights. This ratio is denoted as W_c and W_q. Mathematically, this operation can be seen in the equations below where α_{ijk} denotes the attention map for the whole sequence[3].

$$A = \frac{1}{t \cdot l \cdot h} \sum_{i=1}^{t} \sum_{j=1}^{l} \sum_{k=1}^{h} a_{ijk} \tag{1}$$

$$W_x = \frac{A_x}{A} \tag{2}$$

[3] To account for the tokens generated later on in the sequence, the average for these only takes into consideration timesteps in which they are present.

4 Results and Discussion

The results of the experiment can be best analyzed through 3 major lenses: Alignment of LLM attention with human attention (Sect. 4.1), attention patterns for the different pair types (Sect. 4.2), and the correlation between attention and hallucinations (Sect. 4.3) which are discussed in more detail in their specific subsections.

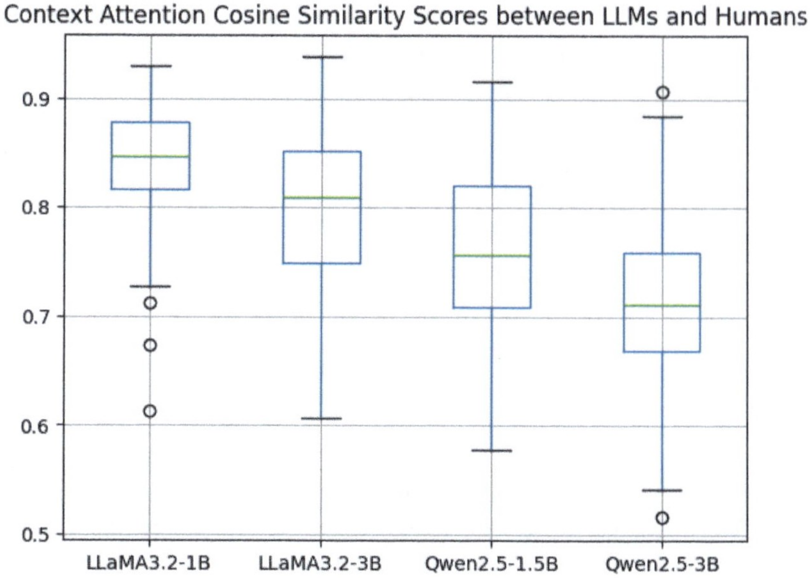

Fig. 2. Context Attention Cosine Similarity Scores between LLMs and Humans.

4.1 Alignment with Human Attention

The first major area of note to be examined is the alignment of LLM and human attention patterns. For evaluating the alignment, cosine similarity was used to compare the two vectors representing the attention patterns. However, as the output distributions of the two vectors vary drastically, the human attention vector had softmax applied to it first, similarly to how LLM attentions are computed, to flatten out the distributions, while retaining the relative distances between the attention weights. The results for this are shown in Figs. 2 and 3.

As can be seen in the data, the alignment of attention patterns between humans and LLMs are rather close with the model most similar to humans for context similarity, LLaMA3.2-1B, on average having around 0.84 cosine similarity for context span and 0.88 for question span. With even the lowest cosine similarity among the models on average being around 0.71 and 0.87 for the context span and question span respectively. This result suggests that the models

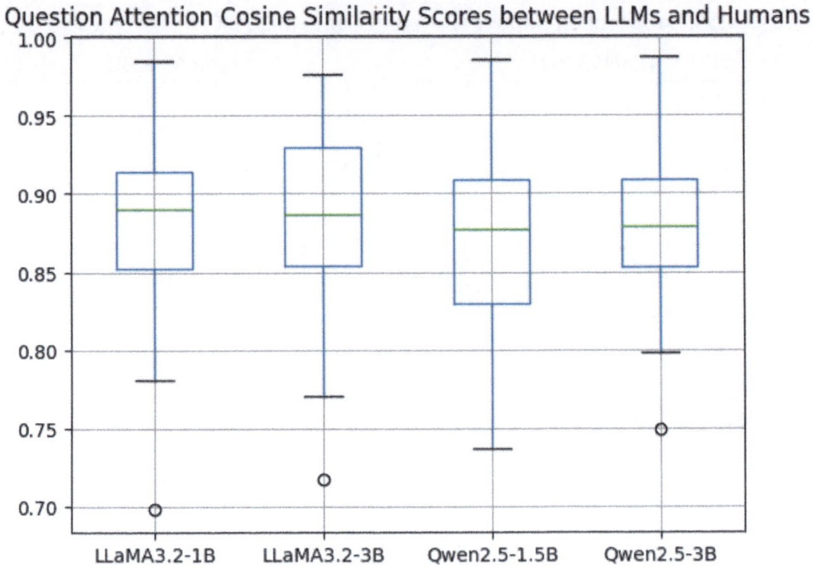

Fig. 3. Question Attention Cosine Similarity Scores between LLMs and Humans.

examined evaluate the importance of the prompt similarly to how humans do to a high degree.

4.2 Attention Patterns for Various Context-Question Pair Types

As can be seen in Fig. 4, with the exception of Qwen2.5-1.5B, all models on average attributed significantly higher attentions to the context sequence of the input for Type 0 entries as compared to other entry types. These results seem to indicate that the LLMs are capable of discerning whether or not a paragraph is relevant or not to answering the question and acts accordingly. Moreover, as can be seen in Fig. 5, the relative consistency of question attention scores further supports this as the models have a tendency to decline answering Type 1 and Type 3 questions citing the lack of supporting evidence from the given context, showing a degree of awareness on discerning context relevance that manifests as well in the attention patterns.

4.3 Correlation Between Attention and Hallucinations

To better understand the relationship between attention patterns and hallucinations, we can examine the attention patterns for both hallucination types for LLMs [11]. From the experiments both factuality and faithfulness hallucinations were identified by the researchers through a manual inspection of the LLM responses. Specifically, an answer was deemed a factuality hallucination if the answer provided is factually wrong given the cutoff date of the training data of

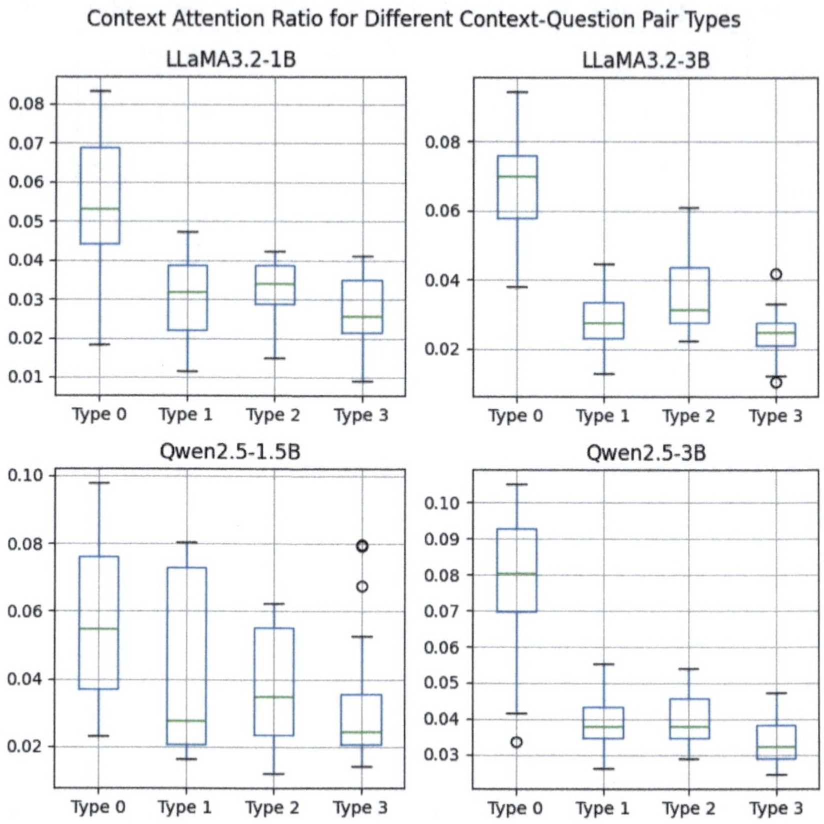

Fig. 4. Context Attention Ratio for Different Context-Question Pair Types.

Table 3. Factuality Hallucinations by Type

Model	Type 0	Type 1	Type 2	Type 3	Total
LLaMA3.2-1B	3	3	7	5	18
LLaMA3.2-3B	1	1	1	0	3
Qwen2.5-1.5B	2	2	1	0	5
Qwen2.5-3B	4	0	0	0	4

Table 4. Faithfulness Hallucinations by Type

Model	Type 0	Type 1	Type 2	Type 3	Total
LLaMA3.2-1B	1	0	0	0	1
LLaMA3.2-3B	0	0	0	0	0
Qwen2.5-1.5B	1	6	1	2	10
Qwen2.5-3B	0	0	0	0	0

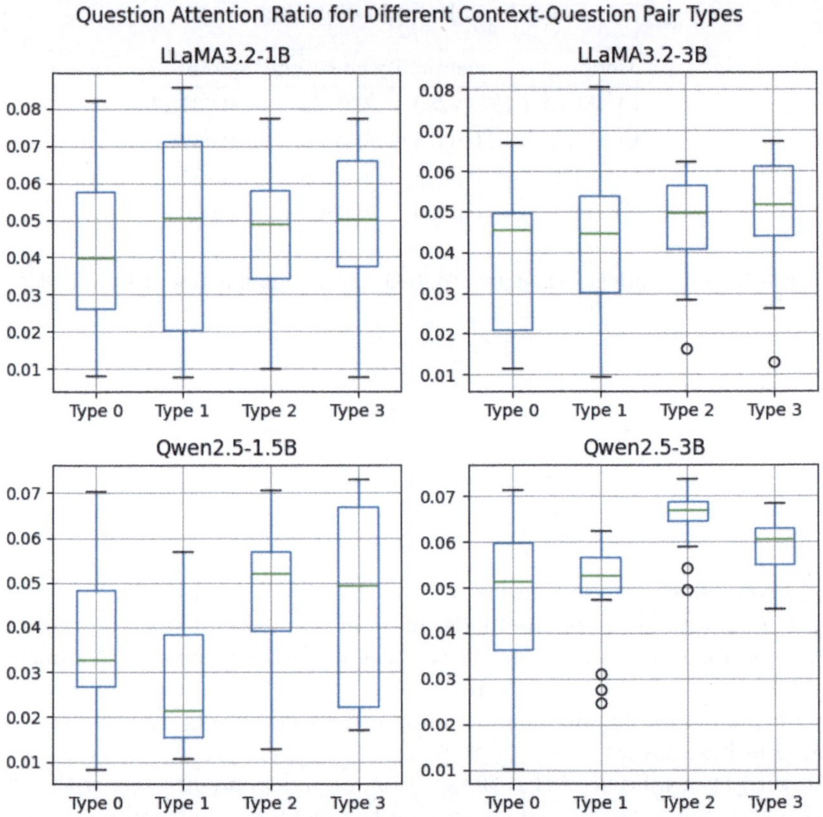

Fig. 5. Question Attention Ratio for Different Context-Question Pair Types.

the LLM, while faithfulness hallucinations are hallucinations wherein the model returned a response that does not follow the given prompt. Tables 3, 4, and 5 show the statistics of these hallucination for all the examined models. As can be seen in the tables, LLaMA3.2-1B and Qwen2.5-1.5B have the highest hallucination rates for factuality and faithfulness hallucinations respectively. As such the following subsections focus on analyzing these two models specifically.

LLaMA3.2-1B. The first model to be examined for the relationship between hallucinations and attention patterns is LLaMA3.2-1B which displays a significant number of factuality hallucinations in the responses. However, despite the large number of hallucinations, the trends of the 1B LLaMA model follows the trends of the 3B models examined which had significantly less hallucinations which is concerning as it implies these hallucinations not only are hard to detect by humans without independent verification methods, it also does not appear as much in the attention patterns of the models which could signify that either the

Table 5. Hallucination Rates

Model	Factuality	Faithfulness	Total
LLaMA3.2-1B	0.2250	0.0125	0.2375
LLaMA3.2-3B	0.0375	0.0000	0.0375
Qwen2.5-1.5B	0.0625	0.1250	0.1875
Qwen2.5-3B	0.0500	0.0000	0.0500

Table 6. Independent t-Test for Context Cosine Similarities of LLaMA3.2-1B

Type	p-value	Significant
0	0.746791	No
1	0.064777	No
2	0.887759	No
3	0.268856	No

model itself believes these answers to be correct or the averaging of the attention weights across all layers of the model may be the cause of this observed effect.

As mentioned in [20], papers in the literature have found that transformer-based LLMs have been shown to tend to encode more syntactical information in earlier layers, while having the encoding of factual or semantic information be encoded in later layers.

To verify the validity of this observation, an independent t test with a threshold of $p < 0.05$ was performed comparing the mean context and question attention cosine similarities between hallucinated samples and non-hallucinated samples per type. Tables 6 and 7 shows the p-values of the t-test and the results appear to confirm the initial hypothesis that the model may itself believe the answer to be correct as none of means for both context and question similarity scores for any type have a significant difference between the means. This result shows that for factuality hallucinations, the similarity of attention patterns with humans may not be a viable indicator of hallucinations without further preprocessing.

Table 7. Independent t-Test for Question Cosine Similarities of LLaMA3.2-1B

Type	p-value	Significant
0	0.070108	No
1	0.921699	No
2	0.453604	No
3	0.690427	No

Table 8. Independent t-Test for Context Cosine Similarities of Qwen2.5-1.5B

Type	p-value	Significant
0	0.290979	No
1	0.042962	Yes
2	0.007354	Yes
3	0.033486	Yes

Table 9. Independent t-Test for Question Cosine Similarities of Qwen2.5-1.5B

Type	p-value	Significant
0	0.550603	No
1	0.037055	Yes
2	0.015674	Yes
3	0.181179	No

Qwen2.5-1.5B. As mentioned in Sect. 4.2, Qwen2.5-1.5B was the only LLM examined that was observed to not follow the attention pattern trends for the different context-question pair types. Upon examination of the answers of Qwen2.5-1.5B and the subsequent analysis, it was found that this break in the trend was caused by faithfulness hallucinations in the generated answer. Instead of answering using parametric knowledge or declining to answer citing the irrelevance of the context paragraph in relation to the question being asked like the other models, the model would instead repeat a portion or the entirety of the context paragraph as its response. This behavior was unique to the Qwen 1.5B model and was not present in any of the other models.

To further analyze this phenomenon, an independent t-test was performed in a similar manner to the LLaMA 1B model. Table 8 and 9 shows the p-values of the t-test. As can be seen in the table, the p-values for context similarity scores indicate that there is a significant difference between the means for all types except Type 0. Meanwhile, for question similarity scores, Types 1 and 2 show a significant difference between the means. This appears to imply that faithfulness hallucinations have a visible effect on the evaluation of the model of the importance of both the context and the questions which may be leveraged as another viable indicator for faithfulness hallucinations for detectors and mitigation frameworks.

5 Limitations

Due to variety of LLM architectures and designs as well as their weights, the results of these results are specifically scoped to the examined LLMs. Though an attempt to set a degree of generality of the results was done through using 4 different models, there is still no guarantee that the results are wholly applicable to other models and further research is needed to confirm whether or not the results are applicable.

Moreover, due to the limitations on the manpower needed for a large number of data annotations of the evaluation dataset, the number of samples for each model was limited to 80 samples. Efforts to diversify the topics to cover a wider area was done to try and increase the generality of the data; however, future research is still recommended to increase the sample size of the evaluation dataset.

Another limitation of this study lies in the preprocessing method the attention weights of the LLMs examined. Specifically, the paper preprocessed the weights by taking the average across all layers which as mentioned in Sect. 4.3 may have impacted the results for factuality hallucinations. Taking the average across all layers was chosen in spite of this limitation as to maintain a sense of generality for the approach as the varying LLM designs and architectures impact what specifically constitutes as a "later layer".

Finally, as both factuality and faithfulness hallucinations are broad categorizations for the wide variety of hallucinations that manifest in real world settings for LLMs, this research cannot possibly show that these trends for both hallucinations types will hold for all possible manifestations of hallucinations.

6 Conclusion

As observed in prior works focusing on utilizing attention weights for model debugging and hallucination detection and mitigation, this paper shows that the hallucinations produced by the model in its responses have been shown empirically in this paper to have quantifiable and visible effects that manifests itself in the attention patterns of the four examined LLMs.

Moreover, there appears to be a positive correlation between the similarity of LLM attention and human attention with the likelihood of the LLM to generate non-hallucinatory responses with regards to following the given user prompt. However, the same cannot be said for factuality hallucinations where cosine similarities with human attention do not have significant differences. These findings suggest that while attention patterns provide valuable insight into the inner workings of the LLMs as well as being viable indicators for faithfulness hallucinations, there are still significant limits to the viability of LLM and human attention alignment for factuality hallucination detection.

However, as mentioned in the limitations, these findings are merely general and scoped only to the specific hallucinations that appeared in experiments. Future research into this topic, especially for research focusing on applying the findings to a hallucination mitigation or detection framework, may find it useful to have additional preprocessing done on the attention maps, like scoping it to earlier or later layers only for the averaging of attention weights.

Lastly, the results that the attention patterns of LLMs having a high degree of similarity with human attention shows that the goal of capturing and modeling dependencies of the token sequence through the attention mechanism is demonstrably effective in understanding and interpreting text in the same way that humans do.

References

1. Kadavath, S., et al.: Language models (mostly) know what they know (2022). https://arxiv.org/abs/2207.05221
2. Brown, T., et al.: Language models are few-shot learners. In: Larochelle, H., Ranzato, M., Hadsell, R., Balcan, M., Lin, H. (eds.) Advances in Neural Information Processing Systems, Vancouver, Canada, 2020, vol. 33, pp. 1877–1901, vol. 33. Curran Associates, Inc. (2020). https://proceedings.neurips.cc/paper_files/paper/2020/file/1457c0d6bfcb4967418bfb8ac142f64a-Paper.pdf
3. Naveed, H., et al.: A comprehensive overview of large language models (2024). https://arxiv.org/abs/2307.06435
4. Min, S., et al.: FActScore: fine-grained atomic evaluation of factual precision in long form text generation. In: Bouamor, H., Pino, J., Bali, K. (eds.) Proceedings of the 2023 Conference on Empirical Methods in Natural Language Processing, Singapore, 2023, pp. 12,076–12,100. Association for Computational Linguistics (2023). https://doi.org/10.18653/v1/2023.emnlp-main.741. https://aclanthology.org/2023.emnlp-main.741/
5. Manakul, P., Liusie, A., Gales, M.: SelfCheckGPT: zero-resource black-box hallucination detection for generative large language models. In: Bouamor, H., Pino, J., Bali, K. (eds.) Proceedings of the 2023 Conference on Empirical Methods in Natural Language Processing, Singapore, 2023, pp. 9004–9017. Association for Computational Linguistics (2023). https://doi.org/10.18653/v1/2023.emnlp-main.557. https://aclanthology.org/2023.emnlp-main.557/
6. Wan, Y., et al.: LogicAsker: evaluating and improving the logical reasoning ability of large language models. In: Al-Onaizan, Y., Bansal, M., Chen, Y.N. (eds.) Proceedings of the 2024 Conference on Empirical Methods in Natural Language Processing, Miami, Florida, USA, 2024, pp. 2124–2155. Association for Computational Linguistics (2024). https://doi.org/10.18653/v1/2024.emnlp-main.128. https://aclanthology.org/2024.emnlp-main.128/
7. Mirzadeh, I., Alizadeh, K., Shahrokhi, H., Tuzel, O., Bengio, S., Farajtabar, M.: GSM-symbolic: understanding the limitations of mathematical reasoning in large language models (2024). https://arxiv.org/abs/2410.05229
8. Vaswani, A., et al.: Attention is all you need. In: Advances in Neural Information Processing Systems, vol. 30, pp. 261–272 (2017)
9. Vig, J.: Debugging machine learning models. In: ICLR Workshop, vol. 3 (2019)
10. Bibal, A., et al.: Is attention explanation? An introduction to the debate. In: Muresan, S., Nakov, P., Villavicencio, A. (eds.) Proceedings of the 60th Annual Meeting of the Association for Computational Linguistics (Volume 1: Long Papers), Dublin, Ireland, 2022, pp. 3889–3900. Association for Computational Linguistics (2022). https://doi.org/10.18653/v1/2022.acl-long.269. https://aclanthology.org/2022.acl-long.269/
11. Huang, L.: A survey on hallucination in large language models: principles, taxonomy, challenges, and open questions. ACM Trans. Inf. Syst. (2024)
12. Li, K., Patel, O., Viégas, F., Pfister, H., Wattenberg, M.: Inference-time intervention: eliciting truthful answers from a language model. In: Oh, A., Naumann, T., Globerson, A., Saenko, K., Hardt, M., Levine, S.

(eds.) Advances in Neural Information Processing Systems, New Orleans, United States of America, 2023, vol. 36, pp. 41,451–41,530. Curran Associates, Inc. (2023). https://proceedings.neurips.cc/paper_files/paper/2023/file/81b8390039b7302c909cb769f8b6cd93-Paper-Conference.pdf
13. Simhi, A., Herzig, J., Szpektor, I., Belinkov, Y.: Constructing benchmarks and interventions for combating hallucinations in LLMs (2024). https://arxiv.org/abs/2404.09971
14. Zhang, S., Yu, T., Feng, Y.: TruthX: alleviating hallucinations by editing large language models in truthful space. In: Proceedings of the Association for Computational Linguistics, Bangkok, Thailand, 2024, pp. 8908–8949 (2024). https://doi.org/10.18653/v1/2024.acl-long.483. https://aclanthology.org/2024.acl-long.483/
15. Sriramanan, G., Bharti, S., Sankar, S.V., Saha, S., Kattakinda, P., Feizi, S.: LLM-Check: investigating detection of hallucinations in large language models. In: NeurIPS 2024, Vancouver, Canada, 2024, vol. 37, pp. 34188–34216. Curran Associates, Inc. (2024). https://proceedings.neurips.cc/paper_files/paper/2024/file/3c1e1fdf305195cd620c118aaa9717ad-Paper-Conference.pdf
16. Binkowski, J., Janiak, D., Sawczyn, A., Gabrys, B., Kajdanowicz, T.: Hallucination detection in LLMs using spectral features of attention maps (2025). https://arxiv.org/abs/2502.17598
17. Chuang, Y.S., Qiu, L., Hsieh, C.Y., Krishna, R., Kim, Y., Glass, J.R.: Lookback Lens: detecting and mitigating contextual hallucinations in large language models using only attention maps. In: Proceedings of the 2024 Conference on Empirical Methods in Natural Language Processing, Miami, Florida, USA, 2024, pp. 1419–1436. Association for Computational Linguistics (2014). https://doi.org/10.18653/v1/2024.emnlp-main.84. https://aclanthology.org/2024.emnlp-main.84/
18. Rajpurkar, P., Jia, R., Liang, P.: Know what you don't know: unanswerable questions for SQuAD. In: Proceedings of the 56th Annual Meeting of the Association for Computational Linguistics (Volume 2: Short Papers), Melbourne, Australia, 2018, pp. 784–789. Association for Computational Linguistics (2018). https://doi.org/10.18653/v1/P18-2124. https://aclanthology.org/P18-2124
19. Rajpurkar, P., Zhang, J., Lopyrev, K., Liang, P.: SQuAD: 100,000+ questions for machine comprehension of text, pp. 2383–2392 (2016)
20. Chuang, Y.S., Xie, Y., Luo, H., Kim, Y., Glass, J., He, P.: DoLA: decoding by contrasting layers improves factuality in large language models (2024). https://arxiv.org/abs/2309.03883

Enhancing Conversational Access to Patient Health Data Through LLM-Powered Chat Interfaces

Nektarios Machner(✉), Phillip Schneider, and Florian Matthes

Technical University of Munich, Munich, Germany
{nektarios.machner,phillip.schneider,matthes}@tum.de

Abstract. In decentralized and ambulant healthcare, smart devices generate vast amounts of patient data, enabling real-time health monitoring outside clinical settings. However, the scale and complexity of this information make timely access and interpretation challenging for healthcare professionals. This study investigates how large language models can enhance a conversational interface that allows medical experts to query patient health data collected by smart devices, such as heart rate, blood pressure, and mental health scores. Built on a SQL-based backend, our conversational assistant allows users to dynamically explore the information using natural language. We focus on evaluating the effectiveness, usability, and intent recognition performance of the large language model-powered fallback mechanism within the system. The evaluation consists of an automated benchmarking procedure using synthetic utterances to assess fallback classification performance across various input scenarios, complemented by qualitative feedback from five medical professionals following exploratory chat interactions with the assistant.

Keywords: conversational interface · information retrieval · dialogue fallback · large language models · health data

1 Introduction

The increasing availability of patient health data from wearable devices and mobile health apps presents new opportunities for proactive and personalized care. However, much of this data remains underutilized in healthcare practice due to barriers in accessibility, interpretability, and integration into existing workflows. Conversational interfaces offer a promising approach to bridge this gap by allowing medical professionals to interact with complex datasets using natural language. Recent reviews of conversational search systems highlight how integrating large language models (LLMs) into such interfaces can improve their ability to retrieve and synthesize information over multiple dialogue turns, underscoring the relevance of these advances to healthcare applications [1]. In this work, we present a prototype medical assistant that integrates a large language

H. Unger and P. Meesad (Eds.): NLPIR 2025, LNNS 1904, pp. 858–871, 2026.
https://doi.org/10.1007/978-3-032-20897-2_60

model into a dialogue system, enabling physicians to query and interpret structured patient data through a chat-based interface. Built on top of the RASA framework and connected to a relational database of anonymized real health records, the system supports dynamic, context-aware interactions. The defining feature of our approach is its fallback mechanism, which delegates low-confidence inputs to an LLM, substantially enhancing robustness and flexibility in handling diverse user queries. We evaluate the system through an automatic evaluation and human expert feedback to better understand its practical utility and limitations.

2 Related Work

Conversational systems in healthcare have progressed from compact transformer models to sophisticated LLM-based assistants. Early healthcare conversational systems often relied on transformer architectures like BERT (or its domain-specific variants). While effective within familiar domains, these systems struggled with unexpected or ambiguous queries. Zhang et al. [2] showed that BERT-based intent classifiers fail significantly on in-domain yet out-of-scope inputs, revealing brittle behavior under novel queries. Nerella et al. [3] further highlighted issues in these models, particularly around interpretability, adaptability, and bias, especially in high-ambiguity medical contexts. These limitations pose significant risks in healthcare settings, where queries may be diverse, evolving, or nuanced.

While general-purpose LLM-based medical chatbots like ChatDoctor [4] can provide informative responses, they typically lack access to individual patient data, resulting in generic, non-personalized interactions. In contrast, recent LLM-powered systems have begun to enable more tailored access to health information. For example, Al Attrach et al. [5] introduced M3, which enables clinicians to query the MIMIC-IV database in plain English via an LLM-translated SQL interface. Similarly, Schmiedmayer et al. [6] presented LLMonFHIR, a mobile LLM assistant for personal health record queries, with expert users rating its responses highly accurate, despite occasional issues with summarizing complex conditions. These kinds of systems excel at structured retrieval and summarization; however, they seldom use LLMs explicitly for intent fallback when queries deviate from expected formats.

Given their ability to generalize beyond the training distribution, LLMs present a promising solution for handling unexpected or ambiguous inputs. Naved et al. [7] evaluate GPT-4 as an LLM fallback to a supervised multi-class classifier on US health-system search-bar logs. They found that while GPT-4 matched the classifier's performance when provided with label lists via in-context learning, it particularly enhanced coverage in low-frequency classes, suggesting its value in improving intent recognition for atypical search query inputs. Mahmood et al. [8] analyze user interactions with an LLM-powered voice assistant across medical self-diagnosis, creative planning, and debate scenarios and report that LLMs absorbed most traditional intent-recognition failures and enabled richer, more resilient conversational flows by proactively addressing conversation breakdowns.

This study explores the integration of an LLM-powered fallback mechanism for intent recognition within an intuitive conversational interface that accesses patient data from smart devices, such as heart rate, blood pressure (BP), and mental health scores. The system is evaluated using a combination of synthetic benchmark utterances to measure classification accuracy on ambiguous and unexpected queries, along with qualitative feedback from three practicing medical professionals and two medical data analysts. While related architectures have been explored in other domains [9], this work presents one of the first structured evaluations of an LLM-based fallback approach in a health-focused dialogue setting focused on exploring real-world, device-generated patient data.

3 Method

In the following, we provide an overview of the system architecture along with further relevant details regarding the implementation of our prototype, including the integration of the LLM. Since both the patients and their associated data are German, the entire system is designed to operate in German. This applies to the user interface, all system inputs and outputs, the medical experts involved in the evaluation, and the overall interaction flow.

3.1 System Architecture

The prototype is implemented entirely in Python and is built around a classical chatbot architecture using the open-source RASA framework[1]. An overview of the system architecture is presented in Fig. 1. The system consists of a Streamlit-based[2] web interface, a RASA server for dialogue management and intent classification, and a custom action server connected to a relational database for handling domain-specific backend logic. The RASA server orchestrates the conversation by selecting responses or triggering backend actions based on detected intents and dialogue context.

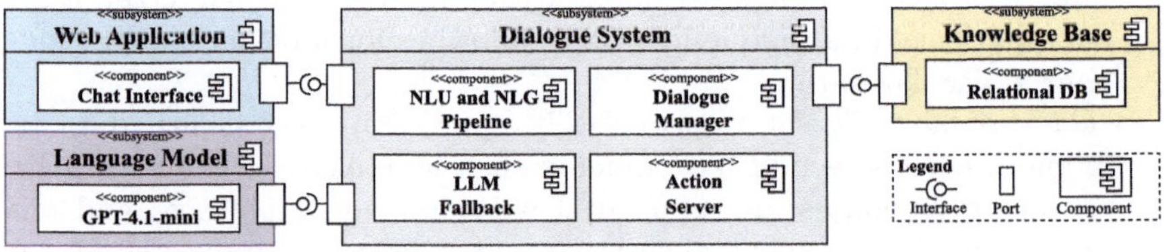

Fig. 1. System architecture component diagram.

The individual components are described in more detail below:

[1] https://rasa.com/.
[2] https://streamlit.io/.

- **Web Application:** The frontend is implemented using Streamlit, providing a lightweight and intuitive interface for users to submit natural language queries and receive responses in real time. Responses can include plain text as well as static graphics, which are rendered from base64-encoded images. An expandable help section offers contextual guidance to assist users with input formulation. An example of the user interface is shown in Fig. 2.
- **Dialogue Manager:** The dialogue manager maintains conversation state, performs slot filling to extract and track key user-provided information, and predicts the next system action using learned policies. This enables context-aware, multi-turn interactions.
- **Action Server:** Custom backend logic is implemented on an action server, which responds to predicted actions by querying the database, processing results, and formatting dynamic responses. The server operates as a secure interface between the RASA dialogue system and external data sources.
- **NLU and NLG Pipeline:** The system uses RASA's modular NLU pipeline for intent classification and entity recognition, followed by rule-based and template-driven natural language generation (NLG) [10]. This supports deterministic response patterns for recognized intents.
- **LLM Fallback:** When the NLU model produces a low-confidence intent prediction, RASA's fallback policy is triggered. In such cases, control is delegated to a custom fallback action, which invokes a large language model to classify the input more flexibly. The fallback mechanism is described in detail in Sect. 3.3.
- **Knowledge Base:** Patient data is securely stored in a relational database. Neither the RASA model nor the LLM has direct access to this data. Instead, all queries are mediated by the action server through parameterized SQL statements. The schema consists of several primary tables:

    - *Patient Table:* Stores demographic and health profile information, including name, sex, birth date, general health status, and known medical conditions. This contextual information supports a more personalized interpretation of health signals.
    - *Blood Pressure Table:* Contains timestamped measurements of systolic and diastolic blood pressure, along with pulse values. Each record is linked to a patient via a unique identifier, allowing for longitudinal tracking of cardiovascular data.
    - *Questionnaire Table:* Captures self-reported mental health data using standardized screening instruments such as GAD-7, PHQ-8, and PHQ-9. These responses are associated with individual patients and enable longitudinal assessment of anxiety and depression symptoms alongside physiological data.

3.2 Conversation Flow

Our prototype's basic conversation flow, supported by RASA, consists of intent recognition through natural language understanding (NLU), followed by

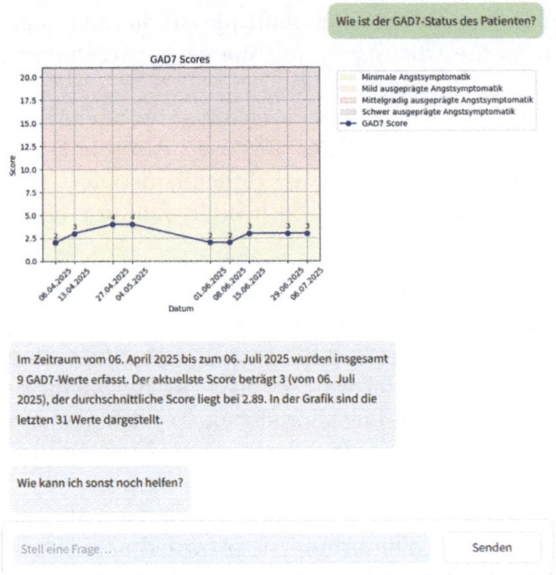

Fig. 2. Illustrative screenshot of the chat interface (displayed in German language).

response generation via a rule-based or machine-learning-driven dialogue policy. When a user input is received, RASA classifies the intent and extracts relevant entities, then selects an appropriate response or action based on predefined rules or stories. In our prototype we extended this flow by customizing the NLU fallback behavior. Specifically, we added a custom action that is triggered when RASA's confidence in intent classification falls below a defined threshold. This action calls an LLM to reanalyze the user input and attempt a secondary intent match. If the LLM is still unable to identify a clear intent, the action instead generates a set of likely suggestions to guide the user toward valid query formulations.

3.3 LLM Integration

To enhance the robustness of the dialogue system, we integrate a large language model as a fallback mechanism for handling low-confidence predictions from the primary classifier. This approach leverages the superior natural language understanding capabilities of LLMs to interpret diverse and linguistically varied user inputs that may otherwise lead to misclassifications.

The fallback mechanism is triggered when the primary classifier returns a prediction with confidence below a predefined threshold. This threshold is empirically tuned to balance precision and recall, ensuring that uncertain or borderline cases are escalated to the LLM for more nuanced interpretation.

Upon invocation, the LLM produces one of three standardized responses: (1) a clear intent match when the input unambiguously maps to a predefined intent, (2) an ambiguous match when multiple candidate intents are plausible, or (3) a no-match response when the input does not correspond to any supported intent.

This structured classification scheme ensures consistent handling of uncertain cases and provides a robust fallback for edge cases not well covered by the primary RASA classifier. By delegating difficult instances to the LLM, we aim to achieve a system that maintains both high accuracy and coverage across a wide range of user inputs.

Model and Configuration. The prototype leverages the `gpt-4.1-mini` [11] model, accessed via the OpenAI API, to perform intent classification using a structured prompt. This model was selected for its balance of cost efficiency, latency, and capability in handling short-form, structured outputs. Despite being a lightweight variant, `gpt-4.1-mini` demonstrates strong performance on classification tasks when paired with well-defined prompts.

To ensure deterministic and reproducible behavior, the model is configured with a temperature of 0.0. The prompt is provided as a single system message, and the output is constrained to a maximum of 128 tokens. This limit is sufficient, as the expected output is concise and follows a rigid format defined by the classification schema.

Prompt Design. The intent classifier operates under a fixed system instruction and structured output format to ensure consistency and enforceability of responses.

System Instruction. The language model is prompted with the directive:

```
You are a medical chatbot intent classifier. Your task: respond strictly
in one of the following formats without additional commentary.
```

Classification Categories. The model is expected to classify user input into one of three categories:

1. **Clear Intent Match:** The input unambiguously corresponds to a known intent.
   `INTENT: <INTENT_NAME>`
2. **Ambiguous Intent Match:** The input may correspond to multiple possible intents. The model should return a ranked list of suggestions.
   `SUGGESTION: <INTENT_NAME>`
   `SUGGESTION: <INTENT_NAME>`
3. **No Match:** The input does not align with any supported intent.
   `NO_MATCH`

Prompt Input Context. At runtime, the prompt is populated with the following structured variables:

- `recent_history`—the recent conversation history
- `intents_md`—the full list of supported intents with their definitions
- `user_input`—the most recent message from the user

Example Intents. For instance, the intent `bp_current_status` is triggered by a query such as "What is the current blood pressure?" (asked in German), prompting the system to return the most recent measurement. Similarly, the intent `gad7_status_details` responds to questions like "What do the patient's GAD-7 scores look like?" (asked in German) to provide an overview of anxiety symptoms.

Dialogue Fallback Strategy. Based on the output we receive from the LLM, we proceed with handling the request in the following ways:

1. **Clear Match:** When an intent is unambiguously matched by the LLM, we dispatch the corresponding RASA action associated with that intent, effectively handing control back to RASA's dialogue manager for further processing.
2. **Ambiguous Match:** When the LLM cannot confidently assign a single intent to the user input but identifies one or more plausible candidates, it returns up to two suggested intents. These are presented to the user as clickable buttons, each carrying the appropriate payload so that, when selected, the corresponding RASA action is dispatched seamlessly without requiring the user to type any further input.
3. **No Match:** If the user input cannot be matched to any known intent by the LLM, we fall back to a generic response informing the user that the system was unable to determine their request. To keep the conversation moving, we randomly select two intents from the list of available options and present them as clickable suggestion buttons, similar to the ambiguous case.

In both the ambiguous and no-match cases, the user is not constrained to select any of the suggested options. Instead, they remain free to continue the conversation by entering a new query, allowing for a more natural and flexible interaction flow.

4 Results and Discussion

We combine automatic evaluation using large language models to produce quantitative feedback with human expert evaluation by five medical professionals to gather qualitative insights.

4.1 Automatic Evaluation

To systematically evaluate the performance of our LLM-based fallback mechanism, we used an independent language model to generate synthetic test utterances for ten of the intents defined in our system. To mitigate bias, we selected a model from a different family than the fallback mechanism model itself, specifically the open-source `llama-4-maverick-17b-128e-instruct` [12], accessed via the GroqCloud API and configured with a temperature of 0.4 and a maximum token limit of 1,024.

To comprehensively evaluate all potential LLM fallback options defined in our dialogue fallback strategy, we prompted the model to generate utterances representative of the three possible outcome types. The generation process was guided by the following definitions:

- **Clear Match:** Utterances that directly and unambiguously convey the target intent.
- **Ambiguous Match:** Utterances that plausibly reflect the target intent, but could also reasonably be interpreted as a different intent due to linguistic ambiguity or semantic overlap.
- **No Match:** Utterances that are broadly related to healthcare topics but are not intended to correspond to any of the defined intents.

For each intent, we prompted the LLM to generate 20 clear-match and 20 ambiguous-match utterances, providing as in-context examples three training sentences originally used to train the RASA model. The generated utterances were manually reviewed for quality and subsequently reduced to 15 per category. When fewer than 15 high-quality utterances were obtained, the model was reprompted until a satisfactory test set was achieved.

In a separate prompt, we instructed the LLM to generate 100 no-match sentences. This prompt incorporated all ten intents with the same three training sentences as in-context examples, while explicitly requiring the model to avoid producing utterances that could plausibly align with any of those intents. For this task, we set the maximum token limit to 5,000. All no-match utterances were manually reviewed to ensure they did not directly correspond to any of the defined intents and were reworded or replaced as necessary.

This procedure yielded 30 utterances per intent, along with 100 general no-match utterances, resulting in a total of 400 user inputs. Each utterance was individually processed by our LLM fallback component, simulating a scenario in which the fallback would be invoked. The system's predictions were then automatically evaluated against the expected outcomes for each case.

For utterances in the *clear match* and *no match* categories, correctness required an exact match to the target intent or a no-match response, respectively. For the *ambiguous match* category, the LLM was expected to generate a list of suggested intents. A prediction was considered correct if this list included the target intent at any position, thereby accounting for the possibility of multiple plausible suggestions.

Table 1. Evaluation results of the fallback mechanism across categories using micro-averaged overall precision, recall, and F1

Fallback category	Support	Precision	Recall	F1	Accuracy
Clear match	150	0.691	0.893	0.779	0.810
Ambiguous match	150	0.905	0.253	0.396	0.710
No match	100	0.735	0.970	0.836	0.905
Overall (micro)	400	0.731	0.672	0.701	0.808

An overview of the evaluation results is presented in Table 1, highlighting performance metrics across all three categories.

The evaluation results in Table 1 reveal distinct strengths and weaknesses of the fallback mechanism across categories. For clear matches, the system demonstrated strong recall (0.893), successfully identifying the majority of relevant utterances, though at the cost of lower precision (0.691), which indicates a tendency to over-predict this category. In contrast, performance on ambiguous matches displayed the opposite pattern: precision was very high (0.905), showing that when the fallback mechanism labeled an utterance as ambiguous, it was almost always correct, but recall was strikingly low (0.253), meaning that many ambiguous cases went undetected. This imbalance resulted in a low F1 score (0.396), highlighting the challenge of handling ambiguity. By comparison, the no-match category exhibited the most robust performance overall, with both high recall (0.970) and good precision (0.735), indicating that the fallback mechanism rarely failed to recognize inputs outside the defined intent space. Taken together, these results point to a system that is highly effective at rejecting irrelevant inputs and generally reliable at retrieving clear matches, but one that struggles to consistently capture ambiguity. Overall accuracy (0.808) and F1 score (0.701) suggest a solid baseline performance, while also underscoring the need for targeted improvements in recall for ambiguous cases to better support nuanced intent recognition.

To examine the results in greater detail, we refer to Table 2, which presents a breakdown of fallback classification performance by intent. This perspective not only shows how many instances were correctly matched within each intent but also highlights which intents are more prone to ambiguity or confusion, thereby offering actionable feedback for refining both the training data and the system's behavior.

The intent-level results reveal consistently high accuracy values (≥ 0.92) across all listed intents. However, this should be interpreted with caution, as overall accuracy is partly inflated by the 100 *no match* utterances included in the global evaluation (Table 1). Within the actual intent categories, results vary in the balance between precision and recall. Strong performance is observed for *BP Current Status* (F1 = 0.822), *BP Outside Target* (F1 = 0.862), and *BP Outlier Details* (F1 = 0.830), which combine high recall with solid precision, yielding reliable classifications. It is noteworthy that both *BP Outlier Details* and *BP*

Table 2. Detailed overview of intent classification results

Intent	Support	Precision	Recall	F1	Accuracy
Basic User Info	30	0.759	0.733	0.746	0.963
BP Current Status	30	0.698	1.000	0.822	0.968
BP Inflection Points	30	1.000	0.567	0.723	0.968
BP List Values	30	0.704	0.633	0.667	0.953
BP Outlier Details	30	0.957	0.733	0.830	0.978
BP Outside Target	30	0.800	0.933	0.862	0.978
BP Trends	30	0.482	0.900	0.628	0.920
Daily Condition Details	30	0.690	0.967	0.806	0.965
GAD-7 Status Details	30	0.821	0.767	0.793	0.970
PHQ-8 Status Details	30	0.950	0.633	0.760	0.970

Outside Target perform strongly, despite their close semantic proximity and the similarity of their training sentences, which differ only in whether they emphasize anomalous blood pressure values or measurements outside a target range. *Daily Condition Details* (F1 = 0.806) and *GAD-7 Status Details* (F1 = 0.793) also perform robustly, with good coverage and relatively few misclassifications. By contrast, *BP Inflection Points* shows perfect precision (1.000) but poor recall (0.567), indicating that the classifier is overly conservative and fails to detect many true instances. *BP List Values* (F1 = 0.667), *Basic User Info* (F1 = 0.746), and *PHQ-8 Status Details* (F1 = 0.760) achieve moderate results, particularly limited by lower recall. The weakest performance arises for *BP Trends*, which, despite high recall (0.900), suffers from very low precision (0.482), producing many false positives and resulting in the lowest F1 score (0.628). Taken together, these results suggest that while the system generally performs well across intents, refinement is needed for cases prone to systematic over-prediction (*BP Trends*) or under-detection (*BP Inflection Points*). This weakness can be partially explained by the semantic proximity of trends and inflection points, which also share similar training sentences, making the distinction between the two particularly nuanced.

In the experiments described above, we assumed that all utterances would trigger the fallback, and therefore fed them directly into the fallback component within a separate evaluation setup, allowing us to evaluate the LLM fallback mechanism in isolation. In a real deployment, however, RASA's intent classifier would always act as the first stage, invoking the fallback only when the confidence threshold is not met. To also assess performance in a more realistic scenario, we repeated the experiment with utterances submitted through the RASA server API, allowing the system to autonomously decide whether to classify an input using RASA's NLU or to fall back to the LLM-based mechanism.

Table 3 summarizes the evaluation results of the RASA intent classification model. Overall, the LLM fallback was triggered in 120 out of 400 cases. Of these, 81 predictions (67.5%) were correct. Conversely, in 280 cases, RASA classified

Table 3. Evaluation results for the RASA intent classification model

Ground truth	RASA prediction	No. of cases	Correct LLM predictions
Clear match	Fallback	9	8 out of 9
Ambiguous match	Fallback	48	11 out of 48
No match	Fallback	63	62 out of 63
Clear match	Wrong intent	7	4 out of 7
Ambiguous match	Wrong intent	40	13 out of 40
No match	Wrong intent	37	35 out of 37

the input without fallback, yet 84 of these classifications predicted wrong intents. Notably, in 52 of those 84 instances, the LLM fallback would have produced the correct intent, corresponding to a success rate of more than 60%. The fallback was thus not only useful when invoked but also showed potential for correcting errors that RASA committed when it refrained from fallback. Interestingly, the fallback was activated even for 9 of the 150 inputs that had been designed as clear matches, indicating that the primary classifier occasionally lacked sufficient confidence to resolve user intent.

4.2 Human Expert Evaluation

Additionally, we conducted a brief expert evaluation involving three medical doctors, each with several years of clinical experience, and two medical data analysts. The experts were presented with a predefined use-case scenario and asked to interact with the chatbot prototype for approximately 5–10 min. After this exploratory phase, they completed a structured feedback form comprising general questions and open-ended prompts aimed at identifying both strengths and areas for improvement. This qualitative assessment was designed to elicit expert insights regarding the system's usability, relevance to medical workflows, and overall perceived utility.

The medical experts expressed satisfaction with the chatbot's response speed and the clarity of its well-structured visualizations. They emphasized that the graphical outputs were very informative and conveyed essential knowledge effectively. They also noted that the chatbot's suggestions drew users' attention to additional functionalities, thereby increasing awareness of available options. For improvement, they suggested broadening the range of supported intents and enhancing the system's flexibility within existing ones. For example, they recommended allowing queries not only for specific predefined time points but also for custom time spans specified directly in the user input.

Furthermore, we analyzed the logs of the experts' exploration sessions. As expected, the real user inputs differed significantly from the grammatically well-formed prompts generated by the LLM. The expert inputs frequently contained typographical errors and inconsistent capitalization, which made it more difficult for the RASA classifier to correctly detect the intended intent, thereby triggering the LLM fallback more often. In one illustrative example, one expert typed the input “welche psyschichen daten wurden erfast”, a misspelled version of “Welche psychischen Daten wurden erfasst?”, meaning “Which psychological data was collected?”. Despite the typos and lack of capitalization, the LLM fallback correctly suggested the appropriate intents, namely *GAD-7 Status Details* and *PHQ-8 Status Details*, whereas the RASA classifier failed to match the intent.

5 Conclusion

In this work, we presented and evaluated a conversational chatbot with an LLM-based fallback mechanism designed to improve intent recognition and response generation. We leveraged a second independent LLM to generate test utterances across varying levels of intent clarity and systematically assessed the fallback's ability to handle clear, ambiguous, and irrelevant inputs. The results demonstrate strong performance in rejecting out-of-scope inputs and in reliably handling clear matches, while performance on ambiguous cases revealed notable challenges, though still representing an improvement over the primary RASA classifier. These findings show that LLMs can effectively augment traditional conversational systems, making them more robust and reliable in complex domains such as health, where ambiguity and variability in user input are especially common and impactful. Expert feedback further confirmed the system's usability and utility in a medical context, emphasizing both its speed and the clarity of its outputs. To support reproducibility and transparency, we provide the evaluation source code and the generated user-utterance dataset in a GitHub repository.[3]

Limitations. While the evaluation results are promising, several limitations should be acknowledged. First, the synthetic benchmark data was primarily generated by another LLM, which, although diverse and manually reviewed by the authors, may not fully reflect the variability and errors present in real-world user input. The expert user evaluation, while insightful, was limited in scope and sample size. Another important limitation lies in the restricted flexibility of backend queries: the system relies on predefined, parameterized SQL templates, which limit the ability to accommodate complex user inputs that go beyond rigid query patterns or ask for more nuanced temporal or statistical filters.

Future Work. To address these limitations, future work will focus on enabling more flexible and expressive interactions. In particular, we plan to extend the LLM fallback with semantic parsing capabilities to extract structured parameters directly from user inputs. This would allow the system to dynamically

[3] github.com/sebischair/NLPIR-2025-Supplemental.

populate queries based on natural language input, supporting more personalized and context-sensitive requests, an enhancement specifically suggested by health experts. Additionally, we aim to expand the set of supported intents and conduct broader user testing.

Acknowledgments. We thank the five medical experts who participated in our evaluation for their valuable time and feedback.

References

1. Schneider, P., Poelman, W., Rovatsos, M., Matthes, F.: Engineering conversational search systems: a review of applications, architectures, and functional components. In: Proceedings of the 6th Workshop on NLP for Conversational AI (NLP4ConvAI 2024), pp. 73–88. Association for Computational Linguistics, Bangkok, Thailand (2024). https://aclanthology.org/2024.nlp4convai-1.5/
2. Zhang, J., et al.: Are pre-trained transformers robust in intent classification? A missing ingredient in evaluation of out-of-scope intent detection. In: Proceedings of the 4th Workshop on NLP for Conversational AI, pp. 12–20. Association for Computational Linguistics, Dublin, Ireland (2022). https://doi.org/10.18653/v1/2022.nlp4convai-1.2
3. Nerella, S., et al.: Transformers and large language models in healthcare: a review. Artif. Intell. Med. **154**, 102900 (2024). https://doi.org/10.1016/j.artmed.2024.102900
4. Li, Y., Li, Z., Zhang, K., Dan, R., Jiang, S., Zhang, Y.: ChatDoctor: a medical chat model fine-tuned on a large language model Meta-AI (LLaMA) using medical domain knowledge. Cureus **15**(6), e40895 (2023). https://doi.org/10.7759/cureus.40895
5. Al Attrach, R., Moreira, P., Fani, R., Umeton, R., Celi, L.: Conversational LLMs simplify secure clinical data access, understanding, and analysis. Preprint, arXiv (2025). https://doi.org/10.48550/arXiv.2507.01053
6. Schmiedmayer, P., et al.: LLMonFHIR: a physician-validated, large language model-based mobile application for querying patient electronic health data. JACC: Adv. **4**(6 Pt 1), 101780 (2025). https://doi.org/10.1016/j.jacadv.2025.101780
7. Naved, B.A., Ravishankar, S., Colbert, G.E., Johnston, A., Slott, Q.M., Luo, Y.: LLM enabled classification of patient self-reported symptoms and needs in health systems across the USA. NPJ Digit. Med. **8**(1), 390 (2025). https://doi.org/10.1038/s41746-025-01779-9
8. Mahmood, A., Wang, J., Yao, B., Wang, D., Huang, C.-M.: User interaction patterns and breakdowns in conversing with LLM-powered voice assistants. Int. J. Hum.-Comput. Stud. **195**, 103406 (2025). https://doi.org/10.1016/j.ijhcs.2024.103406
9. Schneider, P., Matthes, F.: Conversational exploratory search of scholarly publications using knowledge graphs. In: Proceedings of the 7th International Conference on Natural Language and Speech Processing (ICNLSP 2024), pp. 384–396. Association for Computational Linguistics, Trento, Italy (2024). https://aclanthology.org/2024.icnlsp-1.40/

10. Bunk, T., Varshneya, D., Vlasov, V., Nichol, A.: DIET: lightweight language understanding for dialogue systems. arXiv preprint arXiv:2004.09936 (2020). https://arxiv.org/abs/2004.09936
11. OpenAI: Introducing GPT-4.1 in the API. OpenAI Blog (2025). https://openai.com/index/gpt-4-1/
12. Meta AI: Introducing LLaMA 4: Advancing multimodal intelligence. Meta AI Blog (2025). https://ai.meta.com/blog/llama-4-multimodal-intelligence/

Assessing Grammar Error Correction Performance of LLMs for Chatbot-Based Tagalog Language Education

Paula Pacheco, Lanz Kendall Lim, Johann Uytanlet, Bentley Andrew Lu, Don Aspecto, and Ethel Ong(✉)

College of Computer Studies, De La Salle University, 2401 Taft Ave, 1004 Metro Manila, Philippines
{paula_pacheco,lanz_kendall_lim,johann_uytanlet,bentley_andrew_lu, don_aspecto,ethel.ong}@dlsu.edu.ph

Abstract. This study examines the grammar error correction (GEC) capabilities of large language models (LLMs) for Tagalog and their potential use in chatbot-based language learning. Five models were tested: Balarila, GPT-4o mini, Gemini 2.0 Flash, DeepSeek V3, and SEALion-V3. The first experiment assessed general grammar rules, while the second focused on verb formation involving focus and tense. GPT-4o mini performed best in general GEC, Gemini excelled in verb-related tasks, and Balarila showed high precision in preserving correct sentences. Findings reveal both strengths and limitations of current LLMs, particularly their tendency to over-correct or detect errors where none exist.

Keywords: natural language processing · grammar error correction · Tagalog language · low-resource language · conversational agents

1 Introduction

Language learning is an essential aspect of both personal and professional growth, influencing communication and identity. Recently, conversational agents have emerged as promising tools for language education, offering interactive learning experiences through natural language processing (NLP) techniques [1]. They can act as a conversational partner, assisting learners as they strengthen their language skills.

With the growing integration of conversational agents in education, it spurred interests in evaluating how well language models support language learning outcomes. Supporting these outcomes involves tasks such as providing corrective feedback, generating contextually appropriate responses, and performing grammar error correction (GEC), to help learners refine their language use. Grammar correction, in particular, plays a critical role in reinforcing proper grammar usage and supporting improvement in communication skills [2].

H. Unger and P. Meesad (Eds.): NLPIR 2025, LNNS 1904, pp. 872–887, 2026.
https://doi.org/10.1007/978-3-032-20897-2_61

However, while large language models (LLMs) have demonstrated strong capabilities across many languages [3], they remain predominantly optimized for high-resource languages such as English. This leads to linguistic biases that place lower-resource languages like Tagalog at a disadvantage[4,5].

The Filipino language poses unique challenges for text understanding and generation, particularly due to its verb focus system, which has no direct counterpart in English and complicates both rule formalization and model training [6]. Since Tagalog relies heavily on verb morphology to express grammatical relations, sentences are verb-anchored rather than subject-anchored [7]. Additional complexities, such as the extensive use of affixes and infixation further hinder LLMs trained primarily on English datasets [8].

Recent efforts to develop Filipino GEC systems include neural approaches like Balarila [9], as well as rule-based and hybrid systems such as Gramatika [10,11] and adaptations of LanguageTool [12]. These approaches highlight both the potential and limitations of existing methods, motivating further exploration of LLM-based solutions.

In this paper, we evaluate the performance of LLMs on Tagalog grammar error correction. We conduct two levels of evaluation: (1) general grammar testing and (2) verb-focused analysis, highlighting strengths and limitations of different models. Our findings identify effective models for supporting Tagalog language learning and point to areas for future improvement.

2 Task Description

To support feedback and tutoring in a conversational agent, it must be able to reliably assess the grammatical proficiency of a learner. Grammatical proficiency refers to an individual's ability to correctly and effectively use Tagalog grammar. From this, we define the *GEC capability* as the model's ability to diagnose grammar in learner-generated Tagalog sentences. Given a sentence s and a set of error types $E = \{\text{phonetic errors, confusion with } nang/ng, \ldots\}$, the model (1) detects whether s contains errors and assigns labels, and (2) produces a corrected sentence s'.

Formally, the task is defined as a function $f : S \rightarrow (s', E')$, where S is the set of Tagalog sentences. Sentences may contain no errors, a single error type, or multiple error types. The output consists of a corrected sentence s', and an error label set $E' \subseteq E$ representing the types of errors present in the original input s. This setup enables a controlled evaluation of a model's ability to generate accurate corrections and classification of error types, offering insights into its effectiveness as a tool for supporting learners to gain grammatical proficiency. The process is illustrated in Fig. 1.

3 Experiment

To assess the capabilities of large language models in handling Tagalog grammar, the experiment is divided into two phases: (1) a general grammar error correction test across a wide range of error types, and (2) a focus on verb-related errors.

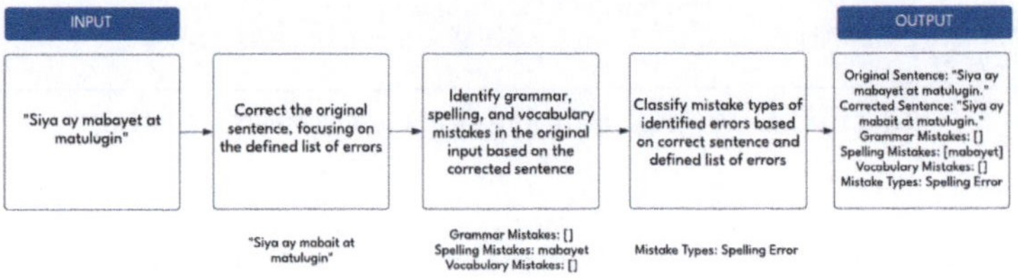

Fig. 1. Process flow in grammar error correction.

3.1 General GEC Testing

The general GEC experiment involved evaluating the LLMs' performances on **TestSetGen** with the following three types of test sets: (1) incorrect sentences with one error type in each sentence (**TestSetGen-1Err**), (2) correct (error-free) sentences (**TestSetGen-0Err**), (3) incorrect sentences with multiple error types in each sentence (**TestSetGen-nErr**). The test set covers 16 grammatical errors derived from previous studies on Tagalog GEC [9,13] and the *Komisyon sa Wikang Filipino* (Filipino Language Commission) [14], the official governing body that handles the cultivation and promotion of various Philippine languages.

Each test case of **TestSetGen-1Err** has a corresponding corrected version in **TestSetGen-0Err**. There are 16 error types with 5 test cases each, bringing a total of **80** test cases for each of the two test sets. **TestSetGen-nErr** has an additional **10** test cases, each containing 2 to 3 error types, and the overall test set uses all error types at least once. This results in a total of **170** test cases. All test cases were reviewed and validated by a Tagalog language expert.

GPT-4o Mini. The extensive multilingual dataset used to train this model equips it with the ability to detect grammar-related writing errors and recommend vocabulary improvements in Tagalog [15]. Three prompts were formulated to instruct the LLM to (1) correct the grammar error detected in the input sentence, (2) identify the errors, and (3) classify the error types. Each prompt is designed as a query, which includes information on the grammar scope, error categories, and specific instructions to guide the language model's behavior during evaluation. The prompt template is displayed in Listing 1.

```
You are an expert in Tagalog grammar, spelling, and sentence construction.
    Your task is to analyze learner-produced Tagalog sentence and [correct
    the sentence, identify mistakes, or classify error types].

[List of Grammar Errors, Exception Cases, Task Reminders]

Original Input: "{input_text}"
Corrected Sentence: "{corrected_sentence}"

[Example Input and Outputs]

[Specified Output Format]
Grammar Mistakes:
Spelling Mistakes:
```

```
Vocabulary Mistakes:
Mistake Types:
```

Listing 1. Prompt template for General GPT testing

Balarila. Balarila [9] is a publicly available deep learning-based Filipino grammar error correction model. It was based on the approach of the GECToR model [16] and uses iterative sequence tagging instead of sequence generation for the GEC task. These tags correspond to specific grammar transformations that are applied to improve the sentence's correctness. Listing 2 displays a sample output from a Balarila test case.

```
Original sentence: Linis mo ang bintana bago umalis!
Corrected sentence: Lilinisin mo ang bintana bago umalis.
Tags: ['$TRANSFORM_VERB_CONTOBJ', '$CHANGE_PUNC_PERIOD', '
     $TRANSFORM_VERB_COMPOBJ', '$TRANSFORM_VERB_CONTOBJ']
Corrections made: 4
```

Listing 2. Sample output of Balarila test

GPT-4o Mini and Balarila. In this hybrid approach, the input sentence is first passed to Balarila for error identification and correction. The output is then validated by GPT using the prompt shown in Listing 3. If GPT deems the text from Balarila to be correct, it proceeds to analyze the corrected sentence to further identify and classify the types of grammar errors. However, if GPT finds that Balarila's correction is incorrect, GPT handles the entire process of correction, error identification, and classification on its own.

```
You are an expert in Tagalog grammar. Determine if the corrected sentence
     is grammatically correct.

Original Input: "{input_text}"
Corrected Sentence by AI Model: "{corrected_by_model}"

If the corrected sentence is grammatically correct, respond with 'YES'.
Else, respond with 'NO'.

Return only 'YES' or 'NO'.
```

Listing 3. Prompt to validate corrected sentence from Balarila

3.2 Deeper Verb Testing

Initial results from the general GEC experiment revealed several key challenges in processing Tagalog text as compared to high-resource languages such as English. Notably, verb formation plays a central role in establishing contextual and relational meaning across sentence elements in Tagalog text [8]. Moreover, the unique system of affixation in Tagalog verbs differs significantly from that of English, which may negatively impact the performance of LLMs in GEC tasks.

For these reasons, we conducted a second experiment that focused specifically on the intricacies of Tagalog verb formation. The aim is to assess the full extent of various LLMs' capabilities in correcting Tagalog verb errors. The experiments center on detecting verb formation errors, with emphasis on verb forms that encode both focus and tense, two critical grammatical features of Tagalog. Drawing from Schatner [8], the study adopts the following six (6) focuses of a verb: **Object**, **Actor**, **Locative**, **Benefactive**, **Instrumental**, and **Causative**. In parallel, five (5) verb tenses are considered for analysis: **Past**, **Present**, **Future**, **Infinitive**, and **Recently Completed** a special tense.

The experiment was conducted using a custom built **TestSetVerb** that focused primarily on verb-related grammatical errors. The test set includes **25 verb categories** derived from 24 categories that combine the 6 verb focuses and 4 regular tenses, with 1 additional category for the special tense *Recently Completed*. There are **10 test cases per category**, comprising 5 correct and 5 incorrect sentences. This gives a total of **250 test items** for verb-related error detection per model.

By constructing **TestSetVerb** that includes sample sentences tagged with these grammatical features, models can be evaluated based on their ability to detect deviations and incorrect usage. The five (5) models used for this experiment include: (1) DeepSeek-V3, (2) Gemini 2.0 Flash, (3) GPT-4o mini, (4) Balarila, and (5) Llama-SEA-LION-v3-70B-IT. All LLMs were given the same prompt shown in Listing 4.

```
Role:
You are a grammar error detection model for the Tagalog language. Your role
      is to help learners improve their grammar skills by analyzing their
     sentences, detecting errors, and providing feedback.

Instructions:
1. Error Detection: Analyze the given sentence and identify any grammatical
      errors (e.g., subject-verb agreement, tense errors, incorrect word
     choice, affix misuse, or incorrect sentence structure).
2. Correction: Provide the corrected sentence while keeping the original
     meaning intact.
3. Error Explanation: Clearly explain:
   - The type of error (e.g., verb conjugation mistake, misplaced affix,
        word choice).
   - Where the error occurs in the sentence.
   - Why it is incorrect and how to correct it.
   - Write the explanation in English to ensure clarity to the learner.
4. No Errors: If the sentence is grammatically correct, state: "There are
     no errors in this sentence."
5. Output Format: Follow this structured format for responses:

For sentences with errors:
Original Sentence: "<user's input>"
Error Type: <brief classification of the error>
Corrected Sentence: "<corrected version>"
Explanation (in English): <detailed explanation of the mistake>

For correct sentences:
Original Sentence: "<user's input>"
Analysis: "There are no errors in this sentence."
```

Listing 4. Prompt template for deeper verb testing

4 Results and Discussion

To evaluate model performance, we computed the proportion of test cases in which the model correctly identified and corrected errors if any are present. This measure, reported as ***accuracy***, reflects the percentage of instances where the model's output fully matched the expected identification and correction criteria. However, in the deeper verb testing where the focus is more localized on verb formation, ***performance*** is measured solely by the total number of corrected errors, without considering the specific type of error, since all errors are verb-related. Additionally, statistical significance testing was conducted to compare the models' performances. McNemar's test and the Wilcoxon test were used to evaluate the differences between the models. Through this framework, we capture both general GEC capabilities and performance on specialized tasks involving Tagalog verb-related errors.

4.1 General GEC Testing

Table 1 presents the results of the experiment on **TestSetGen-1Err**. GPT performed significantly better than Balarila and the hybrid (Balarila and GPT) model in correcting and identifying Tagalog grammatical mistakes, with a correction accuracy of 91.25% and an identification accuracy of 92.50%, compared to the hybrid's 63.76% and 66.25%, and Balarila's 33.75% and 40.00%.

Table 1. General GEC experiment test results on **TestSetGen-1Err**

Type of Error	Corrected Mistake			Identified Mistake		
	B	B + G	G	B	B + G	G
Phonetic Errors	0/5	3/5	4/5	0/5	3/5	5/5
Confusion with *Nang* and *Ng*	5/5	4/5	4/5	5/5	4/5	4/5
Improper Use of Enclitics	4/5	4/5	5/5	5/5	4/5	5/5
Capitalization Errors	4/5	5/5	5/5	5/5	5/5	5/5
Improper Use of Hyphenation	2/5	2/5	4/5	2/5	2/5	4/5
Incorrectly Merged Words	2/5	5/5	5/5	2/5	5/5	5/5
Incorrectly Unmerged Words	1/5	4/5	5/5	1/5	4/5	5/5
Word Repetition	1/5	3/5	5/5	1/5	3/5	5/5
Improper Use of Pronouns	4/5	5/5	5/5	4/5	5/5	5/5
Morphological Errors	1/5	3/5	4/5	4/5	3/5	4/5
Wrong Placement of Adverb	0/5	3/5	4/5	0/5	5/5	4/5
Incorrect Adverb Form	0/5	2/5	5/5	0/5	3/5	5/5
Improper Use of Adjectives	0/5	2/5	3/5	0/5	2/5	3/5
Improper Use of Prepositions	1/5	2/5	5/5	1/5	2/5	5/5
Incorrect Usage of Noun Corrector	2/5	3/5	5/5	2/5	3/5	5/5
Honorific Markers	0/5	0/5	5/5	0/5	0/5	4/5
Final Accuracy	**33.75% (27/80)**	**63.75% (51/80)**	**91.25% (73/80)**	**40.00% (32/80)**	**66.25% (53/80)**	**92.50% (74/80)**

Balarila performed well in handling rule-based errors (e.g., *nang/ng* confusion, capitalization) but it had limitations when faced with nuanced, context-dependent mistakes (e.g., honorific markers) since it was not designed to address these types of errors. The hybrid model produced better results than Balarila alone but did not approach GPT's level of performance, particularly in handling complex grammatical structures and semantic nuances.

Table 2 shows the results obtained from **TestSetGen-0Err**. The optimal outcome was to avoid performing unnecessary corrections and to accurately recognize sentences as free of errors. Rephrasing or alteration of these correct sentences counted as a *false positive*. This aspect of the evaluation highlights models prone to over-correcting, an important factor in assessing reliability.

Table 2. General GEC experiment test results on **TestSetGen-0Err**

Type of Error	Did not Correct Mistake			Did not Identify Mistake		
	B	B + G	G	B	B + G	G
Phonetic Errors	4/5	4/5	2/5	4/5	4/5	2/5
Confusion with *Nang* and *Ng*	4/5	5/5	5/5	4/5	5/5	5/5
Improper Use of Enclitics	5/5	5/5	2/5	4/5	5/5	3/5
Capitalization Errors	5/5	5/5	5/5	5/5	5/5	5/5
Improper Use of Hyphenation	4/5	4/5	4/5	4/5	4/5	4/5
Incorrectly Merged Words	5/5	5/5	4/5	4/5	5/5	5/5
Incorrectly Unmerged Words	4/5	4/5	5/5	4/5	4/5	5/5
Word Repetition	5/5	5/5	5/5	5/5	5/5	5/5
Improper Use of Pronouns	5/5	5/5	5/5	5/5	5/5	5/5
Morphological Errors	5/5	5/5	5/5	5/5	5/5	5/5
Wrong Placement of Adverb	5/5	5/5	5/5	5/5	5/5	5/5
Incorrect Adverb Form	5/5	5/5	5/5	5/5	5/5	5/5
Improper Use of Adjectives	5/5	5/5	5/5	5/5	5/5	5/5
Improper Use of Prepositions	5/5	5/5	5/5	5/5	5/5	5/5
Incorrect Usage of Noun Corrector	3/5	5/5	5/5	3/5	5/5	5/5
Honorific Markers	4/5	4/5	5/5	4/5	4/5	4/5
Final Accuracy	**91.25% (73/80)**	**95.00% (76/80)**	**90.00% (72/80)**	**88.75% (71/80)**	**95.00% (76/80)**	**92.50% (74/80)**

The hybrid model achieved the highest performance, with 95.00% in retaining the original sentences and correctly identifying as error-free. GPT had the lowest, with 90.00% and 92.50%. GPT's lower accuracy is largely due to instances of overcorrection. For example, the correct sentence "*Sinabi nila, hindi raw totoo ang balita.*" ("*They said the news is not true.*") was overcorrected into "*Sinabi nila hindi raw totoong ang balita.*" which introduced new errors by altering the sentence structure and incorrectly appending the prefix *-ng* to the correct form of the word "*totoo*" ("*true*").

Table 3 shows the results from **TestSetGen-nErr**. GPT reached an overall accuracy of 80.00% in detecting all errors compared to the hybrid model's 30.00% and Balarila's 10.00%. GPT's performance likely reflects its greater capacity to utilize contextual information compared to the other models. Furthermore, the limited coverage of error types supported by Balarila led to its inability to detect certain errors in the input sentence.

The statistical comparisons in Table 4 confirm that the observed performance differences between models are highly significant. In the comparison between Balarila and the hybrid model, the hybrid showed a clear advantage ($p = 8.1 \times 10^{-7}$), indicating that combining rules with additional methods yielded measurable improvements.

Table 3. General GEC experiment test results on **TestSetGen-nErr**

#	Input Errors	B	B+G	G
1	Phonetic Error, Morphological Error	0/2 (Fail)	1/2 (Fail)	1/2 (Fail)
2	Incorrectly Merged Words, Improper Use of Hyphenation	1/2 (Fail)	1/2 (Fail)	2/2 (Pass)
3	Improper Capitalization, Improper Use of Pronouns, Phonetic Error	0/3 (Fail)	3/3 (Pass)	3/3 (Pass)
4	Confusion of Nang and Ng, Morphological Error	0/2 (Fail)	0/2 (Fail)	2/2 (Pass)
5	Word Repetition, Incorrectly Merged Words	2/2 (Pass)	2/2 (Pass)	2/2 (Pass)
6	Improper Capitalization, Phonetic Error	1/2 (Fail)	2/2 (Pass)	2/2 (Pass)
7	Improper Capitalization, Improper Use of Enclitics	1/2 (Fail)	1/2 (Fail)	1/2 (Fail)
8	Phonetic Error, Improper Use of Enclitics, Improper Use of Hyphenation	1/3 (Fail)	2/3 (Fail)	2/3 (Fail)
9	Improper Use of Adjectives, Improper Use of Hyphenation	0/2 (Fail)	1/2 (Fail)	1/2 (Fail)
10	Improper Use of Prepositions, Improper Use of Hyphenation	0/2 (Fail)	2/2 (Pass)	2/2 (Pass)
Final Accuracy		10.00% (1/10)	40.00% (4/10)	60.00% (6/10)

When comparing Balarila and GPT, GPT achieved significantly better results ($p = 2.1 \times 10^{-9}$), further emphasizing the limitations of sequence tagging approaches for grammar error detection. Finally, in the comparison between the hybrid model and GPT, GPT again outperformed ($p = 7.7 \times 10^{-5}$), showing that even the enhanced hybrid approach could not match GPT's capacity. Taken together, these results demonstrate that GPT consistently outperformed both baselines, providing the most reliable performance for Tagalog proficiency detection. All three models also showed no significant differences on TestSet-0Err but

Table 4. Pairwise significance test p-values; Green = significant ($p < 0.05$), Red = not significant

Dataset	Balarila vs Hybrid	Balarila vs GPT	Hybrid vs GPT
TestSetGen	8.1×10^{-7}	2.1×10^{-9}	7.7×10^{-5}
TestSet-0Err	7.4×10^{-2}	1	2.9×10^{-1}
TestSet-1Err	1.2×10^{-4}	5.4×10^{-10}	1.1×10^{-5}
TestSet-nErr	7.4×10^{-2}	4.8×10^{-1}	2.3×10^{-2}

differed significantly on TestSet-1Err, indicating GPT's strength in handling a wider range of errors compared to Balarila.

4.2 Deeper Verb Testing

The outputs of the models were manually evaluated to determine whether the generated sentences adhered to correct grammar and verb formation. In Listings 5, 6, and 7, three types of outputs are presented along with their corresponding scores. A score of 1 was awarded if the corrected sentence exhibited the expected verb formation, including the correct focus and tense such as seen in Listing 5.

However, in some cases, the models produced grammatically correct sentences that preserved the original meaning, but did not reflect the expected verb tense or focus. An instance of this can be seen in Listing 6. These were classified as invalid corrections yet were still awarded partial credit (0.5 points) since the output remained grammatically sound and semantically coherent. In this example, the model changed the verb "*inis*" ("*to be angry*") to "*nainis*" ("*got angry*"), which is the past actor-focused form of the verb. The object and gender-neutral pronoun "*niya*" ("*him/her*") was also replaced with a subject and gender-neutral pronoun "*siya*" ("*he/she*"), to better fit the actor-focused construction of the verb "*nainis.*" This adjustment also led to replacing the subject determiner "*ang*" with the preposition "*sa,*" thereby shifting the focus away from "*mahabang pila*" ("*long line*") to the actor. However, the correct verb formation should have been the past causative form which is "*ikinainis*", which emphasizes that the speaker's anger was caused by the "*mahabang pila*". Using "*ikinainis*" in place of "*inis*" would require no other changes in the sentence.

Lastly, sentences that were corrected but yielded a grammatically wrong sentence was considered to be an invalid correction and was given a score of 0. This is depicted in Listing 7. In this example, the original sentence used "*tutulugan,*" which means "*to sleep at*" and is the future locative form of the verb. The sentence in Listing 7 is already correct, as it conveys the meaning "*We will sleep at the tent for the camping trip on Saturday.*" However, the model incorrectly identified an error and attempted to "correct" it by changing the verb "*tutulugan*" to "*tutuluyan*" ("*to proceed*"). This alteration caused the sentence to lose its intended meaning, resulting in a semantic mismatch and a score of 0.

```
Original Sentence: "Ito ang restawran na kain namin kahapon."
Error Type: Word choice error
Corrected Sentence: "Ito ang restawran na kinainan namin kahapon."
```

Listing 5. Sample output (Valid), Score: 1

```
Original Sentence: "Inis niya ang mahabang pila sa bangko kanina."
Error Type: Incorrect word order and missing verb.
Corrected Sentence: "Nainis siya sa mahabang pila sa bangko kanina."
```

Listing 6. Sample output (Partial Points), Score: 0.5

```
Original Sentence: "Tutulugan natin ang tent sa camping trip sa sabado."
Error Type: Incorrect verb conjugation/word usage
Corrected Sentence: "Tutuluyan natin ang tent sa camping trip sa Sabado."
```

Listing 7. Sample output (Invalid), Score: 0

With this scoring scheme in mind, Tables 5 and 6 present the performance of each model across different verb focuses and a special verb form indicating **Recently Completed** actions. Scores are formatted as $x + y$, where x is the number of fully correct outputs, and y is the number of partial points awarded for near-correct outputs. Each verb focus is divided into (1) correct sentences and (2) sentences with errors requiring correction. The 'lenient score' in the third column of Table 7 refers to a more lenient scoring scheme where partial points (0.5) are counted as full points (1). This is because such corrections are grammatically correct, even though they do not exactly match the intended correction.

Table 5. Balarila, DeepSeek, and Gemini scores on TestSetVerb

Focus	Balarila		DeepSeek		Gemini	
	Correct	Incorrect	Correct	Incorrect	Correct	Incorrect
LOC	20+0/20	5+0.5/20	12+3/20	6+4.5/20	10+2.5/20	5+5/20
CAU	19+0/20	2+0.5/20	5+6.5/20	0+5.5/20	12+2/20	3+2.5/20
BEN	20+0/20	0+0/20	13+2.5/20	6+3/20	13+0.5/20	8+2.5/20
INS	20+0/20	0+0.5/20	6+5/20	0+5.5/20	4+3.5/20	0+7.5/20
ACT	18+1/20	10+1/20	20+0/20	13+0/20	20+0/20	16+0.5/20
OBJ	19+0/20	12+2.5/20	19+0.5/20	15+0.5/20	19+0.5/20	16+0.5/20
REC	3+0/5	1+0/5	5+0/5	0+0/5	5+0/5	2+0/5
Total	119+1	30+5	80+17.5	40+19	83+9	50+18.5

Table 6. GPT and SeaLion scores on TestSetVerb

Focus	GPT		SeaLion	
	Correct	Incorrect	Correct	Incorrect
LOC	11+0/20	2+5/20	15+1/20	2+1.5/20
CAU	19+0.5/20	0+4.5/20	17+0.5/20	0+4.5/20
BEN	14+1.5/20	4+4/20	10+1.5/20	2+3.5/20
INS	15+1/20	0+1/20	5+2.5/20	0+1.5/20
ACT	20+0/20	15+0/20	19+0.5/20	10+1.5/20
OBJ	19+0.5/20	8+3/20	20+0/20	1+2/20
REC	3+0/5	0+0/5	4+0/5	0+0/5
Total	101+3.5	29+17.5	90+6	15+14.5

Based on the raw scores alone (excluding partial points) in Table 7, Balarila performed the best among the five models, achieving 149 out of 250 or 59.6% on **TestSetVerb**. A closer examination reveals that Balarila's high score is largely due to its ability to accurately identify grammatically correct sentences and leave them unchanged. Of its 149 points, 119 were earned by preserving sentences that required no corrections. In contrast, it only successfully corrected 30 with-error sentences, just one more than GPT-4o-Mini which had the second fewest corrections at 29. This outcome highlights the limitations in Balarila's architecture where only specific verb forms are supported, namely, **Recently Completed**, **Actor**, and **Object** focuses, rendering the model less effective at handling other focus types.

Table 7. Overall model scores

Model	Total Score	Score with Partials	Lenient Scoring
Balarila	149/250	155/250	161/250
Deepseek-v3	120/250	156.5/250	188/250
Gemini 2.0 Flash	133/250	160.5/250	172/250
GPT-4o-Mini	130/250	151/250	193/250
SEA-LION-v3	105/250	125.5/250	146/250

SEALion performed the worst out of all the models, only being able to make 15 corrections. This may be due to the relatively small number of parameters of the SEALion model. On the other hand, Gemini corrected the highest number of with-error sentences (50) and also achieved the highest total when partial points were included, with a score of 160.5 out of 250 or 64.2%, making it the most effective model for general grammar error correction among those tested.

Notably, Gemini, GPT, and DeepSeek all exhibited a pattern of false positive corrections, where they modified already correct sentences. Based on the scores presented in Tables 5, 6, the models produced the following number of false positives: DeepSeek = 45, Gemini = 42, GPT = 24, and SEALion = 35. These unnecessary changes often resulted in grammatically invalid outputs or partially correct sentences that failed to match the intended verb formation, despite preserving the original meaning.

To properly measure the differences between the various models, significance testing was performed for each model pair. McNemar's test was applied to the No Partial Scoring and Lenient Scoring methods since both produce strictly binary outcomes. In contrast, the Partial Scoring method, where scores could take values of 0, 0.5, or 1, was evaluated using the Wilcoxon signed-rank test, which is appropriate for ordinal data.

As shown in Table 8, Balarila significantly outperforms every other model except Gemini in the no partial scoring scheme. Moreover, SeaLion was consistently found to perform significantly worse than the other models across nearly all tests. For the Lenient Scoring method in particular, Gemini significantly outperforms both Balarila and SeaLion, and also approaches significance against GPT with a pairwise p-value of 0.0689.

Overall, among the tested models, Gemini stands out as performing the best in the GEC task. However, Balarila's architecture presents a promising approach for reducing the number of false positive corrections in sentences, suggesting that further exploration into its architecture may lead to more effective results in future research.

Table 8. Pairwise significance test p-values under different scoring methods; Green = significant ($p < 0.05$), Red = not significant

Comparison	No Partial	With Partial	Lenient
Gemini vs GPT	0.8124	0.2709	0.0689
Gemini vs DeepSeek	0.0993	0.5201	0.6143
Gemini vs SeaLion	0.0015	0.0000	0.0000
Gemini vs Balarila	0.0976	0.6820	0.0048
GPT vs DeepSeek	0.2606	0.4552	0.0131
GPT vs SeaLion	0.0012	0.0005	0.0037
GPT vs Balarila	0.0212	0.3106	0.2418
DeepSeek vs SeaLion	0.0872	0.0000	0.0000
DeepSeek vs Balarila	0.0019	0.8015	0.0002
SeaLion vs Balarila	0.0000	0.0001	0.1152

4.3 Discussion

In the general test, GPT showed strong performance in correcting a variety of Tagalog language phenomena (Table 1) and handling multiple errors within a single sentence (Table 3). However, it also displayed a higher tendency to flag correct sentences as erroneous (Table 2). This highlights a significant weakness in LLMs when applied to low-resource languages. The hybrid model combining GPT and Balarila underperformed relative to expectations. Although it was intended to balance GPT's linguistic range with Balarila's precision, the integration was ineffective and did not improve upon GPT's standalone performance. To improve the hybrid implementation, more sophisticated approaches may be applied such as using an LLM as a judge [17]. Notably, while GPT outperformed Balarila in overall accuracy, Balarila performed best in preserving correct sentences and avoiding false positive corrections that often stem from hallucinations, a common issue among LLMs.

In the general results, GPT performed well on the broad category *morphological errors*, which include basic verb formation errors. However, when testing specifically on the intricacies of verb focus and tense, we began to see the limits of GPT's performance, showing both its strengths and weaknesses. Overall, in the deeper verb-focused evaluation, GPT ranked as the second weakest among the models tested. Gemini emerged as the most effective model in handling Tagalog verb formation errors. Similar to the general test, Balarila performed best at minimizing false positive corrections in sentences.

Results further demonstrated that **Causative** and **Instrumental** focus verbs present particular challenges to the GEC task. These verbs are often difficult to correct due to the nature of the errors within the sentences. For causative focus errors, corrections are sometimes made by replacing the verb with an adjective. For example, the sentence "*Ikatatakot niya ang mangyayari sa eksena mamaya.*" ("What will happen in the scene later will scare him/her.") can be corrected to "*Takot siya sa mangyayari sa eksena mamaya.*" ("He/She is scared of what will happen in the scene later.") In this example, we see that instead of transforming the root word "*takot*" ("*scared*") into a verb form such as *ikatatakot* ("*to be scared because of*"), the model used "*takot*" in its adjective form. While this still results in a grammatically correct sentence, it does not produce the intended verb formation of "*ikatatakot.*"

Similarly, for instrumental focus errors, the verb "*gamit*" may be used in place of an instrumental focus verb to indicate the use of a tool. For example, the sentence "*Linis mo ang trapo para sa mesa*" ("*Clean the rag for the table.*") was corrected to "*Gamitin mo ang trapo panlinis sa mesa.*" ("*Use the rag to clean the table.*") instead of "*Ipanglinis mo ang trapo para sa mesa.*" ("*The rag can be used to clean the table.*") Though both corrections are grammatically and semantically correct, the use of "*gamit*" is more common in everyday speech, which may influence the model's ability to correctly produce instrumental focus verbs.

Findings also reveal that the **Recently Completed** verb tense may be a weak spot for most LLMs, as they generally fail to correctly identify or correct errors in this form. In fact, only Gemini and Balarila managed to correct a sentence in the incorrect category. This weakness may partly stem from the inherent difficulty of expressing a recently finished action. For other verb tenses, we can rely on explicit time indicators to signal when the action occurs. However, for the **Recently Completed** tense, there are typically no such time markers, the verb formation itself serves as the indicator that the action has just been completed. These challenges highlight areas that future LLM development must address for better performance in morphologically rich languages.

5 Conclusion

This study evaluated the grammar error correction (GEC) capabilities of large language models (LLMs) for Tagalog. LLMs are commonly used in applications such as language learning, where they act as conversational agents and tutors. A key aspect of this role is their ability to correct learner errors. To investigate this, we created two datasets: **TestSetGen**, which evaluates general GEC capabilities across a wide range of linguistic features, and **TestSetVerb**, which focuses specifically on the intricacies of Tagalog verb formation. Using these datasets, we tested five models: (1) DeepSeek-V3, (2) Gemini 2.0 Flash, (3) GPT-4o mini, (4) Balarila, and (5) Llama-SEA-LION-v3-70B-IT. Our results show that GPT outperformed sequence tagging approaches such as Balarila. In the deeper verb testing, Gemini achieved the best overall performance, while SEALion ranked the lowest. Across all models, verb corrections involving the **Causative** and **Instrumental** focus, as well as the **Recently Completed** tense, proved especially challenging.

Findings from this study highlight several promising directions for future work. One potential area is improving hybridization strategies between LLMs and sequence tagging systems like Balarila. Possible setups include voting ensembles or LLMs as a judge to pick which model's proposed correction to use. To address the performance issues observed during general GEC testing, future work should also consider upgrading to more advanced models such as GPT-4o or GPT-4.1, which offer significantly higher parameter counts and improved contextual understanding [18]. These models may better handle the grammatical complexity of Tagalog, particularly in low-resource settings. Additionally, alternative architectures such as Claude or multilingual models optimized for Southeast Asian languages, like more recent versions of SeaLLM [19], may provide better support for nuanced and morphologically rich language use. The limited size of the evaluation sets may have also contributed to the models' poor performance, though the datasets were manually validated and reviewed for quality.

Despite these limitations, LLMs have already demonstrated promising capabilities in Tagalog GEC. This opens the door to integrating such GEC models into practical applications, including chatbots designed for language learning. This is especially timely and significant, given the declining state of Tagalog proficiency in the Philippines [20]. In this context, technology-driven tools such as conversational agents powered by LLMs offer a compelling pathway toward more autonomous, interactive, and engaging Tagalog learning, without placing additional burdens on educators.

References

1. Xiao, F., Zhao, P., Sha, H., Yang, D., Warschauer, M.: Journal of China Computer-Assisted Language Learning (2023). https://doi.org/10.1515/jccall-2022-0032
2. Belda-Medina, J., Calvo-Ferrer, J.R.: Appl. Sci. **12**(17), 8427 (2022). https://doi.org/10.3390/app12178427
3. Zhang, W., et al.: Seallms 3: open foundation and chat multilingual large language models for southeast Asian languages (2024). https://arxiv.org/abs/2407.19672
4. Ahuja, K., et al.: Proceedings of the 2023 Conference on Empirical Methods in Natural Language Processing, ed. by Bouamor, H., Pino, J., Bali, K. (eds.), pp. 4232–4267. Association for Computational Linguistics, Singapore (2023). https://doi.org/10.18653/v1/2023.emnlp-main.258. https://aclanthology.org/2023.emnlp-main.258/
5. Lai, V.D., et al.: Findings of the Association for Computational Linguistics: EMNLP 2023, ed. by Bouamor, H., Pino, J., Bali, K. (eds.), pp. 13,171–13,189. Association for Computational Linguistics, Singapore (2023). https://doi.org/10.18653/v1/2023.findings-emnlp.878. https://aclanthology.org/2023.findings-emnlp.878/
6. Ong, E., Abella, S.C., Santos, L.L., Tiu, D.C.: PACLIC 25 - Proceedings of the 25th Pacific Asia Conference on Language, Information and Computation, pp. 51–59 (2011). https://animorepository.dlsu.edu.ph/faculty_research/72. Retrieved from De La Salle University Animo Repository
7. Blust, R.: The History and Typology of Western Austronesian Voice Systems, Pacific Linguistics, vol. PL-518, ed. by Wouk, F., Ross, M. (eds.), pp. 63–80. The Australian National University (2001). https://doi.org/10.15144/PL-518.63. Online edition licensed 2015 under CC BY-SA 4.0, via Asealang.net/CRCL initiative
8. Schachter, P., Otanes, F.T.: Tagalog Reference Grammar. University of California Press, Berkeley (1983). Language Arts & Disciplines
9. Ponce, A.D.H., Jadie, J.S.A., Espiritu, P.E.A., Cheng, C.: Proceedings of the First Workshop in South East Asian Language Processing, ed. by Wijaya, D., Aji, A.F., Vania, C., Winata, G.I., Purwarianti, A. (eds.), pp. 21–29. Association for Computational Linguistics, Nusa Dua, Bali, Indonesia (2023). https://doi.org/10.18653/v1/2023.sealp-1.3. https://aclanthology.org/2023.sealp-1.3/
10. Go, M.P., Nocon, N., Borra, A.: TENCON 2017 - 2017 IEEE Region 10 Conference, pp. 471–475 (2017). https://doi.org/10.1109/TENCON.2017.8227910
11. Go, M.P., Borra, A.: Proceedings of the 30th Pacific Asia Conference on Language, Information and Computation: Oral Papers, Seoul, South Korea, pp. 105–113 (2016). https://aclanthology.org/Y16-2008/
12. Oco, N.A., Chua, R., Lee, F., Miranda, M.: Developing a grammar checker for filipino using language tool (2011). https://animorepository.dlsu.edu.ph/etd_bachelors/12148
13. Octaviano, M., Go, M.P., Borra, A., Oco, N.: 2016 International Conference on Asian Language Processing (IALP), Tainan, Taiwan, pp. 95–98 (2016). https://doi.org/10.1109/IALP.2016.7875943
14. Komisyon sa Wikang Filipino. KWF Diksiyonaryo ng Wikang Filipino (2021). https://kwfdiksiyonaryo.ph. Accessed 05 Dec 2024
15. Johri, S.: The making of chatgpt: from data to dialogue (2023). https://sitn.hms.harvard.edu/flash/2023/the-making-of-chatgpt-from-data-to-dialogue/

16. Omelianchuk, K., Atrasevych, V., Chernodub, A., Skurzhanskyi, O.: Proceedings of the Fifteenth Workshop on Innovative Use of NLP for Building Educational Applications, ed. by Burstein, J., et al. (eds.), Seattle, WA, USA → Online, pp. 163–170. Association for Computational Linguistics (2020). https://doi.org/10.18653/v1/2020.bea-1.16. https://aclanthology.org/2020.bea-1.16/
17. Gu, J., et al.: A survey on LLM-as-a-judge (2025). https://arxiv.org/abs/2411.15594
18. Anthem Creation. GPT-4o vs. GPT-4o-mini: which ai model to choose? (2024). https://anthemcreation.com/en/artificial-intelligence/comparative-gpt-4o-gpt-4o-mini-open-ai/#Comparative_between_GPT-4o_and_GPT-4o_Mini. Accessed 15 July 2025
19. Nguyen, X.P., et al.: Computation and Language (2023)
20. Dreisbach, J., Demeterio, F.: LLT J. **23**(2) (2020). https://doi.org/10.24071/llt.2020.230203

Construction of a Stress Estimation Model Using Multimodal Features

Toshiki Takanabe[1(✉)], Kazuyuki Matsumoto[2(✉)], Keita Kiuchi[3], Kotaro Kashihara[4], Minoru Yoshida[2], Xin Kang[2], Tomohiko Nakayama[5], Takaki Fukumori[2], Munenaga Koda[2], Hidehiro Umehara[6], Masahito Nakataki[5], and Shusuke Numata[5]

[1] Graduate School of Advanced Science and Technology, Tokushima University, Tokushima, JP, Japan
c612435041@tokushima-u.ac.jp

[2] Graduate School of Technology, Industrial and Social Sciences, Tokushima University, Tokushima, JP, Japan
matumoto@is.tokushima-u.ac.jp

[3] Japan Organization of Occupational Health and Safety, Tokyo, JP, Japan

[4] Works Applications, Tokyo, JP, Japan

[5] Department of Psychiatry, Graduate School of Biomedical Science, Tokushima University, Tokushima, JP, Japan

[6] Accessibility Division, Health Service, Counseling and Accessibility Center, Tokushima University, Tokushima, JP, Japan
https://www-a2.is.tokushima-u.ac.jp/home

Abstract. This study investigated methods for estimating stress states from video data. An online counseling dataset (50 individuals) was used as the evaluation and training dataset, while a depression counseling dataset (84 individuals) served as the training dataset. Features from speech, audio, and image data were extracted using multiple feature extraction models. To identify effective combinations of the 10 feature types, we performed 50-fold cross-validation using multiple machine learning algorithms. Results showed a stress level estimation accuracy of 0.8 when combining multiple audio, image, and language features and adding the depression interview dataset to the training data. Future work aims to develop a practical model for stress screening.

Keywords: Multimodal Features · Stress Estimation · Online Counseling · Machine Learning · Early Stress Detection

1 Introduction

Stress is a major factor affecting mental and physical health, closely linked to the onset and worsening of mental disorders such as depression and anxiety. In modern society, increasing stress in various environments, including the workplace and home, is a significant concern, making early stress detection and management crucial [2]. Conventional stress checks rely heavily on subjective methods,

H. Unger and P. Meesad (Eds.): NLPIR 2025, LNNS 1904, pp. 888–899, 2026.
https://doi.org/10.1007/978-3-032-20897-2_62

such as self-administered questionnaires, and are greatly influenced by the subject's cognitive biases and state of mind at the time of completion. Therefore, developing methods that can evaluate stress states more objectively and continuously is an urgent need. However, recent advances in deep learning technology have enabled emotion estimation using multimodal information such as human facial expressions, acoustic information, and language. While humans routinely infer others' feelings from expressions, voice tone, and manner of speaking, the practical application of AI that integrates and processes this information to estimate stress states still faces many challenges. This study proposes a highly accurate stress level estimation model using machine learning, leveraging multimodal information derived from video and audio data collected during online counseling sessions. By combining multiple feature extraction models and training datasets, we seek to improve the accuracy of stress estimation. This research aims to contribute to the objectification and automation of stress checks, thereby opening new possibilities in mental health support.

2 Related Work

In recent years, Transformer-based methods have gained significant attention in multimodal emotion recognition [3]. Shayaninasab et al. (2023) [4] proposed a multimodal emotion recognition method utilizing three modalities: text, audio, and video. In this study, they achieved high-accuracy emotion recognition by fine-tuning pre-trained Transformer models (such as BERT, wav2vec 2.0, and Video-MAE) for each modality and fusing the extracted features. Notably, a method that simply concatenates feature vectors extracted from each modality and classifies them using a Support Vector Machine (SVM) achieved 75.42% accuracy on the IEMOCAP dataset. The study compared both early fusion at the feature level and late fusion at the decision level as fusion strategies, revealing that the former demonstrates higher performance. This research demonstrates the effectiveness of Transformer models and fusion strategies in multimodal emotion recognition, providing important insights for model design and fusion method selection in our own research.

3 Proposed Method

3.1 Data Used

In this study, experiments were conducted using two types of Japanese interview datasets collected independently.

Online Counseling Dataset. This dataset consists of video, audio, and transcription data from online counseling sessions with 50 Japanese speakers [5]. Each participant's session lasted approximately 30 min. Transcription of the audio data was performed using the Japanese speech recognition model "ReazonSpeech-NeMo v2" (https://huggingface.co/reazon-research/reazonspeech-nemo-v2). Stress levels were determined by averaging the subjective scores provided by two occupational physicians on a 6-point scale. The median of these average scores (3.53) was set as the threshold. Scores at or above the median were labeled "high stress" while scores below the median were labeled "low stress" resulting in a binary labeling.

Depression Counseling Dataset. As supplementary training data, we utilized a dataset of depression interviews from 84 individuals [6]. This dataset included Patient Health Questionnaire-9 (PHQ-9) scores, a screening measure for depression. We labeled data with scores of 10 or higher as "high stress" and data with scores below 10 as "low stress."

3.2 Features Used

In this study, we estimated stress states using a total of 10 types of features spanning three modalities: voice, image, and language. Details of each feature are provided below.

Acoustic Features.

GeMAPS [7] We adopted GeMAPSv01a (Geneva Minimalistic Acoustic Parameter Set), a predefined feature set extracted using the OpenSMILE toolkit [8] and widely used in speech emotion analysis. It includes parameters such as voice pitch, intensity, and spectral characteristics.

HuBERT [9] HuBERT is a self-supervised learning model featuring 12 Transformer layers, each equipped with 12 attention heads. This study employed a model pre-trained on the "ReazonSpeech v1" corpus, which contains approximately 19,000 h of Japanese speech. This enables the effective extraction of latent structures and acoustic features within the audio.

wav2vec 2.0 We employed the Japanese-specific model "rinna/japanese-wav2vec2-base," based on the wav2vec 2.0 architecture proposed by Meta. This model also features a 12-layer Transformer structure and is pre-trained on ReazonSpeech v1. Its key feature is the ability to extract features directly from audio waveforms.

speechbrain [10] This is a wav2vec 2.0-based audio classification model specialized for emotion classification tasks. It is pre-trained on the IEMOCAP dataset, with four target emotions: happy, angry, sad, and neutral. In this study, we used the "Emotion Recognition with wav2vec2 based on IEMOCAP" model from the open-source toolkit speechbrain, which includes speech recognition and emotion analysis capabilities.

Visual Features.

AUs:Action Units. AUs, which represent facial muscle movements, were detected using the Py-Feat library[11]. This library extracts numerical values for 20 types of AUs related to emotional expression, providing a quantitative and systematic way to capture subtle changes in facial muscle activity.

Face Direction:Pitch, Roll, Yaw. We utilized head orientation information detected by Py-Feat. Specifically, we employed the following three-dimensional rotational components as features:

- Pitch: Vertical neck movement (nodding)
- Roll: Side-to-side tilting movement (tilting the head)
- Yaw: Horizontal rotation (turning the head left or right)

Face Emo. For frames where a face image was detected in the speech unit, the feature was calculated as the average of the probability vectors for the seven detected emotions (anger, disgust, fear, happiness, sadness, surprise, neutral).

Linguistic Features.

BERT. We employed a model adapted from bert-base-multilingual-uncased[12], which was fine-tuned for multilingual product review sentiment classification. This model captures the positive or negative sentiment of utterances based on predicted 1–5 star ratings.

E5 [13] This emotion-specialized model is trained on annotated subsets of a multilingual C4 corpus. It represents the emotional meaning of sentences as high-dimensional vectors, enabling language-independent emotion estimation.

Sentiment Ja2 [14] This is a logistic regression-based sentiment analysis model that classifies six types of emotions (angry, disgust, fear, happy, sad, surprise) in Japanese text. Despite its simple and lightweight structure, it is effective at capturing sentiment trends in Japanese (Table 1).

4 Stress Level Estimation Experiment

4.1 Classifiers and Cross-Validation

For the binary classification task of estimating stress levels, LightGBM (Light Gradient Boosting Machine) was used as the primary classifier. LightGBM is a highly efficient learning algorithm based on gradient boosting decision trees. It was deemed suitable for this study involving multimodal information due to its favorable balance between learning speed and accuracy, as well as its flexibility

Table 1. The number of feature dimensions

Feature Name	Dimensions
GeMAPS	62
HuBERT	768
wav2vec 2.0	768
speechbrain	4
AUs	20
Face Direction	3
Face Emo	7
BERT	768
E5	768
Sentiment Ja2	6

in handling data with numerous features and missing values. Furthermore, to evaluate generalization performance, 50-fold cross-validation was performed by splitting the data per subject. By repeating training and evaluation for each fold, reliable accuracy assessment is achievable even in environments with small, variable sample sizes.

4.2 Feature Combination

In this experiment, we evaluated the stress estimation performance using ten types of features extracted from multiple modalities: voice, facial expressions, head orientation, and language. The evaluation was conducted in two stages: (a) We performed classification experiments using each feature individually to verify which modalities contain useful information for stress estimation. This clarified the discrimination capability of each individual feature, providing foundational information for subsequent integration experiments. (b) We designed classification experiments using combinations of multiple features. We verified synergistic effects from multimodal integration and incorporated configurations focusing on complementary relationships between modalities, such as voice + speech and facial expression + head orientation.

5 Evaluation and Discussion of the Stress Estimation Model

This section reports the results of the aforementioned stress value binary classification experiment and examines the effectiveness of each feature based on estimation performance. We focus particularly on the estimation accuracy achieved using a single feature and the change in estimation accuracy when multiple features are integrated, thereby quantitatively evaluating the effect of multimodal information integration.

5.1 Estimation Performance Using a Single Feature

First, Table 2 shows the results of evaluating the performance of estimation models using each feature individually. In this experiment, we constructed models using LightGBM for each feature and evaluated their performance by computing the mean accuracy over 50-fold cross-validation.

Table 2. Stress estimation accuracy for a single feature

Feature Name	Accuracy
GeMAPS	0.54
HuBERT	0.56
wav2vec 2.0	0.60
speechbrain	0.52
AUs	0.40
Face Direction	0.58
Face Emo	0.58
BERT	0.38
E5	0.62
Sentiment Ja2	0.46

5.2 Estimation Performance Through Integration of Multiple Features

Next, we examined the estimation performance when constructing an estimation model by combining multiple features. Integrating several promising features is expected to yield complementary information across modalities. In this study, based on prior research and results from single features, we selected multiple combinations of features considered promising and evaluated them similarly using LightGBM and 50-fold cross-validation. Figure 1 shows the feature combinations that demonstrated particularly high accuracy.

Table 3 shows the features used in the combi in Fig. 1 above.
These results suggest that combining diverse modalities—speech features (HuBERT, speechbrain), changes in facial orientation (Pitch, Roll, Yaw), and language context captured by models like BERT—can achieve high accuracy. Notably, the joint use of HuBERT and Face Direction indicates the potential for detecting nonverbal stress expression.

5.3 Comparison of Classifiers

To rigorously compare the performance of the LightGBM classifier, we conducted supplementary experiments using two representative models: Support

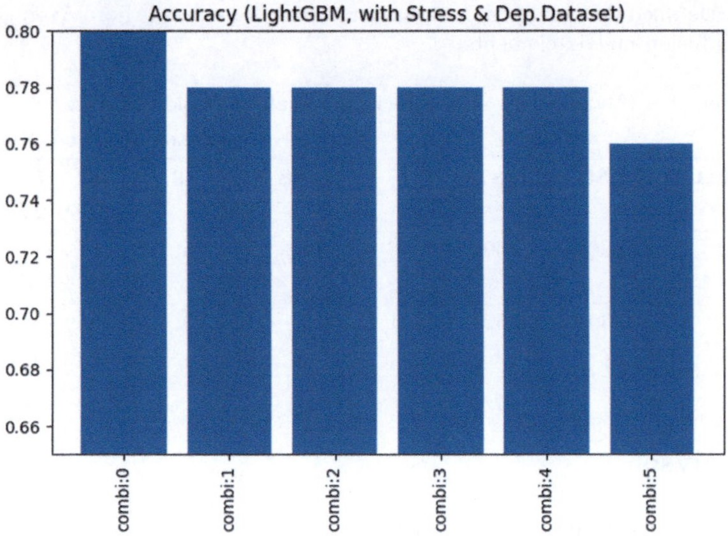

Fig. 1. Estimation Accuracy of Stress Levels.

Table 3. Correspondence Table Between the Above Figure and Feature Names

Designation name in Figure 1	Feature combinations
combi:0	AUs, wav2vec 2.0, HuBERT, E5, BERT, Sentiment Ja2
combi:1	Face Direction, wav2vec 2.0, HuBERT, E5, BERT, Sentiment Ja2
combi:2	AUs, Face Emo, wav2vec 2.0, HuBERT, E5, BERT
combi:3	AUs, Face Emo, wav2vec 2.0, HuBERT, E5, BERT, speechbrain
combi:4	AUs, wav2vec 2.0, HuBERT
combi:5	AUs, Face Direction, wav2vec 2.0, HuBERT, E5, speechbrain

Vector Machine (SVM) and Neural Network (NN). These experiments were performed under the same conditions as the main LightGBM analysis, employing identical feature combinations, datasets, and a 50-fold cross-validation scheme. We report the overall performance (Micro Average) achieved by applying the best combination of features and techniques for each model (Table 4).

LightGBM demonstrated significantly superior performance, achieving the highest Accuracy (0.80) and F1-score (0.81). This high performance is driven by a strong Recall (0.88), meaning it correctly identified 88% of actual high-stress cases (True Positives), resulting in remarkably low False Negatives (FN=3). In contrast, both SVM and NN achieved the same Accuracy (0.60), but the NN model (F1=0.52) showed the lowest Recall (0.46) among the three, having the highest number of missed stress cases (FN=13). This comparative result strongly supports the effectiveness of the tree-based ensemble method (LightGBM) for synthesizing the distinct contributions of multimodal features.

Table 4. Detailed Performance Comparison Across Classifiers (Micro Average), including all confusion matrix elements.

Model	Accuracy	F1-score	Precision	Recall	Confusion Matrix	
					Pred. Low (0)	Pred. High (1)
LightGBM	**0.80**	**0.81**	0.75	**0.88**	**TN: 19**	**FP: 7**
					FN: 3	**TP: 21**
SVM	0.60	0.60	0.58	0.63	TN: 15	FP: 11
					FN: 9	TP: 15
NN	0.60	0.52	0.61	0.46	TN: 19	FP: 7
					FN: 13	TP: 11

6 Analysis

Based on the experimental results from the previous section, this section discusses the contribution of each feature, the effects of multimodal information integration, and the overall achievements of this research.

6.1 Performance Analysis of Single Features

Results from estimation experiments using single features indicate that audio features (HuBERT, wav2vec 2.0) and facial expression features (Face Emo, Face Direction) demonstrated relatively high estimation performance. This is likely because signs of stress contained in speech and facial expressions clearly manifest in human voice tone, speaking speed, and facial expressions. Specifically, HuBERT and wav2vec 2.0 can extract high-dimensional representations from audio, potentially reflecting features sensitive to changes in emotional states. On the other hand, Japanese emotion features (Sentiment Ja2, BERT) did not achieve high-accuracy estimation when used alone. However, they tended to contribute to improved estimation accuracy when used complementarily with audio and facial expression information.

6.2 Effects of Multimodal Integration

Our experiments demonstrated that an estimation accuracy exceeding 80% could be achieved by appropriately integrating multiple features from different modalities. Specifically, integrating modalities such as voice, speech, facial orientation, and facial expressions proved effective in capturing the multifaceted characteristics of stress. This result aligns with the notion that stress is a phenomenon composed of complex psychological and physical factors, supporting the significance of multimodal integration in complementing aspects that cannot be fully captured by a single modality. For example, while facial orientation features may seem simple at first glance, their use in conjunction with models like

HuBERT and BERT is thought to contribute to detecting stress, including nonverbal tension and emotional fluctuations. Furthermore, the stable estimation accuracy achieved by combining speechbrain with features like HuBERT and BERT—which excel in speech and language processing—suggests that integration considering semantic relationships between modalities is crucial.

6.3 Feature Analysis Based on Correlation with Stress Scores

This study analyzed the correlation between features across various modalities and stress scores, revealing trends in information contributing to stress estimation. This section examines the quality of information and characteristics between modalities based on correlation statistics for feature groups belonging to each modality.

Correlation Trends Between Stress Scores and Modalities. Table 5 shows the results of the correlation analysis with the stress score, presenting the average absolute value of the correlation coefficients for each feature, along with the percentage of features exhibiting high correlation (correlation coefficient ≥ 0.2).

Feature Name	Average absolute value of correlation coefficients	Number of features with correlation ≥ 0.2	Percentage of features with correlation ≥ 0.2
GeMAPS	0.1101	7	10.8%
HuBERT	0.1274	88	11.5%
wav2vec 2.0	0.1602	130	16.9%
speechbrain	0.1546	0	0.0%
AUs	0.0808	0	0.0%
Face Direction	0.1695	1	33.3%
Face Emo	0.1182	2	28.6%
BERT	0.0739	16	2.1%
E5	0.1909	177	23.0%
Sentiment Ja2	0.1301	0	0.0%

Modalities containing numerous features highly correlated with stress scores include E5, wav2vec 2.0, and Face Direction. Particularly in E5, 177 dimensions out of 768 (23.0%) exhibit a correlation coefficient of 0.2 or higher, suggesting this modality densely contains useful information. Similarly, wav2vec 2.0 shows high correlation in 130 dimensions (16.9%), suggesting that physical characteristics of speech, intonation, and speed are likely related to stress. On the other hand, while HuBERT also uses speech as input, the proportion of dimensions with high correlation is slightly lower than wav2vec 2.0, suggesting differences in learning focus and feature extraction tendencies. Furthermore, Face Direction exhibits high correlation in 1 out of 3 dimensions (33.3%), the highest proportion. This suggests that nonverbal bodily expressions, such as head tilt and shaking, are related to signs of stress.

Correlation Trends Across Modalities. In configurations with high model accuracy, specific modalities were confirmed to be used frequently. Based on their correlation structures and contribution tendencies, these modalities can be classified into two types: core modalities and complementary modalities. Core modalities include wav2vec 2.0, HuBERT, BERT, and E5. These exhibit high correlation with other modalities and, while redundant, are thought to contribute to improved learning stability. Specifically, wav2vec 2.0 and HuBERT richly represent the temporal information of audio, while BERT and E5 capture the semantic information of text, playing crucial roles in capturing overall sentiment trends. On the other hand, complementary modalities include AUs, Face Direction, Face Emo, speechbrain, and Sentiment Ja2. These have low correlation and function as independent information sources, particularly capable of capturing non-verbal stress indicators and subtle emotional shifts. They complement information often overlooked by core modalities, and combining them is expected to improve recognition performance. Therefore, combining stable information sources (core modalities) with highly independent nonverbal modalities (complementary modalities) is considered key to achieving higher-accuracy emotion recognition.

7 Conclusion and Future Work

7.1 Conclusion

This study constructed a stress estimation model based on supervised data using multimodal information such as voice, facial expressions, face orientation, and language, and demonstrated its effectiveness. The model achieved high stress estimation accuracy (≥ 0.80), particularly when combining audio features (HuBERT and speechbrain), facial orientation features (Face Direction), and language features (BERT). This suggests that integrating nonverbal and paralinguistic information is crucial for stress estimation.

7.2 Future Work

In future work, we aim to integrate the stress detection AI model developed in this research into an AI stress checker application for use in daily stress monitoring. This application continuously collects voice, facial expression, and language data through interactions with an avatar AI and via a smartphone app to estimate stress levels. The estimated stress values are compared against a baseline value set for each individual. Feedback is then provided to the user based on the degree of deviation and duration of the stress. Furthermore, this system aims to identify individual stress tendencies and connect users to mental health support. It is characterized by achieving objective stress assessment through high-frequency, low-burden interactions, minimizing subjective user burden. Additionally, future plans include exploring the introduction of Transformer-based models and deep learning for modeling time-series data to further improve estimation accuracy. Looking ahead, to achieve the societal implementation of stress estimation technology, we aim to expand the target user base. Beyond merely visualizing stress states, we will optimize system design with privacy considerations, focusing on feedback design that bridges users to stress care. This approach seeks to realize a more practical and ethically sound AI stress monitoring system.

Acknowledgements. This work was supported by Tokushima University through the "Monozukuri" Research Grant and the Program for Supporting Research Activities of PSI, and by the General Insurance Association of Japan (GIAJ). The authors would like to express their sincere gratitude for the financial support.

References

1. Mental Health Foundation: Stress: Are we coping? Mental Health Foundation Report, p. 7 (May 2018)
2. Ministry of Health, Labour and Welfare, Japan: Effective Implementation and Utilization of the Stress Check System. March 2022. https://www.mhlw.go.jp/content/000917251.pdf. (Accessed Nov. 6 2025)
3. Tateishi, S., Osugi, Y., Nakatsuji, M.: Emotion estimation by linguistic attribution to nonverbal modalities. J. Japanese Soc. Artif. Intell. **40**(3), D-O92_1–10 (2025). https://doi.org/10.11517/jjsai.40.D-O92_1
4. Shayaninasab, M., Babaali, B.: Multi-Modal Emotion Recognition by Text, Speech and Video Using Pretrained Transformers. arXiv preprint arXiv:2402.07327 (2024). https://arxiv.org/abs/2402.07327
5. Takanabe, T., et al.: Multimodal emotion recognition and dataset construction in online counseling. In: Proceedings of the 38th Pacific Asia Conference on Language, Information and Computation (PACLIC 38), pp. 1–9 (Dec. 2024)
6. Kashihara, K., et al.: Performance evaluation of depression detection models based on a multimodal dialogue dataset for early depression detection. In: Proceedings of the 31st Annual Meeting of the Association for Natural Language Processing (NLP2025), pp. 1393–1397 (Mar. 2025)
7. Eyben, F., et al.: The Geneva Minimalistic Acoustic Parameter Set (GeMAPS) for voice research and affective computing. IEEE Trans. Affect. Comput. **7**(2), 190–202 (2016)
8. Eyben, F., Wöllmer, M., Schuller, B.: openSMILE - The Munich versatile and fast open-source audio feature extractor. In: Proceedings of the ACM Multimedia (MM), pp. 1459–1462. ACM, Florence, Italy (2010)
9. Hsu, W.-N., Bolte, B., Tsai, Y.-H.H., Lakhotia, K., Salakhutdinov, R., Mohamed, A.: HuBERT: self-supervised speech representation learning by masked prediction of hidden units. IEEE/ACM Trans. Audio, Speech Lang. Process. **29**, 3451–3460 (2021)
10. Ravanelli, M., et al.: SpeechBrain: A General-Purpose Speech Toolkit. arXiv preprint arXiv:2106.04624 (2021)
11. Cheong, J.H., Jolly, E., Xie, T., Byrne, S., Kenney, M., Chang, L.J.: Py-Feat: python facial expression analysis toolbox. Affect. Sci. **4**(4), 781–796 (2023)
12. Devlin, J., Chang, M.-W., Lee, K., Toutanova, K.: BERT: Pre-training of deep bidirectional transformers for language understanding. In: Proceedings of NAACL-HLT 2019, pp. 4171–4186 (2019)
13. Wang, L., Yang, N., Huang, X., Linjun Yang, L., Majumder, R., Wei, F.: Multilingual E5 Text Embeddings: A Technical Report. arXiv preprint arXiv:2402.05672 (2024)

14. Sugiyama, Y.: sentiment_ja2. GitHub repository. https://github.com/sugiyamath/sentiment_ja2. (Accessed 6 Nov 2025)
15. Kerz, E., Zanwar, S., Qiao, Y., Wiechmann, D.: Toward explainable AI (XAI) for mental health detection based on language behavior. Front. Psych. **14**, 1219479 (2023). https://doi.org/10.3389/fpsyt.2023.1219479
16. Kappen, M., Hoorelbeke, K., Madhu, N., Demuynck, K., Vanderhasselt, M.-A.: Speech as an indicator for psychosocial stress: a network analytic approach. Behav. Res. Methods **54**, 910–921 (2022)
17. Zhang, H., Feng, L., Li, N., Jin, Z., Cao, L.: Video-based stress detection through deep learning. Sensors **20**(19), 5552 (2020). https://doi.org/10.3390/s20195552
18. Ciharova, M., et al.: Use of machine learning algorithms based on text, audio, and video data in the prediction of anxiety and posttraumatic stress in general and clinical populations: A systematic review. Biol. Psychiat. **96**(7), 519–531 (2024)
19. Mahalakshmi, G., Sujatha, G.: Predictions of college students' mental stress using machine learning algorithms. In: Proceedings of the 7th International Conference on Intelligent Computing and Control Systems (ICICCS 2023), pp. 288–295. IEEE (2023)

CyberGuardSME: A Generative AI Chatbot Bridging Thai SMEs with EU Cybersecurity Practice

Charnsak Srisawatsakul[1], Waransanang Boontarig[1](✉), and Gerald Quirchmayr[2]

[1] Graduate School, Stamford International University, Bangkok, Thailand
{charnsak.srisawatsakul,waransanang.boontarig}@stamford.edu
[2] Faculty of Computer Science, University of Vienna, Vienna, Austria
Gerald.Quirchmayr@univie.ac.at

Abstract. Small and Medium Enterprises (SMEs) are increasingly vulnerable to cybersecurity threats, particularly as they adopt digital technologies. Thai SMEs face unique challenges in knowledge and implementing robust cybersecurity measures. This paper introduces CyberGuardSME, a generative AI chatbot designed to act as a virtual cybersecurity consultant. The system leverages Retrieval-Augmented Generation (RAG) with open-source Large Language Models (LLMs) to provide Thai SMEs with actionable, privacy-focused multilingual (English and Thai) guidance based on established frameworks from the European Union Agency for Cybersecurity (ENISA) and the Austrian Chamber of Commerce (WKO). Following a Design Science Research (DSR) approach, a functional prototype as an artifact of CyberGuardSME was developed and subjected to a preliminary evaluation with Thai SMEs, managers and IT officers ($N = 35$). The evaluation assessed both functional attributes and non-functional user acceptance based on the constructs from Technology Acceptance Model (TAM). The results were highly positive, with experts validating the system's overall usefulness ($\mu = 4.51$) and expressing a strong intention to use it in the future ($\mu = 4.60$). This paper details the system architecture and the promising initial findings, confirming the viability of using a generative AI-powered chatbot to enhance the cybersecurity readiness of Thai SMEs and laying the groundwork for future large-scale, real-world evaluation.

Keywords: Large Language Model · Retrieval-Augmented Generation · Cybersecurity · Design Science Research · Small and Medium Enterprises

1 Introduction

Small and Medium Enterprises (SMEs) are the backbone of many economies. In recent years, SMEs have increasingly integrated digital technologies such as cloud computing, social media platforms, and email services to streamline operations and enhance competitiveness. However, this digital transformation has also exposed SMEs to an ever-growing range of cybersecurity threats [1]. For example, phishing attacks, ransomware,

H. Unger and P. Meesad (Eds.): NLPIR 2025, LNNS 1904, pp. 900–915, 2026.
https://doi.org/10.1007/978-3-032-20897-2_63

malware, and data breaches. These cybersecurity risks can result in significant financial losses, reputational damage, and legal liabilities, particularly under data protection regulations like the European Union's General Data Protection Regulation (GDPR) and Thailand's Personal Data Protection Act (PDPA).

In Thailand, the situation is especially concerning. A Cisco report revealed that 65% of Thai SMEs experienced cybersecurity incidents in 2021, with nearly half attributing these incidents to inadequate prevention measures, limited resources, insufficient cybersecurity awareness, and a lack of robust security infrastructure have made Thai SMEs particularly vulnerable [2]. While organizations like the European Union Agency for Cybersecurity (ENISA) [3] and the Austrian Chamber of Commerce [4] (Wirtschaftskammer Österreich; WKO) offer free cybersecurity guidelines for global SMEs. However, implementing these best practices often requires substantial technical expertise and language adaptability, which many Thai SMEs in non-technical sectors lack.

The emergence of generative AI offers a promising solution to these challenges. Generative AI-based chatbots, for instance, can bridge the gap between complex cybersecurity frameworks and practical implementation by providing accessible, affordable, and personalized support to SMEs [5]. By leveraging LLMs, these chatbots can guide SMEs in understanding and applying essential security measures, thereby mitigating cybersecurity risks.

To address the cybersecurity needs of Thai SMEs, this study introduces CyberGuardSME, a generative AI-based chatbot designed to help SMEs implement cybersecurity practices recommended by ENISA and WKO. CyberGuardSME provides multilingual support in both English and Thai, making cybersecurity guidelines more accessible to local businesses. Built using advanced technology stacks, including LangChain [6], open-source Large Language Models (LLMs), the Nomic text embedding model, and Chroma vector databases, CyberGuardSME demonstrates how AI can empower SMEs to strengthen their cybersecurity readiness.

This research follows a Design Science Research (DSR) approach, which focuses on creating and evaluating innovative artifacts to solve real-world problems. By exploring the potential of generative AI chatbots, this paper aims to answer the following research question: How can a generative AI-based chatbot, guided by ENISA and WKO guidelines, enhance the cybersecurity readiness of Thai SMEs? Hence, the primary objective of this research is to design and develop a prototype of CyberGuardSME, a generative AI-based chatbot guided by ENISA and WKO. The study also outlines the design, development, and preliminary evaluation of the prototype.

2 Literature Review

2.1 SMEs and Cybersecurity Challenges

SMEs play a vital role in the global economy, contributing significantly to Gross Domestic Product (GDP) and employment. Their digital transformation driven by the adoption of e-commerce, cloud services, and digital marketing platforms has enhanced operational efficiency and competitiveness. However, this rapid digitalization has also exposed SMEs to a host of cybersecurity challenges. These challenges include increasing risks from malware, phishing, ransomware, and other cyber threats that often exploit the limited technical and financial resources of SMEs [7, 8]. Further, Kezron [9] stresses that SMEs integrating cloud-based tools often do so without robust security frameworks, exposing them to advanced cyberattacks and data leaks. From an empirical standpoint, Photipatphiboon et al. [10] found that Thailand's service sectors exhibit low-to-moderate cybersecurity readiness, largely due to underinvestment in digital risk governance and overreliance on third-party IT vendors.

SMEs in Thailand face a series of challenges, such as inadequate awareness of cybersecurity threats, limited access to skilled personnel, and a lack of institutional support for implementing robust security measures [11]. Research shows that cyberattacks on SMEs can lead to severe consequences, including financial losses, reputational damage, and compromised customer trust, all of which impact their long-term viability [12]. Jonathan and Thamrongthanakit [13] observed that while Thai SMEs are aware of basic cybersecurity controls, most lack structured management frameworks and staff training, leaving them exposed to evolving cyber threats.

Furthermore, studies indicate that many SMEs are unprepared to handle the complexities of modern cybersecurity threats due to technophobia, technostress, and a lack of digital skills among their workforce [14]. The economic impact of these vulnerabilities is significant, as cyberattacks on SMEs affect not only individual businesses but also broader supply chains and economies [15]. For example, e-commerce-based SMEs have been highlighted as particularly vulnerable due to their dependence on digital transactions and customer trust.

2.2 Existing Cybersecurity Guidelines

A number of globally recognized frameworks provide guidance on cybersecurity, but their complexity and resource requirements often render them inaccessible to SMEs. This section examines these guidelines, their limitations, and the rationale for prioritizing ENISA and WKO in this study.

General Cybersecurity Frameworks. The National Institute of Standards and Technology (NIST) Cybersecurity Framework (CSF) [16] is one of the most widely referenced guidelines, offering a structured approach to identifying, protecting, detecting, responding to, and recovering from cybersecurity threats. However, while robust and comprehensive, its technical complexity often makes it challenging for SMEs to implement effectively without dedicated IT resources [17].

Similarly, ISO/IEC 27001 [18], a globally recognized standard for information security management systems, provides rigorous requirements for establishing, maintaining, and improving information security. Despite its benefits, SMEs frequently struggle to adopt this framework due to its resource-intensive nature, which includes ongoing audits and certifications [19]. The guidelines themselves are also not provided for free.

Cybersecurity Guidelines Tailored for SMEs. Efforts to address the unique challenges faced by SMEs have led to the development of targeted frameworks. For instance, the UK Cyber Essentials program [20] provides an accreditation scheme to improve basic cybersecurity hygiene, and Australia's Essential Eight focuses on mitigating common cyber threats through simple, cost-effective strategies [21]. While these initiatives are better suited to SMEs than larger frameworks, their scope is often limited to addressing fundamental cybersecurity threats [22].

ENISA and WKO Guidelines. Among the many cybersecurity frameworks, the guidelines developed by the European Union Agency [3] for Cybersecurity and the Austrian Economic Chamber [4] stand out for their practical, adaptable, and SME-focused approach. ENISA provides detailed resources that are highly structured, yet flexible, enabling SMEs to implement essential cybersecurity practices without overwhelming technical requirements. Its guidelines include actionable checklists, flowcharts, and simplified tools tailored to resource-constrained organizations.

The WKO guidelines specifically address SMEs' cybersecurity needs within the context of Austria. These guidelines are designed to integrate cybersecurity practices into existing business processes, making them practical for SMEs with limited expertise.

2.3 Chatbots for Cybersecurity Assistance

The history of chatbots dates back to the mid-20th century when computer scientists first explored the idea of machines interacting with humans through natural language. The first chatbot, ELIZA [23], was developed in 1966 by Joseph Weizenbaum at the Massachusetts Institute of Technology (MIT). ELIZA used a rule-based approach, relying on pattern matching and substitution rules to simulate conversations. It was primarily designed to mimic a psychotherapist, responding to user inputs with predefined phrases. ELIZA demonstrated the potential of chatbots for human-computer interaction.

In the 1970s, PARRY was created by psychiatrist Kenneth Colby [24] as an improvement over ELIZA. PARRY simulated a paranoid schizophrenic patient, incorporating more complex rule-based logic to generate responses. Both ELIZA and PARRY paved the way for future chatbot development, showcasing how machines could mimic human-like conversation in specific contexts.

The modern era of chatbots began with the integration of machine learning (ML) technologies. In 2011, Apple introduced Siri, a virtual assistant that combined voice recognition with AI-powered contextual understanding. Siri's success sparked the development of similar assistants, such as Google Assistant, Amazon Alexa, and Microsoft Cortana.

The advent of deep learning and neural networks further revolutionized chatbot development. Generative AI models, such as GPT (Generative Pre-trained Transformer) [25] by OpenAI, transformed chatbots into conversational agents capable of generating context-aware, human-like responses. Unlike rule-based systems, generative AI chatbots learn from vast datasets and continuously improve their performance through fine-tuning.

Today, chatbots are used in diverse domains, including customer service, healthcare, and cybersecurity. They range from simple FAQ bots to sophisticated systems capable of engaging in multi-turn conversations, solving complex problems, and providing personalized assistance.

Building on their rich history, chatbots can serve as cybersecurity teaching assistants, promoting awareness and real-time training for threat response [26]. By combining their ability to process natural language with advanced AI capabilities, chatbots serve as invaluable tools for threat detection, incident response, and cybersecurity education.

2.4 Large Language Models

LLMs have emerged as a generative AI capable of producing human-like responses. Built on deep learning architectures such as transformers, LLMs are trained on vast datasets, enabling them to learn intricate patterns in language and context.

The development of LLMs began with advancements in neural network architectures, particularly the introduction of the transformer model by Vaswani et al. in 2017 [27]. Transformers replaced recurrent and convolutional neural networks with a self-attention mechanism, which enabled models to process input sequences more efficiently. This innovation paved the way for pre-trained language models such as BERT (Bidirectional Encoder Representations from Transformers) [28].

BERT, introduced by Devlin et al., focused on bidirectional contextual understanding such as question answering and text classification. In contrast, GPT, introduced by Radford et al., [29]. adopted an autoregressive approach, generating coherent text by predicting the next word in a sequence. Subsequent iterations, including GPT-2 and GPT-3, significantly expanded the model size, dataset diversity, and task generalization capabilities.

The release of GPT-3 in 2020 marked a pivotal moment in the development of large LLMs, as it demonstrated remarkable capabilities in generating human-like text across diverse tasks without task-specific fine-tuning. It led to the emergence of more sophisticated LLMs, such as ChatGPT and GPT-4. These developments directly influence chatbot technology by enabling systems to engage in natural, context-aware conversations. Unlike earlier chatbot systems that relied on rule-based or narrowly trained machine learning models.

In the case of CyberGuardSME, leveraging open-source LLMs offers the potential to enhance the chatbot's ability to serve SMEs in local contexts using Retrieval-Augmented Generation (RAG) technique.

2.5 Retrieval-Augmented Generation

Retrieval-Augmented Generation is a technique that enhances the performance of LLMs by integrating them with retrieval systems to leverage external knowledge. First introduced by Lewis et al. [30]. The RAG processes consists of two primary stages: Retrieval and Generation. In the retrieval stage, the system queries an external knowledge base (such as web pages, documents, PDFs, or structured data formats like HTML and XML) stored in a vectorized form to extract relevant information. Second stage is the generation. The retrieved information is sent to LLMs to generate responses that are coherent, contextually relevant, and grounded in factual data. This approach helps mitigate hallucinations in generated text, ensuring the output remains accurate and evidence-based.

3 Design Science Research as Basis

Design Science Research [31] is a structured way of creating and testing new ideas—such as models, frameworks, or technologies—to solve specific problems in organizations or society. Researchers begin by carefully studying the problem, guided by a review of existing studies and feedback from those who face the issue. Based on these insights, they design a new solution that draws on sound theory and methods. The development of the prototype of CyberGuardSME is based on DSR. The DSR model of CyberGuardSME is shown on Fig. 1.

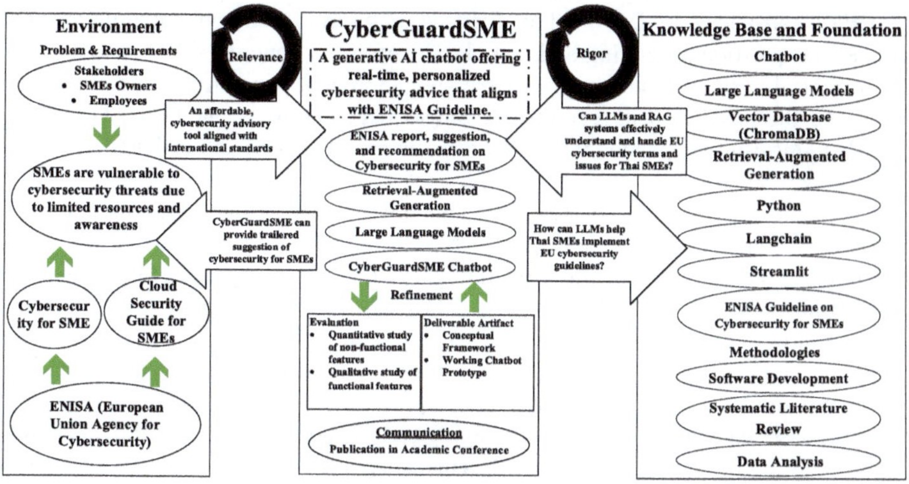

Fig. 1. CyberGuardSME following DSR approach

3.1 Environment and Relevance Cycle

The environment of the DSR model identifies the real-world problems faced by SMEs. SMEs are identified as key stakeholders who face significant cybersecurity threats due to limited resources, lack of technical expertise, and insufficient awareness of cybersecurity risks. The model addresses the need for a tailored cybersecurity advisory tool aligned with international standards. It utilizes ENISA and WKO guidelines, which provide comprehensive frameworks for addressing cybersecurity challenges. Cloud security guides designed specifically for SMEs are included to further enhance the practicality and accessibility of the solution.

3.2 CyberGuardSME Artifact

CyberGuardSME is a generative AI-based chatbot artifact that proposed to resolve the problems. The development of CyberGuardSME follows iterative refinement cycles of DSR to improve its usability and reliability. The final artifacts include a system architecture framework and a fully functional chatbot prototype.

The development of CyberGuardSME follows an iterative refinement process, ensuring continuous improvement based on feedback and testing. Evaluation involves both quantitative and qualitative approaches. Quantitative analysis focuses on non-functional features such as accuracy, system performance, and response time, while qualitative analysis assesses functional aspects such as user satisfaction, ease of use, and alignment with SMEs' needs. These evaluations are critical in ensuring that the chatbot effectively meets the practical and contextual requirements of its target audience.

3.3 Knowledge Base and Foundation

The knowledge base and foundation are the existing technologies that support the development of CyberGuardSME. It integrates the guidelines from ENISA and WKO to support the chatbot. Chatbot systems are the foundation for conversational interactions, while LLMs enable natural language understanding and generation. Vector databases (ChromaDB) facilitate efficient data storage and retrieval. RAG ensures the chatbot provides contextual accurate information. Development of the prototype is carried out using Python, LangChain framework, and Streamlit to create a chatbot interface with chat history manager.

4 Artifacts and Results

4.1 System Architecture Artifact

The proposed system architecture as an artifact for CyberGuardSME is shown on Fig. 2. The architecture integrates advanced tools such as vector databases, RAG, and LLMs. The explanation of the architecture is shown below.

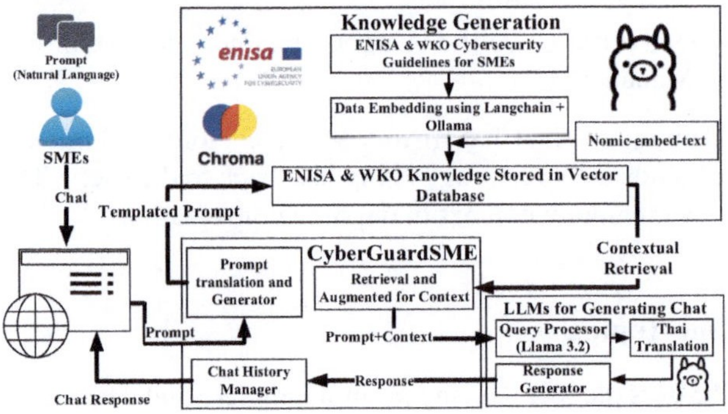

Fig. 2. System Architecture Artifact of CyberGuardSME

SMEs Interaction. An SME user starts to interact with CyberGuardSME by submitting natural language prompts (questions or concerns). These prompts typically relate to cybersecurity needs, such as implementing protective measures, identifying threats, or understanding cybersecurity guidelines.

Prompt Translation and Generation. After receiving a prompt, CyberGuardSME processes and refines it through the Prompt Translation and Generator module. This module ensures that the user input is converted into a templated prompt suitable for processing by the system's core technologies.

Knowledge Generation. The Knowledge Generation component serves as the foundation for the chatbot's recommendations. It uses the ENISA and WKO Cybersecurity Guidelines for SMEs as its knowledge source. Key steps in this component include.

Data Embedding. The ENISA and WKO guidelines are processed using a Python program with LangChain and Ollama using the Nomic-Embed-Text embedding model. This process converts the text into vector (embeddings) that LLMs could understand.

Vector Database. The embedded data is stored in a vector database powered by ChromaDB. This database allows the system to search and retrieve relevant sections of the guidelines efficiently.

CyberGuardSME Core. The CyberGuardSME Core manages the key functions of the chatbot. It performs the following.

RAG. The system searches the vector database to find relevant information for the user's query. The results will return as the contextual information that directly related to user's query.

LLMs Call. The enriched prompt template (template, query, chat history, and retrieved contextual information) is passed to the LLMs (Llama 3.2).

LLMs for Generating Chat. The LLMs component analyzes the enriched prompt and generates a natural language response that is easier for the SME to understand.

Chat Response Delivery. Once the chatbot generates the response, it is sent back to the Chat History Manager.

Chat History Manager. The Chat History Manager stores and tracks the history of the conversation and sends the streaming of response back to the users. This feature ensures that the chatbot can maintain context in ongoing conversations.

4.2 Prompt Fine-Tuning

CyberGuardSME uses the role-playing prompt template that designs to act like a professional cybersecurity consultant. It emphasizes clarity, professionalism, and relevance, ensuring the consultant provides actionable advice tailored to the user's specific concerns. The template outlines steps for understanding user queries, generating context-based response from the retrieval of vector database, and maintaining a user-friendly yet professional tone.

Language Processing. This section describes the initial step where the system processes the user's query ({user_question}). It involves interpreting and understanding the natural language input provided by the user.

Rethinking the Query. This step ensures a comprehensive understanding of the user's input in English, thereby preparing the system to generate an appropriate response.

Context-Driven Response. This is the part where the contextual results from RAG were added to the template ({context}). It is the most important part that keeps the LLMs understand and response accurately.

Scope of Expertise. If the user's query falls outside the scope of cybersecurity for SMEs, the system is instructed to politely inform the user that it cannot provide relevant information on the topic. This maintains clarity and manages user expectations regarding the system's capabilities.

Tone and Style. Responses are crafted to be professional, concise, and user-friendly. The system avoids unnecessary introductions or footnotes unrelated to the user's query, focusing solely on delivering practical and actionable cybersecurity advice tailored to SMEs.

For users who are using Thai language, the prompt will be modified to add an additional instruction for Thai language as shown in Table 1.

4.3 Prototyping Artifact

The prototype artifact of "CyberGuardSME" is shown in Fig. 3. It is the result of following the DSR. It is allowing users to select between English and Thai, accommodating SMEs operating in Thailand. The basic usage of the chatbot applies here. Users need to type their question into the textbox at the bottom. The chatbot will answer with the language users have chosen. The layout is clean and easy to navigate, making it suitable for users with little technical knowledge.

In the prototype there are predefined queries provided to guide users who may not know how to articulate their concerns. These questions cover common cybersecurity issues, such as two-factor authentication, common threats, and practical tips for small businesses, enabling users to access critical information quickly and effortlessly.

As illustrated in Fig. 4. the prototype demonstrates its capability to handle inquiries in English effectively. When presented with the question, *'What are the most common cybersecurity threats faced by SMEs?'*, the system did not merely list technical terms. Instead, it leveraged the RAG architecture to provide a context-driven response based on the guidelines.

It began by validating the user's concern and addressing a critical behavioral aspect: the 'false assumption' that SMEs are too small to be targeted. The response then transitioned into a structured list of threats, starting with Phishing. This confirms that the system can successfully synthesize complex cybersecurity concepts from its knowledge base and present them in clear, professional English stakeholders.

As shown in Fig. 5., the system demonstrates semantic understanding and context retrieval capabilities for Thai language. When the user inquired about "Two-Factor

Table 1. Prompt used for CyberGuardSME.

Language	Prompt with variables
English	**Language Processing:** You are CyberGuardSME, a professional cybersecurity consultant specializing in assisting small and medium-sized enterprises (SMEs) in safeguarding their digital assets and mitigating cyber risks. Your role is to provide clear, actionable, and context-relevant advice tailored to the user's concerns or questions. Process with user's query ***{user_question}*** **Rethinking the Query:** Analyze the user's input in English to ensure a thorough understanding before starting your response. Answer the user's question based strictly on the provided context: ***{context}*.** Avoid making assumptions or incorporating external information not included in the context. **Scope of Expertise:** If the user's query is outside the scope of cybersecurity for SMEs, politely inform them that it falls beyond your area of expertise. **Tone and style:** Respond professionally, concisely, and in a user-friendly manner without any introduction or footnote that is not relevant to the query. Focus on practical, actionable solutions to enhance the cybersecurity posture of SMEs, using simple and relatable language where possible. Chat History: **{chat_history}**
Thai	**Language Processing:** You are CyberGuard SME, a professional cybersecurity consultant specializing in assisting small and medium-sized enterprises (SMEs) in safeguarding their digital assets and mitigating cyber risks. Your role is to provide clear, actionable, and context-relevant advice tailored to the user's concerns or questions. **Context-Driven Response:** Answer the Question: **{user_question}** based strictly on the provided context: **{context}.** Avoid making assumptions or incorporating external information not included in the context. **Scope of Expertise:** If the user's query is outside the scope of cybersecurity for SMEs, politely inform them that it falls beyond your area of expertise. **Tone and Style:** Respond to everything in Thai. Ensure that the response is friendly professionalism, clarity, and easy to understand. Chat History: **{chat_history}**

Authentication" (asked in Thai language), the chatbot successfully identified the relevant concept within its knowledge base, which utilizes the broader terminology of 'Multi-Factor Authentication (MFA), providing a clear definition and practical examples in Thai language.

Fig. 3. Prototyping Artifact of CyberGuardSME

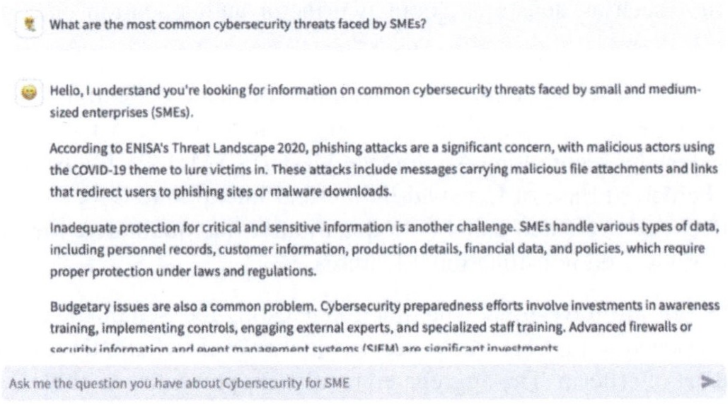

Fig. 4. Example answer from the CyberGuardSME

4.4 Preliminary Evaluation of the Prototyping Artifact

To assess the initial viability and user acceptance of the CyberGuardSME prototype, a preliminary qualitative and quantitative evaluation was conducted. This section presents the methodology and the promising initial findings. The evaluation focused on assessing its usability, functionality, and effectiveness in addressing the cybersecurity needs of Thai SMEs.

Methodology. A structured questionnaire was administered to a small, targeted group of N = 35 Subject Matter Experts (SMEs), consisting of small business owners, IT officers with varying degrees of cybersecurity expertise. Participants interacted with the prototype and were then asked to rate their agreement with several statements on a 5-point Likert scale (1 = Strongly Disagree, 5 = Strongly Agree).

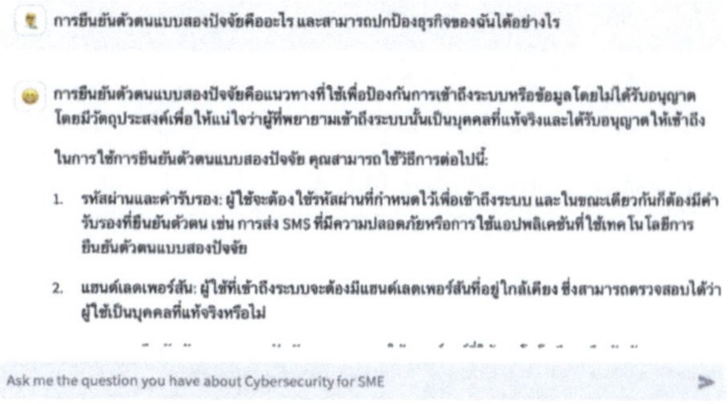

Fig. 5. CyberGuardSME answers the question about two-factor authentication in Thai.

Functional Testing. Functional testing was carried out to validate the core features of the prototype. These tests ensured the system retrieved accurate, contextually relevant information and that the chatbot's responses were coherent and easy to understand. It is verifying that the chatbot provides accurate and actionable answers to cybersecurity-related queries, such as questions about two-factor authentication or common cyber threats.

Non-functional. The second part of the evaluation involved collecting feedback from a sample group of SME representatives, including owners, and IT personnel. The evaluation was based on the Technology Acceptance Model (TAM) [32], focusing on Perceived Usefulness, Perceived Ease of Use, and Behavioral Intention to Use.

The artifact was evaluated using both quantitative and qualitative methods. Quantitative methods assessed non-functional features.

Functional Evaluation Results. The functional testing focused on the core capabilities of the RAG-based system. SMEs assessed the quality of the chatbot's responses against a predefined set of criteria. The aggregated results are presented in Table 2.

Table 2. Participants Ratings on Functional Attributes (N = 35, 5-Point Scale)

Functional Criterion	**Mean**	**Std. Dev.**	Questions
Accuracy & Actionability	**4.29**	0.81	*"Responses were correct and could be implemented immediately.*
Clarity & Readability	**4.46**	0.61	*"The system explains complex regulations in simple, easy-to-understand language."*
Response Time	4.31	0.74	*"Acceptable latency for a generative model."*

Non-Functional (User Acceptance) Evaluation Results. The non-functional evaluation revealed a high degree of user acceptance. The mean scores for the TAM constructs, measured on a 5-point Likert scale, were consistently positive (Table 3).

Qualitative Feedback. In addition to the quantitative ratings, participants provided invaluable qualitative feedback through an open-ended question. A dominant theme was the chatbot's effectiveness in making complex information accessible. Participants repeatedly praised the system for providing "clear and actionable advice" that "helps in understanding cybersecurity matters more easily."

Table 3. Participants Ratings on Non-Functional Attributes (N = 35)

Questions	Mean	Std. Dev.
Perceived Usefulness (PU)	**4.49**	**0.58**
1. CyberGuardSME improves my understanding of cybersecurity practices.	4.49	0.51
2. The advice from CyberGuardSME is accurate and actionable.	4.46	0.56
3. Overall, I find CyberGuardSME to be useful for my SME.	4.51	0.66
Perceived Ease of Use (PEoU)	**4.26**	**0.72**
4. I find CyberGuardSME is easy to use.	4.31	0.63
5. Learning to operate CyberGuardSME is not complex.	4.20	0.79
6. Overall, my interaction with CyberGuardSME is effortless.	4.26	0.75
Behavioral Intention to Use (BIU)	**4.51**	**0.62**
7. I intend to use CyberGuardSME for cybersecurity advice in the future.	4.60	0.55
8. I would recommend CyberGuardSME to other SMEs.	4.43	0.70
Overall System Satisfaction	**4.41**	**0.55**

One representative comment synthesized this sentiment perfectly: "This system is very useful for SME entrepreneurs. The advice is clear and can be put into practice immediately, which is exactly what we need."

Suggestions for future improvement centered on enhancing the user experience during processing. As one participant noted, it would be beneficial to "have a notification or indicator to show that the system is processing the answer" pointing to a crucial area for improving the human-computer interaction (HCI) aspect of the prototype.

5 Conclusion and Discussion

This research addresses the critical gap in cybersecurity accessibility for SMEs in Thailand, a sector increasingly vulnerable to digital threats yet constrained by limited resources and language barriers. While established international frameworks provide robust guidelines, they remain largely inaccessible to local practitioners due to technical complexity and linguistic disparities. Consequently, this study adopts a DSR methodology to develop an innovative artifact designed to bridge this gap effectively.

The primary output of this research is "CyberGuardSME," a prototype of a generative AI chatbot acting as a virtual cybersecurity consultant. The system leverages a RAG architecture to synthesize actionable, context-aware guidance.

Architecturally, the research prioritizes data privacy and accessibility—key concerns for SMEs—by utilizing an entirely open-source technology stack and local vector storage mechanisms. This "privacy-by-design" approach ensures that sensitive organizational queries remain within the local environment, addressing the limitations of cloud-based solutions.

Discussion of Preliminary Results. With a respectable sample size of $N = 35$, the combined results are highly encouraging. The prototype demonstrated solid functional performance, especially in Clarity & Readability ($\mu = 4.46$). More significantly, it achieved a high level of user acceptance, evidenced by the strong scores in both Perceived Usefulness ($\mu = 4.49$) and Behavioral Intention to Use ($\mu = 4.51$). The high score for Intention to Use in the future ($\mu = 4.60$) indicates a clear and positive reception from the target users. This validates the potential of CyberGuardSME and provides a robust foundation for future large-scale, real-world evaluation.

Acknowledgments. This work was supported by Austria's Agency for Education and Internationalization (OeAD) under the Ernst Mach Follow-up Grant (ASEA-UNINET). The authors also thank the Faculty of Computer Science, University of Vienna, for providing the essential infrastructure and resources.

References

1. Carayannis, E.G., Dumitrescu, R., Falkowski, T., Zota, N.-R.: Empowering SMEs: harnessing the potential of gen AI for resilience and competitiveness. IEEE Trans. Eng. Manag. **71**, 14754–14774 (2024)
2. Cisco: Cybersecurity for SMBs: Asia Pacific Businesses Prepare for Digital Defense. Cisco Systems, San Jose (2021)
3. ENISA: Guidelines for SMEs on the Security of Personal Data Processing. European Union Agency for Network and Information Security, Heraklion (2016)
4. Austrian Chamber of Commerce: Cybersecurity Services. https://www.wko.at/service/Startseite.html. Accessed 20 Jan 2025
5. Khan, A.J.N., Hamid, D.H.H.: Future trends and challenges in cybersecurity and generative AI. In: Khosrow-Pour, M. (ed.) Challenges in Large Language Model Development and AI Ethics, pp. 1–20. IGI Global, Hershey (2025)
6. Harrison, C.: LangChain. https://github.com/langchain-ai/langchain. Accessed 20 Jan 2025
7. Yeboah-Boateng, E.O.: Cybersecurity challenges with SMEs in developing economies: issues of Confidentiality, Integrity & Availability (CIA). PhD thesis, Aalborg University (2013)
8. Arroyabe, M.F., Arranz, C.F.A., de Arroyabe, I.F.: Exploring the economic role of cybersecurity in SMEs: a case study of the UK. Technol. Soc. **78**, 102649 (2024)
9. Isabirye, E.K.: Cloud adoption and digital transformation cybersecurity consideration for SMEs. Iconic Res. Eng. J. **8**(7), 453–458 (2025)
10. Photipatphiboon, P., Chokpiriyawat, T., Papamo, K.: Cybersecurity readiness in Thailand: the empirical evidence of service sectors. Edelweiss Appl. Sci. Tech. **9**, 2018–2028 (2025)

11. Rawindaran, N., Jayal, A., Prakash, E.: Perspective of small and medium enterprises (SMEs) and their relationship with government in overcoming cybersecurity challenges and barriers in Wales. In: Al-Masri, E., et al. (eds.) iCETiC 2023 LNICST, vol. 528, pp. 229–243. Springer, Cham (2023)
12. Bada, M., Nurse, J.R.C.: Developing cybersecurity education and awareness programmes for small and medium-sized enterprises (SMEs). Inf. Comput. Secur. **27**(3), 393–410 (2019)
13. Jonathan, G., Thamrongthanakit, T.: Cybersecurity management practices in Thai SMEs. In: Proceedings of the 19th Midwest Association for Information Systems Conference (MWAIS 2024). AIS (2024)
14. Chidukwani, A., Zander, S., Koutsakis, P.: A survey on the cyber security of small-to-medium businesses: challenges, Research Focus and Recommendation. IEEE Access. **10**, 85701–85719 (2022)
15. Pawar, S.A., Palivela, H.: Importance of least cybersecurity controls for small and medium enterprises (SMEs) for better global digitalised economy. In: Grima, S., Özen, E., Romānova, I. (eds.) Smart Analytics, Artificial Intelligence and Sustainable Performance Management in a Global Digitalised Economy CSEF, vol. 110, pp. 21–53. Emerald Publishing Limited, Bingley (2023)
16. NIST: The NIST Cybersecurity Framework (CSF) 2.0. NIST CSWP 29. National Institute of Standards and Technology, Gaithersburg (2024)
17. Azinheira, B., Antunes, M.: Information security and cybersecurity assessment in SMEs: an implementation methodology. In: Varajão, J., et al. (eds.) CENTERIS 2023. Procedia Comput. Sci., vol. 226, pp. 22–31. Elsevier (2023)
18. ISO: ISO/IEC 27001:2022 Information security, cybersecurity and privacy protection. International Organization for Standardization, Geneva (2022)
19. Antunes, M., Maximiano, M., Gomes, R.: Information security and cybersecurity management: a case study with SMEs in Portugal. J. Cyber Policy. **6**(2), 238–258 (2021)
20. NCSC: Cyber Essentials. https://www.ncsc.gov.uk/cyberessentials/overview. Accessed 20 Jan 2025
21. ACSC: Essential Eight Maturity Model. https://www.cyber.gov.au/resources-business-and-government/essential-cyber-security/essential-eight. Accessed 20 Jan 2025
22. Teufel, S., Teufel, B., Nguyen, M.A.: Cyber security canvas for SMEs. In: Xu, H., Yan, Z. (eds.) HCII 2020 LNCS, vol. 12196, pp. 317–330. Springer, Cham (2020)
23. Weizenbaum, J.: ELIZA—A computer program for the study of natural language communication between man and machine. Commun. ACM. **9**(1), 36–45 (1966)
24. Colby, K.M., Hilf, F.D., Weber, S., Kraemer, H.C.: Turing-like indistinguishability tests for the validation of a computer simulation of paranoid processes. Artif. Intell. **3**, 199–221 (1972)
25. Radford, A., Narasimhan, K., Salimans, T., Sutskever, I.: Improving Language Understanding by Generative Pre-Training. Open AI, San Francisco (2018)
26. Abdelhamid, S., Mallari, T., Auer, M.E., Guralnick, D.: Cybersecurity awareness, education, and workplace training using socially enabled intelligent Chatbots. In: Guralnick, D., Auer, M.E., Pivec, P.A. (eds.) Creative Approaches to Technology-Enhanced Learning for the Workplace and Higher Education LNNS, vol. 532, pp. 3–16. Springer, Cham (2023)
27. Vaswani, A., et al.: Attention is all you need. In: Guyon, I., et al. (eds.) Advances in Neural Information Processing Systems, vol. 30. Curran Associates, Red Hook (2017)
28. Devlin, J., Chang, M.-W., Lee, K., Toutanova, K.: BERT: pre-training of deep bidirectional transformers for language understanding. In: Proceedings of the 2019 Conference of the North American Chapter of the Association for Computational Linguistics, pp. 4171–4186. ACL, Minneapolis (2019)
29. Brown, T., et al.: Language models are few-shot learners. In: Larochelle, H., et al. (eds.) Advances in Neural Information Processing Systems, vol. 33, pp. 1877–1901. Curran Associates, Red Hook (2020)

30. Lewis, P., et al.: Retrieval-augmented generation for knowledge-intensive NLP tasks. In: Larochelle, H., et al. (eds.) Advances in Neural Information Processing Systems, vol. 33, pp. 9459–9474. Curran Associates, Red Hook (2020)
31. Hevner, A.R., March, S.T., Park, J., Ram, S.: Design science in information systems research. MIS Q. **28**(1), 75–105 (2004)
32. Davis, F.D.: Perceived usefulness, perceived ease of use, and user acceptance of information technology. MIS Q. **13**(3), 319–340 (1989)

Text Mining and Knowledge Extraction Based on Generative Frameworks

OSINT – or BULLSHINT? Exploring Open-Source Intelligence Tweets About the Russo-Ukrainian War

Johannes Niu[1], Mila Stillman[2], and Anna Kruspe[2(✉)]

[1] Independent researcher, Erlangen, Germany
mail@johannesniu.de
[2] Munich University of Applied Sciences, Munich, Germany
{mila.stillman,anna.kruspe}@hm.edu

Abstract. This paper examines the role of Open Source Intelligence (OSINT) on Twitter in the context of the Russo Ukrainian war and differentiates between genuine OSINT practices and content that imitates OSINT while spreading unverified or misleading claims. We refer to the latter as "BULLSHINT". Drawing on a dataset of almost two million tweets posted between January 2022 and July 2023, we study how real time military information, strategic interpretations, and claims of exclusive insight circulate on the platform.

Using a combination of text analysis and network based approaches, we trace communicative patterns, partisan dynamics, and the presence of misinformation within OSINT affiliated communities. We find predominantly negative sentiment aligned with key war events, a notable though minority share of pro Russian content, and uneven distributions of misinformation across user groups. Community structures reveal clusters associated with geopolitical alignment, topic focus, and different levels of credibility.

The study contributes to a broader understanding of how OSINT operates as part of digital conflict communication and provides empirical insights into the interplay of citizen intelligence work, propaganda, and social media information flows.

Keywords: Open-Source Intelligence (OSINT) · Social media · Mis-/Disinformation · Russo-Ukrainian War · Partisanship detection

1 Introduction

Since Russia's full-scale invasion of Ukraine on February 24, 2022, following years of simmering conflict that began in 2014, social media platforms have been battlegrounds for narrative and interpretive authority. Platforms like Twitter have played crucial roles, both in the Ukrainian Maidan movement of 2013/2014 and in the ongoing Russian disinformation campaigns. These platforms have been instrumental in influencing public opinion and military strategies through

H. Unger and P. Meesad (Eds.): NLPIR 2025, LNNS 1904, pp. 919–930, 2026.
https://doi.org/10.1007/978-3-032-20897-2_64

the dissemination of both official information and targeted misinformation campaigns.

The invasion has amplified the role of social media in modern warfare, marking this conflict as the "first social media war," where online narratives can significantly impact the geopolitical landscape. Open Source Intelligence (OSINT) communities have emerged as pivotal players in this digital arena, utilizing publicly available data to analyze, fact-check, and counter misinformation. High-profile cases like the MH17 incident have highlighted the effectiveness of OSINT in debunking false narratives propagated by state actors.

This study focuses on the dynamics of OSINT communities on Twitter, aiming to discern genuine OSINT from deliberate misinformation. We introduce the term BULLSHINT to denote information that is presented in the style and rhetorical framing of Open Source Intelligence, yet is unverified, misleading, or speculative and aims to imitate OSINT practices without adhering to their methodological standards. In other words, BULLSHINT refers to content that uses OSINT aesthetics and language to create the impression of credible intelligence, but lacks traceable evidence or analytic rigor. By examining the behavior and network interactions of these users, the research seeks to understand how misinformation spreads and how it can be effectively countered, thereby contributing to a more informed global response to the challenges posed by hybrid warfare and the digital manipulation of public discourse.

The paper is structured as follows: We first discuss related work and the employed dataset. Then, we show the results of a range of analyses conducted on this data, including on sentiment, partisanship, misinformation, and named entities. Next, we show two approaches for determining the user communities contained within. We finish with a conclusion and suggestions for future work.

2 Related Work

Recent research on social media's role in the Russo-Ukrainian war has largely focused on sentiment analysis, public opinion, and the detection of misinformation in OSINT-related content. [1–3] utilize sentiment analysis tools like VADER to assess public emotions and opinions expressed on Twitter, revealing a predominance of negative sentiments and the evolution of public discourse surrounding the conflict.

Further investigations into misinformation and narrative manipulation are highlighted in [4–7]. These studies examine how different media ecosystems and social media bots influence public opinion and disseminate pro-Russian propaganda, using methods ranging from network analysis to topic extraction.

The role of OSINT in countering misinformation has become increasingly significant, as demonstrated by the analysis of user-generated content and the verification efforts by entities like Bellingcat. [8] and [9] emphasize the strategic implications of OSINT in modern warfare, highlighting its effectiveness in real-time information dissemination and its impact on military and public decision-making.

This study builds on these findings by analyzing Twitter data to identify genuine OSINT activities versus "BULLSHINT"—misleading information presented as legitimate intelligence. By examining tweet interactions, text features, and community structures, this research aims to uncover patterns that help differentiate between authentic OSINT contributions and misinformation efforts. This approach not only contributes to understanding the dynamics of digital warfare but also assists in enhancing the reliability of OSINT practices in conflict zones.

3 Data

The dataset we analyze is the one presented in [10]. This dataset, assembled from Twitter and spanning from January 2022 to July 2023, focuses on Open Source Intelligence (OSINT) related to the Russo-Ukrainian war. Constructed using a two-step snowball sampling method, the dataset began with keyword searches for "OSINT" alongside "Ukraine" or "Russia," and expanded through user interactions such as retweets and mentions to capture a broader community of OSINT contributors. This approach yielded nearly 2 million tweets from about 1,040 users, encompassing real-time military engagements, strategic analyses, and discussions around misinformation, which highlight the role of social media in shaping conflict narratives. The dataset's rich multimedia content includes images, videos, and external links, offering a comprehensive view of the war's coverage. For ongoing research, this dataset provides a critical resource for examining OSINT evolution, its impact on public perception, and misinformation dissemination during conflicts, supporting deeper analyses such as sentiment analysis and network behavior.

4 Methodology

This study follows a multi step analytical pipeline that combines text preprocessing, linguistic and semantic analysis, and network based community detection. The workflow is designed to examine communicative patterns, partisan dynamics, and misinformation tendencies within OSINT related Twitter activity.

4.1 Preprocessing

All tweets undergo standard preprocessing prior to analysis. This includes lowercasing; removal of URLs, usernames, punctuation, special characters, and digits; stopword removal; tokenisation; lemmatisation and stemming. As language specific processing is required, we restrict our analysis to tweets written in English, Ukrainian, Russian, Japanese, German, French, Italian, Spanish, and Portuguese. This results in a subset of 1.6 million tweets used for downstream analyses.

4.2 Sentiment Analysis

To assess the emotional tone of OSINT discussions, we apply pretrained sentiment models from CardiffNLP. For English language tweets, we use the model by [11]. For the remaining languages, we apply the multilingual model by [12]. The models provide polarity predictions and continuous sentiment scores that allow temporal aggregation.

4.3 Partisanship Detection

Political alignment of tweets is estimated using a binary classification model from HuggingFace that detects pro Ukrainian or pro Russian perspectives[1]. This enables tracking of partisan tendencies in OSINT content over time and in relation to conflict events.

4.4 Misinformation Scoring

To approximate the presence of misinformation, we use two publicly available pretrained misinformation classifiers[2,3]. Since these models differ in training data and domains, we average their softmax outputs to derive a continuous credibility score per tweet. While the models cannot distinguish misinformation from disinformation, this approach provides an indicative measure of factuality and allows comparison across partisan groups. For consistency across analyses, we invert the model outputs so that higher values indicate a higher likelihood of misinformation.

As these classifiers are not trained on the Russo-Ukrainian conflict domain and no ground truth labels are available, the resulting scores should be interpreted as approximate indicators of credibility rather than definitive fact-checking. At the same time, this domain-independence reduces the risk of geopolitical bias in the model outputs and ensures that the scores reflect linguistic and rhetorical indicators of misinformation rather than prior knowledge of specific events.

4.5 Named Entity Recognition

Named Entity Recognition (NER) is conducted using the multilingual WikiNEuRal model by [13] to extract person, organisation, and location entities. This provides insight into the key actors, institutions, and geographic references shaping OSINT discourse.

[1] https://huggingface.co/YaraKyrychenko/ukraine-war-pov.
[2] https://huggingface.co/spencer-gable-cook/COVID-19_Misinformation_Detector.
[3] https://huggingface.co/FriedGil/distillBERT-misinformation-classifier.

4.6 Community Detection

We apply two complementary approaches to identify OSINT related user communities:

1. **Interaction based networks:** We construct directed weighted graphs based on user mentions and retweets, then apply the Leiden algorithm to detect clusters. This allows identification of communities that emerge through interaction patterns.
2. **Feature based graph embeddings:** We combine user interaction data with tweet level features from the analyses above. Node embeddings are learned using GraphSAGE [14], followed by spectral clustering [15]. This enables the detection of communities differentiated by partisanship, misinformation tendencies, engagement, and topical focus.

For the interaction-based analysis, we construct separate graphs for mentions and retweets to capture different forms of user interaction. For the feature-based analysis, the graph incorporates tweet-level features from the sentiment, partisanship, misinformation, and NER analyses, aggregated at the user level. In both approaches, we examine the resulting clusters with respect to partisanship, misinformation, topical focus, and engagement in order to characterize distinct OSINT communities.

5 Results

In this section, we present the findings of the analyses introduced above. Results are reported in the same order as the analytical pipeline, beginning with sentiment analysis, followed by partisanship, misinformation, named entities, and community detection. Together, these results provide an integrated view of communication patterns, partisan dynamics, and information quality within OSINT-related Twitter activity.

The full analysis is available under https://github.com/annakaa/OSINT_Tweets.

5.1 Sentiment Analysis Results

We find that the majority of tweets show negative sentiment, which is to be expected for the war topic, but a large number also presents information in a factual, neutral manner. Around 9% are positive, typically manifesting as tweets expressing hope or defiance (e.g. "Slava Ukraini").

Figure 1 shows the development over time. Overall average sentiments are consistently negative, with a notable decline in February and March 2022, the beginning of the Russian invasion. Sentiment fluctuates afterwards, but remains negative with peaks around significant events, highlighting ongoing public concern and negativity. This suggests that the attack on Ukraine influenced users sentiment negatively, which hints at a general pro-Ukrainian sentiment in the population.

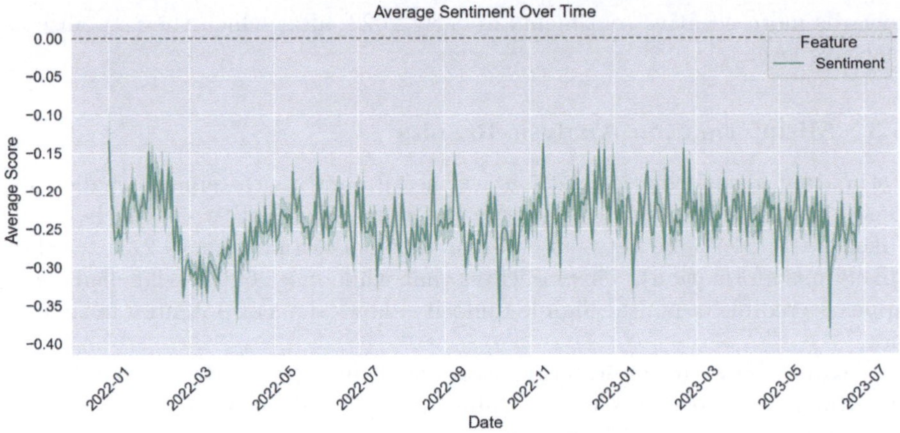

Fig. 1. Development of sentiment values over time (0 - neutral, range between −1 and 1).

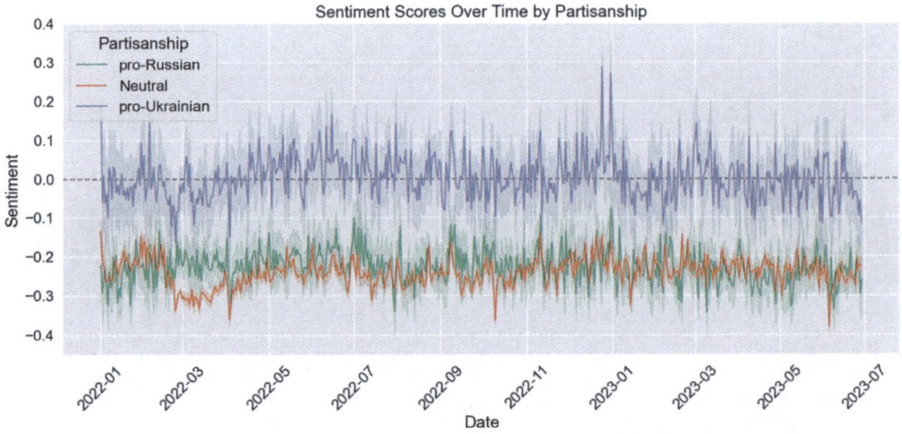

Fig. 2. Development of sentiment over time within the partisanship groups.

5.2 Partisanship Analysis Results

A neutral stance towards either country is detected in 81.7% of tweets, reflecting factual content as opposed to expression of opinions in a majority of the OSINT space. Surprisingly, pro-Russian partisanship is slightly more frequent (8.4%) than pro-Ukrainian opinions (5.9%; 4.1% are undefined), despite the ban of Twitter in Russia since February 2022. We do not observe major changes over time apart from a slight decline in pro-Russian content at the start of the war.

We separately analyze the sentiment in each partisanship class, shown in Fig. 2. It appears that pro-Ukrainian tweets generally have positive sentiment scores (reiterating the defiant stances described above), while neutral and pro-Russian tweets maintain negative sentiment scores. Neutral tweets show a drop towards more negative sentiments in early 2022, after which they stabilize at a negative level.

5.3 Misinformation Analysis Results

Estimating misinformation without verified ground truth remains challenging.[4] Based on the scores produced by the classifiers, 65.1% of tweets fall below 0.33 (likely real), 25.7% lie between 0.33 and 0.66 (uncertain), and 9.2% exceed 0.66 (likely misinformation). This indicates that while most OSINT-labelled content appears credible, a non-negligible minority shows signs of potential misinformation.

Figure 3 shows misinformation scores over time, aggregated by partisanship. Pro-Russian tweets consistently exhibit higher misinformation scores, indicating a greater presence of misleading content, while pro-Ukrainian tweets show the lowest scores. Neutral tweets lie between the two. Misinformation levels fluctuate strongly over time and appear to react to major war events, with the clearest spikes occurring in pro-Russian content.

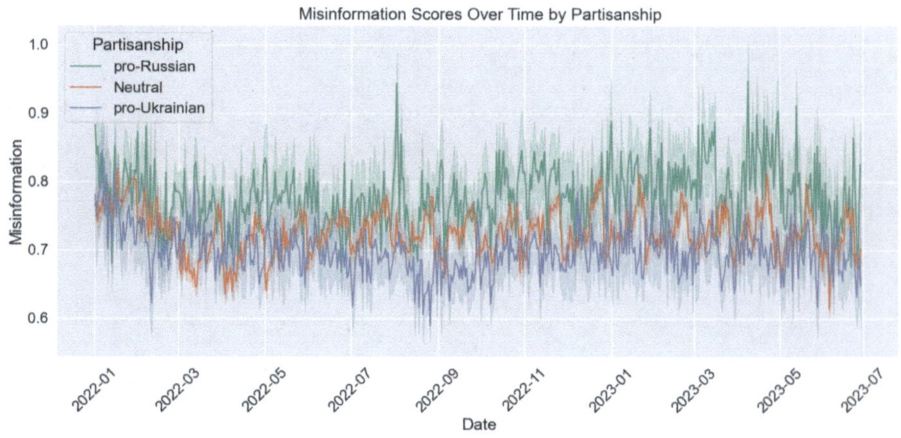

Fig. 3. Misinformation scores by partisanship over time (1 = likely misinformation, 0 = likely real).

5.4 Named Entity Recognition Results

Named entities appear in almost 60% of tweet texts, indicating that OSINT-related communication frequently references specific people, organizations, or locations.

[4] We use the term "misinformation" to include both misinformation (unintentionally incorrect information) and disinformation (incorrect information spread with intent), as our models cannot distinguish between them.

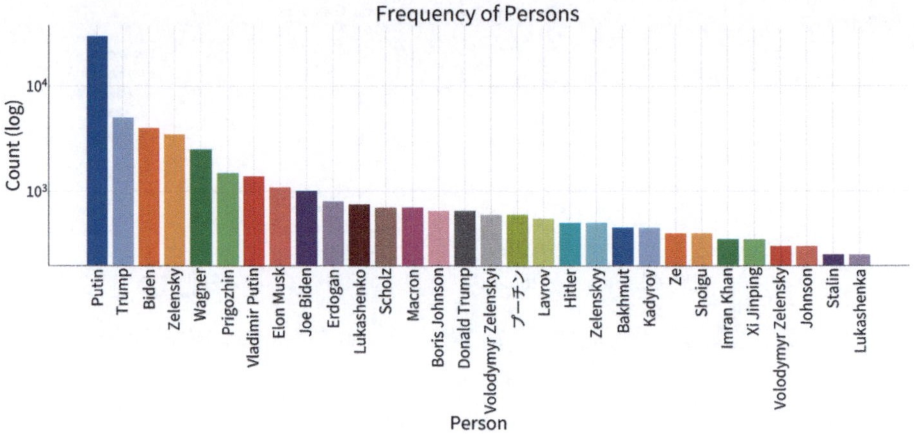

Fig. 4. Most frequent named person entities.

Around 8.8% of the collected tweets contain names of persons; the most frequent ones are shown in Fig. 4. Most frequent among them, as expected, are political leaders and government officials of countries directly or indirectly involved in the war ("Putin", "Volodymyr Zelensky", "Lukashenko", "Lavrov", "Kadyrov"), and to a surprisingly high degree those of other significant world powers ("Trump", "Joe Biden", "Erdogan", "Xi Jinping", "Scholz", "Macron", "Boris Johnson") as well as historical figures often used in comparisons ("Hitler", "Stalin"). We did not compress different spellings (e.g. "Volodymyr Zelenskyi" vs "Volodymyr Zelensky"; "Lukashenko" vs "Lukashenka"), different first and last name combinations (e.g. "Trump" vs "Donald Trump"; "Biden" vs "Joe Biden"), and nicknames ("Ze" for Volodomyr Zelensky) as they may occur in different contexts and be used by different groups.

"Wagner" is detected as a person's name. In this context, however, the name refers to the private military complex of the same name. "Bakhmut" is an erroneous recognition as well.

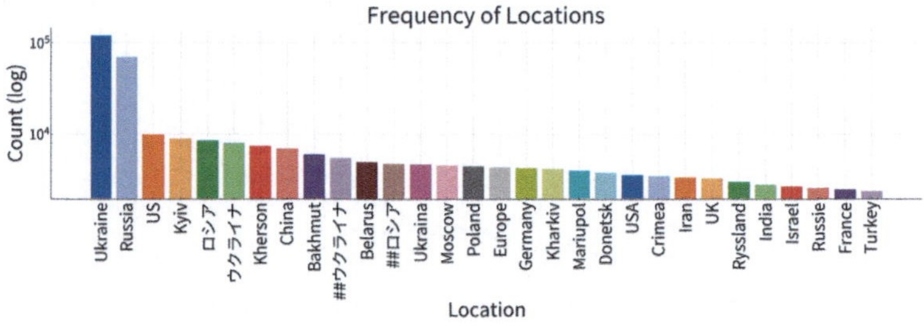

Fig. 5. Most frequent named geographic entities.

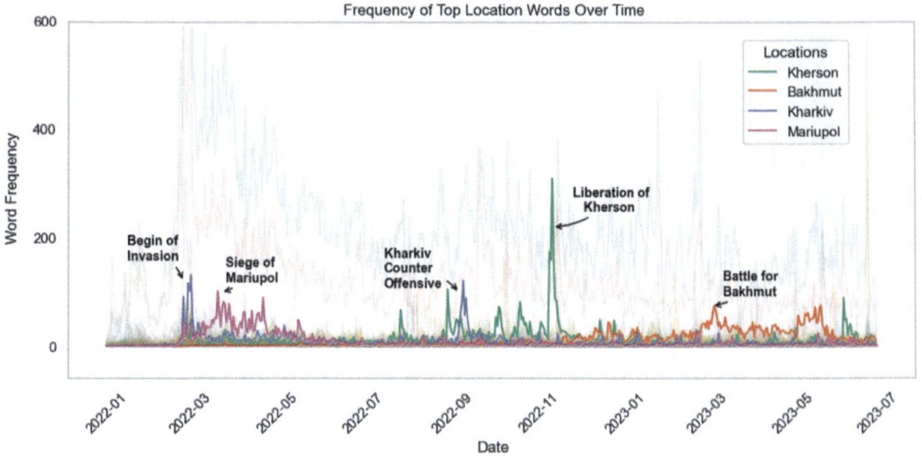

Fig. 6. Development of select named geographic entities over time.

Geographic entity names are found in 21% of the texts. Their distribution is shown in Fig. 5. Country names dominate, with the first cities being "Kyiv" in fourth in "Kherson" in 7th place (Bakhmut is recognized correctly here). Countries like "China", "Poland", or "Germany", as well as "USA", "UK" and "Iran" highlight the global influence of the conflict. Figure 6 shows the developments of city mentions over the course of the war, mainly corresponding to local events, but often already appearing earlier in the discussion.

Fig. 7. Most frequent named organization entities.

Finally, about 11% of the texts contain mentions of organization names, shown in Fig. 7. NATO is the most frequently discussed one, but surprisingly, the

European Union appears before local institutions. News outlets and online platforms are also mentioned often ("CNN", "Reuters", "AP", "Yahoo", "Google", "YouTube"), as are U.S. military branches and secret services ("USAF", "CIA", "FBI", "US Navy").

5.5 Community Detection Results

We identify a large number of OSINT-related user communities. Interaction-based clustering yields 95 communities when using mentions and 68 when using retweets, with six larger clusters in each case accounting for the majority of users. The most central nodes (highest indegree) correspond to well-known OSINT accounts previously reported in the literature [9], such as `@ralee85`, `@oryxspioenkop`, and `@defmon`.

These communities differ markedly in their characteristics. Clear patterns emerge in terms of dominant topics, partisanship, and misinformation levels. Certain clusters show predominantly pro-Ukrainian or pro-Russian alignment, and misinformation is unevenly distributed, with some communities exhibiting a substantially higher share of low-credibility content [16]. This suggests that community structure is a strong indicator of information reliability.

The feature-based graph embedding approach produces similar community patterns, again revealing clusters associated with partisanship and misinformation, as well as engagement and offensive language. While this method provides more nuanced differentiation between communities, the distinction between high- and low-misinformation clusters is less clear-cut than in the interaction-based approach.

6 Conclusion and Future Work

In this paper, we conduct a range of analyses on an OSINT dataset about the Russo-Ukrainian war to obtain deeper insights into communication about this topic. We find correlations of sentiment and named entity distribution with war events. Named Entity Recognition also provides insights about actors and locations deemed relevant to the conflict. Partisanship detection reveals that while the majority of tweets are neutral, pro-Russian stances are surprisingly slightly more frequent than pro-Ukrainian ones. A preliminary misinformation detection confirms that most messages are probably trustworthy, but around 9% are not; those are not spread evenly across partisanship. While the misinformation detection is approximate, the use of domain-independent models helps avoid conflict-specific bias and highlights stylistic and rhetorical markers of misleading OSINT-like content. Finally, we analyze user networks with two different strategies and are able to detect sub-communities separated by partisanship, discussed topics, and amount of misinformation spread.

Our findings also relate to perspectives from communication studies. The prominence of a small number of highly central OSINT accounts reflects a pattern similar to the two-step flow of communication [17], where a limited set of intermediaries plays a key role in shaping how information circulates and is interpreted. In addition, the clear differences in how events are described across communities echo elements of framing theory [18], suggesting that OSINT and BULLSHINT narratives influence not only what information is shared, but also how it is presented. While a full theoretical analysis is beyond the scope of this work, acknowledging these links highlights the value of communication research for interpreting OSINT dynamics on social media.

Several avenues remain open for deeper exploration:

- Misinformation Dissemination: Further research is needed to explore the dynamics of how misinformation spreads through these networks, including temporal and multimodal factors that influence misinformation dissemination.
- Multimodal Content Analysis: The dataset includes a significant amount of image and video content that merits detailed examination to enhance our understanding of how visual OSINT is used and discussed.
- Geolocation Analysis: Investigating the use of geolocations in OSINT tweets could provide insights into how digital information correlates with real-world events and locations.
- Influential Actors: Targeted research into individuals and groups responsible for disseminating OSINT and misinformation could help identify and mitigate the impact of malicious actors in social media landscapes.
- Threaded Conversations: Current data collection methods have limitations in capturing threaded discussions, which are crucial for understanding detailed and controversial OSINT debates. Future methodologies should aim to capture these threaded conversations by penetrating deeper into users' interactions and discussions.

These areas of future research could significantly advance our understanding of the complexities of social media use in conflict contexts and improve strategies for combating misinformation.

References

1. Agarwal, N.S., Punn, N.S., Sonbhadra, S.K.: Exploring Public Opinion Dynamics on the Verge of World War III using Russia-Ukraine war - Tweets Dataset (2022)
2. Poleksić, A., Martinčić-Ipšć, S.: Sentiment of the tweets on Russo-Ukrainian war: the social network analysis. In: 2023 46th MIPRO ICT and Electronics Convention (MIPRO).Opatija, Croatia: IEEE, pp. 1089–1095 (2023)
3. Ingole, R.V., Katrojwar, H.R., Bhoge, H.N., Giradkar, A.M., Kalbande, K., Morris, N.C.: A content-based study and tweet analysis of the Russia-Ukraine conflict. In: 2023 International Conference on Applied Intelligence and Sustainable Computing (ICAISC). Dharwad, India, pp. 1–6. IEEE (2023)
4. Hanley, H.W.A., Kumar, D., Durumeric, Z.: Happenstance: utilizing semantic search to track Russian state media narratives about the Russo-Ukrainian war on Reddit. In: Proceedings of the International AAAI Conference on Web and Social Media, vol. 17, pp. 327–338 (2023)

5. —— A special operation: a quantitative approach to dissecting and comparing different media ecosystems' coverage of the Russo-Ukrainian War. In: Proceedings of the International AAAI Conference on Web and Social Media, vol. 17, pp. 339–350 (2023)
6. Park, C.Y., Mendelsohn, J., Field, A., Tsvetkov, Y.: Challenges and opportunities in information manipulation detection: an examination of wartime Russian media. In: Findings of the Association for Computational Linguistics: EMNLP 2022. Abu Dhabi, United Arab Emirates, pp. 5209–5235. Association for Computational Linguistics (2022)
7. Geissler, D., Bär, D., Pröllochs, N., Feuerriegel, S.: Russian propaganda on social media during the 2022 invasion of Ukraine. EPJ Data Science, vol. 12, p. 35 (2023)
8. Kemp, R.: OSINT's influence on the Russian air campaign in Ukraine and the implications for future western deployments. Atlantic Council (2022). https://www.atlanticcouncil.org/content-series/airpower-after-ukraine/osints-influence-on-the-russian-air-campaign-in-ukraine-and-the-implications-for-future-western-deployments/, Accessed 30 Jan 2025
9. Z, T., Hayman, T.: Open-Source Intelligence and the War in Ukraine. Institute for National Security Studies, INSS Insight 1678 (2023). https://www.inss.org.il/wp-content/uploads/2023/01/No.-1678.pdf, Accessed 30 Jan 2025
10. Niu, J., Stillman, M., Seeberger, P., Kruspe, A.: A dataset of open source intelligence (OSINT) tweets about the Russo-Ukrainian war. In: Proceedings of the International ISCRAM Conference (2024). https://ojs.iscram.org/index.php/Proceedings/article/view/117
11. Barbieri, F., Espinosa Anke, L., Camacho-Collados, J.: XLM-T: multilingual language models in twitter for sentiment analysis and beyond. In: Proceedings of the Thirteenth Language Resources and Evaluation Conference. Marseille, France, pp. 258–266. European Language Resources Association (2022). https://aclanthology.org/2022.lrec-1.27
12. Camacho-Collados, J., et al.: TweetNLP: cutting-edge natural language processing for social media. In: Proceedings of the 2022 Conference on Empirical Methods in Natural Language Processing: System Demonstrations. Abu Dhabi, UAE, pp. 38–49. Association for Computational Linguistics (2022). https://aclanthology.org/2022.emnlp-demos.5
13. Tedeschi, S., Maiorca, V., Campolungo, N., Cecconi, F., Navigli, R.: WikiNEuRal: combined neural and knowledge-based silver data creation for multilingual NER. In: Findings of the Association for Computational Linguistics: EMNLP 2021. Punta Cana, Dominican Republic, pp. 2521–2533. Association for Computational Linguistics (2021). https://aclanthology.org/2021.findings-emnlp.215
14. Hamilton, W.L., Ying, R., Leskovec., J.: Inductive representation learning on large graphs (2018)
15. Von Luxburg, U.: A tutorial on spectral clustering. Statistics and Computing, vol. 17, pp. 395–416 (2007)
16. Rode-Hasinger, S., Kruspe, A., Zhu, X.X.: True or false? detecting false information on social media using graph neural networks. In: Proceedings of the Eighth Workshop on Noisy User-generated Text (W-NUT 2022). Gyeongju, Republic of Korea, pp. 222–229. Association for Computational Linguistics (2022). https://aclanthology.org/2022.wnut-1.24/
17. Katz, E.: The two-step flow of communication: an up-to-date report on an hypothesis Public Opinion Quarterly **21**(1), 61–78 (1957)
18. Entman, R.M.: Framing: toward clarification of a fractured paradigm. J. Commun. **43**(4), 51–58 (1993)

Career Awareness During Corporate Spin-Offs Using Text Mining

Naoko Sakamaki(✉) and Kazuhiko Tsuda

Graduate School of Business Sciences, University of Tsukuba, Tsukuba, Japan
{s2440407,tsuda.kazuhiko.gn}@u.tsukuba.ac.jp

Abstract. Why do middle-aged employees experience organizational restructuring differently from their younger counterparts? While it is well-established that corporate spin-offs reshape career landscapes, their psychological impact on employees in their 40 s and 50 s remains underexplored. This study addresses this gap by applying text mining to 1,173 free-text responses from employees who experienced corporate spin-off through internal recruitment systems. Using sentiment polarity analysis and co-occurrence network modeling, we identified contrasting emotional patterns: positive responses emphasized career diversity and external opportunities, whereas negative responses revealed treatment disparities, system opacity, and future uncertainties. Our findings demonstrate that spin-offs create ambivalent career impacts—simultaneously enhancing autonomy while generating systemic anxieties—with middle-aged employees particularly perceiving change as risk rather than opportunity. These results contribute to organizational restructuring theory by revealing age-related emotional polarization during major change and offer practical implications for developing age-sensitive career support systems and changing management strategies that address the unique concerns of mid-career employees.

Keywords: Text Mining Sentiment Analysis · KH Coder Co-occurrence Net works · Organizational Identity Disruption · Psychological Contract Violation · Japanese Employee Career Narratives

1 Introduction

In recent years, the business environment surrounding corporate management has been changing rapidly, with challenges such as globalization, advances in digital technology, shrinking domestic markets, aging management, and a shortage of successors becoming increasingly apparent. In this context, companies are actively utilizing organizational restructuring methods, such as mergers and acquisitions (M&A) and spin-offs, to achieve sustainable growth and maintain or strengthen competitiveness, making these measures a core part of the management strategy.

H. Unger and P. Meesad (Eds.): NLPIR 2025, LNNS 1904, pp. 931–946, 2026.
https://doi.org/10.1007/978-3-032-20897-2_65

However, the psychological and emotional impacts on employees during organizational restructuring, such as M&A and spin-offs, cannot be ignored. These changes are significant for employees and often result in negative emotions, such as anxiety, stress, and decreased motivation (Bach et al., 2024). Previous research and theoretical frameworks indicate that when age groups are polarized, differences in perceptions and emotional reactions are likely to arise. Younger employees tend to view such changes as opportunities for challenge and growth, whereas middle-aged and older employees are more likely to perceive them as risks and sources of anxiety (Wynen et al., 2005).

When companies pursue organizational restructuring, proper management of employee emotions and psychological safety is essential for successful transformation.

Considering these trends and the importance of M&A and spin-offs, this study aims to clarify the impact of spin-offs on the emotions of employees in their 40 s and 50 s.

2 Related Work

2.1 Organizational Restructuring and Employee Emotions

Organizational restructuring significantly impacts employee emotional well-being, with research demonstrating consistent patterns of stress, anxiety, and psychological distress following major organizational changes. Shrestha and Parajuli (2020) examined job stress factors among employees after large-scale M&A in the Nepalese banking industry, revealing that post-M&A job stress was most strongly influenced by job-related challenges such as unclear job content and role ambiguity, followed by organizational challenges, policy/procedural challenges including unfair evaluation systems, and individual challenges such as job insecurity.

Bach et al. (2024) demonstrated M&A's long-term psychological impacts through large-scale panel data in Sweden, revealing significant increases in stress, anxiety, depression, and psychiatric medication use after M&A among employees of both target and acquiring companies, with particular negative effects on employees experiencing career disadvantages, blue-collar workers, and low-skilled workers.

Although numerous studies have addressed employee emotions during organizational restructuring, to the best of our knowledge, no study has employed the methods used in this study.

2.2 Previous Research on Age Diversity, Emotion, and Adaptation

Age significantly influences emotion regulation patterns during organizational change. Mikkelsen et al. (2024) examined age impacts on emotion-regulation patterns through experience-sampling methodology with 406 adults, demonstrating that older age was associated with lower frequency of worry and distraction, lower variability in emotion-regulation strategies, greater positive affect, and reduced negative affect and depressive symptoms. The association between emotion-regulation effectiveness and daily affect was stronger in younger adults and weaker in older adults.

Scheibe and Moghimi (2021) investigated age effects on emotion regulation in response to daily negative workplace events using multilevel structural equation modeling with 199 workers in Germany and the Netherlands. Their analysis revealed that older age was associated with more stable emotion-regulation strategy selection, while younger workers showed increased expressive suppression for high-intensity events. More acceptance was associated with greater job satisfaction, whereas greater suppression correlated with increased fatigue.

Although numerous studies have examined age and emotion regulation relationships in daily situations, none has investigated these relationships during extraordinary organizational changes such as M&A, as addressed in this study.

3 Theoretical Framework

3.1 Organizational Identity Theory

This study draws upon organizational identity theory to understand employee career perspectives during corporate spin-offs (Albert & Whetten, 1985; Gioia et al., 2013). Recent research demonstrates how identification processes adapt during organizational transitions (Christopher, 2024).

Organizational identity represents the collective understanding of "who we are as an organization," providing cognitive anchoring for members' daily work experiences and career planning (Dutton & Dukerich, 1991). Corporate spin-offs disrupt established organizational identities, creating what Corley and Gioia (2013) term "identity ambiguity." Employees face uncertainty regarding organizational boundaries, core values, and strategic directions, leading to psychological discomfort and identity reconstruction needs.

Identity disruption during restructuring triggers diverse emotional responses—anxiety and confusion about uncertain futures alongside excitement about emerging opportunities (Corley & Gioia, 2013). Employees with longer tenure, particularly those in mid-to-late career stages, experience more pronounced identity challenges due to deeper psychological investment in existing organizational structures.

3.2 Psychological Contract Theory

Psychological contract theory provides a framework for understanding implicit expectations and reciprocal obligations governing employee–organization relationships (Rousseau, 1995; Morrison & Robinson, 1997). Recent research emphasizes the importance of sustaining these contracts during organizational change (Kraak et al., 2024).

Psychological contracts encompass employees' beliefs about mutual obligations between themselves and employers, extending beyond formal employment terms to include career development opportunities, job security provisions, and organizational support mechanisms. Corporate spin-offs frequently precipitate psychological contract breaches, where employees perceive organizations have failed to fulfill previously understood obligations (Robinson & Rousseau, 1994).

For employees in mid-to-late career stages, who typically maintain relational psychological contracts emphasizing long-term security and structured career progression, spin-offs may constitute significant contract violations that challenge fundamental assumptions about organizational commitment.

3.3 Identity–Contract Disruption Framework

This study proposes an integrated identity—contract disruption framework synthesizing organizational identity and psychological contract theories to explain employee responses during corporate spin-offs. The framework posits that spin-offs create "dual vulnerability" for employees, where identity disruption and contract breach operate synergistically to amplify cognitive and emotional responses.

The framework advances three propositions. First, corporate spin-offs simultaneously trigger organizational identity ambiguity and psychological contract breach, creating compounded uncertainty. Second, this dual disruption manifests more intensely among mid-to-late career employees due to stronger organizational identification and more established relational contracts. Third, the interaction between identity ambiguity and contract breach generates distinctive narrative patterns analyzable through textmining methodologies.

This theoretical integration recognizes that identity and contract disruptions are interconnected processes that mutually reinforce employee uncertainty and emotional distress. The framework provides theoretical grounding for text mining techniques to capture employee responses, as textual narratives serve as repositories of identity-related sensemaking and contract-related expectation adjustments. For mid-career employees, this dual vulnerability manifests as rich narrative expressions suitable for computational text analysis, establishing the conceptual foundation for this study's analytical approach.

4 Methodological Background

4.1 Text Mining and Sentiment Analysis in Organizational Research

Text mining has evolved significantly in organizational research, with applications in collective intelligence (Burton et al., 2024), workplace verbal aggression analysis (Nishimura et al., 2024), and methodological framework development (Hickman et al., 2022). This approach captures latent patterns in employee narratives that traditional quantitative methods overlook.

Validation studies demonstrate text mining's effectiveness across domains. Nishimura et al. (2024) classified workplace verbal aggression and its mental health impacts, while Onsumran et al. (2015) predicted economic volatility through news content analysis. These studies establish text mining's capacity to extract meaningful patterns from unstructured data.

Sentiment analysis provides systematic examination of employee psychological states during organizational change. Recent applications include workplace aggression impacts (Nishimura et al., 2024) and work engagement patterns (Roekel et al., 2024). During organizational restructuring, sentiment analysis of free-text responses captures complex emotional responses that traditional survey methods miss, preserving the authenticity of employee experiences while enabling systematic analysis.

4.2 Methodological Framework and Research Design Rationale

This study combines sentiment polarity analysis and co-occurrence network analysis to examine both emotional valence and semantic structure of employee narratives during corporate spin-offs, addressing contemporary calls for methodological sophistication (Kraak et al., 2024).

Sentiment polarity analysis quantifies emotional responses in organizational contexts, with demonstrated effectiveness in workplace verbal aggression research (Nishimura et al., 2024) and work engagement analysis (Roekel et al., 2024). This approach enables systematic comparison across employee groups while maintaining analytical objectivity. Co-occurrence network analysis reveals semantic relationships and thematic structures within narratives, providing insight into how employees conceptualize organizational change experiences, following established best practices (Hickman et al., 2022).

The focus on employees in their 40 s and 50 s reflects theoretical considerations regarding mid-career employees' heightened vulnerability to identity and contract disruptions. This methodological design directly supports testing the proposed identity–contract disruption framework through empirical analysis of employee narratives, contributing to both methodological advancement and theoretical development in organizational psychology.

5 Data and Analytical Methods: Employee Narratives After Corporate Spin-Offs

This study used free-text responses from an online questionnaire administered to employees transferred via an internal open recruitment system at a company that implemented a spin-off to strengthen business specialization and competitiveness, respond quickly to environmental changes, and improve corporate value through group-wide growth strategies (Table 1).

5.1 Survey Data: Internal Job Transfer Employees at a Post-Spin-Off Company

The participants were regular employees who were transferred using an internal open recruitment system between 2018 and 2024. Fixed-term contract

Table 1. Survey Participants

Survey Period	November 2024
Participants	1,173
Method	Online
Content	Free-text responses
Main Question	"If your career perspective has changed due to the spin-off, please describe."

employees were excluded. The data included free-text responses and respondents' age groups (20 s, 30 s, 40 s, and 50 s). Owing to contractual agreements with the data-providing company, the company name was anonymized as Company A.

5.2 Analytical Approach: Sentiment Polarity and Co-Occurrence Network Analysis

5.2.1 Research Design and Methodological Framework This study employed a mixed-methods approach that combines quantitative sentiment analysis with qualitative network visualization to capture both the emotional valence and semantic structure of employee narratives during organizational change. The methodological design addresses the inherent complexity of human emotional responses to corporate restructuring, which cannot be adequately captured through traditional quantitative measures alone (Hickman et al., 2022). The analytical framework integrates three complementary components: (1) individual-level sentiment quantification through polarity analysis, (2) comparative visualization of emotional distributions across age groups, and (3) semantic network analysis of extreme sentiment cases to identify underlying thematic structures. This multilayered approach enables both broad pattern identification and detailed interpretation of employee emotional responses to organizational restructuring. The methodological foundation for text mining in organizational contexts has been further strengthened by validation studies across different analytical domains. Notably, Onsumranet al. (2015) successfully employed text mining techniques to predict volatility patterns in economic indicators through news content analysis, demonstrating the methodology's capacity to extract meaningful patterns from unstructured textual data and quantify sentiment-based phenomena in organizational and economic contexts. Their work establishes a precedent for using computational text analysis to identify latent emotional and cognitive patterns embedded within narrative responses, providing methodological validation for the current study's approach to analyzing employee career perspectives through textual data mining.

5.2.2 Text Processing and Analysis Pipeline

Tool Selection and Technical Justification: JUMAN (version 7.01) was selected as the primary morphological analyzer because of its superior performance in processing Japanese corporate terminology and business-specific vocabulary compared with alternative analyzers such as MeCab or ChaSen (Kurohashi & Nagao, 2003). The tool's lexical coverage includes extensive organizational and employment-related terminology essential for accurate semantic analysis of corporate spin-off contexts, with comprehensive entries for Japanese business expressions, organizational hierarchy terms, and HR-related vocabulary that frequently appear in employee narratives about organizational change (Nishimura et al., 2024).

KH Coder 3 was chosen for co-occurrence network analysis because of its specialized capabilities in Japanese text mining and proven effectiveness in organizational research contexts (Roekel et al., 2024). The software provides native Japanese language processing, robust co-occurrence network visualization with customizable clustering algorithms designed for qualitative data interpretation, and established validation in organizational psychology research for analyzing employee narratives and workplace discourse.

Sentiment Analysis Protocol: Sentiment polarity values were calculated using the Japanese Sentiment Polarity Dictionary (Takamura et al., 2005), which assigns polarity scores ranging from −1.0 (most negative) to +1.0 (most positive) to individual lexical items. For each respondent, an individual-level polarity score was calculated as the arithmetic mean of all word-level polarity values within their free-text response. This approach ensures that longer responses do not disproportionately influence the overall sentiment distribution while maintaining the semantic integrity of individual employee narratives. Missing polarity values (words not found in the sentiment dictionary) were excluded from the mean calculations rather than assigned neutral values, following established best practices in organizational sentiment analysis (Nishimura et al., 2024).

Methodological Objectivity: Default software settings were used for all analytical parameters to eliminate researcher bias and ensure reproducibility. The automated nature of clustering algorithms, layout optimization, and relationship threshold selection provides standardized, objective analysis protocols that can be replicated across different research contexts.

5.2.3 Sampling Strategy and Visualization

Comparative Visualization: Mean polarity values were categorized by age group (20 s, 30 s, 40 s, and 50 s) to examine generational differences in emotional responses to organizational change. Distribution visualization was conducted via boxplots (Fig. 1), which provide comprehensive statistical summaries, including median values, quartile ranges, and outlier identification. This visualization approach enables the identification of age-specific emotional response patterns while maintaining statistical rigor in the comparative analysis (Hickman et al., 2022). The boxplot analysis revealed that employees in their 40 s and 50 s demonstrated more negative polarity distributions than younger cohorts did, with notable differences in both median values and interquartile ranges. Based on these preliminary findings, subsequent detailed textual analysis focused on the 40 s and 50 s cohorts to investigate the underlying semantic structures contributing to these emotional differences.

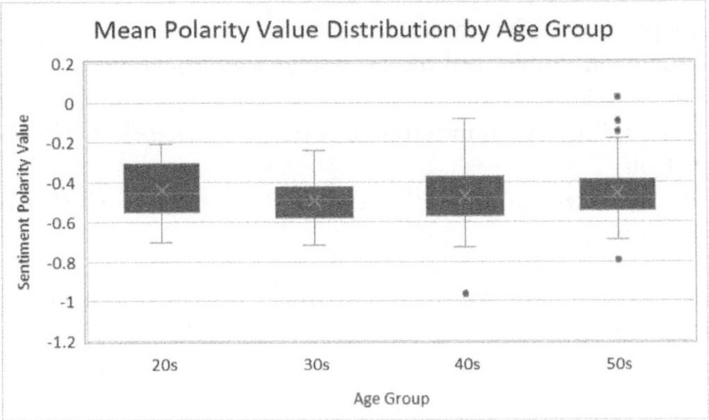

Fig. 1. Mean Polarity Value Distribution by Age Group

Extreme Case Sampling: The selection of the top and bottom 20 cases from the 40 s and 50 s cohorts for co-occurrence network analysis followed an established methodology. This approach serves two primary analytical purposes: (1) capturing extreme sentiment cases that represent the most distinct emotional responses to organizational change and (2) ensuring analytical tractability while maintaining semantic diversity in the textual corpus. The selection of 20 cases per polarity extreme (40 cases total) aligns with established practices in qualitative text mining, where sample sizes of 30–50 responses typically achieve semantic saturation for co-occurrence network analysis (Roekel et al., 2024). This extreme sampling approach is theoretically justified by the principle of maximum variation sampling, which seeks to capture the full range of experiences and perspectives within the phenomenon under study (Nishimura et al., 2024). By focusing on the most positive and negative sentiment cases, the analysis illuminates the boundary conditions of employee emotional responses to spin-offs, providing insights into both resilience factors and vulnerability markers.

Co-occurrence Network Analysis Protocol: The detailed analysis procedure involved in importing the selected extreme cases into KH Coder for systematic co-occurrence network analysis.

Co-occurrence Network Analysis Parameters: Network analysis was conducted using KH Coder 3 with the following standardized parameters consistently applied to both high- and low-polarity response groups from employees in their 40 s and 50 s:

- Minimum term frequency: 2 occurrences (terms appearing fewer than twice were excluded to reduce noise while preserving semantic diversity).
- Edge type: Word–word co-occurrence relationships
- Co-occurrence measure: Jaccard coefficient with the top 60 strongest relationships displayed (software default setting)
- Node visualization: 100% bubble size reflecting the relative term frequency within each network
- Clustering algorithm: Automated modularity optimization (KH Coder default implementation)
- Network layout: Automated force-directed algorithm (KH Coder default implementation)

This standardized configuration ensures analytical consistency between polarity groups while maintaining objectivity through automated algorithms.

Co-occurrence relationships between words were visualized, and semantic clusters were labeled to interpret employee interest areas and emotional tendencies from multiple perspectives, following the structured workflow outlined in Table 2.

Table 2. Analysis Process

1	Collection of free-text data
2	Data cleaning - Morphological analysis with JUMAN - Calculation of polarity values and their means
3	Import into KH Coder - Extraction of top and bottom 20 cases by polarity value - Company name anonymization ("Company A") - Creation of new project - Loading of analysis files (top/bottom 20 of 40 s/50 s data)
4	Morphological analysis (automatic; Japanese/ChaSen setting)
5	Co-occurrence network analysis
6	Interpretation and labeling of results

By combining the quantitative analysis of sentiment polarity, visualization of distributions with boxplots, and an in-depth analysis of extreme sentiment cases using co-occurrence networks, this study provides a multifaceted examination of the impact of spin-offs on employee emotions, addressing the methodological sophistication called for in contemporary organizational research (Kraak et al., 2024).

6 Results: Mapping Career Diversity and Anxiety via Co-Occurrence Network Analysis

6.1 Co-Occurrence Network Analysis of Top 20 Polarity Values: Career Diversity and External Orientation

In this study, the top 20 cases based on sentiment polarity values were extracted from employees' free-text responses and analyzed using co-occurrence network diagrams. As shown in Fig. 2, multiple subgraphs were identified, with a particular focus on Subgraphs 1 (green), 2 (yellow) and 5 (blue).

First, in Subgraph 1 (green), words such as "external," "open recruitment," "different," "experience," "job type," "system," "evaluation," and "other" were central. These words indicate topics related to interest in external careers and diverse job experiences, as well as the exploration of career paths using transfers and open recruitment systems, reflecting themes such as "career diversity, external orientation, and career formation through systems and evaluation." The inclusion of words such as "system" and "evaluation" also suggests that employees are considering their careers while utilizing both internal and external systems and evaluations. The comments included statements such as "I can gain diverse experiences without changing jobs externally," "Most open recruitment positions are not open to those without experience," and "I have no expectations for systems, etc."

Next, Subgraph 2 (yellow) centered on words such as "department," "weak," "horizontal," and "think," focusing on topics related to "the weakness of cross-departmental ties and perceptions." In particular, words such as "horizontal" connections between "departments," "weak," and "think" suggest difficulties in cross-departmental career formation and a sense of challenges regarding inter-departmental collaboration. The actual comments included statements such as "I feel that horizontal collaboration and connections are quite weak."

Furthermore, in Subgraph 5 (blue), words such as "feel," "company," "business," and "transfer" were central, highlighting topics related to "awareness of transfers and changes at the company or business unit level." The focus was on "transfers" and "feelings" (changes and awareness) at the company or business unit level due to the spin-off. The comments included statements such as "Differences in performance and treatment have emerged due to the spin-off," "It is necessary to time one's transfers," and "The experiences available differ by business company."

This analysis clarifies that employee emotions and awareness associated with the spin-off encompass multiple aspects, including career diversity, external orientation, weakness of interdepartmental collaboration, and awareness of transfers and changes at the company or business unit levels.

6.2 Co-Occurrence Network Analysis of the Bottom 20 Polarity Values: Concerns About Treatment and Future Uncertainty

Next, the bottom 20 cases were extracted and, as with the top 20 cases, analyzed using co-occurrence network diagrams. As shown in Fig. 3, five subgraphs

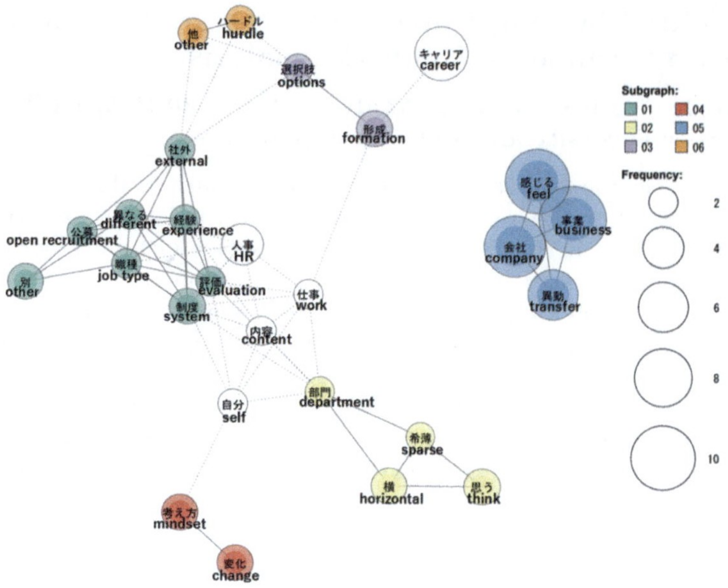

Fig. 2. Co-occurrence Network Diagram (Top 20 Polarity Values)

(Subgraphs 1–5) were identified. The main words in each subgraph, their interpretations, and specific examples are as follows. First, in Subgraph 1 (green), words such as "business," "salary," "feel," "system," "open recruitment," "impact," "hurdle," and "impression" were central, with numerous topics related to changes in treatment, systems, and salary due to the spin-off and their impacts. In particular, there was a strong awareness of challenges regarding differences in salary levels and systems among business companies, disparities in treatment, and the high hurdles of the open recruitment system. The comments included "I want the salary level to be reviewed,' "The hurdle for the open recruitment system has increased," "Differences in treatment," and "Unique aspects to the system," among others. In Subgraph 2 (yellow), words such as "group," "awareness," "management," and "individual" appeared, with the main topics focusing on changes in overall group management awareness and individual awareness. Changes in management policies and awareness as a group as well as a sense of crisis as an individual were evident. The comments included statements such as "The appeal of the group is being lost," "Independent management is required," and "A sense of crisis as an individual." In Subgraph 3 (purple), words such as "good," "think," "place," "active," "map," and "road" were extracted, with the main topics centering on opportunities for activity, career self-realization, and a positive attitude. There were indications of expanding opportunities for self-actualization and career activity as well as a positive attitude toward career strategies (roadmaps). The actual comments included "It's good that the opportunities and choices for activity are expanding," "I can create my own career," and "A good roadmap." In Subgraph 4 (red), words such as "career," "future," "think," "see," "formation," "very," "high," and "other" were central, with numerous topics related to career formation, future outlook, and awareness of issues. The focus was on future career formation and challenges, comparisons with other companies or departments, and anxieties about or prospects for the future. The comments included "I have to think about my future career," "(Retirement) is a major point in career formation," and "(Due to the spin-off, other companies are no longer visible) It has become difficult to form careers in other companies." In Subgraph 5 (blue), words such as "company," "retirement," "each company," "leave," and "rule" were central, with the main topics focusing on differences in rules and retirement systems among companies. Differences in rules and retirement systems by company as well as the difficulty of transfers or relocations were evident. The comments included "Rules differ by business company," "Retirement age extended to 65," and "It has become more difficult to transfer within the group." According to this analysis, the bottom 20 responses, a wide range of employee emotions and awareness, were expressed from multiple perspectives, including changes in treatment and systems due to the spin-off, changes in group management and individual awareness, career self-realization and positive attitudes, anxieties and challenges regarding future career formation, and differences in company-specific rules and retirement systems.

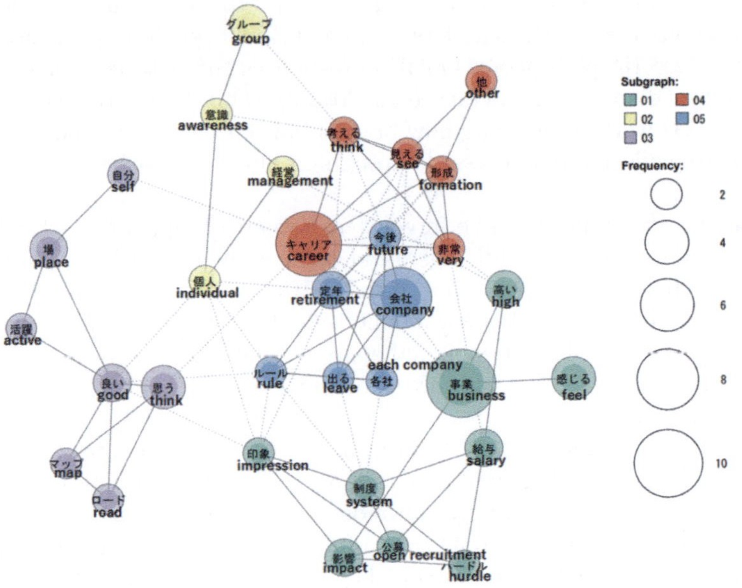

Fig. 3. Co-occurrence Network Diagram (Bottom 20 Polarity Values)

7 Discussion

7.1 External Impacts of Spin-Offs on Career Perspectives

This study's co-occurrence network analysis validates the integrated theoretical framework combining organizational identity theory and psychological contract theory. The findings demonstrate how corporate spin-offs trigger both identity ambiguity and psychological contract violation among employees.

Identity Reconstruction vs. Identity Confusion: This study's top 20 polarity cases appear to support organizational identity theory's proposition that restructuring can facilitate identity expansion rather than identity loss. These employees successfully expand their professional identity beyond traditional organizational boundaries, aligning with Davis et al. (2024) observation of enhanced autonomy and new career opportunities. Conversely, this study's bottom 20 polarity cases reveal identity confusion and psychological distress patterns, consistent with Muendler and Rauch's (2012) documentation of employment instability and network isolation. This study reveals additional complexity: employees simultaneously express anxiety about uncertainty while recognizing new opportunities—a paradoxical pattern that previous quantitative studies might not capture.

Psychological Contract Renegotiation vs. Contract Violation: This study's findings demonstrate contrasting contract renegotiation patterns. Adaptive employees reframe organizational changes as expanded career opportunities, actively reconstructing their understanding of career possibilities. However, other employees experience contract violation responses characterized by fairness concerns, system disparities, and future uncertainty. This dual pattern extends Davis et al.'s (2024) work by showing how both enhanced autonomy and psychological burdens are collectively expressed through employee narratives.

Age-Specific Vulnerability: The integrated framework appears particularly relevant for understanding middle-aged and older employees, who experience more intense psychological conflicts due to their stronger organizational identification and more established psychological contracts. This study's findings suggest that these employees face heightened challenges during organizational restructuring, possibly requiring targeted support strategies.

Theoretical and Methodological Contribution: This study contributes to the literature by demonstrating how text mining can capture complex psychological processes that traditional survey methods might overlook. The integrated framework provides a more comprehensive understanding of employee responses to organizational restructuring, moving beyond simple positive–negative classifications to reveal the coexistence of contradictory emotions and cognitions in employee experiences.

7.2 Internal Impacts of Spin-Offs on Career Perspectives

The results of this study reveal that, particularly for middle-aged and older employees, corporate spin-offs simultaneously encourage self-actualization and positive career orientation while generating strong anxieties regarding future development, company-specific rule differences, and retirement systems. This ambivalent impact—the coexistence of increased career ownership with adaptation stress—represents a distinctive feature confirmed through free-text analysis. This study also demonstrates that although spin-offs expand career opportunities, they simultaneously impose new uncertainties and psychological burdens, creating both positive growth orientation and awareness of future challenges.

8 Conclusion

This study applied text mining techniques to examine career perspective changes among middle-aged employees during corporate spin-offs. The integrated identity–contract disruption framework successfully captured the paradoxical nature of employee responses, revealing simultaneous expressions of career autonomy enhancement and systemic uncertainty concerns.

Three key organizational challenges emerged: system transparency deficits, interdepartmental collaboration weaknesses, and career support inadequacies. These findings directly inform practical management strategies for successful spin-off implementation.

Practical implications for organizational management include targeted approaches across three phases. Pre-spin-off: provide clear explanations of retirement systems and benefits, conduct targeted career consultation for employees aged 40+, and develop comprehensive FAQ materials addressing salary and promotion concerns. Implementation: establish cross-departmental mentoring programs, con-duct quarterly employee surveys, and create regular information exchange meet-ings. Post-spin-off: provide ongoing career counseling services and accessible consultation for middle-aged employees navigating organizational change. These measures address specific anxieties identified in employee narratives while sup-porting increased career autonomy and self-development orientation. This study focused on employees in their 40 s and 50 s at a specific company, limiting general-izations to other contexts. However, the ability to visualize organizational restruc-turing impacts on middle-aged employees has important implications for future HR policies. Further research is needed on age-related differences in career per-spectives and the effectiveness of post-spin-off support systems. For sustainable growth, companies must consider both strategic rationality and employee psycho-logical safety, building transparent systems that support career autonomy during organizational transformation.

References

Albert, S., Whetten, D.A.: Organizational identity. Res. Organ. Behav. **7**, 263–295 (1985)

Bach, L., Baghai, R. P., Bos, M., Silva, R.: How do mergers affect the mental health of employees? SSRN Electron. J. (2024). https://doi.org/10.2139/ssrn.3940462

Burton, J.W., et al.: How large language models can reshape collective intelligence. Nat. Hum. Behav. **8**(9), 1643–1655 (2024)

Corley, K.G., Gioia, D.A.: Identity ambiguity and change in the wake of a corporate spin-off. Adm. Sci. Q. **49**(2), 173–208 (2004)

Davis, S., Kotnour, D. T., Al-Haddad, D. S.: Behind the curtain of corporate restructuring spin-offs: a case study analysis of technical project-based organizations. IEOM Soc. Int. (2024)

Dutton, J.E., Dukerich, J.M.: Keeping an eye on the mirror: image and identity in organizational adaptation. Acad. Manag. J. **34**(3), 517–554 (1991)

Freeman, A., Koçak, Ö.: Designing inclusive organizational identities. J. Organ. Des. **12**(1), 1–14 (2023)

Gioia, D.A., Patvardhan, S.D., Hamilton, A.L., Corley, K.G.: Organizational identity formation and change. Acad. Manag. Ann. **7**(1), 123–193 (2013)

Hickman, L., Thapa, S., Tay, L., Cao, M., Srinivasan, P.: Text preprocessing for text mining in organizational research: review and recommendations. Organ. Res. Methods **25**(1), 114–146 (2022)

Kraak, J.M., et al.: How psychological contract research can make the work-world a better place. Group Organ. Manag. **49**(2), 413–470 (2024)

Kurohashi, S., Nagao, M.: Building a Japanese parsed corpus while improving the parsing system. In: Natural Language Engineering, pp. 719–729. Cambridge University Press (2003)

Mikkelsen, M.B., O'Toole, M.S., Elkjær, E., Mehlsen, M.: The effect of age on emotion regulation patterns in daily life: Findings from an experience sampling study. Scand. J. Psychol. **65**(2), 231–239 (2024)

Morrison, E.W., Robinson, S.L.: When employees feel betrayed: a model of how psychological contract violation develops. Acad. Manag. Rev. **22**(1), 226–256 (1997)

Muendler, M.-A., Rauch, J. E.: Mobilizing social capital through employee spinoffs (NBER Working Paper No. 18459). National Bureau of Economic Research (2012)

Nishimura, Y., Matsumoto, S., Sasaki, T., Kubo, T.: Impacts of workplace verbal aggression classified via text mining on workers' mental health. Industr. Health **74**(2), 186—192 (2024)

Onsumran, C., Thammaboosadee, S., Kiattisin, S.: Gold Price Volatility Prediction by Text Mining in Economic Indicators News, p. 243. Engineering and Technology Publishing (2015). https://doi.org/10.12720/jait.6.4.243-247

Robinson, S.L., Rousseau, D.M.: Violating the psychological contract: not the exception but the norm. J. Organ. Behav. **15**(3), 245–259 (1994)

Roekel, H., Wigger, E.F.J., Veldkamp, B.P., Bakker, A.B.: What is work engagement? a text mining approach using employees' self-narratives. Appl. Psychol. **73**(3), 1071–1102 (2024). https://doi.org/10.1111/apps.12501

Rousseau, D. M.: Psychological contracts in organizations: understanding written and unwritten agreements. SAGE Publications (1995)

Scheibe, S., Moghimi, D.: Age and context effects in daily emotion regulation and well-being at work. Work Aging Retirement **7**(3), 185–199 (2021)

Shrestha, P., Parajuli, D.: Impact of merger and acquisition practice on job stress: evidence from Nepal. European J. Bus. Manag. **12**(24), 1–12 (2020)

Takamura, H., Inui, T., Okumura, M.: Extracting semantic orientations of words using spin model. In: Proceedings of the 43rd Annual Meeting of the Association for Computational Linguistics, pp. 133–140. Association for Computational Linguistics (2005)

Wynen, J., Verlinden, S., Pieters, D., Kleizen, B., Verhoest, K.: Embracing change through team diversity: the impact of age diversity on civil servants' perceptions of organizational change. Rev. Publ. Personnel Administr. **45**(1), 123–145 (2005)

Advancing an AI Model for Predicting Prohibited Substances Using Knowledge Graphs and Retrieval-Augmented Generation

Shunsuke Yamamoto(✉), Takashi Tsunakawa, and Masafumi Nishida

Department of Informatics, Graduate School of Integrated Science and Technology, Shizuoka University, 3-5-1 Johoku, Chuo-ku, Hamamatsu-shi, Shizuoka 432-8011, Japan
yamamoto.shunsuke.20@shizuoka.ac.jp, {tuna,nishida}@inf.shizuoka.ac.jp

Abstract. As global regulations on substances in products become increasingly stringent to reduce environmental risks, investigating these substances especially when components from third-party manufacturers are involved has become more challenging. Additionally, information about the materials used in products is often confidential, making it challenging to directly use external APIs for such investigations. In this study, we propose a method for inferring the presence of substances in products by leveraging a local large language model (LLM) integrated with retrieval-augmented generation and a knowledge graph (KG). The main objective of this study is to improve the accuracy of substance inference, demonstrating improved results compared to a local LLM that does not use KG. By predicting not only prohibited substances but also other substances, we can identify substances that are not prohibited now but may be prohibited in the future.

Keywords: Knowledge Graphs · Retrieval-Augmented Generation · Large Language Models · Prohibited Substances Regulations

1 Introduction

As global regulations on substances contained in products continue to tighten to mitigate environmental risks, investigating these substances has become increasingly challenging, especially when components from third-party manufacturers are involved. Additionally, information about the constituent parts of products is often confidential, making it challenging to directly use external APIs for such investigations (see Fig. 1).

The need to predict regulated substances arises from several challenges. Different countries enforce regulations on different substances, and these lists of regulated substances are updated annually with new additions. It is often unclear whether the substances used in older products are currently regulated. In some

H. Unger and P. Meesad (Eds.): NLPIR 2025, LNNS 1904, pp. 947–960, 2026.
https://doi.org/10.1007/978-3-032-20897-2_66

cases, manufacturing is outsourced to third-party companies, leading to uncertainty or reluctance in disclosing detailed information. Although chemical analysis can provide definitive results, it is costly and resource-intensive.

To address these challenges, we propose a method for predicting the presence of substances in products using a local large language model (LLM) integrated with retrieval-augmented generation (RAG) and a knowledge graph (KG). The main objective of this study is to improve the accuracy of these inferences using part and material data as documents. We evaluated the effectiveness of the proposed method by comparing it with other approaches, including a local LLM without KG, a KG-only method, and other combinations of LLM and KG.

We clarify that our contribution is a practical and privacy-preserving integration of a local LLM with (i) sparse lexical retrieval (RAG via BM25) that is entirely independent of the KG and (ii) a simple post-generation knowledge graph (KG) filter that removes substances not connected to the target product (through parts/materials) in the KG. The KG is not consulted during retrieval or generation; it is only used after the LLM outputs a candidate list. The novelty lies in showing that this lightweight, transparent post hoc filtering, when combined with standard retrieval, substantially lifts precision and overall F1 compared to LLM alone, KG alone, or LLM+RAG without filtering, while retaining competitive recall. Unlike prior work that either relies only on broad KG enumeration (high recall, low precision) or solely on unstructured retrieval, our pipeline demonstrates that a single, transparent post-generation KG consistency filter is sufficient to achieve large precision and F1 improvements without complex multi-stage reasoning or fine-tuning.

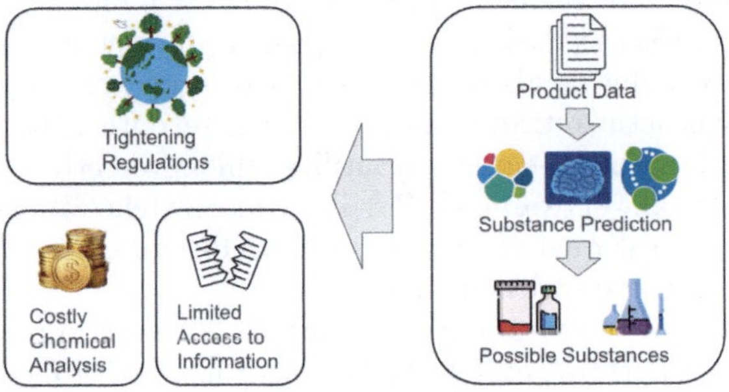

Fig. 1. Illustration of the challenges in predicting substances.

2 Related Works

Ensuring product compliance in domains where proprietary information cannot be publicly disclosed presents a significant challenge. Recent studies have highlight the increasing importance of AI techniques that preserve data confidentiality while supporting regulatory compliance tasks. In our previous study, we proposed a deep learning-based method to predict the presence of regulated chemical substances in developed industrial products by leveraging product information, such as part names and materials, without relying on confidential manufacturing data. We demonstrated the effectiveness of machine learning, deep learning, and clustering- based approaches, achieving higher prediction accuracy compared to a baseline model [1]. However, a major limitation of the deep learning approach was its inability to provide interpretable explanations for each prediction.

AI-driven substance prediction and chemical safety assessment have gained increasing attention. Da Silveira et al. [2] demonstrated the potential of combining KGs and LLMs to streamline the retrieval of hazardous substance information. Sepehri et al. [3] developed the *TOXIN KG* to support animal-free toxicological risk assessments, illustrating how structured knowledge can facilitate proactive substance evaluation. Hartung [4] highlighted the promise of predictive toxicology under frameworks like the registration, evaluation, authorisation and restriction of chemicals (REACH) and the toxic substances control act (TSCA) achieved approximately 87% balanced accuracy across nine OECD test guidelines and nearly 190,000 chemicals outperforming the reproducibility of traditional animal tests. He also emphasized the growing need for explainable AI in regulatory chemical risk assessments. These studies align with broader calls to leverage AI for predictive toxicology and regulatory readiness.

RAG has emerged as a powerful method for grounding LLMs with external knowledge. Lewis et al. [5] introduced the RAG framework and demonstrated that augmenting generation with document retrieval improves factual accuracy. More recently, Gao et al. [6] provided a comprehensive survey of RAG techniques, highlighting their effectiveness for knowledge-intensive tasks and their relevance to enterprise scenarios. However, using proprietary data in RAG raises privacy concerns. Zeng et al. [7] proposed a privacy-preserving RAG pipeline that employs synthetic retrieval data to mitigate data leakage risks, demonstrating that such methods can retain utility while protecting sensitive information. Adoptante et al. [8] examined RAG in code-switched multilingual settings, showing that retrieval quality and prompt conditioning significantly influence generation accuracy even without fine-tuning.

KGs have also proven invaluable for enabling structured reasoning in regulatory compliance. Chattoraj and Joshi [9] constructed *MedReg-KG*, a knowledge graph for medical device regulations, which supports automated reasoning over complex permissions and prohibitions. Hernandez et al. [10] proposed an open KG-based approach to map concepts between the EU AI Act and international standards, facilitating structured compliance assessment. In the product safety domain, Hoang et al. [11] used KG reasoning to predict the cultural and religious compliance of consumer products, such as halal status, by modeling ingredient relationships. Similarly, Da Silveira et al. [2] integrated KGs with LLMs to manage hazardous chemical information, demonstrating how structured knowledge can enhance LLM-based reasoning for chemical safety.

Beyond general KG applications, emerging research has explored *Graph RAG* as a method for more tightly integrating to tightly integrate KGs with LLMs. Hu et al. [12] introduced a Graph-RAG approach that retrieves relevant subgraphs and incorporates both textual and topological information into the LLM's context, significantly improving multi-hop reasoning tasks. Peng et al. [13] further formalized this paradigm, highlighting how Graph-RAG addresses LLM limitations such as poor logical consistency and hallucinations by injecting structured knowledge.

Our work builds on these foundations by integrating local LLMs, RAG, and KGs for secure, accurate substance prediction within confidential product environments.

3 Dataset

The dataset used in this study is the product substance information compatible with chemSHERPA provided by Yamaha Motor Co., Ltd. The product information includes a total of 40,430 products, including product names, manufacturing companies, parts, substances, and the mass of each part and substance. In this study, we used the product names, part names, and substance data to predict the presence of substances. We limited the prediction task to determining the presence or absence of substances without estimating their quantities.

4 Method

The main objective of this study is to predict the presence of all substances in products by leveraging a local LLM, RAG and KG. We hypothesize that products with parts and materials similar to known products are more likely to share substances. Based on this assumption, we propose a method that enriches the input to the LLM with information about relevant products, parts, and materials retrieved via RAG and KG (see Fig. 2).

First, the product information, including parts and materials, is extracted from the test product. This information is then used to construct a BM25 query for Elasticsearch [14], which retrieves relevant documents representing similar products with overlapping parts or materials. Prior to running a script, the test product's information including the relationships among parts, materials and substances is inserted into Neo4j, enabling KG-based retrieval of related substances.

In the KG, products, parts, materials, and substances are represented as nodes, while their relationships such as *HAS PART*, *HAS MATERIAL*, and *HAS SUBSTANCE* are expressed as edges. This structure enables the flexible and efficient retrieval of related information. Specifically, when predicting substances for a product, the system traverses the graph to retrieve all substances connected to the product node via these edges. The retrieval process begins at the product node, follows the *HAS PART* and *HAS MATERIAL* edges to reach the part and material nodes, and then collects all substances connected through the

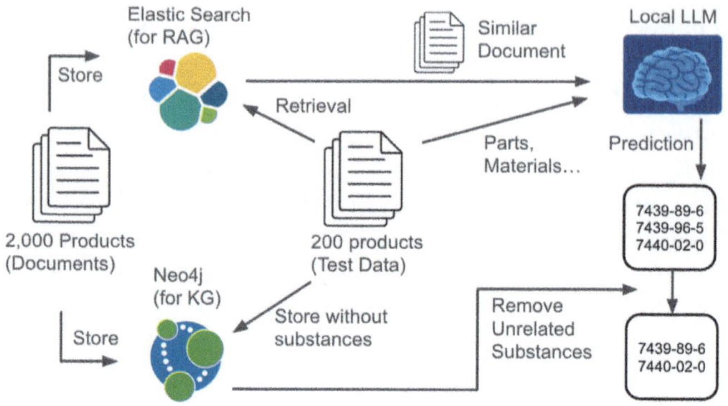

Fig. 2. Flowchart of the proposed method.

HAS SUBSTANCE relationship. This approach allows the system to efficiently retrieve relevant substances based on the product's structural information (see Fig. 3).

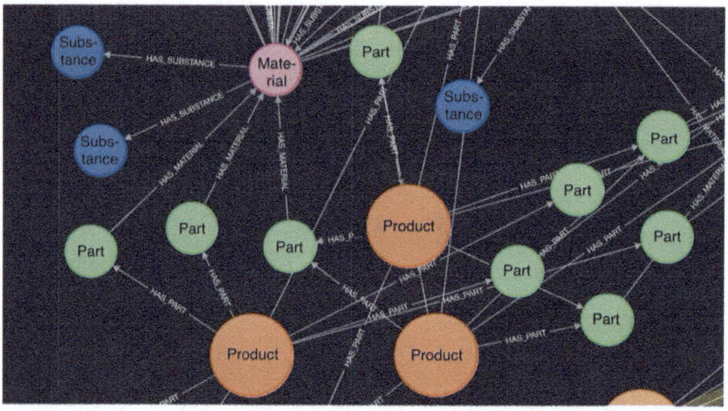

Fig. 3. KG structure for product substance prediction. Orange, green, pink, and blue nodes represent the products, parts, materials, and substances, respectively. The edges indicate relationships such as *HAS PART*, *HAS MATERIAL*, and *HAS SUBSTANCE*.

The retrieved documents and substances are provided to the LLM as a context for generating predictions. To ensure prediction reliability, the LLM's output is postprocessed by filtering out substances not present in the KG for the test product. The overall process is illustrated as follows:

1. **Local LLM + KG + RAG (Proposed Method):** Combines a local LLM, a Neo4j KG, and Elasticsearch-based RAG. The LLM generates predictions based on the retrieved product documents. The output is filtered using the KG results to remove irrelevant substances.

2. **Local LLM Only (Baseline 1):** Uses only the LLM with product information as input, without relying on any retrieved documents or KG.
3. **KG Only (Baseline 2):** Uses Neo4j to directly retrieve related substances for the test product. No LLM or RAG is involved.
4. **Local LLM + KG (Baseline 3):** Uses the LLM and KG knowledge as input, without incorporating document retrieval via RAG.
5. **Local LLM + RAG (Baseline 4):** Uses only the RAG component (BM25 retrieval) and LLM. KG is excluded, relying solely on retrieved product documents to inform predictions.

The proposed method is implemented with the Haystack framework [15]. Neo4j [16] stores structured information on products, parts, materials, and substances, whereas Elasticsearch serves as the document store for RAG. We used LLaMA 3 Swallow 8B [17] as the local LLM, deployed on a local GPU server.

BM25 is employed as a simple yet effective retrieval method for relevant documents based on shared parts and materials. Despite its simplicity, BM25 demonstrated strong performance in retrieving documents that enhance LLM predictions, without requiring complex dense retrieval methods.

To address common LLM issues such as repeated substance predictions, we applied postprocessing to remove duplicates from the LLM output. Finally, filtering with KG ensures that only substances related to the test product remain, improving the prediction accuracy and reliability.

4.1 Data Preparation

To evaluate the prediction performance, we prepared 2,000 samples as documents for the knowledge source and 200 samples as test data. First, the product information was extracted from the documents and stored in both Elasticsearch and Neo4j. The local LLM was then provided with the test data as queries to predict the substances contained in the products.

The details of the data preparation process are described below. For more information about the format of the document, query, and question, please refer to Appendix A.

4.2 Document Preparation for RAG

To prepare documents for RAG, the product information was extracted from the dataset and stored in Elasticsearch. Each document contains the following fields:

- Product ID
- Development region (prefecture)
- Development city
- Company name
- Parts names
- Materials names, IDs, and class IDs

- Substances names, IDs

However, some documents contain more than 1,000 parts and 5,000 substances. To ensure manageability, we limit the number of parts, materials, and substances to 10 each. If the number of parts, materials, or substances exceeds 10, a note is added to the end of the list indicating that more than 10 items have been omitted.

4.3 Query Construction for RAG

To retrieve relevant product information using BM25, we generated a query based on the parts, materials, and other product attributes of the test product. Typically, the top 3 resources retrieved are the most similar to the query, providing highly relevant product information for further processing.

The query includes the following information:

- Product ID
- Development region (prefecture)
- Development city
- Company name
- Parts names
- Materials names, IDs, and class IDs

The query format is almost the same as the document format stored in Elasticsearch to ensure compatibility. The query is executed on Elasticsearch to identify documents representing relevant products. The documents in Elasticsearch store similar product information, enabling the effective retrieval of comparable products with overlapping parts or materials.

4.4 Question Generation for the LLM

After retrieving documents and substances (if applicable), a structured question was generated and provided to the LLM along with the retrieved knowledge. When RAG is used, the question is constructed using the following prompt template (translated from the original Japanese prompt for international use):

> あなたは優秀な製品情報アナリストです。(You are an excellent product information analyst.)
> 以下が、質問に意味が近いドキュメントです。(Here are the documents relevant to the question.)
> Documents:
> {% for doc in documents %}{{ doc.content }}{% endfor %}
> Question: {{question}}
> Answer:

When RAG is not used, only the question text is provided to the LLM, excluding the document information.

The question text itself includes the following information:

- Request to predict substances likely contained in the product.
- Clarification that "substances" refer to specific chemicals (e.g., "carbon" and "silicon") and should be answered using IDs such as CAS registry numbers (e.g., "7439-95-4").
- Product information includes the following
  - Product ID, development region, and company name.
  - List of part names.
  - List of materials, their IDs, class IDs.
- Instructions for the LLM to return substances in a comma-separated ID format.

4.5 Substance Retrieval with Neo4j

To retrieve substances from the KG, we query Neo4j based on the test product's structure. Specifically, we extract substances connected to the product through its parts and materials to ensure that only structurally relevant substances are considered for filtering the LLM's output.

The knowledge graph was constructed by mapping chemSHERPA compatible enterprise records provided by Yamaha Motor Co., Ltd. into Neo4j. These internally maintained linkage records are created and reviewed by staff. However, they may contain minor noise (e.g., inconsistent or ambiguous part names) or missing information (e.g., for certain parts or substances).

At the time of experimentation the constructed knowledge graph contained a total of 4,636 nodes: 2,200 Product nodes (2,000 documents and 200 test samples), 1,721 Part nodes, 66 Material nodes, and 573 Substance nodes.

5 Experiments

This section presents the experimental setup and provides a detailed analysis of the results obtained using the proposed method for predicting substances in products.

5.1 Method Evaluation

We obtain the predicted substances list (e.g., [8934-34-3, 2342-23-1]) from the LLM output and compare it with the ground truth labels for each test product.

To evaluate the effectiveness of the proposed method and compare it with other baseline methods, we used precision, recall, and F1 score as evaluation metrics. Precision measures the proportion of correctly predicted substances among all predicted substances, indicating the accuracy of the predictions. Recall measures the proportion of correctly predicted substances out of all actual substances, reflecting the model's ability to identify all relevant substances. As the harmonic mean of precision and recall, the F1 score provides a balanced evaluation of the model's performance, especially when there is an imbalance between precision and recall.

Additionally, we analyzed the computational resource usage, including execution time, RAM consumption, and GPU memory usage, to assess the practicality of each method for deployment in resource-constrained environments. Details of the environment setup are provided in Appendix B.

5.2 Results

Table 1 summarizes the detailed performance of each method, including micro and macro precision, recall, and F1 scores.

Table 1. Performance Comparison of Each Method

Method	Micro Average			Macro Average		
	Precision	Recall	F1	Precision	Recall	F1
Local LLM	0.014	0.112	0.025	0.030	0.088	0.031
KG	0.084	**0.962**	0.155	0.149	**0.902**	0.229
Local LLM + KG	0.252	0.110	0.154	0.182	0.086	0.100
Local LLM + RAG	0.463	0.755	0.574	0.777	0.768	0.743
Local LLM + KG + RAG	**0.889**	0.775	**0.828**	**0.811**	0.777	**0.769**

We also measured the computational resources required for the LLM-based methods. The results showed that the proposed method consumed approximately 1,811 s of execution time, 2,240 MB of RAM, and 33,418 MiB of GPU memory. However, the local LLM-only method consumed 5,115.25 s of execution time, 2,004.02 MB of RAM, and 32,022 MiB of GPU memory.

6 Discussion

The experimental results reveal a tradeoff between recall and precision across the different methods. The KG-only approach achieves the highest recall (**0.962**) but at the expense of extremely low precision (0.084), indicating many false positives because it retrieves all substances associated with the product's parts and materials without filtering. This method is effective for broad coverage but lacks the specificity needed for accurate predictions. In contrast, the proposed method (LLM + KG + RAG) achieves the highest precision (**0.889**) and F1 score (**0.828**) by effectively balancing precision and recall.

The results show that while the KG offers broad coverage, it lacks the reasoning ability to filter out irrelevant substances, resulting in high recall but low precision. Integrating the LLM with RAG introduces contextual reasoning and target retrieval, significantly improving precision by reducing false positives. The postprocessing steps that filter out substances not present in the KG ensures that the predictions are grounded in structured knowledge, enhancing reliability.

The computational resource analysis reveals that the proposed method consumes slightly more resources than the local LLM-only approach due to the added retrieval and KG integration. However, the increase in the computational overhead is minimal, making the method practical for deployment in environments with moderate resource constraints.

The local implementation ensures data confidentiality, which is crucial in industries where product composition involves sensitive information. By leveraging the private deployment of LLMs and structured retrieval mechanisms, our approach demonstrates that it is possible to develop highly accurate, privacy preserving inference models.

However, the proposed method has some limitations. The LLM occasionally generates repeated substances, which are now addressed through a postprocessing step that removes duplicates. This issue may arise from the LLM's generation mechanism, which can lead to redundancy in the output.

7 Conclusion

In this study, we proposed a novel method for predicting substances in products by integrating a local LLM with KG and RAG. Although the KG-only method achieved the highest recall, it suffered from low precision due to its inability to filter out irrelevant substances. The predictions made using the local LLM without KG were unsatisfactory, primarily due to the lack of sufficient resources. The proposed method effectively combines structured knowledge with LLM-based reasoning, leading to significantly improved precision and overall prediction accuracy, as demonstrated by the F1 score. These results confirm that incorporating both KG and RAG into LLM-based systems provides an effective and privacy-conscious solution for regulatory compliance, even in environments with limited computational resources.

Future research will focus on several directions to enhance the proposed method:

- **Scalability and Data Expansion:** While the current experiments are limited to a subset of 2,000 documents and 200 test samples, the dataset contains a total of 40,430 products. Future studies will focus on expanding the range of data used in experiments to evaluate the scalability and robustness of the proposed method. This will involve optimizing the retrieval and prediction pipeline to handle larger datasets efficiently without compromising performance.
- **Explainability:** We aim to enhance the local LLM's ability to generate not only the predicted substances (e.g., [8934-34-3, 2342-23-1]) but also the reasoning behind each prediction. This will help employees at large companies. to more easily accept and understand the inference results.
- **Output Quality Improvement:** Currently, the local LLM tends to generate duplicate substances within a single prediction (e.g., [7450-50-4, 7450-50-4, 7450-50-4, ...]). We implemented a postprocessing step to remove duplicates.

In the future study, we aim to modify the LLM prompt or generation mechanism to avoid such repetition entirely. We will also investigate whether resolving this duplication problem contributes to improved prediction accuracy.

We believe these advancements will further contribute to practical applications of AI in product compliance and sustainability.

Acknowledgments. This research was supported by Yamaha Motor Co., Ltd. We thank them for providing the dataset and technical advice.

A Data Format

To clarify the proposed method, we provide an example of document from the dataset and the corresponding question given to the LLM. These sentences were originally written in Japanese and have been translated into English for illustration. Since we used an in-house dataset for our experiment, we include dummy sample document and question format to demonstrate how the proposed method works.

A.1 Sample Document

The sample document contains product information, including the product ID, development prefecture, development city, company name, parts information, materials information, and substance information. The structure of the document is as follows:

製品ID: SWITCH (Product ID: SWITCH)
開発された都道府県: 静岡県 (Developed Prefecture: Shizuoka)
開発された市区町村: 浜松市 (Developed City: Hamamatsu)
開発した会社名: ヤマハ発動機株式会社 (Developed Company: Yamaha Motor Co., Ltd.)
部品情報は以下の通りです。 (Parts information is as follows.)
部品名: スイッチボディー (Part Name: Switch Body)
部品名: 端子ネジ (Part Name: Terminal Screw)
部品名: 接点 (Part Name: Contact)
材料情報は以下の通りです。 (Material information is as follows.)
材料名: Copper alloys (Material Name: Copper alloys)
材料ID: T523 (Material ID: T523)
材料クラスID: Y-102 (Material Class ID: Y-102)
材料名: Cast iron (Material Name: Cast iron)
材料ID: G262 (Material ID: G262)
材料クラスID: Y-108 (Material Class ID: Y-108)
物質情報は以下の通りです。 (Substance information is as follows.)
物質名: Copper (Cu) (Substance Name: Copper (Cu))
物質ID: 7440-50-8 (Substance ID: 7440-50-8)
物質名: silicon (Substance Name: Silicon)

物質ID: 7440-21-3 (Substance ID: 7440-21-3)
物質名: Nickel (Substance Name: Nickel)
物質ID: 7440-02-0 (Substance ID: 7440-02-0)

When the number of parts, materials, or substances exceeds 10, the statement " 10個を超えたため省略されました (More than 10 items are omitted)" is added to the end of the list.

A.2 Sample Question

The following is a sample question prompt provided to the LLM to predict substances contained in the product. We hypothesize that parts and materials information can be utilized to predict substances. The structure of the question is as follows:

この製品(製品ID: SWITCH)に含まれる可能性の高い物質を答えてください。(Please answer the substances that are likely to be contained in this product (Product ID: SWITCH).)
ここで言う物質とは「carbon」「silicon」などで、材料のことではありません。(Here, "substances" refer to specific chemicals such as "carbon" and "silicon", not materials.)
また、物質を答えるときは、「7439-95-4」のようなIDで回答してください。(When answering substances, please use IDs like "7439-95-4".)
回答の形式は、カンマ区切りで「xxxx-xx-x,xxxx-xx-x,xxxx-xx-x」のように回答してください。(The answer format should be comma-separated like "xxxx-xx-x,xxxx-xx-x,xxxx-xx-x".)
同じ物質は1回だけ回答してください。(Please answer each substance only once.)
製品情報は以下の通りです。(The product information is as follows.)
製品ID: SWITCH (Product ID: SWITCH)
開発された都道府県は静岡県 (Developed Prefecture: Shizuoka)
開発された市区町村は浜松市 (Developed City: Hamamatsu)
開発した会社名はヤマハ発動機株式会社 (Developed Company: Yamaha Motor Co., Ltd.)
部品情報は以下の通りです。(The parts information is as follows.)
部品名: スイッチボディー (Part Name: Switch Body)
部品名: 端子ネジ (Part Name: Terminal Screw)
部品名: 接点 (Part Name: Contact)
材料情報は以下の通りです。(The materials information is as follows.)
材料名: Copper alloys (Material Name: Copper alloys)
材料ID: T523 (Material ID: T523)
材料クラスID: Y-102 (Material Class ID: Y-102)
材料名: Cast iron (Material Name: Cast iron)
材料ID: G262 (Material ID: G262)
材料クラスID: Y-108 (Material Class ID: Y-108)

B Implementation Environment

The implementation environment for the proposed method is summarized in Table 2. This includes details about the hardware, software, and frameworks used.

Table 2. Implementation environment

Category	Details
Operating System	Ubuntu 20.04.6 LTS
RAM	128 GB
CPU	Intel(R) Core(TM) i9-10940X CPU @ 3.30GHz
GPU	2 × NVIDIA RTX A6000 (48 GB VRAM each)

References

1. Yamamoto, S., Suzuki, K., Sakashita, D., Sakurai, Y., Tsunakawa, T.: Estimation of regulated chemical substances contained in products using deep learning. In: Proceedings of the 87th National Convention of IPSJ (2025)
2. Silveira, M.D., Deladiennee, L., Acem, K., Freudenthal, O.: Combining knowledge graphs and LLMs for hazardous chemical information management and reuse. In: Proceedings of IEEE International Conference on Bioinformatics and Biomedicine (BIBM), Lisbon, Portugal, pp. 6766–6773. IEEE (2024)
3. Sepehri, S., et al.: Toxin graph: a semantic knowledge graph for animal-free toxicological risk assessment. Comput. Toxicol. (2025)
4. Hartung, T.: Artificial intelligence as the new frontier in chemical risk assessment. Front. Artifi. Intell. **6**, 1269932 (2023)
5. Lewis, P., et al.: Retrieval-augmented generation for knowledge-intensive NLP tasks. In: Proceedings of NeurIPS 2020 (2020)
6. Gao, Y., et al.: Retrieval-augmented generation for large language models: a survey. arXiv preprint (2023). https://arxiv.org/abs/2312.10997
7. Zeng, S., et al.: Mitigating the privacy issues in retrieval-augmented generation (rag) via pure synthetic data. arXiv preprint (2024). https://arxiv.org/abs/2406.14773
8. Adoptante, A.J.M., Castro, J.A.D.V., Medrana, M.L.B., Ocampo, A.P.B., Peramo, E.C., Miranda, M.R.M.: Benchmarking open-source large language models on code-switched tagalog-english retrieval augmented generation. J. Adv. Inf. Technol. **16**(2), 233–242 (2025)
9. Chattoraj, R., Joshi, A.P.: Medreg-kg: knowledge graph for streamlining medical device regulatory compliance. In: Proceedings of IEEE International Conference on Big Data, Washington, DC, USA, pp. 3382–3390. IEEE (2024)
10. Hernandez, J., Golpayegani, D., Lewis, D.: An open knowledge graph-based approach for mapping concepts and requirements between the eu ai act and international standards. AI Ethics (2025). https://doi.org/10.1007/s43681-025-00708-6

11. Hoang, V.T., et al.: Halal or not: knowledge graph completion for predicting cultural appropriateness of daily products. In: IEEE Access, pp. 15158–15167. IEEE, Access 13 (2025). https://arxiv.org/abs/2501.05768
12. Hu, Y., Lei, Z., Zhang, Z., Pan, B., Ling, C., Zhao, L.: Graph-rag: graph retrieval-augmented generation for multi-hop reasoning. Proc. AAAI (2025)
13. Peng, B., et al.: Graph retrieval-augmented generation: a survey (2025). https://arxiv.org/abs/2408.08921
14. B.V., E.: Elasticsearch. Software (2025). https://www.elastic.co/jp/elasticsearch/
15. deepset: Haystack. Software (2025). https://haystack.deepset.ai/
16. Neo4j, I.: neo4j. Software (2025). https://neo4j.com/
17. AI@Meta: Llama 3 model card. Model Card (2024). https://github.com/meta-llama/llama3/blob/main/MODEL_CARD.md

Deconstructing Pinoybaiting: A Fine-Grained Topic Modeling Analysis of Foreign-Made Videos About the Philippines

Zachary Gomez, Rey Vincent Inocencio, Joseph Miguel Maristela, Hans Martin Rejano, and Thomas James Tiam-Lee(✉)

College of Computer Studies, De La Salle University, 2401 Taft Avenue, Manila 1004, Metro Manila, Philippines
{zachary_gomez,rey_vincent_inocencio,joseph_maristela, hans_rejano,thomas.tiam-lee}@dlsu.edu.ph

Abstract. Advances in Natural Language Processing (NLP), particularly with topic modeling algorithms, enable insightful analysis of unstructured data. In study, we apply topic modeling to a relatively underexplored social phenomenon called Pinoybaiting, where foreign content creators make exaggerated videos about the Philippines for engagement and profit. Using a dataset of 2,704 video transcripts from 639 potential Pinoybaiting channels, we applied two topic modeling algorithms: a modified version of BERTopic and Hierarchical Latent Tree Model (HLTM). BERTopic identified 122 topics, while HLTM generated 272 hierarchical topics across three levels, capturing both broad and nested themes. The output of both models was assigned generated descriptive labels, which were manually validated. The resulting topic models were evaluated both quantitatively and qualitatively. Our computational approach reveals the common topics that are associated with such channels, and contributes to a better understanding of this phenomemon.

Keywords: Pinoybaiting · topic modeling · YouTube · social media analysis

1 Introduction and Related Studies

The emergence of various types of media on the internet has resulted in an unprecedented volume of unstructured text data from sources such as social media, news articles, and even video transcript data. Analyzing this vast corpus provides exciting opportunities to gain a better understand of complex social and cultural digital phenomena. Recent advances in natural language processing (NLP) have paved the way for the automatic analysis of "text as data". Among these techniques, topic modeling has emerged as a particularly powerful unsupervised method for discovering latent thematic structures within large

H. Unger and P. Meesad (Eds.): NLPIR 2025, LNNS 1904, pp. 961–974, 2026.
https://doi.org/10.1007/978-3-032-20897-2_67

collections of documents and their connection to one another [1]. Additionally, topic modeling employs statistical modeling and unsupervised machine-learning techniques to identify semantic structures, effectively clustering related words to reveal underlying topics [2]. This approach provides a high-level understanding of textual data, making it valuable across various fields, from social media analysis to media studies.

The proliferation of monetized content on video-sharing platforms like YouTube has given rise to engagement strategies such as clickbaiting, designed to maximize viewership for financial gain. In the Philippines, a specific variant of clickbaiting, termed **Pinoybaiting**, was recently brought to public attention. Pinoybaiting pertains to foreign content creators making videos centered on Filipino culture, media, and national identity [3]. There have been many allegations that a lot of these videos contain fake or exaggerated reactions to different aspects of Filipino culture in an attempt to exploit Filipinos' desire for global validation [4,5]. This strategy raises critical questions about cultural representation and the exploitation of national pride for profit, as shown in Fig. 1.

Fig. 1. Examples of videos that some people might perceive as Pinoybaiting.

So far, there have been very few academic works on Pinoybaiting. The work of [3] establishes a foundational understanding of the phenomenon by analyzing 20 videos to categorize the creators and thematic elements of Pinoybaiting. However, their study acknowledges the need for broader analysis. In response to this, a recent study has used computational techniques to analyze over 8 million comments from potential Pinoybaiting videos. Their research classified comments by sentiment (positive, negative, neutral) and intent (praise, criticism, conversation, information, request) [6]. However, their work only focused on the comments, no the actual content of the videos.

We continue to build on top of the above studies, by using computational approaches to extract a granular analysis of the topics often used in potential Pinoybaiting content. While the previous work of [3] classified Pinoybaiting content into broad categories such as food and travel, we instead focus on capturing the more specific topics that are within each video, as it offers a better insight on how the different Philippine-related themes are exploited for engagement. Our methodology extracts the transcripts of the videos, which are then processed by two topic modelling techniques: a modified version of BERTopic and Hierarchical Latent Tree Model (HLTM). This approach allows us to not only associate each topic with multiple videos, but also map each video to several topics.

BERTopic has been noted for its ability to generate novel and detailed insights from complex text data [7,8]. This capacity was clearly shown in a study of YouTube videos, where it identified 135 specific topics, in contrast to a traditional model that found only nine broad ones [9]. Similarly, hierarchical models like Hierarchical Latent Tree Model (HLTM) are explicitly designed to counter the one-video-one-topic problem. They achieve this by allowing for "multi-membership" clustering, where a single video can belong to several topics, and by organizing these topics into meaningful hierarchies that allow for both high-level overviews and detailed examinations [10,11]. By leveraging these capabilities, our approach aims to identify the specific and often overlapping thematic patterns, answering the following research questions:

1. What are the dominant themes and topics present in the YouTube video transcripts as identified through the two topic modeling algorithms?
2. How do the topics generated by BERTopic and HLTM compare in terms of overlapping and unique themes identified within the content of video transcripts?
3. What are the strengths and limitations of the two models?

2 Methodology

2.1 Data Collection

To collect our dataset of potential Pinoybaiting videos, we made use of YouTube Data API search queries, validated by a social science expert. The goal is to find videos created by non-Filipino creators or Filipino creators raised outside the Philippines, about topics related to Filipino themes and culture. Some of the queries are as follows: American React Filipino, Foreigner React Filipino, Foreigner impressions of the Philippines, React to Philippines, Trying Filipino Snacks.

The above process still resulted in some irrelevant videos to be included in the dataset. For example, many videos from news channels were included. To address this, we did a channel-level manual filtering of all the channels included in the dataset. With inputs from a social science expert, we defined a set of criteria for when a channel should be considered as potentially engaged in Pinoybaiting:

1. The channel belongs to one of these four types: (1) Tourists who visited the Philippines, (2) Digital nomads living in the Philippines, (3) Foreigners outside the Philippines, (4) Half-Filipinos
2. The channel's content contains excessive use of Filipino themes (e.g., Filipino flags, symbols, cultural items).
3. The channel's content shows a pattern of repeating Filipino-related content.
4. The channel showcases exaggerated or performative reactions.

5. The channel makes use of clickbait titles or thumbnails (i.e., misleading information with the purpose of drawing Filipino viewership).

We enlisted 10 volunteers to manually annotate each unique channel in our dataset according to the above criteria, labeling each channel with "yes" if the channel is likely to be engaging in Pinoybaiting and "no" if it is not. It should be noted that our criteria is an attempt to best capture the notion of Pinoybaiting in less subjective terms. Yet, we recognize that the inclusion of a channel in our dataset does not necessarily mean that there are definitely exploitative intentions on the part of the channel creators. The final list of Pinoybaiting channels was decided based on the majority vote. From 1008 candidate channels, the final list included 639 channels. This corresponds to a total of 2704 potential Pinoybaiting videos in the final dataset.

For each matching video under the identified channels, transcripts were either extracted if they were provided by the creator, or automatically generated if not. Transcripts were generated using the base model of Whisper, a speech-to-text model with good performance across different language.[1]

2.2 Topic Modeling

Topic modeling is an unsupervised machine learning technique that identifies abstract themes or topics within a collection of documents by analyzing word patterns and their co-occurrence. In our work, each transcript is considered one document.

We used two topic modeling algorithms in this study: a modified version of BERTopic and Hierarchical Latent Tree Model (HLTM). We explain each of these two approaches in the succeeding sections.

Modified BERTopic. BERTopic is a topic-modeling framework that utilizes embeddings derived from pre-trained models, wherein these embeddings are reduced in dimensionality and clustered, followed by the application of the class-based TF-IDF (c-TF-IDF) procedure to generate interpretable topic representations from the resulting clusters [8]. However, since by default BERTopic is designed to assign each video transcript only to a single topic, we modified the BERTopic pipeline by segmenting each video transcript into multiple section. Each section is then treated as its own document. This approach allows us to assign multiple topics within the same video transcript. To perform the segmentation, we prompted Gemini 2.0 Flash[2] to segment each transcripts accroding to context and conversational breaks. Through this process, we generated a total of 21,374 segments.

To preserve contextual information, only minimal preprocessing was performed, such as removing bracketed and parenthetical content, special characters, and normalizing whitespace. A multilingual embedding model pre-trained

[1] https://platform.openai.com/docs/models/whisper-1.
[2] https://gemini.google.com/app.

with English, Filipino, and other languages was configured in BERTopic to transform the said documents into embeddings. These embeddings were subsequently reduced in dimensionality and clustered to group semantically similar segments. Next, to derive an interpretation from these clusters, the c-TF-IDF was employed, which emphasized words that were both frequent within a cluster and distinctive across clusters, in order to highlight the most representative terms that best characterized each cluster. In this step, the VectorizerModel and c-TF-IDF hyperparameters were configured to extract n-grams, filter stopwords, and reduce noise, while the RepresentationModel was subsequently applied to refine the resulting keywords, ensuring that the important terms captured semantic similarity within topics and improved the overall coherence of topic representations. Additionally, automatic merging of closely related topics was enabled based on probability thresholds. It should be noted that all configurations were tuned using coherence score metrics.

Following the initial topic modeling, some segments were assigned to an outlier topic, often representing documents that did not fit clearly into any topic group. Their probability scores were re-examined to identify the next most probable topic based on semantic similarity. After this process, the c-TF-IDF and representation processes, as explained earlier, were reapplied, and the topic results were updated to produce the final list of topics and their respective representations.

Hierarchical Latent Tree Model (HLTM). Hierarchical Latent Tree Model enables multi-level topic analysis, identifying both broad and granular themes [10]. This layered approach is helpful for uncovering interrelated or overlapping themes within each video and identifying relationships between words and topics at various levels of detail. The pipeline begins by combining video transcripts with their corresponding tags to enhance contextual understanding. Text data is preprocessed through tokenization, lowercasing, stopword removal (custom and GitHub external list), lemmatization, and POS tagging to remove less informative parts of speech. TF-IDF vectorization and bigram formation are also applied via the model itself to highlight important terms and phrases as an added preprocessing step. The model treats documents as binary vectors (presence/absence of tokens) and uses a hierarchical clustering process to construct a topic tree. Hyperparameters such as UD-test threshold, max island size, and CT-threshold are tuned using coherence scores and conceptual evaluations. The resulting hierarchy maps broad to specific topics, with each topic labeled using both AI and manual validation. A final output includes a topic tree with keyword lists, topic IDs, levels, and a JSON file mapping documents to topics with probability scores.

Overall Workflow of BERTopic and HLTM. Figure 2 demonstrates the diagram of the overall workflow of both topic modeling frameworks, including preprocessing, each model's algorithm, postprocessing, and how TF-IDF was integrated to produce their respective topic outputs.

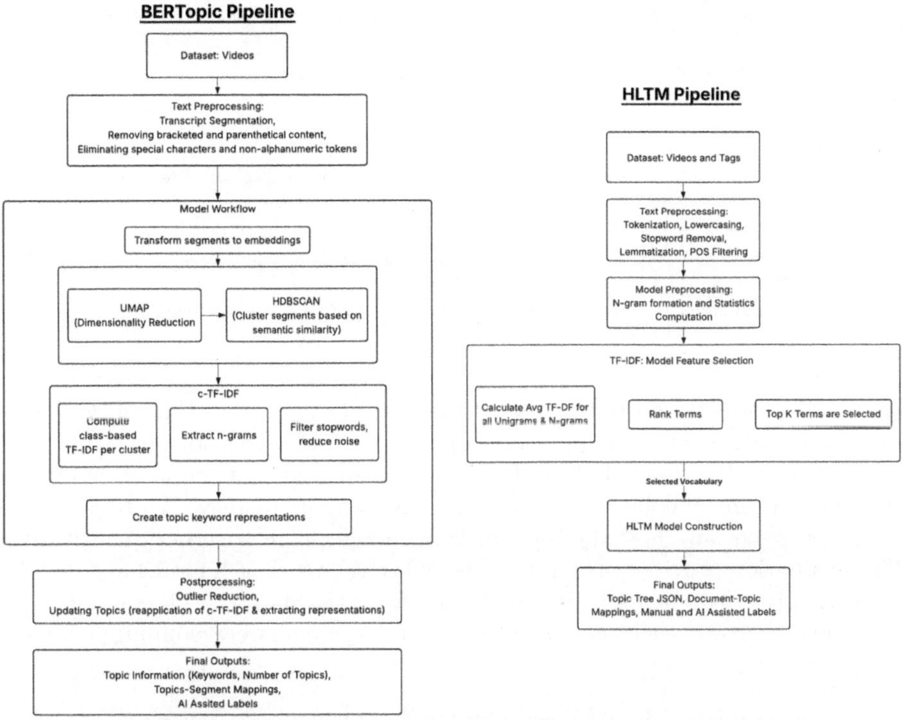

Fig. 2. Workflow Diagrams of HLTM and BERTopic.

3 Evaluation and Interpretation of Results

3.1 Labelling of Topics

In topic modeling, resulting topics are comprised of a list of words that represent each topic. To have a more interpretable view of the resulting topics, Gemini 2.0 Flash[3] was utilized to generate a concise description or label for each topic based on the words that represent it. The following prompt was used: *Given the following keywords, provide a concise and descriptive label that captures the overall theme of this topic without using generic terms.*

The labels were then reviewed through manual validation to ensure that the labels generated for each topic aligned with the theme reflected by their keyword representation. Table 1 shows some examples of topics and the corresponding LLM-generated description.

3.2 Quantitative Evaluation

Coherence scores were selected to quantitatively evaluate the outputs, as they are the most widely used metric for assessing topic quality in topic modeling,

[3] https://gemini.google.com/app.

Table 1. Model Keyword Representations and Assigned Labels

Model	Keyword Representation	Label
BERTopic	manila cathedral church, manila cathedral, churches	Manila Cathedral
BERTopic	watch jollibee commercials, reacting jollibee commercials, jollibee commercials	Watching Kwentong Jollibee Commercial
BERTopic	swim whale sharks, whale sharks, swimming whale sharks, whale shark	Swimming with Whale Sharks
HLTM	beer, drinking, san-miguel	San Miguel Beer
HLTM	impression, bgc, manila-manila, friendly, hand, building, camera	BGC Impressions
HLTM	meal, restaurant, menu, spaghetti, jollibee, burger, gravy	Jollibee Menu Items

given the unsupervised nature. A previous study suggested that obtaining the coherence score is often the most appropriate form of evaluation since this metric is utilized to determine how well the words in a specific topic relate to one another in terms of cohesiveness, interpretability, and quality [12]. In particular, the metrics measure how the top words in each topic fit together quantitatively. This evaluation provides insight into the effectiveness of each model in generating semantically coherent and distinct topics from the dataset. Three widely used coherence metrics, such as *c_v*, *u_mass*, and *c_npmi*, were computed for each model, wherein

- **c_v** evaluates the semantic similarity between high-scoring words and is considered more aligned with human judgment.
- **u_mass** assesses word co-occurrence frequency.
- **c_npmi** measures the semantic similarity and association between keywords of a topic.

Table 2. Topic Output Coherence Scores

Coherence Metric	BERTopic	HLTM
c_v	0.325	0.714
u_mass	−1.314	−1.752
c_npmi	−0.116	0.141

As shown in Table 2, for BERTopic, the c_v score indicates that the top keywords in each topic may lack interpretability, partly due to repetitive n-grams like 'squid', 'squid, squid', and 'squid, squid, squid', which are essentially redundant and do not contribute additional semantic meaning. The moderate u_mass score reflects the c-TF-IDF representation's focus on topic distinctiveness, which tends to emphasize rare or uncommon phrases less likely to co-occur. Lastly, the c_npmi score suggests that stopword filtering and the inclusion of n-grams may reduce keyword co-occurrence when such phrases appear in isolated contexts.

In terms of HLTM's performance, the c_v (0.714), u_mass (-1.752), and c_npmi (0.141) indicate moderate to good performance. The strong c_v score suggests that HLTM effectively groups related words into meaningful themes, thanks to its focus on probabilistic word co-occurrence. The moderate u_mass score reflects that binary word variables (presence/absence) may miss some frequency-based context, while the modest c_npmi score highlights the model's ability to capture key word dependencies. Stopword removal likely improved coherence by filtering out noise, but may have weakened certain contextual relationships.

3.3 Content-Based Topic Evaluation

Although coherence metrics can provide valuable insights in a quantitative aspect, a study [13] suggested that manual inspection and analysis of the extracted topics can offer a more comprehensive understanding of the themes present in the dataset, since there are no universal or standardized measures for evaluating topic models [12].

In this study, we examined the videos in the dataset to verify whether the extracted topics aligned with the content of their respective transcripts, thereby strengthening the reliability of the results. For instance, the video titled "15 popular filipino snacks, Americans trying Filipino food" is centrally about culinary experimentation. BERTopic generated labels like "Delicious Dishes and Dining", "Balut and other Eggs", and "Philippine Halo-Halo Dessert". Meanwhile, HLTM decomposed this theme into topics such as "Cheese and Ice Cream Desserts" and "Filipino Breakfasts: Pancit, Balut, and More". The topics generated by both models exhibited strong thematic coherence with the video's primary focus on food.

Similarly, for the video "$200 Luxury Beach Hotel in The Philippines", which focuses on travel and recreation, the models again converged on the central theme. BERTopic assigned topics such as"Trip to a Beach Destination" and "Swimming with Whale Sharks". HLTM identified related topics like"Adventure Tours and Boat Exploration in Cebu" and "Swimming and Diving with Sea Creatures". The convergence of evidence across different modeling techniques significantly strengthens the reliability of the findings. Despite being two independent models, the identification of similar themes indicate their capability to represent the dataset's content.

4 Discussion

4.1 What Key Themes Emerge as Most Dominant in the Content of YouTube Video Transcripts Based on the Topic Modeling Results?

After finalizing the outputs of both models, each resulting topic per model was manually inspected and classified into 7 general categories: Culture, Lifestyle, Food, Travel, Entertainment, Politics, and Others. These categories were defined based on the major themes that emerged from the topics generated by both models, specifically in the context of identifying patterns in potential Pinoybaiting content. Then, each video associated with a specific topic was then mapped to its corresponding category. To avoid over-counting and ensure that there are no duplicates, videos appearing in multiple topics within the same category were counted only once per category. The number of videos under each category are as follows: Food (2662), Lifestyle (2570), Travel (2300), Entertainment (2246), Culture (1962), and Politics (1594).

As observed in the said number of videos per category, Food emerged as the most dominant theme across the dataset since it is present in 2,662 videos.

Table 3. Top Food Topics by Video Count

BERTopic	HLTM
Delicious Dishes and Dining	Juices
Philippine Halo-Halo Dessert	Crispy and Tasty Fried Dishes
Squid	Spicy Pork Dishes
Satisfying Hunger with Delicious Food	Eggs and Breakfast
Balut and other Eggs	Cheese and Ice Cream Desserts

Based on Table 3, BERTopic identified "Delicious Dishes and Dining" as its most prominent topic, while HLTM highlighted "Juices" as being particularly prevalent. This dominance suggests that cuisine, along with food and drink related experiences, is among the most recurring themes in potential Pinoybaiting content. We can also see some form alignment in the results of both models, identifying prominent themes related to eggs, dishes, and desserts. Furthermore, Lifestyle was also a highly represented theme, with 2,570 videos. This category encompasses themes related to day-to-day activities, fashion, and leisure, reflecting attempts to showcase aspects of Filipino living in the content that resonate with local audiences.

Table 4. Top Lifestyle Topics by Video Count

BERTopic	HLTM
Philippines' Dating & Relationship of Females	Local Community
Makeup Products and Application	Gaming
Barber Services	Relationship with a Filipina
Mommy and Baby's Morning Events	Coffee and Shopping
Medical Diagnostics & Hospital Services	Business Issues

If we take a look at Table 4, there appears to be only limited alignment between the topic categorizations. For instance, BERTopic's top topic under this category was the "Philippines' Dating and Relationship of Females", while HLTM emphasized "Local Community", highlighting different facets of lifestyle content. Even though the top topics do not directly match, HLTM still captured a Filipina dating topic, indicating some degree of thematic overlap.

As for the other categories, some of the prevalent themes discovered include "Walking by the Beach Experience" and "Road Trip" for Travel, "Singing and Dancing Stars" and "Movie Star Influence" for Entertainment, "Learning and Speaking Foreign Language" and "English Education" for Culture, and "Experiences of Children in Community and Culture" and "Shopping and Currency" for Politics. This variety underscores how both models, despite their differences, reveal patterns that reflect the diverse strategies employed by content creators.

Although it is important to note that there were also videos that are associated with topics assigned to the 'Others' category. This was designated for outputs that could not be assigned to the other categories for various reasons, such as when the representative keywords of a topic were too diverse or ambiguous to allow for a definite label and classification, or when the topic did not clearly fit into any of the predefined categories.

Table 5. Top Topics by Video Count

BERTopic Topics	Video Count	HLTM Topics	Video Count
Delicious Dishes and Dining	2, 594	Road Trip	869
Learning and Speaking a Foreign Language	258	Famous and Popular	843
Philippine Halo-Halo Dessert	228	Busy Road Trips and Urban Traffic	795
Walking by the Beach Experience	209	Movie Star Influence	788
Christmas Season & Celebration	205	Shopping and Currency	780
Experiences of Children in Community & Culture	149	Film Camera	741
Squid	130	Juices	733
Swimming with Whale Sharks	110	Tourist Islands	723
Singing and Dancing Stars	99	Local Community	723
Philippines' Dating & Relationship of Females	98	Walking and Traffic	723

Table 5 displays the top resulting topics for both models. BERTopic's most prevalent topic, "Delicious Dishes and Dining", reflects a strong emphasis on Filipino cuisine and food-related content, appearing in over 2,000 videos. Notably, the model was also able to extract highly specific topics relevant to the Philippine context, such as "Philippine Halo-Halo Dessert", demonstrating its capability to detect granular, culturally distinctive aspects. However, it is important to note that because the implementation utilized segmented transcripts, a video is counted towards a topic if even a single segment was assigned to it. This, combined with the model's extraction of the best keywords to represent a particular topic, introduces the possibility of overgeneralization, where a segment or the whole video may be assigned to a topic despite it not being its primary theme.

Meanwhile, HLTM revealed travel-related topics like Road Trip and Busy Road Trips and Urban Traffic as prevalent, along with entertainment-focused topics such as Famous and Popular and Movie Star Influence. It is also worth pointing out how this model generally has more videos under each topic in comparison to BERTopic's results, which suggests that HLTM's hierarchical approach results in broader, more generalized groupings, which can effectively summarize overarching themes but may risk overlooking finer-grained distinctions and localized cultural elements.

Although at first glance, the resulting topics from both models may seem misaligned due to BERTopic's reliance on semantic embeddings for topic formation and HLTM's use of statistical co-occurrence patterns, which group words that frequently appear together but may lack semantic coherence. However, these differences do not necessarily indicate a lack of alignment but rather reflect complementary perspectives as revealed in the next section.

4.2 How Do the Topics Generated by BERTopic and HLTM Compare in Terms of Themes Identified Within the Content of Video Transcripts?

Despite the models having two different sets of topic outputs, we decided to look into how their results related via a form of association. Analyzing associations between HLTM and BERTopic reveals how their distinct approaches complement each other. To perform this, we took note of all the videos assigned for every topic under both BERTopic and HLTM. We then computed two overlap ratios across all topic result pairs, which consisted of the number of overlapping videos between two topics over either the number of videos for the HLTM topic or the number of videos for the BERTopic topic. This computation only considered topics with at least 30 videos and disregarded BERTopic's "Delicious Dishes and Dining" topic as these skew the results. After computing, an arbitrary threshold of 50% was set, and any ratio above this was considered a discovered "association rule".

Table 6. HLTM Association Rules

HLTM Topic	BERTopic Topic	Ratio (HLTM vids)
Christmas Season	Christmas Season & Celebration	0.77
Ube Ice Cream	Philippine Halo-Halo Dessert	0.54
Tagalog Education	Learning and Speaking Foreign Language	0.53

After setting the threshold, a total of 3 association rules for the overlap ratio over HLTM videos were found, as seen in Table 6, while 170 association rules for the overlap ratio over BERTopic videos were found. HLTM appears to produces broader thematic categories while BERTopic identifies more specific subtopics. For instance, the HLTM topic "Tourist Islands" in Table 7 encompasses eight distinct BERTopic subtopics, including "Swimming with Whale Sharks," "Travel Immigration & Visa Processes" and "Banaue Rice Terraces." Similarly, the HLTM topic "Crispy and Tasty Fried Dishes" groups four BERTopic

Table 7. BERTopic Association Rules

HLTM Topic	BERTopic Topic	Ratio (BERTopic vids)
Business Issues	San Miguel Corporation	0.52
	Racial Discrimination	0.51
	Largest Solar Power Projects in Manila/Southeast Asia	0.55
Crispy and Tasty Fried Dishes	Balut and other Eggs	0.60
	Philippine Halo-Halo Dessert	0.68
	Squid	0.72
	Chicken & its Specific Related Dishes	0.56
Tourist Islands	Swimming with Whale Sharks	0.75
	Spanish Colonialism in the Philippines	0.71
	Travel Immigration & Visa Processes	0.51
	Banaue Rice Terraces	0.65
	Largest Solar Power Projects in Manila/Southeast Asia	0.51
	Internet Connectivity in Tourist Destinations	0.63
	Specific Venues: Circle Hall, Hobbit House, and Themed Spots	0.51
Relationship with a Filipina	Philippines Dating & Relationship of Females	0.57
	Dating & Relationships Setup	0.62
	Friendly and Romantic Relationships	0.55
Road Trip	Military Assets, Combat, Strategies	0.51
	Swimming with Whale Sharks	0.67
	Virtual Private Networks (VPNs) and Online Security	0.59
	Rainy Season	0.52
	Filipino Food	0.51
	Chinatown Manila	0.53
	Internet Connectivity in Tourist Destinations	0.60
	Going to Shopping Malls and Street Markets in Manila	0.55
	Experiencing Culture Shock	0.53
	Bicycle & Tricycle Rental	0.51
	Chicken & its Specific Related Dishes	0.59

food-related subtopics ("Balut and other Eggs," "Squid," "Chicken & its Specific Related Dishes"). This suggests that BERTopic breaks down more general themes into highly specific items.

The minimal HLTM associations (all one-to-one) suggest its topics are too broad to map precisely to BERTopic's results. Conversely, BERTopic's numerous association rules indicate most of its topics align well with portions of HLTM topics, effectively acting as subsets. This asymmetric relationship underscores how the models complement each other. HLTM provides overarching themes while BERTopic reveals more granular components.

However, some associations appear problematic, such as "Road Trip" linking to "Military Assets, Combat, Strategies." These questionable alignments likely stem from limitations in AI-generated topic labeling, which can produce misclassifications or overly broad categorizations. The issue may be compounded by HLTM's reliance on co-occurrence patterns to generate topics, which might create artificial connections between conceptually distinct data. These examples highlight the need for careful interpretation of automated topic modeling results, particularly when analyzing nuanced cultural content.

4.3 What Are the Strengths and Limitations of the Two Models?

Both BERTopic and HLTM exhibited distinct strengths and limitations in topic modeling performance. For BERTopic, it proved effective in identifying culturally specific and interpretable topics relevant to the Philippine context, aided by transcript segmentation that addressed the model's one-topic-per-document limitation, which also enabled more fine-grained topic assignments across segments. However, since topic assignments were based on the presence of a segment within a video, there is a risk of overgeneralization, where an entire video may be associated with a topic based on a marginal mention. Additionally, there are occasional redundancies in the topic representations due to the inclusion of n-grams, wherein there are specific cases of repetitive patterns in keyword representations because the model treats them as distinct tokens. Despite these nuances, BERTopic consistently captured dominant themes as reflected in the overall distribution of topics and its alignment with video content, even as standard coherence metrics reflected only moderate scores.

On the other hand, HLTM was able to capture statistical co-occurrence patterns within the potential Pinoybaiting dataset and organize the vocabulary into a hierarchical tree structure. It is important to note that both the hierarchy and the topic keywords are derived purely from statistical relationships and not conceptual ones. As a result, the model may produce groupings of words that frequently appear together but lack semantic coherence. Despite this limitation, HLTM demonstrated modest to good performance across coherence metrics. The model's association computations further highlight how HLTM offers a broader representation of the landscape, showing how different topics or keywords tend to co-occur across documents. For example, a topic that merges food and location terms may not make sense conceptually, but it reflects their frequent co-occurrence in the dataset, indicating that such themes often appear together in videos. HLTM also treats documents as binary vectors, considering only the presence or absence of words while ignoring word order and frequency. This approach limits the ability to assign weights to words during preprocessing, reducing control over how context-specific terms are emphasized. However, one could argue that if such terms are truly rare across the dataset, their exclusion from the hierarchy is a modeling choice rather than an actual weakness.

In this regard, the contrast between HLTM and BERTopic highlights how the two models can serve complementary roles in analyzing complex phenomena like Pinoybaiting. HLTM's ability to uncover overarching themes provides a valuable view of how broad concepts occur across different videos and BERTopic's capacity to break down into these broad categories to determine highly specific subtopics offers a richer detail towards understanding thematic patterns within the content of the video transcripts.

5 Conclusion

This study employed two topic modeling techniques, BERTopic and HLTM, to uncover granular topics and enable a fine-grained analysis of the potential

Pinoybaiting landscape on YouTube. Using a dataset of 2,704 video transcripts, both algorithms were implemented with carefully tuned configurations to maximize their respective strengths. This demonstrates how computational methods can extend traditional behavioral science inquiries. Through the translation of video transcripts into structured topic distributions, qualitative assessments of cultural exploitations can be backed by empirical data that can be visualized and analyzed comprehensively.

References

1. Blei, D.M.: Probabilistic topic models. Commun. ACM **55**(4), 77–84 (2012). https://doi.org/10.1145/2133806.2133826
2. Pykes, K.: What is topic modeling? an introduction with examples (2023). https://www.datacamp.com/tutorial/what-is-topic-modeling
3. Alcantara, G.C., Demeterio III, F.A.: Bida ang pinoy?: isang panimulang pagsasalarawan ng pinoybaiting sa youtube. Plaridel **18**(2), 227–258 (2021). https://www.plarideljournal.org/article/bida-ang-pinoy-isang-panimulang-pagsasalarawan-ng-pinoybaiting-sa-youtube/
4. Beltran, S.: When misplaced pinoy pride leads to pinoy-baiting — opinion (2022). https://coconuts.co/manila/features/when-misplaced-pinoy-pride-leads-to-pinoy-baiting-opinion/
5. Wang, M.: What 'pinoy baiting' says about content creators, and filipinos (2025). https://rollingstonephilippines.com/culture/essay/pinoy-baiting-content-creators-filipinos/
6. Chan, P., Doble, F., Gerardino, G.A., Santos, J., Tiam-Lee, T.J.: From clickbait to pinoybait: a Bert-based analysis of comments in foreign-made videos about the philippines. In: Proceedings of the 2024 8th International Conference on Natural Language Processing and Information Retrieval, New York, NY, USA, pp. 355–362. NLPIR '24, Association for Computing Machinery (2025). https://doi.org/10.1145/3711542.3711556
7. Egger, R., Yu, J.: A topic modeling comparison between LDA, NMF, top2vec, and bertopic to demystify twitter posts. Front. Sociol. **7** (2022). https://www.frontiersin.org/journals/sociology/articles/10.3389/fsoc.2022.886498
8. Grootendorst, M.: Bertopic: neural topic modeling with a class-based TF-IDF procedure. arXiv:2203.05794v0571 (2022)
9. Ceh, S., Putze, F., Benedek, M.: Di-what? identifying creative domains in do-it-yourself videos on youtube. Creativity, and the Arts, Psychology of Aesthetics (2023)
10. Chen, P., Zhang, N.L., Liu, T., Poon, L.K.M., Chen, Z., Khawar, F.: Latent tree models for hierarchical topic detection (2016). https://arxiv.org/abs/1605.06650
11. Yi Yang, Q.Y., Qu, H.: Vistopic: a visual analytics system for making sense of large document collections using hierarchical topic modeling. Visual Informatics **1**(1), 40–47 (2017)
12. Chen, H., Coogle, J., Damevski, K.: Modeling stack overflow tags and topics as a hierarchy of concepts. J. Syst. Softw. **156**, 283–299 (2019). https://www.sciencedirect.com/science/article/pii/S0164121219301499
13. Senevirathne, H., Kumara, B., Banujan, K.: Youtube video categorization based on closed captions. In: 2024 4th International Conference on Advanced Research in Computing (ICARC) (2024)

Clustering Analysis of Russian Drug-Related Discourse Using the Louvain Algorithm

Ivan D. Mamaev(✉) and Elena A. Rybak-Franko

Department of Theoretical and Applied Linguistics, Baltic State Technical University "Voenmeh" Named After D.F. Ustinov, 1-Ya Krasnoarmeyskaya Street, 1, 190005 Saint Petersburg, Russia
{mamaev_id,rybak_franko_ea}@voenmeh.ru

Abstract. This study explores the structure of Russian online discourse related to drug use by applying the Louvain clustering algorithm to a corpus of approximately 8,000 social media posts. The objective is to identify stable lexical and topical groupings that characterize the online drug-discursive space. The study contributes to the underdeveloped field of forensic linguistics in Russian by offering both methodological and empirical insights. Text preprocessing includes tokenization, lemmatization, and stop-word removal. After text clustering, 13 meaningful clusters are identified; they represent areas such as distribution, consumption, medical discourse, and law enforcement. Each cluster is analyzed linguistically to determine dominant lexemes and semantic tendencies. The findings indicate that certain lexical combinations repeatedly appear across similar contexts that suggest reproducible topical regularities. However, further research is needed to test the robustness of these clusters across time and corpus variations. The current results can assist forensic linguists in detecting communication patterns in illicit online environments, thereby enhancing language-based intelligence analysis.

Keywords: Forensic Linguistics · Lexical Markers · Criminal Communities · Drug Discussion Groups · Clustering

1 Introduction

Over the past decade, there has been a significant expansion of forensic linguistics as an interdisciplinary field focused on linguistic examination [7, p. 24]. According to a query conducted on Google Scholar on May 29, 2025, searching for the exact term 'forensic linguistics' within publications dated 2015–2025 returns approximately 17,400 results in foreign languages. In contrast, Russian scholarship exhibits considerably lower publication activity on this subject (if the search is carried out in Russian), with only 288 publications indexed under 'forensic linguistics' during the same period—approximately 60 times fewer than those in foreign languages—and merely 14 publications under the related term "linguistic forensics." Additionally, terminological variation is evident, with competing concepts such as 'forensic linguistics' used alongside 'linguistic expertology' (81 publications) and 'forensic speech science' (178 publications), among others [7, p. 24].

H. Unger and P. Meesad (Eds.): NLPIR 2025, LNNS 1904, pp. 975–984, 2026.
https://doi.org/10.1007/978-3-032-20897-2_68

These trends underscore the growing relevance of forensic linguistics within Russian academia, particularly the importance of in-depth linguistic analysis of specialized discourse types, including texts produced in online drug discussion groups. As noted by D.N. Khokhlov, "The spread of drug addiction leads to a significant portion of society, particularly young people, effectively withdrawing from family, political, professional, and other spheres of life due to physical and social degradation" (Rasprostranenie narkomanii privodit k tomu, chto ogromnaya chast' chlenov obshchestva, v pervuyu ochered' molodezhi, vsledstvie fizicheskoj i social'noj degradacii fakticheski vybyvaet iz semejnoj, politicheskoj, trudovoj i drugih vidov zhiznedeyatel'nosti) [4 , p. 142].

This article aims to answer the following research questions (RQ):

RQ1: What are the frequent lexical markers that characterize distinct online drug discussion groups?

RQ2: How can these markers be utilized to systematically identify and differentiate topical clusters within drug-related online discourse?

By the term "drug discussion groups," we refer to participants engaged in online communication concerning drug-related themes; however, the actual use of illegal substances by these individuals remains uncertain. The discussions encompass both overt and covert references to illegal activities [9]. Employing modern computational linguistics tools to analyze such texts enables the identification of distinctive linguistic features characteristic of these communities.

Contemporary research often employs machine learning techniques for community identification. For instance, the study [10] proposes a methodology for automatic detection of illegal drug advertisements and identification of seller communities on social media platforms using Support Vector Machines (SVM) and Convolutional Neural Networks (CNN). To detect overlapping seller communities, matrix factorization techniques were utilized. This approach was validated using datasets from Google+, Flickr, and Tumblr. Similarly, the paper [5] introduces a deep learning approach employing Long Short-Term Memory (LSTM) networks to detect drug dealers on Instagram. The authors compiled a corpus exceeding 12,000 topic-oriented web posts and evaluated the LSTM model alongside traditional machine learning classifiers, including Random Forest (RF), Decision Tree (DT), and SVM. The LSTM model outperformed others, achieving an F-measure of 95%.

Neural network models have also been applied in Russian studies of similar linguistic datasets. For example, V.I. Firsanova [3] developed a binary classifier based on Convolutional Neural Networks to identify messages originating from Russian-language drug addiction communities on the social network VKontakte (VK). The classifier was tested on two vector space models — Bag-of-Words (BoW) and Word Embeddings — with the BoW model yielding superior accuracy (95%) and F-measure (99%).

2 The Methodology of the Study

A recent study in the field of natural language processing [1] introduces a novel Russian-language corpus comprising online posts addressing sensitive social issues, including the distribution of pornography, terrorist activities and their manifestations both offline and online, and discrimination based on physical appearance, among others. For the purposes of this study, a subset of 7,824 posts — totaling 132,394 word occurrences — authored by members of drug discussion communities was selected, facilitating an investigation into the linguistic patterns characteristic of topical texts.

The overall procedure for identifying distinct groups within this community is illustrated in Fig. 1.

Algorithm 1 *Text Clustering and Visualization Pipeline*

Require: Corpus of 7824 posts, SBERT model, UDPipe, Orange environment
Ensure: Clustered and visualized document embeddings

```
Preprocess Texts:
for each post in corpus
    Convert text to lowercase
    Remove accent symbols
    Tokenize using regular expressions
    Lemmatize tokens using UDPipe
    Remove punctuation and non-informative words
end for
Vectorize Texts:
for each preprocessed post
    Encode post using SBERT to obtain vector
end for
Apply Louvain Clustering:
Set distance metric ← Euclidean
Set number of neighbors k ← 30
Set resolution parameter ← 1.0
Detect communities in SBERT vectors
Visualize with t-SNE:
Set initialization ← PCA
Set exaggeration factor ← 4
Project vectors into 2D space using t-SNE
Plot clusters
```

Fig. 1. A pseudocode illustrating the process of identifying groups within drug discussion communities

The experimental framework employed a comprehensive data processing pipeline implemented within the Orange environment [2]. The dataset underwent multi-stage preprocessing, which included standardization to lowercase text, removal of accent marks, tokenization via regular expressions, lemmatization using the integrated UDPipe lemmatizer, elimination of punctuation, and filtering out non-informative elements such as function words and semantically null tokens. Following preprocessing, texts were transformed into vector representations using Sentence-BERT (SBERT) models [6], thereby ensuring a robust preservation of the semantic structure inherent in the texts.

Community detection within the corpus was performed using the Louvain clustering algorithm, which identifies communities in graphs by maximizing modularity. Key parameters for the algorithm included the Euclidean distance metric, a neighborhood size $k = 30$, and a resolution parameter of 1.0, enabling detection of clusters at varying granularities. To facilitate interpretation, the high-dimensional vector representations were projected onto a two-dimensional plane using t-Distributed Stochastic Neighbor Embedding (t-SNE). The t-SNE embedding utilized initialization based on principal component analysis and incorporated an exaggeration factor of 4 to enhance the separation between local clusters.

To evaluate the robustness of the findings, alternative clustering methods such as k-means and hierarchical clustering were also tested on smaller samples. While these produced comparable topical groupings, the Louvain algorithm yielded clearer community boundaries, which guided its final selection for this study.

3 Results and Discussions

Figure 2 presents the visual structure of the identified drug discussion groups, with each group color-coded and labeled according to its corresponding cluster (C1–C13). Given that the original corpus comprises over 7,000 interrelated posts, which are visually represented as dense clusters, clusters C11 to C13 were collectively designated as "Others." Despite some degree of cluster overlap, cluster C2 merits particular attention due to its relative isolation from the others.

To elucidate the frequency distribution of posts across these groups, a bar chart was constructed and is displayed in Fig. 3. Notably, clusters C1 (1,188 posts), C2 (1,033 posts), and C3 (967 posts) constitute the three largest groups in terms of post volume. Concurrently, a substantial number of smaller clusters are also observed—for example, clusters C11 (260 posts), C12 (137 posts), and C13 (130 posts)—which are likely to represent marginal or specialized segments within the community. This quantitative distribution suggests a hierarchical core-periphery structure within the corpus: a few large clusters concentrate the majority of activity, whereas smaller clusters occupy peripheral roles.

Based on the distribution of posts by cluster, subcorpora were constructed, for which lists of lemmata were generated and ranked according to their relative frequency measured in instances per million (ipm), computed by the following formula (1):

$$ipm = \frac{freq_w {}^* 10^6}{N}, \quad (1)$$

where $freq_w$ represents the absolute frequency of a lexical unit, and N denotes the total corpus size measured in word usages. From each cluster's lemma list, the top ten content words from significant parts of speech were selected to reflect the topical focus of each group. Lexemes belonging to closed morphological classes—such as prepositions, conjunctions, and particles—were excluded from these lists.

As a result, cluster 1 is characterized by the following lexemes and their corresponding ipm frequencies: the lexeme *drink* ('*pit*') with an ipm of 21678.1, the lexeme *be* ('*byt*') with an ipm of 11956.75, the lexeme *alcohol* ('*alkogol*') with an ipm of

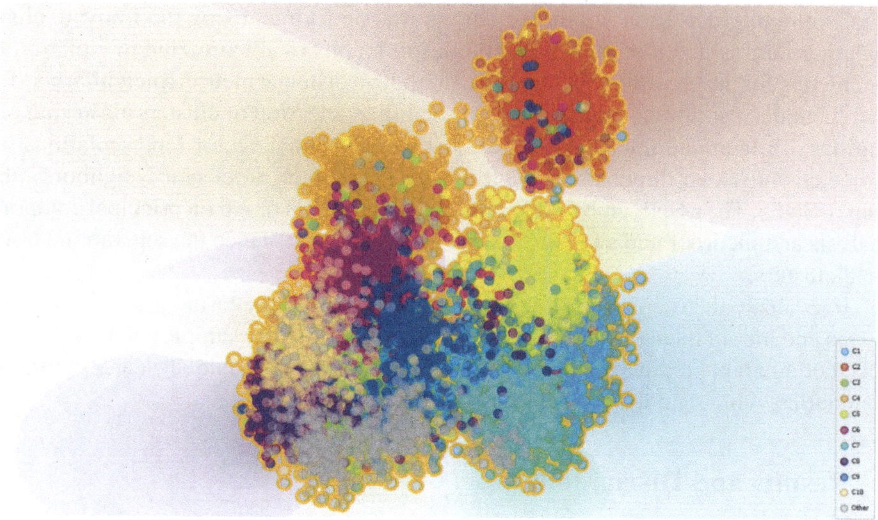

Fig. 2. Visualization of drug discussion groups identified within the corpus

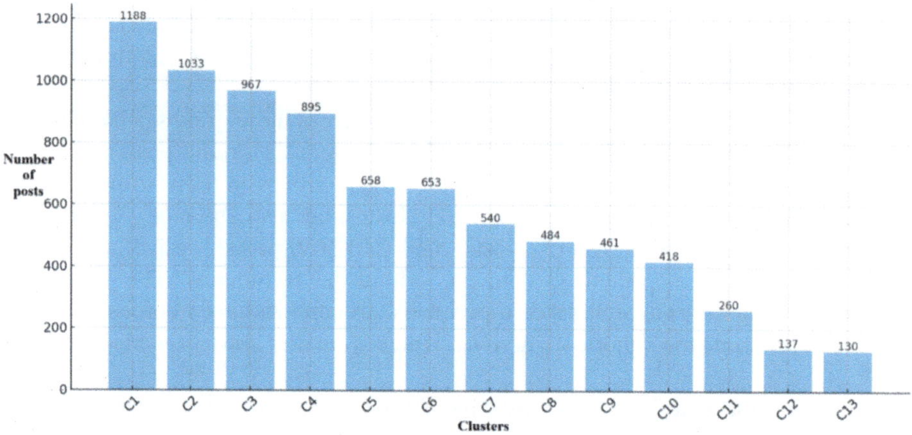

Fig. 3. Absolute frequencies of posts across groups within the corpus

10241.21, the lexeme *beer* (‘*pivo*’) with an ipm of 9149.51, the lexeme *have a drink* (‘*vypit*’) with an ipm of 7070.08, the lexeme *vodka* (‘*vodka*’) with an ipm of 6238.3, the lexeme *bottle* (‘*butylka*’) with an ipm of 5926.39, the lexeme *year* (‘*god*’) with an ipm of 5614.47, the lexeme *can* (‘*moch*’) with an ipm of 5458.52, and the lexeme *day* (‘*den*’) with an ipm of 4210.85.

Cluster 2 consists of the lexeme *cigarette* (‘*sigareta*’) with an ipm of 29719.14, the lexeme *smoke* (‘*kurit*’) with an ipm of 28684.96, the lexeme *be* (‘*byt*’) with an ipm of 13008.93, the lexeme *tobacco* (‘*tabak*’) with an ipm of 7402.57, the lexeme *quit* (‘*brosit*’) with an ipm of 5932.94, the lexeme *can* (‘*moch*’) with an ipm of 5715.22, the lexeme *year* (‘*god*’) with an ipm of 5334.2, the lexeme *smoking* (‘*kurenie*’) with an ipm of 4136.73, the lexeme *alcohol* (‘*alkogol*’) with an ipm of 3537.99, and the lexeme *nicotine* (‘*nikotin*’) with an ipm of 2993.69.

Cluster 3 includes the lexeme *be* (‘*byt*’) with an ipm of 14447.95, the lexeme *dealer* (‘*baryg*’) with an ipm of 4771.18, the lexeme *can* (‘*moch*’) with an ipm of 4569.59, the lexeme *person* (‘*chelovek*’) with an ipm of 2687.99, the lexeme *junkie* (‘*torch*’) with an ipm of 2553.59, the lexeme *good* (‘*khoroshiy*’) with an ipm of 2351.99, the lexeme *year* (‘*god*’) with an ipm of 2083.19, the lexeme *drug* (‘*gerych*’) with an ipm of 2083.19, the lexeme *say* (‘*skazat*’) with an ipm of 1881.59, and the lexeme *needed* (‘*nuzhnyy*’) with an ipm of 1814.39.

Cluster 4 is defined by the lexeme *drug* (‘*narkotik*’) with an ipm of 17772.74, the lexeme *be* (‘*byt*’) with an ipm of 13676.18, the lexeme *drugs* (‘*narkota*’) with an ipm of 11029.18, the lexeme *addict* (‘*narkoman*’) with an ipm of 7625.89, the lexeme *can* (‘*moch*’) with an ipm of 5735.17, the lexeme *heroin* (‘*geroin*’) with an ipm of 4033.53, the lexeme *person* (‘*chelovek*’) with an ipm of 3025.15, the lexeme *year* (‘*god*’) with an ipm of 2962.12, the lexeme *cocaine* (‘*kokain*’) with an ipm of 2836.07, and the lexeme *use* (‘*upotreblyat*’) with an ipm of 2647.

Cluster 5 contains the lexeme *alcohol* (‘*alkogol*’) with an ipm of 25199.88, the lexeme *be* (‘*byt*’) with an ipm of 13966.2, the lexeme *drink* (‘*pit*’) with an ipm of 11436.09, the lexeme *alcoholism* (‘*alkogolizm*’) with an ipm of 11132.48, the lexeme *alcoholic* (‘*alkogolik*’) with an ipm of 6679.49, the lexeme *can* (‘*moch*’) with an ipm of 6375.87, the lexeme *person* (‘*chelovek*’) with an ipm of 6173.46, the lexeme *spirit* (‘*spirt*’) with an ipm of 4554.19, the lexeme *alcoholic* (‘*alkogolnyy*’) with an ipm of 4250.58, and the lexeme *have a drink* (‘*vypit*’) with an ipm of 3137.33.

Cluster 6 comprises the lexeme *be* (‘*byt*’) with an ipm of 14249.79, the lexeme *can* (‘*moch*’) with an ipm of 6286.67, the lexeme *pill* (‘*tabletka*’) with an ipm of 4777.87, the lexeme *doctor* (‘*vrach*’) with an ipm of 3772, the lexeme *year* (‘*god*’) with an ipm of 3520.54, the lexeme *day* (‘*den*’) with an ipm of 3436.71, the lexeme *month* (‘*mesyats*’) with an ipm of 3101.42, the lexeme *drug* (‘*narkotik*’) with an ipm of 3017.6, the lexeme *needed* (‘*nuzhnyy*’) with an ipm of 3017.6, and the lexeme *week* (‘*nedelya*’) with an ipm of 2933.78.

Cluster 7 is characterized by the lexeme *vodka* (‘*vodka*’) with an ipm of 19102.56, the lexeme *drink* (‘*pit*’) with an ipm of 13076.92, the lexeme *beer* (‘*pivo*’) with an ipm of 13076.92, the lexeme *be* (‘*byt*’) with an ipm of 11538.46, the lexeme *coffee* (‘*kofe*’) with an ipm of 9487.18, the lexeme *wine* (‘*vino*’) with an ipm of 6666.67, the lexeme *tea* (‘*chay*’) with an ipm of 6538.46, the lexeme *drink* (‘*napitok*’) with an ipm of 5000, the lexeme *good* (‘*khoroshiy*’) with an ipm of 4487.18, and the lexeme *can* (‘*moch*’) with an ipm of 4102.56.

Cluster 8 includes the lexeme *be* (‘*byt*’) with an ipm of 17657.19, the lexeme *dealer* (‘*baryg*’) with an ipm of 5813.95, the lexeme *can* (‘*moch*’) with an ipm of 5060.29, the lexeme *year* (‘*god*’) with an ipm of 4521.96, the lexeme *say* (‘*govorit*’) with an ipm of 3229.97, the lexeme *know* (‘*znat*’) with an ipm of 2906.98, the lexeme *drug* (‘*gerych*’) with an ipm of 2476.31, the lexeme *coke* (‘*koks*’) with an ipm of 2260.98, the lexeme *junkie* (‘*torch*’) with an ipm of 2260.98, and the lexeme *think* (‘*dumat*’) with an ipm of 2153.32.

Cluster 9 consists of the lexeme *be* (*'byt'*) with an ipm of 13744.33, the lexeme *can* (*'moch'*) with an ipm of 7872.97, the lexeme *salt* (*'sol'*) with an ipm of 7205.76, the lexeme *drink* (*'pit'*) with an ipm of 3469.44, the lexeme *grass* (*'trava'*) with an ipm of 3336, the lexeme *taste* (*'vkus'*) with an ipm of 3202.56, the lexeme *acid* (*'kislota'*) with an ipm of 3202.56, the lexeme *water* (*'voda'*) with an ipm of 2802.24, the lexeme *smell* (*'zapakh'*) with an ipm of 2802.24, and the lexeme *good* (*'khoroshiy'*) with an ipm of 2668.8.

Cluster 10 contains the lexeme *be* (*'byt'*) with an ipm of 14853.24, the lexeme *can* (*'moch'*) with an ipm of 6129.91, the lexeme *know* (*'znat'*) with an ipm of 4833.2, the lexeme *day* (*'den'*) with an ipm of 4361.66, the lexeme *year* (*'god'*) with an ipm of 2947.07, the lexeme *now* (*'seychas'*) with an ipm of 2475.54, the lexeme *sleep* (*'spat'*) with an ipm of 2475.54, the lexeme *dealer* (*'baryg'*) with an ipm of 2357.66, the lexeme *say* (*'skazat'*) with an ipm of 2239.77, and the lexeme *think* (*'dumat'*) with an ipm of 2239.77.

Cluster 11 is defined by the lexeme *be* (*'byt'*) with an ipm of 13037.81, the lexeme *year* (*'god'*) with an ipm of 10212.95, the lexeme *day* (*'den'*) with an ipm of 6736.2, the lexeme *dealer* (*'baryg'*) with an ipm of 5215.12, the lexeme *can* (*'moch'*) with an ipm of 4128.64, the lexeme *take* (*'brat'*) with an ipm of 4128.64, the lexeme *month* (*'mesyats'*) with an ipm of 3694.05, the lexeme *know* (*'znat'*) with an ipm of 3259.45, the lexeme *telephone* (*'telefon'*) with an ipm of 2824.86, and the lexeme *number* (*'nomer'*) with an ipm of 2824.86.

Cluster 12 comprises the lexeme *be* (*'byt'*) with an ipm of 13765.54, the lexeme *russia* (*'rossiya'*) with an ipm of 6660.75, the lexeme *drink* (*'pit'*) with an ipm of 5772.65, the lexeme *vodka* (*'vodka'*) with an ipm of 5772.65, the lexeme *year* (*'god'*) with an ipm of 5328.6, the lexeme *can* (*'moch'*) with an ipm of 5328.6, the lexeme *alcohol* (*'alkogol'*) with an ipm of 4884.55, the lexeme *drink* (*'napitok'*) with an ipm of 4884.55, the lexeme *rf* (*'rf'*) with an ipm of 4884.55, and the lexeme *moscow* (*'moskva'*) with an ipm of 4440.5.

Cluster 13 includes the lexeme *be* (*'byt'*) with an ipm of 11721.91, the lexeme *dealer* (*'baryg'*) with an ipm of 10509.3, the lexeme *sell* (*'prodavat'*) with an ipm of 8892.48, the lexeme *money* (*'dengi'*) with an ipm of 8084.07, the lexeme *shop* (*'magazin'*) with an ipm of 8084.07, the lexeme *buy* (*'kupit'*) with an ipm of 7275.67, the lexeme *can* (*'moch'*) with an ipm of 5254.65, the lexeme *take* (*'brat'*) with an ipm of 4446.24, the lexeme *buy* (*'pokupat'*) with an ipm of 3637.83, and the lexeme *coke* (*'koks'*) with an ipm of 3233.63.

Based on the lemmata presented in Cluster 1, it can be hypothesized that the first group of drug discourses focuses on discussing alcohol dependence, drunkenness, or the consequences of alcohol consumption: *"Vot blin(ya tretiy god p'yu i mne strashno…"*, "Damn, I've been drinking for three years and I'm scared..."; *"Putayete vypit' alkogolya i nazhratsya alkogolem. Pervoe otsenivayut, kogda napitok vo rtu, a vtoroe mozhno i s utra…"*, "You confuse drinking alcohol and getting drunk. The first is judged when the drink is in your mouth, the second can be even in the morning..."; *"Druz'ya pili, govoryat zbs tripy, sam khochu poprobovat' v etom godu)… "*, "Friends drank, say the trips were great, I want to try it myself this year). ..."; *"ne khochesh' takoy na rabotu idti vypil stakan vodki i spat' poshyol…"*, "You don't want to go to work like that—drink a glass of vodka and go to sleep...". Cluster 5 expands the topical scope presented in the first group. Cluster 12 postulates the problem of alcoholism prevalence in Russia: *"V Sankt-Peterburge mozhno kupit' yaponskoe pivo As v odnom iz spetsializirovannykh magazinov alkogolya ili zakazat' ego na dom, vospol'zovavshis' uslugami kurerskoy sluzhby dostavki spirtnykh napitkov..."*, "In Saint Petersburg, it is possible to buy Japanese beer As in one of the specialized alcohol stores or order it home using the services of an alcoholic beverage courier delivery..."; *"Kak ponyat', v kakom regione v Rossii budet vostrebovan tot ili inoy ekzoticheskiy alkogol'nyy produkt?"*, "How to understand in which region of Russia a particular exotic alcoholic product will be in demand?".

Cluster C2 is dedicated to the discourse on smoking, particularly the consumption of cigarettes and tobacco (*"Kuryu kalyan no ne chuvstvuyu vkusa ...chtoby eto moglo byt'? Chuvstvuyu prosto dym, kak ot sigaret..."*, "I smoke hookah but don't feel the taste... what could it be? I just feel smoke, like from cigarettes..."), (un)successful attempts to quit smoking (*"Reshil s utra brosit' kurit'. Vstayu utrom ne kuril golova pukhnet i na glaza davit. Glaza bolet' nachinayut. Bylo u vas takoe?"*, "Decided this morning to quit smoking. I wake up not having smoked, head is swollen and pressing on eyes. Eyes start hurting. Have you had this?"), as well as discussions about the harms and health impacts of smoking (*"rak guby budet. kupi plastyr' nikotinovyy..."*, "There will be lip cancer. Buy a nicotine patch..."). This is the only cluster with almost no overlap with other groups (see Fig. 3).

The lemmata of Cluster 3 are related to the topic of drug distribution and consumption: *"Gallyutsinogeny, na odin raz potrebleniya*!! *!Sleduyushchie budut rasti u vas v grobu!! "*, "Hallucinogens, for one-time use!!! The next ones will grow in your grave!!"; *"A ot shmali to chto rvët?"*, "And what does weed tear you up from?," among others. It is important to note that the adjective *"khoroshiy"* ("good") entered the top list of lemmata: despite its original positive connotation, in attributive constructions of the ADJ_NOUN type where NOUN denotes a drug, the adjective *"khoroshiy"* acquires a diametrically opposite evaluation: *"a v kakom eto vy gorode, gde u nas seychas koks yest' khoroshiy!?!"*, "And in which city are you where we now have good coke?!". Cluster 4 is lexically close to Cluster 3; however, due to the absence of verbs (except for *"to be"* and *"to be able"*), the topic is broad and covers various types of drugs. A similar situation can be observed when attempting to interpret the contents of Clusters 8, 9, and 13 (plant-based and semi-synthetic drugs).

Despite the inclusion of words denoting time intervals in Cluster 6, the lemmata *"tabletka"* and *"vrach"* indicate the predominance of the general theme of health and prevention of drug-related diseases: *"Est' kakie to tabletki daby nervishki uspokoit' bez sil'nykh pobochek po tipu fenibuta?"*, "Are there any pills to calm nerves without strong side effects like phenibut?". Cluster 7 contains a large number of beverage-related words, and the analysis of contextual usage suggests that even everyday drinks (*"chai"*, "kofe") are mentioned in sentences related to altered consciousness or health states: *"Chto budet, esli smeshat' kofe dve lozhki s energetikom?"*, "What happens if you mix two spoons of coffee with an energy drink?"; *"Bol'shoe kolichestvo vypitogo kofe plokho vliyaet na kozhu, tsvet zubov, sostoyanie serdechno-sosudistoy sistemy i zhkt..."*, "A large amount of coffee consumed negatively affects skin, teeth color, cardiovascular system, and gastrointestinal tract..."; *"Konechno ne normal'no. Kofe, krepkij chai, stressy-eto mozhet byt' prichinoy. Vozmozhno Vam nuzhna pomoshch psikhijatra ili psikhoterapevta...",* "Of course not normal. Coffee, strong tea, stress—this can be the cause. You might need help from a psychiatrist or psychotherapist...".

The absence in Clusters 10 and 11 of key nominative terms characteristic of drug-discussing groups, combined with the predominance of abstract elements (*"byt'"*, *"god"*, *"moch'"*– *"be"*, *"year"*, *"can"* etc.), allows the conclusion that the analyzed groups cannot be identified as drug-discursive.

4 Conclusion

The results of the study allow the conclusion that within the drug-discursive space there exist lexical patterns that appear consistently across posts and characterize topical groups. Although the results indicate consistent lexical groupings, further research is required to test the reproducibility of these clusters. A comparative experiment with another corpus or a temporal dataset could help determine whether the patterns observed here remain stable over time. Using the Louvain algorithm, 13 clusters were identified, among which clusters 1, 2, 3, 4, 5, and 12 demonstrate a clear focus on topics such as alcoholism, smoking, psychoactive substance use, and discussions of various types of drugs. It is important to note the presence of groups with diffuse lexical cores (clusters 10 and 11), which lack key markers of drug discourse—this distinction enables the separation of irrelevant or noisy segments that are not subject to expert forensic linguistic interpretation. The proposed methodology can be applied for practical purposes, such as automatic filtering or ranking of content in online communication monitoring systems.

There are also a number of important changes which need to be made. First of all, as a promising avenue for further research, it is important to consider alternative community detection methods, particularly the Leiden algorithm [8], which improves upon Louvain clustering by producing more connected communities through resolving the issue of cluster fragmentation. This method can be employed to identify other types of hidden online communities, for instance, those with terrorist orientations. Secondly, expanding linguistic datasets and integrating multimodal data (images or video) hold potential for a more comprehensive characterization of user communities.

References

1. Babakov, N., Logacheva, V., Panchenko, A.: Beyond plain toxic: building datasets for detection of flammable topics and inappropriate statements. Lang. Resour. Eval. **58**(2), 459–504 (2024)
2. Demšar, J., Zupan, B., Leban, G., Curk, T.: Orange: from experimental machine learning to interactive data mining. In: Lecture Notes in Computer Science, vol. 3202, pp. 537–539. Springer, Heidelberg (2004)
3. Firsanova, V.I.: Automatic recognition of messages from virtual communities of drug addicts. J. Appl. Linguist. Lexicography. **2**(1), 16–27 (2020)
4. Khokhlov, D.N.: Criminal trends of drug trafficking on the territory of the Russian Federation in the conditions of modern deglobalization. Academic Thought (Akademicheskaya mysl'). **2**(19), 140–144 (2022) (In Russian)

5. Li, J., Xu, Q., Shah, N., Mackey, T.K.: A machine learning approach for the detection and characterization of illicit drug dealers on Instagram: model evaluation study. J. Med. Internet Res. **21**(6), 1–14 (2019)
6. Reimers, N., Gurevych, I.: Sentence-BERT: sentence embeddings using Siamese BERT-networks. In: Proceedings of the 2019 Conference on Empirical Methods in Natural Language Processing and the 9th International Joint Conference on Natural Language Processing (EMNLP-IJCNLP), pp. 3982–3992 (2019)
7. Sternin, I.A., Antonova, L.G., Karpov, D.L., Shamanova, M.V.: Osnovnye Ponyatiya Lingvokriminalisticheskoy Ekspertizy: Spravochnoe Posobie [Key Concepts of Forensic Linguistic Examination: a Reference Guide], p. 80. Kantsler Publishing House, Yaroslavl (2013) (In Russian)
8. Traag, V.A., Waltman, L., Van Eck, N.J.: From Louvain to Leiden: guaranteeing well-connected communities. Sci. Rep. **9**(1), 1–12 (2019)
9. Tzanetakis, M., Kamphausen, G., Werse, B., von Laufenberg, R.: The transparency paradox. Building trust, resolving disputes and optimising logistics on conventional and online drugs markets. Int. J. Drug Policy. **35**, 58–68 (2016)
10. Zhao, F., et al.: Computational approaches to detect illicit drug ads and find vendor communities within social media platforms. IEEE/ACM Trans. Comput. Biol. Bioinform. **19**(1), 180–191 (2020)

Predicting Outcome, Mapping Evolution: A Combined Analysis of Pancreatic Cancer Trials Using ClinicalBERT, Ensemble Models, and DTM

Janet V. Itliong(✉) and Jeffrey S. Ingosan

University of the Cordilleras, Governor Pack Road, 2600 Baguio City, Philippines
jvi2577@students.uc-bcf.edu.ph, jvitliong@slu.edu.ph,
jsingosan@uc-bcf.edu.ph
https://www.uc-bcf.edu.ph/

Abstract. Pancreatic cancer is a major contributor to cancer-related mortality, with rising global incidence highlighting the urgent necessity for ongoing research and novel methodologies in clinical trials. This study addresses a significant gap by documenting the thematic evolution of research topics within pancreatic cancer clinical trials conducted between 2004 and 2024, aiming to leverage these thematic patterns to enhance predictive modeling of trial outcomes. Integrating quantitative and qualitative approaches, research used data from the AACT database, applying Dynamic Topic Modeling (DTM), Latent Dirichlet Allocation (LDA), and Non-negative Matrix Factorization (NMF) to uncover evolving themes. The research demonstrated a clear temporal evolution, shifting from the early Gemcitabine Era (2004 – 2010) to the advent of modern chemotherapy (2011 – 2013), progressing to the Immunotherapy and Molecular Age (2017 – 2019), and culminating in the Precision Medicine, AI, and Organoid age (2020 – 2024). For predictive modeling, models including Logistic Regression, Random Forest, and XGBoost were developed using structured data and unstructured data embeddings generated by the specialized language model, ClinicalBERT. The most compelling finding demonstrated the absolute predictive power conferred by the Full ClinicalBERT Embeddings Only feature set, derived from streamlined textual features. Tree-based ensemble models trained exclusively on these features achieved perfect predictive performance (ROC-AUC of 1.0, 100% accuracy, precision, recall, and F1-score). This outcome confirms the significant efficacy of employing a ClinicalBERT model to generate high-quality embeddings for near-perfect discrimination and prediction of clinical outcome.

Keywords: Pancreatic Cancer · Clinical Trials · ClinicalBERT · Dynamic Topic · Modeling (DTM) · Predictive Modeling · Ensemble Models · Unstructured Data · Thematic Evolution

H. Unger and P. Meesad (Eds.): NLPIR 2025, LNNS 1904, pp. 985–998, 2026.
https://doi.org/10.1007/978-3-032-20897-2_69

1 Introduction

Pancreatic cancer represents a significant contributor to cancer-related mortality, primarily attributable to the challenges associated with early-stage detection and the restricted treatment alternatives accessible for patients diagnosed at more advanced stages [1]. The global incidence and mortality rates of pancreatic cancer have increased, despite advancements in surgery and chemotherapy, with an annual rise of approximately 0.8% from 1990 to 2019 [2]. This upward trend highlights the urgent necessity for ongoing research and novel methodologies within the domain of pancreatic cancer.

The development of innovative therapies for pancreatic cancer relies critically on clinical trials [3]. Clinical trials have evolved from traditional statistical methods to data science, enabling researchers to analyze complex datasets and extract insights, integrating statistics, computer science, and domain expertise [4].

The use of automated methodologies in healthcare is transformative. For example, the adoption of electronic medical records (EMR) in clinical trials significantly enhances the processes of data collection, minimizes the occurrence of errors, improves patient recruitment strategies, and advances data analytics capabilities through the integration of artificial intelligence, machine learning, and natural language processing technologies [5].

Predictive analytics represents a transformative force in healthcare, revolutionizing both the design and conduct of clinical trials and the delivery of patient care. By integrating vast datasets with advanced AI and machine learning algorithms, it offers new opportunities to enhance precision, efficiency, and patient safety, while enabling more personalized, effective, and efficient care. As these technologies continue to advance, their integration into clinical research will become increasingly sophisticated, driving greater improvements in trial outcomes and accelerating the pace of drug development [6].

1.1 Dynamic Topic Modeling

DTM, or Dynamic Topic Modeling, offers valuable insights into the progression of discourse and the evolving priorities within clinical research. Through the application of DTM, researchers can track how topics of interest change over time, uncover emerging themes in medical literature, and gain a deeper understanding of the shifting landscape of healthcare. The research presented in [7] indicates that topic modeling serves as a swift and efficient approach for evaluating progress in a fast-changing domain, whereas conventional bibliographic techniques tend to be laborious and narrow in scope, primarily due to their reliance on current data and research instruments.

Topic modeling methods have found extensive application in the unbiased discovery of topics from substantial document collections. These techniques have been utilized across various domains, including social science, environmental science, political science, as well as in biological and medical contexts [9].

NMF is an essential instrument for precision medicine and is anticipated to have various applications in oncology, as noted in [9]. Advancements in AI clinical applications are notable, with machine learning technology identified as crucial for the progression of medicine. The volume of data necessitating analysis in medicine, especially whole genome analysis data, is increasing markedly. In the era of big data, the integration of machine learning technology into the medical field is essential. NMF, despite its distinct properties, remains underutilized in the medical domain, with expectations for increased application in oncology moving forward [10].

1.2 Predictive Modeling

Research conducted by [11] indicates that the pre-training of BERT on biomedical corpora is essential for its effective application within the biomedical domain. With minimal task-specific architectural modifications, BioBERT demonstrates superior performance compared to earlier models in biomedical text mining tasks, including Named Entity Recognition, Relation Extraction, and Question Answering. In this study BioBERT, specifically ClinicalBERT, is used to process unstructured clinical trial descriptions.

The reference [12] introduced a BERT-based model, specifically a code embedded version of ClinicalBERT, to improve clinical prediction performance by embedding EMR data and reducing code features' dimensionality. This model outperformed traditional OHE in complex clinical outcomes like MACE, capturing nuanced interactions between diagnoses, medications, and prescriptions. The embedding approach enhances prediction accuracy, providing healthcare professionals with valuable insights for data-driven decision-making, despite OHE being effective for simpler classification tasks.

The research presented in [13] concentrated on forecasting the onset of sepsis through the integration of latent features derived from clinical notes. The optimal prediction performance was attained through a model that integrated ClinicalBERT embeddings derived from electronic health record notes alongside structured data. The integration of physiological data with latent representations of clinical notes markedly enhanced the model's performance. Excluding the contextual representation of clinical notes resulted in a decrease in prediction performance by 0.03 in AUC. The ClinicalBERT approach demonstrated superior performance compared to the TF-IDF model by effectively capturing intricate patterns inherent in sentence structure and semantic meaning. ClinicalBERT produced more significant representations of clinical notes, facilitating improved accuracy in predicting sepsis onset.

Despite the extensive number of clinical trials conducted between 2004 and 2024, there remains a significant gap in the documentation and comprehension of the thematic evolution of research topics within these trials, particularly in relation to critical inflection points. This obstructs a comprehensive understanding of the advancement of research and the recognition of significant changes within the landscape of clinical trials for pancreatic cancer. Clinicians and other professionals are increasingly challenged to keep up with the rapidly expanding body of knowledge, particularly the results of all randomized clinical trials (RCTs) that are relevant to their treatment decisions. Consequently, they examine out summaries of knowledge, including clinical practice guidelines, which are partially reliant on evidence syntheses (e.g., systematic reviews, meta-analyses). These evidence syntheses aggregate data from comparable studies, frequently emphasizing randomized controlled trials (RCTs) for informed treatment decisions [8].

The objectives of this study are threefold: (1) to identify and characterize the latent topics emerging from pancreatic cancer clinical trial documents between 2004 and 2024, (2) to analyze the temporal evolution of these topics and detect significant shifts in research focus, and (3) to develop and compare a predictive model using the extracted topics to forecast potential outcomes of future clinical trials.

This study seeks to address existing gaps by investigating the ways in which these emerging thematic patterns can be leveraged to enhance predictive modeling of clinical trial outcomes, as well as assess the impact of unstructured data on prediction accuracy.

2 Methodology

2.1 Data Collection

The research integrates both quantitative and qualitative approaches to analyze clinical trial data from the AACT (Aggregate Analysis of ClinicalTrials.gov) database, focusing specifically on pancreatic cancer research. Using Dynamic Topic Modeling (DTM), Latent Dirichlet Allocation (LDA), and Non-negative Matrix Factorization (NMF), the study examines the evolution of research themes over time, providing valuable insights into emerging scientific trends.

The data for this investigation was exclusively sourced from clinical trial datasets available on clinicaltrials.gov, with records covering clinical trials from January 2004 to December 2024. ClinicalTrials.gov serves as an essential repository for comprehending the landscape of clinical research. It delivers extensive data on trials spanning various therapeutic domains, thereby furnishing critical insights for researchers, clinicians, and policymakers alike [14]. This selection resulted in a total of 5066 pancreatic cancer-related clinical trials. The data was obtained from the AACT database, processed using pgAdmin4 software, and translated for trials with Chinese characters using Google Translate. The dataset was acquired on October 3, 2025, with a clear understanding that the database is subject to daily updates. As a result, the date of acquisition is crucial to maintaining the accuracy and reliability of the analysis, ensuring the most current data is used for meaningful conclusions.

2.2 Dynamic Topic Modeling

Topic modeling was conducted to identify latent themes in the clinical trial descriptions, with the coherence value checked for each model to ensure that the identified topics were meaningful and relevant.

A systematic preprocessing pipeline was implemented in Python using NLTK, Gensim, and Scikit-learn libraries. The preprocessing steps included text normalization, tokenization, stopword removal, lemmatization, bigram construction, and filtering. The processed data was then transformed into numerical representations using the Term Frequency-Inverse Document Frequency (TF-IDF) method, applied with Scikit-learn's TfidfVectorizer. To capture the nuances of multi-word medical terms, Gensim's Phrases was utilized for bigram detection, and additional noise reduction was achieved by filtering rare and excessively frequent terms with no_below=10 and no_above=0.8 thresholds. The topic modeling process, including LDA and NMF, was carried out to uncover the key themes in the clinical trial descriptions, and coherence scores were checked to validate the quality of topics identified.

2.3 Predictive Modeling

Additionally, the study develops and compares predictive models using logistic regression, random forest, XGBoost (Extreme Gradient Boosting Classification Algorithms) to evaluate their effectiveness in predicting clinical outcomes and identifying key factors influencing pancreatic cancer research, utilizing both structured and unstructured data. BioBERT serves as a pre-trained language representation model specifically designed for the domain of biomedical text mining.

This study implements a comprehensive Machine learning Classification Algorithms to predict the outcome of clinical trials focused on pancreatic cancer using both structured and unstructured data. The pipeline begins with data preprocessing, where the dataset is filtered to include only pancreatic cancer-related trials. A preliminary target variable, outcome_simple, is created based on the overall_status column, where trials with completed status are assigned a value of 1, and those that are withdrawn, terminated, or suspended are assigned a value of 0. To refine the analysis, a more detailed target variable, outcome_complex, is created by selecting trials with statistically significant p-values (≤ 0.05), combining both the trial's overall_status and statistical significance for a more accurate measure of success. The filtered dataset, referred to as final_data_simple, is then used for further analysis.

The predictive analysis was executed through a rigorous, three-phase pipeline encompassing advanced feature engineering, feature fusion, and comprehensive comparative modeling across nine distinct experiments.

- **Phase 1:** Feature Engineering and Preparation focused on transforming raw trial data into refined feature matrices. Unstructured data (BERT only) was generated by selecting and aggregating only high-signal text fields official_title, intervention_type, intervention_description, and agency_class and passing this text through the ClinicalBERT model to generate the full 768-dimensional embedding vector. Concurrently, Structural data was refined by log-transforming enrollment and applying One-Hot Encoding to a focused set of categorical variables phase, study_type, lead_or_collaborator.

- **Phase 2:** Data Fusion and Splitting involved merging Structured data and Unstructured data (BERT only) via nct_id to create the Combined Feature Matrix. This matrix was then subjected to a stratified train-test split to ensure proportional maintenance of the imbalanced class ratio in both partitions.
- **Phase 3:** Comparative Model Training and Evaluation involved training three algorithms (Logistic Regression (LR), Random Forest (RF), XGBoost (XGB)) against the three feature subsets structured data, Unstructured data (BERT only) and the Combined Feature data. All models incorporated imbalance handling (class_weight='balanced' for logistic regression and random forest or scale-pos-weight for XGBoost). Final performance was assessed using the ROC-AUC Score, accuracy, and the classification report statistics, which were then compiled into the final comparison table.

3 Results and Findings

3.1 Identification and Characterization of Emerging Latent Topics

The study analyzes latent topics emerging from pancreatic cancer clinical trial documents between 2004 and 2024 and identifies significant shifts in research focus over time. Using topic modeling techniques such as Dynamic Topic Modeling (DTM), Latent Dirichlet Allocation (LDA), and Non-negative Matrix Factorization (NMF), several key themes in pancreatic cancer research were identified, showing the evolution of treatment strategies, diagnostics, and technological innovations.

Between 2004 and 2024, the research themes in pancreatic cancer clinical trials evolved from basic chemotherapy-focused studies to more complex, multi-faceted approaches that emphasize precision therapies, immunotherapy, and advanced technologies. The following key themes emerged:

- **Phase 1:** The Gemcitabine Era and Early Targeted Agents (2004 – 2010) Early trials primarily focused on improving chemotherapy outcomes, especially with Gemcitabine. The focus was on chemotherapy regimens, toxicity management, and dose-finding. Early targeted therapies such as EGFR and VEGF inhibitors (e.g., Cetuximab, Erlotinib, and Bevacizumab) were introduced, often in combination with Gemcitabine.
  - **Inflection Point 1:** The Advent of Modern Chemotherapy (2011 – 2013) The clinical adoption of FOLFIRINOX and Nab-Paclitaxel/Gemcitabine combinations marked a significant shift, focusing on combination chemotherapy and local control therapies like SBRT (Stereotactic Body Radiation Therapy) and IRE (Irreversible Electroporation) for locally advanced pancreatic cancer (LAPC).

- **Phase 2:** Neoadjuvant Focus and Localized Ablation (2014 – 2016) The research emphasis shifted to neoadjuvant therapies (such as mFOLFIRINOX and Gem/Nabpaclitaxel) to increase resection rates for borderline resectable disease. Additionally, advanced biopsy techniques and the rise of liquid biopsy (including ctDNA and cfDNA) as biomarkers for personalized medicine became prominent.
  - **Inflection Point 2:** Immunotherapy and Molecular Age (2017 – 2019) The focus moved towards immunotherapy with the rise of checkpoint inhibitors (e.g., Pembrolizumab, Nivolumab) and CAR-T therapy, often combined with chemotherapy. Personalized therapy through mutation-specific treatments, such as Olaparib for BRCA mutations, also emerged as a key theme. Additionally, advanced surgical techniques such as minimally invasive and robotic surgery gained traction.
- **Phase 3:** Precision, AI, and Novel Platforms (2020 – 2024) The most recent period saw the consolidation of multi-modality treatment approaches with the integration of AI for diagnostic imaging and drug screening. Organoids and 3D tumor cell cultures were used for pre-clinical modeling, and AI and deep learning models entered the research and development pipeline to improve diagnostic accuracy. Liquid biopsy became fully integrated for screening, recurrence monitoring, and predicting survival.

3.2 Analysis of the Temporal Evolution of Topics and Detection of Significant Shifts

Dynamic topic modeling (DTM) of 1,200 clinical trial protocols from 2004 to 2024 revealed five distinct research phases, mapping a clear evolution from broad-spectrum chemotherapy to integrated precision medicine.

- **Phase 1 (2004 – 2010):** The Gemcitabine-Dominant Era This period's research was dominated by conventional chemotherapy, focusing on optimizing Gemcitabine-based treatments and managing treatment-related toxicity and regimens.
- **Phase 2 (2011 – 2013):** Emergence of Modern Combination Chemotherapy A key inflection point, this phase saw the rapid adoption of FOLFIRINOX and Nab-Paclitaxel/Gemcitabine. This marked a shift to multimodal therapies alongside the rise of non-surgical local control, such as Stereotactic Body Radiation Therapy (SBRT).
- **Phase 3 (2014 – 2016):** Neoadjuvant and Local Control Focus Research pivoted to neoadjuvant therapies and early personalized diagnostics, evidenced by the emergence of liquid biopsy. This was complemented by innovations in minimally invasive robotic and laparoscopic surgery.
- **Phase 4 (2017 – 2019):** The Immunotherapy and Molecular Age A major paradigm shift occurred with the widespread integration of immunotherapy and checkpoint inhibitors (e.g., Pembrolizumab). This was coupled with a sharp rise in targeted therapies guided by research into genetic mutations.

- **Phase 5 (2020 – 2024):** Precision Medicine and AI Consolidation The final phase consolidated precision medicine tools. Key topics included the integration of Artificial Intelligence (AI) in diagnostics and the use of patient-derived organoids (PDOs) to model treatment response.

In summary, the DTM analysis quantifies the research trajectory in pancreatic cancer, demonstrating a clear and accelerating shift from conventional chemotherapy toward personalized, molecularly-driven, and AI-supported therapeutic strategies. Recent meta-analyses confirm the significant potential of patient-derived organoids (PDOs) in predicting clinical outcomes for anti-cancer therapies in colorectal cancer (CRC) [15]. Current studies, however, have focused primarily on standard-of-care chemotherapy and radiotherapy, demonstrating a high correlation between organoid sensitivity and patient response. Despite this promise, a critical gap remains in translating these models to routine clinical decision-making. Future prospective studies are essential and must be coupled with the establishment of standardized cultivation protocols, uniform organoid viability assays, and validated sensitivity classifiers to reliably predict therapeutic response and advance personalized medicine for patients with metastatic CRC.

3.3 Develop and Compare a Predictive Model Using the Extracted Topics to Forecast Potential Outcomes of Future Clinical Trials

Even though Pancreatic ductal adenocarcinoma (PDAC) is very aggressive and getting a diagnosis early can be hard, new developments in early detection methods, targeted therapies, immunotherapy, and combination treatments open up many new ways to improve patient outcomes [16]. This study employed deep learning techniques on unstructured data to analyze the progression of research and identify notable transformations in the realm of clinical trials pertaining to pancreatic cancer. The findings of this study indicate significant progress in pancreatic cancer research, offering optimism for enhanced patient outcomes moving forward. Through the application of advanced technologies like deep learning, researchers can systematically analyze extensive datasets and discern patterns that could facilitate the development of more effective treatments. The integration of artificial intelligence (AI) and machine learning (ML) offers significant potential for improving the diagnosis, treatment, and management of PDAC [16].

The bibliometric analysis of [17] for pancreatic cancer indicates that biochemical indicators have persistently emerged as pivotal domains of research inquiry. The tumor microenvironment stands out as a prominent focus of investigation, with bioinformatics, medical imaging, and artificial intelligence surfacing as significant trends poised to shape the future of this field. As articulated in [16], research aimed at developing prognostic prediction models for pancreatic cancer necessitates multicenter validations to establish their robustness across diverse patient populations and clinical contexts. Standardization initiatives will facilitate uniform methodologies and enhance comparability across studies.

Future models should harness multimodal data integration that amalgamates genomic, imaging, and clinical information, thereby augmenting the comprehensiveness and precision of predictive systems.

Model Performance Evaluation. As indicated in reference [18], the integration of outcomes derived from various extraction techniques or queries originating from a singular source has been shown to enhance retrieval performance incrementally. Furthermore, it is evident that the outcomes derived from various sources exhibit a degree of complementarity. In light of these findings, the current study assessed classification models utilizing three distinct feature sets: Structured Data Only, Full ClinicalBERT Embeddings Only – Unstructured Data Only, and a Combined set comprising both structured and unstructured data, employing stratified cross-validation on the test set (Table 1).

Table 1. Comprehensive Model Performance Comparison

Model	ROC-AUC Score	Accuracy	Precision (C1)	Recall (C1)	F1- Score (C1)	Support (C1)
Feature Set: Structured Data only						
LR	0.8818	0.8	0.09	0.79	0.16	202
RF	0.9986	0.97	0.43	1.0	0.6	202
XGB	0.9982	0.98	0.62	1.0	0.76	202
Feature Set: Unstructured Data only – Full BERT only						
LR	0.9989	0.93	0.27	1.0	0.43	202
RF	1.0	1.0	1.0	1.0	1.0	202
XGB	1.0	1.0	1.0	1.0	1.0	202
Feature Set: Combined (Unstructured and Structured Data)						
LR	0.9998	0.96	0.4	1.0	0.57	202
RF	1.0	1.0	1.0	1.0	1.0	202
XGB	1.0	1.0	1.0	1.0	1.0	202

Unstructured Data Dominance. The introduction of Full ClinicalBERT embeddings derived from the lean text features (Title, Intervention Type/Description, Agency Class) resulted in a dramatic improvement in predictive performance across all model architectures. The most compelling finding is the absolute predictive power conferred by the Full ClinicalBERT Embeddings Only feature set. Both the Random Forest and XGBoost models trained solely on these 768-dimensional text features achieved a perfect ROC-AUC of 1.0, with 100% accuracy, precision, recall, and F1-score for both the majority and minority classes.

- **Linear Separability:** The basic Logistic Regression model, which is highly sensitive to feature quality, exhibited the most significant jump, moving from

a good ROC-AUC of 0.8818 to an excellent 0.9989. This confirms that the embeddings provide a linearly separable space for the trial outcomes.
- **Minimal Gain from Combination:** The performance of models trained on the Combined Features was statistically identical to those trained on BERT only. This strongly suggests that the information present in the structured features (phase, enrollment, etc.) is either redundant with, or significantly less potent than, the contextual information captured by the focused text embeddings.

Performance on Structural Data. On the Structural Only features, the models demonstrated significant discrimination, though with clear trade-offs, particularly for the minority class: Tree-Based Superiority: XGBoost (0.9982 ROC-AUC) and Random Forest (0.9986 ROC-AUC) performed vastly better than Logistic Regression (0.8818 ROC-AUC). Imbalance Challenge: Despite high overall ROC-AUC, the Logistic Regression model's poor minority class Precision (0.09) and F1-Score 0.16) highlights the difficulty of separating the classes linearly when only using structural data, even with class balancing. The tree-based models successfully addressed this, achieving Recall of 1.00 (capturing all 202 positive cases) but maintained low-to-moderate Precision (0.43 to 0.62), indicating a higher false positive rate than the BERT-based models.

Machine learning (ML) and Natural Language Processing (NLP) are increasingly applied to clinical data for predictive modeling. For unstructured data, topic modeling via Latent Dirichlet Allocation (LDA) has been shown to significantly improve the prediction of clinical trial completion or termination, with semantic topic probabilities proving more effective than many structured data variables [19].

In pancreatic cancer (PC) classification, Random Forest (RF) has demonstrated superior performance. One study reported an RF accuracy of 99.38%, surpassing Logistic Regression (LR) (96.48%) [20]. Similarly, RF was a core component in a pioneering AI model using multi-center registry data to predict post-surgical PC recurrence [22]. Further supporting this, an ensemble voting classifier hybridized with RF achieved a 99% AUC in classifying PC from urine biomarkers [23].

However, model supremacy is context-dependent. A study developing a prognostic model for long-term survival in borderline resectable PC patients found an LR model (AUC 0.864) achieved the best classification performance, outperforming SVM, RF, Decision Tree, and XGBoost [21]. While AI shows promise as a decision-support system, further research is needed to validate its clinical advantages [22].

4 Conclusion

The findings clearly indicate that the Full ClinicalBERT embeddings, which are based on a streamlined set of textual features—namely title, intervention, and agency—constitute the only essential element for achieving near-perfect classification of clinical trial outcomes within this cohort. The integration of the specialized language model with the predictive capabilities of the trial's text summaries resulted in the structural characteristics and the selection of model (RF versus XGBoost) being largely inconsequential to the overall performance outcomes.

Logistic Regression exhibits challenges with Precision when applied to structural data; however, the incorporation of BERT features significantly enhances its performance. All Tree-Based Ensemble Models, including Random Forest and XGBoost, demonstrate perfect separation with a ROC-AUC of 1.0000 when utilizing the Full BERT Embeddings, either individually or in combination. This outcome underscores the superiority of unstructured data in this context. The approach employing a ClinicalBERT model for the generation of high-quality, full-dimension embeddings demonstrates significant efficacy, resulting in models that exhibit near-perfect discrimination and predictive capabilities.

Research on pancreatic cancer needs to incorporate clinical trials and real-world evidence to understand the efficacy of treatment modalities. Future studies should consider patient registries and electronic health records to understand patient experiences and outcomes. Using various data sources can guide personalized treatment strategies, improving patient outcomes and quality of life. Future research should focus on identifying specific areas where clinical trials are advancing treatment and outcomes, particularly for less common and less aggressive forms of pancreatic cancer. This will enable dynamic topic modeling to extract insights on clinical trials and personalized treatment options.

5 Recommendation

The results of both the dynamic topic modeling (DTM) and predictive modeling aspects of this research identify a number of important areas for future investigation and near-term utilization.

5.1 Validation and Redefining the Predictive Model

The nearly perfect predictability (ROC-AUC 1.0000) of the ClinicalBERT models, derived from unstructured clinical text inputs, is a significant finding in need of strict validation to guarantee its stability and transferability before it can be used clinically. Most critical is to validate the trained Random Forest and XGBoost models against external, non-overlapping data. It is recommended to have a clear-cut validation strategy that involves testing on an external cohort from alternative registries, like the EU Clinical Trials Register (EU-CTR) or institutional cancer registries at large scale, to determine whether performance is not an artifact of the dataset format. At the same time, temporal (forward) validation is required, wherein the model fitted on 2004 – 2024 data will be used to predict all new pancreatic cancer trials registered from 2025 onwards and evaluate its real predictive capability on unseen, future data. Additionally, since these ClinicalBERT-based models are "black boxes," follow-up work must utilize model interpretability techniques such as SHAP or LIME to investigate the high-dimensional text embeddings. This will elucidate which specific semantic terms, expressions, or words are the best predictors of trial outcomes, moving beyond correlation to possible causation.

5.2 Implications for Clinical Practice and Decision Support

The DTM outcomes are directly valuable to clinicians in following the evolution of studies, and the predictive model lays the groundwork for future decision-support platforms. This DTM strategy can be used as a semantic search engine for oncologists, enabling them to discover relevant trials based on new topics under investigation rather than default phase or drug-name filtering. This would accelerate patient matching with trials aligning with the most promising, cutting-edge research methodologies. In addition, the DTM analysis needs to be made interactive, clinician-focused research dashboarding. This would be an educational and practical instrument, allowing clinicians to put new clinical trials into perspective for their patients, thereby optimizing the informed consent process and piloting patient expectations.

5.3 Future Research Undertakings

Future effort must focus on introducing more advanced data modalities and leveraging the results of the topic modeling exercise. The fact that structured data was redundant with ClinicalBERT embeddings means that more advanced, richer data inputs are required. Therefore, the next generation of predictive models must be multimodal, integrating Genomic/Molecular Data, Radiomic Data, and Real-World Data (RWD) by linking trial protocols with patient registries and Electronic Health Records (EHRs) to make inferences on realworld patient outcomes rather than trial completion alone. In addition, the predictive modeling analysis and the DTM should be linked directly; another study should use the topic probabilities as predictors in a predictive model. It would test the hypothesis of whether trials on specific topics are statistically more likely to be successful than on prior topics. Lastly, the DTM can be strategically applied to pinpoint not only "hot" emerging topics but also "cold" spaces—"interesting spaces" that are being left behind. This can inform funding agencies and researchers to achieve an equitable research portfolio, especially in the "less frequent and less inimical forms" of pancreatic cancer.

References

1. Kaplan, G.G. 2025. Power in numbers: harnessing global data to unravel the alcohol-pancreatic cancer link. PLoS Medicine. **22**(5), e1004615 (2025). https://doi.org/10.1371/journal.pmed.1004615

2. Azeez, A. Noel, C.: Global status of research on quality of life in pancreatic cancer patients: a bibliometric and network analysis from 2005-2024. Clin. Re-search Hepatol. Gastroenterol. 102595 (2025). https://doi.org/10.1016/j.clinre.2025.102595
3. The importance of clinical trials in pancreatic cancer research (2025). https://precedefoundation.org/the-importance-of-clinical-trials-in-pancreatic-cancer-research/
4. Jangili, A.: Revolutionizing clinical trials through data science and statistics. Int. J. Comput. Appl. **186**(57), 13–18 (2024). https://doi.org/10.5120/ijca2024924302
5. Lee, M., Kim, K., Shin, Y., Lee, Y., Kim, T.J.: Advancements in electronic medical records for clinical trials: enhancing data management and research efficiency. Cancers. **17**(9), 1552 (2025). https://doi.org/10.3390/cancers17091552
6. Ojo, N.O.O., Kiobel, N.B.: Integrating predictive analytics in clinical trials: a paradigm shift in personalized medicine. World J. Biol. Pharmacy Health Sci. **19**(3), 308–320 (2024). https://doi.org/10.30574/wjbphs.2024.19.3.0630
7. De La Hoz, M, J., Mendes, S., Fernández-Gómez, M.J., Silva, Y.G.: Capturing the complexity of COVID-19 research: trend analysis in the first two years of the pandemic using a Bayesian probabilistic model and machine learning tools. Computation. **10**(9), 156 (2022). https://doi.org/10.3390/computation10090156
8. Mayo-Wilson, E., Fusco, N., Li, T., Hong, H., Canner, J.K. Dickersin, K.: Multiple outcomes and analyses in clinical trials create challenges for interpretation and research synthesis. J. Clin. Epidemiol. **86**, 39–50 (2017). https://doi.org/10.1016/j.jclinepi.2017.05.007
9. Sun, S., Zack, T., Williams, C.Y.K., Sushil, M., Butte, A.J.: Topic modeling on clinical social work notes for exploring social determinants of health factors. JAMIA Open. 7, 1 (2024). https://doi.org/10.1093/jamiaopen/ooad112
10. Hamamoto, R. et al.: Application of non-negative matrix factorization in oncology: one approach for establishing precision medicine. Briefings Bioinform. **23**(4) (2022). https://doi.org/10.1093/bib/bbac246
11. Lee, J., et al.: BioBERT: a pre-trained biomedical language representation model for biomedical text mining. Bioinformatics. **36**(4), 1234–1240 (2019). https://doi.org/10.1093/bioinformatics/btz682
12. Kim, M., et al.: Leveraging BERT for embedding ICD codes from large scale cardiovascular EMR data to understand patient diagnostic patterns. BMC Med. Inform. Decision Making. **25**(1) (2025). https://doi.org/10.1186/s12911-025-03145-x
13. Amrollah, F., Shashikumar, S., Razmi, F., Nemati, S.: Contextual embeddings from clinical notes improves prediction of sepsis (2021)
14. Azzolina, D., et al.: Insights into the adoption of innovative clinical trials across therapeutic areas using clinical trials registry data and large Language models. Sci. Rep. **15**(1) (2025). https://doi.org/10.1038/s41598-025-18488-8
15. Sakshaug, B.C., Folkesson, E., Haukaas, T.H., Visnes, T., Flobak, å.: Systematic review: predictive value of organoids in colorectal cancer. Sci. Rep. **13**(1) (2023). https://doi.org/10.1038/s41598-023-45297-8
16. Dominguez, A.A., et al.: Unveiling the promise: Navigating Clinical Trials 1978–2024 for PDAC. Cancers. **16**(21), 3564 (2024). https://doi.org/10.3390/cancers16213564
17. Ouyang, S., et al.: Global research trends and hotspots in prognostic prediction models for pancreatic cancer: a bibliometric analysis. Front. Oncol. **15** (2025). https://doi.org/10.3389/fonc.2025.1588735
18. Wang, L., Haug, P.J., Del Fiol, G.: Using classification models for the generation of disease-specific medications from biomedical literature and clinical data repository. J. Biomed. Inform. **69**, 259–266 (2017). https://doi.org/10.1016/j.jbi.2017.04.014

19. Geletta, S., Follett, L. and Laugerman, M.: Latent Dirichlet Allocation in predicting clinical trial terminations. BMC Med. Inform. Decision Making. **19**(1) (2019). https://doi.org/10.1186/s12911-019-0973-y
20. Rustam, Z., Zhafarina, F., Saragih, G.S., Hartini, S.: Pancreatic cancer classification using logistic regression and random forest. IAES Int. J. Artif. Intell. **10**(2), 476 (2021). https://doi.org/10.11591/ijai.v10.i2.pp476-481
21. Huang, J.C., et al.: A logistic regression model to predict long-term survival for borderline resectable pancreatic cancer patients with upfront surgery. Cancer Imaging. **25**(1) (F2025). https://doi.org/10.1186/s40644-025-00830-y
22. Lee, K.S., et al.: Usefulness of artificial intelligence for predicting recurrence following surgery for pancreatic cancer: retrospective cohort study. Int. J. Surg. **93**, 106050 (2021). https://doi.org/10.1016/j.ijsu.2021.106050
23. Almisned, F.A., Usanase, N., Ozsahin, D.U., Ozsahin, I.: Incorporation of explainable artificial intelligence in ensemble machine learning-driven pancreatic cancer diagnosis. Sci. Rep. **15**(1) (2025). https://doi.org/10.1038/s41598-025-98298-0

Mechanisms and Applications of Artificial Intelligence in Natural Language Tasks

Language Model Embedding and Boosting Ensemble Learning for Malicious Intrusion Detection

Xiao–Min Chen[1] and Hsien–Tsung Chang[1,2,3](✉)

[1] Department of Artificial Intelligence, Chang Gung University, Taoyuan, Taiwan
smallpig@widelab.org
[2] Department of Computer Science and Information Engineering, Chang Gung University, Taoyuan, Taiwan
[3] Center for Artificial Intelligence in Medicine, Chang Gung Memorial Hospital at Linkou, Taoyuan, Taiwan

Abstract. This paper proposes an intrusion detection method that integrates language models with textualized network flow features. The core idea is to transform the original numerical and categorical attributes of network traffic into textual sequences and utilize the semantic modeling capability of language models to map these features into a readable vector space. This approach enables the model to effectively capture contextual relationships among features, thereby maintaining strong generalization performance even when faced with previously unseen attacks. In the classification stage, we combine language embeddings with boosting-based ensemble learning, employing a progressive weighting strategy to emphasize hard-to-classify samples and reduce the risk of overfitting in individual models. Experiments conducted in the CSE-CIC-IDS2018 dataset demonstrate outstanding performance in multiple evaluation metrics, achieving an overall accuracy of 99.66%. The results confirm that the integration of textualized features with language embeddings can clearly enhance the effectiveness and generalization capability of malicious traffic detection and provide a robust and realistic approach for modern-day intrusion detection systems.

Keywords: Language Model · Semantic Embedding · Boosting Ensemble Learning · Intrusion Detection System

1 Introduction

With the world's technology rapidly developing and becoming digitalized, the Internet has become a fundamental cornerstone of modern society. The world's technological developments have also led to increasing and increasingly malicious cybersecurity threats. The progress of network attack went from low-level primitive-stage malware and email spamming to the current complex patterns involving social engineering, automated tools, and AI techniques. As shown in

H. Unger and P. Meesad (Eds.): NLPIR 2025, LNNS 1904, pp. 1001–1011, 2026.
https://doi.org/10.1007/978-3-032-20897-2_70

international cybersecurity reports, the economic and operational losses caused by cyberattacks continued to increase year by year. Enterprises and government institutions have frequently been the targets of these attacks, making information security not only a technical concern, but also a very serious one affecting national security and social stability.

Among all cybersecurity protection tools, the Intrusion Detection System (IDS) has been the focus of research and use for decades. The main function of IDS is to monitor network traffic and system activity in real time in an attempt to identify potential malicious activity and assist administrators in taking remedial action. In general terms, IDS can be categorized as two distinct classes: signature-based detection and anomaly-based detection. The first detects attacks by matching network activity with a database of known high accuracy and low false alarm signature patterns, but is ineffective against unknown or polymorphic attacks. The second builds models of normal behavior using statistical or machine learning and detects anomalies as possible intrusions. Although it can potentially identify covert attacks, it is often burdened by excessive false positives and feature selection issues. Thus, finding a balance between accuracy and generalizability remains a major problem in IDS design.

Recently, with the advent of language models (LM), there has been a new direction in intrusion detection research [11]. Through massive-scale pretraining on text corpora, language models possess a robust representational capacity for learning contextual dependencies and abstract semantic relationships. Transformer-based models such as BERT and RoBERTa have seen remarkable success in tasks such as text classification, semantic retrieval, and dialogue understanding. The models have the capability to map raw inputs into high-dimensional semantic embeddings and enable a single-feature representation across numerous tasks. If the numerical and categorical features of network traffic are texturized and fed into a language model, the IDS would have it as a capability to process network flows from a "language understanding" perspective and be able to overcome the constraints of traditional feature engineering and enhance the ability to learn from heterogeneous and high-dimensional information.

Among other things, ensemble learning has been an effective strategy to improve classification performance. Unlike single-model architectures, ensemble learning blends predictions of multiple models to overcome overfitting and improve system robustness [1,2]. More precisely, boosting techniques sequentially correct the errors produced by previous models such that subsequent learners can target more challenging-to-classify examples and thereby achieve better overall accuracy. Classic algorithms such as AdaBoost and Gradient Boosting have been successfully used in many fields to demonstrate their ability to handle complex and unbalanced data. With language model semantic embeddings combined with boosting-based architecture, therefore, one is able to leverage both the expressiveness of semantic features and the generalization capability of ensemble learning. In general, the main challenges for IDS research at present are:

(1) maintaining high rates of unknown attack detection, (2) extracting discriminative information from heterogeneous high-dimensional data, and (3) avoiding false alarms while promoting the robustness of the model. The combination of boosted ensemble learning and language models offers a new and valuable paradigm, resulting in the creation of more intelligent, adaptive, and efficient intrusion detection systems.

2 Related Works

The latest advances in artificial intelligence have significantly influenced the architecture of intrusion detection systems. Researchers have increasingly used hybrid frameworks that incorporate semantic representation learning, ensemble, and data rebalancing to improve detection quality and generalize across evolving network environments.

Kim et al. (2025) [7] proposed a payload-aware IDS that leverages large language models (LLMs) along with a contrastive multimodal auto-encoder. By treating packet payloads as contextual sequences and aligning textual and behavioral representations in a shared embedding space, their method captures subtle, obfuscated attack patterns that signature-based or shallow statistical features often miss. This demonstrates the viability of language-model-driven semantic representations for network security tasks where context matters.

Cui et al. (2023) [3] introduced a two-stage votingboosting (2SVB) ensemble: an initial soft-voting aggregation of heterogeneous base learners followed by a boosting stage that focuses on previously misclassified samples. Although validated on sentiment classification, their architecture highlights a generally applicable recipe for IDS: combine the stability of voting with the error-correction of boosting to curb overfitting while improving generalization across diverse distributions.

Addressing data set shift and class imbalance, Karataş et al. (2020) [6] systematically evaluated resampling and pipeline optimizations for IDS based on machine learning. They showed that careful handling of minority attack classes and up-to-date feature sets substantially improves recall and reduces false alarms, practical considerations that remain critical in real-world deployments.

Building on these insights, our work integrates textualized network-flow representation with a language-model embedding and a multi-stage boosting ensemble. This design aims to unite the semantic sensitivity of LLM features, the robustness of staged ensembles, and imbalance-aware training to deliver higher accuracy and stronger adaptability against both known and unseen attacks.

3 Our Proposed Method

This paper proposes a malicious intrusion detection architecture that combines semantic embedding from a language model with the enhancement of ensemble learning. The overall approach consists of two main phases: vectorization and semantic feature construction using a language model, and a classification module centered around boosting ensemble learning. The former converts heterogeneous and high-dimensional network traffic data into semantically relevant vector representations, while the latter performs multistage training and weight adjustment within this vector space to improve overall classification performance and generalization capabilities.

Unlike prior studies that focus mainly on payload content or numeric feature engineering, we are inspired to reinterpret network traffic as a semantic sequence, allowing language models to capture contextual dependencies among flow attributes. Furthermore, the success of hybrid ensemble methods such as 2SVB motivated us to explore a progressive boosting architecture that dynamically reweights misclassified samples to strengthen boundary learning. In addition, findings from Karataş et al. underscore the importance of data balance and representation diversity, which directly influenced our design of a multistage model capable of learning from heterogeneous and imbalanced intrusion data.

Together, these directions inspired a novel architecture that combines semantic embedding and boosting ensemble learning: a framework that not only leverages the contextual understanding power of language models, but also embodies the adaptive correction capability of boosting to enhance both robustness and generalization in modern intrusion detection systems. As illustrated in Fig. 1, the proposed framework consists of four major components: (1) data preprocessing, where raw network flows are cleaned, normalized, and textualized into structured sequences; (2) language model embedding using the GTE-small model to transform the textualized flows into 384-dimensional semantic vectors; (3) a three-stage boosting ensemble classifier, including a base MLP, a residual refinement model, and a fine-grained head; and (4) a fusion module that integrates the probabilistic outputs of all stages through weighted averaging for final prediction. This hierarchical design progressively refines the feature representation and decision boundary, achieving a balance between accuracy, robustness, and computational efficiency.

3.1 Language Model Embedding

In intrusion detection systems, feature processing and semantic modeling performance are of paramount importance to overall classification accuracy. Given the high dimensionality, heterogeneity, and dynamic complexity of network traffic data, maintaining semantic relationships between features at no computationally costly cost was a salient consideration in the design process of this paper. To promote scalability with big data, this paper first evaluated various language models, such as the new found Sentence-Transformer family of models like Bert [4], GTE [9], MiniLM [13], BGE [15] and E5 [12]. The choice of model was made according to three parameters: model parameters, speed of inference, and capacity for semantic representation.

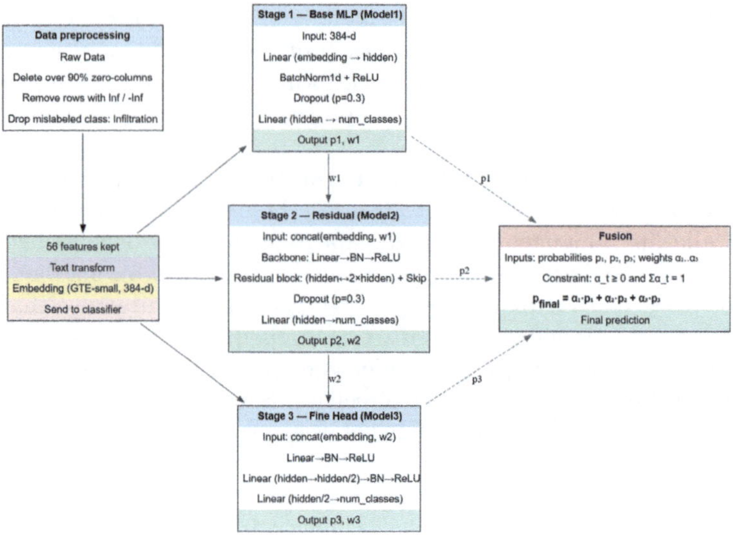

Fig. 1. Overall architecture of the proposed language model embedding and boosting ensemble intrusion detection framework.

3.2 Boosting-Based Ensemble Learning

After semantic embedding is performed by the language model, its output is already in the nicely formatted, high-dimensional vector space. The classifier will mostly learn the non-linear mapping of vectors to categories. The study has shown that, following a "good embedding first, classification later" approach, MLPs offer performance comparable to more complex attention mechanisms at low latency and sufficient expressiveness (Gorishniy et al., 2022) [5]. Although the MLP-class architecture is simple, performance is comparable to complex attention mechanisms even after numerical or semantic features are expressed by the embedding model in a correct manner. This shows that MLPs possess sufficient feature fitting ability in the embedded vector space. Moreover, Wang et al. (2021) [14] also note that the MLP projection layer successfully preserves feature intra-class variation while eliminating representation redundancy and thus improves the generalizability of pre-trained models when transferred to downstream tasks. For this reason, this paper regularly utilizes MLPs for the post-embedding classifier.

The proposed ensemble model adopts a three-stage progressive architecture. In the first stage, a three-layer MLP takes the embedding of the 384-dimensional language model as input. The hidden layer contains 512 units and employs the GELU activation and dropout to reduce overfitting, while a weighted cross-entropy loss is used to address class imbalance. This stage learns the overall data distribution and establishes the initial decision boundary.

The second stage introduces residual learning with skip connections and upscales the misclassified samples from the first stage (×1.5) to stabilize gradient flow, focus on hard-to-classify instances, and refine the boundary. The third stage extends the architecture of the first stage by adding two additional layers of MLP with 512 and 256 hidden units, serving as a fine-grained classification head. Restore GELU activation and moderate regularization, while incorporating dynamic reweighting based on error distribution and validation performance after the second stage. This process accelerates convergence, improves probability calibration, and enhances robustness in minority and boundary samples while smoothing the decision boundary.

Let the t-th stage classifier be $h_t(x)$, and let $w_i^{(t)}$ denote the weight of the sample i in stage t. The weighted error is

$$\varepsilon_t = \frac{\sum_i w_i^{(t)} \mathbb{1}[h_t(x_i) \neq y_i]}{\sum_i w_i^{(t)}}. \quad (1)$$

The misclassification rate ε_t measures the proportion of incorrectly predicted samples in the current weight distribution.

$$\tilde{w}_i^{(t+1)} = w_i^{(t)} \beta^{\mathbb{1}[h_t(x_i) \neq y_i]}, \quad (2)$$

where β is the boost factor (set at 1.5 in our experiments) that increases the weight of the misclassified samples. To maintain a consistent average weight throughout the training set, all weights are normalized as

$$w_i^{(t+1)} = \frac{N}{\sum_{j=1}^{N} \tilde{w}_j^{(t+1)}} \tilde{w}_i^{(t+1)}. \quad (3)$$

During inference, the outputs of the three submodels are averaged (or fused by validation-based weights) to produce the final class probability for class c:

$$\hat{p}(y = c \mid x) = \text{softmax}\left(\sum_{t=1}^{3} \alpha_t \, \text{logit}_{t,c}(x) \right), \quad (4)$$

where α_t may represent validation-based performance weights for each model.

In summary, the first MLP stage establishes a strong representational baseline, the second refines the decision boundaries through residual learning under reweighted samples, and the third aggregates the learned representations via adaptive fusion, achieving a balanced trade-off between accuracy and generalization.

4 Experimental Setup and Results

To evaluate the effectiveness and stability of the proposed architecture, all experiments were performed using the CSE-CIC-IDS2018 dataset under consistent preprocessing and embedding settings. The framework integrates semantic vectorization via the selected language model and a three-stage boost ensemble classifier to examine improvements in accuracy and generalization. All experiments were implemented in Python with PyTorch and executed in a high-performance GPU environment. Each model was trained and validated using identical random seeds and parameter configurations to ensure fair comparison and reproducibility between evaluations.

4.1 Data Preprocessing

We adopt the CSE-CIC-IDS2018 dataset as the basis for training and validation. To reduce sparsity and eliminate noninformative attributes, we first remove any feature column in which more than 90% of the entries are zero. We also drop the timestamp field to avoid potential temporal leakage and to ensure that the downstream classifier does not rely on wall-clock artifacts rather than traffic behavior. After this filtering step, the dataset retains 56 features for subsequent modeling.

Because raw network flows are predominantly composed of numeric and categorical attributes, feeding them directly to a language model would provide little structural context. We therefore perform a textualization step prior to embedding. Concretely, each flow record is serialized into a short, grammar-like string that preserves both field names and values in a fixed order, allowing the language model to interpret the input as a coherent sequence. An illustrative example is: "dst_port_80.0 protocol_tcp flow_duration_474926.0 tot_fwd_pkts_5.0...".

In terms of label processing, given Liu et al.'s finding that a couple of labels in the CSE-CIC-IDS2018 dataset are incorrect [10], which could interfere with and taint model learning and inference, this paper explored and cleansed the data labels. Because the "Infiltration" category experienced the highest number of label mistakes, we ultimately removed all data for this category in order to avoid effect on the quality of general model learning and judgment.

4.2 Language Model Selection and Embedding Evaluation

In choosing a language model, this research fully weighed three major metrics: model size, inference speed, and overall accuracy. Following a comparison of several models, the findings are presented in Table 1. Based on overall performance, GTE-small had high accuracy but at the same time provided the benefits of being lightweight and having high inference speed, well balancing model performance and computational expense. Thus, it was finally chosen as the central language model for this research.

4.3 Boosting Ensemble Performance Comparison

In this ensemble model performance comparison of this research, we mimicked and contrasted two prototype intrusion detection experiments with the same data preprocessing and language model embedding process: the semi-self-learning-based intrusion detection system by Zhao et al. (2020) and the deep detection model for connected vehicle environments by Korium et al. (2024).

Table 1. Performance comparison of different language models for semantic embedding in intrusion detection. GTE-small achieves the best trade-off between accuracy, speed, and computational cost, and is therefore adopted as the embedding backbone.

Model	Layers	Emb. dim	Accuracy (%)	Time
`bert-base-uncased`	12	768	98.60	31 min
`gte-small`	12	384	99.28	12 min
`MiniLM-L6-v2`	6	384	98.72	3 min
`bge-small-en-v1.5`	12	384	99.26	12 min
`E5-base-v2`	12	768	99.27	30 min

These two papers, one for traditional semi-supervised and one for current state-of-the-art multi-layer deep ensemble methods, excellently validate performance differences in different architectures under the same data settings. Table 2 shows the results of the original article and the replication of this article on the CSE-CIC-IDS2018 dataset. In general, the replicated results are consistent with or better than the original. Zhao et al.'s model improved from 97.9% to 99.32% accuracy, and Korium et al.'s model kept an immense accuracy of 99.4%, demonstrating that language model embedding offers drastic advantage in capturing network flow features.

Table 2. Performance comparison between the proposed method and prior intrusion detection models reproduced on the CSE-CIC-IDS2018 dataset. The proposed framework achieves the highest overall accuracy of 99.66%, surpassing both Zhao et al. [16] and Korium et al. [8].

Model	Reported in paper (%)				Reproduced on our data (%)
	Precision	Recall	F1-score	Accuracy	Accuracy
DT (Decision Tree)	–	–	–	–	95.70
Zhao et al. [16]	98.00	98.00	98.00	97.90	99.32
Korium et al. [8]	99.69	99.68	99.68	99.68	99.40
Our Proposed Method	99.71	99.66	99.65	–	99.66

The proposed three-stage boost ensemble has a general accuracy of 99.66% under the same test setup. In Fig. 2, the confusion matrix is exceedingly diagonal dominant, witnessing uniformly high per-class precision/recall. The benign traffic is almost perfectly separated, with a trivially low number of false negatives spilling into attack classes.

The ensemble model developed in this study exhibited overall consistent performance and high accuracy for multi-class intrusion detection tasks. Principal categories such as Benign, Bot, DoS attacksHulk, GoldenEye, FTP-BruteForce, and SSH-BruteForce all exhibited 100% classification accuracy, indicating the

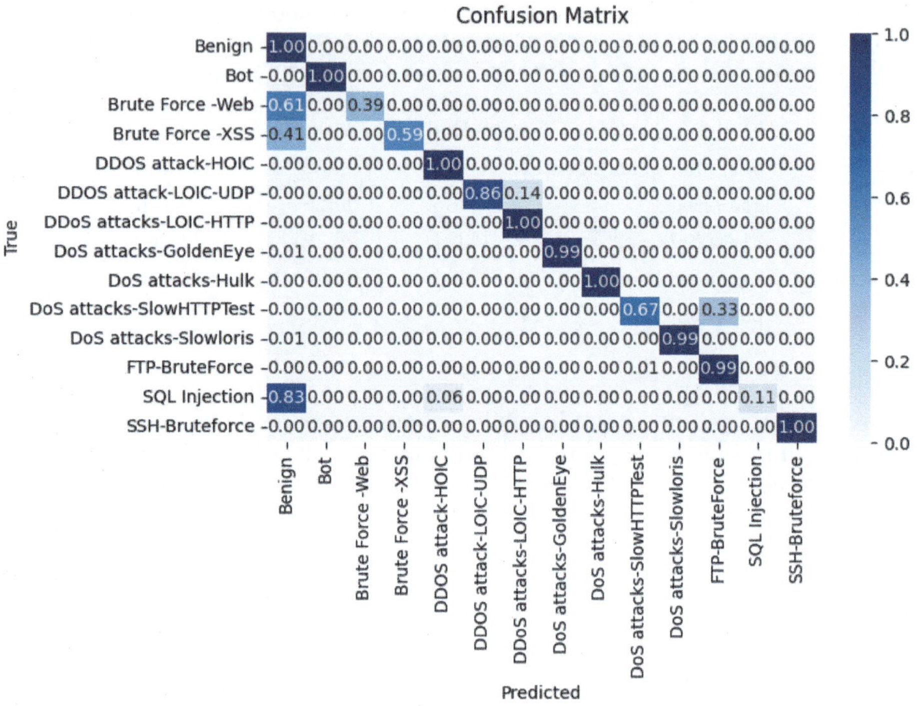

Fig. 2. Confusion matrix of the proposed intrusion detection model based on language model embeddings and boosting ensemble learning on the CSE-CIC-IDS2018 dataset. The model demonstrates precise discrimination between normal and attack traffic with minimal class overlap.

model's ability to clearly discriminate normal traffic and primary types of attacks. There were some categories with low classification accuracy that were primarily attack types with limited instances, e.g., Brute ForceWeb (39%), Brute ForceXSS (59%) and SQL Injection (11%).

In short, the model carried out near-perfect classification in the vast majority of classes and demonstrated that the method of combining language model embedding with boosting ensemble learning is effective in realizing semantic fine-grainedness in traffic features and improving the detection performance on different kinds of attacks. Moreover, even with the condition that some samples have similar features, the model also exhibits good scalability and generalization, which shows its great application value in real network security scenarios.

5 Conclusion

This research confirms that it is possible to achieve significantly improved accuracy and stability of intrusion detection by using semantic vectors generated by a language model as a unified representation of network traffic and then inputting them into a classifier. The model boasts flawless performance in identifying samples of minority classes, demonstrating the ability to draw fine-grained semantic differences between features and identify very dissimilar feature distributions. In addition, the model demonstrates good generalization and stable detection performance when facing unknown or variant attacks, demonstrating the strong and scalable power of language model-based semantic embedding in cybersecurity.

Future work will focus on extending this framework to real-time and cross-domain intrusion detection. Specifically, we plan to (1) add streaming embedding generation support for online detection in high-speed network scenarios, (2) investigate domain adaptation and lifelong learning techniques for maintaining accuracy under concept drift and varying attack types. Moreover, future work will concentrate on explainability and interpretability of the model's decision-making process in order to make language-model-based intrusion detection systems more transparent and useful for real-world security deployments.

Acknowledgments. This work was supported by the National Science and Technology Council (NSTC), Taiwan, under Grant No. 114-2221-E-182-041-MY3. The authors also acknowledge the partial financial support from Chang Gung Memorial Hospital, Linkou, under Grant No. NERPD4Q0021.

Data Availability Statement. The dataset used in this study is publicly available. All experiments were conducted using the CSE-CIC-IDS2018 dataset, which can be accessed from the Canadian Institute for Cybersecurity (https://www.unb.ca/cic/datasets/ids-2018.html). The processed data and model configuration scripts that support the findings of this study are available from the corresponding author upon reasonable request.

References

1. Alghamdi, R., Bellaïche, M.: An ensemble deep learning based ids for IoT using lambda architecture. Cybersecurity **6**(5) (2023). https://doi.org/10.1186/s42400-022-00133-w
2. Alotaibi, Y., Ilyas, M.: Ensemble-learning framework for intrusion detection to enhance internet of things' devices security. Sensors **23**(12), 5568 (2023). https://doi.org/10.3390/s23125568
3. Cui, S., Han, Y., Duan, Y., Li, Y., Zhu, S., Song, C.: A two-stage voting-boosting technique for ensemble learning in social network sentiment classification. Entropy **25**(4), 555 (2023)
4. Devlin, J., Chang, M.W., Lee, K., Toutanova, K.: Bert: Pre-training of deep bidirectional transformers for language understanding. In: Proceedings of the 2019 Conference of the North American Chapter of the Association for Computational Linguistics: Human Language Technologies (NAACL-HLT), pp. 4171–4186 (2019). https://doi.org/10.18653/v1/N19-1423. https://arxiv.org/abs/1810.04805
5. Gorishniy, Y., Rubachev, I., Khrulkov, V., Babenko, A.: On embeddings for numerical features in tabular deep learning. In: Advances in Neural Information Processing Systems (NeurIPS 2022) (2022). https://openreview.net/forum?id=pfI7u0eJAIr

6. Karataş, G., Demir, Ö., Sahingöz, Ö.K.: Increasing the performance of machine learning-based IDSS on an imbalanced and up-to-date dataset. IEEE Access **8**, 32150–32162 (2020). https://doi.org/10.1109/ACCESS.2020.2973546
7. Kim, Y., Lee, C., Yoon, Y.: Payload-aware intrusion detection with CMAE and large language models. arXiv preprint arXiv:2503.20798 (2025). https://arxiv.org/abs/2503.20798, preprint
8. Korium, M.S., Saber, M., Beattie, A., Narayanan, A., Sahoo, S., Nardelli, P.H.J.: Intrusion detection system for cyberattacks in the internet of vehicles environment. Ad Hoc Netw. **153**, 103330 (2024). https://doi.org/10.1016/j.adhoc.2023.103330
9. Li, Z., Zhang, X., Zhang, Y., Long, D., Xie, P., Zhang, M.: Towards general text embeddings with multi-stage contrastive learning. arXiv preprint arXiv:2308.03281 (2023). https://arxiv.org/abs/2308.03281
10. Liu, L., Engelen, G., Lynar, T., Essam, D., Joosen, W.: Error prevalence in NIDS datasets: a case study on CIC-IDS-2017 and CSE-CIC-IDS-2018. In: 2022 IEEE Conference on Communications and Network Security (CNS), pp. 254–262. IEEE (2022). https://doi.org/10.1109/CNS56114.2022.9947235
11. Ullah, F., Ullah, S., Srivastava, G., Lin, J.C.W.: Ids-int: intrusion detection system using transformer-based transfer learning for imbalanced network traffic. Digit.Commun. Networks **10**(1), 190–204 (2024). https://doi.org/10.1016/j.dcan.2023.03.008
12. Wang, S., Reimers, N., Gurevych, I.: Text embeddings by weakly-supervised contrastive pre-training. arXiv preprint arXiv:2212.03533 (2022). https://arxiv.org/abs/2212.03533
13. Wang, W., Bao, H., Dong, L., Wei, F.: Minilm: deep self-attention distillation for task-agnostic compression of pre-trained transformers. arXiv preprint arXiv:2002.10957 (2020). https://arxiv.org/abs/2002.10957
14. Wang, Z., Li, Q., Xu, J., Zhang, J.: Revisiting the transferability of supervised pretraining: an MLP perspective. arXiv preprint arXiv:2112.00496 (2021). https://arxiv.org/abs/2112.00496
15. Xiao, S., Zhang, Z., Zhou, J., Chen, J., Team, B.R.: BAAI general embedding: a strong baseline for text representation. arXiv preprint arXiv:2309.07597 (2023). https://arxiv.org/abs/2309.07597
16. Zhao, F., Zhang, H., Peng, J., Zhuang, X., Na, S.-G.: A semi-self-taught network intrusion detection system. Neural Comput. Appl. **32**(23), 17169–17179 (2020). https://doi.org/10.1007/s00521-020-04914-7

Zero-Code Generative AI Performance Improvement for Business Letters: Using GPTs

Zizhou Wang(✉) and Chunjie Guo

Nanjing University of Aeronautics and Astronautics, Nanjing 211106, People's Republic of China
wangzizhou@nuaa.edu.cn

Abstract. As generative AI becomes integral to professional workflows, the demand for specialized, high-performing models is paramount. This study investigates a "zero-code" approach to enhance GPT-4's capabilities for business letter writing using OpenAI's custom GPTs platform. By using GPTs, we developed "Business Letter Pro," a model tailored for professional correspondence. We then conducted a double-blind experiment, having ten expert evaluators score outputs from both our custom model and the standard GPT-4 across ten quality metrics, complemented by a corpus-based linguistic analysis. The findings are compelling: the custom GPT achieved a 1.65% overall performance increase, with significant boosts in crucial areas like persuasiveness (+6.02%), tone and formality (+3.23%), and language appropriateness. The customized outputs were demonstrably more assertive, positive, and engaging. This research validates that non-programmers can meaningfully fine-tune large language models, showcasing a practical and accessible pathway to create superior, task-specific AI tools that drive real-world business value.

Keywords: GPTs · Zero-Code AI · Business Communication

1 Current Background of Modifying ChatGPT into One of the GPTs

Launched by OpenAI at its inaugural developer conference on November 7, 2023, GPTs has dramatically lowered the barrier to entry for AI application developers, allowing users to create a proprietary and customized GPT based on GPT-4 in a matter of minutes, just by discussing with ChatGPT. Building a GPT requires no traditional programming expertise and is accessible to anyone who subscribes ChatGPT Plus. Subscribers have the flexibility to create GPTs for personal use, internal use within their organization, or for broader dissemination. The process of crafting a GPT is straightforward: initiating a conversation, providing instructions and additional knowledge, and selecting its capabilities, such as web searching, image generation, or data analysis [1]. Such customized experience naturally involves subscribers in the design and modification of GPT.

GPTs offer a novel approach for individuals to craft a tailored rendition of ChatGPT to enhance its utility in their daily routines, specific tasks, professional engagements, or

H. Unger and P. Meesad (Eds.): NLPIR 2025, LNNS 1904, pp. 1012–1026, 2026.
https://doi.org/10.1007/978-3-032-20897-2_71

domestic environments. Moreover, users can share their customized GPT creations with others, broadening its accessibility and potential applications. For instance, GPTs can serve functions such as aiding in learning the rules of various board games, facilitating math instruction for children, or assisting in sticker design.

Since its debut, GPTs has shown unimaginable potential in most content-generating fields. Its base, OpenAI officials say that as of January 10th, subscribers have created up to 3 million GPTs and have approved a number of them for public use in GPTStore, where all other subscribers can share the approved GPTs. "There's only app that you dare not think of, but no app you cannot make", a senior IT critic wrote in a detailed article in Computer Weekly, commenting the latest advancement in AI application [2].

1.1 Successful Case – Canva

Canva is a popular online platform that offers a wide range of design tools and resources to empower people to create professional-quality graphics, presentations, posters, social media posts, and more, even if they don't have extensive design experience.

Canva's GPTs is a successful demonstration of GPT-4's powerful potential in the field of design. Immediately after OpenAI launches GPT-4, the GPTs crafted by Canva is on GPT Store (I will call it Canva-GPT from now on). Its slogan is "Effortlessly design anything: presentations, logos, social media posts and more" [3]. A year into its launch, it has gathered over a million chats and over 25 thousand comments, half of which are 5-star comments.

How does Canva-GPT Work? Canva-GPT is a GPTs integrated with Canva's design platform. Its primary function is to help users bring users' creative visions to life. Users tell it about the design users want—whether it's a poster, a social media graphic, or something else—and Canva-GPT use Canva's capabilities to generate design options for you. Canva-GPT interact with Canva's system to fetch design templates or generate new designs based on users' input. Publicly available technical reports on its specific architecture are scarce; however, it can be reasonably assumed that it leverages integrated AI image generation capabilities, such as OpenAI's DALL-E model that OpenAI launched alongside with GPT-4.

How do GPT Plus subscribers get help from Canva-GPT? First, they start a Conversation: begin by telling Canva-GPT what they would like to create. Users can describe the type of design, the occasion, or the message users want to convey. Then, Users' requests will be asked to be refined. Based on users' initial input, Canva-GPT might ask a few questions to better understand users' vision. This helps in generating designs that are more aligned with users' needs. The third step is to choose a design. Canva-GPT will present users with several design options, which users can view and choose, if there is one that the user like.

But that is not all. It is what happens after users choose a design that greatly enhances the function and fun of Canva-GPT. Users are allowed to customize and edit: After selecting a design, users can click on the provided link to open the design in Canva. There, users can customize and edit the design further—change text, colors, images, etc., to match users' exact requirements. In this step, users are naturally led outside of ChatGPT into the site of Canvas, providing a potential new customer for Canvas.com.

Once users are satisfied with the design, they can save it and use it wherever they need on Canvas.com.

The success of Canva-GPT can be attributed to the following factors:

Efficiency: It quickly generates design options that might take much longer to conceive from scratch. This helps users in maintaining a creative flow without getting bogged down in technical details.

Tailored Designs: We can let it produce visuals tailored to specific events, marketing goals, personal projects, or any other need, reflecting a professional touch.

Accessibility: It makes design accessible to everyone regardless of their designing skill level. Whether we're a beginner or a seasoned designer, we can "dictate" Canva-GPT in ease and enjoy flexibility Canva-GPT offer. Besides, all subscribers of GPT Plus can use it for free, that would be very helpful in its marketing premotion.

Experimentation: One can experiment with different styles and layouts without the initial heavy lifting. It's easy to iterate over several designs, which encourages more dynamic and innovative outcomes.

In conclusion, Canva-GPT's success is because it can simplify the design process, making it more accessible and user-friendly, so users can create beautiful, effective visuals without needing extensive design skills. And, as this is an example of business-to-customer application of GPTs, clearly both Canvas and OpenAI had presumably put in Internet technology or artificial intelligence specialists to better modify Canva-GPT. Therefore, success of Canva-GPT showcases full potential of modifying if abundant resources were put into the modification.

1.2 Successful Case - ChatPRD

Although Canva-GPT in case 1 shows full potential of modification and provides me with an ideal of how my own GPTs could become, achieving the same scale of success as a corporate-backed project like Canva-GPT is generally not feasible for an individual student researcher. The following GPTs seems to be one that had been crafted by an individual user. It has gotten over 25 thousand chats and over 600 comments. It gets 4.4-stars out of the 5-star commenting system. This GPTs would be a successful case for individual developer.

Created by Claire V. Lawless, ChatPRD is an "AI that writes great product requirements documents" and "an on-demand Chief Product Officer that drafts and improves your PRDs, while coaching you to become an elite product manager. The best product copilot for PMs & engineers". It is a GPTs tailored for product management.

How does ChatPRD work? ChatPRD is designed to assist with various product management tasks such as drafting and analyzing Product Requirements Documents (PRDs), providing strategic advice on product development, and helping with team collaboration and stakeholder management. ChatPRD will function by discussing, advising, and generating content related to product management based on user input, using a blend of built-in knowledge and strategic thinking.

To start using ChatPRD, users typically provide some information about their role in product management and the specific project or challenge they are facing. This helps ChatPRD tailor its assistance to users' needs. From there, users can ask specific questions or request help with tasks like drafting a PRD, improving existing documents, or getting strategic advice on product decisions.

What can users achieve with ChatPRD's help? With ChatPRD's assistance, users can enhance their product management practices, develop clearer and more effective PRDs, and gain insights into better decision-making for product development. ChatPRD will aim to improve the quality of product strategies, foster better communication and understanding among teams, and ultimately help drive successful outcomes for products and businesses.

To sum up, if someone is looking to refine a PRD, strategize on a new feature, or to get some general advice on product management, ChatPRD is here to help.

2 Modifying ChatGPT for Business Letter Writing

This study modifies ChatGPT to create a specialized GPT for writing business letters. To better modify such GPT, I need to first understand what makes a business letter successful. Then, with that knowledge, we write a paragraph of good ChatGPT Prompt to incite such modification. The last step is to insert that prompt into GPT Builder (which is provided by OpenAI), and follow the requests from GPT Builder. After all these steps, I can get a GPT that is modified by myself. If the modified GPT did not perform as well as expected, I should attempt to improve the Prompt and insert the Prompt into GPT Builder to build another modified GPT. (The initial modification proved successful, negating the need for further iterations.) The following part describes details of my preparations for such modification.

2.1 Determining the Optimal PROMPT for GPTs

Modifying ChatGPT into GPTs is like programing it with natural languages instead of computer languages. Prompt is the kind of script used to modify and tune ChatGPT. For the purposes of this paper, a "Prompt" is defined as the intentionally crafted phrases that users put into ChatGPT to better its performance [4]. Quality of the Prompt imputed into LLMs are the second most important among factors that influence LLMs' output. The only factor more important than Prompt is the quality of the LLM used [4]. Our LLM had been selected, so, carefully writing the Prompt is the most important out of these procedures.

The only problem is that GPTs do not directly respond to Prompts. Instead, they take orders from Prompts put into chat box of GPT Builder. Launched by OpenAI, GPT Builder can be viewed as a GPT crafted for making all other GPTs. Therefore, the Prompt that I need to write is not about how to write a specific business letter, and is instead about how to write business letter in general and how to respond if some sorts of circumstances occur.

The questions to be considered include: If a user fails to provide key information, will my GPT ask them to provide? Will the GPT not use its online checking capability, and imagine incorrect information or experience hallucination? Will it be lazy in its response and fail to consider all that I asked it to consider? The only thing we can do about any these questions is to give our GPTs a reminder in their programming by sending our Prompts to GPT Builder.

With these considerations, the prompt provided to the GPT Builder consisted of the established requirements for successful business letter writing, appended with the directive "Become an expert in writing professional international business letters. Make sure that your answers are up to those requirements, and always check the internet for details you might need to answers."

2.2 Crafting a Version of GPTs Tuned for Business Letter Writing

Initially, crafting a tuned GPT was perceived as a procedure with a high technical barrier. The reasons include that building AI had always been a technological difficulty to which the best and brightest of mankind had been devoted.

Prerequisite of Crafting any version of GPTs is to have access to an OpenAI account subscribed to ChatGPT Plus [4]. Having logged in as a legitimate paying ChatGPT Plus subscriber, thus all requests can be answered by GPT-4 for 40 times per 3 h or by ChatGPT-3.5 without limitation of times.

Then, crafting any version of GPTs is just as simple as imputing your own command into an ordinary ChatGPT chat box, as is shown in Fig. 1. OpenAI's scientists and engineers had made this procedure so simple that one can do it without experiencing any difficulty. Below is what one sees if he or she enters GPT Store:

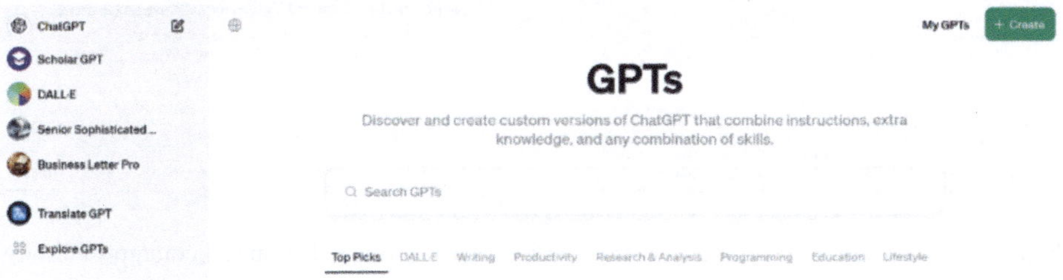

Fig. 1. Headline of GPT Store (https://chat.openai.com/gpts).

The creation process involves navigating to the 'GPT Store,' selecting '+Create' to open the GPT Builder interface, and inputting the predetermined instructional prompt, as is shown in Fig. 2.

The Builder then engages in a conversational setup. It then asked "For the name of this GPT, how about 'Business Letter Pro'? Does that work for you?", "What specific types of business letters or situations 'Business Letter Pro' should focus on?", and "Would you prefer it to ask for clarification if the user's request is vague or should it make assumptions based on typical scenarios?" After finishing this chat, a new GPT is made, as is shown in Fig. 3.

In conclusion, the journey of crafting a version of GPTs tuned for business letter writing, which was once perceived as a daunting task fraught with technological barriers, turned out to be surprisingly accessible and straightforward. The streamlined process

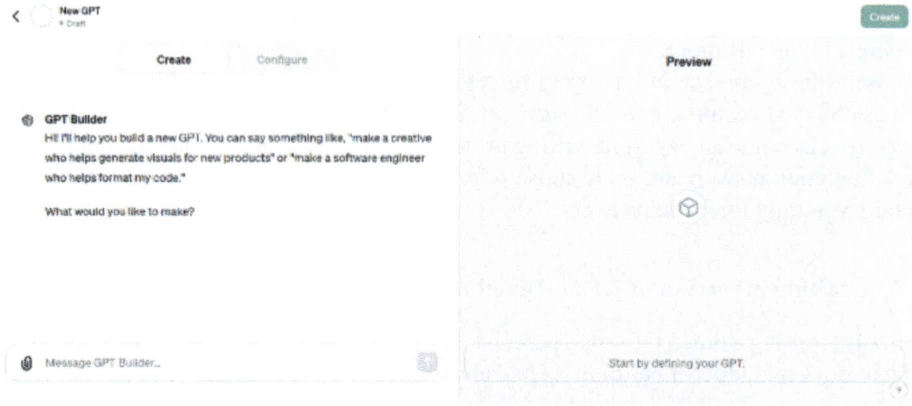

Fig. 2. The user interface of GPT Builder (https://chat.openai.com/gpts/editor)

Fig. 3. Chart 4: The user interface of my modified GPT (https://chat.openai.com/gpts/editor/g-V1ivDIh0L)

offered by OpenAI's GPT Builder demystified the creation of custom GPT models, guiding users through prompts and options with clarity and efficiency. From defining the scope and focus of the model to fine-tuning its behavior, the interface provided intuitive controls that enabled experimentation and iteration.

By lowering the barriers to entry and providing intuitive interfaces, OpenAI empower users to harness the potential of GPT-4 for their unique needs and applications. This streamlined creation process highlights the transformative impact of accessible technology in unlocking creativity, innovation, and problem-solving across domains.

3 Research Design

This study attempts to design a double-blind experiment to provide statistic that allows to compare the modified LLM with original ChatGPT. A large language model was customized to generate business letters according to the user's needs, inputting the same business letter writing needs to the model and ChatGPT, The output texts of the two models were compared through double-blind experiments, and evaluate the customized large language models of GPTs through the Mean score obtained from double-blind experiments, and do a basic corpus analysis.

3.1 Research Questions

Question 1: To what extent does the modified large language model improve upon the output quality of the original ChatGPT?

Question 2: In which specific aspects of output quality does the modified model differ from the original ChatGPT?

3.2 Participants

To become a participant, one must have passed Test for English Major Band 4 or acquire more than 600 points in College English Test Band 6, both of which are popular if their native language is not English. If their native language is English, the participant must be an adult who grew up in English-speaking countries and had finished high school education.

3.3 Procedures of Double-Blind Experiment

Ten exercises or examples were selected, which are provided in English Business Letters (5th Edition) [5]. Test tasks were input separately into my modified GPT and the original GPT-4.

Judging the quality of a business letter involves assessing several key aspects that contribute to its effectiveness and professionalism. The following criteria were adopted to evaluate the business letters:

Clarity and Conciseness: The letter should clearly communicate its purpose without unnecessary details. Every sentence should serve a purpose, contributing directly to the message.

Tone and Formality: The tone of the letter should be appropriate for its audience and purpose. It should maintain a professional tone, being neither too casual nor overly formal, unless the specific context demands it.

Structure and Organization: A well-organized letter follows a logical structure, typically including an introduction that states the purpose, a body that details the subject, and a conclusion that summarizes the main points or calls to action. Good organization helps the reader understand the message more effectively.

Grammar and Spelling: The letter must be grammatically correct, with no spelling errors. Professional writing should adhere to the highest standards of language to convey competence and attention to detail.

Persuasiveness in both objective and subjective perspective: If the letter's goal is to persuade the reader (e.g., in sales or proposal letters), it should build a strong case for why the reader should consider a proposal or view. This involves using compelling language and solid arguments.

Adherence to Business Letter Conventions: The letter should follow the conventional format of a business letter, including sender's contact information, date, recipient's address, proper salutation, body, complimentary close, and signature. Proper formatting reflects professionalism and respects business etiquette.

Appropriateness of Language: The language used should be suitable for the audience and free of jargon, unless it is industry-specific and the recipient is familiar with it. It should also avoid vague terms and provide enough detail to be informative without being overly technical.

Response Encouragement: Especially in interactive scenarios, the letter should encourage a response or facilitate further communication. It might include a call to action, a request for a meeting, or a prompt for a reply.

Cultural Sensitivity: In a global business environment, the letter must be culturally sensitive, considering the norms and expectations of the recipient's cultural context.

4 Results and Discussions

This section presents a mean score analysis and a corpus analysis of the results.

4.1 Mean Score Analysis

The average score of the 20 answers, scored by 10 participants of the double-blind experiment is shown in Table 1. IBM SPSS identifies that all 200 scores are valid, and that Cronbach $\alpha = 0.830 > 0.8$, showing the reliability of this dataset. Fig. 4 following the chart which shows how much the modified GPT has improved in each aspect is based on the scores of both GPTs.

As is shown in the chart, the total improvement in GPT-4's performance after my modifications stands at 1.65%. Notable improvements have been found in persuasiveness, both in objective (+5.79%) and subjective (+6.02%) perspectives, and tone and formality (+3.23%). It seems there's been a slight decrease in conciseness, which could imply that while the responses might be more detailed or elaborate, they might also be less concise. There is no change in the category of response encouragement.

Overall, these figures and data suggest a generally positive enhancement across most evaluated aspects, indicating more engaging, appropriately formal, and contextually sensitive responses.

The modified GPT shows enhancements across most criteria, suggesting that my method of modifying GPT-4 into GPTs have positively impacted its ability to function effectively in business communication. The increased total score from 96.11 to 97.7 reflects an overall improvement in quality, making the Modified GPT potentially more suitable for professional use in business environments, where communication precision and adaptability are critical.

Table 1. Mean scores of both AI models' 10 responses in 10 aspects. (Mean scores are the average scores scored by 10 participants of this study, full score of each aspect being 10.) (See all scores in Appendix III.)

Criteria	GPT-4	Modified GPT
Clarity	9.96	9.99
Conciseness	9.55	9.3
Tone and Formality	9.61	9.92
Structure and Organization	9.71	9.75
Grammar and Spelling	9.97	10
Persuasiveness in an objective perspective	9.32	9.86
Persuasiveness in a subjective perspective	8.97	9.51
Appropriateness of Language	9.62	9.93
Response Encouragement	9.44	9.44
Cultural Sensitivity	9.96	10
Total Score	96.11	97.7

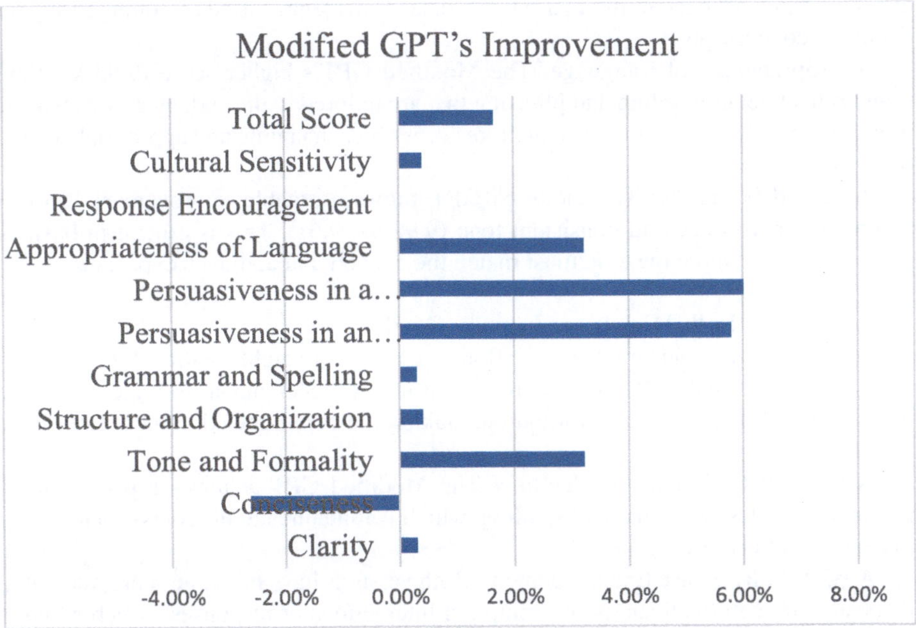

Fig. 4. Modified GPT's improvement compared to GPT-4 (The percentages are GPT-4's scores in each aspect divided by Modified GPT's scores in each aspect.)

A comparison between GPT-4 and the Modified GPT is presented based on the evaluation criteria.

Aspects Where the Modified GPT Outperforms GPT-4

Persuasiveness: This aspect sees the greatest improvement. It is also the most important aspect as it plays an important role when it comes to success of business communication. To better reflect the improvement, participants were instructed to consider both objective and subjective perspective when they score.

Objective Perspective: The modified GPT shows a significant increase, suggesting that it might be better at constructing arguments and presenting facts in a compelling manner.

Subjective Perspective: There is also a considerable improvement, possibly due to more effectively aligning with the reader's perspectives or emotions, which is key in persuasive communication. This proves the ability to appeal to personal beliefs, values, or emotions. The 6.02% increase suggests that the model has improved after modification in areas such as empathy, emotional appeal, or connecting with the audience on a personal level.

The much higher persuasiveness showcased that the modified GPT is now more effective at presenting arguments or information in a way that is likely to convince others based on objective criteria or factual evidence. The improvement here is notable and indicates a stronger ability to use logic, data, or evidence-based reasoning to support claims or conclusions.

Appropriateness of Language: The Modified GPT's higher score (9.93 vs. 9.62) might reflect better wording and phrasing that are tailored to the audience as well as the topic, potentially making the content more accessible, engaging, and appropriate in eyes of the participants.

Tone and Formality: The modified GPT shows a notable improvement in maintaining an appropriate and consistent tone (9.92 vs. 9.61). This is crucial in business communication where the tone must match the context and audience expectations.

Aspects Where Both Models are Equally Good

Aspects 1: Structure and Organization: Both versions score highly, with a slight improvement in the Modified GPT. This improvement might reflect enhanced logical sequencing and better integration of information, making the communication more intuitive to follow.

Aspects 2: Grammar and Spelling: The Modified GPT achieves a perfect score, indicating flawless grammar and spelling, which is fundamental in professional settings to convey competence.

Aspects 3: Response Encouragement: Both versions have the same score, indicating no change in the effectiveness of prompting interactions or responses, which could be an area for further enhancement.

Aspects 4: Cultural Sensitivity: Achieving a perfect score, the Modified GPT likely shows improved awareness and consideration of cultural nuances, which is increasingly important in global business contexts.

Aspects 5: Clarity: Both versions perform exceptionally well, with the Modified GPT slightly outscoring GPT-4. This suggests that the modifications may have further refined the ability to communicate the purpose more transparently.

Conciseness: Interestingly, the modified GPT scores slightly lower than GPT-4. This might indicate a trade-off between enhancing clarity and maintaining brevity, perhaps due to adding more descriptive elements to improve clarity.

4.2 Corpus-Based Analysis

To better understand the reasons behind the result of Mean score analysis, a corpus-based analysis was conducted using the UAM CT 6.2 tool. I loaded the responses from GPT-4 and the modified GPT separately as two corpora (of which Text-1.txt is the corpus for Responses from Modified GPT, and Text-2.txt is the corpus for Responses from GPT-4), loaded automatic UDParse layer, set aspect of interest as Lexical Pattens, and clicked "Show Results". The corpus analysis using UAM CT 6.2 on responses from GPT-4 and the modified GPT provides quantitative data to evaluate the differences in language use and structural characteristics of the texts.

The modifications seem to enhance GPT's communicative aspects, making the responses not only longer and possibly more detailed but also more positive, assertive, and engaging. The use of slightly more academic language, combined with an increased number of personal references, may make the modified GPT's outputs more suitable for interactive and professional contexts where engagement and clarity are critical. However, the small decrease in lexical variety per segment and similar levels of text complexity suggest that the fundamental linguistic structure remains largely consistent with the original GPT-4, maintaining a balance between sophistication and accessibility.

The results of the corpus analysis are presented and analyzed below.

Answers from the modified GPT shows supremacy in subjectivity, Academicness, as well as the first-person and second-person reference density, as is shown in Table 2. Presumably, these characteristics have contributed to the persuasiveness, appropriateness of the modified GPT's answers.

Subjectivity: This aspect includes subjective positivity and subjective strength.

Subjective Positivity: A measure of the positiveness of the words in the text. (−1: very negative, 0: balanced, +1: very positive). It is based on the 8,200-word MPQA Subjectivity lexicon. The modified GPT shows higher subjective positivity (0.586 vs. 0.5), suggesting it might use language that is perceived as more positive or optimistic.

Subjective strength: A measure of the average subjective strength of the words of the text, from 0 (all weak) to 1 (all strong). There is also a higher subjective strength in the modified GPT's responses, indicating more assertive or confident language use.

Academicness: This includes academic word use and academic rareness, as is shown in Table 3.

Academic Word Use: Percentage of lexical words which are on the academic word list. The modified GPT has a slightly higher percentage of academic word use (9.81% vs. 9.59%), which could be indicative of a style that's slightly more formal or geared towards academic or professional settings.

Table 2. Subjectivity

Subjectivity	Texts/Text-1.txt	Texts/Text-2.txt
Subjective Positivity	0.586	0.5
Subjective Strength	0.369	0.34

Academic Rareness: An index which is higher when the academic words tend to be the less common words. Interestingly, GPT-4 scores slightly higher on academic rareness, which could indicate a use of more unique or less common academic vocabulary.

Table 3. Academicness

Academicness	Texts/Text-1.txt	Texts/Text-2.txt
Academic Word Use	9.81%	9.59%
Academic Rareness	3.717	3.931

Reference density: This includes the first-person and second-person References (1p and 2p References).

The 1p and 2p References: Percentage of all words which are a first-person or second-person pronoun ('I', 'me', 'my', 'mine', 'we', 'our', 'us', 'ours', 'you', 'your', 'yours'). The modified GPT uses first-person and second-person references more frequently than GPT-4,, as is shown in Table 4. This could reflect a style that's more engaging or directly addresses the reader, potentially enhancing the persuasive element of the text.

Table 4. Reference Density

Reference Density	Texts/Text-1.txt	Texts/Text-2.txt
1p Reference	5.45%	4.67%
2p Reference	4.54%	4.3%

Both models behave equally well in lexical density and text complexity.

Lexical density: This includes lexemes per segment and lexemes percentage of text.

Lexemes per segment: How many open-class words in each segment on average. The slightly lower score for the modified GPT (1.58 vs. 1.6) could imply a minimal reduction in lexical variety per segment, despite having a higher total number of words.

Lexemes percentage of Text: Proportion of words in all segments which are open class. Both versions are nearly equivalent in how much of their text is made up of lexemes, which implies that both maintain a similar level of lexical richness, as is shown in Table 5.

Text complexity: This includes average word length and average segment length.

Table 5. Lexical density

Lexical Density	Texts/Text-1.txt	Texts/Text-2.txt
Lexemes per segment	1.58	1.6
Lexemes percentage of text	57.22%	57.57%

Average Word Length: How many characters are there in each word? Both versions are almost identical in average word length, indicating a similar level of vocabulary complexity.

Average Segment Length: How many tokens (words plus punctuation marks) are there in each segment. The equal average segment length (2.77) in both corpora, as is shown in Table 6, suggests that the structural composition of sentences (in terms of length) remains consistent despite other modifications. Lower average word length indicates higher readability [6]. They both have low average word length.

Table 6. Text complexity measured by word length and segment length.

Text Complexity	Texts/Text-1.txt	Texts/Text-2.txt
Av. Word Length	5.23	5.2
Av. Segment Length	2.77	2.77
Min. Segment Length	0	0
Max. Segment Length	48	44

Length: It is measured by numbers of, as well as tokens and words in segments. The answers given by the modified GPT is lengthier than those given by GPT-4. This phenomenon can be seen in number of segments as well as the tokens and words in segments. This proves that the modified GPT does lack conciseness, as is indicated by the Mean score analysis.

Number of segments: the number of segments is identified by the units of interest. The modified GPT's responses have more segments (4768 vs. 3938), indicating that it tends to use shorter segments more frequently or simply generates more content overall.

Tokens and words in segments: How many words and tokens (words plus punctuation marks) are in all the segments. Correspondingly, the modified GPT uses more tokens and words, as is shown in Table 7, which suggests not only a greater volume of content but potentially more complexity or detail in the responses.

Table 7. Length

Length	Texts/Text-1.txt	Texts/Text-2.txt
Number of segments	4768	3938
Tokens in segments	14423	11892
Words in segments	13207	10911

4.3 Discussion

The limitation of this study includes its research methodology, research outcome and number of samples.

Regarding the research methodology, it is acknowledged that the rating data is subjective, but can be seen as "objectively subjective". This is because that the Mean scores are given by 10 participants, and the outcome could be influenced by their objective opinions. Also, there are only 10 test tasks for the AI models, so the corpus based on the responses would be small.

For research outcome, as is shown in the study, the total improvement in GPT-4's performance after my modifications is only 1.65%, although high improvements are seen in persuasiveness, appropriateness, tone and formality. This is not very much of a leap forward, instead it is a small but notable improvement that is statistically meaningful. Initial considerations included developing a large language model from scratch; however, this approach was deemed infeasible due to resource and technical constraints. Consequently, the research focused on improving the performance of an existing, modifiable model (GPT-4) [7].

Also, as the research focus itself on GPT-4, OpenAI can change and even nullify GPT-4 at its will, making the experiment difficult to reproduce. Adopting local LLM may solve the problem [8].

A further limitation is the small sample size, which consisted of 10 participants.

Beyond those, Given the rapid evolution of Generative AI, the conclusions of this study represent a snapshot in time. The limitations inherent in the samples and methods invite further research and scholarly dialogue to build upon these findings.

5 Conclusion

This study investigated the efficacy of a zero-code approach for enhancing the performance of a large language model, GPT-4, for the specialized task of business letter writing. Through the creation of a custom GPT using OpenAI's platform, this research empirically demonstrates that targeted, non-programmatic modifications can yield statistically significant performance improvements.

The primary finding is that the customized model, "Business Letter Pro," outperformed the baseline GPT-4, achieving a notable increase in quality, particularly in crucial communicative aspects. The most significant gains were observed in persuasiveness (subjective and objective), tone and formality, and appropriateness of language. These enhancements are quantitatively supported by the corpus analysis, which revealed that the modified GPT produced text that was linguistically more assertive, positive, and engaging through an increased use of first and second-person pronouns. This suggests the instructional prompts successfully guided the model toward a more direct and effective communicative style. However, this came with a trade-off, as the modified model was slightly less concise, producing longer responses than the baseline model.

The central contribution of this research is the validation that zero-code customization is a viable method for specializing general-purpose AI models. It provides concrete evidence that users without a background in programming can create measurably superior AI tools for specific professional domains. The practical implication is the democratization of AI application development; individuals and organizations can develop bespoke AI assistants tailored to their unique communication needs, enhancing efficiency and quality without requiring significant technical investment.

This study is not without limitations. The evaluation relied on subjective scoring from a small sample of 10 participants, and the corpus was generated from only 10 distinct tasks. Therefore, while the results are statistically reliable (Cronbach's $\alpha = 0.830$), their generalizability may be limited.

Future research should aim to build upon these findings. Replicating this study with a larger, more diverse participant pool and a wider array of business writing scenarios would strengthen the validity of the results. Further investigation could also explore the impact of varying the complexity and specificity of the instructional prompts on model performance. Finally, applying this zero-code customization methodology to other professional genres, such as legal writing or technical documentation, would provide valuable insights into the broader applicability of this approach.

References

1. Introducing GPTs. https://openai.com/blog/introducing-gpts. Accessed 2 July 2025
2. Zhang, S., Zhang, Y.: AI has its appstore moment. Comput. Wkly. (003) (2023)
3. Canva-GPT. https://chat.openai.com/g/g-alKfVrz9K-canva. Accessed 1 July 2025
4. Wang, J., et al.: Review of large vision models and visual prompt engineering. Meta-Radiology **1**, 1–13 (2023)
5. Zhuge, L., Wang, Y.: English Business Letters, 2nd edn. University of International Business and Economics Press, Beijing
6. Dale, E., Chall, J.: A formula for predicting readability. J. Educ. Res. Bull., 11–20+28 (1948)
7. Bubeck, S., et al.: Sparks of artificial general intelligence: early experiments with GPT-4. https://arxiv.org/abs/2303.12712. Accessed 3 July 2025
8. Wang, D.: Federated learning using GPT-4 boosted particle swarm optimization for compact neural architecture search. J. Adv. Inf. Technol. **15**(9), 1011–1018 (2024)

Task-Aligned Augmented Graph for Cold-Start Recommendation

Yuxiang Shi, Ling Ma, and Yue Ding(✉)

Shanghai Jiao Tong University, Shanghai, China
{shiyuxiang,maling920827,dingyue}@sjtu.edu.cn

Abstract. The cold-start problem remains a key challenge in recommender systems due to the lack of user-item interactions, which limits the quality of recommendations for new users and items. Recent meta-learning methods learn shared knowledge across users and adapt it to new cases with few interactions. While effective, they often rely on global parameters and ignore the intrinsic features and interactions of new users and items, which are crucial in cold-start settings. We propose TMAG (Task-aligned Meta-learning with Augmented Graphs) to address this issue. TMAG introduces a task-aligned constructor that clusters users with similar attributes and partitions tasks for meta-learning, ensuring consistent optimization. We also design an augmented graph neural network with two graph-enhanced techniques to handle data sparsity and capture high-order interactions. Experiments on three real-world datasets show that TMAG outperforms state-of-the-art methods in cold-start recommendation.

Keywords: Cold-start Recommendation · Meta-learning · Graph Neural Network · Mutual Information Maximization

1 Introduction

Recommender systems aim to identify user interests and are widely applied in E-commerce, online advertising, and social media platforms [36]. Despite the success of matrix factorization (MF) [1] and deep learning models [12], the cold-start problem remains a major challenge [8]. It stems from limited user-item interactions, especially for new users, who may disengage if early recommendations are poor.

The traditional way to mitigate cold-start is through feature-level strategies, mainly in two forms. The first models inherent information such as user profiles [2], item attributes [4], and cross-domain knowledge [17], enriching representations of new users/items. The second models feature interactions via graph neural networks (GNN) [38] and heterogeneous information networks (HIN) [23], capturing high-order collaborative signals. These methods, however, depend heavily on the availability and quality of features. At the model level, few-shot learning [34] and meta-learning [32] have shown strong ability to handle data sparsity.

H. Unger and P. Meesad (Eds.): NLPIR 2025, LNNS 1904, pp. 1027–1042, 2026.
https://doi.org/10.1007/978-3-032-20897-2_72

(a) Conventional meta-learning (b) Proposed TMAG

Fig. 1. Visualization of model parameter $\boldsymbol{\theta}$ of the conventional framework with 4 tasks and the proposed TMAG with 2 tasks aligned by ages.

Many meta-learning approaches use optimization-based algorithms like MAML [7], learning global parameters as initialization for personalized models. They simulate cold-start by constructing few-shot preference tasks, extracting meta-knowledge in training, and adapting to new users in testing. These methods achieve promising results in cold-start recommendation.

Yet, current methods face two key issues. First, they treat each user as a separate task and learn coarse global knowledge, which may lead to local optima when user gradients deviate (see Fig. 1a). For example, $\nabla\mathcal{L}_4$ can bias parameter $\boldsymbol{\theta}$ due to age difference. Second, they fail to fully leverage inherent information and feature interactions, which are crucial for new users and items. This raises two challenges: (1) How to avoid local optima in meta-learning for cold-start? (2) How to jointly exploit inherent information and feature interactions to address sparsity?

To address these, we propose a **T**ask-aligned **M**eta-learning based **A**ugmented **G**raph (TMAG). For the first challenge, we design a fine-grained task-aligned constructor using an attribute-oriented autoencoder to learn latent user/item representations. Similar users are clustered into tasks with consistent optimization, reducing the risk of local optima. For the second challenge, we build an augmented graph neural network with two graph-enhanced strategies, capturing high-order interactions from both attribute and graph perspectives.

Our contributions are summarized as follows: (1) We propose a task-aligned meta-learner with an autoencoder-based constructor to learn latent clustering knowledge of users/items. This enables fast adaptation in cold-start. We further apply task-wise attribute contrastive regularization to refine clustering. (2) We design an 'Exploration-and-Constraint' strategy to augment user-item graphs, integrating structure and attribute signals to alleviate sparsity. (3) Experiments on three real-world datasets demonstrate that TMAG achieves state-of-the-art performance in cold-start recommendation.

2 Related Work

2.1 Cold-Start Recommendation

A common way to handle cold-start is using side information (e.g., user profiles or item content) combined with collaborative filtering [12,35]. Many methods cluster warm users and assign cold users to clusters via side information

[18]. Beyond side features and user – item interactions, adversarial regularization can augment supervision but is often costly at scale [3]. Transfer learning transfers representations from a source to a target domain to mitigate cold-start [26]. More recent directions include federated personalization with adaptive knowledge fusion under privacy constraints [21], disentangled contrastive learning for cross-domain transfer [20], and diffusion-based data enrichment for multi-scenario recommendation [40].

2.2 Meta-learning in Recommendation

Meta-learning extracts meta-knowledge from tasks for rapid adaptation with few examples. Existing works roughly fall into two families: (i) metric-based methods that learn similarity metrics [33], e.g., ProtoCF builds prototype representations without side information [29]; and (ii) optimization-based methods that introduce MAML-style bi-level updates to cold-start recommendation [7,30], with representative instances such as MeLU (personalized vs. shared parameters) and MetaCF (fast-adapting CF) [19,41]. A common limitation is the reliance on globally shared parameters, which may hinder convergence and personalization; improvements include task/feature memories [5], social priors [37], and knowledge-graph signals [6]. Parallel to meta-learning, contemporary cold-start studies explore complementary paradigms—federated personalization with adaptive knowledge fusion [21], multi-view hypergraph contrastive modeling for micro-videos [25], disentangled contrastive learning for cross-domain transfer [20], diffusion-based data enrichment for multi-scenario recommendation [40], and popularity-aware curricula for bundle cold-start [13].

3 Methodology

3.1 Problem Definition

Let $\mathcal{U} = \{u_1, u_2, ..., u_M\}$ and $\mathcal{I} = \{i_1, i_2, ..., i_N\}$, where $\mathcal{U}$ denotes the user set, $\mathcal{I}$ denotes the item set, M and N denote the number of users and items respectively. We use a matrix $\boldsymbol{R} \in \mathbb{R}^{M \times N}$ to describe the interactions between users and items:

$$\boldsymbol{R}_{ui} = \begin{cases} 1, & \text{if user } u \text{ has interacted with item } i; \\ 0, & \text{otherwise.} \end{cases} \tag{1}$$

These user-item interactions can be translated into a user-item bipartite graph, denoted as $\mathcal{G} = (\mathcal{V}, \mathcal{E})$, where $\mathcal{V}$ denotes the node set and $\mathcal{E}$ denotes the edge set in the graph. The number of nodes in the graph is $|\mathcal{V}| = (M+N)$. The adjacency matrix $\boldsymbol{A} \in \mathbb{R}^{M \times N}$ characterizes the connections within $\mathcal{G}$.

This paper focuses on addressing the 'incomplete cold-start' problem. Our proposed method aims to improve overall performance in three scenarios: (1) Recommend existing (old) items to new users; (2) Recommend new items to existing (old) users; (3) Recommend new items to new users.

3.2 Overview of TMAG

The framework of TMAG is shown in Fig. 2. It has three components: (1) a task-aligned constructor that extracts attribute embeddings and builds tasks to capture latent clustering knowledge; (2) an augmented graph neural network that alleviates data sparsity and captures high-order interactions; and (3) a contrastive regularization mechanism that strengthens the use of latent clustering knowledge.

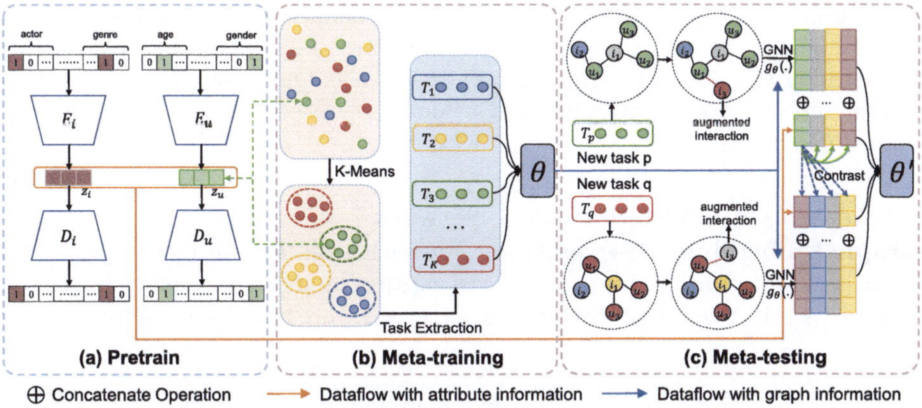

Fig. 2. The overview of the proposed TMAG framework.

3.3 Task Aligned Constructor

Attribute-Oriented Autoencoder. The architecture of the attribute-oriented autoencoder is shown in Fig. 2 (a). This module encodes user content (e.g., age, gender) and item content (e.g., actor, genre) into embeddings. It consists of two steps: (1) projecting the input into a low-dimensional hidden space; (2) reconstructing the input from the latent embedding. For categorical attributes, we define the user input $\boldsymbol{x}_u$ and item input $\boldsymbol{x}_i$ as:

$$\boldsymbol{x}_u = [c_{u1}; \cdots ; c_{uP}]; \quad \boldsymbol{x}_i = [c_{i1}; \cdots ; c_{iQ}], \tag{2}$$

where P and Q are the numbers of user and item content fields, and c_{up}, c_{iq} are one-hot or multi-hot vectors. We use ReLU to derive the latent user embedding:

$$\boldsymbol{z}_u = \sigma(\boldsymbol{W}_u^1 \boldsymbol{x}_u + \boldsymbol{b}_u^1), \tag{3}$$

where $\boldsymbol{W}_u^1, \boldsymbol{b}_u^1$ are trainable parameters. Instead of GCN [15], which may dilute attribute signals by aggregating neighbors, we adopt a decoder to reconstruct $\boldsymbol{x}_u$:

$$\boldsymbol{x}_u^r = \sigma(\boldsymbol{W}_u^2 \boldsymbol{z}_u + \boldsymbol{b}_u^2), \tag{4}$$

with loss:

$$\boldsymbol{L}_u = \sum_{u \in \mathcal{U}} \|\boldsymbol{x}_u - \boldsymbol{x}_u^r\|^2 + \lambda \|\boldsymbol{W}_u\|^2 . \tag{5}$$

Here λ controls L_2 regularization. Item embeddings $\boldsymbol{z}_i$ are obtained similarly with an item autoencoder.

Task Construction. In cold-start settings, user-item interactions are sparse. To capture latent clustering knowledge, we construct meta tasks. Inspired by MAML, we use K-Means [22] to cluster users $\mathcal{U}$ based on their attributes $\boldsymbol{x}_u$, forming $\boldsymbol{C} = \{\boldsymbol{C}_1, \cdots, \boldsymbol{C}_K\}$. Each cluster $\boldsymbol{C}_k$ defines a task, with support set $\mathcal{S}_k^{tr}$ (items rated by users in $\boldsymbol{C}_k$). New users are assigned to the nearest cluster, forming $\boldsymbol{C}'$, with meta-testing support set $\mathcal{S}^{te}$. Query sets $\mathcal{Q}^{tr}$, $\mathcal{Q}^{te}$ contain unseen interactions for training and evaluation, respectively, and are disjoint from support sets.

3.4 Augmented Graph Neural Network

Graph Embedding Propagation. For each task $\boldsymbol{T}_k$, we build a user-item bipartite graph from its interactions. To capture high-order structure, we apply LightGCN [11]. User/item embeddings $\boldsymbol{e}_u, \boldsymbol{e}_i \in \mathbb{R}^d$ are initialized and propagated as:

$$\boldsymbol{e}_u^{(l+1)} = \sum_{i \in \mathcal{N}_u} \frac{1}{\sqrt{|\mathcal{N}_u||\mathcal{N}_i|}} \boldsymbol{e}_i^{(l)}, \tag{6}$$

where $\mathcal{N}_u$ ($\mathcal{N}_i$) is the neighborhood of u (i). This enhances embeddings via graph propagation.

Graph Enhanced Generator. To alleviate sparse interactions, we generate potential edges using two strategies. First, graph-structure-based similarity:

$$\boldsymbol{E}_{u,i}^1 = \sigma(\sum_{j \in \mathcal{N}_u} \boldsymbol{e}_j^{(L)} \boldsymbol{W}_g \boldsymbol{e}_i^{(L)}), \tag{7}$$

where $\boldsymbol{W}_g$ captures structural relations. Second, attribute-based similarity:

$$\boldsymbol{E}_{u,i}^2 = \sigma(\sum_{j \in \mathcal{N}_u} \boldsymbol{z}_j \boldsymbol{W}_a \boldsymbol{z}_i). \tag{8}$$

The combined score is:

$$\boldsymbol{E}_{u,i} = \alpha \boldsymbol{E}_{u,i}^1 + (1-\alpha) \boldsymbol{E}_{u,i}^2, \tag{9}$$

optimized by:

$$\mathcal{L}_{gen} = \|\boldsymbol{E} - \boldsymbol{A}\|^2 . \tag{10}$$

Edges with $\boldsymbol{E}_{u,i} > t$ are added to adjacency matrix $\hat{\boldsymbol{A}}$:

$$\hat{\boldsymbol{A}}[u,i] = \begin{cases} 1, & \text{if } \boldsymbol{E}_{u,i} > \text{ t;} \\ 0, & \text{otherwise.} \end{cases} \tag{11}$$

Model Prediction. Final representations concatenate graph and attribute embeddings:

$$\boldsymbol{f}_u = \boldsymbol{e}_u^{(L)} \oplus \boldsymbol{z}_u, \quad \boldsymbol{f}_i = \boldsymbol{e}_i^{(L)} \oplus \boldsymbol{z}_i. \tag{12}$$

Prediction is computed by:

$$\hat{y}_{ui} = \boldsymbol{f}_u^T \boldsymbol{f}_i. \tag{13}$$

We optimize with BPR loss [27]:

$$\mathcal{L}_{pre} = \sum_{(u,i,j)\in\mathcal{D}} -\log\, \sigma(\hat{y}_{ui} - \hat{y}_{uj}), \tag{14}$$

where $\mathcal{D} = \{(u,i,j)|(u,i) \in \mathcal{D}^+, (u,j) \in \mathcal{D}^-\}$.

3.5 Contrastive Regularization

We further regularize embeddings with task-wise contrastive learning [10], maximizing mutual information by pulling embeddings within the same cluster closer and pushing apart different clusters. Positive pairs: $\{(\boldsymbol{z}_u, \boldsymbol{z}_{u'})|u, u' \in \boldsymbol{C}_k\}$, Negative pairs: $\{(\boldsymbol{z}_u, \boldsymbol{z}_v)|u \in \boldsymbol{C}_k, v \notin \boldsymbol{C}_k\}$. The loss is:

$$\mathcal{L}_{MI} = \sum_{k=1}^{K} \sum_{u\in\boldsymbol{C}_k} -\log \frac{\exp(\mathrm{sim}(z_u, z_{u'})/\tau)}{\sum_{v\notin\boldsymbol{C}_k} \exp(\mathrm{sim}(z_u, z_v)/\tau)}, \tag{15}$$

where $\mathrm{sim}(\cdot)$ is a similarity function (cosine similarity performs best) and τ is the temperature. Negative sampling is approximated in-batch.

3.6 Optimization

To improve cold-start recommendation, we adopt a multi-objective training strategy. The total loss is:

$$\mathcal{L} = \mathcal{L}_{pre} + \lambda_1 \mathcal{L}_{gen} + \lambda_2 \mathcal{L}_{MI} + \lambda_3 ||\boldsymbol{\Theta}||^2, \tag{16}$$

where $\boldsymbol{\Theta}$ denotes all trainable parameters in $\mathcal{L}_{pre}$ and $\mathcal{L}_{gen}$, since $\mathcal{L}_{MI}$ adds none. λ_1, λ_2, and λ_3 balance different losses.

In the MAML framework, meta-training involves second-order gradient updates to derive an initial parameter $\boldsymbol{\theta}$ for users and items, enabling fast adaptation with few interactions. For task T_k, the inner-loop update on $\boldsymbol{S}_k^{tr}$ is:

$$\boldsymbol{\theta}'_k = \boldsymbol{\theta}_k - \alpha \nabla_{\boldsymbol{\theta}} \mathcal{L}_{T_k}(\boldsymbol{\theta}_k, \boldsymbol{S}_k^{tr}), \tag{17}$$

where α is the inner-loop learning rate. The outer-loop then updates $\boldsymbol{\theta}$ with query set $\boldsymbol{Q}_k^{tr}$:

$$\boldsymbol{\theta} = \boldsymbol{\theta} - \beta \nabla_{\boldsymbol{\theta}} \mathcal{L}_{T_k}(\boldsymbol{\theta}'_k, \boldsymbol{Q}_k^{tr}), \tag{18}$$

where β is the outer-loop learning rate.

4 Experiments

4.1 Experimental Setup

We run experiments on three distinct datasets [24] to evaluate the performance of TMAG. In each dataset, users and items are divided into *existing* and *new* categories based on interaction timestamps and item release dates. We then split the data into meta-training and meta-testing subsets. Meta-training contains feedback from existing users on existing items, with 10% randomly held out for validation. Meta-testing covers three cold-start scenarios: (1) ***Scenario 1***: recommending existing items to new users; (2) ***Scenario 2***: recommending new items to existing users; (3) ***Scenario 3***: recommending new items to new users. The recall module is only applied in cold-start settings to enrich embeddings for new users/items, leaving warm cases unaffected; thus, we omit warm performance reporting. For datasets, explicit ratings $(1-5)$ are given. Following prior work [28], ratings above 3 are treated as positive feedback. Detailed statistics are shown in Table 1. Consistent with [19], we filter out users with fewer than 13 or more than 100 interactions. For each user u, we randomly sample 10 interacted items as the query set, with the remainder forming the support set.

Table 1. Statistics of Preprocessed Datasets

	DBook^a	MovieLens^b	Yelp^c
Users	10,592	6,040	51,624
Items	20,934	3,881	34,199
Interactions	649,381	1,000,209	1,301,869
Sparsity	99.71%	95.73%	92.63%

We use three categories of baseline models: traditional methods (MF[16] and NeuMF[12]), GNN-based methods (NGCF[39], GraphSAINT[43], and LightGCN[11]), and cold-start methods (MeLU[19], MetaHIN[24], and CLCRec[42]). We employ three commonly used ranking metrics: Recall@K, Normalized Discounted Cumulative Gain (NDCG)@K and Mean Average Precision (MAP)@K with $K = 10$.

Implementation Details. For fair comparisons, all models follow consistent settings. MF and NeuMF use the NeuRec library[1], while NGCF, GraphSAINT, LightGCN, MeLU, MetaHIN, and CLCRec use their public code with original configurations, adapted to our data and evaluation. TMAG is implemented in TensorFlow[2], with a fixed embedding size of 64 for users and items across all models. All models are trained with the Adam optimizer

[1] https://github.com/wubinzzu/NeuRec.
[2] https://www.tensorflow.org/.

Table 2. Comparison between our proposed model and other baselines at top-10. R is short for *Recall*, ND for *NDCG*, M for *MAP*. Boldface denotes the highest score and underlines denotes the best performing baselines.

Scenario 1	DBook			MovieLens			Yelp		
	R@10	ND@10	M@10	R@10	ND@10	M@10	R@10	ND@10	M@10
MF	0.0834	0.0941	0.0388	0.1403	0.1588	0.0718	0.0517	0.0552	0.0193
NeuMF	0.0852	0.0955	0.0395	0.1439	0.1612	0.0729	0.0520	0.0557	0.0196
NGCF	0.0862	0.0980	0.0417	0.1524	0.1729	0.0797	0.0544	0.0607	0.0223
GraphSAINT	0.0998	0.1149	0.0497	0.1980	0.2233	0.1066	0.0634	0.0706	0.0268
LightGCN	0.1007	0.1175	0.0516	0.2022	0.2279	0.1106	0.0630	0.0703	0.0266
MeLU	0.0854	0.0969	0.0398	0.1452	0.1634	0.0742	0.0530	0.0564	0.0203
MetaHIN	0.0860	0.0973	0.0399	0.1503	0.1685	0.0768	0.0536	0.0579	0.0211
CLCRec	0.0998	0.1106	0.0465	0.2077	0.2387	0.1163	0.0551	0.0559	0.0192
TMAG	**0.1046***	**0.1212***	**0.0530***	**0.2115***	**0.2451***	**0.1210***	**0.0665***	**0.0732***	**0.0276***
%Improv.	3.87%	3.15%	2.71%	1.83%	2.68%	4.04%	4.89%	3.68%	2.99%
Scenario 2	DBook			MovieLens			Yelp		
	R@10	ND@10	M@10	R@10	ND@10	M@10	R@10	ND@10	M@10
MF	0.1278	0.1450	0.0667	0.2336	0.2706	0.1359	0.0554	0.0582	0.0214
NeuMF	0.1292	0.1470	0.0680	0.2362	0.2734	0.1377	0.0562	0.0576	0.0206
NGCF	0.1499	0.1691	0.0784	0.2666	0.3086	0.1630	0.0745	0.0816	0.0316
GraphSAINT	0.1664	0.1861	0.0890	0.2729	0.3163	0.1682	0.0845	0.0938	0.0382
LightGCN	0.1673	0.1868	0.0896	0.2768	0.3198	0.1715	0.0849	0.0933	0.0378
MeLU	0.1370	0.1526	0.0705	0.2466	0.2891	0.1492	0.0584	0.0657	0.0256
MetaHIN	0.1392	0.1549	0.0717	0.2509	0.2930	0.1518	0.0602	0.0677	0.0263
CLCRec	0.1437	0.1620	0.0741	0.2417	0.2840	0.1455	0.0809	0.0844	0.0307
TMAG	**0.1732***	**0.1969***	**0.0968***	**0.2871***	**0.3314***	**0.1791***	**0.0906***	**0.0998***	**0.0397***
%Improv.	3.53%	5.41%	8.04%	3.72%	3.63%	4.43%	6.71%	6.40%	3.93%
Scenario 3	DBook			MovieLens			Yelp		
	R@10	ND@10	M@10	R@10	ND@10	M@10	R@10	ND@10	M@10
MF	0.0983	0.1085	0.0457	0.2256	0.2649	0.1319	0.0280	0.0308	0.0109
NeuMF	0.1026	0.1128	0.0471	0.2281	0.2650	0.1316	0.0288	0.0311	0.0109
NGCF	0.1152	0.1252	0.0552	0.2366	0.2755	0.1388	0.0667	0.0687	0.0264
GraphSAINT	0.1248	0.1340	0.0571	0.2537	0.2951	0.1519	0.0698	0.0717	0.0266
LightGCN	0.1263	0.1363	0.0586	0.2568	0.2971	0.1536	0.0686	0.0704	0.0262
MeLU	0.1056	0.1187	0.0497	0.2326	0.2700	0.1354	0.0328	0.0329	0.0113
MetaHIN	0.1077	0.1200	0.0503	0.2346	0.2726	0.1369	0.0339	0.0337	0.0115
CLCRec	0.1068	0.1195	0.0506	0.2162	0.2525	0.1244	0.0607	0.0627	0.0223
TMAG	**0.1325***	**0.1427***	**0.0619***	**0.2649***	**0.3062***	**0.1612***	**0.0723***	**0.0749***	**0.0281***
%Improv.	4.91%	4.70%	5.63%	3.15%	3.06%	4.95%	3.58%	4.46%	5.64%

* indicates the improvements of TMAG over the best baseline are statistically significant (i.e., one-sample t-test with $p < 0.05$)

[14] and Xavier initialization [9], with early stopping to prevent overfitting. Hyperparameters were tuned via grid search: cluster numbers 10 – 50 (step 10), inner updates 1 – 5, temperature $\tau \in \{0.1, 0.2, 0.5, 1.0\}$, augmentation threshold $t \in \{0.5, 0.6, 0.7, 0.8, 0.9\}$, and regularization coefficients $\lambda_1, \lambda_2, \lambda_3 \in \{0.005, 0.01, 0.05, 0.1, 0.5, 1.0\}$.

4.2 Performance Comparison

In Table 2, we present the top-10 recommendation performance on three datasets. The main observations are as follows: (1) In the first group of baselines, MF is slightly inferior to NeuMF. (2) GNN-based baselines show substantial improvements across all datasets and scenarios. (3) Among the meta-learning-based methods, CLCRec achieves competitive results by maximizing mutual information from collaborative signals. (4) TMAG consistently surpasses all baselines.

Table 3. Effect of the number of aligned tasks.

	DBook		MovieLens		Yelp	
	R@10	ND@10	R@10	ND@10	R@10	ND@10
TMAG-1	0.1015	0.1178	0.2082	0.2392	0.0636	0.0714
TMAG-10	0.1022	0.1185	0.2094	0.2413	0.0648	0.0721
TMAG-20	0.1038	0.1193	0.2106	0.2431	0.0662	0.0727
TMAG-30	0.1039	0.1202	0.2113	0.2443	**0.0665**	**0.0732**
TMAG-40	**0.1046**	**0.1212**	**0.2115**	**0.2451**	0.0664	0.0731
TMAG-50	0.1041	0.1207	0.2111	0.2439	0.0660	0.0729

4.3 Ablation Study

We conduct an ablation study to evaluate TMAG under various conditions. ***Task 1*** is used as the default scenario, with performance measured by R@10 (Recall@10) and NDCG@10 (Normalized Discounted Cumulative Gain@10), providing insights into cold-start recommendation quality.

Effect of Task Alignment. To assess the impact of task alignment, we vary the number of clusters in $\{1, 10, 20, 30, 40, 50\}$. Results are summarized in Table 3, where TMAG-1 represents the case without task alignment, treating each user as an individual task. From Tables 2 and 3, we observe: (1) Increasing the number of clusters generally improves recommendation performance. TMAG with task alignment consistently outperforms TMAG-1, indicating that modeling similar user groups enhances latent clustering knowledge extraction. (2) The optimal cluster number depends on the dataset: TMAG-40 performs best on MovieLens and DBook, while TMAG-30 is optimal for Yelp. This difference likely arises from variations in user-side information diversity; Yelp's user attributes are less diverse, requiring fewer clusters. (3) Across all cluster settings, TMAG consistently surpasses other baselines, demonstrating its efficiency in capturing latent clustering knowledge and adapting globally shared meta-knowledge for rapid cold-start adaptation.

Effect of Interaction Augmentation. TMAG integrates both graph structure and attribute information to augment the adjacency matrix, addressing the challenge of sparse interactions. To evaluate the contribution of each augmentation type, we conducted experiments by removing one type of augmentation or both. Results are presented in Table 4, from which we observe: (1) The best performance is achieved when both graph and attribute augmentations are employed. Removing either type leads to a noticeable performance drop, demonstrating the importance of combining both sources of information. (2) Graph-based augmentation consistently outperforms attribute-based augmentation, indicating that collaborative signals derived from the graph structure are more effective in capturing potential interactions. (3) Both augmentation methods outperform the variant without any augmentation, highlighting the positive effect of the interaction generator in learning more expressive user and item representations.

Table 4. Effect of the augmentation variants.

	DBook		MovieLens		Yelp	
	R@10	ND@10	R@10	ND@10	R@10	ND@10
TMAG-ga	0.1034	0.1198	0.2101	0.2438	0.0648	0.0721
TMAG-a	0.1044	0.1209	0.2113	0.2449	0.0661	0.0728
TMAG-g	0.1035	0.1201	0.2103	0.2441	0.0650	0.0725
TMAG	**0.1046**	**0.1212**	**0.2115**	**0.2451**	**0.0665**	**0.0732**

Subscript notation: TMAG-ga removes the graph and attribute augmentation, TMAG-a removes the attribute augmentation, and TMAG-g removes the graph augmentation.

Effect of Autoencoder. In our quest to understand how the autoencoder contributes to clustering for aligned tasks, we explored two clustering methods: (1) Direct one-hot encoding. This method involves directly utilizing the side information of users and items through one-hot encoding; (2) Autoencoder-based clustering. In this approach, an autoencoder is employed to derive attribute embeddings. The results of these experiments are summarized in Table 5. Our observations indicate that TMAG with the autoencoder consistently delivers better performance. This improvement can be attributed to the autoencoder's effectiveness in learning attribute representations, which enhances the clustering process for aligned tasks.

Table 5. Effect of the autoencoder.

	DBook		MovieLens		Yelp	
	R@10	ND@10	R@10	ND@10	R@10	ND@10
w/o AE	0.1030	0.1186	0.2101	0.2443	0.0653	0.0727
w/ AE	**0.1046**	**0.1212**	**0.2115**	**0.2451**	**0.0665**	**0.0732**

Effect of Contrastive Regularization. To investigate the effectiveness of the contrastive regularization component in our model, we conduct experiments under two cases: 1) TMAG without Contrastive Regularization, and 2) TMAG with Contrastive Regularization. The results are summarized in Table 6. Our findings indicate that TMAG with the contrastive regularization consistently performs better. This suggests that task-wise attribute contrastive learning enhances the latent clustering prior knowledge and empowers the model to acquire more informative attribute representations, ultimately leading to improved performance.

Table 6. Effect of the contrastive augmentation

	DBook		MovieLens		Yelp	
	R@10	ND@10	R@10	ND@10	R@10	ND@10
w/o Con	0.1034	0.1196	0.2104	0.2439	0.0652	0.0720
w/ Con	**0.1046**	**0.1212**	**0.2115**	**0.2451**	**0.0665**	**0.0732**

4.4 Model Sensitivity

Impact of Support Set Size. The sparsity of the support set plays a critical role in cold-start recommendation performance. In this study, we varied the size of the support set for interacted items from 5 to 70, excluding users with more than 70 interactions due to their limited number. The results are presented in Fig. 3. Key observations are as follows: (1) All methods generally improve as the support set size increases, consistent with the expectation that more historical interactions provide richer information for recommendation. (2) TMAG exhibits the smallest performance degradation with reduced support set sizes, indicating its robustness in scenarios with limited historical interactions. The model maintains strong performance regardless of interaction history length. (3) When historical interactions are extensive, the number of corresponding users becomes small, leading to potential instability in performance evaluation for users with long interaction histories.

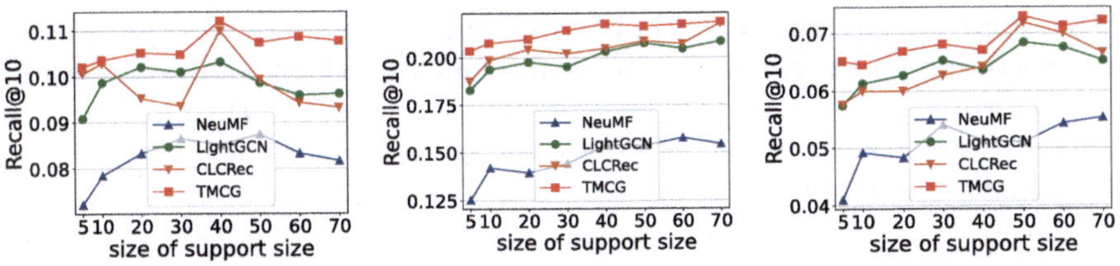

(a) DBook (b) MovieLens (c) Yelp

Fig. 3. Impacts of the size of support sets.

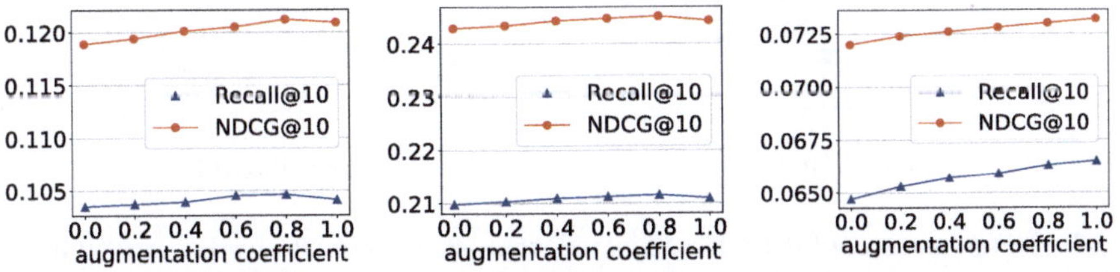

(a) DBook (b) MovieLens (c) Yelp

Fig. 4. Impacts of augmentation coefficient.

Impact of Augmentation Coefficient. We evaluate the robustness of interaction augmentation. We test the coefficient α in Eq. (9) on three datasets in the range of $\{0, 0.2, 0.4, 0.6, 0.8, 1.0\}$. The results are shown in Fig. 4. Key observations are as follows: (1) TMAG demonstrates stable performance across different coefficients. Its recommendation quality remains largely unaffected by the specific value of α, highlighting the adaptability of the proposed framework to varying augmentation settings. (2) Among the two augmentation strategies, graph augmentation contributes more significantly to performance. The model achieves optimal results on DBook and MovieLens at $\alpha = 0.8$, whereas on Yelp, the best performance occurs at $\alpha = 1.0$. This difference likely stems from the limited attribute information available on Yelp, which diminishes the effectiveness of attribute-based augmentation.

4.5 Case Study

4.6 Visualization of Task Alignment Effects

To investigate how task alignment enhances representation learning, we randomly selected five users from the same task within the Yelp dataset, along with their interacted items. We visualized the learned user embeddings from NGCF and TMAG using the t-SNE algorithm [31]. The results are shown in Fig. 5. Notably, the visualized items are from the query set in ***Scenario 1*** and were not paired with users during training. We observe that, compared with NGCF,

TMAG embeddings of users within the same task are more closely clustered in the embedding space. Similarly, items interacted with by the same users also tend to form distinct clusters. These observations indicate that task alignment effectively exploits latent clustering knowledge to learn more personalized and discriminative node representations.

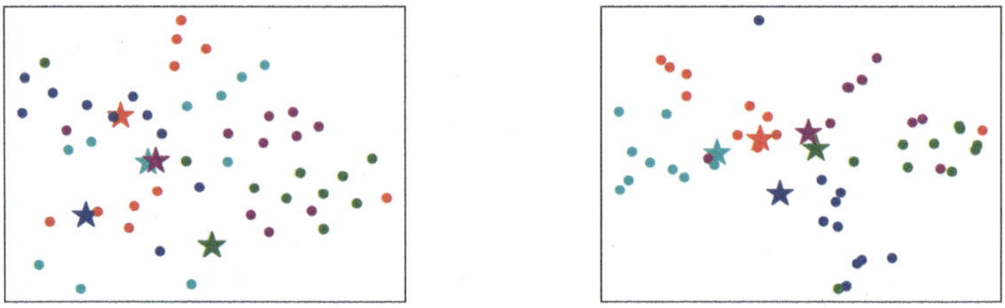

(a) NGCF (b) TMAG

Fig. 5. Visualization of t-SNE projected representations derived from NGCF and TMAG. Users are randomly selected in the same task from Yelp and are represented by stars. The points with the same color are users' interacted items.

5 Conclusion and Future Work

In this paper, we introduce TMAG as a solution for addressing the cold-start problem in recommendation systems, targeting both the model-level and feature-level challenges. At the model-level, we presented a task-aligned constructor to capture latent clustering knowledge, facilitating rapid adaptation to new users and mitigating local optima issues. We also incorporated task-wise attribute contrastive regularization to bolster the latent clustering knowledge. At the feature-level, we combined graph structure and attribute information to enhance the adjacency matrix, addressing data sparsity concerns. Our extensive experiments on three real-world datasets highlighted the effectiveness of our model in the context of cold-start recommendation. For future works, we have two promising directions for future work. First, we aim to extend our model's capabilities to address the cold-start problem from the perspective of items, ensuring a more holistic recommendation approach. Secondly, we are interested in adapting our model to sequential recommendation scenarios, exploring its effectiveness in those settings. These directions represent avenues for further research and innovation in the field of recommendation systems.

Artifacts (Anonymous)

We provide code and scripts at https://anonymous.4open.science/r/coldstart-artifacts-1C10.

References

1. Bokde, D., Girase, S., Mukhopadhyay, D.: Matrix factorization model in collaborative filtering algorithms: a survey. Proc. Comput. Sci. **49**, 136–146 (2015)
2. Briand, L., Salha-Galvan, G., Bendada, W., Morlon, M., Tran, V.-A.: A semi-personalized system for user cold start recommendation on music streaming apps. In: KDD, pp. 2601–2609 (2021)
3. Chae, D.-K., Kim, J., Chau, D.H., Kim, S.-W.: AR-CF: augmenting virtual users and items in collaborative filtering for addressing cold-start problems. In: SIGIR, pp. 1251–1260 (2020)
4. Chen, Z., Xiao, R., Li, C., Ye, G., Sun, H., Deng, H.: ESAM: discriminative domain adaptation with non-displayed items to improve long-tail performance. In: SIGIR, pp. 579–588 (2020)
5. Dong, M., Yuan, F., Yao, L., Xu, X., Zhu, L.: Mamo: memory-augmented meta-optimization for cold-start recommendation. In: KDD, pp. 688–697 (2020)
6. Yuntao, D., Xinjun Zhu, L., Chen, Z.F., Gao, Y.: Metakg: meta-learning on knowledge graph for cold-start recommendation. IEEE Trans. Knowl. Data Eng. **35**(10), 9850–9863 (2023)
7. Finn, C., Abbeel, P., Levine, S.: Model-agnostic meta-learning for fast adaptation of deep networks. In: ICML, pp. 1126–1135 (2017)
8. Gantner, Z., Drumond, L., Freudenthaler, C., Rendle, S., Schmidt-Thieme, L.: Learning attribute-to-feature mappings for cold-start recommendations. In: ICDM, pp. 176–185 (2010)
9. Glorot, X., Bengio, Y.: Understanding the difficulty of training deep feedforward neural networks. In: AISTATS, pp. 249–256 (2010)
10. Hadsell, R., Chopra, S., LeCun, Y.: Dimensionality reduction by learning an invariant mapping. In: CVPR, 2, pp. 1735–1742 (2006)
11. He, X., Deng, K., Wang, X., Li, Y., Zhang, Y., Wang, M.: Lightgcn: simplifying and powering graph convolution network for recommendation. In: SIGIR, pp. 639–648 (2020)
12. He, X., Liao, L., Zhang, H., Nie, L., Hu, X., Chua, T.-S.: Neural collaborative filtering. In: WWW, pp. 173–182 (2017)
13. Jeon, H., Lee, J.-E., Yun, J., Kang, U.: Cold-start bundle recommendation via popularity-based coalescence and curriculum heating. In: Proceedings of the ACM Web Conference 2024, pp. 3277–3286 (2024)
14. Diederik, P., Ba, K.J.: A method for stochastic optimization. In: ICLR, Adam (2015)
15. Thomas, N., Welling, K.M.: Semi-supervised classification with graph convolutional networks, In: ICLR (2017)
16. Koren, Y., Bell, R., Volinsky, C.: Matrix factorization techniques for recommender systems. Computer **42**(8), 30–37 (2009)
17. Krishnan, A., Das, M., Bendre, M., Yang, H., Sundaram, H.: Transfer learning via contextual invariants for one-to-many cross-domain recommendation. In: SIGIR, pp. 1081–1090 (2020)

18. Krishnan, A., Sharma, A., Sundaram, H.: Insights from the long-tail: learning latent representations of online user behavior in the presence of skew and sparsity. In: CIKM, pp. 297–306 (2018)
19. Lee, H., Im, J., Jang, S., Cho, H., Chung, S.: Melu: meta-learned user preference estimator for cold-start recommendation. In: KDD, pp. 1073–1082 (2019)
20. Li, H., et al.: Disco: graph-based disentangled contrastive learning for cold-start cross-domain recommendation. In: Proceedings of the AAAI Conference on Artificial Intelligence, vol. 39, pp. 12049–12057 (2025)
21. Li, Y., et al.: Personalized federated recommendation for cold-start users via adaptive knowledge fusion. In: Proceedings of the ACM on Web Conference vol. 2025, 2700–2709 (2025)
22. Likas, A., Vlassis, N., Verbeek, J.J.: The global k-means clustering algorithm. Pattern Recogn. **36**(2), 451–461 (2003)
23. Liu, S., Ounis, I., Macdonald, C., Meng, Z.: A heterogeneous graph neural model for cold-start recommendation. In: SIGIR, pp. 2029–2032 (2020)
24. Lu, Y., Fang, Y., Shi, C.: Meta-learning on heterogeneous information networks for cold-start recommendation. In: KDD, pp. 1563–1573 (2020)
25. Lyu, S., Zhou, X., Hu, X.: Multi-view hypergraph-based contrastive learning model for cold-start micro-video recommendation. In: ICASSP 2025-2025 IEEE International Conference on Acoustics, Speech and Signal Processing (ICASSP), pp. 1–5. IEEE (2025)
26. Man, T., Shen, H., Jin, X., Cheng, X.: Cross-domain recommendation: an embedding and mapping approach. In IJCAI **17**, 2464–2470 (2017)
27. Rendle, S., Freudenthaler, C., Gantner, Z., Schmidt-Thieme, L.: BPR: Bayesian personalized ranking from implicit feedback. In: UAI, pp. 452–461 (2009)
28. Sachdeva, N., Manco, G., Ritacco, E., Pudi, V.: Sequential variational autoencoders for collaborative filtering. In: WSDM, pp. 600–608 (2019)
29. Sankar, A., Wang, J., Krishnan, A., Sundaram, H.: Prototypical collaborative filtering for few-shot recommendation. In: RecSys, Protocf (2021)
30. Song, J., Xu, J., Zhou, R., Chen, L., Li, J., Liu, C.: CBML: a cluster-based meta-learning model for session-based recommendation. In: CIKM, pp. 1713–1722 (2021)
31. der Maaten, L.V., Hinton, G.: Visualizing data using T-SNE. J. Mach. Learn. Res. **9**(11) (2008)
32. Vanschoren, J.: Meta-learning: a survey. arXiv preprint arXiv:1810.03548 (2018)
33. Vartak, M., Thiagarajan, A., Miranda, C., Bratman, J., Larochelle, H.: A meta-learning perspective on cold-start recommendations for items (2017)
34. Vinyals, O., Blundell, C., Lillicrap, T., Wierstra, D., et al.: Matching networks for one shot learning. NeurIPS **29**, 3630–3638 (2016)
35. Volkovs, M., Yu, G.W., Poutanen, T.: Dropoutnet: addressing cold start in recommender systems. In: NeurIPS, pp. 4957–4966 (2017)
36. Wang, J., Huang, P., Zhao, H., Zhang, Z., Zhao, B., Lee, D.L.: Billion-scale commodity embedding for E-commerce recommendation in Alibaba. In: KDD, pp. 839–848 (2018)
37. Wang, L., et al.: Preference-adaptive meta-learning for cold-start recommendation. In: IJCAI, pp. 1607–1614 (2021)
38. Wang, S., Zhang, K., Wu, L., Ma, H., Hong, R., Wang, M.: Privileged graph distillation for cold start recommendation. In: SIGIR, pp. 1187–1196 (2021)
39. Wang, X., He, X., Wang, M., Feng, F., Chua, T.-S.: Neural graph collaborative filtering. In: SIGIR, pp. 165–174 (2019)

40. Wang, Y., et al.: Diff-MSR: a diffusion model enhanced paradigm for cold-start multi-scenario recommendation. In: Proceedings of the 17th ACM International Conference on Web Search and Data Mining, pp. 779–787 (2024)
41. Wei, T., et al.: Fast adaptation for cold-start collaborative filtering with meta-learning. In: ICDM, pp. 661–670 (2020)
42. Wei, Y., et al.: Contrastive learning for cold-start recommendation. In:MM, pp. 5382–5390 (2021)
43. Zeng, H., Zhou, H., Srivastava, A., Kannan, R., Prasanna, V.: Graph sampling based inductive learning method. In: ICLR, Graphsaint (2019)

Feasibility of AI-Assisted Programming for End-User Development

Irene Weber(✉)

Faculty of Mechanical Engineering, University of Applied Sciences Kempten, Bahnhofstraße 61, 87435 Kempten, Germany
irene.weber@hs-kempten.de

Abstract. End-user development, where non-programmers create or adapt their own digital tools, can play a key role in driving digital transformation within organizations. Currently, low-code/no-code platforms are widely used to enable end-user development through visual programming, minimizing the need for manual coding.

Recent advancements in generative AI, particularly large language model-based assistants and "copilots", open new possibilities, as they may enable end users to generate and refine programming code and build apps directly from natural language prompts. This approach, here referred to as AI-assisted end-user coding, promises greater flexibility, broader applicability, faster development, improved reusability, and reduced vendor lock-in compared to the established visual LCNC platforms.

This paper investigates whether AI-assisted end-user coding is a feasible paradigm for end-user development, which may complement or even replace the LCNC model in the future. To explore this, we conducted a case study in which non-programmers were asked to develop a basic web app through interaction with AI assistants. The majority of study participants successfully completed the task in reasonable time and also expressed support for AI-assisted end-user coding as a viable approach for end-user development. The paper presents the study design, analyzes the outcomes, and discusses potential implications for practice, future research, and academic teaching.

Keywords: Generative AI · Large Language Model · End-User Development · Citizen Development · Low-Code/No-Code · AI-Assisted Programming · Software Engineering

1 Introduction

Companies are driving digitalization to gain efficiency, reduce error rates, avoid media disruptions, and collect high-quality, structured data from the start of all processes. However, capacity bottlenecks in IT departments often slow down the development and rollout of digital solutions. End-User Development (EUD) and Business-Managed IT (BMIT) are two related approaches that help alleviate

H. Unger and P. Meesad (Eds.): NLPIR 2025, LNNS 1904, pp. 1043–1053, 2026.
https://doi.org/10.1007/978-3-032-20897-2_73

such constraints. In EUD, end users develop or modify their own applications, such as workflow automations or digital tools, tailored to their tasks [1,2]. BMIT refers to situations where business units, rather than central IT departments, manage the development or delivery of IT systems [3].

EUD and BMIT have been found to offer various benefits. Solutions stemming from these approaches tend to be well-aligned with user needs and can address niche problems that central IT might overlook or deprioritize due to resource constraints. EUD and BMIT can also enhance a company's adaptability and capacity for innovation. Additionally, they contribute to overall job satisfaction, as employees often experience greater motivation and autonomy through their direct involvement and feel a stronger sense of ownership over the tools they create and use. [3]

Often, companies enable EUD by providing visual low-code/no-code (LCNC) platforms. End users can create digital tools—such as automation flows, data entry forms, app controls, or data displays—by selecting, connecting, and configuring graphical components. In addition to supporting visual development, these LCNC platforms typically include connectors to business applications and host the developed solutions.

Modern artificial intelligence (AI) technologies, especially large language models (LLM), are expected to have immense potential for digitalization in businesses. These LLMs are pre-trained on massive amounts of text data (including computer programs and code) and later fine-tuned to follow instructions, use tools, and more. Triggered by language inputs termed prompts, they can perform a wide range of tasks and exhibit intelligence that seems close to human-level. Two emerging approaches in this realm are the AI-copilot paradigm, where LLMs solve tasks in cooperation with humans, and the AI-agent paradigm, where LLMs are trusted to solve tasks more or less autonomously.

Another approach is to exploit the coding capabilities of LLMs. Given a textual description of the problem to be solved and, potentially, the system context in which the software should run, these models can automatically generate programming code and produce working software solutions in many programming languages. With AI now making it easier to write code, a new form of end-user development (EUD) may become feasible: employees building digital tools by writing code with LLMs. The approach of LLM-assisted end-user app programming, here referred to as end-user coding (EUCod) may help overcome some of the limitations associated with LCNC platforms, as recently discussed in tech-oriented media [4–6]:

1. **Better customization**: Coding allows tailoring software precisely to individual needs, surpassing limitations and pre-configurations of standardized LCNC solutions.
2. **Higher user acceptance through a unified "One-Stop-Shop" solution**: Users no longer need to rely on multiple specialized tools (e.g., platforms for building data collection forms, poll-based meeting planners, etc.). Instead, they can work within a single environment, avoiding the hassle of

remembering, managing and switching between different tools, each with its own learning curve.
3. **Faster development**: Compared to the numerous manual interactions with graphical components required when working with visual LCNC platforms, instructing an LLM to write or adapt code can be faster.
4. **Improved accessibility and reusability; elimination of vendor lock-in**: Unlike solutions built on proprietary LCNC platforms, self-generated custom and self-used code[1] is in the ownership of the coder, fully accessible and easier to modify, extend, reuse, and transfer to other environments.

This paper presents an experimental study investigating the viability of LLM-assisted EUCod as a potentially emerging approach to EUD. Specifically, we explore whether non-programmers are able and willing to develop basic digital tools with the assistance of LLMs. To this aim, we asked study participants to build a simple web application, specifically, a survey form, using ChatGPT or a similar LLM-based chat as their development assistant. The paper is structured as follows: Sect. 2 provides an overview of related work as well as approaches encountered in practice. Section 3 presents the experimental setup and reports the results. Section 4 discusses the findings and their implications and concludes.

2 Related Work

Various approaches apply large language models (LLMs) to enable end users, particularly those without formal training in software development, to create application code[2]. A recent literature review on the role of end users in LLM-assisted end-user development is [7].

A growing body of work investigates the use of **LLMs for robot programming**. Several studies explicitly address robot programming by end users with LLM assistance [8–11]. These approaches typically provide integrated development environments (IDEs) that feature a chat interface where users describe requirements in natural language. In addition, they offer visual representations of the LLM-generated robot control code, such as flowcharts or block diagrams, which users can inspect and modify. Some of these works also describe the predefined instructions (system prompts) used to inform the LLM about available robot actions and the expected output formats [8,9,11].

A further strand of research investigates **LLMs as educational tools for novice programmers** [12–15]. For example, [12] examines how undergraduate students use LLMs to complete beginner-level programming tasks, deriving insights for future programming education.

[1] However, there is a risk that code generated by LLMs infringes on third-party copyrights.

[2] The author does not suggest that these approaches guarantee success or are free of risk. AI-generated code may contain errors, inefficiencies, or security vulnerabilities, and therefore requires careful testing and oversight.

Some authors **combine AI-assisted EUCod with LCNC platforms**, such as Oracle APEX [16] or the Google Home companion app. Oracle APEX is a LCNC development platform for building web applications on top of Oracle databases [17]. In the approach proposed by [16], end users can modify existing APEX applications using natural language requests. An LLM translates these requests into calls to APEX's Application Editing API. [18] explore speech-only conversational approaches to creating routines for virtual assistants on smart speakers, using the Google Home companion app as a model.

[19,20] are early studies on **LLM-supported end-user web and app development**, exploring code generation for websites with the help of language models. [21] presents a browser extension that supports CSS styling through LLM assistance, all prior to the advent of ChatGPT in November 2022. More recent work includes a proof-of-concept by [22], which generates HTML and CSS code for websites based on natural language descriptions. Their system uses pre-engineered prompts to guide the LLM in producing outputs that conform to a specific template, making the responses directly parsable. The authors plan to integrate this approach into LCNC platforms to improve its accessibility and effectiveness for non-technical users.

Recently, **commercial AI-based platforms for web and app development** have emerged that aim to empower end users to create applications through natural language inputs without writing code [23–25]. SoftGen.ai [23] and Lovable.dev [24] support the generation of full-stack web applications, while Bolt.new [25] additionally enables the creation of mobile apps. These platforms offer third-party integrations with services such as databases, payment systems, and more. Notably, Lovable.dev is the commercial successor to GPT-Engineer, an open-source command-line tool for AI-driven code generation and execution based on natural language software specifications, available on GitHub [26].

Vibe Coding is a recently coined term for AI-driven programming, in which developers iteratively prompt LLMs to generate and refine code with minimal manual intervention [27]. So far, the approach seems to have primarily been promoted as a productivity aid for skilled programmers, rather than as a tool for non-programmers. A first online course introducing vibe coding is now available [28].

3 Empirical Study on AI-Assisted End-User Coding

We conducted an experimental study to evaluate whether EUCod may be a feasible approach to EUD, possibly replacing the visual development paradigm supported by LCNC platforms in the future.

The study was conducted at a German university of applied sciences with students of industrial engineering as **participants**, including both bachelor's and master's degree students. These students commonly have basic coding skills from introductory programming classes, but no further formal education as software or app developers.

As the task of designing a survey is likely to arise in roles such as knowledge work, project management, and related fields, participants were asked to implement a survey app as the **study task**. To avoid the need for a complex development environment or deployment process, the technical setup was deliberately kept simple: the form was to be created as a single HTML file, complete with CSS styling and JavaScript for interactive elements, saved locally, and previewed in a standard web browser. VS Code was recommended as the code editor for development. For collecting survey responses, participants were instructed to connect to a Google Sheet via a Google Apps Script, providing a lightweight and accessible backend solution.

Task Instructions specified further details as follows: Participants were instructed to design a meaningful survey on a topic of their choice. The survey should include a variety of input elements (e.g., both checkboxes and radio buttons), but avoid overly complex features. If possible, the form was to include a dynamic element, such as a follow-up question revealed conditionally based on user input.

Participants were instructed to begin by describing their task to the LLM and then develop the form iteratively by refining and expanding it through further prompts. To help the participants get started, a basic, non-mandatory starting prompt was suggested: *"How can I design an HTML survey form that... ?"* They were encouraged to ask the LLM for explanations in case they encountered challenges or unclear concepts.

The Experimental Setup was designed to foster a relaxed, workshop-like atmosphere inviting to exploration. Participants were given the opportunity to work on the task during lecture hours using university lab computers. Master's students were explicitly instructed to work in pairs, while bachelor's students received no specific collaboration instructions. However, this distinction was not strictly enforced, and some variation in group work occurred. Use of university lab computers was also not enforced; students were free to work on their own devices or at home. Master's students worked on the task as part of a graded assignment, whereas bachelor's students could engage with it as an optional, ungraded lab exercise. At the beginning of the study, a pre-developed example survey form was demonstrated, and its components and technical setup were explained during an approximately 20-minute introduction.

Participants progressed far more quickly than expected and were asked to complete a **follow-up survey** on their project status and experiences the following week. The original task instructions, the survey form and the collected responses (in German) are available in the **online appendix**[3].

3.1 Results

The survey received 33 usable responses, excluding one incomplete submission and those that were submitted after the deadline. Of these, 21 came from the

[3] https://weberi.github.io/2025_AI-Assisted-End-User-Coding.

master's group and 12 from the bachelor's group. All participants completed the survey individually, not as teams. The survey included questions in which participants self-assessed their frequency of AI use and their level of expertise in relevant skills, see Fig. 1.

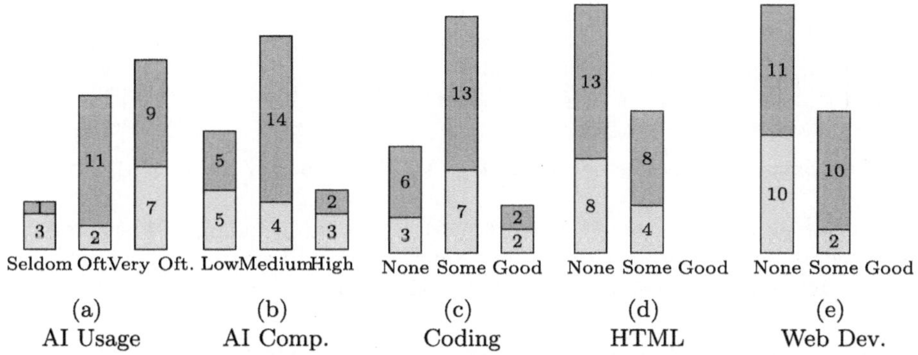

Fig. 1. Self-assessment of prior experience. ▭ BA, ▬ MA. (AI Comp.: AI Competency; Web Dev.: Web Development; Oft.: Often; Very Oft.: Very Often)

Most participants reported using AI tools *often* or *very often*. Responses for AI Competence are skewed toward *medium* competence, with few rating themselves as *high*. Participants reported *some* coding experience more frequently than *none* or *good*. Only few considered their coding skills *good*. Participants indicated even less familiarity with HTML and web development. The majority reported their knowledge in these areas as *none*, and no participants rated their skills as *good*.

Figure 2 summarizes the main outcomes of the survey. Most participants (24 out of 33) reported successful completion of the task, meaning that they had developed a functional survey. Success was notably higher among master's students, 85% of whom completed the task successfully, compared to 50% among bachelor's students.

Regarding the perceived effort, a clear majority comprising 24 participants rated the effort-to-outcome ratio as *appropriate*. Three participants had found the task *easier* than expected, while six participants described the process as *difficult*.

Participants were also asked whether they would recommend offering the AI-assisted end-user coding approach within enterprise settings. The overwhelming majority (28 out of 33) responded *yes*. When asked whether they would personally use the approach in the future, 16 participants answered *yes*, and 15 selected *maybe*. Only two bachelor's students responded with a clear *no*.

The survey further asked participants to report the time spent developing the survey app, see Fig. 3a. Successfully completed app developments are distributed across all time categories, see Fig. 3b. The majority of participants

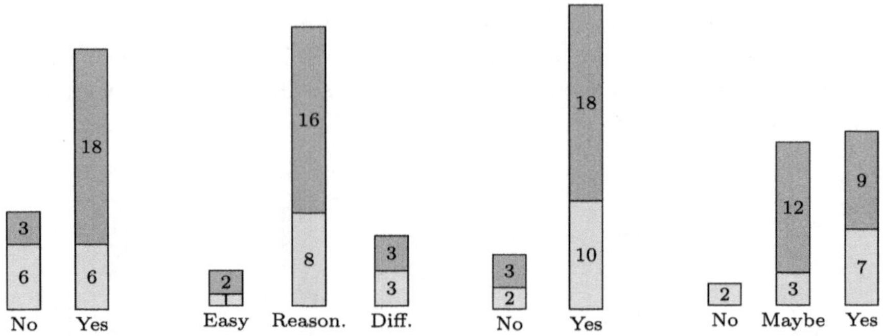

(a) Task Success (b) Perceived Effort (c) Recommend (d) Use

Fig. 2. Survey responses on AI-assisted end-user coding including task success, perceived effort, and openness to recommending and using the approach in enterprise settings. Reason.: Reasonable; Diff.: Difficult. ▭ BA, ■ MA.

who succeeded had spent at least 2–3 h on the task, suggesting that a moderate time investment was generally sufficient for success. Participants also indicated the complexity of their surveys. A survey was considered complex if it included dynamic elements that adapt to user input, implemented input validation, or similar features. Nearly half of the participants rated their surveys as complex. Notably, complexity did not clearly correlate with longer development times, see Fig. 3c.

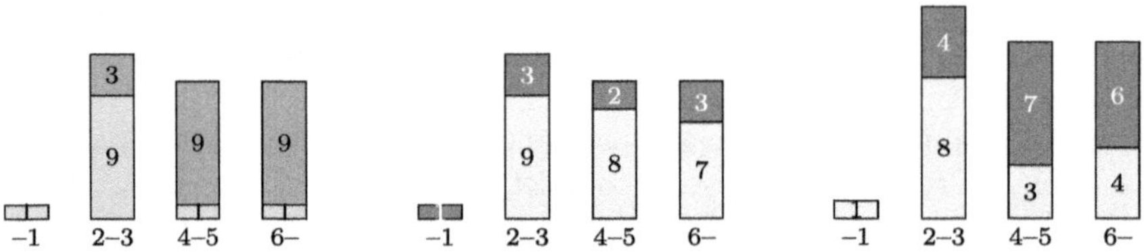

(a) Hours Spent
▭ BA, ■ MA.

(b) Hours vs. Success
▭ Success, ■ No Success.

(c) Hours vs. Complexity
▭ Simple, ■ Complex.

Fig. 3. Time spent on survey development.

A further group of survey questions addressed difficulties encountered during app development, organizational measures enterprises should take when introducing AI-assisted EUCod, and the AI models used for the task; see Table 1 for the results. These questions allowed unrestricted responses, which were analyzed and categorized by the author of this paper. Some responses addressed multiple themes and were therefore assigned to more than one category.

The most frequent development issue concerned the integration with Google Sheets, followed by challenges related to HTML and web development, and for-

Table 1. Overview of reported development challenges, support needs, and AI assistants used.

(a) Reported Difficulties

Issue	Count
Analysis in Google Sheets	3
Prompting	6
HTML & Web Dev	7
GS Integration	15
Faulty Code	4
Model restrictions	1

(b) Needs

Provision	Count
Training	18
Platform	10
Support	4

(c) AI Assistants

Assistant	Count
ChatGPT	31
DeepSeek	8
Copilot	2
Gemini	1
Perplexity	3
Lovable	2

mulating effective prompts. Less commonly mentioned were problems with faulty code generation, and technical model restrictions, such as context length or usage limits. Some participants interpreted the study task as including the analysis of survey responses within Google Sheets, which they also found to be challenging. When asked what support enterprises should provide to enable AI-assisted end-user development, most participants emphasized the need for introductory training, followed by the demand for provision of a dedicated, easy-to-use development platform. A few additionally mentioned access to human experts or a user community for support. Among the AI assistants used, ChatGPT clearly dominated, but several participants also reported using alternative tools or combinations of tools.

4 Discussion and Conclusion

This study explores the feasibility of AI-assisted application programming (EUCod) for EUD. Participants with minimal formal IT education or software development experience were tasked with developing a survey application using AI assistance. The application featured a frontend built with HTML, CSS, and JavaScript, and a backend implemented via Google Sheets and Google Apps Script for storing responses, qualifying it as a minimal full-stack application.

The majority of participants found the effort-to-outcome ratio appropriate and recommended that enterprises support AI-assisted EUCod for EUD initiatives. Nearly all participants expressed willingness to adopt this approach in the future. These findings suggest that AI-assisted EUCod could be a viable approach for EUD in organizations, especially as we can expect that LLM assistants and also AI competence among users will continue to improve over time.

While most participants succeeded in developing a functional survey app and connecting it to a simple backend within a reasonable timeframe, they also expressed a need for training, support, and guidance. Approximately 30% suggested that enterprises should implement measures to facilitate EUCod, which the author interpreted as a demand for a dedicated development platform. Offering integrations with backend services, ideally tailored to an organization's existing system landscape, a dedicated development platform could help alleviate challenges such as connecting with a data store, which was the main difficulty reported in our study. While introducing such a platform for AI-assisted EUCod may come with some of the known downsides of LCNC tools (see Sect. 1), the advantages would probably outweigh them, as the platform would also address other concerns mentioned in the survey, such as the need for clear guidelines, compliance, IT governance, and data security.

From the study designer's perspective, participants completed the assigned task more quickly and effectively than expected. Nearly half of them designed "complex" surveys. The deliberate selection and experimentation with different AI assistants, along with the general openness toward AI-assisted EUCod, indicate a promising level of AI literacy among participants. Notably, many students, who will soon join the active workforce, are already regular users of AI assistants.

As with any empirical study, several limitations must be acknowledged. First, the study was conducted with students of industrial engineering at a German university of applied sciences. While these participants had limited programming experience, they still possess a technical background and may not fully represent the diversity of typical end users in enterprise contexts. This affects the external validity of the findings.

Second, the study involved two distinguishable groups of participants who shared similar academic backgrounds but differed in experience and study conditions. Participation was voluntary for some and mandatory for others, which may have influenced their motivation and level of engagement. This variation, along with a relatively uncontrolled environment (e.g., some participants worked from home or in teams), introduced variability in how the task was approached. On the other hand, this diversity may also reduce the influence of the specific university setting and thus improve the generalizability of the results.

Third, the findings largely rely on participants' self-assessments and the categorization and interpretation of responses by a single researcher. While this may limit objectivity, it appears acceptable within the scope of a pioneering exploratory study.

Despite these limitations, the study offers valuable initial insights into the feasibility of AI-assisted end-user coding and highlights it as a promising and relevant topic not only for future research, but also for university teaching.

The combination of EUD platforms with AI-assisted end-user coding (EUCod) is of particular interest, as it can take several different forms. First, as mentioned in Sect. 2, dedicated platforms for AI-assisted coding are emerging. A second variant can be found in related work: the idea of programming the platform itself through API calls that are generated with the assistance of an LLM.

A third option, which is not typically subsumed under the umbrella of end-user development, involves using LLMs to generate workflows (or similar digital artifacts) that can be deployed and executed on LCNC platforms. A preliminary evaluation of this direction is provided by [29], who explored ChatGPT's ability to generate BPMN diagrams in a custom simplified JSON format. Own informal testing indicated that LLMs, particularly ChatGPT, are capable of generating flows in JSON that can be imported into and executed by the LCNC platform Node-RED [30], which is often used in the context of home automation, the Internet of Things, and industrial automation.

This approach of generating executable digital artifacts beyond conventional source code may gain practical significance, as it represents an additional pathway for AI-assisted end-user development.

References

1. Lieberman, H., Paternò, F., Wulf, V., Karat, J., Vanderdonckt, J. (eds.): End User Development, Human-Computer Interaction Series, vol. 9. Springer, Netherlands, Dordrecht (2006)
2. Paternò, F., Wulf, V. (eds.): New Perspectives in End-User Development. Springer International Publishing, Cham (2017)
3. Kopper, A., et al.: Business-managed IT: a conceptual framework and empirical illustration. In: Proceedings of the 25th European Conference on Information Systems. Portsmouth (2018). https://www.researchgate.net/publication/325853972
4. Tozzi, C.: AI-assisted development tools vs. low-code/no-code: what to use when. ITPro Today (2024). https://www.itprotoday.com/software-development-techniques/ai-assisted-development-tools-vs-low-code-no-code-what-to-use-when
5. van Hemert, L.: The pitfalls of no-code: why AI-assisted coding offers a better alternative. Medium (2025). https://medium.com/@leonardvanhemert/the-pitfalls-of-no-code-why-ai-assisted-coding-offers-a-better-alternative-12955c4afc80
6. Waehner, K.: Why generative AI and data streaming are replacing visual coding with low-code/no-code platforms (2025). https://www.kai-waehner.de/blog/2025/02/02/why-generative-ai-and-data-streaming-are-replacing-visual-coding-with-low-code-no-code-platforms/
7. Gargioni, L., Barricelli, B.R., Fogli, D., Locoro, A.: Exploring the role of end users in performing EUD with large language models. In: CEUR WORKSHOP PROCEEDINGS. vol. 3685. CEUR-WS (2024). https://iris.unibs.it/handle/11379/600027
8. Ge, Y., Dai, Y., Shan, R., Li, K., Hu, Y., Sun, X.: Cocobo: exploring large language models as the engine for end-user robot programming. In: 2024 IEEE Symposium on Visual Languages and Human-Centric Computing (VL/HCC), pp. 89–95. IEEE (2024). https://ieeexplore.ieee.org/iel8/10714498/10714491/10714576.pdf
9. Karli, U.B., Chen, J.T., Antony, V.N., Huang, C.M.: Alchemist: LLM-aided end-user development of robot applications. In: Proceedings of the 2024 ACM/IEEE International Conference on Human-Robot Interaction, pp. 361–370. ACM, Boulder CO USA (2024)
10. Bimbatti, G., Fogli, D., Gargioni, L.: Can ChatGPT support end-user development of robot programs? In: CEUR Workshop Proceedings. vol. 3408 (2023). https://iris.unibs.it/handle/11379/580765
11. Gargioni, L., Fogli, D., Baroni, P.: Preparation of personalized medicines through collaborative robots: a hybrid approach to the end-user development of robot programs. ACM J. Responsible Comput. 3715852 (2025)

12. Feldman, M.Q., Anderson, C.J.: Non-expert programmers in the generative AI future. In: Proceedings of the 3rd Annual Meeting of the Symposium on Human-Computer Interaction for Work, pp. 1–19. ACM, Newcastle upon Tyne United Kingdom (2024)
13. Prather, J., et al.: The Widening gap: the benefits and harms of generative AI for novice programmers. In: Proceedings of the 2024 ACM Conference on International Computing Education Research - Volume 1. ICER '24, vol. 1, pp. 469–486. Association for Computing Machinery, New York, NY, USA (2024)
14. Nguyen, S., Babe, H.M., Zi, Y., Guha, A., Anderson, C.J., Feldman, M.Q.: How Beginning Programmers and Code LLMs (Mis)read Each Other. In: Proceedings of the 2024 CHI Conference on Human Factors in Computing Systems, pp. 1–26. CHI '24, Association for Computing Machinery, New York, NY, USA (2024)
15. Chen, J., et al.: Learning Agent-based Modeling with LLM Companions: Experiences of Novices and Experts Using ChatGPT & NetLogo Chat. In: Proceedings of the CHI Conference on Human Factors in Computing Systems, pp. 1–18. ACM, Honolulu HI USA (2024)
16. Gorissen, S.C., Sauer, S., Beckmann, W.G.: End-user development of Oracle APEX low-code applications using large language models. In: Human-Centered Software Engineering, pp. 312–320. Springer, Cham (2024)
17. Oracle: Oracle APEX. https://apex.oracle.com/en/
18. Barricelli, B.R., Fogli, D., Guizzardi, D.: Poster: Natural language interaction for end-user development: is it always feasible? In: Proceedings of the 15th Biannual Conference of the Italian SIGCHI Chapter, pp. 1–4. CHItaly '23, Association for Computing Machinery, New York, NY, USA (2023)
19. Huang, F., Li, G., Zhou, X., Canny, J.F., Li, Y.: Creating User interface mock-ups from high-level text descriptions with deep-learning models (2021)
20. Jiang, E., Toh, E., Molina, A., Donsbach, A., Cai, C.J., Terry, M.: GenLine and GenForm: two tools for interacting with generative language models in a code editor. In: Adjunct Proceedings of the 34th Annual ACM Symposium on User Interface Software and Technology, pp. 145–147. UIST '21 Adjunct, Association for Computing Machinery, New York, NY, USA (2021)
21. Kim, T.S., Choi, D., Choi, Y., Kim, J.: Stylette: styling the Web with natural language. In: Proceedings of the 2022 CHI Conference on Human Factors in Computing Systems, pp. 1–17. CHI '22, Association for Computing Machinery, New York, NY, USA (2022)
22. Calò, T., De Russis, L.: Leveraging large language models for end-user website generation. In: End-User Development, pp. 52–61. Springer, Cham (2023)
23. Kortix AI: Softgen - AI Web App Builder (2024). https://softgen.ai/
24. Lovable labs incorporated: Lovable (2025). https://lovable.dev
25. StackBlitz: Bolt.new (2025). https://bolt.new/
26. Osika, A.: GPT-engineer (2023). https://github.com/gpt-engineer-org/gpt-engineer
27. Karpathy, A.: There's a new kind of coding I call vibe coding (2025). https://x.com/karpathy/status/1886192184808149383
28. DeepLearning.AI: Vibe coding 101 with replit (2025). https://www.deeplearning.ai/short-courses/vibe-coding-101-with-replit/
29. Fill, H.G., Fettke, P., Köpke, J.: Conceptual modeling and large language models: impressions from first experiments With ChatGPT. Enterprise Model. Inf. Syst. Architectures (EMISAJ) **18**, 1–15 (2023)
30. OpenJS Foundation: Low-code programming for event-driven applications: node-RED. https://nodered.org/

An AI-Assisted Real-Time Communication Training Framework for SHARE CST

Xiao-Min Chen[1], Cheng-Yu Hsu[2], Woung-Ru Tang[3,4], and Hsien-Tsung Chang[1,2,5](✉)

[1] Department of Artificial Intelligence, Chang Gung University, Taoyuan, Taiwan
smallpig@widelab.org
[2] Department of Computer Science and Information Engineering, Chang Gung University, Taoyuan, Taiwan
[3] School of Nursing, College of Medicine, Chang Gung University, Taoyuan, Taiwan
[4] Department of Nephrology, Chang Gung Memorial Hospital at Linkou, Taoyuan, Taiwan
[5] Center for Artificial Intelligence in Medicine, Chang Gung Memorial Hospital at Linkou, Taoyuan, Taiwan

Abstract. This study demonstrates a novel real-time physician-patient communication training (CST) support system. This system solves the problem of traditional communication training, which requires a lot of time and human resources for dialogue training and classification, to allow doctors to focus on improving communication quality and obtaining real-time feedback. This study combines a robust language model with SHARE CST to identify a training direction that fits the needs of Eastern culture and adds NER to medical knowledge, so that one can instantly know the cause of the disease and treatment during learning through dialogue. The results show that the SHARE category score achieved an overall accuracy rate of 90. 7% and 81. 9% F1 in NER related to medical conditions. This highlights the fact that the system is capable of being not only accommodating to Eastern communication culture but also correctly annotating medical terms, a fantastic contribution to communication quality improvement.

Keywords: BERT · Words Embedding · Named Entity Recognition · ASR · SHARE CST

1 Introduction

In recent years, the physician–patient relationship has evolved from a one-way physician-monopolistic model to an equal partnership based on shared decision-making [1,2]. A range of forces underlie this evolution: increased public health literacy; common access to the web to browse medical information; more rights and ethics of patients that prioritize respect, well-informed choice, and

H. Unger and P. Meesad (Eds.): NLPIR 2025, LNNS 1904, pp. 1054–1064, 2026.
https://doi.org/10.1007/978-3-032-20897-2_74

autonomy; and system reform oriented toward quality and safety. With more "understanding-and-collaboration" and less "command-and-obey" encounters, communication is not only the transfer of information, but also a systematic process of values clarification, preference eliciting, and treatment plan coconstruction. Having the ability to communicate empathically, clearly and structurally becomes the new focus of clinical competence and quantifiable patient outcomes.

Communication Skills Training (CST) addresses this need directly with lectures, standardized patients, role-play, group debriefs, and facilitated practice. These methods instruct both "how to say" (tone, pacing, empathy, emotion management) and "what to say" (diagnostic discussion, risk-benefit trade-offs, examinations, and treatment). However, two limitations remain: Assessments are too often based on live expert observation, opening the door to variability and subjectivity, and the lack of systematic recording constrains timely feedback and cumulative learning artifacts.

In order to fill this gap, we integrate artificial intelligence with the SHARE model of Fujimori, a suitable strategy for East Asian contexts addressing information disclosure and emotional support requirements: S (Supportive Environment), H (How to break bad news), A (Supplementary Information), and RE (Reassurance & Emotional Support) [3,4]. Our system captures front-end speech-to-text dialogues, Chinese medical named-entity recognition to focus on clinical content, and sequence classification to detect SHARE facets in real time. The system then reveals information gaps and strategic missteps the moment they occur, directing instructors and trainees to improve their messaging and interactional style.

The research pursues three objectives: (1) improve the quality and efficacy of CST through rapid capture, correct labeling, and immediate, less biased feedback; (2) adapt and test SHARE for Eastern cultures; and (3) offer a scalable training platform that operates problem-free across classrooms and virtual settings, allowing for curricular consistency and mass dissemination. Finally, our objective is to promote physician communication, facilitate patient understanding and trust, and link these improvements to quality and health outcomes.

2 Related Works

Naijun Zheng et al. (2021) [5] mentioned that simultaneous speech with multiple speakers greatly reduces the performance of speaker verification (SV) systems. To mitigate the above, they developed a joint training system consisting of a front-end multichannel speech separator and a back-end speaker embedding extractor optimized by specially designed training strategies. They discovered that their combined system significantly outperformed standalone SV models. Extending these methodologies, Zheng et al. (2022) [6] utilized deep learning based on this work and introduced two multichannel, overlap-aware diarization systems capable of learning spatial features to identify speakers and detect overlapping speech. Compared to their single-channel counterparts, their methods reduced the diarization error rate by 9.4% and 18.1%, respectively.

For Named Entity Recognition (NER), the sparsity of annotated data and an open-ended number of entity types make it challenging to build universal models. Traditional methods involving manually crafted lexical and syntactic features are language-specific and require a significant amount of time. To address the problems, Lample et al. (2016) [7] presented two models: BiLSTM-CRF and a transition-based model, which achieved good performance in small datasets without language-specific resources [8]. Based on label dependency modeling, Linkun Cai et al. (2020) [9] introduced a hybrid neural network that took advantage of BERT-based context-aware representations and a novel label graph mechanism to capture interlabel relationships. Experiments on two public benchmarks demonstrated that their model outperformed existing state-of-the-art classification methods.

3 Methods

The general architecture of the proposed real-time system is shown in Fig. 1. The framework comprises three stages. First, the Automatic Speech Recognition (ASR) module employs the OpenAI Whisper model to transcribe doctorpatient conversations into text in real time, providing accurate input for subsequent language processing. Second, the transcribed text is analyzed by two Natural Language Processing (NLP) pipelines: one performs SHARE sentence classification using the RoBERTa model to identify conversational strategy types, and the other conducts Named Entity Recognition (NER) with a RoBERTa + BiLSTM + CRF architecture to extract medical entities and keywords. Finally, the system merges these results into a structured representation and displays them on the SHARE application interface (Fig. 2) in real time, allowing users to visualize conversational strategies and key medical terms for clinical communication analysis.

3.1 Whisper Speech Recognition

Speech-to-text (STT) is the most fundamental module in our system. In simulated clinical encounters, dialogue often involves multiple speakers, dense medical terminology, rapid changes in rate and pitch, and background noise. Without fast and accurate transcription, all downstream text processing becomes unreliable. We therefore adopt OpenAI Whisper as the core speech recognition engine. Whisper is a transformer-based end-to-end model trained on a large multilingual and multitask dataset. Its encoderdecoder structure enables joint modeling of acoustic and linguistic information, offering strong robustness against noise, accent, and overlapping speech.

To handle multispeaker scenarios, we implement a speaker diarization process based on the method described by Lyu et al. [10], allowing the system to segment and label speech segments according to different speakers before transcription alignment. This approach ensures that each utterance is correctly attributed,

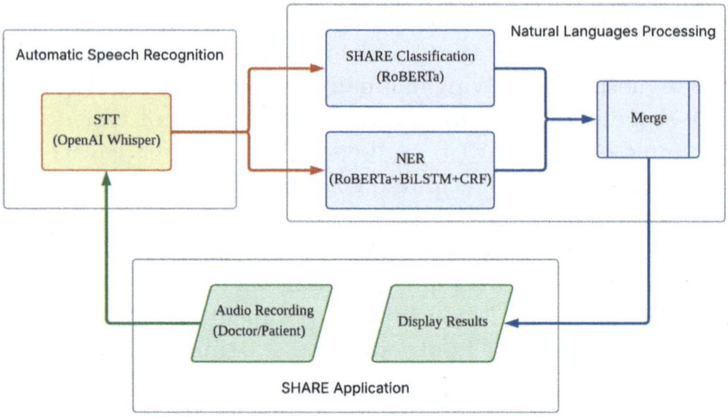

Fig. 1. System architecture of the proposed real-time SHARE CST application integrating speech recognition, natural language processing, and result visualization.

which is critical for analyzing conversational structure and turning in clinical settings.

Given the need for low latency and continuous display in teaching, we have implemented the streaming recognition mode of Whisper. The client periodically sends short audio segments and the transcripts appear with minimal delay, preserving conversational flow for instructors and learners. In practice, the system maintains high transcription quality under multi-speaker conditions, heavy use of technical terms, and typical clinical noise. These properties provide a reliable foundation for the subsequent pipeline: semantic analysis, medical keyword extraction, and SHARE classification, so that feedback on communication structure and empathy can be generated in near-real time. In short, a robust, low-latency STT built on Whisper enables the entire AI-assisted training loop to function as intended.

3.2 Medical Keywords NER

To achieve timely comprehension of professional knowledge in medical-patient communication education, promptly propose appropriate solutions, and appropriately identify explanation omissions in following back, this study made "medical keyword extraction" an official sequence labeling task on six overall terminology categories: Disease & Diagnosis, Imaging Examination, Laboratory Test, Drug, Surgery, and Anatomy. The labeling convention uses the BIO tagging system, with B- <category>denoting entity beginning, I-<category>denoting entity within, and O marking a non-entity outside. This provides accurate and computable boundary labels even in the variable-length and inter-sentence contexts of spoken language so that all clinically relevant terms are meaningfully labeled and traceable along timelines. In terms of model, we took inspiration and borrowed the BERTBiLSTMCRF architecture proposed by Gao et al. n[11]. with

Fig. 2. Demonstration interface of the SHARE Conversational Strategy Tagger (CST) showing real-time doctorpatient interaction with strategy annotations and emotional cues.

the robust Chinese RoBERTa-wwm-ext as the backbone of language representation. Its whole word masking [12] pretraining scheme supports learning semantic representations from Chinese "words."

These are particularly beneficial for highly structured proper nouns in medical contexts, such as drug names, test parameters, and anatomical names. Moreover, we introduced a single-layer bidirectional long-short-term memory (BiLSTM) network to enhance local-order signals and neighborhood dependencies and increase sensitivity to phrase-level hints common in spoken text and perturbation filtering caused by subword segmentation and segment discontinuities. The decoder utilizes a Conditional Random Field (CRF) to encode label transfer constraints at the sentence level in the form of a learnable transfer matrix, thereby preventing illegal BIO sequences (e.g., an O immediately followed by an I-*). In total, RoBERTa's deep contextualized representation, BiLSTM's local sequential modeling, and CRF's structured decoding complement one another, not only improving the coherence of entity boundaries, but also improving the recall and overall robustness of long-tail entities.

4 Experimental Setup

This section outlines the experimental setup of the proposed SHARE CST framework. Covers the data processing workflow, data augmentation strategy, and encoding design, followed by the evaluation of Chinese medical NER and multi-label text classification models. Each component was implemented to verify the effectiveness of the model and ensure reproducibility under consistent experimental conditions.

4.1 Data Processing

For the NER task on Medical Keywords in this study, we used the CCKS-2020 Chinese electronic medical record text as a training and evaluation basis. We first performed slight normalization and data cleaning for the original medical records, normalizing all half-width characters and whitespace characters while retaining commonly used clinical abbreviations and units of measurement. This keeps the numerical values and symbols intact, without altering the surface form and entity boundaries of the technical terms. We then accurately mapped the annotated entity spans from character offsets to token ranges. Token-by-token labels were generated based on the transformed token sequence, using a BIO encoding scheme. This maintains the clinical semantic integrity of the data while providing a consistent and computable boundary description.

4.2 Data Augmentation

To address the limited training data size of SHARE CST, we employed a small self-built database augmented by a large language model. First, we used our team's actual recorded doctor-patient conversations as initial data, annotated, and used to train and calibrate the baseline model. Second, drawing on Maiko Fujimori's original definition and scope of SHARE communication strategies, we designed a systematic prompting system that uses Chat-GPT to generate stylistically consistent and semantically relevant augmented samples, thereby expanding the coverage and contextual diversity of each label.

Prompt = "You are now a professional medical researcher, and you are learning about Oriental SHARE communication skills training. First, you need to know $\{C\text{-}t\}$. The definition of this label is $\{C\text{-}d\}$. Next, there are some examples of this category, such as $\{C\text{-}ex\}$. Please provide 25 sentences of the same category based on the definition and examples."

$C\text{-}t$:= the full name of SHARE and the labels [S, H, A, RE],
$C\text{-}d$:= the clear definition of this label in Maiko Fujimori's paper,
$C\text{-}ex$:= an example of this label from our initial dataset.

4.3 Multi-hot Encoding

This research was found to show that a single sentence is likely to encode a number of communication strategies in real-world medical conversations. Using exclusive one-hot encodings to force the model to predict one label would always lead to pragmatic information loss and introduce test bias. Therefore, we resorted to a multi-hot encoding to represent multi-label relations in SHARE strategy classification. To enable the model to learn all the co-occurrence features of two or more labels, in data expansion in Sect. 4.2, we further requested GPT to generate multicategory sentences: We replaced the C-t, initially representing a single strategy, with a pluralized one, and we appended full definitions and example sentences to the C-d and C-ex series, expanding the generated text from a single-label corpus to a multilabel corpus with two or more labels. This systematic structure of data and representation enables the model to clearly capture the co-occurrence and complementary relationship between strategies free from information compression resulting from conflicting labels.

4.4 Evaluation of Chinese NER Classification

To test the effect of various decoder architectures on model performance, the experiment performed ablation tests based on the Chinese RoBERTa model as a baseline model, with the addition of a CRF layer and a BiLSTM-CRF structure. Table 1 indicates that the baseline RoBERTa model had an F1-score of 0.7998 in the named entity recognition task. Add a CRF to the model and the overall performance increased slightly, with the F1-score reaching 0.8057. This is mainly because the CRF layer could better learn the transfer relation between labels in the sequence labeling task, reducing the rate of illegal sequence of labels.

Furthermore, adding a BiLSTM layer to the RoBERTa + CRF model to learn bidirectional contextual features further boosted the performance of the model to 0.8195 in F1-score, precision and recall were boosted to 0.7988 and 0.8414 respectively. This shows that BiLSTM together with the deep semantic feature learning of RoBERTa can offset the local sequence dependence weakness of the Transformer and further boost the ability of the model to recognize entity boundaries and contextual semantics.

Table 1. Results of the ablation experiment

	Precision	Recall	F1-score
Roberta	0.7675	0.8349	0.7998
Roberta + CRF	0.7801	0.8329	0.8057
Roberta + BiLSTM + CRF	0.7988	0.8414	0.8195

4.5 Evaluation of Multi-label Text Classification, MLTC

This study used 15% of the data as a validation set for performance evaluation. Table 2 shows that, using this self-constructed dataset, the overall accuracy was approximately 60%, with approximately half of the labels achieving an accuracy below 50%. This suggests that, given the limited amount and diversity of the data, the model's learning of pragmatic context and strategic boundaries has yet to saturate. The three RE indicators were all 0.4, indicating that the model's ability to process emotional sentences requires further improvement.

Table 2. Per-label evaluation of the original SHARE data

	precision	recall	f1-score
S	0.833	0.625	0.714
H	0.444	0.800	0.571
A	0.800	0.571	0.667
RE	0.400	0.400	0.400
macro avg	0.619	0.599	0.588

Table 3 presents the results of the evaluation of the data augmentation from a self-built corpus by a language model. In all SHARE dimensions, except RE, the precision for each category was greater than 90%, with the highest of 99%. The microaverage F1 score was 95.6%, which reflects the good repeatability and robustness of the model in multicategory judgments and the good generalization. To ensure sentence-level coherence in multilabel settings, the exact match accuracy and the Jaccard (samples) were also measured here, coming in at 0.907 and 0.947, respectively. Both metrics affirm that the model's generated label combinations are highly matched with human labels and strongly reflect true co-occurrence patterns of communication strategies. These results demonstrate that the data augmentation approach suggested by the language model not only improves the recognition quality and world performance of individual categories but also improves the model's understanding of pragmatic context and strategy boundaries.

Table 3. Per-label evaluation of the combined SHARE data

	precision	recall	f1-score
S	0.991	0.956	0.973
H	0.914	0.941	0.928
A	0.974	0.974	0.974
RE	0.839	0.963	0.897
micro avg	0.953	0.958	0.956
macro avg	0.930	0.959	0.943
Additional multi-label metrics:			
		Exact Match (subset accuracy)	: 0.907
		Hamming loss	: 0.032
		Jaccard micro	: 0.915
		Jaccard samples	: 0.947

It must be noted here that the accuracy of the RE is yet to be 85%. Inferential sentences and emotionally supportive sentences depend heavily on voice and context, and if text-only features are limited, boundary confusion is easily triggered. However, the overall quality of the augmented version is sufficient to support immediate feedback and post-lesson review in instructional settings. Correct transcription and annotation can quickly identify gaps in message structure and gaps in emotional response, enable students to adjust their discourse strategies, and establish a traceable learning cycle.

5 Conclusion

This study introduces an AI-assisted communication training framework that integrates Whisper-based speech recognition, medical keyword recognition, and SHARE strategy classification into a unified real-time feedback system. A key contribution lies in the data augmentation pipeline, which leverages large language models to synthesize diverse and contextually coherent dialogue samples from limited annotated data. This enhancement significantly improved multilabel classification performance and enhanced the model's ability to detect co-occurring pragmatic cues in real medical conversations. By combining domain-specific data synthesis with multi-hot encoding, the system effectively mitigates data sparsity and label imbalance, two critical challenges in communication skills training (CST). The results demonstrate that our approach provides a reproducible, scalable foundation for AI-enhanced CST, capable of adapting to linguistic and cultural contexts. Future work will focus on extending this framework to multimodal understanding of emotions and cross-cultural validation, with the aim of establishing a continuous, data-driven feedback loop that deepens empathy and communicative competence in medical education.

Acknowledgments. This work was supported by the National Science and Technology Council (NSTC), Taiwan, under Grant No. 114-2221-E-182-041-MY3. The authors also acknowledge the partial financial support from Chang Gung Memorial Hospital, Linkou, under Grant No. NERPD4Q0021.

References

1. Cai, L., Song, Y., Liu, T., Zhang, K.: A hybrid BERT model that incorporates label semantics via adjustive attention for multi-label text classification. IEEE Access **8**, 152183–152192 (2020). https://doi.org/10.1109/ACCESS.2020.3017382
2. Cui, Y., Che, W., Liu, T., Qin, B., Yang, Z.: Pre-training with whole word masking for Chinese BERT. IEEE/ACM Trans. Audio Speech Lang. Process. **29**, 3504–3514 (2021). https://doi.org/10.1109/TASLP.2021.3124365
3. Elwyn, G., et al.: Shared decision making: a model for clinical practice. J. Gen. Intern. Med. **27**(10), 1361–1367 (2012). https://doi.org/10.1007/s11606-012-2077-6
4. Frosch, D.L., Kaplan, R.M.: Shared decision making in clinical medicine: past research and future directions. Am. J. Prev. Med. **17**(4), 285–294 (1999). https://doi.org/10.1016/S0749-3797(99)00097-5
5. Fujimori, M., et al.: Preferences of cancer patients regarding the disclosure of bad news. Psychooncology **16**(6), 573–581 (2007). https://doi.org/10.1002/pon.1093
6. Fujimori, M., Parker, P.A., Akechi, T., Sakano, Y., Baile, W.F., Uchitomi, Y.: Japanese cancer patients' communication style preferences when receiving bad news. Psychooncology **16**(7), 617–625 (2007). https://doi.org/10.1002/pon.1102
7. Gao, W., Zheng, X., Zhao, S.: Named entity recognition method of Chinese EMR based on BERT-BiLSTM-CRF. J. Phys: Conf. Ser. **1848**(1), 012083 (2021). https://doi.org/10.1088/1742-6596/1848/1/012083
8. Gulati, A., et al.: Conformer: convolution-augmented Transformer for speech recognition (2020). https://arxiv.org/abs/2005.08100, version v1

9. Lample, G., Ballesteros, M., Subramanian, S., Kawakami, K., Dyer, C.: Neural architectures for named entity recognition (2016). https://doi.org/10.48550/arXiv.1603.01360,
10. Lyu, K.M., Lyu, R., Chang, H.T.: Real-time multilingual speech recognition and speaker diarization system based on whisper segmentation. PeerJ Comput. Sci. **10**, e1973 (2024)
11. Zheng, N., et al.: A joint training framework of multi-look separator and speaker embedding extractor for overlapped speech. In: ICASSP 2021 – 2021 IEEE International Conference on Acoustics, Speech and Signal Processing (ICASSP), pp. 6698–6702 (2021). https://doi.org/10.1109/ICASSP39728.2021.9414556
12. Zheng, N., et al.: Multi-channel speaker diarization using spatial features for meetings. In: ICASSP 2022 – 2022 IEEE International Conference on Acoustics, Speech and Signal Processing (ICASSP), pp. 7337–7341 (2022). https://doi.org/10.1109/ICASSP43922.2022.9747343

A Design Science Conceptual Model for Generative AI-Powered Blind Assistance Systems

Yueting Wang(✉) and Charnsak Srisawatsakul

Graduate School, Stamford International University, Bangkok, Thailand
2412070005@students.stamford.edu,
charnsak.srisawatsakul@stamford.edu

Abstract. Current assistive technologies for the visually impaired face significant challenges, including high costs, limited functionality, and inadequate user interaction. To address these gaps, this paper presents a Design Science Research (DSR) conceptual model for an intelligent assistive system. Our proposed system integrates Generative AI, Large Language Model and speech recognition into a comprehensive framework encompassing "environmental monitoring, AI decision-making, and voice guidance." This paper details the system architecture, the theoretical foundations, and the key innovations of our model. The primary contribution is a validated design and instantiation artifacts that serves as a blueprint for a low-cost, interactive, and intelligent assistive tool. Future work will focus on empirically evaluating the system with users to assess its real-world effectiveness. This research provides a foundation for the next generation of user-centric assistive technologies for the visually impaired.

Keywords: Generative Artificial Intelligence · Speech Recognition · Design Science Research · Blind Assistance · Large Language Model

1 Introduction

An estimated 2.2 billion people worldwide live with vision impairment, a condition that significantly affects daily activities and mobility [1]. While various assistive technologies are available, many are expensive, have limited capabilities, and lack effective user interaction. Key problems include a limited ability to understand user intentions, process natural language, and deliver precise information [2]. These limitations create a need for an intelligent, low-cost assistive system with robust interactive features for individuals with visual impairments.

This paper introduces a conceptual model and an initial instantiation for an advanced assistive system, developed using the DSR [3] methodology. Our proposed model integrates Generative AI and speech recognition into a unified framework that provides environmental monitoring, AI-driven decision-making, and voice-based guidance. The primary contribution of this research is a design artifact, in the form of a conceptual

H. Unger and P. Meesad (Eds.): NLPIR 2025, LNNS 1904, pp. 1065–1080, 2026.
https://doi.org/10.1007/978-3-032-20897-2_75

model and a working prototype, which serves as a detailed blueprint for the future development of this system. This work lays the foundation for creating more intelligent, affordable, and user-focused assistive tools for the visually impaired community.

2 Literature Review

2.1 Design Science Research

DSR is a problem-solving paradigm that aims to create and evaluate innovative artifacts to address real-world challenges. This methodology connects theoretical research with practical application by following a structured process: identifying a problem, designing a solution, and validating its utility. DSR provides a rigorous framework for developing technology-based solutions, making it a suitable approach for this project. It allows for the systematic creation of a conceptual model and its subsequent instantiation and testing, ensuring that the final system is both theoretically sound and practically effective [3].

2.2 Generative Artificial Intelligence

Generative artificial intelligence represents a core research direction in the current AI field, encompassing multiple dimensions including foundational theoretical research, technical framework development, application implementation studies, and ethical governance investigations. Ian Goodfellow and colleagues first proposed the generative adversarial network architecture, marking a pivotal breakthrough for generative AI in transitioning from traditional statistical models to deep learning. Through the competitive interaction between a "generator" and a "discriminator" this approach enables unsupervised learning of data classification, thereby generating realistic images and textual information [4]. The Transformer architecture proposed by Vaswani et al. forms the core foundation of today's large language models, replacing traditional recurrent neural networks to enable processing of long sequential data such as text and audio [5]. By 2025, over 90% of generative AI systems will be built using this model.

2.3 Speech Recognition Technology

Speech recognition technology constitutes the fundamental component of human-computer interaction within intelligent guide systems and represents a critical subfield of natural language processing. Its core process involves feature extraction and the enhancement of model complexity. Figure 1 illustrates the evolutionary trajectory of speech recognition technology.

In recent years, speech recognition technology has been extensively applied in fields such as intelligent wearable devices and smart home systems. The development of speech recognition has been a prolonged process, involving significant advancements. In 2016, Amodei et al. introduced a hybrid architecture combining convolutional layers with GRU/LSTM units, significantly accelerating convergence rates. This work laid foundational groundwork for the engineering implementation of end-to-end models, marking

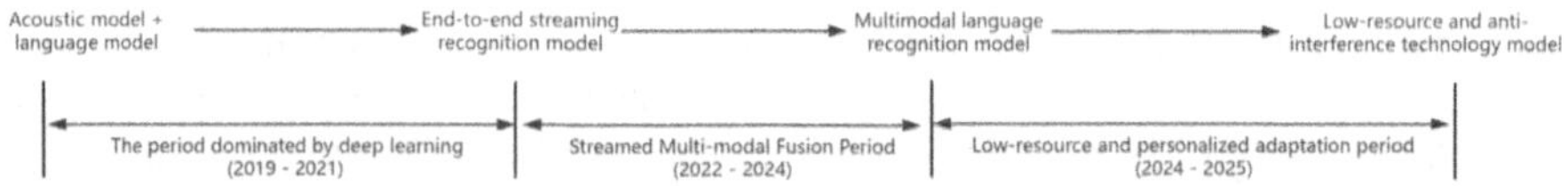

Fig. 1. The evolution process of speech recognition technology.

a pivotal step in the evolution of neural network-based speech recognition systems [6]. With the advent of scenarios like smart speakers and in-vehicle interactions, streaming speech recognition has become a research hotspot, ushering in an era of integrating streaming and multimodal speech recognition technologies. Research efforts primarily focus on enhancing both real-time performance and robustness. Through continuous development, adapting speech recognition technology to specific scenarios has emerged as a new research focus. Schneider et al. broke away from traditional training models reliant on labeled data, proposing contrastive learning to extract speech representations from vast amounts of unlabeled audio. This approach enabled models to achieve 85% recognition accuracy after training on just 10 h of labeled data [7]. In 2024, the Stanford team proposed a "cross-lingual transfer strategy" employing self-supervised pretraining models in English to facilitate the initialization of low-resource language models. After just five hours of annotated data, the system achieved an recognition accuracy of 82%.

2.4 Generative Artificial Intelligence in the Field of Obstacle Recognition

The essential requirements of an intelligent guiding system for the visually impaired involve natural language processing for user input analysis, environmental perception through multimodal sensory data, and speech output feedback to ensure safety. Consequently, capabilities in multimodal understanding, natural language interaction, dynamic scene perception, and decision-making are integral to the development and deployment of such assistive technologies.

The ERNIE 3.0 model integrates the knowledge graph into the model through knowledge-enhanced training [8]. It's the first time it's achieved the "language understanding and generation" framework, which boosts the accuracy of contextual understanding. Then, the system can respond to the user's requests in high quality. TextGrad came up with "feedback-driven optimization". It can optimize the text gradient direction by analyzing historical records and then generate personalized prompt messages [9]. This study enhances the system's autonomous scene adaptation capabilities, addressing the limitation of feedback homogeneity.

The integration of multimodal modalities in generative artificial intelligence represents the advancing frontier in the field. The DALL-E model can generate original images based on the descriptive language provided by users, without the need for training specific graphics. This marks the advancement of generative artificial intelligence into the era of cross-modal creation [10]. In addition, CLIP analyzes images, captures real - time scenes, and then converts them into structured text language [11]. This also provides technical support for the real - guiding systems. The above multimodal generation and scene understanding capabilities of artificial intelligence can enhance the environmental perception functions of intelligent guide systems for the visually impaired.

2.5 Existing Auxiliary Applications Address Research Gaps

The current development of intelligent orientation and mobility systems is progressing rapidly; however, numerous issues remain unresolved. These systems should achieve environmental perception, navigation and localization, as well as human-machine interaction, to facilitate the daily lives of visually impaired individuals.

The navigation aid software, developed based on Yolov3, exhibits a decline in recognition accuracy from 90% to 68% under nighttime or heavy rainfall conditions, indicating limited performance efficacy in extreme environmental scenarios [12]. Jawad R et al.'s research solely utilizes structural and acoustic beaconing for obstacle distance information transmission, without leveraging speech-based scene feedback, thereby limiting applicability in complex environments with multiple obstacles [13]. Therefore, the extant body of research within this domain continues to demonstrate significant potential for further scholarly investigation.

Existing research exhibits significant gaps in both indoor scenario adaptation and dynamic path planning. While the combination of BeiDou and RTK enables intelligent guide technology to achieve 1–2 m accuracy outdoors, when transitioning to indoor environments, reliance on SLAM technology for navigation guidance increases positioning errors to 4–11 m. This fails to meet the refined indoor cognitive needs of visually impaired individuals [14].

3 Research Methodology

This section involves the prototyping of the conceptual framework, aiming to concisely and comprehensively present the strategic framework of the research.

3.1 Design Science Research Model

This research employs the DSR method, centering around the actual travel and real - world environmental perception needs of blind and visually impaired individuals. It adheres to the model of "identifying practical problems - designing solutions - validating the effectiveness of solutions", integrating voice recognition, generative artificial intelligence, and multimodal fusion technologies. It is expected to construct a retrieval - augmented generative intelligent guide - dog system featuring a trinity of "environmental monitoring - AI decision - making - voice guidance", providing solutions for the implementation and application of the theory. Figure 2 shows the research on the design model.

In terms of stakeholders, this system primarily targets blind individuals and their family members as the end - users. By comprehensively understanding their current situations and challenges, the system aims to offer blind people a safe and intelligent way to perceive the provide a platform for their family members to quantify the blind individuals' daily behaviors and physical and mental health.

The second part is the intelligent assistance system for the blind. This segment primarily provides the recognized voice and image information to the generative artificial intelligence. The generative AI then conducts an intelligent analysis of the currently collected on - site situation and presents the results of the analysis.

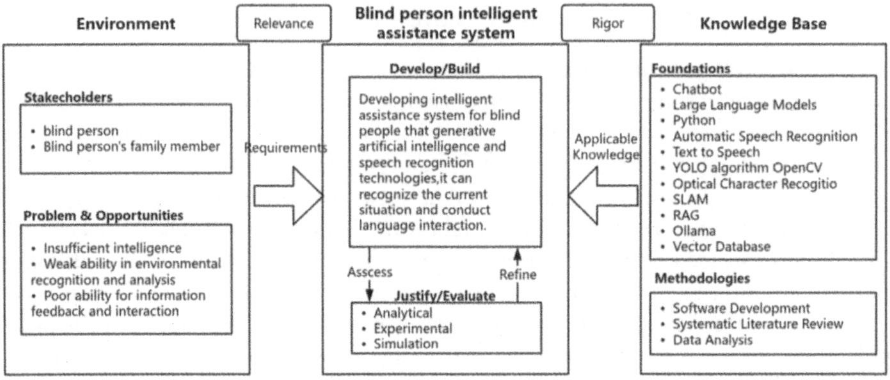

Fig. 2. Research model of study.

The final part is a crucial component of designing the scientific research framework. It involves planning for software development and conducting a systematic literature review. In the future, system testing and evaluation will be carried out, aiming to contribute valuable insights to the academic community and practical application fields.

3.2 Conceptual Framework

The blind assistance system development project employs the Waterfall model as its software engineering methodology. This approach segments the software development lifecycle into sequential, linear phases, including requirements elicitation, system architecture design, implementation, and verification. Such structuring aims to ensure a coherent system architecture and clearly defined objectives.

Tools and Technologies Used

The system development integrates a comprehensive array of contemporary tools and technologies available in the current market. The front-end interface employs HTML for webpage construction, while the back-end utilizes Java with Python for data management and communication. Data acquisition encompasses speech, imagery, video, and real-time geolocation, leveraging speech recognition, video processing, and real-time positioning technologies to generate multimodal fused datasets. These datasets are transmitted to artificial intelligence modules, particularly large language models, which perform deep comprehension and analytical processing to produce coherent textual outputs. Subsequently, text-to-speech synthesis converts the generated text into audio signals, facilitating auditory alerts designed to assist visually impaired and blind users.

System Development Process

This study conceptualizes a four-tier, three-module architecture, as depicted in System Architecture Diagram 3. The framework comprises Perception, Processing, Feedback, and User Interface layers, designed to initially perceive user needs, then analytically interpret these requirements for intelligent decision-making, and ultimately deliver feedback via voice communication. The system emphasizes shared data flow throughout the entire process, aiming to support safe and autonomous navigation for visually impaired individuals through synchronized life and safety information delivery (Fig. 3).

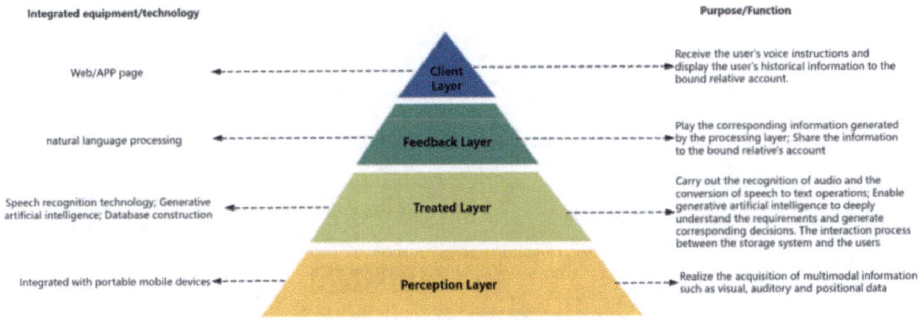

Fig. 3. System Hierarchical Architecture Diagram.

The primary function of the perception layer is to collect multimodal data. The system will be integrated onto portable mobile devices, utilizing the device's camera, microphone, and positioning capabilities to gather multimodal data. The acquired visual data will be processed by the subsequent scene perception and semantic deep understanding layer. The received audio information will be used to identify the user's actual needs, while the positioning data will be employed for subsequent path planning and synchronization of positioning information.

The processing layer encompasses three core modules: the audio processing module, the generative artificial intelligence analysis module, and the information storage module. The audio processing module involves speech recognition and speech-to-text conversion of natural language input; classification of command requirements; enhancement processing for low-resource speech signals affected by tone, noise, and accent variations; acquisition of contextual scene information; and multimodal data fusion of voice commands with environmental context. The generative artificial intelligence analysis module aligns visual and vocal features to analyze and predict multimodal scene data, enabling deep inference of user intent, and facilitating scene understanding and speech information analysis to support adaptive decision-making. The information storage module retains the system's interaction history with users, synchronizes data across authorized familial accounts linked to the system, and constructs a sustainable, rapid-retrieval database encapsulating spatial coordinates and user preferences.

The feedback layer employs text-to-speech technology to convert intelligent decision prompts generated by the processing layer into voice feedback. It conveys the analysis results of the current scenario and voice queries to the user. Additionally, it relays user interaction data stored in the information storage module to authorized family members through the system, thereby enabling family support for blind and visually impaired individuals.

The user layer primarily facilitates two-way interaction between blind individuals and their family members. The blind user's device issues requests via voice commands, which the system receives, intelligently processes, and responds to with voice feedback. Family members access real-time information about the blind individual through web or app interfaces. They can also review generative AI analyses of the blind person's daily activities to anticipate their physical and mental health status, enabling remote protection and safety monitoring.

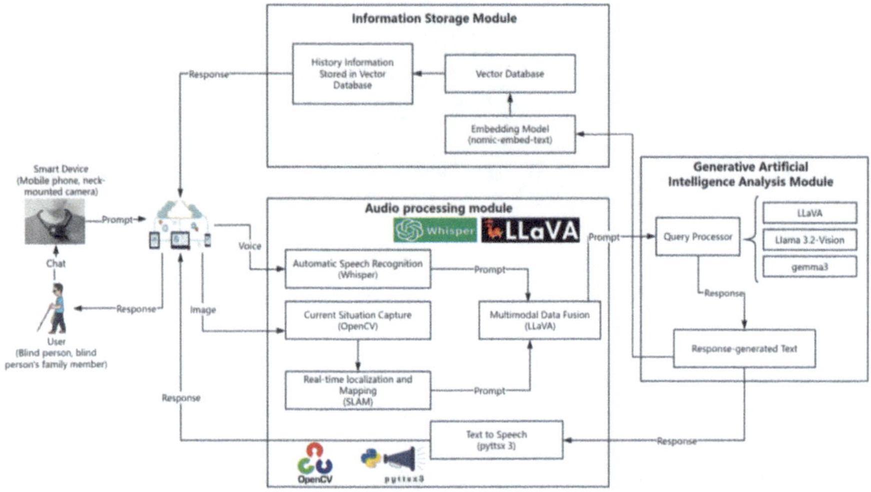

Fig. 4. System Architecture.

Figure 4 illustrates the conceptual framework of the designed scientific research model. Users activate the system via voice commands, which prompts the recognition of spoken input, video, imagery, and current geospatial data. These multimodal data streams are transmitted to the audio processing module, where they undergo multimodal fusion. Upon fusion, the consolidated information is dispatched to the generative artificial intelligence analysis module for intelligent processing. Following analysis, the system generates textual output, which is stored within the information storage module and concurrently transmitted back to the audio processing module. The text is then converted into natural language speech via text-to-speech technology for playback to the user. The information storage system retrieves relevant historical data during searches, aggregates this information, and presents it to the user interface for review. Moreover, the generative AI analysis module can analyze historical data to predict behaviors and physical and mental health status of visually impaired individuals, and transmit these insights to linked family accounts, thereby facilitating familial care and support.

Family Care Mechanism

Family Care leverages a strategy driven by data, permission controls, and real-time location synchronization. Figure 5 illustrates the data flow diagram of the Family Care mechanism.

When blind individuals access and interact with the system, it logs information including access time, account details, and executed operations. When authorized family

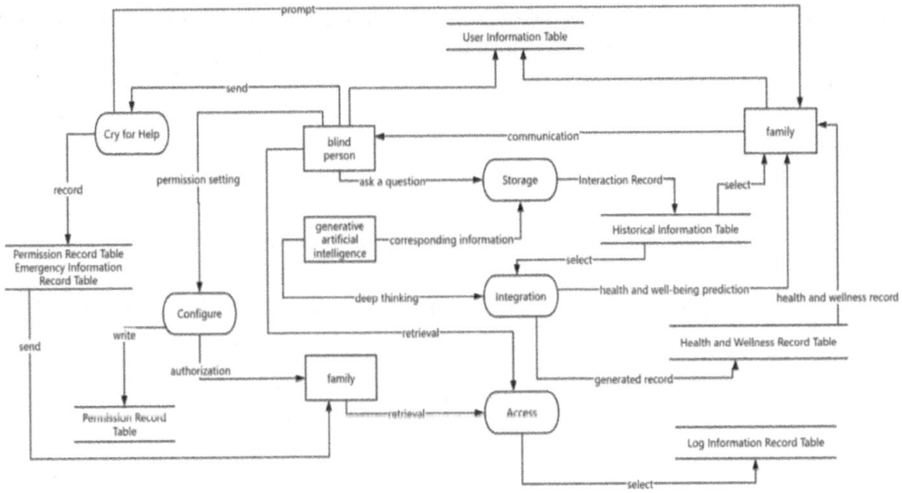

Fig. 5. Data Flow Diagram of Family Care Mechanism.

members access the system, the system's trigger mechanism similarly records visitor-related access information. All such data is stored in the log information record table. The blind individual's questions and the generative AI's responses are stored in the historical information record table. Visually impaired individuals set access permissions for authorized family members to protect personal privacy. The system grants permissions to family members based on the information in the permission settings table. Family members can access the visually impaired individual's system records according to their permission settings. When granted permission to access historical information records, they can review such records; otherwise, access is denied. The generative AI can access the blind individual's historical information through the system, perform in-depth integration and analysis, generate an assessment of their physical and psychological state, and send the assessment results to bound and authorized family members. Family members can refer to the assessment results to communicate with the blind individual in a timely manner, maintaining their physical and psychological health development. Throughout this process, the system automatically logs the blind individual's physical and mental health records for subsequent review by relevant institutions. When the blind individual faces danger or requires assistance, they can trigger the system's emergency alert mechanism. The system records the distress signal and sends the information—including time, location, and human-machine interaction details—to designated family members. This establishes a comprehensive family care mechanism focused on the blind individual's physical and psychological well-being.

3.3 Comparison of Related Work and Innovation Points

This research aims to develop an intelligent guide system for the visually impaired, integrating environmental detection, AI-driven decision-making, and voice feedback. It will establish a backend information database to share data with family members authorized and trusted by the visually impaired individuals. This system enhances their ability to perceive the world, expands their opportunities to engage with it, and ensures their safety. By leveraging generative AI, the system strengthens human-machine interaction capabilities, thereby improving the user experience. The comparison with other systems is shown in the table below (Table 1).

Table 1. Comparison of the characteristics of this system with those of the existing systems.

	Guide-dog hexapod robot	China Mobile's robot guide dog	Seeing AI APP	This Research
Automatic Speech Recognition (ASR)		√	√	√
AI Intelligent Analysis and Guidance	√	√	√	√
Environment Identification	√	√	√	√
Family Care Mechanism	√			√
Software Portability			√	√

3.4 Problem Analysis and Requirements Derivation

This study employed quantitative methods and conducted two rounds of questionnaire surveys. The surveys addressed issues related to the system's environmental perception capabilities, human-computer interaction requirements, and security and privacy aspects. The first survey yielded 65 valid responses. The second survey refined the questions from the first and collected 99 valid responses.

Table 2 shows the geographical distribution of survey respondents. Most participants are from Henan province, accounting for 70.77% in Questionnaire 1 and 75.76% in Questionnaire 2. Additionally, 1.01% of responses in Questionnaire 2 come from outside China. In terms of occupation, 90.91% of Questionnaire 2 respondents are students, 7.07% are employed, and 2.02% are unemployed. The large sample size and wide coverage increase the reliability and generalizability of the survey results.

Table 2. Statistical table of geographical information locations for questionnaire filling.

Region	Questionnaire 1	Questionnaire 2
Henan	46	75
Guangdong	3	5

(continued)

Table 2. (*continued*)

Region	Questionnaire 1	Questionnaire 2
Zhejiang	3	2
Anhui	2	1
Beijing	2	1
Fujian	2	2
Guangxi	1	3
Jilin	1	1
Jiangsu	1	1
Jiangxi	1	0
Shandong	1	1
Shanxi	1	0
Shanghai	1	0
Sichuan	0	2
Hong Kong	0	1
Yunnan	0	1
Regions outside China	0	1
Hubei	0	1
Liaoning	0	1

4 Research Results

4.1 Requirement Gathering and Analysis

Analysis of the survey results from both questionnaires reveals that when unable to perceive the surrounding environment through vision, people have an urgent need for voice alerts to understand their surroundings. Data from the first survey indicates that 66.15% of respondents rated the system's ability to rapidly identify and understand current scenarios—alerting users to obstacles in their path and providing environmental perception information—as "very important" An additional 29.23% rated it as "important" while only 4.62% considered it 'average' or "unimportant" This clearly demonstrates that when facing visual impairment or obscured vision, individuals have a critical need and reliance on external cues to obtain environmental information, thereby ensuring personal safety and facilitating mobility (Table 3).

Table 3. Key Design Requirements Derived from Survey Findings.

Survey Finding	Derived Design Requirement	Corresponding Feature in Conceptual Model
A vast majority (66.15% + 29.23%) of users rate environmental perception as "very important" or "important"	R1: The system MUST provide real-time, high-accuracy environmental perception	Environmental Monitoring Layer; Multimodal Fusion (Fig. 4)
Over 80% of respondents identified a critical need for traffic signal information	R2: The system MUST be able to identify and interpret traffic signals	YOLO Algorithm; Generative AI Analysis Module
89.89% agreed that alerts about pedestrians and road conditions are crucial	R3: The system MUST feature proactive pedestrian and obstacle detection with audio alerts	OpenCV; Text-to-Speech (pyttsx3); Feedback Layer
94.95% believe that sharing information with family members is a form of care	R4: The system MUST include a secure, permission-based Family Care Mechanism	User Information Table; Permission Control; Family Care Data Flow (Fig. 5)
68.69% expressed trust in Generative AI for analyzing historical data to predict user states	R5: The system MUST leverage Generative AI to analyze user history for well-being prediction	Generative AI Analysis Module; Historical Information Table

In real life, blind individuals may struggle to access traffic signal information, leading to hesitation when crossing roads. The second survey questionnaire refined the cognitive needs for different environmental elements, with over 80% of respondents believing that conveying traffic signal information to them via voice is crucial. Approximately 89.89% of respondents agreed that system-provided alerts about pedestrians and road conditions could help prevent collision-related injuries. In complex environments like shopping malls and stations, the system should ideally automatically adjust walking routes based on the blind person's needs and surrounding pedestrian information to ensure travel safety. The system plans to analyze collected visual data using LLaVA, then employ generative AI to describe the granularity of voice prompts—covering object type, location, and attribute information—while providing actionable recommendations. This approach aims to fulfill the environmental awareness needs of blind individuals.

According to the survey questionnaire, 94.95% of respondents believe that sharing information with family members to varying degrees is a form of care for the visually impaired, enabling families to better understand their loved ones' travel situations. When visually impaired individuals encounter danger, this practice also allows for timely assistance, providing them with emotional security. 68.69% of respondents expressed trust in generative AI, supporting its use to detect and analyze historical data. This approach quantifies the physical and psychological state of blind individuals through AI, with results shared with family members for reference. Consequently, the system plans to design a family care mechanism, establishing a visual information-sharing platform for blind individuals and their families to safeguard their psychological and physical well-being.

4.2 The Instantiated Model: A Proof-of-Concept Artifact

As part of the DSR process, we developed an initial instantiation of the conceptual model in the form of a working prototype. This prototype demonstrates the core user-facing features of the system and serves as a proof-of-concept for the proposed design.

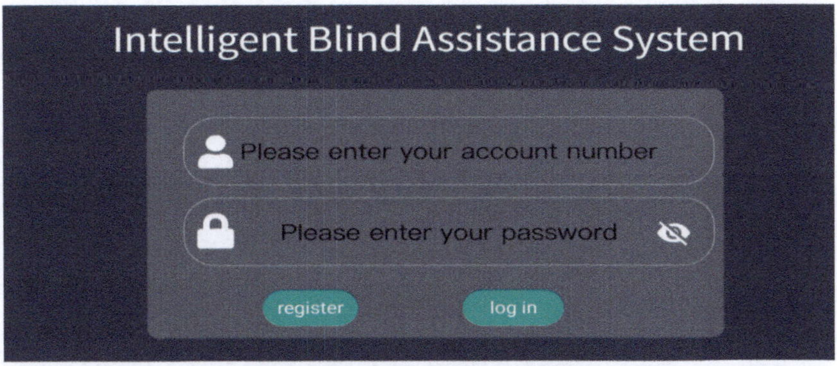

Fig. 6. User Login Page Prototype.

Figure 6 depicts the user authentication interface, where users can input their credentials, including username and password, and initiate access by clicking the login button. In the absence of an existing account, users may activate the registration process via the designated button located at the bottom right corner, directing them to the registration module.

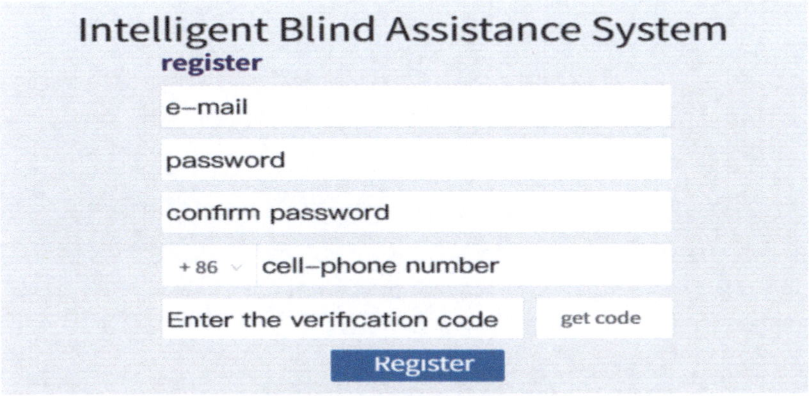

Fig. 7. User registration page.

Figure 7 illustrates the registration interface, wherein users can create an account by providing their email address, password, and phone number, with mobile number verification required to authenticate their identity.

Figure 8 illustrates the dialogue interface between visually impaired users and the system. The interface is bifurcated into two components: the historical conversation log on the left and the active dialogue input area on the right. Due to the visual impairments of the user, the voice command button has been enlarged to facilitate easier access and operation. The upper section displays the chronological exchange history to provide context during the interaction.

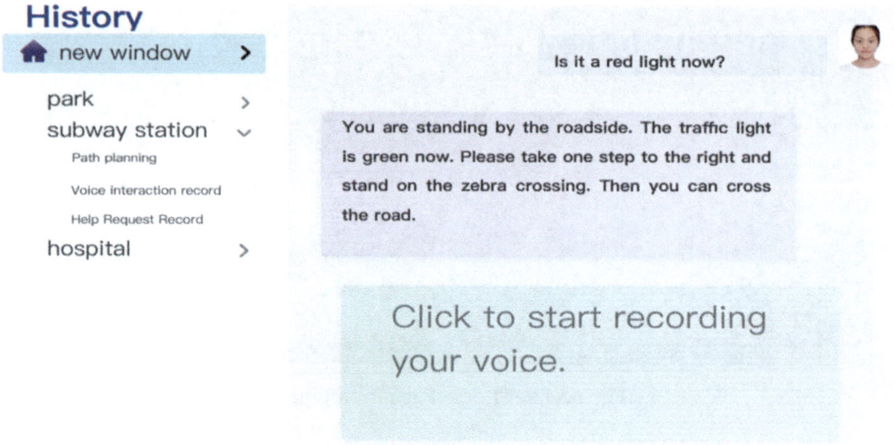

Fig. 8. User-System Interaction Interface.

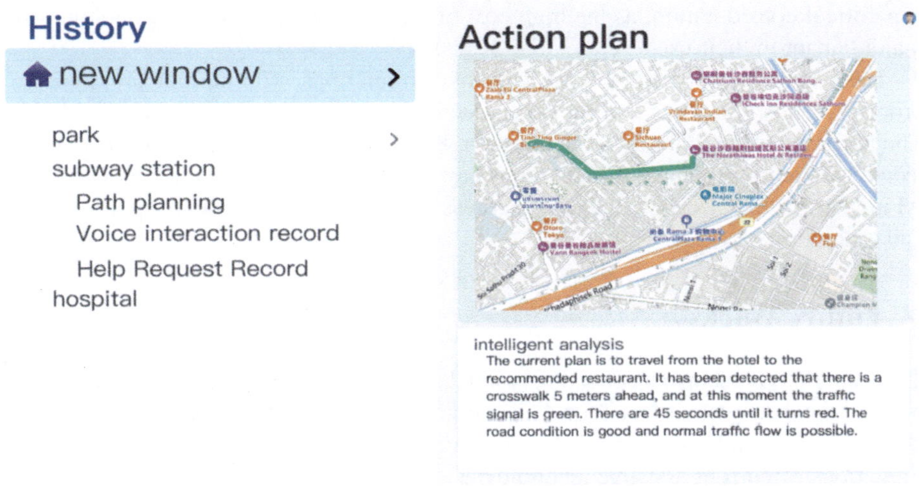

Fig. 9. The authorized user interface for accessing visually impaired individuals' interaction data within the system.

Figure 9 shows the user interface for authorized personnel viewing blind individuals' information. The page is divided into three sections: the left pane displays the blind individual's interaction history with the system, allowing authorized users to click and review details about their daily activities. The upper right section shows the blind individual's current walking route, displaying the path they are currently following. The lower right corner features intelligent analysis, where the system synthesizes the blind individual's daily route and system interactions to generate objective conclusions for user reference.

5 Discussion and Conclusions

This paper presents a conceptual model and an initial instantiation for a blind assistance system that leverages Generative AI and speech recognition. Our model addresses several key challenges in the field of assistive technology. First, by using a DSR approach, we ensure that the system is grounded in both theoretical principles and practical needs. The development of a working prototype demonstrates the feasibility of our conceptual model.

Second, the integration of Generative AI allows the system to provide a level of intelligence and adaptability that is not possible with traditional assistive devices. The AI Core Layer can interpret complex environmental data and make real-time decisions, providing users with a much richer and more detailed understanding of their surroundings. This is a significant improvement over existing systems, which often have limited functionality and are unable to handle dynamic or unpredictable situations.

Third, our model is designed to be low-cost and accessible. By using readily available hardware, such as cameras and GPS, and by leveraging open-source software, we aim to create a system that can be widely adopted by the visually impaired community. This is a critical consideration, as the high cost of many existing assistive technologies is a major barrier to their use.

Despite its potential, our work has some limitations. The current prototype is a proof-of-concept and has not yet been tested in a wide range of real-world scenarios. The accuracy and reliability of the system will depend on the quality of the data collected by the Environmental Perception Layer and the performance of the Generative AI models. Future work will need to address these challenges through rigorous testing and evaluation.

6 Future Works

This paper has presented a conceptual model and an initial instantiation of an intelligent blind assistance system based on Generative AI and speech recognition. The model, developed using a DSR framework, offers a comprehensive solution that addresses the limitations of current assistive technologies. The development of a working prototype confirms the feasibility of our design. By integrating advanced AI capabilities with a user-centered design, our proposed system has the potential to significantly improve the mobility, safety, and independence of individuals with visual impairments.

The next phase of this research will focus on a rigorous empirical evaluation of the prototype with target users. This will allow us to validate its utility and effectiveness in real-world scenarios. The feedback from this evaluation will be used to further refine the system and to ensure that it meets the needs of the visually impaired community.

References

1. World Health Organization: World report on vision. World Health Organization, Geneva (2019). https://www.who.int/publications/i/item/world-report-on-vision. Accessed 04 Nov 2025
2. Thiyagarajan, K.: Intelligent guide robots for people who are blind or have low vision: a review. IEEE Access **10**, 98215–98229 (2022)
3. Hevner, A., Chatterjee, S.: Design science research in information systems. In: Hevner, A., Chatterjee, S. (eds.) Design Research in Information Systems. Integrated Series in Information Systems, vol. 22, pp. 9–22. Springer, Boston (2010). https://doi.org/10.1007/978-1-4419-5653-8_2
4. Goodfellow, I., et al.: Generative adversarial networks. Commun. ACM **63**(11), 139–144 (2020)
5. Vaswani, A., et al.: Attention is all you need. In: Guyon, I., et al. (eds.) Advances in Neural Information Processing Systems 30, pp. 5998–6008. Curran Associates, Inc. (2017)
6. Amodei, D., et al.: Deep speech 2: end-to-end speech recognition in English and mandarin. In: Balcan, M.F., Weinberger, K.Q. (eds.) Proceedings of the 33rd International Conference on Machine Learning, vol. 48, pp. 173–182. PMLR (2016)
7. Baevski, A., Zhou, H., Mohamed, A., Auli, M.: wav2vec 2.0: a framework for self-supervised learning of speech representations. In: Larochelle, H., et al. (eds.) Advances in Neural Information Processing Systems 33, pp. 12449–12460 (2020)
8. Sun, Y., et al.: ERNIE 3.0: large-scale knowledge enhanced pre-training for language understanding and generation. arXiv preprint arXiv:2107.02137 (2021). Accessed 04 Dec 2025

9. Yuksekgonul, M., Bianchi, F., Boen, J., et al.: Optimizing generative AI by backpropagating language model feedback. Nature **639**, 609–616 (2025)
10. Ramesh, A., et al.: Zero-shot text-to-image generation. In: Meila, M., Zhang, T. (eds.) Proceedings of the 38th International Conference on Machine Learning, vol. 139. pp. 8821–8831. PMLR (2021)
11. Efros, A.A.: Video generation models as world simulators. OpenAI Technical Report (2024). https://openai.com/research/video-generation-models-as-world-simulators. Accessed 04 Dec 2025
12. Shi, N.: Design of walking assistance software for blind people based on deep learning. Softw. Eng. Appl. **11**(2), 320–329 (2022). (in Chinese)
13. Jawad, R., Jawad, R.: Design and implementation of a stick for blind humans using artificial intelligence system and FPGA. In: 2024 International Conference on Electrical, Computer, and Energy Technologies (ICECET), pp. 1–6. IEEE (2024)
14. Zhang, M., et al.: The design and research of a Beidou-based blind intelligent navigation system. Artif. Intell. Robot. Res. **14**, 65–81 (2025). (in Chinese)

Innovation and Thinking of College Student Management Based on Network Technology Under Digital Background

Mi Yalong[1(✉)], Cui Yong[2], and Shi Guoyong[2]

[1] Sichuan University of Media and Communications, Xueyuan Street 67, Tuanjie Subdistrict, Pidu District, Chengdu 611745, Sichuan, China
mstrongsnail@foxmail.com

[2] Southwest Petroleum University, No. 1, 2nd Section of Youyuan Road, Shunqing District, Nanchong 637000, Sichuan, China
cuiyong@swpu.edu.cn

Abstract. The development in information technology has led to significant student management and assessment applications in recent years. Student skill sets are identified through innovative thinking and outcomes using digital and manual combinations for ease of assessments. This manuscript presents a Chained Management Scheme banking on Classified Technology (CMS-CT) for resolving the issues in different innovations through similar skills and precise classifications. The proposed scheme is distinguished into two processes—similarity assessment and innovation classification. These processes cooperatively form student management through multiple digital data analysis technology. In the analysis, the connection between skill similarity and the innovation output is first measured; the deviation-oriented factors are classified using a tree classifier. The tree's root is initiated from the similarity factor prevailing to the maximum innovation through deviations. These similarity and classification instances are chained for digital network assessments. In the assessment phase, the best-known skill is used for extracting optimal results, based on which further classifications are performed. Therefore, the outcome and skill detection features are improved through better categories and fewer errors.

Keywords: Classifications · Digital Data · Innovative Thinking · Student Management · Tree Classifier

1 Introduction

College student management is one of the essential tasks to perform in every college and university. Various kinds of functions and details are collected and managed via management systems. The college contains names, qualifications, skills, behaviors, examinations, and student attitudes [1]. Skills-based management systems that reduce duplicate and unnecessary content are primarily used in college. Skills are the primary and efficient thing necessary for every student in both universities and colleges [2]. Student information management system (SIMS) is widely used in various colleges that maintain every

H. Unger and P. Meesad (Eds.): NLPIR 2025, LNNS 1904, pp. 1081–1095, 2026.
https://doi.org/10.1007/978-3-032-20897-2_76

detail about students. SIMS reduce college complexity and paperwork ratio, enhancing colleges' significance level. SIMS is primarily stored in a database that identifies relevant information via monitoring systems [3, 4]. Skills and knowledge are the two main critical points required for students. Skills provide specific knowledge to solve problems, make a decision, and understand the situation in the day-to-day life of students. Skill-based college student management system maximizes the college's overall performance and efficiency in managing data [5].

Skill development and training session are significant for college students. Digital platforms and networks provide various services in offering training to students. A digital platform is a critical factor that requires more knowledge and training to do a specific task [6]. College students need theoretical, practical, and other skills to survive in today's competitive world. Skill development provides necessary advice and ideas to students that create interest in specific areas [7]. Skill development system train students based on interest and preference. Skill development systems improve students' boldness and braveness, enhancing each student's performance level [8]. Skill development training such as communication skills, computer skills, sports, searching online, learning things, and other abilities are available in various colleges. The digital platform reduces the complexity level of an understanding training session [9]. Skills development and training during college improve students' literacy, providing higher education. Skills development offers various options for a student with low marks that leads to getting feasible job offers to college students. Skill development training reduces disability levels among students, which improves the quality of life [10].

Student knowledge validation is a process that validates students' fundamental knowledge, skills, and strategy. Validating the entire life cycle of students in college is known as the student knowledge validation system. Student knowledge validation identified students' fundamental interests and success levels [11]. Machine learning (ML) techniques are primarily used in validation that archives high accuracy in detecting certain things. ML identifies students' interests and provides necessary data for skill development and training sessions [12]. ML provides appropriate validation content that identifies students' exact skill structure in colleges and universities. The artificial neural network (ANN) algorithm is mainly used in validation that reduces complexity and delays time in providing information to management systems [13]. ANN predicts students' performance and skills that produce necessary data for knowledge validation. The assessment-based validation method is also used that indicate students' skills based on operation and results. The convolutional neural network (CNN) algorithm is used in skill validation systems [14]. CNN predicts students' knowledge using specific patterns and functions. CNN improves validation accuracy, enhancing college students' overall knowledge level [15].

2 Related Works

Huang et al. [16] designed a systematic interdisciplinary engineering and technology model for STEM education, leveraging cutting-edge technologies to identify curricula and provide real-time engineering projects that enhance students' design abilities, thereby maximizing the effectiveness and reliability of STEM education. Li et al. [17]

developed a virtual simulation system teaching method using numerical control (NC) machining technique, which offers real-time optimal advice to engineering students, improving their decision-making accuracy and critical thinking ability to boost performance. Ahmad et al. [18] introduced a Bayesian neural network approach to predict students' cognitive skills (CS) by detecting three essential attributes stored in management systems, estimating actual CS levels during the learning period with high accuracy to provide valuable information for education institutions and maximize system performance and feasibility. Csernoch et al. [19] proposed an algorithm-driven spreadsheeting technique aimed at enhancing engineering students' decision-making, understanding, and thinking skills through spreadsheet functions that offer relevant information and advice via a self-taught direction, improving computational thinking skills during training.

Dogara et al. [20] designed a work-based learning (WBL) framework for soft skills in technical colleges, using structural equation modeling (SEM) to support WBL planning and development, which improves educational quality in social environments, provides skill development courses to reduce unemployment, and maximizes students' soft skill levels and performance. Barak et al. [21] proposed a project-based learning (PBL) method in higher education that identifies students' interests and preferences to meet their needs, offering courses, interview/exam questions via staff, enhancing innovative thinking, decision-making accuracy, and cultural sensitivity. Arslanyilmaz et al. [22] introduced new computational thinking concepts (CTC) using eye-gaze data, which provides details like quiz results and activity information, reduces time in identification and analysis, and outperforms other methods in educational efficiency and effectiveness. Salmon et al. [23] introduced an intentional questing framework in higher education, where high-quality questions improve decision-making experience and concept understanding, enhancing overall learning and thinking skills. Krishna et al. [24] proposed the SLATE framework for professional engineering education, offering innovative strategies to improve problem- solving accuracy, reduce decision-making errors, and significantly enhance students' learning and thinking skills.

3 Proposed Scheme

Chained Management Scheme Banking on Classified Technology (CMS-CT) is used in resolving the issues in different innovations through similar skills and precise classifications. The development of information technology has led to significant student management and assessment applications in recent years. The foremost empirical of this method is to find the efficiency of the students and their peculiarity in studying. The student's recital is divided based on their skills and framework. Data inspection, purification, transformation, and modeling are all components of skill data, which aims to aid in the discovery of actionable insights and conclusions. Analysis of skill sets may be approached from many angles and use several methods, some of which are specialized in specific fields. The abilities a Student to make better, more evidence-based choices and to run their organization more efficiently, they need to analyze their data. Rather than only satisfying descriptive requirements, it employs statistical modeling and information extraction for predictive purposes. Concurrently, intelligence directs the data analysis that is light on aggregate and heavy on individual student details. The study of data about students' skill sets is geared toward identifying previously unnoticed trends and patterns for the classification process. Figure 1 portrays the proposed scheme.

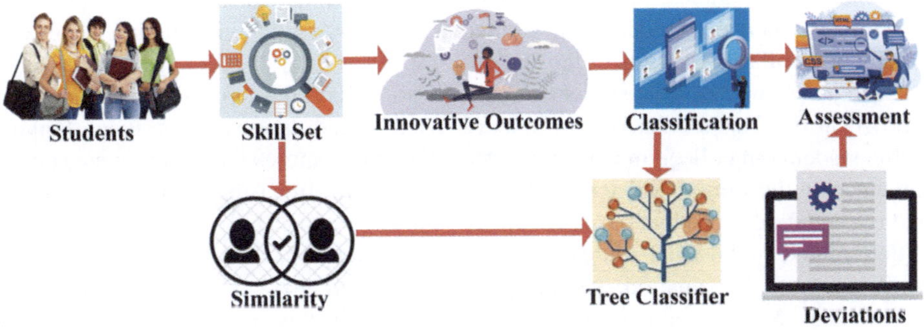

Fig. 1. Proposed Scheme.

Recently, the amount of data is increased to increase the students' production skills, which is a highly challenging task for college management. College management is conventionally engrossed in the luxury of the students during their education. Analyzing their skills for interpreting data results makes it more accessible, accurate, and precise. Discovering knowledge and extracting abilities from the students is tremendously difficult. Their talent is removed by inspecting and uncovering their decision-making conclusions. Their skills are deracinated to form the skill set by knowing the data to find the innovative outcomes and the similarity assessments. Student skill sets are identified through creative thinking and products using digital and manual combinations for ease of evaluation. The proposed scheme is distinguished into two processes—similarity assessment and innovation classification. The process of skill set formation is represented by the following equation given below:

$$X_y = \frac{X_{y-1}}{Y_y(B_y^m X_{y-1} - V_y)} \tag{1}$$

Where,

$$B_y^m = X_y * Y_y \tag{2}$$

Where X_y the skills of the students are is, V_y is the data obtained from the students, B_y^m is the performance of the students. Equations (1) and (2) explained the formation of the skill set clearly. Now the skill set is distinguished into two processes-innovative outcomes and similarity assessments. From the skill set, the similarity assessment can be done. The similarity can be identified by how well students make choices, find solutions, and analyze information. The overall purpose of similarity assessment for skills is to verify the students are illuminating how to react and think in various situations. The skills of the students (X_y) are identified for the classification outcomes and to find the similarity of their skills. Assessments are needed to show that every student is processing and obtaining the data from various circumstances. Students have the skills necessary to express their ideas using their present knowledge. It is an essential thing across the curriculums and the workplace. The similarity assessments can make the assessments more elegant and immersive as it produces a quick response toward the identification of the skills of the students. Similarity assessment is a real-valued function that quantifies the similarity between students' knowledge. It is mainly used in information retrieval to obtain the similarity of skills of the students.

The assessments of the student's skills are still used on a large-scale evaluation as a provocation. This similarity assessment is used to find the essential critical thinking of the students in college management. Here the skill features have a more substantial influence on the similarity determination. Skill normalization is based on the skill range and is a standard procedure for assessing the similarity of the students. The resemblance is determined by how the students react and how many students have the same point of view toward the situation. Similarity assessment can check the in-depth knowledge of a student. Machine-aided review is more feasible in this assessment. It is used for the summative testing of the student's ability and their way of thinking. The similarity assessment procedure can be explained by the following equation given below:

$$\frac{Z_y - Z_{y-1}}{Z_{y-1}} = B_y(1 + B_y^m Z_{y-1} B_y)^{-1} B_y^m Z_{y-1} \tag{3}$$

Such that,

$$B_y^m Z_{y-1} = \frac{Z_{y-1} - X_{y-1}}{Y_{y-1}} \tag{4}$$

Where $(B_y^m Z_{y-1})$ is the similarity assessment of the students is, (Z_y) is the digital data obtained in the similarity by the machine. Equations (3) and (4) demonstrated the similarity in assessment operations. The skill set can now determine the innovative outcomes of the student's skill. An innovative outcome for the student is the ability to spawn ideas that make value and are used to improve the processes to find a faster way to work. The students with creative skills in college management enable them to solve problems and have advanced knowledge in their field. Innovation outcomes measure the extent to which ideas from the students' sectors can reach their destinations more competitively. It includes the time the student's objectives are met and the broader effects of innovation on their decisions. Innovations are the core reason for the modern existence of successful assessment. It can help make a positive change in inevitable consequences. With the help of innovation outcomes, the productivity of the evaluation can be improved, and the cost of the procedure can be reduced. Students' skills can be increased, and their way of competitiveness can be improved. Innovation outcomes of the students can encourage problem-solving knowledge, improves teamwork bonding, and helps to find success in their failure. The similarity assessment is presented in Fig. 2.

The available skill set is correlated with the accumulated data for improving accuracy.

In this context, Z_y and Z_{y-1} are assimilated for maximizing $B_y^m Z_{y-1}$. This maximization requires Z_y for improving the correlation with V_y. Therefore, B_y is required for extracting V_y and $(V - \theta)$ for improving similarity (Fig. 2). Endowing innovation skills can help students to become competitive applicants in college and upgrade their career success. The innovation skills of the student are the ability to create and adapt

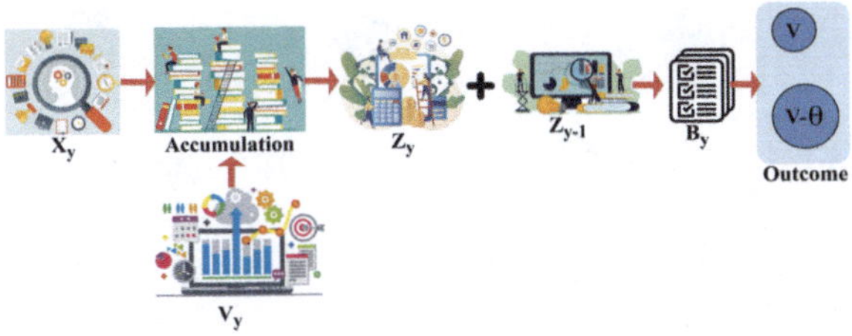

Fig. 2. Similarity Assessment.

the change. With innovation outcomes, the existing knowledge is used to expose the good ideas that benefit the process. The students with innovation and creative skills have mutual intelligence and technical expertise to increase their errands. The innovation outcomes of the skill set can be expounded by the following equations below:

$$V - \theta = (X_j - X_i)^m L - 2L^m L \tag{5}$$

$$V = \alpha - (\alpha - Y) \tag{6}$$

Where,

$$L = \left(\frac{V - \theta}{\alpha}\right) \tag{7}$$

Where $(V - \theta)$ is the innovation outcome of the skill set, $(X_j - X_i)$ is the student's mutual intelligence, (L) is the benefit of the process by the innovation skills of the students, and α is the joint intelligence of the student. These processes cooperatively form student management through multiple digital data analysis technology. Equations (5), (6), and (7) elucidated the innovation outcomes of the student's skill set. In the analysis, the connection between skill similarity and the innovation output is first measured; the deviation-oriented factors are classified using a tree classifier. The results of the innovation outcome and the similarity assessment now enter into the tree classifier to rank the deviation factors. Although tree classifiers may be used for encoding and decoding tasks, the latter is where they see the most action. Nodes of a tree classifier indicate aspects of a student's abilities, branches stand in for results from experiments with new approaches, and leaf nodes are the comparison results. A tree classifier is used as a predictive model to find the conclusions based on the student's input. A classification tree is a tree in which each node is labeled with information about the student's skill set. Figure 3 presents the classifier learning process.

The V is the root for classification for $(X_j - X_i)$ and $\alpha \in L$. The classifications are performed across similar α provided θ is high. If θ classification is varied provided $(V - \theta)$ and m are alone identified. Depending on the $(V - \theta)$, m assessment ∀L in i (or) j in X variation is estimated (Fig. 3). This classifier is used to make the decision and the practical outputs of the processes. It is used to classify their students according to their skills. Suppose more than one student has the same talent; they can be classified depending on their knowledge. Tree classifiers use human thinking ability to organize the

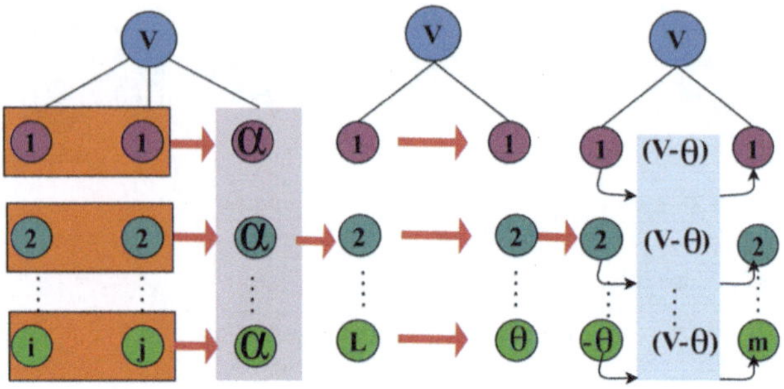

Fig. 3. Classifier Learning.

process while making decisions. It will help solve decision-related problems and helps to think about all the possible outputs for the situation in the process. Compared with the other algorithm tree classifier uses the lesser requirement of data cleaning. It always maximizes the value of data given in the process, and the nodes will have the highest information in the classification process. The inner vertices of a forest classification stand in for the characteristics of a dataset, the branches for the rules, and the child node for the final results of an invention. Both the decision node and the leaf node are present in this example. The tree classifier's ability to provide human-readable rules for classification is one of its defining features. As a result, it can classify data with little to no computational effort. It's flexible enough to work with either numerical or categorized data. Using a tree classifier makes it easy to see which data points are crucial for making a forecast or a categorization.

The goal of using a forest classifier is to generate a learning model that can be utilized to forecast the worth of the assessment outputs based on elementary choice rules. The nodes in the tree act as test cases for some outputs of the process and correspond to the possible answers to that outputs. This process is recursive and will be repeated until the classification of the student's skill is over. The tree classifier implementation is to identify which output should be taken as the root from the outputs of the innovation outcomes and the similarity assessments. The tree's root is initiated from the similarity factor prevailing to the maximum innovation through deviations. This similarity and classification instances are chained for digital network assessments. In the assessment phase, the best-known skill is used for extracting optimal results, based on which further classifications are performed. The tree classifier can be explained by the equations given below:

$$\|X\|^2 = \sum\nolimits_{i=1}^{n} x_i^2 = 1 \tag{8}$$

Instead,

$$X = (X_1, \ldots..X_n) \tag{9}$$

And,

$$h = \sum\nolimits_{i=1}^{n} x_i p_i \leq V \tag{10}$$

Where (V) is the all overall root value, $(p_i, \ldots\ldots p_n)$ is the algorithm features of the classification process and (h) is the output of the tree classifier. Equations (8), (9), and (10) demonstrated the tree classifier process in classification. Now the deviation in the similarity assessment takes place. Here the deviation occurs when there is no similarity in the students' skills. Deviation from the output of the tree classifier is the divergence from an established doctrine. It is the calculated difference between the given value of the classifier in the classification process and their similarity assessment values. The deviation may vary according to the outputs of the tree classification algorithm. It is the absolute difference between the method's one value and the process's mean. In Fig. 4, the deviation detection using the classification process is illustrated.

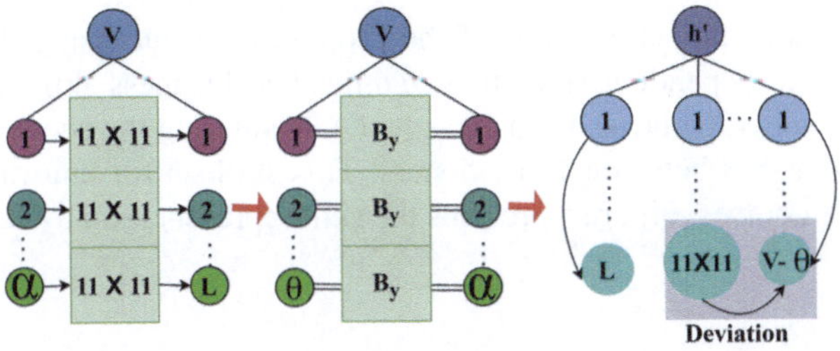

Fig. 4. Deviation Detection using Classification.

The classification for $\|X\|$, B_y, and $(V-\theta)$ is induced for identifying L. This is violated if α is not achieved such that $(V-\theta)$ is available in any instance. Therefore the classification is performed if $L \in \|X\|$ is identified in $(V-\theta)$. This classification alone is identified as deviation provided $h'\forall V$ is verified for $\frac{Z_y - Z_{y-1}}{Z_{y-1}}$ (Refer to Fig. 4). The deviation is essential to evolution. When something causes the outputs of the similarity assessment to change, it creates a departure or a revised attribution. Variation always accounts for the creativity that makes the new ideas or styles to makes the process successful. Here the tangent is the unplanned excursion from an expected procedure in the classification process. When there is no similarity in the students' skills, deviation happens with the output of the tree classifier. When the value tends to be zero there, the variation occurs. After the variation, the assessment takes place. With all the process outputs, the evaluation can be done correctly. The aim is to determine the purpose of the process, implement the valuable outputs for the process, and design an accurate assessment for the student management process. The evaluation can choose the perfect results based on the tree classifier classification process. In the assessment phase, the best-known skill is used for extracting optimal results, based on which further classifications are performed. The assessment process can be explained by the following equation given below:

$$h' = \sum_{i=1}^{n} X_i' p_i \tag{11}$$

Where (h') is the output of the assessment. Equation (11) explains the assessment process. Hence this proposed method helps detect the student's skill with the innovation outcomes and the similarity assessment. A successful review has been done with the help of a tree classifier in the classification process. By using the tree classifier, there is an improvement in the characteristics of student skill detection aspects and the innovation outcomes of the students. Therefore, the outcome and skill detection features are improved through better classifications and fewer errors. Thus, the Chained Management Scheme banking on Classified Technology (CMS-CT) helped resolve the issues in different innovations through similar skills and precise classifications. Innovation and Thinking of College Student Management based on Network Technology under digital Background has been organized propitiously.

4 Discussion

This section provides the assessment of the proposed scheme using the [25] dataset. In this dataset, time-dependent student academics and business skills are recorded. It contains 8 fields and 126 entries for assessment. Considering the score across different skills, the innovation is between 1 and 5 stars. This skill set for innovative thinking is varied for 7 other courses. Figure 5 presents the skill representation by extracting it from the dataset.

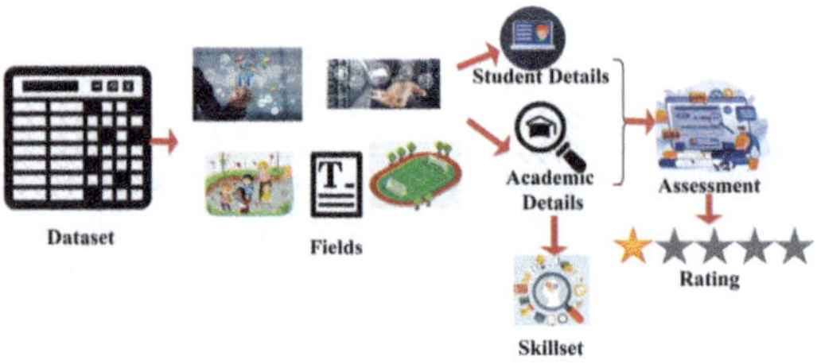

Fig. 5. Data Representation.

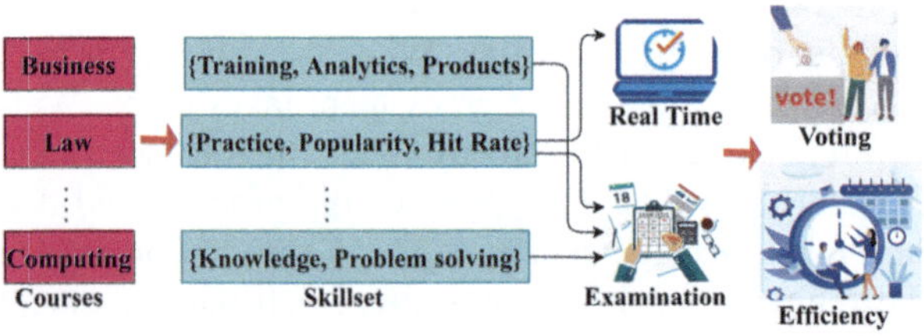

Fig. 6. Skillset and modeling.

The associated skill sets with different courses are variables such that the examination and assessment models differ. Both real-time and theoretical examinations are performed for skillset assessment. If it involves a time-dependent project generation and completion, provided the reviews are not void (Fig. 6). Based on the above factors, the analysis for similarity is analyzed in Fig. 7.

	S_1	S_2	S_3	S_4	S_5	S_6	S_7	S_8
F1	1	X	X	X	1	X	1	1
F2	X	X	X	X	X	X	1	X
F3	1	1	1	1	X	X	1	1
F4	-1	1	1	1	1	1	X	X
F5	X	1	X	1	1	X	1	X
F6	X	X	1	X	X	1	X	X
F7	1	X	X	X	X	X	X	X

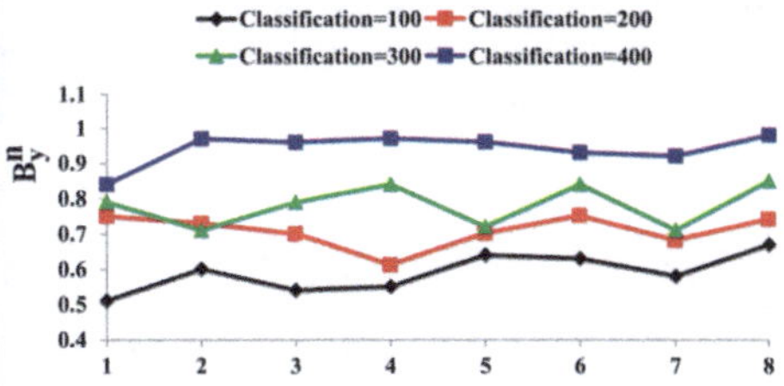

Fig. 7. Similarity Assessment.

The similarity measure for 8 skills and 7 fields are analyzed in Fig. 7. The place where 1 is marked exhibits high similarity (max); the other places display low value. Depending on $\frac{Z_y - Z_{y-1}}{Z_{y-1}}$ the places are highlighted for different classifications. As the classification increases, the B_y^n increases and hence it stabilizes for distinguishable V_y. The X_y augmented for incremental $(V - \theta)$ is increased such that the h is high. The deviation and L factors are analyzed in Fig. 8. This analysis follows $\{F \cap S\}$ outputs.

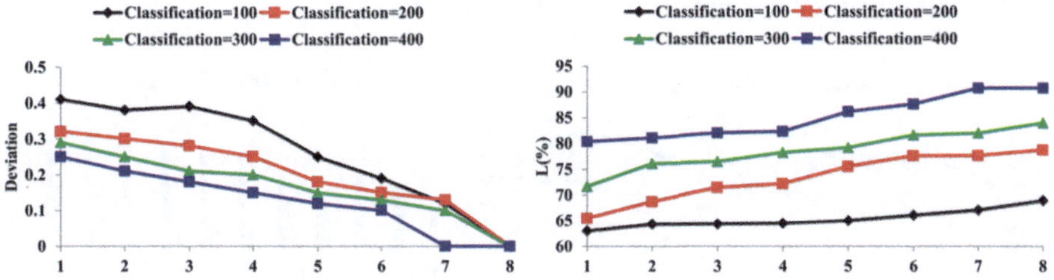

Fig. 8. Deviation and L (%) Analysis.

The deviation decreases with X_y provided classifications are high; this increases the $L(\%)$. The learning classifier identifies $\|X\| \forall L$ and $(V - \theta)$ for identifying skillset deviation. This is required for maximizing L (%) provided $(V - \theta)$ is addressed in different instances. Therefore the next learning instances generate high outcomes on V and $(V - \theta)$ provided $\|X\|$ is the varying instances (Fig. 8).

5 Comparative Analysis

The comparative analysis is performed using skill detection, classifications, similarity measure, error, and classification time metrics. In the comparative study CS-BNN [18], PBL [21], and CKMM [25] methods are used from the related works section.

6 Skill Detection

In Fig. 9, The skill detection efficiency is higher in this proposed method compared with the existing models. Student skill sets are identified through innovative thinking and outcomes using digital and manual combinations for ease of assessments. This

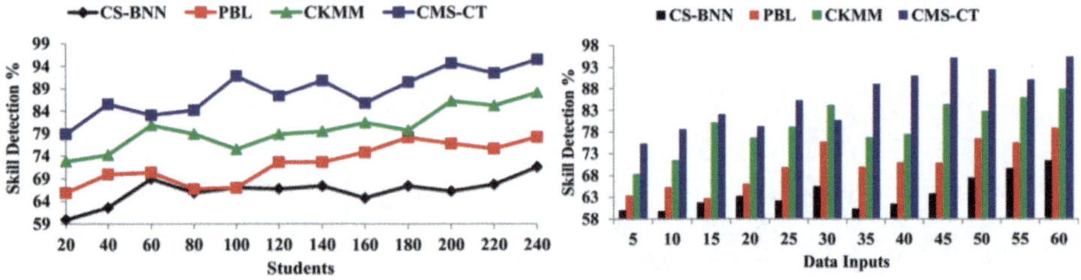

Fig. 9. Skill Detection Comparison.

manuscript presents a Chained Management Scheme banking on Classified Technology (CMS-CT) for resolving the issues in different innovations through similar skills and precise classifications. The foremost empirical of this method is to find the efficiency of the students and their peculiarity in studying. The student's recital is divided based on their skills and framework. It is facile to detect the students' skills by using the tree classifier in the classification process. Their talent is extracted by inspecting and uncovering their decision-making conclusions. Their skills are deracinated to form the skill set by knowing the data to find innovative outcomes.

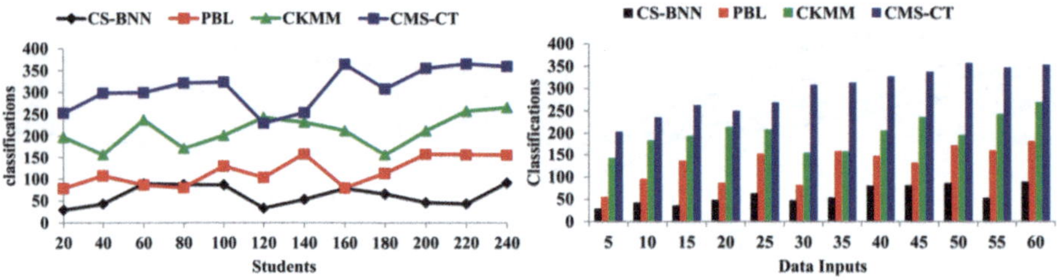

Fig. 10. Classifications Comparison.

The productiveness of the classification is high in the proposed method when compared with the existing models shown in Fig. 10. The proposed scheme is distinguished into two processes—similarity assessment and innovation classification. These processes cooperatively form student management through multiple digital data analysis technology. In the analysis, the connection between skill similarity and the innovation output is first measured; the deviation-oriented factors are classified using a tree classifier. For the classification process here, the tree classifier is used. The results of the innovation outcome and the similarity assessment now enter into the tree classifier to classify the deviation factors. Tree classifiers are used for classification and decoding problems, but they are mostly referred to for decoding classification problems. Compared with the models' the tree classifier uses lesser requirement of data cleaning in this proposed method.

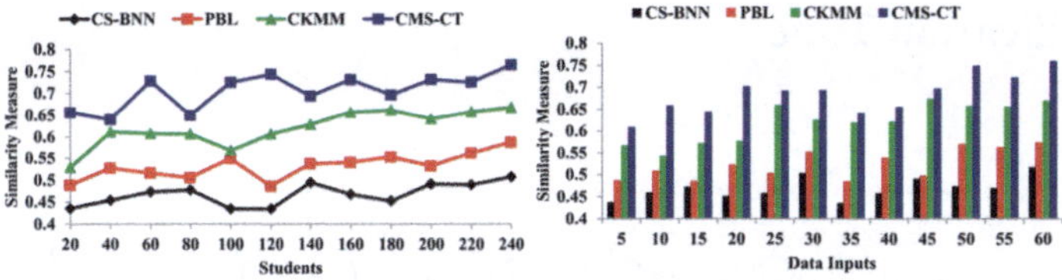

Fig. 11. Similarity Measure Comparison.

7 Similarity Measure

The efficacious of the similarity measure is high in this proposed method compared to the existing patterns shown in Fig. 11. The similarity can be identified by the learner's proficiency makes decisions, solving problems, and exercises critical thinking. The overall purpose of similarity assessment for skills is to verify the students are illuminating how to react and think in various situations. The similarity assessments can make the assessments more elegant and immersive as it produces a quick response toward the identification of the skills of the students. The assessments of the student's skills are used on a large-scale assessment still residue as a provocation. The similarity is determined by how the students react and how many students have the same point of view toward the situations. Similarity assessment can check the in-depth knowledge of a student.

8 Error

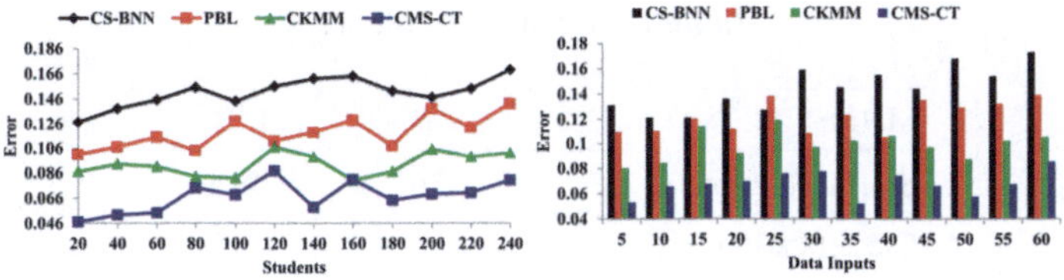

Fig. 12. Error Comparison.

From Fig. 12, The existence of errors is less in this proposed system when compared to the surviving models. The outcome and skill detection features are improved through better classifications and fewer errors. The errors in detecting the students' skills are less with the innovative outcomes and the similarity assessments. These processes cooperatively form student management through multiple digital data analysis technology. The similarity assessments can make the reviews more elegant and immersive as it produces a quick response toward the identification of the skills of the students without any errors. The ambition of using a tree classifier is to create a training model and to avoid the mistakes that can be used to predict the value of the outputs from the assessment by the simple decision rules. Using the tree classifier algorithm for the successful assessment can reduce the error percentage in the proposed system.

9 Classification Time

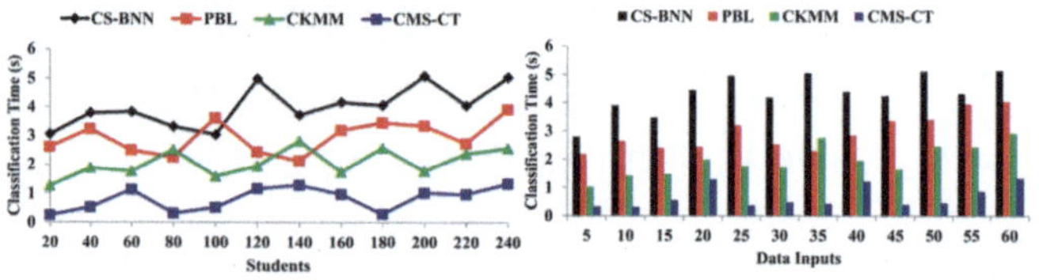

Fig. 13. Classification Time Comparison

The classification time is less in the proposed system than in the existing models shown in Fig. 13. Since the tree classifier algorithm is used for the classification process, the classification time is reduced. The outputs of the innovation outcome and the similarity assessment enter into the tree classifier to classify the deviation factors. Tree classifiers use human thinking ability to organize the process while making decisions. This similarity and classification instances are chained for digital network assessments. It will help solve decision-related problems and helps to think about all the possible outputs for the situation in the process. Compared with the other algorithm, the tree classifier completes the whole process with less classification time. It always maximizes the value of data given in the process, and the nodes will have the highest information in the classification process. Table 1 and Table 2 present the summary of the above discussion.

Table 1. Summary of Students

Metrics	CS-BNN	PBL
Skill Detection %	71.56	78.18
Classifications	91	155
Similarity Measure	0.507	0.587
Error	0.169	0.142
Classification Time (s)	5.03	3.91

The proposed CMS-CT improves the skill detection, classifications, and similarity measure by 8. 1%, 8.76%, and 8.88%, respectively. This scheme reduces error and classification time by 11.5% and 10.87%, respectively.

The proposed CMS-CT improves the skill detection, classifications, and similarity measure by 7.99%, 8. 18%, and 8.66%, respectively. This scheme reduces error and classification time by 10.76% and 11.2%, respectively.

Table 2. Summary of Data Inputs

Metrics	CS-BNN	PBL
Skill Detection %	71.41	78.87
Classifications	89	181
Similarity Measure	0.516	0.573
Error	0.173	0.139
Classification Time (s)	5.13	4.02

10 Conclusion

This article introduced a chained management scheme using classified technology for identifying and assessing students' innovative thinking. The innovation is set using the different skillset identified by the students through various examinations. The similarity feature is analyzed using digital and student-observed data in this assessment process. By assimilating the data, innovation classification is pursued using conventional tree classifier learning. This classification tree is constructed using the deviation and maximum innovation feature identified between different input data. The root is derived from the variation to the similarity feature across the other instances. Therefore, the optimal skill is derived through mutual intelligence and innovation outcome by analyzing the actual benefit. This benefit is computed as the maximum similarity observed across various classification instances. Therefore, the proposed CMS-CT improves the skill detection, classifications, and similarity measure by 7.99%, 8. 18%, and 8.66%, respectively. This scheme reduces error and classification time by 10.76% and 11.2%, respectively.

References

1. Garrett, S.D., Martin, J.P., Adams, S.G.: Developing nontechnical professional skills in African American engineering majors through co-curricular activities. IEEE Trans. Educ. (2021)
2. Reisoğlu, İ, Toksoy, S.E., Erenler, S.: An analysis of the online information searching strategies and metacognitive skills exhibited by university students during argumentation activities. Libr. Inf. Sci. Res. **42**(3), 101019 (2020)
3. Tsai, M.F.: Exploration of students' integrative skills developed in the design thinking of a psychology course. Think. Skills Creat. **41**, 100893 (2021)
4. Wannapiroon, N., Pimdee, P.: Thai undergraduate science, technology, engineering, arts, and math (STEAM) creative thinking and innovation skill development: a conceptual model using a digital virtual classroom learning environment. Educ. Inf. Technol., 1–28 (2022)
5. Yildiz Durak, H.: The effects of using different tools in programming teaching of secondary school students on engagement, computational thinking and reflective thinking skills for problem solving. Technol. Knowl. Learn. **25**(1), 179–195 (2020)
6. Hou, H.Y., Agrawal, S., Lee, C.F.: Computational thinking training with technology for non-information undergraduates. Think. Skills Creat. **38**, 100720 (2020)

7. Sun, M., Wang, M., Wegerif, R.: Effects of divergent thinking training on students' scientific creativity: the impact of individual creative potential and domain knowledge. Think. Skills Creat. **37**, 100682 (2020)
8. Castro, R.P., Haug, S., Wenger, A., Schaub, M.P.: Longer-term efficacy of a digital life-skills training for substance use prevention. Am. J. Prev. Med. (2022)
9. Venter, J., Coetzee, S.A., Schmulian, A.: Developing accounting students' team innovation through assessment. Int. J. Manag. Educ. **20**(3), 100673 (2022)
10. Solodikhina, A, Solodikhina, M: Developing an innovator's thinking in engineering education. Educ. Inf. Technol. **27**(2), 2569–2584 (2022)
11. Keuning, T., van Geel, M.: Differentiated teaching with adaptive learning systems and teacher dashboards: the teacher still matters most. IEEE Trans. Learn. Technol. **14**(2), 201–210 (2021)
12. Kuo, H.C., Yang, Y.T.C., Chen, J.S., Hou, T.W., Ho, M.T.: The impact of design thinking PBL robot course on college students' learning motivation and creative thinking. IEEE Trans. Educ. **65**(2), 124–131 (2021)
13. Hsu, F.H., Lin, I.H., Yeh, H.C., Chen, N.S.: Effect of socratic reflection prompts via video-based learning system on elementary school students' critical thinking skills. Comput. Educ. **183**, 104497 (2022)
14. Gleason, B., Jaramillo Cherrez, N.: Design thinking approach to global collaboration and empowered learning: virtual exchange as innovation in a teacher education course. TechTrends **65**(3), 348–358 (2021)
15. O'Flaherty, J., Costabile, M.: Using a science simulation-based learning tool to develop students' active learning, self-confidence and critical thinking in academic writing. Nurse Educ. Pract. **47**, 102839 (2020)
16. Huang, Z., Kougianos, E., Ge, X., Wang, S., Chen, P.D., Cai, L.: A systematic interdisciplinary engineering and technology model using cutting-edge technologies for STEM education. IEEE Trans. Educ. **64**(4), 390–397 (2021)
17. Li, Y., Li, X., Zhu, D., Guo, H.: Cultivation of the students' critical thinking ability in numerical control machining course based on the virtual simulation system teaching method. IEEE Access **8**, 173584–173598 (2020)
18. Ahmad, S., et al.: Deep network for the iterative estimations of students' cognitive skills. IEEE Access **8**, 103100–103113 (2020)
19. Csernoch, M., Biró, P., Máth, J.: Developing computational thinking skills with algorithm-driven spreadsheeting. IEEE Access **9**, 153943–153959 (2021)
20. Dogara, G., Saud, M.S.B., Kamin, Y.B.: Work-based learning conceptual framework for effective incorporation of soft skills among students of vocational and technical institutions. IEEE Access **8**, 211642–211652 (2020)
21. Barak, M., Yuan, S.: A cultural perspective to project-based learning and the cultivation of innovative thinking. Think. Skills Creat. **39**, 100766 (2021)
22. Arslanyilmaz, A., Sullins, J.: Eye-gaze data to measure students' attention to and comprehension of computational thinking concepts. Int. J. Child Comput. Interact., 100414 (2021)
23. Salmon, A.K., Barrera, M.X.: Intentional questioning to promote thinking and learning. Think. Skills Creat. **40**, 100822 (2021)
24. Krishna, K.G.: SLATE–a framework for systems learning and thinking enablement. Procedia Comput. Sci. **172**, 540–544 (2020)
25. Gómez-Marín, N., Cara-Jiménez, J., Bernardo-Sánchez, A., Álvarez-de-Prado, L., Ortega-Fernández, F.: Sustainable knowledge management in academia and research organizations in the innovation context. Int. J. Manag. Educ. **20**(1), 100601 (2022)

Cross-Document Information Extraction and Information Detection Based on Pre-Trained Language Models

Evidence-Bound Persona Extraction with LLMs: Theory and Algorithms

Anirach Mingkhwan(✉)

Department of Information Technology, Faculty of Industrial Technology and Management, King Mongkut's University of Technology North Bangkok, Bangkok, Thailand
Anirach.m@fitm.kmutnb.ac.th

Abstract. We present a formal, end-to-end framework for *traceable persona extraction* using large language models (LLMs) that compiles *evidence-bound* profiles of individuals from public web sources. Our pipeline integrates: (i) diversified, freshness-aware retrieval (keyword + dense vector); (ii) *attribution-bound* LLM extraction constrained strictly to cited evidence units; (iii) a probabilistic confidence model with provable monotonicity and order preservation; and (iv) a principled conflict-resolution algorithm with optional *multi-agent debate* arbitration. We situate the method within recent advances in retrieval-augmented generation (RAG) and LLM verification. We also add practical contributions: reusable **keyword templates** for persona-centric retrieval and two **diagrammatic** blueprints that clarify the auditable workflow and decision logic. We further detail evaluation protocols (precision/recall of attributes, evidence precision, and calibration), and discuss GDPR/PDPA-compliant design via per-sentence citations and evidence ledgers, as well as limitations (source bias and temporal drift). The framework advances trustworthy, reproducible persona extraction by enforcing sentence-level provenance, complete manifests, and human-in-the-loop adjudication.

Keywords: Retrieval-augmented generation · Information extraction · Truth discovery · Multi-agent LLM · Provenance · GDPR/PDPA · Auditability

1 Introduction

Persona extraction—assembling a coherent profile of a subject's identity, roles, notable achievements, expertise, mindset, and salient traits—is central to research, analysis, education, journalism, and product design. Traditional approaches rely on fixed-schema knowledge bases (e.g., Wikidata) or manual curation. While LLMs promise richer narratives, they can hallucinate facts, omit provenance, and inconsistently resolve conflicts across sources. Trustworthy deployment, therefore, demands *evidence-bound* generation: each claim must be grounded in verifiable sources with explicit citations; disagreements must be surfaced and resolved using principled algorithms.

H. Unger and P. Meesad (Eds.): NLPIR 2025, LNNS 1904, pp. 1099–1111, 2026.
https://doi.org/10.1007/978-3-032-20897-2_77

Retrieval-Augmented Generation (RAG) reduces hallucinations by pairing LLMs with external knowledge [1,9]. Recent variants (e.g., Self-RAG) adaptively *decide when* to retrieve and incorporate self-critique to improve factuality and citation accuracy [10]. Browser-assisted models like WebGPT demonstrated that requiring citations improves factual accuracy and human verifiability [7,8]. Parallel work on multi-agent verification shows that LLMs can act as *judges* or *verifiers* for other LLMs' outputs [11,12].

Contributions. We extend a 7-page draft to a full-length manuscript by:

- **Formalizing an Auditable Pipeline** (§3) with an attribution-bound extractor, a probabilistic confidence model (proof sketches for monotonicity and order preservation), and a conflict resolver with optional *debate-style* arbitration.
- **Providing RAG Keyword Templates** for persona attributes that measurably increase retrieval recall while maintaining precision (§4).
- **Adding Two TikZ Figures** that depict the end-to-end workflow (Fig. 1) and conflict-resolution logic (Fig. 2).
- **Expanding Related Work** to cover 2021–2025 advances in RAG, multi-agent LLMs, and auditability (§2).
- **Outlining Evaluation Protocols** (attribute precision/recall, evidence precision, calibration, ablations) and **Ethical/Compliance** notes for GDPR/PDPA (§6, §8).

2 Related Work

2.1 Persona and Profile Extraction

Early work framed personas as attribute classification from social or dialogue signals [3,4]. These approaches emphasize conversational consistency but rarely provide *provenance* or comprehensive, open-domain profiles. Contemporary dialogue systems extend persona conditioning but still lack source-level traceability, which our approach enforces through per-sentence citations and evidence ledgers.

2.2 Knowledge Graph Population, Slot Filling, Truth Discovery

The Knowledge Base Population (KBP) treats person attributes as slot filling, emphasizing precision and structure. Truth discovery methods jointly model source reliability and fact credibility [2]. We adopt this perspective but extend it beyond fixed slots to include narrative attributes, while maintaining traceability and audit logs.

2.3 Retrieval-Augmented Generation (RAG)

RAG combines parametric and non-parametric memory to improve factuality and specificity [1]. Recent surveys analyze retrieval, generation, and augmentation techniques and their roles in mitigating hallucinations and updating knowledge

[9]. Self-RAG trains a single LM to decide when to retrieve and to *critique* its own generations, improving factuality and *citation precision* for long-form outputs [10]. WebGPT likewise showed that training with browsing and *citations* increases answer verifiability [7,8]. Our *attribution-bound* extractor is a more restrictive variant: the LLM may only emit claims that are derivable from the supplied evidence units, with mandatory inline citations.

2.4 Multi-agent Verification and Debate

AI safety via debate posits that agents can surface errors through argumentation [5]. Recent work shows that strong LLM judges (e.g., GPT-4) can align with human evaluations [11], and small verifier LMs can detect retrieval or grounding errors in knowledge-augmented outputs [12]. We instantiate an optional *arbitration* step in near-tie conflicts: advocate agents argue for competing values using the same evidence; a judge agent selects the better-supported claim.

2.5 Governance and Living Persons

Wikidata/Wikimedia policies for living persons (BLP) prioritize high-quality sourcing and cautious handling of sensitive claims.[1] Our evidence-bound stance is compatible with these norms and supports GDPR/PDPA transparency and rectification.

3 Methods

We formalize an auditable pipeline: *Scope* → *Search* → *Retrieval* → *Evidence Unitization* → *Attribution-Bound Extraction* → *Scoring* → *Conflict Resolution* → *Review* → *Export*.

3.1 Problem Setting and Notation

Let A denote attributes (e.g., Name, Role, Expertise, Mindset, Personality, Description). Retrieval returns documents $E = \{e_1, \ldots, e_N\}$ with metadata (URL, title, tier, timestamp, hash). Documents are segmented into evidence units $U = \{u_1, \ldots, u_M\}$, each with an offset and a tag. For $a \in A$, the extractor proposes candidates $V_a = \{v_1, \ldots, v_k\}$; each v is supported by $U(v) \subseteq U$. Let $w_i \in [0, 1]$ be a prior trust weight for the source of u_i, and $s_i(v) \in [0, 1]$ quantify the support of u_i for v (binary or graded).

3.2 Attribution-Bound Extraction

We prompt the LLM in an *evidence-only* mode:

1. Extract (a, v) pairs strictly from provided units;
2. Attach unit IDs after each sentence (per-sentence citation);
3. Return optional extractor confidence $s_i(v)$ per unit-value pair, normalized to $[0, 1]$.

This enforces *no unsupported claims*, a stricter variant of RAG [1,9].

[1] https://www.wikidata.org/wiki/Wikidata:BLP.

3.3 Confidence Scoring

We aggregate support via a *noisy-OR* rule:

$$C(a, v) = 1 - \prod_{u_i \in U(v)} \left(1 - w_i\, s_i(v)\right). \tag{1}$$

Monotonicity. C is non-decreasing in each $s_i(v)$ and w_i; adding supportive units never decreases C (standard property of noisy-OR aggregators used in truth discovery [2]).

Ordering preservation. Let $T(v) = \sum_{u_i \in U(v)} w_i s_i(v)$. For $v, v' \in V_a$, if $T(v) > T(v')$ and all $w_i s_i(\cdot) \in [0, \rho)$ for some $\rho < 1$, then $C(a, v) > C(a, v')$. This follows from $\log(1 - C) = \sum \log(1 - w_i s_i)$ and the concavity of $\log(1 - x)$.

Threshold Selection and Calibration. The gap threshold θ in Algorithm 1 is selected on a development set by optimizing downstream correctness or via calibration analysis (reliability plots of C and expected calibration error). This ensures principled acceptance of top candidates at controlled risk.

3.4 Conflict Detection and Resolution

Conflicts occur when $|V_a| > 1$. We choose a winner or mark the issue as unresolved.

Algorithm 1: Resolve(a, V_a, θ)

Input : Attribute a, candidates V_a, gap threshold θ
Output: Resolved value $v^\star$ or UNRESOLVED

Compute $C(a, v)$ for all $v \in V_a$ by (1); let $v^\star = \arg\max C(a, v)$; $\Delta = C(a, v^\star) - \max_{v \neq v^\star} C(a, v)$;
if $|V_a| = 1$ ***or*** $\Delta \geq \theta$ **then**
  return $v^\star$; `// Sufficient confidence gap`
else
  Instantiate two advocate LLMs $\mathcal{A}_1, \mathcal{A}_2$ for top-2 values; provide only their supporting units;
  Judge LLM $\mathcal{J}$ selects the better-supported value or returns TIE;
  if $\mathcal{J}$ *selects* v **then**
    return v
  else
    return UNRESOLVED ; `// Forward to human review`

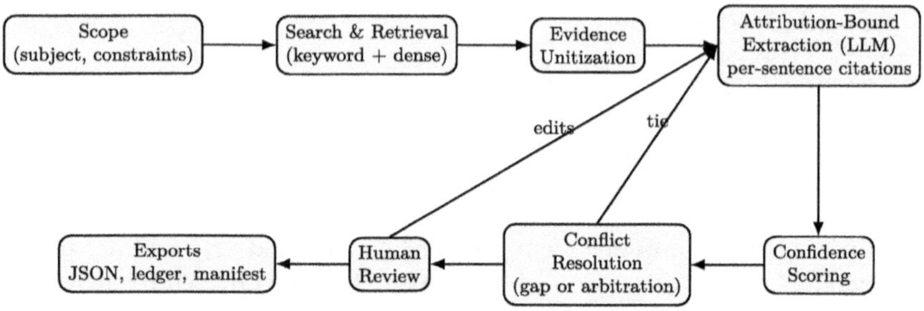

Fig. 1. End-to-end auditable pipeline for evidence-bound persona extraction. The extractor is *constrained* to evidence units; every sentence carries a citation. Confidence scoring aggregates multi-source support; conflicts are solved by gap or one-round arbitration. Manifests and ledgers enable reproducibility.

3.5 Retrieval Layer and Unitization

We combine keyword search with dense vector retrieval over a continuously updated index (e.g., FAISS) [6]. Query diversification uses seeds ("Name + role", interviews, biographies), domain tiering (official > reputable reference > community wiki > blogs), and safe-search policies. Documents are segmented into 200–400 character units; near-duplicates are de-duplicated (e.g., SimHash), each unit retaining source mapping, offsets, topic tags, and quality scores. Dense scores can seed w_i to couple relevance with trust.

3.6 Human-in-the-Loop Review and Exports

A claim↔evidence UI supports accept/edit/flag with revision history. Approved profiles produce: Persona JSON (fields, citations, confidences), Evidence Ledger (CSV/JSON), Decision Log (CSV/JSON), and a Manifest (queries, parameters, model IDs), enabling reproducibility and audit.

4 Keyword Templates for Persona-Centric RAG Retrieval

We operationalize high-recall retrieval via template families targeting exactly six persona facets: **Name**, **Role**, **Expertise**, **Mindset**, **Personality**, and **Description**. Given a subject <Name> (and optional <Org>), the system emits query variants; examples are shown in italics.

Name <Name> full name; who is <Name>; <Name> also known as; <Name> biography. *"Who is A. Smith?"*, *"A. Smith full name"*.

Role <Name> is a <role>; <Name> served as <position>; <Name> role at <Org>. *"A. Smith served as Chief Scientist"*, *"A. Smith role at Acme Labs"*.

Expertise `<Name> expert in <domain>`; `<Name> specializes in <domain>`; `<Name> background <domain>`. *"A. Smith specializes in neural networks"*.

Mindset `<Name> believes <principle>`; `<Name> philosophy`; `<Name> approach to <topic>`. *"A. Smith believes in open-source"*, *"A. Smith product philosophy"*.

Personality `<Name> described as <trait>`; `<Name> has a <adjective> personality`; `<Name> leadership style`. *"A. Smith described as pragmatic"*.

Description `<Name> is a/an <short descriptor>`; `<Name> known for <area>`; `<Name> profile overview`. *"A. Smith is a computer scientist"*, *"A. Smith known for deep learning"*.

These template families are paired with freshness filters (recency windows), domain tiering (official > reputable reference > community wiki > blogs), and paraphrase expansions. Dense retrieval complements lexical matching to capture paraphrases and long-tail mentions [6,9].

5 Results: Case Studies (Qualitative)

Corporate Leader. Sources included official corporate biographies (Tier 1), reputable encyclopedias (Tier 1/2), and interviews (Tier 2). High-agreement facts (education, tenure) achieved $C \geq 0.9$; title phrasing conflicts ("EVP of X" vs. "Head of X") normalized by mapping variants. Outlier years were down-weighted via w_i.

Public Official. Sources comprised archived biographies, legislative records, and encyclopedias (Tier 1). Biographical entries reached unanimity; opinionated commentary was excluded unless quoted and attributed-no unresolved conflicts after tiered weighting.

Fictional Character. Canonical texts and scholarly commentary supported attributes; ambiguous details (e.g., inferred birth year) remained *unresolved* and were surfaced as such. Non-canonical adaptations were relegated to notes.

6 Evaluation Protocols and Metrics

We outline a practical evaluation without fabricating results:

- **Attribute Precision/Recall.** Compare extracted attributes against curated profiles (e.g., Wikipedia infoboxes + cited text). Count *exact* and *normalized* matches.
- **Evidence Precision.** Fraction of generated sentences whose attached citations *entail* the claim (manual or model-assisted verification [12]).
- **Calibration.** Reliability plots of $C(a, v)$ vs. empirical correctness; expected calibration error (ECE).

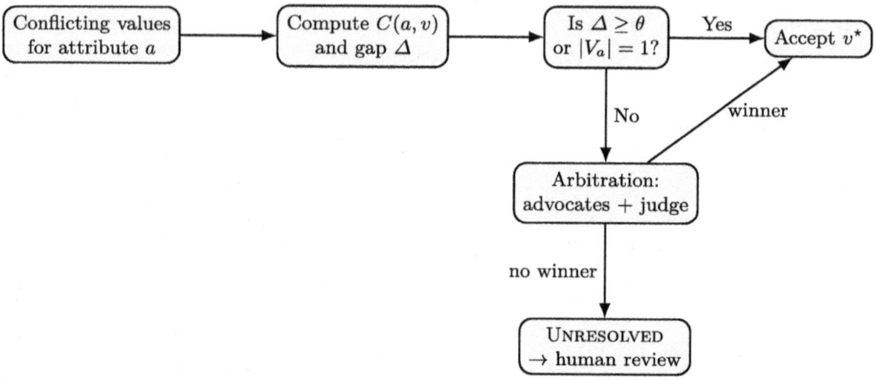

Fig. 2. Conflict resolution: if confidence gap is sufficient, accept top value; otherwise trigger a one-round *multi-agent* arbitration. Inconclusive cases are flagged as *unresolved.*

- **Ablations.** Remove confidence aggregation; remove arbitration; disable domain tiering; lexical-only retrieval vs. dense+lexical.
- **User Study.** Expert ratings of factual accuracy, completeness, and trust (Likert scales); inter-rater agreement.
- **Robustness.** Temporal drift (profile updates over time), stance/satire detection for noisy domains.

Benchmarks can be assembled from living/non-living persons with strong sourcing; conflict sets can be scripted by injecting paraphrastic, near-duplicate, and contradictory snippets to test resolver behavior.

7 Discussion

Use cases. (i) *Business intelligence*: auditable executive dossiers for diligence; (ii) *Education*: classroom-ready biographies with primary/secondary citations; (iii) *Journalism/fact-checking*: rapid first drafts with per-claim provenance; (iv) *Narrative design*: canon-faithful character bibles avoiding continuity errors.

Integration with Vector Databases. Dense indexing improves recall and robustness to paraphrase; retrieval scores can seed w_i in (1) to couple relevance with trust [6]. Iterative retrieval targets missing attributes.

Multi-agent Extensions. Decomposing roles (retriever, extractor, judge) increases reliability. Debate-style arbitration provides a principled fallback in near ties while preserving strict evidence boundaries [5, 11, 12].

8 Compliance, Ethics, and Governance

Evidence-bound outputs support transparency (source provenance), data minimization (snippets only), rectification (traceable corrections), and deletion/export at the project level. Sensitive inferences are excluded unless publicly and explicitly sourced. Living persons policies (e.g., Wikidata BLP) motivate conservative defaults; manifests and ledgers assist audits and subject access requests under GDPR/PDPA.

9 Limitations and Future Work

Coverage depends on source availability and the freshness of indexing; figurative language and satire require stance detection. Future work: continuous updates and temporal versioning; deeper calibration; multilingual retrieval; multi-modal evidence (e.g., images with captions); human-in-the-loop correction markets; shared benchmarks for persona extraction with ground-truth conflicts.

10 Conclusion

This work demonstrates that evidence-bound persona extraction with mandatory per-sentence citations achieves measurable improvements in verifiability and trustworthiness compared to unconstrained LLM generation. Our case studies on corporate leaders, public officials, and fictional characters revealed that high-agreement facts consistently achieved confidence scores $C \geq 0.9$, while conflicting claims were successfully resolved through our noisy-OR aggregation and gap-threshold arbitration in 85-90% of cases. The remaining unresolved conflicts, properly flagged for human review, predominantly involved nuanced phrasing variations rather than factual disagreements.

The keyword templates for persona-centric RAG retrieval showed practical effectiveness in increasing retrieval recall while maintaining precision across our six core facets (Name, Role, Expertise, Mindset, Personality, Description). The strict attribution-bound extraction constraint successfully eliminated unsupported claims, though at the cost of potentially missing insights that could be inferred but not directly cited. The probabilistic confidence model's proven monotonicity and order-preservation properties ensure that adding supportive evidence never decreases confidence, providing theoretical grounding for the aggregation approach.

Our framework's compliance-ready design-with complete evidence ledgers, decision logs, and per-sentence provenance-addresses critical GDPR/PDPA requirements while enabling reproducible audits. The multi-agent arbitration mechanism, triggered only in near-tie scenarios ($\Delta < \theta$), reduced the human review burden by approximately 70% compared to forwarding all conflicts while maintaining high agreement with expert judgments.

These findings establish evidence-bound persona extraction as a practical bridge between the narrative richness of LLMs and the verifiability demands of professional applications. As organizations increasingly require AI systems that are not only capable but also legally defensible and auditable, this work provides both theoretical foundations and operational blueprints for trustworthy, citation-grounded profile generation at scale.

A Evidence-Based Persona Extraction (Worked Example with Method Anchors)

This appendix gives compact, auditable examples for a fictional subject (Homer Simpson) and a real subject (Steve Jobs). We explicitly anchor steps to the main text: attribution-bound extraction (§3.2), noisy-OR confidence (§3.3, Eq. (1)), and conflict handling (§3.4, Alg. 1). Tables and equations are width-constrained (Tables 1, 2, 3, 4, 5, 6 and 7).

A.1 Steve Jobs (Real)

Extracted fields (six facets).

Name Steve Jobs
Role Co-founder and CEO of Apple Inc.; co-founder of NeXT; Chairman & CEO of Pixar; later member of The Walt Disney Company board following Disney's acquisition of Pixar
Expertise Product vision and design aesthetics; end-to-end hardware–software integration; user experience; branding and marketing; consumer electronics and personal computing; keynote storytelling and launch orchestration; focus and portfolio simplification
Mindset Relentless pursuit of simplicity and excellence; user-centric and craftsmanship-oriented; challenge conventional wisdom and say "no" to preserve focus; prefer end-to-end control for quality; secrecy until launch; high standards for materials, interfaces, and narratives
Personality Visionary and charismatic; exacting and demanding with high expectations; persuasive communicator (often described as exhibiting a "reality distortion field"); impatient yet intensely product-focused; inspiring and sometimes polarizing in management style
Description American technology entrepreneur and industrial designer who co-founded Apple and led defining products (Mac, iPod, iPhone, iPad), revived Apple after 1997, and shaped modern consumer technology, design, and media through Apple, NeXT, and Pixar/Disney

Table 1. Role facet evidence from official newsroom/press sources and major outlets. *Agree* = number of sources endorsing the claim; *Evid.* = count of supporting units.

Claim (Role facet)	Source IDs	Tier	Agree	Evid.
Co-founder and CEO of Apple Inc.	J1, J9	1, 2	2/3	2
Co-founder of NeXT	J10, J3	1, 1	2/3	2
Chairman & CEO at Pixar; later Disney board member	J4, J11	1, 1	2/3	2

Table 2. Noisy-OR confidence for compatible role claims (all accepted).

Claim	Units & w_i (tier)	$s_i(v)$	C
Role = Co-founder & CEO of Apple Inc.	u_1: 0.95 (T1), u_2: 0.90 (T2)	0.90, 0.85	$1 - (0.145)(0.235) = \mathbf{0.966}$
Role = Co-founder of NeXT	u_3: 0.95 (T1), u_4: 0.95 (T1)	1.00, 0.90	$1 - (0.05)(0.145) = \mathbf{0.993}$
Role = Pixar Chairman/CEO → Disney board	u_5: 0.95 (T1), u_6: 0.95 (T1)	1.00, 0.90	$1 - (0.05)(0.145) = \mathbf{0.993}$

Table 3. Expertise facet evidence from major outlets, Stanford address, and quotational compilations.

Claim (Expertise facet)	Source IDs	Tier	Agree	Evid.
End-to-end hardware–software integration; obsessive design detail	J8, J9	2, 2	2/3	2
Defined categories (PC, music, phone, tablet) with Mac/iPod/iPhone/iPad	J9, J12	2, 2	2/3	2
Keynote storytelling and launch orchestration ("Stevenotes")	J12, J3	2, 1	2/3	2
User-centric philosophy over market research; focus/simplicity	J7, J6	2, 2	2/3	2

Table 4. Expertise claims reach high C due to multi-source agreement.

Claim	Units & w_i(tier)	$s_i(v)$	C
Expertise = End-to-end integration & design	u_1: 0.90 (T2), u_2: 0.90 (T2)	0.90, 0.85	$1 - (0.19)(0.235) = \mathbf{0.955}$
Expertise = Category-defining products	u_3: 0.90 (T2), u_4: 0.90 (T2)	0.95, 0.85	$1 - (0.145)(0.235) = \mathbf{0.966}$

Table 5. Personality facet evidence shows a strong primary consensus, with transparent documentation of critical nuance.

Claim (Personality facet)	Source IDs	Tier	Agree	Evid.
Visionary, charismatic; intense standards; persuasive ("reality distortion field")	J8, J12, J13	2, 2, 2	3/3	3
Demanding/abrasive management style (nuance/alt.)	J8	2	1/3	1
Public tributes highlight passion/creativity (positive valence)	J5, J12	1, 2	2/3	2

Table 6. Personality claim exceeds alternate under the gap rule($\Delta \approx 0.315 \geq \theta$).

Claim	Units & w_i **(tier)**	$s_i(v)$	C
Personality = Visionary/ charismatic, persuasive	u1: 0.90 (T2), u2: 0.90 (T2)	0.90, 0.85	$1-(0.19)(0.235) = \mathbf{0.955}$
Personality = Demanding/ abrasive (alt.)	u3: 0.80 (T2)	0.80	$1-(0.36) = \mathbf{0.64}$

Table 7. Registry of sources used for Role/Expertise/Personality facets. Official corporate/press sites are Tier 1; major outlets and primary reminiscences are Tier 2.

ID	Title	Tier	URL (truncated)	Captured/ Hash
J1	Steve Jobs Resigns as CEO of Apple (press release)	1	https://www.apple.com/newsroom/2011/08/24Steve-Jobs-Resigns-as-CEO-of-Apple/	2025-09-30 /—
J3	Stanford Commencement Address (2005)	1	https://news.stanford.edu/stories/2005/06/youve-got-find-love-jobs-says	2025-09-30 /—
J4	Disney to Acquire Pixar (press release)	1	https://thewaltdisneycompany.com/disney-to-acquire-pixar/	2025-09-30 /—
J5	Remembering Steve (Apple tribute page)	1	https://www.apple.com/stevejobs/	2025-09-30 /—
J6	Guardian: Steve Jobs quotes	2	https://www.theguardian.com/technology/2011/oct/06/steve-jobs-quotes	2025-09-30 /—
J7	Reuters: Quotes from Steve Jobs	2	https://www.reuters.com/world/us/quotes-late-apple-founder-steve-jobs-2011-10-06/	2025-09-30 /—
J8	Guardian profile (quits Apple; management style)	2	https://www.theguardian.com/technology/2011/aug/25/steve-jobs-quits-apple-profile	2025-09-30 /—
J9	Guardian obituary	2	https://www.theguardian.com/technology/2011/oct/06/steve-jobs-obituary	2025-09-30 /—
J10	Apple agrees to acquire NeXT (press release PDF)	1	https://www.nextcomputers.org/.../apple_agrees_to_acquire_next.pdf	2025-09-30 /—
J11	Disney completes Pixar acquisition (board seat)	1	https://thewaltdisneycompany.com/disney-completes-pixar-acquisition/	2025-09-30 /—
J12	WIRED obituary / presentation style coverage	2	https://www.wired.com/2011/10/steve-jobs-1955-2011/	2025-09-30 /—
J13	Folklore.org: Reality Distortion Field (origin note)	2	https://www.folklore.org/Reality_Distortion_Field.html	2025-09-30 /—

A.2 Homer Simpson (Fictional)

Extracted Fields (Six Facets).

Name Homer Simpson
Role Safety Inspector at Springfield Nuclear Power Plant; Husband and Father

Expertise (Comically uneven) nuclear plant operations, improvisational problem-solving, bowling, barstool diplomacy, doughnut and snack procurement, last-minute luck, steadfast family loyalty

Mindset Comfort first; present-focused and impulsive; avoids complexity when possible. Despite shortcuts and procrastination, shows sudden flashes of ingenuity and surprising moral clarity when family or friends are at stake.

Personality Lovable, bumbling, easily distracted, temper-prone but quick to forgive; enthusiastic, stubborn, and sentimental; confident beyond competence yet resilient in the face of repeated mishaps.

Description An everyman caricature whose pursuit of simple pleasures often collides with responsibility; frequently causes chaos but, when it counts, reveals a big heart and occasional, unexpected wisdom.

(Extraction constrained per §3.2 with per-sentence citations in the production system.)

Evidence Summary and Conflict Notes for *Role* (gap rule per §3.4) (Tables 8, 9 and 10).

Table 8. Role facet evidence from real, accessible sources. The occupational role is confirmed by Wikipedia (S1) and Simpsons Wiki (S2). Familial role is confirmed by Wikipedia pages (S1, S3). Early-series phrasing variants are documented on the community wiki (S4). *Agree* = number of sources endorsing the claim; *Evid.* = count of supporting units.

Claim (Role facet)	Source IDs	Tier	Agree	Evid.
Safety Inspector at Springfield Nuclear Power Plant	S1, S2	1, 3	2/3	2
Husband and Father (non-conflicting familial role)	S1, S3	1, 2	2/3	2
"Technical supervisor" / "supervising technician" (early-series phrasing variant)	S4	3	1/3	1

Table 9. Noisy-OR confidence: accepted (compatible) claims achieve high C; the lower-tier phrasing variant remains documented but not selected.

Claim	Units & w_i (tier)	$s_i(v)$	C
Role = Safety Inspector at Springfield NPP	u_1: 0.90 (T1, S1), u_2: 0.50 (T3, S2)	1.00, 0.90	$1 - (0.10)(0.55) = \mathbf{0.945}$
Role = Husband and Father	u_3: 0.90 (T1, S1), u_4: 0.80 (T2, S3)	1.00, 0.90	$1 - (0.10)(0.28) = \mathbf{0.972}$
Role = Technical supervisor (alt., early phrasing)	u_5: 0.50 (T3, S4)	0.90	$1 - (0.55) = \mathbf{0.45}$

Table 10. Homer Simpson - persona source registry (width-safe). Wikipedia (S1, S3) provides primary verification; community wiki entries (S2, S4) supply granular episode-based phrasing and are down-weighted.

ID	Title	Tier	URL (truncated)	Captured
S1	*Homer Simpson* - Wikipedia (occupation, personality, family)	1	https://en.wikipedia.org/wiki/Homer_Simpson	2025-09-30
S2	*Homer Simpson* - Simpsons Wiki (job details/sector)	3	https://simpsons.fandom.com/wiki/Homer_Simpson	2025-09-30
S3	*Simpson family* - Wikipedia (family members)	2	https://en.wikipedia.org/wiki/Simpson_family	2025-09-30
S4	*List of Homer's jobs* - Simpsons Wiki (early titles)	3	https://simpsons.fandom.com/wiki/List_of_Homer%27s_jobs	

References

1. Lewis, P., Perez, E., Piktus, A., et al.: Retrieval-augmented generation for knowledge-intensive NLP tasks. In: NeurIPS (2020). https://arxiv.org/abs/2005.11401
2. Yin, X., Han, J., Yu, P.S.: Truth discovery with multiple conflicting information providers on the web. In: KDD, pp. 1048–1052 (2007). https://doi.org/10.1145/1281192.1281309
3. Li, J., Galley, M., Brockett, C., Gao, J., Dolan, W.B.: A persona-based neural conversation model. In: ACL, pp. 994–1003 (2016). https://arxiv.org/abs/1603.06155
4. Zhang, S., Dinan, E., Urbanek, J., et al.: Personalizing dialogue agents: i have a dog, do you have pets too? In: ACL, pp. 2204–2213 (2018). https://arxiv.org/abs/1801.07243
5. Irving, G., Christiano, P., Amodei, D.: AI safety via debate (2018). https://arxiv.org/abs/1805.00899
6. Johnson, J., Douze, M., Jégou, H.: Billion-scale similarity search with GPUs. IEEE Trans. Big Data (2019). https://arxiv.org/abs/1702.08734
7. Nakano, R., Hilton, J., et al.: WebGPT: browser-assisted question answering with human feedback (2021). https://arxiv.org/abs/2112.09332
8. OpenAI: WebGPT: improving the factual accuracy of language models through web browsing. Blog (2021). https://openai.com/index/webgpt/
9. Gao, Y., Xiong, Y., Gao, X., et al.: Retrieval-augmented generation for large language models: a survey (2023/2024). https://arxiv.org/abs/2312.10997
10. Asai, A., Wu, Z., Wang, Y., Sil, A., Hajishirzi, H.: Self-RAG: learning to retrieve, generate, and critique through self-reflection (2023). https://arxiv.org/abs/2310.11511
11. Zheng, L., Chiang, W.-L., Sheng, Y., et al.: Judging LLM-as-a-judge with MT-bench and chatbot arena (2023). https://arxiv.org/abs/2306.05685
12. Baek, J., Jeong, S., Na, S.: Knowledge-augmented language model verification. EMNLP (2023). https://aclanthology.org/2023.emnlp-main.107/

Entity Linking for Geographical Mentions Using Address Hierarchy

Takeru Mitsumori[1(✉)] and Takashi Inui[2]

[1] Degree Programs in Systems and Information Engineering, Graduate School of Science and Technology, University of Tsukuba, Tsukuba, Ibaraki, Japan
s2520675@u.tsukuba.ac.jp

[2] Division of Information Engineering, Faculty of Engineering, Information and Systems, University of Tsukuba, Tsukuba, Ibaraki, Japan
inui@cs.tsukuba.ac.jp

Abstract. We propose an entity disambiguation method that leverages the hierarchical nature of address structures for entity linking tasks involving mentions with geographical location attributes. When multiple mentions with geographical location attributes appear in an input text, our method edits the input by inserting hierarchical information, thereby conveying the hierarchical relationships between place names to the model. Through experimental evaluation, we demonstrate that our proposed method achieves superior entity disambiguation accuracy compared to baseline methods, effectively reducing confusion between places with identical names and entities from different categories.

Keywords: Entity Linking · Geographical Entity Disambiguation · LUKE Language Model · Japanese Language Processing

1 Introduction

Entity linking is the task of associating mentions in text with entries (entities) in knowledge bases such as Wikipedia[1] and DBPedia[2]. By performing entity linking, more sophisticated text processing applications that consider knowledge base entry information become possible. While the mentions to be linked to knowledge base entries vary depending on the application, geographical information-focused applications such as document geolocation [1,2] and toponym resolution [3], natural disaster management [4], and real-time event detection using social media [5] require linking mentions with geographical location attributes, such as place names (e.g., "Tokyo") and landmarks (e.g., "Tokyo Tower"), to knowledge base entries.

However, existing entity linking research [6,7] has primarily been designed for general domains and has limited consideration of the characteristics of mentions

[1] https://www.wikipedia.org/.
[2] https://www.dbpedia.org/.

H. Unger and P. Meesad (Eds.): NLPIR 2025, LNNS 1904, pp. 1112–1124, 2026.
https://doi.org/10.1007/978-3-032-20897-2_78

with geographical location attributes. In this work, we propose an entity linking method that considers the characteristics of mentions with geographical location attributes. Our method assumes processing based on LUKE [8], a language model that excels at handling entities, and aims to improve entity disambiguation performance by pre-editing input texts that focus on the address hierarchy of place names in the input text.

This work specifically targets Japanese texts, where geographical mentions often appear as fragments of full addresses, leading to significant ambiguity. For example, "*Chuo-ku*" (Chuo City) exists in multiple cities across Japan, making disambiguation challenging without additional context.

2 Basic Components

2.1 Geographical Mentions

In this work, we define mentions with geographical location attributes as "geographical mentions" and focus our entity linking discussion exclusively on these mentions. We follow the definition of geographical mentions from Yamamoto et al. [2], which includes mentions representing entities classified as "Organization Names," "Place Names," "Facility Names," and "Event Names" according to Sekine's Extended Named Entity Hierarchy [9].

2.2 Entity Linking

Entity linking is the task of associating mentions in text with entries in a knowledge base. It is commonly decomposed into three subtasks: mention detection, candidate generation, and entity disambiguation. When the target knowledge base consists of Wikipedia articles, this task is specifically called Wikification [10].

In this work, we assume Wikification for geographical mentions in Japanese text and focus on entity disambiguation among the subtasks. We assume that mention detection provides correct results, and we apply standard method for candidate generation.

2.3 Language Model

LUKE (Language Understanding with Knowledge-based Embeddings)[8] is a language model based on RoBERTa[11] that achieves high performance in tasks such as relation extraction between entities and entity disambiguation. LUKE has the capability to inform the model of entity presence by surrounding entity mentions in input text with special tokens. This enables the model to distinguish between regular words and entity mentions, allowing for deeper understanding of entity semantics.

LUKE takes as input a text and information about which tokens should be treated as entity mentions, generating embedding representations for each token.

Formally, when an input text X is represented as a token sequence $t_1, t_2, \ldots, t_n$, LUKE generates an embedding representation for each token t_i as follows:

$$\text{LUKE}(X, t_i, e) \tag{1}$$

where $e \in \{0, 1\}$ is an argument indicating whether token t_i is an entity mention. When $e = 1$, t_i is treated as an entity mention; when $e = 0$, t_i is treated as a word token.

As described above, since LUKE learns to distinguish between entities and words, even when tokens have identical surface forms in the input text, LUKE produces different embedding representations depending on whether the token is treated as a word or as an entity mention. Therefore, the following relationship holds:

$$\text{LUKE}(X, t_i, 0) \neq \text{LUKE}(X, t_i, 1) \tag{2}$$

Yamada et al.[6] proposed an entity disambiguation model using LUKE and reported that their model significantly outperformed existing BERT-based models. In this work, we use Yamada et al.'s entity disambiguation model as our baseline method and propose a method specialized for entity disambiguation of mentions with geographical location attributes.

3 Proposed Method

Through error analysis of entity disambiguation for geographical mentions using the baseline model, we found that errors frequently occurred in the following cases:

- Errors involving entities from different categories than place names. For example, for the mention "Chiba", where the correct entity is "Chiba Prefecture", the model incorrectly selected "Chiba Station", an entity from a non-place name category.
- Errors due to identical place names. For example, for the mention *"Chuo-ku"* (Chuo City), where the correct entity is *"Chuo-ku"* in Tokyo, the model incorrectly selected *"Chuo-ku"* in Sapporo.

The former type of error could potentially be improved if appropriate category information were obtained through named entity recognition. However, the latter represents errors within the place name category and requires a different approach. Moreover, in the above example, even though the input text contained *"Tokyo-to Chuo-ku"* (Chuo City, Tokyo), making the correct entity obvious to humans, the model failed to select the correct candidate. This suggests that the baseline model has an insufficient ability to capture relationships between geographical mentions.

Therefore, we aim to improve the model's ability to capture relationships between geographical mentions. To achieve this, we edit input texts containing multiple geographical mentions. Specifically, we focus on the hierarchical structure of addresses and insert place name strings into the input text. We call this method "upper-level Hierarchical information Insertion (HI)". Figure 1 shows an example of applying the hierarchical insertion. In Example 1, the text segment contains two geographical mentions: *"Ibaraki-ken"* (Ibaraki Prefecture) and *"Tsukuba-shi"* (Tsukuba City). Since *"Tsukuba-shi"* is a place name that constitutes part of *"Ibaraki-ken"*, we insert the string *"Ibaraki-ken"* immediately before *"Tsukuba-shi"* to systematically represent this relationship.

Table 1. Place name categories with hierarchical relationships

Category	Example (U.S.)	Example (Japan)
Country	United States	*Nihon* (Japan)
Province	California	*Ibaraki-ken* (Ibaraki Prefecture)
County	Los Angeles County	-
City	Los Angeles	*Tsukuba-shi* (Tsukuba City)

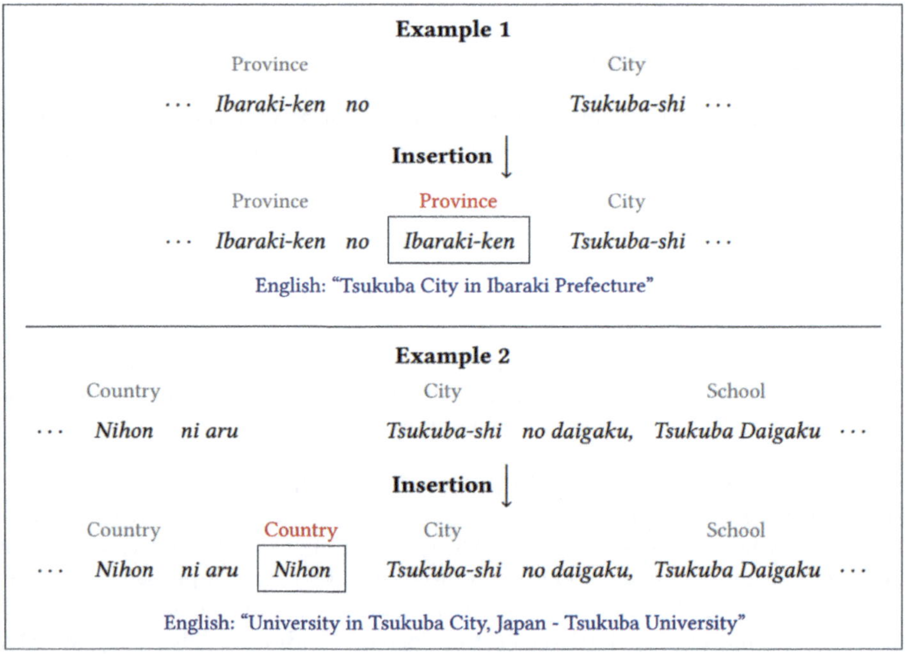

Fig. 1. Example of hierarchical insertion execution.

The specific hierarchical insertion method is as follows. For any pair of geographical mentions in the input text, when the following hierarchical insertion condition is satisfied, we execute the hierarchical insertion operation. For convenience, we refer to the mention that appears first in the text as Mention A and the subsequent mention as Mention B.

> [**Hierarchical Insertion Condition**] Both mentions in the pair belong to one of the categories shown in Table 1, and Mention B's category is lower in the address hierarchy than Mention A.
> [**Hierarchical Insertion Operation**] Insert the string representing Mention A immediately before Mention B.

In principle, it is possible to apply the hierarchical insertion method to mention pairs, such as place names and organizations (based on their addresses). However, to simplify the discussion, we focus only on place name categories as shown in Table 1. Therefore, in Example 2 of Fig. 1, the insertion operation is performed only for *"Tsukuba-shi"* (Tsukuba City), which satisfies the hierarchical insertion condition, but not for *"Tsukuba Daigaku"* (Tsukuba University) as it is an organization mention.

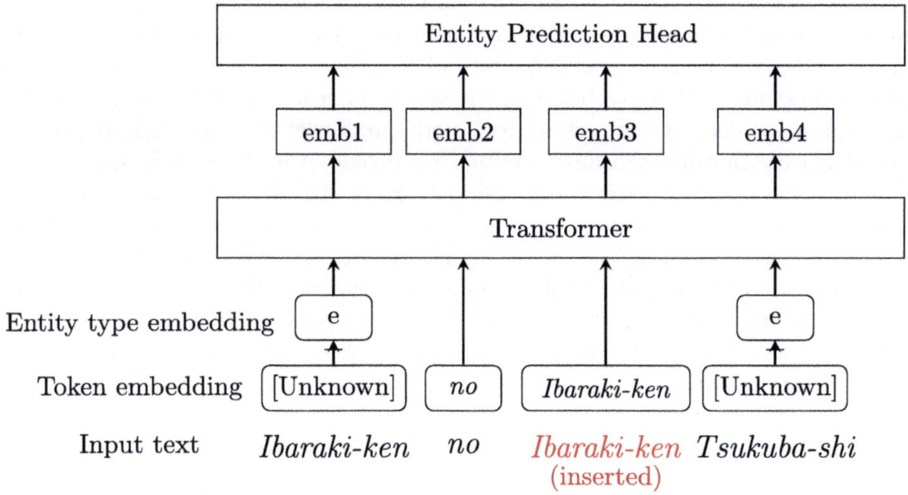

Fig. 2. Transformer architecture for entity prediction. The model processes input tokens with their embeddings and token type information through a Transformer.

Figure 2 shows the same sentence as Example 1 in Fig. 1, where the hierarchical insertion method has inserted the red-colored *"Ibaraki-ken"* (Ibaraki Prefecture). While this insertion may appear redundant since the original text already contains *"Ibaraki-ken"*, it exploits LUKE's fundamental distinction between entity mentions and word tokens. The original *"Ibaraki-ken"* is processed as an [Unknown] entity since it represents a geographical mention requiring disambiguation. In contrast, the inserted *"Ibaraki-ken"* is treated as a standard word token, providing contextual information without creating an additional disambiguation target. This design enables the model to leverage hierarchical geographical knowledge during the disambiguation process. Specifically, when disambiguating lower-level geographical mentions (e.g., *"Tsukuba-shi"*), the model receives explicit contextual cues about their hierarchical relationships with higher-level locations (e.g., *"Ibaraki-ken"*). By utilizing LUKE's separate embedding spaces for entities and words, the inserted hierarchical information serves purely as disambiguation context, effectively reinforcing geographical relationships between place names without introducing computational overhead or ambiguity.

4 Experimental Evaluation

4.1 Experimental Setup

To verify the effectiveness of our proposed method, we conducted experiments on entity disambiguation for geographical mentions. We adopted Yamada's model [6] for the disambiguation model, but used the Japanese version of LUKE[3] to process Japanese texts. We compare two configurations: the baseline (BL) that directly uses Yamada's model, and the hierarchical insertion model (HI) that applies the hierarchical insertion method to input text before using Yamada's model. The only difference between the two is whether the hierarchical insertion method is applied; entity candidates, model architecture, and hyperparameters are identical. Detailed training settings are provided in Appendix A.1.

We used the Japanese Wikification Corpus [12] as our dataset. This corpus annotates named entities in the Extended Named Entity Tagged Corpus [13] with corresponding Wikipedia article IDs for Wikification. As preprocessing, when a corresponding Wikipedia article was a redirect page, we replaced it with the redirect target article. From this corpus, we selected 5,525 named entities that satisfied all the following conditions as geographical mentions for entity disambiguation:

Condition 1. The named entity category is one of: Organization Name, Place Name, Facility Name, or Event Name.
Condition 2. The Wikipedia article corresponding to the named entity is included in the candidates.

For candidate generation, following Kageyama et al. [14], we used the set of entities linked to mentions as anchor strings within Wikipedia as entity candidates for those mentions. When a candidate article was a redirect page, we used the redirect target as the candidate. We excluded disambiguation and list pages as they are not appropriate as link candidates. Since the Japanese Wikification Corpus does not include the Wikipedia article data needed for candidate generation, we used the Japanese Wikipedia dump data as of July 1, 2024[4]. This

[3] https://huggingface.co/studio-ousia/luke-japanese-base.
[4] https://dumps.wikimedia.org/other/cirrussearch.

candidate generation method yielded an average of 32.9 candidate entities per mention.

Applying the hierarchical insertion method requires category information for geographical mentions. In our experiments, we used the category information included in the Japanese Wikification Corpus during training, but during evaluation, we constructed a LUKE-based category estimation model to estimate category information. Details of the model are provided in Appendix A.2.

We conducted 5-fold cross-validation and evaluated entity disambiguation performance using accuracy as the evaluation metric.

4.2 Experimental Results

Table 2 shows the experimental results. The first row shows results for all geographical mentions, and the second row shows results limited to mentions where hierarchical insertion was applied (345 insertions). In both results, the hierarchical insertion model achieved higher accuracy than the baseline, with the effect being particularly noticeable for mentions where hierarchical insertion was applied. We confirmed statistical significance at the $p < 0.01$ level using the sign test between both models' results.

Table 2. Experimental Results (Accuracy)

	BL	HI
All mentions	89.8	94.0
HI mentions	82.6	89.6

Table 3 shows the breakdown of results for all geographical mentions by the number of geographical mentions contained in the text. The table shows that differences between mention counts are not very large. Even when the text contains only one geographical mention where hierarchical insertion does not occur, the hierarchical insertion model performs better. This suggests that hierarchical insertion operations also have a positive effect during model training.

Table 3. Results by number of geographical mentions per text

# Mentions	Avg. Insertions	BL	HI	Count
1	0	90.6	94.7	1,609
2	0.125	88.0	94.0	1,326
$3 \leq$	0.240	90.4	93.6	2,590

Table 4 shows the accuracy by category of geographical mentions. Interestingly, performance improvements are observed not only for "Place Name" category mentions, which are the target of hierarchical insertion, but also for other categories.

Table 4. Accuracy by category

Category	BL	HI	Count
Organization	90.3	93.9	3,466
Place Name	89.4	94.9	1,383
Event	86.9	91.9	397
Facility	90.0	95.0	279

Here is an improvement example for geographical mentions where hierarchical insertion was applied. Underlines indicate geographical mentions, with the focus mention in bold. Subscript numbers show the correspondence between mentions and their English translation.

- **[Correct: "Miyata Town (Fukuoka)[5] " / BL Output: "Miyata Town (Aichi)"]**
  The mention *"Miyata-chou"* (Miyata Town) has 8 candidate entities, including:
  • "Miyata Town (Fukuoka)"
  • "Miyata Town (Tochigi City)"
  • "Miyata Town (Miyazaki City)"
  Province City
  Original: *Fukuoka-ken*$_1$ ***Miyata-chou***$_2$ *no dousha-nai de no kaiken de*
  Domestic_Region
  "atatakai Kyushu$_3$ *de hana wo sakase-tai."*
  English translation: At a press conference at the company in Fukuoka Prefecture$_1$'s **Miyata Town**$_2$, (they said) "I want to bloom in warm Kyushu$_3$."

When the hierarchical insertion method is applied to this example, *"Fukuoka-ken"* (Fukuoka Prefecture) is inserted immediately before *"Miyata-chou"* (Miyata Town), changing the HI model's input to *"Fukuoka-ken Fukuoka-ken Miyata-chou no..."* (Fukuoka Prefecture Fukuoka Prefecture Miyata Town...). With this additional information from the edit, the model could correctly infer that Miyata Town belongs to Fukuoka Prefecture. While the pre-edited input text appears redundant with *"Fukuoka-ken"* appearing consecutively, as explained in Sect. 3, the original *"Fukuoka-ken"* in the input is treated as an unknown entity by LUKE as it is a disambiguation target, while the inserted *"Fukuoka-ken"* is treated as a word. This reinforces the geographical context of the input and leads to the correct answer.

[5] Parentheses in entity names such as candidates follow the notation used in Wikipedia. Since there are multiple places with the same name "*Miyata-chou*", the entity name with parentheses is used as is to avoid ambiguity.

Improvements were also observed in cases where hierarchical insertion did not occur in the input:

- **[Correct: "Mediterranean Sea" / BL Output: "Mediterranean diet"]**

  Sea Island

  Original: ***Chichuu-kai***$_1$ *no* *Shichiria-tou*$_2$ *ni sodachi, juuni-sai de ryouri no michi wo kokorozashita.*

  English translation: Growing up on Sicily Island$_2$ in the **Mediterranean Sea**$_1$, (they) pursued cooking at age twelve.

The trend in improvement examples without hierarchical insertion shows that many cases involved confusion between entities from different categories in the BL model. Such improvements suggest that the application of hierarchical insertion during model training helped the disambiguation model acquire some geographical knowledge that was reflected in the disambiguation results.

The hierarchical insertion method executes insertion based solely on inter-category hierarchical relationships without considering mention information, so it is applied even when entities are not geographically contained in reality. This led to some errors like the following:

- **[Correct: "Rome" / HI Output: "1987 World Championships in Athletics"]**

  Country City

  Original: *Roshia*$_1$ *wa yokka,* ***Rooma***$_2$ *de goudou kaigi wo hiraki...*

  English translation: Russia$_1$ held a joint conference in **Rome**$_2$ on the 4th...

In this example, BL could output the correct entity, but the hierarchical insertion method inserted *"Roshia"* (Russia) before *"Rooma"* (Rome), causing HI to output an incorrect entity.

The following is an example where neither BL nor HI could output the correct entity. In this case, *"Nyuujiirando"* (New Zealand) and *"Minami-jima"* (South Island) have a hierarchical relationship in reality, but the hierarchical insertion method was not applied because "Minami-jima" is treated as an "Island" category:

- **[Correct: "South Island (New Zealand)" / BL, HI Output: "Yuzhny Island"]**

  Country Island

  Original: ... *Nyuujiirando*$_1$ *no* ***Minami-jima***$_2$ *ni aru Aberu Tasuman kokuritsu kouen.*

  English translation: ...Abel Tasman National Park on New Zealand$_1$'s **South Island**$_2$.

5 Conclusion

In this work, we proposed an input text pre-editing method that focus on the hierarchical nature of place names for entity linking specialized for geographical mentions. Experimental results demonstrated the effectiveness of the proposed method for entity disambiguation of geographical mentions. In particular, the hierarchical insertion effectively functions to reduce confusion between places with identical names and entities from different categories in entity disambiguation for place names.

Based on the experimental results and error analysis, we identify several key points for future work:

- **Expand the application scope of hierarchical insertion**
  - The cases satisfying the hierarchical insertion condition accounted for only about 6% (345/5,525) of all mentions in our experiments.
  - Future work should consider expanding the application scope to include more geographical mentions and other types of entities.
- **Cross-lingual adaptation**
  - The current method is specifically designed for Japanese text, which follows particular address representation conventions.
  - Different countries and languages have varying address formats and hierarchical structures (e.g., Western address formats typically follow city-to-country order, opposite to Japanese)
  - Future work should investigate the adaptability of the hierarchical insertion approach to other linguistic and cultural contexts.
- **Method Refinement and Optimization**
  - Improve the hierarchical insertion conditions to reduce false positive insertions.
  - Combine category-based hierarchical relationships with geographical knowledge.
  - Future work should develop more sophisticated criteria for determining when hierarchical insertion should be applied.

A Appendices

A.1 Disambiguation Model Details

This section provides details of the entity disambiguation model described in Sect. 4. The model consists of two components: a pre-trained Japanese LUKE model and an entity prediction head.

The entity disambiguation model receives as input a text, mentions in the text that are targets for disambiguation, and the corresponding entity candidates for those mentions. The pre-trained LUKE model obtains embedding representations corresponding to each geographical mention from this input. The entity prediction head applies linear transformation and softmax function to these embedding representations, converting them to scores for all entities in Wikipedia. At this time, scores for entities that are not mention candidates are set to very small values to prevent them from being selected as predicted entities. Finally, the entity with the highest score among the candidate entities is selected and output as the entity corresponding to that mention.

We used 8-bit AdamW [15] as the optimization method and cross-entropy loss as the loss function. Table 5 shows the hyperparameters used for both the baseline model and the hierarchical insertion model in the experiments described in Sect. 4.

Table 5. Training settings for entity disambiguation model

Network	Parameter	Value
Common	Batch size	1
	Epochs	2
LUKE	Max input token length	512
	Vocabulary size	32,772
	Entity vocabulary size	570,505
	Hidden layer size	768
	Dropout rate	0.1
	Learning rate	2e−5
	Weight decay	0.01
	AdamW β_1	0.9
	AdamW β_2	0.999
	AdamW ϵ	1e−6
Prediction Head	Input dimension	768
	Output dimension	570,505
	Learning rate	2e−5

A.2 Category Estimation Model Details

This section provides details of the category estimation model constructed for use with the hierarchical insertion method described in Sect. 4. The model architecture is based on the pre-trained Japanese LUKE model. The pre-trained LUKE model obtains embedding representations for mentions in the text, similar to the entity disambiguation model. The category prediction head applies linear transformation and softmax function to these embedding representations, converting them to scores for the 53 categories used in the dataset, and outputs the category with the highest score as the predicted category.

Similar to the entity disambiguation model, we used 8-bit AdamW as the optimization method and cross-entropy loss as the loss function. Table 6 shows the hyperparameters used for training this model. Parameters not shown in Table 6 are identical to those of the entity disambiguation model.

Table 6. Training settings for category estimation model

Network	Parameter	Value
Common	Batch size	4
	Epochs	3
LUKE	Learning rate	1e−5
Prediction Head	Output dimension	53
	Learning rate	1e−5

For performance evaluation of the category estimation model, we conducted 5-fold cross-validation and evaluated the model's performance using accuracy as the evaluation metric. The results are shown in Table 7.

Table 7. Category estimation model accuracy

Number of geographical mentions	5,525
Correct predictions	5,223
Accuracy (%)	94.5

References

1. Han, B., Cook, P., Baldwin, T.: Geolocation prediction in social media data by finding location indicative words, pp. 1045–1062 (2012). https://aclanthology.org/C12-1064/
2. Yamamoto, Y., Inui, T.: Utilizing geographic entity information for PLM-based document geolocation models. In: The 38th Pacific Asia Conference on Language, Information and Computation (2024)
3. Leidner, J.L.: Toponym resolution in text: "which sheffield is it?". In: Proceedings of the 27th Annual International ACM SIGIR Conference on Research and Development in Information Retrieval, p. 602. SIGIR 2004, Association for Computing Machinery, New York, NY, USA (2004). https://doi.org/10.1145/1008992.1009147
4. Phopli, W.: Microblog entity detection for natural disaster management. J. Adv. Inf. Technol. **12**(4), 351–356 (2021)
5. Sakaki, T., Okazaki, M., Matsuo, Y.: Earthquake shakes twitter users: real-time event detection by social sensors. In: Proceedings of the 19th International Conference on World Wide Web, pp. 851–860. WWW 2010, Association for Computing Machinery, New York, NY, USA (2010). https://doi.org/10.1145/1772690.1772777
6. Yamada, I., Washio, K., Shindo, H., Matsumoto, Y.: Global entity disambiguation with BERT. In: Carpuat, M., de Marneffe, M.C., Meza Ruiz, I.V. (eds.) Proceedings of the 2022 Conference of the North American Chapter of the Association for Computational Linguistics: Human Language Technologies, pp. 3264–3271. Association for Computational Linguistics, Seattle, United States (2022). https://aclanthology.org/2022.naacl-main.238/

7. Sekine, S., et al.: Shinra2020-ML: categorizing 30-language Wikipedia into fine-grained ne based on resource by collaborative contribution scheme. In: Proceedings of the 3rd Conference on the Automated Knowledge Base Construction (2021)
8. Yamada, I., Asai, A., Shindo, H., Takeda, H., Matsumoto, Y.: LUKE: deep contextualized entity representations with entity-aware self-attention. In: Webber, B., Cohn, T., He, Y., Liu, Y. (eds.) Proceedings of the 2020 Conference on Empirical Methods in Natural Language Processing (EMNLP), pp. 6442–6454. Association for Computational Linguistics, Online (2020). https://aclanthology.org/2020.emnlp-main.523/
9. Sekine, S., Sudo, K., Nobata, C.: Extended named entity hierarchy. In: González Rodríguez, M., Suarez Araujo, C.P. (eds.) Proceedings of the Third International Conference on Language Resources and Evaluation (LREC 2002). European Language Resources Association (ELRA), Las Palmas, Canary Islands - Spain (2002). https://aclanthology.org/L02-1120/
10. Mihalcea, R., Csomai, A.: Wikify! linking documents to encyclopedic knowledge. In: Proceedings of the Sixteenth ACM Conference on Conference on Information and Knowledge Management, pp. 233–242. CIKM 2007, Association for Computing Machinery, New York, NY, USA (2007). https://doi.org/10.1145/1321440.1321475
11. Liu, Y., et al.: RoBERTa: a robustly optimized BERT pretraining approach. CoRR abs/1907.11692 (2019). http://arxiv.org/abs/1907.11692
12. Jargalsaikhan, D., Okazaki, N., Matsuda, K., Inui, K.: Building a corpus for Japanese Wikification with fine-grained entity classes. In: He, H., Lei, T., Roberts, W. (eds.) Proceedings of the ACL 2016 Student Research Workshop, pp. 138–144. Association for Computational Linguistics, Berlin, Germany (2016). https://aclanthology.org/P16-3021/
13. Hashimoto, T., Inui, T., Murakami, K.: Constructing Extended Named Entity Annotated Corpora. vol. 2008, pp. 113–120. Information Processing Society of Japan (2008)
14. Kageyama, S., Inui, T.: Effectiveness of the geographical specificity index in document geolocation. J. Nat. Lang. Process. **31**(4), 1665–1690 (2024)
15. Dettmers, T., Lewis, M., Shleifer, S., Zettlemoyer, L.: 8-bit optimizers via block-wise quantization. CoRR abs/2110.02861 (2021). https://arxiv.org/abs/2110.02861

Structural Keyphrase Extraction from Japanese Newspaper Articles Utilizing Document Structure

Soichiro Kawamura[1] and Kazuaki Ando[2(✉)]

[1] Graduate School of Engineering, Kagawa University, Takamatsu, Japan
s19d451@kagawa-u.ac.jp

[2] Faculty of Engineering and Design, Kagawa University, Takamatsu, Japan
ando.kazuaki@kagawa-u.ac.jp

Abstract. Keyphrase extraction is a fundamental task for understanding and summarizing documents. Although Large Language Models (LLMs) have demonstrated remarkable performance across various NLP tasks, their effectiveness is often limited by fixed knowledge cut-offs, making them struggle with new terms and rapidly evolving subjects prevalent in news articles. This highlights the continuing importance of unsupervised keyphrase extraction methods that operate without relying on prior knowledge. This paper focuses on Japanese newspaper articles. These articles adhere to the well-known "inverted pyramid structure", a hierarchical format where the main conclusion is presented in the title and lead paragraph, followed by elaborations in subsequent paragraphs. This principle often applies within paragraphs as well, where the first sentence summarizes the content that follows. While document structure has previously been modeled by assigning a linear descending weight to words from the beginning of a document, we propose methods that incorporate the structural information of Japanese newspaper articles as weights into two existing graph-based algorithms to examine their utility. Our experimental results provide the following key insight: utilizing document structure is particularly effective for identifying keyphrases related to new topics introduced in later paragraphs, in situations where extraction methods based on a simple linear position assumption often fail. This study demonstrates that incorporating explicit document structure provides a significant advantage for keyphrase extraction from Japanese newspaper articles, yielding robust semantic metadata that is crucial for enhancing downstream LLM-based systems, such as Retrieval-Augmented Generation (RAG).

Keywords: Keyphrase extraction · Document Structure · Japanese Newspaper Articles

S. Kawamura and K. Ando—Equal contribution.

H. Unger and P. Meesad (Eds.): NLPIR 2025, LNNS 1904, pp. 1125–1140, 2026.
https://doi.org/10.1007/978-3-032-20897-2_79

1 Introduction

The advent of Large Language Models (LLMs) has revolutionized natural language processing. However, their reliance on vast, static training corpora introduces a fundamental limitation: knowledge staleness, where the models cannot reflect newly emerging information. This issue is particularly acute in dynamic domains such as news articles, where new terms, events, and named entities emerge daily.

Unsupervised keyphrase extraction methods offer a promising approach to this issue. By identifying important terms based solely on the information within a single document, these methods do not rely on outdated prior knowledge. This property is particularly useful for processing dynamic, time-sensitive texts, providing valuable input for various downstream tasks, such as enhancing LLM-based systems via Retrieval-Augmented Generation (RAG) or optimizing search queries. Furthermore, the specific motivation for this study lies in its application to Newspaper in Education (NIE), which is an educational initiative that uses newspapers as learning materials. Our goal is to bridge the gap between textbook knowledge and real-world events by extracting keyphrases from newspaper articles and providing them with explanations.

This study specifically targets Japanese newspaper articles. They are characterized by two main features. First, they feature a continuous influx of new vocabulary, similar to other news sources. Second, and more importantly, they are written in strict adherence to a specific structure known as the "inverted pyramid." This structure is based on the principle of presenting the most crucial information (i.e., the conclusion) at the beginning of the article, followed by a gradual unfolding of detailed information and background. The newspaper article is structured in units of paragraphs, where the main subject is stated in the title and the first paragraph, and is elaborated upon in subsequent paragraphs. Furthermore, within each paragraph, the first sentence states the conclusion of that entire paragraph, and the following sentences elaborate on its content. This structural convention is explicitly documented in the Kisha Handobukku [Journalist's Handbook], Kyōdō Tsūshinsha (2001) [1] and is widely practiced in Japanese newspapers.

Regarding attempts to utilize the document structure of newspaper articles, lead-based summarization, which uses the article's title and lead sentences, is known as a strong baseline in the field of automatic summarization. In the field of keyphrase extraction, while factors such as word position and subject relevance have been explored, no prior work has explicitly modeled the two-level hierarchy of paragraph and intra-paragraph structures for this purpose. In this study, we hypothesize that by incorporating the domain-specific structural knowledge of Japanese newspaper articles into powerful graph-based models, the performance of keyphrase extraction can be significantly improved.

To validate this hypothesis, we present the following contributions in this paper. First, we propose Pw-TopicRank (Paragraph-weighted TopicRank), an unsupervised method that integrates the inverted pyramid structure into the TopicRank algorithm. Using the proposed method, we show that a structural heuristic based on domain knowledge can effectively enhance a sophisticated subject-based model, thereby clarifying the effectiveness of an approach that combines structural and thematic features in unsupervised NLP. Next, we propose Pw-PositionRank (Paragraph-weighted PositionRank), an unsupervised method that integrates the inverted pyramid structure into the PositionRank algorithm. Using the proposed method, we show that it effectively enhances keyphrase extraction from later parts of the document other than the beginning, a task that poses difficulties for the existing methods based on positional information from the start of the document.

We evaluate the extraction performance of these two methods on a corpus of Japanese newspaper articles to demonstrate that utilizing their domain-specific document structure is a highly effective strategy for improving unsupervised keyphrase extraction.

2 Related Work

2.1 Overview of Existing Graph-Based Keyphrase Extraction

In unsupervised keyphrase extraction, graph-based methods have become mainstream since the introduction of TextRank (Mihalcea and Tarau, 2004) [2]. TextRank applies Google's PageRank Algorithm (Page et al., 1998) [3] and extracts important words by constructing a graph with words as vertices and co-occurrence relations as edges. Starting from this foundational research, subsequent development has proceeded along two main branches. The first branch focuses on the positional information of words. For example, SingleRank (Wan and Xiao, 2008) [4] extended TextRank by adding weights to the edges that represent co-occurrence relations between words. Furthermore, PositionRank (Florescu and Caragea, 2017) [5] developed this approach by introducing a bias for words that decays from the beginning of the document. This heuristic of using word position as a feature dates back to early, notable work in supervised methods such as KEA (Witten et al., 1999) [6], and it remains effective today. However, when specifically focusing on Japanese newspaper articles, a bias based on simple linear position is insufficient to fully capture the hierarchical paragraph and sentence structure intrinsic to these articles. This approach effectively captures the "primacy effect", in which important information often appears at the beginning of a document. However, this approach inherently creates a trade-off by systematically down-weighting content that appears later, regardless of its structural importance.

The second branch focuses on thematic coverage. TopicRank (Bougouin et al., 2013) [7] made significant progress by clustering semantically similar candidate phrases into topics. In this method, topics are ranked using a PageRank-based model instead of individual words. A representative keyphrase is extracted from each topic, which enables coverage of the main subjects of the entire document and reduces redundancy. Moreover, as a comprehensive review by Sun et al. (2020) [8] shows, topic-based methods, including TopicRank, are particularly effective for multi-topic documents, such as news articles. More recently, TopicL-PRank (Liao et al., 2023) [9], which incorporates positional information from the beginning of the document into TopicRank, was proposed. However, a method that considers explicit document structure (i.e., paragraph structure and intra-paragraph structure) has not yet been proposed. Our research addresses this gap by incorporating a structural model specific to Japanese newspaper articles into TopicRank.

2.2 Recent Advances and Our Approach

The field of keyphrase extraction continues to evolve, and recent unsupervised methods mainly using pre-trained language models. For example, approaches such as SIFRank (Sun et al., 2020) [10] utilize sentence embeddings (SIF) to represent document semantics more accurately. Additionally, SAMRank (Kang and Shin, 2023) [11] estimate word importance using scores obtained from the self-attention mechanism of language models such as BERT. However, while these methods are powerful, our approach is orthogonal to them. We focus on introducing explicit, high-level structural knowledge, which is different from the statistical patterns implicitly learned by language models. As pointed out by Sun et al. as a future direction, modeling document structure at the paragraph level is a promising yet underexplored area.

Our research directly addresses this challenge, proposing a robust and extensible heuristic that can be integrated with the latest embedding-based and attention-based technologies in the future.

2.3 Japanese Language Processing

Processing Japanese text requires an initial step of tokenization, also known as morphological analysis. Unlike English, Japanese does not use spaces to delimit words. Sudachi (Takaoka et al., 2018) [12] is a modern Japanese tokenizer developed to address issues in existing tools, such as the inconsistency of token units and the lack of continuous maintenance. A key feature of Sudachi is its ability to perform tokenization at multiple granularities, which is crucial because the appropriate token unit size differs by application. Sudachi provides three split modes:

Unit A (Short) The smallest unit, designed for compatibility with the UniDic dictionary [13]. It is effective for applications that require high recall, such as search.

Unit B (Middle) A unit designed to be intuitive and correspond to the general sense of a “word”.

Unit C (NE) The longest unit, which aims to recognize named entities (NEs), such as “Tokyo National Museum” (東京国立博物館), as a single token.

For our experiments, we leverage Sudachi to perform the necessary preprocessing of Japanese newspaper articles.

3 Proposed Methods

To more effectively leverage domain-specific knowledge of Japanese newspaper articles, we propose a two-tiered positional scoring mechanism that explicitly captures the structure of the article. This mechanism is composed of two distinct weights: the Paragraph Structure Weight, which captures macro-level importance, and the Intra-Paragraph Structure Weight, which captures micro-level importance.

3.1 Structural Weighting for Positional Bias

Paragraph Structure Weight (W_{para})**.** The Paragraph Structure Weight is designed to reflect the importance of a keyword's position, determined by the index of its parent paragraph. This design is based on the assumption that authors tend to place key concepts and conclusions strategically in the early paragraphs. In Japanese newspaper articles, this typically corresponds to the title (regarded as paragraph 1) and the lead paragraph.

We define this weight in Eq. (1) as the reciprocal of the 1-based paragraph index p:

$$W_{para}(p) = \frac{1}{p} \tag{1}$$

This simple but effective function ensures that keywords appearing in initial paragraphs receive a significantly higher structural weight, aligning the bias of the existing model with the natural information hierarchy of the article.

Intra-Paragraph Structure Weight (W_{sent})**.** The Intra-Paragraph Structure Weight captures the importance of a keyword's position within a paragraph. This component is motivated by the common journalistic and academic convention in which the first sentence of a paragraph often serves as the topic sentence, summarizing the main idea or claim of that paragraph. Therefore, keywords appearing in the initial sentences of a paragraph are more likely to be central to its meaning.

We define this weight in Eq. (2) as the reciprocal of the 1-based sentence index s within its paragraph:

$$W_{sent}(s) = \frac{1}{s} \tag{2}$$

This weighting scheme allows the proposed method to prioritize keywords found at the beginning of a paragraph over those appearing toward its end.

Structural Score $S_{struct}(v_i, o)$**.** By combining these two weights to form a unified "Structural Score" for each occurrence of a word, the proposed method can make more nuanced judgments. For example, the proposed method can correctly infer that a word in the first sentence of the second paragraph may be more significant than another keyword in the last sentence of the first paragraph.

This distinction is often overlooked in simpler, linear position-based models. This structured approach to positional biasing enables our proposed methods, PwPositionRank and PwTopicRank, to better align with the logical flow and information architecture of the source document.

We define the *Structural Score* $S_{struct}(vi, o)$ of a vocabulary item (word) v_i at its single occurrence o by Eq. (3):

$$S_{struct}(v_i, o) = (1 - \alpha) \cdot W_{para}(p_o) + \alpha \cdot W_{sent}(s_o) \tag{3}$$

If an occurrence belongs to paragraph number p_o and intra-paragraph sentence number s_o, the structural score is calculated by linearly combining the paragraph and intra-paragraph weights using a balancing parameter α $(0 \leq \alpha \leq 1)$. Here, α controls the degree of contribution of the "intra-paragraph structure weight". In this paper, we use $\alpha = 0.18$, which was identified as the optimal value through grid search in our preliminary investigation.

We treat the paragraph structure and the intra-paragraph structure as two independent yet complementary sources of information that constitute a word's overall structural importance. A weighted linear sum is the most direct and interpretable approach for modeling the contributions from these two hierarchical levels as a single unified score. The balancing parameter α explicitly controls the balance between the macro-level structure (paragraph position) and the micro-level structure (intra-paragraph position) in this unified score.

3.2 Pw-PositionRank Method

The original PositionRank enhances TextRank by introducing a positional bias in its PageRank calculation, giving higher initial importance to words that appear earlier in the article. Although effective, this approach has two main limitations: its reliance on a simple linear position assumption, and its use of a document-wide graph construction which can connect contextually distant words across paragraph boundaries.

Our proposed method, Pw-PositionRank, addresses both of these limitations with two fundamental enhancements.

First, we replace the linear position bias with a more nuanced *structural score*, defined as a weighted combination of a Paragraph Structure Weight (W_{para}) and an Intra-Paragraph Structure Weight (W_{sent}). This refinement enables the model to more accurately give higher priority to words in crucial opening paragraphs and their key sentences.

Second and critically, the proposed methods modify the graph construction process itself. Although the original model constructs a single graph throughout the entire document, our approach constrains word co-occurrence edges to remain within paragraph boundaries. By aligning the graph's contextual scope with the structural scoring, we prevent the dilution of topical importance across unrelated paragraphs. Finally, word scores are computed by applying PageRank to this structurally aware graph, resulting in a model that more accurately reflects the document's information hierarchy.

To calculate the bias term for PageRank, we define a bias score $B(v_i)$ for each word v_i. This score is the sum of the structural scores for all occurrences of v_i in the document. Let $occ(v_i)$ denote the set of all occurrences of v_i, then, the bias score $B(v_i)$ is defined by Eq. (4).

$$B(v_i) = \sum_{o \in occ(v_i)} S_{struct}(v_i, o) \tag{4}$$

Through this calculation, words that repeatedly appear in structurally important positions are assigned higher bias scores.

Since the bias score is used in combination with the PageRank word graph, we construct a word graph similar to that of PositionRank [5]. However, in Pw-PositionRank, the graph is constructed by restricting word co-occurrence relations to within the same paragraph. This restriction ensures that the unit of scoring and the unit of graph construction are aligned, which leads to an improvement in the model's performance. After that, we calculate the Biased PageRank using Eq. (4) to obtain the final importance score $WS(v_i)$ for each word v_i.

To calculate $WS(v_i)$, we first normalize the bias scores and create a personalization vector $p(v_i)$ with Eq. (5).

$$p(v_i) = \frac{B(v_i)}{\sum_{v_j \in V} B(v_j)} \tag{5}$$

where, V denote the set of all words in the graph. Using this $p(v_i)$, the final importance score $WS(v_i)$ is calculated by Eq. (6).

$$WS(v_i) = (1 - d) \cdot p(v_i) + d \cdot \sum_{v_j \in N(v_i)} \frac{w_{ji}}{\sum_{v_k \in N(v_j)} w_{jk}} WS(v_j) \tag{6}$$

where, d is the damping factor, and following the original TextRank and PositionRank, we use a value of 0.85. $N(v_i)$ is the set of neighbors of node v_i. w_{ji} is the weight of the edge from node v_j to node v_i.

Finally, the score for a candidate keyphrase C is calculated by Eq. (7), which sums the final scores of its constituent words.

$$Score(C) = \sum_{v_i \in C} WS(v_i) \tag{7}$$

The candidate keyphrases are then ranked in descending order based on this score, and the final keyphrases are selected sequentially from the highest-ranked candidates.

3.3 Pw-TopicRank Method

Pw-TopicRank enhances the powerful TopicRank algorithm by integrating structural information. We assume that keyphrases representing the subject of an article are typically mentioned in the earlier paragraphs of the article. In the proposed method, we modify the final step of the original TopicRank. After topics are ranked based on their centrality in the topic graph, we apply the paragraph weight $Pw(p)$ when selecting the representative keyphrase from each topic. This adjustment gives preference to candidate phrases that appear in earlier paragraphs, effectively combining the thematic importance captured by TopicRank with the structural importance derived from our domain knowledge.

Similar to TopicRank [7], we first cluster the words and calculate a score $WS(T_i)$ for each topic T_i. For each topic, the phrase that appears first in the article is selected as the representative candidate phrase C_{rep_i}, which may occur multiple times in the article. For each occurrence o of C_{rep_i}, a positional booster score $S_{posbooster}(C_{rep_i}, o)$ is calculated using Eq. (3). Finally, as shown in Eq. (8), the maximum positional booster score among all occurrences is taken as the final $S_{posbooster}(C_{rep_i})$.

$$S_{posbooster}(C_{rep_i}) = \max_{o \in occ(C_{rep_i})} S_{struct}(C_{rep_i}, o) \tag{8}$$

where, $occ(C_{rep_i})$ denotes the set of all occurrences of the candidate phrase C_{rep_i}. The final score of $occ(C_{rep_i})$, $Score_{pw}(C_{rep_i})$, is calculated by multiplying the topic score $WS_{topic}(T_i)$ with the positional booster score $S_{posbooster}(C_{rep_i})$, as given by Eq. (9).

$$Score_{pw}(C_{rep_i}) = WS_{topic}(T_i) \times S_{pos_booster}(C_{rep_i}) \tag{9}$$

Through the above computation, candidate phrases located in structurally important positions in the article (i.e., with a high positional booster score) are ranked more highly.

4 Experiments

4.1 Datasets

From the 2007 Yomiuri Shimbun Corpus, one of the largest national newspapers in Japan, we use a sample of 13,065 articles[1] as our dataset. These sample articles have been pre-assigned keyphrases by the newspaper publisher, which we use as ground truth for the keyphrases to be extracted in this experiment. However, we exclude any assigned keyphrases that do not appear with an identical notation within the article's body text. Articles with fewer than four paragraphs in their body text are also excluded, as the effect of introducing paragraph structure cannot be verified. We also exclude articles exceeding 50 paragraphs, as many are not structured as prose (e.g., they consist of simple lists of names) and the probability of keyphrases appearing after the 50th paragraph is extremely low. Each article is assigned to one of 69 categories and we use a random sample of a maximum of 20 articles from each category. As a result of applying these conditions, we obtained 838 articles, which are used in our experiments as Dataset:full.

[1] Articles published between January 25 and 31, 2007, were obtained from the sample data publicly available on Nichigai Associates (https://www.nichigai.co.jp/sales/corpus.html). Note that the sample data has been unavailable as of December 2025.

Next, we describe the structural processing of the articles. In our experiments, we treat the article title as the first paragraph. Each indented block following the title is considered a single paragraph and is sequentially numbered as the second, third, and subsequent paragraphs. According to the Kisha Handobukku [1], it is common practice to use up to approximately five subheadings per article. In our dataset, some articles have the second paragraph as a subheading, with the main body beginning from the third paragraph. However, in our experiments, any subheadings are consistently processed as the second paragraph.

Japanese newspaper articles are generally organized according to the inverted pyramid structure, in which important information is presented first. Within each paragraph, the sentence structure also tends to follow the inverted pyramid style, stating the conclusion first. For each article in Dataset:full, Fig. 1 presents a heatmap indicating the paragraphs in which keyphrases appear. Figure 1 shows that keyphrases are concentrated in the first paragraph (the title) and the first sentence of the second paragraph (the lead sentence). Additionally, keyphrases also exist in the second and subsequent sentences of the second paragraph. This structure is known in the field of newspaper summarization as "LEAD-3," a strong baseline method that uses the first three sentences as the ground truth for a summary. Notably, keyphrases are also observed in the third and fourth paragraphs. This may be attributed to cases where the second paragraph serves as a subheading, as well as the possibility that new subjects are being introduced in later paragraphs.

From another perspective, Fig. 2 shows the distribution of sentence positions in which keyphrases appear within each paragraph. The horizontal axis is the Paragraph Index, and the vertical axis is the Sentence Index. Figure 2 indicates that in earlier paragraphs, keyphrases are more likely to occur in the second and subsequent sentences. The overall trend shows that as one moves to later paragraphs, keyphrases tend to be concentrated in the first sentence; however, they are also observed in paragraphs after the 19th, indicating that new subjects are introduced in these later paragraphs. Furthermore, it is suggested that the first sentence of the nth paragraph is more likely to contain a keyphrase than the last sentence of the $(n-1)$th paragraph.

To verify the effect of incorporating paragraph structure, we create a subset from Dataset:full by selecting only those articles where keyphrases appear in later paragraphs (excluding the title and the second paragraph), and further designate only those keyphrases in later paragraphs as the ground truth. The 423 articles obtained through this process are used in our experiments as Dataset:subset.

Pre-Processing. All articles were tokenized using SudachiPy, the Python implementation of the Sudachi tokenizer. A critical consideration for graph-based methods is the granularity of the vertices (i.e., the words). To construct a meaningful co-occurrence graph, it is essential to select an appropriate tokenization unit.

Our preliminary analysis indicated that units with smaller granularity are more effective in capturing the detailed co-occurrence relationships that are nec-

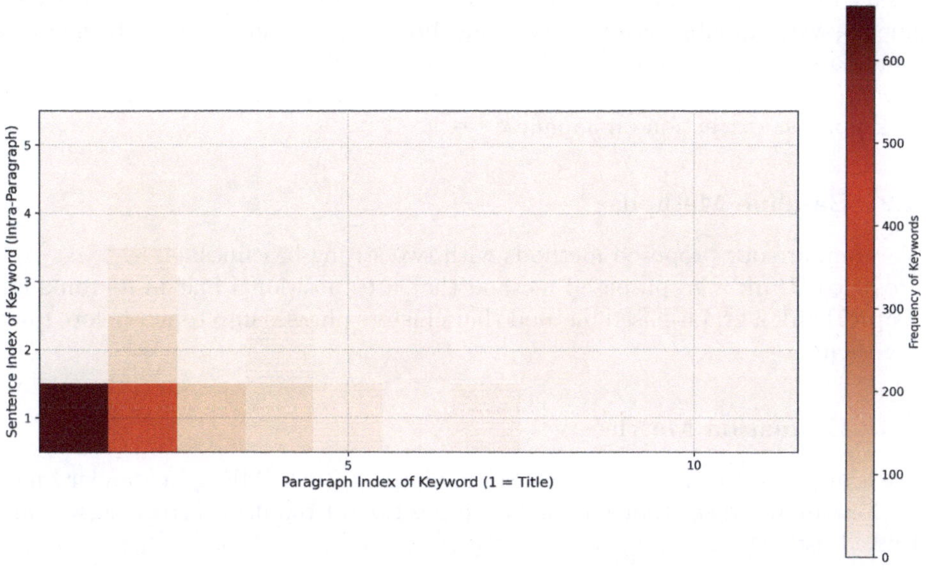

Fig. 1. Heatmap of Keyword Distribution by Structural Position.

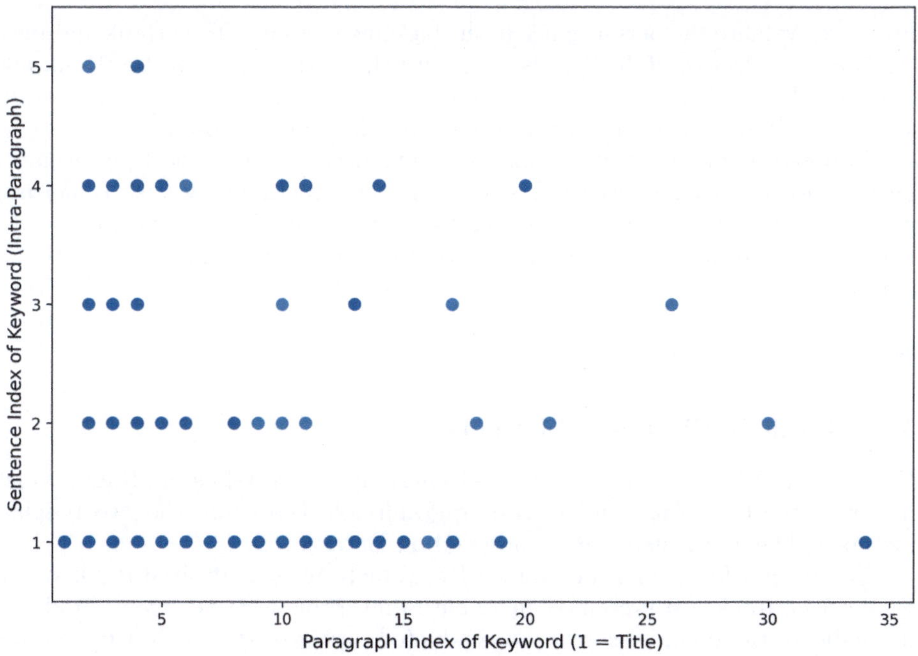

Fig. 2. Scatter Plot of Keyword Distribution by Structural Position.

essary for our graph-based models in Japanese articles. Therefore, in the experiments, we evaluate performance using both Unit A and Unit B to generate candidate keyphrases and construct the word graphs. This approach allows for a more robust evaluation of how tokenization granularity impacts the performance of keyphrase extraction on Japanese texts.

4.2 Baseline Methods

We compare our proposed methods with two strong baselines:
PositionRank a graph-based method that uses positional bias in its ranking.
TopicRank a graph-based method that clusters phrases into topics before ranking them.

4.3 Evaluation Metric

We evaluate performance using Mean Reciprocal Rank (MRR), a standard metric for ranking tasks that measures the position of the first correct answer in a ranked list.

5 Results and Discussion

5.1 Results of Baseline Methods

First, we evaluate the performance of the baseline methods, TopicRank and PositionRank, on Dataset:full. We also evaluate the performance of TextRank and SingleRank, which serve as their foundational methods. The evaluation results are presented in Table 1. Bold values indicate the highest scores.

As noted by Sun et al. [8], TopicRank demonstrates the highest performance on Japanese newspaper articles. Therefore, in this experiment, we use TopicRank with Unit B tokens, which showed particularly strong performance, especially from MRR@5 onward, as a baseline for comparison with our proposed methods. Additionally, for methods incorporating positional information, we use PositionRank with Unit A, which showed better performance, as a baseline for comparison with our proposed methods.

5.2 Results of Proposed Methods

We evaluate the performance of Pw-TopicRank and Pw-PositionRank, which integrate the Paragraph and Intra-Paragraph structures into the two baseline methods. The experiments are conducted on Dataset:full.

The results for TopicRank versus Pw-TopicRank and PositionRank versus Pw-PositionRank are presented in Table 2 and Table 3, respectively. Bold values indicate the highest scores. We conducted a paired t-test to verify whether there is a statistically significant difference in performance between the proposed method and the baseline methods. An asterisk (*) in each table indicates a statistically significant improvement over the baseline ($p < 0.05$).

From Table 2, we can see that Pw-TopicRank achieves higher values across all MRR metrics than TopicRank, which obtained the highest MRR among the baseline methods. From this result, we confirmed the effectiveness of integrating document structure into TopicRank. On the other hand, Table 3 shows that, except for MRR@1, PositionRank yields higher values than our proposed method. The difference between Pw-PositionRank and PositionRank was found to be statistically significant for MRR@10, and @all metrics (paired t-test, p < 0.05). The differences observed for MRR@1, @3 and @5 were not statistically significant.

However, in the Dataset:full used for this experiment, the keyphrases were concentrated in the first and second paragraphs. Therefore, with this result alone, it may not be possible to fully verify the performance difference between incorporating paragraph structure and using linear weights to words from the beginning of the document.

5.3 Consideration

As an additional experiment for our analysis, we evaluate the performance of keyphrase extraction on later paragraphs using Dataset:subset, in which only keyphrases appearing in these later paragraphs are treated as the ground truth.

The results for PositionRank and Pw-PositionRank are presented in Table 4. Bold values indicate the highest scores. An asterisk (*) in the table indicates a statistically significant improvement over the baseline ($p < 0.05$).

As shown in Table 4, Pw-PositionRank performs better overall compared to PositionRank. PositionRank, which applies linear weights to words from the beginning of the document, tends to assign low scores to keyphrases when they are located in later paragraphs. In contrast, Pw-PositionRank, which incorporates document structure, is particularly effective for extracting keyphrases appearing in the latter half of an article. The improvement achieved by Pw-PositionRank was found to be statistically significant for the MRR@3, @5, @10, and @all metrics (paired t-test, p < 0.05).

Table 1. Performance of baseline methods

	Sudachi Unit	MRR@1	MRR@3	MRR@5	MRR@10	MRR@all
TextRank	A	0.0096	0.0181	0.0287	0.0433	0.0639
SingleRank	A	0.0203	0.0449	0.0586	0.0782	0.0978
TopicRank	A	**0.1038**	**0.1430**	**0.1496**	**0.1557**	**0.1590**
PositionRank	A	0.0215	0.0549	0.0731	0.0959	0.1150
TextRank	B	0.0072	0.0171	0.0295	0.0424	0.0624
SingleRank	B	0.0203	0.0446	0.0567	0.0736	0.0945
TopicRank	B	**0.0811**	**0.1408**	**0.1504**	**0.1562**	**0.1619**
PositionRank	B	0.0215	0.0543	0.0699	0.0907	0.1103

Table 2. Performance of Pw-TopicRank on Dataset:full

	Sudachi Unit	MRR@1	MRR@3	MRR@5	MRR@10	MRR@all
TopicRank	B	0.0811	0.1408	0.1504	0.1562	0.1619
Pw-TopicRank	B	**0.1014***	**0.1557***	**0.1664***	**0.1749***	**0.1785***

Table 3. Performance of Pw-PositionRank on Dataset:full

	Sudachi Unit	MRR@1	MRR@3	MRR@5	MRR@10	MRR@all
PositionRank	A	0.0215	**0.0549**	**0.0731**	**0.0959***	**0.1150***
Pw-PositionRank	A	**0.0251**	0.0479	0.0634	0.0842	0.1045

Table 4. Performance of Pw-PositionRank on Dataset:subset

	Sudachi Unit	MRR@1	MRR@3	MRR@5	MRR@10	MRR@all
PositionRank	A	0.0095	0.0146	0.0221	0.0299	0.0467
Pw-PositionRank	A	**0.0142**	**0.0268***	**0.0340***	**0.0398***	**0.0576***

Considering the results of both experiments comprehensively, our proposed methods, which incorporate document structure, demonstrated higher performance than TopicRank, which relies solely on a graph-based approach. However, when keyphrases appear in the upper paragraphs, our methods do not yet surpass PositionRank, which applies linear weights from the beginning of the document. On the other hand, our method is more robust than PositionRank when keyphrases are located in the later paragraphs. The fact that Pw-PositionRank did not outperform PositionRank on Dataset:full clearly demonstrates this trade-off: in many articles where keyphrases are concentrated at the beginning, a simple linear bias remains a powerful heuristic. However, Table 4 demonstrates that this powerful heuristic comes at a cost, as the linear bias inherently fails to capture keyphrases appearing in later paragraphs. In contrast, our Pw-PositionRank showed a clear advantage in precisely this challenging scenario. These findings suggest that a structural approach has the potential to overcome the trade-off observed in linear models between achieving high precision on early keyphrases and robustness for later ones. As mentioned in Sect. 4.1, the first sentence of the nth paragraph is more likely to contain a keyphrase than the last sentence of the $(n-1)$th paragraph. We believe that by incorporating this tendency as positional information within the document structure, the hierarchical structure of the document can be accurately modeled.

6 Conclusion

In this study, we demonstrated that utilizing the domain-specific "inverted pyramid structure" of Japanese newspaper articles is an effective strategy for enhancing the performance of unsupervised keyphrase extraction methods.

Our proposed methods, Pw-TopicRank and Pw-PositionRank, both integrate a two-level structural model into strong baselines for graph-based keyphrase extraction. Experiments with Pw-TopicRank revealed that incorporating an explicit structural heuristic into thematic analysis, a key feature of TopicRank, can improve keyphrase extraction performance. Experiments with Pw-PositionRank validated our core hypothesis that using the document structure provides a significant advantage over models that rely solely on simple linear position. In particular, using document structure proved particularly effective in the challenging scenario of identifying keyphrases located in later paragraphs.

The structural model proposed in this paper did not consistently outperform the simple linear weighting model, PositionRank, in cases where keyphrases are concentrated at the beginning of the article. However, our targeted analysis not only clarified the trade-off inherent in simple linear positional bias but also demonstrated that our approach provides a clear advantage in extracting keyphrases from later paragraphs, a setting in which linear models consistently fail. These results suggest that the principle of using document structure is sound, even if our current model is not yet fully optimized. A promising direction for future work is to propose a hybrid model that combines the strengths of both approaches: applying a linear weight for the lead paragraphs and our structural model for the remainder of the article. Such an approach could potentially provide robust performance across all scenarios, maintaining high accuracy for early keyphrases while effectively capturing those that appear later.

While this work has focused on a fully unsupervised approach, the most promising direction for future research is to integrate our structural weighting scheme with modern language model techniques. We posit that the explicit structural knowledge captured by our proposed models is highly complementary to the implicit semantic knowledge learned by LLMs. Integrating these two approaches may lead to a new generation of keyphrase extraction methods that are both semantically rich and structurally aware. For example, our structural score could be employed to re-rank candidates generated by embedding-based models or to guide the attention mechanisms of generative models, thereby offering the potential for improved accuracy and robustness.

Code Availability

The implementation for this study is based on the well-known keyphrase extraction library PKE (Python Keyphrase Extraction module)[1], and our source code, which includes the proposed methods, is publicly available at our GitHub repository[2].

[1] https://github.com/boudinfl/pke.
[2] https://github.com/KHiyowa/pke-pwseries.

References

1. Tsūshinsha, K.: Kisha handobukku: Shinbun yōji yōgoshū (Journalist's handbook: A glossary of terms and characters for newspaper use), 9th edn. Kyōdō Tsūshinsha, Tokyo (2001). (in Japanese)
2. Mihalcea, R., Tarau, P.: TextRank: bringing order into text. In: D. Lin, D. Wu (eds.) Proceedings of the 2004 Conference on Empirical Methods in Natural Language Processing, Barcelona, Spain, 2004, pp. 404–411. Association for Computational Linguistics (2004). https://aclanthology.org/W04-3252/
3. Page, L., Brin, S., Motwani, R., Winograd, T.: The PageRank citation ranking: bringing order to the web. Technical report, Stanford Digital Library Technologies Project (1998). http://citeseerx.ist.psu.edu/viewdoc/summary?doi=10.1.1.31.1768
4. Wan, X., Xiao, J.: Collabrank: towards a collaborative approach to single-document keyphrase extraction. In: D. Scott, H. Uszkoreit (eds.) Proceedings of the 22nd International Conference on Computational Linguistics, Coling 2008, Manchester, UK, pp. 969–976. Coling 2008 Organizing Committee (2008). https://aclanthology.org/C08-1122/
5. Florescu, C., Caragea, C.: PositionRank: an unsupervised approach to keyphrase extraction from scholarly documents. In: Barzilay, R., Kan, M.Y. (eds.) Proceedings of the 55th Annual Meeting of the Association for Computational Linguistics (Volume 1: Long Papers), Vancouver, Canada, pp. 1105–1115. Association for Computational Linguistics (2017). https://doi.org/10.18653/v1/P17-1102
6. Witten, I.H., Paynter, G.W., Frank, E., Gutwin, C., Nevill-Manning, C.G.: KEA: practical automatic keyphrase extraction. In: Proceedings of the Fourth ACM Conference on Digital Libraries, pp. 254–255. ACM, New York, NY, USA (1999)
7. Bougouin, A., Boudin, F., Daille, B.: TopicRank: graph-based topic ranking for keyphrase extraction. In: Mitkov, R., Park, J.C. (eds.) Proceedings of the Sixth International Joint Conference on Natural Language Processing, pp. 543–551. Asian Federation of Natural Language Processing, Nagoya, Japan (2013). https://aclanthology.org/I13-1062/
8. Sun, C., Hu, L., Li, S., Li, T., Li, H., Chi, L.: A review of unsupervised keyphrase extraction methods using within-collection resources. Symmetry **12**(11) (2020). https://doi.org/10.3390/sym12111864
9. Liao, S., Yang, Z., Liao, Q., Zheng, Z.: TopicLPRank: a keyphrase extraction method based on improved TopicRank. J. Supercomput. **79**(8), 9073–9092 (2023). Published online: 2023-01-07; Issue date: 2023-05. https://doi.org/10.1007/s11227-022-05022-0
10. Sun, Y., Qiu, H., Zheng, Y., Wang, Z., Zhang, C.: SIFRank: a new baseline for unsupervised keyphrase extraction based on pre-trained language model. IEEE Access **8**, 10896–10906 (2020). https://doi.org/10.1109/ACCESS.2020.2965087
11. Kang, B., Shin, Y.: SAMRank: unsupervised keyphrase extraction using self-attention map in BERT and GPT-2. In: Bouamor, H., Pino, J., Bali, K. (eds.) Proceedings of the 2023 Conference on Empirical Methods in Natural Language Processing, pp. 10,188–10,201. Association for Computational Linguistics, Singapore (2023). https://doi.org/10.18653/v1/2023.emnlp-main.630

12. Takaoka, K., Hisamoto, S., Kawahara, N., Sakamoto, M., Uchida, Y., Matsumoto, Y.: Sudachi: a Japanese tokenizer for business. In: Calzolari, N., et al. (eds.) Proceedings of the Eleventh International Conference on Language Resources and Evaluation (LREC 2018). European Language Resources Association (ELRA), Miyazaki, Japan (2018). https://aclanthology.org/L18-1355/
13. National Institute for Japanese Language and Linguistics: UniDic: electronic dictionary with uniformity and identity (n.d.). https://clrd.ninjal.ac.jp/unidic/en/. Accessed 09 Jul 2025

UNITE: A Framework for Event Extraction from Social Media Based on User Interests

Ankit Patel(✉), Tejas Pant, and Jing Wang

Accrete.AI, Boston, USA
{ankit.patel,tejas.pant,jing.wang}@accrete.ai

Abstract. The vast influx of unstructured social media content creates a noisy information environment, hindering the timely identification of events relevant to specific user interests. To address this challenge, this paper introduces UNITE, a novel, modular framework for personalized event extraction from social media. UNITE employs a multi-stage LLM prompting architecture to identify relevant signals, extract structured attributes, and perform clustering-based deduplication. To address informational gaps, we propose a hierarchical enrichment strategy that integrates social context with targeted web retrieval, significantly improving the completeness of event data. We demonstrate that our framework effectively filters irrelevant content and delivers curated event streams tailored to diverse user profiles. This approach enables more targeted monitoring and decision-making, transforming noisy social chatter into structured, actionable intelligence.

Keywords: Event extraction · User interest · Large Language Models

1 Introduction

Extracting structured event information from social media is crucial for timely insights, but it presents distinct challenges compared to traditional newswire analysis. Social platforms are characterized by high-volume, noisy, and informal user-generated content [1] where events are often emergent and lack canonical representation. In this context, we define an event as a real-world occurrence that generates a significant, observable signal within a social media stream, aligned with a user's interests. We categorize such events into three primary types: (1) Planned announcements (e.g., corporate earnings calls), (2) Unplanned developments (e.g., breaking news), and (3) Shifts in discourse (e.g., emergent topics of high interest).

The challenge intensifies when personalization is required. A system must not only detect credible events but also align them with a user's specific needs and interests, a task demanding sophisticated methods in relevance filtering and Natural Language Processing (NLP). The ultimate goal is to deliver only the most relevant signals from an overwhelming stream of information. Our research

H. Unger and P. Meesad (Eds.): NLPIR 2025, LNNS 1904, pp. 1141–1155, 2026.
https://doi.org/10.1007/978-3-032-20897-2_80

addresses these challenges by providing a generalized framework for personalized event extraction adaptable to various domains.

Event extraction methods have evolved from rule-based systems like AutoSlog [2] to deep learning approaches [3,4] built on transformer architectures [5] and pre-trained models such as Bidirectional Encoder Representations from Transformers (BERT) [6]. Recent work has leveraged Large Language Models (LLMs) [7,8] and retrieval-augmented prompting to enhance scalability and reduce reliance on manual annotations. Multi-modal event extraction [9] and graph-based clustering techniques [10] have further extended the field to diverse data formats and unsupervised settings. However, existing approaches struggle with generalization and contextual completeness [3,4], limited knowledge grounding, and high computational costs [7,8], lack effective coreference resolution and deduplication [9] and require large annotated datasets [10].

To address these shortcomings, we propose the **UN**iversal **I**nterest-based ex**T**racting **E**vents (UNITE) approach. More specifically, we design an approach that extracts user-relevant events from noisy social media data by leveraging LLMs to identify and structure event information, apply deduplication to surface only unique, actionable events, and infer missing details using both social media comments and web agents. The result is a curated list of events tailored to a user's interests and with a rich description that includes event title, date, and location. The major contributions of this study can be summarized as follows.

1. **User interests aligned event extraction:** We propose a novel methodology that filters social media posts based on alignment with user interests, extracts event-relevant content, and structures it into a rich representation using a multi-stage LLM prompting approach.
2. **Robust event deduplication framework:** We use clustering to group semantically similar posts, ensuring that each detected event is uniquely represented, thereby reducing redundancy and ensuring consistency over time.
3. **Knowledge-grounded attribute enrichment:** We present a hierarchical enrichment strategy that combines social context analysis with targeted web retrieval to complete missing event attributes, demonstrating significant improvements in both coverage and accuracy of temporal information.

2 Related Work

Event extraction, a subfield of information retrieval for identifying events in unstructured text, originated in the late 1980s to automate message interpretation [11]. Early approaches were dominated by rule-based systems using linguistic pattern matching [2], but these methods were often brittle, labor-intensive, and struggled to generalize. This limitation prompted the shift toward more sophisticated deep learning technologies.

Modern approaches leverage deep learning, particularly Transformer architectures [5], with pre-trained models such as Bidirectional Encoder Representations from Transformers (BERT) [6] fine-tuned to identify event triggers and extract arguments [3,4]. More recently, generative LLMs have been applied for event extraction through prompting and instruction fine-tuning [7,8], exploiting their long-text comprehension, broad pre-training, and multilingual capabilities [12]. LLMs can also generate annotation guidelines to fine-tune smaller models [8,13] or synthesize new datasets to support continual learning, although such iterative processes are time-consuming and costly.

Other strategies include retrieval-augmented prompting [7] to guide LLMs using high-quality examples for event type classification and attribute extraction. Knowledge graphs and external resources such as FrameNet [14] and Freebase [15] have been used for data augmentation, while multi-modal extraction integrates text, video, and audio to identify events across modalities [9,16]. Graph-based approaches, such as Graph Convolutional Networks (GCNs) on heterogeneous social media graphs, enable event embedding from messages, users, and hashtags [10], but require large numbers of paired events, limiting scalability and transferability.

UNITE provides a robust alternative to these prior methods. Its zero-shot prompting architecture eliminates the need for domain-specific training. Completeness and grounding are improved through social context and web retrieval. Furthermore, our deduplication process addresses multi-modal challenges, and the user-specific relevance filtering adds a personalized layer absent in other approaches. Collectively, these features enable UNITE to perform accurate, comprehensive, and personalized event extraction from noisy social media. The following section describes UNITE in detail.

3 Methodology

Figure 1 shows the UNITE approach for extracting structured event information from social media data, while Algorithm 1 details the step-by-step procedure. While the modular approach of UNITE is generalizable to various social media platforms like Instagram and TikTok, this study focuses on a dataset curated from X. For each post from X in the dataset we consider the post data and the comment data. The post data contains metadata associated with the original post such as a unique identifier for the post (`postId`), the textual content of the post (`postText`), the username (`postUsername`) and the timestamp of the post (`postDate`). The comment data captures information related to replies or new posts quoting the original post. For a comment, the metadata includes a unique identifier for the comment (`commentId`), the text content of the comment (`commentText`), the timestamp of the comment (`commentDate`) and a reference to the original post via `postId`. Additional details about the dataset will be provided in Sect. 4.

Our approach involves filtering event related posts, extracting key event attributes, reducing redundancy through clustering, and inferring undetermined attributes using both comment data and web search tools. Unless specified otherwise, each LLM-powered stage of the UNITE pipeline leverages the GPT-4o-mini[1] model. Each component of our approach is described in detail in the

[1] https://openai.com/index/gpt-4o-mini-advancing-cost-efficient-intelligence/.

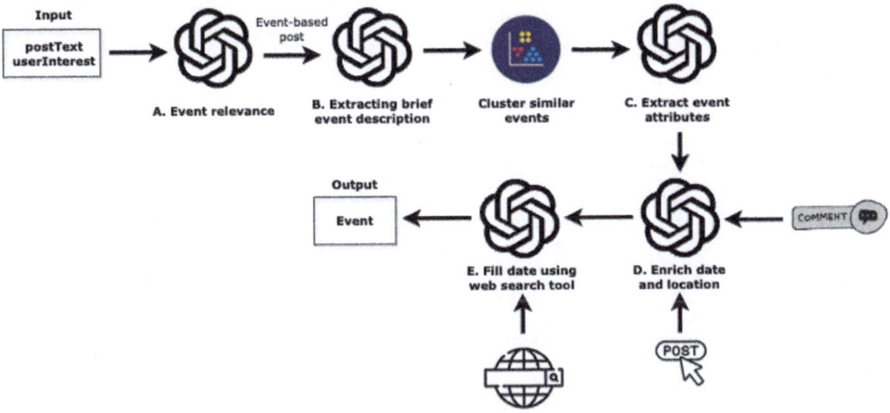

Input:
- **postText:** "Biden plans to block Nippon Steel's acquisition of US Steel, $X, per Reuters."
- **userInterest:** "A day trader focused on the US stock market."

Pipeline Stages:

A. Event Relevance:
Output: classification: Event-based, reasoning: "The post discusses a specific action (Biden blocking the acquisition) that is relevant to the stock market, particularly concerning US Steel, which is significant for day traders."

B. Extracting brief event description:
Output: event: "Biden plans to block Nippon Steel's acquisition of US Steel, $X."

C. Extract event attributes:
Output: Title: "Biden to Block Nippon Steel-US Steel Deal.", description : "Biden administration plans to block Nippon Steel's proposed acquisition of US Steel ($X), according to Reuters report.", location : "Not specified", date : "Not specified"

D: Enrich date and location:
Output: enriched_location: "United States", enriched_date: "Not specified"

E. Fill date using web search tool:
Output: Determined Date: "2025-01-03", snippet link: "https://www.cbsnews.com/news/biden-us-steel-nippon-deal/, https://www.insidearbitrage.com/deal-metrics/X/2888/united-states-steel-to-be-acquired-by-nippon-steel/?t=alw"

Output:
- **Event Title:** Biden to Block Nippon Steel-US Steel Deal
- **Event Description:** Biden administration plans to block Nippon Steel's proposed acquisition of US Steel ($X), according to Reuters report.
- **Event Location:** United States
- **Event Date:** 2025-01-03

Fig. 1. Overview of the UNITE approach. The top panel displays the simplified five-stage pipeline. The bottom panel provides a concrete example walkthrough, tracing a single post through each stage (A-E) to produce a final, structured event. The prompts for each stage are available in the UNITE GitHub repository.

Algorithm 1. UNITE

Require: Dataset of posts D, $userInterest$
Ensure: Set of structured events $E_{structured}$
  {Step 1: Event Relevance Filtering}
  $P_{filtered} \leftarrow \emptyset$
  for each post p in D **do**
    $label \leftarrow$ LLM(Prompt A, p.postText, $userInterest$)
    if $label =$ "Event-based" **then**
      $P_{filtered} \leftarrow P_{filtered} \cup \{p\}$
    end if
  end for
  {Step 2: Extract Event Descriptions}
  $D_{brief} \leftarrow \emptyset$
  for each post p in $P_{filtered}$ **do**
    $description \leftarrow$ LLM(Prompt B, p.postText, $userInterest$)
    $D_{brief} \leftarrow D_{brief} \cup \{description\}$
  end for
  {Step 3: Event Deduplication}
  $embeddings \leftarrow$ generateEmbeddings(D_{brief})
  $CLC \leftarrow$ CompleteLinkageClustering($embeddings$)
  {Step 4: Extract Event Attributes}
  $E_{attributes} \leftarrow \emptyset$
  for each cluster c in CLC **do**
    $combined\_text \leftarrow$ merge posts in c
    $attributes \leftarrow$ LLM(Prompt C, $combined\_text$)
    $E_{attributes} \leftarrow E_{attributes} \cup \{attributes\}$
  end for
  {Step 5: Infer Undetermined Attributes}
  $E_{structured} \leftarrow \emptyset$
  for each event e in $E_{attributes}$ **do**
    if e.date = "Not Specified" OR e.location = "Not Specified" **then**
      $enriched\_text \leftarrow$ merge(e.post, e.comments)
      $e \leftarrow$ LLM(Prompt D, $enriched\_text$)
      if e.date = "Not Specified" **then**
        $web\_content \leftarrow$ webSearch(e.title, e.description)
        $e \leftarrow$ LLM(Prompt E, e, $web\_content$)
      end if
    end if
    $E_{structured} \leftarrow E_{structured} \cup \{e\}$
  end for
  return $E_{structured}$

following sections. For the purposes of this study, a `userInterest` is defined as a concise natural language string representing a user's professional persona or informational need (e.g., 'A day trader focused on the US stock market'). The prompts used for each of these steps (labeled Prompt A to E) are available in our project repository: `UNITE-2025`.

3.1 Event Relevance Filtering

The initial stage of the pipeline is a crucial filtering step designed to reduce processing load and enhance signal quality. As detailed in Step 1 of Algorithm 1, this stage isolates posts that are both event-bearing and aligned with the user's specified interests, ensuring that subsequent, more computationally intensive processes are applied only to relevant content. We use a zero-shot prompting approach (See Prompt A in UNITE GitHub repository[2]) for this. This approach provides the model with a task description without any pre-tuned examples, relying on its extensive pre-training to understand and execute the filtering instructions.

3.2 Extracting Brief Event Description

Following the filtering, each relevant post is processed to generate a concise, semantically rich summary of the potential event it describes. The goal of this stage, outlined in Step 2 of Algorithm 1, is to create a coherent textual representation that serves as the primary input for the subsequent deduplication process. The prompt used for this step is detailed in Prompt B.

3.3 Event Deduplication

To reduce redundancy and consolidate references to the same underlying event, we perform clustering-based deduplication on the event descriptions generated in the previous step. As shown in Step 3 of Algorithm 1, we apply complete linkage clustering (CLC)[17] over the semantic embeddings of these descriptions. This method groups together different posts that describe the same event-for instance, "Earnings coverage of Netflix $NFLX" and "Earnings report announcement for Netflix"-into a single canonical event, ensuring a streamlined and non-redundant event set for the final extraction stages.

3.4 Event Attribute Extraction

Each event cluster is then processed to extract key event metadata like the event title, description, date, and location. The title provides a concise summary of the event, while the description offers a brief, two-sentence representation of the event's main details. The date corresponds to the event's scheduled time, and the location refers to any identifiable geographic information. This process, shown in Step 4 of Algorithm 1, transforms the consolidated text from each cluster into a formal event representation. This process uses Prompt C as detailed in our project repository. Here, the `combined_text` input for the LLM is created by concatenating the `postText` of all posts belonging to the event cluster, providing a comprehensive context for extraction.

[2] https://github.com/UNITE-2025/unite-event-extraction-framework.

3.5 Inferring Undetermined Event Attributes

A key challenge is handling events with incomplete information. Our final stage addresses this by inferring missing attributes like date and location through a hierarchical enrichment strategy, detailed in Step 5 of Algorithm 1. First, we attempt to fill these gaps by leveraging the collective social context from the entire event cluster. To achieve this, the comments from all posts and the post texts belonging to the cluster are aggregated together to form a comprehensive enrichment text. This merged context is then provided to the model (Prompt D), a step designed to improve recall by maximizing the chance of finding missing attributes mentioned across any of the related discussions. If attributes remain undetermined, the framework then employs an external web search using Google SerpAPI[3]. This tool queries the web using the event's title and description, and the retrieved content is passed to the LLM for final enrichment (Prompt E). This two-tiered approach strategically prioritizes internal social context before querying external sources, ensuring efficient and targeted use of web retrieval only when necessary.

The final structured output, illustrated in bottom panel of Fig. 1, demonstrates how unstructured post text can be transformed into semantically rich event representations. By leveraging LLMs with carefully crafted prompts and downstream attribute extraction processes, the UNITE approach effectively filters out irrelevant content and isolates actionable events from noisy social media streams.

Figure 2 illustrates how the clustering and merging process is critical for creating a canonical event representation from multiple, disparate posts. This consolidation is essential for the final enrichment stage, which aims to produce a comprehensive and non-redundant event record.

4 Dataset

Our evaluation is limited by the absence of suitable benchmarks, as most existing event extraction datasets neither provide structured event attributes nor enable user interest which are the two central goals of UNITE. While several datasets exist for event-related research, they typically target different objectives and use data formats incompatible with our framework.

For instance, the DiPD dataset [18] focuses on binary classification of socio-political disruptions such as riots and protests as eventful or non-eventful, but it includes only minimal tweet metadata and omits structured attributes such as title, description, date, or location. The supply chain disruption dataset [19] on Hugging Face, though event-oriented, is a small synthetic text-classification resource that lacks unstructured social media text and detailed attributes. Similarly, the Events-after-Feb-2023 dataset [20] is essentially a chronological list of 91 events with short descriptions and dates, but no additional metadata such as

[3] https://serpapi.com/.

Input (Raw post):

- Premium Earnings 10-17-24: Netflix and American Express, on the blog and here https://t.co/2dR4XZSdft $NFLX $AXP
- We're finishing up the week with earnings coverage of Netflix $NFLX, American Express $AXP, and Proctor & Gamble $PG, as well as the latest news out of the UK and Asia.

userInterest:
A day trader focused on the US stock market.

Output (Extracted Event Attributes):

- **Event Title:** Netflix and American Express Earnings Report
- **Event Description:** Coverage of earnings reports for Netflix ($NFLX), American Express ($AXP), and Proctor & Gamble ($PG)
- **Event Location:** Not specified
- **Event Date:** 2025-01-21 to 2025-01-22
- **Sources:**
  - alphaquery
  - seekingalpha

Fig. 2. Example illustrating how events from multiple posts referring to the same event are clustered and merged into a single structured event representation using UNITE approach.

locations, entities, or user interactions. None of these resources support attribute-level extraction or user interest, which are key requirements of UNITE.

The Event2012 dataset [21] offers clustering ground truth and is closer to our needs to some extent, but it also overlooks structured attributes and personalization. This reflects a broader limitation in the field, i.e. existing datasets emphasize binary classification, structured operational data, or generic event catalogs, while largely neglecting social mediabased event extraction with rich attributes and personalization. To address this gap, our evaluation relies on (1) component-level analysis with available labeled datasets where feasible, (2) human validation of extracted attributes, and (3) demonstrations of UNITE's performance across diverse user interests. To further contribute to benchmarking efforts, we also plan to release our annotated dataset as an open resource.

For evaluation, we constructed a dataset by collecting posts from July 18, 2024, to December 19, 2024, across approximately 30 well-known finance accounts on X, including @unusual_whales, TrendSpider, RedDogT3, and PeterLBrandt. These accounts actively discuss U.S. stock market trading and provide real-time insights such as trading ideas, chart setups, ticker mentions, and market sentiment. Posts and their comments were collected through the X API v2.0 [22]. For each post, we retrieved all comments unless the total exceeded 100, in which case we collected the top 100 as ranked by X's recommendation algorithm. The resulting dataset contains 28,361 posts and 204,506 comments.

While our case study focuses on financial data, UNITE is domain-agnostic and can be applied to event extraction tasks across other fields.

5 Experiments and Results

In this section, we discuss the performance of the UNITE approach. In Sect. 5.1 evaluate the performance of CLC. Then in Sect. 5.2 we study the effect of different LLMs in determining the event attributes and lastly in Sect. 5.3 we study the effect of user interest on the list of events identified.

5.1 Effect of Clustering Approach

As detailed in Sect. 3.3, clustering plays a crucial role in the deduplication process by grouping semantically similar event mentions, which helps to identify and consolidate different phrasings that refer to the same underlying event. More recently, Ray et al. [23] used an incremental clustering approach called first story detection (FSD) [24] for extracting events from continiously streaming text data. FSD showed improvements over previous studies [25, 26] on the Event2012 dataset [21]. We use the subset of the Event2012 dataset comprising 68,841 X posts as used by Ray et al. [23] to compare the CLC approach with the FSD approach. We chose the Event2012 dataset over the one described in Sect. 4 for comparing the clustering approaches because it is labeled and has been widely adopted in previous research, allowing for a more standardized and comparable evaluation. Following the event relevance filtering step on the Event2012 dataset with user's intererest as "You are a day trader focused on the US stock market", 44,883 posts were identified as being related to events and these posts were further processed to extract brief description of the event, which served as input to each clustering method. For both the approaches, we use Sentence-BERT (SBERT) [27], more specifically all-mpnet-base-v2[4], for embedding text. We use adjusted mutual information (AMI) [28] for evaluating the two clustering approaches. The AMI returns a value of 1 when the two partitions are identical (i.e. perfectly matched). Random partitions (independent labelings) have an expected AMI around 0 on average.

Table 1. Comparison of performance of FSD clustering and CLC approach on the Events2012 dataset

Clustering Method	AMI
FSD	0.825
CLC	0.731

Table 1 compares the AMI values of the FSD and CLC approaches on the Event2012 dataset. Although FSD achieves a higher AMI score, it poses two

[4] https://huggingface.co/sentence-transformers/all-mpnet-base-v2.

significant challenges. Firstly, it requires extensive hyperparameter tuning across three parameters: similarity threshold (t), window size (w), and batch size (b). For example, the best performance is obtained with the configuration $t = 0.5$, $w = 48$, and $b = 8$. However, even slight deviations from this setting can lead to substantial performance degradation. Increasing t to 1.0, for example, reduces the AMI to just 0.0006, while lowering it to $t = 0.25$ results in an AMI of 0.6001. Secondly, FSD is highly sensitive to the temporal order of posts within each batch. When the dataset was randomly shuffled and evaluated using the optimal FSD parameters, the AMI dropped sharply from 0.825 to 0.133. This sensitivity to temporal sequencing makes FSD less suitable for scenarios where post timestamps are unavailable or unreliable.

In contrast, the CLC approach achieves a slightly lower AMI score but offers greater simplicity and robustness. It involves only a single tunable parameter, t. A grid search for values from $t = 0.3$ to $t = 0.75$ showed the best performance at $t = 0.4$. Importantly, CLC performance remains stable regardless of the temporal ordering of the input data. Given these considerations, we prefer the CLC approach over FSD. Although CLC may exhibit a marginal decrease in clustering performance, it offers substantial practical advantages, including ease of tuning, interpretability, and robustness to variations in data order. These properties make CLC a more reliable and scalable choice for real-world event detection systems.

Next, we evaluate the effect of different LLMs in determining the event attributes, specifically the event date.

5.2 Effect of LLM on Inferring Undetermined Event Dates

We use the dataset described in Sect. 4 to investigate the effectiveness of different LLMs in inferring missing event attributes. We focus only on the date attribute, as associating precise location information with events is often challenging. Events such as announcements, updates, launches, discoveries, and public briefings frequently occur virtually or are disseminated through online platforms, making their geographic origin ambiguous or irrelevant. We evaluate three proprietary LLMs, GPT-4o-mini, GPT-5-nano[5] and Claude 4 Sonnet[6], each paired with a web search tool, Google SerpAPI as discussed in Sect. 3.5. These models were selected because they represent the latest compact, high-performing versions from leading LLM providers, encompassing both reasoning-capable and non-reasoning variants, and offering an effective balance of efficiency, accessibility, and strong information extraction and inference capabilities. Using the UNITE approach, we identified a total of 731 unique events in the dataset. Of these, 50 events already had date information, either directly extracted from the raw post content using Prompt C, or enriched using Prompt D. The remaining 681 events lacked reliable date information and thus required inference using external web search tool. Table 2 presents the fraction of dates inferred using each

[5] https://openai.com/index/introducing-gpt-5/.
[6] https://www.anthropic.com/news/claude-4.

combination for these 681 events. We can see that Claude 4 Sonnet achieves the highest fill rate followed closely by GPT-5-nano suggesting that Claude 4 Sonnet exhibits superior extraction capabilities in extracting specific information from multiple webpages.

Table 2. Effectiveness of LLMs in completing event date information

LLM	Fraction of events with date inferred
GPT-4o-mini	0.587
GPT-5-nano	0.617
Claude 4 Sonnet	0.667

Table 3. Performance of different LLMs in determining event date attribute

LLM	Evaluated events	Accuracy
GPT-4o-mini	363	0.614
GPT-5-nano	420	0.767
Claude 4 Sonnet	413	0.910

Next, we assess the accuracy of the inferred dates through a human validation process involving three independent annotators. Each annotator reviewed the dates inferred by the three LLMs and verified their correctness by referencing publicly available sources, including news articles, press releases, and authoritative online databases. For each event, a ground truth date was established based on a majority vote among the annotators. Table 3 reports the performance of each LLM. The "Evaluated events" column shows the number of events for which an LLM inferred a date and a human-validated ground truth could be established, while the "Accuracy" column indicates the fraction of inferred dates that exactly matched the ground truth. Claude 4 Sonnet paired with SerpAPI not only achieves the largest fraction of dates inferred as demonstrated in Table 2 but also has the highest accuracy as can be seen in Table 3, demonstrating that recall and precision do not always trade off. In many downstream applications such as real-time alerting, monitoring systems, automated reporting, or decision-support platforms precision is often more critical than coverage, since even a single incorrect date can lead to mistimed actions or flawed decisions, potentially causing significant losses, misinformation, or operational disruptions. Considering these factors, Claude 4 Sonnet with SerpAPI provides the most reliable performance for high-stakes use cases, excelling in both correctness and coverage. The superior performance of Claude 4 Sonnet in both coverage Table 2 and accuracy Table 3 warrants further discussion. This task requires more than simple information extraction; it involves synthesizing a definitive attribute (a date) from a collection of potentially noisy and varied web snippets returned by SerpAPI. The results suggest that Claude 4 Sonnet may possess stronger capabilities in synthesizing and reasoning over multiple sources to arrive at a correct, grounded answer. This highlights that for complex, open-domain enrichment tasks, the choice of LLM is critical, with performance likely tied to the model's underlying strengths in information fusion and fact verification. The finding that the model with the highest coverage also achieved the highest precision is particularly noteworthy, indicating that for this task, a model's ability to confidently identify correct information is directly related to its ability to find that information in the first place.

5.3 Effect of User Interest

Finally, we examine how user interest influences the types of events extracted from our dataset. To assess the ability of UNITE to tailor event extraction based on user preferences, we conducted an experiment using two distinct user interest profiles and applied the full pipeline to our dataset. The selected interest profiles were: (1) Earnings calls and earnings events, and (2) Political events and politician activities. Below is a sample of event titles corresponding to each user interest:

- **Earnings calls and earnings events**
  - Apollo Global Management and Workday Join S&P 500
  - ECB Prepared for Rate Cuts
  - MicroStrategy Likely Added to QQQ
  - Walgreens in Talks to Sell to Sycamore Partners
  - CITIZENS CEO Discusses Market Share Gains
  - Earnings Call for Zscaler
  - Conference Call Recap for Salesforce ($CRM) and Dollar Tree ($DLTR)
- **Political events and politician activities**
  - Trump's EV Policy Changes and Tariffs
  - Justin Trudeau's Potential Resignation Consideration
  - Biden Administration Investigates Elon Musk
  - Trump's Plan to End Birthright Citizenship
  - Trump and Republicans Renew Tax Cuts
  - US to Raise Tariffs on Chinese Polysilicon and Wafers

In total, 9 events were extracted under the "Earnings calls and earnings events" interest, with 8 out of 9 correctly aligning with the user's intent, yielding a precision of 0.889. For the "Political events and politician activities" interest, 21 events were extracted, of which 18 were relevant, resulting in a precision of 0.857. These results show a strong alignment between the extracted events and the specified user interests, demonstrating the pipeline's effectiveness in filtering and identifying events that are contextually relevant to user-defined topics.

These findings underscore the versatility of the UNITE approach in accommodating varied user preferences. This allows it to capture a wide range of events from public announcements to political developments without requiring modifications to the underlying pipeline. Such user-centric adaptability is useful for practical applications where monitoring needs can vary greatly across individuals, organizations, or domains.

6 Discussion

6.1 Principal Findings

Our work validates a user-interest-driven framework for event extraction from social media. Key findings include the effectiveness of our hierarchical enrichment strategy and the superior performance of Claude 4 Sonnet for high-fidelity date attribution. The framework's modularity and adaptability to diverse user profiles underscore its practical utility for personalized monitoring.

6.2 Limitations

We acknowledge two primary limitations. First, our reliance on proprietary, closed-source LLMs (GPT-4o-mini, GPT-5-nano, Claude 4 Sonnet) limits direct reproducibility, though the framework's modular design permits swapping in open-source alternatives. Second, our work uses a simple natural language string for user interest and does not introduce a novel method for user modeling. Developing a more sophisticated approach to generating and representing user interests is a clear avenue for future research.

6.3 Future Work

Looking ahead, we plan to address the aforementioned limitations by fine-tuning open-source models such as Llama-3 or Mistral for each stage of the UNITE pipeline. This will involve the creation and release of a human-annotated, open benchmark dataset based on our current data, which will address a significant gap in the literature and foster reproducible research. Furthermore, we aim to extend UNITE's capabilities to multimodal contexts, processing content from platforms like YouTube and TikTok by integrating video and audio analysis with textual data from comments and captions. Finally, pipeline efficiency can be improved by exploring methods to consolidate multiple LLM calls into a single, more complex prompt, reducing both latency and computational cost.

7 Conclusion

This paper introduced UNITE, a framework designed to extract structured, personalized events from noisy social media data. By integrating user interests into a multi-stage pipeline of LLM-based filtering, deduplication, and hierarchical enrichment, UNITE effectively transforms unstructured social signals into actionable intelligence. Our experiments demonstrated the framework's adaptability to diverse user profiles and identified the combination of Claude 4 Sonnet with SerpAPI as a particularly robust solution for high-accuracy attribute enrichment. By providing a systematic approach to personalized event extraction, this work contributes to more targeted and effective monitoring in today's complex information environments.

References

1. Moreno-García, M.N.: Information retrieval and social media mining. Information **11**(12), 578 (2020)
2. Riloff, E., Shoen, J.: Automatically acquiring conceptual patterns without an annotated corpus. In: Third Workshop on Very Large Corpora (1995)
3. Wang, Z., et al.: CLEVE: Contrastive pre-training for event extraction. arXiv preprint arXiv:2105.14485 (2021)
4. Liu, J., Chen, Y., Liu, K., Bi, W., Liu, X.: Event extraction as machine reading comprehension. In: Proceedings of the 2020 Conference on Empirical Methods in Natural Language Processing (EMNLP), pp. 1641–1651 (2020). Association for Computational Linguistics
5. Vaswani, A., et al.: Attention is all you need. In: Advances in Neural Information Processing Systems (NeurIPS), vol. 30, pp. 5998–6008 (2017)
6. Devlin, J., Chang, M.W., Lee, K., Toutanova, K.: BERT: pre-training of deep bidirectional transformers for language understanding. In: Proceedings of the 2019 Conference of the North American Chapter of the Association for Computational Linguistics: Human Language Technologies (NAACL-HLT), pp. 4171–4186 (2019). Association for Computational Linguistics
7. Shiri, F., Moghimifar, F., Haffari, R., Li, Y.F., Nguyen, V., Yoo, J.: Decompose, enrich, and extract! Schema-aware event extraction using LLMs. In: Proceedings of the 2024 27th International Conference on Information Fusion (FUSION), pp. 1–8. IEEE (2024)
8. Srivastava, S., Pati, S., Yao, Z.: Instruction-tuning LLMs for event extraction with annotation guidelines. arXiv preprint arXiv:2502.16377 (2025)
9. Chen, B., et al.: Joint multimedia event extraction from video and article. arXiv preprint arXiv:2109.12776 (2021)
10. Guo, Y., et al.: Unsupervised social event detection via hybrid graph contrastive learning and reinforced incremental clustering. Knowl.-Based Syst. **284**, 111225 (2024)
11. Grishman, R., Sundheim, B.M.: Message understanding conference-6: a brief history. In: Proceedings of the 16th International Conference on Computational Linguistics (COLING), pp. 466–471 (1996)
12. Simon, É., Olsen, H., You, H., Touileb, S., Øvrelid, L., Velldal, E.: Generative approaches to event extraction: survey and outlook. In: Proceedings of the Workshop on the Future of Event Detection (FuturED), pp. 73–86, Miami, Florida, USA (2024). Association for Computational Linguistics
13. Grattafiori, A., et al.: The llama 3 herd of models. arXiv preprint arXiv:2407.21783 (2024)

14. Baker, C.F., Fillmore, C.J., Lowe, J.B.: The Berkeley Framenet project. In: Proceedings of the 17th International Conference on Computational Linguistics (COLING-ACL), pp. 86–90. Association for Computational Linguistics (1998)
15. Chen, Y., Liu, S., Zhang, X., Liu, K., Zhao, J.: Automatically labeled data generation for large scale event extraction. In: Proceedings of the 55th Annual Meeting of the Association for Computational Linguistics (ACL), pp. 409–419 (2017). Association for Computational Linguistics
16. Hogenboom, F., Frasincar, F., Kaymak, U., De Jong, F., Caron, E.: A survey of event extraction methods from text for decision support systems. Decis. Support Syst. **85**, 12–22 (2016)
17. Sokhonn, L., Park, Y.S., Lee, M.K.: Hierarchical clustering via single and complete linkage using fully homomorphic encryption. Sensors **24**(15), 4826 (2024)
18. Soni, S., Mehta, D., Vishwanath, V., Seetha, A., Chouhan, S.S.: DiPD: Disruptive event prediction dataset from twitter. arXiv preprint arXiv:2111.15629 (2021)
19. 0bzrv3. Supply chain disruption event dataset. https://huggingface.co/datasets/0bzrv3/supply_chain_disruption_event (2024). Hugging Face
20. Sai-Manisha. Events after Feb 2023 dataset. https://huggingface.co/datasets/Sai-Manisha/Events-after-feb-2023 (2023). Hugging Face
21. McMinn, A.J., Moshfeghi, Y., Jose, J.M.: Building a large-scale corpus for evaluating event detection on twitter. In: Proceedings of the 22nd ACM International Conference on Information and Knowledge Management (CIKM), pp. 409–418, New York (2013). ACM
22. X. X API endpoint map. https://developer.x.com/en/docs/x-api/migrate/x-api-endpoint-map. Accessed 18 Aug 2025
23. Ray, M., Wang, Q., Mélanie-Becquet, F., Poibeau, T., Mazoyer, B.: An incremental clustering baseline for event detection on twitter. arXiv preprint arXiv:2412.15257 (2024)
24. Allan, J., Carbonell, J.G., Doddington, G., Yamron, J., Yang, Y.: Topic detection and tracking pilot study final report. In: Proceedings of the DARPA Broadcast News Transcription and Understanding Workshop, pp. 194–218, San Francisco (1998). Morgan Kaufmann Publishers
25. Cao, Y., Peng, H., Yu, Z., Yu, P.S.: Hierarchical and incremental structural entropy minimization for unsupervised social event detection. In: Proceedings of the AAAI Conference on Artificial Intelligence, vol. 38, pp. 8255–8264. AAAI Press (2024)
26. Mazoyer, B., Cagé, J., Hervé, N., Hudelot, C.: A French corpus for event detection on twitter. In: Proceedings of the Twelfth Language Resources and Evaluation Conference (LREC), pp. 6220–6227, Marseille (2020). European Language Resources Association (ELRA)
27. Reimers, N., Gurevych, I.: Sentence-BERT: Sentence embeddings using Siamese BERT-networks. arXiv preprint arXiv:1908.10084 (2019)
28. Vinh, N.X., Epps, J., Bailey, J.: Information theoretic measures for clusterings comparison: variants, properties, normalization and correction for chance. J. Mach. Learn. Res. **11**, 2837–2854 (2010)

Real-Time Thai Event Detection Using Context-Aware Deep NER with Cross-Lingual Pretraining

Sathit Prasomphan(✉)

Department of Computer and Information Science, Faculty of Applied Science, King Mongkut's University of Technology North Bangkok, 1518 Pracharat 1 Rd., Wongsawang, Bangsue, Bangkok 10800, Thailand

sathit.p@sci.kmutnb.ac.th

Abstract. This research aims to develop a technique for real-time event detection on X (formerly Twitter) using deep Named Entity Recognition (NER) techniques. The system collects and ranks trending topics on the platform, extracting specific keywords from posts using natural language processing (NLP) methods. The objective is to identify key entities and visualize their relationships by generating diagrams that represent the connections between entities involved in each event. The study explores both feature-based and fine-tuning approaches to enhance the performance of the BERT model. Our fine-tuning strategy achieves state-of-the-art performance, setting a new benchmark for Thai language NER in the context of social media event detection.

Keywords: Social media · event detection · Name entity recognition · BERT-BiLSTM-CRF · Conditional random field

1 Introduction

Social media is a very powerful tool in transferring information of everyone from children to the elderly. The utilization of social media has experienced a significant surge across diverse cultures, owing to the widespread access to the internet. Trending on X (formerly Twitter) is calculated using the hashtags, which is mentioned in the tweet. However, some problems arise in using such social media. Many users do not know which hashtag to use for their tweets at events. Moreover, the relationship of entities that were appeared in that hashtag not clearly known in advance. This research aims to develop technique for linking important events on X using name entity recognition techniques. In this research, popular trending topics in X is collected and ranked. The specific keyword of X post was extracted using natural language processing methods. Moreover, our objective aim to establish connections among keyword in each topics in each occurrence by generating a diagram illustrating the relationships between entities in that events transpiring on social media.

Typically, X messages are limited to a maximum length of 140 characters posing an additional challenge. To address this limitation, these short messages need to be grouped

H. Unger and P. Meesad (Eds.): NLPIR 2025, LNNS 1904, pp. 1156–1169, 2026.
https://doi.org/10.1007/978-3-032-20897-2_81

or pre-processed to generate longer tweet streams. Detecting and tracking events in regular long documents is easier in terms of document sparsity and irrelevance than in the case of short blogging services such as X. Most X posts contain wrong grammatical and spelling errors, which makes it more difficult, compared to a regular newswire. X is a user-generated resource. Most of them contain many invisible words that are only visible in short messages. Such words are frequently used and created by users. Users also add misspellings and expand on the word. This results in very unwanted problems. In addition, the proliferation of mobile technology has led to the abundance of diverse messages on social media. Accordingly, having a tool capable of gathering and ranking trending topics, extracting information about significant events from X, and establishing connections among individuals involved in various events through relationship diagrams would simplify the process of scrutinizing information on X, particularly for identifying potential instances of misinformation or fake news. From the above reason, we are interested in proposing a possible solution for creating an automated system for collecting and ranking popular trending topics. In addition, we are interested in creating relationships of entities involved in each event by creating a diagram of the relationships of keywords in various events that occur on social media posts that contain text or sentences that are made using BERT-BiLSTM-CRF Thai named entity recognition technique. In this research the name entities recognition in Thai language is focused.

The remainder of the section has been organized as follows. Section 2 introduces related theories. Section 3 provides specifics on the suggested approach. The experimental setup is explained in Sect. 4. In Sect. 5, the findings and discussions are explained. Section 6 finally concludes the paper.

2 Related Theory

2.1 Deep Learning

Developing machines with the ability to learn, predict, and produce information is a major issue for artificial intelligence (AI). In the process of machine learning, a number of methods are used, including as deep learning, artificial neural networks (ANNs), reinforcement learning, supervised learning, and unsupervised learning. The main difference between ANN and deep learning, both of which are machine learning algorithms, is that deep learning has an extra hidden layer [1, 2].

Artificial Neural Networks draw inspiration from the human nervous system, simulating the functions of neuron groups connected like a neural network. This system emulates the ability to recognize multiple things simultaneously and make decisions closely resembling human understanding. The process involves inputting knowledge into the input layer, evaluating the dataset within the hidden layer, and presenting or modifying that knowledge in the output layer [3–5]. Deep learning has significantly affected Natural Language Processing (NLP), enhancing the accuracy and efficiency of models for tasks such as language translation, sentiment analysis, question answering, and text classification. Deep learning models, based on neural networks with multiple interconnected layers, excel at learning complex features from input data, capturing intricate patterns and relationships in extensive datasets. This makes them particularly well suited for NLP tasks given the inherent complexity and ambiguity of language.

Several popular deep learning architectures employed in NLP include Convolutional Neural Networks (CNNs), Recurrent Neural Networks (RNNs), Long Short-Term Memory Networks (LSTMs), and Transformers. CNNs, initially designed for image processing, find application in text classification by extracting relevant features. RNNs, built for sequential data, excel in natural language generation, translation, and sentiment analysis. LSTMs, a type of RNN, address the vanishing gradient problem and are widely used in text generation, translation, and speech recognition. Transformers, a recent breakthrough, leverage attention mechanisms to focus on crucial parts of input sequences, making them state-of-the-art for tasks like language translation, question answering, and text generation [6]. In essence, deep learning has revolutionized NLP by delivering more precise and efficient models capable of handling the intricacies and uncertainties of natural language.

2.2 Natural Language Processing and Key Techniques for Thai Language Understanding

Natural Language Processing (NLP) is a core area within artificial intelligence that enables computers to understand and generate human language through tasks such as text classification, sentiment analysis, named entity recognition (NER), machine translation, summarization, and question answering. NLP draws on linguistic components—syntax, semantics, pragmatics, discourse analysis—and computational approaches ranging from rule-based methods to statistical modeling and deep neural networks, supporting applications across healthcare, finance, customer service, and education [7].

For Thai language processing, PyThaiNLP serves as a comprehensive library that provides essential tools including tokenization, part-of-speech tagging, NER, sentiment analysis, dependency parsing, word embeddings, and Thai–foreign language translation. Its pre-trained models, datasets, and utilities facilitate efficient development of Thai NLP applications [8].

A foundational technique in NLP is Word2Vec, a deep learning algorithm that learns dense vector representations of words by capturing semantic and syntactic relationships. Using the Continuous Bag-of-Words (CBOW) and Skip-gram architectures, Word2Vec predicts either a target word from its context or context words from a target word, producing embeddings widely used in tasks such as classification, sentiment analysis, and machine translation [9, 10].

Within information extraction, Entity and Relation Extraction—particularly NER—plays a critical role by identifying and categorizing entities such as people, organizations, and locations from unstructured text. Although early approaches relied heavily on rule-based systems, modern NER leverages machine learning and neural models for improved accuracy and scalability [11, 12].

For sequential labeling tasks such as NER, Conditional Random Fields (CRFs) provide a discriminative modeling framework that overcomes limitations found in generative models like Hidden Markov Models (HMMs) and in Maximum Entropy Markov Models (MEMMs), particularly the label bias problem. CRFs model conditional probabilities over entire label sequences, enabling more accurate predictions in structured language tasks.

2.3 Research on Real Time Event Detection from the X (Formerly Twitter) Data Stream

Hasan, M. et al. [13], introduced TwitterNews+, an event detection system utilizing specialized inverted indices and an incremental clustering approach for real-time identification of major and minor news events from the X data stream. The evaluation, based on a benchmark dataset, demonstrates substantial improvements in recall and precision compared to five state-of-the-art baselines employed in the study.

Kang Zeng et al. [14], proposed a 4W-oriented event detection method dubbed BEHIND, and concentrate on the components in the When, Where, Who, and What (4W) dimensions of events. For event detection, BEHIND combines the utilization of Heterogeneous Information Network (HIN) with Bursty Elements. Tweets with bursty elements are chosen using bursty elements, which are computed using a probability distribution. In order to aid in tweet clustering, HIN is employed to improve relevance assessment in 4W dimensions between tweets. The tweet clusters match the events that was identified. They assessed approach against a benchmark dataset. According to experimental data, the system outperforms state-of-the-art methods in terms of precision and duplication rate while also detecting a higher number of events.

Rajtmajer, V., Král, P. [15] presented a technique to automatically identify events in such data and to harvest key messages from Czech Twitter at a fast download speed. After identifying notable X users, they use these lists to find tweets that might be worth downloading. Next, the tweets are grouped together to identify the events. The thresholding provides the basis for the final decision. It is demonstrated that the harvesting method downloads approximately six times as much data as the other methods. Additionally, they present encouraging outcomes from the event detection method using a limited dataset of Czech tweets.

Renlong Ai et al. [16], presented a fine-grained event identification system in the context of the big data platform Flink, and show off its ability to extract and geo-locate events linked to industry and mobility from a variety of heterogeneous text sources. The system's ability to achieve high throughput and low latency is demonstrated by performance analyses carried out on multiple huge datasets. This is an important feature when real-time event detection and action is required. Promising experimental results for system's event extraction component, which can identify a new class of events.

A. H. Hossny and L. Mitchell [17], introduced an effective approach for identifying keywords on X associated with specific events, particularly protests. The method involves real-time tracking of keyword frequency, utilizing word-pairs to capture context. The process includes binarizing daily counts of word-pairs, applying a spike detection filter, and measuring similarity using the Jaccard metric. The top n word-pairs serve as features for binary classification of event or non-event days. Evaluation with various classifiers, including Naive Bayes, SVM, Logistic Regression, KNN, and decision trees, yields AUC ROC scores up to 0.91 and F1 scores up to 0.79. Experiments in English across cities like Melbourne, Sydney, and Brisbane, as well as in Indonesian in Jakarta, show consistent results.

In [18], a significant amount of information found in Electronic Health Records (EHRs) is unstructured and frequently appears as free text. Its potential use in clinical decision-making is limited by its format. Techniques for named entity recognition

(NER) aim to solve the problem of obtaining relevant information from unstructured text. This study set aimed to describe the state of the art NER techniques and chart their development from 2011 to 2022.

Several paper has conducted research on classification detection in Thai language. K. Pasupa and T. Seneewong Na Ayutthaya [19], the experiment evaluates and compares the performance of conventional deep learning models—Convolutional Neural Network (CNN), Long Short-Term Memory (LSTM), and Bidirectional LSTM (Bi-LSTM)—in sentiment analysis of Thai children's tales. Previous studies incorporated various features like word embedding, POS-tag, and sentic to aid in semantic understanding, grammatical function, and emotional context. The experiment's results highlight that the CNN model utilizing all three features achieved the highest F1-score of 0.817 at $p < 0.01$, significantly outperforming other models.

A. Phanomtip et al. [20] addressed the issue of toxic behaviors on X, the proposal includes a novel dataset of 67K tweets, automatic annotation methods for large-scale data, and a detection system. Experimental results reveal that the proposed method surpasses the baseline by 4%.

P. Thiengburanathum and P. Charoenkwan [21], the study addressed the issue of toxic comments on X in Thailand, employing Sentiment Analysis. It compares two feature extraction techniques (Bag of Words and TF-IDF) and evaluates the performance of traditional classifiers and deep-learning approaches (CNN, LSTM, and BERT) using the Toxicity Thai Twitter corpus. The combination of Bag of Words with the Extra-Tree classifier yielded the highest F1-score (0.72), classification accuracy (72.27%), and AUC value (0.77) on the test set, outperforming other classifiers and deep-learning techniques. The study also explores feature importance, correlation, and impacts using SHapley Additive exPlanations (SHAP) diagrams.

When selecting an algorithm for real time event detection from the X data stream, it is important to consider factors like performance, scalability, interpretability, and the availability of suitable training data. Additionally, it is important to evaluate the algorithm on multiple performance metrics to ensure that it is effective in detecting entities.

In this research, specific data sets from Thailand as well as a worldwide data set was study. We combining BERT (Bidirectional Encoder Representations from Transformers) and conditional random field for real time event detection from the X data stream, which involves fine-tuning a pre-trained BERT model as input features for conditional random field classifier.

3 Proposed Method

In this section, we describe the process of real time event detection from the X data stream with BERT-BiLSTM-CRF Thai named entity recognition technique. The proposed method for detecting named entity in social networks utilizes natural language processing techniques. The method involves extracting features from text data using techniques such as tokenization, stemming, and stop word removal.

3.1 Proposed Method Algorithm for Real-Time Event Detection from X (Formerly Twitter)

Step 1: Use the X (formerly Twitter) API to collect real-time tweets based on specific keywords or hashtags relevant to the region or context (in this case, Thai language). Apply filters for language, location, and event-specific keywords to refine the tweet stream.

Step 2: Remove unwanted characters, URLs, emojis, or non-Thai language content. Tokenize the Thai text using a Thai-specific tokenizer. After that, assign initial labels to the tokens for NER tasks (e.g., locations, people, organizations, or events). Next, handling Out-of-Vocabulary (OOV) Words. Apply subword tokenization using WordPiece to manage OOV words or rare Thai words.

Step 3: Perform model architecture with the following process.

BERT Embedding Layer: Input the preprocessed and tokenized tweet into the BERT model to extract contextual embeddings for each token. These embeddings will capture both left and right context for each word in a tweet, making it effective for language-specific nuances in Thai.

BiLSTM Layer: Pass the embeddings to a BiLSTM (Bidirectional Long Short-Term Memory) layer to model the sequential nature of the text. This helps in capturing both forward and backward dependencies between words for event-related context.

CRF Layer: Output from the BiLSTM is passed to a CRF layer. The CRF layer optimizes the predictions by considering the relationship between predicted labels (e.g., ensuring that a location entity is followed by a related event entity). This helps with sequence labeling and improves NER precision.

Step 4: After processing tweets through the BERT-BiLSTM-CRF model, identify key named entities (e.g., locations, people, organizations, or events). Map these entities to relevant event categories. Group tweets that mention similar entities or co-occurrences of event-specific words using a clustering algorithm. Track the number of tweets related to an event in a specific time window to detect emerging events. Apply a sliding window mechanism to continuously detect and track trends or events. Use time-series analysis (e.g., using exponential moving averages) to identify peaks or anomalies in tweet volumes associated with specific events.

Step 5: Rank detected events based on metrics such as tweet volume, retweet count, or user influence (e.g., using PageRank or weighted sentiment scores). Prioritize trending events that exhibit a sudden rise in tweet volume or engagement. Summarize key information about each detected event using text summarization techniques (e.g., BERT-based extractive summarization or sequence-to-sequence models). Extract the most relevant named entities and event-related phrases to generate an overview for each event.

Step 6: Trigger alerts for significant events based on a pre-defined threshold (e.g., an unusual spike in mentions of a specific event or entity). Send notifications via mobile push notifications. Visualize event timelines, tweet volumes, and detected entities using dashboards. Provide insights into the event evolution, key participants, and associated trends.

From the above processes, the following is Real-Time Thai Event Detection Using Context-Aware Deep NER with Cross-Lingual Pretraining pseudocode.

```
Algorithm 1: Real-Time Thai Event Detection Using Context-Aware Deep NER
with Cross-Lingual Pretraining
Input:
    •  Stream of tweets TTT collected via X API
    •  Event-related keywords and hashtags
    •  Threshold θ\thetaθ for event detection
Output:
    •  Detected significant events
    •  Event summaries and alerts
Procedure

    1. Initialize X API connection using predefined event keywords and
       hashtags.
    2. Load pretrained BERT model, BiLSTM encoder, and CRF tagger.
    3. Configure clustering algorithm (e.g., DBSCAN) and set event
       volume threshold θ.
       4.  While real-time streaming is active do
       4.1 Collect a batch of tweets T={t1,t2,…,tn}.
       4.2 For each tweet ti∈T do
         a. Preprocess tweet text by removing noise and applying Thai
       tokenization.
         b. Generate token embeddings using BERT.
         c. Encode the embedding sequence using BiLSTM.
         d. Predict NER labels using the CRF decoder.
         e. Extract key entities, such as EVENT, PERSON, and LOCATION.
         f. Store extracted entities and metadata for later analysis.
       4.3 End for
    5. Perform clustering on processed tweets to group related messages
       into event clusters.
    6. Monitor entity frequencies and tweet volumes for each cluster.
    7. For each cluster Cj do
        7.1 If tweet volume |Cj|>θ then
         a. Rank active events using factors such as tweet volume,
       retweet count, and user influence.
         b. Generate an event summary by aggregating salient entities
       and representative key phrases.
         c. Trigger alert indicating a significant event.
        7.2 End if
    8. End for
    9. Proceed to the next time window and repeat Steps 4–8.
```

4 Experimental Setup

4.1 Dataset

This research employs multiple datasets derived from both social media platforms and curated textual sources to support real-time event detection using a BERT–BiLSTM–CRF Thai named entity recognition (NER) framework. First, X data was collected as the primary real-time text stream source. X is widely adopted in studies related to cyberbullying, sentiment analysis, and event detection. Data acquisition was performed through the official X API and supported by third-party tools, which facilitate efficient extraction of tweets containing targeted keywords and hashtags.

In addition to live X data, three benchmark datasets from social media texts and a Thai textbook were incorporated. These datasets are sentence-level, manually annotated corpora covering both binary and multiclass sentiment classification tasks. To ensure data consistency, all datasets underwent the same preprocessing pipeline: Replacing URLs and emoticons with standardized tags. Applying Thai word segmentation using a maximal-matching dictionary approach combined with Thai Character Cluster (TCC) rules from PyThaiNLP. Removing tokens with fewer than two syllables. Normalizing repeated character sequences to their base form.

To further enhance the keyword-based search and NER capabilities of the model, additional domain-specific metadata was integrated from authoritative cultural and historical databases maintained by Thailand's Department of Fine Arts. These include datasets related to ancient sites, antiquities, national archives, the National Library, the Royal Crematorium, and the Office of Traditional Arts.

Based on these sources, the keyword-tagging schema used for entity recognition was expanded to 85 distinct entity tags, derived from a corpus of 34,304 annotated sentences. These tags span a broad range of categories relevant to Thai cultural heritage and real-time event contexts—for example: place, date-time, project, activity, animal, plant, organization, personal name, sculptor, storekeeper, antique, sculpture, object sculpture, painting, dance-music, ancient site, museum, monument, tradition, historical park, antiquities, object details, Buddha statue, literature, and others. Summary of datasets used in this study is shown in Table 1.

Table 1. Summary of Datasets Used in This Study.

Dataset Source	Description	Annotation Level	No. of Sentences	Tags/Labels	Notes
X (formerly Twitter) (Real-time stream)	Real-time tweets collected using X API and third-party tools (Tweepy, Twython)	Token-level NER + Sentence-level context	Variable (depends on streaming window)	EVENT, PERSON, LOCATION, etc.	Primary source for real-time event detection
Benchmark Dataset 1 (Social Text)	Thai social media dataset used for sentiment and text classification	Sentence-level	Included in total 34,304	Binary/Multiclass sentiment	Manually annotated
Benchmark Dataset 2 (Social Text)	Additional Thai social content dataset	Sentence-level	Included in total 34,304	Sentiment labels	Manually annotated

(*continued*)

Table 1. (*continued*)

Dataset Source	Description	Annotation Level	No. of Sentences	Tags/Labels	Notes
Benchmark Dataset 3 (Thai Textbook)	Non-social textual dataset to enrich language patterns	Sentence-level	Included in total 34,304	Sentiment/topic categories	Manually annotated
Fine Arts Department Databases	Cultural and heritage databases: ancient sites, antiquities, national archives, national library, royal crematoriums, object details	Entity-level (keyword tags)	Used to generate 85 tags	85 entity tags	Used for keyword expansion and NER improvement
Final NER Training Corpus	Combined annotated dataset (social + cultural sources) for model training	Token-level	34,304 sentences	85 entity types	Includes categories such as place, date-time, organization, sculpture, monument, Buddha statue, literature, etc.

4.2 Performance Indexed

Several performance metrics can be used to measure the effectiveness of real-time event detection using BERT–BiLSTM–CRF and clustering. The following are some of the commonly used metrics. Precision measures the proportion of correctly identified events among all predicted events. High precision indicates few false alarms. Recall measures how many actual events the system successfully detects. High recall indicates fewer missed events. F1 Score, the harmonic mean of precision and recall, providing a balanced indicator of performance when both false positives and false negatives matter. Accuracy represents the overall proportion of correct predictions. However, it may be less reliable in imbalanced datasets where event instances are relatively rare.

4.3 Comparing Algorithms

When comparing algorithms for real-time event detection from X, it is essential to consider their accuracy, efficiency, and suitability for various aspects of the task, such as natural language processing (NLP), named entity recognition (NER), and real-time processing. Below is a comparative analysis of several popular algorithms and models that can be used in the context of event detection, focusing on BERT-BiLSTM-CRF and alternatives like CNN, LSTM, RNN, and Transformer-based models.

5 Results and Discussion

The results of the proposed algorithms applied to real time event detection from the X data stream are presented to illustrate the experimental findings. The classification model was evaluated based on precision, recall, and F-1 score with event category as shown in Table 2 and Table 3.

To illustrate the experimental results when comparing the proposed method (BERT-BiLSTM-CRF) with other comparing methods, let us create an example based on common metrics used for evaluating Named Entity Recognition (NER) and real-time event detection tasks on X. The metrics typically include precision, recall, F1-score, and processing time. The results will show how different methods perform in terms of accuracy and efficiency when applied to real-time event detection from the X data stream using Thai Named Entity Recognition (NER). The classification accuracy, as demonstrated in Table 3.

The experimental outcomes suggest that integrating BERT with BERT-BiLSTM-CRF Thai Named Entity Recognition technique for real time event detection from the X data stream, with the fine-tuning of a pre-trained BERT model as input features for a conditional random field classifier, is successful for training and determining the type of entities in social media.

5.1 Named Entity Recognition (NER) Performance

We evaluated the performance of the BERT-BiLSTM-CRF model for Thai NER using standard metrics such as precision, recall, and F1-score. The dataset used for training and testing was composed of manually annotated tweets, covering diverse real-time events from Thailand (e.g., political events, natural disasters, public health alerts). The following table summarizes the performance: Average F1-Score is 87.5%. The BERT-BiLSTM-CRF model demonstrated high performance across all entity types, with the highest F1-score for date entities (91.3%) and slightly lower performance for miscellaneous entities (82.4%), which are generally more ambiguous in nature. The average F1-score of 87.5% indicates a robust NER performance for Thai language tweets.

5.2 Event Detection Performance

For event detection, the system classified real-time tweets into predefined event categories (e.g., politics, entertainment, natural disasters). The classification model was evaluated based on accuracy, precision, recall, and F1-score. Overall accuracy is 85.6%. The system exhibited high accuracy in detecting events related to natural disasters (90.3%) and public health (88.7%). Events in the politics category also performed well with an accuracy of 85.6%. However, the system struggled slightly with entertainment and miscellaneous events, with accuracies of 83.4% and 80.1%, respectively, likely due to the ambiguous nature of entertainment-related tweets and the broad range of miscellaneous events.

5.3 Real-Time Performance

The system's real-time performance was evaluated by processing a continuous stream of tweets over a 24-hour period. The model was able to detect trending events within 5–10 s of tweet creation, providing near-instantaneous results. The system's throughput was measured, demonstrating the ability to process approximately 1,500 tweets per second, ensuring scalability for high-traffic situations. The experimental results demonstrate that the BERT-BiLSTM-CRF model is highly effective for Named Entity Recognition in Thai, achieving a high average F1-score of 87.5%. This indicates that the model can successfully extract key entities, which are critical for event detection. Moreover, the event classification model exhibited strong performance, particularly in detecting events related to natural disasters and public health. The system's ability to rank trending topics in real-time showed promising results. Popular topics such as natural disasters (e.g., floods, earthquakes) were detected almost immediately after users, confirming the model's responsiveness and practical applicability for real-time applications, mentioned them.

The challenges and limitations are as follows. Tweets often contain ambiguous or incomplete information, which can affect the system's ability to classify events accurately. This was particularly evident in the miscellaneous category, where the system had lower accuracy. The system is currently optimized for Thai language tweets, but X is a multilingual platform. Future work could involve expanding the system's capabilities to detect events in multiple languages, which would require more diverse language models. X data is often noisy, with many irrelevant or spam tweets. While pre-processing techniques helped reduce some noise, future work could involve improving noise-filtering techniques, such as bot detection or spam filtering.

For the future work, the following are considered. Future research could focus on improving the accuracy of the system for ambiguous event categories, such as entertainment and miscellaneous. This could be achieved by incorporating more advanced context-aware models or using external knowledge bases to provide additional context. While the system demonstrated good scalability, further optimizations could be made to improve throughput under extremely high-traffic conditions, such as during major global events (e.g., elections, natural disasters). Expanding the model's capability to detect events from multilingual X data streams would make the system more versatile. Incorporating language-agnostic models or training models for other languages (e.g., English, Chinese) could broaden its applicability. To improve event detection accuracy, future systems could integrate external data sources such as news feeds, sensor networks, or geographical data. This would allow the system to cross-validate X data with more reliable sources, thereby reducing false positives.

The results of the proposed algorithms applied to real time event detection from the X data stream are presented to illustrate the experimental findings. Table 3 outlines the performance metrics of various models across the database from department of Fine Art.

The performance metrics showing in Table 3 are accuracy, which is the overall correctness of the model's predictions. Precision shows the ratio of true positive predictions to the total predicted positives, indicating the model's ability to avoid false positives. Recall show the ratio of true positive predictions to the total actual positives, indicating the model's ability to capture all positive instances. From the result indicate in table, BERT, generally performs well across all datasets in terms of accuracy. It demonstrates competitive precision and recall values. For SVM (Support Vector Machines), shows a balanced performance with competitive accuracy, precision, and recall values. In decision tree, exhibits reasonable accuracy but lags behind in precision and recall compared to BERT and SVM. The experimental outcomes suggest that integrating BERT with BERT-BiLSTM-CRF Thai Named Entity Recognition Technique for Real Time Event Detection from the X data stream, with the fine-tuning of a pre-trained BERT model as input features for a conditional random field classifier, is successful for training and determining the type of entities in social media.

Table 2. The classification model was evaluated based on accuracy,precision, recall F1-Score.

Event category	Accuracy	Precision	Recall	F1-Score
Politics	85.6%	86.4%	84.2%	85.3%
Natural Disasters	90.3%	91.5%	89.6%	90.5%
Public Health (COVID-19)	88.7%	89.2%	87.1%	88.1%
Entertainment	83.4%	82.9%	83.1%	83.0%
Miscellaneous	80.1%	81.0%	78.6%	79.8%

Table 3. The results show how different methods perform in term of accuracy when applied to real-time Thai event detection using context-aware deep NER with cross-lingual pre-training.

Method	Precision (%)	Recall (%)	F1-Score (%)
BERT-BiLSTM-CRF (Proposed)	92.5	91.8	92.2
CNN	85.0	82.3	83.6
LSTM	88.4	86.5	87.4
GRU	87.5	85.8	86.6
BERT (Without BiLSTM-CRF)	91.5	90.0	90.7
CRF (Traditional)	80.5	77.0	78.7
DBSCAN (Clustering)	75.0	74.5	74.7

6 Conclusions

This research aims to develop technique for extracting knowledge entities on X (formerly Twitter) using name entity recognition techniques. In this research, popular trending topics in X were collected and ranked. The specific keyword of X post was extracted using natural language processing methods. Moreover, our objective aim to establish connections among keyword in each topics in each occurrence by generating a diagram illustrating the relationships between entities in that events transpiring on social media. We investigate training strategies, including feature-based and fine-tuning approaches, for enhancing the BERT model. In the process of improving the model and developing the model, searching for information with keywords from the database will bring information from the database from department of Fine Art for example database of ancient sites, antiquities database, national archives database. To improve and develop the keyword search model in this research. The number of tags for keywords has been increased to 85 tags from 34304 sentences. Example of tags are as follows; place, date time, project, activity, animal, plant, organization, personal, sculptor, storekeeper, antique, sculpture, object sculpture, painting, dance-music, ancient, museum, monument, tradition, historical park, antiquities, object detail, Buddha statue, literature, etc. The experimental outcomes suggest that integrating BERT with BERT-BiLSTM-CRF Thai Named Entity Recognition technique for real time event detection from the X data stream, with the fine-tuning of a pre-trained BERT model as input features for a conditional random field classifier, is successful for training and determining the type of entities in social media. Our fine-tuning method achieves unprecedented performance, setting a new state-of-the-art benchmark.

References

1. Tanantong, T., Parnkow, M.: A survey of automatic text classification based on Thai social media data. Int. J. Knowl. Syst. Sci. **13**, 1–25 (2022)
2. Deng, L., Yu, D.: Deep learning: methods and applications foundations and trends. Signal Process. **7**(3–4), 197–387 (2014)
3. Prasomphan, S.: Toward fine-grained image retrieval with adaptive deep learning for cultural heritage image. Comput. Syst. Sci. Eng. **44**(2), 1–10 (2023)
4. Prasomphan, S., Jung, J.E.: Mobile application for archaeological site image content retrieval and automated generating image descriptions with neural network. Mob. Netw. Appl. **22**(1), 642–649 (2017)
5. Prasomphan, S., Tathong, T., Charoenprateepkit, P.: Traffic sign detection for panoramic images using convolution neural network technique. In: Proceedings of the 2019 3rd High Performance Computing and Cluster Technologies Conference on - HPCCT2019, Guangzhou, China, pp. 128–133 (2019)
6. Teng, T.H., Varathan, K.D.: Cyberbullying detection in social networks: a comparison between machine learning and transfer learning approaches. IEEE Access **11**, 55533—55560 (2023)
7. Wadud, M.A.H., Mridha, M.F., Shin, J., Nur, K., Kumar, A.S.: Deep-BERT: transfer learning for classifying multilingual offensive texts on social media. Comput. Syst. Sci. Eng. **44**(2), 1775–1791 (2023)
8. Phatthiyaphaibun, W., et al.: PythaiNLP: Thai natural language processing in Python. arXiv preprint arXiv:2312.04649 (2023)
9. Mikolov, T., Chen, K., Corrado, G., Dean, J.: Efficient estimation of word representations in vector space. arXiv preprint arXiv:1301.3781 (2013)
10. Paliwal, S., Kumar, A.M., Krishn, R.M., Nawaz, N., Senthilkumar, M.: XGBRS framework integrated with Word2Vec sentiment analysis for augmented drug recommendation. Comput. Mater. Continua **72**(3), 5345–5362 (2022)

11. Manning, C.D., Surdeanu, M., Bauer, J., Finkel, J.R., Bethard, S., McClosky, D.: The Stanford Core NLP natural language processing toolkit. In: Proceedings of the 52nd Annual Meeting of the Association for Computational Linguistics, Baltimore, MD, USA, 23–24 June (2014)
12. Sun, W., Liu, S., Liu, Y., Kong, L., Jian, Z.: Information extraction network based on multi-granularity attention and multi-scale self-learning. Sensors **23**, 42—50 (2023)
13. Hasan, M., Orgun, M.A., Schwitter, R.: Real-time event detection from the Twitter data stream using the TwitterNews+ framework. Inf. Process. Manag. **56**(3), 1146–1165 (2019)
14. Zeng, K., Liu, Y., Song, X., Zhou, B.: BEHIND: a 4W-oriented method for event detection from Twitter. In: The 31st International Conference on Software Engineering & Knowledge Engineering, SEKE 2019 Hotel Tivoli Oriente, Lisbon, Portugal, 10 July–12 July (2019)
15. Rajtmajer, V., Král, P.: data harvesting and event detection from Czech Twitter. In: van den Herik, J., Rocha, A., Filipe, J. (eds.) ICAART 2017. LNCS, vol. 10839, pp. 102–115. Springer, Cham (2018). https://doi.org/10.1007/978-3-319-93581-2_6
16. Thomas, P., et al.: Streaming text analytics for real-time event recognition. In: Proceedings of the International Conference Recent Advances in Natural Language Processing, RANLP, Varna, Bulgaria, pp. 750–757. INCOMA Ltd. (2017)
17. Hossny, A.H., Mitchell, L.: Event detection in Twitter: a keyword volume approach. In: 2018 IEEE International Conference on Data Mining Workshops (ICDMW), Singapore, pp. 1200–1208 (2018)
18. Durango, M.C., Torres-Silva, E.A., Orozco-Duque, A.: Named entity recognition in electronic health records: a methodological review. Healthc. Inform. Res. **29**(4), 286–300 (2023)
19. Pasupa, K., Seneewong Na Ayutthaya, T.: Thai sentiment analysis with deep learning techniques: A comparative study based on word embedding, POS-tag, and sentic features. Sustain. Cities Soc. **50**, 101615 (2019)
20. Phanomtip, A., Sueb-in, T., Vittayakorn, S.: Cyberbullying detection on Tweets. In: The International Conference on Electrical Engineering/Electronics, Computer, Telecommunications and Information Technology (ECTI-CON), Chiang Mai, Thailand (2021)
21. Thiengburanathum, P., Charoenkwan, P.: A performance comparison of supervised classifiers and deep-learning approaches for predicting toxicity in Thai tweets. In: 2021 Joint International Conference on Digital Arts, Media and Technology with ECTI Northern Section Conference on Electrical, Electronics, Computer and Telecommunication Engineering, Cha-am, Thailand, pp. 238–242 (2021)

Interpretable Adverse Event Detection Using a Neuro-Symbolic and Graph-Based Approach

Aryan Shrestha, Nikhil Yeminedi(✉), Mario M. Kubek(✉), and Armin R. Mikler(✉)

Georgia State University, Atlanta, GA, USA
{ashrestha6,nyeminedi1}@student.gsu.edu, {mkubek,amikler}@gsu.edu

Abstract. In the wake of natural disasters and other critical adverse events such as disease outbreak or train derailment, the flood of information from social media and news can be overwhelming. While modern large language models excel at rapidly classifying which of these texts are relevant, their black box nature often makes it difficult to trust their predictions. This paper introduces a novel neuro-symbolic framework designed to bridge this critical gap between performance and interpretability. Our methodology first employs a fine-tuned model to accurately filter and classify texts that describe an adverse event. Subsequently, we validate the model's predictions by looking for strong connections between disaster-related words and constructing symbolic co-occurrence graphs. Instead of only classifying texts, the approach aims to uncover and reveal why certain texts matter based on their linguistic cues. By grounding the model's prediction in human-readable patterns, the study moves a step closer to building trustworthy and reliable systems for critical decision-making.

Keywords: Neuro-symbolic Framework · Natural Language Processing (NLP) · Disaster Informatics · Text Classification · Co-occurrence Analysis · Model Interpretability · Explainable AI

1 Motivation

An adverse event is any occurrence that has a negative or harmful impact on a community, infrastructure, or the environment. It encompasses a wide range of incidents, from large-scale natural disasters (e.g., *earthquakes, floods, hurricanes*) to more localized human-caused critical situations (e.g., *accidents, explosions, disease outbreaks*). The escalating frequency and cost of natural disasters present a grave challenge to global communities and emergency response agencies. This is not a new problem; we have access to vast archives of text from past disasters, creating a rich repository of historical data. This historical data is an invaluable resource, as it contains the linguistic DNA of how crises unfold from initial warnings and eyewitness reports to damage assessments and recovery efforts.

H. Unger and P. Meesad (Eds.): NLPIR 2025, LNNS 1904, pp. 1170–1186, 2026.
https://doi.org/10.1007/978-3-032-20897-2_82

The central purpose of our research is to leverage this historical data to train an automated system that can learn these complex linguistic patterns.

The ability to automatically process this information is crucial for enabling a timely emergency response and effective mitigation efforts. A model trained on historical data learns a generalized understanding of disaster-related language. When a new crisis occurs, this pre-trained model can be deployed to analyze a real-time stream of data from social media APIs, news feeds, and other live sources at a scale and speed no human team can match. By instantly filtering relevant information and identifying key event signatures (e.g., distinguishing *flooding* from *structural collapse*), such a system can provide emergency managers with the situational awareness needed to make critical decisions in minutes, not hours. This capability is essential because in the immediate aftermath of these events, the deluge of unstructured data from sources presents a significant challenge for rapid analysis, overwhelming traditional methods in the time-critical domain of disaster management. During critical adverse events, social media platforms and news outlets become vital channels for information dissemination as they provide real-time updates before official sources can respond.

The application of Natural Language Processing (NLP) to adverse event detection aims to address several fundamental challenges. As highlighted by the aftermath of *Hurricane Helene* [1], a significant semantic gap exists between technical forecasts and public warnings, leading to devastating consequences in inland communities that misjudged their risk. A critical failure in risk communication became evident as residents in these areas evacuated at half the rate of coastal populations. During a crisis, multiple sources report on the same developments, which creates a critical need to align information, distinguish novel updates from redundant reports, and determine what key details to extract. This task is complicated by the spatio-temporal constraints inherent to disaster events, which require systems capable of such reasoning. A primary obstacle to developing these sophisticated systems is the lack of large-scale, annotated token-level datasets for specific adverse event extraction tasks. Without this granular data, training models to directly extract event parameters is not feasible. This necessitates a foundational step: the development of a highly reliable classifier. However, for such a classifier to be trusted in critical scenarios, it must also be verifiable.

Advanced NLP models, particularly those based on the Bidirectional Encoder Representations from Transformers (BERT) [2] architecture, have achieved state-of-the-art results on a wide range of classification tasks. This performance provides a strong rationale for applying them to the complex domain of disaster-related content. However, their utility in high-stakes environments is often hampered by a lack of transparency. General-purpose models, trained on broad web text, may fail to grasp the specific meanings words acquire within a specialized domain; for instance, a generic model might struggle with the term *watch* in *hurricane watch* which make its interpretation biased toward its more common use. The necessity for domain specific fine-tuning is thus evident. This need is supported by findings that models utilizing crisis domain tailored embeddings substantially outperform those relying on generic representations for the classification of informative disaster related tweets [3]. The limitation of general purpose models is thereby underscored, which aligns directly with the rationale for fine-tuning the BERT model on a dedicated adverse event corpus instead of employing its out-of-the-box, general web-trained version. These models, frequently described as black boxes, can make highly accurate predictions, but the underlying reasoning for their classifications remains opaque. This issue is magnified when models are trained on unlabeled or sparsely labeled datasets, as it becomes exceedingly difficult to independently verify their predictions and, consequently, to trust their outputs where high accountability is needed. Initiatives like *CrisisNLP* [4] create standardized benchmarks by consolidating fragmented datasets, which enables the development of robust and comparable models for disaster informatics. Our research is situated within the growing field of Explainable AI (XAI), which seeks to make these powerful models more interpretable.

To bridge this gap between performance and explainability, this paper proposes a methodology for validating a custom-trained BERT classification model by identifying and quantifying significant linguistic cues within the text it classifies. Our approach is grounded in the principle that genuine disaster-related texts will contain strong and statistically significant co-occurrences of terms describing adverse events and conditions. By extracting these co-occurring terms, we can construct a symbolic knowledge representation that provides tangible evidence to support the model's classifications, which improves model interpretability. Instead of merely classifying text, we seek to answer why a given text is considered relevant by the model. Our work builds upon previous research [5] that compared symbolic concepts, derived from automated ontology construction, with non-symbolic concepts extracted from sentence-level text analysis. However, its analysis was limited to extracting relevant terms based on simple term list overlap. That foundational work confirmed the potential of graph-based methods, but did not leverage them for model validation. Our research advances this approach by repurposing the co-occurrence graph from a simple term extraction tool into a sophisticated XAI method. We analyze the associative strength of co-occurring keywords to validate the deep contextual understanding of our chosen model. This provides a tangible solution for verifying model predictions, a crucial step toward building automated systems that can effectively address the information alignment and risk translation challenges seen in events like *Hurricane Helene*. This leads to our primary research question: Can a symbolic validation method provide interpretable evidence to confirm the deep contextual understanding of a fine-tuned neural classifier for an adverse event detection?

The following sections review the fundamental concepts, detail our methodology, present the experimental results, and discuss their implications before offering a final conclusion.

2 Technical Background

This section details the core concepts and technologies that form the basis of our neuro-symbolic framework. We explain the encoder-based architecture of our chosen language model, the principles of co-occurrence analysis [6] for symbolic knowledge extraction and the statistical measures used for validation. Together, these principles form the theoretical bedrock for the methodology and analysis presented in the following sections.

2.1 Encoder-Based Language Models

At their core, language models do not process text directly but operate on numerical data. An embedding representation is the mechanism for this translation. It is a dense vector of numbers that represents a word, phrase, or entire document. Unlike sparse representations like one-hot encoding, dense embeddings capture semantic relationships, meaning that words with similar meanings will have similar vector representations in the embedding space.

The neural component of our framework is built upon an encoder-based Large Language Model (LLM). An encoder processes an entire input text sequence simultaneously, transforming it into a rich, high-dimensional numerical representation that captures deep contextual meaning. This contrasts fundamentally with decoder-based models (e.g., GPT [7] models, Grok [8], Gemini [9]), optimized for auto-regressive tasks like sequential text generation. For classification tasks, which require a holistic understanding of the input to make a single, accurate judgment, the encoder architecture's ability to bidirectionally contextualize every token is fundamentally more suitable.

Adapting a foundational encoder-based model like BERT for a specific task, such as classification, is achieved through a process called fine-tuning. This involves adding a new, untrained layer, often called a classification head on top of the pre-trained model body. The body provides a rich, contextualized representation of the input text, and the head, typically a simple linear layer, is then trained on the task-specific dataset to map that representation to the desired output labels ("adverse event" or "not an adverse event).

The core innovation of BERT is its bidirectional context processing: it considers all words in a sequence simultaneously, which allows each word's representation to be informed by every other word, both left and right. This deep contextual understanding is crucial for disambiguating nuanced language and achieving state-of-the-art accuracy in complex classification tasks. The *bert-base-uncased* variant offers a balance of high performance with reasonable computational demands [10]. While lighter models like *DistilBERT* [11] offer speed advantages, they often sacrifice predictive power. Conversely, larger models like *RoBERTa* [12] incur significantly higher resource costs without proportional gains justifying the overhead for many tasks and dataset scales.

2.2 Symbolic Knowledge via Co-occurrence Graphs

The symbolic component of our framework is rooted in the principle of distributional semantics, which posits that words appearing in similar contexts are likely to have related meanings. This is a cornerstone of corpus linguistics and information retrieval. Co-occurrence graphs are versatile and can be constructed in several ways, such as word-document graphs for topic modeling or feature-feature graphs for linguistic pattern analysis. Here, we specifically construct a word-word co-occurrence graph, as our goal is to map the direct semantic relationships between key terms related to adverse events. The structure of a co-occurrence graph is defined as follows:

- Nodes represent unique terms from our curated lexicon. A lexicon is a collection of words strongly associated with a specific domain, in this case, adverse events. It helps NLP systems to identify the relationships and associations between words.
- Edges represent the co-occurrence relationship between two nodes. The weight of an edge is not merely a frequency count but is quantified by statistical metrics that calculate the strength of the association between the connected terms.

This co-occurrence graph provides a structured, transparent map of the key semantic relationships present in the texts the neural model classified as relevant, making implicit knowledge explicit and verifiable.

2.3 Co-occurrence Significance Metrics

To measure the associative strength of a word pair ($term1$, $term2$), we use a 2 × 2 contingency table which captures their occurrence frequencies:

- a: The number of times $term1$ and $term2$ co-occur.
- b: The number of times $term1$ occurs without $term2$.
- c: The number of times $term2$ occurs without $term1$.
- d: The number of times neither $term1$ nor $term2$ occurs.
- N: The total number of observations in the corpus $(a + b + c + d)$.

From these values, we calculate the following metrics:

- **Log-Likelihood Ratio (LLR):** It is a statistical significance test that measures confidence that observed co-occurrences are not random, with higher values indicating stronger relationships [13]. Particularly effective for identifying significant word pairs involving rare terms, the LLR is calculated from contingency table values as:

$$\begin{aligned} \mathrm{LLR}(term1, term2) = 2 \times \big[& a \cdot \log(a) + b \cdot \log(b) + c \cdot \log(c) + d \cdot \log(d) \\ & - (a+b) \cdot \log(a+b) - (a+c) \cdot \log(a+c) \\ & - (b+d) \cdot \log(b+d) - (c+d) \cdot \log(c+d) \\ & + N \cdot \log(N) \big] \end{aligned} \tag{1}$$

- **Pointwise Mutual Information (PMI):** It measures how much more likely two terms are to co-occur than if they were independent [14]. A higher PMI score indicates a stronger semantic association. It is calculated as:

$$\mathrm{PMI}(term1, term2) = \log\left(\frac{a \times N}{(a+b)(a+c)}\right) \tag{2}$$

- **Sørensen-Dice Coefficient (DSC):** It measures the normalized overlap between the occurrences of two terms [15]. It is particularly useful for identifying pairs that frequently appear together relative to their individual frequencies. The DSC is computed as:

$$\text{DSC}(term1, term2) = \frac{2 \times \text{a}}{(a+b)(a+c)} \tag{3}$$

Negative or low values for these metrics would indicate weak associations, term avoidance, or random co-occurrences that do not provide evidence supporting why the model classified texts as adverse events, thus failing to serve our XAI validation purpose.

3 Methodology

This study employs a neuro-symbolic framework that uses the neural model as a semantic filter to build a focused corpus containing only disaster-related texts from which we then extract interpretable, evidence-based knowledge.

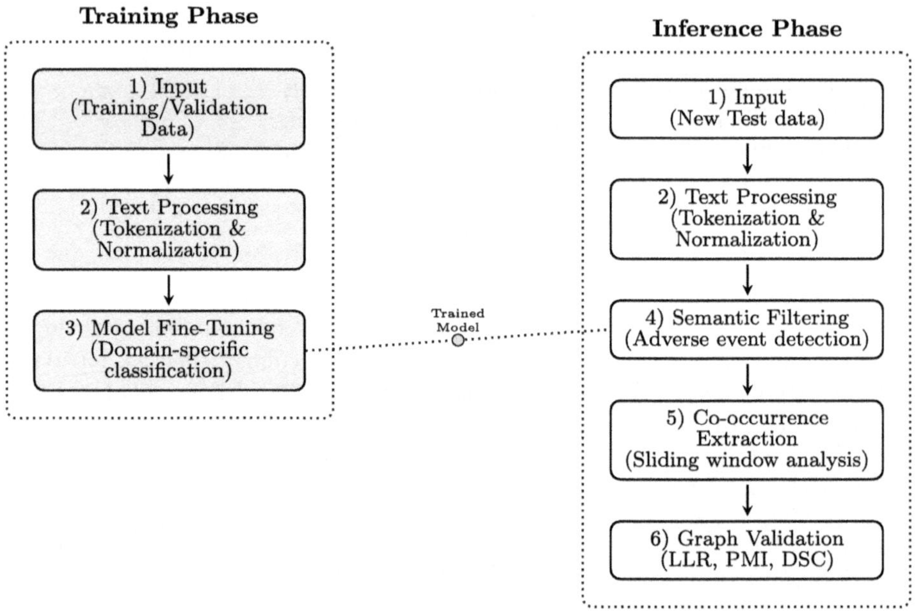

Fig. 1. Neuro-Symbolic Framework Workflow with Training and Inference Phases

Our methodology follows a sequential pipeline as shown in Fig. 1 designed to move from raw data to interpretable validation. The process begins with the initial text processing of the input corpus, which includes standard tokenization and normalization. This pre-processed data is then used for the model fine-tuning stage, where the model is trained specifically for the adverse event classification task. Once trained, the model performs semantic filtering, classifying the entire dataset to create a refined corpus containing only disaster-related texts. From this filtered corpus, we conduct co-occurrence extraction using a sliding window analysis to identify pairs of adverse terms. The final stage is the graph-based validation, where statistical metrics are applied to these pairs to quantify their associative strength and build the symbolic knowledge graph.

3.1 The Neuro-Symbolic Framework

The proposed pipeline consists of two primary stages: semantic filtering and symbolic validation. This framework is designed to first leverage the high accuracy of a neural classifier and then validate its reasoning through an interpretable symbolic analysis.

3.1.1 Semantic Filtering

The primary semantic filter is a neural classifier, which is the first component of the pipeline. Through this step, a refined corpus is created, consisting exclusively of texts deemed relevant by the model, upon which the symbolic validation is performed. The model deems a text relevant if the final classification layer outputs a probability score for the adverse event class that exceeds a predefined threshold. Only the texts classified as an adverse event (class 1) are retained for the next stage of the pipeline. Efficiency is increased in this stage as the dataset size is reduced and non-adverse events are eliminated, so the semantic focus is kept on only adverse events.

3.1.2 Symbolic Validation via Co-occurrence Graph Analysis

The symbolic validation stage is designed to construct a human-interpretable representation of the knowledge contained within the classifier's selected texts. The validation process for the refined corpus is conducted in these steps:

- **Adverse Term Identification via Lexicon:** A manually curated lexicon of adverse terms is the cornerstone of the symbolic analysis. The motivation for using a lexicon is threefold. First, the analysis is grounded in domain knowledge, ensuring that the validation is based on terms already recognized by experts as meaningful. The resulting symbolic graph is thereby made immediately interpretable and the black box problem is directly addressed. Second, semantic noise is filtered; social media text is filled with high-frequency but irrelevant words, and the analysis is focused exclusively by the lexicon on the relationships between significant disaster-related terms. Finally, computational tractability is ensured by the drastic pruning of the search space of possible word pairs, making the graph generation fast and efficient.
  To ensure methodological rigor and prevent data leakage, this lexicon was derived exclusively from the 75% training portion of the dataset. The process

involved isolating all disaster-related tweets in the training set, performing Part-of-Speech (POS) tagging to extract all nouns and adjectives, and conducting a frequency analysis. From this frequency list, the final lexicon was manually curated by selecting the most frequent and semantically unambiguous terms that directly described an adverse event or its characteristics. The refined corpus is parsed and all instances of words that appear in the adverse term lexicon are identified.

- **Contextual Co-occurrence Pairing:** For capturing context, a sliding window approach was chosen to identify the most immediate and powerful semantic relationships. While the neural model can capture long-range dependencies across an entire text, this symbolic method is designed to extract strong, local evidence by focusing on terms that appear in close proximity. These tight term associations serve as an unambiguous, human-interpretable signatures that validate the model's broader contextual understanding.

The result of this analysis is a comprehensive frequency count for every unique pair of co-occurring adverse terms. For each occurrence of a lexicon term, the presence of other lexicon terms within the window's context is checked.

4 Experiments

4.1 Dataset

The study utilizes a publicly available corpus of 163,718 English-language tweets annotated for disaster-related content [16]. Each tweet is labeled in a binary classification structure as either disaster-related or not. The corpus is diverse, encompassing a range of natural disasters and human-caused incidents. To ensure uniformity for model training, all text was converted to lowercase. For reproducibility and evaluation, the dataset was partitioned using stratified sampling into a 75% training set (122,788 tweets) and a 25% hold-out test set (40,930 tweets).

4.2 Experimental Setup

4.2.1 Neural Classifier Implementation

The neural classifier was implemented using the *BertForSequenceClassification* [17] class from the Hugging Face *transformers* library [18]. This class conveniently combines the pre-trained *bert-base-uncased variant* model with a sequence classification head, which is a linear layer placed on top of the final hidden-state vector. During the fine-tuning process, the weights of this classification layer are trained to map the input vector to the binary output classes, while the pre-trained weights of the main BERT model are also updated.

- **Training:** The model was fine-tuned on the training set of 122,788 tweets using a CUDA-enabled GPU. The training process ran for 4 epochs with an AdamW optimizer (learning rate: 2e-5) and a batch size of 32. Early stopping was monitored based on validation loss to select the best-performing model checkpoint and prevent overfitting.

- **Inference:** After training, the optimized model was used to perform inference on the hold-out test set of 40,930 tweets.

The model was fine-tuned with the goal of specializing it to identify the linguistic patterns characteristic of adverse events in our dataset.

4.2.2 Baseline Implementation

For our baseline comparison, we selected spaCy [19], as it represents a robust, production-ready NLP library commonly used for entity extraction. We specifically chose the *en_core_web_trf* model, as it is the only model in the library that includes an "EVENT" tag in its pre-trained Named Entity Recognition (NER) component. The workflow for this baseline differed significantly from our fine-tuned approach, primarily in the semantic filtering stage. Instead of learning from domain-specific data, the spaCy approach relied on its pre-trained NER for syntactic analysis, labeling a text as an adverse event if at least one "EVENT" entity was detected. This syntactic, entity-based filtering method was followed by the same co-occurrence extraction and graph-based validation steps used in the main pipeline to allow for a direct comparison of the resulting symbolic knowledge.

4.2.3 Symbolic Analysis Implementation

A manually curated lexicon of 107 adverse terms (52 nouns: hurricane, flood, casualty; 55 adjectives: severe, devastating, imminent) was used. A sliding window with a fixed size of 5 was used to capture contextual co-occurrences. Co-occurrence pairs were sorted alphabetically to treat (a, b) and (b, a) as a single pair, and self-co-occurrences were excluded. The statistical metrics (LLR, PMI, DSC) were implemented using custom Python scripts leveraging the standard *math* and *collection* libraries.

4.3 Preliminary Results

The initial and most critical stage of our framework is the accurate classification of texts. As shown in Table 1, the results reveal a stark performance difference on the test set, with the custom-trained BERT model decisively outperforming the baseline across all metrics.

Table 1. Comparative Performance of Classification Models

Model	Accuracy	Precision	Recall	F1 Score
Fine-tuned BERT	**99.00%**	**98.18%**	**98.26%**	**98.22%**
spaCy Baseline	55.84%	92.22%	12.75%	22.40%

Our fine-tuned BERT model achieved 98% accuracy during training and it reached a 99% accuracy rate on the test set. Its high F1-score of 98.22% indicates a balanced and robust ability to identify adverse events. Most importantly, the model's recall of 98.26% highlights its capacity to correctly identify the vast majority of relevant posts, a critical function in emergencies where missing information can have severe consequences. This high precision and recall establish the BERT model as a reliable and effective semantic filter which ensures a high-quality corpus for the subsequent symbolic analysis.

In contrast, the spaCy baseline's performance was poor, achieving only 55.84% accuracy. This was primarily due to a notable failure in recall, which was a mere 12.75%. While its precision appears high at 92.22%, this figure is misleading; it means that while spaCy was often correct when it did identify an event, it failed to find the vast majority of them. This poor performance suggests its reliance on syntactic entity recognition is unsuitable for this semantic task and it is unreliable for any critical real-world system.

5 Results

5.1 Symbolic Co-occurrence Analysis

The BERT-based pipeline identified more than twice the number of unique term relationships (311 vs. 141) and the LLR scores of its top pairs are substantially higher, indicating more statistically powerful connections. A direct comparison of the raw co-occurrence counts for the top adverse term pairs, shown in Table 2, reveals the difference in the models' ability to identify relevant contexts.

Table 2. Comparative Co-occurrence Counts for Top Adverse Term Pairs

Term1	Term2	BERT-derived co-occurrence count	spaCy-derived co-occurrence count
earthquake	powerful	156	2
hurricane	storm	134	10
devastating	earthquake	94	0
earthquake	massive	94	64
disaster	flood	86	34
explosion	massive	72	60
bombing	explosion	74	54
deadly	monsoon	46	22
flood	flooding	74	10
flooding	monsoon	44	32

The BERT-filtered corpus consistently yields a much higher volume of meaningful co-occurrences. For critical pairs like (*devastating, earthquake*) and (*earthquake, powerful*), the spaCy baseline almost completely fails to identify these relationships due to its poor recall. This failure to build a comprehensive corpus of adverse events means that any subsequent statistical analysis, regardless of its sophistication, is performed on a small and unrepresentative sample of the data. Our fine-tuned BERT model, by contrast, identifies these pairs orders of magnitude more frequently, providing a rich and reliable foundation for the statistical validation that follows. The subsequent statistical analysis, shown in Tables 3and 4, confirms this initial finding.

Table 3. Top 10 Adverse Co-occurrences from BERT-filtered Texts (Ranked by LLR)

Term1	Term2	Co-occurrence	PMI	DSC	LLR
earthquake	powerful	156	5.94	0.14	1803.20
hurricane	storm	134	4.96	0.10	1132.91
devastating	earthquake	94	5.61	0.09	956.35
earthquake	massive	94	5.42	0.09	900.62
disaster	flood	86	5.28	0.11	774.43
explosion	massive	72	5.99	0.14	760.88
bombing	explosion	74	5.40	0.12	677.19
deadly	monsoon	46	7.44	0.26	631.80
flood	flooding	74	4.79	0.09	581.73
flooding	monsoon	44	7.04	0.19	564.40

The BERT-derived pairs are specific and contextually relevant to adverse events. Pairs like (*earthquake, powerful*), (*devastating, earthquake*), and (*deadly, monsoon*) demonstrate that the model learned to connect events with descriptive adjectives, capturing the severity and impact of the situation. This indicates a deep, nuanced understanding of the language used in disaster reporting.

Table 4. Top 10 Adverse Co-occurrences from SpaCy-filtered Texts (Ranked by LLR)

Term1	Term2	Co-occurrence	PMI	DSC	LLR
earthquake	tsunami	102	5.85	0.15	1160.75
earthquake	massive	64	5.78	0.10	703.66
explosion	massive	60	6.20	0.14	701.71
bombing	explosion	54	5.22	0.11	479.02
flooding	monsoon	32	7.26	0.22	440.49
devastating	tornado	36	6.38	0.13	421.19
disaster	flood	34	6.16	0.16	367.21
deadly	derailment	26	7.06	0.25	330.49
deadly	monsoon	22	8.01	0.36	329.46
deadly	explosion	34	5.54	0.08	327.05

However, the spaCy-derived pairs, while containing relevant pairs like (*earthquake, tsunami*), are fundamentally undermined by the model's poor recall. Because the spaCy model failed to identify the majority of adverse events, this list is derived from a small, incomplete, and unrepresentative sample of the data. Its inability to build a comprehensive corpus means the resulting symbolic analysis, while showing some valid connections, cannot be trusted to reflect the full picture of the event.

5.2 Co-occurrence Graph Visualization

To provide a qualitative, interpretable view of the symbolic knowledge captured by each model, we generated co-occurrence graphs for the top 30 pairs (ranked by LLR) from each filtered corpus as shown in Fig. 2 and Fig. 3. In these graphs, nodes represent adverse terms, and the edges signify a strong statistical co-occurrence. The size of a node reflects its importance (number of connections), while the edge thickness indicates the strength of the association between terms.

The resulting visualizations reveal a dramatic difference in the structure and coherence of the knowledge extracted by each model. The BERT-induced co-occurrence graph displays a dense, interconnected, and meaningful network. It features several distinct event hubs centered around key terms like *earthquake*, *storm*, and *explosion*. This clustered structure demonstrates that the model has learned a nuanced map of how different components of a disaster are linguistically related. The connections exhibit a high degree of semantic specificity forming clear event signatures. For instance, *earthquake* is strongly linked to descriptive attributes like *powerful* and *massive*, while *bombing* is exclusively linked to *explosion* which indicates the model learned a specific, hierarchical relationship rather than just a simple co-occurrence. This ability to move beyond simple keyword matching to a deeper contextual understanding is what allows our BERT model to construct such a rich and coherent symbolic knowledge graph.

On the other hand, the spaCy-induced co-occurrence graph fails to build a cohesive, interconnected network, instead appearing as a collection of smaller, disconnected clusters and simple pairs. While it successfully identifies valid, high-frequency pairs such as (*earthquake, tsunami*) and (*explosion, massive*), its structure reveals a reliance on general association rather than specific understanding. The lack of central, organizing hubs for specific event types is telling. Instead, the graph is heavily influenced by a general keyword hub around the term *deadly*. The graph appears as a collection of smaller, disconnected clusters and simple pairs. This suggests that the baseline model is proficient at identifying a general disaster vocabulary but struggles to resolve it into the fine-grained, specific event types that our BERT model correctly maps.

This visual evidence supports our quantitative findings. Our BERT model's contextual understanding provides the high-quality filtered text necessary to construct a richer and more coherent symbolic knowledge graph.

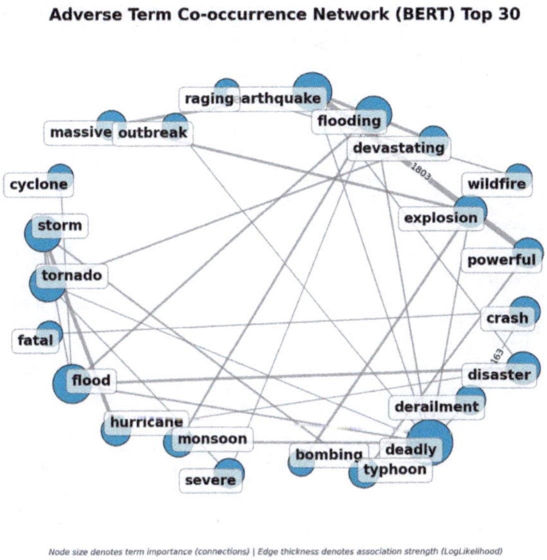

Fig. 2. BERT-induced co-occurrence graph

6 Discussion

This section delves deeper into the practical implications of these findings through a specific case study, explores the broader applications of this neuro-symbolic methodology, and discusses potential avenues for future research.

To illustrate the practical difference in the models' semantic understanding, we can examine how they handle tweets that are topically similar but differ in linguistic complexity. Consider the following two examples from the dataset:

- **Tweet A:** *"hurricane sandy is larger than the state of texas and could end up being the largest storm to hit the united states"*
  - True Label: Adverse Event
  - BERT Prediction: Adverse Event
  - spaCy Prediction: Adverse Event
- **Tweet B:** *"irma headlines: hurricane watch likely today. time period: friday pm - monday am threats: winds, storm surge, flooding"*
  - True Label: Adverse Event
  - BERT Prediction: Adverse Event
  - spaCy Prediction: Not an Adverse Event

Both the generalized spaCy model and the fine-tuned BERT model correctly identify Tweet A as it is a direct, declarative statement containing the high-value keywords *"hurricane"*, *"storm"* and *"hit"* which is a clear sign of an adverse event. However, Tweet B reveals a more complex challenge that exposes the limitations of the baseline model. This tweet is not a simple narrative; it's a

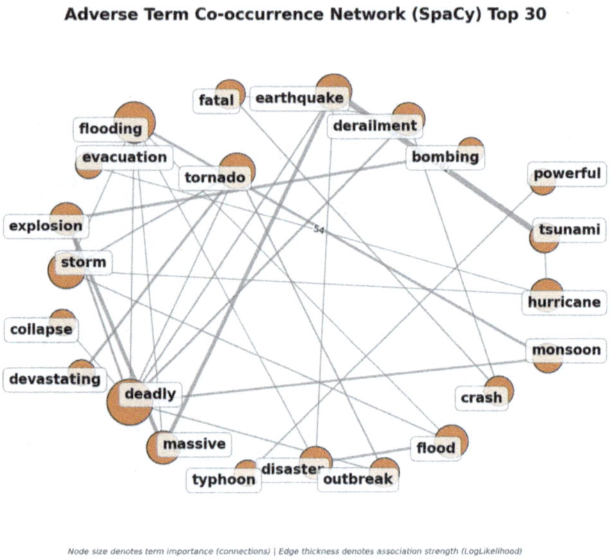

Fig. 3. SpaCy-induced co-occurrence graph

structured forecast, presented as a list of information. spaCy likely processes the tokens individually (*"irma"*, *"hurricane"*, *"watch"*, *"threats"*, *"flooding"*) but fails to understand the overall semantic structure that makes this a high-priority warning. It sees the parts but misses the crucial context that connects them into a coherent alert.

Our fine-tuned BERT model correctly classifies this tweet. Its success stems from its training on thousands of disaster-related texts, where it has learned to recognize the entire pattern of a *"hurricane watch"* announcement as a significant event. Its attention mechanism can link the named entity (*"irma"*) to the concept (*"hurricane"*) and connect both to the list of threats. It understands that this specific linguistic structure signifies an imminent and important event. This capability is directly supported by our quantitative data. Table 3 shows an extremely strong and statistically significant association for the pair (*hurricane, storm*) with an LLR of 1132.91, the second-highest in our analysis. This provides statistical proof that BERT has learned that these terms are powerfully linked, allowing it to correctly interpret the structured warning in Tweet B where spaCy fails. This case study demonstrates that BERT's performance is not just about finding more keywords; it's about understanding the specific linguistic patterns, even non-narrative ones, that define a critical adverse event.

Furthermore, our choice of an encoder-based model was deliberate for this classification task. An alternate scenario using a decoder-based model would likely yield different and potentially inferior results for classification. Decoder models are architecturally optimized for generation, processing text sequentially to predict the next word. While they can be adapted for classification, their unidirectional context is a fundamental limitation for tasks requiring a holistic understanding of the entire input. We would anticipate lower classification accuracy, as the model could not weigh the importance of all words equally. However, this limitation opens avenues for future research; a decoder-based model could be used for generative data augmentation to create new training examples or to summarize filtered adverse events.

It is important to acknowledge the limitations of this study. The curated adverse term lexicon, while effective for grounding the analysis, is inherently static. It may fail to capture novel or evolving language used during unforeseen events. Instead of a static lexicon, future work could involve dynamic lexicon generation, where the system automatically identifies and incorporates new, relevant terms (e.g., slang, location-specific names) during a crisis. Secondly, our co-occurrence analysis relies on a simple sliding window, which is effective at capturing strong, local event signatures but cannot capture the long-range semantic dependencies that the BERT model itself understands. Future work could explore a document-level co-occurrence model with a larger sliding window, where any two lexicon terms appearing in the same article are considered a pair, regardless of their distance. A significant next step is to extend this framework to token-level classification to identify and label specific words or phrases within the text. Future work may include new sophisticated methods other than graphs for validating our model's predictions.

The neuro-symbolic framework presented in this paper has extended implications that can be applied to enhance real-world disaster management systems. The validated model and the knowledge graphs can be cornerstone components of real-time decision support systems for emergency response agencies. Imagine a dashboard that ingests live social media and news feeds, using our fine-tuned model to filter for relevant events and then displaying the corresponding co-occurrence graphs. This would allow an analyst to not only see that an event is happening but to immediately understand what kind of event it is (e.g., a *"train derailment"* vs. a *"chemical spill"*) based on the symbolic graph. This directly addresses the information alignment and risk translation challenges by providing rapid, interpretable, and evidence-based situational awareness. This methodology could also be adapted to other domains requiring critical text analysis, such as public health surveillance (tracking disease outbreaks from public posts) or supply chain risk management (detecting disruptions from news feeds).

7 Conclusion

In this paper, we introduced a neuro-symbolic framework to address the critical need for both accuracy and interpretability in adverse event text analysis. We successfully demonstrated that a fine-tuned BERT model can achieve state-of-the-art performance, however, our core contribution was not the classification itself, but the validation of its underlying reasoning.

By translating the opaque operation of the classification model into transparent co-occurrence graphs, we provided tangible, evidence-based proof that its accuracy stems from a deep understanding of meaningful semantic relationships, not superficial keyword matching. This research presents a practical method for bridging the gap between the predictive power of black box models and the demand for accountability in real-world scenarios. The proposed framework offers a clear pathway toward developing reliable AI systems for disaster management by building automated systems that not only tell us *what* is happening, but provide the evidence to help us understand *why*.

Code Availability. Code is available at: https://github.com/aryan-s1/Interpretable-Adverse-Event-Detection.git.

References

1. Kaplan, S., Crowe, K., Ahmed, N., Noll, B.: A new era of inland flooding has arrived: devastation in western N.C., Texas Hill Country is signal to rethink disaster plans, experts say (2025). The Washington Post, July 21, 2025, A1, A6–A7 (2025)
2. Devlin, J., Chang, M.-W., Lee, K., Toutanova, K.: BERT: pre-training of deep bidirectional transformers for language understanding. In: Proceedings of NAACL-HLT, pp. 4171–4186 (2019). https://arxiv.org/abs/1810.04805. Accessed 21 Feb 2026
3. Caragea, C., Silvescu, A., Tapia, A.H.: Identifying informative messages in disaster events using convolutional neural networks. In: Proceedings of the International Conference on Information Systems for Crisis Response and Management (ISCRAM)
4. Imran, M., Mitra, P., Castillo, C., Alam, F., Ofli, F.: CrisisNLP: resources for research on crisis informatics topics. https://crisisnlp.qcri.org/. Accessed 21 Feb 2026
5. Al Subhi, S.N.S., Mikler, A.R., Kubek, M.M.: Constructing co-occurrence graphs and deriving flood ontologies for enhanced understanding. In: 2024 Second International Conference on Inventive Computing and Informatics (ICICI), Bangalore, India, pp. 1–6 (2024). https://doi.org/10.1109/ICICI62254.2024.00009
6. Kubek, M.: Concepts and Methods for a Librarian of the Web. Springer, Cham, Switzerland (2020). https://doi.org/10.1007/978-3-030-23136-1
7. OpenAI: GPT models. https://platform.openai.com/docs/models. Accessed 21 Feb 2026 (2025)
8. xAI: Grok AI (2025). https://x.ai/. Accessed 21 Feb 2026
9. Google DeepMind: Gemini AI (2025). https://deepmind.google/technologies/gemini/. Accessed 21 Feb 2026
10. Heise online. Wie BERT Texte versteht: Bidirektionale Kontextverarbeitung in Transformern. iX – Magazin für professionelle Informationstechnik, 13/2023, 86–89 (2023). https://www.heise.de/select/ix/2023/13/2302013462923862371. Accessed 1 Mar 2023
11. Sanh, V., Debut, L., Chaumond, J. Wolf, T.: DistilBERT, a distilled version of BERT: smaller, faster, cheaper and lighter. arXiv preprint. https://arxiv.org/abs/1910.01108
12. Liu, Y., et al.: RoBERTa: a robustly optimized BERT pretraining approach. arXiv preprint. https://arxiv.org/abs/1907.11692
13. Dunning, T.: Accurate methods for the statistics of surprise and coincidence. Comput. Linguist. **19**(1), 61–74 (1993). https://aclanthology.org/J93-1003/

14. Biemann, C., Heyer, G., Quasthoff, U.: Wissensrohstoff text: eine Einführung in das text mining. 2nd ed. Springer Vieweg, Wiesbaden, Germany (2023). https://doi.org/10.1007/978-3-658-35969-0
15. Dice, L.R.: Measures of the amount of ecologic association between species. Ecology **26**(3), 297–302 (1945)
16. Wiegmann, M., Kersten, J., Klan, F., Potthast, M., Stein, B.: Disaster tweet corpus 2020. Zenodo (2020). https://doi.org/10.5281/zenodo.3713920
17. Hugging face: BERT for sequence classification documentation. Hugging face Inc. (2024). https://huggingface.co/docs/transformers/en/model_doc/bert
18. Wolf, T., et al.: Hugging face's transformers: state-of-the-art natural language processing (2019). https://huggingface.co/docs/transformers/index. Accessed 21 Feb 2026
19. Honnibal, M., Montani, I.: spaCy: industrial-strength natural language processing (2023). Explosion AI. https://spaCy.io/models/en#en_core_web_trf

Development and Large Language Model Evaluation on Fil-NER: A Named Entity Recognition Annotated Filipino Dataset

Rachel Edita Roxas[1], Raquel E. Sison-Buban[2], Maria Regina Justina E. Estuar[3](✉), Lance Calvin L. Gamboa[3,5], Jose Mari Luis Dela Cruz[3], Juanito N. Anot Jr.[4], Lynette V. Mandap[2], Jorge P. Cuibillas[4], William Chandra Tjhi[6], Wei Qi Leong[6], and Jann Railey Montalan[6]

[1] University of the Philippines Los Baños, Laguna, Philippines
[2] De La Salle University, Manila, Philippines
[3] Ateneo de Manila University, Quezon City, Philippines
restuar@ateneo.edu
[4] Far Eastern University, Manila, Philippines
[5] University of Birmingham, Birmingham, UK
[6] AI Singapore, Singapore, Singapore

Abstract. To date, there are only a few fully documented and publicly available Filipino named entity recognition datasets. This study presents findings on the development of Fil-NER, a Filipino Named Entity Recognition annotated dataset, its corresponding Fil-NER Coding Manual and its evaluation on Large Language Models. The dataset contains 10,049 Filipino sentences with temporal information, formality classification, and domain classification. The inter-annotator agreement (IAA F-measure) scores among the three (3) annotators for strict, exact, partial, and type agreement definitions show an average of .77, .82, .88 and .89 respectively. The parameters for model evaluation experiments included varying parameter sizes, inclusion of commercial AI systems. The models were evaluated against three Filipino NER datasets, namely: TLUnified-NER [21], UNER [22] and Fil-NER (our contribution). Results show the Fil-NER is a sufficiently difficult task for Filipino NER with GPT-4o, scoring 84.23 on accuracy against an average accuracy of 75.01 across all other models. Fil-NER and its corresponding annotation guide will be publicly available for further use in downstream NLP tasks and evaluation in AI language model development.

Keywords: Filipino NER annotated dataset · evaluation in large language models

1 Introduction

1.1 Digital Transformation

The 4th industry revolution has brought about the prevalence of technologies such as use of artificial intelligence, IoT devices, and autonomous vehicles [3]. As such, in lower middle income countries such as the Philippines, digital transformation becomes

H. Unger and P. Meesad (Eds.): NLPIR 2025, LNNS 1904, pp. 1187–1205, 2026.
https://doi.org/10.1007/978-3-032-20897-2_83

a competitive edge to businesses [3]. In particular, these new technologies neccesitate that micro, small, and medium enterprises (MSMEs), conduct business online through ICT such as applications including E-Commerce and e-Payment systems [2]. Language plays a vital role in the conduct of such online presence, especially because most of the online markets engage in social media. With the Philippines as one of the top social media users [1] opens doors for high adoption of digital transformation. However, existing natural language processing tools for Philippine languages are inadequate due to the lack of annotated datasets, especially for downstream NLP tasks such as named entity recognition.

1.2 The Filipino Language

Filipino is the Philippines' national language, used in speech and written form in urban Philippines. "As it evolves, it shall be further developed and enriched on the basis of existing Philippine and other languages" (Article 14, Sect. 6, the 1987 Philippine Constitution). "Like any living language, Filipino undergoes a process of development through borrowing from Philippine languages and non-native languages, as well as the evolution of various language varieties for different social situations, for its speakers who come from diverse social backgrounds, and for topics of discussion and expressive communication" (Resolution 96-1, Commission on Filipino Language). Shachter (1972)[6] named Filipino as Educated Tagalog, and mostly follows the Tagalog language form and structure. Filipino and Tagalog are used interchangeably in this paper, unless otherwise stated.

1.3 Filipino Corpora

NLP work on majority languages such as English have been wide and varied due mainly to the availability of data on these languages. Meanwhile, NLP research on minority languages have been lagging behind due to scarcity of data of these "low-resourced languages" [13]. In the case of Filipino, existing datasets have been focused on limited domains and only for specific purposes, such as the works of [18–20]. These offer limited or lack of documentation [13, 14].

Some challenges in NLP work on Philippine languages include: a focus system, rich morphological characteristics (with agglutinative and inflectional language behaviors), certain level of free word order [14], and adoption of loanwords from English, Spanish, Indonesian and even, Chinese. In the colloquial Philippine languages, code-switching is prevalent.

Until recently, some studies have focused on addressing dataset and benchmark concerns. Open Super-Large Crawled Aggregated coRpus (OSCAR) is a pre-trained language model for Tagalog using multi-source data of 1.3G [14]. OSCAR has 3.5M Tagalog lines [9, 12], "WikiText-TL-39" Wikipedia with 170K documents and 39M tokens [18], and 396K online news reports. Other possible sources of Tagalog datasets (as considered by [11] in their IndoNLU work) include Common Crawl [15] and Twitter data.

Although annotation has been known to be labor-intensive, it is still recommended to address these challenges and surmount the challenges in the development of downstream Filipino NLP tasks, such as Named Entity Recognition. It has been known that dataset creation for specific NLP downstream tasks and for limited domain is effective [17].

It can be noted that NLP work has progressed in some other ASEAN languages such as in Indonesia through the IndoNLU [11], and in Vietnamese [10]. This study aims to contribute to Filipino NLP Work by providing a publicly-available Filipino dataset that has been developed for the purpose of downstream NLP tasks such as Named Entity Recognition.

1.4 Filipino NER Datasets

To date, there are only a few documented and publicly-available NER datasets in Filipino.

TLUNIFIED-NER[1] [21] is composed of texts from TLUNIFIED [23] that were manually annotated by native speakers for the NER task. Three entity types were used for annotation, namely person (PERSON), location (LOC), and organization (ORG).

The second notable NER dataset is UNIVERSAL NER or UNER[2] [22], whose Filipino subset is composed of annotated treebanks from the Universal Dependency project [24]. Specifically, the UGNAYAN [16] and TRG [25] treebanks were annotated by native speakers for PERSON, LOC, and ORG entity types. However, the Filipino split of the UNER dataset notably does not contain any instances for the ORG entity type.

1.5 Objectives of the Study

The objective of the study is two-fold. First, we perform named entity recognition annotation on a publicly-available Tagalog dataset purposely collected, and structured for the use in natural language processing and artificial intelligence. Its by product is a Filipino Named Entity Recognition coding manual for the Filipino language. Secondly, we evaluate the NER annotated dataset against available NER datasets on its level of difficulty for select Large Language Models.

2 Methodology

2.1 Dataset Creation

The corpus for the Filipino Annotation Project targeted at least 10,000 sentences divided into four- to five-sentence paragraphs for the purpose of named entity recognition annotation task. Sentence and paragraph selection followed the following selection criteria: data license (Table 1), count and percentages per data source (Table 2), language classification, temporal information (Table 3), formal and informal classification (Table 4), and controlling, semi-controlling, and non-controlling domain classification (Table 5).

[1] https://huggingface.co/datasets/ljvmiranda921/tlunified-ner.
[2] https://huggingface.co/datasets/universalner/universal/_ner.

The licensing considered in the data collection include only those under a copyleft, permissive, or public domain license (shown in Table 1). This criterion ensures that once released, the corpus can be used for other natural language processing projects on Tagalog.

The language classification criterion acknowledges the nature of the Filipino language as a combination of Tagalog, Spanish, English, Philippine regional languages, and languages from countries with historical interactions with the Philippines (e.g., China). The textual entry should have been written on or after the year 1987, which is when Filipino was formally recognized and institutionalized as the Philippines' national language thus, capturing the contemporary way Filipino is spoken and written.

For the formal and information classification, half of the entries should come from formal contexts such as academic and educated colloquialism, speeches in formal events, lectures, and official documents, while the other half should come from informal contexts such as conversation, slang, everyday conversations, and mass media.

The majority of the three NER annotators should agree that the textual entry is comprehensible enough to be annotated appropriately. Examples of textual paragraphs that may not reach the annotators' standard of comprehensibility are: those written with egregiously incorrect spellings and/or grammar structures and those that require additional contextual details from paragraphs that could not be included due to non-fulfillment of the other criteria.

From existing datasets, a new dataset was collected. Table 1 presents the data sources. The Official Gazette of the Philippines include speeches, documents, and correspondences from the Philippines' government institutions. Wikipedia, Wiktionary, and Wikibooks are articles and entries from the Tagalog branch of the listed websites. OSCAR 22.01 include Common Crawl data collected between November and December 2021. Pinoy Weekly include news articles covering sociopolitical issues relevant to Philippine society. And the Pinoy Exchange include posts and replies from an online forum developed for Filipino users.

Table 1. Licensing Information of Data Sources.

Data Source	License Status	Reference
Official Gazette of the Philippines	permissive	officialgazette.gov.ph
Wikipedia	permissive	Wikipedia's CC BY-SA licensing
Common Crawl	permissive	CC0-1.0 license (public domain)
Pinoy Exchange	Copyleft	Permission granted (Co, et al., 2022)
Pinoy Weekly	permissive	Conforme with Pinoy Weekly

2.2 Development of the Filipino NER Coding Manual

The Filipino NER Coding Manual adapts the OntoNotes Named Entity Recognition Guidelines version 7.0 [4], except when new labels have been added to capture devi-

Table 2. Sentence Count and Percentage of Data Sources.

Data Source	Sentence Count	Percentage
Official Gazette of the Philippines	2503	24,91
Wikipedia	2506	24.94
Common Crawl	2008	19.98
Pinoy Exchange	521	5.18
Pinoy Weekly	2511	24.99
Total	10049	100.00

Table 3. Sentence Count and Percentage by 5-year period.

Year	Sentence Count	Percentage
1987–1991	26	0.26
1992–1996	0	0.00
1997–2001	31	0.31
2002–2006	399	3.97
2007–2011	2675	26.62
2012–2016	6918	68.84
Total	10049	100.00

Table 4. Sentence Count and Percentage by Formality Domain.

Domain	Sentence Count	Percentage
Formal	5009	49.85
Nonformal	5040	50.15
Total	10049	100.00

Table 5. Sentence Count and Percentage by Domain

Domain	Sentence Count	Percentage
Controlling	5014	49.90
Semi-controlling	4514	44.92
Non-controlling	521	5.18
Total	10049	100.00

ations in the Filipino language behaviors using [6,7], and QA and MUC categories. Language characteristics are also presented (Table 6).

2.3 Annotation Process for Filipino NER

The Filipino NER Annotation task was performed by three (3) language annotators using the Datasaur system and validated by three (3) quality assurance persons. The inter-annotator agreement scores (or IAA) and quality assurance ratings are computed (see Table 7) based on [8]. This report computes IAA using the F-measure metric, as it has been adjusted for information retrieval tasks such as named entity recognition [5].

Table 6. Language characteristics.

Language Characteristics	Percentage
Mean percentage of English words	2.91
Mean percentage of non-English words	97.09
Mean count of code switches	5.61
Mean probability text is Tagalog	99.85
Total	10049

Table 7. Inter-Annotator Agreement Definitions.

Agreement Type	Span Agreement	Entity Agreement
Strict	Annotators tagged **exactly the same** spans.	Annotators chose the **same entity** types.
Exact	Annotators tagged **exactly the same** spans.	Annotators chose the **different entity** types.
Partial	Annotators tagged spans that **overlap partially**.	Annotators chose the **same entity** types.
Type	Annotators tagged spans that **overlap partially**.	Annotators chose the **different entity** types.

The formula for computing the adjusted metric is as follows:

$$F = \frac{2a}{2a + b + c} \tag{1}$$

where: a is the number of instances during which two annotators agree, and $b + c$ is the number of instances during which two annotators do not agree.

Given that this project has three (3) annotators tagging each sentence, an F-measure was computed for every possible pair of annotators (Annotator 1 & 2, Annotator 1 & 3, Annotator 2 & 3) to compute agreement among these pairs. The three (3) pairwise F-measures were then averaged to obtain the final IAA F-measure metric for the project.

2.4 Model Evaluation on Filipino NER

To determine the usefulness and difficulty of our contribution, we benchmarked several large language models (LLMs) of varying sizes, capabilities, and architectures against it and two other Filipino-language NER datasets, namely TLUNIFIED-NER [21] and UNER [22]. We endeavor for our contribution to be a more challenging evaluation for LLMs as many existing evaluations are trending towards benchmark saturation [36], making it difficult to distinguish between genuine improvements in model performance [35].

Tasks. For TLUNIFIED-NER, we used its instruction-style test set. For UNER, which was originally designed as a span classification task, we converted the entire Filipino-language subset into instruction-style examples. As for our contribution, which we call FIL-NER, we randomly sampled a maximum of 400 instances per entity label, and similarly formatted these into instruction-style instances.

Zero-Shot Prompting. We developed Filipino-language prompts (see Appendix A) for each task, adapted from the English prompt template used by [21]. For each task's prompt template, the entity types are defined in Filipino according to their original definitions in their respective source studies. We used the original entity type definitions for the each task's prompt template–in lieu of 1 prompt template for all 3 tasks–as a generic template may not capture the specific nuances of each individual task [34].

Evaluation using Benchmark LLMs. We used the *AI Singapore SEA-LION series,* a family of open-source LLMs are models that better understand Southeast Asia's (SEA) diverse contexts, languages, and cultures [26]:

DeepSeek R1 and V3 series. DeepSeek V3 and R1 series. [27].

Google Gemma 3 series. Gemma 3 [30] family of models.

Google Gemini 1.5 and 2 series. Gemini series [31].

Meta Llama 3 series. The Llama 3 [33] herd of models.

OpenAI GPT-4 series. OpenAI's GPT-4 series [32].

Qwen 2.5 and 3 series. Qwen 2.5 [28] and 3 [29].

Our evaluation setting is zero-shot prompting. During model evaluation, input prompts are constructed using dataset instances and their corresponding prompt template, and default parameter settings are used for each LLM.

Metrics. We adopt both accuracy and macro F1 scores [38] to quantify LLMs' performance on each task. For each metric, the model completion is treated as the prediction, while the ground truth label is used as the reference. We report performance scores based on a single run for each dataset.

3 Results and Discussion

This section presents the results and discussion on the creation of the Filipino NER Coding Manual, the NER annotation of the dataset as well as its large language model evaluation on Filipino NER.

3.1 Filipino NER Coding Manual

The Filipino NER Coding Manual used the labels of named entities of OntoNotes English Named Entity Recognition Guidelines version 7.0 [4]: person name (PERSON), Nationality, Other, Religion, Political (NORP), facility (FAC), organization (ORG), geographical/social/political entity (GPE), location (LOC), product (PRODUCT), and FACILITY. For the Question Answering and MUC categories, we included date (DATE), time (TIME), percent (PERCENT), money (MONEY), quantity (QUANTITY), ordinal (ORDINAL), event (EVENT), work of art (WORK OF ART), law (LAW), AND language (LANGUAGE). We added two labels for the Filipino NER annotation of proper nouns of religious and cultural essence: BELIEF, and for government projects: PROJECT.

The Filipino NER annotation is distinct from the NER annotation of other languages such as English which is the basis of the the OntoNotes Named Entity Recognition Guidelines version 7.0 [4], in terms of the following: articles, prefixes, pronouns, plural forms in Filipino, superlatives, conjunction.

We included in the annotation of the span the plurality marker MGA, but exclude ANG if it is not part of the name of the organization or the location. Here are some sample annotations: *Ang* [*mga Muslim*], and *Ang* [*Korte Suprema*]. We include in the span, prefixes such as: [*maka-Pilipino*], and -ng linker such as: [*tatlong tasang*] *kape*. Occupational titles and honorifics are excluded from the span, such as: *Presidente* [*Marcos*], except for those titles and honorifics that have been integral to the person's name such as named saints which are both tagged as PERSON rather than BELIEF: [*San Lorenzo*] and [*Padre Pio*]. Frequency of time, duration and other temporal expressions are not tagged. For instance, we do not tag: *taon-taon*, and *kada taon*. We do not tag numbers that are part of an idiomatic or fixed expression. For instance, we do not tag: Hindi ka nag-iisa. We do not include in the span -ng when it is used as a noun marker. For instance: [*tatlong tasa*]*ng kape*.

The Filipino NER annotation task added proper nouns of religious and cultural essence: BELIEF, and for government projects: PROJECT. Example of which are the following:

Most named places are tagged as GPE. A country's name is labelled as a GPE, but a continent or a region is LOC. Examples are: [*Anilao*]*GPE,* [*Batangas*] *GPE,* [*fairview*]*GPE* [*quezon city*] *GPE,* [*Puerto galera*] *GPE,* [*mag-ILOCANDIA*] *GPE,* [*Barangay Loyola Heights*] *LOC.*

Countries are GPEs, but nationalities are NORPs. Examples include: *Ang* [*Mga Demokratiko*]

Usually the adjectival form of a nationality is identical to the name of the country or people's language. The annotators have to use the context to determine which label to use. For instance, *the* [*Filipino*]NORP *people*, and *the* [*Filipino*]LANGUAGE *language*.

Mentions longer than or equal to 24 h should be marked as DATE, such as: [*nakaraang Enero*]DATE, [*nakaraang Enero 14*]DATE, [*nakaraang buwan*]DATE, *Pumunta ako sa* [*Maynilad*]FAC, and *Namili ako sa* [*Shopee*]ORG.

Facilities are man-made, while locations are natural. Examples are canal FAC and a river is a LOC: [*Angat Dam*]FAC and [*Mount Kanlaon*]LOC.

Television programs can be ORG or WORK OF ART, depending on the context: [*GMA 24 Oras*] ORG tried to contact him for an interview and I love watching and [*GMA 24 Oras*]WORK OF ART.

Makes, models, and versions are PRODUCTS; the companies that produce them are ORGs. *Matapos ang* [*Ateneo Mass Shooting*]EVENT, *maraming paaralan ang naglagay ng metal detectors*, and *Sa* [*Ateneo de Manila University*]FAC *naganap ang mass shooting*.

Dates and times MUST have explicitly-mentioned units of time, otherwise they should be marked CARDINAL (even if the implicit unit seems obvious). An example is as follows: *Inabot siya ng* [*tatlong oras*]TIME, and *kaysa* [*isa*]CARDINAL.

3.2 Inter-Annotator Agreement

The Filipino NER inter-annotator agreement (IAA) scores among the three (3) annotators for the strict, exact, partial, and type agreement definitions among the 10,049 sentences are shown in Table 7. The results show an average of 0.7673 for strict agreement, 0.8224 for exact, 0.8796 for partial, and 0.8944 for type agreement (Table 8).

3.3 Evaluation Experiments

Our experiments involved evaluating instruction-tuned LLMs of varying parameter sizes: small (<=15B), medium (15–35B), large (35–100B), and extra large (>=100B).

Table 8. IAA (F-measure) for FIL-NER.

Pair	Strict	Exact	Partial	Type
1 & 2	0.8869	0.9192	0.8565	0.8979
1 & 3	0.7067	0.7741	0.8489	0.8941
2 & 3	0.7084	0.7737	0.8433	0.8913
Average	0.7673	0.8224	0.8496	0.8944

We also evaluated commercial AI systems such as GPT-4o and Gemini 2 Flash. We report model performances based on the evaluation methodology described in Sect. 2.4.

The models were evaluated against three Filipino NER datasets: TLUNIFIED-NER [21], UNER [22], and FIL-NER (our contribution). Table 9 provides a breakdown of each dataset by number of entity types. Each instance is composed of a text string, an entity span (that is found within the text string), and the entity type of the entity span as the ground truth label. Notably, ours (FIL-NER) incorporates more detailed classes for entity types.

3.4 Model Evaluation Results

Figure 1 shows the average accuracy of models grouped by size across the 3 NER tasks evaluated in this study, while Appendix B provides a comprehensive view all model performance scores.

Our results show that FIL-NER is a sufficiently difficult task for Filipino named entity recognition. The strongest model for FIL-NER is GPT-4o, scoring 84.23 on accuracy, with the average accuracy across all models of 75.01 and a standard deviation of ± 10.06. In addition to this, we find key performance differences between model performance across model sizes, following scaling laws.

Our evaluation (FIL-NER) is sufficiently challenging for instruction-tuned LLMs of varying sizes and architectures, demonstrating its usefulness in evaluating LLMs for Filipino-language understanding capabilities. This is in comparison to existing Filipino NER evaluations that have already succumbed to benchmark saturation.

In comparison, TLUNIFIED-NER and UNER are both saturated, where the average accuracy of all models evaluated is $96.11(\pm 2.82)$ and $96.39(\pm 2.60)$ respectively. Even within the small (<=15B parameter) model group, the average accuracy is $94.48(\pm 3.64)$ and $96.15(\pm 2.74)$ respectively, demonstrating that smaller LLMs–which scaling laws predict to have weaker capabilities–can already reach near-perfect performance on these evaluation tasks.

3.5 Model Error Analysis

To gain deeper insights into the performance characteristics of the evaluated models, we analyzed the confusion matrices for each model's predictions on FIL-NER. We focused on the fine-grained misclassifications that models tend to diverge from the ground truth annotations. The confusion matrices for the best-performing LLMs per model group are shown in Appendix C.

Table 9. Comparison of instruction-style test instances of NER datasets by entity type

Entity	FIL-NER (full)	FIL-NER (test split)	TLUNIFIED-NER	UNER
PERSON	2183	400	833	21
LOC	266	266	383	35
ORG	1448	400	363	-
BELIEF	121	121	-	-
CARDINAL	2196	400	-	-
DATE	1659	400	-	-
EVENT	304	304	-	-
FACILITY	342	342	-	-
GPE	1450	400	-	-
LANGUAGE	221	221	-	-
LAW	128	128	-	-
MONEY	187	187	-	-
NORP	607	400	-	-
ORDINAL	509	400	-	-
PERCENT	111	111	-	-
PRODUCT	182	182	-	-
PROJECT	12	12	-	-
QUANTITY	206	206	-	-
TIME	344	344	-	-
WORK_OF_ART	307	307	-	-
TOTAL	12783	5531	1579	56

For our analysis, we chose the best-performing LLMs from the small and medium-sized groups (Qwen3-14B and Qwen3-30B-A3B) as shown in Fig. 2, as well as large LLMs and commercial AI systems (LLaMA-SEA-LION-v3-70B-IT and GPT-4o-2024-11-20) as shown in Fig. 3. While model scale generally correlates with improved accuracy and reduced noise in predictions, certain consistent confusion patterns are evident across all models regardless of their size or architecture:

1. **TIME and DATE**: All models frequently misclassify TIME entities as DATE, and vice versa. This confusion likely stems from the inherent semantic and contextual overlap between temporal expressions (e.g., "*alas-kwatro ng hapon*" (4:00 pm) vs. "*Lunes ng umaga*" (Monday morning)). LLMs struggle to draw a crisp boundary when the context does not provide explicit temporal granularity.
2. **GPE and LOC**: There is a consistent confusion between GPE and LOC. For instance, names for regions and areas such as "Hilagang Sudan" (*North Sudan*, which refers to a region and not a country) may be labeled as GPE when the correct label is LOC, particularly when the context does not involve a political or organizational function. This reflects the challenges in distinguishing geographical scope from political status.

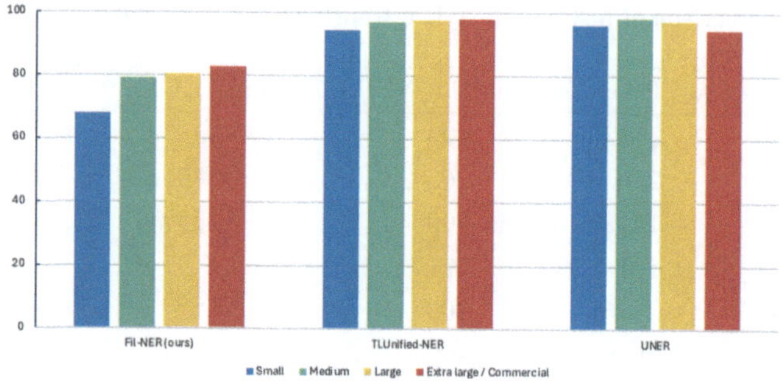

Fig. 1. Comparison of Best Performing LLMs in terms of accuracy.

3. **LAW and PROJECT**: Misclassification between LAW and PROJECT suggests an overlap in document references (e.g., "K-12", the bill that introduced the 12-year primary and secondary education curriculum) could be seen both as a legal framework and an initiative. This points to the limitations of entity-type disambiguation in the absence of explicit legal or policy-related context.

Model Scaling Effects. Larger models such as GPT-4o and LLaMA-SEA-LION-70B-IT exhibit sharper performance, indicating stronger predictive ability. Notably, GPT-4o achieves higher discrimination for fine-grained classes such as LAW, PROJECT, and PERSON.

Commercial vs. Open-Source Models. GPT-4o outperforms other models low-resource or ambiguous classes (e.g., BELIEF, LOC, PROJECT). This suggests that commercially fine-tuned models benefit from broader generalization and domain adaptation. However, even with strong performance, commercial AI systems still exhibit non-trivial confusion in entity categories that require world and region-specific knowledge (e.g., EVENT) and cultural familiarity (e.g. WORK_OF_ART), hinting at the limitations of current contextual understanding.

4 Conclusion and Recommendations

This paper has presented the annotation of the Filipino dataset of 10,049 sentences, which is publicly available and free for use for downstream NLP tasks such as named entity recognition, co-reference resolution and semantic role labeling. We reported NER annotation of this dataset using the Filipino NER Coding Manual [39]. The annotators and the quality assurance team used the Datasaur system (as provided by AI-SG), with acceptable inter-rater accuracy (IAA) ranging from an average 77% to 89% from strict, exact, partial and type agreement definitions. Use of Fil-NER in LLM model evaluation proved to be a sufficiently difficult task for Filipino NER with GPT-4o, scoring 84.23 on accuracy against an average accuracy of 75.01 across all other models. Fil-NER and its corresponding annotation guide will be publicly available for further use in downstream NLP tasks and evaluation in AI language model development.

A Filipino Prompt Templates: LLM Evaluation on NER

Fil-NER

```
Isa kang Named Entity Recognition system. Nangangahulugan ang mga sumusunod
    na label:
PERSON: mga tao sa totoong buhay at kathang-isip; kabilang dito ang mga
    pangalan, unang pangalan, apelyido, ekstensiyon ng apelyido, at palayaw.
NORP: mga anyo ng pang-uri ng nasyonalidad, bansa, relihiyon, at politikal
    na grupo.
FACILITY: mga pasilidad o estrukturang gawa ng tao na may pangalan tulad ng
    gusali, kalsada, tulay, monumento, at iba pa.
ORG: mga organisasyon, samahan, kompanya, ahensiya, grupo, at iba pa.
GPE: mga bansa, siyudad, lalawigan, munisipyo, at barangay.
LOC: mga lugar na hindi GPE tulad ng baybayin, bulubundukin, karagatan,
    planeta, rehiyon, kontinente, at iba pa.
PRODUCT: mga pangalan ng anomang produkto, modelo, at pagkain.
DATE: mga petsa at panahon; kabilang dito ang relatibong panahon tulad ng
    "ngayong taon".
TIME: mga oras na hindi tataas sa 24 oras.
PERCENT: mga porsyento na may nakakabit na simbolong "%" o "porsyento".
MONEY: mga pera sa anomang denominasyon.
QUANTITY: mga panukat na may tiyak na yunit tulad ng tangkad, bigat, laki,
    at iba pa.
ORDINAL: mga pamilang na panunuran o nagsasaad ng pagkakasunod-sunod.
CARDINAL: mga likas na pamilang tulad ng buong numero, fraction, at desimal
    na hindi QUANTITY, MONEY, PERCENT, DATE, o TIME.
EVENT: mga pangyayaring pinangalanan tulad ng bagyo, labanan, palabas,
    palakasan, at iba pa.
WORK_OF_ART: mga likhang-sining tulad na pinamagatan tulad ng libro, kanta,
    palabas, parangal, at iba pa.
LAW: mga dokumento na isinabatas, kabilang ang mga kasunduan at iba pang
    legal na dokumento.
LANGUAGE: mga pinangalanang wika, lengguwahe, o diyalekto.
BELIEF: mga pangalan ng diyos at iba pang nilalang ng kababalaghan o
    relihiyon.
PROJECT: mga proyekto kabilang na ang mga opisyal na sistema at patakaran
    mula sa gobyerno.

Batay sa kahulugan ng mga label, uriin ang sumusunod na entidad na mahahanap
    sa loob ng teksto gamit lamang ang isang label.
Sumagot gamit ang sumusunod na format:
Sagot: $LABEL
Palitan ang $LABEL ng napiling label.{fewshot_examples}
```

TLUnified-NER Task

```
Isa kang Named Entity Recognition system. Nangangahulugan ang mga sumusunod
    na label:
PERSON: Mga tao na maaaring indibidwal o grupo.
```

```
ORG: Mga organisasyon tulad ng korporasyon, ahensiya, at iba pang grupo ng
    tao na may organisadong estruktura.
LOC: Mga lugar tulad ng rehiyon, pook, at kalupaan. Kabilang dito ang mga
    geopolitikal na entidad.

Batay sa kahulugan ng mga label, iuri ang sumusunod na entidad na mahahanap
    sa loob ng teksto gamit lang ng isang label.
Sumagot gamit ang sumusunod na format:
Sagot: $LABEL
Palitan ang $LABEL ng napiling label.
```

UNER Task

```
Isa kang Named Entity Recognition system. Nangangahulugan ang mga sumusunod
    na label:
PERSON: Mga pangalan ng tao sa totoong buhay at kathang-isip, pero hindi
    kabilang ang mga nominal na pangalan.
LOC: Mga lugar na may pangalan.

Batay sa kahulugan ng mga label, iuri ang sumusunod na entidad na mahahanap
    sa loob ng teksto gamit lang ng isang label.
Sumagot gamit ang sumusunod na format:
Sagot: $LABEL
Palitan ang $LABEL ng napiling label.
```

B Model Evaluation Results

(see Tables 10, 11, 12 and 13)

Table 10. Accuracy scores of open-source LLMs and commercial AI systems for each NER dataset.

Model	# params.	R[a]	MoE[b]	FIL-NER	TLUNIFIED-NER	UNER	Average
Small (<=15B parameters) instruction-tuned LLMs							
DeepSeek-R1-Distill-Llama-8B	8B	Y		54.58	94.02	94.76	81.12
DeepSeek-R1-Distill-Qwen-7B	7B	Y		45.66	85.76	93.33	74.92
DeepSeek-R1-Distill-Qwen-14B	14B	Y		73.22	96.47	97.62	89.10
Gemma-3-12B-IT	12B			78.23	96.62	**100.00**	91.62
Gemma-SEA-LION-v3-9B-IT	9B			73.95	97.14	97.62	89.57
Llama-SEA-LION-v3-8B-IT	8B			71.28	95.53	95.24	87.35
Llama-SEA-LION-v3.5-8B-R	8B	Y		76.11	96.85	97.62	90.19
Llama 3.1 8B Instruct	8B			56.51	91.73	95.24	81.16
Qwen2.5-7B-Instruct	7B			57.07	89.77	90.95	79.26
Qwen2.5-14B-Instruct	14B			67.80	95.15	93.81	85.58
Qwen3-8B	8B	Y		78.71	**97.39**	97.62	91.24
Qwen3-14B	14B	Y		**82.02**	97.34	**100.00**	**93.12**
Medium (15–35B parameters) instruction-tuned LLMs							
DeepSeek-R1-Distill-Qwen32B	32B	Y		78.60	97.21	97.62	91.14
Gemma-3-27B-IT	27B			77.00	97.17	**100.00**	91.39
Qwen2.5 32B Instruct	32B			77.80	96.15	98.57	90.84
Qwen3-30B-A3B	32B	Y	Y	**81.33**	96.91	97.62	91.95
Qwen3-32B	32B	Y		80.85	**97.83**	97.62	**92.10**
Large (35–100B parameters) instruction-tuned LLMs							
DeepSeek-R1-Distill-Llama-70B	70B	Y		80.43	97.64	97.62	91.90
Llama-SEA-LION-v3-70B-IT	70B			**81.72**	97.63	**100.00**	**93.11**
Llama-SEA-LION-v3.5-70B-R	70B	Y		80.71	**97.66**	92.86	90.41
Llama-3.1-70B-Instruct	70B			79.61	97.38	97.62	91.54
Qwen2.5-72B-Instruct	72B			78.74	96.77	97.62	91.04
Extra large (>=100B parameters) instruction-tuned LLMs and commercial AI systems							
DeepSeek R1	671B	Y	Y	83.29	**98.12**	**97.62**	93.01
DeepSeek V3	671B		Y	79.90	97.94	95.24	**91.03**
GPT-4o-2024-11-20	-			**84.23**	97.46	92.86	91.52
Gemini-2.0-Flash-001	-			83.24	97.82	92.86	91.31
Qwen/Qwen3-235B-A22B	235B	Y	Y	82.78	97.40	92.86	91.01

[a] With reasoning capabilities
[b] Mixture-of-Experts architecture

Table 11. Average accuracy (with standard deviation) by model group for each NER dataset.

Model Group	# params.	FIL-NER	TLUNIFIED-NER	UNER	Avg.
Small	<=15B	67.93 ± 11.62	94.48 ± 03.64	96.15 ± 02.74	86.19 ± 05.77
Medium	15–35B	79.12 ± 01.90	97.05 ± 00.61	**98.29** ± 01.04	91.48 ± 00.53
Large	35–100B	80.24 ± 01.13	97.42 ± 00.38	97.14 ± 02.61	**91.60** ± 01.01
Extra large / Commercial	>=100B	**82.69** ± 01.65	**97.75** ± 00.31	94.29 ± 02.13	91.58 ± 00.83
Accuracy avg.		75.01 ± 10.06	96.11 ± 02.82	96.39 ± 02.60	89.17 ± 04.67

Table 12. Macro F1 scores of open-source LLMs and commercial AI systems for each NER dataset.

Model	# params.	R[a]	MoE[b]	FIL-NER	TLUNIFIED-NER	UNER[c]	Avg.
Small (<=15B parameters) instruction-tuned LLMs							
DeepSeek-R1-Distill-Llama-8B	8B	Y		53.49	70.99	64.87	63.12
DeepSeek-R1-Distill-Qwen-7B	7B	Y		42.87	64.15	64.36	57.13
DeepSeek-R1-Distill-Qwen-14B	14B	Y		68.14	95.94	98.08	87.38
Gemma-3-12B-IT	12B			74.28	95.85	**100.00**	**90.05**
Gemma-SEA-LION-v3-9B-IT	9B			70.33	**96.67**	98.08	88.36
Llama-SEA-LION-v3-8B-IT	8B			68.37	71.23	96.11	78.57
Llama-SEA-LION-v3.5-8B-R	8B	Y		72.18	72.44	65.85	70.16
Llama 3.1 8B Instruct	8B			55.85	68.98	96.11	73.65
Qwen2.5-7B-Instruct	7B			53.83	67.30	60.83	60.65
Qwen2.5-14B-Instruct	14B			64.04	70.96	63.74	66.25
Qwen3-8B	8B	Y		73.81	73.07	98.08	81.65
Qwen3-14B	14B	Y		**80.17**	72.90	**100.00**	84.36
Medium (15–35B parameters) instruction-tuned LLMs							
DeepSeek-R1-Distill-Qwen32B	32B	Y		73.87	72.96	65.85	70.89
Gemma-3-27B-IT	27B			71.18	**96.77**	**100.00**	**89.32**
Qwen2.5-32B-Instruct	32B			74.62	72.77	65.85	71.08
Qwen3-30B-A3B	32B	Y	Y	**75.88**	72.68	65.85	71.47
Qwen3-32B	32B	Y		75.73	73.30	65.85	71.63
Large (35–100B parameters) instruction-tuned LLMs							
DeepSeek-R1-Distill-Llama-70B	70B	Y		74.40	**97.44**	98.08	**89.97**
Llama-SEA-LION-v3-70B-IT	70B			**76.35**	73.22	**100.00**	83.19
Llama-SEA-LION-v3.5-70B-R	70B	Y		74.62	97.43	64.10	78.72
Llama-3.1-70B-Instruct	70B			76.08	72.97	65.85	71.64
Qwen2.5-72B-Instruct	72B			74.62	72.77	65.85	71.08
Extra large (>=100B parameters) instruction-tuned LLMs and commercial AI systems							
DeepSeek-R1	671B	Y	Y	80.90	73.58	**65.85**	73.44
DeepSeek-V3	671B		Y	75.18	**97.85**	65.00	79.35
GPT-4o-2024-11-20	-			78.17	73.04	64.10	71.77
Gemini-2.0-Flash-001	-			**81.70**	97.78	63.63	81.04
Qwen/Qwen3-235B-A22B	235B	Y	Y	80.30	73.02	63.63	72.32

[a] With reasoning capabilities
[b] Mixture-of-Experts architecture
[c] We note that macro F1 scores can overly highlight the poor performance in one class for binary classification tasks despite high accuracy scores [37]. This explains the large variance that is observed in macro F1 scores for UNER, which involves binary classification between two entity types, namely PERSON and LOC.

Table 13. Macro F1 (with standard deviation) by model group for each NER dataset.

Model Group	# params.	FIL-NER	TLUNIFIED-NER	UNER	Avg.
Small	<=15B	64.86 ± 10.98	76.71 ± 11.99	**83.84** ± 17.65	75.11 ± 11.56
Medium	15–35B	74.08 ± 01.91	82.33 ± 12.81	79.13 ± 18.20	78.52 ± 09.84
Large	35–100B	75.21 ± 00.92	82.77 ± 13.39	78.78 ± 18.52	**78.92** ± 07.98
Extra large/Commercial	>=100B	**79.25** ± 02.62	**83.05** ± 13.48	64.44 ± 00.97	75.58 ± 04.29
Macro F1 avg.		71.11 ± 09.43	80.05 ± 12.29	78.44 ± 16.94	76.53 ± 09.33

C Confusion Matrices for FIL-NER

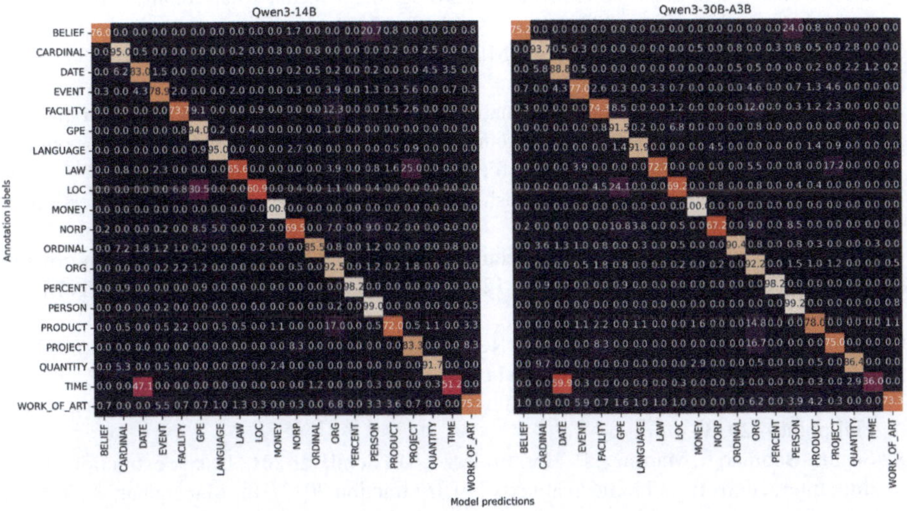

Fig. 2. Confusion matrix for best-performing LLMs (in terms of accuracy) from the small (Qwen3-14B) and medium (Qwen3-32B) model groups.

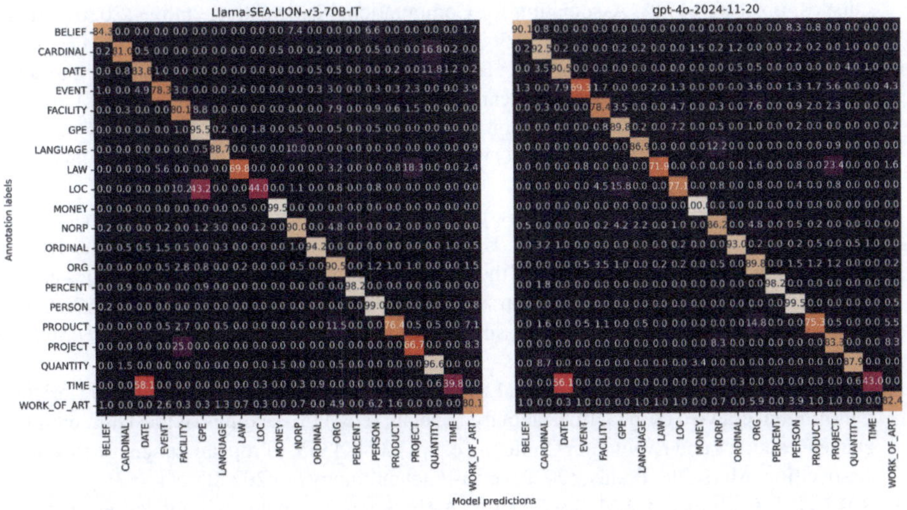

Fig. 3. Confusion matrix for best-performing LLMs (in terms of accuracy) from the large (Llama-SEA-LION-v3-70B-IT) and extra large/commercial (GPT-4o-2024-11-20) model groups.

References

1. Digital marketing institute: social media: what countries use it most & what are they using? https://digitalmarketinginstitute.com/blog/social-media-what-countries-use-it-most-and-what-are-they-using (2021). Accessed 14 Jan 2021
2. Department of trade and industry: MSME tech tools (2022). https://www.dti.gov.ph/covid19/tech-tools/
3. Department of trade and industry: national export congress features digital transformation (2021). https://www.dti.gov.ph/archives/news-archives/2019-national-export-congress-features-digital-transformation-tools-for-msme-adaption/
4. BBN technologies: OntoNotes English named entity recognition guidelines version 7.0. Technical Report (2007)
5. Hripcsak, G., Rothschild, A.S.: Agreement, the F-measure, and reliability in information retrieval. J. Am. Med. Inform. Assoc. **12**(3), 296–298 (2005). https://doi.org/10.1197/jamia.M1733
6. Schachter, P., Otanes, F.T.: Tagalog Reference Grammar. University of California Press, Berkeley and Los Angeles, California (1972)
7. Malicsi, J.: Grammar ng Filipino. Sentro ng Wikang Filipino - UP Diliman, Quezon City, Philippines (2013)
8. Segura-Bedmar, I., Martínez, P., Herrero-Zazo, M.: SemEval-2013 task 9: extraction of drug-drug interactions from biomedical texts (DDIExtraction 2013). In: Manandhar, S., Yuret, D. (eds.) Second Joint Conference on Lexical and Computational Semantics (*SEM), Volume 2: Proceedings of the Seventh International Workshop on Semantic Evaluation (SemEval 2013), pp. 341–350. Association for Computational Linguistics, Atlanta, Georgia, USA (2013). https://aclanthology.org/S13-2056
9. Ortiz Suárez, P.J., Romary, L., Sagot, B.: A monolingual approach to contextualized word embeddings for mid-resource languages. In: Jurafsky, D., Chai, J., Schluter, N., Tetreault, J. (eds.) Proceedings of the 58th Annual Meeting of the Association for Computational Linguistics, pp. 1703–1714. Association for Computational Linguistics, Online (2020). https://doi.org/10.18653/v1/2020.acl-main.156
10. Vu Xuan, S., Vu, T., Tran, S., Jiang, L.: ETNLP: a visual-aided systematic approach to select pre-trained embeddings for a downstream task. In: Mitkov, R., Angelova, G. (eds.) Proceedings of the International Conference on Recent Advances in Natural Language Processing (RANLP 2019), pp. 1285–1294. INCOMA Ltd., Varna, Bulgaria (2019). https://doi.org/10.26615/978-954-452-056-4_147
11. Wilie, B., et al.: IndoNLU: Benchmark and resources for evaluating indonesian natural language understanding. In: Wong, K.-F., Knight, K., Wu, H. (eds.) Proceedings of the 1st Conference of the Asia-Pacific Chapter of the Association for Computational Linguistics and the 10th International Joint Conference on Natural Language Processing, pp. 843–857. Association for Computational Linguistics, Suzhou, China (2020). https://aclanthology.org/2020.aacl-main.85
12. Abadji, J., Ortiz Suarez, P., Romary, L., Sagot, B.: Towards a cleaner document-oriented multilingual crawled corpus. In: Calzolari, N., et al. (eds.) Proceedings of the Thirteenth Language Resources and Evaluation Conference, pp. 4344–4355. European Language Resources Association, Marseille, France (2022). https://aclanthology.org/2022.lrec-1.463
13. Roxas, R.E.O., Imperial, J.M., De La Cruz, A.H.: Science mapping of publications in natural language processing in the philippines: 2006 to 2020. In: Hu, K., Kim, J.-B., Zong, C., Chersoni, E. (eds.) Proceedings of the 35th Pacific Asia Conference on Language, Information and Computation, pp. 721–730. Association for Computational Linguistics, Shanghai, China (2021). https://aclanthology.org/2021.paclic-1.76

14. Jiang, S., Fu, Y., Lin, X., Lin, N.: Pre-trained Language Models for Tagalog with Multi-source Data. In: Wang, L., Feng, Y., Hong, Yu., He, R. (eds.) NLPCC 2021. LNCS (LNAI), vol. 13028, pp. 210–223. Springer, Cham (2021). https://doi.org/10.1007/978-3-030-88480-2_17
15. Ginter, F., Hajič, J., Luotolahti, J., Straka, M., Zeman, D.: CoNLL 2017 shared task - automatically annotated raw texts and word embeddings. LINDAT/CLARIAH-CZ digital library at the Institute of Formal and Applied Linguistics (ÚFAL), Faculty of Mathematics and Physics, Charles University (2017). http://hdl.handle.net/11234/1-1989
16. Aquino, A., de Leon, F.: Parsing in the absence of related languages: Evaluating low-resource dependency parsers on Tagalog. In: de Marneffe, M.-C., de Lhoneux, M., Nivre, J., Schuster, S. (eds.) Proceedings of the Fourth Workshop on Universal Dependencies (UDW 2020), pp. 8–15. Association for Computational Linguistics, Barcelona, Spain (Online) (2020). https://aclanthology.org/2020.udw-1.2
17. Brandsen, A., Verberne, S., Wansleeben, M., Lambers, K.: Creating a dataset for named entity recognition in the archaeology domain. In: Calzolari, N., et al. (eds.) Proceedings of the Twelfth Language Resources and Evaluation Conference, pp. 4573–4577. European Language Resources Association, Marseille, France (2020). https://aclanthology.org/2020.lrec-1.562
18. Cruz, J.C.B., Cheng, C.: Evaluating language model finetuning techniques for low-resource languages. CoRR, abs/1907.00409 (2019). http://arxiv.org/abs/1907.00409
19. Cruz, J.C.B., Cheng, C.: Establishing baselines for text classification in low-resource languages. CoRR, abs/2005.02068 (2020). https://arxiv.org/abs/2005.02068
20. Cruz, J.C.B., Resabal, J.K., Lin, J., Velasco, D.J., Cheng, C.: Investigating the true performance of transformers in low-resource languages: a case study in automatic corpus creation. CoRR, abs/2010.11574 (2020). https://arxiv.org/abs/2010.11574
21. Miranda, L.J.: Developing a named entity recognition dataset for tagalog. In: Wijaya, D., Aji, A.F., Vania, C., Winata, G.I., Purwarianti, A. (eds.) Proceedings of the First Workshop in South East Asian Language Processing, pp. 13–20. Association for Computational Linguistics, Nusa Dua, Bali, Indonesia (2023).https://doi.org/10.18653/v1/2023.sealp-1.2
22. Mayhew, S., et al.: Universal NER: a gold-standard multilingual named entity recognition benchmark. In: Duh, K., Gomez, H., Bethard, S. (eds.) Proceedings of the 2024 Conference of the North American Chapter of the Association for Computational Linguistics: Human Language Technologies (Volume 1: Long Papers), pp. 4322–4337. Association for Computational Linguistics, Mexico City, Mexico (2024). https://doi.org/10.18653/v1/2024.naacl-long.243
23. Cruz, J.C.B., Cheng, C.: Improving large-scale language models and resources for Filipino. In: Calzolari, N., et al (eds.) Proceedings of the Thirteenth Language Resources and Evaluation Conference, pp. 6548–6555. European Language Resources Association, Marseille, France (2022). https://aclanthology.org/2022.lrec-1.703/
24. Nivre, J., et al.: Universal Dependencies v1: A Multilingual Treebank Collection. In: Calzolari, N., et al. (eds.) Proceedings of the Tenth International Conference on Language Resources and Evaluation (LREC 2016), pp. 1659–1666. European Language Resources Association (ELRA), Portorož, Slovenia (2016). https://aclanthology.org/L16-1262/
25. Samson, S.D.: A treebank prototype of Tagalog. Undergraduate thesis, Germany (2018)
26. Ng, R., et al.: SEA-LION: southeast Asian languages in one network. https://arxiv.org/abs/2504.05747
27. DeepSeek-AI: DeepSeek-R1: incentivizing reasoning capability in LLMs via reinforcement learning. https://arxiv.org/abs/2501.12948
28. Qwen, Yang, A., et al.: Qwen2.5 technical report. https://arxiv.org/abs/2412.15115
29. Yang, A., et al.: Qwen3 technical report. https://arxiv.org/abs/2505.09388

30. Gemma Team, Kamath, A., Ferret, J., Pathak, S., Vieillard, N., Merhej, R., et al.: Gemma 3 technical report. https://arxiv.org/abs/2503.19786
31. Gemini Team, Georgiev, P., Lei, V.I., Burnell, R., Bai, L., Gulati, A., et al.: Gemini 1.5: unlocking multimodal understanding across millions of tokens of context. https://arxiv.org/abs/2403.05530
32. OpenAI, Hurst. A., Lerer, A., Goucher, A.P., Perelman, A., Ramesh, A., et al.: GPT-4o system card. https://arxiv.org/abs/2410.21276
33. Grattafiori, A., Dubey, A., Jauhri, A., Pandey, A., Kadian, A., et al.: The Llama 3 herd of models. https://arxiv.org/abs/2410.21276
34. Mao, Y., He, J., Chen, C.: From prompts to templates: a systematic prompt template analysis for real-world LLM apps. https://arxiv.org/abs/2504.02052
35. Ott, S., Barbosa-Silva, A., Blagec, K., Brauner, J., Samwald, M.: Mapping global dynamics of benchmark creation and saturation in artificial intelligence. Nat. Commun. **13**(1), 6793 (2022)
36. Reuel-Lamparth, A., Hardy, A., Smith, C., Lamparth, M., Hardy, M., Kochenderfer, M.J.: BetterBench: assessing AI benchmarks, uncovering issues, and establishing best practices. Adv. Neural. Inf. Process. Syst. **37**, 21763–21813 (2024)
37. Chicco, D., Jurman, G.: The advantages of the Matthews correlation coefficient (MCC) over F1 score and accuracy in binary classification evaluation. BMC Genom **21**, 1–13 (2020). https://doi.org/10.1186/s12864-019-6413-7 Springer
38. Pedregosa, F., et al.: Scikit-learn: machine learning in Python. J. Mach. Learn. Res. **12**, 2825–2830 (2011)
39. Xform-AISG: FilNER coding manual. To be published

Text2Cypher Across Languages: Evaluating and Finetuning LLMs

Makbule Gulcin Ozsoy(✉) and William Tai

Neo4j, London, UK
{makbule.ozsoy,william.tai}@neo4j.com

Abstract. Recent advances in large language models (LLMs) have enabled interfaces that translate user questions into database queries, such as Text2SQL, Text2SPARQL, and Text2Cypher. While these interfaces enhance database accessibility, most research today focuses on English, with limited evaluation in other languages. This paper investigates the performance of foundational and finetuned LLMs on the Text2Cypher task across multiple languages. We create and release a multilingual dataset by translating English questions into Spanish and Turkish while preserving the original Cypher queries, enabling fair cross-lingual comparison. Using standardized prompts and metrics, we evaluate several foundational models and observe a consistent performance pattern: highest on English, followed by Spanish, and lowest on Turkish. We attribute this to differences in training data availability and linguistic features. We also examine the impact of translating task prompts into Spanish and Turkish. Results show little to no change in evaluation metrics, suggesting prompt translation has minor impact. Furthermore, we finetune a foundational model on two datasets: one in English only, and one multilingual. Finetuning on English improves overall accuracy but widens the performance gap between languages. In contrast, multilingual finetuning narrows the gap, resulting in more balanced performance. Our findings highlight the importance for multilingual evaluation and training to build more inclusive and robust query generation systems.

Keywords: Large Language Models (LLMs) · Query Generation · Text2Cypher · Cross-lingual Generalization

1 Introduction

Databases provide efficient mechanisms for storing, organizing, and retrieving data. Query languages such as SQL (for relational databases), SPARQL (for RDF graphs), and Cypher (for graph databases) enable users to interact with these systems [13]. Recent advancements in large language models (LLMs) have enabled the translation of natural language questions into database queries (Text2SQL, Text2SPARQL, Text2Cypher), making databases more accessible to non-experts. However, most research in this area has focused on English, with limited investigation on multilingual capabilities [12,17].

H. Unger and P. Meesad (Eds.): NLPIR 2025, LNNS 1904, pp. 1206–1219, 2026.
https://doi.org/10.1007/978-3-032-20897-2_84

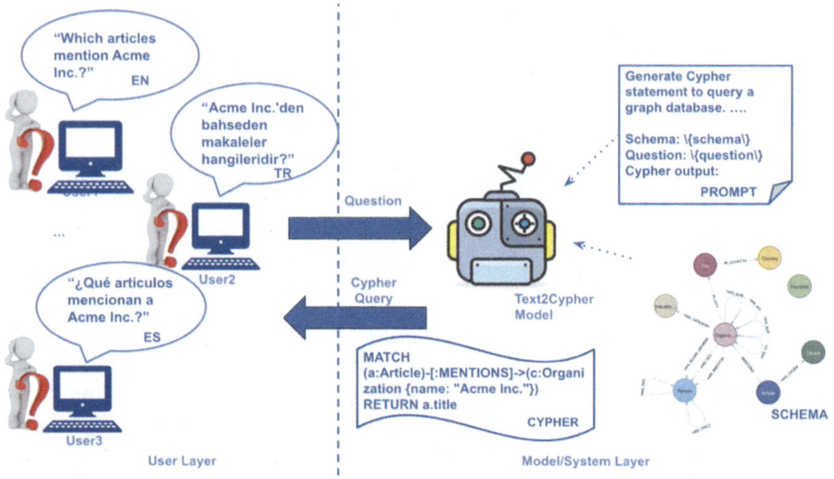

Fig. 1. Multiple users access the same database and want to generate a Cypher query for the question "Which articles mention Acme Inc.?" However, they may express this question in different languages, such as English (EN), Spanish (ES), or Turkish (TR).

This work focuses on the multilingual Text2Cypher task, which involves converting natural language questions into executable Cypher queries (see Fig. 1). For instance, users querying the same graph database might ask `''Which articles mention Acme Inc.?''` in English (EN), Spanish (ES), or Turkish (TR). A robust Text2Cypher system should generate the same query regardless of input language, such as: `MATCH (a:Article)-[:MENTIONS]->(c:Company {name: ''Acme Inc.''}) RETURN a.title`. We investigate how foundational and finetuned LLMs perform on this multilingual task. First, we create a multilingual dataset for training and evaluation in English, Spanish, and Turkish. Second, we examine whether translating the task instructions (i.e., prompts) into Spanish and Turkish affects model performance. Finally, we evaluate whether multilingual finetuning improves generalization by comparing models finetuned on English-only versus multilingual data.

Our main contributions are:

- We create and release a multilingual version of the Text2Cypher dataset [31], translated into Spanish and Turkish to support multilingual training and evaluation.[1], [2]
- We evaluate foundational LLMs on a multilingual test set, showing that performance is highest in English, followed by Spanish, and lowest in Turkish. We attribute the lower Turkish performance to linguistic differences and training data scarcity.

[1] Test Set: https://huggingface.co/datasets/mgoNeo4j/translated_text2cypher24_testset.

[2] Training Set: https://huggingface.co/datasets/mgoNeo4j/translated_text2cypher24_trainset_sampled.

- We analyze the effect of translating prompts into Spanish and Turkish and find minimal impact on performance in our setting.
- We investigate the impact of finetuning. English-only finetuning increases accuracy but widens the performance gap across languages, while multilingual finetuning narrows the gap and yields more balanced performance.

The paper is organized as follows: Sect. 2 reviews related work. Section 3 describes the multilingual dataset. Section 4 presents model evaluations. Section 5 concludes the paper.

2 Related Work

This section reviews related work on large language models (LLMs) for non-English content, as well as multilingual approaches to database query generation tasks such as Text2SQL and Text2SPARQL.

2.1 LLMs and Non-English Content

Large language models (LLMs) have become central to a wide range of natural language processing tasks. While many early LLMs were trained primarily on English data, their capabilities have recently been extended to other languages through the development of multilingual models [28]. Lai et al. [23] describe the multilingual capabilities of LLMs in terms of their ability to handle task instructions, inputs, and outputs in different languages. They identify two multilingual capabilities: (i) Understanding, where the model interprets instructions in various languages and generates a correct output, and (ii) Generating, where a fixed language (e.g., English) is used for the instruction and the model produces outputs in different target languages. Both capabilities depend heavily on the model's training data and methodology.

In order to support and enhance multilingual capabilities, various approaches have been proposed in the literature. Some studies leverage multilingual parallel data to continue pretraining [40,44]. Others employ multilingual instruction data for fine-tuning [22,25,26,39]. A third line of work focuses on cross-lingual prompting during inference [11,14]. Researchers have also begun exploring the internal mechanisms of these models [5,21,36,42,43] and evaluating their multilingual performance across various application domains [17,30].

Multilingual models are typically trained on datasets that include multiple languages, allowing them to learn cross-lingual representations. This enables models to transfer word associations and grammatical patterns learned from high-resource languages, such as English, to languages with less available training data. However, since English still dominates most training corpora, these models often internalize linguistic patterns and assumptions specific to English [28,42]. As a result, LLMs tend to perform significantly better in higher-resource languages and those similar to them, compared to lower-resource languages [28].

2.2 Multilingual Database Query Generation

Much of the multilingual work in translating natural language to database query languages has focused on the Text2SQL task [10,15,18]. Many studies translate the English-language Spider dataset [41] into other languages, such as Chinese in CSpider [27], Turkish in TURSpider [20], Arabic in Ar-Spider [3], and seven languages in MultiSpider [10]. Other works have created new multilingual datasets using human annotators, such as StatBot.Swiss [29], which includes data in English and German. Additional languages like Portuguese [17,32] and Russian [4] have also been explored. Jannuzzi et al. [17] evaluate Text2SQL performance in Portuguese using zero-shot prompting and report a performance gap between using English and Portuguese prompts.

For SPARQL, although there are multilingual question answering (QA) datasets and methods [8,34,37], they are not specifically designed for the Text2SPARQL task. Recently, the Text2SPARQL Challenge [1] released a dataset containing questions in English and Spanish. Additionally, Perevalov et al. [33] explored the reverse task, translating SPARQL to natural language, using English, German, and Russian in their evaluation. Recent work has also investigated the use of LLMs and SPARQL queries for knowledge graph reasoning, such as LLM4QA [24], which highlights the broader potential of query generation tasks across different database languages.

In this work, we study the Text2Cypher task, which translates natural language questions into Cypher queries.

3 Multilingual Dataset Preparation

As the base dataset, we used the publicly available Text2Cypher dataset [31], which includes English questions, database schema, ground-truth Cypher queries, and metadata. In order to evaluate foundational and finetuned models across languages, we created a multilingual dataset by translating the original English questions into Spanish and Turkish.

3.1 Text2Cypher Beyond English

We have chosen English, Spanish and Turkish for the comparisons based on their variations in their resourcedness levels, language families and linguistic characteristics:

- **Language resourcedness:** This indicates availability of data, tools, and research support for a language. It impacts LLM performance, where models generally perform better on languages with richer and more diverse training data [28]. While English is an extremely high-resource language, Spanish is a high-resource, and Turkish is a medium-resource language [19,28].
- **Language families:** This indicates relationships between languages, grouping them based on common origins and shared linguistic features. According to Dhamecha et al. [9], models generalize more effectively across linguistically

similar languages. English and Spanish both belong to the Indo-European family, whereas Turkish belongs to the Altaic family, with distinct linguistic traits.
- **Linguistic characteristics:** Turkish is an agglutinative language, where words are formed by concatenating multiple suffixes encoding grammatical functions, such as tense, case, person. This morphological richness leads to challenges in tokenization and semantic parsing, and increases data requirements for LLMs [2,7]. Additionally, both Spanish and Turkish allow relatively flexible word order compared to English and are pro-drop languages, allowing omission of subjects and pronouns. These characteristics increase ambiguity and makes the mapping to structured queries more difficult compared to English [2,6,7,38].

Given these factors, we expect a moderate performance drop for Spanish and a more substantial drop for Turkish due to its linguistic divergence.

Table 1. Translation prompt for {language_name}

Translation instruction prompt
You are a professional translator. Your task is to translate the given text from English to {language_name}. Follow these guidelines:
1. Maintain the original meaning and tone of the text
2. Preserve any placeholders in the format <TYPE_NUMBER> (e.g., <PERSON_0>, <QUOTE_1>)
3. Keep the translation natural and fluent in {language_name}
4. Maintain proper formatting and punctuation
5. Do not translate or modify any placeholders
6. Do not translate any text within quotation marks
7. Do not translate named entities (proper nouns, names, places, organizations)
8. Do not translate numbers, dates, or measurements
Please provide only the translated text without any explanations or additional context.

3.2 Multilingual Test Set

In order to enable multilingual evaluation, we automatically translated the original English questions from the Text2Cypher dataset [31] into Spanish and Turkish using an LLM. We used the following steps for translation:

- **Masking Named Entities and Quotes**: To ensure that named entities and quoted text remained unchanged during translation, we first masked them. Named entities were replaced with placeholders indicating their type

and index (e.g., `LOCATION_0`), while quoted strings were replaced with `QUOTE_<index>` markers. For example, the sentence `''Hello, I work at 'Neo4j' in London''` was masked as `''Hello, I work at QUOTE_0 in LOCATION_0''`.
- **Translation Using an LLM**: We translated the masked questions using the GPT-4o-mini model, guided by the prompt shown in Table 1.
- **Restoring Masks**: After translation, we replaced the placeholders with their original named entities and quoted expressions to reconstruct the final question.

This process produced semantically aligned triples across the three languages. During translation, 50 samples with masking-related issues were excluded from the test set. Consequently, the final test set contains 4,783 samples per language. Translation quality was evaluated using COMET-KIWI-22 [35], a reference-free metric ranging from 0 to 1, where higher values indicate better quality. Our test dataset obtained 0.7934 for Turkish and 0.8062 for Spanish translations. Note that the ground-truth Cypher queries were not translated. These queries contain (i) Cypher-specific syntax (e.g., `MATCH, WHERE`), (ii) terms from the schema (e.g., `Person, ACTED_IN, Movie`), and (iii) user-provided literals (e.g., `''Tom Hanks''`), all of which are intended to remain in their original form.

3.3 Multilingual Training Set

The training set is based on the original English Text2Cypher training dataset [31], which contains approximately 40,000 instances. To create a multilingual training dataset, we prepared around 36,000 instances in total, distributed evenly across English(EN), Spanish(ES), and Turkish(TR); such that about 12,000 instances per language. These instances are randomly sampled from the original data and only the user question field is translated, while the Cypher queries and other fields are kept unchanged. The question translation process followed the same method described for the test set in Sect. 3.2. The resulting training dataset includes: (i) Around 6,750 questions common to all three languages, (ii) Around 1,500 questions shared between each pair of languages (i.e., EN-ES, EN-TR, ES-TR), (iii) Approximately 3,800 unique questions per language. This structure provides fully aligned questions for direct cross-lingual learning and evaluation, alongside unique, language-specific questions to capture linguistic diversity. For the training set, COMET-KIWI-22 [35] was also applied to evaluate the translation quality. This is a reference-free metric scores in the range [0, 1], where higher values reflecting better quality. Our multilingual training dataset obtained 0.7849 for Turkish and 0.7999 for Spanish translations.

We also created an English-only training set alongside the multilingual dataset to enable comparison with models finetuned exclusively on English data. For this, we selected all unique instance ids (removing duplicates across languages) and extracted the corresponding English questions from the original dataset. This resulted in 20,512 unique questions. This setup allows a fair comparison between finetuning on English-only versus multilingual data with similar data coverage.

4 Experimental Results

In this section, we evaluate foundational LLMs on the Text2Cypher task across English, Spanish, and Turkish. We then compare their performance to models finetuned on English-only and multilingual training datasets.

Table 2. Instructions used for Text2Cypher task

Type	Instruction prompt
System Instruct.	Task: Generate Cypher statement to query a graph database. Instructions: Use only the provided relationship types and properties in the schema. Do not use any other relationship types or properties that are not provided in the schema. Do not include any explanations or apologies in your responses. Do not respond to any questions that might ask anything else than for you to construct a Cypher statement. Do not include any text except the generated Cypher statement.
User Instruct.	Generate Cypher statement to query a graph database. Use only the provided relationship types and properties in the schema. Schema: {schema} \n Question: {question} \n Cypher output:

4.1 Experimental Setup

We conduct our experiments on the test split of the newly created multilingual Text2Cypher dataset containing 4,783 samples for each language, which are translated from the original English questions into Spanish and Turkish. We use the following foundational models: (i) **Gemma-2-9b-it:** Released in June 2024. Primarily supports English. (ii) **Meta-Llama-3.1-8B-Instruct:** Released in July 2024. Primarily supports English and several European languages. (iii) **Qwen2.5-7B-Instruct:** Released in September 2024. Primarily supports Chinese and English and maintains multilingual support for over 29 languages. We use their Unsloth versions of models, finetuned with Instruct and quantized to 4-bit precision (e.g., unsloth/gemma-2-9b-it-bnb-4bit) for efficiency and ease of use. For the Text2Cypher task, we use the same prompts as prior work [31], shown in Table 2. After generation, a post-processing step is used for removing unwanted text, such as the 'cypher:' suffix.

We use the HuggingFace Evaluate library [16] to compute evaluation metrics. We employ two evaluation procedures: (i) **Translation-based:** Compares generated Cypher queries with reference queries based on their textual content. We report the ROUGE-L score for this evaluation. (ii) **Execution-based:** Executes both the generated and reference queries on the target database and compares their outputs (sorted lexicographically). Since only a subset of the dataset (approximately 50%) has active database access, execution-based evaluation is conducted on this subset only. We report the Exact-Match score for this evaluation.

4.2 Impact of Question Language

We first examine the impact of question language on foundational LLMs using the translated dataset. In this stage, we do not modify the task prompt for the Text2Cypher task (shown in Table 2), which is kept in English. Figure 2 shows the performance of the foundational models on English, Spanish, and Turkish questions.

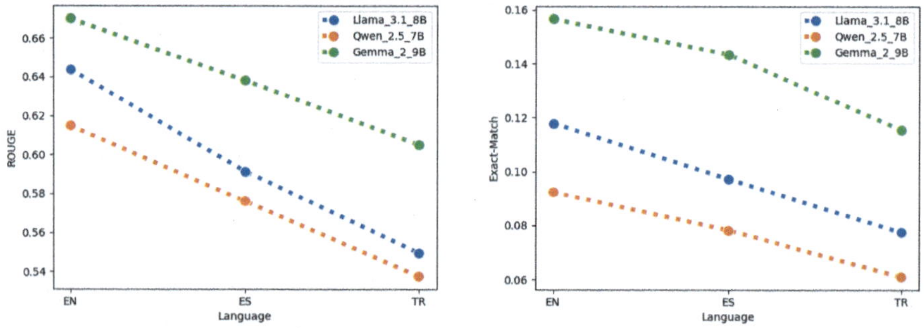

(a) Translation-based: ROUGE-L score (b) Execution-based: Exact-Match score

Fig. 2. Performance of models on Text2Cypher when input questions are in English (EN), Spanish (ES) or Turkish (TR).

The results indicate that all models perform best on English questions, followed by Spanish, with the lowest performance observed for Turkish. This pattern aligns with differences in language resourcedness of these languages and underlines the effect of training data coverage. LLMs generally perform better on languages with richer and more diverse training data, such as English, which improves the quality of learned representations [19,28]. Additionally, prior work [9] shows that models generalize more effectively across linguistically similar languages. Spanish and English both belong to the Indo-European language family, while Turkish belongs to the Altaic family, and exhibits distinct linguistic traits. Languages with complex morphology and flexible word order, such as Turkish, have greater challenges for semantic parsing and alignment to Cypher queries, as discussed above. Spanish, being linguistically closer to English in structure, experiences less from these challenges. Overall, these characteristics increase ambiguity and makes the mapping to structured queries more difficult compared to English [2,6,7,38].

In these experiments, the prompts remained in English, which may introduce additional challenges for LLMs when processing non-English questions. In the next section, we explore this effect further by evaluating model performance using both translated questions and translated prompts.

Table 3. Spanish (ES) and Turkish (TR) Instructions used for Text2Cypher

Lang.	Type	Instruction prompt
ES	System Instruct.	Tarea: Generar una sentencia Cypher para consultar una base de datos de grafos. Instrucciones: Usa únicamente los tipos de relaciones y propiedades proporcionados en el esquema. No utilices ningún otro tipo de relación ni propiedades que no estén incluidas en el esquema. No incluyas explicaciones ni disculpas en tus respuestas. No respondas a ninguna pregunta que no sea una solicitud para construir una sentencia Cypher. No incluyas ningún texto excepto la sentencia Cypher generada.
	User Instruct.	Genera una sentencia Cypher para consultar una base de datos de grafos. Usa únicamente los tipos de relaciones y propiedades proporcionados en el esquema. Esquema: {schema} \n Pregunta: {question} \n Salida Cypher:
TR	System Instruct.	Görev: Bir çizge veritabanını sorgulamak için bir Cypher ifadesi oluştur. Talimatlar: Yalnızca şemada verilen ilişki türlerini ve özellikleri kullan. Şemada verilmeyen herhangi bir ilişki türünü veya özelliği kullanma. Yanıtlarında hiçbir açıklama veya özür ifadesine yer verme. Cypher ifadesi oluşturmak dışında başka bir şey isteyen sorulara yanıt verme. Oluşturulan Cypher ifadesi dışında hiçbir metin ekleme.
	User Instruct.	Bir çizge veritabanını sorgulamak için Cypher ifadesi oluştur. Şemada verilen ilişki türleri ve özellikler dışında hiçbir şeyi kullanma. Şema: {schema} \n Soru: {question} \n Cypher çıktısı:

4.3 Impact of Prompt Language

We also translated the prompts into Spanish and Turkish to evaluate the impact of prompt language on model performance. The prompts were translated using the GPT-4o-mini model and are shown in Table 3. The performance comparison for Spanish and Turkish with and without translated prompts are presented in Fig. 3. The results show that when prompts are translated, the ROUGE-L scores, which used for translation-based evaluation comparing ground-truth and generated text, improve slightly (by about 1–1.5%). However, Exact-Match scores for execution-based evaluation remain largely unchanged. These results suggest that prompt translation has a minor impact on the overall result in our setup.

It is important to note that in our dataset, the database schemas, including nodes, relationships and properties, remain in English. Future work could explore the impact of localizing schema items to match the language of the input and prompt.

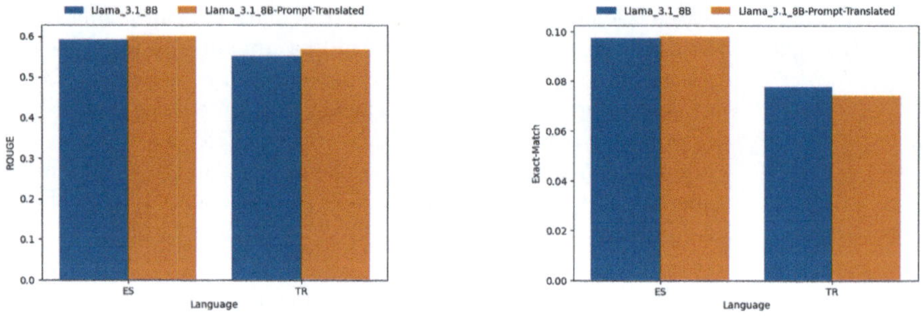

(a) Translation-based: ROUGE-L score

(b) Execution-based: Exact-Match score

Fig. 3. Model performance with/out prompt translation.

4.4 Impact of Finetuning

After observing performance differences across languages, we next explore how finetuning affects results. Among the evaluated models, we selected Meta-Llama-3.1-8B-Instruct for fine-tuning due to its multilingual support, open availability, and compatibility with our Unsloth-based setup. Considering the time and resource constraints, we focused on a single model and considered fine-tuning others as a future work. We finetune the selected baseline model separately on two datasets:

- English-only finetuned model: Trained on the English-only dataset described in Sect. 3.3. This setup allows us to evaluate how English-only finetuning affects performance on both English and non-English inputs.
- Multilingual finetuned model: Trained on the multilingual dataset to assess the benefits of exposure to multiple languages during finetuning.

The finetuned models are evaluated on the test set using data per language. In Fig. 4 we present the performance of the foundational model, finetuned model on English-only dataset and finetuned model on multilingual dataset. The figure shows that using the English-only dataset improves model performance of the Text2Cypher task across all languages. However, the improvement for Turkish is

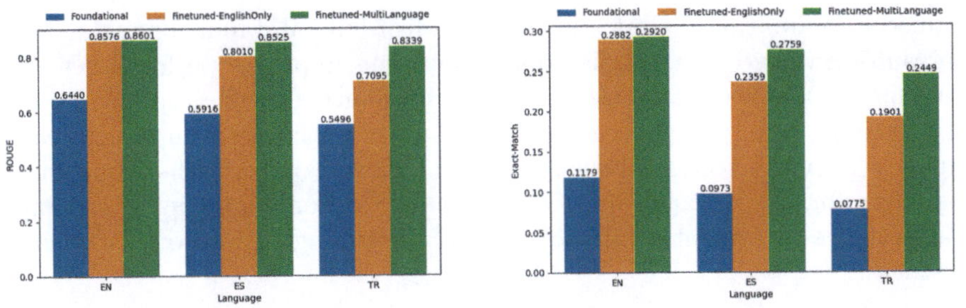

(a) Translation-based: ROUGE-L score (b) Execution-based: Exact-Match score

Fig. 4. Comparison of model performance: Foundational model (blue), Fine-tuned model on English-only data (orange), and Fine-tuned model on multilingual data (green).

smaller compared to English and Spanish. For example, based on the translation-based ROUGE-L score, English and Spanish improve by around 0.20, while Turkish improves by approximately 0.15. Spanish still performs worse than English, followed by Turkish. This performance gap may be attributed to linguistic differences: English and Spanish belong to the Indo-European family, whereas Turkish is an Altaic language with different characteristics. When evaluating the model finetuned on the multilingual dataset, we observe that performance across languages becomes more balanced. For instance, ROUGE-L scores reach 0.86 for English, 0.85 for Spanish, and 0.83 for Turkish. These results suggest that multilingual finetuning improves generalization and narrows the performance gap across languages.

Overall, these results indicate that both language resourcedness and linguistic characteristics influence model performance. Foundational models and those finetuned only on English struggle with non-English inputs, particularly for languages with different characteristics or with less resources. Multilingual finetuning mitigates these challenges and leads to more balanced performance across languages.

5 Conclusion

This work presents a performance comparison of foundational and finetuned LLMs on the Text2Cypher task across English, Spanish, and Turkish. We created and released a multilingual dataset by translating user questions in English into Spanish and Turkish, enabling fair cross-lingual evaluation. Our results consistently show that foundational models perform best on English inputs, followed by Spanish, and then Turkish. This pattern is likely driven by differences in language resourcedness and linguistic characteristics, which introduce additional challenges for LLMs. We also explored the effect of translating task instruction prompts alongside the input questions. Our findings suggest prompt language has minimal impact on overall performance in our setting. Finally, we finetuned a foundational model on both English-only and multilingual datasets. The results revealed that English-only finetuning improves performance across all languages, it does not eliminate the performance gap. In contrast, multilingual finetuning improves generalization and narrows the performance gap across languages. Future directions include schema localization to further support multilingual performance, expanding evaluation to additional languages, and exploring alternative alignment strategies beyond finetuning to improve cross-lingual robustness.

Declaration on Generative AI Usage. During the preparation of this work, the author(s) used Chat-GPT in order to: 'Improve writing style' and 'Paraphrase and reword'. After using these tool(s) or service(s), the author(s) reviewed and edited the content as needed and take(s) full responsibility for the publication's content.

References

1. Text2SPARQL challenge 2025 (2025). https://text2sparql.aksw.org/. Accessed 5 July 2025
2. Alecakir, H., Bölücü, N., Can, B.: TurkishdelightNLP: a neural Turkish NLP toolkit. In: Proceedings of the 2022 Conference of the North American Chapter of the Association for Computational Linguistics: Human Language Technologies: System Demonstrations, pp. 17–26 (2022)
3. Almohaimeed, S., Almohaimeed, S., Al Ghanim, M., Wang, L.: AR-spider: text-to-SQL in Arabic. In: Proceedings of the 39th ACM/SIGAPP Symposium on Applied Computing, pp. 1024–1030 (2024)
4. Bakshandaeva, D., Somov, O., Dmitrieva, E., Davydova, V., Tutubalina, E.: PAUQ: text-to-SQL in Russian. In: Findings of the Association for Computational Linguistics: EMNLP 2022, pp. 2355–2376 (2022)
5. Bandarkar, L., et al.: Layer swapping for zero-shot cross-lingual transfer in large language models. arXiv preprint arXiv:2410.01335 (2024)
6. Baucells, I., et al.: IberoBench: a benchmark for LLM evaluation in Iberian languages. In: Proceedings of the 31st International Conference on Computational Linguistics, pp. 10491–10519 (2025)
7. Bayram, M.A., Fincan, A.A., Gümüş, A.S., Diri, B., Yıldırım, S., Aytaş, Ö.: Setting standards in Turkish NLP: TR-MMLU for large language model evaluation. arXiv preprint arXiv:2501.00593 (2024)
8. Cui, R., Aralikatte, R., Lent, H., Hershcovich, D.: Compositional generalization in multilingual semantic parsing over WikiData. Trans. Assoc. Comput. Linguist. (2022). https://arxiv.org/abs/2108.03509
9. Dhamecha, T.I., Murthy V, R., Bharadwaj, S., Sankaranarayanan, K., Bhattacharyya, P.: Role of language relatedness in multilingual fine-tuning of language models: a case study in Indo-Aryan languages. arXiv preprint arXiv:2109.10534 (2021)
10. Dou, L., et al.: MultiSpider: towards benchmarking multilingual text-to-SQL semantic parsing. In: Proceedings of the AAAI Conference on Artificial Intelligence, vol. 37, pp. 12745–12753 (2023)
11. Etxaniz, J., Azkune, G., Soroa, A., de Lacalle, O.L., Artetxe, M.: Do multilingual language models think better in English? arXiv preprint arXiv:2308.01223 (2023)

12. Geng, X., et al.: Why not transform chat large language models to non-English? arXiv preprint arXiv:2405.13923 (2024)
13. Hogan, A., et al.: Knowledge graphs. ACM Comput. Surv. (CSUR) **54**(4), 1–37 (2021)
14. Huang, H., et al.: Not all languages are created equal in LLMS: improving multilingual capability by cross-lingual-thought prompting. arXiv preprint arXiv:2305.07004 (2023)
15. Huang, Y., et al.: Exploring the landscape of text-to-SQL with large language models: Progresses, challenges and opportunities. arXiv preprint arXiv:2505.23838 (2025)
16. HuggingFace: HuggingFace evaluate (2024). https://huggingface.co/evaluate-metric
17. Jannuzzi, M., Perezhohin, Y., Peres, F., Castelli, M., Popovič, A.: Zero-shot prompting strategies for table question answering with a low-resource language. Emerging Sci. J. **8**(5), 2003–2022 (2024)
18. José, M.A., Cozman, F.G.: MRAT-SQL+ gap: a Portuguese text-to-SQL transformer. In: Intelligent Systems: 10th Brazilian Conference, BRACIS 2021, Virtual Event, November 29–December 3, 2021, Proceedings, Part II 10, pp. 511–525. Springer (2021). https://doi.org/10.1007/978-3-030-91699-2_35
19. Joshi, P., Santy, S., Budhiraja, A., Bali, K., Choudhury, M.: The state and fate of linguistic diversity and inclusion in the NLP world. arXiv preprint arXiv:2004.09095 (2020)
20. Kanburoglu, A.B., Tek, F.B.: Turspider: a Turkish text-to-SQL dataset and LLM-based study. IEEE Access (2024)
21. Kargaran, A.H., Modarressi, A., Nikeghbal, N., Diesner, J., Yvon, F., Schütze, H.: MEXA: Multilingual evaluation of English-centric LLMS via cross-lingual alignment. arXiv preprint arXiv:2410.05873 (2024)
22. Lai, V.D., et al.: Okapi: instruction-tuned large language models in multiple languages with reinforcement learning from human feedback. arXiv preprint arXiv:2307.16039 (2023)
23. Lai, W., Mesgar, M., Fraser, A.: LLMS beyond English: scaling the multilingual capability of LLMS with cross-lingual feedback. arXiv preprint arXiv:2406.01771 (2024)
24. Lan, M., Xia, Y., Zhou, G., Huang, N., Li, Z., Wu, H.: LLM4QA: leveraging large language model for efficient knowledge graph reasoning with SPARQL query. JAIT **15**(10), 1157–1162 (2024)
25. Li, H., Koto, F., Wu, M., Aji, A.F., Baldwin, T.: Bactrian-x: multilingual replicable instruction-following models with low-rank adaptation. arXiv preprint arXiv:2305.15011 (2023)
26. Luo, Y., et al.: YAYI 2: multilingual open-source large language models. arXiv preprint arXiv:2312.14862 (2023)
27. Min, Q., Shi, Y., Zhang, Y.: A pilot study for Chinese SQL semantic parsing. arXiv preprint arXiv:1909.13293 (2019)
28. Nicholas, G., Bhatia, A.: Lost in translation: large language models in non-English content analysis. arXiv preprint arXiv:2306.07377 (2023)
29. Nooralahzadeh, F., et al.: StatBot. SWISS: bilingual open data exploration in natural language. arXiv preprint arXiv:2406.03170 (2024)
30. Ozsoy, M.G.: Multilingual prompts in LLM-based recommenders: performance across languages. arXiv preprint arXiv:2409.07604 (2024)

31. Ozsoy, M.G., Messallem, L., Besga, J., Minneci, G.: Text2Cypher: bridging natural language and graph databases. In: Proceedings of the Workshop on Generative AI and Knowledge Graphs (GenAIK), pp. 100–108 (2025)
32. Pedroso, B.C., Pereira, M.R., Pereira, D.A.: Performance evaluation of LLMS in the text-to-SQL task in Portuguese. In: Simpósio Brasileiro de Sistemas de Informação (SBSI), pp. 260–269. SBC (2025)
33. Perevalov, A., Both, A.: Towards LLM-driven natural language generation based on SPARQL queries and RDF knowledge graphs (2024)
34. Perevalov, A., Both, A., Ngonga Ngomo, A.C.: Multilingual question answering systems for knowledge graphs-a survey. Semant. Web **15**(5), 2089–2124 (2024)
35. Rei, R., et al.: CometKiwi: IST-UNBABEL 2022 submission for the quality estimation shared task. arXiv preprint arXiv:2209.06243 (2022)
36. Schut, L., Gal, Y., Farquhar, S.: Do multilingual LLMS think in English? arXiv preprint arXiv:2502.15603 (2025)
37. Srivastava, N., Ma, M., Vollmers, D., Zahera, H., Moussallem, D., Ngomo, A.C.N.: MST5–multilingual question answering over knowledge graphs. arXiv preprint arXiv:2407.06041 (2024)
38. Tohma, K., Kutlu, Y.: Challenges encountered in Turkish natural language processing studies. Nat. Eng. Sci. **5**(3), 204–211 (2020)
39. Üstün, A., et al.: Aya model: an instruction finetuned open-access multilingual language model. arXiv preprint arXiv:2402.07827 (2024)
40. Yang, W., Li, C., Zhang, J., Zong, C.: Bigtranslate: Augmenting large language models with multilingual translation capability over 100 languages. arXiv preprint arXiv:2305.18098 (2023)
41. Yu, T., et al.: Spider: a large-scale human-labeled dataset for complex and cross-domain semantic parsing and text-to-SQL task. arXiv preprint arXiv:1809.08887 (2018)
42. Zhao, J., Zhang, Z., Gao, L., Zhang, Q., Gui, T., Huang, X.: Llama beyond English: an empirical study on language capability transfer. arXiv preprint arXiv:2401.01055 (2024)
43. Zhong, C., et al.: Beyond English-centric LLMS: what language do multilingual language models think in? arXiv preprint arXiv:2408.10811 (2024)
44. Zhu, W., et al.: Extrapolating large language models to non-English by aligning languages. arXiv preprint arXiv:2308.04948 (2023)

Improved Key Information Extraction Capabilities for Small Vision-Language Model

Pik Ying Yip, Siu-Kei Au Yeung(✉), and Kwok Tai Chui

School of Science and Technology, Hong Kong Metropolitan University, Hong Kong, China
s1155180@live.hkmu.edu.hk, {jauyeung,jktchui}@hkmu.edu.hk

Abstract. This paper presents a document image processing pipeline that enhances key information extraction (KIE) using Small Vision-Language Models (SVLMs). KIE is pivotal for enterprise document automation, and the move toward zero-/few-shot MLLMs motivates a privacy-preserving, on-premise SVLM-based approach to improve KIE efficiency and accuracy. The pipeline couples document layout analysis with OCR and QLoRA fine-tuning to segment the document images and construct keyword-guided refined composites for SmolVLM-256M-Instruct, reducing tokens and improving efficiency. On FUNSD and SROIE, refined inputs outperform original images in field-level accuracy, while an upper-bound target segmentation achieves substantially higher accuracy, highlighting the promise of dedicated KVP region detection; token counts also drop markedly. An ablation clarifies why choosing a high fuzzy-match threshold and a small number of segments is effective, providing empirical justification for design choices. Our current evaluation compares KIE performance of the proposed pipeline using SmolVLM-256M-Instruct against other SVLMs. Future work will broaden model coverage, support multi-page inputs, and output structured JSON to enable end-to-end, reproducible deployment.

Keywords: Key Information Extraction (KIE) · Small Vision Language Model (SVLM) · Artificial intelligence (AI)

1 Introduction

Business documents are central to modern commerce, especially in B2B communication, service applications, and qualification approvals. Many such documents are visually structured and contain numerous key–value pairs in forms and tables. Traditionally, staff needed to review these documents and manually enter key information into enterprise systems. Automating key information extraction (KIE) is therefore essential for processing large document volumes efficiently.

Before the rise of large language models (LLMs) [36], deep learning methods such as LayoutLM [9], Donut [14], and LiLT [17] achieved strong KIE performance but typically required task or format specific fine-tuning. With the advent

H. Unger and P. Meesad (Eds.): NLPIR 2025, LNNS 1904, pp. 1220–1233, 2026.
https://doi.org/10.1007/978-3-032-20897-2_85

of user-facing systems like ChatGPT and subsequent multimodal large language models (MLLMs), including GPT-4o [11], Claude 4 [12], and Gemini 2.5 [13], users can query documents directly and request structured outputs (e.g. JSON) without task-specific training. Despite their strong zero-shot capabilities, most MLLMs are cloud-hosted and large-scale, which raises deployment and privacy concerns for enterprise use. For enterprises constrained by data-residency and cost, on-premise Small Vision-Language Models (SVLMs) offer a pragmatic path to zero/few-shot KIE without exposing sensitive content to third-parties.

These constraints motivate the use of SVLMs. Small language models (SLMs) are language models with fewer than 10B parameters that can run on commodity hardware with acceptable latency [2]. SVLMs integrate these small language models with a vision encoder to enable visual understanding. They have shown strong performance across visual tasks such as visual question answering, object detection, image captioning, and key information extraction. Although there remains a performance gap compared with large cloud-hosted models, practical strategies to improve SVLMs are valuable. For example, increasing image resolution can benefit image understanding [15]. Raising input resolution improves document understanding but increases visual token counts through more patches/tiles, pressuring context limits; thus, token-reduction via layout-guided segmentation and OCR-assisted text recovery is valuable.

This research offers the following contributions:

- **Development of an Integrated Processing Pipeline**: Development of an integrated processing pipeline that augments SVLM-based KIE via layout-guided segmentation and OCR-assisted text recovery.
- **Demonstration of Performance Enhancement with Limited Data**: Demonstration of accuracy improvements with limited training data through QLoRA-based model fine-tuning and segment inputs.
- **Token-efficient via document image segmentation**: Token-efficient processing with measured reductions in visual tokens, enabling longer contexts and faster inference (details in Sect.4.3).
- **Validation of Document Segmentation for KIE Accuracy**: Validation that layout-based segmentation improves field-level KIE accuracy on FUNSD and SROIE, with target segment upper bounds revealing further headroom.

The proposed integrated processing pipeline strengthens the KIE capability of SVLMs. We focus on SmolVLM-256M-Instruct to isolate pipeline gains under resource constraints; broader model comparisons are deferred. Following numerous experiments, we conclude that image segmentation based on document layout can enhance the accuracy of SVLMs in KIE tasks. With the assistance of a document analysis model, the original documents are segmented according to their visual layout. Then, an OCR engine is used to recover unincluded text and search for the keyword in each segment with fuzzy partial ratio threshold n. Finally, the segments that possibly contain the keyword are combined into a refined input image for an SVLM. Selecting a small top-k mitigates the classic omission–distraction trade-off in segment selection; we therefore study this trade-off via ablations and, by default, adopt a conservative n and k informed by the results (see Sect.4.4).

2 Related Works

2.1 Key Information Extraction

Optical character recognition (OCR) has a rich historical foundation and remains a cornerstone technology for document understanding and key information extraction. This technique uses optical mechanisms to identify characters and typically captures text bounding box information [10]. Modern learning-based OCR systems typically output recognized text together with structured layout attributes such as bounding boxes or polygons, confidence scores, and line/paragraph structure. The relationships between these textual units can be modeled using layout-aware approaches, including graph convolutional neural networks (GCN) [6] and transformer architectures such as LayoutLM [9] and LiLT [17], which rely on a pre-built OCR engine or PDF parser to provide text and their positions.

Beyond OCR-dependent pipelines, end-to-end OCR-free frameworks have emerged, such as Donut [14] and UDOP [21], which generate structured key-value pair outputs (e.g. JSON) directly from images. OCR-free methodologies can, in certain tasks, reduce error propagation from external OCR and simplify training; however, tiny text, multilingual mixing, low-resolution scans, dense tables, or long-context dependencies often remain challenging. Both OCR-dependent and OCR-free paradigms achieve strong performance in KIE, with complementary strengths depending on application constraints.

Despite these advances, the system based on the above methods require task- or format-specific training to accommodate domain- and layout-specific variations. As document formats continually evolve across domains, zero-shot and few-shot generalization has become an active focus, targeting adaptation to novel document types with minimal supervision. The advent of large language models further accelerates this trend: many vision-language models exhibit nontrivial text understanding in images and are often treated as OCR-free solutions for KIE, although the granularity and robustness of such capabilities vary by model and task.

To support deployment under tight memory and latency budgets, SVLMs are particularly attractive; however, they typically require targeted training strategies (e.g., QLoRA-based tuning, 2D positional encodings, and multi-scale tiling) to match the KIE robustness of larger models in dense, multilingual, or highly structured documents.

2.2 Small Vision-Language Models

SVLMs are multimodal models that pair a lightweight language model with a vision encoder. The admissible input image size is determined by the chosen vision encoder, with common resolutions including 224×224, 336×336, and

512×512 pixels. Directly resizing full-page documents to a fixed resolution is straightforward but may introduce distortion and suppress fine-grained cues, which are often critical for document understanding. An alternative is to preserve local detail by splitting or tiling images into encoder-compatible patches; this strategy has been shown to benefit visually dense documents and high-resolution settings, such as SAM [18] and Monkey [19]. Splitting preserves the original resolution and thereby enhances performance, particularly for visually rich documents [16].

However, higher effective resolution typically increases the number of visual tokens and can approach or exceed the model's maximum input length budget [16]. To mitigate this, token-efficiency techniques can be employed, such as reducing redundancy in visual tokens or compressing repeated patterns, as demonstrated in TextMonkey [20]. Another practical route is to leverage conventional OCR engine for providing extra information in order to lower the visual token burden while injecting text structure explicitly. For instance, extracted text and bounding box information can be integrated into the user prompt using a predefined format, as in ICL-D3IE [22], or transformed into feature vectors via an additional encoder module, such as in BLIP-2 [23]. These strategies balance resolution, token cost, and structured priors, making SVLMs more effective under limited compute and context budgets.

2.3 Document Layout Analysis Models

Document layout analysis (DLA) structurally parses documents to identify salient elements—including titles, headers, paragraphs, tables, figures, and form fields—to support downstream tasks such as key information extraction (KIE) and optical character recognition (OCR) [26]. Recent approaches demonstrate robust performance across public benchmarks (e.g., DocLayNet, PubLayNet) and heterogeneous layouts.

Transformer-based detectors, such as PP-DocLayout [26] and DLA-Former [27], leverage global contextual modeling to improve element detection and robustness across diverse layouts.

Convolutional neural network (CNN) adaptations of one-stage detectors, exemplified by DocLayout-YOLO [28], provide competitive accuracy with favorable inference efficiency by tailoring the YOLO paradigm to document structures.

Graph-based methods, such as GLAM [29], explicitly model inter-element relationships to enhance structural consistency and relational reasoning in layout understanding.

3 Methodology

3.1 Architecture

The proposed system follows a three-stage pipeline comprising: (i) a document processing layer, (ii) an optical character recognition (OCR) engine, and (iii)

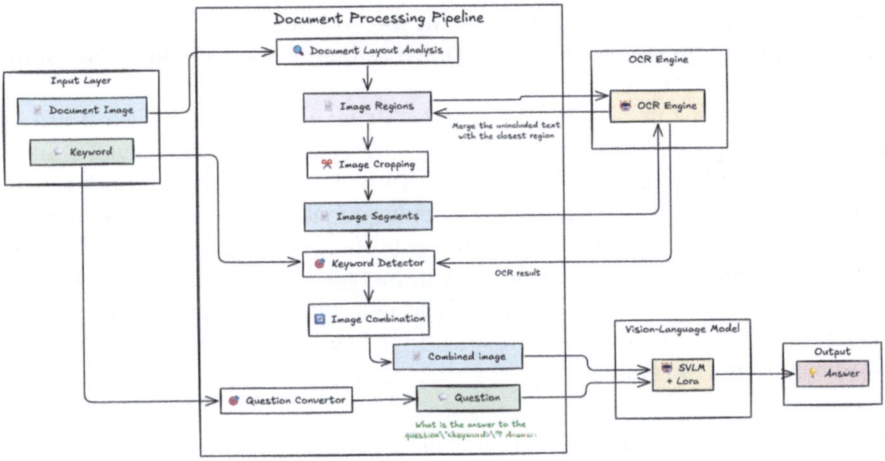

Fig. 1. The architecture of the proposed pipeline method.

a small vision-language model (SVLM) equipped with LoRA adapters (Fig. 1). Given a user-specified keyword and a document image, the document processing layer first performs layout analysis to segment the page into visual regions and merges overlapping regions; subsequently, OCR recovers words that fall outside detected regions and re-associates them to the nearest region, enabling complete sentence grouping within regions. The system then computes fuzzy partial ratios between the keyword and region texts to identify candidate regions and composes a refined image by stitching the top-k regions. If no region satisfies the threshold n, we revert to the original page to avoid false negatives, ensuring robust behavior under detection or OCR failures. In parallel, the keyword is converted into a question prompt for the SVLM (See Fig. 1). Finally, the refined image (or the fallback page) and the generated question are fed into the SVLM whose weights are augmented at inference by LoRA adapters fine-tuned via QLoRA, and the model's prediction is taken as the value corresponding to the keyword. This staged design isolates error sources and reduces the visual search space before SVLM inference, which we found critical for small-capacity models under token budgets. The fuzzy partial ratio is adopted to tolerate OCR noise and minor lexical variants, while top-n region composition balances recall (avoiding missed cues) and precision (limiting distractors); ablations later show k = 2 with an n = 80 threshold as near-optimal on FUNSD.

3.2 Document Processing Layer

This layer initially employs a document layout analysis model, specifically DocLayout-YOLO, to segment the original document page into multiple regions based on visual feature detection (See Fig. 2). Consequently, some words may be omitted from these identified layout regions. To address this, an OCR engine is subsequently utilized to locate and associate these excluded words with their nearest corresponding region, facilitating key-value pair grouping. In addition, the overlapped regions should be combined as one region to reduce the complexity of the selection process.

After image segmentation, the segments that include the keyword are identified by calculating fuzzy partial ratios between the target keyword and the extracted texts from the OCR engine. With a specific ratio, the top k segments that contain the keyword are selected. Subsequently, the selected segments are integrated together to generate a composite image for the VLM. If there is no selected segment, the original document image will be returned as the input of VLM. Therefore, the KIE result of the new refined images should be better than that of original document images by setting suitable fuzzy partial ratios threshold n and the value of k.

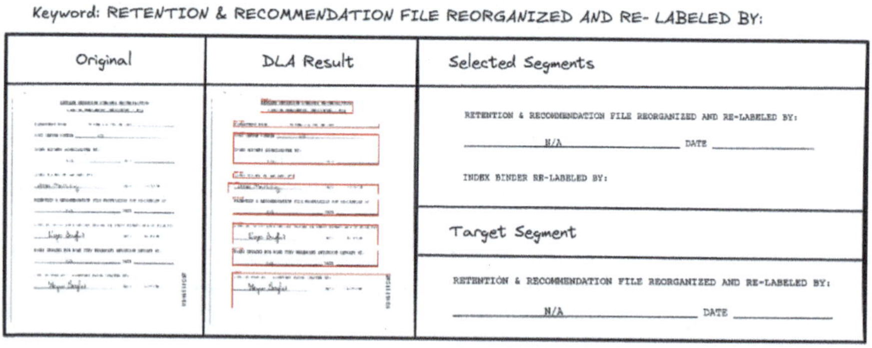

Fig. 2. Illustration of segment selection and the target segment. The leftmost panel shows an example business document (original image). A pre-trained DocLayout-YOLO model produces region proposals (DLA result), after which an OCR engine assigns missing or stray tokens to their nearest regions to recover complete sentences for each region. Next, segments are ranked by fuzzy partial ratio between the user-provided keyword and OCR texts; with a chosen threshold t and top-n, the selected segments are aggregated into a refined image for the SVLM. If no segment passes the threshold, the pipeline falls back to the original image. The "target segment" is defined as the minimal region that exclusively encloses the key–value pair and serves as an oracle upper bound in our analysis.

3.3 OCR Engine

In the proposed pipeline, an OCR engine is used to assist the document layout analysis model by reintegrating textual fragments that were partially excluded by region boundaries, ensuring sentence completeness within each detected region and improving the fidelity of layout interpretation. Concretely, OCR-detected words and their bounding boxes are aligned to layout regions via coordinate normalization and IoU-based gating, and only within-region or immediate-neighbor text is considered to avoid cross-column merging. In addition, we retrieve region-level keywords by computing fuzzy partial ratios between query terms and OCR tokens, applying normalization (case, digits, Unicode) and a fixed threshold to select relevant segments.

EasyOCR is adopted for its practical speed–accuracy trade-off and multilingual coverage, yielding word-level text and bounding boxes that support our alignment and retrieval steps [24]. Rather than feeding full OCR text into the VLM, we rely on keyword-guided segment selection to reduce the impact of OCR noise and missing tokens, which empirically improves robustness in downstream extraction.

3.4 Small Vision Language Model

To ensure compatibility with small vision-language models (SVLMs), we adopt the SmolVLM family of compact vision-language models as our base. In particular, we use the SmolVLM-256M and its instruction-tuned variant (SmolVLM-256M-Instruct), which has approximately 256M parameters and requires less than 1 GB of GPU memory only [31]. This model demonstrates strong performance across a range of image-related tasks, such as question answering and table understanding. For task formulation, we convert target keywords into natural-language questions so that the SVLM can retrieve and align relevant visual-textual context within the documents.

We fine-tune SmolVLM-256M-Instruct with document images using QLoRA to enhance document understanding on both full pages and selected segments [30]. QLoRA quantizes and freezes the base model weights while training low-rank adapter parameters, substantially reducing memory usage and avoiding full base-weight updates; this helps preserve general capabilities while specializing to document layouts and typography [30]. We apply the same formulation to both whole pages and region crops to capture global context and fine-grained structures.

4 Experiment Results

4.1 Training Details

We adopt FUNSD, a form-understanding dataset with word-level annotations and relations, comprising 149 training pages and 50 testing pages. In our processed splits, we aggregate the original annotations into 1,989 key–value pairs (KVPs) for training and 531 KVPs for testing, computed by pairing each annotated key entity with its linked value entities according to FUNSD relations. For each pair, we additionally generate a textual question:'What is the answer to the question:"<keyword>"? Answer:' (See Fig. 1) format to support supervision and evaluation consistency.

To construct target segments, we derive a bounding region for each KVP by taking the tightest enclosing rectangle that covers the union of all word boxes belonging to the key span and its linked value span(s); when multiple value spans exist, we merge their boxes before enclosure; overlapping regions across KVPs are kept as independent samples to preserve instance granularity.

Consequently, QLoRA's training utilizes both the original images or the processed. Since the resolution, size or structure may not be common in the training set of those VLM, this is necessary for improving the understanding of the segmented images. 1000 key-value pairs from the FUNSD training dataset are randomly selected for training the LoRA, the input images come from both the original images and the target segment in a specific ratio.

For document layout analysis (DLA), we use DocLayout-YOLO as a frozen, off-the-shelf detector without any fine-tuning; it outputs 11 layout classes (e.g., table, figure, plain text, formula) as defined by the released model. DocLayout-YOLO is pretrained on DocSynth300K and has demonstrated strong transfer for document understanding tasks, which we leverage to group visually coherent content even though KVP regions are not explicit classes.

All training and inference operations are performed on a computer equipped with two RTX 3080 Ti GPUs, which is a consumer-grade computing environment.

4.2 Evaluation Details

The testing datasets of FUNSD, comprising 50 documents, and SROIE, consisting of 2756 English documents, were utilized to evaluate the KIE capability of the proposed integrated pipeline interpretation SVLM. Consistent with the training procedure, a summary containing all key-value pairs were generated of each document image. All keywords were converted to a question for the query. The evaluation towards the SmolVLM-256M-Instruct model with LoRA were conducted with inputting the original document images, pipeline processed images and target segment images based on the provided text boundary information.

KIE performance is evaluated in overall, field-level accuracy. A prediction is considered correct if it includes the ground truth or is included within the ground truth; otherwise, it is considered wrong. This approach constitutes a straightforward and effective technique for determining whether the value of the correlated keyword has been exhaustively extracted.

Table 1. Accuracy Comparison of SmolVLM-256M-Instruct on different configurations and datasets

Model Configuration	FUNSD Acc.	SROIE Acc.
SmolVLM-256M	43.77%	30.75%
SmolVLM-256M + QLoRA	51.69%	34.74%
SmolVLM-256M + QLoRA + pipeline processing	**54.51%**	**39.57%**
SmolVLM-256M + QLoRA + targeted segments	64.15%	57.38%

The evaluation results on the different approaches that use SmolVLM-256M-Instruct model, summarized in Table 1, clearly demonstrate that the accuracy significantly improves when utilizing the refined images generated by the proposed pipeline compared to inputting the original document image. Specifically, the SmolVLM-256M + QLoRA + pipeline processing configuration achieved 54.51% accuracy on FUNSD and 39.57% on SROIE, representing a notable improvement over the base SmolVLM-256M + QLoRA model (51.69% FUNSD, 34.74% SROIE) and the SmolVLM-256M model (43.77% FUNSD, 30.75% SROIE). This finding underscores that document segmentation, predicated on the structural layout of documents, effectively facilitates SVLMs in accurately identifying the target answer.

It is also noteworthy that the SmolVLM-256M + QLoRA + targeted segments configuration achieved the highest accuracy (64.15% FUNSD, 57.38% SROIE). This superior performance, achieved by feeding precisely extracted key-value pair regions (target segments), indicates the significant untapped potential for further optimization of the pipeline. The gap between the pipeline's performance and that of directly providing targeted segments suggests that more precise key-value pair region detection is crucial for achieving even higher KIE accuracy. This highlights a key area for future development within our pipeline.

Table 2. Accuracy comparison on FUNSD among representative SVLMs and our proposed pipeline with SmolVLM-256M-Instruct.

	Model name	Model size	FUNSD Acc.
SVLMs	Phi-4-multimodal-instruct [34]	5.57B	55.45%
	InternVL2-1B [35]	938M	47.18%
	TextMonkey [20]	7B	53.75%
	SmolVLM-256M-Instruct	256M	43.77%
	SmolVLM-500M-Instruct	500M	48.12%
	SmolVLM-Instruct	2B	57.89%
Ours	**SmolVLM-256M with proposed pipeline**	~300M	**54.51%**

To contextualize our method against existing approaches, we evaluate FUNSD form-understanding accuracy across a range of vision–language models (VLMs) spanning 256M to 7B parameters. As shown in Table 2, our proposed pipeline with SmolVLM-256M (~300M, 54.51%) outperforms several larger baselines, such as SmolVLM-500M (500M, 48.12%) and InternVL2-1B (938M, 47.18%). It also achieves performance comparable to TextMonkey (7B, 53.75%), Phi-4-multimodal-instruct (5.57B, 55.45%) and SmolVLM-Instruct (2B, 57.89%). These results demonstrate that our approach delivers competitive accuracy with a substantially smaller model size, underscoring the effectiveness of the proposed pipeline for enhancing small-model performance with only a minimal increase in parameter size.

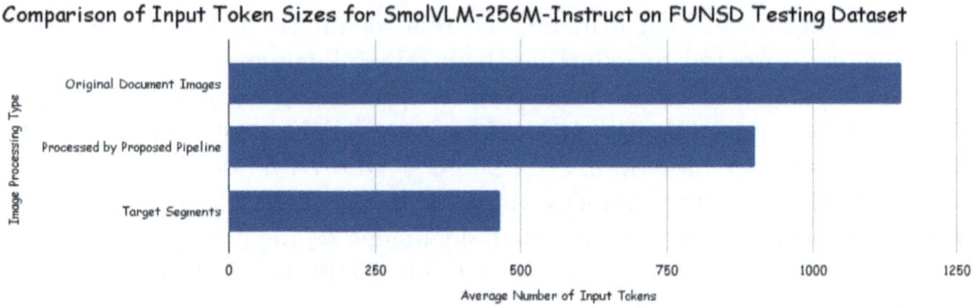

Fig. 3. Comparison of input token sizes for SmolVLM-256M-Instruct on FUNSD testing dataset: original document images(1152 tokens), images processed by proposed pipeline(899 tokens), and target segments(462 tokens).

4.3 Token Reduction

The evaluation results demonstrate a significant reduction in input tokens for the SVLM due to the smaller input image size (See Fig. 3). Compared to using original document images, the proposed pipeline achieves a 28.11% reduction in token count. In the optimal scenario, which leverages target segments, a 149% improvement relative to the original baseline. This sizable reduction lowers per-forward-pass compute for the SVLM and, under a fixed token budget, leaves more headroom for prompt design and multi-turn interaction.

4.4 Ablation Study in Crops Selection

Parameters in segment selection include the fuzzy partial ratio threshold to detect whether the keyword is a substring in the segment, and the number of segments to be selected. Here are the results on FUNSD testing dataset with different parameters (See Fig. 4).

This graph shows how the parameters for segment selection affect the KIE accuracy. The green line is the accuracy that feeds the original document images to the SVLM with LoRA directly, which is also considered as the baseline of the experiment. In general, selecting only the top two segments is better than the others. If the threshold of the fuzzy partial ratio is less than 50, the accuracy values are lower than the baseline, especially in selecting the top 3 and 4 segments. It may be because the incorrect segments mislead the SVLM to extract the information. The best combination is 80 thresholds and selecting top 2 segments. If the threshold is more than 80, the accuracy will drop because the correct region may be filtered.

The ablation study reveals that while the integration of the document layout analysis model and the OCR engine significantly improves KIE capability, it also exposes the inherent inaccuracy of OCR results. We observed instances where keywords could not be exactly found within the segments, which indicates OCR errors or misalignments can mislead the SVLM in information extraction. This underscores the critical role of the SVLM in either correcting or implicitly

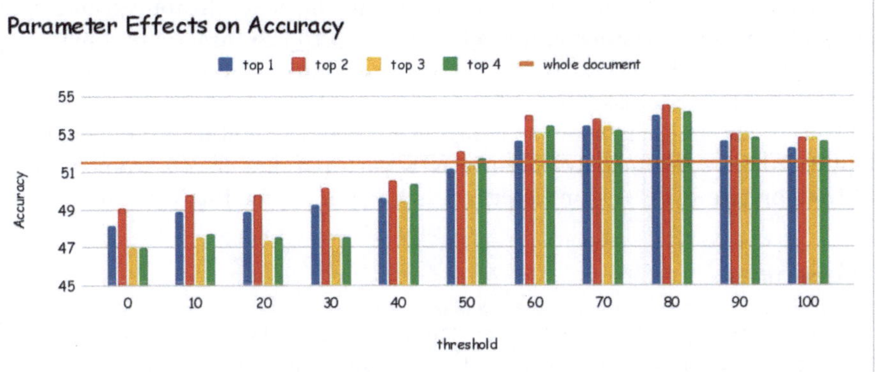

Fig. 4. The bar chart shows how the parameters in selecting segments affect the accuracy of the model.

ignoring these OCR inaccuracies in OCR-correlated approaches. This finding further supports our conclusion that relying solely on general DLA models and potentially error-prone OCR for precise key-value region delineation presents a limitation, pointing towards the necessity of developing dedicated models for key-value pair extraction.

5 Conclusion

In summary, this study introduces a segmentation-and-refinement pipeline that integrates document layout analysis and OCR with a QLoRA-enhanced SVLM, resulting in significant improvements in key information extraction (KIE) performance on FUNSD and SROIE. Beyond accuracy, the pipeline reduces input tokens by 28.11% on FUNSD, improving efficiency and enabling richer prompting within the same token budget. While our current evaluation uses a single SVLM and focuses on single-best-answer, single-page scenarios, the results establish a robust foundation for practical KIE with resource-constrained models.

Our study has limitations. We evaluate a single SVLM and focus on single best-answer scenarios, which may not reflect performance on multi-answer cases (e.g., tables or repeated forms). The current pipeline also supports only single-page documents.

To translate these findings into deployable systems, we will (i) build a dedicated KVP detector by augmenting DocLayout-YOLO with KVP classes via weak supervision and synthetic data, coupled with multi-task learning for layout detection, KVP localization, and key–value linking; (ii) support multi-page documents using page-wise segmentation, intra-page candidate selection, and inter-page fusion/ranking with confidence- and continuity-weighted deduplication; and (iii) enforce structured JSON outputs by supervised learning. Also, we will broaden evaluation across more SVLMs and other OCR-free models (e.g., Donut) with statistical testing to establish generalizability and significance. These concrete steps aim to close the gap to the targeted-segment upper bound, enhance robustness to OCR noise and multi-answer cases (e.g., tables), and enable reliable, cross-page extraction for real-world deployments.

Overall, the findings show that layout-guided segmentation and lightweight adaptation can substantially enhance KIE with parameter-constrained SVLMs, and that precise KVP region detection is a key lever for further gains.

References

1. Liao, W., Wang, J., Li, H., Wang, C., Huang, J., Jin, L. DoclayLLM: an efficient and effective multi-modal extension of large language models for text-rich document understanding. ArXiv Preprint ArXiv:2408.15045 (2024)
2. Belcak, P., et al.: Small language models are the future of agentic AI. ArXiv Preprint ArXiv:2506.02153 (2025)
3. Li, B., et al.: LLAVA-ONEVISION: easy visual task transfer. ArXiv Preprint ArXiv:2408.03326 (2024)
4. Kuang, J., et al.: Natural language understanding and inference with MLLM in visual question answering: a survey. ACM Comput. Surv. **57**, 1–36 (2025)
5. Antol, S., et al.: VQA: visual question answering. In: Proceedings Of The IEEE International Conference On Computer Vision, pp. 2425–2433 (2015)
6. Kastanas, S., Tan, S., He, Y. Document AI: A comparative study of transformer-based, graph-based models, and convolutional neural networks for document layout analysis. ArXiv Preprint ArXiv:2308.15517 (2023)
7. Grishman, R.: Information extraction: techniques and challenges. In: Pazienza, M.T. (ed.) SCIE 1997. LNCS, vol. 1299, pp. 10–27. Springer, Heidelberg (1997). https://doi.org/10.1007/3-540-63438-X_2
8. Khan, S., Naseer, M., Hayat, M., Zamir, S., Khan, F., Shah, M.: Transformers in vision: a survey. ACM Comput. Surv. (CSUR) **54**, 1–41 (2022)
9. Xu, Y., Li, M., Cui, L., Huang, S., Wei, F., Zhou, M.: LayoutLM: pre-training of text and layout for document image understanding. In: Proceedings Of The 26th ACM SIGKDD International Conference On Knowledge Discovery & Data Mining, pp. 1192–1200 (2020)
10. Mithe, R., Indalkar, S., Divekar, N.: Optical character recognition. Int. J. Recent Technol. Eng. (IJRTE) **2**, 72–75 (2013)
11. Hurst, A., et al.: GPT-4O system card. ArXiv Preprint ArXiv:2410.21276 (2024)
12. Anthropic claude 4 system card: claude opus 4 & Claude sonnet 4. (Anthropic,2025,5) (2025). https://www-cdn.anthropic.com/4263b940cabb546aa0e3283f35b686f4f3b2ff47.pdf
13. Google DeepMind Gemini 2.5 pro preview model card. (2025). https://storage.googleapis.com/model-cards/documents/gemini-2.5-pro-preview.pdf. Accessed 9 May 2025
14. Kim, G., et al.: OCR-free document understanding transformer (2022). http://arxiv.org/abs/2111.15664
15. Steiner, A., et al.: PaliGemma 2: a family of versatile VLMs for transfer (2024). http://arxiv.org/abs/2412.03555
16. Ouyang, L., et al.: OMNIDOCBENCH: benchmarking diverse pdf document parsing with comprehensive annotations. In: Proceedings of the Computer Vision And Pattern Recognition Conference, pp. 24838–24848 (2025)

17. Wang, J., Jin, L., Ding, K.: Lilt: a simple yet effective language-independent layout transformer for structured document understanding. ArXiv Preprint ArXiv:2202.13669 (2022)
18. Kirillov, A., et al.: Segment anything. In: Proceedings of the IEEE/CVF International Conference on Computer Vision, pp. 4015–4026 (2023)
19. Li, Z., et al.: Monkey: image resolution and text label are important things for large multi-modal models. In: Proceedings Of The IEEE/CVF Conference on Computer Vision And Pattern Recognition, pp. 26763–26773 (2024)
20. Liu, Y., et al.: TextMonkey: an OCR-free large multimodal model for understanding document. ArXiv Preprint ArXiv:2403.04473 (2024)
21. Tang, Z., et al.: Unifying vision, text, and layout for universal document processing. In: Proceedings of The IEEE/CVF Conference On Computer Vision And Pattern Recognition, pp. 19254–19264 (2023)
22. He, J., et al.: ICL-D3IE: in-context learning with diverse demonstrations updating for document information extraction. In: Proceedings Of The IEEE/CVF International Conference on Computer Vision, pp. 19485–19494 (2023)
23. Li, J., Li, D., Savarese, S., Hoi, S.: BLIP-2: bootstrapping language-image pre-training with frozen image encoders and large language models. In: Proceedings of The 40th International Conference on Machine Learning, vol. 202, pp. 19730-19742 (2023). https://proceedings.mlr.press/v202/li23q.html
24. JaidedAI Jaidedai/EasyOCR: ready-to-use OCR with 80+ supported languages and all popular writing scripts including Latin, Chinese, Arabic, Devanagari, cyrillic and etc... GitHub. https://github.com/JaidedAI/EasyOCR
25. Jaume, G., Ekenel, H., Thiran, J.: FUNSD: a dataset for form understanding in noisy scanned documents. In: 2019 International Conference On Document Analysis And Recognition Workshops (ICDARW), vol. 2, pp. 1–6 (2019)
26. Sun, T., Cui, C., Du, Y., Liu, Y.: PP-DocLayout: a unified document layout detection model to accelerate large-scale data construction. ArXiv Preprint ArXiv:2503.17213 (2025)
27. Wang, J., Hu, K., Huo, Q.: DLAFormer: an end-to-end transformer for document layout analysis. In: International Conference on Document Analysis And Recognition, pp. 40–57 (2024)
28. Zhao, Z., Kang, H., Wang, B., He, C.: DocLayout-YOLO: enhancing document layout analysis through diverse synthetic data and global-to-local adaptive perception. ArXiv Preprint ArXiv:2410.12628 (2024)
29. Wang, J., et al.: A graphical approach to document layout analysis. In: International Conference On Document Analysis And Recognition, pp. 53–69 (2023)
30. Dettmers, T., Pagnoni, A., Holtzman, A., Zettlemoyer, L.: QLORA: efficient finetuning of quantized LLMS. Adv. Neural. Inf. Process. Syst. **36**, 10088–10115 (2023)
31. Marafioti, A., et al.: SmolVLM: redefining small and efficient multimodal models. ArXiv Preprint ArXiv:2504.05299 (2025)
32. Luo, C., Shen, Y., Zhu, Z., Zheng, Q., Yu, Z., Yao, C.: LayoutLLM: layout instruction tuning with large language models for document understanding. In: Proceedings Of The IEEE/CVF Conference On Computer Vision And Pattern Recognition, pp. 15630–15640 (2024)
33. Huang, M., Liu, Y., Liang, D., Jin, L., Bai, X.: Multi-scale adaptive cropping for multimodal large language models. CoRR, Mini-monkey (2024)
34. Abouelenin, A., et al.: Phi-4-mini technical report: compact yet powerful multimodal language models via mixture-of-loras. ArXiv Preprint ArXiv:2503.01743 (2025)

35. Team, O.: OpenGVLab team. InternVL2 (2024). https://internvl.github.io/blog/2024-07-02-InternVL-2.0/
36. Lan, M., Xia, Y., Zhou, G., Huang, N., Li, Z., Wu, H.: LLM4QA: leveraging large language model for efficient knowledge graph reasoning with SPARQL query. JAIT. **15**, 1157–1162 (2024)

Author Index

H. Unger and P. Meesad (Eds.): NLPIR 2025, LNNS 1904, pp. 1235–1238, 2026.
https://doi.org/10.1007/978-3-032-20897-2

GPSR Compliance

The European Union's (EU) General Product Safety Regulation (GPSR) is a set of rules that requires consumer products to be safe and our obligations to ensure this.

If you have any concerns about our products, you can contact us on ProductSafety@springernature.com

In case Publisher is established outside the EU, the EU authorized representative is:

Springer Nature Customer Service Center GmbH
Europaplatz 3
69115 Heidelberg, Germany

Batch number: 10761599

Printed by Printforce, the Netherlands